AF704324

Energy & Pollution Control Opportunities to the Year 2000

ENERGY & POLLUTION CONTROL OPPORTUNITIES TO THE YEAR 2000

Published by
THE FAIRMONT PRESS, INC.
700 Indian Trail
Lilburn, GA 30247

ENERGY & POLLUTION CONTROL OPPORTUNITIES TO THE YEAR 2000

Library of Congress Catalog Card No. 93-73196

Published by
THE FAIRMONT PRESS, INC.
700 Indian Trail
Lilburn, GA 30247

Printed in the United States of America

10 9 8 7 6 5 4 3 2 1

ISBN 0-88173-189-7 FP

ISBN 0-13-300062-1 PH

Distributed by PTR Prentice Hall
Prentice-Hall, Inc.
A Paramount Communications Company
Englewood Cliffs, NJ 07632

Prentice-Hall International (UK) Limited, London
Prentice-Hall of Australia Pty. Limited, Sydney
Prentice-Hall Canada Inc., Toronto
Prentice-Hall Hispanoamericana, S.A., Mexico
Prentice-Hall of India Private Limited, New Delhi
Prentice-Hall of Japan, Inc., Tokyo
Simon & Schuster Asia Pte. Ltd., Singapore
Editora Prentice-Hall do Brasil, Ltda., Rio de Janeiro

CONTENTS

SECTION 6: HVAC CONTROLS 443

SECTION 7: COMPETITIVE POWER TECHNOLOGIES 567

INTRODUCTION

The energy industry continues to evolve and create new opportunities for energy engineers and managers. The Energy Policy Act (EPAct), integration of energy and environmental issues and meeting regulations are the buzz words of the '90s.

The Energy Policy Act is the most comprehensive legislation enacted since the passage of the National Energy Conservation Policy Act of 1978. States will need to upgrade their energy codes, inefficient equipment and lighting products will be phased out, integrated resource planning, which encourages utilities to implement demand-side management programs, will be mandated through the EPAct. The sweeping scope of the EPAct also opens up utility transmission lines for use by new non-utility power producers, encourages conservation in federal buildings and establishes a comprehensive research and development program for efficient technologies and alternative fuels.

The integration of energy and environmental concerns is best illustrated through the Environmental Protection Agency's Green Lights Program and the newly announced Adjustable Speed Drives Program. It is now recognized that pollution prevention can be achieved by reducing energy consumption. How can a mere light bulb play a role in slowing the far-reaching effects of global warming? The less electricity utilities produce, the less carbon dioxide they generate. By replacing an incandescent light bulb with an energy efficient compact fluorescent, 300 pounds of carbon dioxide a year would be saved.

The integration of energy and environmental concerns is also important in recycling, reclamation and reuse of refrigerants to reduce CFCs, thus minimizing ozone depletion. The "sick building syndrome" is another example of how energy and the environment go hand-in-hand. By improving indoor air quality through innovative filtration systems we can save energy and eliminate the poor quality of indoor air.

In order to determine how the energy industry is evolving, the Association of Energy Engineers conducted a comprehensive survey of 8000 energy engineers and managers; 1500 individuals responded. The survey found that many companies have done an outstanding job in applying energy management technologies and reducing energy costs as indicated below:

Cost Savings Since Program Started	***Respondents***
Less than $100,000	22.3%
$100,000 – $500,000	23.3%
$500,000 – $1,000,000	13.8%
$1,000,000 – $2,000,000	10.6%
$2,000,000 – $5,000,000	13.3%
Over $5,000,000	16.6%

The survey also found that the issues which will have a high impact on an organization's business during the next five years are increased environmental regulations (70.6%), price of oil, gas and electricity (67.4%), reduced reliability of electricity (45.7%), CFC phaseout (43.5%), and availability of gas (41.6%).

Energy awareness is up dramatically in most companies: 46.6% indicated energy awareness increased during the last year; however, 16% felt energy awareness was up significantly. Environmental issues of most concern to companies are CFC phaseout (30.6%), "sick building syndrome" (24.3%), emission control to meet Clean Air Act (24.3%), and hazardous and solid waste (20.8%).

We congratulate our cosponsors of the WEEC and ETE — The Alliance To Save Energy, Electric Power Research Institute, General Services Administration, Georgia Office of Energy Resources, National Registry of Environmental Professionals, *Pollution Engineering* magazine, U.S. Department of Energy: State Energy Programs, U.S. Environmental Protection Agency: Green Lights Program, U.S. Environmental Protection Agency: Region IV — for supporting this program and playing a major role in fostering technology transfer.

We also congratulate the Association of Energy Engineers and its 8000 members for providing the leadership for the development of the energy and environmental industry.

For 16 years the World Energy Engineering Congress has provided the essential forum for the energy industry. The sharing of information is important to the continued growth of the energy engineering profession. AEE is proud to play a major role in sponsoring this vital conference.

Albert Thumann
Executive Director
Association of Energy Engineers

CONTRIBUTORS

Agudo, Michael E., *Naval Surface Warfare Center*
Alnajjar, Rogeh K., *Aaron Environmental Company, Inc.*
Anderson, Donald J., *EnerSys, Inc. (EnChill)*
Ausmus, Guy H., *Inland Steel Company*
Bandy, R.L., *Public Service Co. of Colorado*
Beach, Bobby G., *Enercon Services, Inc.*
Beckwith, William, *Indoor Environment & Energy*
Berish, Cory, *U.S. Environmental Protection Agency, Region IV*
Broberg, Frederick N., *H.F. Lenz Company*
Burch, Gary D., *U.S. Department of Energy*
Burke, David J., *Dean Witter Reynolds*
Campbell, Jeffrey S., *Chicago Board of Trade*
Case, Michael P., *U.S. Army Construction Engineering Research Lab*
Clunie, Robert M., *Applied Engineering, Inc.*
Coe, William F., *Southern Electrical International*
Coleman, Thomas N., *University of Tennessee*
Conlan, Paul E., *A&C Enercom*
Courtright, Henry A., *Electric Power Research Institute*
Czeschin, Laurie, *Cooper Lighting*
Daellenbach, Keith K., *Pacific Northwest Laboratory*
Dalton, John C., *Reed Consulting Group*
Damberger, Thomas A., *Kaiser Permanente*
de Leon, Armando J., *City of Austin, TX*
DeBaillie, Lee P., *U.S. Army Construction Engineering Research Lab*
DeGroat, Kevin, *Meridan Corporation*
Delurey, Michael W., *Naval Surface Warfare Center*
DePaull, Thomas K..
Depies, Chris, *Digisonix, Inc.*
Dineen, Susan, *Digisonix, Inc.*
Dixon, Douglas R., *Pacific Northwest Laboratory*
Doubek, David, *Cooper Lighting*
Dunham, Eric, *General Services Administration*
Dybowski, Jaroslaw, *Polish Power Grid Company*
Elliott, R. Neal, *American Council for an Energy-Efficient Economy*
Feldman, Roger D., *McDermott, Will & Emery*
Fetters, John L., *AT&T Network Systems*
Fiorino, Donald P., *Texas Instruments, Inc.*
Foster, Scott R., *Lorusso & Lord*
Gage, Cynthia, *U.S. Environmental Protection Agency*
Geers, James R., *PLM Technologies*
Gerkin, Carl D., *Storage Technology Corporation*
Gignilliat, Henry, *U.S. Army Installation Support & Activity Center*
Gillespie, Adrian H., *Army Forces Command*
Ginsberg, Mark, *U.S. Department of Energy*
Gracias, Cecillio E., *New York State Energy Office*
Green, Joanta H., *Bechtel Group, Inc.*
Gregoire, Robert, *Applied Engineering, Inc.*
Griswold, George H., *Southern Company Services, Inc.*
Gross, Robert G., *Hydroscience, Inc.*
Groves, David G., *Stanford University*
Guibord, Russell P., *Applied Engineering, Inc.*
Haakenstad, Mathew D., *Electric Power Software*
Hansen, Shirley J., *Hansen Associates, Inc*
Hardin, William B., *Springs Industries*
Hardy, Thomas C., *Wright Laboratory, U.S. Air Force*
Hedden, Barry K., *Baltimore Gas and Electric Company*
Hendrickson, Paul L., *Pacific Northwest Laboratory*
Herrera, Salvador, *Fund for Electric Energy Efficiency*
Higgins, John E., *Reed Consulting Group*
Hille, Ewaryst, *Polish Foundation for Energy Efficiency*
Hoffstatter, Lloyd S., *New York State Energy Office*
Holmes, William A., *Holmes Energy Services, Inc.*
Homan, Kelly, *U.S. Army Construction Engineering Labs*
Hoshide, Robert K., *Rockwell International / Rocketdyne Division*
Hutchins, Paul F., *Reynolds, Smith and Hills, Inc.*
Hyre, Regi A., *Oklahoma State University*
Iden, Mark K., *Defense Fuel Supply Center*
Jendrucko, Richard J., *University of Tennessee*
Kane, Christopher, *ER3, Inc.*
Kazachki, Georgi, *Acurex Engineronmental Corporation*
Kelly, G. Robert, *Naval Surface Warfare center*
Kiang, C.S., *Georgia Institute of Technology*
Knutson, Kent S., *Resource Data International, Inc.*
Kumar, Dinesh, *DynCorp*
Lacki, Thomas R., *Oklahoma State University*
Larranaga, F.J., *Refineria Isla (Curazao) SA*
Laventure, George C., *Tyndall Air Force Base*
LeBlanc, William J., *Barakat & Chamberlin, Inc.*
Ledbetter, Marc, *American Council for an Energy-Efficient Economy*
Lewis, W. Michael, *W.M. Lewis & Associates, Inc.*
Lindsay, Bruce B., *University of Wisconsin*
Lowe, Mick, *Digisonix, Inc.*
Magana, Rogelio, *Fund for Electric Energy Efficiency*
Mauch, James C., *Vista Environmental Information, Inc.*
McRae, Marjorie R., *Barakat & Chamberlin, Inc.*
Meyers, S. David, *RMT, Inc.*
Miller, Richard K., *Richard K. Miller & Associates*
Mitchell, Frederick G., *Energy Investment, Inc.*
Moshage, Ralph E., *U.S. Army Construction Engineering Research Lab*
Nadas, Denis, *New York State Energy Office*
Nemeth, Robert J., *U.S. Army Construction Engineering Research Lab*
Newcomb, Ellen C., *The Trane Company*
Nezhad, Hameed G., *Moorhead State University*
Nicholson, J. Patrick, *N-Viro Energy Systems, Ltd.*
Noble, Eric C., *New York State Energy Office*
Norton, William E., *Westinghouse Remediation Services, Inc.*
Oster, John, *Lockwood Greene*
Owan, Ransome E., *New Jersey Transit*
Packard, Cristina P., *Baltimore Gas and Electric Co.*
Parfomak, Paul, *Barakat & Chamberlin, Inc.*
Penney, Robert A., *Washington State Energy Office*
Peters, Jane S., *Barakat & Chamberlin, Inc.*
Petze, John D., *Teletrol Systems, Inc.*
Peyton, Joe C., *Southern Company Services, Inc.*
Polaski, Joseph A., *Measuring & Monitoring Services, Inc.*
Preston, George V., *Phillips Lighting Company*
Price, Robert O., *CH2M Hill*
Prue, Virginia W., *Kaiser Permanente*
Pulfrey, Robert J., *Environmental Science & Engineering, Inc.*
Reddy, Shylender, *ACR Engineering, Inc.*
Robertson, C., *Consultant*
Robertson, J. Wayne, *Black Gold Corporation*
Rodriguez, Ignacio, *Fund for Electric Energy Efficiency*
Rojeski, Peter, *North Carolina AT State University*
Rose, Patricia H., *U.S. Department of Energy*
Rouleau, Jeffrey, *Westinghouse Remediation Services, Inc.*
Rowley, Stephen D., *Department of the Army*
Running, Richard M., *Electric Power Software*
Said, Salem A., *Aaron Environmental Company, Inc.*
Salavani, Reza, *Tyndall Air Force Base*
Saleem, Jwamir M., *Aaron Environmental Company, Inc.*
Savoie, Martin J., *U.S. Army Construction Engineering Research Lab*
Schiess, Klaus, *KSEngineers*
Scott, Roy R., *Arde Barinco, Inc.*
Seaburn, Gerald S., *Law Companies*
Seiner, Yan, *Hydroscience, Inc.*
Selgrade, Edward L., *Tillinghsat, Collins & Graham*
Selmon, Eric H., *Stanford University*
Shay, Colin, *Meridan Corporation*
Sherif, S.A., *University of Florida*
Simpson, Phill, *Furman University*
Smith, Lee L., *New York State Energy Office*
Smith, Mark D., *Tyndall Air Force Base*
Smith, Loretta A., *American Council for an Energy-Efficient Economy*
Smith, Garyl D., *Naval Air Weapons Station*
Sohn, Chang W., *U.S. Army Construction Engineering Research Lab*
Sparts, Gary, *Reynolds, Smith and Hills, Inc.*
Stebbins, Wayne L., *Hoechst Celanese Corporation*
Stein, Jay, *E-Cube, Inc.*
Sullivan, William E., *Enercon Services, Inc.*
Svard, Carl D., *Integrated Planning & Engineering, Inc.*
Swalla, Dana R., *University of Florida*
Thomas, Todd M., *University of Tennessee*
Tiller, Jeffrey S., *Southface Energy Institute, Inc.*
Tolman, A. Jean, *Arnall Golden & Gregory*
Tormey, David, *EML Research, Inc.*
Traudt, Jon E., *Air Pro, Inc.*
Troncoso, Ricardo, *ACR Engineering, Inc.*
Tucker, R. Arnold, *Osram Sylvania, Inc.*
Turner, Stephen C., *Florida Atlantic University*
Turner, Wayne C., *Oklahoma State University*
Vaillencourt, Richard R., *Energy Investment, Inc.*
van Kemenade, Johan, *Phillips Lighting Company*
Varno, Edward J., *Larsen Engineers*
Venkatesan, V., *Refineria Isla (Curazao) SA*
Vera, Ivan A., *Decision Analysis Corporation of Virginia*
Voigt, Thomas C., *CSU Bakersfield*
Wade John W., *Reynolds, Smith & Hills, Inc.*
Walker, Terri C., *Richard K. Miller & Associates*
Warren, Robert H., *Commercial Energy Systems Management Corporation*
Warren, Carlos S., *Reynolds, Smith & Hills, Inc.*
Wayte, Patrick W., *Vanderbilt University*
White, Sam C., *RMT, Inc.*
Wiles, Charles F., *Metro Environmental Co.*
Willis, Reginald A., *Logical Technology, Inc.*
Wise, Steve, *Digisonix, Inc.*
Wolcott, David, *RCG / Hagler, Bailly, Inc.*
Wolpert, Jack S., *E-Cube, Inc.*
Ziming, Yang, *City of Austin, TX*

ACKNOWLEDGEMENTS

Appreciation is expressed to all those who have contributed their expertise to this volume, to the conference chairmen for their contribution to the 16th World Energy Engineering Congress and 3rd Environmental Technology Expo, and to the officers of the Association of Energy Engineers for their help in bringing about this important conference.

The outstanding technical program of the 16th WEEC can be attributed to the efforts of the 1993 Advisory Board, a distinguished group of energy managers, engineers, consultants, producers and manufacturers:

Richard Aspenson
Consultant

Keith Davidson
Gas Research Institute

Douglas A. Decker
Johnson Controls, Inc.

Shirley Hansen, Ph.D.
Hansen Associates, Inc.

Jon R. Haviland, P.E., CEM
Criterium Haviland Engineers

MaryAnne Lauderdale, P.E.
Energy Investment, Inc.

Dilip Limaye
Synergic Research Corporation

Konstantin K. Lobodovsky
Consultant

William H. Mashburn, P.E.
Virginia Polytechnic Institute

Malcom Maze
Abbott Laboratories

Harvey Morris
Independent Power Producers

Martin A. Mozzo, Jr., P.E.
Kenetech Energy Management Inc.

Graham B. Parker
Battelle Pacific Northwest Laboratories

Patricia Rose
U.S. Department of Energy

Frank Santangelo, P.E., CEM
Consultant

Ron Smith
General Motors

Walter P. Smith, Jr.
Energy Technology Services Consulting

Albert Thumann, P.E., CEM
Association of Energy Engineers

Wayne C. Turner, Ph.D., P.E., CEM
Oklahoma State University

The following organizations have given outstanding support in promoting principles and practices to help achieve energy savings, greater productivity, and lower operating costs.

CORPORATE SUSTAINING MEMBERS

Association of Energy Engineers

Abbott Laboratories
Aircon Energy Inc.
Alken-Murray Corporation
Alliance To Save Energy
American Institute of Plant Engineers
Association of Physical Plant Administrators of Universities and Colleges
Bailey Controls Company
Baskin-Robbins, USA, Co.
Battelle Pacific Northwest Laboratories
California Energy Commission
City of Houston
Cooper Lighting
CP National/Trident Energy Group
Datastream Systems Incorporated
Department of Corrections/Office of Capital Management
Destec Energy
Diesel & Gas Turbine Publications
DMC Services
Energy Initiatives, Inc.
Energy Investment Inc.
Energy Management Associates, Inc.
Energy Management Specialists Inc.
Engineer's Digest
Enron Gas Marketing, Inc.
Falcon Seaboard Power Corporation
Ferreira Service Inc.
Florida Power & Light Company
General Services Administration
HaVAC Control Systems, Inc.
Hydra-Co Operations, Inc.
IBM Corporation
Indiana University of Pennsylvania
International Energy Society
H.F. Lenz Company
Los Angeles Dept. of Water & Power
Luis Malheiro da Silva, Lda
MAMAC Systems, Inc.
Montana Power Company
Mission Energy Company
NALMCO (intN'l. Assoc. of Lighting Mgmt. Companies)
National Electrical Contractors Association
National Wood Energy Association
Naval Energy & Environmental Support Activity
NEOS Corporation
New York State Electric & Gas
Niagara Mohawk Power Corporation
Northern Illinois Gas Company
Northern States Power Company
Northstar Energy Corporation
Old Dominion Electric Cooperative
Philadelphia Thermal Energy Corporation
Pima County Facilities Management
PSI Energy
Public Service Conservation Resources Corporation
Public Service Electric & Gas Company
Rocky Mountain Institute
Sam Young Mechanical Engineering Co., Ltd.
Shanghai Society of Energy Research
SCT Yarns, Inc.
Southwire Company
SSI Services Inc.
State Industries, Inc.
Stone & Webster Engineering Company
Sverdrup Corporation
Synergic Resources Corporation
Tennessee Valley Authority
Texas A&M University
3 M Company
Trend Control Systems
Turlock Irrigation
United Assoc. of Journeymen Apprentices of the Plumbing & Pipefitting Industry in the US & Canada
United States Air Force
Universal Energy Control, Inc.
University of Mass/Amherst
University of Missouri
US Army Construction Engineering Research Laboratory
US Army Logistics Evaluation Agency
US Dept of Energy/Chicago
US Dept of Energy/New Mexico
US Dept of Energy/New York
Utah Association Municipal Power Systems
V.E.I. Inc
Verle A. Williams & Associates Inc.
Western Area Power Administration/California
Western Area Power Administration/Colorado

SECTION 1
ENVIRONMENTAL MANAGEMENT

Chapter 1

Clean Air Act Amendments Overview and Update

W.C. Turner, R.A. Hyre, T.R. Lacki

ABSTRACT

The Clean Air Act Amendments of 1990 will be one of the most (if not the most) industry affecting set of regulations to date. Likely, its final impact will dwarf even RCRA. This paper summarizes the 1990 Amendments with an emphasis on how industry will be impacted. The paper ends by presenting a list of management suggestions. These ideas should help industry prepare for and incorporate the regulations while minimizing negative profit (cost) interruptions.

INTRODUCTION

Purpose

The total impact on industry of the 1990 Clean Air Act Amendments (CAAA) will be significant and broad, affecting practically every manufacturing plant in the country. Cost implications in most cases will be expensive and, in some cases, industry may find continued existence to be impossible. For example, most if not all manufacturing plants will have to obtain air permits. As stated in Title V (see below), there will be a fee associated with these permits.

Fortunately, there are provisions in the act that will help industry prepare for these changes and minimize costs. For example, small business assistance centers are being established to help small businesses understand and meet the regulations. Also, early reductions of some Hazardous Air Pollutants (HAP'S) will "buy time" for those industries astute enough to seize the opportunities.

The purpose of this paper is to help industry prepare for and react to the changes coming because of the CAAA. This is accomplished by first summarizing the CAAA with an emphasis on those sections impacting industry. Then, management suggestions (MS) are made. These suggestions are intended to serve as guidelines for industry to minimize cost and other negative impacts of the CAAA. Finally, some thoughts for the future are given.

We (industry, academia, researchers, EPA, state environmental agencies, environmental advocate groups, etc.) need to realize that all breath the same air and use it as a resource in our manufacturing endeavors. This is not a "me vs they"; it is (or should be) a group attack on what we all recognize. Air (and the environment in general) is a resource we have borrowed from our kids. We use it to accomplish our objectives and then return it for their later use. To return that air unused for any societal improving activity is almost as big a sin as returning it in very poor condition. We must compete economically; but we must protect the environment while competing.

CAAA SUMMARY[1]

Introduction

The 1990 amendments consist of eleven titles as shown in Table One.

Table One - Titles of the CAA 1990 Amendments

Title I -	Nonattainment Provisions
Title II -	Provisions Relating to Mobil Sources
Title III -	Hazardous Air Pollutants
Title IV -	Acid Deposition Control
Title V -	Permit Provisions
Title VI -	Stratospheric Ozone Protection
Title VII -	Federal Enforcement Provisions
Title VIII -	Miscellaneous Provisions
Title IX -	Clean Air Research
Title X -	Disadvantaged Business Concerns
Title XI -	Clean Air Employment Transition Assistance

The eleven titles of the 1990 amendments will fit into the six titles of the Clean Air Act. The CAA previously consisted of three titles. The 1990 amendments have added three new titles dealing with acid deposition, permits, and stratospheric ozone protection, as shown in Figure One. Following is a discussion on the highlights of each of the CAA amendment titles, and the probable impact on industry.

Title I - Provisions for Attaining and Maintenance of National Ambient Air Quality Standards (NAAQS)

The entire nation is divided into air quality regions. Regions that meet or exceed NAAQS are classified as "attainment" areas, areas that do not meet NAAQs are classified as "nonattainment". The pollutants being monitored for NAAQs purposes are, ozone (smog),

1 This part of the paper draws heavily from Mozingo, Carl A. and Hyre, Regi A. "Clean Air Act Amendments of 1990" Proceedings World Energy Engineering Congress, Atlanta, GA, October 1992.

EVOLUTION OF CAA		
CAA 1977	**1990 Amendments**	**CAA 1990**
Title I-Nonattainment **Title II-Mobile Sources** **Title III-General**	**→ 11 Titles →**	**Title I-Nonattainment** **Title II-Mobil Sources** **Title III-General** **Title IV-Acid Deposition** **Title V-Permits** **Title VI-Stratospheric Ozone Protection**

FIGURE 1: THE NEW CLEAN AIR ACT

carbon monoxide (CO), particulates, nitrogen oxides (NO_x), sulfur dioxide (SO_2), and lead. Some nonattainment areas are further categorized based on the severity, and type of pollution. CO and particulate nonattainment areas are broken into two categories, moderate (reach attainment by 1995) or serious (reach attain-ment by 2000). Ozone areas are as:

Marginal - attainment by 1993
Moderate - attainment by 1996
Serious - attainment by 1999
Severe - attainment by 2005, 2007 (Chicago, Houston, and New York)
Extreme - attainment by 2010 (Los Angeles)

Ozone Nonattainment: The amended act seeks a continuous improvement in ozone non-attainment areas. This is accomplished by requiring all nonattainment areas (except marginal areas) to have a 15% reduction in Volatile Organic Compounds (VOCs)[2] by 1996, and a 3% annual reduction thereafter.

Service stations in nonattainment areas will use vapor recovery devices on pumps to reduce VOC emissions. These recovery devices will be required at stations that dispense 10,000 or more gallons per month.

The definition of a "major" source has changed. Previously it was any source that emitted 100 TPY of VOC or NO_x. The new definition of a "major" source is based on the type of nonattainment area where the source is located. Major sources are:

Extreme - 10 TPY of VOC or NO_x
Severe - 25 TPY of VOC or NO_x
Serious - 50 TPY of VOC or NO_x
Moderate - 100 TPY of VOC or NO_x
Marginal - 100 TPY of VOC or NO_x

This will have a large impact on industry - especially those in the more tightly controlled areas.

The use of Reasonably Available Control Technology (RACT)[3] on NO_x sources is now required in certain areas and will be applied to "major" sources. EPA is charged with developing a set of generic RACTs for specific industrial source categories. RACTs are to be issued in the form of Control Technology Guidelines (CTGs). CTGs, in essence, have the same authority of a federally mandated rule. To date there are about thirty CTGs. Under the new amendments, EPA is charged with developing CTGs for eleven new specific industrial sources. Table Two shows when RACT and BACT is required.

Table Two - Attainment and nonattainment Requirements

Nonattainment
Old Sources - RACT
New Sources - LAER Emission Offsets
Attainment PSD (BACT)
New Sources must meet NSPS - Emission Based

The effectiveness of NO_x emission reductions is to be determined. Depending on EPAs findings the NO_x control measures could change. Furthermore, if it can show that no net air quality improvement results from using these controls then sources can petition EPA for exemption. Another measure that affects both ozone and CO nonattainment areas is vehicle Inspection & Maintenance (I&M) programs. I&M programs that already exist must be upgraded, and in most ozone nonattainment areas where I&M programs do not exist they must be implemented.

Areas can also be in nonattainment for particulate matter less than 10 microns in size (PM_{10}) or for not meeting SO_2, and CO standards. At present there are 41 CO Nonattainment areas and 70 PM_{10} Nonattainment areas.

State Implementation Plans (SIPs) must provide for attainment and maintenance of NAAQSs in areas within the state. If EPA finds that the SIP is not effective or is inadequate, then it can develop a Federal

2 VOC's are generally agreed to be ozone precursors; so control of VOC emissions will reduce ozone.

3 See the Appendix for a description of RACT and other acronyms.

Implementation Plan (FIP) for use in the state. FIPs can be whole plans or partial plans designed to meet specific needs. If for some reason a state makes an inadequate effort towards implementing this title or meeting NAAQSs, the act has provided certain sanctions that can be applied. Cutoff or restriction of federal highway funds, or a requirement that new industry offset emissions by a 2 to 1 ratio, are sanctions EPA has available. This is a much stronger Federal presence than before.

Title I Impact on Industry: The classification of the Air Quality Control Region (AQCR) in which a facility is located is the major factor for determining the impact of Title I. Since Title I mainly deals with nonattainment problems, facilities in attainment areas are not directly affected. Facilities in attainment areas need to be aware of any problems that arise in their AQCR. If an attainment area is reclassified as non-attainment, a company could have to drastically alter its operating procedures. Basically the idea is to strive for attainment, and once achieved, work to keep it. Industries in nonattainment areas must meet all Title I requirements and should be communicating with the state air regulating board. Facilities in non attainment areas should conduct comprehensive emissions inventories.

Title II - Provisions Relating to Mobile Sources

The amendments take two primary approaches to reducing mobile source emissions: reducing tailpipe emissions and using "clean" fuels. Additionally, there are provisions for fleet vehicles. An annual, centralized Inspection and Maintenance (I&M) program is required in nonattainment areas classified as "serious" and above.

The traditional approach to mobile source emissions has been to reduce, as much as possible, tailpipe emissions. By 1994 all automobile manufacturers are required to reduce tailpipe emissions on 40% of the cars produced. All must meet the new standards by 1998. These emissions consist of Non-methane-hydrocarbons (NMHCs), Carbon Monoxide (CO), and oxides of Nitrogen (NO_xs).

The "clean" fuels approach to emission reductions consists of reformulated gasolines and oxygenated fuels. Reformulated gasolines are initially for use in nine ozone nonattainment areas (Los Angeles, San Diego, New York, Houston, Chicago, Milwaukee, Baltimore, Philadelphia, and Hartford) starting in 1995. Other nonattainment areas can opt into this program. Reformulated gasolines after January 1, 1995 must meet the following fuel content specifications:

1. 2% average oxygen.
2. 1% average benzene
3. Use detergents.
4. Contain no heavy metals.
5. Cannot increase NO_x emissions.

Oxygenated fuels will be required in CO Nonattainment areas during the four winter months (the time of year prone to high CO levels). EPA can allow a shorter oxygenated fuel period if it will not cause any problems with the CO standard. Starting in 1992, a minimum oxygen content of 2.7% was required in fuels sold in CO nonattainment areas.

Under the CAAA fleet vehicles in ozone nonattainment areas classified as, serious, severe, or extreme, and one CO nonattainment area (Denver), will face new requirements. A fleet vehicle is defined as belonging to a group of 10 or more vehicles capable of being centrally fueled, with the exception of vehicles normally garaged at personal residences at night. Generally these fleets must meet standards requiring emissions to be about 60%-70% less than non fleet vehicles. This will often be met with alternative fuels such as compressed natural gas and electricity.

Title II impact on Industry: The auto and oil industries will be greatly affected by Title II. It is the auto manufacturers who will design and produce cars to meet the new standards. Automakers will have to build Low Emission Vehicles (LEV's) for sale in affected areas and for use by companies that have fleet vehicles. The oil industry will be responsible for developing and producing reformulated and oxygenated fuels. Overall, Title II will force automakers and oil refiners to redesign their products so that they will comply with the amendments. Industries with fleets in nonattainment areas should be preparing for the fleet rules.

Title III - Hazardous Air Pollutants (NESHAPS)

National Emission Standards for Hazardous Air Pollutants (NESHAPS) exist for 8 substances. Any industry using these substances, must meet the NESHAP.

The basic approach the amendments have taken to control toxic emissions is to first apply the maximum available control technology (MACT) to a source, then determine if there is any "residual risk" above what the MACT can control. In Title III, Congress has mandated that 189 additional air pollutants are to have NESHAPs developed for them. This is significant since to that date, EPA had developed only 8 NESHAPS.

Types of Hazards: Three classes of hazardous air pollutant sources have been targeted for regulation, these are.

1. Routine everyday stack & fugitive emissions for major sources (oil refineries, manufacturing plants, etc.)
2. Routine everyday stack & fugitive emissions from small (area) sources (gas stations, dry cleaners etc.)
3. Accidental, catastrophic releases that are life threatening.

A Title III major source is defined as any source emitting 10 tons per year of a single listed pollutant, or a total of 25 tons per year of a combination of any listed pollutants. A list of major source categories is to be developed by EPA.

Emission reduction standards for major sources will be based on the Maximum Available Control Technology (MACT) for each major source category. For new major sources, MACT will be the most restrictive available control. For existing sources MACT will be the control technologies used in the cleanest 12% of a source category.

After MACT rules have been set for major source categories EPA is to determine if adequate protection is provided, or if there is any residual risk posed. EPA and other federal agencies will determine acceptable residual risk levels. More to the point, if the MACT provides sufficient protection against listed hazardous air pollutants then all is well. Its when there is still not enough protection against unwanted emissions, after installation of MACT, that the area of risk between what is desired and what is possible needs to be evaluated. If this area is determined to be acceptable by EPA then no further controls are needed, if the risk is not acceptable then further steps will have to be taken. Each major source that is subject to the new NESHAPs standards is required to apply for an operating permit under Title V of the 1990 amendments.

"Area sources" are those small industries that each generate small amounts of pollution but together generate a large amount. Area sources are subject to the same NESHAPS list as major sources. EPA is to develop a list of categories and subcategories for area sources. Enough categories are to be identified to assure reduction of 90% of the 30 most serious toxic air pollutants emitted from area sources. MACT rules will be established for area sources in the same manner as for major sources.

EPA has developed a list of extremely hazardous substances (EHSs). EHSs are pollutants that would likely cause death, injury, or require evacuation of the immediate area if an accidental release were to occur.

If the threshold amount of a listed EHS is exceeded at a source, then the facility has to meet certain requirements. An engineering analysis of the facility will be done to determine any public health hazards, also, risk management plans (RMP) must be developed. These studies and plans are to be made available to the public. EPA is directed to develop plans to deal with facilities handling EHSs. In addition, reporting requirements will increase beyond the current SARA Title III requirements. RMP requirements are developed in 40 CFR and will be due soon.

A new board designed to investigate non-transportation chemical accidents, called the Chemical Safety Board, is established. This board will consist of five members. The board is to determine the cause of accidents and to recommend ways to avoid similar accidents in the future.

Title III Impact on Industry: Title III is the title that will likely have the biggest impact on industry. Some of the impacts are summarized below:

1) If an industry generates 10 TPY of a single listed substance or 25 TPY of any combination, it is a "major" industry and will have to meet all requirements including applying for a permit.

2) The Hazardous Organic NESHAP (HON) has been issued. This rule would cover about 150 of the 189 chemicals Congress added. Synthetic Organic Chemical Manufacturing Industries (SOCMI) are covered as are some non SOCMI's.

3) Risk Management Plans (RMP) regulations have been finalized and appear in 40 CFR. Companies with more than a threshold quantity (TQ) of these chemicals must do complete Risk Assessment plans on a phased in schedule.

4) Early reduction provisions are allowed in the CAAA. Basically, any company that reduces a listed HAP before compliance dates by 90% "buys" an additional 6 years for compliance dates. This could be very profitable as 6 years is enough for technology to significantly mature.

5) Process Safety Management plans are required by OSHA for all plants that have chemicals listed in Appendix A in amounts greater than threshold quantities (listed). The number of chemicals covered is greater than EPA's Risk Management Plan list but the efforts are similar. EPA (RMP) is more concerned about releases to the environment while OSHA (PSM) is more concerned about exposure to employees.

Title IV - Acid Deposition Control

Title IV deals with reducing SO_2 and NO_x emissions from electric utility power plants. Utilities will be able to emit set levels (allowances) equal to a permitted emission level. Basically, an allowance is a marketable permit to emit one ton. If a utility reduces its emission below its permitted emission level, then it will accumulate whatever allowances it does not use. These allowances can be saved for future use, or given to another plant owned by the same company, or sold outright to another utility. If allowances are sold to another facility it would allow the purchasing facility to emit above its own permitted emission level, yet remain in compliance with the regulations. The idea is to reduce emissions overall in the most cost effective manner.

Sources emitting amounts of SO_2 and NO_x over what their allowances are valued (both permitted and purchased allowances) will be required to pay an excess emission fee of $2,000 per ton. Sources that over emit and become subject to fees must submit a plan showing how it will correct the problem. Also, continuous emission monitoring equipment will be required at affected facilities. Recordkeeping and reporting requirements have been changed to better keep track of emissions.

This is a novel approach that will be interesting to watch. Most utilities have several professionals presently working on Title IV, so no more will be said in this article.

Title IV Impact On Industry: Title IV directly targets emissions from coal fired electric utilities. The regulations that will come out of this provision will require many utilities to refit to come into compliance. As a result, industry can look for higher energy costs.

Title V - Permit Provisions

A significant addition to the CAA is the provisions for a federal permit program (Title V) to be overseen by EPA. The new permit program is similar to the National Pollution Discharge Elimination System (NPDES) permit required by the Clean Water Act (CWA) for discharges to surface water.

The responsibility for implementing and maintaining Title V permit programs is delegated to the state, to be integrated into the existing SIP. The program must be in place nationwide by November 1994. If a state does not develop or adequately enforce a permit program then EPA can move in and develop its own. States are allowed to develop and add more measures as long as they are more stringent than EPAs.

Title V permit provisions are directed at significant contributors to air emissions. Sources that emit more than 100 tons per year (major source) and smaller sources in more polluted areas (according to nonattainment classification, extreme, serious, etc.) are required to have a permit. Sources that emit Title III air toxics of 10 tons or more of any single listed hazardous air pollutant or 25 tons or more of any combination of listed hazardous air pollutants, are require to obtain a permit. New sources subject to NSPSs are also required to obtain permits. Thus, most industrial plants will eventually be forced to obtain a Title V permit.

Sources have a year after the programs start to apply for a permit. If a permit is applied for in a reasonable amount of time, the source can continue operations during the review and issuance process. After the first year of the program, permit applications must be acted on by the state within 18 months. The state is to send draft permit applications to EPA for approval who has the authority to reject permits. If EPA objects to a permit application, the state has 90 days to revise it. If the state does not revise the permit, then EPA can revise it or deny it.

If, after a source receives its permit, it decides to modify its operations, no formal permit revision will be required if the following requirements are met:

1. The changes are not defined as modifications according to existing statutes.

2. An increase in emissions will not occur because of the changes.

3. Seven days written notice to state and EPA, prior to the changes.

To fund this program states are allowed to collect a permit application fee of not less than $25 per ton of emitted regulated air pollutants, except CO. This amount cannot exceed $100,000 because no fees will be assessed for emissions above 4,000 tons a year. If a source exceeds its permitted emissions, then a fee of not less than $10,000 a day can be imposed.

While processing permit applications prior to issuance, the state is required to allow for public comment and hearing. Any person can petition EPA to issue an objection to a permit application. EPA has 60 days to act on the petition. Objections in these petitions need to have been raised during the public comment period in order for EPA to act on them. Once a permit is issued, a citizen petition cannot delay its use.

Title V Impact on Industry: Initially the responsibility for this program lies with the states. After a permit system has been implemented then industry can apply for a permit. A system like that of the National Pollutant Discharge Elimination (NPDES) under the Clean Water Act (CWA) can be expected. Industry will have to build on existing record keeping and reporting structures used for other environmental permits. Since this permit requirement will be new for most and changed for all industries, there will be a big learning curve that must be climbed.

Title VI - Stratospheric Ozone Protection

Title VI provisions are intended to prevent deterioration in the stratospheric ozone. This title contains many "do or die" provisions. It sets firm deadlines for the phase out of production of certain substances. These substances are chlorofluorocarbons (CFCs), halon, carbon tetrachloride, and methyl chloroform. The intent is to phase out the use and production of these substances and force industry to develop new and more environmentally safe substitutes.

There are two classes of substances to be phased out and regulated. Class I includes CFCs, halons, methyl chloroform, and carbon tetrachloride. Class I substances are to be phased out by the year 2000, except for methyl chloroform which has until 2002. Class IIs principle substance is hydrochlorofluorocarbons (HCFCs). In 2015 production of HCFCs, except for certain uses, is to be banned. HCFCs can continue to be used as a refrigerant until 2020. HCFC production will be completely banned in 2030.

The most noticeable changes is with the recovery, recycling, and disposal of Class I and II substances. The servicing of air conditioners and refrigerators is tightly regulated. Since July 1, 1992 no direct venting of refrigerants during service (including car air conditioners) has been allowed. Cars produced in model year 1994 and thereafter will be required to use refrigerants that will not deplete the ozone.

Title VI Impact on Industry: Replacements for CFCs used in industry will have to be found. These provisions force new development by gradually banning all use of CFCs and HCFC's. Manufacturers of CFCs are responding by developing new products. This will translate into higher costs for these new products. The recycling requirement for all CFC's used in air conditioning is already dramatically affecting industry. Also phase out studies are being conducted including using non CFC refrigerates, revitalizing absorption systems and going to ammonia systems.

Title VII Federal Enforcement

This title basically increases federal control and punishment for violations of SIP, permits, or any other provisions of the new act. The EPA can use Title VIII to go after states that allow SIP or permit violations.

Criminal penalties can be up to $250,000 and five years in prison for individuals or $500,000 for corporations. Basically, criminal penalties can result from any knowing violation of the revised act or SIPs, or knowingly endangering others.

The amendments do deal with individuals involved in violations. Employees that knowingly release air toxics that have the potential of doing severe bodily harm to others can be fined $250,000 a day and 15 years in prison. For corporations the fine can be $1,000,000. When dealing with unknowing violations the new act places responsibility on employers or senior personnel rather than actual operators.

Field Citations: Government officials now have the ability to write on-site citations for minor violations. These "tickets" can be up to $5,000 a day for each violation. Facilities that have been given a citation have the right to a hearing.

Other Provisions: Citizen suits regarding violations or noncompliance can be brought against facilities to force them to obtain a permit and comply with the SIP. In cases where the violation has been repeated, the suit can also cover past violations. Another innovative provision is the Award Program for individuals providing information on violations - in effect a "bounty" program.

Title VII Impact On Industry: Title VII gives the federal government broad powers to enforce these amendments. Individual and corporate liability seems to be clearly defined. One provision that has the potential to cause trouble is the "bounty" provision. If industry complies with these amendments and conducts business in an environmentally responsible manner, these provisions probably will not be used against them. One way for industry to respond to environmental concerns as a whole is to make use of environmental audits to provide a "picture" of their environmental operations. (See managements suggestions - emission inventory).

Titles VIII thru XI

Title VIII contains miscellaneous provisions, ranging from studying Class I PSD areas, to analysis of costs and benefits. Title IX deals with clean air research into problems such as acid rain and the health effects of both long and short term air pollution. Title X contains provisions for disadvantaged business concerns. Title X requires EPA to make sure that no less than 10% of total federal funding for clean air research goes to those disadvantaged concerns. Disadvantaged business concerns are those of which women or minorities own a majority interest. Title XI contains provisions for clean air employment transition assistance. These provisions set up a program dealing with employees displaced because of compliance with the Clean Air Act.

SUMMARY

The Clean Air Act Amendments of 1990 set ambitious goals and deadlines for the various provisions. How some of these provisions will be transferred into actual regulations, and how effective these regulations will be, remains to be seen. Only after several years and after the regulations are tested in both the courts and the work environment, will the true impact of the 1990 amendments be known. These amendments have the potential for a far reaching economical impact, not just now, but for years in the future. Most of these provisions and their resulting regulations will translate into expenses for industry. Ultimately, it will be up to those in industry to implement the CAA 1990 in an environmentally responsible manner.

It is impossible for this article (or any other) to be completely up to date. At this very minute, new regulations are being developed and implemented. Industry must keep up with the CAAA and this will be no simple task.

MANAGEMENT SUGGESTIONS

The goal of this section is to provide a set of management suggestions that will help industry to respond to the CAAA in a cost effective manner. Getting into and staying in compliance will be costly and have a negative impact on profits. However, there is no other alternative short of ceasing to exist. Thus, industry will comply; the goal is to do it cost effectively.

Obviously, this is a rapidly changing and individualistic list. Some will be helpful to any industry but all will not apply. Management must be selective. Also, there are other ideas not developed here. This is a start, not a final product.

Management Suggestion 1: IF YOU ARE GOING TO BE AFFECTED BY TITLE III, GO AFTER EARLY REDUCTIONS NOW. Early Reductions buy an additional 6 years and the technology will develop rapidly in those 6 years. For more information, see the Tuesday, December 29, 1992 Federal Register.

Management Suggestion 2: ASSIGN THE COMPLETE CLEAN AIR ACT COMPLIANCE SET OF DUTIES TO ONE PERSON (GROUP) IN YOUR PLANT. The Hazardous Organic NESHAP (HON) alone is approximately two inches thick and it's just begun!!! (Federal Register December 31, 1992). Just knowing what the regulations require is going to be a full time job for some plants. Then, research and vendor contacts will have to be done just to identify alternatives. Someone must be in charge.

Management Suggestion 3: CONDUCT EMISSIONS INVENTORIES NOW. To understand whether a plant complies or not, requires that management knows what all emissions are and what levels are allowed. MS 2 identifies the requirements; but sometimes knowing present emissions is also quite difficult. Management should identify all possible emitters in the plant and conduct complete inventories of all emissions. Measurements must be taken and often fairly complex calculations performed. These should be done now to ensure the proper data is available when needed.

Management Suggestion 4: CONDUCT COMPREHENSIVE CFC REPLACEMENT STUDIES. Almost all industry presently use CFC's as refrigerants. First, these refrigerants are extremely expensive and will become more so leading to an eventual phase out. Alternatives must be found. Prudent management will be conducting replacement studies and will not be bound by existing equipment constraints. For example, some plants might be able to reactivate absorption chillers that were mothballed in recent years due to high energy costs. Absorption systems do not use CFC's; but they do carry an energy penalty. Similarly, the once very popular centralized ammonia system will likely become more popular again. The American Society of Heating, Refrigeration and Air Conditioning Engineers (ASHRAE) and others have conducted numerous studies on CFC replacements including absorption and ammonia systems. This would be a good place to start; but again, new equipment and/or reactivating old are alternatives that should be considered along with different refrigerants. Going to HCFC replacements only buys time; they will also be phased out eventually.

Management Suggestion 5: CONDUCT LEAK DETECTION AND REPAIR STUDIES FOR REFRIGERANTS AND RECYCLE THEM DURING ANY MAINTENANCE ACTIVITY. CFC's are extremely expensive and rapidly becoming more so. Leak detection repair is presently very cost effective. Recycling is absolutely required for any maintenance work involving possible rejection into the atmosphere.

Management Suggestion 6: CONDUCT CFC REPLACEMENT AND EMISSION MINIMIZATION STUDIES FOR ANY CFC'S USED IN DEGREASING. Vapor degreasers are still very popular in industry and likely there will always be some. Frequently, these degreasers use solvents that are Ozone depleters (e.g. 1,1,1,-Trichloroethane). Recently much has been done toward

1) Reducing emissions by better control, equipment redesign, rack redesign, etc.
2) Using industrial strength surfactants (parts washers) as alternatives to vapor degreasers.
3) Using semiaqueous solvents (frequently citrus or terpine based products) as alternatives.

All of these should be considered in a proper management program. There does not seem to be any universal solution; so each plant should study all alternatives and several chemicals within each alternative to find the best for that plant.

Management Suggestion 7: STUDY THE STATE'S AIR PERMIT PROGRAM CAREFULLY AND BE PREPARED TO RESPOND AS NEEDED. Each state will likely have a slightly different program so management must be alert as to what that state requires. Fines are significant; traveling "air cops" will be looking for noncompliant companies; and the "bounty" program will become a money making endeavor for environmental groups and disgruntled employees. Simply stated, industry will be watched very carefully by a host of parties. Management will have to be up to date.

Management Suggestion 8: BECOME FAMILIAR WITH CIVIL AS WELL AS REGULATORY LIABILITY ISSUES. Potential fines were discussed in the main body of this report. They will be significant and early on there will be some fines. Cost effectiveness requires being in compliance. Also, however, management must become more familiar with civil liability issues where all have a responsibility not to harm neighbors.

If a plant does harm a neighbor, then they are liable for that damage and could even have their doors closed due to being a "nuisance". This has long been recognized as a problem in dealing with solid or hazardous waste and surface or ground water contamination. The authors predict the CAAA will draw much more attention to air contamination which will lead to more civil suits involving air.

Management Suggestion 9: IMPLEMENT RISK MANAGEMENT PLANS. The Clean Air Act Amendments require risk management plans (RMP) for any industry with threshold quantities of certain chemicals. These plans entail rather comprehensive "fault tree" or "what if" types of analysis. They are not simple and their execution will require expenditure of time and monies. However, the savings could be truly significant if accidental releases are avoided or minimized. The authors also contend that the methods employed could be used on other chemicals especially those with a propensity to escape into the air and could create significant environmental and/or health problems.

Management Suggestion 10: IMPLEMENT FLEET VEHICLE STUDIES. Plants in certain ozone nonattainment areas will be required to examine alternative fuels. As a result, the technology is rapidly maturing in several areas and state tax credits are being offered. For example, several states offer help through tax credits and even rebates (oil overcharge funds) to convert to compressed natural gas (CNG) or electric vehicles. There are locations where these conversions are quite cost effective today. Numerous fleets in Oklahoma, Texas, and other gas producing states are converting or have converted to CNG with significant savings. Self Help Gas has made gas inexpensive in many areas. CNG and Electric conversions should be studied by all companies with fleets. Some will find the conversion cost effective today. A check with the local energy office will determine what types of state aid is available. Studies are being conducted as to the true environmental and health impacts of these alternative fuels. Management should stay alert as to these results.

SUMMARY AND CONCLUSIONS

One goal of this paper was to identify the direction and present status of the Clean Air Act. This was partially achieved but before the print is dry, the information is out of date. Periodically, this will have to be revised.

Then the paper developed a list of 10 management suggestions intended to help minimize cost to industry. Some of the suggestions should be applicable to any industry and are worthy of consideration. Obviously, this list is incomplete.

Hopefully, the list will plant some ideas that will grow, prosper and lead to other prodegy. The authors hope that the reader will identify other ideas and will share those with other managements and with the authors. Good luck!!! (You will need it).

References

Blattner, J. Wray, Esq., The Clean Air Act Compliance Handbook, Executive Enterprises, New York, NY 1989.

Burke, Robert L. Permitting For Clean Air A Guide To Permitting Under Title V of The Clean Air Act Amendments of 1990, Air and Waste Mgmt. Assoc., Pittsburg, PA, 1992.

Graves, Michael D.; Jones, Randolph L.; Livingood, Mathew G.; Bentley, Royce H.; The Historical Perspective on the Clean Air Act; Hall, Estill, Gable, Golden, & Nelson, P. C., Tulsa, OK, Oklahoma State University, Engineering Extension, Stillwater, OK; 1991.

Graves, Michael D.; Jones, Randolph L.; Livingood, Mathew G.; Bentley, Royce H.; The (New) Clean Air Amendments; Hall, Estill, Gable, Golden, & Nelson, P. C., Tulsa, OK, Oklahoma State University, Engineering Extension, Stillwater, OK; 1991.

Doughty, Dennis G.; The New Clean Air Act from the States Perspective; Oklahoma State Dept. of Health, Air Quality Service, Oklahoma City, OK; Oklahoma State University, Engineering Extension, Stillwater, OK; 1991.

Hosford, R. Blair, "Clean Air Act Title V Knocking on Your Door", Pollution Engineering, Jan. 15, 1993.

Findley, Roger W.; Farber, Daniel A.; Environmental Law in a Nutshell; West Publishing Co., St. Paul Minn., 1988.

Quarles, John; Lewis, William H.; The NEW Clean Air Act, A Guide to the Clean Air Program as Amended in 1990; Morgan, Lewis & Bockius, Washington, D.C.; 1990.

Note: Glossary is available from one of the speakers

Chapter 2

Regulation of Hazardous Air Pollutants Emitted from Fossil-Fired Boilers

P.L. Hendrickson, K.K. Daellenbach

ABSTRACT

The changes made in §112 of the Clean Air Act by the 1990 Amendments to the Act will affect the regulation of hazardous air pollutants (HAPs) emitted by fossil-fired boilers. The 1990 Amendments designated 189 chemicals/compounds as HAPs. Major and area sources of these pollutants in categories designated by the Environmental Protection Agency (EPA) will be subject to emission standards set by EPA. Industrial and institutional/commercial boilers are two such categories of HAPs designated by EPA for which emission standards will be issued. Fossil-fired boilers can emit a variety of HAPs. All or a portion of such emissions that exceed designated thresholds are likely to be regulated. Source owners have the option of participating in the early reduction program before final standards are issued.

STATUTORY BACKGROUND

Hazardous air pollutants (HAPs) were a significant concern of Congress when it amended the Clean Air Act (CAA) in 1990. According to industry estimates, more than 2.4 billion pounds of HAPs were emitted to the atmosphere in 1988 [1, p. 30]. HAPs can cause a variety of adverse health effects including cancer, reproductive effects, birth defects, and respiratory illness [1, p. 30].

Title III of the 1990 CAA Amendments completely replaced §112 which covers HAPs. Before 1990, regulation of HAPs at the federal level was relatively limited. The original §112, enacted as part of the CAA in 1970, required the Environmental Protection Agency (EPA) to identify HAPs "which may reasonably be anticipated to result in an increase in mortality or an increase in serious irreversible, or incapacitating reversible, illness." Once a HAP was identified, EPA had one year to issue a final emission standard that provided "an ample margin of safety" to the public. By 1990, EPA had issued only seven national emission standards for hazardous air pollutants (NESHAPS). The standards cover mercury, beryllium, radionuclides, asbestos, vinyl chloride, benzene, and inorganic arsenic. These standards apply to both new and existing sources, are still in effect, and are codified in Part 61 of Title 40 of the Code of Federal Regulations (CFR).

1990 CLEAN AIR ACT AMENDMENTS

Congress took a different approach to the regulation of HAPs in the 1990 CAA Amendments. Section 112(b) of the CAA now designates 189 chemicals/compounds, including heavy metals and organic compounds, as HAPs. Section 112(a) of the CAA defines a "major source" of HAPs to be a stationary source or group of stationary sources that emits 10 or more tons per year of any single HAP or 25 or more tons per year of any combination of HAPs. "Area sources" are defined to be stationary sources of HAPs that are not major sources.

Section 112(c) of the CAA directs EPA to issue a list of categories and subcategories of major and area sources of HAPs. The initial list includes industrial boilers and institutional/commercial boilers as major source categories [57 Federal Register (FR) 31591, July 16, 1992]. Emission standards for the designated categories and subcategories of major and area sources are to be issued according to the schedule in §112(e) of the CAA. EPA's draft schedule for issuing emission standards for categories of HAP-emitting sources was issued September 24, 1992 (57 FR 44147).[a] According to the draft schedule, emission standards for the industrial and institutional/commercial boilers categories will be issued by November 15, 2000. This is the latest date authorized by §112(e)(1) of the CAA. If EPA does not issue final emission standards by the scheduled due date, the states are to set case-by-case emission limits for every major source that would have been subject to the emission standard set by EPA [CAA, §112(j)]. These state emission limits will be enforced through the Title V operating permit program (see below).

Section 112(d) of the CAA directs EPA to set the emission standards for the categories of major and area sources it identifies under §112(c). The standards are to require the maximum degree of reduction in emissions of the HAPs, taking into consideration the cost of achieving such emission reduction and any non-air quality health and environmental impacts and energy requirements. This standard is commonly referred to as the maximum achievable control technology (MACT) standard. Under §112(d)(3) of the CAA, EPA is to set emission standards requiring new sources to reach at least the level of emission control achieved in practice by the best-controlled similar source. A new source is defined at §112(a)(4) of the CAA as a stationary source for which construction started after EPA issued the emission standard applicable to the source. For existing sources, EPA is to set an emission standard that requires reaching the level of emission control achieved by the best performing 12% of existing sources or the average emission limitation achieved by the best performing five sources in the category [CAA, §112(d)(3)]. Owners of industrial and institutional/commercial boilers (as well as other owners of HAP sources) will want to get involved in EPA's standard-setting process as early as possible to make their views known on an appropriate emission standard under the preceding Congressional guidance.

Because the final emission standards for industrial and institutional/commercial boilers will not be issued for approximately seven years, sources in this category have ample opportunity to participate in the early reduction program authorized by §112(i)(5) of the CAA. Under this program, sources that achieve 90% reduction in HAP emissions (95% for particulates) before the emission standard for the source is first proposed can obtain a permit authorizing the source to operate at the 90%/95% level of control for up to six years after the compliance date of the applicable §112(d) emission standard. The compliance date for existing sources can be up to three years after an emission standard is issued [CAA, §112(i)(3)(A)]. Consequently, a source participating in the early reduction program can potentially operate at the 90%/95% level for up to nine years after the §112(d) MACT standard is issued. Under EPA's draft schedule for issuing emission standards, existing industrial or institutional/commercial boilers whose owners participate in the early reduction program can thus potentially operate at the 90%/95% level until November 15, 2009. A source owner who is considering participating in the early reduction program should compare the present value cost of meeting the early reduction program requirements with the likely MACT emission reduction requirements. Final regulations for the early reduction program were issued December 29, 1992 (57 FR 61970). The regulations will be codified at 40 CFR Part 63, Subpart D. Procedures for participating in the program are included in the final regulations.

Many sources of HAPs will need to have an operating permit issued by EPA or a state permitting authority that has been delegated authority by EPA. Specifically, any major source in an EPA-designated category of major sources or area source in a designated category of area sources will need to have an operating permit [40 CFR 70.3(a)(3)]. The operating permit program was added to the CAA by Title V of the 1990 CAA Amendments. Final regulations for the program were issued by EPA on July 21, 1992 (57 FR 32250) and are codified at 40 CFR Part 70.

An existing source of HAPs that is modified can potentially become subject to a MACT standard even before the standard is issued. After the effective date of a state's operating permit program, a major source of HAPs may not be modified unless EPA or the state determines that the MACT standard for that category of existing sources will be met [CAA, §112(g)(2)(A)]. If the MACT standard has not yet been issued, an appropriate emission standard is to be determined on a case-by-case basis. A modification is defined at §112(a)(5) of the CAA as "any physical change in, or change in the method of operation of, a major source which increases the actual emission of any HAP emitted by such source by more than a de minimis amount or which results in the emission of any HAP not previously emitted by more than a de minimis amount." State operating permit programs are likely to become effective beginning in 1995 [CAA, §502(d)]. Owners of sources of HAPs that will be subject to a §112(d) MACT standard will want to consider carefully the implications of any changes to their

(a) EPA's most recent Regulatory Agenda indicates that a final schedule will be issued in October 1993 (58 FR 25079, April 26, 1993).

facility that constitute a modification once the operating permit program becomes effective in their state.

Under §112(l) of the CAA, states can apply to EPA for delegated authority to implement and enforce the requirements in §112 of the CAA. States can apply for complete or partial delegation of EPA authority. State emission standards established under this delegated authority must be at least as stringent as EPA's standards. EPA issued a proposed rule on May 19, 1993 (58 FR 29226) to provide guidance to the states regarding EPA approval of rules or programs that states can implement and enforce in place of certain requirements under §112 of the CAA.

All of the preceding discussion covers regulation of HAPs under the federal CAA. Owners of sources of HAPs must also be aware that many states regulate the emission of HAPs under state laws. Under §116 of the CAA, states can regulate stationary sources of air pollution provided the state requirements are at least as stringent as federal requirements. States have been particularly active in regulating HAP sources because the federal program has been so slow to develop. State programs vary widely in their scope and application. Some states regulate more than the 189 chemicals/compounds designated in §112(b) of the CAA and apply their program to sources that would not be major sources under §112 of the CAA. A common approach to setting standards is for the state to establish maximum acceptable ambient concentrations of particular pollutants [2, §17.01]. Using this standard as a basis, the state regulatory authority establishes emission limits to be imposed on individual sources.

APPLICATION TO FOSSIL-FIRED BOILERS

A common industrial process with the potential to emit HAPs is operation of fossil-fired boilers. Thirty-seven of the 189 HAPs designated in §112(b) of the CAA have been detected in flue gas from fossil power plants [3, p. 10]. These 37 HAPs are listed in Table 1. Any of these 37 HAPs could also be present in the emissions from fossil-fired boilers.

There is considerable uncertainty regarding what particular HAPs are emitted from fossil-fired boilers and their rate of emission. Reasons for this uncertainty include 1) lack of universal quality assurance procedures and methodologies during sampling and testing, 2) wide variability in the trace elements found in coal, and 3) variations in boiler design and controls. Also, emissions data for coal-fired utility boilers are largely based on calculations from test burns under carefully controlled conditions [4].

TABLE 1. HAZARDOUS AIR POLLUTANTS FOUND IN FOSSIL POWER PLANT FLUE GASES

Acetaldehyde
Antimony compounds
Arsenic compounds
Benzo-a-pyrene
Benzene
Beryllium compounds
Biphenyl
Bis-(2-ethylhexyl)-phthalate
Cadmium compounds
Carbon disulfide
Carbon tetrachloride
Carbonyl sulfide
Chlorine
Chlorobenzene
Chloroform
Chromium compounds
Cobalt compounds
Dibenzofurans
1,4-Dichlorobenzene (p)
Formaldehyde
Hexachlorobenzene
Hydrochloric acid
Hydrofluoric acid
Lead compounds
Manganese compounds
Mercury compounds
Naphthalene
Nickel compounds
Pentachlorophenol
Phenol
Phosphorus
Selenium compounds
2,3,7,8-Tetrachlorodibenzo-p-dioxin
Tetrachloroethylene
Toluene
Trichloroethylene
2,4,5-Trichlorophenol

Tables 2, 3, and 4 present average emission factors for oil-fired, stoker-fired, and pulverized coal-fired boilers, respectively. The data in the tables should be considered with the limitations noted above in mind. Much of the data in Tables 2, 3, and 4 are from a 1986 report prepared for EPA by the Radian Corporation [5,6]. Because the emission factors for each boiler type are generally

TABLE 2. AVERAGE EMISSION FACTORS FOR OIL-FIRED BOILERS

Hazardous Air Pollutant	Average Emission Factors ($lbm/10^{12}$ Btu of Heat Input)	
	Residual Oil	Distillate Oil
Arsenic	19	4.2
Beryllium	4.2	2.5
Cadmium	15.7	10.5
Chromium	21	48
Copper	280	280
Mercury	3.2	3.0
Manganese	26	14
Nickel	1260	170
Polycyclic Organic Matter	8.4	22.5
Formaldehyde	405	405

independent of capacity (Btu/hr), the emission factors could apply to fossil-fired boilers used for residential, commercial/institutional, industrial, or utility applications. The emission factors indicate the level of emissions for a currently operating plant (pounds of HAPs per 10^{12} Btu of fuel burned) in various operating modes including no control of emissions, i.e., "uncontrolled," as well as selected emission control equipment for the coal-fired boilers. Stoker-fired boilers typically have steam capacities less than 100,000 pounds of steam per hour. Pulverized coal (PC) boilers generally have capacities greater than 100,000 pounds of steam per hour [7]. Average emission factors for stoker and PC coal boilers are shown in Tables 3 and 4, respectively.

The quality of coal varies greatly among geographical areas in the United States. Bituminous, lignite, and anthracite coal are included in Table 4. These coal types have higher heating values (HHV) of 13,770 Btu/lbm; 7,194 Btu/lbm; and 12,700 Btu/lbm, respectively. Tables 3 and 4 break out average emission factor values for these coal types.

Controlling the emissions shown in Tables 2, 3, and 4 to a MACT emission standard could involve significant expenditures. Emission control equipment to meet MACT standards could include 1) cold-side electrostatic precipitators, 2) fabric filters (conventional and pulse-jet), 3) spray dry and wet lime/limestone flue gas desulfurization scrubbers, 4) low NO_x burners, 5) post-combustion NO_x systems, or 6) emerging technologies such as alumina adsorbents for mercury removal [8]. Once again, the quality and quantity of data concerning the effectiveness of emission control equipment is limited. Furthermore, certain of these control technologies may require that a sorbent or other combustion-modifying chemicals be added to the furnace or flue gases. The effect of such additions on the nature and quantity of HAP emissions is not fully known [4].

SUMMARY

The federal program for regulating emission of HAPs was significantly altered by the 1990 Amendments to the Clean Air Act. Over the next seven years, EPA will be issuing emission standards for a number of source categories of HAPs. These standards will require the application of MACT to HAP emissions. Two common source categories which will be subject to MACT standards are industrial and institutional/commercial boilers. Under EPA's proposed schedule, the MACT standards for these categories are to be issued by the end of the year 2000.

Owners of industrial and institutional/commercial boilers can take several steps before the final emission standards are issued. First, they can participate in EPA's standard setting process. Second, they can elect to participate in EPA's early reduction program to delay the time when compliance with the MACT standard is required. Third, they should carefully consider any planned modifications to a facility that might subject the facility to a MACT standard set by a state in advance of the EPA-set standard.

TABLE 3. AVERAGE EMISSION FACTORS FOR STOKER-FIRED BOILERS AND SELECTED EMISSION CONTROL EQUIPMENT

Hazardous Air Pollutant	Average Emission Factors (lbm/10^{12} Btu of Heat Input) for Stoker-Fired Boilers								
	Uncontrolled Emissions			Multicyclone			Electrostatic Precipitator		
	Bitm.	Lign.	Anth.	Bitm.	Lign.	Anth.	Bitm.	Lign.	Anth.
Arsenic	650	1,320	250	320	650	125	80	165	30
Beryllium	75	120	45	30	45	20	6	10	4
Cadmium	50	55	15	30	35	8	15	15	3
Chromium	1,255	1,505	2,975	725	925	1,720	360	460	850
Copper	905	1,590	990	535	940	585	135	240	150
Mercury	15	20	20	15	20	20	10	15	15
Manganese	2,170	11,800	2,230	600	3,290	630	250	1,380	260
Nickel	1,030	930	1,795	520	470	910	280	250	490
Uranium-238[a]	NA[b]	NA	NA	NA	NA	NA	15	15	15
Thorium-232[a]	NA	NA	NA	NA	NA	NA	15	15	15
Polycyclic Organic Matter (POM)	NA	NA	NA	30	30	30	30	30	30
Formaldehyde	710	710	710	NA	NA	NA	NA	NA	NA
Lead	NA	NA	NA	1,410	220	1,420	50	5	NA

Notes:
(a) Emissions for uranium-238 and thorium-232 are measured in picoCuries per 10^6 Btu of heat input.
(b) NA indicates that emission factor information was not found.

TABLE 4. AVERAGE EMISSION FACTORS FOR PULVERIZED COAL BOILERS AND SELECTED EMISSION CONTROL EQUIPMENT

	Average Emissions Factors (lbm/10^{12} Btu of Heat Input) for Pulverized Coal Boilers											
	Uncontrolled Emissions			Multicyclone			Electrostatic Precipitator			Flue Gas Desulf. (Scrubber)		
Hazardous Air Pollutant	Bitm.	Lign.	Anth.	Bitm	Lign.	Anth.	Bitm.	Lign.	Anth.	Bitm.	Lign.	Anth.
Arsenic	1,010	2,060	395	500	1,010	195	105	215	40	50	95	20
Beryllium	80	130	50	50	85	30	3	5	2	0.1	0.2	0.1
Cadmium	55	65	15	40	45	10	15	15	3	1	1	0.3
Chromium	1,295	1,550	3,070	750	890	1,775	370	440	875	110	130	250
Copper	710	1,250	780	420	740	460	140	100	150	25	40	25
Mercury	15	20	20	15	20	20	10	15	15	4	5	5
Manganese	1,895	10,330	1,950	880	4,815	910	410	2,230	420	35	195	40
Nickel	1,065	660	1,850	540	485	940	290	260	500	40	35	70
Uranium-238[a]	NA[b]	NA	NA	NA	NA	NA	295	295	295	75	75	75
Thorium-232[a]	NA	NA	NA	NA	NA	NA	170	170	170	35	35	35
Polycyclic Organic Matter (POM)	NA	NA	NA	20	20	20	20	20	20	20	20	20
Formaldehyde	110	110	110	NA	NA	NA	NA	NA	NA	NA	NA	NA
Lead	315	NA	NA	130	155	NA	55	5	90	20	NA	NA

Notes:
(a) Emissions for uranium-238 and thorium-232 are measured in picoCuries per 10^6 Btu of heat input.
(b) NA indicates that emissions factor information was not found.

ACKNOWLEDGMENT

Preparation of this paper was supported by the U.S. Department of Energy (DOE) under Contract DE-AC06-76RLO 1830. Pacific Northwest Laboratory is operated for DOE by Battelle Memorial Institute.

REFERENCES

1. U.S. Environmental Protection Agency. January 1991. Implementation Strategies for the Clean Air Act Amendments of 1990. EPA Office of Air and Radiation, Washington, D.C.

2. D. R. Selmi and K. A. Manaster. 1992. State Environmental Law. Clark Boardman Callaghan, Deerfield, Illinois.

3. D. Boutacoff. March 1991. "New Focus on Air Toxics." EPRI Journal. 16(2):4-13. Electric Power Research Institute, Palo Alto, California.

4. T. D. Brown, C. E. Schmidt, and A. S. Radziwon. November 1991. "Comprehensive Assessment of Toxic Emissions from Coal-Fired Power Plants." CONF-911139-3. Proceedings of the Managing Hazardous Air Pollutants: State of the Art Conference, Washington, D.C.

5. R. C. Mead, B. K. Post, and G. W. Brooks. July 1986. Summary of Trace Emissions from and Recommendations of Risk Assessment Methodologies for Coal and Oil Combustion Sources. DCN No. 86-203-024-41-10. Radian Corporation, Research Triangle Park, North Carolina.

6. U.S. Environmental Protection Agency (EPA). May 1992. Air Chief CD-ROM User's Manual, Version 2.0 Beta. EPA-450/4-92-015. Research Triangle Park, North Carolina.

7. E. A. Avallone and T. Baumeister III. 1978. Mark's Standard Handbook for Mechanical Engineers, Ninth Edition. McGraw-Hill Book Co., New York.

8. W. Chow, M. J. Miller, J. Fortune, G. Behrens, and E. Rubin. January 1991. "Managing Air Toxics Under the New Clean Air Act Amendments." Power Engineering 95(1):30-34.

Chapter 3

Phase-Out of CFC Usage in Industrial Refrigeration Equipment

W.E. Sullivan, B.G. Beach

ABSTRACT

After January 1, 1996, with limited exceptions, refrigerants containing chlorofluorocarbons (CFCs) will no longer be produced in the United States. A production phase-out of hydrochlorofluorocarbons (HCFCs) will begin January 1, 2003 resulting in a complete ban on production by January 1, 2030. Federal regulations placed into effect at various times in 1992 and 1993 make it illegal to knowingly vent CFCs or HCFCs to the environment and require users of CFCs and HCFCs to meet strict EPA regulations concerning servicing, maintenance and disposal of refrigeration and air conditioning equipment. This is all in response to a significant environmental threat, depletion of the earth's ozone layer and will have an immediate impact on companies who use CFCs in their air conditioning and refrigeration systems. Following is a brief summary of current regulations applying to industrial refrigeration and air conditioning systems, their impact on users of CFCs and HCFCs and a recommended course of action. Servicing of motor vehicle air conditioners, restrictions on non-essential products and labeling requirements for ozone-depleting substances are outside the scope of this paper and will not be discussed.

DEFINITIONS

Appliance: Any device which contains and uses a class I or class II substance as a refrigerant and which is used for household or commercial purposes, including any air conditioner, refrigerator, chiller or freezer.

Class I Refrigerant: Refrigerant composed of fully halogenated chloroflourocarbons (CFCs). Commonly used refrigerants included in this category are R-11, R-12, R-111, R-114, R-500 and R-502. Class I refrigerants are commonly used in chiller systems.

Class II Refrigerant: Refrigerant composed of partially halogenated chlorine containing halocarbons (HCFCs). R-22 and R-123 are examples of class II refrigerants. Because of it's high volumetric efficiency, R-22 has been widely applied for air conditioning units and refrigeration systems.

Commercial Refrigeration: Refrigeration appliances utilized in the retail food and cold storage warehouse sectors.

Montreal Protocol: Formally known as the Montreal Protocol on Substances that Deplete the Ozone Layer, represents an international agreement to impose limits on the production and consumption of certain ozone depleting substances. The original Protocol was agreed to by the United States and 22 other countries in 1987. Parties to the Protocol met again in London in 1990 and in Copenhagen in 1992 to approve amendments to the original agreement. The agreement is now signed by over 92 nations who represent over 90% of the world's consumption of CFCs and halons.

Reclaim: to reprocess refrigerant to at least the purity specified in ARI Standard 700 and to verify this purity using analytical methodology prescribed in ARI 700. Reclamation typically involves processes and procedures only available at a reprocessing or manufacturing plant.

Recover: to remove refrigerant in any condition from an appliance without necessarily testing or processing it in any way.

Recycle: to extract refrigerant from an appliance and clean the refrigerant for reuse by oil separation and other mechanical means without chemical analysis. Recycling can typically be implemented at the field job site.

Small Appliance: Certain products manufactured, charged and hermetically sealed in a factory with five pounds or less of refrigerant. Typical of this group are home refrigerators, freezers, air conditioners and heat pumps.

BACKGROUND & PHASE-OUT SCHEDULE

CFC refrigerants released to the environment during the use, maintenance and disposal of refrigeration and air conditioning appliances have contributed significantly to the accelerated depletion of the earth's protective ozone layer. In 1990, the Federal Government responded to this challenge by adopting sweeping changes to the Clean Air Act of 1977. Title VI of the Clean Air Act, originally promulgated in the Montreal Protocol of 1987, established a phase-out schedule for substances known to deplete the stratospheric ozone layer. Starting in 1991 the production of CFCs would be limited to a fixed percentage of 1986 production levels ending with a complete production phase-out by the year 2000. A complete production phase-out of HCFCs, which are much less damaging to the ozone layer, would be implemented by the year 2030. However, in response to renegotiation of the Montreal Protocol in Copenhagen in November of 1992 and petitions by several prominent environmental groups the EPA accelerated it's phase-out schedule to require a complete production phase-out for CFCs by January 1, 1996. The proposed phase-out schedule for HCFCs was also accelerated on a compound-specific basis. HCFCs with the greatest ozone depletion potential will be phased-out as early as January 1, 2003. Proposed phase-out schedules issued in March of 1993 for CFCs and HCFCs are included as Table 1 and Table 2.

OVERVIEW OF PROPOSED REGULATIONS

Authority for enforcement of the Clean Air Act is the responsibility of the Environmental Protection Agency. The EPA has instituted regulations under Title 40 Part 82 of the Code of Federal Regulations. Production phase-out schedules for Class I and Class II substances are issued under Subpart A. We have already discussed the phase-out schedule. Regulations for recycling and emission reduction are issued under Subpart F. Following is a brief summary of the required practices issued under Subpart F for recycling and emission reduction dated April 23, 1993. The EPA has made certain exceptions for small appliances and motor vehicle air conditioners (MVACS). Persons should refer to Subpart F for clarification of those exceptions and for specific regulations.

- Effective June 14, 1993, no person performing maintenance, service, repair or disposal activities on appliances shall knowingly release into the environment any class I or class II substance used as a refrigerant in such equipment.
- Effective July 13, 1993, all persons opening appliances except for MVACs for maintenance, service or repair must evacuate the refrigerant to a system receiver, recovery or recycling machine properly certified in accordance with 40 CFR 82 to the level of evacuation as specified by Subpart F.
- Effective July 13, 1993, all persons opening appliances except for small appliances and MVACs for maintenance, service, repair or disposal must have at least one piece of certified, self-contained recovery equipment available at their place of business.
- Effective November 15, 1993, recovery and recycling equipment must be certified by an approved testing organization to meet the requirements of Subpart F.
- Effective July 13, 1993, persons who take the final step in the disposal process of an appliance must recover remaining refrigerant and evacuate the appliance to the levels specified by Subpart F or verify that the refrigerant has been evacuated.
- Effective July 13, 1993, owners of commercial refrigeration and industrial process refrigeration must maintain leakage rates below 35% of the total charge during a twelve month period or develop a one-year retrofit or replacement plan for the leaking equipment within 30 days. The limit for leakage from appliances not considered as commercial refrigeration or industrial process refrigeration containing more than 50 pounds of refrigerant is 15% of the appliances total charge.
- Effective November 14, 1994, persons who maintain, service, or repair appliances except MVACs, and persons who dispose of appliances, except for small appliances, room air conditioners, and MVACs, must be certified by an approved technician certification program.

INCREASED FEDERAL TAXES

In 1991, under 26 CFR 52 the Internal Revenue Service issued final regulations pertaining to the tax on ozone depleting chemicals (ODCs). The regulations relate to the environmental taxes imposed by chapter 38 of the Internal Revenue code. The amount of the tax is determined by the weight in pounds of the ODC and the ODC's ozone depletion potential. The tax on refrigerants containing CFCs will increase yearly to encourage a rapid changeover to environmentally safe refrigerants.

IMPACT ON USERS OF CFCs

The recently imposed federal regulations will have an immediate impact on users of CFCs. As production levels drop to zero in 1996, users of CFCs will find it increasingly difficult to procure adequate quantities of CFCs to maintain refrigeration and cooling equipment. Prices for CFC refrigerants have already gone up significantly. At the time of this writing the price for a pound of R-12 had increased to almost eight dollars. A special tax on the sale of CFC refrigerants will more than triple between 1991 and 1999. As tax levels go up and availability goes down, the situation will get much worse.

IMPACT ON USERS OF HCFCs

The phase-out schedule for commonly used HCFCs will not be implemented for several years. HCFCs are still being recommended by the Environmental Protection Agency as an alternative to CFCs. For many smaller applications, equipment manufacturers have not developed equipment to utilize completely non-ozone depleting substances to take the place of HCFCs. HCFC usage will continue until cost effective alternatives are developed.

RECOMMENDED COURSE OF ACTION

There are several steps that companies can take to minimize the impact of increased refrigerant costs and start the transition to environmentally safe refrigerants. The EPA's Stratospheric Ozone Protection Action Guide recommends the following course of action:

- Designate a Facility Refrigerant Manager who is responsible for tracking regulatory and technical developments. This is extremely important as the regulatory environment is constantly changing, new environmentally safe refrigerants are still being developed and pricing for CFC refrigerants may change drastically within a very short period of time.
- Conduct an inventory of refrigeration and air conditioning equipment and refrigerants. Refrigeration equipment should be categorized by type and quantity of refrigerant used. Refrigerant inventories should be established and monitored.
- Perform a review of maintenance practices and procedures to determine the extent of refrigerant recovery, recycling and reclaiming efforts. This effort should coincide with a review of refrigerant usage to determine compliance with federal limits on refrigerant losses and required maintenance practices as outlined by subpart F to 40CFR82.
- Develop a plan for phasing-out CFCs. Companies which use CFCs in air conditioning or refrigeration systems should prepare now to minimize the economic impact of substantially higher CFC prices resulting from limited production and increased federal taxes. Decisions to replace or retrofit existing equipment should be based on equipment condition, maintenance history and expected life. Priority should be given to equipment with the largest inventories of CFC refrigerants.

ALTERNATIVES TO CFCs

There are currently a limited number of acceptable substitutes for CFC based refrigerants. Acceptable substitutes must provide an acceptable level of ozone depletion potential, flammability, toxicity, exposure potential, and their use in refrigeration and air conditioning equipment must be economically and technically feasible. Evaluating alternatives could become a significant challenge. Fortunately, the 1990 Amendments to the Clean Air Act require the EPA to evaluate substitutes for ozone depleting substances. Alternatives are reviewed through the EPA's Significant Alternatives Policy Program, referred to as "SNAP". SNAP will provide a list of acceptable and unacceptable alternative refrigerants, including a list of substitutes for which requests are pending, updated quarterly. Substitutes that pose a greater overall risk to human health and the environment relative to other alternatives will be restricted. Tables 3 and 4 are reproductions of the SNAP list made available by an EPA fact sheet in February of 1993. They provide a list of proposed acceptable refrigerants suitable for chiller and commercial refrigeration retrofits and new equipment.

ALTERNATIVES TO HCFCs

The EPA list of recommended refrigerants still includes several HCFCs as alternatives to CFCs in refrigeration and air conditioning equipment. At the time of this writing, SNAP has not yet published a list of environmentally safe substitutes for HCFCs. For many smaller applications, equipment manufacturers have not developed equipment to utilize completely non-ozone depleting substances to take the place of HCFCs.

RETROFIT VS. NEW EQUIPMENT

Once the decision has been made to replace a CFC refrigerant with a more environmentally friendly substitute the decision will have to be made to retrofit or replace the existing equipment. For large chiller systems which still have a considerable useful life the best solution may be to retrofit the machine for use with one of the new environmentally safe refrigerants. For a retrofit to be successful, it is important that the thermodynamic properties of the new refrigerant closely approximate that of the refrigerant being replaced. Figure 1 compares the thermodynamic properties of HFC-134A with those of CFC-12, a commonly used Class I refrigerant. HFC-134A is a chlorine-free non ozone depleting fluorinated refrigerant designed to replace CFC-12. It was originally designed for automotive air conditioning systems and domestic refrigeration. It is acceptable as a CFC-12 substitute for medium and high temperature applications.

DESIGN CONSIDERATIONS

Original equipment manufacturers should be consulted for special design requirements before considering a retrofit. Following is a sample of design considerations which must be evaluated for a successful changeout:

- Compatibility of lubricants
- Expansion valve sizing
- Filter dryer type

- Verification of safety control setpoints
- Operating setpoint changes
- Pressure regulator and pilot operated solenoid valve sizing
- Compatibility of seals and gaskets

CONCLUSION

There may be doubts about global warming and the need to control ozone depleting substances but there can be no doubt about the impact of recent federal legislation on the use of CFCs. CFCs are rapidly becoming very expensive and will soon become very scarce. Persons who choose to continue using CFCs will have to pay the price including the additional costs associated with certification of equipment and technicians. The time to develop a plan is now.

REFERENCES

1. 40 CFR 82, EPA STRATOSPHERIC PROTECTION REGULATIONS

2. 29 CFR 52, IRS REGULATIONS ON ENVIRONMENTAL TAXES FOR CERTAIN CHEMICALS AND HAZARDOUS WASTE

3. COPELAND, APPLICATION ENGINEERING BULLETIN AE-1295, DECEMBER 1992

4. EPA STRATOSPHERIC OZONE PROTECTION ACTION GUIDE, EPA-430-F-93-003, COOLING AND REFRIGERATING WITHOUT CFCs

5. EPA STRATOSPHERIC OZONE PROTECTION FACT SHEET, EPA-430-F-93-004, SHORT LIST OF ALTERNATIVE REFRIGERANTS

6. EPA STRATOSPHERIC OZONE PROTECTION RESOURCES, EPA-430-F-93-005, AIR CONDITIONING AND REFRIGERATION

7. TRANE AIR CONDITIONING MANUAL, COPYRIGHT 1965

8. ASHRAE HANDBOOK FOR REFRIGERATION SYSTEMS AND APPLICATIONS, COPYRIGHT 1990

TABLE 1
PHASE OUT SCHEDULE FOR CFC REFRIGERANTS

PERCENT (FOR YEAR BEGINNING JANUARY 1) [1]		
1994	1995	1996
25%	15%	0%

[1] Allowed percent of baseline production and consumption

TABLE 2
PHASE OUT SCHEDULE FOR HCFC REFRIGERANTS

DATE	AFFECTED COMPOUND	RESTRICTION
Jan 1, 2003	HCFC-141b	Ban on production and consumption
Jan 1, 2010	HCFC-22, HCFC-142b	Production and consumption frozen at baseline levels
	HCFC-22, HCFC-142b	Ban on use of virgin chemical [1]
Jan 1, 2015	all other HCFCs	Production and consumption frozen at baseline levels
	all other HCFCs	Ban on use of virgin chemical [1]
Jan 1, 2020	HCFC-22, HCFC-142b	Ban on production and consumption
Jan 1, 2030	all other HCFCs	Ban on production and consumption

[1] Unless used as feedstock or refrigerant in appliances manufactured prior to January 1, 2010

TABLE 3
PROPOSED ACCEPTABLE REFRIGERANT ALTERNATIVES FOR CHILLERS

Applications	Alternatives for Retrofitting Equipment			Alternatives for New Equipment				
	HCFC 123	HCFC 22/ HFC 152a HCFC 124 blends	HFC 134a	Ammonia systems*	HCFC 123	HCFC 22	HFC 134a	Lithium/ bromide/ water absorption blends
CFC 11 Centrifugal	•			•	•	•	•	•
CFC 12 Centrifugal			•	•	•	•	•	•
CFC 12 Reciprocating			•			•	•	
CFC 500 Centrifugal		•	•	•		•	•	

*ASHRAE Standard 15-1992 sets special requirements for systems using ammonia-containing refrigerants.

TABLE 4
PROPOSED ACCEPTABLE REFRIGERANT ALTERNATIVES FOR REFRIGERATION EQUIPMENT

Applications	Alternatives for Retrofitting Equipment				Alternatives for New Equipment			
	HCFC 22	HCFC 22/ HFC 152a HCFC 124 blends	HCFC 22/ propane/ HFC 125 blends	HFC 134a	Ammonia systems*	HCFC 22	HCFC 22/ propane/ HFC 125 blends	HFC 134a
CFC 12 Cold Storage Warehouses	●	●		●	●	●		●
CFC 12 Commercial Ice Machines		●			●	●		●
CFC 12 Industrial Process Refrigeration	●	●		●	●	●		●
CFC 12 Refrigerated Transport		●		●		●		●
CFC 12 Retail Food	●	●		●	●	●		●
CFC 12 Vending Machines	●	●		●		●		●
CFC 500 Refrigerated Transport		●	●	●		●	●	●
CFC 502 Cold Storage Warehouses	●		●		●	●	●	
CFC 502 Industrial Process Refrigeration	●		●	●	●	●	●	●
CFC 502 Refrigerated Transport	●		●	●		●	●	●
CFC 502 Retail Food	●		●		●	●	●	
CFC 502 Commercial Ice Machines					●	●		

*ASHRAE Standard 15-1992 sets special equipment room requirements for systems using ammonia-containing refrigerants.

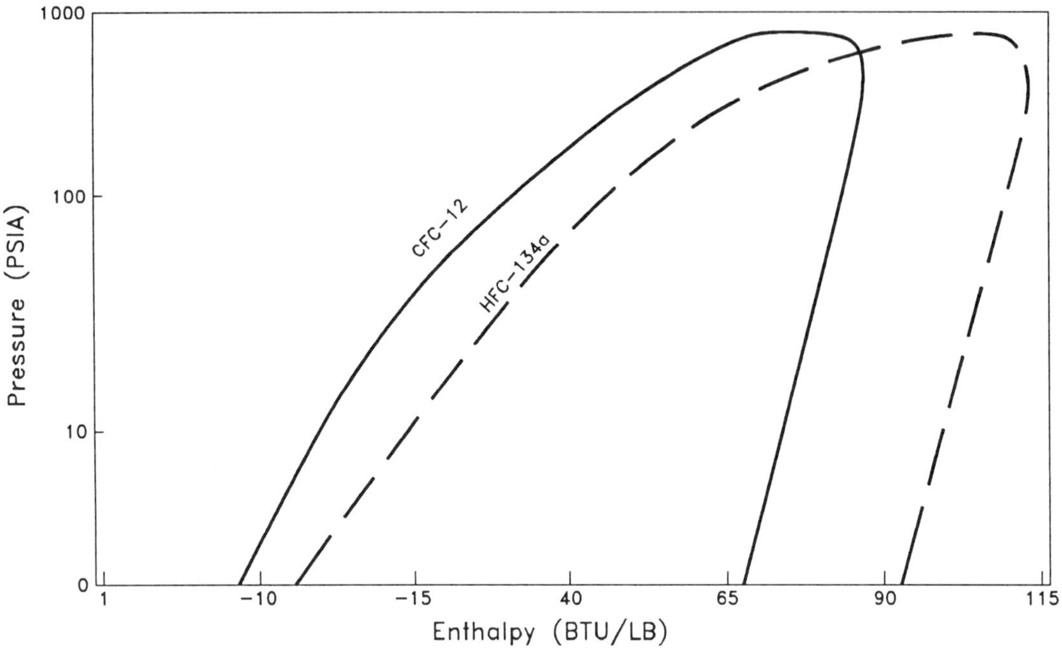

FIGURE 1
PRESSURE - ENTHALPY DIAGRAM
HFC-134a and CFC-12

Chapter 4
RCRA Regulatory Developments

A.J. Tolman

ABSTRACT

The nine years following the extensive 1984 amendments to the Resource Conservation and Recovery Act (RCRA) have witnessed a constant stream of RCRA rule developments, largely as a result of statutory mandates and deadlines imposed by Congress. Oftentimes EPA missed these deadlines as it struggled to match its resources to the requirements imposed by both new and ongoing programs. A number of lawsuits by environmental interest groups ensued, resulting in court-imposed deadlines for rule issuance.

While the past two years have seen more of the same, the RCRA program has also had to react during 1992 and 1993 to the unexpected impact of major judicial decisions that have undercut significant principles and provisions of EPA's RCRA regulations. This paper discusses these and other important RCRA developments, including current judicial and EPA trends in enforcing RCRA.

LAND DISPOSAL RESTRICTIONS

EPA's "Third-Third" Rule

One of the most important RCRA developments occurring during the past year has involved EPA's implementation of the final portion of the "land ban" imposed by the 1984 Amendments to the Resource Conservation and Recovery Act (RCRA).[1] The so-called land ban prohibited land disposal of hazardous wastes unless they were first treated to applicable treatment standards to be established by EPA.[2] In a progression of events that was extremely unsettling to American companies, portions of EPA's "third-third" land disposal restrictions implementing the land ban on characteristic hazardous waste--the largest volume of hazardous wastes generated by U.S. industry--were overturned by a federal appeals court in late 1992, and then partially reinstated by EPA on an emergency basis in May of 1993.

Under a schedule imposed by RCRA, EPA first imposed land ban restrictions ("LDRs") on untreated wastes containing dioxin (F020-23 and F026-28) and solvents (F001-05), effective in November of 1986.[3] Next, the "California list" wastes were banned from land disposal effective in July of 1987, unless they were first treated to the EPA-promulgated standards.[4] Finally, all other hazardous wastes were to be ranked and dealt with one-third at a time.[5] EPA imposed LDRs on the highest priority wastes in August 1988 and on the second group in June of 1989.[6]

The "third-third" of the ranked wastes included all "characteristic" hazardous wastes, which are deemed hazardous based on their exhibiting the characteristic of reactivity, corrosivity, ignitability, or toxicity. They comprise the largest volume (more than 50%) of the hazardous wastes generated each year in the United States. These wastes became subject to EPA's third-third LDRs in May of 1990.[7]

Both industry representatives and several environmental organizations challenged various aspects of EPA's third-third LDRs in a lawsuit filed in the D.C. Circuit Court of Appeals (the court designated by RCRA to determine the validity of EPA rules promulgated under RCRA.[8]) The industry petitioners challenged provisions of the rule requiring treatment beyond the point at which a characteristic waste no longer exhibits the hazardous characteristic. The environmental groups challenged provisions of the LDRs allowing deactivation by dilution, as opposed to treatment with specified technologies, prior to land disposal. They also challenged EPA's authority to allow underground injection of untreated, formerly characteristic hazardous wastes, or placement of such wastes in surface impoundments regulated under the Clean Water Act.

The D.C. Circuit Court of Appeals issued its decision in September of 1992.[9] Dealing a major setback to industry, the appeals court ruled that EPA has the authority to regulate a hazardous waste "beyond the removal of the attribute" that caused the waste to be classified as hazardous. However, the decision also caused considerable consternation on EPA's part. The court invalidated portions of EPA's third-third LDR, indicating that EPA must go further than the rule presently requires in certain cases to reduce the amount of hazardous constituents present.

The appeals court ruled that dilution of characteristic hazardous wastes *may* constitute applicable treatment if no hazardous constituents remain at levels that would endanger human health or the environment. According to the court, however, EPA had generally not shown that the dilution standard would accomplish this result. The court found the existing deactivation, or dilution, standard to be "permissible only in the case of corrosive wastes," and then only if EPA demonstrates that the deactivated wastes do not still contain hazardous constituents posing a threat to human health or the environment. Accordingly, the court "vacated" (voided) the dilution treatment standard for ignitable wastes until EPA can "identify the ignitable wastes that include, after dilution, sufficiently high levels of hazardous constituents to pose a risk to human health or the environment," and propose an appropriate treatment method. The court also vacated EPA's dilution treatment standard for reactive wastes because it did not demonstrate that dilution would eliminate the risk of a violent reaction.

Other portions of the rule were not vacated (so they remained in effect) but were "remanded" to the Agency to be corrected through future rulemaking. For instance, the court decided that diluting wastes in Clean Water Act (CWA) facilities is acceptable only if the toxicity of waste discharged from an impoundment is minimized or eliminated consistent with RCRA provisions. Underground injection may be used for hazardous waste disposal only if the hazardous characteristics have been eliminated and health and environmental dangers minimized. This will require EPA to promulgate methods for integrating RCRA and CWA/SDWA requirements, including a method for demonstrating that the waste treated in a CWA treatment system has been subjected to the equivalent of treatment required under RCRA.

Invalidation of much of EPA's third-third LDR was potentially very disruptive to industry because it meant that no treatment standard existed that would allow legal land disposal of large volumes of hazardous wastes generated by industry. The next several months following the court's decision saw extensive judicial maneuverings, as proponents on both sides of the case sought reversal or refinement of the decision. The court denied EPA's request for rehearing but stayed the decision until January 5, 1993. Before that date, the Chemical Manufacturers' Association petitioned the Supreme Court for review of the case, further delaying the issuance of the court's mandate.

On April 26 of this year, the Supreme Court refused to review the appeals court's decision, setting the stage for EPA's emergency issuance of a rule governing treatment of characteristic hazardous wastes. The "interim final rule" was issued as expected, on an emergency basis, effective May 10, 1990.[10] The rule is intentionally narrow in scope, addressing only the treatment standards for ignitable and corrosive wastes that were vacated (as opposed to remanded) by the D.C. Circuit Court of Appeals.

The revised rule retains the deactivation requirement for certain ignitable and corrosive wastes, but adds numerical treatment standards for about 200 regulated hazardous constituents contained in the waste. EPA was unable, in the emergency rulemaking, to establish levels of treatment standards "at which threats to human health and the environment are minimized." According to EPA, it will continue to evaluate various approaches for setting such minimization of threat levels and propose them in future rulemakings. In issuing the emergency rule, EPA granted a 90-day national capacity variance to allow industry time to implement the treatment technologies.

The remanded standards, not addressed by the interim final rule, cover many more facilities and hazardous waste sources than do the vacated and revised standards. For instance, the interim standards affect about 75 facilities that treat wastes on site and about 275 facilities that receive wastes from off site. For both categories, incineration is the most common form of treatment. The interim final rule also affects about 100 Class V injection wells. By contrast, more than 6 billion gallons of hazardous waste are disposed of yearly in Class I injection wells, which are covered by the still effective remanded standards.

Treatment Standards for Soil and Debris

Both contaminated soils and contaminated debris present unique problems under the LDR program because they are regulated as hazardous wastes under the controversial "mixture" rule (discussed below). Separate treatment standards have not been established for wastes contained in other materials. Therefore, the treatment standard for soils and debris is based on the hazardous waste they contain rather than the soil-waste matrix.[11] Applying the incineration treatment standard to soil or debris poses obvious and significant technical problems. It also hinders the initiation of voluntary cleanup operations.

Recognizing this problem, EPA has announced its intention to issue a separate set of treatment standards for hazardous soils based on the use of alternative technologies such as soil washing, thermal desorption and biodegradation.[12] A final hazardous debris rule, containing treatment standards for debris contaminated with listed or characteristic hazardous wastes that are prohibited under the land disposal ban, was published on August 18, 1992, with an effective date of November 16, 1992.[13]

The final contaminated debris rule provides that such debris must be treated prior to land disposal by using one or more of three treatment technologies: extraction, destruction, or immobilization. Moreover, hazardous debris that is treated by extraction or destruction, and no longer exhibits a hazardous characteristic, will no longer be subject to regulation as a hazardous waste. The rule establishes a new waste management unit --a "containment building"--that will allow storage of non-liquid hazardous wastes in buildings for up to 90 days without such storage being considered land disposal.

EPA estimated that the debris rule would save industry an estimated $600 million annually by reducing unneeded treatment for low-risk debris and by exempting treated debris from further RCRA regulation. The debris rule is an incomplete measure, however, as long as the companion rule designating treatment and disposal methods for contaminated soil remains unissued. The contaminated soil rule has been delayed due to the controversy surrounding the "third-third" land disposal rules. Even the contaminated debris rules may be changed as a result of the final resolution of these regulations.

Capacity Variances For Soil and Debris

The 1984 RCRA Amendments gave EPA limited authority to grant an extension of up to two years of the land disposal restrictions for specific hazardous wastes based on a lack of currently available alternative treatment or disposal capacity. EPA can also grant a one-year extension, renewable once, that is a "case-by-case capacity variance" based on an individual company's demonstration of need and certain other prerequisites.[14]

In the final rule establishing LDRs for third-third wastes, a two-year "national capacity variance," extending the LDR deadline to May 8, 1992, was granted for contaminated soil whose best demonstrated available technology (BDAT) is incineration, retorting or vitrification.[15] The variance did not include soils contaminated with listed solvents, dioxin, or "California list" wastes.[16] The national capacity variance expired on May 8, 1992. In a rule published in October of last year, EPA granted a case-by-case variance to these same contaminated soils.[17] EPA effectively extended the national capacity variance, because the rule did not require individual variance applications. Rather, it required companies desiring coverage by the extension to meet certain record keeping requirements. This variance expired on May 8, 1993, and EPA declined to grant any further extension of the ostensible case-by-case variance for the contaminated soils.[18]

A similar process was followed for certain debris contaminated with hazardous waste, except that EPA was willing to extend the variance to the maximum allowable by law. Having already employed the two-year national capacity variance and a one-year ostensible case-by-case capacity variance,[19] EPA recently extended the case-by-case capacity variance for contaminated debris to May 8, 1994.[20] Although EPA confirmed that a lack of treatment capacity for hazardous debris continues to exist, it concluded that "some capacity may exist, at least for some forms of debris."[21] Therefore, to qualify for the extension, generators of such waste are required to submit a report to EPA demonstrating a good-faith effort to locate treatment capacity.[22]

No further variance for contaminated soils appears forthcoming, and, under RCRA's provisions, no further variance or extension of the LDR effective date can be given for contaminated debris.

"No Migration" Treatment Variances

The RCRA land ban program provides for a variance from land disposal restrictions if the subject land disposal facility submits a petition to EPA demonstrating that there will be "no migration" of hazardous constituents from a particular hazardous waste disposal unit for as long as the waste remains hazardous.[23] In a proposal published August 11, 1992, EPA set forth its interpretation of the "no migration" variance.[24] EPA's proposed rule interprets "no migration" as not exceeding health-based levels at the boundary of the land disposal unit in any environmental medium, that is, in groundwater, surface water, soil or air. The proposal also includes new procedural and substantive requirements for petitioning EPA and demonstrating "no migration," along with EPA's criteria for evaluating such petitions.[25]

HAZARDOUS WASTE IDENTIFICATION

The Mixture and Derived From Rules

In late 1991, the U.S. Court of Appeals for the District of Columbia Circuit found EPA's mixture and derived from rules to be invalid on procedural grounds.[26] As a result, in early 1992 EPA issued the mixture and derived from rules on an emergency basis as an "interim final rule," effective on February 18, 1992.[27] They are still in effect today.

The so-called "mixture rule" classifies as hazardous any solid waste containing a hazardous constituent listed in 40 CFR §§ 260.20 and 260.22.[28] The "derived from rule" defines as hazardous all residues generated from treating or disposing of hazardous wastes, regardless of whether they show hazardous characteristics.[29] The mixture and derived from rules have required millions of tons of arguably contaminated soil, wastewater and incinerator ash to be handled and disposed as hazardous.

These rules have been central to EPA's hazardous waste program for over a decade. EPA originally proposed RCRA regulations governing management of hazardous wastes at the end of 1978. The final rules, containing both the mixture and derived-from provisions, were issued in May 1980.[30] A number of lawsuits were filed challenging the RCRA regulations. Although most were settled, several that could not be resolved were consolidated and ruled on by the appeals court in its December 1991 decision.

In its lawsuit, Shell Oil had claimed that the mixture rule was not included for public comment in the proposed rules and was therefore not promulgated in accordance with the Administrative Procedure Act (APA). The derived from rule was also missing from the proposed regulations. The court noted that the APA does not require that a final rule mirror the draft version, but it must at least be a "logical outgrowth" of the version issued for public comment. The court found that both the mixture and derived from rules had dramatically changed the proposal and were not logical outgrowths of the proposed regulations. Therefore, proper notice of the rules had not been given, and they were ruled invalid.

The court's decision threw EPA's RCRA program into turmoil, because the elimination of these rules could cause up to forty percent of all RCRA "hazardous" waste to be classified as non-hazardous. EPA quickly asked the appeals court to reconsider its decision and to clarify that its decision would have no impact on previous regulatory and enforcement actions that were based on violations of the mixture and derived from rules. Although the court denied EPA's request for clarification, EPA immediately pronounced that the court's decision was *not* retroactive. In the meantime, EPA issued its "interim final rule," reinstating both key provisions.[31] This rule will remain in effect indefinitely while EPA searches for an effective alternative.[32]

In the first decision on the issue following the December 1991 invalidation of the mixture and derived from rules by the D.C. Court of Appeals, another federal court held in June of 1992 that EPA may *not* rely on the RCRA mixture rule to support criminal and civil enforcement actions initiated before the rule's invalidation.[33] The Eighth Circuit Court of Appeals ordered a new trial for an aircraft maintenance company and its owner because the jury may have based its decision to convict them for unlawful disposal of hazardous waste on the invalidated rule.

In its opinion, the Eighth Circuit held that the December 1991 decision invalidating the mixture and derived from rules *should apply retroactively*. The court found that the reinstatement and current validity of the rules did not alter the fact that they were previously invalid. The Eighth Circuit's decision appeared to have the potential to jeopardize years of EPA enforcement actions, both civil and criminal, that were based on violations of these rules. The decision's effect was limited, however, by the fact that many convictions and enforcement actions were based on the state RCRA laws. The 45 states authorized by EPA to administer RCRA are required to have laws at least as stringent as the federal program. Thus the programs in the RCRA authorized states were not immediately affected by the court's ruling and, both before and after reinstatement of the federal rule, continued enforcement as usual.

The New Hazardous Waste Identification Rules

On May 20, 1992, EPA proposed a new Hazardous Waste Identification Rule (HWIR) to replace the mixture and derived from rules.[34] The proposed HWIR outlined two approaches--Concentration Based Exemption Criteria ("CBEC") and the Expanded Characteristics Option ("ECHO")--for excluding low-risk wastes from stringent federal hazardous waste controls. CBEC would have exempted wastes falling below a certain toxicity threshold, and ECHO would have made a set of toxicity characteristics the sole factor in determining whether a waste is deemed hazardous. While industry showed some support for the CBEC option, environmental groups, several states, and the Association of State and Territorial Solid Waste Management Officials indicated strong opposition to either of the HWIR options. Responding to the flood of criticism, EPA withdrew the HWIR on October 30, 1992.[35]

After EPA withdrew its controversial HWIR from consideration, it convened a series of meetings ("roundtable discussions") with representatives of business and industry, environmental groups, and other interested parties aimed at reaching a consensus on an approach to categorizing contaminated media. A workgroup that is focusing on options for capturing wastes not currently regulated under RCRA Subtitle C was, as of June 1993, leaning toward the use of a concentration-based trigger, that is, identifying waste as hazardous by using the presence of specific chemical constituents at levels greater than a designated concentration.[36] Another workgroup, that is focusing on contaminated media, has developed a state versus national tiered approach whereby contaminated media falling below a certain risk assessment level (a "bright line") would, as low-risk wastes, be managed under requirements set by the state agency. Contaminated media falling above the bright line would come under a national technology-based system aimed at achieving risk reduction.[37] The hazardous waste discussion group has several more months of work ahead to iron out the details before their efforts can take the form of a proposed rule.

THE CAMU RULE

In early 1993, EPA issued a final RCRA rule that allows hazardous waste facilities subject to RCRA corrective action more flexibility in the use of alternative cleanup technologies for wastes that are handled as part of an overall site remediation project.[38] The Corrective Action Management Unit (CAMU) rule, signed on January 19 and effective April 19, 1993, was separated out from a larger RCRA corrective action rule, proposed in 1990.[39]

The CAMU rule allows non-contiguous areas of contamination at the same site to be designated as a single waste management unit. This means that wastes can be moved for treatment within the site and handled within the total corrective action management unit without being subject to RCRA land disposal regulations. The rule also provides for a "temporary" waste management unit ("TU") that may operate for up to 180 days under limited waste management standards. The CAMU rule recognizes that the remediation of existing contamination is inherently different from the management of as-generated industrial wastes, and that applying strict Subtitle C requirements to remediation wastes does not always result in implementation of the best remedies.

The rule imposes predictable limitations. Wastes transported to the facility from an outside source will not be considered remediation wastes for the facility and cannot be managed in a CAMU or TU at the facility. Neither can wastes that are excavated at the facility, transported to an off-site treatment facility, and returned to the facility be considered remediation wastes manageable in a CAMU. Finally, a CAMU designation is not automatic upon request. The regulatory agency determines its applicability following consideration of the individual characteristics of the site and certain decision criteria contained in the rule.

EPA has estimated that CAMUs will be used at approximately 1500 RCRA facilities nationwide to remediate roughly 6,000 solid waste management units (SWMUs) EPA also expects the CAMU rule to result in a remediation cost savings of $16.6 billion to $26.6 billion, largely due to the decrease in the use of off-site incineration and disposal.

On May 14, 1993, the Environmental Defense Fund, Hazardous Waste Treatment Council and Natural Resources Defense Council sued EPA over the CAMU rule. The groups alleged that the rule's standards are not protective enough and that the rule does not provide for adequate public involvement.

Used Oil Will Not Be Listed As A Hazardous Waste

In mid-1992, EPA issued as a final rule that it will not list "used oil destined for disposal" as a hazardous waste under RCRA.[40] This maintained the current regulatory status of used oil. However, used oil destined for disposal will continue to be subject to management as a hazardous waste if it is a "characteristic" hazardous waste, that is, if the toxicity characteristic contaminants in 40 CFR § 261.24 exceed the specified maximum levels.

EPA later issued a final determination not to list recycled used oil as a hazardous waste.[41] EPA also issued management standards for handling used oil that are intended to promote recycling while safeguarding health and the environment. The standards, which are promulgated under Section 3014(a) of RCRA, are codified in a new Part 279 of 40 CFR.[42] The standards apply to all used oil generators, transporters, processors, re-refiners, and off-specification used oil burners. Although generators comprise the largest segment of the used oil recycling industry, the most stringent standards apply to used oil processors and re-refiners.

RCRA ENFORCEMENT TRENDS

Developments In RCRA Criminal Liability

Every principal environmental statute, including RCRA, contains provisions for criminal liability. Traditionally, culpability (such as wrongful intent, willful misconduct) has been a necessary element of a criminal offense and must be proven to establish guilt for a crime. Congress and the courts appear willing to deal more harshly with wrongdoers when their actions violate statutes dealing with protection of human health and safety. In addition, it is assumed that companies that are highly regulated under RCRA should and must have knowledge of the statute's extensive regulatory requirements.

A "knowing" violation or failure to comply is a crime under nearly all of the federal environmental statutes, including RCRA. The courts have interpreted the "knowledge" requirement liberally, so that one can be charged with knowing commission of violation if the violator *should* have known or if the knowledge can be implied by circumstantial evidence. RCRA examples follow.

Section 3008(d)(1) of RCRA[43] imposes criminal liability on anyone who "knowingly transports...any hazardous waste...to a facility which does not have a permit." Section 3008(d)(2)[44] imposes criminal liability on anyone who "knowingly treats, stores, or disposes of any hazardous waste...without a permit..." There is no question that the government prosecutor must prove that the defendant knew he performed the act of transport, treatment, storage or disposal. The question that has perplexed the courts is whether the prosecutor must also prove that the defendant knew that the act was a violation, that is, that the facility lacked a required RCRA permit.

In decisions issued this past year, two federal appellate courts came to opposite conclusions about whether, in order to obtain a criminal conviction under RCRA, prosecutors must prove that a defendant knew that a waste facility lacked a RCRA permit. The Ninth Circuit Court of Appeals held that a waste generator must know that a receiving facility does not have a permit to be convicted of knowingly transporting hazardous waste to an unpermitted facility.[45] The Sixth Circuit Court of Appeals held that the government need not prove that the defendant knew he lacked a required RCRA permit when he treated, stored or disposed of hazardous waste.[46]

The Sixth Circuit concluded that the word "knowingly" in §3002(d)(2) modifies only "treats," "stores" and "disposes of." It does not also modify "without a permit." Therefore, knowledge of the lack of a permit is not required. The court supported its conclusion on policy grounds; if a defendant could claim ignorance of the permit requirement and avoid prosecution, the force of RCRA's statutory scheme would be significantly weakened.

The Ninth Circuit case dealt with Section 3009(d)(1) of RCRA. In holding that the government under that provision must prove knowledge of the lack of a RCRA permit, the Ninth Circuit focused on the difference between the defendant knowing about the permit status of his own site and that of another site to which he sends his waste. The court thought that the government should not have to prove that a defendant knew that his own site lacked a permit. On the other hand, the government should have to prove that a defendant knew that a facility to which he was transporting his waste lacked a permit because he could have been deceived by the receiving facility concerning its permit status.

In a case involving illegal transportation of waste, the U.S. Court of Appeals for the Eleventh Circuit affirmed the criminal conviction of a Georgia man for knowing violations of RCRA.[47] The Eleventh Circuit held that, in a criminal prosecution for illegal transportation and storage of hazardous waste, the government does not have to prove that the defendant knew that a chemical had been identified as a "hazardous waste"; it only has to prove that he knew that the chemical(s) had the potential to harm others or the environment.

In the Eleventh Circuit case, the defendant was charged with transporting drums of chemicals from one place of business to another business he owned. The drums were stored for a time at the defendant's warehouse, which was not permitted as a RCRA storage facility. Later, other individuals dumped the drums in a wooded area. The discarded drums were eventually discovered and reported to EPA, which removed and disposed of them at a cost exceeding $200,000. A number of the drums contained chemicals classified as hazardous under RCRA due to their ignitability.

The defendant was convicted for transporting hazardous waste to unpermitted facilities in violation of Section 3008(d)(1), and for storing hazardous waste without a permit in violation of Section 3008(d)(2)(A). In upholding the conviction, the appeals court held that the government only had to prove that the defendant knew of "the general hazardous character" of the chemical.

Summary of EPA's 1992 Enforcement Efforts

EPA recently released its "Enforcement Accomplishments Report for FY 1992."[48] This report details enforcement actions by Region and by environmental program ("media"). According to the report, civil judicial and administrative penalties assessed by the Agency totaled $78,733,341 for Fiscal Year 1992 (October 1991 through September 1992). These numbers include enforcement actions in all media including hazardous waste management. Civil penalties assessed under RCRA totaled $13,586,122.

Criminal enforcement is the fastest growing component of EPA's enforcement effort. By the end of FY 1992, EPA had increased the number of criminal agents by nearly 40 % over the FY 1988 level. The enforcement report bears out EPA's increased emphasis on criminal prosecutions. In the criminal area, RCRA penalties for 1992 totaled $22.9 million. Criminal fines for FY 1992 total $66.9 million for all media (before deducting suspended sentences). This is a four-fold increase from FY 1991 and the highest amount ever in one year. In fact, 59% of all criminal fines in EPA's history were assessed in FY 1992. Ninety-four years of incarceration were imposed (before suspension), a significant increase from the 75 years imposed in FY 1991.

It is important to note that EPA is well into implementing its "cross-program/multi-media" enforcement initiative. This means that, wherever appropriate, EPA is attempting to bring enforcement actions against facilities that address violations in more than one program or environmental medium. Toward that end, EPA is developing cross-program/multi-media training courses for its compliance inspectors and its technical and legal staffs. These courses are also being made available to state environmental programs and legal personnel. In the near future, some facilities will face the daunting prospect of inspections from state/EPA teams comprised of representatives from several different enforcement programs.

Notes

1. 42 U.S.C. §§ 6901-6992k.

2. 42 U.S.C. § 6924.

3. 42 U.S.C. § 6924(e) and 40 CFR §§ 268.30 and 268.31.

4. 42 U.S.C. § 6924(d) and 40 CFR § 268.32.

5. 42 U.S.C. § 6924(g).

6. *See* 40 CFR §§ 268.33 and 268.34.

7. 42 U.S.C. § 6924(g)(4)(C) and 40 CFR § 268.35.

8. *See* 42 U.S.C. § 6976(a)(1).

9. *Chemical Waste Management v. EPA,* 976 F.2d (D.C. Cir. 1992).

10. 58 FR 29860 (May 24, 1993).

11. 57 FR 47773 (October 20, 1992).

12. *See* 56 FR 55160, 551-72-77 (October 24, 1991).

13. 57 FR 37194-37282 (August 18, 1992).

14. 42 U.S.C. § 6924(h).

15. *See* 55 FR 22520 (June 1, 1990).

16. *Id.*

17. 57 FR 47772 (October 20, 1992).

18. *See* 58 FR 28510 (May 14, 1993).

19. *See* 57 FR 20766 (May 15, 1992).

20. 58 FR 28506 (May 14, 1993).

21. *Id.*

22. 58 FR 28510 (May 14, 1993).

23. *See* 42 U.S.C. § 6924(i) and 40 CFR § 268.6.

24. 57 FR 35940 (August 11, 1992).

25. EPA also announced the availability of a draft guidance manual entitled "No Migration Variances to the Hazardous Waste Land Disposal Prohibitions: A Guidance Manual for Petitioners," U.S. EPA, Draft, July 1992. *Id.*

26. *Shell Oil v. EPA*, 950 F.2d 741 (D.C. Cir. 1991).

27. 57 FR 7628 (March 3, 1992).

28. 40 CFR § 261.3(a)(2)(iv).

29. 40 CFR § 261.3(c).

30. 45 FR 33095 (May 19, 1980).

31. 57 FR 7628 (March 3, 1992).

32. The rules were reinstated with an April 28, 1993, expiration date. Last fall, however, EPA lifted the sunset provision to allow the rules to remain in effect indefinitely. 57 FR 49278 (October 30, 1992).

33. *United States v. Goodner Bros. Aircraft*, 966 F.2d 380 (8th Cir. 1992); *cert. denied*, 113 S. Ct. 967 (U.S. 1993).

34. 57 FR 21450 (May 20, 1992).

35. 57 FR 49280 (October 30, 1992).

36. Environmental Policy Alert, June 9, 1993, at 12-13.

37. Environmental Policy Alert, May 26, 1993, at 16-17.

38. 58 FR 8658 (February 16, 1993).

39. 55 FR 30798 (July 27, 1990).

40. 57 FR 21524 (May 20, 1992).

41. 57 FR 41566 (Sept. 10, 1992).

42. *Id.*

43. 42 U.S.C. § 6928(d)(1).

44. 42 U.S.C. § 3008(d)(2).

45. *United States v. Speach*, 968 F.2d 795 (9th Cir. 1992).

46. *United States v. Dean*, 969 F.2d 187 (6th Cir. 1992).

47. *United States v. Goldsmith*, 978 F.2d 643 (11th Cir., 1992).

48. U.S. EPA, Office of Enforcement, EPA 230-R-93-001, April 1993.

Chapter 5

Corrective Action Management Units (CAMUs)

R.J. Pulfrey

ABSTRACT

The U.S. Environmental Protection Agency has promulgated final regulations effective April 19, 1993 on Corrective Action Management Units (CAMUs) as part of the process to finalize the RCRA Corrective Action Subpart S rules. The CAMUs are a key element in the remediation of wastes under RCRA authority continued in Sections 3004 (permits), 3008 (compliance), and 7003 (imminent danger), and for handling and remediating RCRA hazardous wastes under Superfund (CERCLA) authority. The final rules on CAMUs are fully intended to aid owner/operators undergoing remediation by providing greater flexibility, more opportunity for onsite treatment, greater use of innovative technologies and less use of offsite incineration and/or land disposal. In issuing these rules regarding CAMUs, the agency recognized the impact of Land Disposal Requirements (LDRs) in limiting viable options for the remedial selection process, and has adopted a decision criteria for the CAMUs based on technical and engineering solutions. The CAMU rules are most applicable to management of soils from both Solid Waste Management Units (SWMU) and regulated units as well as offsite soils/sediments affected by past practices.

The criteria in deciding the appropriateness of the CAMU is based primarily on technical considerations, reduced exposure, expediting remediation, enhancing long-term effectiveness, public acceptance, protectiveness of human health and environment and cost-effectiveness factors. The USEPA in issuing these regulations consider these rules as less stringent leaving the decision making to the regional offices. However, in doing so, fully authorized Hazardous and Solid Waste Amendment states may choose to not adopt these regulations.

BACKGROUND AND PURPOSE

A mechanism was needed by the U.S. Environmental Protection Agency (USEPA) for a more expedient cleanup of hazardous wastes or hazardous constituents in soil, groundwater, surface water or sediment at Resource, Conservation and Recovery Act (RCRA) facilities undergoing corrective action. In 1990, USEPA had proposed RCRA corrective action regulations for those facilities conducting investigations, evaluating potential remedies, and selecting and implementing corrective action remedies. These proposed RCRA corrective action regulations were modeled after the revised National Contingency Plan, and were also known as the new Subpart S regulations to Title 40 Code of Federal Regulations (CFR) Part 264. A remedial waste management unit known as the Corrective Action Management Unit (CAMU) was introduced as part of these proposed RCRA corrective action requirements for the sole purpose of managing RCRA wastes in a single land-based unit. As proposed, the area encompassed by this CAMU unit would be exempt from the current in-force land disposal restrictions (LDRs) and minimum technology requirements (MTRs) to provide expeditious and cost-effective remediation of hazardous wastes and constituents.

USEPA announced on October 22, 1992 in the Federal Register the availability of a report entitled "Supplemental Information on Corrective Action Management Units" that presented and compared the regulatory, remedial, and cost aspects of three regulatory options for the proposed CAMU: (1) eliminating the CAMUs, (2) enacting CAMU as proposed in 1990, or (3) implementing a more flexible expanded approach to a CAMU. The conclusion reached in the report was that the more flexible, expanded CAMU would accomplish the intended purpose of a more expeditious and cost-effective remediation by (1) promoting the use of more innovative technologies depending less on incineration, (2) reducing the use of less effective technologies (*in situ* treatment), (3) increasing waste treatment at the facility, thereby decreasing the practice of transporting wastes for offsite incineration, (4) allowing more hazardous waste to be excavated and remediated *ex situ*, and (5) decreasing the use of capping waste in place without treatment. According to the report, the estimated cost saving of using the expanded CAMU approach in lieu of a no CAMU approach was $17-27 billion. Given the pressure on USEPA to expedite remediation, the expanded CAMU option was adopted and incorporated into the final CAMU rules issued on February 16, 1993 and became effective April 16, 1993.

INTRODUCTION

USEPA has issued final regulations establishing a regulatory procedure for the sole purpose of remediating past releases of hazardous wastes and constituents at facilities governed by Subtitle C of RCRA. The new management mechanism under the final regulations is the CAMU. The new regulations define the CAMU as an area within a facility that is designated by the USEPA Regional Administrator (under Title 40 CFR Part 264 Subpart S) for the sole purpose of implementing correction action requirements for facilities under permit or under interim status. As defined in the rules, "A CAMU shall be used for the management of remediation wastes pursuant to implementing such corrective action requirements at the facility." Thus as stated, the corrective action requirements (issued in the Federal Register, V.58, No.29, p.8683), under which this rule was promulgated, pertain only to past environmental releases, not wastes that are currently being generated from ongoing industrial processes.

TECHNICAL AND ENGINEERING APPROACH

The largest change in philosophy and approach between the proposed CAMU rules and the final rule can be found in the definition of a CAMU. The proposed rule defined a CAMU as a contiguous area within the facility which is contaminated by hazardous waste (including hazardous constituents like groundwater). However, the final rule does not specify CAMUs as being contiguous areas of contamination nor does the final rule specify a CAMU needs to be restricted to areas solely which have been contaminated, that is, CAMU boundaries will not be set according to the presence of contamination. The final rule definition does specify that CAMUs will be used for purposes of managing only remediation waste and, in that respect, can only be used for corrective action under interim status 3008(h) Administrative Orders or at permitted facilities [under Section 3004(u) or 3004(v)].

In essence, the definition of a CAMU relates to the function and purpose a CAMU will serve, rather than to the location and contiguous nature of the wastes to be remediated. The expanded, flexible rules imply an engineering approach to managing wastes based on technical considerations. By specifying how and where wastes are to be managed, USEPA is looking for a sound, well-engineered, technical approach to remediation of wastes. This approach is reflected in the decision criteria for designating CAMUs.

CRITERIA OF A CAMU

The owner/operator of the RCRA facility must demonstrate to the Regional Administrator (RA) and the public that the following criteria for CAMU designation have been met:

- CAMU must result in expediting the remediation;
- CAMU must facilitate a reliable, efficient, and cost-effective remedy;
- CAMU must facilitate use of treatment technologies (long-term containment structures are discouraged) resulting in a reduction of toxicity, mobility, or volume (with no preference to one over another);
- CAMU must minimize area of contamination after closure (consolidation);
- If waste remains in CAMU after closure, it must be managed to minimize future releases, in essence, forcing a decision as to the long-term reliability and effectiveness of the remedy; and
- CAMU must be managed in a way to be protective of human health and the environment and preclude minimizing short-term risks due to implementation.

If the CAMU is approved by the regional administrator (RA), it will be incorporated into a RCRA decision document called the Statement of Basis for public review.

These criteria, although needing sufficient proof to support the decision, do not require a quantitative risk assessment to prove whether or not the CAMU will be protective of human health and the environment, or a detailed cost/benefit analysis to prove that the remedy is cost effective. A qualitative analysis is sufficient to receive approval. However, the Statement of Basis will be reviewed by the public for comment. In this sense, the rationale and analysis in arriving at the conclusion must be sufficient to support the Statement of Basis decision.

ANALYSIS OF CAMU REGULATIONS

A critical component in the definition of the CAMU is defining the term "facility," which, as defined in the new

CAMU regulations for purposes of implementing RCRA corrective action, means all contiguous property under the control of the owner/operator seeking a permit under Subtitle C of RCRA Section 3004(u) or applying to interim status facilities implementing corrective action under RCRA Section 3008(h). This definition is thus restricted to those facilities managing waste generated as a result of implementation of corrective action. The final CAMU regulations also define remediation wastes as "...all solid and hazardous wastes, and all media (includes groundwater, surface, soil, and sediment) including debris that contains listed hazardous wastes or which themselves exhibit a hazardous waste characteristic, that are managed at a facility for the purpose of implementing correction action (under permit or interim status) as long as those remediation wastes have originated from within the facility boundary." Otherwise, excavating or extracting remediation wastes and placing them into a land-based unit either at the facility or offsite that has not been designated a CAMU will need to meet LDRs and MTRs. The CAMU rule also applies to RCRA-regulated units such as landfills, surface impoundments, and waste piles that received hazardous waste after July 26, 1982. Therefore, waste moved from a regulated unit for remediation purposes can be conducted without meeting LDR and MTR restrictions if moved to a designated CAMU. Likewise, moving and consolidating wastes within a designated CAMU will not be subject to LDR or MTR requirements, or moving remediation wastes from one CAMU to another or from a CAMU onsite treatment unit such as a bioremediation tank will not trigger additional Subtitle C RCRA requirements.

Under Subtitle C regulatory rules for remediation of RCRA waste the LDRs and MTRs imposed a distinct disincentive to remediating waste. Essentially, the LDR imposed treatment standards prior to remediated wastes being returned to a land-based unit such as a waste pile, surface impoundment, etc. In addition, the MTR imposed criteria that land-based management units such as waste piles and surface impoundments were required to meet prior to receiving wastes. More often than not, the LDR standards imposed treatment by incineration, a very expensive method of remediation and substantial offsite transportation costs.

However, the newly enacted regulatory purpose of the CAMU is to establish a separate regulatory framework for managing and remediating past releases of RCRA hazardous waste, including provisions made for waste released beyond the facility boundary. Thus in those instances where contaminated groundwater, surface water, sediment, soil (and debris) outside the facility boundary can be proven to be associated with a release from within the facility, or facility hazardous constituents have migrated beyond the facility boundary, the wastes can be managed under the CAMU regulatory framework. Wastes generated as part of a RCRA Facility Investigation (RFI) are also considered remediation wastes under this CAMU rule, for the facility is undergoing corrective action. What the CAMU regulations do, in effect, is provide an incentive to excavate and remediate waste *ex situ* at the facility. For instance, a source of contaminated soil that is excavated within a designed and approved CAMU and stored would not have to meet the full RCRA Subtitle C regulations of LDR and MTR prior to treatment. Without a CAMU, at a minimum, the excavated soil would have to meet LDR standards, since it would become a regulated waste pile, and a permit or interim status is then required under RCRA regulations. Applying for a permit is a very arduous time-consuming paperwork task. Meeting the LDR and permit regulations have proven to hinder timely cleanup of wastes under the current RCRA Subtitle C program. These RCRA Subtitle C regulations were intended to apply to as generated hazardous wastes, that is, wastes generated as part of a facility's operation, but the CAMU regulations are applied to past releases. Thus, the overall purpose of the CAMU is to allow onsite treatment in a clearly defined management area in order to bring about a reduction in toxicity, mobility, volume and land area of waste regardless of where the contamination exists.

Although the CAMU regulations constitute a critical management component in RCRA Subpart S Corrective Action regulations, the applicable regulations have not been finalized to determine which cleanup standards wastes must meet in order to be considered "clean." Therefore, although the CAMU regulations explicitly state how to manage remediation wastes, no standards are as yet available to say how clean the waste must be before final disposal. The only exceptions are the groundwater protection standards which have been established under the corrective action process of a permit or interim status.

PROCESS

A CAMU designation for a facility can be acquired by the owner/operator two ways: (1) under a Section 3008(h) administrative order for interim status facilities, or (2) under a permit. USEPA can also use the CAMU regulations to remedy an eminent danger situation under Section 7003. Final approval in designating a CAMU is authorized to the USEPA RA or, in the case of a state authorized by USEPA to administer RCRA corrective action, the state hazardous waste program director. The owner/operator requesting the CAMU designation must specify:

- The areal extent and configuration of the CAMU (being a land-based unit, it must be designated by actual physical boundaries on map, preferably a facility map with SWMUs located);
- The specific design, operating, and maintenance requirements of managing the waste for treatment, storage (including containment), and disposal; and
- Groundwater monitoring, closure, post-closure and financial assurance requirements.

To obtain a CAMU designation under interim status, a modification to a 3008(h) Administrative Order is possible, as it is in a permit modification. These modifications are also possible early in the corrective action process, should interim measures need to be implemented providing for an expedient reduction in risk to human health and the environment prior to a final remedy.

To receive a CAMU designation under permit, two procedures are possible. The owner/operator may request a Class III permit modification under 40 CFR 270.42, or the permit modification may be processed under an USEPA-initiated permit modification under 40 CFR 270.41. If, however, the owner/operator disagrees with the USEPA decision on how the CAMU will be designated, an administrative appeal may be filed to contest the decision with the provisions not becoming effective until a decision is rendered. Because the CAMU was first approved by the RA, the appeal will be decided by the USEPA Environmental Appeals Board.

However, this CAMU process does not preclude the owner/operator of an interim status facility from seeking another mechanism (such as the closure plan approval process) or allowing an accelerated cleanup to proceed outside the control of an enforcement order.

As a result of this new final rule on CAMUs, one or more CAMUs may be designated for a facility and may also incorporate RCRA-regulated units. Under RCRA rules, regulated units are land-based units such as landfills, surface impoundments, and waste piles that received hazardous waste after July 26, 1982. They would continue to be regulated under provisions of 40 CFR Parts 264 and 265, but existing units can be subject to the less stringent CAMU standards if two criteria are met:

1. The regulated unit is closed or closing; and

2. Inclusion of the units will facilitate effective, protective and reliable remedial actions for the facility.

Nevertheless, the requirements for groundwater monitoring, closure, post-closure, and financial assurance for regulated units as stipulated in 40 CFR Parts 264 and 265 would still apply. Placing a tank or any other non-land-based unit in a CAMU is not considered part of the CAMU and therefore must comply with all applicable Subtitle C standards, thus maintaining a separate regulatory identity (Subpart T Standards for Tanks). Under the new CAMU rules, the USEPA designates such a unit as a temporary unit or TU.

PROGRAMS AFFECTED

Besides affecting the RCRA program, the CAMU regulations are applicable to those instances where RCRA hazardous wastes are managed at a CERCLA site. The CAMU regulations for RCRA waste become applicable or relevant and appropriate requirements (ARARs). As typical with all ARARs, the CAMU requirements for RCRA hazardous wastes and constituents would be incorporated into the CERCLA Record of Decision (ROD), the equivalent to the RCRA Statement of Basis. The incorporation of the CAMU into the CERCLA program is also expected to provide expeditious and cost-effective remedies in remediating CERCLA sites since it precludes LDR and MTR requirements that have been a substantial roadblock to remediating waste to date at CERCLA sites.

The CAMU program is applicable to all state programs except in those states that are authorized for the Hazardous and Solid Waste Amendment (HSWA) corrective action program, which applies to about 15 states. The CAMU rules do not apply in those states because the regulations are less restrictive for then-existing corrective action requirements previously adopted by HSWA-authorized states. In the effort to provide flexibility and expeditious cleanups under their hazardous waste program, states have a number of options. One option is to modify the program by submitting to USEPA according to the state implementation program. Another is using an authority comparable to Section 7003 or authority under a state Superfund program.

By adopting this more flexible, expanded approach to CAMU regulations, it represents, for the first time, a less restrictive program, which is a very large change in philosophy indeed.

Chapter 6

Billion-Dollar Environmental Projects Update

T.C. Walker, R.K. Miller

ABSTRACT

Twenty-five (25) environmental projects in the United States with projected expenditures of $1 billion or more have been identified. These projects include those in the feasibility stage and in active construction.

REMEDIATION

Eight (8) remediation projects with projected expenditures of $1 billion or more have been identified within the United States. They are:

• The U.S. Army's Aberdeen Proving Ground (MD) remediation project is in the assessment phase. The site is so contaminated that it is potentially uncleanable and could become a national sacrifice zone.

• Boston Harbor Cleanup, Island and Boston Harbor, Boston, MA, is estimated to cost approximately $6.1 billion. To comply with the 1972 Clean Water Act requirements, The Massachusetts Water Resources Authority will spend about $6.1 billion to clean up Boston Harbor. To make the harbor safe for fishing and swimming requires an additional $2 billion. A $1.2 billion wastewater treatment plant is under construction on Deer Island to process the 500 million gallons of waste being dumped into the harbor every day.

• Fernald Environmental Management Project (Fernald, OH) is estimated to cost $2.9 billion. During its thirty-six of years of operation, the Feed Materials Production Center (FMPC), recently-renamed the Fernald Environmental Management Project (FEMP), generated large quantities of radioactive, hazardous, and mixed wastes. After preliminary characterization studies in 1986 and the cessation of production in 1989, environmental remediation at the site is now in full swing. In September 1992, DoE announced the selection of Fluor Daniel Environmental Restoration Management Corporation (Irvine, CA) as the prime contractor at Fernald, with a five year, $2.2 billion contract. Flour subcontractors include Jacobs Engineering Group, Inc. (Pasadena, CA), Halliburton NUS Environmental Corporation (Houston, TX), and Nuclear Fuel Services.

• Hanford Reservation (Richland, WA) is a 560 mile site in the Columbia River Basin. Approximately one half of the entire DoE program, between $160 billion to $200 billion by 2018, will be for the Hanford Reservation site. Kaiser Engineers Hanford Co. (a subsidiary of ICF Kaiser Engineers) is completing a five year, $570 million remediation project at Hanford, and is competing for the five year follow-on contract valued at $500 million.

• The Jefferson Proving Ground (Madison, IN) remediation project is in the assessment phase. It is so contaminated that is potentially uncleanable and could become a national sacrifice zone.

• The Oak Ridge National Laboratory (Oak Ridge, TN) Y-12 Plant occupies an 811 acre site in Bear Creek Valley, approximately two miles from downtown Oak Ridge, TN. The site is drained by Bear Creek and East Fork Poplar Creek. The Y-12 Plant, built in 1943 as part of the Manhattan Project, was established to separate uranium isotopes by the electromagnetic process. When the process was discontinued after World War II, the Y-12 Plant's role changed to manufacturing and developmental engineering. The Y-12 Plant contains many facilities that have been used for treating, storing, or disposing of hazardous and radioactive waste, and hazardous substances. Examples of this include landfills, incinerators, drum storage areas, aboveground storage tanks, underground storage tanks, surface impoundments, and treatment facilities. The Oak Ridge reservation, which include the Y-12 Plant, was placed on the National Priorities List in 1989. By 2015, the Department of Energy plans to have completed the East Fork Poplar Creek cleanup and completed remediation of major contaminated areas at the Y-12 Plant.

• McClellan Air Force Base (near Sacramento, CA) is the U.S. Air Force's highest ranked Superfund site because of extensive groundwater contamination on site and off. There are approximately 240 different waste sites on the base, with some contamination as deep as $400 feet. Current efforts, including an $8 million pump and treat system, have helped reverse polluted groundwater plumes. McClellan AFB has formed a partnership with federal and state environmental agencies, and is to be the center for an innovative test of steam vapor extraction technology. The effort is expected to be partly funded by private sector waste generators, such as Monsanto Chemical Co. and IBM, which can share in the results but do not have to worry about liability for failure.

• Rocky Flats (Jefferson County, CO) is a U.S. Department of Energy's nuclear weapon plant which was operated by Rockwell International Corp. from 1975 until 1989, when it was replaced by EG&G. The annual site cleanup budget has risen from $50 million in fiscal 1989 to $280 million in 1993, and the project will cost between $1 billion and $1.5 billion in the next decade. Rocky Flats will have a budget of $1.85 billion through 1995.

RADIOACTIVE WASTE MANAGEMENT AND DISPOSAL

Four (4) radioactive waste management and disposal projects with projected expenditures of $1 billion or more have been identified within the United States. They are:

• Savannah River Waste Disposal Facility (Savannah River near Aiken, SC): The Savannah River Site (SRS), located on 192,000 acres along the Savannah River, produces nuclear materials, primarily tritium and plutonium, for national defense. The primary mission of SRS is to support national security as a major source of reactor-produced material. The number of environmental restoration issues at SRS is commensurate with the size of the installation. SRS has an active program that addresses issues in the following areas:

1. Identification of waste sites.
2. Investigation to confirm and quantify contamination.
3. Technology development and demonstration to conduct cleanup.
4. Installation of postclosure environmental monitoring.
5. Remedial design and cleanup to restore natural resources.
6. Decontamination and decommissioning of surplus facilities.

• Waste Isolation Pilot Project is a $2 billion waste isolation pilot project was built by the Department of Energy near Carlsbad, NM. The plant consists of caverns, 2,150 ft. below the earth's service, for the burial of plutonium-contaminated waste.

• In Yucca Mountain, NV, the Department of Energy has implemented a $6.3 billion study program to determine whether to bury spent nuclear fuel. In April, DoE advertised a $100 million contract for driving 14 miles of exploratory tunnels within the remote, faulted mountain in desert 100 miles northwest of Las Vegas. DoE plans to make a decision by 2001 whether to bury up to 70,000 metric tons of spent nuclear fuel at Yucca Mountain, mostly from civilian nuclear power units. The existing 114 units in the U.S. currently store more than 20,000 metric tons of spent fuel within temporary on-site pools.

• West Valley Nuclear Project (West Valley, NY): The only commercial nuclear fuel reprocessing facility to ever operate in the United States is located on approximately 200 acres of the 3,345 acre Western New York Nuclear Service Center (WNYNSC) near West Valley, NY. The WNYNSC is owned by New York State through the New York State Energy Research and Development Authority (NYSERDA). From 1966 to 1972, the plant chemically reprocessed approximately 640 metric tons of spent nuclear fuel to recover useable uranium and plutonium. Nearly 600,000 gallons of liquid, high-level nuclear waste (HLW), a byproduct of reprocessing, was stored in an underground tank contained within a concrete fault. In 1976, the operator of the facility notified NYSERDA of its intention to cease operations.

WATER AND WASTEWATER MANAGEMENT

Twelve (12) water and wastewater management projects with projected expenditures of $1 billion or more have been identified within the United States. They are:

• Alaska-to-California Subsea Pipeline: In response to California's on-going acute water shortages, some cursory work on engineering feasibility and cost to construct a subsea pipeline to transfer water from Alaska to southern California were conducted by the U.S. Department of Interior's Bureau of Reclamation in 1975 and by Flour Daniel Corporation in 1991. Flour estimated that a 2,000 mile version of the pipeline would cost approximately $150 billion. Interest in subsea pipeline must be classified as long-term. A study by the Office of Technology Assessment concluded that there is no current or near-term demand for expensive water from Alaska. However, the possibility that such water might be eventually be needed cannot be completely dismissed. No one knows what California's water demands might be 50 years or more from now nor the relative costs of the options available at that time for meeting those demands.

• Central Arizona Project (CAP) is a water resource development and management project under construction by the Bureau of Reclamation for the State of Arizona. Authorized in September 1968, the project will bring an average of 1.5 million acre-feet (MAF) of Colorado River water into central and southern Arizona each year. The CAP is a supplemental water supply. It will help Arizona produce its two million acre-foot per year groundwater overdraft, and also allow it to use its full 2.8 MAF allocation of Colorado River water. Prior to the CAP, the state used only about one million acre-feet of Colorado River water annually, all in areas adjacent to the river. In 1989 dollars, the estimated federal cost of constructing the CAP, if all authorized features are built, is $3.4 billion. (Non-federal entities will contribute an additional $389.7 million.) Federal expenditures are expected to exceed $2.2 billion by October 1, 1989.

• Chicago Tunnel and Reservoir Plan: The Metropolitan Water Reclamation District of Greater Chicago's (MWRDGC) solution to the problem of waterways pollution and flooding due to Combined Sewer Overflows (CSO) is the Tunnel and Reservoir Plan (TARP). TARP is a multi-purpose urban public works project with a total cost of $3.7 billion. TARP Phase I is a pollution control project estimated at $2.5 billion with $1.7 billion worth of projects either completed or under construction. A total of 109 miles of tunnels ranging in diameter from nine ft. to 33 ft., three dewatering pumping stations and associated appurtenances are key components of TARP Phase I. TARP Phase II, which is primarily for flood control, is estimated at $1.2 billion. The Tunnel and Reservoir Project is ongoing. About 51 of the 131 miles of planned tunnels have been completed and placed into service. Another 25 miles are under construction. The first of three huge terminal reservoirs is under construction. These reservoirs are a joint project of the Water Reclamation District and the Army Corps of Engineers. When completed, these reservoirs will further increase the capacity of the TARP system by 41 billion gallons, depositing the combined sewer overflow into reservoirs via the tunnel system where it can be held until the storms subside. This overflow will then be pumped out and conveyed to the treatment plants. The completion of this entire project, along with the sewer upgrades in the local municipalities, will eliminate pollution and flooding problems in the Cook County combined sewered area. It will relieve each local municipality of the burden of designing, building and operating its own system to capture and treat wastewater in order to comply with federal regulations. The District provides a regional approach to wastewater collection and treatment which is considerably more efficient and cost-effective for the communities within our jurisdiction.

• Domenigoni Valley Reservoir (Fresno, CA) has been proposed by the Metropolitan Water District. The reservoir would include two mile dams to store enough water for a continuous drought of two years and cost $1.35 billion.

• Hyperion Wastewater Treatment Plant Expansion (Los Angeles, CA) is undergoing a $1 billion expansion to increase the capacity from 425 million gallons per day to 450 million gallons per day by 1998.

• Las Vegas Water Pipeline is a $1.5 billion water pipeline, to be completed by 2000. The Las Vegas Water District plans to use the facility to pump water from four counties north of the city to accommodate city growth.

• Milwaukee's Deep Tunnel Project: The Water Pollution Abatement Program (WPAP) progress has been impressive on the Inline Storage System, commonly referred to as the "deep tunnels." Some 300 ft. below street level, mining was completed on the North Shore leg of the tunnels and work continued on the Kinnickinnic/Lake Michigan leg. With a projected final cost of $2.3 billion, the Program is within 8.5% of the original cost estimate established at the first bond sale in 1983. The Water Pollution Abatement Program will meet the requirements mandated by the United States Clean Water Act. When it is finished, Milwaukee will have a sewerage system second to none in the world. It will protect public health, preserve Milwaukee's water resources, and give Greater Milwaukee a unique infrastructure advantage over major metropolitan communities in the intense competition for new business and industry.

• New York's Third Water Tunnel is a $5 billion project by the Bureau of Water Supply and the City of New York. The first-stage tunnel runs 13.7 miles into the northern part of the city. Later stages will expand the system approximately 70 miles through the city.

• San Diego Wastewater Expansion and Upgrade: San Diego's city council approved a scaled-down clean water plan to cut the cost of court-ordered cleanup in half. Original estimates were $2.5 billion. The consumers alternative plan will cost $1.3 billion through 2003, reflecting San Diego's actual sewer capacity needs. The move has put a number of sewerage design and construction contracts on hold. Under the revised plan, the 180 million gallon per day Point Loma wastewater treatment plant will not implement secondary treatment, but will be improved to provide advanced primary treatment handling 240 mgd. This eliminates the need for a separate 41-to-55 mgd secondary treatment plant in South Bay and four proposed water reclamation plants. One plant, already under design, will be constructed, along with a sludge processing facility. After 2003, two more reclamation plants would be built. The alternative plan will produce half as much reclaimed water as the original plan, which could delay approval. Four design contracts for reclamation plants and connecting pipelines are on hold, along with one for a portion of design of the San Diego River outfall. The city has also suspended bids for design work in the South Bay. The alternative plan supports extension of the existing nine ft. dia. Point Loma outfall, just repaired after its February rupture, and construction of a new parallel outfall. The city awarded Morrison Knudsen Corporation and Manson Construction Company a $54.8 million, design-build contract for the 12,500 ft., 12 ft. dia. extension.

• San Francisco Sewage Treatment Upgrade: San Francisco, CA has almost completed a $1.3 billion upgrade of the sewage treatment system. The project includes storage boxes, treatment plants, and pumping stations.

• Water Diversion Scheme, California, is a $3.8 billion scheme to divert water from British Columbia to drought-stricken California is being proposed by Canadian and U.S. firms. The joint venture, which sought participation by the Metropolitan Water District of Southern California, filed plans for the project in September of 1991.

• Water Pipeline Network Expansion (Southern California) is a $1 billion expansion for the 750 mile California Water Pipeline Network will extend the pipeline 95 miles.

AIR POLLUTION CONTROL

One (1) air pollution control project with projected expenditures of $1 billion or more has been identified:

• Navajo Generating Station (Arizona): In 1992, Navajo Generating Station reached the first settlement agreement under a Clean Air Act provision designed to improve visibility at national parks. Sulfur dioxide emissions from the 2409 MW coal-fired plant will be reduced 90% before August 1999. It will cost $2 billion to install pollution control equipment on the plant's three smokestacks. The plant, operated by Salt River Project, is 12 miles from the northern end of the Grand Canyon. Environmentalists have considered it a major reason for poor visibility in the area.

Chapter 7

Site-Specific Criterion at a Textile Dyeing and Finishing Plant

R. Gross

ABSTRACT

The State of South Carolina recently adopted EPA's water quality criteria as standards to comply with the 1987 amendments to the Clean Water Act. South Carolina's regulations allow for a site specific standard which is less stringent than the EPA criterion if justified through biological testing. This paper describes the steps a textile dyeing and finishing facility took to obtain South Carolina's first site specific water quality standard.

The textile facility was placed on South Carolina's §304(L) toxic discharger list because the discharge resulted in stream concentrations higher than EPA's water quality criteria for copper and zinc. Periodic instream biological assessments had shown adverse impact to the receiving stream during periods of low flow. These adverse impacts supported the State's position that copper and zinc were causing a condition of toxicity.

The paper describes the biological studies conducted as part of the §304(L) evaluation and the surprising results that followed. The paper also describes the control measures taken by the textile plant in order to achieve compliance with the state's water quality standards, to include the development of a hydrographic controlled release (HCR) discharge system and the construction of an upstream reservoir to augment stream flow during periods of naturally low flow.

INTRODUCTION

This paper discusses the activities leading up to a site specific criterion for copper at a textile dyeing and finishing plant in South Carolina. The site specific criterion approach was necessary because there was no other alternative available including relocation of discharge, increased level of treatment, waste minimization, land disposal, or joining a municipal sewer system. Each of these alternatives was considered and ruled out for good reasons.

This is the first site specific criterion granted in South Carolina.

BACKGROUND

As a matter of background information, Dixie Yarns is headquartered in Chattanooga, Tennessee and owns several plants throughout the United States. Dixie acquired the Ti-Caro Division in 1988, and one of the plants is located in Jefferson, South Carolina.

The Caro-Knit plant in Jefferson knits, bleaches, dyes, and finishes high quality fabric goods. The fabric produced at Jefferson becomes golf and rugby shirts and other fashionable apparel of L. L. Bean, Lands End, Polo, and other well known quality brands. The high quality nature of the fabric requires that dye colors be consistent and without flaw.

The Caro-Knit plant has had a wastewater treatment facility since it was constructed in the 1950s. The original treatment plant was upgraded in the mid 1970s to comply with the EPA Textile Effluent Guidelines Best Practicable Control Technology (BPT). The upgraded system included separation and pretreatment of metal bearing wastes, extended aeration, clarification, and effluent polishing prior to discharge.

In 1988 and 1989 Dixie Yarns spent roughly one million dollars improving the wastewater treatment plant at Caro-Knit. Clarifiers were rebuilt, the aeration basin was enlarged, a mixer/flocculator was added, the old pretreatment clarifier was converted to a sludge thickener, sludge storage basins were cleaned and taken out of service, and the effluent polishing pond was converted to an emergency effluent storage basin.

The receiving stream is South Fork Creek which is a small tributary to Lynches River. The flow in the South Fork Creek is highly variable depending on rainfall and season of the year. Flows can vary from near flood conditions to periods of no flow during hot dry summer months.

The early NPDES permits issued to Caro-Knit contained a limit on copper based on a water quality criterion of 0.1 mg/l. The limit varied according to the flow in the receiving stream with a maximum concentration of 1.5 mg/l in the effluent. The plant was able to consistently meet this copper limit with only rare exceptions. All other parameters of the NPDES permit have consistently been well below the permitted values.

Previous NPDES permits have required biological monitoring. Tests included acute evaluations using bluegill sunfish as the test species. The plant was also required to conduct instream biological assessments during the summer months. The results of those tests showed no toxicity to bluegill. The instream biota were affected during periods of low flow, but recovered quickly when stream flows increased.

In 1988 South Carolina developed the Section 304 (L) list of discharges that were impacting receiving streams as a result of the discharge of toxic pollutants. When South Carolina reviewed the data from Caro-Knit plant, it found that the instream concentration of copper exceeded the EPA Water Quality Criterion of 0.0065 mg/l. South Carolina also noted that the instream assessments showed biological impact. They therefore concluded that the impact was caused by the discharge of copper and placed the Caro-Knit plant on the Section 304 (L) list. The EPA criteria were adopted in 1989 as Water Quality Standards for South Carolina.

Dixie Yarns and its consultant, Hydroscience, argued vigorously that the metals in the Caro-Knit effluent were organically bound and were not the ionic species upon which the water quality criteria were based. Limited data were available to convince the State that Caro-Knit should not be on the 304(L) list. There was enough information, however, that the State did allow Caro-Knit to pursue a site specific criterion based on biological testing.

Dixie Yarns felt that the metals were not the cause of toxicity because they are contained in the dyes and are actually part of the dye molecule. As EPA notes in its water quality criteria book (the Gold Book):

> "EPA believes that a measurement such as 'acid-soluble' would provide a more scientifically correct basis upon which to establish criteria for metals. The criteria were developed on this basis. However, at this time, no EPA approved methods for such a measurement area available to implement the criteria through the regulatory programs of the Agency and the States. The Agency is considering development and approval for a measurement such as acid-soluble. Until available, however, EPA recommends applying the criteria using the total recoverable method. This has two impacts: (1) certain species of some metals cannot be analyzed directly because the total recoverable metal does not distinguish between individual oxidation states, and (2) these criteria may be overly protective when based on the total recoverable method.

Dixie Yarns further felt that the cause of toxicity was most likely the salts (sodium chloride and sodium sulfate) used in the dyeing process. These salts would account for the impact to the receiving stream and the quick recovery once stream flow dilutes them to acceptable levels. Limited testing with *Ceriodaphnia dubia* showed that the toxicity of the effluent was comparable to the toxicity of clean water with similar levels of salt.

When the effluent polishing pond was taken out of service in 1989, it was drained and cleaned. During this process a most unusual discovery was made. In the thin layer of sludge on the bottom of the polishing pond, bluegill sunfish (bream) had made beds and were living and reproducing in 100% treated effluent. Recognizing the opportunity at hand, Caro-Knit called Shealy Environmental Services who came to the site and collected fish from the polishing pond for analysis. The results showed that the fish contained no measurable levels of contaminant and were eatable. The only reason that the fish were not larger was that there was insufficient food and no predator species to thin the population. This reproducing fish population in the effluent polishing pond provided and early positive sign that the effluent was not very toxic to aquatic organisms.

The toxicity reduction evaluation involved biological testing with fathead minnows and *Ceriodaphnia dubia*. The plan was developed in conjunction with Hydroscience and Shealy Environmental Services who performed the toxicity tests and the wet chemistry. The fathead minnow larval growth study showed no adverse effect from the Caro-Knit effluent at 100% concentration.

The toxicity testing with *Ceriodaphnia* was conducted under a variety of circumstances. The existing data on Caro-Knit effluent with *Ceriodaphnia* shoed variable toxicity results ranging from a chronic value of 26.5% to 61.2%. Several tests were conducted with effluent spiked with copper.

During this testing period, the Caro-Knit plant was required to meet low concentrations of copper due to the low flow in the receiving stream. In order to achieve compliance with these low limits, the Caro-Knit plant used various polymers to lower the copper concentration. Toxicity tests performed during this time showed that while the copper concentration decreased, the effluent toxicity increased. Further tests showed that the polymers used to lower the copper were toxic. The toxicity tests were suspended, therefore, until the higher permit values prevailed and polymers were no longer needed. The tests were resumed in the fall and a correlation was demonstrated between toxicity and hardness. A final report was prepared showing the toxicity results.

CONCLUSIONS

The results showed some variability, but generally showed that an instream waste concentration of 33% of more would not be expected to have a toxic effect. Based on an instream waste concentration of 33% and an effluent copper concentration of 1.0 mg/l, a site criterion of 0.6

mg/l has been requested and conceptually approved.

IMPLEMENTATION

About the same time that the Caro-Knit toxicity issue arose, the Town of Jefferson was notified by the State that it needed to enlarge its water supply capacity. The Town has agreed to construct a reservoir funded by Dixie Yarns at the headwater of South Fork Creek. This reservoir will provide a constant release of water to South Fork Creek sufficient to provide a 2:1 dilution ratio of the Caro-Knit effluent. The reservoir will also serve as a backup water supply for the Town.

The former treated effluent polishing pond at Caro-Knit will be used as an emergency effluent storage basin. The effluent discharge system will be operated in the hydrographic controlled release mode, storing effluent when needed and releasing it when the stream flow allows it.

Chapter 8

Industrial Wastewater Toxicity Case Histories

S.C. White, S.D. Myers

INTRODUCTION

Biomonitoring is becoming a common NPDES permit requirement for many dischargers of industrial wastewater. The permits require one of several different bioassay tests to determine if an effluent is in compliance with water quality-based whole effluent toxicity requirements. When bioassay results indicate unacceptable toxicity in an effluent, then a Toxicity Reduction Evaluation (TRE) is the tool used to identify the problem, demonstrate the remedy, and bring the discharge back into compliance.

After you have determined that a toxicity problem exists, your first actions are to take common sense steps to improve the situation (e.g., good housekeeping, control chemical spills and usage, and improve waste treatment operations). If toxicity persists, then the facility is a candidate for a TRE. EPA has issued guidance documents that define many of the step-by-step procedures of TREs and the recommended test protocols. The flow diagram in Figure 1 illustrates the many steps and potential paths that can be followed while conducting these investigations.

OBJECTIVE

Being able to conduct bioassays is not the only skill needed to carry out an effective TRE, and EPA's guidance documents are not "cookbooks" that automatically guide the investigator to a solution. A good bioassay laboratory must interact closely with an experienced wastewater treatment engineer to be truly effective in a TRE. This paper presents three case histories to illustrate how the waste treatment specialist and bioassay laboratory need to act in concert to interpret results, know when to follow the guidance and when to deviate from it, and how to improvise when necessary.

CASE 1 - A "TEXTBOOK" TRE

This example is called a textbook case because it followed the main logic flow of the TRE diagram (Figure 1) through all three phases of Toxicity Identification and included both the Treatability Evaluation and Source Investigation. This is presented first to emphasize the contrast among the three cases.

This industrial production facility generated a number of waste streams that contained organics and metals. The waste treatment plant included two treatment trains. The first used precipitation with lime as pretreatment ahead of the activated sludge process. The second was chemical precipitation, flocculation, and filtration for metals removal. After treatment, both main streams were combined for discharge. This combined stream exhibited toxicity.

The Phase I (Characterization) bioassay testing repeatedly and clearly implicated metals as the toxic influence. The fraction treated with EDTA and the high pH Filtration characterization tests were the only ones that produced positive results. After concluding that metals were the problem, laboratory analyses (Phase II - Identification) quantified the presence of copper, nickel, and zinc in concentrations that supported bioassay results. The presence of three metals thought to contribute to the problem complicated the Phase III Confirmation tests.

EPA guidance documents outline more than six different approaches available to confirm the identification of toxic causative agents. Collaboration between the wastewater treatment engineer and the bioassay laboratory were needed to select the proper confirmation approaches for this case. Two methods, Spiking and Correlation, were selected as the

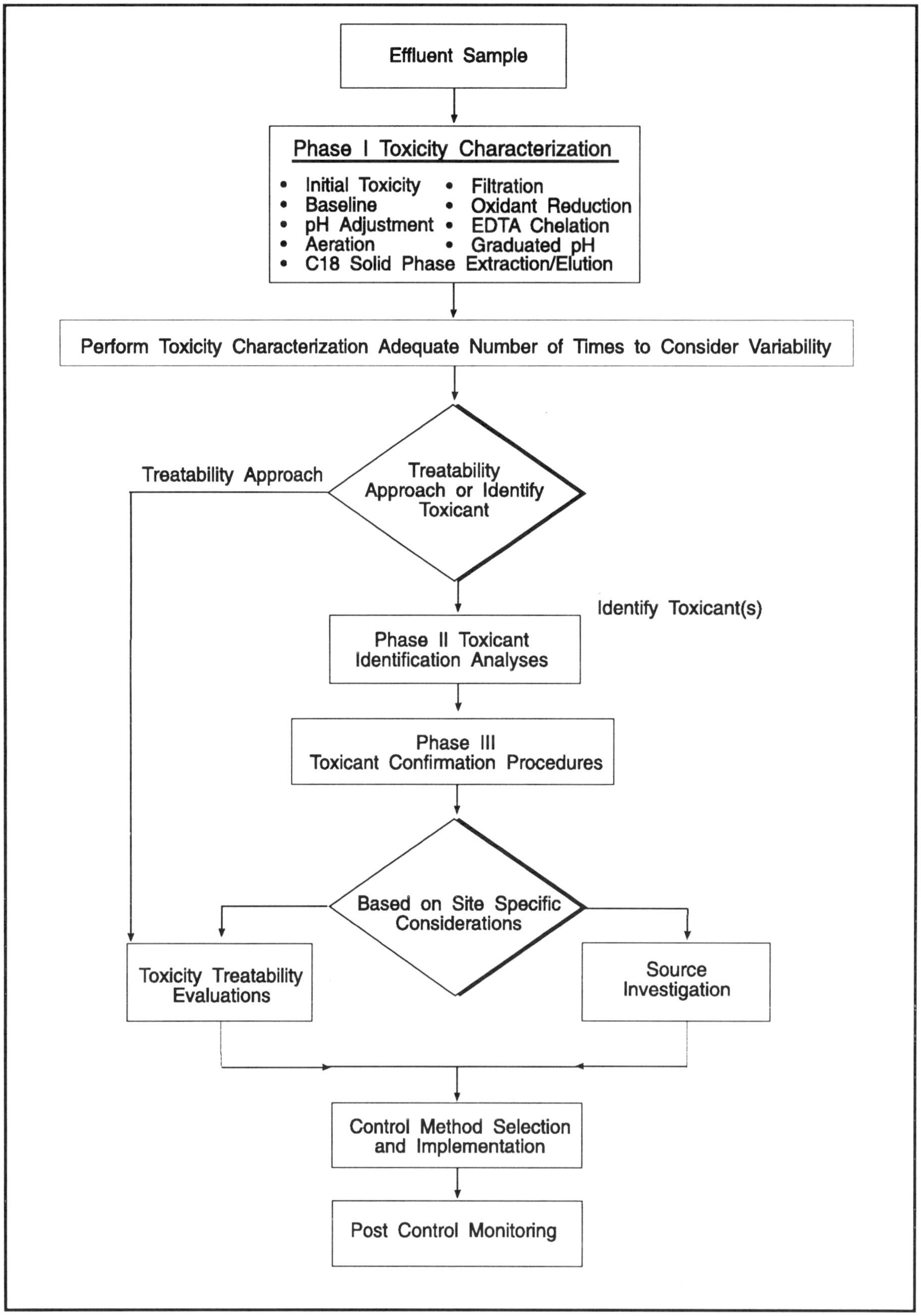

FIGURE 1
TIE FLOWCHART

most applicable means of confirmation. After the engineer removed the suspected metals (and toxicity) by bench-top chemical treatment, the bioassay laboratory "spiked" the treated wastes with the three metals and demonstrated that the toxicity was again similar to the untreated effluent. The correlation approach required that a complex matrix of combinations of the three metals, at varying concentrations, be subjected to bioassays to establish a correlation between toxic response and the matrix conditions. These results also confirmed that all three metals were indeed contributing to the effluent's toxic response and that improved removal of all three would be required to achieve permit compliance.

After completion of the Confirmation Phase of a TIE and the identity of the toxicants are known, a choice is usually made to either eliminate or modify the source of the problem, or to determine how to change the treatment processes to remove the toxic impact. Based on the waste treatment engineer's experience and knowledge of this facility, a rare decision to pursue both options was recommended and two key alterations were made to reduce the effluent's toxicity.

The chemical precipitation system treating the majority of metals produced a sludge that was dewatered, generating a filtrate stream containing metals and some residual organics. That filtrate was being routed to the parallel biological system, but the pretreatment prior to the activated sludge system was not operated at a sufficiently high pH to remove metals from the filtrate stream and they subsequently "passed through" the activated sludge process. Therefore, a key modification was to re-route this toxicity source, the filtrate, back into the head end of the chemical treatment system. The second key modification, a treatability process improvement, was to elevate the pH set point for chemical precipitation to greater than 9.5 to improve removal of the resulting higher metals loading.

In summary, even a textbook TRE can require wastewater engineering expertise to devise bench-top systems that permit valid confirmation experiments and to effectively apply the toxicity test results to achieve compliance in the full-scale system.

CASE 2 - PRACTICAL TREATMENT

The only similarity of this project to the "textbook" case described above is that the bioassay laboratory was again able to repeatedly and clearly demonstrate the nature of the toxicity present in the wastewater during the Phase I Characterizations. But that finding actually precluded continuing with Phases II and III of the TIE and left no choice but to follow the option of progressing directly to the Treatability Evaluation. Without a practical knowledge of available treatment technologies and their capabilities, this investigation would not have been successful. This project clearly demonstrates the necessity of close cooperation between waste treatment engineers and bioassay labs in TREs.

The effluent from the wastewater treatment facility did not meet permit limits for chronic toxicity. A TRE was begun and all characterization fractions indicated in Figure 1 were examined. In three sets of Phase I tests, filtration consistently removed the toxicity. Subsequent microscopic examination of the Ceriodaphnia from unfiltered tests that failed confirmed that the organisms had indeed ingested large quantities of particulates that could have resulted in test failures. Therefore, the Phase I testing implicated "solids", or TSS, as the wastewater component causing the problem. However, the bacterial or algal biomass that typically constitutes effluent TSS in these discharges is not thought to have a toxic effect. It was suspected that some of the effluent TSS was from a source other than biomass. The production facility in this case discharged wastewaters into an aerated lagoon. The treatment process was not as effective as desired for the removal of BOD and TSS. Subsequent analysis of the effluent demonstrated a strong correlation between TSS and BOD.

These findings led to preliminary treatability evaluations using bench-top tests aimed at removing both TSS and BOD components. Based on waste treatment experience, two candidate treatment technologies (filtration and biological degradation) were considered to address the analytical findings discussed above. Filtration can remove effluent TSS and a higher level of biological degradation may be effective on the residual BOD.

Two filtration technologies were tested: sand media and diatomaceous earth media. Sand filtration is a less expensive technology, but is not as effective in removing fine particulates as diatomaceous earth (DE) filters. For these practical reasons, both technologies were screened. Sand filtration lowered the toxicity impact, but did not totally remove it. The DE media was more effective and allowed the treated effluent to pass the bioassay, but the cost of such a system would have been prohibitive when the cost of sludge disposal was considered. A technology that was not as costly as DE filtration, but more effective than sand filtration was needed.

For those reasons, activated sludge was also screened in a bench-top test and it was demonstrated that further degradation of the insoluble BOD was possible and that this process also lowered effluent TSS. These results indicated that a relatively insoluble organic material from the production process was able to transit the lagoon system without being solubilized or degraded. Based on all these screening results, it appeared that an upgraded treatment facility more capable of solids and BOD removal could successfully address the problem.

In this case study, the logic flow diagram was quickly interrupted following Phase I Characterizations and was redirected towards the more practical, direct method of treatability demonstration. Interaction between the bioassay laboratory and engineer in the interpretation of characterization findings and actual wastewater data allowed selection of appropriate bench-top screening tests. A large scale activated sludge pilot plant is now being operated to demonstrate compliance capability and to develop design criteria for the upgraded system.

CASE 3 - IMPROVISATION

This project had no similarity to the textbook case example and is the best illustration of how laboratory and engineer must work together like detectives investigating a mystery. When you follow the guidance documents and reach a "dead end," sometimes you have to improvise based on experience.

In this third case, the industrial wastewater has been discharged into a small stream for many years and previous in-stream studies indicated that the discharge did not have a deleterious impact on aquatic life in the stream. However, when a bioassay was performed on the effluent, it failed, and a TRE was begun.

The Phase I Characterization tests illustrated in Figure 1 were performed, as they were for the other two cases described above, but the answer was "no answer." Repeated tests indicated that the toxicity was consistent and did not vary significantly, but none of the fractions tested could be linked to the toxicity of the effluent. At this point, the engineer and laboratory resorted to a more conventional approach and analyzed the effluent for a full array of both conventional wastewater parameters and priority pollutants.

Results of those analyses were then compared to literature values on the toxicity of individual constituents. None of the priority pollutants were implicated in this search and all the parameters of the NPDES permit were well within limits. However, the total dissolved solids (TDS) level, which was not normally monitored, was noted by the bioassay laboratory. When maintaining a Ceriodaphnia culture in the laboratory for use in these tests, the laboratory must document that the cultures are sufficiently sensitive to a "reference toxicant." That toxicant is typically sodium chloride, or salt. Our bioassay laboratory and engineer determined that, according to the literature values and our own toxicant tests, the level of TDS in the wastewater was probably responsible for the toxic response.

If just salts in the wastewater, not toxic organics nor heavy metals, were responsible for the toxic response, how could this be demonstrated? Again, waste treatment experience was called on to identify a method to remove salts while leaving any other potentially toxic materials in the effluent sample. An ionic resin was selected that removed sodium and chloride, and bioassays were conducted on this modified effluent. The modified effluent passed the bioassay tests.

SUMMARY

As an organization, we have placed our Bioassay Laboratory and Wastewater Treatability Laboratory as parallel functions within the Wastewater Process Engineering Department. With the increasing emphasis on toxicity controls and compliance, we believe the closer these two capabilities are merged, the more efficient we become in problem solving for our clients.

Chapter 9

Design of Effluent Diffuser Systems

Y. Seiner

INTRODUCTION

Most municipal and industrial wastewater treatment facilities discharge their effluent to a receiving waterbody. The degree of treatment required is determined by federal and state regulations. These regulations typically include primary, secondary, and tertiary treatment standards, various guidelines categorized by industry, and acceptable water quality standards in receiving waterbodies. The regulations are typically promulgated by federal environmental agencies, with enforcement delegated to individual state environmental regulatory agencies.

The water quality standards have been rising, and the definition of acceptable impact on the receiving waterbody has been getting narrower and narrower. One of the more recent restrictions has been the implementation of mixing zone requirements, where a discharger may have a very limited length of river run to achieve complete mixing.

The primary method used in the past for discharging wastewater to a waterbody is a single port (opening) discharge. This would consist of an open ended pipe extending into the waterbody. When the flow through the pipe was sufficiently large, and the effluent was of a different color than the receiving water, the resulting streak could at times be seen extending for miles downstream. Besides the aesthetic considerations, such effluent, if toxic, could pose a hazard to wildlife. The concept of mixing zones was introduced to remedy this problem.

A mixing zone is a box drawn around the discharge. The United States Environmental Protection Agency (USEPA) gives recommendations for the specific guidelines for mixing zones in receiving waters. Each individual state formulates their own specific requirements accordingly. The regulatory mixing zones are usually too small to allow the use of single port diffusers.

A mixing zone can be applied to wastewater parameters that are listed under the limits of water quality criteria but are not found to be toxic at any concentration. For example, a state regulatory agency may allow dissolved oxygen to fall below 4 mg/l, or for pH to exceeded 7.5 at a given discharge. However, this paper will focus on mixing zones as they apply toxicity.

The use of mixing zones has led to the use of multi-port diffusers. A multi-port diffuser consists of a number of ports located along the main discharge pipe. The size, location, number, and orientation of the diffuser ports will be dependent upon the requirements for achieving complete mixing of the discharge throughout the receiving waterbody. As a result, the successful design and implementation of a multi-port diffuser depends heavily on computer simulation and careful analysis of the diffuser geometry, location in the receiving waterbody, and the flow conditions under which discharge takes place.

Figures 1 and 2 show a typical outfall before and after diffuser installation. Prior to intallation, the plume is bank attached and concentrations remain high. After diffuser concentration, the plume dissipates much quicker, inspite of the flow in the river for Figure 2 being much lower than Figure 1.

The emphasis on improving water quality through the use of better diffusers has fostered an increased interest in

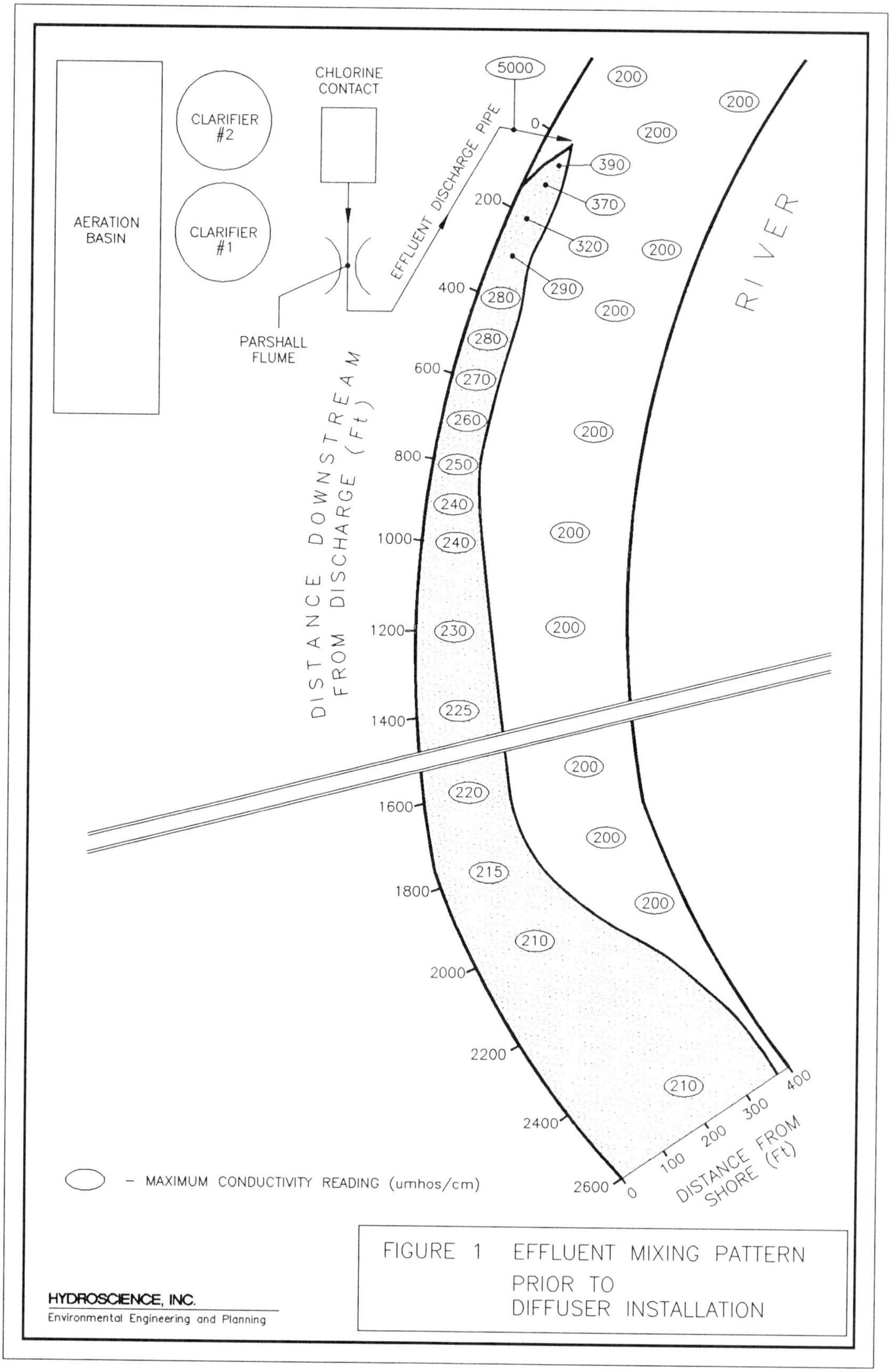

FIGURE 1 EFFLUENT MIXING PATTERN PRIOR TO DIFFUSER INSTALLATION

analysis of the fluid mechanics and the interaction between the discharge and the receiving waterbody. When the discharge dynamics are properly analyzed the impact on the receiving waterbody can be greatly decreased by modifying the design of the discharge structure. The objective is to adequately mix the wastewater with the receiving waterbody to achieve an acceptable dilution level within a desired distance from the discharge point, and to reduce the possibility of the creation of highly concentrated regions of wastewater in the receiving waterbody. Therefore, it is necessary to evaluate the existing hydrodynamics of the receiving waterbody in relation to the particular characteristics of the diffuser design.

The following sections describe (1) the regulatory evolution of discharge requirements which have influenced the development of fluid mechanics analysis of receiving waterbodies, (2) the USEPA guidelines for achieving a mixed system, and (3) the specific computer model used to determine the design parameters for the diffuser structure.

REGULATORY HISTORY

Over the past forty years several landmark environmental acts have been enacted by the federal government. These acts have guided and mandated a large portion of the individual state regulations relating to water quality and wastewater treatment. In 1948 the Water Pollution Control Act provided federal funding and technical assistance to all states to aid in monitoring and control of pollutant discharges. The Federal Water Pollution Control Act of 1956 provided construction grants for municipal wastewater facilities and also formed the basis for the present-day enforcement structure.

In 1965 the Water Quality Act required states to set their water quality standards individually. The Federal Water Pollution Control Act was amended in 1972 to include the National Pollutant Discharge Elimination System (NPDES). The NPDES system is designed to eliminate all pollutants in all discharges at a gradual rate, by the enforcement of permit violations, and the development of the Best Practical Treatment (BPT) and the Best Available Treatment (BAT) effluent pollutant limitations. The Clean Water Act of 1977 included BAT requirements for toxic substances which directly influenced the development of diffused discharge systems.

Throughout the development of all of these acts the USEPA issued guidelines and recommendations from which all states determined their specific pollutant discharge limitations. The impact of specific pollutants and toxicity on the receiving waterbody is gauged against these guidelines.

The Federal Water Pollution Control Act strives to reduce the impacts of the discharge pollutants on receiving waterbodies. The development and application of diffuser systems adheres to this objective.

USEPA GUIDELINES

The United States Environmental Protection Agency has issued specific guidelines on what constitutes a completely mixed zone and how to calculate the dimensions of the zone in reference to the level of toxicity of the wastewater. In accordance with the Water Quality Control Act states are required to develop their own standards for the level of acceptable toxicity. State standards must be at least as stringent as USEPA

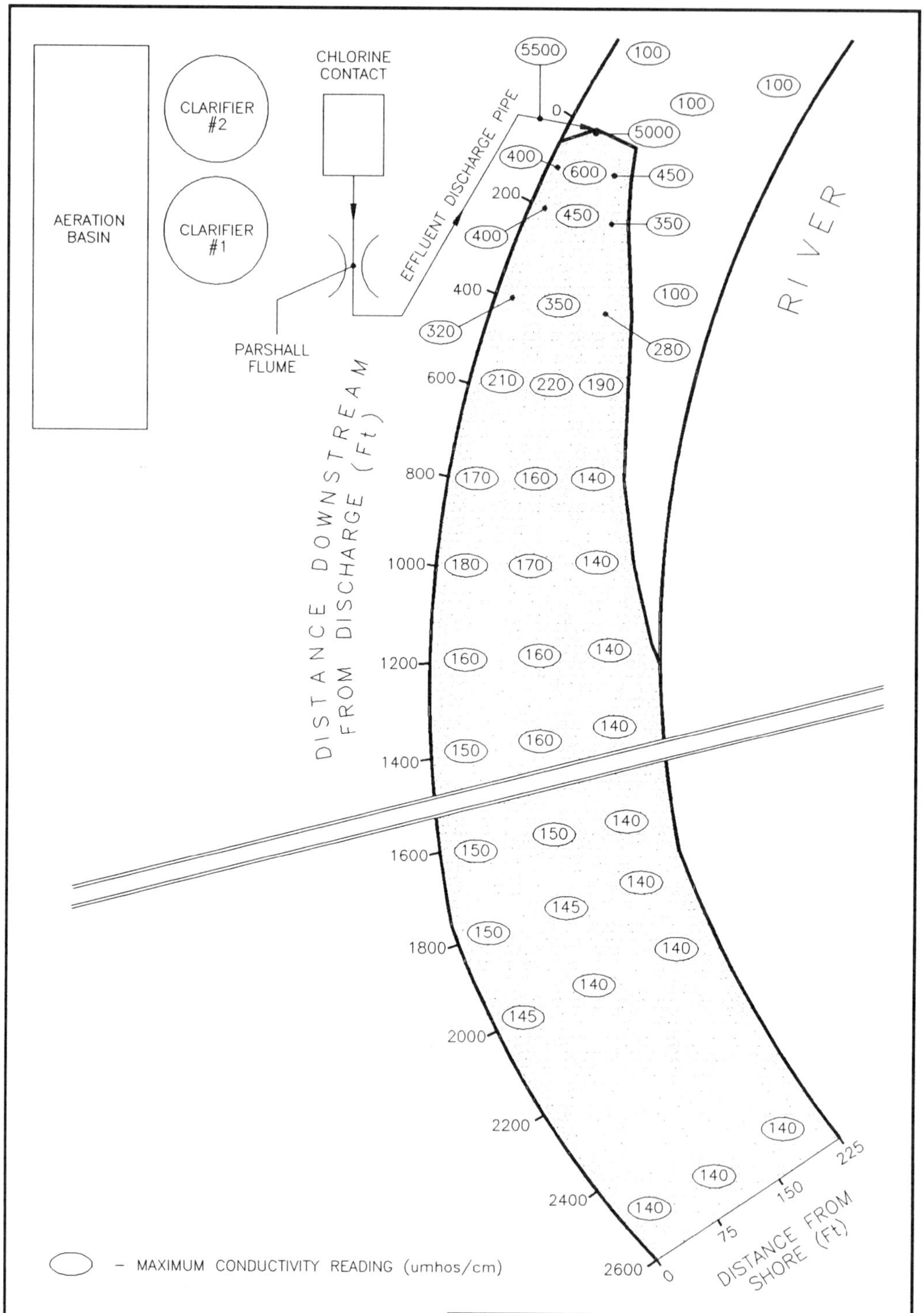

FIGURE 2 EFFLUENT MIXING PATTERN FOLLOWING DIFFUSER INSTALLATION

requirements. Therefore, regulations will vary from state to state. The method for determining the toxicity for a discharged pollutant will also vary from state to state.

The most common practice is to analyze the effect of the various concentrations of the wastewater on the reproduction rate of living organisms. Typically invertebrates are used as test organisms. Ceriodaphnea (Water Fleas) are commonly used in one battery of tests. Toxicity is categorized by two types, acute and chronic.

Acute toxicity is defined as that which causes lethal effects to a test organism. Chronic toxicity effects the reproduction of the test organism. The determination of a waste to be acutely or chronically toxic is dependant upon the test method the individual state regulatory agency has chosen to use as a standard.

For chronic toxicity, the threshold level of reduction in reproduction which indicates a particular waste to be toxic is determined by each state regulatory agency. Therefore, the concentration of a particular wastewater at which it is determined as toxic can vary from state to state. Once the concentration is determined the USEPA guidelines can be applied to determine the dimensions of the Mixing Zone unless specified differently by the individual state regulatory agency.

The USEPA and state guidelines can be used to determine the dimensions of a mixing zone. The mixing zone is a regulatory concept defined as the volume surrounding the point of discharge where water quality criteria limits can be exceeded. It is also referred to as an allocated impact zone. Acute toxicity levels must controlled within the mixing zone to avoid fish kills, but chronic effects can be tolerated.

The water quality criteria apply at the boundary of the Mixing Zone, not within the zone itself. In defining the Mixing Zone the USEPA suggests the area be as small as possible and not interfere with aquatic establishments. The effect of the actual discharge plume in relation to the recommended Mixing Zone is very critical. The discharge plume is defined as the physical characteristics of the wastewater as it moves through the waterbody and is observed as a continuous mass. These characteristics can be used as a primary indicator of how well the wastewater will be mixed with the waterbody through the Mixing Zone.

There are two criteria used by the EPA within the Mixing Zone to define the extents of regions of acute and chronic toxicity.

The first of these criteria is the Criterion Maximum Concentration. The CMC is defined as 0.3 x LC_{50} of the most sensitive of at least three test spoecies. The CMC is intended to prevent lethal or acute effects in the aquatic biota. It describes the condition under which lethality will not occur if the duration of the CMC exposure is less than one hour. The CMC defines the boundary of acute toxicity from the point of discharge.

The second of these criteria is the Criterion Continuous Concentration, or CCC. The CCC is defined as the chronic toxicity limit to the most sensitive of at least three test species. The CCC defines the boundary of chronic toxicity.

The CMC is the more stringent criterion because of the possible adverse impacts to the receiving waterbody due to acute toxicity effects. The CCC is the less restrictive criterion and must be met at the edge of the Mixing Zone. The design criteria are spelled out in the Technical

Summary below. Many states require that the water depth used be equal to the 7Q10 flow, which is the lowest water level occurring in the specific waterbody for a sustained period of seven days and is expected to recur once every ten years.

State regulatory agencies may choose to develop standards for toxic dilutions or they may evaluate each individual discharge and apply specific standards and guidelines. If these standards are available they can be used in conjunction with computer models to determine the most effective diffuser system required to achieve these criteria. If there are no individual state standards then the USEPA guidelines should first be used in evaluating the particular application. After the data is produced then the application should be further evaluated by the individual state regulatory agency.

COMPUTER MODELING FOR TOXIC MIXING ZONES

The specific computer model used by Hydroscience, Inc. to determine the most effective diffuser design is the CORMIX2 Mixing Zone Expert System. This model was developed by Dr. G.H. Jirka of Cornell University. The model uses volume flux, momentum flux, and buoyancy flux equations to predict the type of plume which will be produced by various diffuser structures. This model takes into account the physical characteristics of both the wastewater and the receiving waterbody. By simulating the development of the plume through the waterbody the degree of mixing can be modeled. Therefore, to achieve the CMC and CCC requirements the proper plume must be produced.

The CORMIX2 modeling system was primarily designed for modeling the discharge of wastewater facilities where both the effluent flow and stream flow is fairly constant. Therefore, there are limited guidelines for design of effluent diffusers as applied to variable-flow industrial dischargers. This model can also be used to effectively model industrial applications if 1) the limitations and basis of the model are understood and considered, and 2) the range of design conditions is limited to that which will occur in the greatest frequency. Hydroscience, Inc. has complied a methodology for evaluating industrial wastewater and designing diffuser systems, through the use of the CORMIX2 model, in compliance with regulatory effluent standards.

The effluent diffuser can serve many useful purposes for the industrial discharger. Not only can the impacts of toxicity be reduced through a diffuser but also the impacts relating to color concentration and temperature. To accurately model a proposed diffuser design, extensive data must be acquired due to the large number of variables required for input to the model. The first group of variables relates to the wastewater characteristics, i.e. temperature, density, flow, etc. The second group of variables relates to the receiving waterbody characteristics, i.e. stream channel dimensions, 7Q10 flow, temperature, etc.

The effectiveness of a diffuser is, to a large extent, a function of the nozzle velocity at the point of discharge. Therefore the type of nozzle used is critical for an effective diffuser, and the model must be calibrated for the particular nozzle used. Typically, laboratory testing is required to establish calibration curves that can be used in the model.

The remaining variables required relate to toxicity. The CMC and CCC concentrations must be entered. These are obtained by conducting laboratory tests of test organisms to determine acute and chronic toxicity limits. When these concentrations are entered into the model the output shows the location in the receiving waterbody where the dilution percentage is achieved, and the model indicates whether the diffuser meets the EPA criteria.

The model simulates the dilution of conservative substances. Although the model was designed for toxic modeling it may also be used in the same manner for modeling of color concentrations. All that is required is the dilution percentage desired and the model will determine the distance at which it is obtained. The difficulty is determining what dilution is acceptable, since the human eye response to color is logarithmic rather than linear. Typically, empirical and field studies are required to determine the dilution required for acceptable color levels.

Presently restrictions enforced by regulatory agencies relating to the color of the wastewater are based upon the methods and scales set forth by the American Dye Manufacturers Institute (ADMI), and described in Standard Methods.

The ADMI color can be used with the model as a target dilution. ADMI color numbers do not exhibit the properties of conservative substances. A red color, for example, is more easily visible in a typical river than a brown color.

Hydroscience, Inc. has developed a methodology for using CORMIX with ADMI color for modeling stream response to discharges with high ADMI color.

SUMMARY

Once all the input variables are defined various design options can be run through the model until the criteria for all areas of concern are satisfied. As previously stated the model uses the equations of fluid mechanics. The use of these equations results in a prediction of the discharge plume characteristics. Pilot tests by the author of the model have shown that the theories are relatively accurate under practical conditions. Therefore, the major objective of modeling a diffuser system is to define the theoretical dynamics of the discharge plume as accurately as possible and allow for some variations in the actual field applications.

APPENDIX I

TECHNICAL SUMMARY

Industrial discharge flows can vary widely during the course of the year due to manufacturing requirements and seasonal product variations. A typical diffuser is designed to accommodate a very narrow range of discharge rates. The EPA requires that discharge velocities exceed 10 fps, and the shallow water discharges found in most low-lying areas preclude discharge velocities in excess of 20 fps. If the receiving waterbody is sufficiently shallow, it is possible for the diffuser jet to create a visible jet on the surface of the waterbody.

With these parameters, it is easy to see that flows cannot vary more than 2:1 from minimum to maximum. However, in industry, it is not unusual to find discharge ratios of 3:1 or even greater. While it is possible to install equalization basins, this option is usually cost-prohibitive. A better solution requires the careful design of the diffuser in such a way as to provide the required velocity over a much greater range of flows than is normally possible.

An additional problem with low discharge velocities from diffusers is the possibility of silting in the main diffuser pipe, and the resulting decrease in capacity. What is needed is a check valve at the diffuser nozzle to prevent this. However, flap valves commonly used for this purpose decrease jet velocity and are prone to clogging, jamming, and mechanical failure.

In 1990, Hydroscience approached Red Valve Corp. about modifying their Tide-Flex Check Valves for use with diffusers. Red Valve complied, and named the new series of valves Hydro-Flex. These check valves are unique in that they employ a duck-bill design rather than a flapper to prevent backflow. These valves also have the ability to maintain a more constant jet velocity over a wider range of pipe velocities than with other designs.

As the flow in the riser pipe increases, the duck-bill opens wider to allow greater flow. The relationship is not linear, but still allows the diffuser to maintain jet velocities between 10 and 20 fps for a range of flows exceeding a 3:1 ratio.

Hydroscience has compiled design data for Red Valve Hydro-Flex valves as used in diffuser design. These data indicate that a diffuser can operate over a much greater range of flow rates than is possible with open pipe designs. In addition, the valves act to keep foreign materials from flowing into the diffuser pipe, and will pass relatively large solids that flow down through the outfall pipe.

DESIGN CRITERIA

The EPA has set forth the following requirements:

I. The discharge velocity from a diffuser must exceed 3 m/sec, or approximately 10 fps.

II. The CMC must be met within 10% of the distance from the edge of the outfall structure to the edge of the regulatory mixing zone.

III. The CMC must be met within a distance of 50 times the discharge length scale in any spatial direction. The discharge length scale is defined as $L_Q = A^{1/2}$, where A is the cross-sectional area of any discharge outlet.

IV. The CMC must be met within a distance of 5 times the local water depth in any horizontal direction. The local water depth is defined as the natural water depth prevailing under mixing zone design conditions (i.e. 7Q10 for rivers). This restriction, while it does not prevent locating the discharge in very shallow waters, requires careful design to reduce surface and bottom concentrations.

In addition to these criteria, individual states have adopted additional restrictions on the design of diffusers and size of mixing zones. South Carolina, North Carolina, and Georgia do not have a legislated mixing zone, opting rather for a case-by-case basis. However, Hydroscience has researched the existing regulations by other states, and uses one additional criterion.

V. If the discharge is to a river, the width of a diffuser shall not exceed 33% of the width of the receiving waterbody.

DESIGN AND INSTALLATION CONSIDERATIONS

Diffusers are subject to plugging from foreign objects carried by the wastewater stream, silting during periods of low discharge flow, and most seriously, being washed away during high river flows.

The likelihood of plugging from foreign matter carried in the wastewater stream can be reduced by using a bar screen or basket strainer where the wastewater enters the outfall pipe. This screen should be smaller than the largest particle that can pass through the diffuser. If an object becomes lodged in a diffuser, the flow is reduced and the back pressure rises. Under these conditions, the Hydro-Flex check valve opens to allow the object to pass through the diffuser.

Silting is a problem for diffusers where the discharge velocity falls below the ambient river velocity. Under these conditions, silt and sand can be carried back through the riser pipes, eventually rendering the diffuser useless. Flap valves can reduce this problem, however, they are prone to mechanical failure through corrosion and jamming. In addition, they can be damaged by objects being carried by the river. Hydro-Flex valves close with reduced flow and do not have metal parts that can corrode.

Diffusers can be washed away or severely damaged by objects being carried by the river. For this reason, the Hydroscience design approach is to bury diffusers where possible entirely in the river bed. Only the check valves protrude above the river bottom. These check valves are made of flexible rubber, so that if an object does lodge against the diffuser, the rubber nozzles will flex to avoid damage.

Chapter 10

Can Hazardous Materials Regulations be Met in a Cost-Effective Manner?

R.A. Willis

ABSTRACT

Many managers have stated, "to truly comply with the myriad of hazardous materials regulations in the United States it will cost many dollars." Another prevailing attitude is: "To manage these materials, like I manage other inventories, will cost many dollars." There is some truth in both statements. However, the additional truth is, if you are not truly managing these materials and their uses, you are probably spending thousands of dollars needlessly.

Through electronic data management, (i.e., gathering of information, sound decision making, distribution of information, manipulation of data gathered and distributed, and distribution of responsibility via computer) many costs can be avoided and reduced. One company reduced their on-hand inventory by over 10%. Another developed a hazardous materials handling plan that received a 147% Return on Investment Analysis (ROI) through this approach. This approach requires an up front commitment of time and money, but the payoff is real.

Let's explore how this works. Today, typically, hazardous materials' managers are required to retrieve information from file cabinets, other departments, several computer systems, or hire consultants to do all of the above, etc. This tends to be very expensive and time consuming and the accuracy can be highly questionable for several reasons. First, for example, computer systems not designed to produce the required reports, are generally not designed to collect the proper information. Second, people who perform this task once a year, are generally slow and unless retrained, are not necessarily accurate. Third, people hired for a single task are not always as dedicated as a person who has the ongoing responsibility.

Well, enough sour grapes. What is required? Simply, put - good, sound data collection into a computer system designed to use the information for reporting and other management purposes. The information can be retrieved in several ways. Data can be manually input by the user or Materials Department. Or the data can be input via bar code entry by the user or the Materials Department. Or the data can be electronically input by passing information from other systems if they are already in use. Of course all three methods will require some manual setup, however the third alternative can reduce that amount by passing data electronically, say from an MRP system.

Once the system is set up, then the data can be added on a daily, weekly, monthly, semiannual, or annual basis. However, in some reporting cases, the more often information is input, the more accurate the report will be. For example, a SARA Tier I or Tier II report based on daily input will be far more accurate than a SARA Tier I or Tier II report, based on annual totals.

Tracking material quantities is only part of the requirement. The manager who understands managing hazardous materials, realizes that the task is information management. Information management requires the management of the 6 W's. The what, why, who, when, where, and how much must be readily available for reports, studies, training, etc. This type of information is not only required for hazardous materials/chemicals, but also for hazardous wastes. This type of information tracking has been called "cradle to grave" tracking.

Although a 147% ROI or a 10% reduction may not be available to you and your employer, hazardous materials' management is required if the cost savings and cost avoidance available to you and your employer are going to be achieved. Cost savings can come from reduced material purchases, wiser material purchases, reduced material handling costs, and reduced waste and container disposal costs. Cost avoidance can come from reduced lost time from injury and illness, a more confident work force, reduced liability, reduced medical, insurance and worker's compensation costs, and reduced material spoilage and waste.

The reductions in these areas can only be achieved, if and only if, accurate and current data is available for study. Questions to be asked while doing such studies are typically the 6 W's. What are we doing? Why are we doing it? Where are we doing it? Are we doing it anywhere else? What are we using in each situation? Why are we using it? Is a less costly or less hazardous material being used in one location? Can the more expensive and/or the more hazardous be eliminated? What quantities are we using? In what quantities are we buying it? Are we using the material anywhere else for a different process? Are we buying it in different quantities for different uses? Could we consolidate the purchases? Could we buy in bulk and dispense it in the required quantities? Or could we buy in smaller quantities and reduce spoilage? Can we track expiration dates? If one use location runs out of material, can they readily identify another internal use location that they could contact to obtain the material?

Additional questions may be: What is the true cost of using this material? I.e., does the use of this material require special handling equipment? Special Personal protective equipment? Special disposal practices? Do the containers the material is received in require special handling equipment and practices? Special disposal practices? Would another material eliminate these costs? Why did we choose this material? Was it a "super salesman" that convinced someone to use it? When we started using it, was it the only thing available? Does the process require this material? Can the process be changed to reduce or eliminate these costs? Who do I discuss these findings with to make final determinations?

Even if you do not need to achieve the cost savings and cost avoidances available, the regulations have began to require hazardous materials to be managed. U.S. OSHA requires every employer to have a plan. Either a "Written Hazard Communication Plan," required for every employer except laboratories or a "Chemical Hygiene Plan" for laboratories.

The good manager also recognizes that "accurate reporting," while being required and important, may be the wrong end objective. The proper end objective for "Hazardous Materials' Management" is the elimination of the need to report, as long as, it is cost effective. Then if elimination is not possible or not cost effective, "accurate reporting" should always be the objective. Electronic data collection and distribution can be a cost effective way to have the answers readily available to many of these questions. Using the computer to retrieve data to derive these answers can save much time and heartache.

Plus, through the use of the computer, responsibility can be distributed along with information. It is my observation that every employee is to be responsible for hazardous materials. The responsibilities differ based on their position. For example, receiving personnel should be responsible for assuring the materials are labeled correctly and are handled safety when they are received. Storage personnel should be responsible for assuring the materials are stored properly. Dispensing and distribution people should be responsible for assuring the materials are delivered properly and safely. The using personnel should be responsible for using the materials in a manner that is safe to the environment and to human life. The waste disposal personnel should be responsible for proper collection and disposal. Shipping personnel should be responsible for assuring these materials are shipped properly. All of the preceding should have the responsibility to provide accurate information about their activities. Hazardous Materials' Professionals (HMP) should have the responsibility of providing accurate, current and timely information to all employees to assist them in meeting their responsibilities. Also, the HMP should have the responsibility to then provide accurate reports to the regulators. With the information in a system that is designed to readily provide the needed information, the responsibility is much easier to accept. Once responsibility is accepted, generally compliance and the cutting of costs are much easier to accomplish.

Realizing that we probably cannot eliminate every hazardous material, the good manager then looks for ways to reduce workload and costs required for training and reporting. The computer, of course, is the ultimate tool for such task.

Why is it the ultimate tool for such a task? There are several reasons. First, it is generally available 24 hours a day at less cost than a staff of people. While there may be some downtime to back up files, in my experience, there is less downtime to do this than sick time and vacation of staff, so to take care of all the responsibilities previously mentioned, the computer generally is less expensive than the comparable staff operating in a manual (paper) mode. Second, it is a tool that many people can use at the same time. So, why is that important? Because, as you look at hazardous materials' in general, many disciplines have a need to know something about them. What I mean, for example, is that environmental personnel, hygiene personnel, safety personnel, fire safety personnel, medical personnel, maintenance personnel, production personnel, janitorial personnel, material handling personnel, purchasing personnel, etc. have needs to access the various bits of data from a system that is collecting the correct data about hazardous materials. Third, manipulation of the information is generally far easier and less expensive than retrieving the data from paper files and either using calculators or pencils to complete the required calculations. Or retrieving all locations that need to be air sampled, that have flammable materials and need to be checked to see if they are stored properly, that are making use of a chemical that may require epidemiological studies, that are producing regulated wastes to be able to determine that they be properly disposed.

Third, the data collected may be used to make smarter purchases. To purchase the least cost quantity. Which may include everything from monthly quantities to "just in time" quantities, but reducing handling by your materials' personnel. Changing materials and/or processes to allow for larger quantity purchases, which usually provide lower prices. Or to prevent over purchase. Yet, being able to determine that waste disposal can be accomplished without being cost prohibitive.

Fourth, for those materials that must be handled by your materials personnel, a plan can be developed to create greater efficiency in making those deliveries to use locations and in quantities, that reduce spoilage and waste. Another part is the acquisition of the proper tools for delivering these quantities, i.e., containers, vehicles, etc.

With the appropriate data available, such a plan can be developed and implemented that will allow for the purchase of proper quantities delivered directly to the user or to a central dispensing location. Those being delivered directly to the user are typically unique materials that have a minimum of uses within the facility, quantities may be large or small. Materials delivered to a central dispensing location are typically higher use in multiple locations, so that "just in time" is accomplished internally. Large quantity materials are generally purchased in large quantities and delivered by tank truck or liqui-bins. If delivered by tank truck, the truck driver will off-load the material eliminating the need for your materials' personnel to be involved, except for verification of delivery. Liqui-bins will need to be offloaded, set in a designated location in the dispensing facility, and dispensed from into permanently labeled transportable containers to the user on some predetermined basis. This reduces the use of drums. The importance of reducing drums is the simple elimination of handling the drums through the facility and then disposal of the drum. This may not sound like much, but there are significant dollars involved in this handling. There is also typically spoilage and waste through clingage, improper placement of the pump, and oversupply. Plus, drums tend to be storage containers in work areas that go unaccounted for in on hand figures.

A final area of cost avoidance that cannot be overlooked is that of reducing or eliminating potential fines, legal costs, and imprisonment.

Chapter 11

ASTM Standards for Environmental Site Assessments

J.C. Mauch

I. INTRODUCTION

ASTM (formerly the American Society of Testing and Materials) is a private, nonprofit organization established for the purpose of shepherding the development of standard practices in industry. ASTM Subcommittee E.50.02 was organized in 1989 to publish a standard for environmental site assessments in commercial real estate transactions.

Two standards (E.1927 and E. 1928) for environmental due diligence were published by ASTM on May 13, 1993. This paper will discuss some of the issues raised by the Subcommittee regarding environmental due diligence, and it will outline the new standards.

II. ASTM PROCESS

The philosophy of the ASTM organization is based on a highly representative process that safeguards and ensures a consensus result. In order to achieve that objective, members of each committee are categorized as "Users", "Producers" or "Other". The Users are members of the industry who consume the products and services for which the standard is being established and the Producers are the manufacturers and service providers of those products and services. One of the purposes of categorizing members in this way is to safeguard against a Producer dominated standards process. In fact, under ASTM rules, Users must constitute a majority of voting members on any committee. If there are more Producers than Users on a committee, the number of Producers that exceeds the number of Users will not have voting privileges.

Subcommittee E.50.02 on environmental assessments in commercial real estate transactions was organized into seven task groups, each representing a different aspect of the standard. For example, the task groups included legal, drafting, technical, prior use, government records research, transaction screen and FannieMae. Each task group met and discussed the proposed standard from their vantage point and submitted draft language for inclusion in the final document. Each draft of the standard was then reviewed by each task group and comments were reported back to an Executive task group formed within the Subcommittee.

In keeping with the consensus philosophy, a unanimous ballot is required to approve a standard within ASTM. The ballot options are affirm, affirm with comments, negative, or abstain. All affirm with comments or negative ballots must be addressed in writing by the full Subcommittee. If a negative is found persuasive, the draft must be changed and a new draft sent out for reballoting. If the negative is found non-persuasive, it must be voted down by a majority of the Subcommittee with specific reasons recorded. This process typically results in many rounds of balloting and reballoting to achieve the consensus required.

III. THE PROPOSED STANDARD

The Innocent Landowner Defense under CERCLA or Superfund requires that a purchaser conduct "all appropriate inquiry into the previous ownership and uses of the property consistent with good commercial or customary practice." The charter of the ASTM Subcommittee E.50.02 was to identify good commercial and customary practice for "all appropriate inquiry" or due diligence in commercial real estate transactions and incorporate that practice in the standard.

The first issue faced by the Subcommittee was whether commercial practice for environmental assessments was the same for every transaction. Specifically, certain User groups within the Subcommittee felt that a Phase I assessment can be cost prohibitive in smaller transactions and that something less than a full Phase I better reflects commercial practice for those transactions.

A related issue was whether commercial practice would require that an environmental professional conduct the Phase I. Proponents of the small transaction standard claimed that much of the environmental due diligence in small transactions is performed by the purchaser or by a lender's staff. The Subcommittee's solution to this issue is two standards: the Transaction Screen and the Phase I Assessment.

The Transaction Screen Process

The Transaction Screen is a property characterization study designed to help the User determine whether a Phase I Assessment is necessary. Consistent with the concerns of small transaction Users, the Transaction Screen was designed to be performed by nonprofessionals, such as the property owner, a bank employee or someone they hire.

The scope of a Transaction Screen includes the following steps:

- A User questionnaire regarding prior and current uses of the property.
- A government records review for hazardous waste activity on or in proximity to the property.
- A site visit with a checklist that parallels the topics covered in the questionnaire.

At the conclusion of the Transaction Screen Process, the User should conclude either 1) that there is no indication of Recognized Environmental Conditions on the property and therefore no further inquiry is required, or 2) that there are indications of Recognized Environmental Conditions and therefore further inquiry is necessary. The questionnaire is designed to elicit a "Yes", "No" or "Don't know" response. For any "Yes" or "Don't know" responses, the User must either convert that response to a "No" through additional inquiry, or obtain a complete Phase I Assessment.

The Phase I Environmental Site Assessment

The second standard practice produced by the Subcommittee is for a Phase I Environmental Site Assessment of commercial real estate. Since the charter of the Subcommittee was to codify customary practice, the ASTM Phase I Assessment standard reflects a typical scope of work performed in the industry today. One exception to that observation is a limitation imposed by the Subcommittee's charter. The ASTM assessment standard only addresses hazards covered by CERCLA. Therefore, certain non-CERCLA matters which may be customary Phase I practice today were not included in the standard, but are discussed in a detailed appendix to the standard.

The scope of the ASTM Phase I includes the following steps:

- Research of historical property records.
- Research of government databases.
- Interviews with key site managers, occupants and local agency officials.
- Documentation and report preparation.

Under the standard, a Phase I Assessment must be conducted by an "Environmental Professional." Although the standard does not establish specific qualifications, an Environmental Professional is defined as:

> A person possessing sufficient training and experience necessary to conduct a Site Reconnaissance, Interviews and other activities in accordance with standard practice E.50.02.2, and from the information generated by such activities, having the ability to develop conclusions regarding Recognized Environmental Conditions in connection with the property in question.

"Recognized Environmental Conditions" is defined as part of the required conclusion in both the Transaction Screen and Phase I report. The required Phase I report under the standard must include documentation of each of the steps taken and a summary of findings and a conclusion. The conclusion must state one of the following:

> This Assessment has revealed no evidence of Recognized Environmental Conditions in connection with the Property.
>
> OR
>
> This Assessment has revealed no evidence of Recognized Environmental Conditions in connection with the Property except for the following: [list of exceptions here].

"Recognized Environmental Conditions" is defined as follows:

> The presence or likely presence of any Hazardous Substances or Petroleum Products on a Property under conditions

> that indicate an existing release, a past release or a material threat of a release of any Hazardous Substances or Petroleum Products into structures on the Property or into the ground, groundwater or surface water of the Property. The term includes Hazardous Substances or Petroleum Products even under conditions in compliance with laws. The term is not intended to include de minimis conditions that generally do not present a material risk of harm to public health or the environment and that generally would not be the subject of an enforcement action if brought to the attention of appropriate governmental agencies.

Under the standard, no testing and no recommendations are required. Although subject to some debate, it was generally recognized by the Subcommittee that testing is more appropriately a Phase II activity. Many Users on the Subcommittee also felt strongly that recommendations should only be included in the report if the User specifically requests them to be them included in the report. This was based on a concern that written recommendations could come back to haunt the User should the User choose not to follow the recommendations for business reasons.

IV. OTHER ISSUES

Non-CERCLA Matters

As mentioned above, the charter for the ASTM Subcommittee was to address hazardous substances covered by CERCLA. With one exception, items which are customarily addressed in Phase I assessments but not regulated under CERCLA were excluded from the standard. The exception is petroleum products. Although petroleum is an excluded substance under CERCLA, it is such a common problem, particularly with underground storage tanks, that the Subcommittee included petroleum issues in the Phase I Assessment standard.

On the other hand, asbestos, radon, lead based paint and wetlands are not included under the standard, although in appropriate circumstances, they are customarily included in Phase I Assessments. In order to recognize the importance of these issues in the Phase I process, the standard includes a detailed Appendix which discusses each issue and their relevance to a comprehensive environmental assessment of real estate.

Other Due Diligence Factors

The Innocent Landowner Defense specifically lists five factors that a court will consider when determining whether a purchaser has satisfied the due diligence requirements of the Defense:

- Any specialized knowledge or experience.
- Relationship or purchase price to value.
- Commonly known or reasonably ascertainable information.
- Obviousness of the contamination.
- Ability to detect through appropriate inspection.

Although the last three factors are taken into account by the standards, the first two factors are not. Therefore the User should be aware that it has a responsibility to consider any specialized knowledge or experience it may have, and whether the purchase price reflects an inordinate discount, in order to satisfy the Innocent Landowner Defense.

Use of Prior Assessments

Many Users on the Subcommittee felt strongly that Phase I site assessments should become part of the property's historical documentation, much like title. Moreover, they felt that prior assessments could be relied upon in a subsequent transfer of the property. This was desirable as much to save money on the subsequent site assessment as it was to accurately document the history of the property. In response, the ASTM standard states that a User can rely on a prior assessment as long as it meets or exceeds the requirements of the ASTM standard for Phase I assessments and as long as conditions likely to affect the environmental impairment of the property have not changed materially since the prior assessment was conducted. If specific issues have changed, those issues should be addressed and updated to the time of the current transaction. For example, if more than 90 days has elapsed since the last assessment, the government records review should be updated to the date of the current transaction.

V. CONCLUSION

The ASTM standard practices for the Transaction Screen and for the Phase I Assessment represent consensus standards resulting from input from Users and Producers across the country. In the past, ASTM standards have been given great deference by courts and legislative bodies in most states. Accordingly, it is assumed that the ASTM standard will be adopted by courts when considering whether a party has satisfied the due diligence requirements under the Innocent Landowner Defense of CERCLA. They will also provide the first uniform national guide for environmental due diligence for both the user and producer communities.

Chapter 12

Preparation of Liquid Fuels for Use at Cement Kilns from HazardousWaste Slurries Solid and Semi-Solid

R.R. Scott

INTRODUCTION

A major substitute energy source is presently in use by Portland Cement kilns in the United States, Belgium, France, England, Norway, Netherlands, Mexico, Japan and a growing list of industrialized countries. At the present time, approximately three fourths of the organic chemical waste generated in the U.S. is disposed in this manner. The equivalent amount of coal that does not need to be burned would fill a line of coal cars 85 miles long [Ref. 1].

Disposal of hazardous waste materials, specifically organic solvents such as methyl ethyl ketone, methylene chloride, cyclohexane, and others by thermal treatment was dictated as the sole route of disposal in the United States following the land ban rules initiated in 1987.

Liquid wastes or low viscosity slurries containing a small amount of solids can be burned as fuel through direct injection into rotary kilns with little or no pre-processing required. However, if the wastes consist of lumpy solids or if the slurry is unpumbable, then the waste must be processed before liquid injection into rotary kilns. These wastes typically are deposited in 200 liter drum containers by the generator. If the contents of the drums can be processed into a liquid, pumpable fuel with adequate energy content, this liquid fuel can be supplied to cement kilns.

The operating temperatures of cement kilns exceed those required by the United States EPA to cause a destruction and removal efficiency (DRE) of 99.99 percent. The initial temperature existing in the end of the kiln where the waste is incinerated is approximately 3,000 deg. F. The vapor residence time in a typical cement kiln exceeds 6 to 9 seconds [Ref.1]. The unburned ash including any heavy metals are captured in the cement solids which are, of course,

eventually solidified into non-leachable concrete.
Cement kiln incineration offers the greatest release from liability to the generator since 100% of the generated waste is destroyed or converted to a non-hazardous state.

From an environmental standpoint, there is the closest possible total destruction of the organic solvents and the safe dilution and solidification of the unburned ash. Thermal energy that would not be put to use in classical incineration contributes significantly to the energy requirements of the cement kiln.

There is a whole range of materials including dried or solidified resins, dirt, pigments, fibrous materials, and other trash which may become sufficiently contaminated with solvent so as to make it unacceptable for disposal in sanitary or hazardous waste landfills. The handling of these materials in particular will be discussed in detail in this report.

GENERATION, COLLECTION, AND TRANSPORTATION
Waste solvents that are generated in sufficient quantities can be held in storage tanks and pumped into tanker trucks for transportation to TSD facilities. However, most larger generators of waste solvents have installed their own solvent recovery equipment. Hence, most waste solvents and solvent/solids mixtures are deposited in 55 gallon drums or square material handling drums approximately 350 gallons in capacity.

These are picked-up by licensed transportors and assembled at the TSD facility. An assay and a permanent record of the content of each drum are maintained by the TSD facility.

DRUM HANDLING
Each drum of generated hazardous waste must be treated within a specific short time table. Hence, it is important that drums are arranged and labeled so that they can be processed in the required time frame. Typically, a drum conveyor is used to move drums from the receiving and holding area to the processing area.

The first step is the removal of the top head of the drum. The operator of the drum deheading and drum dumping device will need to make a decision as to whether or not the drum is immediately ready for fuel blending and fine grinding or whether the contents of that drum should be subjected to a rough shredding process first. In some cases, the contents of the drum will not readily fall out of the drum even if the drum is totally inverted. Special loosening devices are employed to make the contents of the drum easier to remove. [Ref. 2]

The inversion and dumping of drums is best done using mechanical devices. Whether the drum will be dumped directly into a fuel blending tank or a rough shredder will depend upon the nature of the contents.

ROUGH SHREDDING OF SOLID MATERIALS

High torque, low speed shredding equipment exists for the handling of large particles greater than 2-6" in diameter. These machines are often called shredders and are capable of reducing particle down below 2-3" [Ref. 3]. Shredding machines are capable of ripping and shredding rubbery materials, plastic materials, wood, paper, rags, dried sludges, and some thin gauged metallic items.

FUEL BLENDING/MIXING VESSEL

Following pre-shredding, the contents of drums should be added to a mixing tank which already contains some waste solvent, waste oil or other fuel mixture. A typical mixing/blending tank is 500-2,000 gallons in capacity. These sizes are utilized because they are easier to handle from a mixing and fine grinding standpoint. The mixing tank should be a cylindrical vertical vessel with a dished or coned bottom to promote good mixing. There should be a bottom outlet in the center of the vessel which will be used for the removal of foreign objects which settle out in the bottom of the vessel. The most widely used mixer type is known as a high speed disperser. This machine utilizes a variable speed, belt drive shaft running at 1,000 - 2,000 RPM to power a wide, flat, circular blade with a serrated edge.

FINE GRINDING

In order to grind particles fine enough to prevent burner nozzle plugging, there must be size reduction to a level 1/8" or smaller in most cases. High speed rotor/stator equipment is utilized to this final step. After solids are added to a quantity of diluting solvent in the mixing tank, a typical practice is to use a grinder pump to recirculate the contents of the mixing tank through itself several times. By recirculating through the grinder pump more than once, finer grinding can be attained while not requiring that 100% of the particles be ground to 1/8" or less on the first pass. This allows the machinery to be resistant to plugging.

Fine grinders use a high speed, typically 900-3600 RPM impeller, that is meshed into a stationary liner with some sort of ripping or shredding teeth. These teeth do the grinding. The teeth are typically meshed at a tolerance of 0.060 - 0.030 inches. The finer the toothed/tooth tolerance, the finer the grinding will be [Ref. 4].

Clearly, one of the most persistent problems in the grinding of hazardous waste solids finely, is the presence of tramp metal such as nuts, bolts, heavy wire, cans, tools, and other objects. Fine grinding machinery cannot grind up this type of material. Hardened metal objects almost always jam the fine grinding machines. Sometimes, failure of machine parts can occur. Usually, only jamming of the machine occurs. Any measure taken to prevent tramp metal material from entering fine grinding machinery is worthwhile since the effort will save on down time and possible repair cost.

STORAGE AND HOLDING

Following preparation of a fuel blend, the batch is normally transferred into a holding vessel. If the fuel blend has to be stored over several hours, it is suggested there be installed some sort of blending mixer such as an axial flow turbine to prevent resettling of solids. In addition, some facilities have installed additional fine grinders in a recirculation loop in order to provide additional finer grinding and also to combat reagglomeration of solids. Reagglomeration often occurs with rubbery or plastic type materials which tend to adhere to each other.

TRANSPORTATION TO THE KILN

The usual method of transport from the TSDF to the cement kiln is via tanker truck or rail tank car. If the transport time exceeds 12 hours, there can be a problem with settling. Solids are often a significantly higher density than the organic solvent carrying liquid. Several steps can be taken to increase stability. First, the transportation and waiting time should be minimized, if possible. This is sometimes feasible with good advance planning when fuel blends are transported via tanker truck. However, with rail transportation, it is assumed that a long time will exist between shipment and arrival at the kiln. Some tanker trucks and rail cars have been equipped with recirculating pumps which draw off the bottom and return a flow to the top. A recent development is the fully agitated tanker truck and rail car. These designs can prevent settling or even re-suspended settled solids [Ref. 5].

A more permanent solution is to prepare a more stable fuel blend. There are two available means to increase the stability of any solid suspension in liquid. If one uses Stokes' law to illustrate the theory, we can see that the settling rate will decrease if the individual particles are made smaller or if the viscosity of the dispersion is increased. Finer grinding utilizing fine grinding machinery with close tolerances is one method to obtain a more stable suspension.

CONCLUSION

A wide variety of solid hazardous waste materials can be incorporated into liquid fuel blends for use at cement kilns. This accepted technology yields major benefits to the generator, processor, and the cement kiln. Specialized equipment is required for handling and dumping drums of solids and semi-solid materials. Rough shredding equipment is required to condition the waste materials so that it can be diluted, mixed, dispersed, and ground fine enough to prepare a stable, pumpable suspension that will not clog cement kiln burner nozzles. By utilizing this type of system, a TSDF can significantly reduce the cost of incineration for a wide variety of materials [Ref. 6 Expanding The Limits Of Solid Fuel Processing For Kilns, E.I. Digest, March 1992, Environmental Information, Ltd., Minneapolis, MN.] The cement kiln method of incineration is very attractive to generators since it represents a complete (cradle to grave) destruction of organic solvents and also renders harmless all remaining unburned ash.

Chapter 13

An Overview of the SO_2 Allowance Market

J.S. Campbell

The Clean Air Act Amendments of 1990 establish an innovative new system that uses tradable emission "allowances" to lower the amount of sulfur dioxide (SO_2) emitted by coal-fired power plants.

HOW THE ALLOWANCE SYSTEM WORKS

Instead of establishing absolute levels that an individual plant can emit with across-the-board emission cuts as previous legislation has dictated, the new system provides for reduction of aggregate SO_2 emissions.

According to the 1990 legislation, power plants emitting SO_2 are issued a predetermined number of emission "allowances." Allowances can be thought of as certificates which permit the holder to emit one ton of SO_2 during a designated year. Therefore, a power plant holding 150,000, SO_2 emission allowances is allowed to emit 150,000 tons of SO_2 during that year. If the power plant were to emit more than 150,000 tons, for example 155,000, it must purchase 5,000 additional allowances sometime during that year or face a fine of $10 million ($2,000 for each excess ton it emits) and forfeit 5,000 allowances the following year.

If the power plant were to emit less than 150,000 tons of SO_2, for example 143,000, it has two choices. First, the 7,000 additional emission allowances can be sold to another power plant needing them, or they can be "banked" or saved for use in future years.

The major benefit of this plan is that by using a market system, total SO_2 emitted into the atmosphere will be reduced at the lowest possible total cost.

For example, suppose power plant A, because of better facilities or technology, can decrease its SO_2 emissions at a lower cost than the market price of allowances. Plant A can install the necessary scrubbing equipment, and generate revenue by selling any excess SO_2 allowances it holds above what it emits. Plant B, on the other hand, because it can purchase additional allowances at a lower cost than it can install scrubbing equipment will buy A's excess allowances. Thus the goal of reducing total SO_2 emissions is accomplished at the lowest possible cost to the ratepayer.

Overall, by the year 2000 annual SO_2 emissions will be cut to about 9 million tons. This is less than half of the 1980 level of 19 million tons. The clean air allowance program will be implemented in two phases.

Phase I takes effect in 1995 and requires the 110 most pollution-intensive power plants, located in 21 states, to reduce their total SO_2 emissions to about 60% of their aggregate 1980 level.

Phase II becomes effective in the year 2000 and affects about 800, less pollution-intensive plants. Phase II cuts total annual SO_2 emissions to less than half the 1980 level.

Additional allowances can become

available through an "opt-in" provision. (Under this provision smaller power plants and other industries that emit SO_2, who are not included under the 1990 Clean Air Act, can "opt-in" to the allowance trading program.) These opt-in polluters are given allowances equal to their average emission level during the years 1985-87. If their emission level is reduced below the amount of allowances they hold, the excess allowances can be sold.

The Environmental Protection Agency (EPA) is charged with the administration of the program. The legislation specifically states, however, that the EPA can contract to have a non-governmental organization conduct the market activities associated with the program.

For more information about the Acid Rain Program, the EPA publishes a series of fact sheets containing information about the Program. Call the Acid Rain Hotline at (617) 674-7377 or the EPA Public Information Center (PIC) at (202) 260-2080.

THE CHICAGO BOARD OF TRADE

A safe and efficient marketplace is critical to the success of a market-oriented pollution program. Price discovery and risk management are both needed for the electric industry to make least cost compliance decisions.

As the world's oldest and largest futures exchange, the Chicago Board of Trade (CBOT) enables businesses and individuals who have a price risk inherent to their industry to transfer that risk to investors who are willing to accept it. Futures markets also provide useful price information that the world looks to as a benchmark in determining the value of a particular commodity or financial instrument at any given time.

The Chicago Board of Trade has the resources and expertise necessary and, therefore, intends to operate an integrated market for SO_2 allowances.

CBOT Objectives:

Specifically the Chicago Board of Trade intends to

- Administer the official EPA allowance auction and direct sales.
- Conduct periodic, private, cash allowance auctions.
- Operate an ongoing cash allowance market.
- Establish and operate an active allowance futures and options market.

I. Administration of Allowance Auction and Sale

In order to provide a liquid market, the provisions of the 1990 Clean Air Act require that 2.8 percent of the emission allowances allocated each year be withheld in a Special Reserve. Therefore, if a power plant were eligible for 100,000 allowances, it would in actuality receive only 97,200 with the remaining 2,800 withheld in the Special Reserve.

A predetermined number of allowances from this Special Reserve will be available for direct sales at a set price of $1500 per allowance. Independent power producers will have access priority to direct sales.

The remainder of the allowances in the special reserve will be sold at open auction each year. Proceeds from direct sales and the annual auction will be redistributed pro-rata to those who forfeit them. Allowance auctions are held no later than March 31 of every year beginning in 1993.

September 25, 1992 the EPA announced that it had selected the CBOT to administrator of the annual allowance auctions and direct sales. Specifically, the CBOT has been selected to administer the following functions: Collect bids and payments; Tabulate the auction; Publish the results of the auction; Collect direct sale applications and payments. Policy decisions, money disbursements, and allowance transfers will remain with EPA. The first EPA auction was conducted March 29, 1993 by the CBOT.

II. Conduct Periodic, Private, Cash Allowance Auctions.

An efficient cash market is essential to ensure that program participants can readily buy and sell allowances as needed. Utilities need to be able to cheaply and easily obtain allowances to cover unexpected increases in emissions and sell excess allowances. The CBOT plans to provide two forums for cash market trading of Emission Allowances through periodic private auctions and its new electronic trading system -- Project A.

In addition to the official EPA auction, the CBOT will conduct periodic private auctions.

The private auctions will complement the annual EPA auction by providing a centralized allowance pricing mechanism reflecting buying and selling that is open to all participants.

The CBOT auctions will let buyers and sellers place bids and offers with complete anonymity, Bids and offers can be placed confidentially through CBOT members, thus letting participants choose from many brokers that will compete for customer business.

A single, market clearing price will be calculated for each emission allowance series (year usable). Using this system, no bidder pays more than his bid price (some will pay less), and no seller will receive less than his offer price (some will receive more).

Market clearing prices, as well as bid and offer information (but not identities) from CBOT private auctions will be disseminated through the CBOT's extensive price reporting network.

III. Operate an Ongoing Cash Allowance Market.

The CBOT plans to provide an ongoing forum for cash market trading of Emission Allowances through its new electronic trading system -- Project A. The Project A system will provide:

- A centralized low-cost trading system.
- Trades that are guaranteed by the Board of Trade Clearing Corporation.
- Easy access and wide dissemination of bid, offer, and other trade information.
- An efficient pay and collect system.
- Full audit services
- A reliable transfer and tracking system.

III. Establish and Operate an Active Allowance Futures and Options Market

Risk Management - Hedging

The creation of a SO_2 allowance market brings with it risk. In order to comply with the new legislation, a power plant can either purchase SO_2 allowances or reduce emissions. Reducing SO_2 emission involves either the installation of expensive pollution control equipment or switching to low-sulfur coal. Without the ability to establish future prices for SO_2 allowances, power plants are unable to make informed decisions on how to best comply.

For example, suppose management at power plant A determines that it is economically better for it to comply by installing scrubbers thus generating excess allowances that can be sold. Once that decision is made, plant A is at substantial risk that the price of allowances will fall, thus decreasing the value of its excess allowances and jeopardizing its rate of return.

On the other hand, power plant B might decide that it is more feasible to purchase allowances rather than installing expensive pollution control equipment. In this instance plant B is exposed to the risk that in the future, allowance prices will rise.

Had plant A and B been able to establish or "lock-in" a future price for SO_2 allowances when the initial decision was made, price risk would have been eliminated.

The Chicago Board of Trade has almost 150 years experience in helping the agricultural and financial industries manage price risk. Futures markets allow both buyers and sellers to establish a price for a commodity several months or even years before the actual transaction will occur thus eliminating the risk of price fluctuations. Using futures markets to protect against adverse price fluctuations is known as *hedging*.

Hedging can also be accomplished through the use of options on futures. Purchasing an option is similar to purchasing insurance. The buyer of an option pays the option seller a premium. In exchange for the premium, the buyer is granted the right to buy or sell a futures contract at a predetermined price if he chooses to exercise the option. Therefore, the option buyer is protected against a price move against him, but is still able to take advantage of a possible favorable price move.

The CBOT proposes to establish both a futures and options market for SO_2 allowances. These markets will allow buyers such as power plant B in the above example, and sellers like plant A, along with other market participants, to come together and negotiate the price of contracts for the future delivery of SO_2 allowances. At the same time these market participant can transfer their price risk to speculators who are willing to assume it. Establishing a future price for allowances, enables power plants to make economically sound decisions as to how best comply with the Clean Air legislation.

Price Discovery

Futures markets create a centralized marketplace where buyers and sellers come together to trade. Through this process of bidding and offering, prices are discovered. The Clean Air futures market will play a critical role to utilities, by signaling the maximum cost they should be willing to accept to reduce their emissions.

In deciding whether to install pollution reduction technology, or purchase allowances, utilities must determine the value of SO_2 allowances. The prices established in the futures market will indicate this value. Once a decision has been made, utilities can protect themselves against adverse price fluctuations by hedging with futures.

For More Information

• **CBOT** For more information about futures markets, hedging, Project A, and the emission allowance auction & direct sales procedures, cash market, and futures and options market, contact the CBOT Education & Marketing Services Department at (312) 341-7955.

• **EPA** For more information about the Acid Rain Program, the EPA publishes a series of fact sheets containing information about the Program. Call the Acid Rain Hotline at (617) 674-7377 or the EPA Public Information Center (PIC) at (202) 260-2080.

Chapter 14

Management of Storm Water Runoff from Uncontrolled Hazardous Waste Sites

W.E. Norton

ABSTRACT

Storm water runoff from uncontrolled hazardous waste sites presents special challenges to property owners, regulators, engineers, and contractors. The property owner must establish positive control and prove to the regulators that the storm water is not carrying pollutants off the site. The regulators must assure themselves that the interests of the public are protected and that the appropriate (but sometimes confusing) regulations are followed. Engineering firms and remediation contractors working to restore the site must develop plans to route the storm water through or around the project to prevent off site water from being affected by on site contamination. If required, they must secure their own permits for treatment and/or discharge to local streams or sewer systems of water that does accumulate. The first and most obvious step is for engineers and contractors to prevent run-on storm water from entering the contaminated area. The next is to prevent not only rain from water falling on the contaminated area but also ground water from entering and mixing with the contaminants if at all possible. Recognizing that this is not always possible, they must provide a means of properly containing and disposing of contaminated water should mixing occur. This paper presents a discussion of the intricacies of the regulations and a summary of accepted practices for solving the problems of storm water runoff from uncontrolled hazardous waste sites.

INTRODUCTION

One example of the success of the program to restore the environment has been the slow but steady improvement of the quality of our nation's surface waters. The principal cause for that improvement has been the establishment of control over the major point sources of pollution such as municipal sewers and industrial waste waters. With this success has come the realization that non point sources such as storm water runoff also have a significant detrimental effect on the quality of our nation's waters. The control of storm water runoff is part of an overall program designed to control this source.

This paper is directed towards examining the requirements for the control of storm water runoff from uncontrolled hazardous waste sites. The regulations are complex and vary from state to state but there are general requirements that apply nation-wide. The first part of this paper discusses the relevant federal, state, and local regulations that apply. The second part discusses the closure of uncontrolled hazardous waste sites and the problems that can be encountered with storm water. The third section addresses controlling erosion and collecting the water. The final section discusses the treatment of contaminated water.

The following laws and amendments are referred to by their acronyms in the text of this paper and are defined here for clarity:

- CERCLA - The Comprehensive Environmental Response, Compensation, and Liability Act of 1976 amended by the Superfund Amendments and Re-authorization act of 1986. The requirements of both of these are referred to as CERCLA throughout this paper.

- RCRA - The Resource Conservation and Recovery Act of 1976 as amended by the Hazardous and Solid Waste Amendment of 1984. The requirements of both are collectively referred to as RCRA throughout this paper.
- CWA - The Clean Water Act . The CWA is the oldest and most often amended of the major pollution statutes (almost annually). The reference to the CWA in this paper refers to the original act and its amendments including the Federal Water Pollution Control Amendments (FWPCA) of 1972 and the Water Quality Act (WQA) of 1987.

THE REGULATORY ENVIRONMENT

In discussing storm water runoff from uncontrolled hazardous waste sites, it is necessary to define what is meant by the term "uncontrolled hazardous waste site". In general terms, a hazardous waste site is a place where hazardous waste is used, treated, stored or disposed. A controlled hazardous waste site is a site where all of the controls required by the regulations are present and maintained. An uncontrolled hazardous waste site is a site which does not have such controls. Thus a hazardous waste landfill where a double liner system is installed, runoff and leachate collection provided, and ground water monitored is a controlled hazardous waste site. A site where hazardous waste has been discarded without the appropriate controls is not. Uncontrolled sites exist in violation of law, and it is required that these sites be brought into compliance with the regulations through a formal cleanup process. The regulations that created and control this process are based on the requirements found in RCRA, CERCLA, and CWA.

Upon the implementation of RCRA, the treatment, storage, or disposal of hazardous waste without a RCRA permit was prohibited. There were specific exemptions made in the law, but for the most part all facilities were included. The requirement was phased in by allowing facilities to operate in an interim status until the application process could be completed. RCRA established the rules that control hazardous wastes and hazardous waste sites including accidental releases.

The primary objective of CERCLA was to establish the authority and the process to clean up uncontrolled hazardous wastes sites. Its enactment closed the loop with RCRA to create a comprehensive set of rules. A major issue in the cleanup of hazardous waste sites had been the responsibility, liability and financing of the work. CERCLA provided that cleanup is performed first and arguing about who pays for it comes later. It also provided a great deal of guidance about how to conduct cleanups.

The two laws work together. RCRA provides the guidance for the management of active facilities that handle both solid and hazardous wastes. It also requires cradle-to-grave accountability for hazardous waste through a detailed system of records. CERCLA provides direction and authority to clean up hazardous waste sites regardless of status; however, CERCLA is not normally implemented on active facilities where RCRA applies since the owner can be held responsible. RCRA sites vastly outnumber CERCLA sites, and most of the cleanup of environmental problems being done today are under the requirements of RCRA.

RCRA contains its own corrective action program intended to require owners and operators to clean up their own facilities. The permitting process is the driving force. If the waste is not cleaned up, the permit is not issued and the facility is shut down. For the owner there is no escape since CERCLA provides the authority for the regulators to enter the process, clean up the waste, and hold the owner liable for the costs. The RCRA corrective action process closely corresponds with the CERCLA process although the individual steps have different names. A significant difference is that while CERCLA uses the concept of reportable quantities, RCRA does not. Under RCRA, all releases of hazardous materials must be addressed.

The CWA also impacts the remediation of hazardous waste sites. The CWA's roots go back to 1948, but prior to 1972 the act and its numerous amendments simply affirmed the federal policy to recognize, preserve and protect the primary responsibilities and rights of the states in controlling water pollution. It was not until 1972 and the passage of the FWPCA that the CWA as it is thought of today was established. The 1972 law set ambitious goals and required the Environmental Protection Agency (EPA) to establish federal limits on the amounts of specific pollutants that could be released by municipal sewage treatment plants and industrial facilities. These effluent limitations were based on the level of cleanup that could be achieved by existing technologies for waste water treatment. The limits were written into National Pollution Discharge Elimination System (NPDES) permits and became conditions of the permits. The act made it illegal for municipalities and industrial facilities to discharge pollutants without an NPDES permit or in violation of the permit requirements.

The system does allow an industry to discharge to a Publicly Owned Treatment Works (POTW) without an NPDES permit as long as the facility obtains an indirect discharge permit from the POTW. It then becomes the responsibility of the POTW to meet the requirements of the NPDES permit. In either case, the discharge is made through a permitted facility.

The change in the CWA brought about in 1972 was focused on reducing pollutants from point sources. The reason for this focus was that in 1972 these sources were not adequately controlled and represented pressing environmental problems. The discharges were easily identified as responsible for massive pollution and serious degradation of national water quality. However, as pollution control measures were developed for these point sources, it became evident that more diffuse sources such as agricultural and urban storm water runoff were also major sources of water quality problems.

To remedy this situation, the CWA was amended again in 1987 (WQA) to require the EPA to address storm water discharges. The WQA required that storm water discharges associated with industrial activities and municipalities be required to meet appropriate water quality standards. In 1990, the EPA issued the implementing regulations, requiring that large urban areas and selected industries that discharge storm water directly into receiving streams apply for an NPDES permit. As with RCRA, the program is being phased in starting with the industries and cities that have the greatest impact on our nation's waters. In addition to meeting quality standards, the regulations also required the reporting of any spill of hazardous materials that occurred within the last three years.

The storm water regulations took another step toward the development of a system that requires industries and cities to be responsible for cleaning up their sewage and their storm water runoff. The principal enforcement mechanism of the system is the synergistic effect of CERCLA, RCRA and the CWA as they impose permit requirements for operations.

CLOSING AN UNCONTROLLED HAZARDOUS WASTE SITE

RCRA regulates solid and hazardous wastes; the CWA regulates waste water and storm water discharges; and CERCLA provides the enforcement mechanisms to cleanup hazardous waste sites when things go wrong. Uncontrolled hazardous waste sites exist in violation of the law and must be brought into compliance. The normal way to do this is by closing the site either under CERCLA or RCRA.

A critical decision that must be made early on is whether or not immediate action is required. This decision is made by the regulators based on the urgency of the threat to human health and the environment. This is done by a risk assessment that weighs the threat against the cost and effectiveness of immediate action. If the threat is not an immediate danger, the owner is usually allowed time to engineer an appropriate response.

RCRA requires that any facility that treats, stores or disposes of hazardous waste be considered a Treatment, Storage, and Disposal facility (TSD) subject to 40 Code Of Federal Regulations (CFR) 264 or 265, unless excluded. This means that they require a RCRA permit to operate and must meet the standards specified for TSD facilities. As part of the permitting process, uncontrolled hazardous waste sites on a RCRA facility have to be reported, and the owner has to submit a closure plan with the permit application specifying how the problem will be remediated. Once the plan is approved, the permit is issued with the closure plan as a condition of the permit (40 CFR 270.14(b)(13)).

The primary purpose of the closure plan is to insure that a hazardous waste site is closed in a manner that protects the public and the environment and controls, minimizes or eliminates the post-closure escape of hazardous waste, hazardous constituents, leachate, contaminated storm water runoff, or waste decomposition products to the ground water or atmosphere (40 CFR 264/265.111). The closure plan has to provide for either removing or decontaminating all hazardous waste and waste residue (clean closure) or closing the site with the hazardous waste in place (dirty closure) and post-closure monitoring similar to that of a hazardous waste landfill.

One of the greatest risks for contaminants moving off-site is provided by water. Water provides two general opportunities for hazardous wastes to migrate. One is flow over the ground (surface water) and the other is flow within the ground (ground water). Ground water is usually the most important path because the contaminants are either buried or have penetrated into the ground. The contaminants that contact the ground water may travel dissolved in it or as a separate liquid phase carried along with it. The natural filtering action of the soil usually inhibits particulate matter from migrating so that contaminants adsorbed on the soil particles stay in place. Surface water can furnish a rapid and direct transport route as waste may move with it, or it can erode the soil particles mobilizing the adsorbed contaminants. In addition, surface water can infiltrate the site, raising the water table and increasing ground water flow.

For the most part, the behavior of surface water is well known, and controlled hazardous waste sites are constructed deliberately with the intention of preventing surface water from either carrying the waste off site or from infiltrating into the soil. However, some infiltration does occur resulting in ground water that moves through the waste and extracts the leachable contaminants. This leachate must be captured and treated. Controlled hazardous waste sites have impervious caps, liners and leachate collection systems in order to do just that.

The control of surface water at an uncontrolled hazardous waste site is just as important. During remediation, it is assumed that any surface water that comes in contact with the hazardous waste is contaminated and must also be treated as hazardous. For this reason, the precipitation that falls within the hazardous waste area and is allowed to contact the hazardous waste is collected and (usually) treated prior to discharge. Once the site is properly closed and the rain no longer comes in contact with waste, it can be discharged without treatment, but it is still subject to monitoring and regulation. Thus, a properly constructed cap serves the important purposes of preventing infiltration and keeping the runoff from contacting the wastes.

Some feel that it is generally advantageous for a facility owner or operator to remediate a site under CERCLA rather than RCRA. CERCLA allows for site specific, performance based standards that are technologically feasible and economically cost effective. Once the long and involved CERCLA process has been completed, the final decision allows the remediator to proceed without further encumbrances. The EPA has determined that since the CERCLA process involves "essential compliance" with all known requirements, further permits are not required.

Under CERCLA, the method of handling the storm water is determined considering all of the requirements and restrictions and is written into the Record of Decision (ROD). The approval of the ROD then becomes the approval of the disposal procedure, whether it be treatment and discharge to a receiving stream or collection and transport to a permitted treatment and disposal facility.

This is not the case with RCRA. The RCRA closure plan may consider all of the requirements, but the method of compliance is the permitting process, and no automatic exemptions are given. Regulators are generally not satisfied with the differences in the processes and work is being done in some states to try to resolve them. However, it is vitally important to understand what the governing rules will be before remedial action is started.

An example may help to illustrate the point. With RCRA, the regulators that approve closure plans are often not the same regulators that permit water treatment plants and coordination between them is not automatic. The regulators that permit water treatment plants may be accustomed to a deliberate process where the application can take months for approval. Expecting to set up a water treatment facility and start discharging without a permit simply because a closure plan has been approved can lead to problems. In addition, contemplating site remediation without the concurrence of the responsible regulatory agency is risky. It is also unnecessary since most state agencies are happy to cooperate with cleaning up waste sites and will be as helpful as they can.

An additional permitting requirement has been placed on remediation projects of over five acres. The CWA now requires that storm water runoff for all construction projects over five acres be permitted, and remediation is considered construction. In most states, this requirement is being met by the issuance of a state-wide permit under which contractors can discharge. The permit covers all construction over five acres and sets general standards for water quality. The key to the permit is the requirement for owners to use Best Management Practices (BMP's) in the development of their pollution prevention plans. Until recently, it was difficult to find exactly what was meant by BMP's, but the EPA recently published an excellent guide which is listed in the references to this paper. This requirement is new and its impact has not yet been generally felt, but its net effect on water quality should be positive.

CONTROLLING EROSION AND COLLECTING THE WATER

It has been previously stated that the storm water that accumulates on the job site is presumed to contain hazardous constituents if it is allowed to come in contact with the hazardous waste. This is somewhat of a universal assumption and not unreasonable, especially if the remediation effort requires excavating the waste in order to remove or treat it. Once uncovered, the waste is in its most vulnerable condition, and if there is any potential for migration off site, it will most likely occur then. In such cases, the water that contacts the hazardous waste is collected and one of the following actions taken:

- The water is treated on site to the appropriate water quality standard and discharged to a receiving stream.
- The water is treated on site to the appropriate pre-treatment standard and discharged to a designated POTW.
- The water is collected and transported to a RCRA permitted facility for treatment before discharge.
- The water is not treated on site but either discharged to a receiving stream or allowed to infiltrate into the ground with only monitoring for water quality required.

Discharging without treatment seems to be allowed only infrequently although examples of sites where it has been done are not hard to find. It is now the exception to the rule.

Collected contaminated water can be transported to commercial RCRA facilities permitted to treat hazardous waste containing water. These facilities are few and the transportation distances may be considerable, but they do provide an alternative to on site treatment. They are not cheap, and costs for transportation and treatment can vary considerably and can cost as much as one dollar per gallon.

It is obviously to everyone's advantage to limit the amount of storm water that contacts the hazardous wastes. One way this can be done is by keeping the area designated as the zone of contamination as small as is practical considering the work space needed. The larger it is, the more contaminated water it will collect. It must also be isolated hydraulically from the rest of the site by the use of berms and other structures to keep outside storm water from running on. That way, the only contaminated storm water that has to be dealt with is that which falls on the contaminated portion of the site. The runoff that cannot be diverted should be deliberately routed through the project with great caution to cause as few problems as possible. If it cannot be controlled because the quantity is too great, then contingencies for site flooding will have to be considered. It is also important to remember that if the project is going to significantly change the local drainage pattern that the neighbors must be considered. Increasing or significantly altering local storm water flow because of a hazardous waste clean up project can be a serious public relations and liability issue.

Erosion is a major problem on any construction site, and especially so on a hazardous waste site. It is visual evidence that soil and contaminants are migrating. In order to prevent erosion, it is necessary to keep large quantities of fast moving water from developing. The EPA manual of BMP's has an excellent discussion on the subject and is intended to help in controlling erosion as a major source of non point source pollution. Some suggestions extracted from that and other sources are listed below:

- Use check dams to reduce the velocity of the water.
- Use temporary impoundment's and catch basins to slow the rate of discharge.
- Use intercept ditches and berms to divert the water to where it can be collected and controlled.
- Protect soil surfaces by compaction or sealing where necessary.
- Plan and design the drainage structures using sound engineering principles.

It should be remembered that when large quantities of water need to be controlled quickly, such as during storms, it needs to be done deliberately before nature takes over. For permanent facilities, twenty-five and one-hundred year storms are often used to establish design criteria (40 CFR 264.18/264.301). Short duration remediation projects often use storms of shorter periods such as three or ten years.

In addition to erosion, contamination of storm water by contact with the hazardous waste is also an issue. There are some techniques that are widely used in the industry that can help prevent the generation of excess contaminated storm water. These are:

- Line the site to prevent storm water from contacting the contaminated soil.
- Grade the site to move the water to where it can be collected and pumped easily.
- Limit the amount of bare soil exposed at any time.
- Cover and seal the site before each rain.
- Keep all non-active surfaces sealed.

However, liners are not the solution to all situations. One only need think of the small army it takes to cover a baseball diamond during a rain shower to visualize the challenge. The larger the area to be lined, the more complex the management problems. Liners can limit the amount of contact water and can help control costs. On small sites they can greatly reduce the quantity of contaminated water.

TREATING CONTAMINATED WATER

When all else fails, the storm water has to be treated. Fortunately, water treatment technology has been around for centuries, and wherever people have gathered, the necessity to prepare safe drinking water and dispose of raw sewage has been a priority. An abundance of literature exists on the subject, and one only need to look to get almost any question answered. The 1993 Public Works Manual listed in the references is one of the best resources available because it not only discusses available technology but also gives the names of contractors and suppliers that are usually left out of technical discussions.

Drinking water preparation and sewage treatment evolved differently and are driven by different requirements and conditions, but each has items of interest to our discussion. The classic means of treating drinking water for human consumption has four steps:

- Coagulation and flocculation
- Sedimentation
- Filtration
- Disinfection

Systems that use this process exist that range in size from huge city water treatment plants to small field units. There are other ways to treat drinking water though, and systems that use distillation, reverse osmosis, etc., are available commercially to fit almost any need. Industry has even devised systems that can provide safe drinking water in the most hostile of environments whether on a contaminated battlefield or under the ocean.

Sewage treatment has evolved on a somewhat different line. The conventional way of referring to the steps in the sewage treatment process is as follows:

- Primary treatment
- Secondary treatment
- Tertiary treatment

Primary treatment is the use of screens and comminutors followed by grit chambers and sedimentation basins to remove debris, sand, and "grit" from sewage. Secondary treatment adds the oxidation of the dissolved organics, the removal and further treatment of the derived sludge, and the disinfection of the effluent. Nearly all secondary treatment plants use biological methods as their major treatment process although there are other means. The vast majority of sewage treatment plants provide only secondary treatment prior to discharge. Additionally, waste water treatment technologies use an extremely wide variety of options for solids removal from flocculation and settlement to aeration and flotation.

Tertiary treatment is the further processing of effluent to remove the undesirable constituents left after secondary treatment. Tertiary treatment may take the form of solids removal by multi-media filters or micro strainers; carbon filters or columns to absorb nitrogen or organics; ion exchangers; or combinations of processes required for specific purposes.

All of these processes are transportable to the field. In considering a water treatment process for a remediation project, there are many possible options and variations. A very reliable and versatile process that is applicable in most cases is as follows:

- Pre-treatment by screening out debris and settling out sand and grit.
- Clarification by coagulation/sedimentation using chemical coagulants.
- Filtration through a sand, multi-media filters, or other type of filter.
- Adsorption using activated carbon.
- Disinfection if required.

This process is relatively thorough in its treatment capability, can be easily adapted to use in the field, can be modularized if steps are not needed, and consists of well-developed technology. Whatever the process selected for treating the contaminated storm water, it is vital to do the following:

- Know what the contaminants are present.
- Develop the processes needed to remove them.
- Conduct a thorough treatability study.
- Verify the scale up from the lab to the field.
- Study and understand the economics.
- Make the decision.

CONCLUSION

Storm water runoff from an uncontrolled hazardous waste site can be a serious problem if not properly managed. Sound engineering practices, and a deliberate planning process can detect most of the problems early and can prevent disasters. The regulatory climate is proceeding in the direction of increasingly complex and stringent requirements. It is prudent for all concerned to stay updated.

REFERENCES

Institute of Hazardous Materials Management, Handbook On Hazardous Materials Management, 1990.

James T. Maughan, Ecological Assessment of Hazardous Waste Sites, New York: 1992.

O'Brien & Gere Engineers, Inc., Hazardous Waste Site Remediation - The Engineer's Perspective, tech. ed. by Robert Bellandi, New York: 1988.

Office of Federal Register National Archives & Records Administration, Code of Federal Regulations - Protection of Environment, parts 260-299, 1992.

The 1993 Public Works Manual, vol. 124, no. 5, New Jersey: 1993.

Travis P. Wagner CHMM, The Complete Guide To The Hazardous Waste Regulations, 2d ed., New York: 1990.

United States Environmental Protection Agency, Office of Wastewater Enforcement and Compliance, Developing Pollution Prevention Plans and Best Management Practices, 1992.

United States Environmental Protection Agency, Office of Solid Waste/Permits and State Programs Division, Office of Ombudsman and the Association of State and Territorial Solid Waste Management Officials, Resource Conservation and Recovery Act Manual, 1990.

Chapter 15

Storm Water Pollution Prevention Plans for Industrial Activities

G.E. Seaburn

INTRODUCTION

Storm water discharges have been identified as a significant source of water pollution in numerous nationwide studies on water quality. To address this problem, the Clean Water Act Amendments of 1987 required EPA to publish regulations to control storm water discharges under the National Pollutant Discharge Elimination System (NPDES). EPA published storm water regulations on November 16, 1990, which require certain discharges of storm water to waters of the United States to apply for NPDES permits. According to the final rule, facilities with a "storm water discharge associated with industrial activity" are required to apply for a storm water permit. EPA has defined this phrase in terms of eleven categories of industrial activity that include: 1) facilities subject to storm water effluent limitations guidelines, new source performance standards, or toxic pollutant effluent standards under 40 CFR Chapter N; 2) heavy manufacturing facilities; 3) mining and oil and gas operations with contaminated storm water discharges; 4) hazardous waste treatment, storage, or disposal facilities; 5) landfills, land application sites, and open dumps; 6) recycling facilities; 7) steam electric generating facilities; 8) transportation facilities, including airports; 9) sewage treatment plants; 10) construction operations disturbing five or more acres; and 11) other industrial facilities where materials are exposed to storm water.

The storm water regulations present three permit application options for storm water discharges associated with industrial activity. The first option is to submit an individual application consisting of Forms 1 and 2F. A second option is to participate in a group application. The third option is to file a Notice of Intent to be covered under a general permit in accordance with the requirements of an issued general permit.

Regardless of the permit application option a facility selects, the resulting storm water discharge permit will most likely contain a requirement to develop and implement a Storm Water Pollution Prevention Plan (SWPPP).

NPDES permits are issued by the state for states that have been delegated NPDES permitting authority or by EPA for states that have not been delegated NPDES authority. Therefore, the specific EPA general permit requirements discussed in this paper apply only to facilities located in one of the twelve non-delegated states or territories. EPA expects, however, that the federal general permit will be used as a model by NPDES - authorized states tailored to meet the state-specific conditions even though storm water permit requirements vary from state to state depending on water quality conditions and permitting priorities for the permitting authority, EPA expects that most NPDES storm water discharge permits will contain Storm Water Pollution Prevention Plan requirements similar to those discussed in this paper.

A Storm Water Pollution Prevention Plan is a written document designed to be site specific, to eliminate or minimize pollutants in storm water discharges to the maximum extent possible, and to promote best management practices for the control of storm water pollutants. The plan is developed in five phases: 1) planning and organization; 2) site evaluation; 3) best management practices; 4) implementation; and 5) evaluation and monitoring. This paper will outline the elements that each of these phases of work.

PLANNING AN ORGANIZATION

Pollution Prevention Team

Each plan should identify a specific individual or a team of individuals within the facility who will be responsible for developing and implementing the Storm Water Pollution Prevention Plan. Commonly, the facility or plant manager is designated as the team leader and will delegate responsibilities to others, as necessary. Some of the responsibilities that may be delegated include: 1) implementation of the physical features of the plan; 2) training; 3) inspections; 4) record keeping; and 5) plan revisions and update.

Consistency with Other Plans

The SWPPP must consider other contingency plans that the facility is subject to; these may include the Preparedness, Prevention and Contingency Plan (40 CFR 264 and 265), the Spill Prevention Controls and Countermeasures requirement (40 CFR 112), and the National Pollutant Discharge Elimination System Toxic Organic Management Plan (40 CFR 413, 433, 469) and the Occupational Safety and Health Administration Emergency Plan (29 CFR 1910).

SITE EVALUATION PLAN

Develop a Site Map

The facility site map must include:

- An outline of the drainage area of each storm water outfall
- Location of any existing structural control measures used to reduce pollutants in storm water runoff.
- Surface water bodies
- Locations where significant materials are exposed to precipitation
- Location where major spills or leaks have occurred
- Location of each of the following activities (where exposed to storm water);
  - Fueling stations
  - Vehicle and equipment maintenance and/or cleaning areas
  - Loading/unloading areas
  - Treatment, storage, and waste disposal areas
  - Liquid storage tanks
  - Processing areas
  - Storage areas

Develop a Drainage Map

For each area of the facility that generates storm water discharges associated with an industrial activity, identify the direction of storm water flow including the types of pollutants expected to be included in the runoff, including erosion of sediments.

Inventory of Exposed Materials

The plan must include the location on the site map and a narrative description of 1) significant materials that have been handled, treated, stored, or disposed in a manner to allow exposure to storm water for the last three years; 2) method and location of on site storage and disposal; 3) materials management practices employed to minimize contact of these materials with storm water runoff over the last three years; 4) existing structural and non-structural control measures to reduce pollutants and storm water runoff; and 5) any treatment of storm water runoff.

Identify Past Spills and Leaks

Prepare a list of spills and leaks above reportable quantities of toxic and hazardous pollutants that have been exposed to storm water over the past three years.

Certify Non-Storm Water Discharges

The plan must provide a procedure to inspect all storm water outfalls for dry weather flows. If flows are present, an investigation of the source of flow by testing -- visual inspection from the outfall to the source, smoke test, dye test, or TV monitoring -- are required. A certification by a "responsible corporate officer" is required and includes: identification of potential storm water discharges, a description of the results of any tests and/or evaluation of the presence of non-storm water discharges, the evaluation criteria or test method used, the date of testing and/or evaluation, the on-site drainage points that were directly observed during the test and/or evaluation.

Description of Any Existing Data

Any existing storm water data describing pollutants in the storm water must be included with the plan.

Risk Identification and Summary of Potential Pollution Sources

The plan must provide an evaluation of the level of risk to storm water from each pollutant source. The list should be prioritized and used in developing an implementation schedule for allocating resources in the subsequent phases of work.

BEST MANAGEMENT PRACTICES

Best Management Practices are measures used to prevent or mitigate pollution from any type of activity. Best Management Practices are very broad classes of measures and may include processes, procedures, schedules of activities, prohibition of practices, and other management practices to prevent or reduce water pollution. In essence, they are anything a plant manager, department foreman, environmental specialist, consultant, or employee may identify as a method, short of actual treatment, to curb water pollution. They may be inexpensive or costly. BMPs can be just about anything that does the job of preventing toxic and hazardous substances from entering the environment.

There are generally four classes of BMPs:

- Prevent and minimize exposure
- Contain or divert flow
- Recycle or reuse the resource
- Treatment

Baseline BMPs

Each facility covered by the permit must develop a description of storm water management controls appropriate for the facility and implement such controls. These controls may include: good housekeeping, preventive maintenance, visual inspections, spill prevention and response, sediment and erosion prevention, management of runoff, employee training, recordkeeping and reporting.

Additional BMPs

More intensive and often times more costly BMPs may be required for special applications. These may include structural changes to control storm water exposure such as, diversion structures, retention ponds, or sediment basins, for example. They may also include structural changes to the manufacturing process, or materials handling procedures to eliminate exposure to storm water runoff.

IMPLEMENTATION PHASE

The plan requires the development of a schedule for implementing each of the elements of the SWPPP. It requires the assignment of responsibilities to individuals to implementing the various elements of the plan. It also requires the development of a commitment from management to support the implementation schedule and strategy for the plan.

The plan requires the development of an employee training program to inform all employees of the plan and the implementation schedule. The training can be formal or informal, but must be documented and must address each element of the SWPPP.

THE EVALUATION PHASE

Qualified and designated personnel must conduct site compliance evaluations at intervals as specified in the plan, and not less than once a year. These evaluations must be documented in a report. The report must summarize the inspection and any follow up actions, the date and the personnel conducting the inspection, and any incidents of non-compliance or certified that the facility is in compliance with the plan. Any changes specified by the evaluation must be implemented within twelve weeks of the inspection.

Each report serves as the documentation of the plan and must be kept with the plan at all times.

Each SWPPP must be amended whenever there is a change in design, construction, operations, or maintenance at the facility with the potential for impacting storm water runoff.

GENERAL REQUIREMENTS

Deadlines

See TABLE 1.

Deadlines for state issued permit may differ from these dates.

Signatures and Certifications

The SWPPP, all reports, certifications, and other information submitted to authorities must be signed by a "responsible corporate officer," a general partner, or a proprietor. A responsible corporate officer is a president, secretary, treasurer, or vice president in charge of a principle business function, or a manager employing more than 250 people, or having gross sales in excess of $25 million dollars with authority assigned or delegated by corporate policy.

If the facility is subject to SARA III, Section 313 regulations, then the plan must also be reviewed and certified by a registered professional engineer for the application of good engineering practice. The professional engineer must also re-certify the plan every three years.

Plan Location

The SWPPP must be maintained on-site at all times and available for inspection by regulatory authorities. Regulators and municipal operators may request copies for review at any time. Plan and records submitted to the Director are subject to public access.

SPECIAL REQUIREMENTS

Discharges to Municipal System

Industrial facilities discharging storm water to a municipal storm water storm sewer system (MS 4)are subject to any compliance requirements established by the MS 4 operators, as part of their NPDES permit.

Special Requirement for SARA Title III, Section 313 Reporting Facilities

Section 313 of SARA requires operators of manufacturing facilities that handle toxic chemicals in amounts exceeding threshold levels listed in 40 CFR 372.25 to report to the government on an annual basis. Because these facilities handle large amounts of toxic chemicals, EPA concluded that they have an increased potential to degrade the water quality of receding streams. To address this risk, EPA established specific control requirements in its general permit. In particular, these requirements apply to Section 313 facilities that report "water priority chemicals" for any of over 200 chemicals that have been identified by EPA as specifically toxic to water ecosystems. These facilities have more stringent requirements to meet as part of the SWPPP.

Salt Storage Piles

The plan requires that all salt storage piles must be covered or enclosed to prevent exposure to precipitation.

SUMMARY

The SWPPP is the principle management tool established by EPA in designated states to initially reduce or minimize storm water pollutants in runoff. The SWPPP is designed to be site specific, to minimize the specific pollutant sources, and to promote implementation of Best Management Practices throughout the facility. This initial "first cut" attempt to control pollutants in storm water, if properly prepared and implemented, will significantly improve the quality of the storm water runoff from industrial facilities and may preclude subsequent regulations.

TABLE 1

EPA BASELINE GENERAL PERMIT REQUIREMENTS
Schedule for Plan Development and Implementation
Part IV.A.

Type of Facility	Deadline for Plan Completion	Deadline for Plan Compliance
Facilities discharging storm water associated with industrial activity on or before October 1, 1992	April 1, 1993	October 1, 1993
Facilities beginning to discharge storm water after October 1, 1992, but on or before December 31, 1992	60 days after commencement of discharge	60 days after commencement of discharge
Facilities beginning to discharge storm water associated with industrial activity on or after January 1, 1993	48 hours prior to commencement of discharge (upon submittal of NOI)	48 hours prior to commencement of discharge (upon submittal of NOI)
Oil and gas exploration, production, processing or treatment operations discharging a reportable quantity release in storm water after October 1, 1992	60 days after release	60 days after release
Industrial facilities rejected or denied from the group application process	365 days after date of rejection or denial	545 days after of rejection or denial

Note: The Director may grant a written extension for plan preparation and compliance for new dischargers (after October 1, 1992) upon showing of good cause.

Chapter 16

Storm Water Best Management Practices That Work

J. Oster

1.0 INTRODUCTION

As of September 9, 1992, permits for storm water discharges associated with industrial activity require regulated facilities to prepare and implement a Storm Water Pollution Prevention Plan (SWPPP). The SWPPP emphasizes Best Management Practices (BMPs) and Spill Prevention Control and Countermeasures (SPCC) to minimize pollutant discharge. Several years before the storm water regulations were finalized, industries were already developing SPCC plans under the 1973 CFR 112 rules. Although these rules regulated the management of oil, many facilities voluntarily expanded the regulated materials to include toxic and hazardous chemicals. As a result, these facilities can extensively incorporate these SPCC plans into the SWPPP.

In addition to using BMPs established under SPCC regulations, there are a large number of specific practices that can be implemented by facilities to minimize pollutants discharged as a result of storm event. This paper will present pollution prevention methods that have been successfully implemented under the SPCC regulations as well as other accepted practices that can be used to meet the SWPPP requirements.

BMPs can best be divided into two general categories: activity-specific and site-specific. The first category, activity-specific BMPs, relates to industrial activities that may contaminate storm water. These activities commonly include fueling, vehicle maintenance, loading and unloading practices, liquid storage in above-ground tanks, outside storage of raw materials, or finished products and salt storage. The second and last category, site-specific BMPs, relates to prevention practices that prevent on-site contaminants from mixing with storm water. The remainder of this paper discusses these two general categories in detail.

2.0 ACTIVITY SPECIFIC BMPS

There are many industrial activities which could be discussed; however, within the time allotted we will limit our discussions to the following:

- Fueling Stations
- Loading and Unloading Materials
- Liquid Storage in Above-ground Tanks
- Outside Storage of Raw Materials

2.1 Fueling Stations

Fuel overflow during storage tanks filling is a major source of spills and storm water contamination. Overflow situations can be prevented by either the implementation of strict operating procedures or installation of equipment which shuts off flow, restricts flow, or sounds an alarm when the tank is near full.

Federal regulations now require overfill prevention equipment be installed in all underground storage tanks (USTS) installed after December 1989. For USTS installed before December 1989, overfill prevention equipment is required by 1998. The best overfill prevention systems include the following elements:

- A high-level alarm, which warns the operator of an impending overfill condition.

- An automatic shut-off device which prevents overfilling by shutting off delivery pump or valve.

- Interlocking of the unloading process and the overfill prevention system such that tank filling cannot take place if the overfill system is inoperative.

- Bypass prevention so that the overfill prevention system cannot be overridden.

In many cases, systems that include all of the above can be prohibitively expensive for many UST owners. Therefore, as a minimum, we recommend the following systems be installed:

- On tanks which are filled by gravity flow from the delivery vehicle, the vent line should be equipped with a float vent valve or similar device.

- On tanks which receive product under pressure from the delivery vehicle, the tank should be equipped with a combined audible/visual high level alarm mounted in a location that is both visible and audible to the delivery vehicle operator.

In addition to the shut-off devices discussed above, we recommend that the following operating practices be followed:

- Provide tight connections between the hose and fill pipe.

- Inspect all hoses periodically.

- Inspection of tank ullage before product delivery to ensure that receiving tank has sufficient capacity.

- Proper identification of stored products and container capacities.

- Proper training of all operators who perform loading or unloading operations.

Another source of fuel spills and storm water contamination is the topping off of vehicle fuel tanks. Under normal conditions, gas pumps are automatically shut off when the vehicle tank is almost full to prevent spills. Trying to completely fill the tanks or topping off the tank often results in overfilling the tanks and spilling fuel. This practice should be discouraged.

2.2 Loading and Unloading Materials

Considering the large amounts of petroleum and hazardous substances that can be handled daily through truck loading/unloading docks or terminals, the potential for spills and for leaks to cause storm water contamination is significant. For tank truck loading/unloading, a diked area should be used. The diked enclosure should be designed to hold at least the maximum capacity of any single tank truck to be loaded or unloaded. The diked enclosure is best constructed of concrete and includes a graded paved slab with a storm water drainage system.

The high risk storm water that collects within the dike area is best isolated from the low risk storm water (outside the diked area) by a normally closed manual drain valve. Storm water that collects inside the diked enclosure is potentially contaminated and we recommend that it be tested prior to release. If the storm water is determined to be contaminated above acceptable limits, treatment prior to discharge will be required. Although we do not recommend this practice, storm water that tests "clean" from these areas can be drained directly into "clean" storm water systems. However, routine discharging to "clean" storm water systems, runs a higher risk of polluting uncontaminated storm water unless strict testing procedures are adhered to. Additionally testing can be a very expensive, depending on the pollutants to be analyzed.

Loading areas also can employ a quick drainage or diversionary system to divert large spills to a holding tank, dike, catchment basin, or containment area. These containment systems will also collect storm water, which must be managed similar to the tank truck area previously discussed. An alternative to managing storm water collected in the spill containment areas or vessels is to cover the loading/unloading docks and prevent exposure to rainfall.

2.3 Liquid Storage in Above-Ground Tanks

Spillage from above-ground tanks can occur from several causes, such as: overfilling, leaks from tank walls/floor, leaks from valves, piping and/or pumps and catastrophic tank failure. Secondary containment such as concrete dikes is preferred for containment of spill materials. We recommend that all pumps used to transfer liquids from the tanks be placed within the tank's secondary containment structure. Storm water which collects inside the containment is potentially contaminated and we recommend that it be tested and discharged to the plant's pretreatment facility or sanitary sewer. This will ensure that even small amounts of contamination will not be discharged to a surface water. If the storm water collected within the containment structure is contaminated above the "clean" standard for discharge, the water will require treatment prior to discharge.

2.4 Outside Storage of Raw Materials

In some cases, raw material storage exposed to rain and/or runoff can pollute storm water. A good example of this is coal pile storm water run-off, which may become polluted by contaminants such as oil, grease, suspended solids, copper and/or nickel. The most effective way to avoid contaminated storm water from outside material storage areas is to prevent storm water run-on or rain from coming into contact with these materials. This can be done by -

- Storing the materials indoors.
- Covering the area with a roof.
- Covering the material with a temporary covering made of polyethylene, polyurethane, polypropylene, or Hyplon.
- Minimizing storm water run-on by enclosing the area or building a berm around the area.

3.0 SITE-SPECIFIC BMP ACTIVITIES

The site-specific BMP consist of activities that prevents pollutants from your site from mixing with storm water. These BMPs can be grouped into the following categories:

- Flow diversion practices.
- Exposure minimization practices.
- Mitigative practices.
- Preventive practices.
- Sediment and erosion prevention practices.
- Infiltration practices.

Each one of the above categories is discussed below:

3.1 Flow Diversion Practices

Flow diversion is the use of a structure (conveyances, diversion dikes, graded pavement) to divert flow. Conveyances can either be used to channel clean storm water away from industrial areas or carry contaminated storm water to treatment facilities. The second type of structure are diversion dikes or berms, which are used to block runoff from passing beyond a certain point. The third and last structure is graded and/or graded paved areas that are used to direct runoff away from industrial activity areas.

For most large industrial sites, conveyance systems work well and generally should be designed to handle only low risk "clean" storm water runoff from roads, parking areas, roofs, and landscaped areas. Flows from high risk areas such as tank farms, loading/unloading areas, material storage areas, and process areas should be conveyed via a separate system. In using a conveyance system for either high risk or low risk storm water, consideration needs to be given to how occasional on-site spills can readily be recovered prior to discharge. For high risk storm water, the conveyance structure is usually a closed pipe system that can be isolated from spills of hazardous chemicals or petroleum by a manually operated valve. In the unlikely event that hazardous materials or petroleum enters a closed pipe storm water system, the discharge receptor (usually a local sanitary waste water treatment plant) should be protected by an on-site waste water pretreatment system.

On the other hand, low risk storm water conveyance systems are usually open channels and very susceptible to liquid hazardous/petroleum spills. Ideally, the industrial site should include an on-line detention pond that can intercept a spill before being discharged to a surface water.

3.2 Exposure Minimization Practices

By eliminating or minimizing the possibility of storm water coming into contact with pollutants, industrial plants can reduce contamination of storm water discharges associated with industrial activity. Completely eliminating the exposure of materials to storm water is not always possible; however, there are some simple and inexpensive BMPs that a facility can use. Containing spills with diking, curbing, drip pans, catch basins and sumps are commonly used.

Another practice which is gaining greater acceptance in chemical plants is double containment piping systems with leak detention. This is a very expensive practice; however, environmental risks may be high enough to justify the expense.

3.3 Mitigative Practices

Mitigation is considered a second line of defense after a spill has occurred, and simply involves cleaning or recovery of the spills before it potentially impacts the environment. Because spills cannot be avoided at industrial sites, it is necessary to plan for these events and have response procedures designed for each hazardous material used on-site. Mitigative practices that have been successfully used for spill clean-up include manual clean-up (sweeping, shoveling), mechanical clean-up (excavation or vacuuming), and physical/chemical clean-up (gel and sorbents).

3.4 Preventive Practices

Preventive Practices are additional measures that can be taken at an industrial site to limit or prevent the exposure of storm water runoff to contaminants.

There are a number of preventive measures that can be implemented to prevent exposure of storm water runoff to pollutants. These measures include the following:

- Dust Control for Material Handling
- Vehicle Washing
- Area Control

The first measure, "dust control" has been successfully used to prevent pollutants from entering storm water discharges by reducing the amount of dust discharged by industrial activities. Types of controls used include; baghouses, cyclone, and filter systems. The purpose of dust control is to contain dust that would otherwise cause storm water pollution if allowed to settle outside the manufacturing building.

The second preventative practices is "vehicle washing" and is used to remove materials that accumulate on vehicles before being transported across the industrial site. This is an appropriate practice for sites where vehicles come into contact with raw materials that are hazardous or toxic.

The third and last preventative practice is "area control" and is simply maintaining good housekeeping standards in areas where toxic or hazardous materials are managed. By maintaining a high level of cleanliness in material management areas the risk of accumulated materials being picked up and transported to unprotected areas outside is reduced.

3.5 Sediment and Erosion Prevention Practices

Sediment and Erosion Preventive Practices are used to prevent soil and/or rock materials from being transported off-site and deposited into a nearby surface water. The major benefit of sediment and erosion control is to protect aquatic life in surface waters which are adversely affected by high suspended solids loading. Additionally, the exposure of surface water to toxic and/or hazardous materials that adhere to soil particles (due to previous spills and/or past management practices) is reduced.

There are several common ways to control sediment and erosion at an industrial site. These methods are as follows:

- Leave as much vegetation on-site as possible.
- Divert storm water flows away from exposed areas.
- Prevent sediments from moving off-site by adding crushed rock and/or curbing.
- Provide drainage ways for heavy rainfalls.
- Slow down runoff flowing across the site.

3.6 Infiltration Practices

Infiltration practices are surface or subsurface measures that allow for quick infiltration of storm water runoff. Rapid infiltration may be possible if the structure or soils used in those practices are very porous. Infiltration practices offer an advantage over other practices in that they provide some treatment of runoff, recharge groundwater and can reduce the velocity of the runoff so that it will not cause damaging erosion. This option, although rarely used on industrial sites, may have application for sites with porous soils, less than 5 acres, and the depth from bottom of the infiltration area to groundwater of at least 3 feet.

4.0 CONCLUSION

The BMPs discussed in this paper are measures that have been successfully used to reduce pollutants at the source before they have the opportunity to contaminate storm-water runoff. Also discussed are BMPs that can be used to direct storm water away from exposed materials or to direct contaminated storm water to treatment locations. The storm water BMPs that are used at your specific site must be selected on a case-by-case basis, depending on site topography, soil conditions, chemicals and petroleum products used on-site, activities on-site, existing soil contamination, and the amount of space available. With an understanding of your specific site conditions, BMPs can be successfully selected based on practices that have already been implemented by other facilities under SPCC regulations.

Chapter 17

Energy, The Environment and the Economy: An Educational Pilot Program

P. Rojeski, P. Rose

ENERGY CONSERVATION IN THE 90's

Under the administration of President Clinton, energy conservation, environmental protection and economic growth are top priorities. In his "Vision for Change for America" the President describes bold conservation initiatives the likes of which have not been attempted since the Carter Administration's response to the OPEC oil embargo of 1973. During the Carter Administration, rising energy prices and the limited availability of gasoline were the two primary driving forces. Under the Clinton-Gore Administration the environment and the economic competitiveness of the United States (particularly as required for the reduction of the national debt) are the primary driving forces. The details of several of the President's major conservation and environmental protection objectives follow. Of primary concern is the number of highly trained engineers and scientists required for the attainment of these goals.

Increase the Efficiency of Federal Buildings and Facilities

Federal investment in energy efficiency improvements currently runs around $150 million per year. This new initiative will increase spending to almost $500 million per year by 1996. Over 700 energy managers will be trained in 1994, and over 2,000 per year in 1995-1998. Outside energy audit teams will review 600 Federal sites in 1994 and 1,000 sites per year in 1995-1998.(1)

DOE/Increase funding for Renewable Energy and Energy Conservation Programs

The Energy Policy Act of 192 contains new responsibilities for the Federal Government, including: (1) establishment of new energy efficiency standards; (2) authorization for enhanced research programs; (3) new demonstration/commercialization programs for renewable energy and energy conservation. This initiative increases funding in these areas, reaching an increase of $500 million in 1997, for a four-year increase of $1.3 billion. (1)

EPA/Environmental Technology

The proposal would increase funding for environmental engineering and technology development by $36 million for 1994, and a total of $626 million through 1998, and a total of $1.85 billion over nine years. (1)

Expand EPA's Voluntary "Green" Programs

The EPA launched its "Green Lights" program two years ago to encourage Fortune 500 companies to convert to more energy-efficient lighting, which will reduce electricity generation and greenhouse gas emissions. Estimated outlays over four years 1994-1997 - $69 million; 1997 - $25 million.

The cost-effective implementation of this monumental program will require a large pool of highly trained engineers whose interdisciplinary training includes avoiding some of the failures of previous conservation efforts.

CONSERVATION MISCUES OF THE 70's

As shown in Figure 1, the United States has made strides in improving its energy efficiency in terms of BTU's per unit of GNP. Two areas of success have been the improvements in automobile fuel economy resulting from the Corporate Average Fuel Economy (CAFE) standards and improvements in the efficiency rating of air conditioning units, electric motors, and lighting. The progress made in these areas resulted from equipment efficiency standards which were promoted by industry and government.

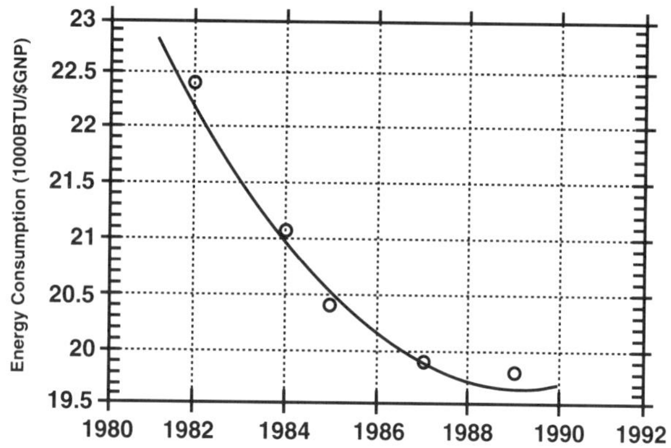

FIGURE 1: NATIONAL ENERGY EFFICIENCY (2)

Not all funding spent on conservation programs was effectively utilized, however. Three main reasons for this ineffectiveness included:

(a) the lack of training and organization of the maintenance staffs;

(b) the failure by engineers to consider the impact that conservation programs might have on the health and productivity of the building occupants; and

(c) management's failure to monitor conservation progress and provide incentives for achievement.

Maintaining a building at peak operating efficiency requires manpower and training. Too often organizations embarked on energy conservation programs with inadequate maintenance staffs. Improved energy efficiency often requires the installation of sophisticated, computerized monitoring and control systems. Many maintenance organizations simply did not have the staff support required to maintain these control systems and they were either abandoned or switched to the manual mode.

The near panic resulting from energy shortages, and the lack of interdisciplinary training of energy engineers resulted in some projects which had detrimental effects on the building occupants. In fact, by 1985, energy conservation became synonymous with uncomfortable, unhealthy working conditions. Sick building syndrome and building related illness are but two examples. This was unfortunate because, if implemented properly, conservation can result in savings through improved efficiency with no degradation of occupant health, comfort or productivity. For that matter, energy efficiency can often result in enhanced occupant comfort and health.

Since energy management programs require additional effort on the part of the maintenance staff, and cooperation/participation from the building occupants, the success of the program often depends on the incentives provided by management. But energy costs (in a building) are a small fraction of the salary costs for personnel. In hospitals, for example, energy costs are only three to seven percent of total operating costs. Therefore it is often difficult to get management's commitment for these programs, especially if they believe that the result will be a less comfortable building. An excellent example of management leadership was the Houston, Texas School district. In the district, the Superintendent returned a fraction of the annual savings to each school. This encouraged the schools to strive for high energy savings and provided a mechanism whereby monthly progress was reviewed. Similarly, the Philadelphia Schools implemented an incentive plan by returning up to 40% of the savings back to the schools with the best energy savings records. This approach, first implemented in the 1982-83 school year, continues today and has earned the district over $60 million - $9.5 just this last year. Most organizations, however, failed to provide incentives or monitoring, and the maintenance staffs often resented the perceived intrusion by outside consultants. These attitudes hampered the success of many conservation efforts.

ENERGY ENGINEERS FOR THE 90's

The energy conservation challenge of the 1990's will require a significant pool of talented engineers. Billions of dollars are to be spent over the next four years to:

(a) increase energy efficiency by reducing waste;

(b) improve worker health, safety and productivity by improving the indoor environment;

(c) reduce global environmental pollution; and

(d) enhance the economic competitiveness of U.S. business and industry.

The strategic planning and implementation of this program will require the education of thousands of engineers who understand building systems and technology, worker health, safety and productivity, and energy and maintenance management. In order to avoid the mistakes of the 1970's, these engineers must also be skilled in communicating conservation ideas to non-technical people, especially upper-level administrators. It is many of these individuals who might have had negative experiences with the conservation efforts of the 1970's.

Educational Trends

For the five years following the OPEC oil embargo, many engineering schools were active in energy conservation, especially solar and other renewable energy resources. Since the early 1980's, most of these programs have been downsized or abandoned. Also, very few engineering departments currently teach heating and air conditioning, lighting, electrical power distribution, energy conservation and/or alternative energy systems. Of course, thermodynamics and energy conversion are still available at most mechanical engineering departments and courses in solar and renewable energy systems are generally available as electives.

Programs in energy and the environment appear to receive the most emphasis at the graduate level. At the graduate level, there are currently 92 energy-related programs being offered at universities across the country. (None of these programs is at historically black universities or minority institutions.) A summary of the types of programs follows in Table 1.

TABLE 1: ENERGY-RELATED GRADUATE PROGRAMS (3)

PROGRAM TYPES AND NUMBERS	
Architecture	8
Basic Sciences	4
Engineering	23
Environment	7
Public Policy & Law	6
Transportation	6
Interdisciplinary	
(Degree Granting)	15
(Non-Degree Granting)	20

	92

Of the 23 engineering programs, only four appear to be comprehensive building conservation programs. These four programs graduate approximately 50 students per year.

Architectural Engineering and Energy Conservation

While graduate programs in energy conservation can produce qualified energy engineers and researchers, a basic undergraduate-level knowledge of building design, building sciences, economics and industrial psychology/hygiene can also prepare students for careers as energy engineers. As shown in Figure 2, residential and commercial energy use accounts for approximately 42% of the total energy consumption in the United States. If industrial lighting and space air conditioning were included, the total would approach 50%.

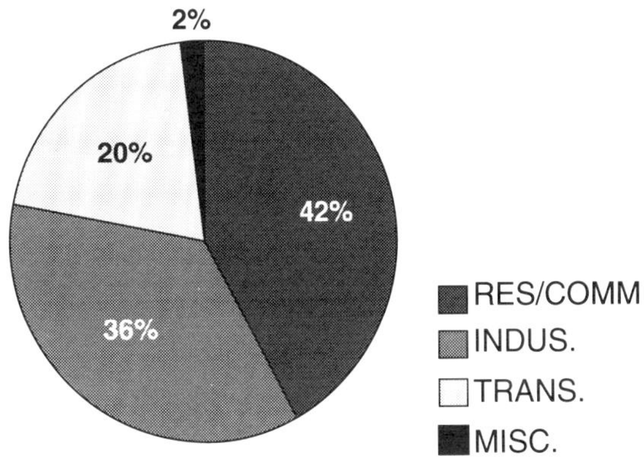

FIGURE 2: U.S. ANNUAL ENERGY USE (2)

The courses taught in Architectural Engineering programs prepare graduates to understand the energy implications of building design, heating and air conditioning systems design (including indoor air quality), and energy efficient lighting design. Architectural Engineering students who specialize in building systems analysis take additional courses in this area. Elective courses are also available in industrial psychology and industrial hygiene, both of which are important courses for energy engineers.

There are currently twelve Accreditation Board for Engineering and Technology-accredited Architectural Engineering programs in the United States. As reported by Singh (4), each of these programs typically includes at least one course in building design, and one course in heating and air conditioning. Additional courses include lighting design, and electrical power systems design. Most departments also offer elective courses in advanced heating and air conditioning systems, building energy conservation, and alternative energy systems. These Architectural Engineering programs provide the basic concepts and analysis skills required for an energy conservation engineer.

It is estimated that annually 10 students with an interest in energy conservation graduate from these architectural Engineering programs.

Architectural Engineering at North Carolina A&T State University

North Carolina Agricultural and Technical State University has one of the twelve ABET-accredited Architectural Engineering programs. The A.E. program is based on a five-year, 160 semester credit-hour curriculum. The curriculum requires three courses in building design, two courses in heating and air conditioning systems, one course in building illumination and one course in electrical power distribution. Elective courses in energy areas include the following:

* Energy and the Environment: A Global Perspective;

* Energy Conservation in Buildings;

* Advanced Heating and Air Conditioning Systems Design;

* Energy and Maintenance Management;

* Energy Management Planning; and

* Heating and Air Conditioning Systems Simulation.

These courses are taught by Peter Rojeski, Jr., PhD, PE and Harmohindar Singh, PhD, PE. Both of these instructors have extensive consulting experience in energy conservation planning, analysis and retrofit in public buildings and industrial facilities.

For the past several years, an average of five students per year graduated from the Architectural Engineering Department and pursued careers in an energy-related field.

DOE/NCA&T PROGRAM ON ENERGY AND THE ENVIRONMENT

This unique Department of Energy Program will accomplish several goals for the Department and the Office of Energy Efficiency and Renewable Energy. The particular goal for the Office of Energy Efficiency and Renewable Energy is to build a cadre of entry level, technically trained individuals who will be able to bring energy efficiency and renewable energy technologies from the research arena to the marketplace. Course material has been and will continue to be developed for the students in this program. With their formal training and their DOE co-op work experience, the students will enter the work force with a broader understanding of the role of the energy engineer, the mission of the Department of Energy, Energy Efficiency and Renewable Energy, and the place this technical work has in the global picture of energy, the environment and the economy.

The year-one program objective included the identification of at least 11 students (who are potentially interested in energy careers) from architectural engineering, and industrial engineering. Participation in the program includes:

(a) two required energy related courses and a course in energy economics;

(b) a summer work experience at the Department of Energy;

(c) trips to two national labs and attendance at two national energy conferences;

(d) involvement in building energy research projects;

(e) a one-year, graduate student cooperative education period with the Department of Energy;

(f) student stipends; and

(g) faculty development support.

To participate in the program, students must first enroll in the course "Energy and the Environment: A Global Perspective" in the Spring semester. They apply to the DOE program during the Spring semester and, if accepted, work in Washington, D.C. for the summer.

This summer, eight students of the eleven who were selected by the Department entered the program. They began work in six program areas within the Office of Energy Efficiency and Renewable Energy. These areas are: the Federal Energy Management Program (FEMP), the Geothermal Energy Division, the Office of Planning and Analysis, the Office of Codes and Standards, the Office of Industrial Programs, and (the Facilities and Systems Coordination Division. Examples of the summer projects, on which they report not only to the Department but also to Professors Rojeski and Singh, are a district heating anaylsis of federally owned buildings in the District of Columbia, an energy analysis of buildings planned for the National Renewable Energy Laboratory, and a computer anaylsis of the compliance status of refrigerator/freezer manufacturers. After the summer work experience, students decide whether they wish to continue with the program. If so, they receive a $1,500 stipend for each of the next two semesters. They must also enroll in the two courses: "Building Energy

Conservation" and "Energy and Economics."

Upon successful completion of their Senior year, (with a minimum overall grade point of 2.6), the students co-op, on a full time basis, for one year with DOE. Assignments during the one-year assignment can be at headquarters, one of the national laboratories, at one of the Support or Operations offices or one of the Department's contractor sites. At the end of that year, they either become permanent DOE employees or pursue careers elsewhere. (Should the students decide not to follow through with the one year co-op, they must return the $3,000 stipend which they received.)

Recruiting the Students

At NCA&T there are several government-sponsored and industry-sponsored intern and co-op programs which all compete for students. In order to avoid some of the problems experienced by other programs, the strategy of this program includes the following high priority items:

(a) involvement of several key faculty who teach specialized courses, attend conferences with the students, and interact with DOE personnel throughout the program; and

(b) involvement of key DOE personnel who select the co-op students, identify mentors and work supervisors, and monitor the students' progress throughout the program.

Recruiting of students for the program was made easier because of an active student ASHRAE chapter at NCA&T. ASHRAE (the American Society of Heating, Refrigerating and Air-Conditioning Engineers) has had a very active student chapter for at least five years and six international Society Presidents have visited the campus. Due to their participation in ASHRAE-related activities, the students were aware of the issues of conservation, global warming, ozone depletion, etc. Students who applied for the DOE program also received travel support to attend the ASHRAE Annual Meeting in Chicago in January. The students will also attend the Association of Energy Engineers (AEE) meeting in Atlanta in October.

Twenty students enrolled in the introductory pre-requisite course, "Energy and the Environment: A Global Perspective," and fifteen of these applied for the DOE program. The final selection of eleven students was made by representatives of DOE based on the student's application, college transcript, a research paper, and a personal interview.

Energy and the Environment: A Global Perspective

In order for the program to be successful, the first course had to be one that captured the students' interest while educating them about career opportunities as energy engineers. Although the students were aware of conservation and environmental protection, the depth of their knowledge of the subject was very limited. This course is being taught as a fast-paced, current events course with a number of expert guest lecturers, including:

Mr. Frank Stewart, Deputy Assistant Secretary for Technical and Financial Assistance, US DOE;

Mr. Richard Charles, past-President, ASHRAE;

Dr. Joel Levine, Senior Research Scientist, NASA Langley, Virginia;

Dr. Joseph Roop, Staff Economist, Battelle Pacific Northwest Laboratories;

Mr. Ron Gibson, Vice President of Planning, Duke Power Company;

Mr. Donald Harrow, Vice President, Government Relations, Piedmont Natural Gas;

Dr. Shoou Yuh Chang, Hazardous Waste Specialist, NCA&T State University.

The students also had assigned readings from five references:

Earth in the Balance by Al Gore;

Trashing the Planet by Dixie Lee Ray;

Cool Energy, Renewable Solutions to Environmental Problems by Michael Brower;

Energy and the Environment - 1991 Proceedings of the 1991 International Symposium in Espoo, Finland, ASHRAE; and

Energy and Environmental Strategies for the 1990's, Association of Energy Engineers.

The course also includes three research papers and a conservation project which involves an analysis and retrofit design of the NCA&T campus utility system.

BUILDING A SUCCESSFUL PROGRAM

The key to maintaining the students' keen interest in energy conservation will be the success of their summer experience at DOE. NCA&T students have worked at DOE facilities before, and those with good experiences came back with a very high level of enthusiasm. Those students had assignments that were interesting and which involved interaction with other students, engineers, and scientists. Successful students were those who felt they had the opportunity to make a contribution and to achieve a worthwhile objective during their summer experience. Great care was taken by DOE personnel in the selection of Project Directors and Mentors for the students. Special Wednesday "Brown Bag" lunch time seminars and a variety of local field trips were conducted for the students. The students were also in contact with NCA&T faculty in an effort to minimize any problems that might have occurred.

While eight students may not seem like a significant number, one must keep in mind that only about one hundred engineering students graduate each year with degree concentrations in energy conservation. Similarly, at the graduate level, only about 50 students graduate each year with extensive training in building sciences and conservation. Certainly a much larger number of energy engineers will be required in order to meet the goals established by President Clinton and the Energy Policy Act of 1992. This project is designed to educate the students required to meet this demand.

The first year of the program will serve as a pilot. If all participants feel that it could become successful, then there is the possibility that it could be expanded to other schools and departments. The model could be applied, for example, to industrial energy conservation in the manufacturing sector, or to alternative fuels in the transportation sector.

REFERENCES

(1) Clinton, Bill, A Vision of Change For America, US Government Printing Office, February 17, 1993.

(2) Energy Facts 1990, Energy Information Administration, U.S. Department of Energy.

(3) Directory of Energy-Related Graduate Programs in U.S. Universities, 1992. The Energy Foundation, Berkeley, CA.

(4) ASHRAE Insights, American Society of Heating, Refrigerating and Air-Conditioning Engineers, Inc., November, 1992.

Chapter 18

Survey Analysis of Waste Generation and Minimization Opportunities in Small to Medium-Sized Manufacturing Facilities

R.J. Jendrucko, T.N. Coleman, T.M. Thomas

ABSTRACT

The EPA-supported University of Tennessee Waste Minimization Assessment Center has completed waste assessments for 40 regional small- to medium-sized manufacturing plants. The results demonstrate that a typical plant has on the order of 10 significant waste streams to manage. Losses of volatile compounds (including water) were found to constitute the waste stream class of greatest magnitude. Among attractive measures for waste minimization, it was most common to recommend actions related to simple process changes and improved housekeeping.

Over a five year period of operation (1988 - 1993), the U.S. Environmental Protection Agency-funded University of Tennessee Waste Minimization Assessment Center (WMAC) has provided technical assistance in the performance of waste assessments for 40 regional manufacturing plants. This program was focused on small- to medium-sized firms in a 150-mile radius of Knoxville, Tennessee with limits of $75 million in sales, 500 employees and limited in-plant waste minimization experience. Clients were identified by random telephone solicitation. Selected plants were required to have at least one hazardous or non-hazardous waste stream of major concern. The distribution of client characteristics is depicted in the data in Table 1 below.

TABLE 1 CLIENT CHARACTERISTICS

Number of Clients Served*	SIC	Industry Class	Average Sales ($ million)
8	34	Fabricated metal products, except for machinery and transportation equipment	21
6	35	Industrial and commercial machinery and computer equipment	25**
6	37	Transportation equipment	21
4	33	Primary metals industry	62**
4	36	Electronic and other electrical equipment and components, except computer equipment	31
3	27	Printing, publishing, and allied industries	25
3	28	Chemicals and allied products	52
2	39	Miscellaneous manufacturing industries	37
2	26	Paper and allied products	63
2	32	Stone, clay, glass, and concrete products	22
1	25	Furniture and fixtures	4
1	20	Food and kindred products	39
1	23	Apparel and other finished products made from fabrics and similar materials	6
1	30	Rubber and miscellaneous plastic products	15

*The numbers of clients in column 1 total more than 40 owing to the fact that several companies have multiple SIC's

**Unusually high sale figures for two clients were omitted

Due to the primarily random solicitation process used, it is believed that the distribution of clients served reflects that for industries in the WMAC service area. Included were from one to eight manufacturers in each of 14 among the 20 Standard Industrial Classification (SIC) categories. The highest frequency of industries served were for fabricated metal products (SIC 34), industrial and commercial machinery (SIC 35) and transportation equipment (SIC 37). Average plant size for each SIC reflected in gross sales varied widely from about $4 million to $62 million excluding two exceptional cases. Individual plant annual sales in all classifications varied from $4 million to $350 million and the average for all 40 clients served was $44 million per year.

Client firms selected for program participation were scheduled for two to four successive plant visits over a period of weeks in order to collect historical data on waste generation and to perform a detailed study of waste generating processes. Quantitative data were obtained from hazardous waste and raw material shipping records, mass balancing, and when necessary, estimations based on plant employee experience.

In general, the types of hazardous or other regulated waste identified for the 40 clients fell into one of the major categories listed in Table 2.

TABLE 2. HAZARDOUS WASTE CATEGORIES IDENTIFIED

General Waste Category	Common Sources of wastes
Spent Solvent	Spent cleaning solvents (vapor degreasing, dip tanks)
Volatile Organic Compound (VOC) Emissions	Solvents from open cleaning tanks and mixing containers (paint and other coatings); curing of coatings
Distillation Bottoms	On-site distillation of solvents
Other Hazardous Liquids	Off specification paint and coatings; spent plating liquids, waste oils; spent ink
Hazardous Sludges	Water curtain paint booth liquid waste; waste water treatment sludges with hazardous characteristics
Hazardous Solids	Incinerator ash; plating bath and other dip tank solids; paint residues; floor coverings and paint booth air filters; cleaning brushes and rags

In addition, it was routinely found that one or more non-hazardous waste streams were important for most clients. The most common non-hazardous waste streams identified typically included one or more of those listed in Table 3.

For the 40 clients served, the distribution of identified actual waste streams leaving client plants is summarized in Table 4 below.

TABLE 3. NON-HAZARDOUS WASTE CATEGORIES IDENTIFIED

General Waste Category	Common Sources of wastes
General Landfill Waste	Cardboard, paper, plastic, and foam packaging waste; office paper; floor clean-up waste; cafeteria waste; wooden pallets
Spent Oils	Hydraulic machine cylinders; machine lubrication; part and die lubrication; fork lift oil changes; protective oils (metal surfaces)
Scrap Metal	Steel, iron, aluminum, copper, and brass from machining operations; defective metal parts
Non-Hazardous Process Wastes	Oily sludges from machine clean out; spent stones from polishing/finishing operations; waste water treatment sludges
Process Waste water	Aqueous washers, rinse and wash tanks; fume scrubbers; on-site treatment system effluent

As is indicated, among the plants served there were found to be an average of about 4 hazardous and 2 significant non-hazardous waste streams. In addition, there were an average of about 3 regulated air emissions and one waste water stream in the plants served. From these data it is noted that a typical small- to medium-sized plant has on the order of 10 significant waste streams to manage.

TABLE 4. FREQUENCY OF WASTE STREAM OCCURRENCE

Waste Type	Min.	Max.	Avg.	Total
Hazardous	0	13	3.9	157
Non-Hazardous	1	4	2.4	95
Air emissions	0	11	3.0	122
Waste water	0	2	1.0	39

The frequency distribution of waste streams of particular types varied significantly as illustrated in the histogram of Figure 1. From this histogram it is evident that air emissions of VOCs ranks as the major waste stream in a typical small- to medium-sized manufacturing plant. Conversely, the generation rates of spent solvents (usually after in-house recovery efforts) and distillation bottoms are encountered at a significantly lower frequency.

During the waste assessment for each client served, a major concern was the identification of the precise point(s) of origin of the principal waste streams. Relevant to the summary data presented in Table 4 above, it was found that for hazardous and otherwise regulated wastes, on average there were 2 identifiable discrete point sources while for non-hazardous streams there were found to be on average 5 point sources for each major stream. The distribution of the identified point sources among the major categories of wastes generated is given in Table 5 below.

From these findings it is clear that there were an average of 38 significant point waste sources of all types for the 40 clients served. This result clearly demonstrates the very demanding effort that is required to characterize waste generation and to formulate a comprehensive waste management plan in a typical small- to medium-sized plant.

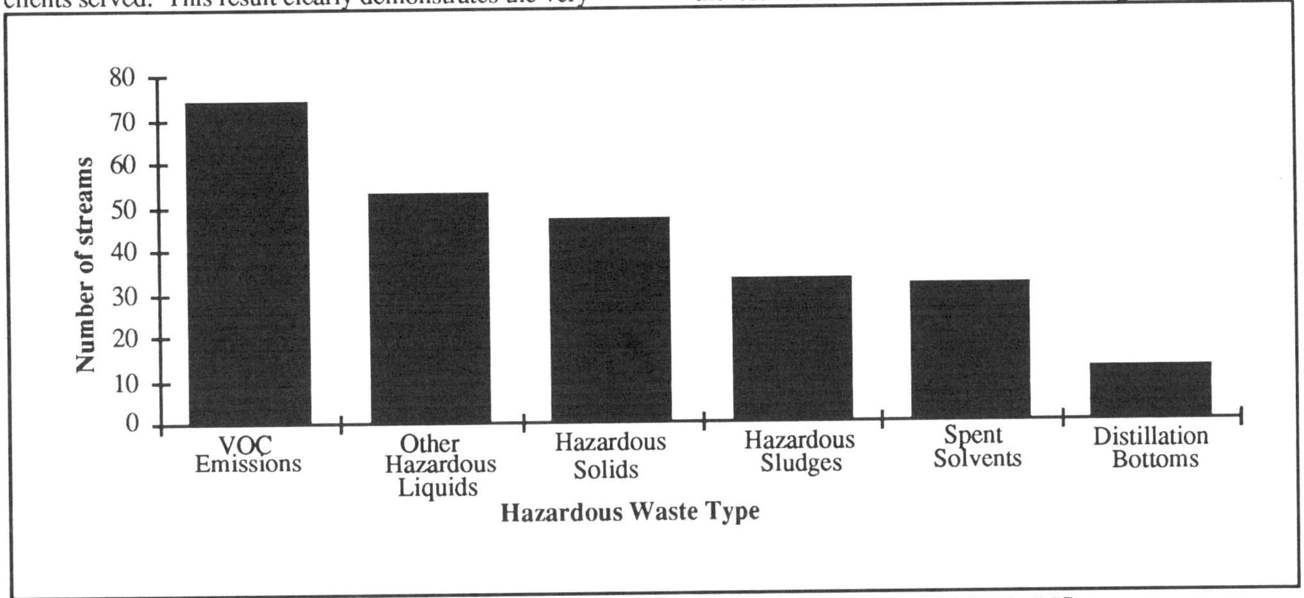

FIGURE 1. DISTRIBUTION OF FREQUENCY OF OCCURRENCE OF WASTE STREAMS

TABLE 5. FREQUENCY OF OCCURRENCE OF WASTE STREAM POINT SOURCES

Waste Type	Min.	Max.	Avg.	Total
Hazardous	0	25	8.4	335
Non-Hazardous	2	30	10.5	421
Air emissions	0	31	8.1	323
Waste water	0	47	11.1	445

In the increasingly important area of water use and discharge, it was found that for 35 among the 40 plants served for which detailed water use data were available, 26 plants or 65% have on-site waste water treatment systems. For both treated and untreated waste water, plants averaged about 23 million gallons per year of waste water discharged. With the rapidly escalating cost of make-up water and municipal sewer charges, it is becoming imperative for many plants to investigate options for in-plant recycling of process or treated water.

There are many factors which provide motivation for plants to pursue pollution prevention. Among these are:

1. Reduction in raw materials costs
2. Reduction in waste management costs
3. Improvement of plant and community environmental conditions
4. Reduction in liability risks

As is readily apparent, most of these factors have an economic component resulting in a clear incentive to reduce the production of all wastes to the extent possible. It is particularly noteworthy that for the 40 clients served the estimated total cost of waste management excluding make-up raw materials purchases ranged from \$13,600/yr to \$1,520,000/yr with an average of \$205,000/yr. Waste management costs, at least at the upper span of the indicated range, was found to be considered significant by most of the plants served. Thus, there was a strong incentive to identify and implement cost effective waste reduction measures.

A major component of each assessment was the identification of a set of specific recommendations for waste reduction. Although each recommendation was specialized for each client's application, most fell into one of the following action categories:

1. Simple procedural changes and housekeeping measures
2. Materials recycling/recovery
3. Major Process changes/material substitution
4. Improved on-site waste treatment to reduce waste volume shipped off-site
5. Worker training to reduce process waste

For each plant served the number of recommendations ranged from 1 to 8 with an average of 4.5 measures being specified per plant. For all 40 plants a total of 181 recommendations were made. The distribution of the total recommendations among the five categories listed above is illustrated in the histogram of Figure 2.

As indicated in Figure 2, most commonly, recommendations fell into the first category involving simple procedural changes and housekeeping measures. This result reinforces the view that substantial waste reduction can still be achieved by giving attention to

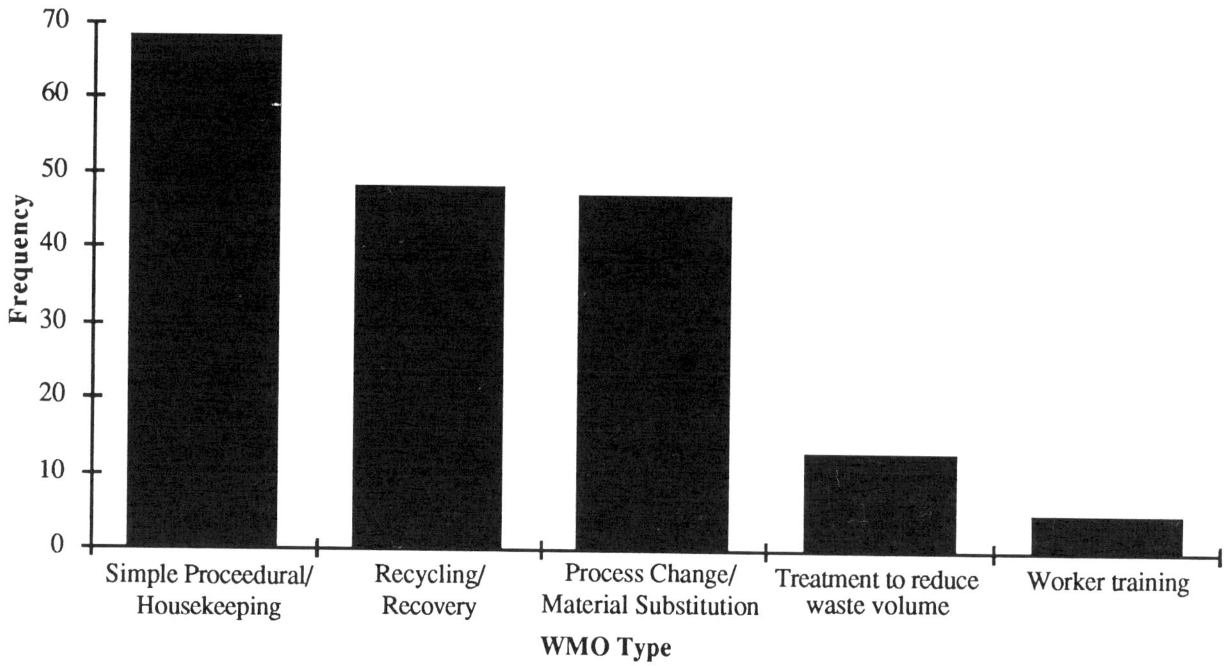

FIGURE 2 DISTRIBUTION OF WMO TYPES

often obvious measures such as covering all solvent containers to reduce evaporative losses.

There were also numerous recommendations in category 2, materials recycling and recovery (48 total measures) and category 3, major procedural changes and materials substitutions (47 total measures). In the area of recycling materials, frequently specified for recycling were hydrocarbon solvents, metal shavings, and process water. Procedural changes recommended most commonly involved alternative surface coating processes. Recommendations for material substitution were frequent in the area of parts cleaning and most often involved the suggested use of a non-hazardous (often aqueous) cleaning agent as a substitute for a hydrocarbon-based cleaner.

The most frequently recommended specific measures among the 40 clients served are listed in Table 6 below. As discussed above, the economic aspects of measures considered for waste minimization are of prime importance to industrial plants. Of specific relevance are the measure capital cost, annual operating cost savings and payback period. These indices are included in Table 6 for the most frequently recommended measures. For all of the 181 recommended measures, implementation costs ranged from $0 to $2.2 million (average of $69,000), while annual cost savings were estimated to range form $17 to $2.7 million (average of $60,000), with paybacks computed to range up to 9.2 years (average of 1.4 years). It is common in industry to consider measures requiring capital expense and having paybacks exceeding about two years to be

TABLE 6: MOST FREQUENTLY RECOMMENDED MEASURES AMONG THE 40 CLIENTS SERVED

Measure	Total Number of Recommendations	Average Annual Cost Savings	Average Annual Payback (yr)
Cover tanks and containers to reduce evaporative losses	15	$14,140	0.2
Treat and reuse or reuse already treated waste water	10	$59,030	3.4
Utilize a sludge drying oven to reduce the volume of waste shipped off-site	11	$30,890	2.0
Convert solvent-based painting to electrostatic powder coating	8	$188,180	2.7
Recover (distillation) solvents on-site	8	$11,490	2.4
Convert conventional painting operations to electrostatic painting	6	$95,900	1.2

unattractive. However, in the area of pollution prevention measures with substantially longer payback periods may be deemed acceptable due to concern with banned chemicals, changing regulations and future financial liability risks.

With further reference to Table 6 and the above additional program summary data, it is apparent that the potential cost savings for viable waste reduction measures is substantial and is often of an equal or greater order of magnitude than utility costs for plant operations. The relationship of measure projected payback period to annual cost savings is depicted in Figure 3.

4. Simple, easily implemented measures can be commonly recommended and tend to associate with the shortest payback periods.
5. Measures resulting in the most significant reductions in waste amounts usually require substantial capital expense and associate with longer payback periods.

Finally, it is concluded that waste reduction programs in manufacturing plants should always begin with a thorough in-plant assessment. Only in this way can it be ascertained that all economically significant waste streams have been identified and that the actual costs associated with waste generation are quantified. Having accomplished these objectives, economically attractive waste minimization opportunities can be readily identified and reliably evaluated.

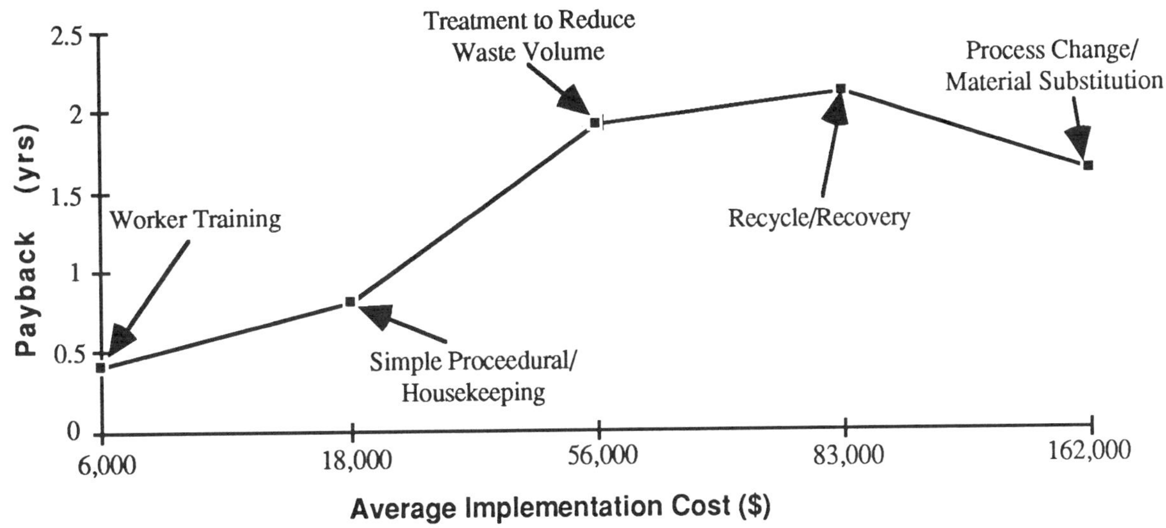

FIGURE 3 PAYBACK VS IMPLEMENTATION

From Figure 3 it is clear that for the recommended measures payback periods tend to lengthen as implementation costs increase. This result most likely relates to the fact that while significant waste reduction-related savings can often be achieved with minimal cost (for simple measures), large potential savings usually require major process changes resulting in substantial capital expense.

In summary, the data presented here lead to the following conclusions:

1. In a typical small- to medium-sized manufacturing plant there are numerous (both hazardous and non-hazardous) waste streams of significant concern.
2. The evaporative loss of volatile chemicals including water represents the most commonly encountered waste stream in plants.
3. There are typically many identifiable measures for waste minimization which have payback periods of two years or less.

Chapter 19

How to Transform Liabilities in Used Oils and Other Waste Fluids into Valuable Assets and Free Energy On-Site

J.W. Robertson, P.W. Wayte

ABSTRACT

Companies must learn to redefine the challenge represented by common waste fluids they generate: used crankcase oil, brake fluid, antifreeze, solvents, parts washer cleaners, gear lube and automatic transmission fluid. The risks of traditional waste management analysis have become greater because of the regulatory and legal climate in which business now operates and because of improved pollution-detection technology. Companies must recognize that the discipline of environmentally-sound waste fluid management can strengthen the bottom-line. This new perspective is driven by current management concepts, by increased costs and liabilities connected to wastes and by the value-adding potential of new, non-polluting technologies.

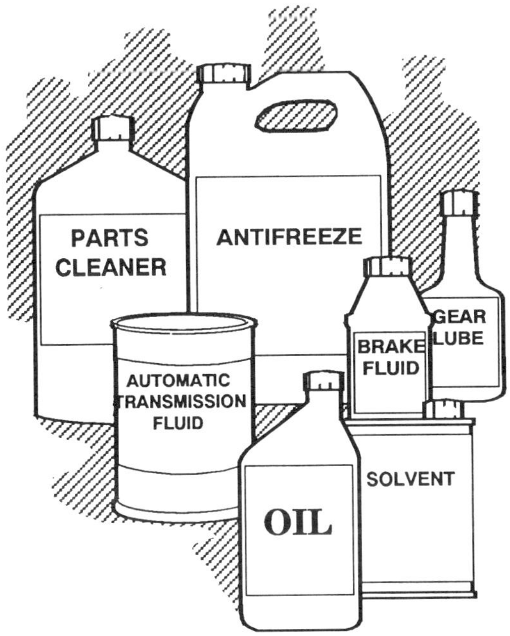

Technologies are now available for use on-site to enable businesses and government facilities to end or minimize their liability and costs associated with waste fluids and to turn the fluids into assets that provide free energy or significant savings over present off-site hauling. This paper provides an understanding of these technologies and the critical features energy and environmental engineers will want to know about and specify.

This presentation is most valuable for businesses that generate used lubricants. It will also be valuable for any business with waste fluids whose flash point is below 450° F including such non-petroleum oils as cottonseed, vegetable, peanut, soybean, etc.

INTRODUCTION

Current management theory–Deming, JIT, world-class manufacturing and TQM–leaped from theory to reality for us just after we doubled our sales three years ago. **[1, 2]** Like any red-blooded American manager, I dreamed of a much bigger building. I asked our Board of Directors to help me locate a new facility. Fortunately, we had a Board member from Northern Telecom who warned that I didn't need a larger building, I needed to rethink my perspective and ask different questions. From a modern management perspective, my space limitation could be an asset rather than a liability, depending on how well I used it. Instead of asking how to get more space, I should be asking what changes in the use of the current space could reinforce our management goals and strengthen the bottom-line.

He was right. I had been managing our production like a man trying to take a bath without closing the drain. No matter what I did, the tub would never fill up with water and the water bill was horrendous. All of us in the company needed to think and act differently. We discovered that the discipline of limited space helped us to focus on eliminating waste and do the things that enhance the bottom line: to design the product for ease and quickness of production, to keep inventories small, to receive components just-in-time, to organize production into tight efficient u-cells for increases in productivity.

When it comes to managing potential environmental problems, many of us face the same need to change our analysis of the situation. We constantly complain that environmental regulation is hurting the economy and hurting our profits. We have a separate out-of-the-way environmental department that we tolerate and whose mission it is to keep up with EPA regulations and to fill out our permits. We continue to manage our environmental affairs as a specialty with no real line-management responsibility. We view the waste products of our businesses as problems to be disposed of. We need to ask how they can be transformed into valuable resources. When that is not possible, we need to examine options for modifying our process so that the production of these wastes is eliminated or reduced.

Near-sighted Management

Traditional management theory has held that waste generated from daily operations is an expensive nuisance in the near term because of safety precautions and removal costs associated with waste disposal. Less attention is paid to the costly long-run liability if waste is disposed of incorrectly, or if it later creates an environmental hazard, regardless of the company's handling of it.

This traditional thinking springs from an obsession with the short-term bottom-line. Such thinking ignores the fact that the proper management of waste yields **long-run** cost savings, and hence, an improved bottom-line.

The risks of traditional waste management analysis are great because of the regulatory and legal climate in which business now operates and because of improved pollution-detection technology.

Strict Legislation Makes Mistakes Costly

Environmental laws **can** seriously affect your profits. For example, many companies continue to generate used lubricants and have them hauled off. If your hauler goes out of business three years from now, and leaves a mess, your company will be paying to clean it up, even though you are innocent of any wrong-doing. That is the way CERCLA (Superfund) or RCRA, if necessary, work. Those clean-up costs will directly reduce your profits since insurance companies don't cover this kind of off-site clean-up.

How much can it hurt? Let's say your company makes $1 million of net profit off $60 million of sales. Several years from now, one of your locations ends up paying their share for a clean-up at the usual rate of $100 per gallon or $500,000 for the 5,000 gallons that were hauled off this year. That represents the profits from $30 million of sales or some 50% of total sales. You know what you went through for that 50%. You know how many people and how much time and money you had to invest in advertising, new product development, sales and marketing to generate those sales revenues.

The damage to your image can be as bad as the clean-up costs. One major corporation asked us to help them develop an on-site alternative to managing their used oil after they received a chilling phone call from a local television station. The conversation went like this:

"The folks who pick up your used oil have just spilled 8,000 gallons of it into the river. What are you going to do about it?" The representative tried to explain to the reporter that his company simply turned the used oil over to this EPA-licensed hauler. The oil didn't belong to them anymore and there wasn't anything they could do at that point. Will you have to deal with such a call in the future? It doesn't have to happen.

The Court System and Cleanup

If you ask your attorneys to explain why judges are holding so many innocent businesses responsible for cleaning up other people's mistakes, they will spend hours taking you through each case individually, explaining the legal theory under which the Judge held this way in that particular circumstance. But if you stand back and look at the hundreds of cases affecting businesses like yours, you'll notice a common thread. Where there is pollution, the Judge usually finds a way to get it cleaned up and to get the clean-up paid for without public expenditures.

As a Potentially Responsible Party (PRP) in the Peak Oil Superfund site protested, "It's not fair to me. We didn't have any control over what Peak Oil did after they left my business with the oil. So why am I being asked to pay? I thought Peak was a good business, I thought they were helping." [3]

New Detection Capabilities Increase Risk
In 1987, we began looking for a building in which to house our operations. We soon discovered one that was about the right size, unoccupied and in a desirable location. The city's property records revealed that it was owned by GE. We contacted them about leasing or purchasing it only to find out that the property had been contaminated with waste transformer oil containing PCBs. What could have been a valuable property site now represented a major liability. At the time the facility was contaminated, PCBs were not recognized as a serious problem.

What does this mean for how we manage today? Daily we learn more about the harm caused by different kinds of waste. The Gas Research Institute's *Environmental Trends and Issues at the Research Horizon* cites one critical trend as the "capability to detect and analyze potentially toxic substances in the environment at lower and lower concentrations." [4] Detection levels for many substances are now at parts per billion. The subsequent reports of pollution and media attention to health consequences greatly increase a company's potential liability.

Credit-worthiness and Insurability Affected
Banks and lenders take this issue seriously. Before they approve your loans, they now want to know: "What are your potential environmental liabilities and what are you doing about them?" They are beginning to recognize one type of modern bad credit risk — a company that produces potential environmental liabilities is needlessly losing money and setting itself and its lender up for a financial catastrophe. Because of this, many loan agreements now have very strict environmental stipulations (See Exhibit A).

As stated by B.E. Dishman of NationsBank, "Banking is about risk management. In recent years, banks' ability to manage their risks have been significantly impaired by court interpretations of current environmental legislation. As a result, banks have been forced to deny credit to companies which do not carefully manage their wastes. "

Insurance companies, too, understand risk management and the seriousness of waste liability, especially off-site. That's why most of the ways your company can be held liable for its waste when it is hauled off-site are clearly not covered by your liability insurance policy. Special environmental insurance to cover off-site liability is also not available at any price.

This is especially true for such smaller operations as small service stations, truck and car dealers, service garages, and others. Basically stated, "smaller operations of this type are simply unable to buy insurance coverages or obtain a rider to their existing policy that would properly protect them for the contingenices involved in disposing of their waste products." [5]

End-of-the-Pipe vs. Source Management
In attempting to respond to this rapidly-changing environment, many businesses have taken a "band-aid" approach. All too often, businesses continue to view the waste products as problems to dispose of, rather than analyzing how the wastes are produced to unearth ways to eliminate problems at their source. They favor end-of-the-pipeline management as opposed to management at the source.

Outside counsel can function as a technical knowledge base for many businesses, but too many consultants, engineers and attorneys are paid to advise your business with the end-of-the-pipe solutions in mind or simply to defend the lawsuit. For this reason, they may not be focusing on finding innovative ways to solve problems at the source.

The answer to this dilemma is to ask the question differently. Instead of asking "How do we dispose of this waste," ask, "How do we turn our waste into an asset or, failing that, how can we generate less of it?" If there is no obvious way, how can we bring together an internal task force to work on the solution. The result could even be a new product or service for others in the industry.

The Environmental Bottom-Line for Business
There is another powerful reason to integrate environmental management into our total management approach — the additional profits that transforming and eliminating waste and pollution can generate. Here are three examples.

Company A mines and sells copper and emits the SO_2 by-product to the atmosphere. The area surrounding Company A looks like a nightmare vision of Mars, with barren, eroded red clay. Company B mines and sells copper and captures and sells the sulphuric acid. EPA regulations aside, which company is going to make more money and win the competitive contest? Which company is going to complain when EPA cracks down? Which company do you want next door to your business and in your community?

Another example is the transformation of used oil into free energy on-site. One technology releases zero hydrocarbon emissions during combustion; the other emits considerable unburned hydrocarbons. EPA allows either at this time. On that basis, some clients will choose the polluting technology that costs several hundred dollars less. Yet, over the ten-year life cycle of the equipment, how many thousands of dollars of lost energy savings do the unburned hydrocarbons and their insulating soot represent?

A third example involves Northern Telecom and other electronics manufacturers. The Montreal Protocol, signed by some 93 countries in June 1990, stipulated the elimination of CFC-113 by the end of the decade. In 1988, Northern Telecom's CFC-113 use stood at roughly 1 million kilograms. By December of 1991, CFC-113 solvents had been completely eliminated from all operations, well ahead of schedule. Northern Telecom "spent approximately $1 million, but realized savings in the vicinity of $4 million by not purchasing CFC-113 solvents, by not having to pay U.S. CFC taxes, and by reducing waste disposal costs." **[6]**

As these three examples indicate, doing the right thing for the environment often also means more profits and more savings for your company. So, what are the basic steps in a good integrated environmental management approach?

THE NEW MANAGEMENT APPROACH

The Basic Steps

(1) Look at every step in the process that creates or provides your ultimate product or service.

(2) Determine if that step or action actually adds any value to the product or service.

(3) Change it or eliminate it if it doesn't add value.

(4) Note the steps that add value, but create a waste liability.

(5) Figure out a way to change that waste liability into a valuable asset.

(6) If you can't make it a valuable asset, figure out how to eliminate or minimize waste production.

Any system that produces a waste liability at the end of its processing cycle should be viewed as defective. Businesses today must act as swiftly to correct this type of defect as they would to a defect in their product or service that threatened the financial integrity of the company.

Application of these Concepts to Used Motor Oil

To apply the new management approach to your waste fluids, go through the six steps described above.

(1) Look at every step in the process that creates or provides your ultimate product or service.

You select used motor oil to examine first.

(2) Determine if that step or action actually adds any value to the product or service.

You generate waste oil each time you do an oil change on vehicles. If these are your own vehicles, eliminating this step would cause excessive wear on the engines; your company would risk the expense of having to replace them much sooner. If you are changing oil as a service for others, it's part of how your company makes a living. Either way, changing oil does add value.

(3) Change it or eliminate it if it doesn't add value.

You skip this step since you concluded in step two that oil changes add value.

(4) Note the steps that add value, but create a waste liability.

Does the waste oil generated by your oil changes create a waste liability? The answer would certainly be "yes" if you took the quick-and-dirty route and illegally dumped the oil down a storm sewer. As Washington attorney Mark Atlas points out, "with EPA sanctions running at $25,000 a day plus penalties, fines can quickly hit the stratosphere." **[7]**

Your company, like most responsible businesses, does not dump the used oil but has it hauled off on a regular basis by an EPA-approved hauler. Does this create a waste liability? Under Superfund and RCRA laws, your company, as the generator, is responsible for the oil. Your liability extends from "cradle to grave," including what happens to the oil after it leaves your site.

Once a site is designated as a Superfund site, all those businesses who contributed to the hazard (i.e., used the hauler's service), can be held liable. In fact, a surprising number of Superfund sites involve waste oil. EPA has found that, "of the 445 National Priorities List (NPL) facilities having documented Records of Decision, 185 (42%) have had used oils co-disposed with other hazardous or industrial solid waste." **[8]** Thousands of businesses and government entities are having to pay to clean up for someone else's mistakes. Considering this, you properly conclude that oil changes create a serious waste liability.

(5) Figure out a way to change that waste liability into a valuable asset.

Under RCRA, "burning the used oil on-site for energy recovery" is defined as one form of "recycling." **[9]** The energy recovery is usually for space heating or hot water. Recycling on-site can save you thousands of dollars in energy bills and end your "cradle to grave" liability. In effect, it becomes an insurance policy that pays for itself. Payback for the equipment is usually from one to three years. That means a 33% to 100% return on investment. The Used Oil Management Table summarizes the various alternatives.

Used Oil Management Table

	DUMPED	HAULED AWAY	RECYCLED
POTENTIAL LIABILITY	$25,000 per day	"cradle to grave"	none
COST PER GAL.	none	$.05 to $2.00	none
ENERGY SAVINGS	none	none	$1.00/gal
ENVIRONMENT	damaged	at risk	safe on-site energy recovery

TECHNOLOGY

The key to good on-site management of used lubricants has been new technology. The old technology for burning used oil on-site was dirty and unreliable. Old technology generally consisted of a "box" unit with a target wall that burned out periodically (Fig. 1). Incomplete combustion left unburned hydrocarbons as soot in the heat-exchanger. Since soot acts as an insulator, it reduces energy recovery. Hydrocarbon emissions also make these units vulnerable as EPA begins to examine and regulate emissions. Certain unburned hydrocarbons (PAHs) can cause cancer.

Old Technology

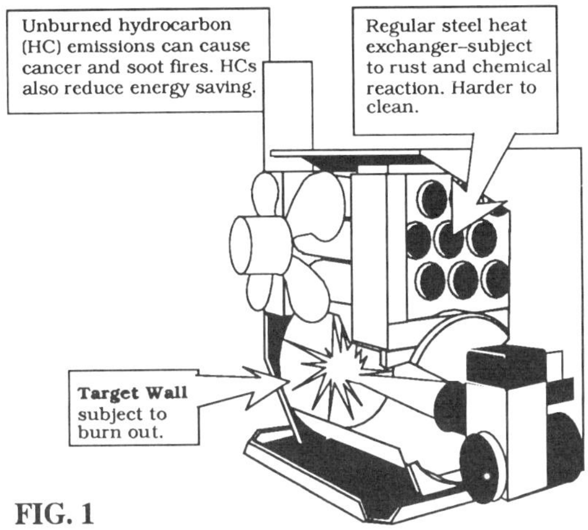

FIG. 1

New advanced technology enables you to burn used motor oil and other lubricants with zero ppm hydrocarbon emissions, zero percent carbon monoxide and the lead emissions of one **unleaded** gasoline car. Here's what to look for as indicators of state-of-the art technology (Fig. 2).

New Technology

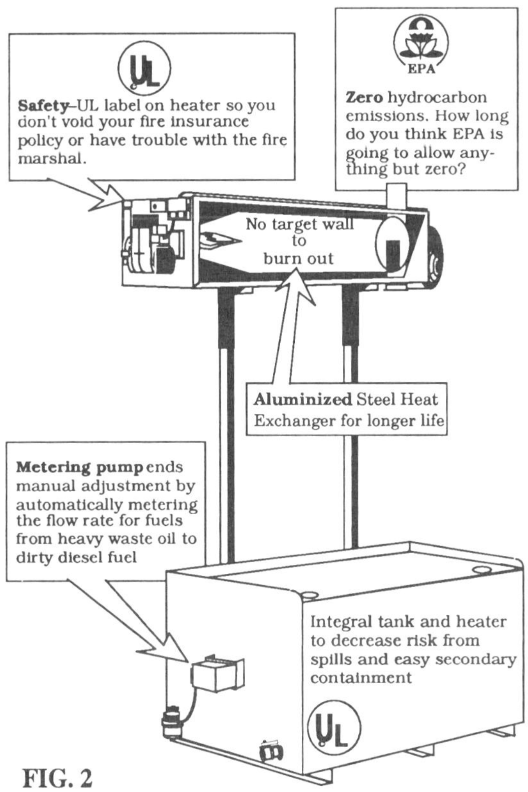

FIG. 2

Specifications for Modern, On-Site Used Oil Recycling Equipment

a. Zero hydrocarbon emissions, as documented by independent test lab.
b. UL listing of heater and used oil storage tank for safety and insurance.
c. Aluminized steel heat exchanger - two metals protect better against chemical attack during high temperature burning.
d. Viscosity-insensitive metering pump to automatically maintain the same fuel rate without manual adjustment
e. Heat exchanger back wall should be 36" or greater from the flame front.

(6) If you can't make it a valuable asset, figure out how to eliminate or minimize the waste.

If, for some reason, you cannot transform your used oil into space heat, you should consider the options for minimizing the oil changes that generate the used oil. You might extend your maintenance schedule through the use of synthetic oil or better filtration.

In both steps (5) and (6), you call upon the collective wisdom of your whole staff in searching for value-adding alternatives.

Other Common Waste Liquids
Advanced technology enables these used oil recycling systems to safely burn, for energy recovery, other common waste fluids whose flash points are below 450° F: brake fluid, gear lube, automatic transmission fluid, certain transformer oils, vegetable oils including cottonseed, peanut, soybean etc.

Special parts washer cleaners are now available that reduce off-site hauling liability. They are non-chlorinated and do not contain benzene, xylene and toluene. They are classified as non-hazardous with a flashpoint above 140° F. EPA's used oil management rules allow these special cleaners to be added to your used oil tank and burned on-site. The parts washer itself should have filtration on it to decrease the annual volume of cleaner used.

Another type of parts washer design uses hot soapy water to clean. Select a model whose design separates the oil and dirt from the water so the water can be reused or legally disposed of through your sewer system.

Anti-freeze and paint solvents can be filtered or distilled on-site and recycled or reused. The anti-freeze will require an additive package before it can be returned to protect the cooling system. The recycled paint solvent is usually suitable for a lower grade use, such as paint gun cleaning. A small amount of residue typically will remain. In evaluating different products, you will want to find out whether the residue produced will test hazardous under TCLP (Toxicity Characteristic Leachate Procedure). Disposal costs for residue that tests hazardous will be much higher.

Application of the six management steps outlined earlier by your total staff can produce other changes in maintenance procedures and innovations to substantially reduce the volume of other waste liquids you generate. For example, something as simple as more frequent use of a drip tray has significantly reduced the volume of waste liquid generated at some facilities.

HOW BUSINESSES AND GOVERNMENT CAN SAVE MORE MONEY AND HELP STOP DIY DUMPING

Oil and Water Don't Mix
Approximately 1.4 billion gallons of used oil are generated in the United States annually. Some 200 million gallons are generated by the do-it-yourself (DIY) oil changer. Only about 10% of this oil is recaptured to be used for energy or to be re-refined. Where does the rest end up? Ultimately, in our water supply. **One gallon of used oil can contaminate one million gallons of fresh drinking water.**

The Answer: DIY Collection Sites
A system of DIY oil collection centers is currently being employed by many cities and states as a means of curbing this proliferation of "backyard oil dumping." A simple, detailed manual for cities and counties on how to set up a collection center is available from the authors of this paper. **[10]** The manual also discusses the conditions under which a Superfund exemption is available for those who accept DIY oil.

Water suppliers have, in this sense, acted in a very rational way. They decided that DIY oil dumping could definitely hurt them and took steps to prevent future costs and potential Superfund sites. However, most of this collected oil is subsequently being hauled away from these collection centers. By allowing this to occur, water suppliers have fallen into the same trap as the business that manages its waste without considering the basic steps.

Here is a scenario:
A waste fluid hauler has a leak in his truck, or has an accident. The result is that oil ends up going down the storm sewer, into the ground water or into a stream.

At this point, there is no obvious method to determine the total damage to the city, but clearly the environment has been affected by the accident. There is no way to clean up or really even identify the waste, which will undoubtedly run off into the sewers (where it can foul sewage treatment processes) or migrate into nearby groundwater over time. If the city in question is near an ocean they will worry that the waste will end up off their shores eventually. If the city is inland, they will fear for their rivers and streams. The inability to clean up such spills is as frustrating as it is alarming, and need not have occurred.

One Step Further

Indeed, there is no practical way to eliminate the generation of used oil, nor is there a way to prevent some 50% of the population from changing their oil on their own. So, the state or local government must find money for a used oil collection center. They also have to pay to have the oil hauled away. Further, if there are any accidental spills or any illegal dumping by the hauler, they risk contaminating the very area that they were trying so hard to protect. In fact, the number of waste oil spills reported to EPA each year averages over 700. **[11]**

From both an economic and environmental perspective, therefore, the real cost, of setting up a DIY collection center and having the used oil hauled away could be much higher.

The alternative to hauling is to employ on-site used oil management. The used oil collected at these collection sites is no different than used oil that is generated at a service station. Waste-oil heaters could be used to burn the used oil on-site. Local government programs that set up DIY centers should place them in buildings that can benefit from heat during the winter and hot water in the summer. This would provide heat to local buildings, saving heating dollars and hauling fees. It will also eliminate accidental spills within the community and the possibility of creating additional Superfund sites.

CONCLUSION

Used Oil Management Made Profitable and Easy

Management and engineering professionals must view any company system or process that produces a waste liability as defective. They must act to transform that waste into an asset, eliminate or minimize it. That is the only way to maximize income or savings and minimize risks. New management concepts, lowered pollution detection levels and the banking, insurance and regulatory climate all mandate this approach.

New on-site technology is now available to enable businesses and government to transform many of their common waste liquids, including used oil, gear lube, brake fluid, automatic transmission fluid, parts washer cleaners, anti-freeze, vegetable oils, etc. from expensive waste liabilities into valuable assets.

[**Note:** The authors of this paper are continuing their research in technology and environmental management. We welcome your comments on this paper and your reports of how you have transformed wastes into assets or eliminated or minimized wastes in your work. Also, if your community needs help establishing DIY used oil collection centers, we will be glad to assist. Please contact Wayne Robertson or Patrick Wayte by phone at 615-251-0680; by fax at 615-251-0682; or in writing at 240 Great Circle Road, #344, Nashville, TN 37228-1707.]

EXHIBIT A

EXCERPTS FROM LOAN AGREEMENT

Environmental Law Compliance

The conduct of Borrower's business operations will not violate any federal laws, rules or ordinances for environmental protection, including, but not limited to, the following: Clean Air Act, 42 U.S.C. § 7401 et. seq; Federal Water Pollution Control Act, 33 U.S.C. § 1251 et. seq; Solid Waste Disposal Act, 42 U.S.C. § 6901 et. seq.; Comprehensive Environmental Response, Compensation and Liability Act ("CERCLA" or "SUPERFUND"), 42 U.S.C. § 9601 et. seq.; National Environmental Policy Act, 42 U.S.C. § 4321 et. seq.; regulations of the Environmental Protection Agency and any applicable local or state law, rule, regulation or rule of common law and any judicial interpretation thereof relating primarily to the environment or Hazardous Materials and Borrower will not use or permit any other party to use any Hazardous Materials at any of Borrower's places of business except such materials as are incidental to Borrower's normal course of business, maintenance and repairs. Borrower agrees to permit Bank, its agents, contractors and employees to enter and inspect any of Borrower's places of business at any reasonable times upon three (3) days prior notice for the purposes of conducting an environmental investigation and audit (including taking physical samples) to insure that Borrower is complying with htis covenant. Borrower shall provide Bank, its agents, contractors, employees and representatives with access to and copies of any and all data and documents relating to or dealing with any Hazardous Materials used, generated, manufactured, stored or disposed of by Borrower's business operations within five (5) days of the request therefor. Hazardous Materials include all materials defined as hazardous wastes or substances under any local, state or federal environmental laws, rules or regulations, and petroleum, petroleum products, oil and asbestos.

Notification

Borrower shall immediately advise Bank in writing of (i) any and all enforcement, cleanup, remedial, removal, or other governmental or regulatory actions instituted, completed or threatened pursuant to any applicable federal, state, or local laws, ordinances or regulations relating to any Hazardous Materials affecting Borrower's business operations; and (ii) all claims made or threatened by any third party against Borrower relating to damages, contribution, cost recovery compensation, loss or injury resulting from any Hazardous Materials. Borrower shall immediately notify Bank of any remedial action taken by Borrower with respect to Borrower's business operations.

Indemnification

Borrower shall indemnify, defend and hold Bank and its successors and assigns harmless from and against any and all claims, demands, suits, losses, damages, assessments, fines, penalties, costs or other expenses (including attorneys' fees and court costs) arising from or in any way related to actual or threatened damage to the environment, agency costs of investigation, personal injury or death, or property damage, due to a release or alleged release of Hazardous Materials, arising from Borrower's business operations or in the surface or ground water arising from Borrower's business operations, or gaseous emissions arising from Borrower's business operations or any other conditions existing or arising from Borrower's business operations resulting from the use or existence of Hazardous Materials, whether such claim proves to be true or false.

REFERENCES

1 Charlene B. Adair-Heeley, *The Human Side of Just-In-TIme*, 1991.

2 Hiroyuki Hirano, *JIT Factory Revolution*, 1988.

3 *FuelLine*, June 1993, p. 25.

4 Gas Research Institute, *Environmental Trends and Issues at the Research Horizon*, GRI-92/0156, 1992.

5 Correspondence from Robert L. Coleman, Marsh and McClennan 1990, May 9 1990.

6 Stephan Schmidheiny, *Changing Course: A Global Perspective on Development and the Environment*, p. 230.

7 *Environment Today*, April 1991.

8 *Hazardous Waste Management System; General: Identification and Listing of Hazardous Waste; Used Oil*, EPA/OSW-FR-91-023.

9 EPA *Used Oil Management Standards*, 40 CFR § 261.6(a)(4), September 1992.

10 Leonard Morgan, *Used Motor Oil Management Made Easy: Paid for with Energy Savings*, 1993.

11 EPA correspondence from Don Clay in response to questions concerning waste oil spills from Congressmen Bob Clement and Jim Cooper, April 1992.

Chapter 20

Alternative Aviation Fuels: Near-Term Opportunities to Enhance Energy Independence and Improve Regional Air Quality

R.O. Price

ABSTRACT

The aviation transportation sector, both commercial jet and general aviation operations, exerts an increasingly important, but largely unacknowledged impact upon regional air quality. Emissions from these operations, which currently are uncontrolled, offset many of the emission reductions which are being achieved in the ground transportation sector. The regional air quality implications of aviation transportation will increase as the disparity between these uncontrolled emissions and those of the rapidly improving ground transportation sector widen. This paper discusses the magnitude of current petroleum-fueled aviation operations from a regional air emissions perspective. Implementation of alternative aviation fuels provide a significant opportunity to improve regional air quality and enhance the potential for energy independence.

-o-

Like many highly urbanized areas, Los Angeles, Orange, Riverside, and San Bernardino Counties, which collectively comprise Southern California's South Coast Air Basin (Basin), are plagued with serious air quality problems. In the Basin's case, these problems are the most severe in the United States. The primary malefactor in this situation is the Basin's heavy reliance on an automotive transportation system fed almost exclusively by petroleum-based fuels. To deal with this pressing issue, innovative solutions have been proposed, including the mandatory phased introduction of low emission vehicles (LEVs) into ground transportation fleets [1] and the trading of mobile source emission reduction credits through the early retirement of older, dirtier automobiles [2].

These efforts are laudable, and some progress in improving regional air quality appears to be forthcoming. The aviation transportation sector, however, particularly commercial jet operations, also exerts an increasingly important, but largely unacknowledged impact upon urban regional air quality. Emissions from these operations, which currently are uncontrolled, have the potential to offset many of the emission reductions which are being achieved in the ground transportation sector.

AVIATION AND REGIONAL AIR QUALITY

Despite a few notable allusions to the contrary [3], on a national scale, aircraft emissions historically have been dismissed as insignificant contributors to adverse air quality impacts. This is due largely to the fact that aviation sources account for only a small portion of total emissions from all sources [4]. However, is this necessarily true on a regional and local scale?

Currently, the Basin's five major commercial airports (Los Angeles, Ontario, Santa Ana, Burbank, and Long Beach) host one of the most intense areas of commercial air traffic activity in the country. For example, according to the Southern California Association of Governments (SCAG), in 1987 domestic and international commercial jet air carrier operations at these airports totaled 791,311 landing/takeoff (LTO) cycles, for a daily average of 2168 [5]. An LTO cycle incorporates all of the normal aircraft flight and ground operation modes which impact Basin air quality, including:

- descent/approach from 3,000 feet;
- touchdown;
- landing run;
- taxi in;
- idle and shutdown;
- startup and idle;

- checkout;
- taxi out;
- takeoff; and
- climbout to 3,000 feet.

The LTO operational description is of importance to Basin air quality, since an inversion layer frequently exists at an altitude of approximately 3000 feet over much of the Basin, a situation which can serve to trap and exacerbate aircraft LTO emissions. The question then becomes whether or not this level of commercial aviation activity, and its associated emissions, is of a magnitude which can exert a potentially significant negative impact on regional air quality.

As shown in Table 1, SCAG has compiled LTO emission data on various commercial aircraft/engine combinations [6]. These data take into account aircraft-specific variations in time-in-mode for the LTO segments of approach, taxi/idle (arrival), idle/taxi (departure), takeoff, and climb. Taking a simple average of these data yields carbon monoxide (CO), unburned hydrocarbon (HC), and oxides of nitrogen (NOx) emissions of approximately 64.4, 34.4, and 49.5 pounds per LTO, respectively.

TABLE 1
1987 SOUTH COAST AIR BASIN
DAILY COMMERCIAL AIRCRAFT OPERATIONS
AVERAGE EMISSION RATES (lb/LTO Cycle)

AIRCRAFT/ENGINE	NOx	CO	HC
A300/GE CF6-50	56.64	50.98	19.58
B707/JT3D-7	22.65	163.75	130.81
B727/JT8D-17	25.29	29.75	6.71
B737/JT8D-17	16.86	19.84	4.47
B737-300/CFM56-3	14.69	14.33	0.80
B747-100/GE CF6-50	113.29	101.96	39.17
B747-200/JT9D-7	77.20	166.98	61.05
B747-SP/RB211-524	118.57	43.34	6.47
B757-200/PW 2037	34.96	14.64	1.43
B767-200/GE CF6-80	50.51	21.48	2.23
BAE146/ALF502R5	7.47	12.82	1.54
DC10-10/GE CF6-6	61.00	58.41	20.07
DC10-30/GE CF6-50	84.96	76.47	29.38
DC10-40/JT9D-7	57.90	125.24	45.79
DC8-63/JT3D-7	22.65	163.75	130.81
DC8-70/CFM56-2	35.90	30.71	1.39
DC9-30/JT8D-17	16.86	19.84	4.47
DC9-80/JT8D-217	23.07	8.23	2.50
L1011-1/RB211-22	59.85	132.16	89.44
L1011-5/RB211-524	88.92	32.50	88.92
AVERAGE	**49.46**	**64.36**	**34.35**

Source: SCAG 1989. [6]

Combining these average aircraft emission data with those on historic daily operations yields an estimate of the contribution made to the Basin's 1987 emissions inventory by commercial aircraft LTOs; in this case, 1.40×10^5, 7.45×10^4, and 1.07×10^5 pounds of CO, HC, and NOx per day, respectively. At this point, it is instructive to consider what this level of commercial aircraft activity and associated emissions means in terms of equivalent individual passenger vehicle (automotive) emissions, both now and in the future.

Aircraft vs. Autos

The South Coast Air Quality Management District (SCAQMD) provides guidance for estimating automotive emissions in support of California Environmental Quality Act (CEQA) Environmental Impact Report (EIR) documents [7]. Drawing upon information provided by this guidance, we can estimate a Basin-specific emission rate for an "average" present-day individual passenger automotive vehicle, i.e. 0.15, 0.02, and 0.02 pounds per day of CO, HC, and NOx, respectively (Table 2).

TABLE 2
CURRENT SOUTH COAST AIR BASIN
DAILY PASSENGER VEHICLE
AVERAGE EMISSION RATES

		(Trips/Miles per trip)	
COUNTY		WORK	NONWORK
Los Angeles		0.95/9.6	4.35/5.6
Orange		1.07/10.9	4.32/5.6
Riverside		0.90/17.7	4.48/7.8
San Bernardino		0.91/13.0	4.79/7.0
Regional Average		0.96/13.0	4.48/6.5
TRIP BREAKDOWN		39.27%	60.73%
REG. AVG. VMT		5.11	3.95
TOTAL AVG. VMT		9.06 miles @ 25.4 mph	
EMISSION FACTORS	CO	ROC	NOx
(g/mile)	7.65	0.98	1.1
(g/vehicle)	69.33	8.88	9.97
(lb/vehicle)	0.15	0.02	0.02

Source: SCAQMD 1992 [7]

By combining these automotive emissions data with those previously estimated for the commercial aviation sector, we may roughly approximate jet emissions in terms of passenger vehicle equivalents. As shown in Table 3, 1987 daily emissions of NOx, CO, and HC from combustion of jet fuel during typical LTO cycles, are estimated to have been equivalent to those from approximately 4.88×10^6, 9.13×10^5, and 3.8×10^6 automobiles, respectively.

TABLE 3
1987 SOUTH COAST AIR BASIN
DAILY COMMERCIAL AIRCRAFT EMISSIONS (lb)

AIRPORT	LTOs	NOx	CO	HC
Los Angeles (LAX)	1,603	79,305	103,191	55,078
Long Beach (LGB)	41	2,040	2,655	1,417
Ontario (ONT)	203	10,042	13,067	6,974
Santa Ana (SNA)	170	8,405	10,936	5,837
Burbank (BUR)	150	7,439	9,680	5,167
TOTAL	2,168	107,231	139,529	74,473

PASSENGER VEHICLE EQUIVALENTS

NOx	4.88×10^6
CO	9.13×10^5
HC	3.80×10^6

Source: SCAG 1992 [5]

For perspective, consider that this level of daily commercial aviation emissions is approximately 1.13, 7.4, and 16.9 times higher than the baseline NOx, CO, and HC emissions estimated, respectively, for stationary powerplants operating within the Basin during this same period [8]. It is also constructive to consider that, under proposed Rule 1601, the SCAQMD had originally proposed to target a scant 51,000 nontransit fleet cars and light trucks for mandatory conversion to clean alternative fuels [9].

Finally, considered from a slightly different perspective, the level of commercial aviation generated NOx is nearly 5 times higher than that of all pre-1972 vehicles estimated to be present and on the road in the Basin, which are candidates for emission trading early retirement [2]. This gives rise to yet another observation. Over time, the ground transportation fleet will continue to benefit from overall emission reductions as newer, cleaner vehicles are introduced (Table 4).

Although the aviation sector may show minor emission reductions, due primarily to improvements in turbine engine combustor technology, the penetration into the overall aircraft fleet likely will be slow as a result of high costs and lengthy certification requirements. Concurrent with this generally static emissions scenario, commercial aviation activity is forecast to increase. For example by 2010, SCAG projects an increase over 1990 level of approximately 53% in domestic and international commercial air carrier operations within the Basin.

The net result will be an increased disparity in aircraft/automotive emissions. For example, combining the projected 2010 automotive and aircraft emissions data, yields NOx, CO, and HC jet transport emissions equivalent to approximately 2.5×10^7, 5.8×10^6, and 6.3×10^7 2010-model year automobiles, respectively (Table 5). Note that none of these approximations take into account the emissions contributed by turbine-powered air taxi, helicopter, general aviation, or military aircraft operations within the Basin. A similar situation exists for gasoline-fueled, piston engine general aviation aircraft.

TABLE 4
2010 SOUTH COAST AIR BASIN
DAILY PASSENGER VEHICLE
AVERAGE EMISSION RATES

COUNTY	WORK (Trips/Miles per trip)	NONWORK (Trips/Miles per trip)	
Los Angeles	0.96/10.8	4.34/6.3	
Orange	1.09/11.6	4.36/6.5	
Riverside	0.89/17.0	4.35/9.6	
San Bernardino	0.89/13.6	4.59/7.9	
Regional Average	0.96/13.2	4.41/7.6	
TRIP BREAKDOWN	39.39%	60.61%	
REG. AVG. VMT	5.2	4.6	
TOTAL AVG. VMT	9.8 miles @ 23.2 mph		
EMISSION FACTORS	CO	ROC	NOx
(g/mile)	1.49	0.07	0.27
(g/vehicle)	14.65	0.72	2.62
(lb/vehicle)	0.03	0.002	0.01

Source: SCAQMD 1992 [7]

TABLE 5
2010 SOUTH COAST AIR BASIN
DAILY COMMERCIAL AIRCRAFT EMISSIONS (lb)

AIRPORT	LTOs	NOx	CO	HC
Los Angeles (LAX)	1,484	73,393	95,498	50,972
Long Beach (LGB)	183	9,049	11,775	6,285
Ontario (ONT)	566	27,997	36,429	19,444
Santa Ana (SNA)	379	18,721	24,360	13,002
Burbank (BUR)	283	13,983	18,194	9,711
TOTAL	2,894	143,143	186,257	99,414

PASSENGER VEHICLE EQUIVALENTS

NOx	2.48×10^7
CO	5.77×10^6
HC	6.26×10^7

Source: SCAG 1992 [5]

General aviation is a vital and important segment of personal and business transportation. For example, in California approximately 72 percent of all flying is business related, while the figure nationally approaches 85 percent. The sharp fuel price rises of recent years have adversely affected the general aviation industry by curtailing not only discretionary, but also business and commercial flying. For example, between 1977 and 1982, the average retail price of aviation gasoline increased more than 400 percent from approximately $0.48 per gallon to more than $2.00 per gallon -- a level where it has remained. In addition, supplies of aviation gasoline are in jeopardy and future shortages loom, since aviation gasoline represents a limited, specialty market to oil refiners.

By utilizing data on general aviation operations provided by the California Department of Transportation [10] and piston engine aircraft emission rates from the EPA [11], repetition of the above comparisons for general aviation shows that in 1987 these operations contributed NO_x, CO, and HC emissions equivalent to 1.9×10^4, 1.4×10^6, and 2.4×10^5 automobiles, respectively (Table 6). Like its larger commercial aviation brethren, the regional air quality implications of the general aviation sector will increase as the disparity between these uncontrolled emissions and those of the rapidly improving ground transportation sector widen (Table 7).

TABLE 6
1987 SOUTH COAST AIR BASIN
DAILY GENERAL AVIATION EMISSIONS (lb)

COUNTY	LTOs	NOx	CO	HC
Los Angeles	5,920	230	121,505	2,702
Orange	1,571	54	28,795	640
Riverside	1,756	64	34,090	758
San Bernardino	1,950	68	36,208	804
TOTAL	11,207	417	220,598	4,904

PASSENGER VEHICLE EQUIVALENTS

NOx	1.89×10^4
CO	1.44×10^6
HC	2.45×10^5

Source: CALTRANS 1988 [11]

Clearly, the aviation transportation sector not only is a significant consumer of petroleum-based fuel, but also represents a significant and expanding source of uncontrolled annual emissions within the Basin. With momentum building at the state and federal levels for analysis and implementation of alternative fuels and low emission vehicles for air quality improvement within the automotive sector, the time appears ripe to conduct similar investigations in the aviation sector.

TABLE 7
2000 SOUTH COAST AIR BASIN
DAILY GENERAL AVIATION EMISSIONS (lb)

COUNTY	LTOs	NOx	CO	HC
Los Angeles	5802	233	122,764	2,731
Orange	1729	60	31,850	707
Riverside	1966	73	38,405	854
San Bernardino	1615	82	43,420	965
TOTAL	11,112	448	236,439	5,257

PASSENGER VEHICLE EQUIVALENTS

NOx	4.48×10^4
CO	7.88×10^6
HC	5.26×10^6

Source: CALTRANS 1988 [11]; SCAQMD 1992 [7]

ALTERNATIVE AVIATION FUEL ISSUES

Is there a market for alternative commercial aviation fuels? CEC performed a preliminary assessment of the potential for methanol as a commercial jet fuel in California nearly a decade ago [12]. This analysis was widely distributed and reviewed within the established aviation industry. Despite the passage of time, the study's basic conclusions remain unchallenged. A few of these conclusions are:

- Intrastate commercial airlines represent California's largest "captive fleet";
- On typical intrastate flights, the methanol weight penalty (resulting from its lower per pound Btu content relative to jet fuel) does not significantly increase fuel consumption; and
- Present airline operation and refueling practices could accommodate methanol.

Barriers to Development

Despite the attractive potential environmental benefits of alternative aviation fuels, development of both piston and turbine engine alternatives face a number of obstacles in the form of resource, technical development, investment, regulatory and marketing impediments [13]:

- Availability;
- Distribution;
- Compatibility;
- Economics;
- Energy density;
- Handling;

- Safety; and
- Quality control.

Availability and supply affect both conventional and alternative fuels. For example, refiners are reluctant to gear up for low-volume fuel production in the absence of a large existing or perceived market, while manufacturers are unwilling to commit resources to develop aircraft/engine designs for specialty fuels which have no long-term production commitment.

Alternative aviation fuel market penetration depends on its ability to use the existing fuel distribution system. An alternative which is compatible with an existing or developing fuel distribution system will enjoy a significant advantage over one which requires complex and expensive storage and handling facilities. A substitute fuel also must be compatible with current aircraft engine/fuel systems.

The fuel/direct operating cost ratios of civil aircraft have increased during the past two decades from approximately 0.25 to over 0.60 [14]. The price of jet fuel rose 40% between August 1989 and January 1990 [15]. The cost effectiveness of alternative aviation fuels is, therefore, a key factor in the future viability of aviation in general, and the airline industry in particular. Consequently, a substitute fuel must compete cost-effectively with conventional fuel.

Energy density is another important consideration when screening alternative aviation fuels. Aircraft turbine engines are heat engines, transforming heat released during combustion into useful mechanical work. One result is that aircraft range is, more or less, proportional to fuel-energy density expressed in BTUs per pound or gallon. Lower energy density also can exact a penalty in payload and range.

Studies of alternative aviation fuels are concerned primarily with both quantity and quality as they affect availability, handling, performance and overall economy in terms of both energy and costs.

A prime consideration for any aircraft fuel is handling ease and safety. Fuel quality control translates as a need for an American Society for Testing and Materials (ASTM) or similar technical specification or standard.

Jet Fuel Alternatives

The use of alternatives such as methyl and ethyl alcohols (methanol and ethanol), either as stand-alone substitute fuels or in combination with conventional jet fuel, presents not only an opportunity for reductions in dependence on imported petroleum, but also significant potential as environmentally attractive options to conventional turbine aircraft fuel.

For example, methanol is considered to be a "near-term" alternative to conventional petroleum-based fuels in the automotive sector. The California Energy Commission (CEC) has demonstrated that methanol is an environmentally attractive alternative to conventional turbine fuel for stationary peaking turbines, with engine/fuel system conversion relatively straightforward [16].

Somewhat more limited testing has been conducted with methanol in aircraft turbine engines. For example, early in 1983 General Electric performed an altitude simulation test of methanol in a combustor segment of one of its CF 680 aircraft turbines for the National Aeronautics Association (NAA) [17]. The test established that methanol as an aircraft turbine fuel would produce low NOx emissions, little smoke and operating temperatures lower than with Jet A. Furthermore, with its lower vapor pressure, methanol could diminish the impact of evaporative emissions from aircraft fuel storage and transfer.

Recent ground-based turbine engine performance tests using ethanol in various percentage mixtures with conventional jet fuel, indicate some potentially attractive reductions in exhaust emissions. For example, Alabama Aviation and Technical College recently conducted a series of "probe tests" of various alternative aircraft fuels, including a 10% ethanol/90% Jet A fuel mixture in a Pratt & Whitney PT-6 turbine engine. These tests resulted in the following observations relative to conventional jet fuel: reduced odor; 0 ppm (parts per million) HC; and 0% CO at rated power under simulated cruise conditions [18]. Although ethanol has a lower energy content (BTUs per gallon) than Jet A, no appreciable loss of power was detected. All temperatures were observed to remain within acceptable limits.

Somewhat more complex and intensive investigations of ethanol/jet fuel blends were conducted recently at Southern Illinois University (SIU) [19]. Although these tests were unable to duplicate the dramatic reductions in HC and CO emission observed in the Alabama tests, the SIU investigation did result in the following observations relative to conventional jet fuel: dramatic reductions in soot (particulate) formation; reduced CO_2 production; and substantial reductions in NOx and nitric oxide (NO) formation.

Piston Engine Fuel Alternatives

Methanol already has received considerable attention as an

alternative aviation fuel. For example, a Supplemental Type Certificate (STC) already has been obtained from the FAA for piston engine applications of methanol, based on a significant amount of flight testing by Gordon Cooper and William Paynter. Similar efforts by Dr. Max Shauck of Baylor have focused on the viability of ethanol as a piston engine fuel; efforts which recently culminated with the first trans-Atlantic flight using an alternative fuel and Dr. Shauck's receipt of the prestigious Harmon Trophy.

Preliminary investigations by the author and others [18] indicate that liquified petroleum gas (LPG) displays a significant potential as an alternative fuel for general aviation aircraft in California, nationally, and internationally. The principal advantages for this LPG application include, but are not necessarily limited to:

- low fuel costs;
- established technical fuel specification;
- current and projected future surplus availability;
- declining traditional LPG markets;
- existing national supply system infrastructure;
- wholesale/retail locations proximate to airports;
- decreased maintenance costs;
- high octane resulting in good performance potential;
- low emissions; and
- low-cost, off-the-shelf conversion technology.

In addition, LPG is equal to or superior to gasoline in the following key aviation fuel areas:

- preignition and deposit ignition;
- vapor lock;
- icing;
- cold start;
- hot restart;
- fuel safety;
- valve sticking and wear;
- material incompatibility and corrosion;
- maldistribution;
- spark plug operation; and
- fuel storage stability.

Regarding economics, LPG 1993 California supplies are expected to be at wholesale prices of from \$0.25-0.31 per gallon (\$0.33-0.42 per gallon gasoline equivalent) for fleet applications. For the year 2000, LPG is projected to be available at wholesale prices of from \$0.31-0.39 per gallon (\$0.41-0.52 per gallon gasoline equivalent) for fleet applications [20]. Note that these costs are based on comparisons with unleaded automotive gasoline. LPG will fare even better when compared with more expensive aviation gasoline.

AVIATION FUEL DEVELOPMENT PLAN

Additional development work, including wider industry and public agency involvement, is needed to establish the performance, cost, and emissions characteristics of any of these alternative aviation fuels before commercial applications can commence [21]. In this light, the author has structured a proposal to the SCAQMD, to conduct a unique research, demonstration, test, and evaluation (RDT&E) project to concurrently:

- flight test and demonstrate the potential for the clean alternative fuel LPG to replace conventional aviation gasoline in aircraft piston engines; and

- flight test and quantify the potential for regional air quality improvement through the use of alternative fuel(s) in the turbine engine, commercial aviation sector.

The proposed effort will develop the data necessary to recertify a conventional single engine aircraft for operation on LPG. Concurrently, a series of various alternative fuel/fuel blend performance and emission "probe" tests, under actual LTO cycle flight conditions, will be conducted using a small aircraft turbine engine (Microturbo TRS-18) mounted on the piston engine test aircraft.

It is anticipated that such an approach can garner technical data at significantly lowered engineering, development, test, and logistic costs associated with a flight test program of this magnitude relative to a program focused on a large commercial or corporate jet aircraft.

REFERENCES

[1] Price, R. 1989. Mandated fleet conversions to alternative clean transportation fuels: environmental policy considerations. 24th Intersociety Energy Conversion and Engineering Conference. Washington, DC. August.

[2] ARB (Air Resources Board). 1992. Mobile source emission reduction credits -- a concept paper on the generation of emission reduction credits from mobile sources. California Environmental Protection Agency. July.

[3] Fay, J.A. 1970. Air pollution from future giant jetports. 63rd Annual Meeting of the Air Pollution Control Association. St. Louis, MO. June 14-18.

[4] Naugle, D.F. and D.L. Fox. 1981. Aircraft and air pollution. Environmental Science and Technology, Vol. 15, Number 4. April.

[5] SCAG (Southern California Association of Governments). 1992. Personal communication with T. Merwin, Aviation Planning. November 30.

[6] SCAG. 1989. Personal communication with W.D. Farrell, Aviation Planning. July 7.

[7] SCAQMD (South Coast Air Quality Management District). 1992. CEQA Air Quality Handbook, Final Draft. September.

[8] SCAQMD. 1990. Draft Air Quality Management Plan 1991 Revision; Appendix IV-D, Energy Conservation Measures and Energy Analysis for Control Strategies. December.

[9] SCAQMD. 1990. Proposed Rule 1601 -- Fleet Conversion for Passenger, Light-Duty, and Medium-Duty Vehicles (PR1601A). February 23.

[10] CALTRANS (California Department of Transportation). 1988. The California Aviation System Plan. Element II -- Forecasts. Division of Aeronautics. August.

[11] EPA (U.S. Environmental Protection Agency). 1986. Compilation of air pollutant emission factors. AP-42. Volume II: mobile sources. September.

[12] Price, R. 1981. Methanol as a commercial aviation fuel in California -- a preliminary assessment. California Energy Commission. Technology Assessments Office Staff Report. August.

[13] GAMA (General Aircraft Manufacturers Association). 1981. Crystal balling the next twenty years. Presentation of T.J. Smith at NASA Lewis technical workshop on aviation gasolines and future alternatives. February 3.

[14] Goodger, E. and R. Vere. 1985. Aviation fuels technology. MacMillan Publishers Ltd. London.

[15] McKenna, J.T. 1990. High fuel prices, weak traffic dim prospects for first quarter. Aviation Week & Space Technology. January 8.

[16] CEC 1986. Methanol clean coal stationary demonstration project final report. P500-86-005. February.

[17] Schauffler, P. 1982. Alcohol fuels for aviation. Energy Efficiency in the Eighties -- proceedings of the ninth energy technology conference, February 16-18, 1982. Government Institutes, Inc. June.

[18] Eveleth, E.L. 1989. Ethanol and methanol in intermittent combustion engines. Published in Future Fuels for General Aviation. American Society for Testing and Materials. ASTM STP 1048.

[19] SIU (Southern Illinois University). 1992. Combustion properties of ethanol blended turbine fuels. Aviation Technologies Division.

[20] CEC (California Energy Commission). 1991. AB 234 Report -- cost & availability of low-emission motor vehicles and fuels. Staff Draft Report. P500-91-009. August.

[21] CEC 1989. California's Energy Agenda -- environmental challenges and energy opportunities. California Energy Commission final 1989 Biennial Report. P106-89-001. July.

Chapter 21

Impact of Driving Trends in Automobile Fuel Efficiency

I.A. Vera

ABSTRACT

This paper presents long-term projections of the gap between "on-road" fuel efficiency and "tested" fuel efficiency for automobiles in the U.S. The projections are generated based on the analysis of three important trends in driving patterns that affect fuel efficiency: 1) increasing urban share of vehicle miles traveled (vmt), 2) increasing average highway speed, and 3) increasing level of urban highway congestion. The projections are developed assuming a logistic growth in the driving trends and taking into consideration the decomposition of urban and rural driving into city and highway driving according to road types.

This study shows that the automobile fuel efficiency gap in the U.S. will increase to 21.2% in 2010 from a base gap of 15.2% in 1990, when a logistic growth in driving trends is assumed.

INTRODUCTION

This paper presents long-term projections of the difference between on-road and tested fuel efficiency for automobiles. The projections are based on the analysis of important trends in driving patterns that affect fuel economy. These trends include the increase in urban share driving, highway speeds, and urban congestion. The projections are developed for the period 1990 through 2010.

BACKGROUND

A difference exists between automotive fuel economy as measured by the Environmental Protection Agency (EPA) under controlled laboratory conditions and the actual fuel efficiency observed under real "on-road" conditions. Public and private organizations such as the Department of Energy (DOE), the Environmental Protection Agency (EPA), Ford Motor Company, General Motors Corporation, and Mitsubishi Motors Corporation have conducted independent research on fuel economy, in the past, confirming this difference.[1] The fuel efficiency gap is defined as the difference between on-road fuel economy and EPA tested fuel economy.[2] When fuel economy is expressed in terms of miles per gallons (MPG), the gap is formulated as:

$$GAP = \frac{EPA\ Test\ MPG - On\text{-}Road\ MPG}{EPA\ Test\ MPG} \quad (1)$$

On-road fuel efficiency depends on several determinants which can be classified into technological factors, driver behavior and habits, driving trends, and road and climate conditions. Furthermore, the magnitude of the gap between tested fuel efficiency and on-road fuel efficiency depends on the specific procedures and conditions used during the test and the closeness of the formulations used to represent real driving conditions.

EPA fuel economy estimates for city and highway driving are published every year for each new model available in the U.S.[3] These MPG estimates are obtained based on vehicle tests performed under controlled laboratory conditions and then adjusted downwards to reflect actual driving conditions. Separate tests are used to generate the city and highway MPG estimates.

The EPA city fuel economy estimates are based on a test that simulates a 7.5 mile, stop-and-go trip with an average speed of 20 mph. The trip lasts 23 minutes and has 18 stops. About 18 percent of the time is spent idling, such as waiting for traffic lights or in rush hour traffic. Two types of engine starts are used: a cold start and a hot start.

The cold start is similar to starting the car in the morning after it has been parked all night. The hot start is similar to restarting a vehicle after it has been warmed up, driven and stopped for a short time.

The EPA highway fuel economy estimates represent a mixture of "non-city" driving. Segments corresponding to different kinds of rural roads and interstate highways are included. The test simulates a 10-mile trip and averages 48 mph. The test is run from hot start and has little idling time and no stops.

EPA adjusts these laboratory fuel economy estimates downwards to reflect actual driving on the road conditions. In the *1992 Gas Mileage Guide: EPA Fuel Economy Estimates* the city estimates are lowered by 10 percent and the highway estimates by 22 percent from the laboratory test results. These adjustment factors represent the EPA estimates of the fuel efficiency gap for both city and highway driving.

Fuel economy can also be represented by a composite number that combines city and highway fuel economies. EPA computes composite fuel economies using the following formulation:

$$\textit{EPA Composite MPG} = \frac{1}{\frac{0.55}{MPGc} + \frac{0.45}{MPGh}} \qquad (2)$$

where:
$MPGc$ = Miles per gallon for city driving
$MPGh$ = Miles per gallon for highway driving

EPA's composite formulation is developed based on 55% city driving and 45% highway driving. This formulation, combined with the EPA city and highway fuel efficiency gaps, leads to a base composite MPG gap for all new vehicles of about 15 percent.

Previous attempts at estimating the base fuel efficiency gap have been made. In 1978, McNutt et al., measured the gap for model year 1974 through model year 1977 cars. The resulting estimates of the gap were between 6 and 9 percent.[4] In 1984, Hellman and Murrel estimated a composite MPG gap of 15 percent.[5] More recently in 1992, Oak Ridge National Laboratory (ORNL) reported composite gap estimates that apply to all automobiles and light trucks in operation.[6] The ORNL base composite gap estimate for all automobiles in operation pre-1974 to 1989 was 15.2 percent. For this analysis, ORNL used EPA tested fuel economy data which was verified by the National Highway Safety Administration (NHTSA). These data were compared against on-road fuel economy data from (1) the Federal Highway Administration (FHWA) *Highway Statistics 1989*, (2) the Department of Energy, Energy Information Administration, *1988 Residential Transportation Energy Consumption Survey* (RTECS), and (3) the Bureau of the Census, 1987 Census of Transportation, *Truck Inventory and Use Survey* (TIUS).

Very few attempts to forecast trends in the fuel economy gap are available. In 1989, Westbrook and Patterson analyzed trends in driving patterns and produced forecasts of the fuel economy gap for the year 2010.[7] Their results indicated a composite gap of 29.7 percent for automobiles for the year 2010. This combined fuel efficient gap corresponded to a city fuel efficiency gap of 23.5 percent and a highway fuel efficiency gap of 30.5 percent. The Department of Energy (DOE) and the Energy Information Administration (EIA) in the 1990 National Energy Strategy (NES) projected the fuel efficiency gap to reach 30 percent by 2030 in the NES reference case.[8] The projected gap for the High Conservation and the Very High Conservation cases of NES were 25 and 20 percent respectively. Also, EIA in the *Annual Energy Outlook 1992* (AEO) projected the fuel efficiency gap to increase from 20 percent in 1990 to 25 percent in 2010. An ongoing effort by DOE's Office of Transportation Technologies in conjunction with the University of Tennessee is focused on forecasting the fuel efficiency gap for automobiles and light duty trucks through 2010.

APPROACH

This paper presents projections generated taking into consideration the decomposition of urban and rural driving into city and highway driving according to road types.

The decomposition is based on road types. Thus, vmt driving on roads identified as "interstate" and "other freeways and expressways" in urban areas are considered part of the highway driving share. Other road types located in urban areas are considered part of the city driving share. In addition, vmt driving on roads defined as "minor collectors" and "local" in rural areas are classified as city driving while the rest of the road types in rural areas are considered highway driving. Although this road classification does not exactly replicate reality, it is a closer representation of the actual city/highway driving composition.

Figure 1 presents the percentages of city and highway driving for both urban and rural areas according to the road type classification described above for 1990. Approximately 63 percent of total 1990 vmt consisted of driving in urban areas and 37 percent in rural areas. About 68 percent of the urban vmt is considered city driving and 32 percent highway driving. In rural areas, 17 percent is considered city driving and 83 percent highway driving.

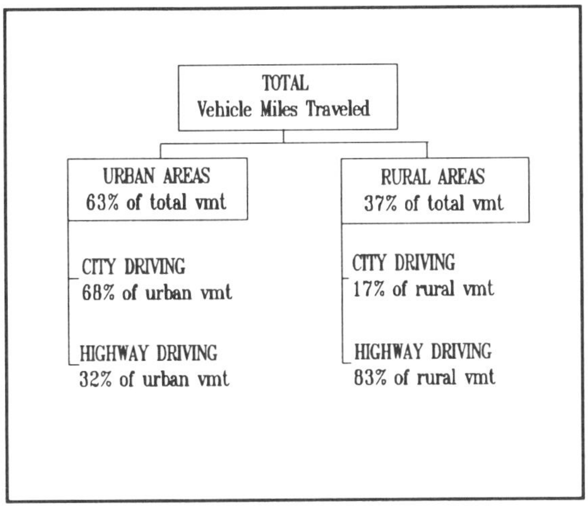

FIGURE 1: CITY/HIGHWAY DRIVING SHARE FOR VEHICLES IN 1990

This composition represents **overall** city and highway driving shares for 1990 of:

City Share:	49.1 %
Highway Share:	50.9 %

These adjusted city and highway shares are the bases for the calculations of the fuel efficiency gap projections in this study. The impact on fuel efficiency, from each of the three factors considered in this study, is affected by these adjusted shares.

The projections are generated based on the analysis of three important trends in driving patterns that affect fuel efficiency. These factors are:

1) increasing urban share of vehicle miles traveled,
2) increasing average highway speed, and
3) increasing level of urban highway congestion.

Forecasts for each of these factors are developed based on logistic growth scenarios.

Increasing Urban Share Driving

A review of the data from the last few decades on vmt for both automobiles and light duty trucks reflects a continuous increase in the share of urban driving.[9] For automobiles the urban share increased from 45.4 percent in 1953 to 62.9 in 1990. This represents a 38.5 percent increase in 37 years, or an average annual rate of increase of 0.88 percent.

Westbrook and Patterson investigated the reasons for this increase in urban share by analyzing the data for the period from 1975 through 1985.[10] Their results indicated that the major reasons for this increase are the larger fraction of travel in urban roads and a larger fraction of roads being classified as urban. Population shifts to urban areas and driving shifts within metropolitan areas account for the larger fraction in urban driving which was estimated to be the cause for 58 percent of the increase in urban share. The other 42 percent increase was determined to be the consequence of the reclassification of roads from rural to urban. Any area reclassified by the U.S. Bureau of the Census from rural to urban results in the reclassification of all roads (regardless of the type) as urban.

In this study, it is assumed that population shifts to urban areas will continue and that future land developments will continue forcing the reclassification of rural areas into urban areas. Therefore, it is expected that the increase in urban shares observed in the past will continue through the forecasting period. In particular, it is assumed that this increasing trend will follow a logistic path. This logistic trend implies that the rate of increase from year to year will eventually diminish, granting an eventual leveling off of the increase in urban share.

Figure 2 shows the historical urban share of automobile vmt driving from 1972 through 1990 and a logistic curve fitted to the historical period and extended through the year 2010. The logistic share values are developed based on a logistic functional form originally formulated by Fisher and Pry[11] and defined by:

$$f_t^U = \frac{f_\infty^U}{1 + e^{-(\alpha+\beta t)}} \tag{3}$$

where:

f_t^U is the urban share in year t,

f_∞^U is the urban share asymptotic limit,

α and β are parameters of the logistic curve defined by

$$\alpha = \ln[f_0^U \,/\, (f_\infty^U - f_0^U)], \tag{4}$$

$$\beta = (1/h^U)\ln[(f_\infty^U + f_0^U) \,/\, f_0^U], \tag{5}$$

f_0^U is the base year urban share, and

h^U is the halving factor for the logistic curve. The halving factor is the time required from the base year for the urban share to reach the midpoint between its base year value and its asymptotic limit.

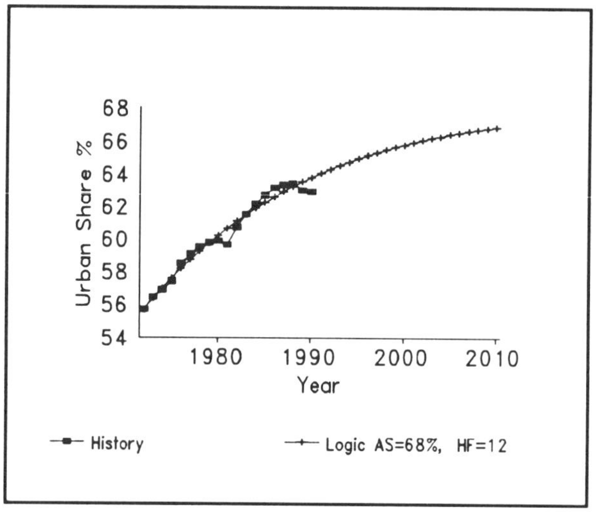

FIGURE 2: AUTO URBAN SHARE VMT DRIVING

The logistic curve in Figure 2 represents the curve that best fits the historical data on urban share for the 1972-1990 period. This curve is generated by assuming two logistic parameters and by selecting a base share year. These two parameters are the asymptotic limit and the halving factor. The asymptotic limit represents an upper limit to the growth of the urban share. The halving factor is a measurement of the time needed for the share to reach this upper limit. The values for both parameters are specific to the best fit curve and they are determined using an iterative approach which minimizes the sum of the squares of the difference between the historical shares and the logistic estimated shares.

The logistic approach generates projections for the urban share of automotive vmt driving of 66.8 percent by the year 2010. Using these projections and assuming that the base composite fuel efficiency gap for automobiles is 15.2 (as estimated by ORNL in 1992) the increase in the gap for the forecasting period can then be estimated.

Table 1 summarizes the impact of the adjusted logistic city share growth on the composite fuel efficiency gap for automobiles. The adjusted logistic city share projection for the year 2010 becomes 51.1 percent. The adjusted logistic forecasts of city share increase are translated into a fuel efficiency gap of 15.90 percent by the year 2010. This represents an increase of about 0.70 percentage points over the base gap of 15.2 percent.

TABLE 1: FUEL EFFICIENCY GAP PROJECTIONS FOR AUTOMOBILES
(ASSUMING A LOGISTIC GROWTH IN CITY SHARE)

	1990	1995	2000	2005	2010
City Share (%)	49.1	50.1	50.5	50.8	51.1
Base Gap	15.20	15.20	15.20	15.20	15.20
Gap Forecast	15.20	15.56	15.73	15.82	15.90
Change	0.00	0.36	0.53	0.62	0.70

Sources: Base Gap from ORNL 1992, Urban Share Forecasts based on Fisher & Pry Logistic Function

Increasing Highway Speeds

The level of speed of a vehicle is one of the relevant factors that affects its fuel efficiency. Specifically, it has been determined that speeds over 45 mph decrease fuel efficiency for most vehicles. Furthermore, EPA estimates that traveling at 65 mph as compared to 55 mph lowers fuel economy over 15 percent.[12] ORNL's 1992 *Transportation Energy Data Book* presents the findings of a fuel economy study performed by the Federal Highway Administration in 1984.[13] This study concluded that, on average, vehicles experience fuel efficiency losses of about 17.8 percent when their speed is increased from 55 mph to 65 mph. This is equivalent to a reduction of 1.78 percent for each mile per hour increase over speed ranging from 55 mph to 65 mph.

Average highway speeds in the United States have shown an increasing trend for several years with few exceptions. Figure 3 presents average highway speeds in mph for the last 45 years. The data in this figure indicate two different increasing trend periods. The first period from 1945 through 1973 corresponds to the largest rate of increase on highway speeds. During these years, highway speed increased at an annual rate of 1.13 percent. In 1973, average highway speed suddenly dropped from about 66 mph to about 55 mph. This sudden drop corresponds to the implementation of the nationwide 55 mph speed limit. After 1974, the increasing trend has continued at a more moderate rate. In the 1974-1990 period the annual rate of speed increase has been 0.15 percent. A closer look at the post-1973 period indicates that through the rest of the 1970s, the average speed remained fairly constant between 55 and 56 mph; and, through the 1980s, the annual rate of increase was 0.34 percent.

The percentage exceeding the speed limit is far from

homogeneous. Significant differences exist across states, highway types, and location for rural or urban areas. For instance, in 1990 the percentage of vehicles exceeding the 55 mph limit in urban interstate highways in New York was 82.5 as compared to 68.2 in California and only 33.7 in South Dakota.[14]

The estimation of the overall impact of speed trends in fuel economy is dependent on the specific data type selected to measure this trend and on the methodology used to forecast this trend. This study uses an aggregated approach based on the nationwide average highway speed for all vehicles and highway types. Average speeds post-1980 are used as the basis to generate forecasts.

As Figure 3 illustrates, average highway speed is influenced by regulatory policies such as the implementation of the nationwide speed limit in 1973-1974. Other factors affecting speed might include safety and environmental regulations, gasoline prices, oil shortages, income fluctuations, etc. Although a methodology to forecast speed trends which includes all relevant factors is desirable, a simplified approach based on historical trends is assumed in this study. The approach implies that the increase in highway speed will continue following a logistic path. This approach assumes that the rate of increase from year to year will eventually diminish granting a leveling off of the increase in highway speeds.

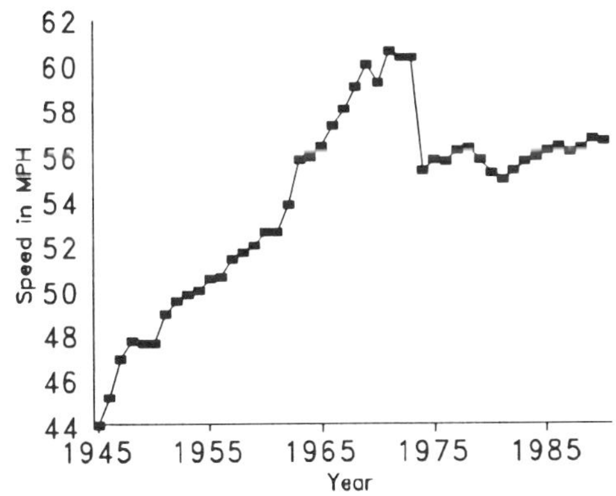

FIGURE 3: AVERAGE HIGHWAY SPEED

Figure 4 shows the historical average highway speeds and a logistic curve fitted to the 1980-1990 period and extended through the year 2010. The logistic speed forecast values are developed based on the logistic functional form originally formulated by Fisher and Pry, described previously, and defined by Equations 3, 4 and 5. This logistic approach generates projections for average highway speeds for all vehicles of 59.22 mph in 2010.

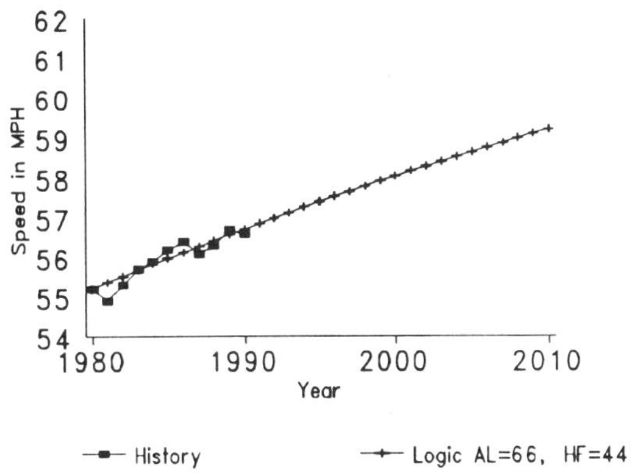

FIGURE 4: AVERAGE HIGHWAY SPEED (HISTORY AND LOGISTIC FORECAST)

The impact of increasing highway speeds on the fuel efficiency gap can then be estimated by using these speed projections. Table 2 lists the fuel efficiency gap projections for automobiles assuming a logistic growth rate of average highway speeds. The increase in the fuel efficiency gap is developed using a base gap of 15.2 percent, and assuming that the fuel efficiency in highway driving is diminished by 1.78 percent for each mile per hour increase in the speed, as determined by FHWA.

Table 2 summarizes the impact of increasing highway speeds on the composite fuel efficiency gap for automobiles using the logistic approach. Based on the forecasts generated for this factor the fuel efficiency gap is expected to grow to 17.02 percent by the year 2010.

TABLE 2: FUEL EFFICIENCY GAP PROJECTIONS FOR AUTOMOBILES
(ASSUMING A LOGISTIC GROWTH OF AVERAGE HIGHWAY SPEED)

	1990	1995	2000	2005	2010
Highway Speed, mph	56.60	57.41	58.06	58.66	59.22
Base Gap	15.20	15.20	15.20	15.20	15.20
Gap Forecast	15.39	15.89	16.31	16.67	17.02
Change	0.19	0.69	1.11	1.47	1.82

Sources: Base Gap from ORNL 1992, Highway Speed Forecasts based on Fisher & Pry Logistic Function.

Increasing Urban Highway Congestion

Congestion is a primary issue of the domestic transportation system. Urban congestion has increased in the last decades in most metropolitan areas as expansion and improvement of the transportation system lagged behind the rapid growth of travel demand.

The Federal Highway Administration (FHWA) classifies the two major causes of urban road congestion as recurring congestion and non-recurring congestion. Recurring congestion is that congestion which is the consequence of inadequate road capacity, reduction of through-put lanes, narrowing of lane widths, physical barriers, inadequate traffic light synchronization, and other similar causes. FHWA estimates that recurrent congestion accounts for 40 percent of all urban road congestion. Non-recurring congestion is that congestion resulting from disabled vehicles and accidents. FHWA estimates that disablement account for 55 percent of overall urban congestion, with the remaining 5 percent due to accidents.

One of the most important road types within urban areas in which congestion takes place is urban freeways. In 1990, 32 percent of the total vehicle miles of travel in urban areas corresponded to freeways, while freeways comprised only 5.7 percent of the urban roadway mileage.[14] The increase in urban congestion can be further analyzed by considering the increase in urban vmt as compared to the increase in urban lane miles. Data corresponding to the period 1975-1987 indicate that urban vmt demand growth rate is over 4 times the rate of new urban lane capacity growth. This corresponds to an increase in the average urban through-put (urban vmt per mile) of 38.9 percent.

Differing methodologies have been developed recently to measure the extent and duration of freeway congestion in urban areas.[15 16] Hanks and Lomax of the Texas Transportation Institute (TTI) have developed congestion indices for 39 urban areas.

Few attempts to forecast urban congestion and its effect on fuel economy are available.[17] Lindley's projections of congestion statistics for the year 2005 take into account factors including time delays, wasted fuel, and user cost.

The projections generated in this study utilize the wasted fuel values developed by Lindley as the basis to measure the impact of urban congestion on the fuel efficiency gap. This study further assumes that the amount of wasted fuel due to congestion will increase to the year 2010 following a logistic trend.

Figure 5 is a plot of the logistic projections of wasted fuel for the period 1984 through 2010. The projections are for both automobiles and light duty trucks. The logistic values are developed based on the logistic functional form formulated by Fisher and Pry as described previously. The wasted fuel values developed by Lindley for 1984, 1987, and 2005 are also presented in Figure 5. The impact of congestion on the fuel efficiency gap is then estimated based on the projections of wasted fuel and logistic projections of urban vmt driving. This approach indicates that the amount of wasted fuel for both automobiles and light duty trucks will be about 13.7 billion gallons by the year 2010. The projections of wasted fuel for just automobiles is then estimated at around 9.2 billion gallons assuming that the light duty truck vmt driving share will increase to 33 percent by the year 2010.

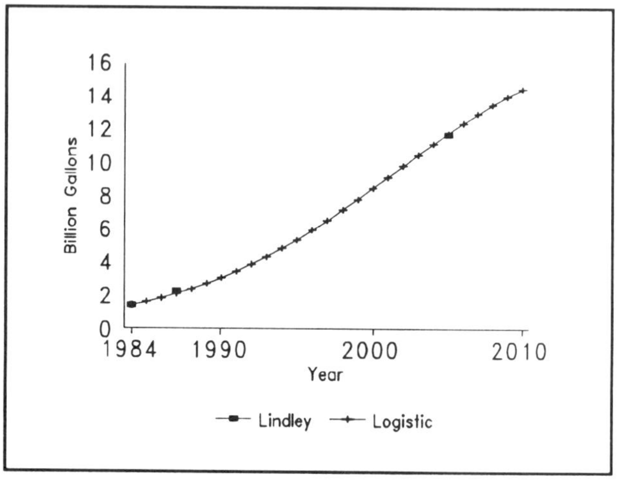

FIGURE 5: WASTED FUEL PROJECTIONS (DUE TO URBAN FREEWAY CONGESTION)

Table 3 presents the projections of wasted fuel for automobiles due to traffic delays and the increase in the fuel efficiency gap due to this factor. Based on this approach the fuel efficiency gap will be 18.66 percent in the year 2010.

TABLE 3: FUEL EFFICIENCY GAP PROJECTIONS FOR AUTOMOBILES

(BASED ON A LOGISTIC INCREASING CONGESTION TREND)

	1990	1995	2000	2005	2010
Wasted Fuel (mil. gal.)	2252	3865	5788	7764	9164
Base Gap	15.20	15.20	15.20	15.20	15.20
Gap Forecast	15.69	16.37	17.34	18.20	18.66
Change	0.49	1.17	2.14	3.00	3.46

Overall Degradation Factor Forecast
Figure 6 summarizes the projections of the fuel efficiency gap using assumptions of logistic growth and adjusted city/highway shares for automobiles. The overall results are listed in Table 4.

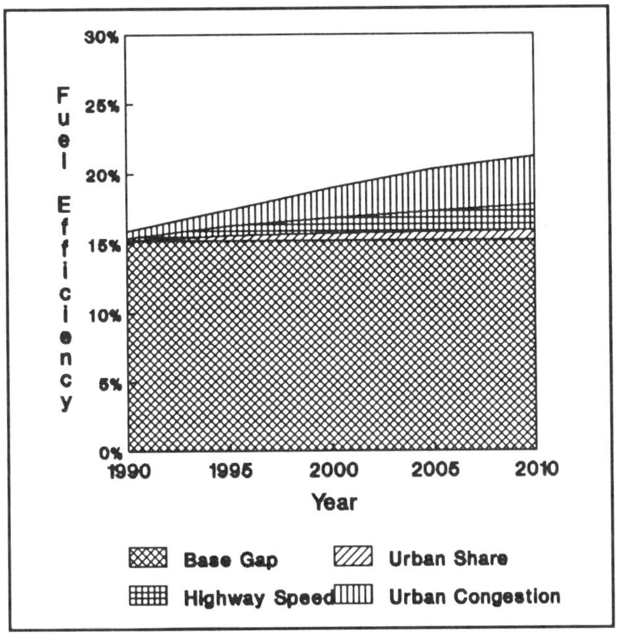

FIGURE 6: OVERALL FUEL EFFICIENCY GAP PROJECTIONS

As illustrated in Table 4, the overall fuel efficiency gap for automobiles is expected to increase from a base of 15.2 to 21.18 by the year 2010 assuming a logistic trend in the three driving trends considered.

TABLE 4: TOTAL FUEL EFFICIENCY GAP PROJECTIONS FOR AUTOMOBILES

	1990	1995	2000	2005	2010
Base Gap	15.20	15.20	15.20	15.20	15.20
Gap Forecast	15.87	17.42	18.98	20.29	21.18
Change	0.67	2.22	3.78	5.09	5.98

CONCLUSIONS
The gap between "on-road" and "tested" fuel efficiency have been determined for automobiles for the period from 1990 through 2010. The projections indicate that the gap for automobiles will increase from a base gap of 15.2 to 21.2 by 2010. The projections were generated assuming logistic growth trends. It is important to realize that if the fuel efficiency testing procedures as defined by EPA were to be modified the efficiency gap projections would vary. Moreover the gap would be affected if the EPA composite fuel efficiency formulation were updated to better represent on-road fuel efficiency estimates. Any such changes would affect the fuel efficiency gap estimates provided in this study. All fuel efficiency gap projections presented in this study are clearly dependent upon specific assumptions regarding relevant factors. Changing assumptions or introducing additional factors will result in different values for the fuel efficiency gap.

ACKNOWLEDGEMENTS
This study was performed with support from the Energy Demand Analysis Branch of the Energy Information Administration (EIA), Contract No. DE-AC01-92EI21946.

The author acknowledges the contribution of John Maples from the Transportation Center of the University of Tennessee in this research. In particular, John Maples provided valuable information and input data necessary for the understanding of the base fuel efficiency gap determination. The concerns about the specific factors affecting the gap and analyzed in this report (i.e. urban share, highway speed and urban highway congestion) were originally identified by Westbrook and Patterson in their 1989 SAE paper entitled "Changing Driving Patterns and Their Effect on Fuel Economy " and presented in the 1989 SAE Government Industry meeting in Washington, D.C. These factors are presently being analyzed in an ongoing effort by the Office of Transportation Technologies of the Department of Energy in conjunction with the University of Tennessee. This ongoing effort allowed further understanding of the fuel efficiency gap.

Finally, the author acknowledges the support from Decision Analysis Corporation of Virginia (DAC) and the contributions from Bill Townsend and Liisa Padgett.

REFERENCES
[1]Davis, S. and Morris, M., Oak Ridge National Laboratory, *Transportation Energy Data Book: Edition 12*, ORNL-6710, (Edition 12 of ORNL-5198), p. 3-9, March 1992.

[2]Westbrook, F. and Patterson, P., "Changing Driving Patterns and Their Effect on Fuel Economy," presented May 2, 1989 at the 1989 SAE Government/Industry Meeting, Washington, D.C.

[3]DOE/EPA, *Gas Mileage Guide: EPA Fuel Economy Estimates*, DOE/CE-0019/10.

[4]SAE 780037

[5]SAE 840496

[6]Davis, S. and Morris, M., Oak Ridge National Laboratory, *Transportation Energy Data Book: Edition 12*, ORNL-6710, (Edition 12 of ORNL-5198), p. 3-9, March 1992; Maples, John D., and Philip D. Patterson, "The Fuel Economy Gap for All Automobiles and Light Trucks in Operation," Draft, Washington, D.C., 1991.

[7]Westbrook, F. and Patterson, P., "Changing Driving Patterns and Their Effect on Fuel Economy," presented May 2, 1989 at the 1989 SAE Government/Industry Meeting, Washington, D.C.

[8]EIA, *Energy Consumption and Conservation Potential: Supporting Analysis for the National Energy Strategy*, SR/NES/90-02, Service Report, p. 89, Washington, D.C., December 1990.

[9]Data on vmt is published annually by the U.S. Department of Transportation, Federal Highway Administration, in *Highway Statistics*.

[10]Westbrook, F. and Patterson, P., "Changing Driving Patterns and Their Effect on Fuel Economy," presented May 2, 1989 at the 1989 SAE Government/Industry Meeting, Washington, D.C.

[11]Fisher, J.G. and Pry, R.M., "A Simple Substitution Model of Technology Change." *Technological Forecasting and Social Change*, Vol. 3, pp. 75-88, 1971.

[12]DOE/EPA, *1992 Gas Milage: EPA Fuel Economy Estimates*, DOE/CE-019/10, October 1991.

[13]Davis, S. and Morris, M., Oak Ridge National Laboratory, *Transportation Energy Data Book: Edition 12*, ORNL-6710, (Edition 12 of ORNL--5198), Table 3.42, p. 3-66, March 1992. 1984 data from U.S. Department of Transportation, Federal Highway Administration, *Fuel Consumption and Emission Values for Traffic Models*, Washington, D.C., May 1985.

[14]U.S. DOT, FHA, *Highway Statistics 1990.*

[15]ibid.

[16]Cottrell, P., "Measurement of the Extent and Duration of Freeway Congestion in Urbanized Areas," ITE 61st Annual Meeting, Milwaukee, Wisconsin, September 1991.

[17]Hanks, J. and Lomax, T., *Roadway Congestion in Major Urban Areas: 1982 to 1987*, Texas Transportation Institute, Research Report 1131-2, College Station, Texas, October 1989.

[18]Lindley, J., "Urban Freeway Congestion Problems and Solutions: An Update," ITE Journal, December 1989, pp. 21-23; Feng, An, "Automobile Fuel Economy and Traffic Congestion," Dissertation for PhD in Applied Physics, University of Michigan, 1992; Westbrook, F. and Patterson, P., "Changing Driving Patterns and Their Effect on Fuel Economy," presented May 2, 1989 at the 1989 SAE Government/Industry Meeting, Washington, D.C.

Chapter 22

Resource Conservation and Recovery—A Worldwide Call to Action

J.P. Nicholson

Abstract: The World Health Organization reports that over 14,000 children die daily, mostly from malnutrition. Concurrently, the United States of America wastes over 200,000,000 tons annually of bio-organic materials, such as domestic wastewater sludge, food processing wastes, yard wastes, animal manures, etc. Moreover, an additional 50,000,000 tons of alkaline by-products are wasted annually. Our worldwide challenge must be to convert natural waste resources into safe, valuable, and environmentally friendly products, to be used as components in sustainable food production.

This spring, I read the book The Crisis Years, 1960-1963, by Michael Beschloss. This excellent historical work reveals the world's terrible proximity to thermonuclear war during the building of the Berlin Wall, and during the Cuban Missile Crisis. Two men of peace, John Kennedy and Nikita Khrushchev, were hard pressed to keep their respective militant leaders from actions that easily could have triggered a worldwide nuclear calamity. How can man be so ignorant?

Unfortunately and tragically, it is not hard to understand. Mankind can be terribly ignorant. How else can one explain today's world leadership's abject failure to recognize the perilous dangers inherent with the wanton destruction of the ozone layer, with the oncoming global warming crisis, with the destruction of the rain forests, with the terrible pollution of our air by acid rain and unnecessary combustion processes such as incinerators; and pollution of our lakes and streams by uncontrolled point and non-point source discharge, the pollution of our ground waters by irresponsible land use disposal programs, and foolish use of chemical fertilizers and pesticides. Mankind must recognize the immensity of the problem now, and must act now to prevent the destruction of civilization!

At the 1991 Searching for Success Conference, sponsored by the National Environmental Awards Council, an organization of respected environmentally active public interest groups such as the Sierra Club, Natural Resource Defense Council, National Wildlife Federation, National Audubon Society, Institute of Self-Reliance, and many others, under the leadership of Renew America, the film The Blue Planet, prepared by the Smithsonian Institute, and the National Air and Space Administration (NASA), was a conference highlight. I cannot recommend any film more highly! The film demonstrates very effectively the world crisis caused by our failure to recognize the critical damage being done by mankind's abuse of the ecology and environment of this planet. The film is shown daily at the National Air and Space Museum in Washington, D.C.

We, in the U.S.A., are the leaders in this terrible environmental, ecological, and

[1] J. Patrick Nicholson, Chief Executive, N-Viro Energy Systems Ltd. Toledo, OH 43606

sociological abuse. A great American President said, "The buck stops here". Well, the buck must stop here and now. We, this generation of leaders, have the watch! It is our responsibility to meet this challenge now - not pass the crisis on to our children, as we have so tragically done in the U.S.A., with our national debt. The U.S.A.. is the biggest worldwide user of resources, and as such, we as a government, first must meet our own responsibilities before we seek to advise others. However, as individuals, we must speak out today so that all nations may see this critical problem and immense opportunity.

Yes, you heard me right. This ecological and environmental problem can, indeed, be converted to a magnificent opportunity. Science and technology exist today to convert and to recycle waste resources into critically necessary products, that in and of themselves, can greatly benefit the world. We can make use of wastes in economical and safe scientific ways to reduce chemical dependency, to improve our worldwide ecology and environment, and to help feed the world.

Mankind does not need to look to the future to see the folly of its actions, or more precisely, its inactions. The World Health Organization, and other respected public health institutions, are dedicated to forcing so-called intelligent industrial nations to recognize the terrible third world devastation caused primarily by food shortages. Is society responding to this terrible human tragedy? Just barely! Malnutrition is the major contributing cause in the deaths of over 14,000 children per day.

Organic by-product materials, such as domestic wastewater residuals, agricultural by-products, food processing by-products, and alkaline by-product wastes, such as fluidized bed residue sulphur scrubbing residue, cement and lime kiln dusts, and certain fly ashes, have tremendous value as organic and aglime products in third world countries that desperately need these materials to cost effectively utilize either chemical or alternative farming technologies. Does it make any sense for New York City or other large municipalities to ask industry to spend hundreds of millions of dollars for pretreatment programs, and then spend as much as $1,000.00 per dry ton to destroy these resources. Until third world countries can treat and develop their own organic residuals resources, the by-product resources of industrialized nations are desperately needed in many parts of the world. The U.S. government should strongly support actions that safely and economically can aid in generating great increases in food supplies throughout the world.

Agricultural scientists are unanimous in their concern for the problems of soil erosion, chemical dependency, and organic, mineral, and microbial deficiencies in soils throughout the world. Many, many of the organic wastes that we now relegate to the oceans, to incineration, to landfills, or otherwise destroy are desperately needed in our worldwide effort to economically provide food for civilization without destroying the soil, the surface waters, and the ground waters near agricultural land.

Sludges are extremely beneficial, however, toxic metals and organics are a concern. Pathogens are a concern. Today, technology exists, either in pre-treatment or in fixation, to reduce toxic concerns to scientifically acceptable levels, and to pasteurize these wastes to eliminate pathogens, while concurrently providing for the survival of microflora, i.e. indigenous soil micro-organisms so necessary to plant growth. The importance of organics, minerals, and microbes to all soils, whether chemically or organically fertilized, is well known.

These technologies make possible the safe use of many organic wastes so desperately important to both industrial and developing nations. These technologies exist today! Why is society so slow in providing the necessary technology transfer? American industry spends hundreds of millions of dollars annually to pretreat sludges to reduce toxicity concerns. Why is our society so slow in taking advantage of this effort through worldwide beneficial utilization? We must bridge the gap between the over supply of organic by-products and the critical demand for low cost worldwide food production.

In a recent speech to the Institution of Water and Environmental Management in England, the Prince of Wales stated, "(Incineration) is seems to me, violates the most profound ecological principle of all, which is to 'close the loop', minimizing resource use and energy wastage at every stage of every economic process. There is no reason at all, surely, why the various beneficial uses: agricultural, forestry, land reclamation, horticulture, amenity land, should not absorb most of U.K. sludge production by the end of the decade. It will only require a leap of the imagination on the part of producers and users, and a guarantee of microbiological and toxicological safety, for an embarrassing sludge to be turned into an embarrassment of riches. Once again, new technology has a part to play. And there are already examples such as the Simon-N-Viro process to show what is actually possible."

If mankind uses its organic wastes, and if that use safely provides critically needed nutrients, organics, minerals, and microbes to our agricultural soils while concurrently reducing our chemical dependency, are we not achieving two critical goals with a single effort? Two for the price of one! Can anything make more sense?

Alternative agriculture is a practice that is being recommended by the Rodale Institute, and other concerned agricultural citizens in the U.S.A. Alternative agriculture is a back-to-the-future approach, in which low input sustainable agriculture (LISA) and/or organic farming methods are used. These efforts to encourage economic and environmental agricultural practices have been constrained by the tremendous political clout of the chemical industry.

In 1989, the Board of Agriculture of the National Research Council published their thorough study of Alternative Agriculture. The following quotes are from that important report.

"Well-managed alternative farming systems nearly always use less synthetic chemical pesticides, fertilizers, and antibiotics per unit o production than comparable conventional farms. Reduced use of these inputs lowers production costs and lessens agriculture's potential for adverse environmental and health effects without necessarily decreasing - and in some cases increasing - per acre crop yields and the productivity of livestock management systems."

"Many federal policies discourage adoption of alternative practices and systems by economically penalizing those who adopt rotations, apply certain soil conservation systems, or attempt to reduce pesticide applications. Federal programs often tolerate and sometimes encourage unrealistically high yield goals, inefficient fertilizer and pesticide use, and unsustainable use of land and water. Many farmers in these programs manage their farms to maximize present and future program benefits, sometimes at the expense of environmental quality."

"Fertilizers and pesticides are often applied at rates that cannot be justified economically without consideration of present or future farm program payments."

"Government policies that discourage the adoption of alternative practices must be reformed."

"Ultimately, farmers will be the ones to decide. However, significant adoption of alternative practices will not occur until economic incentives change. This change will require fundamental reforms in agricultural programs and policies. If these conditions are met, today's alternative farming practices could become tomorrow's conventional practices, with significant benefits for farmers, the economy, and the environment."

One of the foremost proponents of alternative agriculture in the U.S.A. is Richard Thompson, of Boone, Iowa. His farm is highlighted in the Alternative Agriculture report. For many years, Dick Thompson's primary fertilizer has been a blend of Boone, Iowa, wastewater sludge and cement kiln dust, our N-Viro Soil technology.

We believe in a reasonable and scientific blending of traditional and alternative agricultural practices.

Agricultural science, not government subsidies, must dictate safe and economical agricultural practices, particularly in third world countries where subsidies are not available, where by-product organic materials are readily available, and where the high cost of chemical fertilizers and pesticides creates a huge economic burdens.

We fully recognize and respect the important scientific and technological contributions of the chemical industry to food and fiber production in the world. We also understand the immediate environmental problem and future soil problems that are the direct results of over dependence on chemical fertilizers and pesticides. Sensible alternative agriculture practices can reduce chemical dependency by increasing the effectiveness of chemicals. World food and fiber production policies must be driven by good science, sound economics, and reasonable respect for the environment, not by political power!

DOES IT MAKE ANY SENSE TO DESTROY ORGANIC WASTES THROUGH OCEAN DUMPING AND LANDFILLS, CONCURRENTLY CREATING SERIOUS AND UNNECESSARY ECOLOGICAL AND ENVIRONMENTAL PROBLEMS, WHEN TECHNOLOGY EXISTS TO USE THESE RESOURCES TO REPLACE OR REDUCE THE NEED FOR COSTLY AND DANGEROUS CHEMICALS, ESPECIALLY WHEN THESE ORGANIC PRODUCTS AR SO DESPERATELY NEEDED IN THIRD WORLD COUNTRIES?

DOES IT MAKE ANY SENSE TO INCINERATE ORGANIC WASTES, UNNECESSARILY USING ENERGY, UNNECESSARILY ADDING CO^2 TO THE ATMOSPHERE, UNNECESSARILY POLLUTING THE AIR, AND UNNECESSARILY CREATING REAL PUBLIC CONCERNS IN THE LOCAL COMMUNITY, WHEN TECHNOLOGY EXISTS THAT CAN SAFELY CONVERT THOSE ORGANIC WASTES TO CRITICALLY IMPORTANT ORGANIC FERTILIZER PRODUCTS?

We salute the environmental and public interest groups that are providing the worldwide leadership to make civilization realize that waste management is best accomplished through a scientific combination of waste reduction, waste recycling, and waste reuse.

It is right to reuse! Seventy-five years ago, oil refineries had a major disposal problem. That waste was then called "refinery bottoms" or "sludge". Today, that waste is known as asphalt, and its importance in road and roofing construction is well documented. Hundreds of additional products, such as plastics, resins, polymers, etc., have been derived from refinery waste by the oil industry. These development required great efforts, research, risk, investment, and commitment, all important aspects of the free enterprise system. It is amazing what business can accomplish without government interference! Japan has clearly demonstrated the immense value of government and business pulling together.

Unfortunately, a great many of our most valuable organic wastes are generated by government agencies. We know the constraints and limitations of public agencies. It takes real leadership to look beyond today. U.S. municipal wastewater leaders are beginning to recognize the importance of recycling and real beneficial reuse.

They are making large public investments in converting organic wastewater residuals into products with real value, such as Milorganite (Milwaukee's pelletization product), composts, and N-Viro Soil (our own organic aglime product). We particularly wish to salute the Middlesex, New Jersey, County

Utility Authority, and its Executive Director, Fred Kurtz, who resisted the strong incinerator lobby, and chose to replace ocean dumping with the conversion of their 200,000 annual tons of sludge to an organic aglime product using by-product alkaline materials. Here again, two problems (i.e. the management of organic wastes, and the management of alkaline by-products such as kiln dusts or sulphur scrubbing residues), were converted into a major opportunity - the production of an organic aglime agricultural and topsoil product (N-Viro Soil).

Dr. Anson Bertrand, Director of Science and Education, U.S. Department of Agriculture, told a 1980 conference that the annual need for aglime, in the U.S.A. alone, exceeded 90,000,000 tons. By combining organic waste with alkaline by-products, an organic aglime is being produced with amazing soil and fertilizer value. The potential worldwide use of such a product, particularly on acid soils caused by acid rain, chemical fertilizers, or natural causes, is unlimited. The combination of microbes, nutrients, and organics contained in wastewater sludge with the calcium carbonate and minerals of alkaline by-products creates a low cost product with high performance ag and soil value.

The conversion of waste material to valuable products makes sense. In 1976, the Congress of the U.S.A. passed the Resource Conservation and Recovery Act, which strongly supported recycling and beneficial utilization policies. In 1977, Congress passed the Clean Water Act, which instituted America's pre-treatment program, a program implemented to ensure the safe utilization of wastewater sludges.

Today, fifteen years later, the government of the U.S.A. is just beginning to implement programs and regulations that support and encourage recycling and beneficial utilization! Why the delay?

Let's face the truth! Resource conservation and recovery, recycling and beneficial utilization, make sense to everyone except the industries who do not benefit from such concepts. The disposal industry, the incinerator industry, the chemical fertilizer industry, the paper industry, and many other entrenched private interests, have done and will continue to do everything possible to discredit and stop reduction, recycling, and reuse policies and programs in the U.S.A. and throughout the world. Many of these industries do not worry about the world tomorrow. They worry about their profitability today! Our challenge is to find a safer and better way to meet our responsibilities, now and in the future!

The implementation of resource conservation and recovery policies and practices requires real courage and commitment. Such commitment is not unusual for you as leaders of the wastewater industry. We know from personal experience the value of your efforts. We live on Lake Erie. Twenty years ago Lake Erie was called a dead lake. Today, Lake Erie is a swimmer's and fisherman's paradise, thanks to dedicated wastewater point source discharge clean-up efforts by our wastewater leaders and by our government.

During the past two years, the Bush administration, particularly the EPA, under Administrator Reilly, have begun serious efforts to encourage meaningful resource conservation and recovery efforts. We strongly applaud the leadership by U.S. EPA to develop a "clean sludge protocol", whereby sludges or sludge products with very low toxic levels, that are pasteurized or sterilized, will be classified as products with no further regulatory requirements. This protocol, which will dramatically increase community perception and awareness of reuse opportunities, is long over due! The international ramifications are obvious!

On October 31, 1992, our firm, N-Viro Energy Systems Ltd., was on of a small number of Americans honored by the President of the United States in the Rose Garden of the White House. We received one of the first

"President's Environment and Conservation Challenge Award Citations for Excellence in Innovative Technology". In 1990, U.S. EPA recognized our organic aglime process and product as the "#1 Sludge Use Technology in the U.S.A." Similar recognition was given to N-Viro Soil by the National Environmental Awards Council, and by the Friends of the United Nations Environment Programme. Winning awards does not make our product perfect, but it does suggest significant demonstrated credibility. The selection committees for the above awards include such groups as the Sierra Club, World Wildlife Fund, Natural Resources Defense Council, the Soil and Water Conservation Society, Renew America, and other dedicated and respected environmental and conservation organizations. Soil and agricultural professionals in New England and in Northwestern Ohio have developed documented evidence that use of N-Viro Soil has positively reduced farm input costs, and increased farm yield at a value exceeding $40.00 per acre. The importance of such a material, particularly for third world countries, cannot be overstated!

In his 1968 tragic pursuit of the Presidency of the United States, Senator Robert Kennedy concluded his speeches by quoting the poet George Bernard Shaw, "Some people see things as they are and ask 'Why?'" I dream of things that have never been and ask, 'Why not!'"

I dream of a bridge between waste generators of this world and the food and fiber producers of this world. I dream that over that bridge will pass science, technology, understanding, communication and cooperation. I dream that because of that bridge we will greatly reduce starvation and the dangers that face the world now, and that will face our children and our children's children in the future. Moreover, because of that bridge, each of us will enjoy the ultimate satisfaction of knowing that we did our best!

The buck stops here!

Chapter 23

Georgia Tech Experience and Plan in Environmental Science, Technology and Policy

C.S. Kiang

I. Introduction

"Since 1900, the number of people inhabiting the earth has multiplied more than three times. The world economy has expanded 20 times. The consumption of fossil fuels has grown by a factor of 30, and industrial production has increased by a factor of 50; four fifths of that increase has occurred since 1950. This scale of development has produced a world with new realities, realities that have not yet been reflected in human behavior, economics, politics or institutions." (MacNeill, 1989)

Mankind is developing technologies and economies that are changing the physical and biological environment on a regional and global scale more rapidly than ever before. The challenge mankind faces is to develop the capacity to manage these environmental changes effectively, and in a manner that facilitates opportunities for economic development. The world market for environmental technologies is enormous and growing. For example, the Fact Sheet recently published by the Clearinghouse on State International Policies (January, 1992) has shown that the world market for pollution control products is estimated at 228 billion U.S. dollars for 1990 and 370 billion for 2,000.

In the U.S. market, the consulting firm of Alex Brown & Sons, for example, estimated the environmental industry to be in excess of $100 billion in 1989 and growing 25 to 30 percent annually. The same year, The San Diego market research firm EnvironQuest estimated revenues for environmental service firms alone at $120 billion, and growing at the same rate. Naturally, the expenditures and jobs created by those purchasing these services and hardware are much higher. The U.S. EPA estimates that the new Clean Air Act of 1990 will account for a 20 percent added industry growth. The reauthorization in 1991 of the Clean Water Act and the Resource Conservation and Recovery Act will likely have similar accelerating impacts on the industry's growth. Other market research reports project, through the year 2000, a 15 percent and 20 percent annual growth in the market for environmental instrumentation and analytical services, respectively.

Most of the markets noted are attributable to environmental services, pollution control, and facility construction. However, the development and construction of INNOVATIVE TECHNOLOGY, funded both by the public and private sectors, will help drive market competitiveness. For example, the Department of Energy is expected to spend $150 billion on environmental restoration over the next two decades. Of that amount, 10 to 15 percent will be spent on research and development. Similarly, the Department of Defense will be spending about $2 billion annually for remediation and research and development activities.

It has been estimated that perhaps one million of the more than 100 million workers in U.S. today are engaged in occupations related to environment protection. Of these one million jobs, only 2 percent, or 20,000 jobs are in the state of Georgia. Since Georgia contains more than 2% of the U.S. population, environmental employment as a portion of total employment, is lower in Georgia than in the U.S. as a whole. Either Georgia consumes proportionally fewer environmental products and services or the state is a net importer of environmental products and service. Either way, if the environmental market is projected to grow at a faster rate than the economy as a whole, this projection certainly presents a challenge for Georgia's economic competitive advantage regionally and nationally, and perhaps globally, in future years.

The Georgia Institute of Technology is committed to meet these global and local challenges. To assure the industrial growth with appropriate consideration given to the environment and limited natural resources, the mission statement of the Office of Environmental Science, Technology and Policy of Georgia Tech is to establish the Georgia Institute of Technology as a premier technology university via the development of a solid and far-reaching education, research and service program for the interface of the future environment and technology. A overview of the Environmental Science, Technology and Policy program will be presented in the section II.

The Southern Oxidants Study will be presented in Section III as a case study of a successful strategic alliance of universities, state and federal agencies and industries to study the mechanisms involved in the formation and accumulation of ground level ozone air pollution and to understand the relationships among the air quality, transportation and energy use in the Southern United States.

The 1996 Olympic games and the Georgia Tech campus (the site of 1996 Olympic Village) provide a unique opportunity for the development of a model University Supercity to demonstrate integration of innovative technologies, effective operations, and human activities in the infrastructure, buildings and transportation adopting the principles of optimization of energy use and environmental quality.

The purpose of this paper is to introduce the Georgia Tech experience and plan in environmental science, technology and policy and to provide a building ground for the future partners in collaborative education, research and economic development environmental programs.

II. Environmental Science Technology and Policy Program at Georgia Tech

Background

Over the years, individuals from various academic and research units of Georgia Tech have jointly and separately pursued many research initiatives and educational programs broadly defined as environmental science, engineering and policy. Recently, these interactions on projects, curricula, proposals, and state and regional public service activities have highlighted the benefits and need to more formally organize a combined effort in the area of environmental management of science and technology - the interface of environment and technology.

The instances of mutual interaction among the faculty and students have identified a capability that is perhaps unique and potentially unequalled in the U.S. Pooling of resources in basic understanding through science and the ability to provide engineering solutions, offer a necessary capability to address environmental issues. On January 27, 1989, an Environmental Science and Technology Advisory Council was established to develop an inventory and to evaluate active and planned programs, to include all aspects of environmental science and engineering in education, research, and service at Georgia Tech, and to provide recommendations for coordination of planning and execution of environmental education, research and service.

The first task of the advisory council was to develop an inventory of the research capabilities and education programs at Tech. The results of that activities have shown an outstanding and diverse environmental community consisting of nearly every school within the five colleges of Georgia Tech, more than half the Georgia Tech Research Institute Laboratories and several Centers. Over two hundred faculty indicated their involvement in various environmental areas (see Table 1). This capability has been summarized in an extensive inventory depicting the faculty, educational programs, research project experience, facilities and equipment. The inventory indicates a very high level of excellence and significant potential for Georgia Tech to become a major, national and international focal point for environmental science, technology and policy.

Georgia Tech's increased involvement in the arena of environmental study is evidenced by the growth of degree programs in the School of Earth and Atmospheric Sciences, Civil Engineering, and Public Policy, as well as increasing research, technical assistance, and training opportunities across various academic and research units on campus. An Office of Environmental Science, Technology and Policy was established on January 2, 1991 to maximize the coordination of environmental activities at Georgia Tech, and to develop an international leadership position through holistic application of technology based upon sound science, engineering and policy.

Strategic Plan

On February 3, 1992, a strategic plan for environmental science, technology and policy at Georgia Tech was developed.

The relationship between the environmental science, technology and policy program with the other educational and research components at Georgia Tech can be illustrated in Figure 1. Detailed analyses on the external environment of opportunities and challenges, and internal environmental strengths and weaknesses were performed. Some of the general objectives and tasks are summarized below.

General Objectives

1. To develop a basic infrastructure for the linkage of scientific discovery, technology innovation and economic development in environmental science, technology, policy and management.
2. To develop a basic infrastructure for technology transfer and knowledge transfer in environmental

science, technology, policy and management.
3. To develop mechanisms for continuous assessment and dynamic strategic planning in environmental science, technology, policy and management.
4. To develop a strong strategic alliance with state, industry, and universities in environmental science, technology, policy and management.
5. To develop regional, national and global networks in environmental science, technology, policy and management.
6. To develop human resources for the 21st century in environmental science, technology, policy and management.

Tasks

1. To develop mechanisms for assessment and a strategic plan in technology and environment interface.
2. To develop In-Reach programs
   A. To coordinate and develop a strong and integrated research capability in bio-geo-chemical cycles.
   B. To coordinate and develop a comprehensive advanced sensors center.
   C. To coordinate and develop an environmental simulation center.
   D. To coordinate and develop a multi-media information management center.
   E. To coordinate and develop a center of innovative environmental technologies.
   F. To develop educational programs in environmental science, technology and policy.
3. To develop capabilities in social and economic assessment
   A. To develop a critical mass for an environmental policy program.
   B. To develop an environmental economic analysis capability for the management of technology.
4. To develop Out-Reach programs
   A. To develop a coordinated office of technology transfer in environment.
   B. To develop global environmental technology parks.
5. To demonstrate integration of innovative technologies, effective operations, and human activities in the infrastructure, buildings and transportation of a major university city and optimization of energy use and environmental quality. (1996 Olympic Village)

Current Research Programs

A brief summary of selected on-going research and out-reach programs at Georgia Tech is listed as follow:

1. Georgia Environmental Technology Consortium

An umbrella organization of the member universities of the Georgia Research Alliance (Clark-Atlanta University, Emory University, Georgia Institute of Technology, Georgia State University, Medical College of Georgia and University of Georgia), the Georgia Environmental Technology Consortium (GETC), was established at July 1, 1991. GETC will promote cross-disciplinary environmental research and facilitate environmental technology transfer. New telecommunication, environmental technology and genetic technology are three new initiatives proposed by the Governor of Georgia to link the state's economic development and research.

2. Southern Oxidants Studies

Will be described in the following section.

3. Center for Environmental Science and Technology in Pulp and Paper (CETPPI)

The U.S. Pulp and Paper Industry is a multi-billion dollar component of the U.S. manufacturing infrastructure that is especially important to Georgia and other Southeastern states. The focus of CETPPI, collaboration with the Institute of Paper Science and Technology, is to work with Industry to develop and transfer the technologies necessary to prevent environmental problems while, at the same time, helping to keep the U.S. Pulp and Paper industry economically competitive in the global market.

4. Hazardous Substance Research Center/South and Southwest

Hazardous Substance Research Center/South and Southwest is a US EPA Exploratory Research Center with Georgia Tech, Rice University and Louisiana State University as collaborative member institutions. The research programs to be conducted at Georgia Tech with the Center are to be focused on hazardous substances; transport, physiochemical and biochemical processes associated with contaminated sediments; use of sensors in detecting hazardous substances; and training/technology transfer activities associated with hazardous substances.

5. Technical Assistance, Technical Training and Technology Transfer Programs

The Environmental Science and Technology Laboratory of the Georgia Tech Research Institute is working closely with the small and medium industries to reduce risk to the environment, workers and the community by technical assistance to achieve voluntary regulatory compliance. More than 120 training courses are provided to the representatives from industry, government, education institution and consultant firms. Technology transfer occurs through newsletter and engineering extension stations.

6. Global Environmental Technology Parks

Georgia Tech is planning to establish a global network of environmental technology parks to market the innovative technologies for economic development and environmental improvement. The Electricite de France in Paris, France, which is establishing an environmental technology park at 30 miles west of the capital city has contracted with Georgia Tech to be the exclusive U.S. university core-partner. Negotiation with other countries is underway. Plasma Torch, Pulse Combustion, Advanced Analytical Instrumentation and Super-Critical Fluid Separation are some examples of the innovative technologies to be developed at Georgia Tech.

III. A Case Study: The Southern Oxidants Study (SOS)

The southern region of the U.S. has persistent ground level ozone pollution problems in the summer months. Since 1980 ozone levels have not diminished, despite the best efforts of local, state, and federal agencies, coupled with the expenditure of billions of dollars in ozone reduction measures (Chameides,et.al., 1988). However, economic and population growth have been significant in this same time period.

In June, 1988 a multidisciplinary workshop entitled "Atmospheric Photochemical Oxidants: A Southern Perspective" was held on the campus of Georgia Institute of Technology. The 60 Workshop attendees included national experts on all aspects of the photochemical oxidants problem and representatives from all sectors of the research community (e.g., universities, government laboratories, state-regulatory agencies, and key industrial organizations). Recognizing the severity and unique characteristics of the urban and rural ozone problem in the south, the Workshop attendees recommended that a five-year multi-institutional research and monitoring program be undertaken to elucidate the formation, fate, and effects of photochemical oxidants in the South and develop strategies for their management. This research has been named the Southern Oxidants Study (SOS).

General Goals and Scientific Objectives of SOS

SOS has two general goals:

1. Gather the data and refine the numerical models needed to understand the formation of ozone and other photochemical oxidants and their human health, forestry, and agricultural effects in the southern United States; and
2. On the basis of understanding developed from research in Goal 1, evaluate alternative strategies for the management of oxidants and their effect on public health and vegetation effects in the southern states.

Based on these two goals, five specific research and monitoring objectives for SOS have been identified. These are:

1. To establish a network of monitoring sites in various areas of the southern United States and thereby determine the spatial and temporal trends in the chemical meteorology and chemical climatology of ozone and anthropogenic and natural ozone precursors.
2. To determine and quantify, through chemical, emission-rate and meteorological measurements and modeling and other data-interpretive activities, the roles of natural hydrocarbons, anthropogenic hydrocarbons, and nitrogen oxides in the formation of ozone and other photochemical oxidants in the atmosphere in various urban and nonurban areas of the southern United States.
3. To improve emissions factors, methodologies for the development of emissions inventories and improve inventories of natural and anthropogenic VOC and NO_x emissions that lead to formation of ozone in the atmosphere in the southern United States; and to improve understanding of dose-response relationships for the effect of ozone on forests, crops, and engineering materials.
4. To characterize the ozone exposure of humans pursuing different styles of outdoor life in various urban and non-urban areas of the southern United States.
5. To evaluate alternate strategies for management of ozone and other photochemical oxidants and their injurious effects in various urban and non-urban areas of the southern United States.

Scientific Research Programs

To meet the objectives and goals listed above, the university effort in SOS has been apportioned into four closely related scientific research programs.

The four SOS Programs are briefly described below:

1. The Southeast Regional Oxidant Network (SERON) consists of a hierarchy of air-quality monitoring sites situated in mostly rural locations of the southern United States. The network and its implementation is largely designed to address questions related to the regional nature of the southern oxidant problem and will fulfill the first, and partially fulfill the second and fifth objectives of SOS listed above.
2. The Southern Oxidants Research Program on Ozone Non-Attainment (SORP-ONA) consists of a series of intensive field sampling operations in southern cities combined with diagnostic and prognostic model analyses designed to focus on the ozone nonattainment problem in the South. The Program is designed to fulfill the remainder of the second objective and contribute to the fifth objective of SOS.

3. The Southern Oxidants Research Program on Emissions and Effects (SORP-EE) consists of a combination of laboratory studies and field measurements coupled with emission-inventory reviews, designed to better quantify the emissions of natural and anthropogenic ozone precursors and the effects of ozone on living organisms in the South. It addresses in part the second, third, and fourth objectives of SOS.
4. The Southern Oxidants Research Program on Hydrocarbon Measurement Technology and Standards (SORP-HCMTS)will be a collaborative program with the International Committee on Hydrocarbon Intercomparison Study, and will design and oversee the implementation of procedures for the evaluation and standardization of hydrocarbon sampling, storage, and analysis. As such, SORP-HCMTS plays a supportive role to the other three SOS Programs.

IV. A Unique Opportunity: Super-University City Demonstration for the 1996 Olympics

The Olympic games and the Georgia Tech campus provide a unique opportunity for the development, demonstration, and use of environmentally benign technologies for energy, transportation, architectural design, waste management, and water treatment. The development of environmentally benign technologies are important to the world at large in both the global and micro environments. The Olympics provides a "special event" with particular characteristics such as high volumes of people, vehicles, new and special construction, and media attention; however, the facilities being constructed for the Olympics will remain following the event. The innovative technologies developed in the proposed research projects will provide fruitful opportunities for continued research. The technologies developed by Georgia Tech for the Olympic demonstration projects can be replicated at other institutions and be extrapolated as a tool for establishing sound energy and environmental policy.

The Olympics offers a singular opportunity to evaluate energy and transportation technologies which are more environmentally benign than those currently in use. Transportation will be the largest user of energy. Large numbers of people will have to be moved from place to place during the Olympics. Increased tourism throughout the Southeast will escalate energy usage by transportation even more. The use of substitute fuels and their impact on petroleum energy consumption will be assessed. The effect on the environment of alternate transportation fuels is an area which will be targeted for investigation. The fuels will be examined for their airborne combustion products and their effect on storm water runoff.

The energy cycle includes energy generation, distribution, and usage. Each of these aspects should be environmentally benign. Energy generation can have significant impact on outdoor air quality. Ideally energy generation and usage should be regulated by the chemical and meteorological environments, and not only by energy cost. Forecasting and monitoring technology can be developed which will base energy demand on environmental impact/chemical loads; thereby, reducing energy costs and environmental impact.

Innovative energy supply-side (generation) technologies targeted for inclusion in the Georgia Tech Olympic initiative include cogeneration, solar, photovoltaics, fuel cells,, waste energy utilization, and clean coal technology. Each of these technologies have the potential to generate energy more efficiently and more environmentally-friendly than commonly used technologies. The cost, efficiency, and environmental benefits of cogeneration in high-density population areas, such as the Georgia Tech campus/Olympic Village, has been demonstrated in several European cities. The integration of passive solar energy in the design of new Olympic Village facilities is a well-established and environmentally benign technology.

The Power distribution system for the Georgia Tech campus will undergo significant modifications to accommodate the Olympic facility requirements. These modifications can have significant impacts on energy usage and environmental impact. The development of advanced technology for the management of the distribution system will be targeted for inclusion in the Georgia Tech Olympic initiative. Integrated supply and demand-side "smart meters" will be investigated for their potential for improving the efficiency of larger scale grid systems. Advanced energy management systems can result in significantly improved energy distribution systems in an environmentally benign manner by optimizing operation and control of building mechanical systems. The innovative system would efficiently and effectively match supply and demand for power throughout the Georgia Tech campus.

The technology development for the use (demand-side) of least amount of the energy while producing the best possible indoor environments will be targeted for development and demonstration at the Olympics by the development of "smart buildings", efficient and effective air cleaner systems, improved ventilation technology, and lighting control systems. The "smart buildings" will be controlled by artificial intelligence sensor control systems. The control systems can be integrated into new and existing Georgia Tech facilities. The control system is proposed to be used in residential as well as in commercial and institutional buildings. To demonstrate the residential applications, a model energy/environmental house would be constructed and located on the Georgia Tech campus. The model house will combine both the "Good Cents" approach to energy conservation and advanced indoor air betterment technology.

Cold storage technologies are a potential means of reducing costs by shifting cooling loads to off-peak hours. Advancement of these technologies will be targeted for development and evaluation.

Lighting is a major source of energy consumption in buildings and a major contributor to cooling loads. Effective and efficient lighting also is a major factor in human perception of the indoor environment. A project exploring the performance of both user-controlled and automatic lighting control systems is proposed. Lighting performance will be based upon photometric assessment, estimated power loads, and human response data. The development of an advanced campus-wide electrical management system can reduce utility costs by matching electrical supply and demand.

The Olympic construction project is one of the largest and most complex US building projects. It provides an exceptional occasion to develop and evaluate innovative technologies and processes for improved planning, design, and facility management. These innovative technologies will allow for the melding of design and construction for optimum support of the needs and activities of the building users. Energy efficiency, quality of construction, indoor air quality, and occupant comfort and productivity will be optimized. Space usage, movement patterns, participant-interactions and communication in both new and existing individual buildings as well as throughout the entire Olympic event will be evaluated.

The existing Georgia Tech facilities offer an opportunity to serve as a benchmark for university and other large, multibuilding facilities for energy usage and indoor air quality. Traditionally these large facilities, in particular universities, are very inefficient energy users and have poor indoor air quality due to the perception that "doing it right" is too expensive. The development of the benchmark of the Georgia Tech facilities will develop innovative technologies and policies that will demonstrate the cost effectiveness of "doing it right". The benchmark will be established through operation and maintenance management, retrofit, and training and education policies development and implementation.

The Georgia Department of Transportation is currently developing a regional Advanced Traffic Management System (ATMS) that will provide, in a very rudimentary way, surveillance of freeway usage and some simple means of conveying to motorists system conditions so that they can modify their routes, if desired. Very little attention, however, is being given to in-vehicle information units that can convey this information to drivers in a more efficient way and can provide in-vehicle navigational aid. Demonstrations of such systems are currently underway or planned for Orlando, Los Angeles, and Chicago. Very little attention is being given to the energy or environmental impacts of such systems. A pilot project during the Olympics, with specific focus on energy and environmental considerations, would be very important. Similar evaluations could be undertaken with regard to telecommunications substitution for transportation.

Georgia Tech is planning to organize an International Conference to invite the experts around the world to define the possible Supercity University model for the 21st century. This event will give us the opportunity to work together for the future.

IN CONCLUSION, GEORGIA TECH THINKS AND LINKS GLOBALLY AND ACTS LOCALLY AND REGIONALLY

References

Chameides, W.L., R.W. Lindsay, J. Richardson and C.S. Kiang, Science 241, 1473-1475, 1988.

MacNeill, J., Scientific American; 261, 154-165, 1989.

Chapter 24

Patent Laws: Help or Hindrance in Solving Environmental Problems?

S.R. Foster

What is a Patent?

The Founders of our Republic saw fit to include in the Constitution of the United States provision for the establishment of the American Patent system, "The Congress shall have the power... to promote the progress of...useful arts, by securing for limited times to...inventors the exclusive right to their... discoveries." George Washington sought implementation of the Constitutional Provision and the first Patent Act was passed in 1790.

Patents have been said to represent a contract between the government and an inventor, or inventors. The government provides a "monopoly" for a limited number of years in exchange for the inventor disclosing his idea by way of publication of his patent. Upon expiration of the limited time, the invention so disclosed may be used by anyone wishing to use it.

A patent does not provide the inventor (or his assignee) with the right to make, use, and sell that which is patented. The patent enables the patent holder only to prevent others from using the invention he has patented. For example, a patent holder may be prevented from utilizing his invention because of another's broader patent, while at the same time being able to prevent others from using the invention he has patented.

With a few exceptions, the effect of patents is limited to the countries in which they are issued. That is, a United States patent is effective only in the United States.

How do I get a Patent?

The patenting process begins with the preparation and submission to the United States Patent and Trademark Office of a patent application, filed in the name of the inventor or inventors. A patent application includes a "specification", in which the invention is described, usually drawings, in which the invention is pictured, and "claims", which are definitions of the invention.

After filing your patent

application, a Patent Office examiner will examine the application. The examination includes a search of the prior art to determine whether the invention, or something close to it, has previously been patented or otherwise published.

To be patentable, your invention must be new, useful and unobvious. "New", or "novel", means not shown in the prior art. If your invention is shown in the prior art, your invention is said to be "anticipated". "Unobvious" means your invention is "new" and, further, there is no prior art close enough to your invention to render your invention obvious. Most patentability battles are fought over the obviousness issue. Quite often, a search reveals that an invention is not shown in the prior art, and therefore is "new", but that there is prior art which is close to the invention. The question then becomes whether the prior art is so close as to render the invention "obvious" and, therefore, unpatentable.

The requirement for "usefulness" is seldom a point of contention.

If I obtain a patent, how do I use it?

As noted above, your patent enables you to prevent others from making, using and selling the invention defined by your patent claims, that is, the definitions of the invention in your patent. If a patent holder learns of an infringement of his patent, he may elect (1) to do nothing; one is not obligated to enforce a patent, so long as he has not contracted to do so; (2) sue the infringer in Federal District Court; or (3) try to work something out with the infringer; that is, "license" the infringer. A "license" is an agreement by which the patent holder(licensor) permits another(licensee) to utilize the patented invention, and the licensee agrees, in return, to pay money to the licensor, usually a percentage of sales.

Whether one can obtain a patent, and how one elects to use a patent, can depend, at least to some extent, on the country involved. Patent rights gained in one country do not necessarily lead to similar rights in other countries. Each country has its standards for patentability and practices peculiar to that country, as well as statutory differences. Some countries do not permit patenting of certain categories of inventions, such as medicaments, some aspects of bio-tech, and the like. Some countries have "working" requirements, which means that if a patented invention is not utilized in that country within a prescribed term, the patent is subject to revocation and/or compulsory licensing to others.

There has been for several years, and there still is, a strong movement underway to "harmonize" the patent laws of nations. Progress has been, and is being made, but the finish line is nowhere in sight.

How the patent system benefits industry.

The American patent system spurs development by offering to small companies and individuals the perception of opportunity to make big dollars. Stories abound of individuals or small companies patenting a product that sells like hot cakes. Larger companies usually don't look upon the patent system as a vehicle by which to strike it rich, but they do appreciate the patent system as a vehicle by which they are able to build a patent fence around certain technology and keep the competition away for a period calculated to provide a good head start.

The Polaroid Company has done an excellent job of patenting its consumer products so as to keep competition at bay, much to the consternation of Eastman Kodak. In a field known for flagrant copying, Gillette has managed to be the sole seller of a razor system, known as "Sensor", for about five years now, holding some thirty or so patents on the Sensor Razor system, perhaps a couple of dozen other patents on alternative embodiments not sold by Gillette, and a continuing stream of patents on improvements, whether commercial or not. Those who traditionally have copied Gillette products are having a hard time with the Gillette Sensor.

In the long run, all patented technology is made available to the public. Though patents may be obtained with the thought of keeping use of the inventive technology from competitive hands, the use of the patent system insures that, in due course, the technology will be available to the competition.

One who is blocked by a patent has some options; (1) stop making, using and/or selling the patented items; (2) work out a license agreement; (3) attack the patent; and (4) design around the patent. The latter alternative provides an answer by which the one accused of infringement can avoid the accusation and, perhaps, assume a dominant position over the accuser. Thus, the patent system provides incentives which generate the rapid development of new concepts, new products and processes.

<u>How the patent system might impede industry.</u>

The term "monopoly" may not be entirely correct in reference to patents, but in general, it is descriptive of what the patent holder obtains. As is generally recognized in anti-trust law, monopolies are, or can be, very bad things. Monopolies can stifle competition, drive prices up, force acceptance of shady workmanship, and the like.

A United States patent holder is under no obligation to make, use and/or sell that which he has patented. He has a right to prevent others from making, using and selling the product. If a patent holder does elect to make and sell the patented product, he may select a price without regard for the ability of purchasers to pay. He may disregard complaints and treat his customers with indifference. After all, if they want or need the product, the patent holder is the sole source.

Aside from this rather sinister scenario, the more common anti-competitive occurrences include the late discovery of a blocking patent, rendering an R & D program non-productive profit-wise. Even if a product is believed to be non-infringing, fear of suit under a patent which is "close", or possibly could be interpreted in such a manner as to be infringed, could lead to a decision to forego introduction of a product. Infringement usually is not a clear-cut issue. Arguments usually can be made for and against infringement. In short, what patents cover is not always apparent.

The "Doctrine of Equivalency" introduces uncertainty. The Doctrine holds that even if there is not "literal" infringement of a patent claim, there still may be infringement if the accused device does substantially the same thing, in substantially the same manner, to obtain substantially the same result. There have been many cases decided under the Doctrine, many rules set forth judicially, but the matter remains uncertain.

The unpredictability of court cases generally, and patent cases in particular, can lead an accused to back away, even though he may be convinced that he is in the right.

Finally, costs of patent trials and the possibility of treble damages can dampen the enthusiasm of the most ardent proponent of a good fight. Because of the cost factor, many would-be patent litigants are turning to arbitration, which is faster, cheaper by far, and probably productive of as high a quality of "justice" as the court system.

Hypothetical Case

ABC Corporation has invented an inexpensive smoke-stack scrubber capable of removing objectionable matter from emissions more efficiently than current scrubbers. Scientists have opined that in view of the new inexpensive scrubber, acid rain should be virtually eliminated in short order. ABC Corp. is not in the scrubber business. The invention was the brain child of particularly innovative employees who were always working on things not job-related, which was allowed by ABC, within limits.

XYZ Corporation is currently in the scrubber business. They have the capacity to produce scrubbers around the clock, three shifts. They have the floor space, manpower, resources and flexibility to start and carry through with a crash program designed to have the product on the market in three months' time. A study by XYZ's patent counsel has concluded that a license under the ABC patent is the only prudent way to proceed.

XYZ approaches ABC, requesting an exclusive license, but willing to take a non-exclusive license, if necessary. Mr. B.H. Scrooge is CEO of ABC, a family-owned corporation. Mr. Scrooge is not interested in granting licenses. Mr. Scrooge opines that acid rain is God's way of punishing the Earth for its conduct and lack of morals and such things should not be tampered with. Mr. Scrooge concludes by saying anyone who infringes his scrubber patent he will see in court.

XYZ convenes its senior officers and its patent counsel. The options appear to be (1) forget the new scrubber; keep selling the current product; (2) attack the patent by way of (a) re-examination, wherein XYZ would ask the Patent Office to take another look at the patent, or (b) file a Declaratory Judgement Action in Federal District Court, seeking a holding of invalidity. XYZ patent counsel says the patent is a strong one and not likely to be found invalid in a D.J. Action and almost certainly not likely to be found unpatentable in a reexamination proceeding; (3) make and sell the scrubber and hope for the best in court. Patent counsel reminds the meeting of the treble damages prospect plus the real possibility of an injunction shutting down the XYZ production line; (4) design around the patent. The officers of XYZ decide to give top priority to a "design-around" team of its best scientists and engineers. It is recognized that nothing may come of this work, and/or that the project could take years, but it appears to be the only solution.

The above hypothetical case, though improbable, is, of course, possible. In reality, Scrooge probably would jump at the chance to license the scrubber technology for a royalty of 10%, or so, of sales. However, accepting the hypothetical, the ending of the story could be "happy" if the XYZ team is successful in designing its own improved scrubber which has advantages over ABC's and which is, nonetheless, still relatively inexpensive. Further, XYZ probably is only one of several competitors engaged in an R & D effort to design around the patent. If there are six companies in the business to a substantial degree, in all probability there would be six R & D efforts underway. The ending could be different, obviously, if the XYZ team and all other teams similarly engaged meet with no success after years of struggling.

Conclusion

George Washington viewed the American Patent system as contributing to the advancement of agriculture, commerce and manufactures.

Thomas Jefferson opined that the 1790 Patent Act gave "a spring to inventions beyond my conception."

Abraham Lincoln said "the Patent System added the fuel of interest to the fire of genius."

Franklin D. Roosevelt stated that "The American Patent System has promoted countless applications of the arts and services to the needs and well-being of our people."

Aside from the glowing tribute paid the American patent system by the above and a host of other important personages, many today concede that the patent system has contributed to the rapid advancement of bio-technology, electronics, computer science, and everyday consumer products.

It is difficult to assess the general impact of patents on environmental technology because no one keeps records on such things. I would submit, however, that there is no reason to believe that environmental technology so differs from other technology as to stand apart with respect to the benefits of the patent system. As shown by the above hypothetical case, it is possible that in a specific instance patent rights might be used to stop, or at least slow, progress. But it still appears to be the case that in the long run, the patent system must be regarded as a help to the solution of environmental problems, rather than a hindrance. The system holds out the prospect of a protected market for a limited time, enabling the patent owner to reap the benefits of his innovative step forward, and enabling the public to amass a reservoir of public domain technology.

Chapter 25

Evaluating the Performance of Active Noise Control Systems in Industrial Applications

S. Dineen, C. Depies, M. Lowe, S. Wise

ABSTRACT

Active sound cancellation technology is increasingly being used to quiet industrial air-moving devices. Engineers and designers are implementing active or combination active/passive technology to control sound quality in the workplace and the acoustical environment in residential areas near industrial facilities. Sound level measurements made before and after the installation of active systems have proven that significant improvements in sound quality can be obtained even if there is little or no change in the dBA readings. Noise produced by centrifugal and vane-axial fans, pumps and blowers, commonly used for ventilation and material movement in industry, are frequently dominated by high amplitude, tonal noise at low frequencies. Because the A-weighting network, which is commonly used for industrial noise measurements, de-emphasizes low frequencies, its single number rating can be misleading in terms of judging the overall subjective sound quality in impacted areas and assessing the effectiveness of noise control measures. Analysis of frequency spectrum characteristics provide the most effective means of assessing sound quality and determining mitigative measures for achieving suitable background sound levels. Examples will be presented from a number of field installations.

INTRODUCTION

Large air moving devices in industrial settings generate substantial amounts of noise which can cause problems for workers and surrounding neighborhoods. The noise is typically produced by centrifugal and vane-axial fans, vacuum pumps, compressors or blowers and is primarily tonal. Levels can be as high as 140 dB.

Traditional solutions have used passive silencing techniques to eliminate the intake and discharge noise from these devices. Passive silencers can be divided into two types: reactive (or chamber or resonator) silencers and dissipative (or absorptive) silencers. Passive silencers are more effective at mid and high frequencies and, when designed to attenuate low-frequency noise, can be very large and restrictive to flow.

Unlike passive techniques which reduce the sound level by mechanically interfering with the sound wave, active noise control systems produce a sound wave which is the inverse or mirror-image of the noise. This sound wave, which is of equal amplitude and frequency, but of opposite phase, cancels the noise. The basic configuration of an active noise control system is shown in Figure 1. A fan or pump generates waves of sound pressure, recognized by the ear as noise, which travel down the duct or pipe. The upstream or input microphone converts the pressure wave to an electrical signal which is sent to the controller.

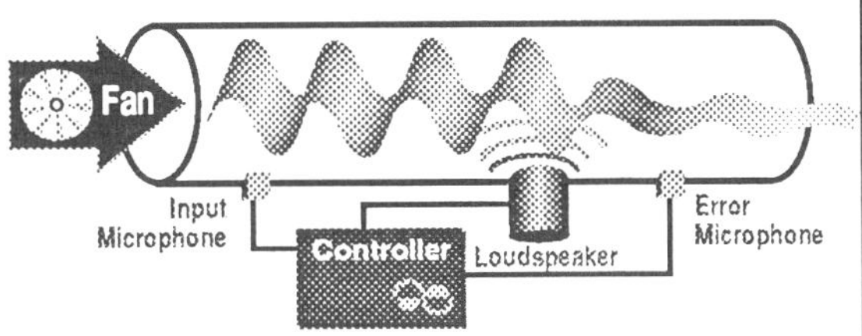

Figure 1

The controller calculates the cancelling signal which is an exact opposite or mirror image of what the incoming signal will look like when it reaches the downstream location of the cancellation loudspeaker. The output of the loudspeaker interferes with the propagating noise and creates nullifying effect. An error microphone downstream from the loudspeaker monitors the residual acoustic pressure after cancellation and allows the controller to adjust itself for optimal results.

The controller precisely estimates the delay and amplitude changes from the input microphone to the loudspeaker, including delays in the microphones, loudspeaker and electronics. This task is known in the signal processing literature as system identification, which is accomplished with adaptive filtering methods. The particular method is filtered recursive LMS, which is described in detail elsewhere (Eriksson, L.J. and Allie, M. C., Proc. Int. Symp. Circuits and Systems, Espoo, Finland, vol. 3, pg. 2387, 1988).

The best results so far have been obtained by cancelling plane waves travelling down ducts or pipes. Typical applications include air pollution control equipment, material handling fans and rotary pumps and blowers. Gas flow is high, typically 25 m/sec (5000 fpm), and there are often suspended solids in the flow stream. Temperatures are often elevated above ambient, and the gas stream is sometimes wet and/or corrosive. These ducts are unlined due to the nature of the gas stream, creating a large standing wave structure. Such issues brought about the development of a range of high-power loudspeakers. Normally two or four loudspeaker modules are used, both to provide the required output power and for redundancy.

Protection of the loudspeakers and microphones from a wet, dirty stack environment is essential for long term performance. A protective membrane material is used to allow transmission of the sound energy into and out of the stack while maintaining a clean, dry environment to ensure long component life.

Figure 2

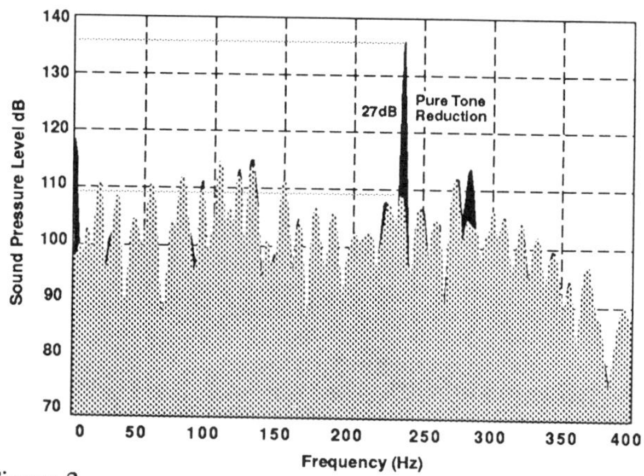

Figure 3

POLLUTION CONTROL EQUIPMENT

Figure 2 shows an active noise control system silencing an industrial incinerator. The noise source is a radial-blade centrifugal fan which pulls high velocity air through a boiler and wet scrubber system. The treated air is discharged through a 20-inch diameter exhaust stack. The stack effluent is hot (180° F), fully saturated and corrosive. Loudspeaker and microphone modules are mounted on the fiberglass duct above the roofline for easy installation and servicing and are protected from the harsh stack envrionment by air-tight membranes.

Figure 3 shows performance of the system. The 235 Hz blade pass tone was reduced by 27 dB in the stack. Nighttime sound measurements at the property line confirmed that the tone was no longer audible.

Another example of active noise control used to attenuate noise from air pollution control equipment is shown in Figure 4. The plant is located in an established residential community. The manufacturing process requires in-plant ventilation for dust control. Air is drawn from in-plant hoods and filtered through baghouses located on the roof of the building. The radial-blade centrifugal fans which draw the air produce a pure tone at the blade pass frequency. Shown is one of eleven fans at this site silenced with active noise control.

Narrowband measurements shown in Figure 5 were taken at a location approximately a quarter mile from the source. The sound energy is concentrated at the primary frequency and harmonics of the fan. The discrete tone was attenuated 25 dB at the stack discharge and a 21 dB reduction was achieved in the far field.

Figure 4

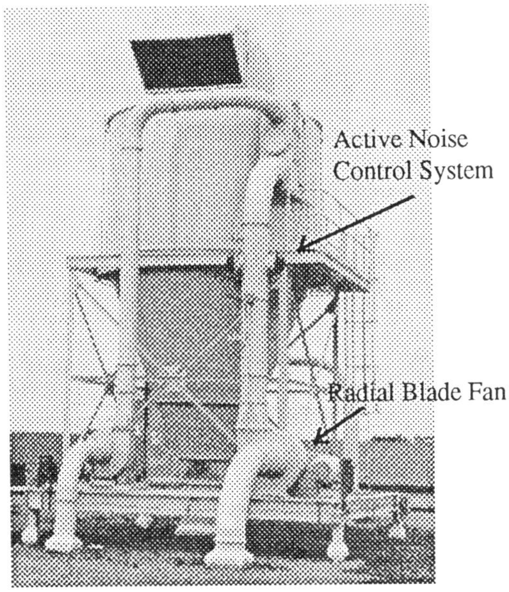

Figure 6

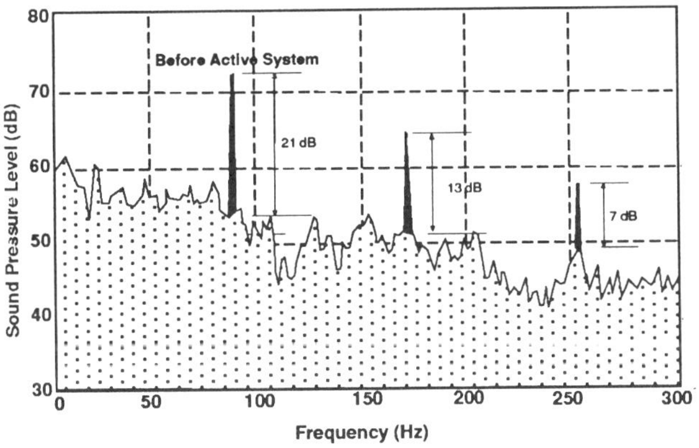

Figure 5

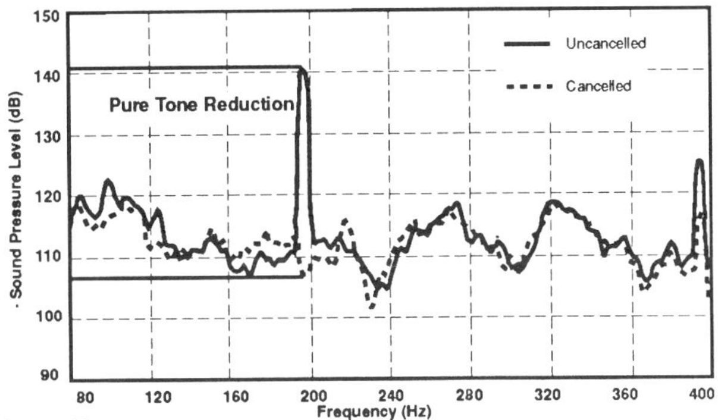

Figure 7

MATERIAL HANDLING FANS

Figure 6 shows a radial-blade material handling fan which pulls material from paper processing stations throughout the plant. The paper scrap discharges into the cyclone for recovery and recycling. A very loud pure tone at 198 Hz is generated by the rotation of the fan blades.

The active noise control equipment is located between the fan and the cyclone. All components are out of the flow stream, eliminating the possibility of clogging.

The noise spectrum, shown in Figure 7 shows a 34 dB reduction in the duct. A far-field measurement, approximately 150 ft. from the cyclone, measured a 21 dB reduction.

EVALUATING PERFORMANCE

The most common noise rating scheme used for industrial and community noise regulations is the A-weighted sound level. A-weighting curves de-emphasize measured sound levels at low frequencies in order to approximate the response of the human ear to low sound pressure levels. The overall A-weighted sound level (dBA) is then the logarithmic sum of the adjusted sound levels. As with most single-number rating schemes, there are inherent shortcomings in using the A-weighting scale. One problem is that high amplitude, low frequency noise, such as discrete tones, may not be discernible in a dBA measurement, even though the noise may be clearly audible and annoying. Consequently, the reduction of such noise is also disregarded in a dBA measurement. Another limitation with the A-weighting scale is that it is based on the 40 phon equal loudness contour and should not be used for sound levels greater than 55 dB. Unweighted or "linear" scale provides accurate sound measurement information and would be more appropriate for use in octave band, third octave or narrowband sound level analyses. Figure 8 shows the comparison between the A-weighting noise descriptor and linear sound levels.

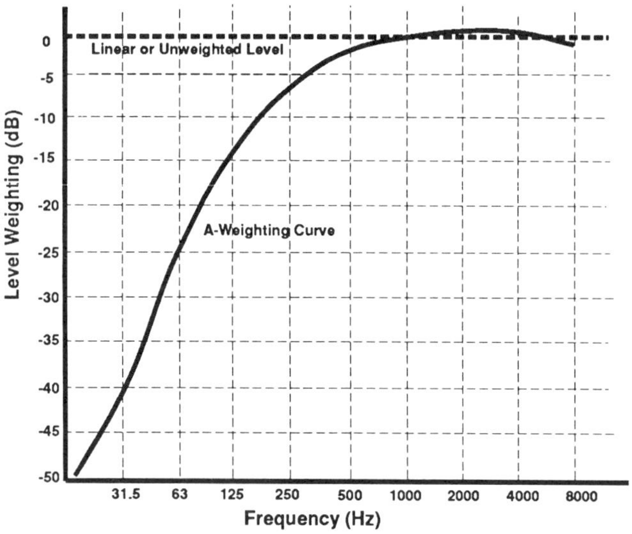

Figure 8

Evaluation of industrial noise control issues must take into account both the level and quality of noise in the location(s) of concern. As mentioned, high amplitude, low frequency tones may annoy residents in the vicinity of industrial facilities and may result in complaints, even though the noise does not exceed the community noise regulations. For example, the 85 Hertz tone produced by the rooftop centrifugal fan in Figure 4 propagated into surrounding neighborhoods and disturbed many residents. The attenuation of the tone by the active sound cancellation system (i.e., 25 dB at the discharge and 21 dB in the far field) improved the sound quality by removing the tone and complaints by the residents stopped. However, the A-weighting scale applies a 22 dB penalty at 85 Hz and the dBA value remained unchanged, consequently under-representing the sound quality improvement provided to the community. In this case, and with other industrial applications that involve low frequency repetitive noise, dBA measurements are not adequate for evaluation and/or mitigation purposes.

High-pressure industrial fans, rotary blowers, industrial engines and compressors produce sound signatures that are dominated by low-frequency pure tones which are related to the operating speed of the equipment. As has been shown by the examples presented, the low-frequency tones are often as much as 30 dB louder than the broadband noise components and, due to their substantial wavelengths and amplitudes, tend to propagate great distances. Because the tones and subsequent harmonics are low frequency, traditional noise control methodologies, such as passive absorptive silencers and enclosures, are often ineffective. The attenuation of passive silencers typically drops significantly at frequencies below 250 Hertz. Improvements in the low-frequency attenuation of passive silencers usually entail substantial increases in the size and/or pressure drop of the silencers. Also, conventional silencers cause air turbulence which creates noise of its own. Alternately, active control can attenuate low-frequency tones by up to 30 dB and broadband noise by 10 to 15 dB, without adding excessive size or pressure losses to the system.

CONCLUSION

Increasingly, active sound cancellation technology is being used to control noise from industrial equipment. Its application is proving to be more effective for attenuating low frequency noise that the passive techniques traditionally employed.

Communities and industrial and commercial facilities are becoming more sensitive to the quality of the environment in which they work and live. It is critical that there be continued advancement in the industry's overall ability to assess the impact of noise on the receiver. Current use of simplified rating schemes such as dBA level provide inadequate information as to the quality of the noise being generated by rotating equipment. A detailed noise survey using unweighted third octave or narrow band frequency analysis should be used in evaluating the impact of industrial noise and the reduction afforded by noise control techniques.

Chapter 26

Infrared High Temperature Thermal Desorption—An Alternative to Mobile Incineration

J.F. Rouleau

ABSTRACT

Mobile incineration faces many barriers to implementation as an on-site remediation technology. The permitting, construction, monitoring, and operation of on-site incineration systems is expensive, complex, and time-consuming. Also, public perception of on-site incineration is strongly negative. This negative bias is principally associated with the potential that off-gas products of incomplete combustion contain harmful hazardous constituents.

Over the last several years, the development of technologies to substitute for on-site incineration has been encouraged by the United States Environmental Protection Agency (USEPA). Thermal desorption is emerging as a preferred treatment for soil affected by volatile and semi-volatile organic compounds and volatile metals. Relatively simple in concept, thermal desorption heats the soil to volatilize contaminants from a soil matrix into a vapor phase. The volatilized contaminants are recovered from the off-gas stream without reaction or the production of harmful air emissions. This process eliminates most permitting issues and the negative public perception associated with other processes such as mobile hazardous waste incinerators.

Thermal desorption processes may be considered in two categories: "low temperature" and "high temperature". Numerous companies have developed and are utilizing low temperature thermal desorption units in the field, principally to treat soils affected by "low" boiling point (less than about 600 °F) compounds. However, very few companies have developed units to treat compounds with high boiling points such as polychlorinated biphenyls (which boil at temperatures as high as 950 °F) and polynuclear aromatic hydrocarbons (which boil at temperatures in excess of 800 °F).

Westinghouse has developed an electrically powered Thermal Desorption Unit (TDU) to handle high and low boiling point contaminants. The TDU heats the contaminant molecules above their boiling point to desorb the contaminants into a vapor phase. The unit operates in an oxygen deficient atmosphere to prevent oxidization of the contaminants. The volatilized contaminants and soil moisture are condensed and collected within the system. Through the utilization of infrared heating and an alloy belt feed system, the process can be precisely controlled.

This paper will present a review of the high temperature thermal desorption technologies available today. Review of the Westinghouse TDU will focus on the design, development, and permitting hurdles that the unit has faced.

INTRODUCTION

To date, there are approximately 25 companies that process affected soils with mobile thermal processes. The majority of the companies utilize mobile incineration type processes. This mobile incineration generally heats the contaminated material, usually inside an externally fired rotary drum, to volatilize the contaminants. The off-gas is then incinerated in a secondary unit. There are two inherent problems with this configuration. The first potential problem is the creation of large amounts of dust carry-over through the tumbling of the soil in the rotary drum. The second potential problem is the products of incomplete combustion in the secondary unit. This second problem created the general negative public perception of mobile incineration.

Despite these problems, numerous companies are processing contaminated materials through mobile thermal units. Most of these companies process soils at relatively low temperatures, less than 500 °F. A few companies process materials at temperatures as high as 1000 °F. However, most of these processes utilize an incineration principle rather than a true desorption principle. Nonetheless, a few companies process soil with equipment that is configured as true desorption processes.

PRINCIPLES OF OPERATION

Thermal desorption is based on the principle of desorption, the process of transferring molecules from the surface of a particle into a fluid (i.e., gas, vapor, or liquid). Desorption occurs when the energy level (related to temperature and concentration) of a molecule at the surface of a particle becomes higher than the energy level of a molecule in the fluid phase.

The TDU operation involves heating the contaminant molecules (and soil surface) above the boiling point of the contaminants present to desorb the contaminants into a vapor phase. The TDU operates at varying temperatures, depending upon the contaminants to be volatilized. Volatilized materials in the vapor phase are then condensed and collected so that no significant air emissions are generated. Temperatures in the heating chamber can be raised to between 400 °F and 1000 °F. This unit operates at just below atmospheric pressure in an oxygen deficient atmosphere. Through the utilization of infrared heating rods and an alloy belt feed system, the process can be exactly controlled and easily adjusted to meet various process requirements.

The thermal desorption process is a simple technology that avoids the negative features of on-site incineration while offering the following benefits:

- it precludes concerns about air emissions, making the technology community friendly;
- permitting requirements are minimized (there is approximately zero process water and approximately 50 cfm process gaseous discharge);
- overall costs of soil detoxification are significantly lower relative to incineration;
- the process is very controlled with well-defined operating parameters; and
- simple treatment by-products are produced (a clean soil which can be re-inserted at the site and a concentrated contaminant which is usually transported off site for final disposition).

In addition, the Westinghouse thermal desorption process offers the following positive characteristics as compared to rotary-type desorption processes:

- electric heating is used;
- no combustion or oxidization takes place;
- known residence time and temperature for each soil particle;
- controlled bed thickness of soil in the unit; and
- limited potential for fines carryover since there is no tumbling of the soils in the purge gas and the purge gas flow rate is very low.

The principle of operation for the TDU is as follows.

- Contaminated soil is fed into the sealed Primary Unit that operates below atmospheric pressure in an oxygen deficient atmosphere. The temperature of the contaminated soil is raised above the boiling point of the contaminants present. This causes the contaminants to volatilize (a change in state from liquid phase to vapor phase).

- The vaporized contaminants are captured in a sweep gas (the water, in contaminated soil, vaporizes prior to contaminants because of the waters lower boiling point) in the Primary Unit and transferred to the Quench/Scrubber Unit.

- The vaporized contaminants and water are collected and condensed in the Quench/Scrubber Unit.

- The condensed contaminants and water are transferred to and processed by the Water Treatment Unit. This Unit consists of an oil/water separator and an air cooling tower. The oil/water separator removes the condensed contaminants from the water. The water is cooled and recycled with a portion of the water being utilized to re-moisturize the treated soil to the original moisture level.

- The concentrated contaminants are collected and transferred off site for final disposition.

The Table 1 highlights the full scale process capabilities of the Westinghouse TDU.

TABLE 1 - PROCESS SPECIFICATIONS

Specification	Value	Units
Total Energy Input:	3×10^6	watt
Soil Layer Thickness on the Belt:	0.5 - 2.0	inches
Residence Time:	5 to 60	minutes
Moisture Content:	5 - 20	% by weight
Flow Rate:	1- 20	ton/hr

The key to successful thermal treatment processes is the careful control of residence time and soil temperature. Control of these parameters maximizes the throughput, minimizes energy use, and produces a treated soil that consistently meets regulatory requirements. Since the Westinghouse process utilizes a belt system and highly controllable heater rods, the residence time and temperature are exactly known, even for the finest particle. In addition, carryover of fines is minimized since there is no active tumbling of the soils in the off-gas and the off-gas flow rate is very low. **Figure 1** shows a diagram of the process flow of the TDU.

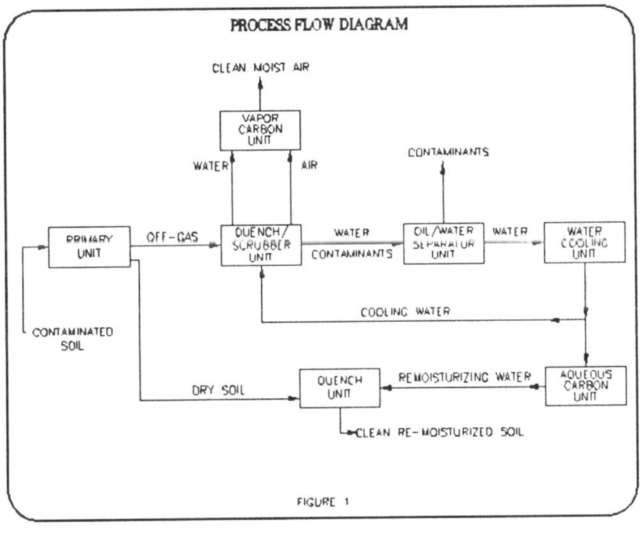

FIGURE 1

This innovative Westinghouse technology is supported by complete in-house capabilities to perform bench scale treatability testing and associated analyses. Westinghouse tailors bench scale testing programs to determine the efficiency and feasibility of thermal desorption in decontaminating site soils and meeting cleanup standards, regulatory criteria, or other treatment goals.

The TDU consists of the following major mechanical subsystems and components:

- Primary Unit;
- Quench/Scrubber System;
- Water Treatment System;
- Treated Soil Handing System; and
- Control System.

The above subsystems are designed to be transported to a hazardous waste site on eight trailers, assembled into one unit, used to thermally treat solids, disassembled into subsystems again, and moved to the next site. The following sections describe the operation of each of the major mechanical subsystems.

Figure 2 illustrates a schematic of the process showing all major components.

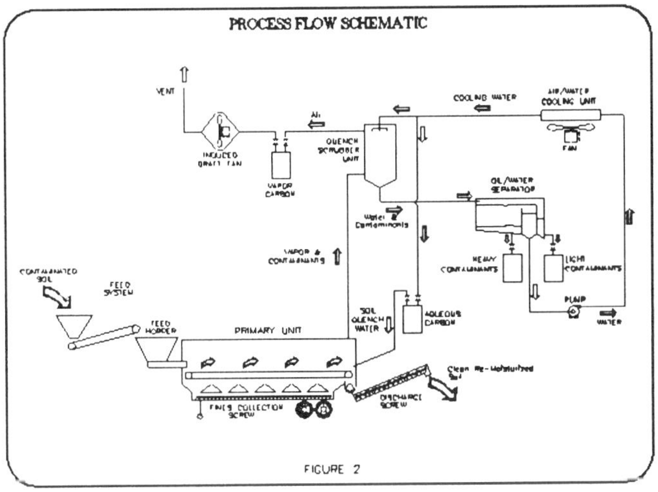

FIGURE 2

Feed System

The feed module is composed of the Inclined Belt Conveyor and Feed Hopper. The pre-sifted contaminated material is placed into the hopper on the Inclined Belt Conveyor where it is weighed by a belt scale. The material is dropped into the Feed Hopper which deposits the material onto a metal alloy belt. This belt transports the material through the Primary Chamber.

Primary Unit

The Primary Unit, shown in **Figure 3**, is a single chamber that is designed to travel over the road. A feed module and a discharge module support the Primary Unit. The specifications for this Unit are shown in **Table 2**.

PRIMARY UNIT

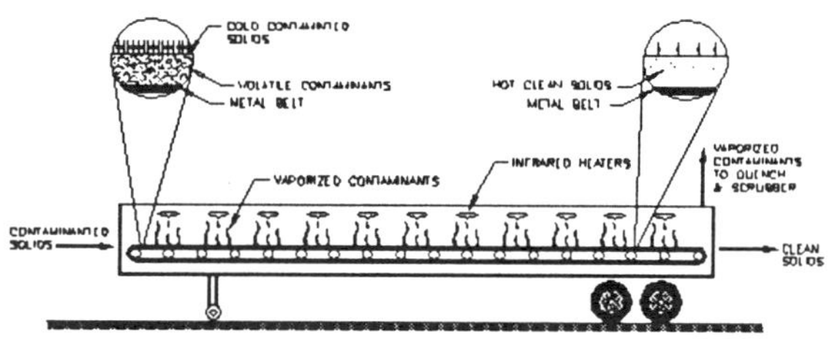

FIGURE 3

TABLE 2 - PRIMARY UNIT SPECIFICATIONS

Specification	Value	Units
Width:	10.0	feet
Length:	61.0	feet
Height:	13.5	feet
Belt Surface Area:	355.3	$feet^2$
Belt Width:	6.8	feet
Total Belt Length:	130.9	feet
Effective Belt Heating Length:	52.0	feet
Internal Height:	34.0	inches
Internal Width:	7.8	feet
Number IR Heater Bars:	108	
Maximum Loading per Heater Bar:	45	$Watt/inch^2$
Total Energy Input	3×10^6	watt

The Primary Unit is a self-contained unit that contains the alloy belt and the silicon heater bars. The soil travels the length of the Primary Unit on the alloy belt where infrared heat volatilizes the contaminants from the affected soil. The belt is supported by externally mounted bearings and is powered by a chain-driven roller system mounted on the Primary Unit.

Six soil lifters are mounted on the Primary Unit just above the belt. The soil lifters turn the material layer on the belt as it moves through the Primary Unit. This mixing of the material on the belt allows for increased contaminant removal.

The processed feed material drops off the end of the belt in the Discharge Module where it is cooled with water sprays then discharged by the Material Soil Handling System.

The floor of the Primary Unit consists of a system of hoppers and screw conveyors to collect and remove treated soil that has fallen from the belt. This soil is then sent to an outlet conveyor.

The sweep gas for the Primary Unit is provided by volatilized water contained in the incoming contaminated material. The sweep gas flow, within the chamber, is from the feed end toward the discharge end, co-current to material movement.

Off Gas System

Exhaust gases flow from the Primary Unit into a Quench Unit where the hot waste gases are cooled with water sprays upon entering the counter-current packed column scrubber. The quenching process provides a significant amount of gas cleaning because over 99.5% of the gas is condensed.

The gas flow in the scrubber is from the bottom to the top and the water flow in the scrubber is from the top to the bottom (counter-current). This process produces water droplets that collect any fine solids or condensed vapors. The atomized droplets flow to a de-entrainment section (mist eliminator) where liquid droplets are removed from the air. The condensed water and contaminants are gravity fed to the Water Treatment System. The minor amount of off gas is then pulled through the induced draft fan and sent through a vapor phase carbon absorption bed and vented.

Water Treatment System

This system removes the contaminants from the water and cools the water before returning it to the quench unit. The water and contaminants enter the Oil/Water Separator that removes the contaminants through a gravity separation process. The majority of the treated outlet water is sent to an Air/Water Cooler that cools the water from approximately 210 °F to approximately 130 °F. The chilled water is then pumped back to the Quench/Scrubber System.

The remainder of the water (approximately 5%) is sent through a carbon absorption bed and returned to the Primary Unit to cool and re-moisturize the treated soil before it is discharged by the Treated Soil Handling System.

The liquid phase contaminants are removed from the Oil/Water Separator and discharged into drums. The drums are then properly labeled and disposed.

Treated Soil Handling System

The treated soil drops off the belt at the discharge end of the Primary Unit and is sprayed with cooling water to quench the soil. This quenching cools and re-moisturizes the soil to its original moisture content. This quenching also eliminates any potential dust problems. The soil then falls into an auger that conveys it out of the side of Primary Unit onto an enclosed inclined screw conveyor. The conveyor deposits the soil into dump trucks or into product bins for testing and final disposition.

Control System

The TDU is controlled by a computer automated control system which is mounted inside a refrigerated box trailer. A Motor Control Center (MCC) is mounted inside to the trailer and provides switching capability for all process motors. The operators' monitors provide the means for communication between the operators and the control system.

The necessary critical control parameters are monitored to ensure that the process operates within acceptable ranges. Whenever the monitored value of a parameter exceeds a limit value, an alarm condition is generated. The Programmable Logic Controls (PLC) are programmed to respond to alarm conditions with appropriate actions. An alarm condition may result in shutdown of some portion of the process, or may disallow certain interlocked devices to operate, or may activate emergency measures.

PROCESS CAPABILITIES

The Westinghouse TDU is designed to treat material contaminated with PCBs, PAHs, dioxins, heavy metals (mercury), and VOCs. The incoming material should have no more than 5 percent VOCs by weight. The moisture content of the inlet material is recommended to be between 5 and 20 percent moisture by weight. At 5% moisture, additional water must be added because the natural moisture does not allow enough steam for the sweep gas. A moisture content above 20% creates an inefficiency problem with the Primary Unit. However, if the moisture content is above 20% a pre-drying system can be utilized to pre-treat the material. Typically, a pre-screening process is required to screen the material to approximately 2 inches in diameter.

MOBILITY

The TDU is a completely mobile unit with the majority of components mounted on over-the-road trailers. The Primary Unit is the only special haul due to the weight and height. The TDU travels to the site on seven (7) trailers and the Primary Unit. The system is designed to be driven up to the site, set-up, connected to the water and electrical supply, and energized. The set-up time at the site is estimated to be ten (10) days. The TDU requires a flat surface with dimensions no less than 150 feet by 100 feet. The TDU requires a clean water supply of 20 gpm and a power supply of 3 megawatts. The Power Trailer of the TDU has a transformer; therefore, all that is required is access to a utility line carrying 3 megawatts of power.

Figure 4 illustrates the typical site layout for the Westinghouse TDU.

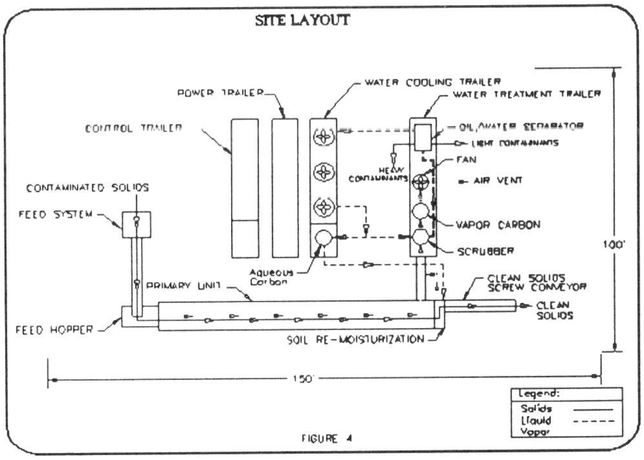

FIGURE 4

RESIDUALS

The Westinghouse TDU is a contaminant separation and concentration process that removes the contaminant from the soil. The contaminant is vaporized from the soil, collected and condensed, and placed into drums for testing and final disposition. No byproducts are created in the thermal desorption process and no contaminants are destroyed in the process.

STATUS OF DEVELOPMENT

Westinghouse has the capability of conducting:

- bench scale treatability testing;
- pilot scale testing; and
- full scale remedial projects.

The bench-scale treatability testing is conducted to determine the technical feasibility of thermal desorption processes for decontaminating hazardous soils. Westinghouse has completed several bench scale treatability studies and is extremely confident in the ability of the bench scale testing program to accurately determine fundamental parameters of the thermal desorption process. Our full-scale equipment has been used to verify performance predictions based on the results obtained in bench-scale testing.

The full-scale equipment can be mobilized quickly and economically to provide this on-site proof of performance pilot testing.

The full scale equipment is currently undergoing some modifications to improve capabilities. These modifications should be completed in late 1993. After completion, additional testing will be conducted by Westinghouse to verify the performance of the TDU.

To date, the Westinghouse TDU full scale unit has successfully completed a series of pilot demonstration tests using a PCB surrogate.

COST OF OPERATION AND EQUIPMENT

Westinghouse understands that for an innovative technology to be competitive in the real world, it must not only perform to the stated requirements, but it must also be cost competitive. The TDU has been designed to be set-up at the site in ten (10) working days. The cost for the set-up at most sites should be less than $100,000. The unit is designed to run with no less than an 80% up-time, processing approximately 10 tons per hour. The cost for the treatment, excluding any pre-treatment, should be approximately between $150 and $250 per ton, with costs being driven by the following parameters:

- volume of material treated;
- moisture content;
- material gradation;
- types of contaminants; and
- contaminant concentration.

CLOSURE

Thermal desorption is emerging as a preferred treatment for CERCLA and RCRA sites. This process removes the contamination from soil affected by volatile and semi-volatile organic compounds and volatile metals. Thermal desorption processes generally heat the contaminated soil then condense and collect the contaminants from the off-gas. The clean soil is then placed back in the excavation and the contaminants are properly disposed.

Westinghouse, like many other companies, has developed the capability to process affected soil utilizing a mobile thermal desorption process. The Westinghouse system differs from those in use by other companies in that it utilizes an off-gas collection system rather than an off-gas incineration system. The use of an off-gas collection system eliminates the majority of the public concern with the potential products of incomplete combustion from an incineration process.

SECTION 2
CFCs & INDOOR AIR QUALITY

Chapter 27

Update on Refrigerant Issues: Accelerated Phase-Out, Safety, and Refrigerant Management

B.B. Lindsay

ABSTRACT

Chlorofluorocarbons (CFCs) have been identified as environmentally damaging compounds that destroy the ozone layer and threaten the existence of life on the planet. Additionally, CFCs are believed to contribute to global warming, with even greater environmental consequences. Public response to these concerns has led to international agreements to limit production and eventually phase out production altogether. The impact of complying with these new agreements and subsequent legislation and regulation is just now being understood by commercial building owners, who use CFC compounds as refrigerant fluids to air condition facilities.

The phase out of CFC production mandates certain actions by facility managers. These will require investment in new equipment, increased maintenance costs, and extensive planning to avoid shortages and building operation disruption. Unfortunately, there is no single "right" answer to the CFC problem. The regulations are changing, new policies are evolving, and new equipment and refrigerant fluids are being developed, which may render today's solution inappropriate in the future. The building owner must evaluate the costs to comply, the costs of not complying, and the risks of numerous alternatives.

BRIEF HISTORY OF CFC AND HCFC PHASEOUT

In 1974, Professors Rowland and Molina published a theory that CFCs deplete the ozone layer. The chlorine in CFC was released in the stratosphere, where it attacked ozone molecules and disrupted the natural equilibrium of oxygen/ozone formation. The long atmospheric life of the CFC compounds permitted these chemicals to make the long 4-7 year transport to the stratosphere without reacting with other chemicals. By 1978, the United States EPA banned the use of CFCs in aerosols. In retrospect, it was only a matter of time before more restrictions were placed on CFCs.

In 1985, public awareness of the CFC issue was ignited by reports of an "ozone hole" over the Antarctic. Public perception, fueled by the extensive press coverage, demanded immediate reaction by elected officials. By 1987, 43 nations had agreed to restrict production of CFCs to 50% of 1986 levels as part of the Montreal Protocol. This represented an even greater than 50% reduction, as CFC use was accelerating rapidly. Users of CFCs would have to convert to alternative compounds that were less damaging to the ozone layer, such as hydrochlorofluorocarbons (HCFCs) and hydrofluorocarbons (HFCs). The HVAC&R industry identified HCFC-22, already in extensive use, as the preferred replacement for a majority of CFC refrigerants.

Scientific studies continued to examine the build up of chlorine in the upper stratosphere and the corresponding depletion of ozone. Atmospheric models indicated that chlorine levels would not return to pre-ozone hole conditions until late in the next century and that more drastic reductions in the production of chlorine compounds was required. In 1990, the United Nations Environmental Program (UNEP) met in London to revise the Montreal Protocol. Over ninety nations agreed to phase out the production of CFCs by the year 2000. Further, the United States Congress amended the Clean Air Act in 1990 to address CFCs. Title VI called for the phaseout of ozone-depleting substances by 2000 and for mandatory recovery and recycling.

In the following years, scientific studies have indicated an even greater concentration of chlorine in the upper atmosphere than had been anticipated. Ozone depletion over the northern hemisphere was discovered. And the Antarctic ozone hole was growing. In 1992, the UNEP meeting in Copenhagen further accelerated the Montreal Protocol. The phase out of CFC production was moved to January 1, 1996. But the four year acceleration would have only minimal impact, and it was deemed prudent to target other chlorine-containing compounds, including HCFCs. The revised Montreal Protocol called for the phase out of HCFCs by the year 2030. The chemicals which had been offered as the preferred replacements for CFCs were now themselves facing withdrawal within 37 years, and possibly much sooner.

The U.S. Environmental Protection Agency (EPA) has announced its proposal to comply with the revised Montreal Protocol. Table 1 summarizes the phase out dates of numerous compounds.

TABLE 1

EPA PROPOSED PHASE OUT SCHEDULE

1996	Cap Production
2003	0%--HCFC 141b
2010	0%--HCFC-142b 0%--HCFC-22 in new products
2020	0%--HCFC-22 0%--HCFC-123 in new products
2030	0%--HCFC-123

It should be noted that various parties will apply pressure to further accelerate the phaseout of chlorine-containing refrigerants. If chlorine levels continue to increase in the upper atmosphere, HCFCs will be targeted. This will be increasingly likely if alternatives become widely available.

A more problematic issue facing the industry is global warming. Regulatory attention to date has been focused on ozone depletion, but the Clinton administration is anticipated to target greenhouse gases and carbon dioxide emissions in the future. CFC, HCFC, and HFC compounds are greenhouse gases and regulations are likely to impact their use.

CFCS AND THE HVAC INDUSTRY

CFCs can be considered the engine of the refrigeration industry. These miracle compounds were developed in the 1930's by General Motors to replace less desirable fluids used in household refrigerators. Ultimately, CFC-11, CFC-12, CFC-113, CFC-114, and CFC-115 were formulated for application in a variety of refrigeration and comfort conditioning equipment. Prior refrigerants were toxic, flammable, inefficient, or expensive, limiting their widespread use. The CFCs were safe, efficient, and cheap.

CFCs were used as refrigerants in automobiles and household refrigerators (R-12), supermarket refrigeration systems (R-12 and R-115), and in large central chilled water systems (R-11, R-12, R-113, and R-114). CFCs were also used as blowing agents to make foam insulation. 38% of the total CFC production in 1985 was used for HVAC&R needs.

Within the commercial building sector, CFCs are used predominantly in chillers. There are approximately 80,000 chillers in the United States, representing 38,000,000 tons of capacity. This large inventory of existing chillers represents approximately 15.2 billion square feet of commercial floor space. Table 2 summarizes the estimated distribution.

TABLE 2

ESTIMATED DISTRIBUTION OF CHILLERS IN THE UNITED STATES

Centrifugal chillers	70%	
CFC-11		67%
CFC-12		12%
CFC-113, 114, 115*		1%
Non-CFC**		20%
Reciprocating and screw chillers***	30%	

* Also includes R-500 and other blends
** Includes HCFC-22, HCFC-123, HFC-134a, and others
*** Use primarily HCFC-22 and R-717

Chillers are initially charged with 2-3 pounds of refrigerant per ton of cooling capacity. A 500-ton chiller will have 1000-1500 pounds of refrigerant. These fluids will be consumed over time, due to leaks, contamination, purging, and venting. Historically, the HVAC&R industry has required 25% of the CFC production to service existing equipment. That was a minor expense 10 years ago, when

CFCs were cheap and available. However, the phase out of CFC production in 1996 mandates radical changes in chiller maintenance. CFC refrigerants will be available in only very limited supply, mostly through recovery and reclamation, and will be exorbitantly expensive. Furthermore, environmental regulations will dictate acceptable service practices to minimize the release of CFCs to the atmosphere.

The HVAC&R industry must address three major issues:

(1) What must be done to comply with the Clean Air Act?

(2) If CFCs are no longer going to be produced, what management practices are required to continue to provide cooling to tenants and processes?

(3) What air conditioning and refrigeration equipment should be purchased in the future?

COMPLIANCE WITH TITLE VI, CLEAN AIR ACT

The U.S. Clean Air Act was amended in 1990 with provisions to protect the stratospheric ozone layer. In addition to the phaseout of CFC production by January 1, 2000, the legislation prohibited the intentional venting of ozone-depleting chemicals and required their recovery, recycling, and reclamation for stationary and mobile equipment. It also restricted the sales of compounds. The U.S. Environmental Protection Agency (EPA) has established regulations on recovery equipment, certification of technicians, labeling, repair, and reclamation, and additional or revised regulations are pending.

One key element on compliance must be stressed: owners are permitted to continue using CFCs and HCFCs beyond the phaseout dates. Owners must ensure that all service is conducted by certified technicians, using EPA-approved recovery devices. Failure to recover refrigerants may expose owners to a $25,000 fine per occurrence. Owners must fix leaks if the annual loss exceeds more than 15% of the refrigerant charge (35% for industrial uses or commercial refrigeration).

It is suggested that owners install "containment" systems, such as high efficiency purges, relief valves, and brazed fittings, on chillers to minimize losses, but these are not required by law (yet). Much of the equipment improvements will be motivated by the higher cost of refrigerant service fluids and uncertainty in their availability. A list of possible containment options is included in Table 3.

TABLE 3

CONTAINMENT OPTIONS FOR LOW AND HIGH PRESSURE CHILLERS

Leak testing/repair
Fittings and valves
Refrigerant oil monitors
High efficiency purge
Refrigerant monitors
Pressurization
Recovery systems

Government agencies and HVAC manufacturers are suggesting that owners convert or replace chillers to utilize non-ozone depleting refrigerants. This is not required by law; owners can continue to use CFCs. Conversion is not a simple task. Materials and lubricants need to be replaced, and modifications to compressors and heat exchanger tubes are required to retain capacity and efficiency. Additionally, some of the new refrigerants require major changes to the equipment room to comply with safety codes. Conversions should cost approximately 50%-60% of a new installation, but site-specific conditions will ultimately dictate costs.

Replacement is an attractive option to some owners, if they have sufficient capital to invest. In addition to the myriad of chillers using ozone-friendly refrigerants (HCFC-22, HCFC-123, and HFC-134a), owners may consider ammonia, absorption, and cool storage. There is no ideal refrigerant choice; each fluid faces compromise. The HCFCs are scheduled for phaseout in either 2020 or 2030, since they contain chlorine (less than CFCs) and deplete ozone. HFCs have high global warming potential and may face future restrictions or costs. Ammonia has toxicity and flammability concerns. Absorption is inefficient and requires larger cooling towers, increasing energy consumption and contributing to global warming.

Faced with the uncertainties of regulatory restrictions and the continuing development of improved refrigerants and cooling technology, many building owners are electing to manage their existing refrigerants and defer the conversion/replacement decision. Chillers are tightened to reduce losses, and selected units are converted or replaced to provide refrigerants to service the remaining equipment. The refrigerant fluids can be stored at "refrigerant banks". Ultimately, all of the chillers will be converted, replaced, or abandoned. Figure 1 shows the CFC management plan at a Texas Instruments facility. This approach minimizes capital investment with acceptable levels of risk. However, it should be noted that the facility is relying on its current inventory of refrigerant to meet its future needs. Owners that assume that reclaimed refrigerant will be available may

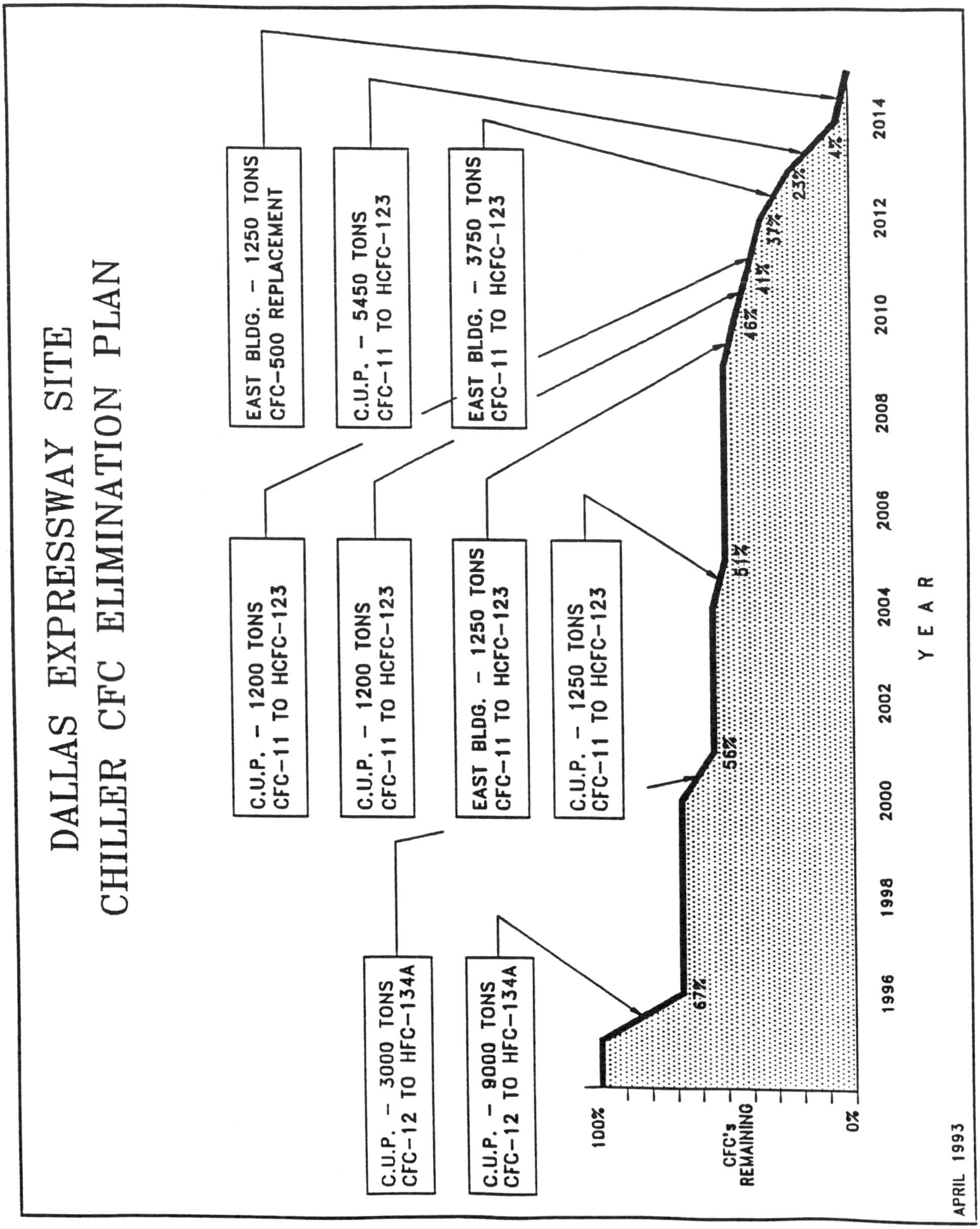

FIGURE 1

CFC REFRIGERANT MANAGEMENT PLAN

be disappointed. Also, owners planning a conversion or replacement in the future may find it difficult to procure the equipment or contractor services.

One benefit of the CFC/HCFC phaseout is the development of more efficient air conditioning systems. Manufacturers are currently offering chillers that require less energy and simultaneously reduce the emissions of refrigerants. One manufacturer is marketing an HCFC-123 chiller that achieves an efficiency of 0.55 kW/ton or a COP of 6.4 at ARI standard conditions. It is anticipated that further improvements will be available.

REFRIGERANT SAFETY

One of the key reasons for the development of CFCs was safety. In order to be widely accepted, the refrigerant fluids had to be non-toxic, non-flammable, efficient, and inexpensive. Additionally, the fluids had to be stable and operate reliably for long periods. Today's new replacement refrigerants must meet even higher expectations for safety and subsequent liability. Extensive testing must be conducted prior to commercial introduction, requiring 3-7 years. Perhaps the most important aspects of this testing are flammability and toxicity.

The refrigeration industry has been aware of the environmental problems with CFCs for nearly 20 years and has been pursuing alternatives. Two compounds, HCFC-123 and HFC-134a, have been developed as alternatives to CFC-11 and CFC-12 respectively. Both chemicals have undergone extensive testing. The results of the toxicity tests on HCFC-123 have caused concern within the industry and some confusion.

Laboratory mice, exposed to HCFC-123 at 5,000 ppm for 30 hours per week for two years, simulating exposure levels for maintenance workers, developed non-malignant tumors in liver and reproductive organs. In order to safeguard employees and technicians, allowable exposure levels (AEL) were established for HCFC-123 at 10 ppm time-weighted average (TWA) and special requirements for equipment room design were adopted by ASHRAE. Some members of the HVAC industry, concerned with the liability issue, decided to withdraw sales and/or service of equipment using HCFC-123. The U.S. Environmental Protection Agency and some manufacturers have conducted tests to determine if equipment with HCFC-123 could achieve the required safety levels and demonstrated levels below 0.6 ppm. Currently, Dupont and Allied Signal are offering HCFC-123 for sale, Trane and York are shipping chillers charged with HCFC-123, and the EPA has provided assurances that HCFC-123 is safe.

Laboratory mice, exposed to HFC-134a at 50,000 ppm, also developed non-malignant tumors. The exposure levels have been established as 1,000 ppm for HFC-134a, comparable to R-11 and R-12.

Several other refrigerants are available to replace CFCs:

- Ammonia exhibits zero ozone-depletion and zero global warming potentials. It is, however, toxic and flammable. Special equipment room requirements are mandated and maintenance requirements would suggest judicious application in commercial buildings.
- Blends of HFC-32 are being considered as replacements for HCFC-22. However, HFC-32 is flammable and special service practices will have to be developed to address the separation of multiple fluid blends.
- Water can be used as a refrigerant in an absorption chiller. While water is non-toxic, safety issues with the fuel or steam used to operate the chiller must be considered.

ASHRAE has published Standard 32-1992 to facilitate the safety classification of refrigerants and Standard 15-1992 to design equipment rooms. All of the refrigerants are considered safe as long as they comply with these standards. Table 4 lists various refrigerants and their safety classification and exposure levels.

TABLE 4

SAFETY CLASSIFICATION AND EXPOSURE LEVELS OF REFRIGERANTS

Refrigerant	Chemical Formula	ODP	GWP (CO_2)	GWP (CFC-11)	Safety Group	Exposure
CFC-11*	CCl_3F	1.0	1500	1.0	A1	1000 TLV
CFC-12*	CCl_2F_2	1.0	4500	2.09	A1	1000 TLV
CFC-113	CCl_2FCClF_2	0.8	2100	1.4	A1	
CFC-114	$CClF_2CClF_2$	0.7	5500	3.67	A1	
CFC-115	$CClF_2CF_3$	0.4	7400	4.93	A1	
HCFC-22*	$CHClF_2$	0.05	510	0.43	A1	1000 TLV
HCFC-123*	CF_3CHCl_2	0.02	29	0.02	B1	10 AEL
HCFC-124	$CHClFCF_3$	0.02	150	0.12	--	500 AEL
HCFC-141b	CH_3CCl_2F	0.10	150	0.1	--	
HCFC-142b	CH_3CClF_2	0.06	540	0.36	A2	
HFC-32	CH_2F_2	0	--	0.15	--	1000 TLV
HFC-125	CHF_2CF_3	0	860	1.00	--	1000 AEL
HFC-134a*	CF_3CH_2F	0	420	0.34	A1	1000 AEL
HFC-143a	CH_3CF_3	0	1000	1.1	--	1000 AEL
HFC-152a	CH_3CHF_2	0	47	0.04	A2	1000 AEL
R-717*	NH_3	0	0	0	B2	25 TLV
R-718*	H_2O	0	0	0	A1	
HC-290	C_3H_8	0	0	0	A3	1000 PEL

*dominant choice

SUMMARY

CFCs and HCFCs are being phased out on an accelerated schedule. There is a need to act immediately, but owners should not panic. Alternatives are available that are safe and efficient. Containment can extend the life of existing systems with minimal risk. Planned conversions or replacements of chillers will ultimately be necessary, but their original CFC refrigerant charge can be used to service remaining units. Financial and management investment is required.

Chapter 28

Performance and Efficiency Evaluations of 11 Non-CFC Refrigerants

G.S. Kazachki, C.L. Gage

ABSTRACT

As a result of the mandated phaseout of ozone-depleting compounds, researchers have been reevaluating a variety of chemicals as potential replacement refrigerants. In order to reduce the number of experimental tests, 11 non-chlorofluorocarbon (CFC) potential refrigerants are theoretically evaluated in this work. An analytical method using a limited set of property data is used as a tool. The performance of these refrigerants is determined for a variety of vapor-compression cycles which are thermodynamic analogies to existing systems. Through separation of the basic vapor-compression cycle into elemental cycles representing only throttling losses or only compression superheating losses, important information is drawn about the design implications of subcooling or superheating the refrigerant. The role of internal heat exchange between the liquid refrigerant before the expansion valve and refrigerant vapor after the evaporator for reducing throttling losses with resulting improvement in the overall cycle efficiency is evaluated for all refrigerants. It is shown through this analysis that alternative, as well as standard, refrigerants fall into two general performance classes. These classes result in important performance and efficiency information which can be used to develop a vapor-compression system design tailored to the refrigerant class.

INTRODUCTION

In the United States, the phaseout of ozone depleting substances is governed by the Montreal Protocol and the Clean Air Act Amendments of 1990. These regulations set into motion the identification of substances which have properties similar to those of the chlorinated refrigerants, but do not deplete the stratospheric ozone layer. The primary alternatives which were identified include fluorinated methanes, ethanes, and propanes (designated HFC or FC); hydrocarbons (designated HC); and fluorinated ethers (designated HFE).

The phaseout refrigerants analyzed and compared in this work include: CFC11, CFC12, HCFC22, CFC114, HCFC123, and R502, an azeotrope of HCFC22 and CFC115. Also evaluated are some of the proposed alternatives including HFC32, HFC125, HFC134, HFC134a, HFC143a, HFC152a, FC218, HC290 (propane), and HFE134. Table 1 presents some of the properties of these 15 refrigerants.

METHOD OF EVALUATION

The analytical expressions for cooling capacity, work, cycle efficiency, temperature after isentropic compression, and efficiency and capacity enhancement from internal heat exchange are used for the thermodynamic evaluation of refrigerants in various cycles. These expressions use limited property data including vapor heat capacity, liquid heat capacity, latent heat of vaporization, and thermal expansion coefficient. The complete derivations have been presented in other works [1,2] and have been applied to the evaluation of both pure and azeotropic refrigerants[3,4,5]. One aspect of the concept is that the refrigerant cycle should be designed around the properties of the refrigerants, and that the refrigerants should be compared using cycles which are appropriate for each refrigerant's properties. Analysis of a variety of refrigerants indicates that they can be divided into two general classes based on the sign of a dimensionless characteristic parameter, κ, which is equal to $(\beta\lambda/c_p - 1)$. Various cycle configurations will be applicable to the refrigerant depending on the value of κ. In the theoretical analysis, the cycle with throttling is applicable to both classes of refrigerants. This cycle can be used to quantify the losses associated with the isenthalpic throttling process between condensation and evaporation while leaving the remainder of the cycle as ideal.

Refrigerants with positive values of κ have a negative

slope of the tangent line to the saturated vapor curve of the two-phase dome in the temperature-entropy diagram. During isentropic compression of saturated vapor, these refrigerants will become superheated. Cycles which apply to this class of refrigerants are the cycle with superheating and the cycle with throttling and superheating. (The cycle with superheating can be used to quantify the losses associated with the compression superheating process while leaving the remainder of the cycle ideal.) Refrigerants with negative values of κ have a positive slope to the tangent line of the saturated vapor curve of the two-phase dome in the temperature-entropy diagram. During isentropic compression of saturated vapor, the process will enter the two-phase region resulting in wet compression. Cycles which apply to this class of refrigerants are the cycle with throttling and dry compression and the cycle with throttling and dry compression provided by internal heat exchange. In these cycles, the minimum amount of superheat is applied to the saturated vapor at the evaporator pressure in order to bring the compressed vapor just to the point of saturated vapor at the condenser pressure. A second cycle which is applicable to both classes of refrigerants is the cycle with internal heat exchange where the condensed liquid is used to superheat the vapor before the compression process.

In order to quantify the efficiency of a given cycle relative to ideal, the individual coefficients of performance (COP) of the cycles have been divided by the COP of the Carnot cycle. The following expressions will be used in the subsequent figures:

$EOC_T = COP_T/COP_C$ = the efficiency of the cycle with throttling;

EOC_{TS} = the efficiency of the cycle with throttling and superheating;

EOC_{TD} = the efficiency of the cycle with throttling and dry compression;

EOC^{IX}_{TD} = the efficiency of the cycle with throttling and dry compression provided by internal heat exchange;

EOC^{IX} = the efficiency of the cycle with internal heat exchange;

$EE^{IX} = EOC^{IX}/EOC_{TS}$ = the efficiency enhancement from internal heat exchange relative to the cycle with throttling and superheating for refrigerants with positive κ, or

$= EOC^{IX}/EOC_{TD}$ = the efficiency enhancement relative to the cycle with throttling and dry compression for refrigerants with negative κ;

VC_{TS} = the volumetric capacity for the cycle with throttling and superheating;

VC_{TD} = the volumetric capacity for the cycle with throttling and dry compression;

VC^{IX} = the volumetric capacity for the cycle with internal heat exchange; and

$VCE^{IX} = VC^{IX}/VC_{TS}$ = the volumetric capacity enhancement from internal heat exchange relative to the cycle with throttling and superheating for refrigerants with positive κ, or

$= VC^{IX}/VC_{TD}$ = the volumetric capacity enhancement relative to the cycle with throttling and dry compression for refrigerants with negative κ.

RESULTS AND DISCUSSION

The analysis of the 15 refrigerants is performed in a wide range of temperature conditions. Results are presented for temperature conditions corresponding to the specific applications of each refrigerant (e.g., low temperature systems such as supermarkets, refrigerator/freezers, chillers, air conditioners, and heat pumps). Correspondingly, refrigerants have been classified as low, medium, or high temperature refrigerants. While a comprehensive set of graphs has been generated for all refrigerants, this work presents a complete set only for the low temperature refrigerants. For the medium and high temperature refrigerants, selected graphs have been chosen as highlights. In all figures, open symbols indicate refrigerants with chlorine while solid symbols indicate non-chlorine containing refrigerants.

From earlier work, it has been concluded that the relationship between performance and κ has the following trends as κ decreases: throttling losses increase, superheating losses decrease, discharge temperature decreases, required minimum superheat increases, and the efficiency and volumetric capacity enhancements from internal heat exchange increase[5]. Figures 1 and 2 show the values of the characteristic parameter, κ, for the 15 refrigerants.

Figures 3 through 11 show the results for the seven low temperature refrigerants evaluated at 43°C condensing and in the evaporating range -45°C to 7°C which are applicable conditions for low temperature systems, refrigerator/freezers, supermarkets, heat pumps, and chillers. One of the refrigerants, FC218, has a negative κ, and as a result has the highest throttling losses (Figure 3). For refrigerants with positive κ, the losses associated with superheating from the compression process are shown in Figure 4. Comparing Figures 3 and 4 it can be noted that the magnitude of superheating losses is smaller than that of the throttling losses. In addition, as expected, refrigerants with the greatest superheating losses (lowest efficiencies in Figure 4) have the smallest throttling losses (highest efficiencies in Figure 3). Results are shown in Figure 5 for the basic vapor compression cycle which includes both throttling and superheating losses. For FC218, which has negative κ, the basic cycle is throttling with dry compression, and this is also shown in Figure 5 along with the cycle with throttling and dry compression provided by internal heat exchange ($FC218_{IX}$). In the basic cycle, nothing outperforms HCFC22, while both HFC32 and HC290 (propane) outperform R502. At -45°C, HFC32 is 3 percent poorer than HCFC22 but 13 percent better than R502, while propane is approximately 8 percent poorer than HCFC22 but 7 percent better than R502. The poorest performer is FC218. Even when the dry compression is provided by internal heat exchange, the additional capacity does not bring the performance of FC218 up to that of the other alternatives. The volumetric capacities in the basic cycles are shown in Figure 6. The capacity of HFC143a is identical to that of R502 and slightly lower than that of HCFC22. HFC32 has the highest capacity and, thus, will require the smallest displacement compressor. Another important characteristic to monitor in compressors is the temperature after compression which is shown in Figure 7. Also shown is the minimum superheat required for FC218 to avoid wet compression. It can be seen that HFC32, the refrigerant with the largest κ, has the highest isentropic compression temperatures. This refrigerant could have compressor reliability problems at some operating conditions.

For refrigerants with small or negative κ, the application of internal heat exchange to subcool the liquid from the condenser by superheating the vapor from the evaporator can significantly improve efficiency. Figure 8 shows the efficiency enhancement from the application of internal heat exchange to superheat the vapor to 7°C. Similar values were calculated for the volumetric capacity enhancement. As expected, FC218 gains the most from internal heat exchange. Results are not presented for this refrigerant at either -7°C or +7°C because superheating the vapor to 7°C is insufficient to provide the minimum superheat required to avoid wet compression as evidenced in Figure 7. It is important to note in Figure 8 that the refrigerants with the largest κs have efficiency enhancements less than 1; that is, the application of internal heat exchange can be detrimental to the performance of some refrigerants. This observation strongly supports the assertion that the cycle should be tailored to the properties of the refrigerant in order to get the best performance from the system.

The efficiency, volumetric capacity, and temperature after isentropic compression for the cycle with internal heat exchange are shown in Figures 9 through 11. In this cycle, the performance of propane is equivalent to that of HCFC22 at low evaporating conditions, while due to the relevant effects of internal heat exchange, HFC32 now performs about 2 percent more poorly than R502. The performance of FC218 has been significantly improved; however, its performance is still 16 to 18 percent lower than that of R502. Volumetric capacities are almost identical for several refrigerants in this cycle including HCFC22, R502, HFC143a, and HFC125. Isentropic compression temperatures in the cycle with internal heat exchange are significantly higher particularly for HFC32 and HCFC22.

Highlight results for the medium temperature refrigerant CFC12 and its potential alternatives are shown in Figures 12 through 17. Since these refrigerants have positive κ, Figures 12 and 13 show the performance in the basic cycle with throttling and superheating. The best alternative is R152a which performs about 3 percent better than CFC12. Volumetric capacities of both HFC152a and HFC134a are close to those of CFC12. The efficiency enhancement from internal heat exchange is shown in Figure 14 where the vapor is superheated from the evaporating temperature to 7°C. (Thus, there is no superheat from internal heat exchange at the 7°C evaporating condition.) Gains are less than 6 percent for three refrigerants; however, for HFC152a there is practically no noticeable effect of internal heat exchange. The improvement for both HFC134a and HFC134, as shown in Figure 15, brings their performances to within 1 percent of those of CFC12, while the ranking of the volumetric capacities remains unchanged from the basic cycle. The final graph in this set shows that the temperatures after isentropic compression for all alternatives will be acceptable.

Highlight results for high temperature refrigerants are shown in Figures 18 through 22. As seen in Figure 2, two of the refrigerants, CFC114 and HCFC123, have negative κ; thus, their basic cycle is the cycle with throttling and dry compression compared to the basic cycle of throttling and superheating for refrigerants with positive κ. The minimum superheat required to avoid wet compression ranged from 7.5C° at 40°C condensing to 19C° at 110°C condensing for CFC114 and from 1C° at 40°C condensing to 7.5C° at 110°C condensing for HCFC123. Although minimum superheat is needed, compressor inefficiencies in actual operation of CFC114 systems tend to superheat the vapor without requiring a design to provide the superheat. Since wet compression is not observed for CFC114, it would not be expected for HCFC123 which requires less superheat.

The efficiencies for the refrigerants in their respective cycles are shown in Figure 18. At low condensing temperatures which represent chiller conditions, the performance of HCFC123 is within a few percent of that of CFC11, while HFE134 is about 6 percent better than CFC114. Also shown in this graph are the efficiencies for refrigerants with negative κ in the cycle with dry compression provided by internal heat exchange ($CFC114_{IX}$ and $HCFC123_{IX}$). Comparing the broken lines with the dashed lines for a given refrigerant shows that the improvement caused by providing the minimum superheat with internal heat exchange is not as great for HCFC123 as it is for CFC114. Volumetric capacities for the refrigerants are shown in Figure 19. At 40°C condensing, the volumetric capacity of HCFC123 is 18 percent lower than that of CFC11 indicating that a larger compressor would be required; while the capacity of HFE134 is 10 percent higher than that of CFC114 for its cycle with dry compression and 4 percent higher for its cycle with dry compression provided by internal heat exchange. Figure 20 shows the efficiency enhancement from internal heat exchange where the vapor is superheated from 4°C evaporating to a temperature which is 5C° below the condensing condition. As expected, the refrigerant which gains the most is CFC114 which has the most negative κ. From Figures 21 and 22, the efficiency of HCFC123 now exceeds CFC11 while the capacity still remains lower, and the efficiency of HFE134 is now poorer than CFC114 while their volumetric capacities are nearly identical.

CONCLUSIONS

The following are based on a theoretical analysis of thermodynamic properties and do not consider other issues (e.g., flammability):

1. For low temperature applications, HFC32 deserves technical consideration as a replacement for R502; however, internal heat exchange should not be used. Smaller displacement compressors would be required, and care should be taken to protect the compressor from potentially high discharge temperatures. If discharge temperatures are a concern, HFC143a could be considered, but internal heat exchange will be needed for this refrigerant in order to bring its performance close to that of R502. Compressor size is unchanged for HFC143a relative to that for R502.

2. For systems using HCFC22, HC290 (propane) deserves technical consideration as a replacement. Internal heat exchange is recommended with this refrigerant. Across a range of applications, performance should closely match that of HCFC22, while compressor sizes need be only slightly larger. Propane should not have a problem with high discharge temperatures.

3. For medium temperature applications, HFC152a, HFC134a, and HFC134 all deserve technical consideration as replacements for CFC12. HFC152a will not need internal heat exchange, while both HFC134a and HFC134 will require internal heat exchange in order to bring their performance up to match that of CFC12. Compressor displacement for HFC152a and HFC134a will be similar to that for CFC12, while HFC134 will require a larger displacement compressor.

4. For high temperature applications using CFC114, HFE134 deserves technical consideration as a replacement. A smaller displacement compressor will be required, and performance should be improved without the addition of internal heat exchange.

5. For high temperature applications using CFC11, HCFC123 is already being successfully used as a replacement. Compressor sizes have been changed, as expected.

NOMENCLATURE

Symbols: β - logarithmic mean thermal expansion coefficient of saturated vapor between evaporating and condensing conditions; c_p - logarithmic mean specific heat of saturated vapor between evaporating and condensing conditions; λ - logarithmic mean latent heat of vaporization between evaporating and condensing conditions; p_{CR} - critical pressure; Tc - condensing temperature; T_{CR} - critical temperature; and T_{nbp} - normal boiling point at atmospheric pressure.

REFERENCES

1. Kazachki, G.S. "Derivation of Dimensionless Parameters for Thermodynamic Evaluation of Refrigerants in Vapor-Compression Cycles," Proceedings of the XVIIIth International Congress of Refrigeration, August 10-17, 1991, Montreal, Canada.

2. Kazachki, G.S. "Criteria for Evaluation of the Efficiency Improvements in Vapor Compression Cycles Using Internal Heat Exchange," Proceedings of the 16th World Energy Engineering and 2nd Environmental Technology Congress, October 26-28, 1993, Atlanta, GA.

3. Kazachki, G.S. and C.L. Gage, "Thermodynamic Evaluation of Five Alternative Refrigerants in Vapour Compression Cycles," Proceedings of the XVIIIth International Congress of Refrigeration, August 10-17, 1991, Montreal, Canada.

4. Kazachki, G.S. and C.L. Gage, "Thermodynamic Evaluation of Predicted Fluorinated Ether, Ethane, and Propane Azeotropes," Proceedings of the 1992 International Refrigeration Conference - Energy Efficiency and New Refrigerants, Vol. II, July 14-17, West Lafayette, IN.

5. Gage, C.L. and G.S. Kazachki, "Performance and Efficiency Evaluations of New Fluorinated Ethers, Propanes, and Butanes," Proceedings of the 16th World Energy Engineering and 2nd Environmental Technology Congress, October 26-28, 1993, Atlanta, GA.

TABLE 1: REFRIGERANTS AND THEIR BASIC PROPERTY DATA
(Italics indicate refrigerants which contain chlorine.)

Chemical Code	Chemical Formula	Mol. Weight	T_{nbp}	T_{CR}	p_{CR}
		kg/kg-mol	°C	°C	kPa
HFC32	CH_2F_2	52.00	-52.0	78.4	5814
HFC125	CHF_2-CF_3	120.03	-48.6	66.3	3631
HFC143a	CH_3-CF_3	84.04	-47.6	73.1	3760
R502	*HCFC22/CFC115*	*111.63*	*-45.4*	*82.2*	*4074*
HC290	CH_3-CH_2-CH_3	44.10	-42.1	96.8	4257
HCFC22	*$CHClF_2$*	*86.47*	*-40.8*	*96.2*	*4990*
FC218	CF_3-CF_2-CF_3	188.00	-36.7	71.9	2680
CFC12	*CCl_2F_2*	*120.93*	*-29.8*	*112.0*	*4116*
HFC134a	CH_2F-CF_3	102.03	-26.1	101.6	4056
HFC152a	CHF_2-CH_3	66.05	-24.2	113.3	4520
HFC134	CHF_2CHF_2	102.03	-19.8	118.9	4562
CFC114	*$CClF_2$-$CClF_2$*	*170.94*	*3.6*	*145.7*	*3261*
HFE134	CHF_2-O-CHF_2	118.03	6.3	147.1	4228
CFC11	*CCl_3F*	*137.38*	*23.8*	*198.0*	*4409*
HCFC123	*$CHCl_2$-CF_3*	*152.93*	*27.9*	*183.8*	*3674*

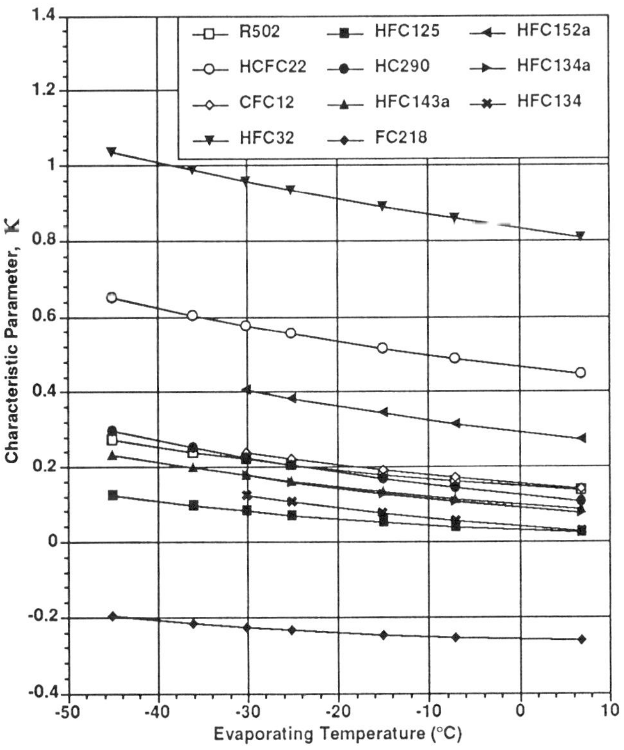

Figure 1: κ for Low and Medium Temperature Refrigerants at 43°C Condensing Temperature.

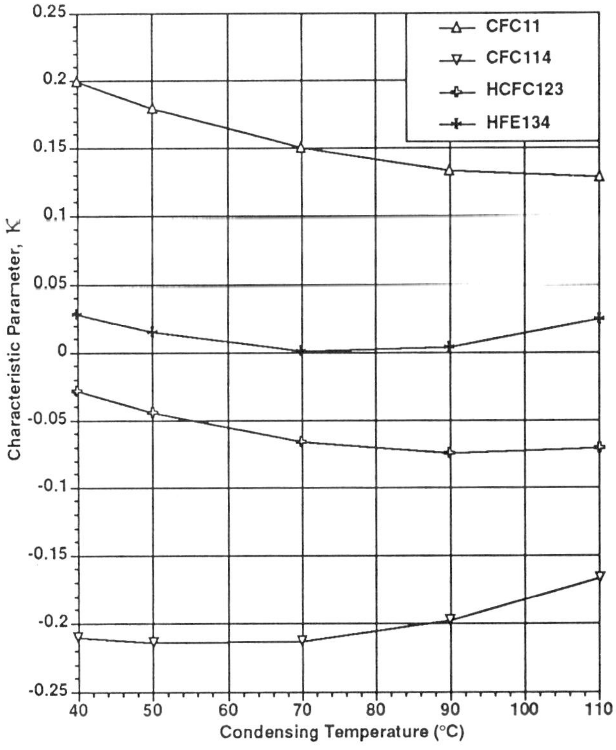

Figure 2: κ for High Temperature Refrigerants at 4°C Evaporating Temperature.

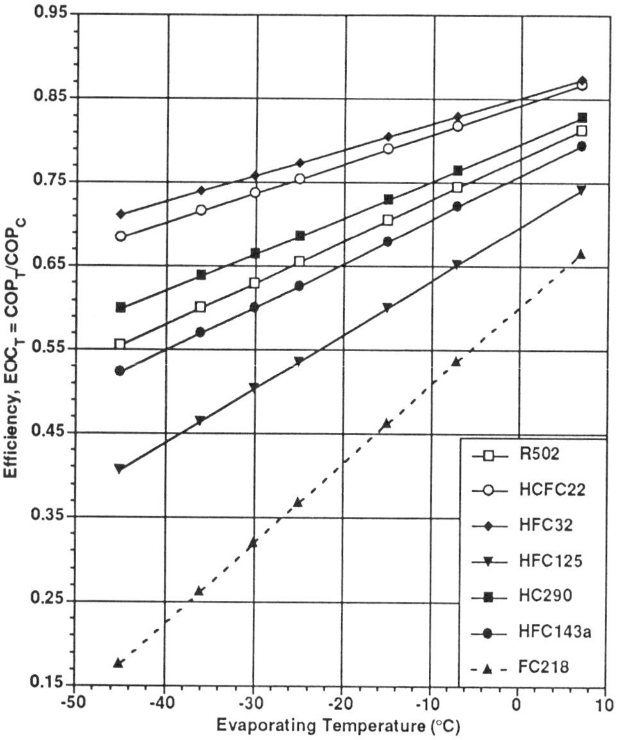

Figure 3: Efficiency of the Cycle with Throttling at 43°C Condensing Temperature. (Dashed line is $\kappa < 0$.)

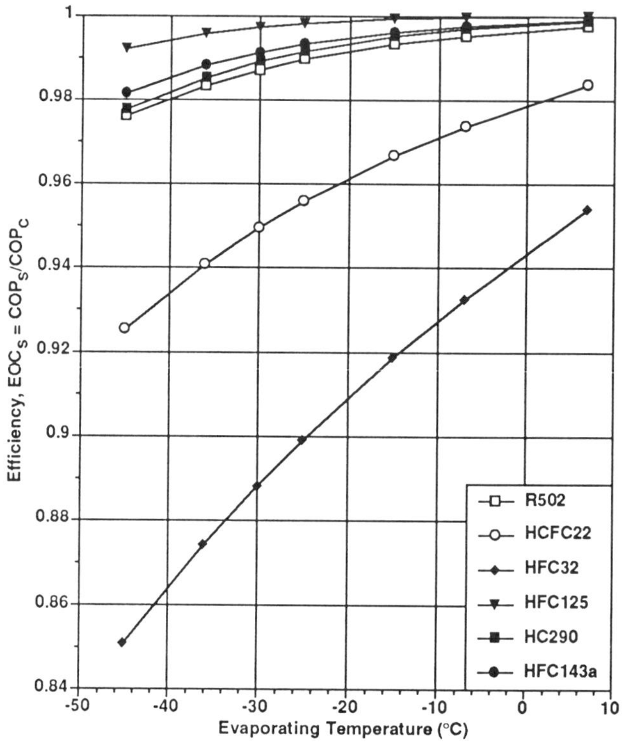

Figure 4: Efficiency of the Cycle with Superheating at 43°C Condensing Temperature.

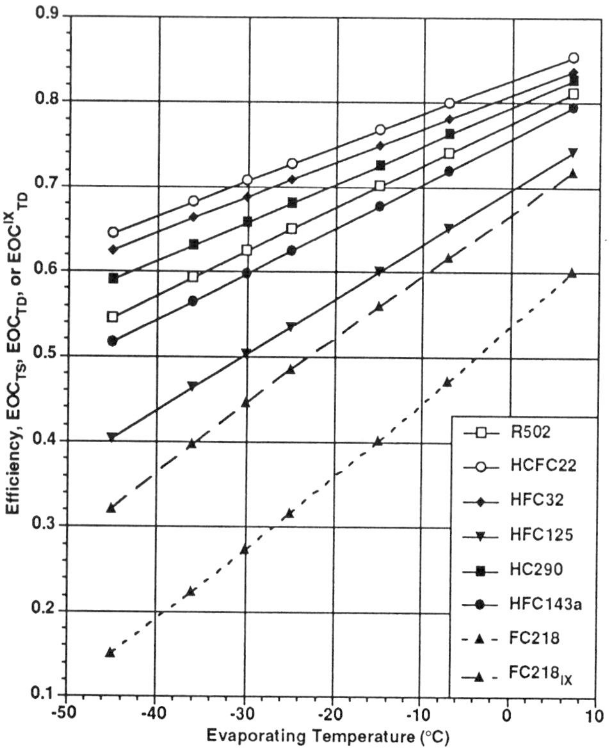

Figure 5: Efficiency of the Cycle with Throttling and Superheating, the Cycle with Dry Compression, and the Cycle with Dry Compression from Internal Heat Exchange at 43°C Condensing Temperature. (Dashed lines are $\kappa < 0$.)

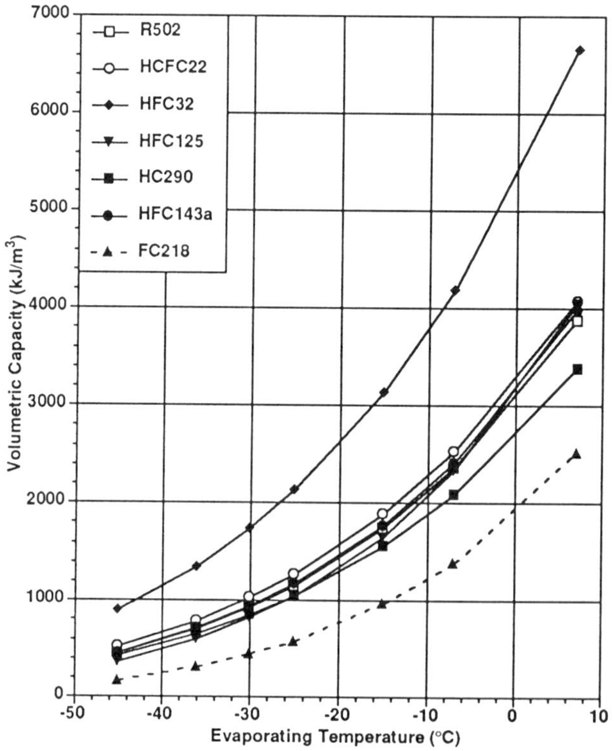

Figure 6: Volumetric Capacity for Cycle with Throttling and Superheating or Dry Compression at 43°C Condensing Temperature. (Dashed line is $\kappa < 0$.)

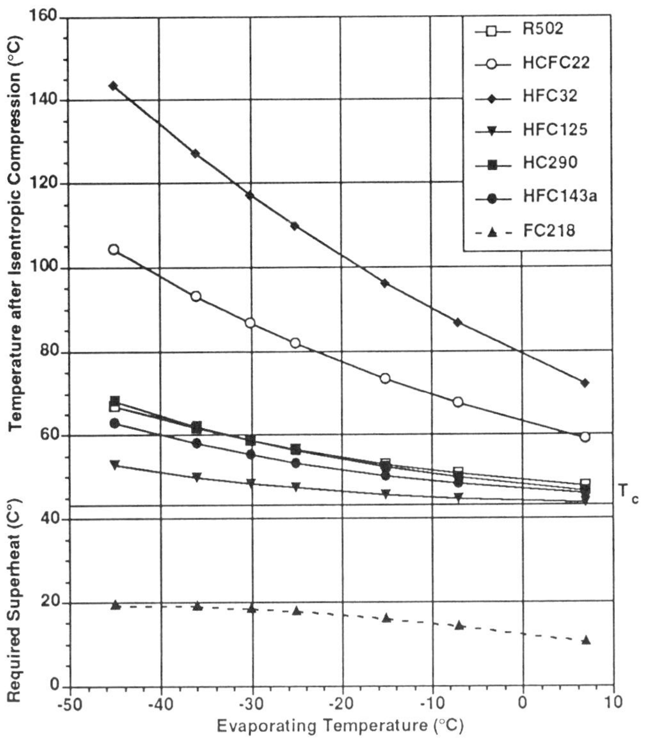

Figure 7: Temperature after Isentropic Compression for the Cycle with Throttling and Superheating and Required Superheat for the Cycle with Dry Compression at 43°C Condensing Temperature. (Dashed line is $\kappa < 0$.)

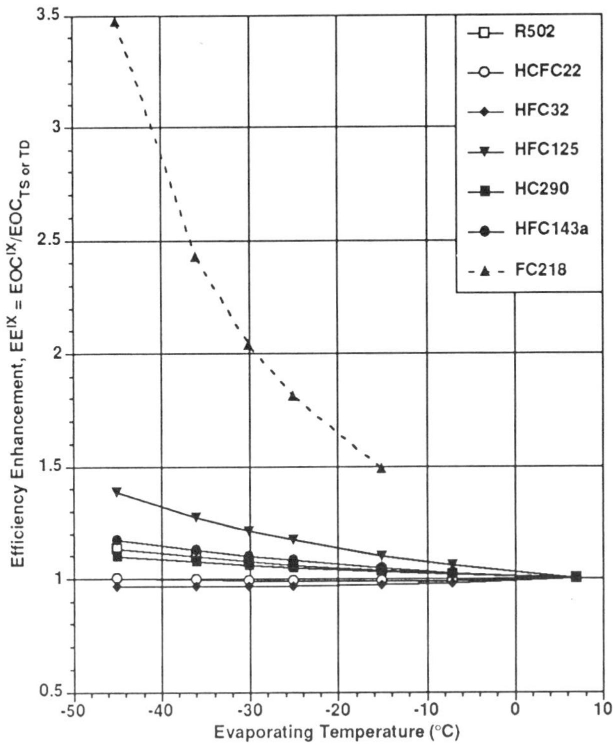

Figure 8: Efficiency Enhancement for the Cycle with Internal Heat Exchange at 43°C Condensing Temperature. Vapor Heated to 7°C. (Dashed line is $\kappa < 0$.)

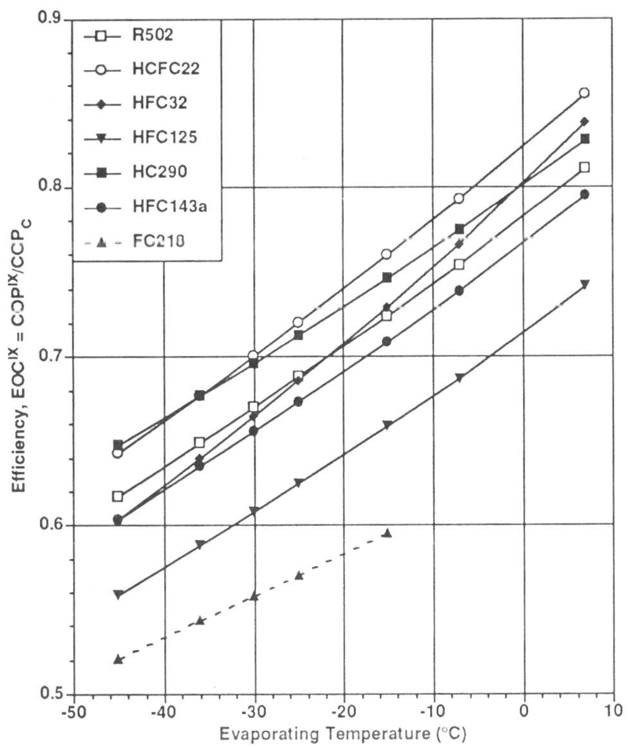

Figure 9: Efficiency of the Cycle with Internal Heat Exchange at 43°C Condensing Temperature. Vapor Heated to 7°C. (Dashed line is $\kappa < 0$.)

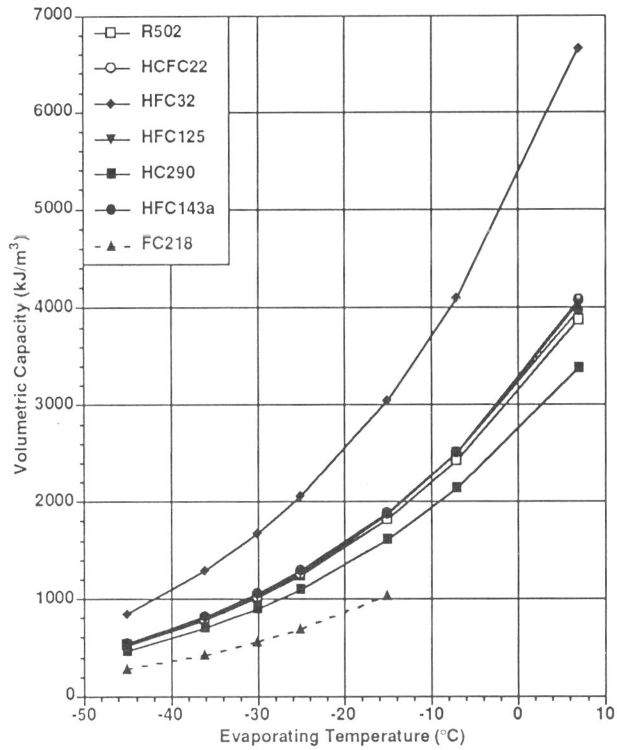

Figure 10: Cycle with Internal Heat Exchange at 43°C Condensing Temperature. Vapor Heated to 7°C. (Dashed line is $\kappa < 0$.)

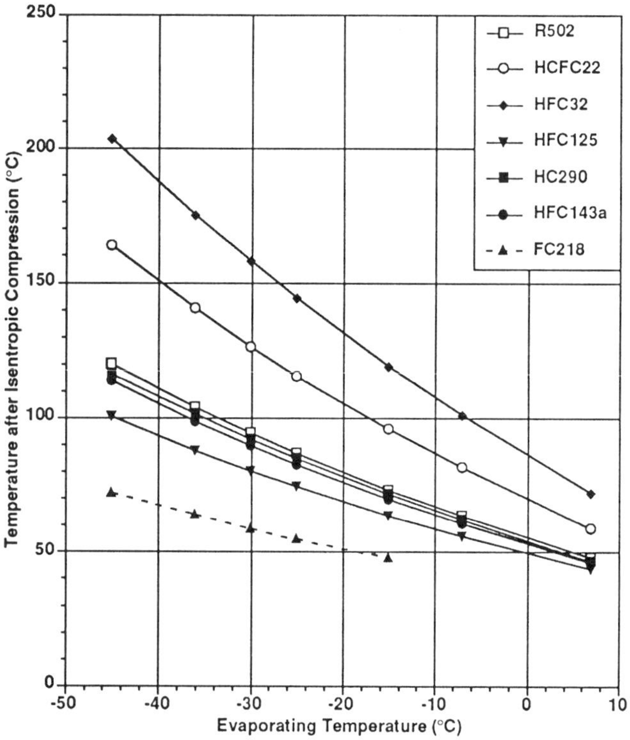

Figure 11: Temperature after Isentropic Compression for the Cycle with Internal Heat Exchange at 43°C Condensing Temperature. Vapor Heated to 7°C. (Dashed line is $\kappa < 0$.)

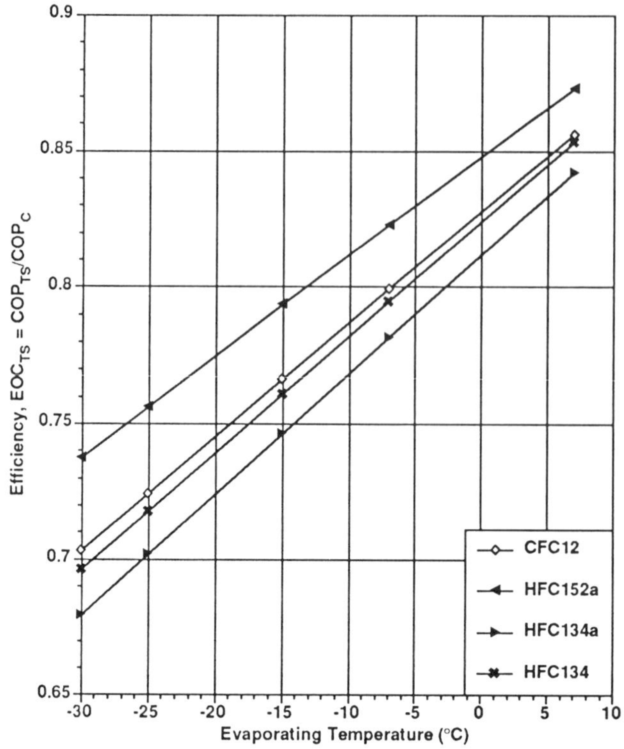

Figure 12: Efficiency of the Cycle with Throttling and Superheating at 43°C Condensing Temperature.

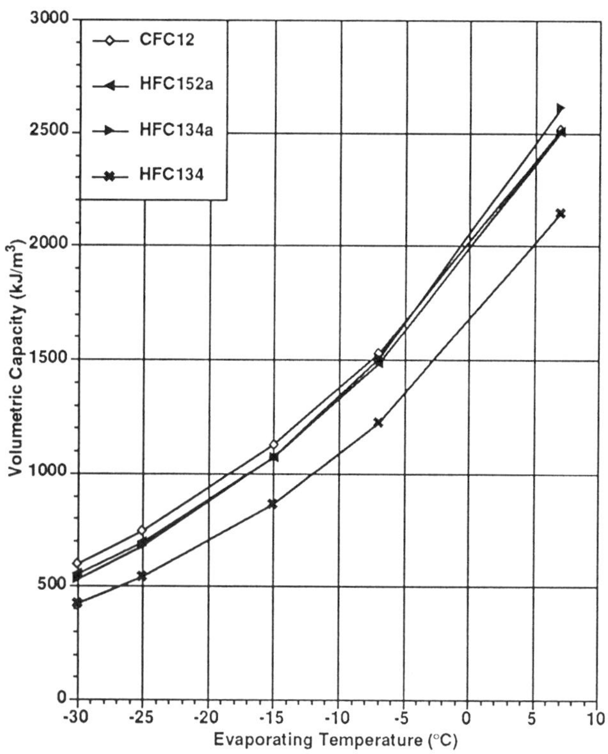

Figue 13: Volumetric Capacity for Cycle with Throttling and Superheating at 43°C Condensing Temperature.

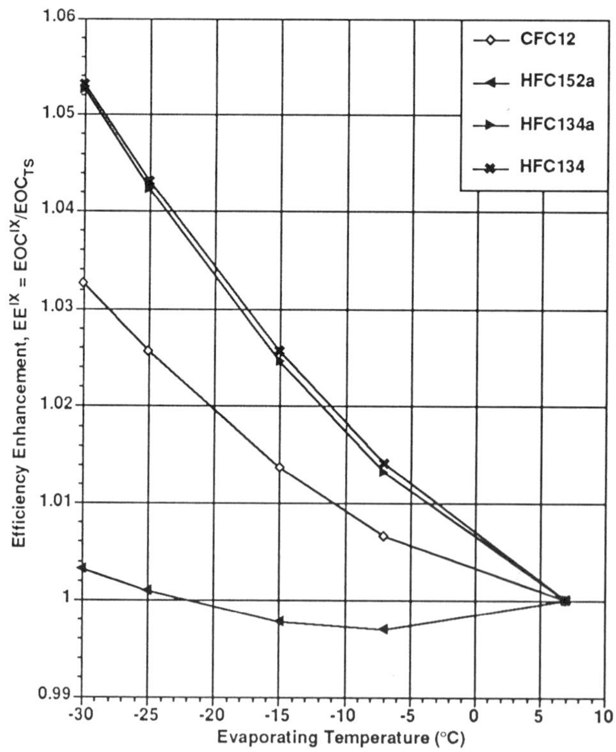

Figure 14: Efficiency Enhancement of the Cycle with Internal Heat Exchange at 43°C Condensing Temperature. Vapor Heated to 7°C.

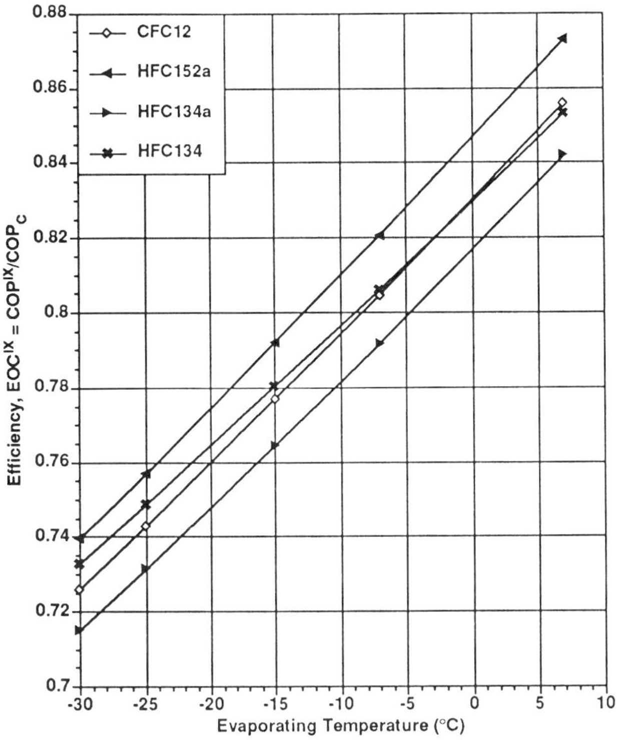

Figure 15: Efficiency of the Cycle with Internal Heat Exchange at 43°C Condensing Temperature. Vapor Heated to 7°C.

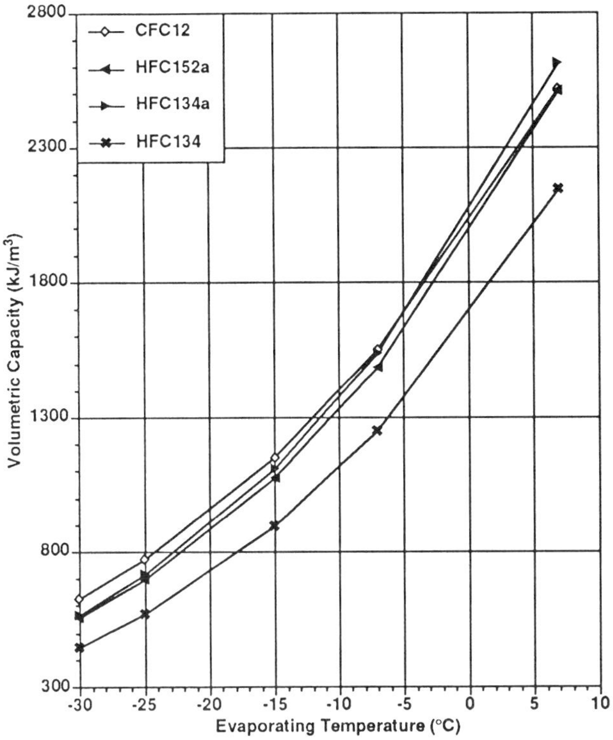

Figure 16: Volumetric Capacity of the Cycle with Internal Heat Exchange at 43°C Condensing Temperature. Vapor Heated to 7°C.

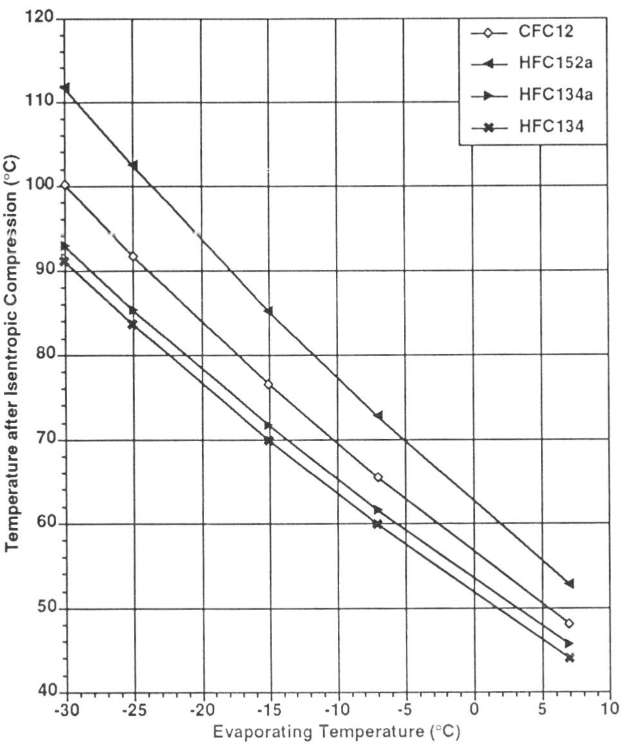

Figure 17: Temperature after Isentropic Compression for the Cycle with Internal Heat Exchange at 43°C Condensing Temperature. Vapor Heated to 7°C.

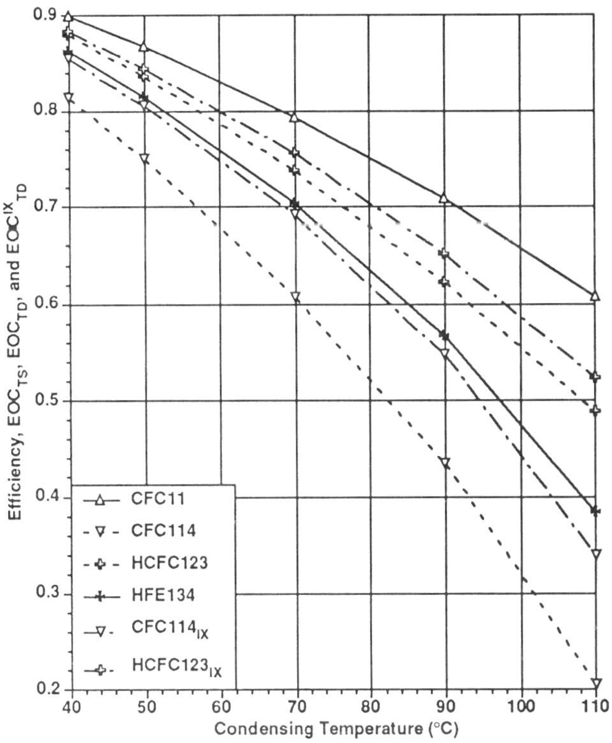

Figure 18: Efficiency of the Cycle with Throttling and Superheating, the Cycle with Dry Compression, and the Cycle with Dry Compression from Internal Heat Exchange at 4°C Evaporating Temperature. (Dashed lines are $\kappa < 0$.)

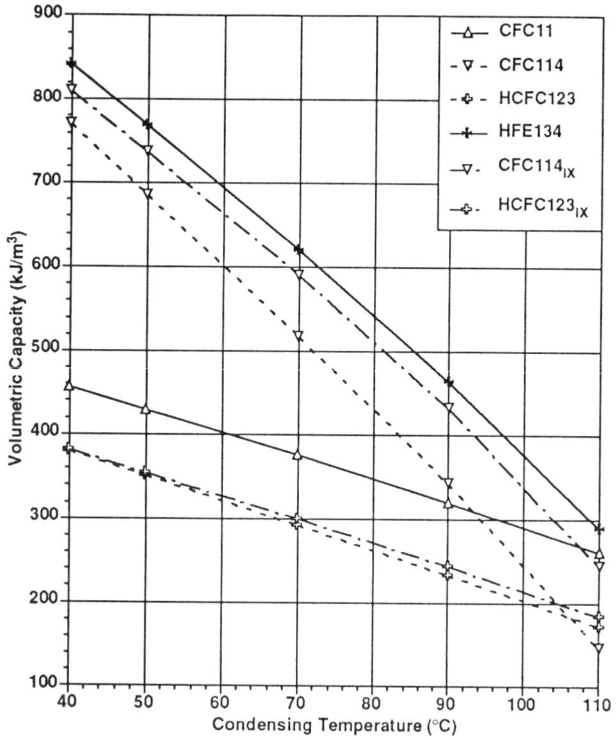

Figure 19: Volumetric Capacity for the Cycle with Throttling and Superheating, the Cycle with Dry Compression, and the Cycle with Dry Compression from Internal Heat Exchange at 4°C Evaporating Temperature. (Dashed lines are K< 0.)

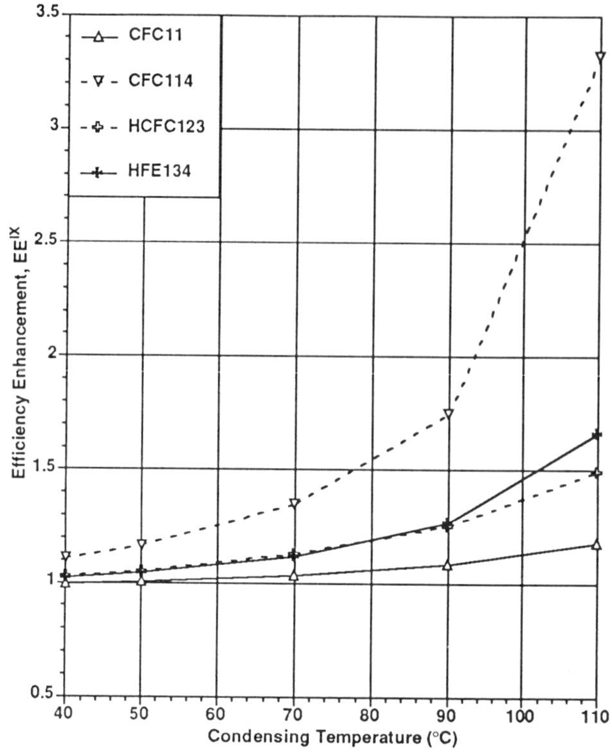

Figure 20: Efficiency Enhancement for the Cycle with Internal Heat Exchange at 4°C Evaporating Temperature. Vapor Heated to Tc - 5C°. (Dashed lines are K< 0.)

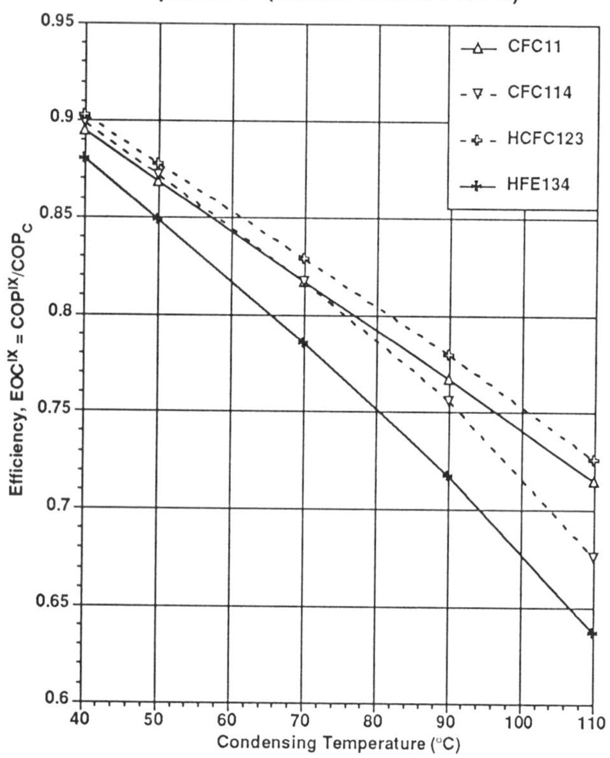

Figure 21: Efficiency of the Cycle with Internal Heat Exchange at 4°C Evaporating Temperature. Vapor Heated to Tc - 5C°. (Dashed lines are K< 0.)

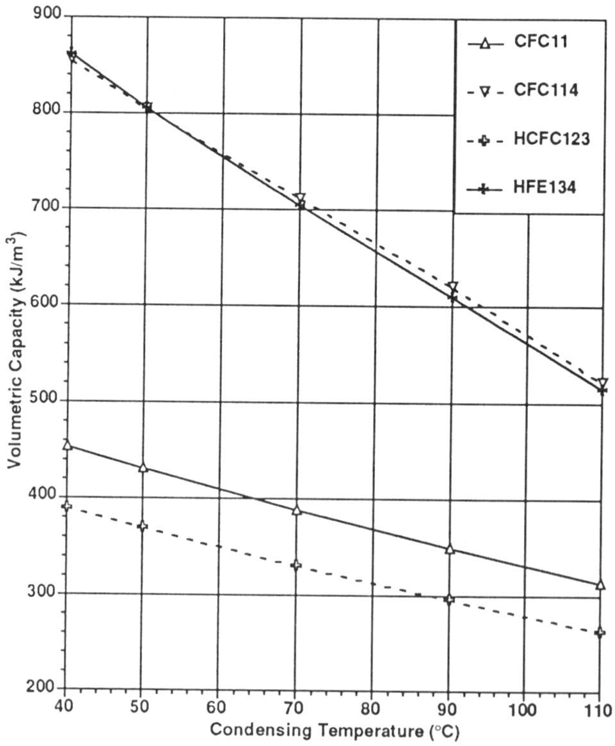

Figure 22: Volumetric Capacity of the Cycle with Internal Heat Exchange at 4°C Evaporating Temperature. Vapor Heated to Tc - 5C°. (Dashed lines are K< 0.)

Chapter 29

Criteria for Evaluation of the Efficiency Improvements in Vapor Compression Cycles Using Internal Heat Exchange

G.S. Kazachki

ABSTRACT

Significant number of chemical compounds has been in a process of evaluation as alternatives to the existing refrigerants since the Montreal Protocol. The energy efficiency and performance characteristics of the vapor compression cycle with each of these refrigerants depend on the cycle design and the organization of the heat fluxes in the system. This is specifically important when evaluating the internal heat exchange between the refrigerant vapor after the evaporator and the liquid refrigerant after the condenser. From the industry experience, refrigerant R22 (chlorodifluoromethane) is known to show a detrimental effect from the internal heat exchange, R12 shows a gain, and a refrigerant such as R114 gains substantially from the internal heat exchange. At the same time the isentropic compression of the first refrigerant ends at a very high temperature often causing problems with the compressor reliability, the second refrigerant is isentropically compressed at very acceptable temperatures, and the third refrigerant needs superheating of the vapor before the compression in order to avoid wet compression which in reciprocating compressors may be critical. These three compounds exemplify the three types of refrigerants used by industry. For a preliminary evaluation of alternative non-CFC refrigerants and for computer simulations of vapor compression cycles and systems, an analytical method for quantifying the effect of the internal heat exchange in vapor compression cycles is developed. For illustration, the suggested method is applied on several representative refrigerants for low, medium, and high temperature applications. A single criterion has been identified as a correlating factor for the thermodynamic performance of a given chemical compound as a refrigerant in the vapor compression cycle and as a predictor of the effect of the internal heat exchange on the cycle efficiency with this refrigerant.

INTRODUCTION

During the last 25 years a large amount of research was devoted to the working fluid in thermodynamic systems. The initial goal was to identify the most efficient refrigerant for each application of refrigeration systems, heat pump systems, or systems for utilizing renewable energy sources. An urgent complementary goal arose during the last decade: the identification of refrigerants which do not destroy the stratospheric ozone. The global warming impact, both direct and indirect, also became an important factor in the last three years. In several iterative steps both the existing chemical compounds had to be evaluated, and new chemicals had to be designed, synthesized, and evaluated. Most of the newly synthesized potential refrigerants as well as some of the existing ones have thermophysical properties resulting in unexpected and undesired performance in the cycles and in the real systems. The efficiencies of both the cycle and the system with these refrigerants in the way they have been traditionally designed and used are very low. Wet compression is likely to take place with such refrigerants with potential for damaging or even destroying the compressor in the case of reciprocating compressors. Developing efficient cycles and designing efficient systems tailored according to the refrigerant properties require specific tools in terms of thermodynamic methods and criteria for cycle and system analysis, evaluation, and design recommendations.

CONCEPT

Three cases of a basic vapor compression cycle (also called simple vapor compression cycle), consisting of isentropic compression of saturated vapor, isenthalpic throttling of saturated liquid, and isobaric evaporation and condensation, are presented in the entropy-temperature diagram in Fig. 1:

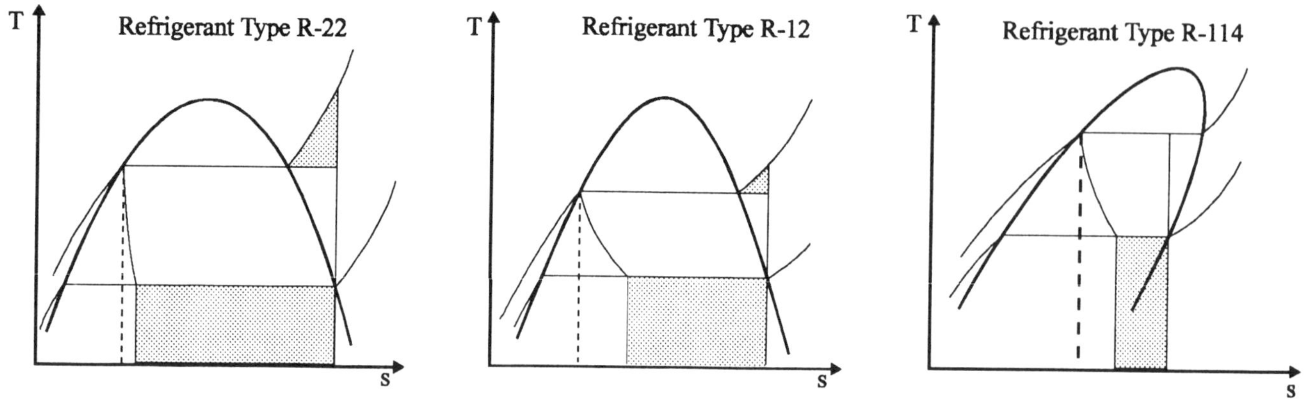

A) B) C)

FIG. 1: BASIC/SIMPLE VAPOR-COMPRESSION CYCLE IN ENTROPY, TEMPERATURE DIAGRAM:
A) CYCLE WITH REFRIGERANT TYPE R22;
B) CYCLE WITH REFRIGERANT TYPE R12;
C) CYCLE WITH REFRIGERANT TYPE R114.

a) a cycle with a refrigerant type R22 (chlorodifluoromethane),

b) a cycle with a refrigerant type R12 (dichlorodifluoromethane), and

c) a cycle with a refrigerant type R114 (dichlorotetrafluoroethane).

The hatched area under the evaporating temperature is equivalent to the cooling capacity of the cycle. The hatched area above the condensing temperature is equivalent to the increase of the cycle work due to the losses from superheating during the compression process. Each of these cycles has specific features which appear also in real systems.

The cycle with refrigerant type R22 is characterized by low throttling losses, high temperature at the end of isentropic compression, and comparatively high losses from the superheat in the compression process. The overall efficiency of the basic/simple cycle with this type of refrigerant is high. A vapor compression system (either refrigeration or any other) implementing as close as possible the basic/simple cycle with refrigerant type R22 is highly desirable. Vapor superheating between the evaporator outlet and the compressor inlet should be avoided. Only mechanical subcooling of the liquid refrigerant before the expansion valve is recommendable. Problems with high discharge temperature should be expected.

Higher throttling losses, lower discharge temperature, and acceptable losses from superheating are inherent to type R12 refrigerants. The cooling capacity and the efficiency of the basic/simple cycle with this type of refrigerant is lower than type R22. This cycle requires improvements which need to be implemented in the real system in order to achieve an acceptable cycle and system efficiency. Both the cycle and the system benefit from the subcooling of the liquid refrigerant before the expansion valve since this reduces the throttling losses in the expansion device. Certain superheat of the refrigerant vapor improves compressor efficiency. The resulting discharge temperature is within acceptable limits. The liquid subcooling and the vapor superheating can be combined through internal heat exchange. As opposed to type R22 refrigerants, cycles with type R12 refrigerants benefit from the internal heat exchange. This was noticed soon after R12 became a commonly used refrigerant, and internal heat exchangers have been part of R12 systems ever since.

Significant throttling losses, which wastes most of the available latent heat of vaporization, and very low specific cooling capacity are inherent to the type R114 refrigerants. The isentropic compression of saturated refrigerant vapors ends in the dome of wet vapor and is called wet compression. Such compression is unacceptable in reciprocating compressors. Therefore, certain minimum suction vapor superheating is needed in order to avoid the wet compression. The minimum suction vapor superheat is defined as superheat which is needed in order for the isentropic compression to end on the vapor saturation curve at condensing pressure. In this case dry compression occurs. The basic vapor compression cycle with type R114 refrigerants has low efficiency. It is clear that this cycle needs improvements: liquid refrigerant subcooling for reducing the throttling losses and suction vapor superheating for avoiding wet compression and improving compressor efficiency. Both can be combined through internal heat exchange.

METHOD

In the discussion above, the refrigerants are divided into three groups by analogy and qualitatively. It is clear that there should be a method and criteria for a quantitative analysis of a vapor compression cycle and for a quantitative basis of the system design as a function of the refrigerant properties. Such a method was derived and described in [1]. The method quantifies the contribution of the refrigerant thermophysical properties into the basic processes of the vapor compression cycle by the efficiency of primitive cycles which differ from the ideal cycle only by the processes being assessed. Thus, the following parameters have been defined and analytically derived from basic thermodynamic principles:

EOC_T - efficiency of cycle with throttling;

EOC_S - efficiency of cycle with superheating;

EOC_{TS} - efficiency of cycle with throttling and superheating;

EOC^{IX} - efficiency of cycle with throttling, superheating, and internal heat exchange;

EE^{IX} - efficiency enhancement of internal heat exchange;

q_v, q_v^{IX} - volumetric capacity of cycle without and with internal heat exchange; and

VCE^{IX} - volumetric capacity enhancement of internal heat exchange.

All these parameters apply for refrigerants with dry isentropic compression of saturated vapors, characterized by a positive value of the parameter $\varkappa$. The following parameters characterizing vapor compression cycles for refrigerants with negative $\varkappa$ are defined here for the first time in the theory of the vapor compression cycles:

EOC_{TD} - efficiency of cycle with throttling and dry compression (cycle with the minimum suction superheat which is needed in order to achieve dry compression);

EOC_{TD}^{IX} - efficiency of cycle with throttling and dry compression achieved through internal heat exchange utilizing the heat of the minimum suction superheating;

EE_{TD}^{IX} - efficiency enhancement of the minimum internal heat exchange related to the efficiency of cycle with throttling and dry compression;

EE^{IX} - efficiency enhancement of the internal heat exchange related to the efficiency of cycle with throttling and dry compression;

q_{vTD} - volumetric capacity of cycle with throttling and dry compression;

q_{vTD}^{IX} - volumetric capacity of cycle with throttling, dry compression, and minimum internal heat exchange;

VCE_{TD}^{IX} - volumetric capacity enhancement of the internal heat exchange in a cycle with throttling, dry compression, and minimum internal heat exchange related to the volumetric capacity of cycle with throttling and dry compression; and

VCE^{IX} - volumetric capacity enhancement of internal heat exchange in a cycle with throttling and internal heat exchange related to the volumetric capacity of cycle with throttling and dry compression.

EE^{IX}, VCE^{IX}, EE_{TD}^{IX}, and VCE_{TD}^{IX} are the criteria for evaluation of the efficiency improvements in vapor compression cycles using internal heat exchange. Compared with the original equations in [1], which resulted from some simplifying assumptions, more accurate analytical expressions were derived for the efficiency enhancement of the internal heat exchange and for the volumetric capacity enhancement of the internal heat exchange for refrigerants with positive $\varkappa$. Accurate analytical expressions were derived for first time for the newly defined parameters related to cycles with negative $\varkappa$. Because of the very theoretical character of the work, neither the derivations nor the final analytical expressions are reported here. However, the results from their employment in the cycle analysis with a wide variety of refrigerants are reported in the next section.

RESULTS AND DISCUSSION

The described method was applied on evaluation of eight low-temperature, eight middle-temperature, and 13 low-pressure refrigerants. The results for the low-temperature refrigerants are presented in Table 1, the results for the middle-temperature refrigerants in Table 2, and the results for the low-pressure refrigerants in Table 3. The refrigerants in each table are positioned in descending order of their parameter $\varkappa$. In such an order the type R22 refrigerants are at the top of each table followed by type R12 refrigerants and then by type R114 refrigerants. The qualitative description of the characteristics of the three types of refrigerants by their performance in a basic vapor compression cycle in the Concept Section can be quantified now. With few exceptions, the following observations are apparent. Going through the tables from the top to the bottom, the throttling losses, represented by EOC_T, become more significant, while the losses from the superheat during the compression process, represented by EOC_S, become smaller. An additional important observation is the smaller value of the EOC_T compared to EOC_S suggesting that the main losses in the vapor compression cycle are encountered during the throttling process and that the throttling losses are determining the overall efficiency of the vapor compression cycle. These losses are substantially smaller in cycles with type R22 refrigerants. The only disturbing circumstance connected with this type of refrigerant is the high temperature at the end of isentropic compression which appears as a high discharge temperature in the compressor. The real vapor compression systems achieve their highest efficiency with these refrigerants without a need for complicating the system with any improvements. They are inherently and thermodynamically the most efficient. As soon as the high discharge temperatures are successfully handled by using advanced construction materials, lubricants, and proper compressor design, these are the preferred refrigerants. This type of refrigerant is to be looked for among the existing chemical compounds. This type of chemical is to be synthesized for use as refrigerants in vapor compression cycles. Until such chemical compounds are designed, refrigerant blends of the same thermodynamic type are to be identified and mixed. This is the right direction, or at least this is the preferable direction in searching for alternative refrigerants in vapor compression systems particularly with reciprocating compressors. Connecting these observations with the value of the parameter $\varkappa$ (the last column in the tables), a very well defined relationship is seen. A conclusion can be drawn that this parameter can be used as a main criterion for searching and screening chemical compounds. Since the value of $\varkappa$ depends not only on the thermodynamic nature of the refrigerant but also on the temperature envelope of the cycle, i.e. on the position of the cycle on the refrigerant dome as well as on the span of the cycle, $\varkappa$ will not have the same value for all refrigeration (heat pump, or any other) applications. However, a narrow range of recommended values of $\varkappa$ can be identified and specified for each application. For instance, in the low temperature range of supermarket applications, which has been and still is the hot topic of the alternative refrigerants research, the value for $\varkappa$ recommended by the author, is in the range between 200 and 500. Refrigerants with $\varkappa$ close to 500 will have the highest efficiency with a simple arrangement of the system. At higher condensing and lower evaporating temperatures, the compressor discharge temperature will be the critical issue to be concerned about and considered. Refrigerants with $\varkappa$ close to 200 will have lower efficiency, will generally not create problems with the discharge temperature, and will benefit from the internal heat exchange, which is already a step towards complicating the system. The quantitative characteristics for refrigerants of type R22 in the temperature range of this application area is $\varkappa > 500$. Typical representatives of this type of refrigerant are ammonia and R32.

After type R22 refrigerants, the next group encountered in the tables is the type R12 refrigerants. For this type of refrigerant EE^{IX} is greater than 1.0, while for type R22 refrigerants EE^{IX} is less than 1.0 indicating a detrimental effect from the application of internal heat exchange relative to the basic vapor compression cycle. The range of $\varkappa$ between 0 and 500 is the quantitative description for type R12 refrigerants. Besides R12, typical representatives of this group are R502 and R134a.

Proceeding down the tables, along with the decreasing overall efficiency of the basic vapor compression cycle and decreasing temperature at the end of the isentropic compression, the point is reached when no more superheat during the compression is encountered. Theoretically this should happen for $\varkappa = 0$. Below this point suction vapor superheating is needed for achieving dry compression. This occurs for refrigerants with $\varkappa < 0$. As can be observed the amount of required minimum superheat increases with decreasing $\varkappa$. This is the quantitative characteristic for refrigerants of type R114. Both $EE_{TD}{}^{IX}$ and EE^{IX} increase down the table; and, in fact, EE^{IX} is largest for the R114 type refrigerants. Most of recently synthesized hydrofluorocarbons chemical compounds are typical representatives of this group. Cycle analysis and design recommendations for vapor compression systems can be drawn from the tables.

ADDENDUM

The efficiencies of the various cycles and the efficiency enhancement from internal heat exchange in the vapor compression cycle were discussed exclusively so far. This is because of the strong relationship between these cycle characteristics and the refrigerant parameter $\varkappa$ whose values can be used not only as a criterion for refrigerant classification but also as a criterion determining the necessity for or the benefit from internal heat exchange. The volumetric capacity enhancement is another important cycle characteristic which benefits from the internal heat exchange in a similar way. However, the volumetric capacity itself is not thermodynamically related to $\varkappa$ as much as to other refrigerant properties and is skipped from this discussion in order to keep a straight and simple line of analysis.

CONCLUSIONS

1. EE^{IX} for refrigerants with positive $\varkappa$ and $EE_{TD}{}^{IX}$ or EE^{IX} for refrigerants with negative $\varkappa$ are the criteria for a quantitative evaluation of the efficiency improvements in vapor compression cycles using internal heat exchange. Exact expressions for these parameters as a function of basic thermodynamic properties have been derived and successfully used in the evaluation of existing and newly synthesized compounds as alternative refrigerants.

2. The parameter $\varkappa$ has been identified as a single property characteristic determining the cycle efficiency as a function of the thermodynamic nature of the refrigerants in vapor compression cycles. It has also been identified as a criterion for predicting the effect of internal heat exchange in vapor compression systems.

NOMENCLATURE

Symbols: c''_{pm} - logarithmic mean specific heat of saturated vapor between evaporating and condensing temperatures; $COP=q_E/w$ - coefficient of performance of cycle; $COP_C=T_E/(T_C-T_E)$ - coefficient of performance of Carnot cycle; $EE^{IX}=EOC^{IX}/EOC$ - efficiency enhancement of cycle with internal heat exchange; $EE_{TD}{}^{IX}=EOC_{TD}{}^{IX}/EOC_{TD}$ efficiency enhancement of cycle with minimum suction-vapor superheating provided by internal heat exchange; $EOC=COP/COP_C$ - efficiency of cycle; l_{Em} - logarithmic mean latent heat of vaporization between evaporating and condensing temperatures; T_C, T_E - condensing and evaporating temperatures; T_1, T_{1min} - suction and minimum-superheated suction temperatures; T_2 - discharge temperature at the end of isentropic compression of saturated vapor; $T_2{}^{IX}$ - discharge temperature at the end of isentropic compression of vapor after the internal heat exchanger; $\Delta T^v{}_{min}$ - minimum suction-vapor superheating for achieving saturated vapor at the end of the isentropic compression; q_E - cooling capacity; q_v - volumetric capacity; VCE^{IX} - volumetric capacity enhancement from internal heat exchange; and w - thermodynamic work.

Superscripts: IX - parameter related to a cycle with internal heat exchange.

Subscripts: T - throttling, S - superheating, TS - throttling and superheating, and TD - throttling and dry compression.

Greek: β''_m - logarithmic mean thermal expansion coefficient of saturated vapor; $\varkappa''_m= [(\beta''_m l_{Em}/c''_{pm}) - 1]$ - efficiency criterion for refrigerant performance in vapor compression cycles.

REFERENCES

1. Kazachki, G.S., "Derivation of Dimensionless Parameters for Thermodynamic Evaluation of Refrigerants in Vapor Compression Cycles," Proceedings of the 18th International Congress of Refrigeration, Commission B2, Montreal, 1991.

TABLE 1: VAPOR-COMPRESSION CYCLE EFFICIENCY CHARACTERISTICS AT -30°C EVAPORATING, 43°C CONDENSING, AND 7°C SUCTION TEMPERATURE AFTER THE INTERNAL HEAT EXCHANGE WITH LOW-TEMPERATURE REFRIGERANTS IN THE ORDER OF THEIR K_M''S.

	Refrigerant	EOC_T	EOC_S	EOC_{TS}	T_2	ΔT^V_{MIN}	EOC_{TD}	EOC_{TD}^{IX}	EE_{TD}^{IX}	EOC^{IX}	EE^{IX}	T_2^{IX}	$\kappa_m''.10^3$
		-	-	-	°C	K	-	-	-	-	-	°C	-
1.	R32	0.758	0.888	0.688	117.2	-	-	-	-	0.665	0.966	157.7	955.8
2.	R22	0.736	0.950	0.707	86.4	-	-	-	-	0.700	0.991	126.2	574.7
3.	R290	0.663	0.989	0.658	58.5	-	-	-	-	0.696	1.058	92.1	224.7
4.	R502	0.630	0.987	0.624	58.6	-	-	-	-	0.670	1.074	94.3	219.8
5.	R143a	0.601	0.991	0.597	55.2	-	-	-	-	0.656	1.098	89.6	175.7
6.	R125	0.503	0.997	0.502	48.3	-	-	-	-	0.608	1.211	80.0	81.0
7.	E125	0.556	-	-	-	5.3	0.537	0.573	1.067	0.670	1.248	72.4	-67.8
8.	R218	0.320	-	-	-	18.6	0.274	0.445	1.626	0.558	2.037	58.8	-228.3

TABLE 2: VAPOR-COMPRESSION CYCLE EFFICIENCY CHARACTERISTICS AT -30°C EVAPORATING, 43°C CONDENSING, AND 7°C SUCTION TEMPERATURE AFTER THE INTERNAL HEAT EXCHANGE WITH MIDDLE-TEMPERATURE REFRIGERANTS IN ORDER OF THEIR K_M''S.

	Refrigerant	EOC_T	EOC_S	EOC_{TS}	T_2	ΔT^V_{MIN}	EOC_{TD}	EOC_{TD}^{IX}	EE_{TD}^{IX}	EOC^{IX}	EE^{IX}	T_2^{IX}	$\kappa_m''.10^3$
		-	-	-	°C	K	-	-	-	-	-	°C	-
1.	R152a	0.752	0.976	0.737	73.2	-	-	-	-	0.739	1.003	111.7	404.9
2.	R12	0.709	0.989	0.703	60.8	-	-	-	-	0.726	1.033	100.1	238.5
3.	R134a	0.683	0.994	0.679	55.7	-	-	-	-	0.715	1.053	92.8	175.6
4.	R134	0.697	0.997	0.696	52.1	-	-	-	-	0.733	1.053	91.0	121.6
5.	E143a	0.679	0.998	0.678	49.3	-	-	-	-	0.725	1.070	86.7	86.7
6.	R245cb	0.569	-	-	-	11.1	0.535	0.601	1.123	0.678	1.267	68.5	-149.1
7.	R227ca	0.518	-	-	-	16.8	0.469	0.580	1.235	0.655	1.397	62.9	-226.3
8.	R227ea	0.529	-	-	-	17.7	0.474	0.597	1.258	0.672	1.417	62.0	-238.3

TABLE 3: VAPOR-COMPRESSION CYCLE EFFICIENCY CHARACTERISTICS AT 4°C EVAPORATING, 40°C CONDENSING, AND 35°C SUCTION TEMPERATURE AFTER THE INTERNAL HEAT EXCHANGE WITH LOW-PRESSURE REFRIGERANTS IN THE ORDER OF THEIR K_M''S.

	Refrigerant	EOC_T	EOC_S	EOC_{TS}	T_2	ΔT^V_{MIN}	EOC_{TD}	EOC_{TD}^{IX}	EE_{TD}^{IX}	EOC^{IX}	EE^{IX}	T_2^{IX}	$\kappa_m''.10^3$
		-	-	-	°C	K	-	-	-	-	-	°C	-
1.	R11	0.902	0.998	0.900	47.4	-	-	-	-	0.895	0.995	80.5	198.9
2.	E134	0.862	1.000	0.862	41.0	-	-	-	-	0.880	1.020	72.5	28.6
3.	R123	0.883	-	-	-	1.0	0.879	0.883	1.005	0.902	1.026	71.2	-28.4
4.	R245fa	0.861	-	-	-	1.4	0.856	0.862	1.007	0.886	1.035	69.9	-39.4
5.	R245ca	0.864	-	-	-	3.8	0.850	0.867	1.019	0.896	1.054	67.5	-105.8
6.	R254cb	0.842	-	-	-	5.3	0.821	0.848	1.032	0.887	1.080	65.6	-146.0
7.	R236ea	0.837	-	-	-	7.0	0.810	0.847	1.045	0.890	1.099	64.1	-193.1
8.	R236ca	0.848	-	-	-	7.5	0.819	0.857	1.047	0.898	1.097	63.8	-208.1
9.	R114	0.844	-	-	-	7.5	0.813	0.855	1.051	0.898	1.105	63.9	-210.2
10.	R236cb	0.806	-	-	-	11.8	0.754	0.833	1.105	0.892	1.183	58.9	-323.8
11.	R347ccb	0.809	-	-	-	13.8	0.749	0.843	1.126	0.902	1.204	57.2	-384.5
12.	R338ccb	0.818	-	-	-	14.2	0.759	0.851	1.121	0.906	1.194	57.0	-395.5
13.	R329cca	0.786	-	-	-	15.1	0.721	0.829	1.149	0.891	1.235	55.7	-414.0

Chapter 30

CFC Issues from a User's Point of View

C.W. Sohn, K.O. Homan, H. Gignilliat

ABSTRACT

The phaseout of chlorofluorocarbon (CFC) refrigerants presents the U.S. Army with the challenge faced by all air conditioning and refrigeration (AC/R) equipment users, to meet the CFC regulatory requirements in a timely and economical manner in the context of rapidly developing technologies. In response to this phaseout, the U.S. Army Construction Engineering Research Laboratories (USACERL), with support from the U.S. Army Engineering and Housing Support Center (USAEHSC), is working to develop options and technologies necessary to meet the challenge. As the first step, USACERL characterized the Army's AC/R equipment inventory in terms of capacity, refrigerant type, and equipment age. The baseline data was used to project the required budget for the Army to comply with current CFC-related regulations. In this paper, the methodologies for establishing baseline data and developing cost models are presented along with a discussion of the compliance options available.

INTRODUCTION

Due to the stratospheric ozone layer depletion potential of the chlorofluorocarbon (CFC) refrigerants, CFC production and use have been prohibited by U.S. law (U.S. Clean Air Act Amendments of 1990) as well as by international protocols (Montreal Protocol of 1987, Copenhagen Amendment of 1992). Since the Army's air conditioning and refrigeration (AC/R) systems are primarily vapor compression units equipped with CFC or HCFC refrigerants, the Army is faced with the same difficult challenge other CFC/HCFC users face. As of 1993, the development of alternative refrigerants and AC/R systems by the industry is rapid; however, the ultimate alternatives are yet to be established. In this changing environment, the U.S. Army and private sector AC/R system users are trying to meet CFC regulatory requirements [EPA, 1993] with a timely and cost-effective solution. To accomplish this transition, the determination of baseline usage data and careful evaluation of available options are essential. The U.S. Army Construction Engineering Research Laboratories (USACERL) has established baseline technical information on the Army's current refrigerant use in AC/R systems based on the analysis of detailed data from three site visits and Army-wide general inventory data. The feasibility of options available to the Army in the current market are under study. Options include keeping existing CFC AC/R systems in operation until the end of their useful service lives but with recovery and recycling of CFCs, determining optimal phaseout schedules for existing CFC equipment through conversion or replacement, and deployment of alternative AC/R systems.

THE ARMY'S AC/R SYSTEM INVENTORY

Anticipating accelerated changes in CFC-related regulations, USACERL, in coordination with the U. S. Army Engineering and Housing Support Center (USAEHSC), completed a study to establish baseline technical information on the Army's AC/R system inventory [Sohn et al., 1992]. The objective was to characterize the Army's AC/R inventory in terms of equipment type, refrigerant use and age. Since a field study of the 164 Army installations is not practical, the information was generated through an Army-wide equipment inventory report and detailed field studies of three representative installations.

The overall inventory data was obtained from the Redbook, the common name for the Facilities Engineering and Housing Annual Summary of Operations [EHSC, 1989].

This document provides the previous year's operating costs and workload performance data for all Real Property Maintenance Activities (RPMA) work done at Army installations, activities, and communities. The Redbook includes 13 categories of data related to air conditioning, cold storage, and refrigeration. The categories separate the AC/R systems by size. For each category, the installed capacity of the units (in tons or hp) is listed. Air conditioning systems, refrigeration systems, and cold storage systems are in separate categories. Data is provided for each of the 164 Army installations.

The detailed information for an individual piece of AC/R equipment, such as the type and amount of refrigerant used and the age, is available only through field inspection of the installed unit. The three Army installations selected for field study are representative of many Army installations. Two of the installations have a large number of personnel staffed and trained at the installation (on-post). The facilities on these installations are suggestive of a small city (population of 20 to 30 thousand) and include dining halls, barracks, training centers, office buildings, schools, hospitals, industrial shops, and family housing units. Air conditioning is supplied by central energy plants as well as many unitary systems. The third installation is involved primarily in the production, maintenance, and storage of various types of ammunition, missiles, and vehicles. The installation is highly industrialized and has many large factory-type buildings. Except for office and administrative areas, most of the floor area in these buildings is not air conditioned. For this reason, the majority of air conditioning systems are packaged or split units located along the perimeter or on the roof of the large industrial buildings. Only a few personnel live onpost in family housing units.

Site study data was obtained by physically examining the equipment and recording the information from the nameplate. The data sought included capacity of the unit (tons or hp), type of refrigerant used, the refrigerant charge, and the unit's age. Although all of this information was generally not on the equipment nameplate, the model number provided a means to obtain it through equipment catalogs or company representatives. This method proved very successful for air conditioning units of all sizes. For small refrigeration units (such as those in dining facilities), information was sparse. For a number of these units, an estimate had to be made.

Analysis of Inventory Data

The equipment examined during the site studies confirmed a number of general observations. Units with positive displacement compressors i.e., unitary air conditioners, use HCFC-22 almost exclusively and are most common in sizes less than 100 tons. Smaller buildings, from offices to family housing units, are cooled by these types of units. Centrifugal chillers are used for large cooling loads from 100 tons to more than 8000 tons. The most commonly used refrigerant in these chillers is CFC-11, although there are machines using CFC-12, CFC-113, CFC-114, CFC/HCFC-500, and HCFC-22. Large cold storage and refrigeration systems are custom designed for a specific application and use either CFC-12 or HCFC/CFC-502 depending on the desired temperature. Small refrigeration units (less than 5 hp) are usually a standardized design. These systems use CFC-12, HCFC-22, or HCFC/CFC-502 depending again on the desired temperature. So one can expect to find CFC refrigerants in chillers of over 100 tons and in refrigeration and cold storage systems.

To use the collected information to predict the type and amount of refrigerants being used in the Army's AC/R equipment, a two-step approach was adopted. First, the total pounds of refrigerant was estimated based on the capacities given in the Redbook and, second, the amount of each type of refrigerant was estimated using calculated distribution data from the site studies.

The first step requires the determination of the ratio of pounds of refrigerant to capacity. This ratio can then be used to estimate total refrigerant use in the following manner:

Pounds of refrigerant = tons capacity x (lb/ton capacity)

The pounds/ton ratio was determined from data gathered during the site studies. The data used was the capacity and refrigerant charge of equipment obtained from equipment nameplates or original equipment manufacturers. The ratio for the air conditioning equipment was determined by scaling the ratios of each size category by the percentage of that equipment type found at the three sites studied. The result is an overall ratio of 2 lbs/ton. This agrees with rules of thumb used in the industry [Salas and Salas, 1992]. Thus, for air conditioning equipment, the total pounds of refrigerant in use are estimated as:

Total lbs = 880,542 tons x 2.0 lb/ton = 1,761,000 lb

The ratio for the refrigeration and cool storage equipment was determined by dividing the total refrigeration quantity by the total refrigeration capacity at the three sites. The estimated ratio is 5.9 lb/hp. Multiplying the total reported refrigeration capacity for all U.S. Army installations as reported in the Redbook yields the estimate of refrigerant use:

Total lb of refrigerant= 126,105 hp x 5.9 lb/hp = 744,000 lb

The second step required the determination of the refrigerant type distribution for the air conditioning group

Table 1. Estimation of Army Wide Refrigerant Usage

	Air Conditioning [1,761,000 lb]		Refrigeration [744,000 lb]		
Refrigerant	**Fraction**	**Estimated**	**Fraction**	**Estimated**	**Estimated Totals**
CFC-11	0.262	461,382			461,382
CFC-12	0.024	42,264	0.755	561,720	603,984
HCFC-22	0.612	1,077,732	0.062	46,128	1,123,860
CFC-113	0.003	5,283			5,283
R-500	0.098	172,578			172,578
R-502	0.001	1,761	0.183	136,152	137,913
			Total lb of Refrigerant		2,505,000

and the refrigeration group. From this, the quantity of each type of refrigerant in the U.S. Army AC/R equipment inventory was calculated. The results are shown in Table 1. The overall quantity of refrigerants is 2.51 million pounds. Of this quantity, 1.29 million pounds (55.4 percent) are CFC based, as illustrated in Figure 1.

The total AC equipment capacities for the U.S. Army are shown in Table 2. Clearly, the chillers of more than 100 ton capacity are the majority of the installed capacity. These are the machines using the CFC refrigerants. Table 3 compares the percentages of AC equipment used at the three installations studied with the overall Army as reported in the Redbook. The comparison supports the assumption that the three installations studied are, in fact, representative of Army installations overall.

Chiller population estimates supplied by the Air Conditioning and Refrigeration Institute [Denny, 1991] are approximately 80 percent, 10 percent, 5 percent, 4 percent, 1 percent, and less than 1 percent, for units using CFC-11, CFC-12, CFC/HCFC-500, HCFC-22, CFC-114, and CFC-113, respectively, based on a total estimated population of 73,000 units. Along with estimates of average unit capacities provided by ARI, the refrigerant distribution for large chillers was estimated. Table 4 shows this information and the distribution obtained from the site surveys. The comparison is quite favorable considering the small number of installations surveyed, but the shortcomings of a small sample size can be seen, for example, in the proportion of R-500 (24.3 percent) in the final estimate of refrigerant quantities. The substantial

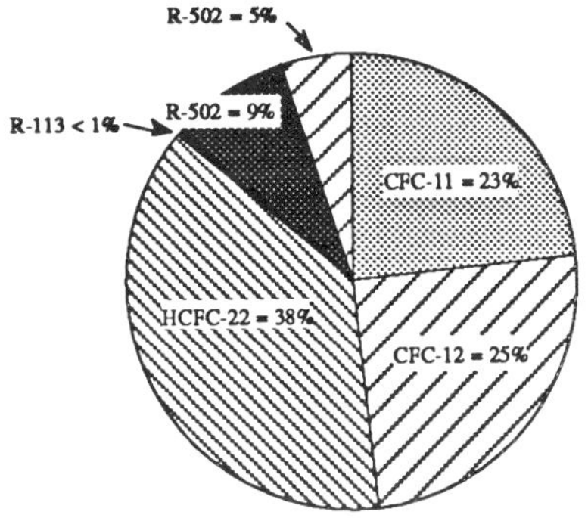

Figure 1. Total Estimated Refrigerant Use in the U.S. Army.

Table 2. Air Conditioning Equipment Use

Category	Fraction	Capacity
Air Cond Plants (>100 Tons)	0.581	511,200
Air Cond Plants (26-100 Tons)	0.083	73,166
Air Cond Plants (5-25 Tons	0.094	82,662
Air Cond Plants (<5 Tons)	0.228	200,780
Heat Pump (>5 Tons)	0.005	4,265
Heat Pump (<5 Tons)	0.010	8,469

Table 3. Equipment Use at Surveyed Sites vs Overall Army

Category	Site Surveys (%)	FY 89 Redbook (%)
Air Cond Plants (>100 Tons)	56.2	58.1
Air Cond Plants (26-100 Tons)	19.4	8.3
Air Cond Plants (5-25 Tons	7.5	9.4
Air Cond Plants (<5 Tons)	12.2	22.8
Heat Pump (>5 Tons)	0.0	0.5
Heat Pump (<5 Tons)	4.8	1.0

Table 4. Large Chiller Refrigerant Use

Refrigerant	Site Studies (%)	ARI (%)
CFC-11	65.3	65.6
CFC-12	3.2	11.9
HCFC-22	6.6	11.9
CFC-113	0.6	0.0
CFC-114	0.0	0.9
R-500	24.3	9.7

fraction of R-500 is a result of three large centrifugal chillers located at one of the sites studied. Regardless of the actual refrigerant usage distributions, most chillers currently use CFC refrigerants i.e., CFC-11, CFC-12, and CFC/HCFC-500.

The total refrigeration capacities and the fractions are shown in Table 5. A large portion of the capacity is household type refrigerators which although using CFC refrigerants need not be replaced.

Absorption units were excluded from the estimate since most large chillers are based on the vapor compression cycle (centrifugal and positive displacement compressors). If the number of absorption units observed at the surveyed sites was, in fact, typical of the Army as a whole, then the amount of CFCs estimated to be used by the Army would be reduced by 11 percent. This means that the CFC use estimated in this study represents an "outside estimate" of Army use. It should be recognized that absorption units contain no CFCs or HCFCs and therefore would not be candidates for retrofit or replacement.

A composite age distribution of AC/R equipment was obtained using data from the three site studies. The equipment types were divided into four categories. The age data is listed in Table 6 and illustrated in Figure 2. The first category — air conditioning units with capacities greater than 100 tons — has a large majority of relatively new units. This suggests that it may be feasible to retrofit or keep many of the units in operation. Units of less than 100 tons are generally less than ten years old. Similarly, family housing equipment can remain in use since these machines also use only HCFC-22. The bottom bar in Figure 2 shows

Table 5. Refrigeration Capacities and Fractions

Category	Capacity (hp)	Fraction
Large Refrig. Units (>5hp)	417	0.106
Small Refrig. Units(<5hp)	496	0.126
Household Refrigerators	3,014	0.768

Table 6. Age Distribution of AC/R Equipment (Composite of Three Site Studies)

Age	AC >100 tn	AC <100 tn	Family Housing	Refrig.
0-5 yrs	14	501	853	1166
6-10 yrs	4	328	1150	483
11-15 yrs	3	141	2155	102
15+ yrs.	3	174	193	37

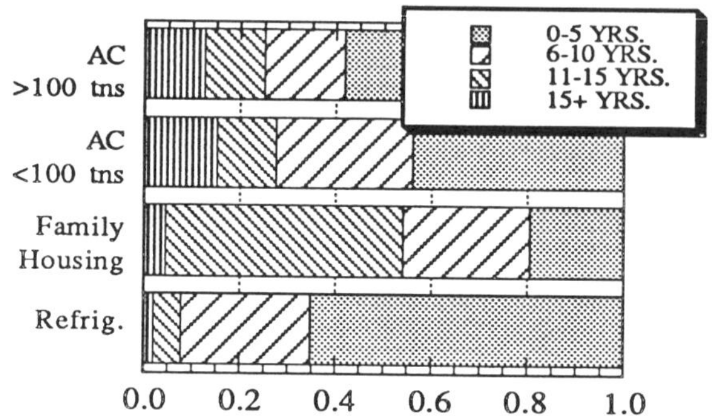

Figure 2. Age Distribution of AC/R Equipment at Three Selected Installations.

the age of refrigeration equipment as found in the USACERL study. Once again, much of the equipment is shown to be relatively new; an important note, since many of these machines use CFCs.

Compliance Cost Estimates Based on Inventory Data

The cost of eliminating the use of CFCs in AC/R equipment was estimated for the planning of retrofit/replacement budgets and schedules as well as to identify specific areas for further study. The estimate included converting at least part of the existing inventory with CFC refrigerants for the use of non-CFC refrigerants, since much of the inventory is expected to be quite new.

The projected retrofit and replacement costs are based on the inventory study presented earlier. The approach is illustrated in Figure 3. The capacity data reported in the Redbook, along with refrigerant and equipment usage data from the CFC inventory study, are used to estimate (1) the quantities of refrigerants used by air conditioning equipment and (2) the capacities of categories of refrigeration equipment. With average equipment size data (and the lb/ton ratio for chillers), the number of units using each type of refrigerant can be estimated. Finally, this estimate can be used with estimates of retrofit and replacement costs obtained from equipment manufacturers to estimate the compliance cost. Note that this is equivalent

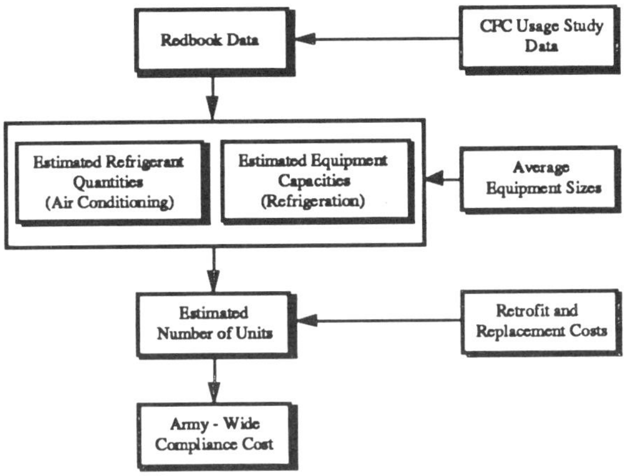

Figure 3. Methodology for Estimate of Compliance Cost.

to estimating the cost per unit capacity, thereby calculating the total cost based on total equipment capacity. However, the estimate of the number of units of any one type is helpful for appreciation of the size of the inventory.

For air conditioning, recall that only the chillers, units of greater than 100 ton capacity, use CFC refrigerants. Thus, the CFC refrigerants used by AC equipment are assumed to be only in the chillers. The retrofit and replacement options illustrated in Figure 4 show that it may be possible to retrofit units using CFC-11, CFC-12, or CFC/HCFC-500. Units using CFC-113 need to be replaced since the physical properties of CFC-113 as compared to the alternative refrigerants make a retrofit impossible. Units using HCFC-22 can be left as is for now.

Two air conditioning equipment manufacturers were contacted for estimates of the cost to replace units typical of those used in the U.S. Army [Bilsland, 1993; Wendl, 1993]. Table 7 summarizes the cost estimates for the alternatives outlined in Figure 4. Note that the replacement costs are for equipment costs only. To account for installation costs, these figures are multiplied by a factor of 1.5. That is, the installation cost is assumed to be 50 percent of the equipment costs.

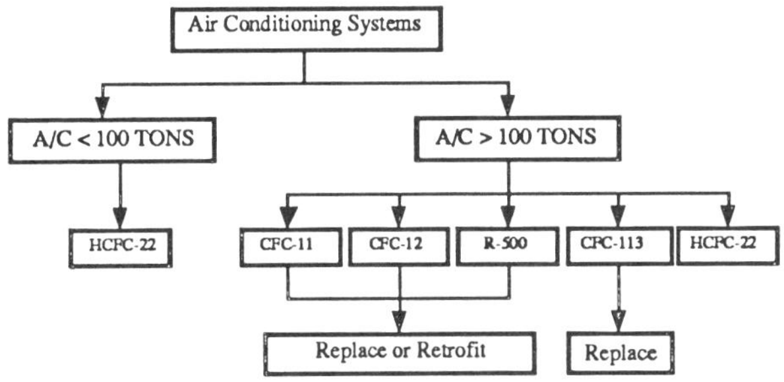

Figure 4. Retrofit and Replacement Options for Refrigerants Used in AC Systems.

The inventory data reported in the Redbook and that obtained in the site studies is the basis for the estimate of retrofit and replacement costs of refrigeration equipment using CFC refrigerants. The fraction of the total refrigeration capacity for each category (greater than 5 hp, less than 5 hp, household refrigerator) was shown earlier. Household refrigerators have been separated from the other refrigeration equipment since they need not be considered for retrofit or replacement, even though they use CFC-12. Most often they are used until they stop functioning properly and then replaced. Refrigerators using R-134a will soon be available and can be used to replace existing units as they become inoperable.

Retrofit and replacement options considered for refrigeration equipment are shown in Figure 5. Based on discussions with equipment suppliers, it may be possible to retrofit units using CFC-12. Replacement of units using R-502 will be necessary. Replacement prices obtained from a local equipment supplier are shown in Table 8. Retrofit costs have been estimated as one-third to one-half of replacement cost.

The estimated cost of compliance with present CFC-phaseout schedules for AC/R equipment located on U.S. Army installations can be determined using the information presented. However, a necessary part of the calculation is an estimation of the fraction of the equipment that can be retrofitted. Based on observations made during

Table 7. Estimated Retrofit and Replacment Costs for Average-Size Chillers

Existing Unit		Replacement		Retrofit	
Size	**Refrig.**	**Cost***	**Refrig.**	**Cost***	**Refrig.**
550 tons	CFC-11	$160 - $180	HCFC-123/HFC-134a	$35	HCFC-123
800 tons	CFC-12	$230 - $250	HCFC-123/HFC-134a	$65	HFC-134a
220 tons	CFC-113	$50 - $100	HCFC-123/HFC-134a	†	
1300 tons	R-500	$250 - $400	HCFC-123/HFC-134a	$85	HFC-134a

*** Cost is given in thousands.**
† Retrofit of this unit is not recommended.

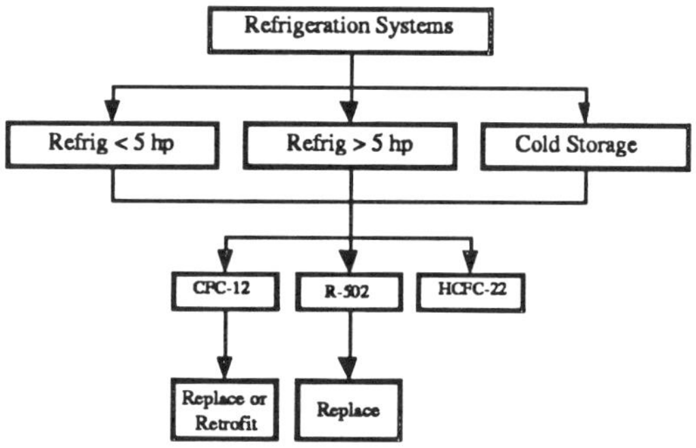

Figure 5. Retrofit and Replacment Options for Refrigerants Used in Refrigeration Systems.

Table 8. Estimated Replacement Costs for Average Size Refrigeration Units

2.5 hp	7.5 hp	35 hp	Explanation
$1,500	$3,500	$20,000	basic system
+ $500	+ $500		misc. parts
$2,000	$4,000		
x 1.25	x 1.25		new refrig. increase
$2,500	$5,000		
+ $2,000	+ $2,500	x 1.5	installation
$4,500	$7,500	$30,000	total cost

the site study and presented earlier, a significant portion of the existing inventory is expected to be new enough that conversion would be economically attractive. Realizing the uncertainty in this and to provide limits for the estimate, the cost is calculated for the range of replacement percentages, 0 to 100 percent. At 100 percent replacement, none of the equipment using CFC refrigerants is retrofitted and, conversely, at 0 percent replacement all CFC-based equipment (those using refrigerants that can be retrofitted) is converted. Retrofit and replacement cost vs. replacement percentage is shown in Figure 6. It is reasonable to expect that some of the equipment will be suitable for retrofit, with the majority requiring replacement. The final cost of compliance will depend on retrofit and replacement strategy, market conditions, and the actual condition of the AC/R equipment inventory.

This methodology was also used to estimate the costs for two installations that had conducted independent cost studies. The comparisons verify the acceptability of the estimates obtained for estimating overall Army costs. For both installations, the cost is within the bounds of the cost estimated by USACERL. A detailed error analysis of the cost estimate is given in Homan and Sohn [1993].

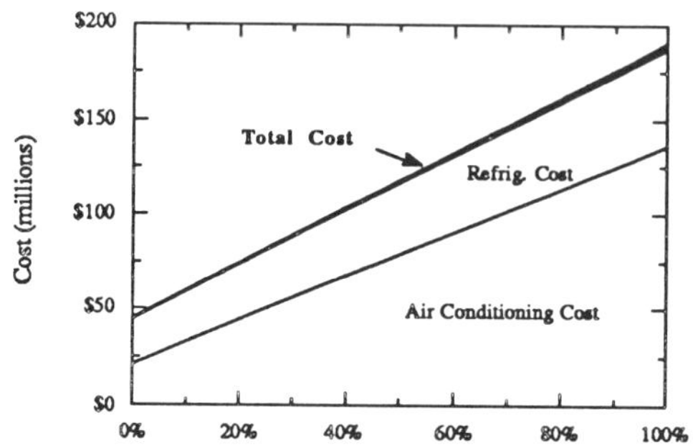

Figure 6. Estimated Retrofit and Replacement Costs vs. Replacement Percentage.

ASSESSMENT OF AVAILABLE OPTIONS

Faced with the production phaseout of CFC refrigerants by January 1, 1996, the AC/R system users are presented with the following options:

(1) maintain and run existing CFC equipment as is
(2) retrofit existing systems with non-CFC refrigerants
(3) replace existing systems with non-CFC systems.

The basic choices are obvious, but the correct choice for any given situation is not as clear. In the next section, the important considerations and questions pertaining to each of the three choices are discussed. These observations are then used to define a number of further studies necessary for selection of the appropriate option. The choice between the options will ideally be made for each CFC-based unit, although it is expected that generalizations will be possible.

Available Options

At least one major AC equipment manufacturer is recommending system users keep equipment with CFCs running as is. This option will probably be most attractive for equipment that is nearing the end of its useful life. Holding off on the purchase of new non-CFC equipment may allow time for more attractive options to make their way into the marketplace. Probably the biggest drawback to the continued use of CFC refrigerants is the increasing cost. As taxes on CFC refrigerants increase and the available supply goes down, the cost of the CFC refrigerants is expected to increase significantly. In the past, a rule of thumb for the refrigerant leakage rate was 5 times the full charge during a 25-year life [Salas and Salas, 1992; Hummel et al., 1991]. Reduction of leakage rates will be necessary not only for economic considerations, but, more importantly, to satisfy statutory requirements. The EPA Final Ruling requires annual leakage rates be less than 15 percent of total charge for comfort cooling (air conditioning) and less than 35 percent of total charge for industrial refrigeration equipment [EPA, 1993]. Therefore, improved techniques for leak detection are an important consideration. USACERL has just completed a market survey on detection equipment which identified 58 refrigerant-detector manufacturers and obtained information on 139 detectors [Tapscott et al., 1993].

Due to the high cost of replacement CFC refrigerant, consideration of a recovery/recycling/reclamation program is logical. This requires the purchase of equipment and training of personnel in its use. Tapscott and Tapscott [1991] provides an exhaustive list of equipment manufacturers. This will serve as an excellent reference for deciding what is available and where it can be obtained.

If running the equipment "as is" is to be successful, a refrigerant management system must be put into place. The system will be necessary not only to meet EPA requirements on inventory reporting by users of systems with over 50 pounds of refrigerant charge [EPA, 1993], but to effectively use refrigerant that is being recovered and recycled. Such a system would also help identify machines that are leaking excessively by maintaining a record of service frequency.

Retrofitting CFC equipment for the use of non-CFC refrigerants is probably the most attractive option for relatively new equipment. In the ideal situation, a retrofit would consist of replacing the CFC refrigerant with a drop-in substitute; however, no drop-in refrigerants have been established at this point. Fortunately, some equipment manufactured in the last few years has been designed to be compatible with CFC replacements. However, this is not true for all equipment, and retrofit costs can be expected to vary considerably depending on make, model, and configuration. A number of engineering considerations must be an essential part of the decision. Mechanical configuration (compressor and vessel), material compatibilities (oil, seals, gaskets and insulation) as well as efficiency and capacity issues must be addressed. Long-term compatibility of replacement refrigerants and the expected life of the retrofit must also be considered. Finally, now that HCFCs are scheduled to be phased out by the year 2030 under the Copenhagen Amendment [Korte, 1993], they may serve only as interim solutions until permanent CFC-alternative systems are fully established in the market.

The final option to be considered is the replacement of the existing CFC equipment with new, non-CFC-based systems. Although this may be appropriate for some systems, replacement of an entire inventory is cost prohibitive. As was noted in the consideration of retrofitting, the ultimate replacement refrigerants remain to be established. Thus, even the purchase of new equipment is not a guarantee of freedom from future phaseout concerns. The replacement option, however, could be an excellent opportunity for overall improvement in the efficiency of the cooling system. This could include the incorporation of an alternative cooling system such as absorption, ammonia (vapor compression), and desiccant.

Further Studies Necessary For Selection of Options

As each option is discussed, it becomes clear that asking the questions is easier than obtaining the answers. However, some of the questions will need to be answered for the correct decisions to be made. A number of studies suggested by the earlier discussion will be a necessary part of an informed decision.

To make appropriate choices, the factors that must considered for option selection must be identified and prioritized. A necessary element of this would be an economic analysis of the options. The previous discussion of the available options will need to be developed more fully. The goal must be a method for ranking the choices among the many options available. Part of the difficulty is that alternative technologies are under rapid development and information may not be available on general retrofit guidelines and the degree of success for retrofits.

Track records from field testing provide necessary information for ensuring the successful operation of retrofitted equipment. Questions about efficiency and capacity loss can be answered by short-term field testing. However, questions about expected life and materials compatibility must be answered with the technical assistance of original equipment manufacturers, since long-term tests would be required for the user. The imminent CFC phaseout precludes waiting for long-term test results.

If replacement of CFC equipment is a logical option, it may be a good opportunity for economical implementation of an alternative air conditioning and refrigeration system. The alternative technologies will be more attractive since the marginal (differential) cost for the unit will be the important factor. As an example, storage cooling systems could be considered where electricity demand charges are significant. Since cool storage systems decouple the instantaneous load from the chillers, the capacity of the installed chillers could be reduced as compared to a conventional system. Where waste heat or idling boilers are available, absorption systems could be used to provide cooling. The need for equipment changes could thus prove to be an excellent time to reduce operating costs and improve reliability and energy efficiency.

CONCLUSIONS

The U.S. Army is a large user of air conditioning and refrigeration systems with over one million tons of cooling capacity in operation. Baseline technical information has been collected for the Army's AC/R equipment inventory based on site studies of three typical Army installations and information compiled in the Redbook. The overall quantity of all types of refrigerants is estimated to be 2.5 million lb. Of this, 1.39 million lb (55.4 percent) are CFC-based. In addition, more than half of the Army's AC/R equipment is expected to be less than 5 years old.

A number of options are available for the Army to comply with current EPA regulations designed to protect the ozone: (1) using existing CFC equipment until the end of its life, with tight leakage control and refrigerant recycling; (2) replacing CFC equipment with AC/R systems using CFC-alternative refrigerants, and (3) retrofitting existing CFC equipment to accept non-CFC refrigerants. In addition, an excellent opportunity exists for economical implementation of other types of AC/R systems such as storage cooling systems. Most importantly, a well planned strategy is urgently needed to make the most of rapidly changing AC/R technology in the most timely and cost-effective manner.

REFERENCES

Bilsland, R. (chlorofluorocarbon Co. Sales Representative), personal communication, Jan. 1993.

Denny, R.J. (Air Conditioning and Refrigeration Institute), personal communication, Aug. 1991.

Facilities Engineering and Housing Annual Summary of Operations, Fiscal Year 1989 (U.S. Army Engineering and Housing Support Center [USAEHSC], Fort Belvoir, VA).

Homan, K.O., and C.W. Sohn, Estimated Retrofit and Replacement Costs of Chlorofluorocarbon Uses in Army Facility Air Conditioning and Refrigeration Equipment, draft Technical Report (USACERL, 1993).

Hummel, K.E., T.P. Nelson, and P.A. Thompson, "Survey of the Use and Emissions of Chlorofluorocarbons from Large Chillers," *ASHRAE Transactions*, Vol 97, pt 2(1991).

Korte, R., "Editor's Page," *Heating/Piping/Air Conditioning*, Vol. 65, No. 4 (April 1993), p. 7.

Salas, C.E., and M. Salas, *Guide to Refrigeration CFCs*, (Fairmont Press, Inc., 1992), p. 6.

Sohn, C.W., K.O. Homan, and B.J. Sliwinski, "Chlorofluorocarbon Uses in Army Facility Air-Conditioning and Refrigeration," USACERL Technical Report FE-93/14/ADA263634, U.S. Army Construction Engineering Research Laboratories [USACERL], December 1992.

Tapscott, R. E., C. W. Sohn, and F. Holcomb, Halocarbon Refrigerant Detection Methods, draft Technical Report (USACERL, 1993).

Tapscott, R. E., and M. K. Tapscott, *Refrigerant Recovery, Recycling, and Reclamation - Meeting the U. S. Clean Air Act* (GlobeTech, Inc., January 1991).

U. S. Environmental Protection Agency, 40 CFC Part 82, "Protection of Stratospheric Ozone: Refrigerant Recycling," *Federal Register*, Vol. 58, No. 92 (Friday, May 14, 1993).

Wendl, D. (Trane Co. Sales Representative), personal communication, Jan. 1993.

Chapter 31

Control of Indoor Air Pressure for Protection of Health, Preservation of Buildings and Conservation of Energy

J.E. Traudt

ABSTRACT

ASHRAE Standard 62-1989, "Ventilation for Acceptable Indoor Air Quality" provides valuable recommendations for providing acceptable indoor air quality without wasting energy.[1]

This paper provides supplemental recommendations on the use of ventilation and air filtration systems to control indoor air pressure and help achieve all four of the following goals:

1. **Protect health** by providing fresh air while minimizing entry of air pollutants and maximizing removal air pollutants released indoors.[2, 3]
2. **Preserve buildings** by reducing or eliminating condensation problems in attics and walls.
3. **Conserve energy** by:
   - Minimizing the entry of pollutants so that less ventilation is required to maintain adequate indoor air quality.
   - Preventing or minimizing moisture accumulation in walls and attics so the insulation in those locations will remain dry and effective.
   - Making it safe to have a building as tight and heavily insulated as anyone desires.
4. **Maintain or improve comfort** by delivering clean fresh air at any desired temperature and velocity.

Special emphasis is given to controlling indoor air pressure to minimize entry of odors, moisture, microbes, radon, and other pollutants through joints, cracks, and pores in walls and floors that contact the soil.

INTRODUCTION

Excessive depressurization of rooms can cause a variety of serious problems:

- Waste of energy if depressurization causes air leakage to be greater than necessary for protection of health during use of heating or air conditioning systems.
- Increased entry of dust, pollen, odors, microbes, etc., from outdoors through unfiltered openings. The American Lung Association estimates that 60,000 Americans — most with lung or heart disease — die prematurely each year from exposure to particles alone.
- Increased entry of odors, moisture, microbes, and radon from the soil. The American Lung Association, Centers for Disease Control, the National Academy of Science, the Surgeon General, the American Public Health Association, and the American Medical Association have all identified radon as a serious national health problem. The EPA estimates that radon may cause more than 20,000 new lung cancer cases in the USA each year.
- Backdrafting (reverse flow) of "natural draft" chimneys.
- Excessive moisture penetration of walls and attics can occur when the outdoor vapor pressure exceeds the indoor air pressure.[4]
- Insulation and other materials within walls and attics may become damp if large quantities of moist outside air are drawn indoors through walls and attics during use of air conditioners. Damp wood can be attacked by fungi. Fungal odors can cause distress to occupants.[5]
- Reduction of exhaust system capacity.
- Increased difficulty in opening exterior doors which swing outward. This is a frequent problem in restaurants with high capacity exhaust systems.

Typical Problems Caused by Excessive Pressurization of Rooms

- Waste of energy if pressurization causes air leakage to be greater than necessary for protection of health during use of heating or air conditioning systems.
- Condensation and moisture damage within attics and outer walls may occur if large quantities of moist indoor air are forced out through walls and attics during cold weather. Damp insulation is less effective. Damp wood can be attacked by fungi. Damp materials can become breeding grounds for microbes that are dangerous to human health.

VENTILATION AND FILTRATION

During the heating and air conditioning seasons, the amount of fresh air required for protection of health, and thus the cost of operating a HVAC system, can be minimized only by minimizing the amount of pollutants that are:

a) entering from outside;

b) released inside; and

c) entering from the soil.

A heating, ventilating, and air conditioning (HVAC) system can deliver fresh air after removing bacteria, pollen, dust and many other impurities that are often present in outdoor air. Air exhaust systems can help to remove air pollutants at their source and prevent them from spreading throughout a building.

Use of a HVAC system can make it unnecessary to have permanent air leaks that often allow entry of unfiltered outdoor air that is too cold, too humid, too hot, or too contaminated with dust, pollen, and microbes.

Use of a modern HVAC system and the elimination of undesirable air leaks makes it possible to prevent excessive depressurization or excessive pressurization of rooms.

What Can Cause Excessive Pressurization and Depressurization of Rooms?

The air pressure in each room may result from the simultaneous action of:[6]

- An unbalanced forced-air circulation system.
- Fresh air supply systems.
- Air exhaust systems.
- Thermal buoyancy ("stack effect") in heated rooms during cold weather.
- Wind pressures causing air to leak in on the upwind side.
- Wind pressures causing air to leak out on the downwind side.

Figure 1 shows how even moderate wind speeds can create significant levels of positive pressure on upwind walls and negative pressure on downwind walls.

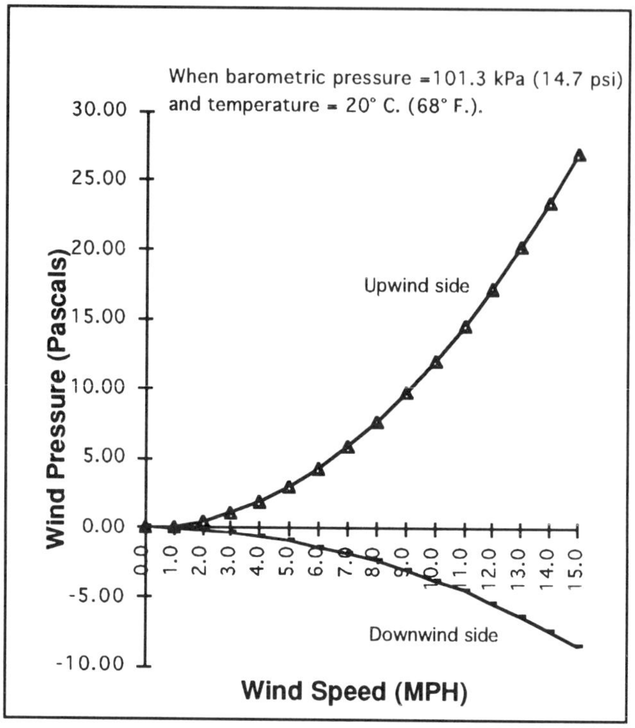

Figure 1: Wind Pressures on a Building

During one study of twelve ordinary homes in Florida, the HVAC systems were found to be so poorly balanced that their operation caused an average of 58% of an air change per hour (ACH) with interior doors open and 99% of an ACH with interior doors closed.[7] These ACH levels were much higher than the 35% ACH recommended by ASHRAE Standard 62-1989 for acceptable indoor air quality in homes.

HVAC system balance can be tested by using an air pressure differential gage to compare the air pressure in a closed room with the air pressure in a hallway or another room.

PRESERVATION OF BUILDINGS

Preservation of buildings requires keeping building materials from becoming wet enough to be attacked by fungi, mold, and other microorganisms. In some buildings, condensation from moist air in attics and walls has enabled fungi and other microorganisms to seriously damage wood and other materials.

Sealing of undesirable air leaks and proper control of indoor air pressure can minimize the amount of moist air entering attics and walls from the inside during cold weather and from the outside during warm humid weather.

AIR LEAKAGE THROUGH SHELL OF BUILDING

Air can be considered to be an incompressible fluid under the conditions expected during use of a blower door to test the air leakage characteristics of buildings.

Therefore, the air flow through an opening is described by Bernoulli's equation:

$$P = .5 * D * V^2$$

where

P = Pressure difference across the opening.

D = Density of air.

V = Velocity of air through the opening

Solving for the velocity:

$$V = (2 * P/D)^{.5}$$

However, Kirschoff showed that the velocity is reduced due to flow contraction and this can be accounted for by introducing a discharge coefficient (K = .61):

$$V = K*(2 * P/D)^{.5} = .61*(2 * P/D)^{.5}$$

The air tightness of a structure is experimentally determined by measuring the flow at several pressures and fitting this data to a flow equation. It has been shown that the equation with the best fit to the pressure and flow data is as follows:

$$Q = C * P^N$$

where

Q = Flow rate

C = Flow coefficient

P = Pressure differential

N = Flow exponent

Once C and N are determined by a log-linearized least squares curve-fit procedure, the flow can be predicted at any other pressure.

The air tightness of a structure can be described as an Equivalent Leakage Area (ELA) by using the previous equations:

$$ELA = Q/V = C * P^{N-.5}/(.61 * 2/D)^{.5})$$

This area is not the exact cross sectional area of all cracks in the structure, but it is the area of a sharp edged hole in a thin wall that would give the same air leakage.

Experimental air leakage data can be fitted to an equation to predict the flow at any pressure and the leakage can be characterized by an equivalent leakage area.

Example:

A blower door was used to estimate leakage characteristics of a building. Figure 2 provides an estimate of the total air leakage per 100 square inches of leaks in the building.

The "Passive" portion of Figure 3 illustrates the wild fluctuations in ventilation rates that are typical of the nearly 100 million homes in the USA.[8] The "Controlled" portion illustrates how adequate ventilation can always be provided in a building that is tight enough to minimize excessive air leakage at all times.

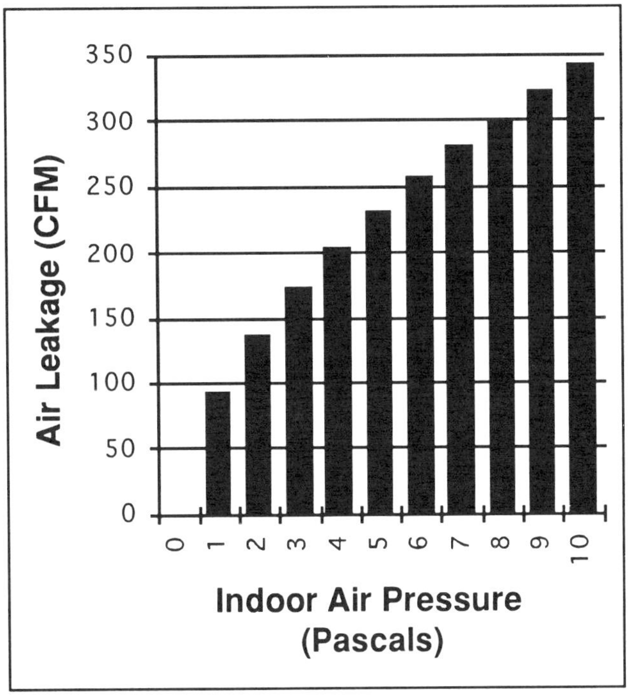

Figure 2: Air Leakage per 100 Square Inches of Leaks in Shell of Building

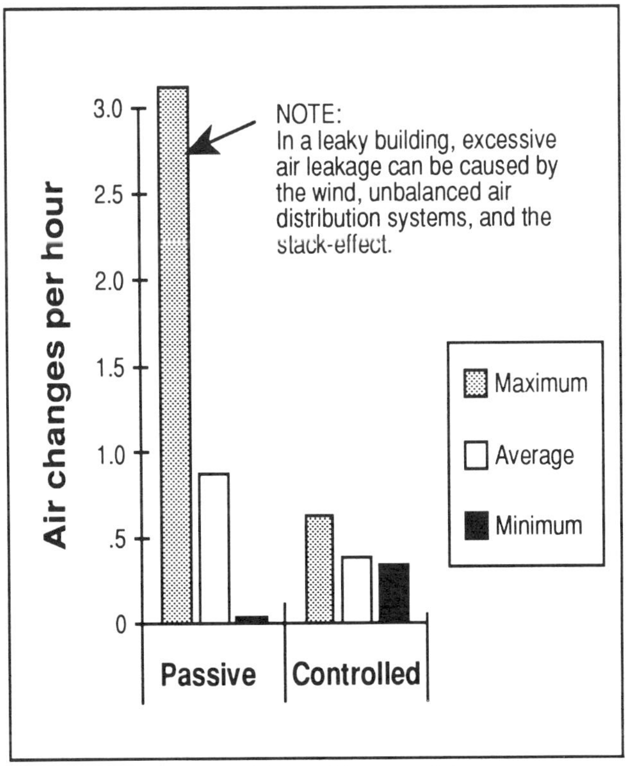

Figure 3: Typical Annual Range of the Ventilation Rate in Houses (when doors and windows are closed).

AVOIDING DEPRESSURIZATION OF ROOMS WITH WALLS OR FLOORS IN CONTACT WITH THE SOIL

Depressurization of rooms with floors and walls in contact with the soil has been identified by the EPA as the leading cause for entry of radon gas from the soil through joints, cracks, and pores.[8] Along with radon, depressurization of the rooms often causes the entry of other undesirable things from the soil, including:

- Odors
- Moisture
- Fungi, bacteria, mold spores, and other microbes.
- Fumes from any insecticides applied to nearby soil on the exterior.
- Fumes from any long-life termite insecticides that have been placed under the slab.

Use of HVAC systems can cause severe depressurization of lower rooms. During a recent study in the State New York State, 80% of the homes with forced-air furnaces had air handlers that were depressurizing the basements.[9] Furnaces, water heaters, and any other combustion system operating in depressurized rooms are less likely to operate safely.

To avoid depressurization or to maintain a very slight positive indoor air pressure (relative to the pressure of gasses in the soil), there is no reliable substitute for actually using the pressure of gasses in the soil as a reference.[10]

An air pressure differential gage (see Figures 4 & 5) or air pressure difference indicator (see Figure 6) can provide a timely warning when indoor air pressure becomes lower than the pressure of gasses in the soil.

Prevention of depressurization can be done manually or by use of an automatic system. When air is being removed from a room, makeup air can be provided directly from outside or from a room that can be safely depressurized.

If prevention of depressurization does not lower the indoor radon level as much as desired, it may be due to diffusion through floors and walls. Concrete is a porous solid. The concrete curing method can change the coefficients of both permeability and diffusion by more than one order of magnitude.[11]

The flow rate of gas through concrete appears to be inversely proportional to the thickness of the slab and directly proportional to the pressure differential.[12]

To counteract entry of radon by diffusion, about 4 Pascals (0.015" w.c.) of indoor air pressure relative to the soil is usually adequate.[13]

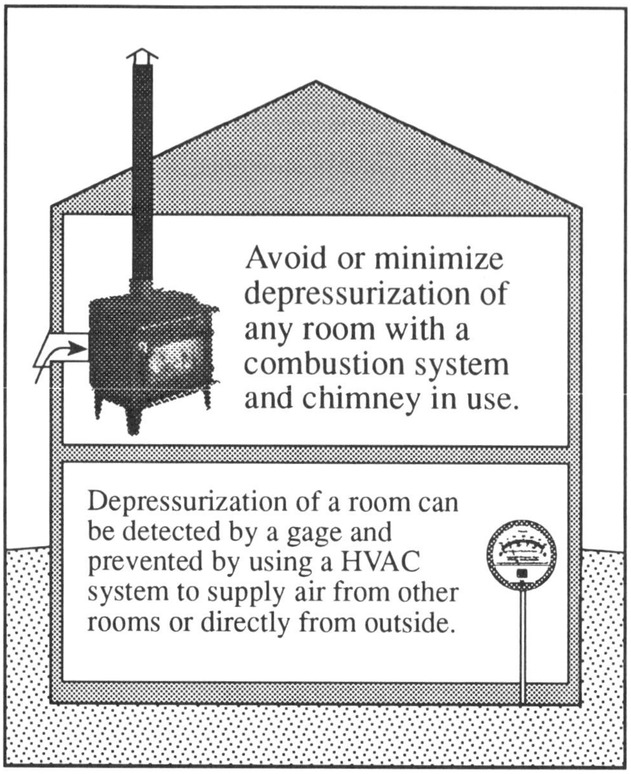

Figure 4: Use of Indoor Air Pressure Gage

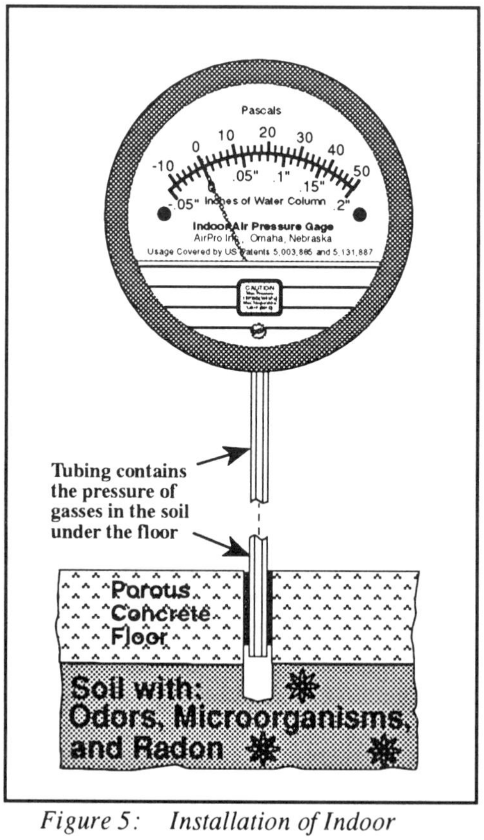

Figure 5: Installation of Indoor Air Pressure Gage

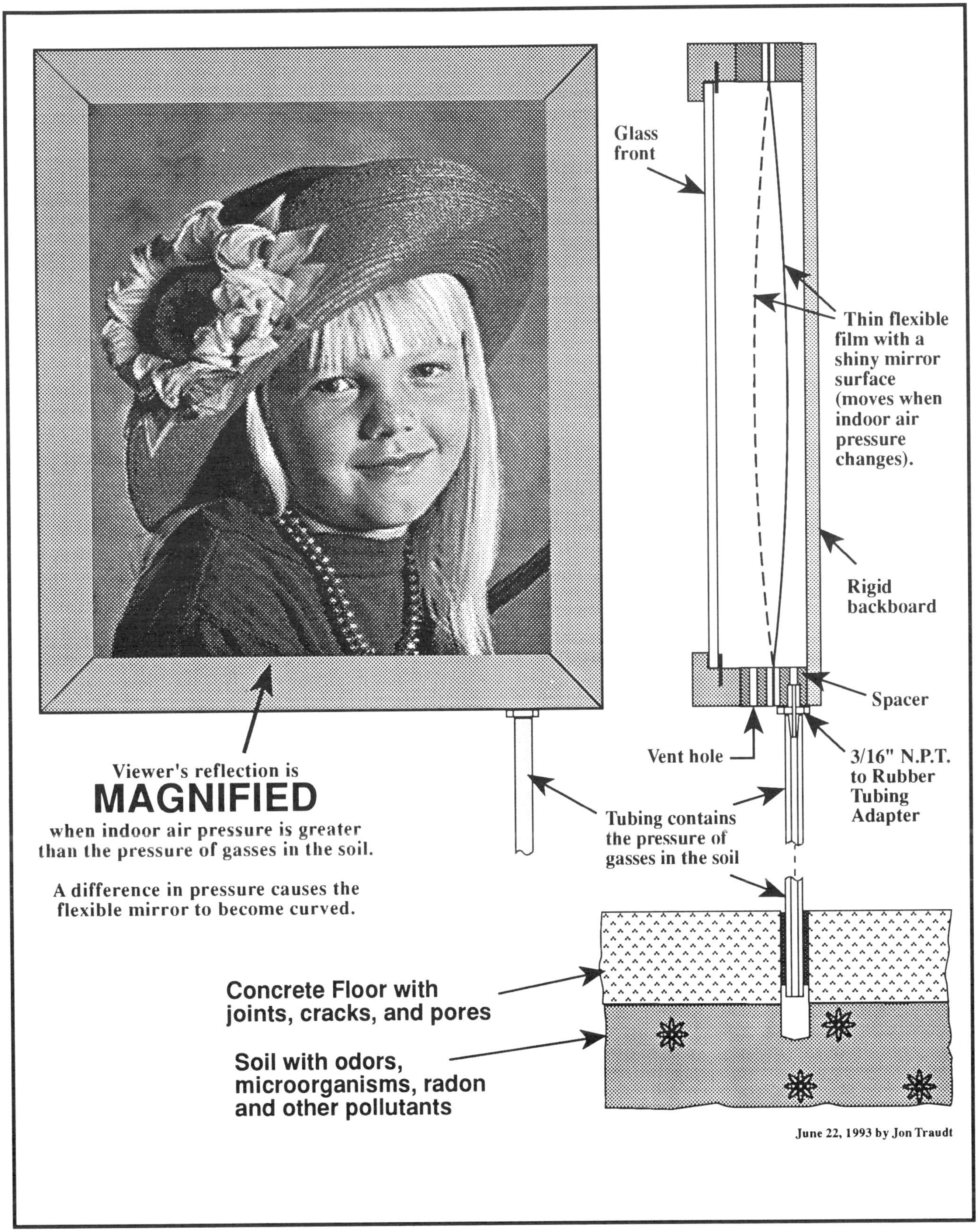

Figure 6: *Flexible Film Mirror (used to detect when indoor air pressure is less than the pressure of radon and other gasses in the soil)*

DETECTING PROBLEMS

No man-made system is 100% reliable. Activities and the number of people in a building may change from time-to-time. Therefore, it is useful to have some way to detect when HVAC systems need to be adjusted or repaired. Commercially available instruments for monitoring indoor air quality and pressure include:

- Carbon dioxide detector (signals when additional ventilation is required).
- Indoor air pressure gage (indicates when indoor air pressure is too high or too low).
- Smoke detector (indicates when products of combustion are present).
- Heat detector (indicates when hot combustion gasses are flowing indoors instead of up through a chimney).
- Carbon monoxide detector (indicates when indoor levels begin to approach dangerous levels).

Problem detection systems should only be used as a "second line of defense". Damage to health can occur from prolonged inhalation of pollutants even if their level remains too low to trigger an immediate alarm. Prevention of problems should always be the "first line of defense".

PREVENTING PROBLEMS

A group of EPA researchers used sophisticated equipment to study exposure of householders to indoor and outdoor air pollutants. The researchers were very surprised: ***No matter how clean or dirty the city, indoor air pollution levels were much worse than outdoor levels.***[2]

The EPA and the American Lung Association both agree that *indoor air pollution* is one of the biggest health threats facing Americans today. Many buildings have polluted outdoor air leaking in through unfiltered openings. Once indoors, the air usually becomes even more polluted by the addition of gasses and particles from materials used indoors. The American Lung Association cites recent studies indicating that airborne pollutants account for about 80% of the pollutants absorbed by a typical person in the USA. Some Medical Doctors believe that air pollution levels commonly found in houses are strong enough to damage the human immune system. The average American is at home 70% of the time (inhaling about 5.6 million times each year). Even very low levels of some common air pollutants in any building where a person spends many hours can gradually cause health problems.

Indoor air pollution can cause symptoms such as headache, lethargy, nausea, or eye irritation, and may show neurological or behavioral effects, such as impaired manual dexterity or reduced ability to perceive time intervals accurately.

Rooms with Chimneys in Operation

Natural-draft chimneys sometimes fail to operate safely when used in a room that is depressurized. Backdrafting is most likely to occur when wind speed is too low to create suction at the top of chimneys by the Venturi effect. Whenever possible, use a power-vented or sealed-combustion system. Compared to combustion systems that use natural-draft chimneys, these modern systems are much safer to use and some of them are much more efficient.

Passive combustion air inlets are usually inadequate to provide sufficient air flow to relieve negative pressure caused by large or numerous exhaust systems or an unbalanced forced air circulation system. When a passive combustion air inlet is open on the downwind side of a building, it may become an air outlet and thus increase combustion room depressurization.

A furnace that uses oil or natural gas fuel can be equipped with a heat detector to shut-off fuel flow when chimney backdrafting or flue gas spillage continues for more than one minute.

Rooms with Exhaust Systems in Operation

Operating of an exhaust system in a room with excessive depressurization can operate prevent the system from operating at full capacity. A fresh air supply system, preferably with an air filter, can provide the necessary makeup air and prevent excessive depressurization of rooms.

Rooms with floors or walls that do not contact the soil

During the winter, maintain a neutral or only slightly negative indoor air pressure to help:

- Prevent the waste of energy that results if excessive depressurization of rooms causes excessive ventilation by pulling too much air in through openings.
- Prevent the waste of energy that results if insulation becomes wet and less effective due to condensation of moisture from too much moist air leaking out through cold attics and walls.
- Help prevent damage to wood and other materials in attics and walls (moisture from moist outgoing air can condense within attics and walls of buildings during cold weather).

During the summer, maintain a neutral or only slightly positive indoor air pressure to help:

- Prevent the waste of energy that results if too much air leaks out during operation of an air conditioner.
- Prevent moisture damage to wood and other materials in attics and walls when an air conditioner is used during humid weather. Moisture from humid air within walls and ceiling materials will condense if the air becomes cooled below the dew point. In some buildings, the condensed moisture has accumulated enough to enable growth of fungi and other microbes.
- Reduce entry of dust, pollen, microorganisms, and other pollutants that can enter from outdoors through unfiltered openings.

CASE HISTORIES

Christopher Smith's House

Christopher Smith was eight years old, living in an ordinary leaky house, and suffering from severe allergies. A fresh air supply system was installed and set according to ASHRAE Standard 62-1989 recommendations. Incoming air is now filtered at point-of-entry. Then the house was tightened to minimize entry of unfiltered air. Several benefits have been noticed by Christopher and his family:

- Christopher feels much better now that he is exposed to allergens and other air pollutants.
- Reduction of allergens entering the home has reduced Christopher's daily exposure and substantially reduced the frequency of allergy symptoms occurrence.
- The home interior is easier to keep clean since the incoming air is now filtered as it enters.
- Utility bills decreased after the house was tightened.
- Prevention of basement depressurization has reduced radon levels below the 4 picoCuries per liter "Action Level" set by EPA.

Ron Wilhelm's House

In Lincoln, Nebraska, a leaky ten year old house nearly killed the Ron Wilhelm family, including three children.

One evening, a fire in the living room fireplace burned down to embers, and the wind became calm. During the night, carbon monoxide filled the house. Early the next morning, after the family woke-up, all three children fell unconscious. A slightly higher concentration of, or longer exposure to, carbon monoxide might have resulted in fatalities.

The Wilhelm house is leakier than many other homes. A blower door test of the 2,000 square foot house indicated 2,600 CFM at 50 Pascals. This is about 60% leakier than many houses that have been professionally weatherized.

A carbon monoxide (CO) detector was installed to provide a warning in case a chimney backdrafted again. The CO alarm has sounded several times each year. When the wind is low, a strongly burning fire in the open fireplace can depressurize the house enough to backdraft the furnace. After the fire in the fireplace has burned down to embers, operation of the furnace can depressurize the living room enough to backdraft the fireplace chimney.

This story illustrates why a leaky building can be much more dangerous than is commonly assumed. Users of combustion sysems with chimneys should not rely on air leaks to provide make-up air and prevent excessive depressurization.

My House

Substantial reductions in the amount of energy required for heating and air conditioning have occurred in homes after installation of ventilation and indoor air pressure control systems followed by sealing leaks to minimize excessive air leakage.

After installing a ventilation system and sealing many air leaks five years ago, the heating costs in my house in Omaha have averaged about $100 per year. The minimum ventilation rate was raised to about 35% of an air change per hour as recommended by ASHRAE Standard 62-1989. Total area of heated floor space is 1,800 square feet. All walls have 3.5" of fiberglass insulation. The natural gas furnace is rated at 90% efficiency. Omaha typically has 6,400 heating degree days per winter. Part of the savings were due to minimizing excessive air leakage. Some of the savings were due to reducing the amount of moisture condensing in the insulation within the walls and attics.

The original goal in my 1,800 square foot home in Omaha, Nebraska was to reduce radon from the initial 24 pCi/L to less than 4 pCi/L. This goal was achieved by tightening the house and avoiding depressurization of rooms with floors and walls in contact with the soil.[14]

Phone Company Office

Failure to detect excessive depressurization can be dangerous to human health.

An attic ventilator had been turned on during the summer to exhaust hot air. The fresh air inlet (a screened opening) to boiler room became clogged by cottonwood tree seeds during the fall. Continued operation of the attic exhaust system during the winter, along with the clogged fresh air inlet in the boiler room, caused the indoor air pressure to become so low that the boiler exhaust fumes could not flow up the chimney, but instead were drawn through the office area. The exhaust fumes were not concentrated enough to cause fatalities, however, inhalation of fumes from combustion systems have been found to weaken the immune system and cause headaches, eye irritation, respiratory distress, etc.

Use of a simple indoor air pressure gage could have provided a timely warning that indoor air pressure was dangerously low.

CONCLUSIONS

A tight and heavily insulated building with a modern HVAC system can offer a highly desirable combination of comfort, good indoor air quality, and minimum energy usage.

REFERENCES

1. ASHRAE Standard 62-1989. The American Society of Heating, Refrigerating, and Air-Conditioning Engineers, Inc., 1791 Tullie Circle, NE Atlanta, GA 30329. (Cost $40. Discount available to members).
2. Total Exposure Assessment Methodology (TEAM) Study: Summary and Analysis, Volume 1 (Washington, D. C.: U. S. Environmental Protection Agency, L. A. Wallace, 1987).
3. "The Usage of Filtration as Fulfillment of Acceptable Indoor and Optimal Energy Management", World Environmental and Energy Congress, 1992. H. E. Burroughs,
4. "Fundamentals of Simultaneous Heat and Moisture Transfer Between the Building Envelope and the Conditioned Space Air", ASHRAE Transactions, Vol. 96, Pt. 2. 1990. S. P. W. Wong, and S. K. Wang.
5. "Microbial Contaminants, Human Allergies, Building Deterioration: Remedial Requirements", World Environmental and Energy Congress, 1992. W. Rhodes and Y. O. Gilyard.
6. ASHRAE Fundamentals Handbook. The American Society of Heating, Refrigerating, and Air-Conditioning Engineers, Inc., 1791 Tullie Circle, NE Atlanta, GA 30329
7. "Infiltration Rates and Pressure Differences in Florida Homes Caused by Closed Interior Doors When the Central Air Handler is On", James B. Cummings of Florida Solar Energy Center and John J. Tooley, Jr. of Natural Florida Retrofit.
8. "The Inside Story. A Guide to Indoor Air Quality". United States Environmental Protection Agency, September 1988. EPA/400/1-88/004.
9. "Air Leakage and the Effects of Mechanical Ventilation: Measurements in Fifty Newer Homes in New York State". SYN TR 93-516, Prepared for The New York State Energy Research and Development Authority, April 1993. Peter R. Strunk and Dr. Larry Kinney.
10. "Pressure Controlled Fresh Air Supply Ventilation System, Soil Gas Pressure as a Reference, and Method of Use", U.S. Patents 5,003,865 issued 1991 and 5,131,887 issued 1992. Jon Traudt. Canadian patents applied for.
11. "Pore Structure and Permeability of Cementitious Materials". Published by Materials Research Society, Research Center of the Cement Industry, Tannenstrasse 2,400 Dusseldorf 30, Germany. Volume 137, Page 346, 1989. Herbert Graf and Max J. Setzer.
12. Handbook of Concrete Engineering. A Nostrand Reinhold book. Edited by Mark Fintel.
13. "Radon Reduction Techniques for Detached Houses," Page 154, EPA/625/5-87/019.
14. "Ventilation Rate and Indoor Air Pressure Control. A Systems Approach to Minimizing Indoor Levels of Radon and Other Dangerous Air Pollutants", 1992. Jon Traudt. Grand Prize — Practical Design, 1992 Innovative Radon Mitigation Design Competition sponsored by EPA, AEE, and EEMI.

Chapter 32

Indoor Air Quality and Energy: A Common Sense Approach

C.F. Wiles

ABSTRACT

The reason to improve indoor air quality is the same reason that money is spent on energy... to satisfy the occupants. Hence, it makes sense to use occupancy as the common denominator for indoor air quality and energy in determining when to ventilate and when not to ventilate. Some refer to this technique as Demand Controlled Ventilation. New technology has allowed this author to analyze well over a million data points within the last year. This article shows that some conclusions previously reached after much calculation are now easily substantiated through extended time monitoring of the indoor environment of occupied spaces using carbon dioxide and temperatures as an index.

INTRODUCTION

In our present-day efforts to "down size" and "right size" corporations, buildings become retrofitted to meet current needs. Walls get changed and more heat and pollution sources (i.e. people, computers, etc.) are shoved into areas that were not originally designed for these new conditions.

ASHRAE Standard 62-1989 provides for two different ways of determining ventilation for buildings. The first way, used in the design and commissioning of a building, is referred to as the Ventilation Rate. The second way, used in the determining of the indoor air quality of an occupied building, is referred to as the Indoor Air Quality Procedure.[1]

This paper will elaborate on the use of extended real time monitoring using the Indoor Air Quality Procedure as the first step in investigating indoor air quality, even when excessive amounts of known contaminants are to be remediated.

It will also elaborate on using non-interactive, non-dispersive, infrared carbon dioxide controllers and Demand Controlled Ventilation as the best way to <u>maximize energy savings while providing acceptable indoor air quality</u>.

THE CARBON DIOXIDE PREMISE

Carbon dioxide has become the surrogate of choice for several reasons. It is a common, inert gas; it degrades slowly; it does not readily interact with other gases; and it is not easily removed without purging. Hence, carbon dioxide becomes a very good index of the ventilation, or the lack of ventilation, within a building area. Keep in mind that we are not necessarily concerned about the health effects of carbon dioxide; we are merely using it as a surrogate or an index.

The premise is simple... if the carbon dioxide levels are high, it can be assumed that anything and everything is rolling over too many times. The HVAC system is simply recirculating contaminants... be it carbon dioxide, perfume, cleaning compounds, or whatever.

THE VENTILATION PREMISE

Inadequate ventilation can cause many things... hence, it accounts for approximately 52% of NIOSH's explanation in 1987 for Sick Building Syndrome.[2]

Consider this... just as there are those who are sensitive to very low levels of one irritant, others may be sensitive to combinations of low level irritants... irritants too low to measure. Toxicologists refer to those with this type of condition as having Multiple Chemical Sensitivities.[3] If the levels of these unknown irritants or chemicals are so low that they cannot be measured, common sense should tell us to increase the building's ventilation to meet the recommendations of the National Standard.

ACCEPTABLE LIMITS

The American Society of Heating, Refrigeration and Air-Conditioning Engineers, Inc., under ASHRAE Standard 62-1989, has established the maximum acceptable indoor air quality limit of carbon dioxide as 1000 ppm. [1]

The California Energy Commission established a maximum acceptable indoor air quality limit of 800 ppm of carbon dioxide for new and retrofitted buildings, effective January 1, 1993.[4]

The United States Air Force, in publication AL-TR-1992-0016, Guide for Indoor Air Quality Surveys, states that "...concentrations above 600 parts per million (ppm) are the cause of some specific IAQ irritations..." [5]

Air Force experts further state that "In our experience, between 15% and 33% of the population will have symptoms when the level is between 600 and 800 ppm. Roughly one-third to one-half become symptomatic between 800 and 1000 ppm, and virtually everyone will have some or all the symptoms when the level is above 1,500 ppm." [5]

Interestingly enough, this same U.S.A.F. report goes on to say that "ASHRAE 62-1989 points out that a concentration of 1,000 ppm CO_2 is not considered a health risk. Many IAQ researchers unfortunately interpret 'no health risk' to mean the same as 'no physiological effect,' and waste their resources looking for a mystery contaminant that is causing discomfort in workers." [5]

In this author's opinion, unless substantial carbon panels are used as part of the IAQ remediation process, the U.S.A.F. report and quoted comments parallel his experience. Carbon collection panels (adsorption filters) remove many of the odors (triggers) that are the underlying cause of the physiological effects, but do not negate the need for adequate ventilation.

NEW TECHNOLOGY

With new technology providing us with non-interactive, non-dispersive infrared carbon dioxide monitors and controllers that have less than 100 ppm drift per year, we can now use extended real time monitoring of this accepted surrogate, along with other pertinent information such as temperature and humidity, to create a record of what is taking place in the building's envelope within 1 ppm.

THE NEED TO MONITOR

Proving Ventilation

The building management may believe that they have adequate ventilation, but can they prove it? Most buildings that have been in service for a few years have become retrofitted to satisfy changing needs. Walls are moved, return and supply grilles are changed and additional people and their computers get shoved into areas that were not originally designed for these new uses. The initial blueprints are now the "as builts" and the "Ventilation Rate" that was used to commission the building is no longer valid. Some buildings open the fresh air louvers a fixed percentage and assume that will take care of the IAQ problem... not energy efficient. Again, just because one thinks that ventilation is adequate does not prove it.

Creating A Record

The need for proving acceptable ventilation and keeping records of the ventilation within buildings is going to become more and more important as the issue of Indoor Air Quality becomes more prominent.

Tenant / Owner Relationships

Ventilation monitoring can be done to entice new tenants by proving acceptable IAQ. Leases can be manipulated or renegotiated on the basis of the building's ventilation records.

Augmenting An Air Balance

A properly balanced air distribution system does not always mean that the occupants of an area are getting the proper air exchange. Take the example of the person who wants to be able to see the entire office, hence the desk is positioned in the far corner. The supply and return grilles are not normally located in the extremes of the room. Since the occupant sits away from the grilles, the air exchange rate that he gets is much diluted.

Checking VAV boxes

Large rooms, common to core designed buildings, with occupants loads that fluctuate may have VAV boxes that are malfunctioning. Extended real time monitoring is an excellent way of checking the performance of VAV boxes.

EXTENDED REAL TIME MONITORING

How Long To Monitor

Experience has shown that seven (7) days is a good length of time for extended real time monitoring. During the period of a week, most of the normal activities within the building are going to be carried out. Some people will have been off sick or on leave. Others will have been out on sales calls or running business-related errands and meetings will have been held, whether in the conference room or at the water fountain. In addition, the weekend will allow us to experience the lowest levels possible. Flat lines in the unoccupied hours make outside (baseline) carbon dioxide monitoring unnecessary in most cases. In short, unlike "grab samples," which usually are impossible to duplicate, a week of extended real time monitoring will capture most of the normal occupant activities.

Why Not Short Term Monitoring ?

Why isn't the 8-hour monitoring acceptable in the graph (Figure #1) done by the author in 1992 indicating the carbon dioxide in a doctor's waiting room in Phoenix, Arizona? Because it really doesn't give enough information. Perhaps this was one of the two days that the office had scheduled in pre-schoolers for shots. The end result might be recommendations based on one of the busiest days of the year. Monitoring conditions should be able to be duplicated as much as possible should a later verification be necessary.

How Often To Monitor

Again, experience has shown that recording data every five (5) minutes allows collection of pertinent data without a lot of unnecessary repetition. Most meaningful occupancy and HVAC conditions will occur within a five (5) minute span.

Results of Monitoring

Extended real time monitoring of the building's indoor air can do several things.

1) When used as the "first step" investigative tool, it creates a "baseline" for the building by recording the ventilation during the monitoring period.

2) If the results of the newly created "baseline" are good, it becomes a permanent record for the building should questions of indoor air quality become an issue in the near future.

3) If the results of the newly created "baseline" are not acceptable, it becomes useful in making recommendations

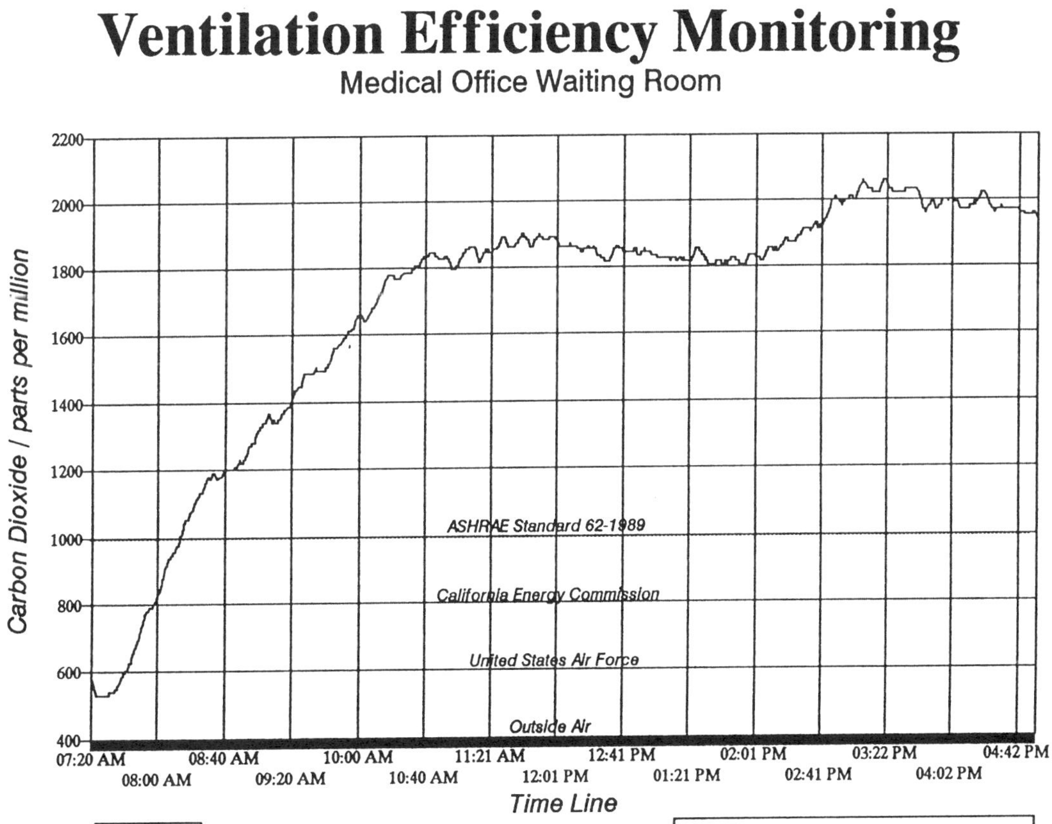

FIGURE #1

SOURCE: Metro Environmental Company

for change. It should be noted here that the ASHRAE 62-1989 Ventilation Rate Procedure may need to be used if the HVAC system requires extensive changes. After changes are made to improve the indoor air quality via improved or increased ventilation, another ventilation monitoring needs to be done to verify that the ventilation complies with the ASHRAE Indoor Air Quality Procedure.

COSTS

IAQ Costs

Much has been written about the cost of energy versus the cost of IAQ. Absenteeism, low productivity and possible litigation all have to be weighed against the cost of HVAC. A lot of the IAQ costs are theoretical whereas the costs associated with HVAC are very real. But, the Air Force has figured that the actual savings of running an HVAC system at 10% outside air versus 20% outside air would cause losses due to infections at 10.5 times the possible energy savings.[6]

HVAC Costs

Energy becomes a concern with the addition of any fresh air into the indoor environment. HVAC simply costs energy dollars. At a minimum, the cost of filtration is a concern. And if we add fresh air during periods in which the outside air temperatures are extreme, costs to condition that air become even more exaggerated. Heat Recovery Ventilators offer some savings of the necessary costs associated with additional ventilation.

A SIMPLE IAQ / ENERGY CASE STUDY

Given the costs involved with any fresh air added to the building, doesn't it make common sense to ventilate for Indoor Air Quality only when you need to? Take the example of the following three graphs (Figures #2, #3 and #4) of a 1500 square foot area occupied by approximately 7 persons servicing banking customers.

Figure #2 is a 5-day monitoring of a banking establishment in which the complaints of Indoor Air Quality necessitated the relocation of part of the employees pending a site investigation, after the employer's decision to clean the ducts failed to mitigate the problem. Note that a ventilation monitoring was done as part of the investigation and that the carbon dioxide exceeded the ASHRAE maximum limits. Recommendations included adding fresh air to the area; there were no provisions for fresh air in the building's ceiling-mounted heat pumps.

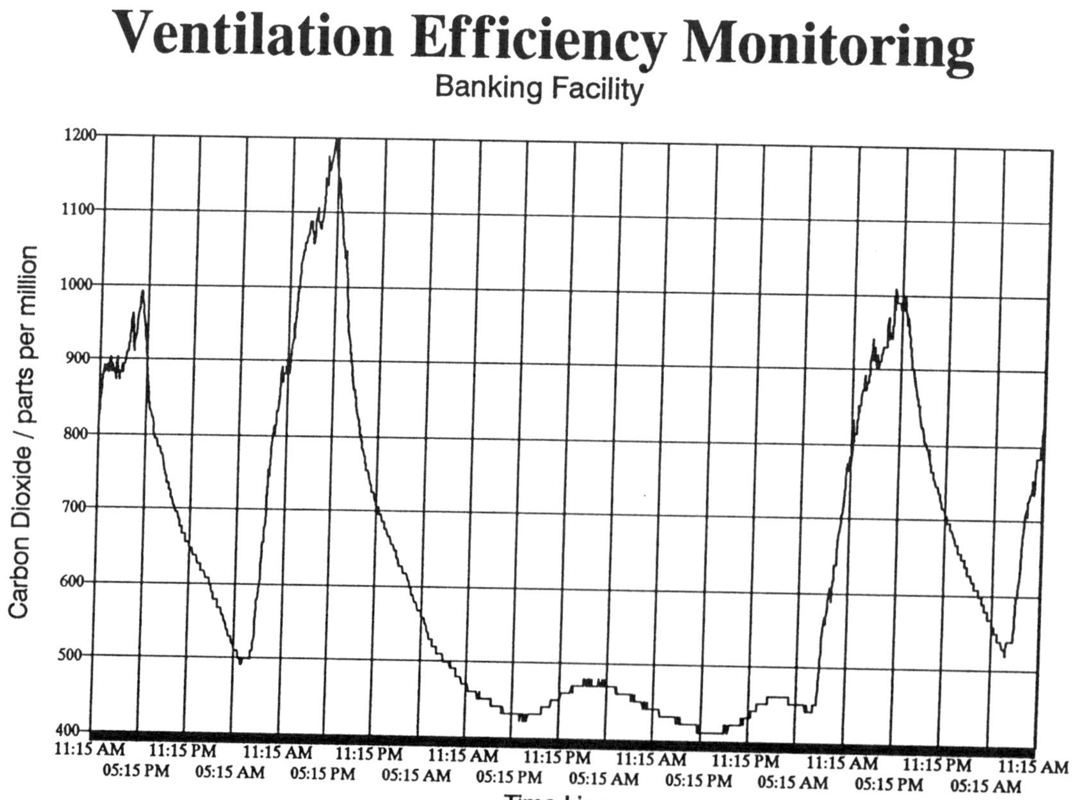

FIGURE #2

SOURCE: Metro Environmental Co.

Ventilation Efficiency Monitoring

Banking Facility

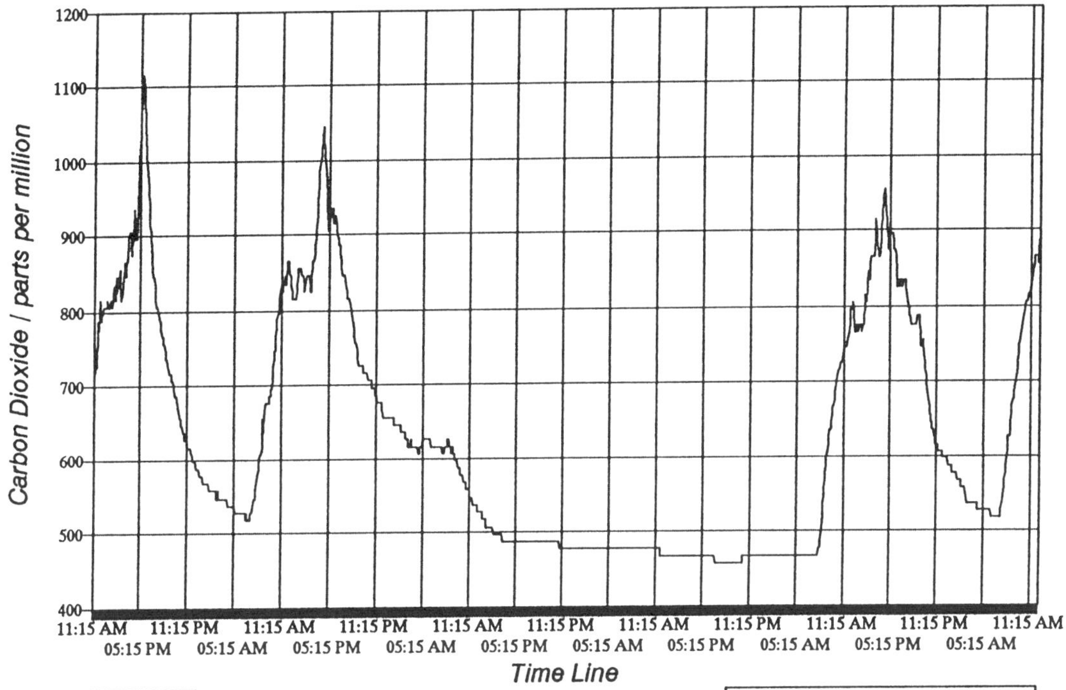

FIGURE #3

SOURCE: Metro Environmental Co.

Ventilation Efficiency Monitoring

Banking Facility

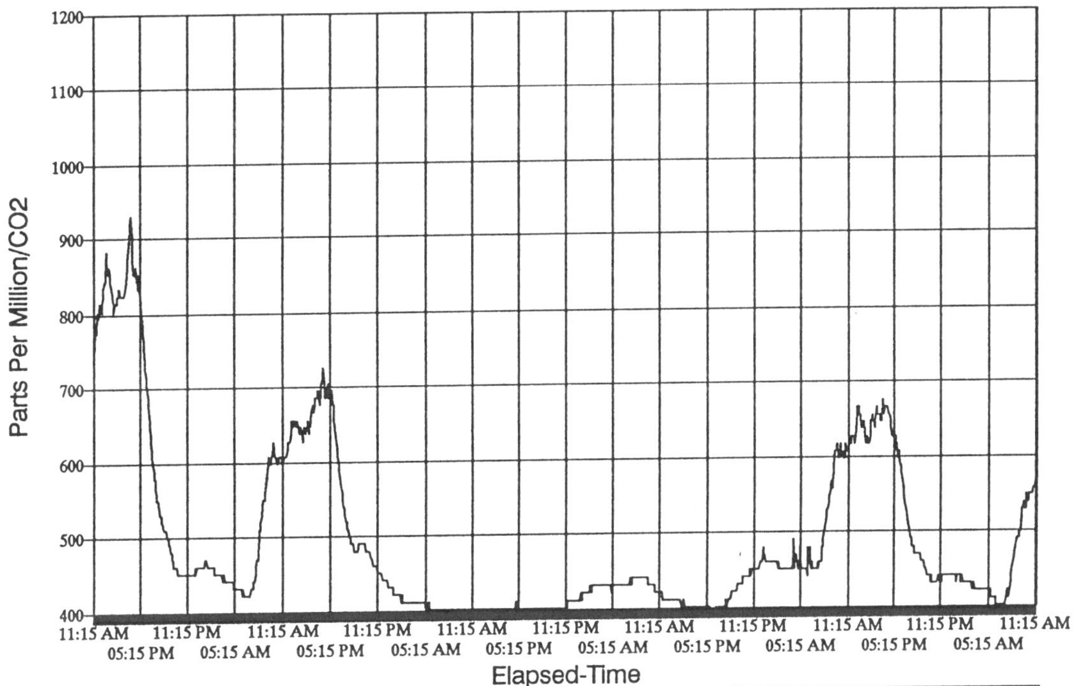

FIGURE #4

SOURCE: Metro Environmental Co.

Figure #3 is a 5-day monitoring of the same area after the employer's "in-house" personnel, who were very HVAC knowledgeable, added fresh air by bringing in outside air via flexible duct to the filter plenum of the ceiling-mounted heat pumps. The logic was that the passive fresh air would mix with the return air when the doors were used by customers. Wrong.

Figure #4 shows monitoring results, again for 5 days, after fans were installed to power the fresh air into the building... Note that during the first day of the graphing, the power was not yet hooked up until after 3 p.m., but the main entrance door (located 30 feet from the monitoring station) was wedged fully open all day. This entrance door being open accounted for little drop in the CO_2. The second day (Friday) is the first full day that power is provided to the fresh air fans. Note the overkill ! Now we have a problem of too much fresh air, hence too much energy cost.

There is no need to run complicated calculations comparing IAQ and Energy. Without the powered ventilation, there is poor Indoor Air Quality. With powered ventilation, it is too costly to condition the environment. Controlling the power ventilation run time to when the occupancy load needs ventilation is the only common sense answer to this building's Indoor Air Quality and Energy problem.

By the way, the employees were allowed to return to the building immediately after the graphing shown in Figure #4. Previous microbial testing, done at the beginning of the first monitoring period using an Andersen Sampler, indicated no microbial concerns. Besides ventilation, carbon adsorption panels containing 300 grams per square foot of 60% activated carbon were installed in the filter racks. There have been no IAQ complaints since the ventilation and carbon panels were added.

IAQ AND ENERGY SUMMARY

Occupancy is the common denominator. When ventilating for Indoor Air Quality, use a carbon dioxide controller to ventilate only when needed... for only as long as necessary to keep the building in compliance with the ASHRAE Standard 62-1989 and to maximize energy savings.

CO_2 controllers are energy efficient in occupied areas. Controllers are available from several companies at a nominal cost. Located on a wall, much like a thermostat, they can be utilized in simplistic methods, such as turning on an exhaust fan when needed. Located in an aspiration box on the side of a large return plenum, they can be incorporated into Direct Digital Control systems, bringing Indoor Air Quality to these often complex computerized energy management systems.

One of the most important side benefits of using carbon dioxide controllers is the ability to maintain records of adequate ventilation for future use.

FINAL THOUGHT

When this author was first introduced to carbon dioxide controllers, there was an immediate feeling of relief... for here was the solution to a common problem: The chief engineer in the basement was reluctant to "open up to outside air" because it may destroy his budget... and maybe his job. Meanwhile, the building manager was insisting that "more" fresh air be brought in because IAQ complaints may cause litigation ... and cost him his job. Here, for the first time, was an answer to both the IAQ and the Energy problem... an instrument... an inanimate, accurate instrument that let both the building engineer and the building manager off the hook... it said when to open up and for how long !

REFERENCES

(1) ASHRAE Standard 62-1989 and ANSI / ASHRAE Standard 62a-1990, Ventilation for Acceptable Indoor Air Quality, American Society of Heating, Refrigerating and Air-Conditioning Engineers, Inc., Atlanta, GA (May, 1992)

(2) Guidance for Indoor Air Quality Investigations, Hazard Evaluations and Technical Assistance Branch, Division of Surveillance, Hazard Evaluations and Field Studies, National Institute for Occupational Safety and Health, Cincinnati, OH (January, 1987)

(3) Workers with Multiple Chemical Sensitivities, Mark R. Cullen, M.D. Occupational Medicine, Hanley & Belfus, Philadelphia, PA (1987)

(4) California Energy Efficiency Standards for Residential and Nonresidential Buildings, Section 121 (c), California Energy Commission, Sacramento, CA (1992)

(5) Guide for Indoor Air Quality Surveys, (AL-TR-1992-0016), Occupational and Environmental Health Directorate, Armstrong Laboratory, Air Force Systems Command, Brooks AFB, TX (May, 1992)

(6) Recommended Carbon Dioxide and Relative Humidity Levels for Maintaining Acceptable Indoor Air Quality, (AFOEHL 90-169CA00111KGA), AF Occupational and Environmental Health Laboratory (AFSC), Human Systems Division, Brooks AFB, TX (October, 1990)

SECTION 3
GLOBAL WARMING

Chapter 33

Energy Conservation Technologies for Global Warming Mitigation

H.A. Courtright

INTRODUCTION

The conservation of energy through the efficiency improvement of existing end-uses and the development of new technologies to replace less efficient systems is an important component of the overall effort to reduce greenhouse gases which may contribute to global climate change. Even though uncertainties exist on the degree and causes of global warming, efficiency improvements in end-use applications remain in the best interest of utilities, their customers and society because efficiency improvements not only reduce environmental exposures but also contribute to industrial productivity, business cost reductions and consumer savings in energy costs.

THE ELECTRIC POWER RESEARCH INSTITUTE

The mission of the Electric Power Research Institute (EPRI) is to discover, develop, and deliver advances in science and technology for the benefit of member utilities, their customers, and society. Funded through annual membership dues from some 700 member utilities, EPRI's work covers a wide range of technologies related to the generation, delivery, and use of electricity, with special attention paid to cost-effectiveness and environmental concerns. More than 350 EPRI scientists and engineers manager over 1600 ongoing projects. EPRI's Customer Systems Division (CSD), manages research and development to help utilities address the "wants and needs" of their customer through electric technologies and services.

Research conducted by EPRI on end-use impacts to global warming is focused in three areas:

1) Efficiency improvements to existing electric applications to reduce kilowatt-hour use.

2) New electrotechnologies with high end-use efficiencies to reduce primary energy use compared to existing electric uses and fossil-fuel applications.

3) Alternative technologies to chlorofluorocarbon (CFC) refrigerants.

EQUIPMENT EFFICIENCY IMPROVEMENTS

Air Source Heat Pumps - EPRI's work to significantly improve heat pump efficiencies resulted in the development of the EPRI-Carrier Hydrotech 2000 heat pump which provided the technology platform for the development of other advanced heat pumps. The development of electronically commutated permanent magnet compressors and blower drives has resulted in significant improvement of heating season performance factors (HSPF) and seasonal energy efficiency ratios (SEER), with HSPFs exceeding 9 and SEERs of 14. In addition, the integration of water heating into the heat pump design effectively increases HSPF to 11 and SEER to 20, far exceeding government efficiency objectives. EPRI's present work with Nordyne Corporation has produced a product line designed for residential new construction and competitive with other heating systems.

Ground Source Heat Pumps - utilizing the ground as a natural heat source increases heat pump efficiency in both the heating and cooling modes. Ground source makes heat pump applications attractive in northern locations, thus providing a highly efficiency heating source which uses considerably less primary energy than the most-efficient gas-fired pulse combustion furnace. Integrated water heating significantly increases overall energy efficiencies.

Heat Pump Water Heaters (HPWH) - EPRI's research has developed a HPWH prototype which is 250% more efficient than resistance water heating and also compares favorably to gas water heating in total primary energy use.

Laundry Use - Advanced versions of the two equipment components of laundries, washers and dryers, are being developed with manufacturers. A high efficiency washer project with Maytag is targeted for a 50% reduction in overall energy use, a 60% reduction water use and a 20% reduction in dryer need through better moisture extraction. A microwave dryer prototype has demonstrated efficiencies in the 76 - 80% range vs. 68% for standard electric dryers while reducing fabric drying time up to 20%. Fifteen prototypes of the dryers will be tested in commercial and residential applications.

Super Efficient Refrigerator Program (SERP) - As an associate member of this 25 member utility consortium, EPRI is participating in the development of a residential refrigerator which is 25 - 50% more efficient than 1993 models, is non-CFC, and preempts DOE's 1998 standards. EPRI will be participating in prototype evaluation and conducting field performance evaluations.

Commercial Water Loop Heat Pumps (WLHP) - EPRI research has resulted in the recent introduction of advanced water loop heat pumps which have cooling energy efficiency rating (EER) levels of greater than 15 under the Air-Conditioning and Refrigeration Institutes' 320 standard and heating coefficients of 4.5. This compares to EERs of 11.5 and COPs of 3.7 for standard WLHP designs. The efficiency improvements are achieved by using thermostatic expansion valves for optimum superheat control, micro-electronic controls integrated to a building energy management systems, and advanced compressors, fans, motors and evaporators.

Motors - EPRI's research program is focusing on two key approaches to energy savings in motors. First, efficiencies in motor component design is likely to result in efficiency improvements of 8-12% over the next decade. Second, research into effectively matching motors to the load through adjustable speed drive (ASDs). ASD's offer the best technology for reducing electrical use by motors. The potential for reducing the power requirements of constant speed motor-driven systems range from 20% to 50%, thus reducing kilowatt-hour use and emissions. The application of ASD-controlled motors to replace non-electric process drives, such as industrial steam turbines and internal combustion engine drive, could reduce primary energy use and CO_2 emissions through more efficient power generation and overall improved process control.

ELECTROTECHNOLOGY DEVELOPMENT

Electricity's efficiency advantage, due to its high controllability and unique abilities at the point of use, has resulted in the development of new technologies which reduce primary energy use compared to existing fossil-fuel applications.

To assure the best use of resources and the most responsible environmental policy, the entire energy-use cycle and its environmental consequences must be considered, from extraction through conversion, delivery, and use. When seen in this comprehensive view, it becomes obvious that many electricity uses are both uniquely valuable to society and the most efficient choices in resource use and environmental protection. The following are examples of primary energy and emission reductions through electrification.

Freeze Concentration - This process is used to separate and remove a liquid - water, for example -- from a fluid mixture. The technology is already being used worldwide for the concentration of fruit juices and is also being applied in processing beer, wine, vinegar, and coffee. One of the most promising applications for the technology is in the concentration of dairy products. In the process, water is frozen out of milk and the ice crystals are skimmed off, leaving a concentrated milk product that can be packaged directly or used in ice cream production. As shown in FIGURE 1, freezing requires a total of 144 Btu/lb of water, or 114 Btu/lb with heat recovery, while evaporating or boiling off water requires considerably more energy -- in this case, a net of 700 Btu/lb with heat recovery. As a result, the electric end-use is so energy-efficient that it needs only half as much primary energy, even including the power plant's energy conversion losses. An most important, consumer tests of reconstituted skim milk have indicated consumers prefer the reconstituted product to regular skim milk. This is because the freeze concentration process results in a creamier product, more like whole milk.

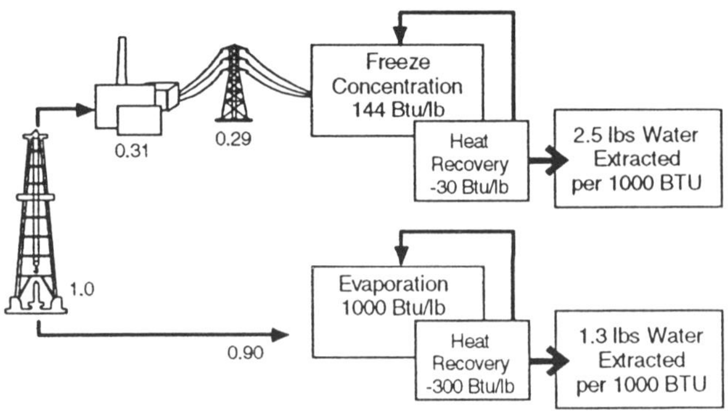

FIGURE 1 - ELECTRIC FREEZE CONCENTRATION IS TWICE AS EFFICIENT

Infrared Drying and Curing - This technology is an alternative to conventional gas convection ovens for setting finishes on a wide variety of products such as painted car bodies and home appliances. IR drying uses only half the total energy resources of the conventional gas-fired convection heating process. IR processing has many other advantages, including faster heating and space savings low maintenance, and adaptability to virtually any heat-treatable coating. IR radiation passes through the coating's surface and dries it from the inside, automatically avoiding the premature surface drying which often can only be eliminated in convection processes by slowing the drying process. IR also lowers CO, CO_2 and VOC's emissions at the factory, although total NO_X and SO_X may be slightly higher at the power plant.

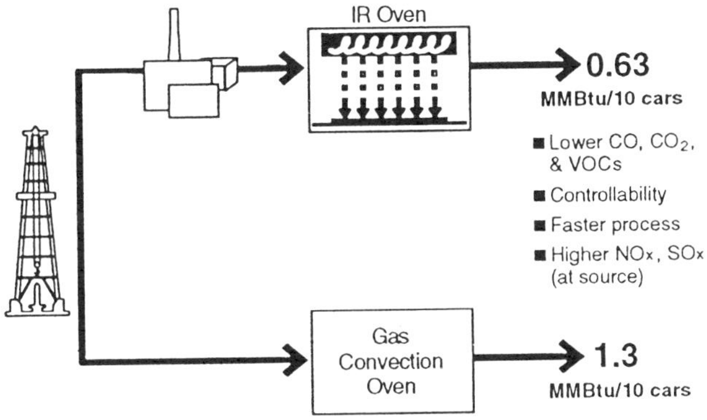

FIGURE 2 - INFRARED DRYING AND CURING

Transportation - the development of electric vehicles (EV) has been accelerated by the need to reduce tailpipe emissions to solve regional air quality problems, however, EVs can also provide in substantial reductions in primary energy consumption and CO_2 emissions. Present day internal combustion engines vehicles (ICEV) emit 60% to over 100% more CO_2/mile than their EV counterparts. The primary advantage of the EV is gained through drive power systems efficiencies of 90% as compared to ICEV drive train of 13.7% to 15.8%[1]. FIGURE 3 shows that EVs are more than 60% more energy efficient than an ICEV and (Table 1) details the emissions reduction from EVs. The benefit of EVs may increase to 2.0 - 2.6 times in favor of EVs as battery capabilities are improved. The development of advanced batteries by the United States Advance Battery Consortium (USABC), of which EPRI is a member, is a critical component in the success of commercially viable EVs having the performance and range needed by consumers.

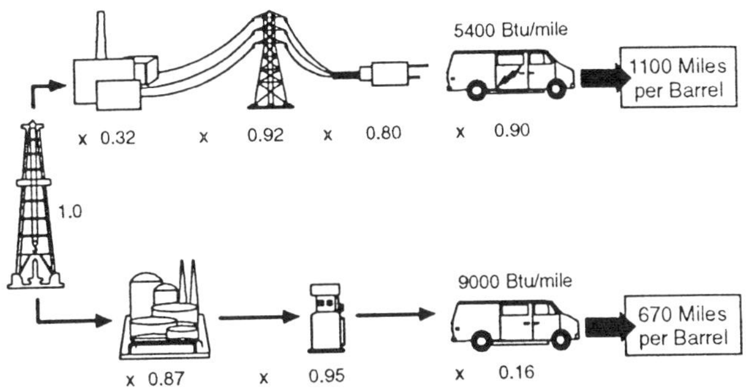

FIGURE 3 - ELECTRIC VEHICLES ARE 60% MORE ENERGY EFFICIENT THAN GASOLINE VANS

Table 1 compares gasoline-powered van emissions based on 1989 U. S. emission standards with EV emissions under a post 1995 U.S. generation mix scenario. The EV emissions take into account all emissions directly or indirectly associated with electric power production.

TABLE 1 - FLEET MINIVAN EMISSIONS (grams per mile)

	Gasoline Powered Van	*Electric Van*
VOCs	1.0	0.01
NO_x	1.8	0.3
CO	10.0	0.05
CO_2	690	320

CFC REPLACEMENTS

EPRI is conducting research with several organization to identify non-CFC refrigerants with acceptable properties for electric heat pump and chillers. A critical component in the evaluation of CFC phase-out is the trade-off of indirect global warming potential (GWP) from changes in equipment efficiencies compared to direct GWP from CFCs. It is important that efficiency is not significantly reduced in equipment redesign to handle non-CFC refrigerants. Efficiency improvement with new refrigerants is anticipated as well as reduction of refrigerant losses, through improved gaskets and refrigerant recovery techniques. It is expected that electric vapor-compression chillers will continue to provide lower GWP than direct-fired absorption machines.

TOTAL BENEFITS

When the estimated benefits of electrification are added to those due to efficiency improvements in existing uses, the results are dramatic. The potential savings in energy and CO_2 emissions are almost as large for new electricity uses as those due to improved existing uses -- so that combining both strategies would almost double the savings.

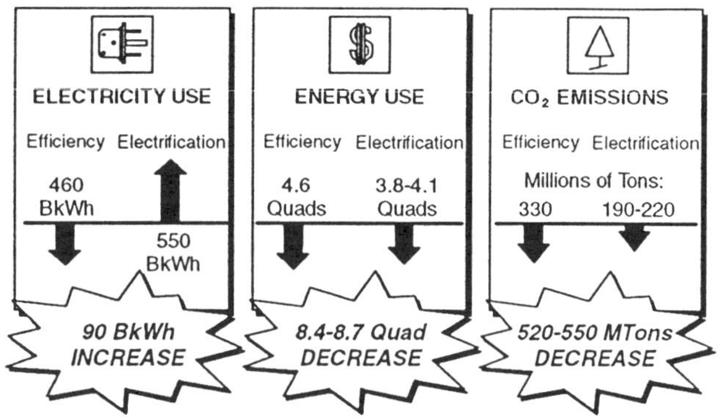

FIGURE 4 - IMPACT OF NEW AND EXISTING ELECTROTECHNOLOGIES - 2010

In addition, development of additional new electrotechnologies could result in significant reductions in primary energy use as well as enhance the quality of life. Electrification is a crucial part of our nation's energy solutions.

References

1 The Impact of Electric Vehicles on CO_2 Emissions, Idaho National Engineering Laboratory, Arthur D. Little, May 1992.

2 Comparing Emissions of End-Use Technologies, EPRI Technical Report TR101574, December 1992

3 Saving Energy and Reducing CO_2 with Electricity: Estimates of Potential, EPRI CU-7440, September 1991.

Chapter 34

Fuel Cells for Energy Generation at Medical Centers

T.A. Damberger, V.W. Prue

ABSTRACT

Traditionally, electrical and thermal energy is produced by a combustion process. Coal, fuel oil and natural gas are common fuels used for electrical generation, while hydro-electric and nuclear are non-combustion processes. All fossil fuels release their stored energy and air pollution simultaneously when burned in a combustion process. To reduce or eliminate air pollution, the process of combustion to release stored energy must be changed to another type of process. Extracting pollution-free energy from fossil fuels can be accomplished through the electrochemical reaction of a fuel cell. This non-combustion process provides the foundation from which pollution-free energy emerges to fulfill our perpetual energy needs without further compromising the environment.

INTRODUCTION

Investigation, research and demonstration of new and emerging technologies at Kaiser Permanente, Southern California Region, has resulted in more efficient uses of our energy resources. It has also brought space-age technology down to earth to help solve health care related issues. Some of our investigations resulted in multiple fuel cell demonstration projects at our medical centers. We are assisting in the process to bring fuel cells toward commercialization with such organizations as Gas Research Institute, U.S. Department of Energy, Institute of Gas Technology, Electric Power Research Institute and many local gas and electric utilities.

BACKGROUND OF ENERGY GENERATION

Throughout history and perhaps as early as one million B.C., fire provided man with the necessary energy to cook, heat and light his environment. Fire was revered as mystical and supernatural by early man. It is one of the essential elements upon which all civilized societies were founded. Use of wood as a primary fuel resulted in denuded forests and barren landscapes. Man found a way, by using fire, to unleash the solar energy stored in the trees of the forests.

Later, about the first century A.D., Hero of Alexandria described a steam boiler system. While the descriptions were found in his writings, they appeared to lack practical application. Early in the 17th century, however, the first steam engine was developed. Further refinements, such as the reciprocating engine, followed in the 18th century. These developments ignited the Industrial Revolution. Steam and the internal combustion engine eventually replaced animal "horsepower" to plow the fields. Fossil fuels became an irreplaceable source of energy. This has led to the present dependence upon fossil fuels.

By the mid 1700s, experiments with electricity and its transport system began in Europe. Early pioneers, including Benjamin Franklin, were able to get electricity to flow through wire several miles long. At that time electricity was more of a curiosity with no known practical application. In 1832, Samuel Morse developed the first crude telegraph which was a landmark in transport of electricity. Its purpose was for communications which he finally demonstrated in May of 1844, with the message, "What hath God wrought?" After the Civil War, in 1879, Charles Brush demonstrated his invention called a dynamo. It produced electricity for an arc light in the City of Cleveland.[1]

Electrical generation and lighting demonstrations evolved rapidly from that point forward. In September of 1882, Thomas Edison closed a switch that started electricity flowing to factories in downtown New York City.[2] Steam engines were used to generate the first electricity. Electricity was used to power only incandescent lights in early factories. Edison was one of the first of many to commercialize electrical generation and transmission of electric energy to consumers.

Edison and other early pioneers did not realize their efforts were to become the foundation for the electric power production industry. Electrical energy became a fundamental element contributing to the success and stability of industrialized societies. It is vital to all nations for sustained economic growth and security.

THE HIDDEN COST OF ENERGY

The cost we currently pay for energy does not reflect all associated costs for that energy. We know the cost of electricity consists of the costs to build generation facilities and distribution systems to bring power to the point of use. It also includes the costs to operate and maintain these systems and generation facilities. There are other hidden costs not reflected in the actual cost of this energy. Generally, these costs relate to pollution created from the production of energy.

Understanding and estimating societal costs are at best difficult but includes the expense society must face when cleaning up the environment. However, a recent study by Jane V. Hall, an economist at California State University in Fullerton, appeared in the February 14, 1992 issue of the journal **Science**.[3] The study indicates that about 1,600 Greater Los Angeles area residents die each year because of air pollution, which costs the economy about $10 billion annually. On November 30, 1992, at a fuel cell conference in Tucson, Arizona, Thomas Gross, from the United States Department of Energy, in his keynote address stated that in 1991, $45 billion was spent on health-related air pollution illnesses nationwide.

Health risks from polluted air are starting to be accepted as an actual cost for energy. Some of these costs are starting to manifest themselves in the form of higher energy costs. We are now experiencing an incremental cost increase due to required pollution controls that reduce NO_x, SO_x and particulate emissions.

Additionally, new laws to control carbon dioxide emissions, the purported cause of the greenhouse effect, and the so called Btu or carbon tax may be forthcoming here in the United States. It is estimated that a tax of $28 per ton on carbon-containing fuels would stabilize the emissions of carbon dioxide. This is because it would ultimately motivate our society to become more energy efficient in order to reduce costs and pollution.

Twenty-nine states now require utility companies to include environmental impacts in their planning and resource acquisition procedures. This is why many utilities are now using demand side management as an option to offset the need for siting new central power plants. According to the California Energy Commission and U.S. Environmental Protection Agency, our greatest resource is energy conservation.

Sooner or later we will have to pay the real cost of energy. Paying this cost will mean gasoline, diesel fuel, natural gas, electricity and home heating oil will consume more of our financial resources. Choosing clean, energy efficient equipment will become critical design elements for our buildings and transportation.

THE ENERGY CONVERSION CYCLE

Conversion of energy from one form to another is a dynamic process. It is also the First Law of Thermodynamics as postulated by Nicolas Carnot, "energy can be neither created nor destroyed but only converted from one form to another".[4] Radiant energy from the sun is converted to mechanical energy of wind, waves and, through photosynthesis, into the latent energy stored in plants. This latent energy, over the millennium, has formed what we now call fossil fuels. It could be said that fossil fuels originated from biologically-stored solar energy.

The carbon in fossil fuels is itself nothing more than atmospheric CO_2 converted into organic form by plants millions of years ago. Oil, natural gas and coal are commonly known as fossil fuels. They release their stored energy by a combustion or burning process. We use the heat energy from fire for production of steam to power a steam engine which generates electricity. Burning any fossil fuel, including wood, results in environmental pollution. Nitrogen oxides (NO_x), sulfur oxides (SO_x), hydrocarbons (HC), carbon dioxide (CO_2), carbon monoxide (CO) and particulate matter (PM) are common contaminants produced by burning fossil fuels. Ozone is formed when volatile organic compounds (VOC) are mixed with NO_x in the presence of sunlight.

Modern central power plants have many of the same working characteristics of those plants built in Thomas Edison's era. As mentioned earlier, steam generation at a conventional power plant requires a fossil fuel to be burned in a combustion chamber. A heat exchanger extracts thermal energy to produce high pressure steam which then powers a steam/gas turbine. Other types of power plants use reciprocating heat engines to convert chemical energy (from natural gas, fuel oil or coal gas) into mechanical energy of a rotating shaft. Modern fossil fuel power plants typically have inherent efficiency losses at each stage of the process. These energy losses result in an overall efficiency of less than 35 percent. One of the challenges for scientists and engineers is to develop new energy conversion systems that provide better utilization of the fuel source, thus increasing system efficiencies.

THE ELECTROCHEMICAL PROCESS; A BATTERY

Batteries have been in use since the early 1800s. Batteries work on the sub-microscopic level of molecules and electrons. When we speak of an electrochemical process, a battery comes to mind. This process is the same today as it was in the 1800s.

A battery is a device that stores electrical energy in chemical form. As a battery is charged or discharged its chemical composition changes electrochemically. They are known as primary (disposable) and secondary (rechargeable) batteries. Batteries are quiet and have no moving parts. It is a clean process that does not create pollution while generating electricity. Simply stated, batteries are chemical engines that push electrons around. Electricity is essentially electrons in motion.

Each cell is comprised of active materials that are electron deficient. Electron deficient cells are equivalent to the positive pole on a battery. The other active material is electron rich and is equivalent to the negative pole on a battery. The flow of electrons from cell to cell is from the negative pole to the positive pole (from cathode to anode). The negative pole can be called the cathode and the positive pole the anode. During the recharge cycle, flow of energy is reversed, electrons flow from the anode to the cathode.

The battery industry continues to expand both in applications and in new electrochemical systems. Battery configurations continue to change and new configurations are being developed. Further, the internal cell structure, including electrode design and composition, is being modified to enhance capacity. Enhancing capacity results in better energy efficiency while reducing cost. Energy density in batteries has increased six to ten times compared to the density of traditional lead-acid batteries of a few years ago.

The uses of a rechargeable batteries continues to expand as their specific energies and volumetric energy densities increase. Applications require not only that the battery system be capable of releasing electrical energy from its chemically-stored form on demand. It must also be capable of consistently returning electrical energy back into the same initially active chemically-stored form with a high degree of efficiency.

THE ELECTROCHEMICAL PROCESS; A FUEL CELL

Imagine a battery that is fully charged at all times and never runs out of electricity. Essentially, the fuel cell is that battery. With a constant fuel supply, it continually converts chemical energy from hydrogen and oxygen electrochemically into virtually emission-free electricity and thermal energy. A fuel cell operates with a fuel and an oxidant, using two electrodes and an electrolyte. It shares many of the same operational characteristics of a battery. The fuel cell has no moving parts so that operation is virtually silent, like a battery.

The electrochemical conversion generates electrical and thermal energy without a combustion process. Unlike the internal combustion of a heat engine, fuel cells are not subject to limits of the intrinsic thermodynamic cycle.

The first fuel cell was devised by Sir William Grove in 1839. He assembled an electrolyte composed of water and sulfuric acid between two platinum electrodes. From that point forward fuel cell research was, for the most part, discontinued until the early 1960s. Two scientists from Cambridge University developed a 6-kilowatt fuel cell using hydrogen and oxygen. This interest in using fuel cells for transportation never went beyond the demonstration stage. Many engineering refinements of the fuel cell by the National Aeronautics and Space Administration (NASA) resulted in a very reliable power supply for use in space craft. Fuel cells were used in applications aboard Gemini, Apollo, and in the ongoing space shuttle program.

At the heart of the fuel cell is a of pair of electrodes, similar to a battery. The negative electrode (anode) uses hydrogen and the positive electrode (cathode) uses oxygen, both separated by an electrolyte. Chemical reaction within the cell is a result of combining hydrogen and oxygen ions (oxygen from air and hydrogen from natural gas or other energy sources) interacting in an exothermic, or heat-releasing reaction.

Typically, natural gas is fed to a reformer after being mixed with steam. Hydrogen from a hydrocarbon molecule, must be separated from the carbon through a process called reforming. Once reformed, the hydrogen is ionized. Hydrogen ions flow through the electrolyte toward the positively charged cathode. Hydrogen ions combine with oxygen at the cathode, producing heat, water and small amounts of carbon dioxide from the reformer. The electron flow produces electric current.

Operating a fuel cell results in almost no polluting emissions and is considered environmentally benign. In the Southern California area, fuel cells are exempt from air quality permits. Fuel cells operate with a high electrical efficiency, with the combined thermal and electrical production exceeding 80 percent. Electrical power generation is very high quality that can run either grid-synchronized, grid-connected or entirely independent of the grid.

Internally, fuel cells look much like a plate and frame heat exchanger. Figure 1. below reflects the operational principles of a fuel cell. A fuel cell has attributes, such as the electrochemical reaction and structure, much like those of a lead-acid battery.

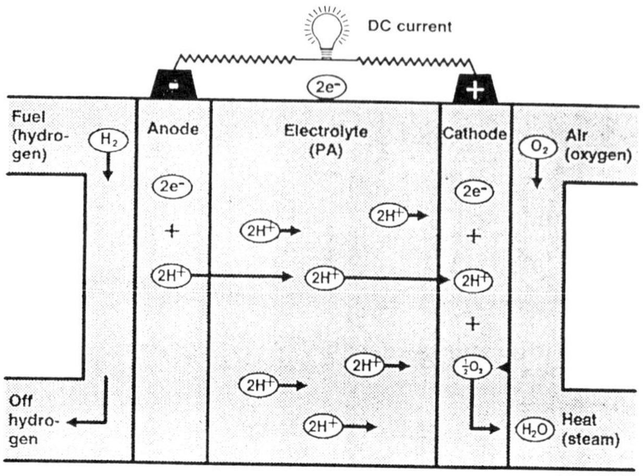

FIGURE 1. PRINCIPLES OF PAFC

TYPES OF FUEL CELLS

Development of the fuel cell technology is advancing on many fronts. Optimistically, they appear to be a very promising technology. There are several types of fuel cells, each with different operating characteristics and at various phases of development. For purposes of discussion and in terms of Kaiser Permanente's needs, phosphoric acid, molten carbonate and solid oxide fuel cells are the only types discussed in this paper. Table 1. lists operating efficiencies and temperatures of each type of fuel cell.

TABLE 1. FUEL CELL OPERATING EFFICIENCIES

Type of Fuel Cell	Electrolyte	Electrical Efficiency	Operating Temp.	Combine Efficienc
Phosphoric Acid	Phosphoric Acid	35-45%	375 F	75-82%
Molten Carbonate	Potassium Carbonate or Lithium Carbonate	50-60%	1200 F	75-85%
Solid Oxide	Zirconium Oxide or yttria-stabilized zirconia	45-55%	1800 F	75-85%

The phosphoric acid fuel cell (PAFC) is by far the most commercially ready product as of this writing.

It is followed closely by the molten carbonate fuel cell (MCFC) which may have market penetration prior to the year 2000. Solid oxide fuel cells (SOFC) are the least developed of the three types currently considered for hospital use.

A PAFC operates at better than 40 percent electrical efficiency (8542 Btu/KWh without heat recovery) with about a 375° Fahrenheit working temperature. A reformer supplies high temperature to the fuel with a catalyst to convert or shift the fuel into hydrogen. A reformer combusts unreacted fuel and uses the thermal energy as a heat source as shown in Figure 2.

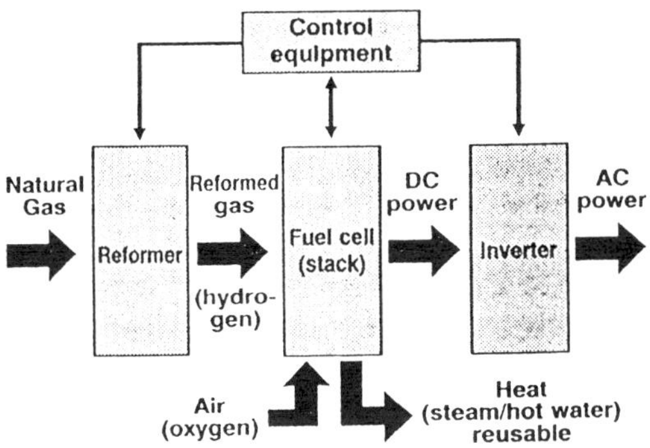

FIGURE 2. PHOSPHORIC ACID FUEL CELL SYSTEM

The discovery of phosphoric acid and its applications have an interesting history. In 1669 it was discovered that phosphorus promoted combustion eventually resulting in phosphorus pentoxide. During the first half of the nineteenth century, the value of phosphates for fertilizers, matches, detergents, animal feed supplements, water softening, rustproofing metals, flavoring in cola soft drinks and flame retardants became evident.

Aside from its acidic behavior, phosphoric acid is relatively unreactive at room temperature and is sometimes substituted for sulfuric acid due to its lack of oxidizing nature. It is manufactured by two processes, the wet process produced directly from phosphatic ores or the furnace process which produces a higher grade acid. The higher grade phosphoric acid is typically used in fuel cells.

The MCFC, uses carbonate as its electrolyte and has a working temperature of about 1200° Fahrenheit. Carbonate separates the cathode and anode plates which operate at about 60 percent electrical efficiency (6300 Btu/KWh without heat recovery). The MCFC has a load following capability that can operate over a wide power range from about 25 percent to 125 percent of rating depending on requirements. Carbonate fuel cells are ideally suited to operate with a wide range of hydrocarbons which consist of hydrogen-rich fuels such as:

- methanol
- natural gas
- carbon monoxide
- methane-rich fuel gas
- propane
- butane
- coal gas

The MCFC has a high overall efficiency and is not contaminated by carbon monoxide or carbon dioxide in the fuel or air streams. The molten carbonate technology may eventually replace centralized power generating stations. The MCFC can be used with a bottoming cycle for additional electricity. This configuration would make full use of steam production.

SOFCs have a high power density and contain no corrosive solutions. Working temperature is about 1800° Fahrenheit. The SOFC has many attractive features that could catapult it into a lead technology. As an example, some manufacturers claim that their units are exceptionally compact and lightweight, offering more flexibility in size and siting of units. This is especially important when space is limited or there are competing demands for space.

The electrochemical reaction in an SOFC employs a semi-conductive electrolyte ceramic such as yttria-stabilized zirconia which operates at about 50 percent electrical efficiency (6800 Btu/KWh without heat recovery). SOFCs process fuel directly without reforming when the electrolyte reaches an operating temperature of 1800° Fahrenheit. At this elevated temperature there is reduced component resistance resulting in an increased electrode reaction rate. Like the MCFC, SOFCs can use any hydrogen-rich fuel. PAFC and MCFC use a liquid electrolyte for ion exchange within the cell. Use of a solid electrolyte reduces corrosion and eliminates problems associated with liquids (evaporation, corrosion and liquid migration from one cell to another).

ENVIRONMENTAL BENEFITS OF FUEL CELLS

Unlike a battery fuel cells generate power continuously, depending only on a steady supply of chemical energy found in hydrocarbons. Unlike the conventional power plant, the fuel cell as an electrochemical engine has no moving parts or combustion process. As such fuel cells are expected to have lower maintenance costs. Unconstrained by the thermodynamic limits of combustion or limitations of the Carnot cycle, friction, lubrication and mechanical energy of a rotating shaft, fuel cells use 40 percent less fuel than conventional systems. Fuel cells are capable of attaining very high efficiencies quietly and with virtually no emissions.

Fuel cells are a key technology for future electrical and thermal energy generation. They are an environmentally clean technology without the traditional pollution found in a burning or combustion process. Impacts of fuel cell technology on reducing emissions directly impacts the reduction of financial, environmental and societal costs of providing energy services. The electrochemical process operates on the hydrogen component (hydrocarbon) of natural gas, methanol or LPG. These fuels are also sulphur-free, which means they produce less pollution than coal and oil. Efficiencies of the electrochemical process in fuel cells can result in substantial reduction of greenhouse emissions and in turn lead toward conservation of natural resources. Reduced fossil fuel consumption rates may help extend proven reserves by decades.

The impacts of fuel cell technology on ground level air pollution are especially dramatic. NO_x emissions are reduced by up to 90 percent as compared with conventional combustion technologies. As an example, NO_x emissions at our operating installations, are lower than that found in the ambient air.

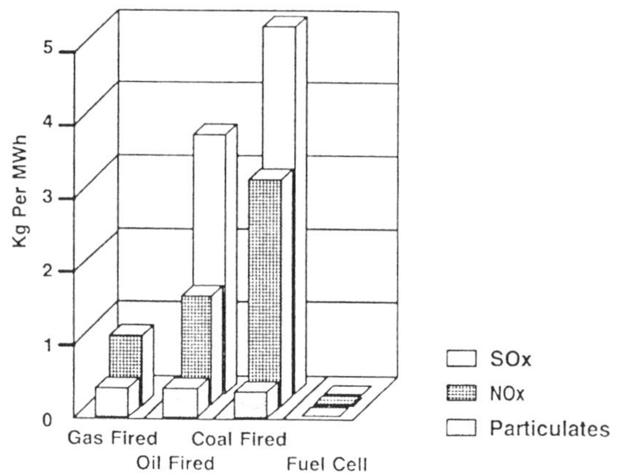

FIGURE 3. POWER STATION EXHAUST EMISSIONS

Figure 3. shows power station exhaust emissions by type of fuel used. Natural gas-fired electrical generation is a clean process when compared to a coal-fired power plant. As demonstrated, fuel cells are the cleanest and most efficient of known power generating technologies utilizing fossil fuels.

Hydrogen is considered to be an important clean, renewable energy source for the future. Fuel cell technology may be viewed as one of the early links in a renewable hydrogen cycle fuel chain. Unlocking the power of hydrogen for use in fuel cells is a transition technology toward a hydrogen-based society.

FUEL CELLS AND THE KAISER PLAN

The Energy Policy Act of 1992 supports continued development of fuel cells for use in transportation, stationary and multiple end-use applications. Budgetary expenditures for 1993 are at $51.5 million and $56 million for fiscal year 1994. The U.S. Secretary of Energy will administer early demonstrations of various fuel cell applications.

In 1989, the Southern California Gas Company and Kaiser Permanente, Southern California Region, entered into an agreement to assist in the large-scale commercialization effort of fuel cells. Three 200-kilowatt phosphoric acid fuel cells manufactured by ONSI Corporation have been installed in two medical centers. The first fuel cell was installed in January 1993 at our Anaheim Medical Center and became operational in April 1993. Additionally, two other units were installed in August 1993 at the Kaiser Permanente Medical Center in Riverside. Both medical centers use all electrical and thermal energy production from the units. Kaiser Permanente's Northern California Region recently signed an agreement with Sacramento Municipal Utility District (SMUD) to install another 200-kilowatt ONSI fuel cell at their South Sacramento Medical Center. They expect to have this unit operational in December 1993. These "partnerships" are demonstrating proof of technology while assisting in the commercialization effort. Reliability of the installed PAFC units are exceeding 90 percent.

In December of 1991 San Diego Gas & Electric (SDG&E) and Kaiser Permanente began discussions that lead to an agreement to install and demonstrate a 250-kilowatt MCFC. The MCFC, manufactured by M-C Power, is scheduled for installation at our San

Diego Medical Center late 1994. Project participants include the U.S. Department of Energy, M-C Power Corporation, Gas Research Institute, SDG&E, EPRI, Bechtel Corporation, Stewart & Stevenson Services, Institute of Gas Technology and Kaiser Permanente. The budget is currently $16.1 million for this "proof of concept" demonstration project.

The MCFC, powered by natural gas, may use a hydrogen-rich fuel produced by another demonstration project, a plasma arc torch. Plasma pyrolyzes (high temperature conversion) medical waste into a hydrogen-rich fuel and glassy slag. Department of Energy, M-C Power and Mason & Hanger National, Inc. are reviewing the gas stream produced by a plasma arc torch system to determine if further gas treatment would be necessary prior to use in the fuel cell. If the two technologies merge sometime in the future, converted medical waste may provide carbon monoxide and hydrogen for fuel cell operation.

Currently, we are reviewing locations for possible installation of a pre-production 2-megawatt MCFC. Installation may be as early as 1998. We are targeting our large medical centers in the Los Angeles Department of Water & Power service area as potential host sites. The selected facility needs to be able to make full use of all electrical and thermal energy produced by the fuel cell. It must also have space available approximately the size of a standard tennis court to house a 2-megawatt MCFC.

Connected load of one of our typical medical centers is generally between 2 to 4 megawatts. All thermal energy requirements for a medical center may be met with the energy produced from a multi-megawatt MCFC. Steam production from the MCFC can be used for domestic hot water, space heating, steam sterilization and absorption chilling. Steam absorption chilling will provide a majority of the thermal load. A MCFC can operate over a wide power range, from a part load of 25 percent to 125 percent of rating depending on our specific operating requirements.

When both the MCFC and PAFCs are fully operational, Kaiser Permanente will gain working knowledge of fuel cells as well as realize cost savings. We also help advance this technology and conserve natural resources. In addition, we receive emissions credits from the local air quality district.

CONCLUSION

The development of steam boiler systems, reciprocating engines, turbines, electricity and electrical distribution systems transformed our industrialized societies. Along with development of our energy infrastructure has come environmental pollution and the hidden costs of energy. We now face dependence upon fossil fuels and those who produce them, as well as the costs related to pollution created from the production of that energy.

Improvements in energy conversion systems have resulted in a cleaner means of generating electricity. The fuel cell is a bridge from a combustion process to electrochemical reaction. This bridge is a conduit to "pollution free" energy. Batteries are one form of electrical generation that do not create pollution while generating power. Fuel cells are much like batteries. They electrochemically convert fuel into electricity and heat as long as there is a constant fuel supply.

PAFC, MCFC an SOFC are three types of fuel cells. All have high electrical efficiency while producing no polluting emissions. Fuel cells have the advantage of having no moving parts, thus reducing maintenance and noise. As they do not use the combustion process, they use 40 percent less fuel than conventional systems. Fuel cells, since they use hydrogen, are looked to as an early link in a renewable hydrogen fuel chain.

In cooperation with public and private groups, Kaiser Permanente is assisting in bringing fuel cell technology toward commercialization. Our total connected fuel cell capacity will exceed one megawatt by the end of 1994. Installation of fuel cells at several of our medical centers will provide Kaiser Permanente with a dynamic working knowledge of various types of fuel cells. We gain operational experience while having additional energy security. Locations using these technologies will have near-zero emissions. Risk of failure and exposure to potential failure is minimized due to shared risk and the number of demonstration projects. This total experience with fuel cells will provide a solid foundation upon which Kaiser Permanente senior management can make sound decisions regarding our energy future.

REFERENCES

[1]Rudolph, Richard and Scott, Ridley, Power Struggle, New York, 1986, p. 24.

[2]Myers, William A., Iron Men and Copper Wires, Glendale, California, 1986, p. 12.

[3]Hall, Jane V., and Winer, Arthur, "Valuing The Health Benefits of Clean Air," Science, Volume 255 (February 14, 1992), pp. 812-817.

[4]Turner, Wayne C., Energy Management Handbook, New York, 1982, p. 31.

Chapter 35

A Methodology and Model for Costing of Environmental Impacts in the Austin Energy Conservation/DSM Process

Y. Ziming, A.J. de Leon

INTRODUCTION

There are two major parts to this paper. The first discusses a methodology and model for costing of environmental impacts due to generation. The second describes how the City of Austin incorporates these impacts and associated costs in an Integrated Resource Planning (IRP) process.

The methodology and model for costing of environmental impact

One goal of utility planning and regulation is to develop resources with the lowest social cost. The social cost of a resource is its internal cost (the utility's revenue requirements, paid by ratepayers) added to its external cost (the cost associated with the resource, not paid by the ratepayers), such as the cost of environmental damage caused by generation. The external cost is considered an environmental externality cost.

It is unlikely that a single model can be used to estimate the full range of environmental externalities associated with an energy resource. The types of impacts produced by a power plant are too diverse for any existing model to capture and evaluate. Most empirical studies have shown however, that the preponderance of environmental costs associated with fossil fuel plants are created by atmospheric pollution. Consequently, models that estimate these effects are most useful in evaluating environmental costs.

Four basic methods have been used for estimating environmental costs. They are (1) estimating the costs that utilities accept for mitigating emission, such as the costs of substituting low- for higher-pollutant-causing fuels, (2) assessing pollution control costs, i.e., the costs of additional scrubbing equipment, (3) assessing the costs for cleaning up negative impacts, and (4) attempting to measure and forecast the value of damage caused by emissions.

The first two methods measure only what the market currently values for a different fuel or piece of equipment. Differences in prices include many other items beside the "valuation" of lower pollution, for example, supply, current political situation, the price of complements, and substitutes.

The third method, the costs for cleaning up the impacts of pollution, is probably closer to the actual. However, it does not encompass all pollution costs. Costs for clean-up are normally expended only when the environmental impacts do not meet basic standards and regulations.

The fourth method measures the actual cost of pollution. Economic values of environmental externalities are those placed by society on damages, including residual environmental impacts, caused by generation. This method is what the Environmental and Conservation Services Department (ECSD) used in building a model to estimate environmental damage due to generation. There are three categories of environmental impacts: local, long-range, and global impacts.

Local environmental damage is much easier to estimate than global costing, though by no means easy. The current City of Austin environmental impact model addresses the local impacts as its primary focus.

An ideal model should convert user-specified information on the resource and its location into estimates of environmental costs. The full range of effects from primary emissions to changes in economic value associated with environmental risks must be clearly and explicitly provided. All the linkages in that chain must be accounted for.

First, it estimates the emissions based on plant type, control technologies, and fuel characteristics. The output of this stage displays the pollutants that are emitted from the stack.

The next stage uses the first stage output to estimate and predict the transport, transformation, and dispersal of pollutants. Some disperse according to prevailing winds, size, and weight. Others, such as sulfates and ozone, are the result of transformations of the stack emissions. During this stage, all transformations, and the path taken by the primary and secondary pollutants, are traced, and exposures are estimated. The output of this stage consists of atmospheric concentrations at various distances from the plant(s).

The exposures calculated in the previous stage are then used to estimate the response of the affected population. This requires estimates of the population size exposed at various distances from the source. "Populations" can be human populations or those of crops, forests, buildings, etc. The result of this stage is an estimate of the physical response of the population, for instance, increase mortality and morbidity, yield reduction of crops, etc., as a result of exposures to pollutants.

Finally, the last stage of the model assigns monetary values to the physical impacts. This is accomplished by multiplying a user-specified unit value by the incremental physical change or environmental risk. This stage gives the financial value of the physical responses.

In the United States several studies of environmental externality costs exist: Professor Ottinger and his colleagues from Pace University (sponsored by New York State Energy Research & Development Authority)[13], Shepard C. Buchanen of Bonneville Power Administration[1]; Mary Lou Munts of the Wisconsin Commission (sponsored by the National Association of Regulatory Utility Commissioners), and The California Energy Commission with help of Regional Economic Research Inc[2].

In Germany, Olav Hohmeyer of Franhaufer Institute has also performed studies[9]. Uwe Fritsche of OKO Institute has also developed software for this purpose.

Most externality work has been done in the North, Southeast, and abroad where the population density, flora, and fauna differ substantially from the South and Southwest.

The Environmental Externality Cost Model (EECM) has been developed to meet these requirements[17]. This model calculates pollutant emission, predicts air quality, estimates environmental damage, and estimates environmental externality costs due to generation. Beside environmental externality costs of local effect, this model evaluates the long-range effects of acid deposition and the value of global effects due to greenhouse gasses, including carbon dioxide.

Two Parts of EECM:
Air Quality Model and Environmental Cost Model

The air quality dispersion model is based on the ISCLT model—The Industrial Source Complex Dispersion Model (Long-Term)[6]. It is a FORTRAN-based program, developed by the EPA for nationwide use. This model was modified by the staff of ECSD so that it is now possible, in a single run, to perform the following tasks: estimate the emission of five pollutants (SO_2, NO_x, TSP, CO, CO_2); predict long term average concentration of six pollutants (SO_2, NO_x, TSP, CO, SO_3, O_3); and estimate acid deposition.

Features of ISCLT

The ISCLT Model is an advanced model, designed to predict annual or seasonal average concentrations of atmospheric pollutants at ground level with grid or discrete receptors and Cartesian or polar coordinates.

Predictions are based on the steady state Gaussian plume hypothesis, Briggs plume rise formulations, Pasquill-Gifford dispersion coefficient approximations, and exponential pollutant decay. The general form of the Gaussian dispersion equation is[15]:

$$C(x,y,0,h) = \frac{Q}{2\pi u \sigma_y \sigma_z} \exp^{(-\frac{y^2}{2\sigma_y^2})} \exp^{(-\frac{h^2}{2\sigma_z^2})}$$

Equation 1

Where
- C is the ground level concentration ($\mu g/m^3$)
- Q is the source's emission rate (g/s)
- U is the average windspeed at stack height (m/s)
- σ_y, σ_z are the standard deviations of the concentration distributions in the crosswind and vertical directions respectively (m)
- h is the effective stack height, which is the sum of stack height and plume rise (m)
- x is the distance downwind from the stack (m)
- y is the crosswind distance from plume centerline (m)

This equation calculates the concentration of pollutants at a receptor from different sources. Atmospheric concentration at each receptor is determined by adding corresponding values from individual sources.

The ISCLT Model's Abilities

- Polar or Cartesian coordinate system
- Rural and urban option
- Plume rise due to momentum and buoyancy
- Building wake effects
- Stack-tip downwash
- Effects of gravitational settling and dry deposition
- Point, line, volume and area sources
- Variation with height of wind speed
- Terrain-adjustment
- Time-dependent exponential decay of pollutants
- The Pasquill method for buoyancy-induced dispersion
- Capability to treat height of receptor above ground
- Accounts for multiple reflections of the plume from the ground surface and at the top of the surface mixing layer

With these features the ISCLT Model not only gives more reasonable predictions of pollutant concentration than other models, but makes it easier to modify these for further applications.

Enhancement of ISCLT

The ISCLT Model was enhanced by the staff of ECSD in December, 1991. The most important enhancement and modification is in the number of predicted pollutant concentrations, which was increased from one to six. Four primary pollutants are studied per run: sulfur dioxide, nitrogen oxide, total suspend particulate (TSP), carbon monoxide and two secondary pollutants are also determined: sulfate and ozone. Concentrations of acid deposition can also be estimated.

The secondary pollutant, sulfate, is transformed from sulfur dioxide. The concentration of sulfate is the difference between the concentrations of sulfur dioxide with and without decay. It is then adjusted by molecular weights. Since the molecular weights of sulfur dioxide and sulfate are 64 and 80 respectively, the adjustment coefficient is 1.25. The transformation of sulfate from sulfur dioxide as a function of time can be expressed in the following equation:

$$C(SO_3) = 1.25 * C(SO_2) * (1 - e^{-(\frac{t}{\tau})})$$

Equation 2

Where t is the elapsed time since a package of SO_2 plume was emitted from stack (s)

τ is the half-live decay time (s)

$C(SO_3)$ is the concentration of sulfate ($\mu g/m^3$)

$C(SO_2) * e^{-(\frac{t}{\tau})}$ is the predicted concentration of sulfur dioxide ($\mu g/m^3$)

The rate of change per second, the reciprocal of τ, is a chemical kinetic coefficient. According to the Boardman Study, sulfur dioxide turns into sulfate at the rate of 0.5% to 0.7% per hour[1].

Ozone is produced in the presence of nitrogen oxides, reactive non-methane hydrocarbon, and sunshine. Ozone concentration can be approximated as a function of the nitrogen dioxide concentration with an adjustment coefficient of 1.043. Stephens indicates that at 10 pphm (parts per hundred million) NO_2 we will generate about 2.7 pphm (0.027 ppm) ozone. According to Boardman, the rate of change is about 5% per hour[1].

Apart from the concentration of pollutants, the model also estimates acid deposition. When sulfur dioxide and sulfate reach ground level, the pollutants come into contact with people, plants, animals, material, and soil. Part of the pollutants are absorbed. In addition, wet precipitation absorbs another part.

The model uses an absorption coefficient of 5%, to estimate acid deposition.

Three files are used to input data. The first contains the power plant operating parameters and seven sets of control parameters for running the ISCLT subroutine. The other two are meteorological data files, one provides STability ARray (STAR) information and the other provides miscellaneous data.

Since this model was developed to estimate emissions from power plants instead of pollutant emission rates (g/s), the input data is power plant generation (GWh) and the emission factor by plant (g/kWh) for each pollutant.

Nine files are produced by the enhanced air quality model, seven of them are interface files, which transfer information to the cost model. Of these, six are concentration ($\mu g/m^3$) distribution files for six pollutants at ground level, and one is acid deposition (tons/km^2-yr) within 300 km^2 in 11x11 grids. Another interface file transfers all miscellaneous messages about generation, pollutant emission, and economic information. The summary file gives two categories of printed output:

1) Emission rates of all primary pollutants (sulfur dioxide, nitrogen oxide, total suspend particulate, carbon monoxide, and carbon dioxide) in tons per year
2) Yearly average concentration of four primary pollutants (sulfur dioxide, nitrogen oxide, total suspend particulate, and carbon monoxide), two secondary pollutants, (sulfate and ozone) at ground level, and acid deposition with 11x11 grid

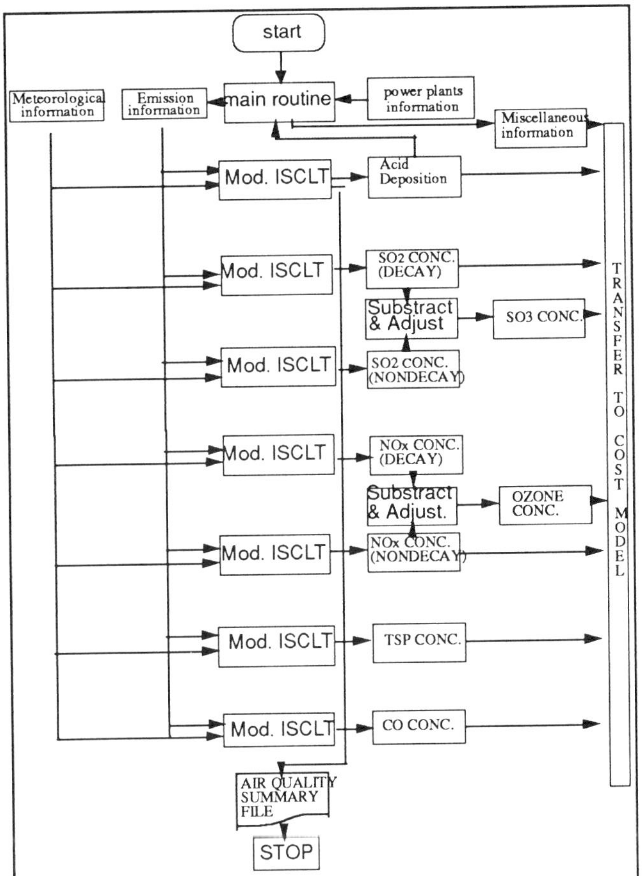

FIGURE 1 - FLOWCHART OF THE AIR QUALITY MODEL

Environmental Cost Model

Air pollution has long been known to have adverse effects on health, plants, and materials. The physical response to a particular pollutant at any time or place around a generation plant is dependent upon the size of the population exposed, the pollutant level, and the response coefficient. It can be expressed with dose-response functions. Dose-response relationships generally fall into four categories: health, vegetation or agriculture, materials, and visibility or aesthetics.

Several functions for environmental damage are optional with this program, so that user can choose one suitable for his/her situation.

EECM is also a FORTRAN-based model that estimates the physical response to a pollutant and assigns it an economic value. Namely, it estimates environmental damage and evaluates the environmental externality cost of local, long-range, and global effects.

Human health effects

Human health effects are expressed in physical and biological terms, such as mortality- or morbidity-pollutant concentration functions. The environmental damage relating the ith physical effect to changes in the dose by j^{th} pollutant is expressed in the following equation:

$$ED_i = \sum_{jk} f_{ij}(AQ_{jk}) * P_k$$

Equation 3

Using the damage function method, the value that society attaches to an air quality change (EC) can be represented as follows:

$$EC = \sum_{i} ED_i * V_i = \sum_{ijk} V_i * f_{ij}(AQ_{jk}) * P_k$$

Equation 4

Where **ED_i** is the ith damage effect, such as annual death rate (mortality), amount of illness cases (morbidity)

EC is the environmental damage in dollar value

$f_{ij}(X_{jk})$ is the dose-response function relating the ith physical effect to changes in the dose by j^{th} pollutant

X_{jk} is the annual average concentration of jth pollutant at k^{th} unit in mg/m^3

P_k is the population density at the kth unit

V_i is the dollar value of the ith type of physical and biological damage, for instance a social value for a change in risk per capita

Whittemore and Korn used the logit model to estimate the probability of an asthma attack. Lave and Seskin estimated a dose-response function where the dependent variable is the death rate in various cities around the United States.

A typical S-shaped damage function shows the damage corresponding from exposure to a specific pollutant[8]. The middle portion of this function is linear, and most concentrations are in this section. The dose-response functions for human health effects from air pollution are generally found to be linear in the range of concentration observed across the United States (see Lave and Seskin (1978) and Mendelsohn and Orcutt (1979))[12]. Laboratory experiments at two and three orders of magnitude above current levels suggest nonlinear characteristics only at those levels. The assumption about linearity of a damage function is most valid in the middle, quasi-linear portion of the curve where data points can be readily interpolated. Therefore, **Equation 4** can be rewritten as follows:

$$EC = \sum_{ijk} V_i * F_{ij} * ((X_{jk} - X_{ij0}) * P_{jk}$$

Equation 5

Where **F_{ij}** is the dose-response coefficient, i.e. slope of middle portion of linear section of dose-response function

X_{ij0} is threshold concentration of jth pollutant to i^{th} physical effect

Agriculture damage

The National Crop Loss Assessment Network (NCLAN) has studied several pollutants and generally concluded that ozone, alone or in conjunction with sulfur dioxide and /or nitrogen dioxide, is responsible for up to 90% of air pollution-related crop damage in the United States.

The dose-response relationship of agriculture damage can be expressed in many functional forms. For different crops, linear and non-linear functions have been studied to fit the yield/pollution data and estimate the dose-response relationships. Ozone's dose-response relationship for barley, kidney bean, corn, cotton, sorghum, soybean, winter wheat, and tomato is given by the Weibull response model with the following form:

$$Y = a \exp[-(x/b)^c] + e$$

Equation 6

Where **Y** is Yield

x is an ozone dose or exposure

b is an ozone concentration when yield is 0.37

c is a dimensionless shape parameter

e is a random error term

Rowe, et al. conducted a comprehensive assessments of California agricultural products, such as dry beans, cotton, and grapes. A linear dose-response functions is fit to estimate crop damage.

Material damage

Material damage has been associated with numerous air pollutants, SO_2 (including acidic deposition), particulates, and ozone. To evaluate material losses, the dose-response function $f_{ij}(X_{ij})$ can be expressed by a linear relationship. The material damage cost due to sulfur dioxide is found by the following expression:

Lower Bound

$$EC = 0.0342 * (X_{SO_2}) + 9.913 \text{ for } X_{SO_2} < 100 \mu g/m^3$$

Equation 7

$$EC = 0.0342 * (X_{SO_2}) + 31.223 \text{ for } X_{SO_2} > 100 \mu g/m^3$$

Equation 8

Upper Bound

$$EC = 0.754 * (X_{SO_2}) - 3.554$$

Equation 9

Where **X_{SO_2}** is concentrations of sulfur dioxide, ozone, and particulate in $\mu g/m^3$

EC is per capita damages in 1983 dollars

There are similar relationships for evaluating crop losses due to ozone and particulates.

Visibility losses

The impairment of atmospheric visibility is caused primarily by small particles whose size is smaller than 2 μm, that scatter and absorb light. Larger particles and colored gases, such as nitrogen dioxide, also reduce visibility, but to a lesser degree. The cumulative percentage of such small particles is usually about 5%.

If the change in the fine-particle mass is known, then the change in visibility in kilometer per year is[7]:

$$\Delta L_v = -\frac{L_o * \Delta M_f}{(1300/L_o) + \Delta M_f}$$

Equation 10

Where L_o is the initial visibility condition in (km/yr)
ΔM_f is the mass of fine particulate in $\mu g/m^3$

The total environmental damage in persons-km per year is a summation of visibility losses around all of the receptor units, weighted by the population density at these points. Environmental cost in dollars, is the product of environmental damage and willingness-to-pay for person-km year of visibility loss.

The dose-response coefficients are highly uncertain. The uncertainty arises from several factors. The spatial correlation of pollutants and other unrelated factors allow the findings to represent a false association between mortality and air pollution. Furthermore, human-health effects are correlated more closely with lifetime exposure patterns rather than current ones, indicating that the consequences of current exposure are latent rather than immediate. The dose-response coefficients for morbidity are even less reliable, large-scale collections of morbidity statistics are rare and imperfect for the purpose of studying dose-response functions. Small-scale morbidity studies have generally failed to account for area-wide variables which confound the pollution effects. The dose-response information for crops is among the most reliable. Scientists can measure the sensitivity of crops to different pollutants in controlled experiments. One of the measures to estimate these uncertainties and dispersion is the standard deviation function.

The total environmental cost is the algebraic sum of individual environmental costs of damage and effect, as shown in **Equation 11**. The total standard deviation is the geometric sum of the standard deviations of individual damage and effect. This can be expressed by the following formula:

$$ST = \sqrt{\sum (ST_i)^2}$$

Equation 11

Where ST is Total standard deviation
ST_i is Standard deviation of individual damage and effect

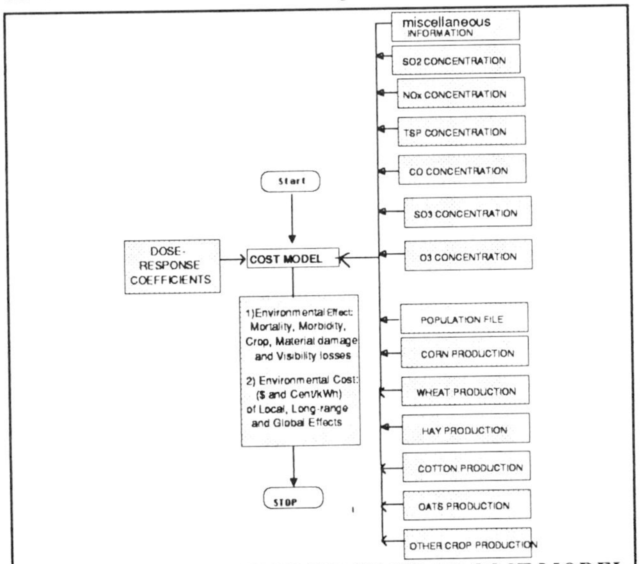

FIGURE 2 - INPUT FILES DIAGRAM OF COST MODEL

Figure 2 shows fifteen input files needed to run the Cost Model, seven of them were produced by the Air Quality Simulation Model; the rest were produced beforehand.

Two types of data are output by this cost model, they are:
1) Environmental damage, namely, mortality (deaths per year), morbidity (cases per year), crop losses (bushel per year), material damage (dollars per year), and visibility losses (per sons-km per year).
2) Total environmental cost both in dollars term and in cent per kWh.

In addition, the cost model also produces the costs of long-range and global effects.

INCORPORATION OF ENVIRONMENTAL EXTERNALITIES INTO THE UTILITY PLANNING PROCESS

Consideration of environmental externalities is becoming an increasingly important factor in utility planning. There has been increased attention by regulatory bodies directed toward taking fuller account of environmental costs. Twenty-six jurisdictions and utility regulatory commissions have taken action to incorporate external environmental costs into utility resource planning, cost-effectiveness evaluation of large DSM programs, and least-cost planning[26]. The City of Austin's Electric Utility Department has also been required to incorporate these environmental externality costs in its planning process.

The City of Austin has four fossil-fired plants, a share of the South Texas Nuclear Project (STP) providing a capacity of 400MW, and a small photovoltaic operation. The four fossil-fired plants are: Holly Street Power Plant with four gas (oil) units and a capacity of 540MW; Decker Creek Power Plant with four gas (or oil) units and a capacity of 710MW; Seaholm Power Plant with a capacity of 100MW; Co-ownership, with the Lower Colorado River Authority, of the Fayette Power Project located in La Grange, Texas where three coal units with steam turbines produce 1,140MW of which Austin's share is 570MW.

Environmental impact of DSM

The City of Austin has shown a strong commitment and investment in energy efficiency[5]. Savings of 553MW were to be achieved by implementing a variety of energy conservation programs and constructing renewable energy projects. In the evaluation of different conservation programs, resources are best utilized when investments in energy efficiency services added to environmental damages are minimized. The assessment of environmental impact is a necessary part of the cost-effectiveness evaluation of DSM program. The following is our study in this respect.

Air quality

The Austin Energy Conservation Program has been studied with this model. The concentration of all of pollutants are increasing with an increase of emissions by year.

Figures 3 and 4 show emission trends for nitrogen oxides and carbon dioxide, that serve as an example of model output. It can be seen that emissions of NO_2 and CO_2 rise as generation increases between 1991 and 2010. Table 1 gives generation savings and avoided emission owing to DSM program.

Table 2 shows the concentration of NO_x. The origin of X and Y coordinates is set at Fayette Power Plant, each element represents an annual average concentration at ground level of square area (30*30 km^2) with the center at the crossing of vertical and horizontal coordinates. The ten highest concentrations were marked with the length of an underline. The downtown area of Austin is 97 km to the West, and 39 km to the North of the Fayette Plant. Prevailing southeasterly winds push air pollution from the Fayette Power Plant toward Austin. In addition, Decker and Holly plants are located in Austin. Their NO_x emission greatly contributes to the highest concentration of NO_x at ground level in Austin area. Another high concentration area is north of the Fayette Power Plant.

The receptor site at (-90.0, 30.0) is close to downtown Austin. The concentration of pollutants at this site is listed in Table 3. It can be seen, that DSM reduces average annual concentrations of pollutants at ground level. In 2000 and in 2010 it decreases by 11~24% and 12~16%. Therefore, it can be deduced, that DSM leads to some improvement of air quality in Austin. Table 3 shows the improvement of air quality due to DSM.

Environmental damage and cost

Despite uncertainties, the cost of environmental damage caused by electric generation can be estimated. Increases in mortality and pollution related illnesses, such as bronchitis, croup, pneumonia, acute illness, and lower respiratory are caused by various pollutants: sulfur dioxide, sulfate, nitrogen oxides, particulate, carbon monoxide, and ozone have been evaluated. Yield reduction of different crops, material and visibility losses were calculated as well. The evaluated environmental costs of local, long-range, and global effects in 2000 are shown in Figure 5. These are about 0.28 ¢/kWh, 0.14 ¢/kWh, and 1.6 ¢/kWh in 2000 without DSM, respectively.

In order to evaluate environmental benefits of the Austin Energy Conservation Program it should quantify future benefits; discount rates are used to assess the value of economic benefits and costs that may occur. The environmental cost of local effects is mainly driven by human health effects. The Nature Resources Defense Council contends that the discount rate for health risks should be zero because a life in the future is as valuable as a life in the present.

The environmental cost of global effect and long-range effect result from material damage. The discount rate used is 8%.

From Table 4 it can be seen that the environmental benefits due to DSM are $182.5M, $0.86M, and $21.9M due to global effect, long-range effect, and local effect respectively.

The City of Austin invests approximately $10M in DSM programs yearly, and takes the same discount rate as environment cost, hence the present value of total investment for 19 years is $100.6M. Provided we only take the environmental benefits into account, the coverage ratio is 2.05.

It can be concluded that DSM leads to improved air quality, decreased environmental damage and cost, and incurs lower true cost to build and operate than generation. Consequently DSM program in City of Austin are cost-effective even if evaluated only in respect of environmental benefit.

Quantification of Environmental Costs and IRP

These days most planners try to include environmental externality cost into costs of energy in their least cost resource selection process.

There are three basic approaches for incorporating environmental externalities into resource planning: qualitative treatment during the resource planning process; direct quantification of environmental costs and impacts as part of integrated resource planning; use of a percentage adder/substractor which is applied to capital cost of supply- or demand-side resources during the planning process[3]. This chapter will show some work of City of Austin in this respect. The figure in Table 5 can be served as an example of direct quantification.

Total social costs of generation with different resources in Table 5 could reflect explicitly competitive of these resources in planning process.

As for internal cost the cheapest fuel is nuclear power, when compared to coal and natural gas. In contrast operations & maintenance expense and depreciation for nuclear power is the highest.

For the coal-fuel Fayette plant the external cost due to global, long-range, local, and other effects — mainly from water pollution and waste heat — are higher than that of the gas-fueled Decker plant. Routine operation of nuclear power plant incurs low external cost, but an accident would result in disastrous costs. Since it is difficult to estimate the probability of such an accident and its consequences, the externality cost of nuclear generation is only a very rough estimate.

The city of Austin is evaluating the future value of photovoltaic (PV) generation for utility applications. PV generation is an attractive option for the future due to low maintenance and external costs, and coincidence with the utility peak periods. There is also a potential for greater cost effectiveness as PV prices decline. The environmental benefits of a 2MW PV system was evaluated. During a 15 year period, 247.4 GWh of electricity will be generated with a PV system instead of fossil-fired units. It leads to environmental benefits at $4.09M, i.e. 1.66 ¢/kWh. The PV system would reduce energy losses in the transmission system during the summer peak hour. The benefits from PV range from 50-$58/kW. Moreover decentralized PV benefits would double the benefits because of transmission savings. Therefore, with the consideration of external costs and the decline of cell prices this renewable resource will be even more competitive in the future.

One important development of Austin's IRP process is the consideration of environmental impacts due to generation and savings due to demand-side options. Figure 5 shows that the largest global environmental cost is due to global warming. The predominate concern for the utility industry with regard to global environmental impacts will be CO_2. The policy considerations and variations in global warming will be addressed by Austin with its IRP optimization matrix. Long-range and local effects are mainly caused by SO_2 and NO_x, their emissions can be estimated. This will allow policymaker to make decisions, and for the investment alternatives to be viewed with associated sensitivities to these factors.

Austin's IRP process produces a matrix of packages with supply- and demand-side investments. These packages are optimal options given various sets of different factors, such as rates/costs, reliability, financial integrity, economics, risk, flexibility, and environment. The environmental factors are emission levels, change of air quality, environmental damage, and damage costs, as caused by local, long-range, and global effects. An example of that optimization matrix is attached in Table 6.

The goal is to provide necessary information that can be compared for all resources (supply- and demand-side, externalities-environmental and economic impacts) so that trade-offs can be weighed in the decision-making process.

Cost-benefit assessment of control measure

Title IV, added by the 1990 Amendments of Clean Air Act establishes a new allowance program designed to reduce emissions of sulfur dioxides ("SO_2") and nitrogen oxide ("NO_x") associated from utilities.

Scrubbers would be used to control particles or gases. The simplest venturi scrubber spray tower is commonly a vertical tower with a number of spray nozzles inside. After consulting with experts, we assume that the efficiency for removing both particles and sulfur dioxide gas approaches 90%.

The scrubber is expected to be on-line in 2003 without DSM and in 2004 in the case which includes DSM. Flue gases from one of two existing units, and all of the new installed coal-fired units in the Fayette Power plant will be treated by the scrubber. In this scenario emission rates for these units are 0.341 g/kWh for SO_2 and 0.026 g/kWh for particulate instead 3.405 g/kWh and 0.263 g/kWh respectively.

After running the cost model, it was determined that the cost of long-range effect are 0.225 ¢/kWh and 0.0222 ¢/kWh before and after installation of scrubber respectively, those of local effect are 0.373 ¢/kWh and 0.123 ¢/kWh before and after installation of scrubber respectively. The estimated capital cost of the scrubber is $70M~$80M. Cost for maintenance and operation is $1.25/MWh, and the discount rate of the benefit is 8.5%.

It can be estimated that in 2009 the cumulative net benefit is $11M which means, all of the capital investment for the scrubber will be paid back in seven years without DSM.

In the case of DSM, the cumulative net benefit is $18M in 2009. It means all of capital investment of scrubber will be paid back in six years.

It can be concluded, that installation of scrubber is an effective measure for controlling air pollution to meet the requirement of Clean Air Act, and it is an economical one as well.

SUMMARY

1) A methodology and Environmental Externality Model system, consisted of enhanced air quality model and cost model, has been developed. Four classes of information are given by this model system, they are emission of pollutants, air quality, environmental damage, and environmental cost. This methodology and computer model system would quantify the environmental impacts of the city's energy conservation programs both by pollutants' amount and dollar value.

2) The DSM program was evaluated with this model. It concluded that from 1991 to 2010 during the 20 years' period, the cumulative avoided generation is 11,843.1 GWh. During the same period 21,251.07 tons of nitrogen oxides, 4,464.73 tons of carbon monoxide, and 9.865M tons of carbon dioxide are avoided. If excluding year 1993, during a 19 year period 3,087.57 tons of sulfur dioxide, and 311.01 tons of particulate are avoided.

The DSM program leads to improved air quality, it reduces concentration of most pollutants by more than 5%, in some cases, up to 30%. Because of the DSM program environmental benefits are $182.5M, $0.86M, and $21.9M due to global, long-range, and local effects respectively. In conclusion, the DSM program is an effective component of the IRP process in the City of Austin.

3) The predominate concern with incorporating environmental externalities into the utility planning processe will be CO_2, SO_2, and NO_x. The policy considerations and variations in global warming valuation will be addressed by the City of Austin with its IRP optimization matrix.

4) The scrubber is a cost-effective measure to prevent air pollution. The investment can be paid back in 6-7 years.

REFERENCES

[1]Bonneville Power Administration. ECO Northwest, Final Report, "Economic Analysis of the Environmental Effects of the Coal-Fire Electric Generator at Boardman, Oregon,"December 29, 1990.

[2]California Energy Commission and Regional Economic Research, Inc." Valuing the Environmental Impacts of Alternative Energy Resources," May, 1991.

[3]Cohen, S.D., et al. "Environmental Externalities: What State Regulators are Doing." The Electricity Journal. Vol. 3, No. 6, June 1990.

[4]de Leon, Armando J., and Yang Ziming. "Evaluation of the Environmental Costs of Nuclear Electricity Production by South Texas Project," City of Austin, ECSD, April, 1993.

[5]ECSD, City of Austin. "The Technical Development to Create an Integrated Resource Planning (IRP) Process: The IRP Development Plan for the City of Austin," May 1990.

[6]EPA Office of Air and Radiation, Office of Air Quality Planning and Standards. "Industrial Source Complex (ISC) Dispersion Model, User's Guide," Second Edition (Revised) Volume I., December, 1987.

[7]Halvorsen, Robert. Benefit-Cost Analysis of Air-Pollution Control, (Lexing-

ton, MA: 1981).
[8]Hershaft, Alex. Air Pollution Damage Functions, Environmental Sciences and Technology, Vol. 10, No. 10 (October, 1976), pp. 992-5.
[9]Hohmeyer, Olav. Social Costs of Energy Consumptions. (Springer-Verlag: Berlin).
[10]Megdal, Lori McCreary, and Yang Ziming. "Environmental Impacts of Austin Energy Conservation Program: Phase I—Air Pollution," City of Austin, ECSD, April, 1992.
[11]Megdal, Lori McCreary, and Yang Ziming. "Incorporating Environmental Impacts into Integrated Resource Planning: The Use of an Optimization Matrix for IRP and the Development of Costing Model for Local Environmental Impacts." City of Austin, ECSD, 1991.
[12]Mendelsohn, R. and G. Orcutt. "An Empirical Analysis of Air Pollution Dose Response Curves", Journal of Environmental Economics and Management (June, 1979), p. 85.
[13]Ottinger, Richard L., Nicholas A. Robinson, and Shepard C. Buchanan, et al. Draft Final Report, "Environmental Externality Costs from Electric Utility Operations." Pace University Center for Environmental Legal Studies, February 28, 1990.
[14]Ottinger, Richard L. "Getting at the True Cost of Electric Power," The Electricity Journal, Vol. 3, No. 6, July 1990.
[15]Perkins, Henry C. Air-Pollution, (New York: McGraw-Hill, 1974). Texas Agriculture Statistics Service, "Texas Agriculture Facts", SM-04-90, USPS 396-750, February 23, 1990.
[16]Ziming, Yang, Armando J. de Leon, and Lori McCreary. "Environmental Impacts of Austin Energy Conservation Program: Phase II—Water Pollution and Others." City of Austin, ECSD, April, 1993.
[17]Ziming, Yang, Armando J. de Leon, and Lori McCreary. "Users's Guide to the Environmental Externality Cost Model, " City of Austin ECSD, April 1993.

Acknowledgements

The work in this paper was initiated and supported by Lori Medgal; we appreciate her support and patience. Her work is a vital part of our study. This work was supported by the City of Austin, Environmental Conservation and Services, we wish to thank everyone who helped with this paper.

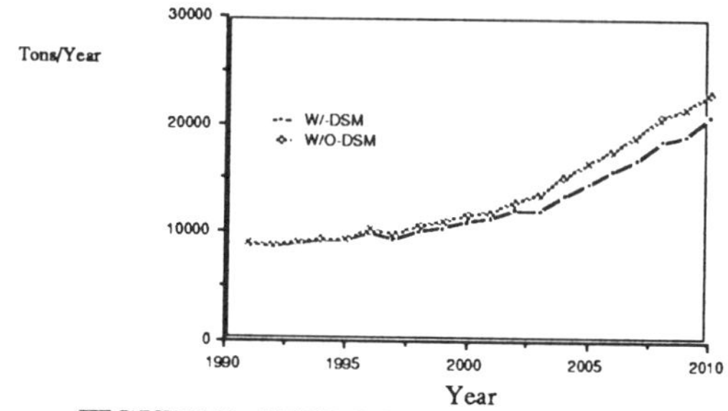

FIGURE 3 - NITROGEN OXIDE EMISSION

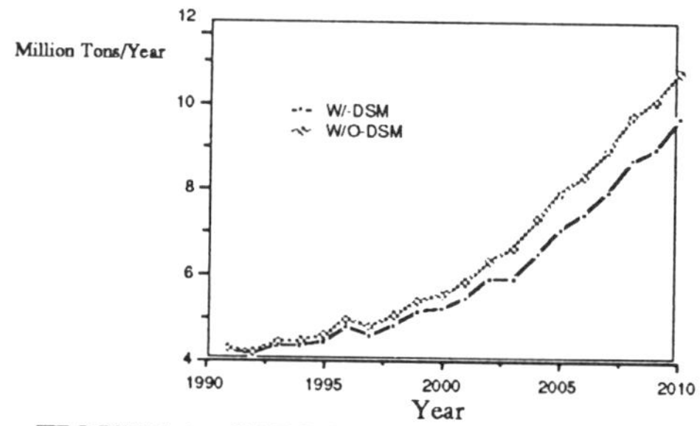

FIGURE 4 - CARBON DIOXIDE EMISSION

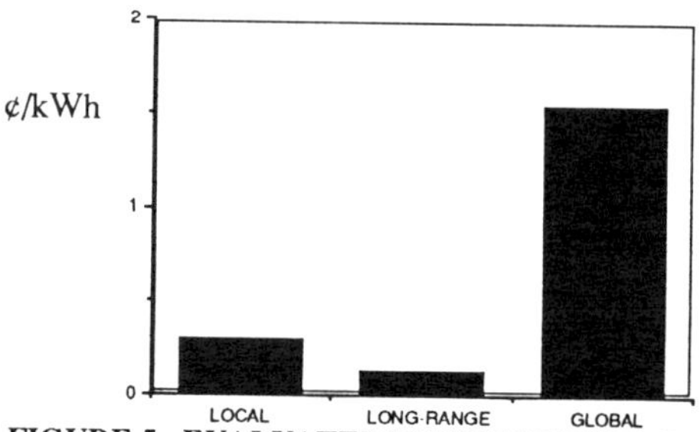

FIGURE 5 - EVALUATED ENVIRONMENTAL COSTS OF LOCAL, LONG-RANGE, & GLOBAL EFFECT

TABLE 1 - AVOIDED GENERATION & EMISSION OF WHOLE SYSTEM WITH DSM PROGRAM[10]

Year	Saving	SO_2	ACID	NO_x	TSP	CO	CO_2
	GWh	T/Y	T/Y	T/Y	T/Y	T/Y	1000*TY
1992	96.5	40.09	35.31	119.65	4.37	20.82	59.319
1993	126.1	120.82	106.47	158.5	10.72	31.07	85.004
1994	172.5	141.02	124.28	224.02	12.93	46.09	113.911
1995	222.8	165.67	146.01	272.13	15.49	58.17	145.3
1996	275.9	75.34	66.36	309.74	9.74	76.2	166.379
1997	326.2	109.17	96.19	415.99	12.99	91.16	198.908
1998	381.1	152.89	134.71	519.02	17.09	114.23	235.518
1999	439.7	65.3	57.51	574.86	11.77	152.57	261.122
2000	506.9	38.99	34.3	634.78	11.42	239.71	300.824
2001	585.9	86.19	75.89	685.59	16.65	335.62	355.425
2002	665.2	71.84	63.23	741.1	17.52	446.6	404.517
TOTAL	3798.8	1067.32	940.26	4655.38	140.69	1612.24	2326.23

TABLE 2 - ANNUAL AVERAGE GROUND LEVEL CONCENTRATION OF NO_2 IN 2000 WITHOUT CONSERVATION PROGRAM ($\mu g/m^3$)[10]

X(KM)	-150.0	-120.0	-90.0	-60.0	-30.0	0.0	30.0
150.0	.136432	.208267	.298028	.218352	.171060	.176939	.115016
120.0	.139208	.230979	.388680	.223410	.176969	.192152	.115153
90.0	.127006	.254471	.564225	.215225	.166677	.216672	.104228
60.0	.101813	.208929	.912524	.146360	.153205	.255801	.093235
30.0	.091506	.143803	.472993	.173099	.137101	.304293	.066171
0.0	.093241	.168294	.225317	.126537	.105646	.046487	.054454
-30.0	.092037	.121343	.135840	.093795	.107527	.131342	.066368
-60.0	.076487	.093384	.105855	.095168	.099170	.110500	.063801

TABLE 3 - IMPROVEMENT OF AIR QUALITY DUE TO DSM[10]

Pollutant	SO_2	NO_x	TSP	CO	SO_3	O_3
In Year 2000						
W/O DSM	0.046954	0.472993	0.009750	0.109869	0.002353	0.025502
W/ DSM	0.046634	0.391112	0.008668	0.083168	0.002344	0.022532
REDUCTION %	0.68	17.31	11.10	24.30	0.38	11.65
In Year 2010						
W/O DSM	0.038857	0.447184	0.008342	0.127337	0.002	0.042952
W/ DSM	0.037245	0.370661	0.007165	0.080769	0.001915	0.0374
REDUCTION %	4.15	17.11	14.11	36.57	4.25	12.93

TABLE 4 - ENVIRONMENTAL BENEFIT DUE TO CONSERVATION PROGRAM (DISCOUNT ADJUSTED)[10]

Year	Global $	Long-Range $	Local $
1992	1,098,500	24,429	158,390
1993	1,457,528	68,375	246,806
1994	1,808,540	73,914	354,614
1995	2,135,996	80,419	432,402
1996	2,264,695	33,810	473,250
1997	2,506,915	45,394	628,015
1998	2,748,450	58,898	807,102
1999	2,821,511	23,249	940,734
2000	3,009,738	12,816	1,253,194
2001	3,292,611	26,322	1,612,830
2002	3,469,800	20,293	1,979,768
Subtotal (1992-2002)	26,614,284	467,919	8,887,105
2003			
2004	5,832,461	59,514	1,212,046
2005	5,746,717	59,630	1,445,712
2006	5,648,098	56,544	1,841,980
2007	5,883,403	60,118	1,651,856
2008	5,748,420	60,084	1,777,916
2009	5,591,914	56,350	2,128,446
2010	4,983,283	38,735	2,919,000
Subtotal (04-10)	39,434,296	390,975	12,976,956
Total (excl 93)	66,048,580	858,894	21,864,061

* In case of DSM installation of scrubber will be deferred for one year.

**Total saving generation from 1992-10 including 1993 is 11843.1 GWh

TABLE 5 - ECONOMIC ANALYSIS OF ELECTRICITY PRODUCTION (¢/kWh)

PLANT	FAYETTE	DECKER	HOLLY	STP[4]	PV
FUEL	COAL	GAS	GAS	NUCLEAR	SOLAR
INTERNAL EXPENSE					
O & M EXPENSE	0.325	0.33	3.35	1.795	2.71
FUEL EXPENSE	1.500	2~3	2~3	0.5	0.0
DEPRECIATION	0.232	0.394	0.291	0.822	22.63**
TOTAL (excl. transm.)	2.06	2.72 ~ 3.72	5.64 ~ 6.64	3.12	25.34
ENVIRONMENTAL EXTERNAL COST					
GLOBAL	1.858	1.141 ~ 1.17	1.14		
LONG-RANGE	0.2251	0.00018~0.0002	0.000195~0.0002		
LOCAL	0.34 ~ 0.41	0.106 ~ 0.23	0.1541 ~ 0.202		
OTHER[16]	0.126	0.0201	0.10	1.55~ 2.43*	
TOTAL	2.584	~ 1.344	~ 1.418	1.55 ~ 2.43	

*Most parts of the external cost is environmental risk of accident, the cost at 2.43 ¢/kWh is cited from Prof. Ottinger's work[13].
** It is from 300 kW experimental installation.
*** Because of low-capacity factor, the Holly plant is operating with an extraordinarily high O&M expense.

TABLE 6 - EXAMPLE OF AN IRP OPTIMIZATION MATRIX[11]

Package 1	Package 2	Package 3	Package 4
Optimized at CO_2 valued at \$20/ton, minimize rates and minimum DSM to each customer group.	Optimized for rates with no global costs considered.	Optimized for minimum revenue requirements, no global costs, and DSM addresses minimum equity across customer group.	Optimized for rates with no global costs, only solar energy options.
Natural Gas Plant 2002 Natural Gas Plant 2008	Natural Gas Plant 1999 Coal Plant 2004 Natural Gas Plant 2008	Natural Gas Plant 2001 Natural Gas Plant 2006	Solar Plant 2002 Solar Plant 2008
DSM investment of \$130M, wavy depending on closeness of plant date, with \$40M single-family, \$35M multi-family, \$50M industrial. Min \$10M to low-income.	DSM investment of \$100M, extremely wavy, depending on closeness to plant date	DSM investment of \$100M, slight wave, with \$30M single-family, \$25M multi-family, \$35M commercial, and \$10M industrial	DSM investment of \$70M, much off-front until solar energy options fall
Economic Impact: \$7M	Economic Impact: \$5M	Economic Impact: \$10M	Economic Impact: \$2M
CO_2: 20 M tons SO_2: 6,502 tons NO_x: 14,245 tons	CO_2: 55M tons SO_2: 17,880 tons NO_x: 34,000 tons	CO_2: 18M tons SO_2: 5,852 tons NO_x: 13,400 tons	CO_2: 15M tons SO_2: 4,888 tons NO_x: 10,300 tons
Average rate: 10¢/kWh Average Summer Bill: \$140 Rate Jumps: 2, slight	Average rate: 6¢/kWh Average Summer Bill: \$140 Rate Jumps: 3, slight-mod	Average Rate: 13¢/kWh Average Summer Bill: \$110 Rate Jumps: 2, slight	Average Rate: 14¢/kWh Average Summer Bill: \$160 Rate Jumps: 2, mod-slight

Chapter 36

A "What if" Program for Creating Energy Efficient Scenarios

J.S. Tiller, C. Berish

INTRODUCTION

Throughout the United States, policy makers at national, regional, state and local levels are making decisions about what types of energy efficient programs to implement. In recent years, many of these programs have emerged from environmental agencies, largely as a result of the evolving view that many of our environmental problems are due to the use of energy. Thus, programs to reduce pollution should include energy efficiency as a key element.

Many of the decision makers who have responsibility for program development lack an extensive background in energy efficient technologies. They are usually not aware of the broad array of energy saving options available and of the potential costs and savings of each option.

These decision makers typically consider a range of program options and must select which ones to assign highest priority. They have limits of both time and program budget. For example, a Pollution Prevention Program Officer might be pondering the following types of programs:

1. A region-wide publicity program to run articles in newspapers on the importance of conserving home energy use to save money and protect the environment.

2. A one-on-one outreach campaign with small businesses to encourage energy efficient lighting, HVAC improvements, recycling and use of recycled materials, and water conservation.

3. Coordinating an effort to seek more stringent building energy codes and better enforcement.

4. Encouraging electric and natural gas utilities to pursue more aggressive Demand Side Management Programs.

5. Educational programs run through state energy offices to promote regular automotive maintenance and proper tire inflation.

6. A energy conservation loan fund for industrial plants.

On which of the diverse programs should the program officer concentrate? The critical variable in most cases is the expected reduction in air pollutant emissions per dollar of program cost. There are few ready tools that a program officer can use in comparing the different programs.

Numerous national studies have investigated alternative futures for reduction of greenhouse gas emissions. For example, ICF Resources, Inc. [1] performed a detailed analysis of the cost of technological measures to reduce U.S. greenhouse gas emissions by 2010. This study evaluated a wide variety of energy savings measures; however, the project did not seek to develop a tool for decision makers to use in selecting from among a wide variety of specifically targeted programs.

* The views in this paper are solely those of the authors and do not reflect official U.S. Environmental Protection Agency policy.

The U.S. Office of Technology Assessment also performed a detailed investigation of alternative greenhouse gas futures [2]. This study linked energy use and the economy in an attempt to determine the overall impacts of various futures on American society. While the study is quite informative, it does not provide a basis to determine which specific types of policies or actions would be most successful.

As a consequence, the Region IV Office of the U.S. Environmental Protection Agency has contracted with the Southface Energy Institute to develop a user-friendly computer model that creates energy efficient scenarios for their specific region of interest -- whether local, state, region or nation.

THE COMPUTER MODEL

Introduction

The computer model is actually a series of Lotus 123^{tm} spreadsheets. As people in all walks of life have become familiar with spreadsheets and since Lotus is one of the most popular, Southface and EPA felt it would be a convenient format for the model to enhance its use. In addition, Lotus 123 has the ability to generate graphics easily.

There are five spreadsheet submodels:

Residential
Commercial
Industrial
Automotive Transportation
Executive Summary

The Executive Summary reads in data from the other spreadsheets and can generate summary graphics. The first four spreadsheets model energy use in four sectors of the economy. These sectors were targeted as ones in which regional and state environmental and energy agencies could have the greatest impact.

The Residential Submodel

Each of the submodels require site specific data. Examples are shown in the References [3,4,5 and 6.] Among the data needs for the residential submodel are:

- Climatic data
- Current energy consumption data
- Current environmental emissions data
- Characteristics of the existing building stock
- Characteristics of typical new homes

Using the above data, the submodel develops energy consumption estimates for a typical existing home. These estimates are extrapolated to the region as a whole to predict regional energy totals. The model compares the predicted consumption with actual and calculates correction factors to mesh predictions with reality.

In preparing an energy efficient scenario, the submodel requests data for projections of future market penetrations for energy saving technologies in residences. Users seeking a quicker view of the possibilities can request that the program develop its own projections.

The submodel employs the market penetration rates to estimate energy use in the residential sector for the energy efficient scenario. It includes an implementation schedule by which the submodel phases in the implementation of the energy efficient technologies.

After finding the future energy use for both base case and energy efficient scenarios, the submodel can provide a number of charts and graphs of the results. As with the other sector submodels, the final residential submodel serves as an input into the executive summary submodel.

Commercial Buildings Submodel

The submodel for the commercial sector is quite similar in design to that for residential. The technologies and some of the procedures differ, but in general the logic is the same.

Industrial Submodel

The industrial submodel is much less complex than the commercial and residential submodels. It divides the industrial sector into the following activities:

Space Heating and Lighting
Steam
Machine Drive
Direct Process Heat
Electrolysis
Mobile Engines

Rather than inputting data on the market penetration of specific technologies, the user types in percentage savings in future energy use for each of the above activities. The model then estimates the savings by fuel type, which are used to predict reductions in air emissions.

Automotive Submodel

The automotive submodel asks the user to make assumptions about automotive miles traveled and average automotive efficiency in miles per gallon. The future scenarios hinge on penetration rates for energy efficient measures and use of alternative fuels. Among the measures are selection of more efficient automobiles, better maintenance, and use of other modes of transit for commuting and travel. The submodel does not ask for implementation rates of any specific energy efficient feature in new cars; instead it asks for projections of future average fuel efficiency.

The submodel also considers development of both alcohol fuels and electric cars. It allows the user to project future efficiencies for both technologies. Of course, it estimates the air pollutant emissions resulting from these technologies, as do all of the submodels.

TABLE 1: ASSUMPTIONS ABOUT RESIDENTIAL PENETRATION RATES

	Low	Moderate	High
Energy efficient measures (insulation, infiltration control, energy efficient lighting, water conservation)	.3	.6	.9
New equipment (heating and cooling system, water heater, refrigerator)	.1	.3	.5
Average savings in new construction over standard homes	.05	.10	.15

TYPICAL SCENARIOS

The CO2 Reduction Scenario Generator will allow the user to input virtually any of the substantial data required to run the model. However, many users would prefer to select the default data and employ the model to examine broader issues, such as the range of possible futures in each of the different sectors.

Figures 1, 2, 3, and 4 show the types of charts available from the model. Figure 1 is a typical set of low, moderate and high impact scenarios for the residential sector. The user can generate scenarios for a region or state. For example, the residential scenarios shown in Figure 1 are based the market penetration rates shown in Table 1.

Figures 2, and 3 show similar findings for other sectors. The model is currrently being finalized with default data for EPA Regions 3, 4 and 5. Any user can quickly generate scenarios if they do not wish to change the data.

Figure 4 shows the contribution of energy savings from each of the technologies employed in the residential sector. Similar graphics can be generated automatically for carbon dioxide savings by measure. Of course, the actual level of savings is a factor of the assumptions the user inserts in the model.

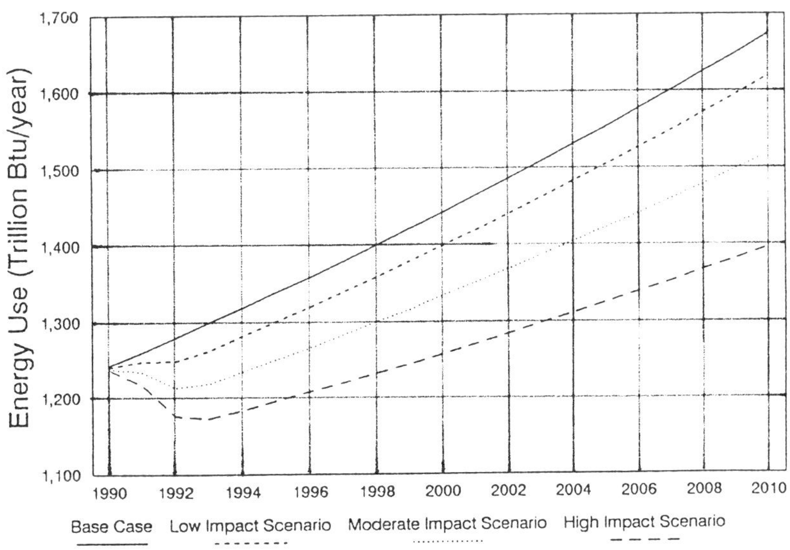

Figure 1: Residential Energy Use Scenarios for Region 3

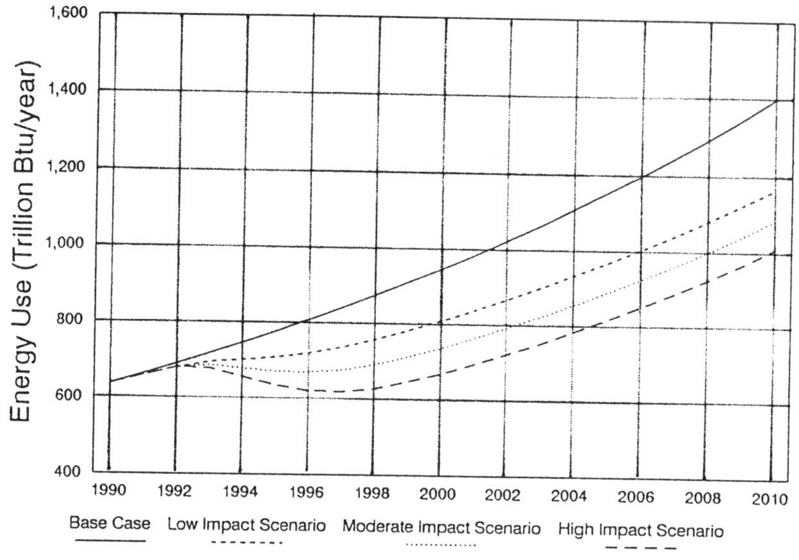

Figure 2: Commercial Energy Use Scenarios -- Region 3

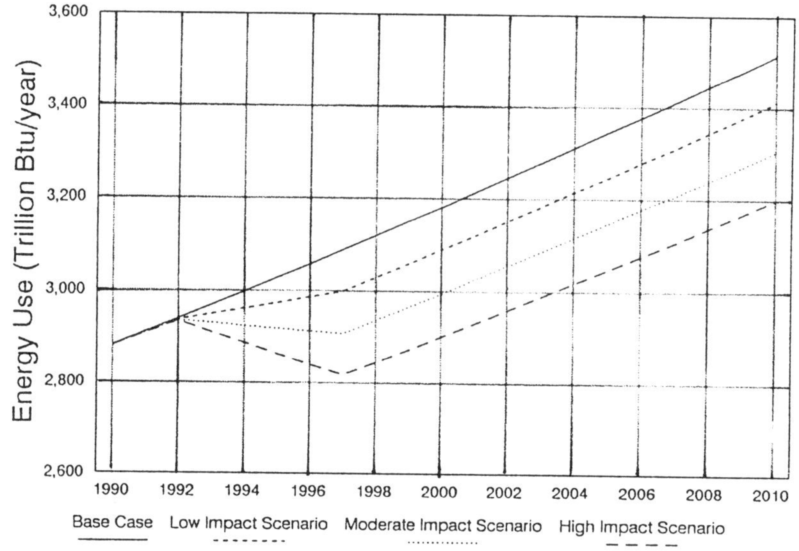

Figure 3: Industrial Energy Use Scenarios -- Region 3

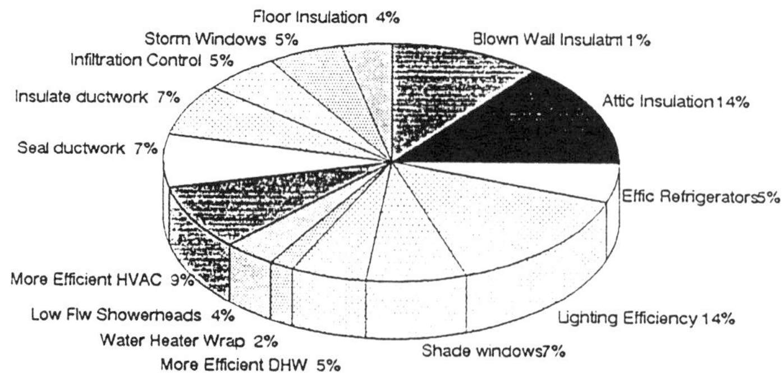

Figure 4: Residential Energy Savings by Measure - Region 3

(56 trillion Btu/year total)

EVALUATION OF PROGRAM OPTIONS

The Introduction contained a list of program options that might be considered in reducing future energy use and consequent air pollutant emissions. The model can provide estimates of the benefits of the various policies. The projected costs of the sample programs are shown in Table 2. While the costs vary widely, it is not clear which type of program will generate the most savings per dollar of investment.

The *residential publicity program* seeks to influence homeowners to undertake specific energy conservation measures in their homes by placing at least four articles in 2,500 newspapers throughout the region. As shown in Table 3, the goal of the program is to encourage at least .5% of the readers to install the measures appropriate for their homes.

The small business outreach campaign is as a door-to-door effort seeking to sign businesses up as energy and water efficient enterprises. Not only will they reap the financial benefits of energy efficiency, but they will be featured in a public relations campaign. The goal is to have 100 small businesses sign up for the program. At least 40% are expected to install energy and water efficient measures.

The *building codes enhancement project* will provide both technical assistance and regulatory encouragement to states to strengthen their state energy code for buildings. In addition, the project will investigate compliance related issues and seek to achieve a higher standard of enforcement of energy code provisions.

The *demand side management assistance program* is similar to the building codes project. It has as a goal to provide technical assistance to 16 utilities in the region. Through the presence of EPA officials and consultants, the project has a goal to encourage both new and existing building owners to implement numerous energy efficient options. The anticipated impact rate for the demand side management program is quite moderate — 1% of existing homes, 5% of new homes, .5% of commercial buildings, and 2.5% of new commercial buildings. Many utilities have experienced much higher long term participation.

The automotive efficient program attempts to secure savings in gasoline through an educational program run through state energy offices. The program will produce and distribute 100,000 brochures and 50,000 automotive efficiency kits -- including tire pressure gauges. The program goal is that 15% of the 50,000 customers receiving the kits will undertake the targeted measures.

TABLE 2:
PROJECTED COSTS OF THE PROGRAM OPTIONS

	Cost for Program
Residential Publicity Program	$56,000
Small Business Outreach Campaign	$150,000
Building Codes Enhancement Program	$160,000
Demand Side Management Assistance	$315,000
Automotive Energy Efficiency Education Campaign	$347,000
Loans for Industrial Energy Efficiency Measures	892,000

The final program being considered is an *industrial energy efficiency loan program*. The program is similar in procedure to the U.S. Department of Energy's Institutional Conservation Program (ICP) with the exception that the funds are loaned rather than granted. The ICP requires a matching grant which would not be stipulated in this program. It is anticipated that the loans would revolve in 5-year cycles. The administrative costs would include 15% of the loan amount.

Table 4 estimates the expected energy and environmental savings from each program and compares these results to the cost of undertaking the program. Despite its projected low impact rate, the demand side management program fares quite well, as does the residential publicity program. The most effective program is that for helping develop more efficient energy codes and implement more stringent enforcement procedures.

CONCLUSIONS

The model developed provides users who are knowledgeable of Lotus 123[tm] a hands-on tool for developing their own scenarios for using energy conservation measures to reduce emissions of carbon dioxide and other atmospheric pollutants. The graphic output will show the potential success of different programs of action that

TABLE 3: PROJECTED IMPACT OF ENERGY EFFICIENCY PROGRAMS

Program	Population	Fraction Impacted	Projected Impact
Residential Publicity Program Articles in 2,500 newspapers (multiple articles)	5,000,000	0.005	Conservation measures installed when appropriate include attic insulation (70%), wall insulation (25%), infiltration control measures (70%), hot water conservation, energy efficient lighting (40%), duct sealing (60%), programmable thermostats (20%)
Small Business Outreach Campaign (100 small businesses)	100	0.400	Conservation measures installed when appropriate, energy efficient lighting (70%), attic insulation (50%), conservation measures (60%), appliance efficiency (40%) duct sealing (25%), infiltration control (40%)
Enhanced Building Codes	24,000 (residential)	0.350	Include R-30 vs. R-23 attic insulation, R-15 vs. R-12 walls, R-19 vs. R-7 floors, double paned vs. .5 single/.5 double windows, infiltration control (.35 air changes/hour or lower), seal ducts (100 cfm of duct leakage on average), hot water conservation
	2,400 (commercial)	0.250	Includes insulation, HVAC, window, hot water, appliance and lighting measures
Demand Side Management 16 utilities			
Existing homes	5,000,000	0.010	Same measures as PR campaign
New homes	24,000	0.050	Same measures as Building Code, with efficient HVAC
Existing commercial	500,000	0.005	Same impact as small business campaign
New commercial	2,400	0.025	20% reduction in energy use
Automotive Energy Education 100,000 brochures/ 50,000 efficiency kits	100,000	0.150	10% improvement in automotive efficiency
Loans for Industrial Energy Measures (4 year payback or less) $5,000,000 worth of loans in 10 industrial plants	10	1.000	$1,000,000 annual savings on energy bills -- lighting, motor efficiency, boiler and process improvements

TABLE 4: PROJECTED BENEFITS OF THE PROGRAM OPTIONS

	Residential Public Relations	Small Business Outreach Campaign	Building Codes	Demand Side Management	Automotive Energy Education	Industrial Loans
Billion Btu Saved/year	25.0	7.5	147.0	604.9	79.2	182.8
$ Saved/Year (thousands)	$250	$82	$1,491	$6,118	$833	$1,000
CO2 Saved (thousand tons)	3.6	1.0	21.1	87.0	7.9	20.2
Years Effective*	10	10	10	10	5	12
Multiplier Effect**	1.10	1.10	5.00	1.14	1.10	2.50
Cumulative TBtu Saved	0.28	0.08	7.35	6.92	0.44	5.48
Cumulative $ Million Saved	$2.75	$0.90	$74.55	$69.98	$4.58	$30.00
Cumulative CO2 Saved (1,000 tons)	40	11	1,057	995	44	607
Cost of Program	$56,000	$150,000	$160,000	$315,000	$347,000	$892,000
Cumulative Savings on Energy Bills ($) Achieved per $ of Program Cost	49	6	466	222	13	34
Cumulative CO2 Reduction per $ of Program Cost (tons/$)	.7	.07	6.6	3.2	.13	.68

would influence implementation of a variety of energy saving technologies.

Several examples of energy conservation program options were presented in this article. Interestingly, funding spent to provide technical information and pressure to those involved in utility demand side management programs and state energy code development and enforcement appear to achieve the greatest savings per dollar of taxpayer money invested. There is obviously a considerable risk of this type of program. However, the success of most energy efficiency programs — other than direct installation of energy measures — has a considerable degree of uncertainty.

The residential publicity campaign runs a strong third in the list of candidate measures. In examining the results, one finds that all of the programs have a high success ratio in dollars of energy saved per dollar invested.

REFERENCES

[1] ICF Resources, Inc. "Preliminary Technology Cost Estimates of Measures Available to Reduce U.S. Greenhouse Gas Emissions by 2010." U.S. Environmental Protection Agency. August, 1990.

[2] Office of Technology Assessment. "Changing by Degrees, Steps to Reduce Greenhouse Gases." U.S. Congress. February, 1991.

[3] Energy Information Administration, U.S. Department of Energy, "State Energy Data Report, May, 1992."

[4] U.S. Department of Commerce. "Statistical Abstract of the United States, 1992."

[5] Energy Information Administration, U.S. Department of Energy, "Commercial Buildings Energy Consumption and Expenditures 1989." (DOE/EIA-0318(89)). April, 1992.

[6] New England Policy Council. "Power to Spare, A Plan for Increasing New England's Competitiveness Through Energy Efficiency." July, 1987.

SECTION 4
ENERGY MANAGEMENT STRATEGIES

Chapter 37

Energy Conservation Projects— Putting It All Together

E.J. Varno

ABSTRACT

Planning and effectively executing an Energy Conservation Program is a challenge that faces industrial managers in the 1990's. Many technical and management concepts are emerging in the energy management field. Though often times confusing once understood they can be used easily to create a successful project. That is, one which reduces production costs and creates a positive cash flow.

Like a jigsaw puzzle the task of putting it all together can prove to be overly challenging unless one can see the big picture. The industrial energy manager needs to begin by understanding how **Demand Side Management** is being practiced by the local utility. To effectively participate in discussions with upper management a working knowledge of the economics of energy conservation is essential. **Return on Investment, Payback Period** and **Cash Flow** are the measuring sticks primarily used.

Many project managers shy away from energy retrofit projects for fear of lack luster results. **Performance Contracting** is an effective risk management tool which is based upon the Pay For Performance concept. Using a performance contract with an Energy Service Company (ESCO) can transfer the risk of an equipment retrofit project and virtually guarantee reduced electric consumption.

ENERGY CONSERVATION PROJECTS - PUTTING IT ALL TOGETHER

Energy, A Taxing Subject

Unfortunately, time does not stand still. If President Clinton's wishes come true the days of expensive energy may be soon coming to an end and the era of sky-high energy costs will begin.

The BTU Tax, The Energy Tax, whatever its called it would have the same effect...higher priced power. Logical reasoning would lead one to conclude those businesses with energy efficient equipment will be more competitive. Those with inefficient manufacturing equipment will be more inclined toward replacing old equipment with energy efficient equivalents.

This inclination may be fraught with fears over how to go about the task and whether investing in energy efficient equipment can meet the economic standards set by company policy. Those facility managers who put it all together make a contribution to the profitability of the company and enjoy the challenge of coordinating the project.

It's not that hard if you understand the big picture.

Start at the Top

It can begin with a memo or even a knock on the door. Your energy conservation plan needs the active endorsement of the top executive of your company. Unless energy conservation is a formal company objective success will be difficult.

There also needs to be a person responsible for developing and implementing the plan. The person can be technical, financial, or administrative. Whatever their background it is important that they work well with people. As with any plan resources will be needed to do the job. These resources must be identified, quantified and communicated to the proper people if the company's balance sheet is sensitive to debt and internal financing is to be used. A more formal presentation of project finances will be required if bank, off-balance-sheet or ESCO financing is desired.

Once the program is underway periodic assessment reports are an effective way to retain program support. Communicating progress to top management, the local utility, and fellow employees will help keep the enthusiasm level high and momentum up. Your public relations department may want to issue a press release now and then to inform the community about your company's energy conservation achievements. A carefully designed communications strategy will make the program flow easier.

Remember: Audits don't save energy, people do.

DSM = Free Cash

It may sound odd but many local utilities are paying their customers big dollars to use less energy. The fact of the matter is that it is less expensive for the local power company to pay you not to use energy than it costs to construct new power plants. This an economic reality driving the DSM industry.

There is a logical history to DSM. As energy production costs increased in the last twenty years government regulatory agencies began to require investor held utilities to look for the most cost efficient way to provide power to their customers. These efforts became known as Least Cost Planning. Demand Side Management or DSM is one of several methods included a utility's Least Cost Planning program.

Although utility efforts to reduce energy consumption have been around since the days of Jimmy Carter the industrial and institutional focus has only emerged in the past several years. A nationwide spotlight was placed on DSM when the Energy Act of 1992 was signed into law. It mandated investor owned utilities to institute DSM programs. Air quality concerns, dependance on foreign oil, the environmental and economic impacts of building new power generating facilities are other driving forces behind DSM. A national energy consumption tax may soon be added to this list.

Conservation of energy using energy efficient equipment is the primary focus of utility DSM Programs. Although not always part of DSM programs, Load Management is another effective way industries use to cut energy costs. By shifting to off peak consumption a company may qualify for lower a service rate during designated hours. The savings can be substantial.

DSM and Your Bottom Line

The real goal of an energy conservation plan is to become more competitive; reducing the cost of production is the objective. With a successful strategy a positive cash flow will result. This is cash that goes right to the bottom line.

Simply because there is money being spent up front convincing those who control company resources to invest in energy efficient equipment may be the biggest challenge to a facility manager wanting to pursue an energy conservation program. Understanding the benefits and the critical components of investment decision making will enable you to prepare proposals in financial terms that are meaningful to those responsible for managing company finances.
A simplified explanation of the most useful measures follows.

RETURN ON INVESTMENT

This is the measure of how well capital available for investment is being managed. The simple formula is:

$$\frac{\textbf{SAVINGS DERIVED}}{\textbf{CAPITAL INVESTED}} = \textbf{ROI}$$

Savings Derived include such items as Net Annual Energy Savings, Annual Savings or Expenses not Energy Related, One Time Savings or Costs, and Useful Life of the Equipment. **Capital Invested** includes all the project costs including principal, interest, any offsetting utility rebates and tax incentives. **ROI** is a broad measure of the cost effectiveness that banks, investment institutions and financial mangers consider important. Return on Investment is often used to set project priorities from an investment perspective. When used in comparison to the Cost of Capital it reveals unprofitable projects and helps to highlight those which are good investments.

CASH FLOW

There is a popular saying in financial circles which goes, "Happiness is a Positive Cash Flow." The sooner actual dollars begin returning after an energy conservation measure is installed the better. Often this cash is shared among the company, the financing source and the ESCO. Sharing the cash flow benefits is a matter of negotiation that takes place during the planning stages of a project. Elements that are considered in the negotiation are the Pay Back Period, how much and when the Principal and Interest payments are due, the useable life of the equipment, who actually provided the investment capital and how much risk and effort was put into the project by each player.

The practical politics of the effort is who gets what percentage of the cash when and how. Remembering the party who advanced the cash must be receive enough to meet the repayment schedule division of the remainder is decided through negotiation. A successful Shared Savings Agreement will be a win -win situation for all parties.

Everyone must benefit from the savings in a way that meets their particular organization's needs.

PAYBACK PERIOD

The most frequently asked question about energy conservation projects is "What's the Payback Period." This is a measure expressed in time which offers a snapshot of how long it will take a project to pay for itself with the energy saved. The formula is:

$$\frac{\text{INVESTMENT IN \$}}{\text{SAVINGS DERIVED}} = \text{SIMPLE PAYBACK}$$

Although the formula seems uncomplicated the reality is the analysis is a bit more complex. The costs which determine the **INVESTMENT** factor not only need to include the dollars spent on equipment but the interest needed to borrow the money. When predicting **SAVINGS DERIVED** soft costs need to be factored in. These include items such as engineering services and changes in O & M costs.

Be aware...the most insidious of all the factors and the most difficult to predict is the cost of energy. Each facility has its own factors which need to be discussed and planned for in the shared savings plan.

As you discuss projects with vendors and consultants and begin to assess the desirability of a project it is important to remember most business managers will not be fond of any projects which exceed a 2.5 year payback.

If the payback period is within acceptable limits and the economics are otherwise favorable **COST OF DELAY** needs to be considered. Although complex algorithms can be developed to assess this matter the basic thrust of this assessment is that "Every minute the current equipment remains on line represents money that can be saved". Cost of Delay can be stressed when there is reluctance to undertake a project.

When the project is final, the equipment operating and long term cost savings are evaluated it is important to understand that actual dollar savings on the electrical bill may not equal that which was originally predicted. Estimates may vary from reality because the actual per unit cost of electricity can vary throughout the evaluation period. If this is the case then an assessment of **COST AVOIDANCE** may be desirable. This is a way to determine what the cost of operation would have been if the efficient equipment had not been installed.

Since dollars saved are, in reality, based on power consumption Cost Avoidance Studies assess the actual reduction in electrical consumption. This requires a monitoring plan to be developed during the project planning stage. A Monitoring Plan involves a pre test of the existing equipment prior to replacement and methodical testing of the new equipment over a period of time.

To assure accuracy the Monitoring Plan must take into account Utility rate schedules and require meter readings which reflect the actual operation of the equipment. Algorithms can get complex in Cost Avoidance. This is particularly true in facilities with changing production schedules or with systems which have many intervening environmental variables such as refrigeration or waste treatment.

Working with ESCOs to Guarantee Results

When searching for technical assistance in developing an energy conservation program it may be best to work with ESCO's which guarantee kilowatt reduction results rather than predict dollars saved. Many lessons have been learned on this topic as the energy conservation industry has matured. One of the most talked about is the problems and conflicts resulting from **SHARED SAVINGS AGREEMENTS.** Periodic changes in energy prices makes prediction of dollars saved difficult at best. Therefore, prediction of kilowatts saved lends itself to a more accurate assessment of the quality of service provided.

If ESCO compensation is to be based on performance, and it should, then it stands to reason that a **PAY FOR PERFORMANCE** agreement is desirable. Assuming an effective Monitoring Plan is developed two possibilities arise. First, The ESCO can predict the actual reduction in electrical consumption to be gained from the installation of the equipment. Payment is based upon meeting the prediction. Second, payment can be based on a set fee per kilowatt saved. This has incentives built into the very nature of the agreement. Creative thinking can produce variations. Pay for Performance concept is fast becoming the agreement of preference in the energy conservation industry.

Putting It All Together

Planning and effectively executing an energy conservation project is an objective that motivates facility managers who are concerned about being competitive. The lure of positive cash flows is often met with the challenge of convincing company officials to invest in energy efficient equipment. Knowing the big picture will generate convincing reasons why these projects make good business sense.

Energy efficient retrofits result in reduced production costs and add to the bottom line. Knowing the facility and the production processes, assessing the economics of a program, selecting a performance partner to provide technical services and building incentives into your agreements are all essential elements to a successful energy conservation project. Put it all together and make it happen.

Chapter 38
Gas Management Strategies, Post-636

G.H. Ausmus

ABSTRACT
Inland Steel is one of the largest natural gas consumers in the United States. This paper does not discuss our broader acquisition strategy for natural gas, but instead focuses on the impacts and countermeasures associated with FERC order 636 and resultant compliance tariffs. The paper concludes with action steps taken by Inland to meet the 636 challenge.

OVERALL ENERGY STRATEGY - A BTU IS A BTU.
Before discussing the narrow topic of 636, I want to make one global point. I call it the zeroeth law of Natural Gas Supply. That law, simply put: **BTU is a BTU is a BTU.** At the risk of belaboring the obvious, natural gas users are in the business of energy management. Anything we do relative to 636 should be in the context of energy management. I continually remind myself. "You have alternatives, you just have to find them."

BTU's, over the long haul, are fungible. Coal, Gas, Oil, Conservation. Economically, there is no difference between these alternatives. OPEC forgot that in the late 70's, and gas producers are forgetting it now.

Finally, look at market elasticity. Energy markets exhibit short term inelasticity, but long term, they are almost perfectly elastic. Think about that before you do any 10 year fixed price deal.

636 ACQUISITION, OR MANAGEMENT?
The title of this session is a misnomer. Why?

This session probes the subject of natural gas acquisition. 636 has its greatest impact on transport and management of gas supply. That is a fine point, but one worth making. When acquiring gas ie, buying the physical commodity, we deal with issues of terms, hedging, basis, counterparty risk, etc. 636 has indirect impact on these issues. 636 more directly impacts where you buy your gas, how you move it, and how you manage the administration and consumption fluctuation. 636 impacts your supply management and logistics.

That's wonderful wordplay, but what does it mean practically?

Practically, it means we have to keep your eye on the goal, which is to provide our plant(s) with gas service. Not gas supply, gas service. We haven't had to do that prior to 636. The right amount, the right price, the right time, all the time. So, we look at the 636 related elements of our gas service portfolio, including the "gas environment" in which our company resides. Some elements of that environment:

- Our Inter/Intra Day Load Swings.
- Neighboring Inter/Intra Day Load Swings, especially IPP's and cogens.
- Operational flow orders.
- LDC actions and attitudes
- Inland's attitude towards risk and towards administrative tasks not incident on the core business.

Describe the elements listed above with clarity, and you'll be halfway to designing a gas service that works. Now, the details.

OUR LOAD PROFILE
636 will, depending on the LDC, give a much narrower operating window for mismatch of gas transport and

consumption. Inland will be competitively advantaged if we can stay in that window. We begin with an understanding of load forecasting. Gas loads can be segregated into a constant baseload, predictable variance, and random variance. Gas supply to a constant load is a trivial problem. Predictable load variances should be quantified and scheduled into your gas deliveries, unless your LDC is going to store on the cheap. Unpredictable load fluctuations can be dealt with through collaboration with other end users (cheapest), intraday nominations (still undeveloped and largely best efforts), fuel switching, scheduling, or no-notice service (NNS). Most pipes will only sell no-notice to the LDC's, and it's expensive. Inland has developed a gas load predictive technique, that, when certain circumstances are met, can eliminate the need for NNS[1].

To minimize the costs of effecting NNS, you should be aware of the profound positive effects load aggregation has on individual load variances. Process load users especially benefit, in that the covariant component of their unpredictable load swings are often small relative to the whole. LDC's are cursed with a load that is almost entirely covariant over their respective service territories. That is expected, as their throughput is weather driven, and weather is a variable that effects everyone in the same way, though not the same degree.

Process users, in contrast, have loads that are largely driven by product throughput. Inland believes that the LDC's could gain much through a incentive load management plan for their process customers. My personal experience in Northern Indiana revealed enough scheduling and fuel switching opportunities to eliminate the need for NNS among a group of industrial gas customers, on an LDC with almost no storage.

I'll editorialize and say that NNS is a natural function of an LDC, and they should be able to do it cheaper than anyone, but rate based costing and conventional system management paradigms have stood in the way of several reasonable schemes, advanced by (who else?) Customers of the LDC and Gas Marketers.

OUR NEIGHBOR'S LOAD PROFILE

The ability of our neighbors (other large end users that are served by the LDC or are proximate on the Interstate pipeline grid) to predict their load may become an issue to us. Especially natural gas fired combustion turbine (CT) generation[2]. If you are going to have a CT behind your LDC you should understand where that unit sits in the electrical power dispatch queue. Modern CT's with high fuel suction pressures have very adverse impacts on line pack when turned on or off. If you don't believe this, conduct a thought experiment: Draw a control volume around the CT. Do an energy balance on the unit. Gas in equals power out. No storage, no ramping, just on or off. Now, take a 200 MW unit and dispatch it. At 8,000 BTU/KWH conversion, that's instantaneous load change of 38,000 MMBTU/D. That's a big hole in the pipe, anyway you look at it.

We should assure ourselves that this kind of load is a) properly planned, preferably near storage and away from us, and b) properly costed, so we aren't carrying the costs of a low load factor end user. Recently, Selkirk Cogeneration Partners was allowed to contract for transport on an MFV, rather than SFV basis on Iroquois pipeline. Why this heresy to the gods of 636? Selkirk's economics flopped when the high demand charges of an SFV rate design conflicted with the revenue created by their anticipated dispatch queue position, sales price and run time[3]. Who pays the rest?

Other large end users behind your LDC will be stirring the soup, based upon fuel switch capability, management of energy, and process. Fortunately for us, the market should allocate the cost of their energy management (or mismanagement) swiftly and surely, leading to a better system for all.

WHAT WILL AN OPERATIONAL FLOW ORDER (OFO) DO?

An OFO is a cry for help, by the pipeline. They scream out: "My flowing inventory is high (low). You must over (under) deliver your nomination to help me - now!" How will Inland react in any given OFO instance? It's too situational to tell in advance. Our strategy is one of compliance and diversity. That's what we plan for under 636. Compliance, all of the time. If we plan for that, manage for that, we shouldn't be far from where we have to be in the event of an OFO. Now, planning for compliance involves contingency planning. We look to 1) cooperation of end users behind our citygate, 2) our own storage arrangements, 3) Market centers for buying/shedding gas, and 4) Curtailment or fuel switch. Importantly, we do not look to over contracting, reserve margins, new physical plant or asset redundancies. The marginal cost of excess reserve is prohibitive, in our view.

WHAT'S YOUR LDC'S DOING ABOUT 636?:

You should be aware of their plans, and be prepared to suggest, challenge, or scream your head off, as the case may be. Challenge them to be creative, as we have. If they are a combined gas/electric utility, ask why they can't make a baseload unit fuel switchable (you can store BTU's in a coal pile, and convert them back into gas by switching off gas and on to coal). Has your LDC suggested that you get paid for voluntary curtailment? Have they asked you to nominate your gas volumes in concert with their (the LDC's) aggregate needs, and pay you for the trouble? Have they asked to help you smooth your load profile? Or are they building storage facilities and pipes willy nilly? Any problem in gas transport can probably be cured with more money and assets in the ground, but why add assets to a grid that is essentially complete?

WHAT'S YOUR COMPANY'S ATTITUDE TOWARD RISK AND ADMINISTRATIVE EXPENSE?

636 will present a riskier world for gas users. There is no bundled merchant gas service that supersedes all other classes of transportation. All parties contracted to create a "gas service" are in a sense "best efforts", ie, they can only perform if the parties upstream of their service perform also. Companies that are very risk averse may want to contract a rebundled service from a gas marketer or the LDC (a rate based gas marketer, if you think about it). A marketer's larger presence in the market, combined with their natural supply and demand diversity should make a complete interrupt remote, as they represent a portfolio effect. Risk seeking firms will have ample opportunities for substantial savings (return) in the capacity reassignment game.

Administrative costs are another issue. The current monthly business cycle of deals, nominations, etc. should expand to a daily business cycle, as parties jockey to match their gas acquisition to their actual usage. In our view, the paperwork for an end user should go up about 30 fold. Further, monthly cashouts will require a timely evaluation of a shipper's imbalance each and every day. In this era of temporaries, contract building maintenance, and outsourcing, you should investigate the use of an outside firm that specializes in gas administration. Those close to this line of work know its a real paper monster, and its only going to get bigger.

WHAT HAS INLAND DONE?

A BTU is a BTU is a BTU

We're switching about 25% of our current gas load to coal. Gas is no longer a value fuel, in our estimation, when full absorbed costs are accurately factored.

Our Load Profile

We're modeling it much more closely, and working to develop unbiased estimates of gas use. We're working closely with other end users on our LDC to capture load aggregation benefits.

Our Neighbor's Load Profile

In working with our end user group, we're sharing our load predicting techniques, in the hope that they can schedule their gas more accurately. Given the state of generating capacity at Nipsco, we don't anticipate any combustion turbine generation in our immediate area soon.

OFO's and Contingencies

Fortunately, we will be close to the Northern Indiana Gas Company (NiGas) hub, which will be the market center for the Midwest. With that and the bag of tricks possessed by ourselves and our neighbors, we are confident of meeting all but the most severe OFO conditions. We will purchase the majority of our gas on term arrangements, and transport it via firm and reassigned capacity with very limited recall rights. We are actively negotiating gas exchanges with nearby LDC's to effect supply flexibility anywhere we can get it.

LDC relations

We discuss the status of their proposed 636 tariff almost every time we talk to them. We are, nonetheless, balancing our gas through the services of a marketer.

Risk management and administrative expense

We carry all commodity risk on our own book, and trade the NYMEX actively. We do this to 1) avoid excessive premiums for fixed price gas, and 2) Keep our vendors from exposure to price risk. 99 out of 100 "force majeure" events are price majeure events and we all know it. Why put your vendors in a position where they want to non perform?

Administration of the supply has been outsourced to a natural gas marketer. This has allowed us to trade and wheel our gas with a great deal more flexibility and accounting accuracy. Long term supply and transport arrangements are still contracted by Inland.

REFERENCES

1. As yet unpublished.

2. "Increased Gas-Firing Creates Load Management problems", *Natural Gas Intelligence,* June, 14,1993, p. 6.

3. "Congen Plant Gets First Exception to Order 636 Rate Design Rule", *The Energy Report,* June 21, 1993, p. 382.

Chapter 39

Vojany Station Reconstruction, Repowering and Expansion Assessment

W.F. Coe, G.H. Griswold, J.C. Peyton

ABSTRACT

Recent European community and state specific environmental guideline, legislative, and regulatory activities have led power producers to assess their currently installed generation technologies with regard to clean air compliance strategies. For the application to older generation facilities, the prudence of linking facility rehabilitation with the addition of environmental control systems to repowering options is warranted. Similarly, ongoing privatization efforts emphasize the necessity for sound economic decisions of site specific technological applications that consider maintaining or enhancing thermal efficiencies.

Southern Electric International (SEI) has conducted such a feasibility study assessment evaluating the reconstruction, repowering and possible expansion of the Slovensky Energeticky Podnik (SEP) 1320 MWe Vojany Station in the Republic of Slovakia. Alternatives included such technologies as atmospheric fluid bed combustion, furnace and duct sorbent injection, low NO_x burners, gas re-burn, selective catalytic reduction, selective non-catalytic reduction, state of the art precipitators and baghouses, wet and dry flue gas desulfurization systems and repowering technologies. In addition, new, turn of the century environmentally plausible and efficient electric power technologies were addressed.

INTRODUCTION

The Vojany Power Station is owned and operated by the Slovak Power Enterprise (Slovensky Energeticky Podnik-SEP).

EVO (Vojany Power Station), the major fossil fuel station of SEP, is located in Vojany, Slovakia, on the Laborec River near the Ukraine border approximately 80 km east of the city of Kosice. The station has two groups of generating blocks, EVO I and EVO II, each having six blocks of 110 MWe each. EVO I is fueled with anthracite T-coal from the Don-Bass region of Ukraine and Russia, and EVO II is fired with either natural gas or heavy oil.

The EVO I blocks were completed in 1967. EVO I emission control devices were originally installed as 84 percent efficient mechanical separators only, with no SOx or NOx abatement. Blocks 1 and 2 of EVO I were recently upgraded (1991 and 1992), respectively, to include electrostatic precipitators rated at 98.5 percent efficiency. In the current configuration, only Blocks 1 and 2 are able to operate near compliance with dust emission standards and none of the EVO I blocks will meet the end of 1997 standards for NOx and SOx. Each of the six EVO I boilers are supported in a structure 27.0 meters wide and 45.0 meters deep.

EVO II was completed in 1973. The six 110 MWe blocks were originally designed to hard burn coal but were reconstructed to burn oil. Both coal and oil became unavailable and finally the blocks were redesigned and reconstructed to burn gas. The boiler can now burn either gas or oil. As the turbines were designed for 565°C-535°C superheat-reheat and the boilers as reconfigured achieve 535°C-535°C, the resultant generation capability is therefore 95 MWe per block. No emission controls exist and compliance with the end of 1997 regulations will require NOx abatement and possibly gas clean-up if sour gas is used for fuel. If the option to burn oil continues, particulate collection equipment could also be required as well as SOx treatment. The blocks have approximately 50,000 operating hours with an estimated remaining life of approximately 150,000 - 200,000 operating hours before major repairs can be anticipated. Each of the EVO II boilers are supported in a structure and space approximately equal to EVO I. Space is available for addition of gas turbines and boiler/duct modifications on the ground level base slab.

DISCUSSION

The feasibility study conducted by Southern Electric International (SEI) for Slovensky Energeticky Podnik (SEP) evaluated the reconstruction, repowering, and possible expansion of SEP's Vojany Station (EVO). The study was conducted by SEI personnel in conjunction with EVO staff, SEP personnel, and selected consultants knowledgeable about SEP, the Slovak Republic, and the range of technical issues addressed herein. Previous SEP evaluations, studies, station records, and technical data were reviewed and utilized as part of the feasibility analysis.

The premise for determining feasibility of:

- Reconstruction of EVO is to enable the blocks (units) to meet current and future emission standards with the most feasible and economical technology;

- Repowering portions of EVO is to meet emissions standards and concurrently maintain or attain additional more efficient generation;

- Expansion of EVO by adding EVO III is to enable SEP and Slovakia to reduce the import of electric power, to become a reliable exporter of power to neighboring utilities and countries, and to investigate potential equity participation in EVO III by SEI.

The general intent of the study, as applicable to EVO I and II, was to review immediate requirements for EVO to meet the targeted emissions deadline of 1997 to evaluate solutions to meet the new emissions limits through equipment modifications, and to develop benefit/cost parameters associated with installing the necessary retrofitting equipment. Additional generation as a result of modifications to EVO I and II and possible greenfield construction of EVO III was reviewed for applicable technology in the time frame outlined by SEP.

The study results recommend a time-phased implementation for the recommended solutions as shown by appropriate project schedules. Specific recommendations are compiled in the following categories.

1. EVO I - Blocks 1 and 2

   Reconstruct with a "dry/semi-dry" scrubber, baghouse, and NO_X control devices.

2. EVO I - Blocks 3 through 6

   Reconstruct/repower with circulating fluid bed boilers and electrostatic precipitators.

3. EVO II - Blocks 7 through 12

   Add low NO_X burners only; provide pre combustion gas scrubbing if sulfur (SO_X) in gas or oil is proven excessive. Provide baghouse filters if solids in oil are proven excessive.

4. EVO - General Reconstruction

   Rebuild support systems in the infrastructure of station as required to support the above modifications.

5. EVO III - Options

   A. Secure cost competitive contracts of tenure for fuel (including coal, gas, and oil).

   B. Add generation expansion as required by existing and forecasted load demand with technology dependent on long-term/spot fuel prices.

Feasibility discussions were initiated by visit of SEI personnel with SEP personnel both in Bratislava and Vojany, (Eastern) Slovakia. Study scope was discussed and modified as required, data was provided, and station walkdowns ensued. The study began with a review of SEP's Basic Technological and Economical Parameters of Production and Maintenance as well as maintenance schedules, and reconstruction plans for the EVO I and II stations. These reviews included discussions with EVO management and technical staff on technology options, existing station limitations, emerging emissions laws and critical path scenarios toward resolutions. The review included EVO's position in the SEP system; variables of concern such as electric load demand, fuel price, stability of all fuel supplies, raw water supply; and construction programs planned and currently underway.

As noted in the scope and reconstruction plans, it was determined and confirmed that EVO cannot meet the future requirements for controlled emissions from either EVO I Blocks 1 and 2 with (recently) new electrostatic precipitators, Blocks 3 - 6, with mechanical separators, or from EVO II as it is currently configured.

The then proposed CSFR law (No. 309.91 Zb, July 1991) established limits for emissions which were not totally concurrent with EVO design requirements or ensuing EEC guidelines. As these were both emerging in implementation and interpretation the assumption was made by the SEI/SEP study team that more stringent requirements would eventually be issued and technology assessments should therefore include prudent considerations for greater removal capability by the components controlling SO_X, NO_X, dust, opacity, particulates, and CO. That is, monies should be allocated for present technologies to ensure compliance but not to the mutual exclusion of expanded compliance.

In addition:

- Each EVO blocks' design thermal capacity rating is very close to a mandated step function decrease in emissions allowable at 300 MW thermal by guideline standards. Therefore, any increase in generation or decrease in efficiency of individual blocks would result in a second category of more stringent emission limits.

- EVO I is currently designed for assisted condensing cooling at low water occurrence on the Laborec River. Additional water usage for ash disposal of sludge or limestone/ash waste, power augmentation of combustion turbines, addition of EVO III, or other water intensive technologies imposes strict water conservation on the EVO site for full capacity generation or a new source of long term, continuous raw water must be secured.

- The main gas supply line, pressure reducing station, and possibly source supply pressure is presently sized for only four blocks of EVO II. If added generation capacity is gas dependent (combustion turbines, gas duct burners, gas reburn, etc.) additional gas pipelines or possibly boosted supply pressure must be installed to meet projected expansion plans.

- Fire protection, stacks, rail facilities, ash ponds, cooling tower fill, limestone handling facilities, crane facilities, industrial water, condensate polishers, boiler acid cleaning, coal/gas/oil contracts improvement, warehousing of major spares, ash marketing, and scrap disposal are items which were assessed for changes or enhancements to support SEP/EVO goals.

- Numerous other existing conditions, normal to any electric generating station of the age and important geographic locale of Vojany, have been identified which will need upgrading for EVO and SEP to meet 1997 emission limit standards and growth projections.

As EVO represents a range of 25-30 percent of SEP's electric production capability and as it is located at the eastern end of the SEP system, it is necessary that EVO be prudently reconstructed and maintained as a valuable asset to SEP.

Factors affecting justification for reconstruction are: Recovery of load demand, assumptions that imports of power can be reversed to export without violation of existing contracts, and production costs lowered or maintained in relation to power sales agreements. Expansion by construction of EVO III, if necessary, can be justified on the same basis as reconstruction or repowering.

Upgrading the functional support systems for all EVO blocks is cost justified based on outages, reduced operating costs, and projected increased revenues.

ASSUMPTIONS

- Recommendations are based upon industry accepted, mature technology, guaranteed by suppliers.

- Forecasted reliability of existing EVO blocks will continue to be maintained by equipment replacement and maintenance according to Vojany supplied equipment/material condition assessment checklists.

- All future solids rejects are assumed to be acceptable from an ecological standpoint (sulfites, sulfates, ash).

- All fuels, water, reducing agents, catalyzers, lime, limestone, etc., and, in general, all products and services, necessary for the reliable and efficient operation of the modified/new blocks shall be deemed available for SEP procurement.

- Preferred solutions shall be those which target maximum local supplies, having in view that a portion of U.S.A. supply is to be considered in order to adhere to the U.S. Trade and Development Program (presently TDA) guidelines.

Bases for the recommendations are: EVO is a vital component of SEP's generation mix and will remain 25-35 additional years. In order for EVO to remain as such, it was recommended:

- First, that emission standards be met by modification of equipment with some capability of margin at minimal evaluated expense in anticipation of future more stringent standards.

- Secondly, it was recommended that SEP/EVO efforts be directed to alternative less expensive fuel sources including imported and domestic coal, oil, and gas obtained on competitive bid bases.

- Third, it was recommended that as enhanced retrofit systems are added, production costs and staff be reduced and/or reassigned to drive down fixed costs.

Specific recommendations were:

EVO I BLOCKS 1 AND 2

SO_X - A spray dryer dry scrubber should be added to control SO_X. A single vessel to handle gas from Blocks 1 and 2 (220 MWeg total) would be added to accommodate space restrictions. Sizing of the components would be for full gas flow at 91 percent removal. Vessel location would require demolition of the existing concrete ducts and induced draft fan changes.

NO_X - A phased approach (targeting economy) should be initiated immediately to test various NO_X control technologies. Due to the design of the existing boilers (double lower arch with fines/overfire air combination) and the "low volatile" anthracite fuel, the resultant boiler configuration is a high (thermal) NO_X producing device without significant turndown control below 85 percent rating.

Following detailed boiler stoichiometry balancing and tuning, a first test would be planned following completion of the NH_3 injection. Gas reburn by installation of gas guns (ignitors) strategically located downstream of the combustion zone would be extensively investigated. Extension of the existing startup gas piping would supply the gas reburn ignitors. Up to 18 percent thermal input would be from the new gas guns. Note: This would increase generation cost.

A second test should begin immediately to evaluate urea/ammonia injection for NO_X reduction via spray pipe into the post combustion zones (SNCR). An adequate supply of NH_3 would have to be secured and tests would have to be sufficiently extensive to determine long range effects on components, ash, operations, etc. One block only should be tested before committing the second block to this scenario.

A third battery of tests should be planned which would be a combination of Tests 1 and 2.

Although these technologies have previously been proven on specific blocks elsewhere, each case is a "customized" application. In addition, these technologies have limited experience with anthracite fuel and the Vojany boiler design.

Following an analysis of the test results, a determination of effectiveness would be evident and a future course of action would be planned.

Should neither test system prove sufficiently effective for NO_X reduction, bidders would be invited to address application of low NO_X burners. Should there be no manufacturers willing to guarantee against liquidated damages, flame stability on the present low volatile anthracite, excessive slagging, and increased carbon carryover (unburned coal in the ash), then bidders would be invited to bid on the application of SCR (selective catalytic reduction) systems for each boiler.

Dust - In conjunction with the new dry SO_X system, NO_X abatement equipment, and existing electrostatic precipitator, it would be necessary to add a baghouse to collect particulate not captured in the precipitator. The action of unreacted limestone on the filter bags would enhance sulfur capture.

Support Systems

Systems necessary to support the above recommended equipment would be required.

- Lime system complete with storage, handling, compressed air, ash loading dust control system, cyclone filters, switchgear, controls, pumps, pipe, delivery rail facilities.
- Waste disposal system for the baghouse including all of the above plus disposal equipment.
- CEMS (Continuous Emissions Monitoring System)
- NH_3 injection system complete with storage, vaporizer, etc.
- Gas duct replacement
- Gas duct heaters for reheat to avoid acid dew point condensation
- Natural gas supply system expansion
- Controls additions

It was further recommended that the existing mills be retained for cost savings but that the mill exhauster fans be reworked or replaced for enhanced control and containment of fugitive dust.

EVO I BLOCKS 3 THROUGH 6

It was recommended that the pulverized coal boilers be replaced with circulating fluidized bed (CFB) boilers designed specifically for multi-fuel application with the existing anthracite as the primary fuel. Replacement would be staged, starting with Block 6. It was recommended that the boilers be purchased from a single vendor also contracted to work within the existing structures as much as possible. The CFB technology would be sufficient to comply with emission regulations for NO_X and SO_X and CO to avoid future compliance issues. The CFB's would also be sufficiently versatile to handle a variety of fuels—anthracite, lignite, brown coal. A dust collection electrostatic-precipitator would be required to replace the existing mechanical separators.

Support systems would include:

- Structural Modifications
- Limestone Handling System
- Ash Handling System
- New ID Fans
- New Pulverizers
- Controls System
- CEMS
- Reassessment and/or Reconstruction of Existing Electrical Components (switchgear, motor starters, termination cabinets, etc.)

It was recommended that incentives for co-manufacturing by vendors be considered for some major component parts of the boiler and duct. Fabrication at Skoda, Brno, Kosice, or closer would be practical.

EVO II BLOCKS 21 THROUGH 26 (7 THROUGH 12)

It was recommended that boiler emissions tests be performed on two separate blocks of EVO II to specifically determine emissions of NO_X, SO_X, CO, and particulate. It was further recommended that the tests be performed while burning gas on one block and oil on the other block to establish baseline data for emission abatement equipment design. Initial analytical assessment anticipated that only NO_X related equipment would be required for burning gas.

EVO GENERAL RECONSTRUCTION

The following systems were recommended to be reconstructed on a phased basis to support reconstruction of EVO I and II.

- Rail Modifications
- Fire Protection Upgrade
- Gas Reducing Station Capacity Increase
- Condensate Polishing
- Station Air Compressors/Dryers
- Industrial Water Plant
- Penthouse Pressurization
- Cooling Tower Drift Eliminator Upgrade
- Full Conversion of New Ash Pond to Dry System
- Conversion of Cooling Tower Blowdown to Ash Disposal Area Dust Suppression Water
- Limestone Silo Erection
- EVO Roof Crane Permanency
- Water Recycle
- Additional Warehouse for Transformers
- Switchyard/Station Service
- Controls/Block (Unit) Performance Monitoring

EVO III OPTIONS

Power plant design is a complex process of engineering and economic optimization. The final design of a power system is at best a compromise based on prioritization of specific parameters and assumptions in the optimization process: capital cost, expected O&M costs, fixed and variable costs, space, operating life, environmental restrictions, efficiency, fuel costs, reliability, availability, schedule, risk assessment, credit rating, interest rates, insurability, locale, infrastructure, short and long term goals, and business purpose.

Environmental demands take precedence over efficiency but economics of competition demand efficiency. In the case of Vojany, fuel flexibility is a critical issue as well as driving production and maintenance costs to the lowest point. Competition between stations within SEP and blocks for economic dispatch and between generation or import of electricity drives efficiency at Vojany to a parallel priority with environmental concerns.

Recent technological advances in gas turbines in a combined cycle and circulating fluidized bed boilers have replaced the pulverized coal fired boiler/steam turbine cycle (especially where low fuel volatility is a possibility) as the primary choice for power generation in many applications. Technology of the future such as Pressurized Fluidized Bed Combustion Combined Cycle (PFBC-CC) and Integrated Gasification Combined Cycle (IGCC) are on the horizon as practical, reliable, and efficient methods of generation but at present are just past the test/prototype or demonstration phase.

Each power project is sufficiently different due to fuel, finance, locale, and demand such that designs for large combined stations like Vojany become a custom or specific design. The design must obviously comply with the anticipated end of 1997 standards but be sufficiently flexible to meet emerging standards not yet legislated (CO, VOC's, opacity). The entire electric generation production thus becomes an integrated mechanical/power/chemical facility producing kV, MW_e, MW_{th} clean gas, and inert ash, economically on demand of dispatch.

Thus, EVO III station options included at the time of the assessment were:

- Simple cycle combustion turbine
- Combined cycle plant (GT-CC)
- Pulverized coal conventional plant
- Pulverized coal supercritical plant
- Circulating fluidized bed conventional plant

- Pressurized fluidized bed plant (PFBC-CC)
- IGCC plant
- ALWR Nuclear plant

A construction schedule for EVO III was not finalized during 1992 due to load demand/growth, reconstruction schedule of EVO I and II, fuel costs, and investigations to secure/assure long term load/debt. Therefore, it was recommended the choice of technology to be applied be postponed to a schedule point, 48 months prior to evaluated need, as dictated by load demand, cost of fuel, and the cost of imported power.

CONCLUSION

In conclusion, we previously have stated some of the critical issues and options, technological and otherwise, that transform perceived short term problems with upcoming environmental regulatory guidelines into challenges which, if executed with due diligence, will lead to long term opportunities in the European community.

Feasibility studies, such as the SEI/SEP Vojany assessment, are a major actor in an intricate cast of "foundation entities." These entities are required to effectively secure adequate financing along with the present trend of equity participation and financial investment guarantees which allow power producers in the sector to meet environmental concerns. If necessary and if evaluated technically possible, the power facilities' thermal efficiencies can concurrently be improved (or maintained) as part of the total rehabilitation investment undertaken. Thus, such feasibility studies become more than a technological screening tool - they become a foundation conerstone in the eventual realization of the power producers goals and the infrastructure they serve and are part of.

Chapter 40

Implementing Demand-Side Management Through Integrated Resource Planning in Poland

J. Dybowski, E. Hille, D. Wolcott

ABSTRACT

Many of the technical and policy disciplines of integrated resource planning (IRP) have been conceived, developed and tested in progressive regulatory jurisdictions throughout the United States over the last decade. The practical effect has been a rapid increase in utility investments in demand-side management (DSM) resources. Through the support of multilateral and bilateral agencies, the U.S. model of IRP and DSM is now being transferred to formerly communist countries in Eastern Europe. This paper describes the activities of the Poland DSM Project, sponsored by the U.S. Agency for International Development, which is providing training and technical assistance on IRP and DSM to Poland. The program's tasks include work with the energy ministry and utilities to enact IRP into law, document the DSM potential in Poland, and develop and implement DSM and end-use load research programs. What makes this activity interesting is that the results will be combined with supply-side and tariff studies to produce an integrated resource plan that will be the basis for substantial multi-lateral development bank power sector loans. This activity is challenging because Poland has different energy infrastructures, pricing systems, and regulatory institutions and it is also experiencing wrenching political and economic transitions. The paper focuses on how the transfer of the U.S. IRP/DSM model will fail without sensitive recognition of conditions in Poland and appropriate adaptation.

INTRODUCTION

The Energy Sector Restructuring, Privatization, and Regulatory Reform Program of the Government of Poland, which is currently underway, aims to:

- liberalize energy prices and eliminate subsidies as far as possible;
- promote enterprise competition through the elimination of barriers to market entry in all productive sectors which are not "natural monopolies;" and
- transform the governance, ownership and control, and regulation of state-owned enterprises through the adoption of a market economy and privatization.

These reforms are expected to revitalize the energy sector, enhance its long-term financial viability, and encourage economically optimal investment and consumption decisions. The program is being conducted with the assistance of the World Bank, the United States Agency for International Development (USAID), and other bilateral agencies.

In order to guide the future development of Poland's power sector, the Energy Sector Program includes an integrated resource planning (IRP) component through which both supply-side and demand-side resources are being identified to produce a mix of energy resources which yield the least cost to society. Several teams are contributing to the development of the integrated resource plan in a two phase program. Under Phase I, A USAID Electricity Transfer Pricing and Retail Tariff Study was completed that formulated an electricity pricing strategy. Austrian and French teams have completed a Supply-Side Development Plan. This paper describes the efforts of another USAID team to identify and assess demand-side management (DSM) resources (the Poland DSM Project). As these Phase I supply-side and demand-side activities are completed in mid-1993, Phase II will commence with the formulation of the integrated resource plan using the results of these studies. The integrated resource plan will then be the basis for power sector loans from multilateral development banks (MDBs) such as the US$ 1 billion loan that is set for appraisal by the World Bank between late 1993 and mid-1994.

The transfer of utility planning concepts (e.g., IRP and DSM) from one country to another is challenging because each country has different energy infrastructures, pricing systems, and regulatory institutions. Perhaps nowhere is the challenge more daunting than in the formerly communist countries of Eastern and Central Europe and the former Soviet Union which are experiencing wrenching political and economic transitions. The Poland DSM Project is prominent in this regard as a precedent and model for MDB financing of DSM investments in the post-communist world.

THE POLAND DSM PROJECT

Under the Energy Sector Program, the Polish power system has evolved from a nationalized, vertically-integrated utility to a disaggregated utility system with 34 generation companies, a single transmission company (the Polskie Sieci Elektroenercetyczne (PSE) or Polish Power Grid Company), and 33 local electric distribution companies. Current proposals call for an aggregation of generation companies to four and distribution companies to twelve. Proposed power transfer pricing and retail tariffs would be based on cost-plus principles designed to yield acceptable profit margins for each entity. Although there are no direct government subsidies to any of the Polish power entities, the widespread undervaluation of assets allows retail tariffs to be set at levels ranging from US$ 0.03 to 0.06 per kWh (plus relevant demand charges) depending on the tariff class, below the expected long-run marginal costs.

Total installed public generation capacity is approximately 29,600 MW, compared to peak demand of around 22,000 MW. Approximately 57% of generation is based on hard coal and 43% on lignite. Peak demand occurs in the winter on a seasonal basis and in the evenings on a diurnal basis. As a result of the continued recession in Poland, electricity sales in the country have fallen by 16% over the last three years, but now appear to have stabilized around 96,000 GWh annually. Because of the decline in sales, new generation capacity is not immediately required, although the Supply Plan calls for the construction of 360 MW of gas turbines between 1996 and 2000. Numerous existing coal plants are candidates for repowering, including environmental controls, and the Supply Plan expects that 2,865 MW of such capacity will be rehabilitated between 1993 and 2000. Load forecasts vary widely reflecting the uncertainty of overall economic trends.

It may seem irrational for Polish utilities to try to sell less of their product, especially with a 35% reserve margin and declining to flat sales. However, the fundamental institutional changes resulting from the Energy Sector Program present a unique opportunity to develop a lasting framework for optimal power system planning and operation. For example, the current reserve margin is not that much more than the 20% level currently deemed prudent for Poland. As the economy improves, load will eventually increase, especially if the rate of electrification increases as Poland's economy modernizes. Also, if old, polluting generating units are shut down instead of being repowered, further capacity may be needed. Because it is uncertain when these events will happen, an institutional framework that encourages IRP should be developed now so that cost-effective DSM resources will be available when they are needed.

There are several other reasons why Poland should be interested in IRP and DSM. First, If IRP is implemented in Poland to account for societal costs, then environmental costs of the severe pollution damage will weigh heavily against the coal-based power plants which account for virtually all of electricity generation. DSM will offer immediate environmental benefits.

Secondly, by adopting DSM, electric consumers will forego energy costs that would have otherwise been incurred, allowing them to price their products more competitively. Overall energy intensity in Poland is two to three times higher per unit of gross national product compared to Western European countries. Clearly, there is considerable scope and justification for DSM programs which emphasize energy efficiency that will increase the competitiveness of the Polish economy.

Thirdly, whereas utilities which rely solely on conventional supply-side resources such as power plants often view themselves as commodity producers, utilities which tap the potential of DSM perceive themselves as service providers, i.e., they are in the business of meeting consumer needs rather than simply producing kilowatt hours. Adopting a customer service approach is an important step as utilities leave the central planning mentality behind.

Finally, a DSM industry will create a new labor market for high skilled energy service workers. The creation of jobs will be an important benefit to Poland's economy which is facing substantial unemployment from the post-communist transition, made worse as coal mines and old power plants are shut down. An energy services industry will also contribute to an economy that needs to move from a heavily industrialized past to a more service oriented future.

To deliver these potential benefits, the Poland DSM Project has the following objectives:

- To establish an institutional and regulatory framework in Poland that will support the development of integrated resource planning;
- To identify DSM resources in Poland which should be considered within the context of an integrated resource plan;
- To evaluate the costs and benefits of these resources to support their appraisal in MDB power sector loans; and
- To assist Polish authorities to develop a lasting institutional capability within the government, the utilities, and the private sector to design, implement, and evaluate DSM programs.

The Poland DSM Project is being carried out through the following activities which simultaneously address the need for institutional development and DSM data collection and assessment at the national level, and the need to develop capabilities at the local electric distribution companies for the accelerated implementation of DSM and end-use load research programs.

Institutional Development

The legislative and regulatory framework under which the power sector operates in Poland must ensure that IRP is the most profitable plan for the utility sector as well as the least-cost plan for its customers and society at large. National energy legislation originally drafted in May 1992 may provide an opportunity to create a system that encourages the use of DSM when justified. Through the Poland DSM Project, the Polish Ministry of Industry and Trade and its Energy Restructuring Group (an international advisory body) are reviewing proposals for different legislative and regulatory mechanisms that encourage the implementation of IRP, covering the following topics:

Development of an Integrated Resource Plan. The draft Energy Law requires that a comprehensive energy plan be developed by the Ministry of Industry and Trade every two years. It appears that this plan is intended to support government policy with subsequent plans to be developed by a proposed Energy Regulatory Agency (ERA) that will address specific actions by energy producers and consumers. ERA may be given the responsibility for overseeing the development of a detailed integrated resource plan for the electric sector which would include the following components:

(1) Identifying the objectives of the plan (e.g., reliable service, minimal environmental effects, low cost of environmental controls, meeting peak demand in a cost-effective manner, and minimal bills for electricity customers); (2) Developing one or more load forecasts for different scenarios; (3) Determining the levels of capacity expected for each year of the plan; (4) Identifying needed resources (e.g., fuels, generating capacity, transmission and distribution capacity, a manageable load shape, and demand reductions) needed to bridge the gap between expected loads and capacities; (5) Evaluating all of the electric resources in a consistent fashion and selecting the most promising resources for fashioning an effective, flexible, and responsive plan, integrating methods of supplying power with methods for controlling and moderating demand; (6) Constructing scenarios that project the selected mixes of resources against possible economic, environmental, and social circumstances; (7) Evaluating the economic and technical characteristics of each mix of resources and analyzing the uncertainties associated with the various scenarios; (8) Screening the alternatives to eliminate those that are not suitable and rank-ordering alternative courses of action; (9) Testing each alternative for cost-effectiveness from a variety of perspectives (e.g., the utilities, ratepayers of different classes, and society at large); (10) Reevaluating the alternatives considering economic, environmental, and societal factors and selecting a plan for implementation, one that most nearly satisfies all the objectives of the plan; (11) Developing and implementing a plan of action; and (12) Monitoring and evaluating the operation of the utilities under the plan and revising the plan if necessary.

Authority for PSE to Implement the Integrated Resource Plan. In order for IRP to function in Poland, it might be appropriate that the PSE be given the authority to implement the plan, under ERA's regulatory oversight (although it may be determined that the distribution companies or others could also play this role). As the monopoly wholesale supplier, PSE could assume the "obligation to serve" by taking the lead responsibility for acquiring both supply-side and demand-side resources according to the plan. "Acquiring" means arranging, directly or indirectly, for resources to be made available to the Polish electric system. PSE could acquire resources through contracting with generating companies for power, contracting with distribution companies for DSM resources, or requesting competitive bids for supply-side and demand-side resources from utility sector entities and independent suppliers. To deliver the DSM resource, distribution companies could contract with customers and energy service companies. Distribution companies could also build and operate their own generating facilities or contract directly with generators. Presumably, however, those facilities and transactions would have been identified as cost-effective resources to be acquired under the integrated resource plan.

Recovery by the Distribution Companies of DSM Costs through Electric Rates. There are two customary mechanisms for recovering the costs of DSM in electric rates: expensing and ratebasing. When the cost of DSM is expensed, the utility recovers 100% of the cost of the program during a short period of time, typically one year. When the cost of DSM is ratebased, the cost is capitalized (listed as a utility investment) for a period that approximates the useful life of the DSM measure and then the full cost, plus an allowed rate of return, is amortized and partially recovered each year until the full cost is recovered. In both cases, the costs are recovered from electric customers through rates. Whichever method is used in Poland, the distribution companies should have the authority to invest in DSM and recover the costs of those investments in electric rates.

Discussions with representatives of the Polish electric industry revealed that the negative impact that DSM expenditures could have on distribution company revenues and profits may be a substantial deterrent to DSM implementation. In the U.S., utilities are increasingly able to recover as program costs the lost revenues from reduced consumption caused by the implementation of DSM. Therefore, the distribution companies could also be allowed to recover revenue losses from qualifying DSM expenditures (those expenditures associated with DSM programs that meet minimum qualifications or performance criteria).

The costs of DSM need not be shared equally among all customers of the distribution companies. For example, it may be decided that the costs of industrial DSM programs should only be charged to industrial customers. Those distributions companies with a higher share of their load from industry would then have more of the costs of DSM charged through the industrial rates.

Ability of the Distribution Companies to Earn Financial Incentives to Implement DSM. For many years, state regulatory commissions in the U.S. exhorted electric utilities to increase their DSM expenditures without providing the utilities any financial incentives to do so. In fact, for most utilities, DSM was undesirable because it resulted in lower revenues and profits. This situation has changed. Many states now provide positive financial incentives to utilities to invest in DSM.

These incentives take many forms. For example, some states set specific performance targets (such as for cost-effectiveness and program participation rates), and then offer financial rewards through subsequent rate increases. Other states allow utilities to ratebase their expenditures and then earn a higher than normal rate of return on that investment. Other states provide utilities a share of the total resource savings achieved by their DSM programs. Experimentation continues with many alternative approaches to rewarding utilities for acquiring cost-effective DSM resources.

Regulators generally believe that such rewards are appropriate to compensate utilities for the increased risks of doing something new and relatively untested, such as implementing DSM programs. For the same reason, the distribution companies in Poland should be provided with financial incentives for implementing DSM consistent with the integrated resource plan.

Without cost recovery and financial incentives, DSM (particularly DSM that focuses on reductions in energy use) is unconventional and antithetical to the normal way that utilities do business. For many utilities throughout the world, DSM is the opposite of what they think they should be doing. Poland will probably be no different than the rest of the world in this respect. Even if IRP is adopted in the Energy Law, there will likely be resistance to DSM activities, at least until the distribution companies can profit from DSM programs. Nonetheless, failure by any distribution company to implement the DSM recommended by the integrated resource plan will raise system-wide costs, even for those distribution companies that implement DSM programs.

It is difficult to judge how receptive each of the distribution companies will be to implementing the integrated resource plan's recommended DSM activities. It may be necessary to institute a mechanism that will dissuade distribution companies from refusing to participate. A possible mechanism would be a surcharge that could be added to the wholesale electric rates charged to the non-participating utility. The surcharge could be set equal to an estimate of the additional costs imposed on the national system as a result of the utility's failure to participate.

Regulatory incentive approaches in the U.S. provide a comparable mechanism in the form of a penalty for utilities that implement DSM programs that fail to meet pre-established performance criteria. Such an approach is probably premature in Poland; however, the surcharge for non-participation may be appropriate. Nonetheless, assessing a surcharge is reminiscent of the old central planning mentality and thus, may unnecessarily raise opposition to DSM by the distribution companies.

Provision for Public Involvement. Public involvement is an essential element of IRP. Ratepayers, electric utilities, generating companies, DSM equipment manufacturers and vendors, environmental interests, and the organizations that represent them should be involved throughout the process of developing and implementing a plan. Although public involvement will surely slow the process initially, both the process and resulting plans will be better for it, and the political support for the plans will be broader and stronger in the long run. Public hearings, solicitation of written comments, and procedures for adjudicating public disputes with the plan could be provided. Thus, a commitment to broad public participation could be made explicit in the Energy Law.

DSM Assessment

Through the Poland DSM Project, PSE has conducted a national assessment of DSM potential to help identify promising aggregate DSM resources. This work addressed:

- How electricity is used in Poland by predominate end-use in each customer class;
- The economic DSM potential based on a "DSM supply curve" that shows how much savings can be achieved by DSM technologies at what cost;
- The achievable DSM potential based on benefit/cost analyses of candidate DSM programs that were defined; and
- The likely market penetration of DSM in Poland based on an assessment of DSM impacts and Polish economic and energy consumption trends.

The results of the DSM assessment are providing a means for PSE to identify investments in DSM resources that could be included in the power sector loan from the World Bank. For this purpose, the Poland DSM Project is using a family of models developed by the Electric Power Research Institute (EPRI) in the United States to perform the following functions:

Development of End-Use Load Shapes. The project documented where and how system and customer load shape changes are required to meet PSE's strategic objectives. It is a new type of load forecasting "from the bottom up" in which end-use load research is used to characterize customer consumption patterns and future requirements. The Polish Foundation for Energy Efficiency (FEWE) and the Polish Institute for Power Engineering were involved in data collection which turned out to be a very difficult process. All pre-1989 data are generally suspect, as they were collected to serve political, rather than technical purposes. Additionally, very little end-use data are available in Poland.

DSM Technology/Measure Screening. PSE used EPRI's DSManager software model to undertake an economic assessment of particular DSM technologies. Fifty eight candidate measures were "screened" or ranked for consideration in DSM program development, including two innovative rate measures (time-of-use and interruptible rates). From this screening, twenty five measures (excluding the rate measures) were considered to be cost-effective resulting in an economic potential of 25,400 GWh per year or roughly 26 percent of current consumption. The DSM supply curve and list of cost-effective measures are presented in the following figure.

DSM Program Assessment. PSE also used DSManager to define utility DSM programs that could be undertaken to implement the cost-effective measures identified in the technology screening process described above. In addition to the installed capital and operating costs of the DSM measures, the utility's program administration costs (overhead, marketing, implementation, and evaluation) were also included. Likely market penetration rates were also modeled. Then benefit/cost analyses of the programs were performed using the total resource cost and societal cost tests. Finally, the energy and demand impacts of cost-effective DSM programs were calculated.

The following table presents a summary of the achievable energy and demand savings in Poland in 2000 by end-use and sector. The achievable energy savings potential is 1,738 GWh per year, equivalent to 8 percent of incremental energy requirements. The achievable demand savings potential is 554 MW, equivalent to 17 percent of incremental demand requirements, which would displace all of the planned capacity additions through 2000 called for in the Supply Plan. This DSM program would be equivalent to the 18th largest power plant in Poland in energy terms and the 14th largest in capacity terms.

Integrated Planning Analysis. PSE will use EPRI's IRP-Manager software model (formerly called LMSTM) to integrate the results of the DSM assessment with those of the Supply Plan and Tariff Study. This process includes: (1) preparation of long-range load forecasts; (2) resource planning to determine the type and timing of supply-side and DSM resources that cost-effectively meet the demand forecast requirements; (3) production costing to establish a cost-optimized mix of fuel and other operating costs to meet forecasted load at particular times; (4) financial analyses to determine the impact of resource selections on the utility's income statement, balance sheet, and source and use of funds; and (5) rate analyses to project the impact of the resource plan on utility revenue requirements

Energy Conservation Supply Curve for Poland

Measure	Cost of Saved Energy	Savings
1 Replace 17.5 kW hi–use motors with hi–eff. motors upon rewind	$0.007 per kWh	336 GWh
2 Replace 120 kW hi–use motors with hi–eff. motors upon rewind	$0.008 per kWh	170 GWh
3 Replace 3.5 kW hi–use motors with hi–eff. motors upon rewind	$0.010 per kWh	257 GWh
4 Retrofit 120 kW hi–use motors with ASDs	$0.013 per kWh	368 GWh
5 Replace 120 kW low–use motors with hi–eff. motors upon rewind	$0.016 per kWh	350 GWh
6 Replace small refigerators (approx. 150 ℓ) upon retirement with 1990 US average	$0.020 per kWh	557 GWh
7 Replace large refigerators (approx. 230 ℓ) upon retirement with 1990 US average	$0.021 per kWh	1,496 GWh
8 Replace 17.5 kW low–use motors with hi–eff. motors upon rewind	$0.021 per kWh	518 GWh
9 Retrofit 17.5 kW hi–use motors with ASDs	$0.023 per kWh	537 GWh
10 Wrap water heaters	$0.024 per kWh	116 GWh
11 Industrial maintenance	$0.024 per kWh	2,552 GWh
12 Retrofit mercury vapor streetlights with high pressure sodium	$0.025 per kWh	314 GWh
13 Optimize production processes through process monitoring	$0.026 per kWh	1,531 GWh
14 Retrofit industrial mercury vapor lights with high pressure sodium	$0.027 per kWh	299 GWh
15 Replace 3.5 kW low–use motors with high efficiency motors upon rewind	$0.029 per kWh	441 GWh
16 Retrofit hi–use commercial incandescent lamps with CFLs	$0.031 per kWh	924 GWh
17 Retrofit 120 kW low–use motors with ASDs	$0.032 per kWh	605 GWh
18 Retrofit low–use commercial incandescent lamps with CFLs	$0.033 per kWh	1,037 GWh
19 Replace hi–use comm. fluorescent lamps & ballasts with T–8 systems upon retirement	$0.036 per kWh	532 GWh
20 Motor downsizing	$0.037 per kWh	271 GWh
21 Retrofit hi–use residential incandescent bulbs with CFLs	$0.038 per kWh	2,850 GWh
22 Retrofit hi–use industrial fluorescent lamps & ballasts with T–8 systems	$0.040 per kWh	888 GWh
23 Occupancy sensors	$0.042 per kWh	440 GWh
24 Retrofit low–use residential incandescent lamps with CFLs upon retirement	$0.043 per kWh	7,632 GWh
25 Retrofit high wattage industrial incandescent lamps with metal halide	$0.046 per kWh	378 GWh
	Total	25400 GWh

Industrial rate programs have not been included in this screening analysis.

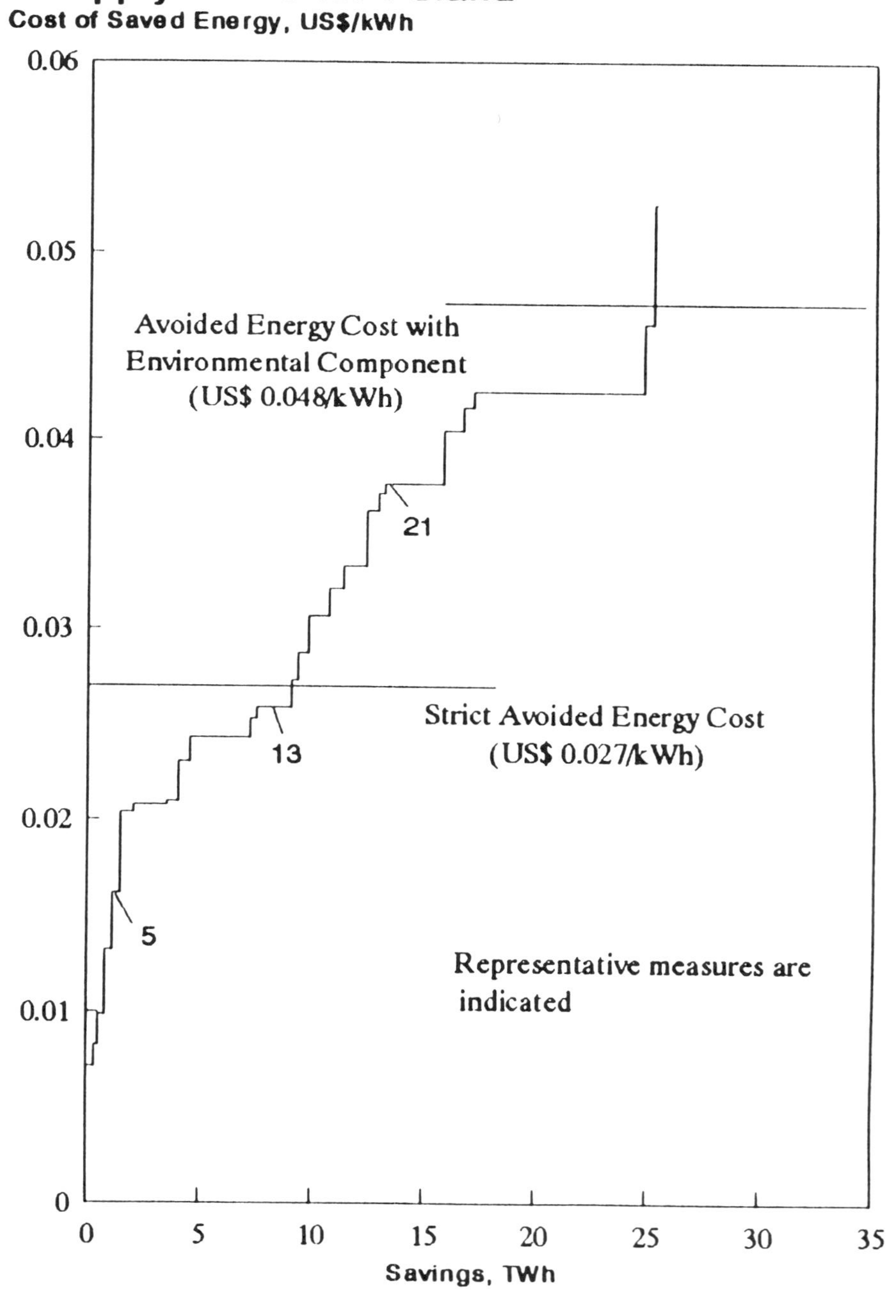

Summary of Achievable Energy and Demand Savings in Poland in the Year 2000

PROGRAM	Energy (GWh)	Demand (MW)	Share of Incremental Energy (1995-2000)*	Share of Incremental Demand (1995-2000)**
Residential Lighting	267	48	1.2%	1.5%
Residential Refrigeration	65	8	0.3%	0.2%
Residential Water Heating	11	2	0.1%	0.1%
RESIDENTIAL SUBTOTAL	**343**	**58**	**1.6%**	**1.8%**
Commercial Lighting	225	36.0	1.1%	1.3%
COMMERCIAL SUBTOTAL	**225**	**36.0**	**1.1%**	**1.3%**
Industrial Motors	401	64	1.8%	2.0%
Drives	68	10	0.3%	0.3%
Motor Downsizing	7	1	0.0%	0.0%
Industrial Process	171	28	0.8%	0.9%
Industrial Lighting	256	44	1.2%	1.4%
Industrial Rates	59	258	0.0%	7.9%
INDUSTRIAL SUBTOTAL	**962**	**405**	**4.4%**	**12.5%**
Street Lighting	208	53	1.0%	1.6%
PUBLIC SECTOR SUBTOTAL	**208**	**53**	**1.0%**	**1.6%**
TOTAL SAVINGS	**1,738**	**554**	**8.1%**	**17.2%**

Note: totals may not add because they have been rounded.
* Incremental energy growth is calculated from the Electric Power Tariff Study: Polish Power Grid Company, Draft Report, February 1993.
**In the modeling exercise, incremental demand growth between 1995 and 2000 is calculated as 3,253 MW.

and customer electricity rates. The IRP-Manager results will be ported to the WASP model which the World Bank is using to appraise the power sector loan for Poland.

DSM and End-Use Load Research Pilot Program Design

The impacts of DSM programs are difficult to assess prior to implementation because of the many factors which influence program effectiveness. Pilot programs provide opportunities to demonstrate whether proposed programs work as expected. Further information can also be gained to reduce uncertainties regarding program impacts and improve program delivery. This task entailed work to identify sectors and end-uses where pilot programs would be most appropriate in Poland. The work was performed with Zaklad Energetyczny Gliwice (Z.E. Gliwice), the largest electric distribution company in Poland, serving 15% of the country's load. It is located in the industrial heartland of Upper Silesia in southern Poland. A pilot industrial DSM program was designed to include incentives that ensure program participation by utility customers, a marketing approach to solicit program participation, and a program evaluation plan. Alternative program delivery mechanisms were explored through focus groups with the utility, its customers, and vendors of DSM products and services that will participate in program implementation.

This work also focused on developing capabilities at Z.E. Gliwice to understand how its customers use electricity, a critical component in DSM program design and evaluation. A pilot end-use load research program was designed to include an analytical framework and load monitoring plan. In addition to whole-premise and end-use load monitoring, the load research program encompasses customer surveys and statistical analyses of billing data. To support the design of the load research program, spot monitoring of utility customers' facilities was conducted using Rustrak Ranger power monitors. The spot monitoring effort was conducted in a range of customer facilities including single family and apartment residences, a school, a hospital, and a number of industrial facilities including a coal mine, a steel mill, and a combined heat and power plant.

CONCLUSION

Integrated resource planning and the demand-side management activities it engenders is principally an American concoction. As such, it relies heavily on government regulation of the utility sector for its implementation. It is also a phenomenon which is relatively new on the scene, and therefore untested in terms of its long-term impacts on the utility industry. For all these reasons, some people have criticized the adoption of IRP and DSM as a key activity in restructuring the utility sectors of post-communist countries such as Poland.

Clearly, a major challenge to the transfer of the U.S. IRP/DSM model is to be sensitive to the specific situation in Poland and to adapt the approach accordingly. In this regard, a number of interesting issues confront the implementation of IRP and DSM in Poland.

Foremost among these is the fact that in restructuring its electric system, Poland has adopted a variation of the "U.K. model." Five years ago, the United Kingdom started a radical restructuring of its electric system. From a nationalized, vertically-integrated utility, the U.K. now has a largely privatized, disaggregated utility system with separate generation, transmission, and distribution companies. This restructuring was undertaken with the primary goal of introducing competition and economic efficiency which would benefit consumers with lower rates.

It is still unclear the extent to which this grand experiment has succeeded in accomplishing its goal. It has become apparent, however, that the U.K. system provides significant disincentives to the development of DSM as an electric resource. With tariffs currently based on cost-plus and rate cap principles, all enterprises in the system are motivated to maximize their profits through increasing electric sales. There are no mechanisms to compensate or reward utilities for undertaking DSM with the result that very little DSM is currently being implemented in the U.K.

To the extent that DSM is a least-cost electric resource, this situation poses a real problem for countries like Poland that have adopted a disaggregated utility industry along the lines of the U.K. model. If the problem is not fixed, Poland may be condemned to a utility planning and resource acquisition process that is seriously flawed from society's point of view. The fix could be accomplished through incorporation in the Energy Law of the institutional mechanisms discussed above such as an integrated resource planning process, the provision of rate incentives for DSM, or the use of a bidding process to acquire electric resources competitively.

A related issue facing the implementation of DSM in Poland is the lack of a regulatory framework. In Poland, the entire nature of the electric utility industry and its relationship to government is in a state of flux. Existing decision making structures may not be able to implement the types of changes required by IRP, unless the Energy Law and subsequent rules promulgated by the Energy Regulatory Agency allow and encourage it.

A politically charged issue is the perception that DSM will contribute to unemployment in the utility and coal mining industries. Coal mining has been hit hard by the economic transformation in Poland, both as a result of the removal of production subsidies as well as by the fall in demand.

The mining workforce has been cut by one-third since 1990 and 22 of 70 mines are slated for closure in the next two years. In the utility industry, a curious holdover from the communist years is that a substantial portion of employee pay is based on sales volume. In both cases, Poles are asking why DSM should be considered a resource option when it may accelerate unemployment and reduce wages.

Another issue is the uncertainty of industrial facility survival. In the highly unstable economic environment of Poland, a large number of industrial and commercial facilities are being closed, many of them permanently. This situation creates a difficult challenge to the successful implementation of DSM programs in which special precautions have to be taken to avoid a high rate of loss or non-performance of utility DSM investments.

Another issue is that many DSM measures that are uncommon or nonexistent in U.S. utilities because of their high labor costs may be economic in Poland, due to the relatively low labor costs. The prevailing labor wages in Poland may allow much more customization for each DSM investment, perhaps involving more direct utility installations. The programs will also probably focus more on no cost/low cost and operations and maintenance measures, compared to high technology solutions.

Finally, while the downward spiral of Poland's economy over the last three years is generally considered to be over, a number of problems still remain such as currency instability and uncertainty in business transactions. These problems have generally discouraged aggressive market development and related business activity by international companies that have the capital which Poland so desperately needs to rehabilitate and modernize its economy after forty years of communism. Of course, this problem is not unique to Poland, but is also shared by the other post-communist countries of Eastern and Central Europe and the former Soviet Union. In that context, the Poland DSM Project should provide a useful model to these other countries on how an electric system can be restructured to obtain the benefits of demand-side management through integrated resource planning.

ACKNOWLEDGMENTS

The work described in this paper was performed by David Wolcott of RCG/Hagler, Bailly, Inc. and Ewaryst Hille of the Polish Foundation for Energy Efficiency under contract to the U.S. Agency for International Development's Bureau for Europe. Jaroslaw Dybowski of the Polish Power Grid Company served as the principal Polish counterpart for this work.

Chapter 41

Achieving Environmental Benefits with Industrial DSM Programs in Mexico

S. Herrera, R. Magana, I. Rodriguez, D. Wolcott

ABSTRACT

This paper presents the results of a preliminary assessment of industrial demand-side management (DSM) in Mexico. The first step of the assessment was to characterize the patterns of electricity consumption in Mexico's industrial sector in order to identify the most important end-uses that could be targeted for energy efficiency and load management. Various DSM measures are described to explain how energy and demand savings can be captured using new technologies, new techniques, and improved operation and maintenance procedures. This initial effort provides estimates of the preliminary economic potential of various DSM measures in Mexico. The second step of the assessment was to conduct a much more detailed analysis of the costs and benefits of alternative DSM programs to identify those most attractive for implementation in the Mexican industrial sector. Using DSManager, a sophisticated computer model developed by the Electric Power Research Institute (EPRI), estimates were derived of the costs per kWh and kW of implementing various industrial DSM programs. The magnitude of the achievable reduction in energy and demand through 2010 was estimated for a variety of industrial end-uses including motors, lighting, and air conditioning.

INTRODUCTION

Mexico's electricity consumption is expected to grow at an average annual rate of about 6 percent between 1992 and 2000, with a slightly higher growth rate expected in the industrial sector. To meet this demand, the Comisión Federal de Electricidad (CFE), Mexico's national utility, plans to build almost 14,000 MW of additional generating capacity, which is to come on line between 1993 and 2000 at an estimated cost of about U.S. $34 billion (including additional transmission and distribution investments). In the year 2000, such a capacity expansion would represent a 78 percent increase over 1990.

Consequently, CFE faces the prospect in the 1990s of substantial increases in external indebtedness. This is an unattractive option for a utility that has already accumulated nearly a billion dollars in external debt as of the end of 1991. Significant growth in CFE indebtedness would represent a setback for a nation that has worked diligently over several years to contain and reduce nearly unmanageable levels of external public sector debt. Additionally, there are large competing capital requirements in Mexico for other public services, such as roads, water and sewage systems.

It is in this context that alternatives to the conventional approach to capacity expansion are being considered in Mexico such as private power generation. Mexico is also beginning to explore a complementary alternative: demand-side management (DSM) which is end-use energy-efficiency in the utility sector. Demand-side management programs involve the identification, use, and evaluation of demand-side resources such as energy efficiency, the reduction of kilowatt hours (kWh) of energy consumption and load management, the reduction of kilowatts (kW) of power demanded or the displacement of demand to off-peak times. DSM encompasses a broad range of measures to encourage consumers to modify their consumption. Tariffs can be designed to stimulate a shift in consumption to off-peak periods. End-use energy efficiency can reduce both energy and peak power demand. Direct load control can likewise limit peak power demand. Low cost/no cost opportunities abound; often it is a matter of providing the right incentives or information to utility consumers. No sacrifice in economic growth or comfort is required. Besides yielding economic benefits, DSM also provides environmental benefits in terms of avoided power plant emissions and reduced effluent discharges.

To assist Mexico in meeting increased electricity demand at the least cost and in an environmentally sound way, the U. S. Agency for International Development, in collaboration with CFE and Mexico's Comisión Nacional para el Ahorro de Energía (CONAE) of the Secretaría de Energía, Minas e Industria Paraestatal, sponsored a project to initiate the analyses required for the development of DSM in Mexico's industrial sector. The industrial sector was chosen for first priority analysis given its importance in accounting for 56 percent of electricity consumption in Mexico.

The goals of the project were to improve the productivity of Mexico's industrial sector while achieving financial, economic, and environmental benefits in utility operations. The work so far has involved an industrial DSM assessment (documentation of the economic and achievable DSM potential, possible DSM impacts, and associated environmental benefits) and the design of an industrial DSM pilot program.

This paper presents the results of the industrial DSM assessment. The first step of the assessment was to characterize the composition of the industrial sector in Mexico and review patterns of electricity consumption by subsector in order to identify the most important end-uses that could be targeted for energy efficiency and load management. Various DSM measures are described in detail to explain how energy and demand savings can be captured using new technologies, new techniques, and improved operation and maintenance procedures. This initial effort provides estimates of the preliminary economic potential of various DSM measures in Mexico and sketches out the magnitude of possible DSM impacts on loads in 2010.

The second step of the assessment was to conduct a much more detailed analysis of the costs and benefits of alternative DSM programs to identify those most attractive for implementation in the Mexican industrial sector. Using DSManager, a sophisticated computer model developed by the Electric Power Research Institute (EPRI) and now used widely by utilities throughout the U.S., estimates were derived of the costs per kWh and kW of implementing various industrial DSM programs. The magnitude of the achievable reduction in energy and demand over the next 17 years was also estimated for a variety of industrial end uses including motors, lighting, and air conditioning.

PRELIMINARY ANALYSIS OF INDUSTRIAL DSM IN MEXICO

The analysis sketches out the economic potential for DSM in the Mexican industrial sector. The study team collected data on end-use consumption and interviewed representatives from CFE and several manufacturing firms. Because Mexican data were scarce, the team used U.S. information on technologies, applications and implementation methodologies and applied them to the appropriate Mexican consumption data to produce rough estimates of the economic potential for various measures. This analysis was based on 1990 Mexican electricity consumption with DSM penetration rate estimates for 2010.

The study team reviewed a variety of measures that have proven attractive in many parts of the U.S. These measures fall into five broad categories: motors and drives, lighting, cooling (air conditioning), low/no cost measures, and innovative rates.

Motors and Drives

Mexican industry uses more electricity to run motors and drives than for any other purpose, representing about 58 percent of total industrial consumption. As in the U.S., these end-uses offer substantial opportunities for energy savings by employing energy-efficient technologies and strategies. The assessment scopes out the technical feasibility and potential impact of four measures that could reduce motor and drive energy consumption in Mexican industry. These measures include:

High-Efficiency Motors. These motors can cut energy consumption with improved designs and high-quality materials. Efficient motors run cooler, last longer, and out-perform standard motors of the same size with increased energy efficiency of over 10 percent in the smaller size ranges.

Downsizing. For a number of technical reasons, underloaded motors are particularly inefficient. Better matching motor size to the load (downsizing) is therefore an important option to improve motor end-use efficiency. Such a measure could improve efficiency 2.4 percent per application, on average, in Mexico.

Adjustable-Speed Drives. Drives offer opportunities for motor-related savings. Adjustable-speed drives (ASDs) can change the speed of electric motors by modulating the voltage supplied to the motor, thereby better adapting motor performance to fluctuations in load. ASDs can produce energy savings in the 24 to 52 percent range.

Cogged V and Synchronous Belts. Improved belts can also boost efficiency by reducing slip. Cogged V-belts, which have notches formed into the face of the belt to reduce slip and dissipate heat, offer efficiency improvements of between 1.6 and 3.8 percent over conventional V-belts. Synchronous belts, which resemble timing belts, are yet a more specialized cogged V-belt that combine the benefits of cogged belts with special grooved sheaves for added reduction in slip. These belts can improve motor system efficiency 4 to 10 percent.

Lighting

Lighting is another important end-use application in Mexican industry, representing 6 percent of consumption. Numerous efficient lighting technologies exist which offer energy savings for both interior and exterior lighting applications, but these technologies vary widely in cost, performance, technical feasibility, and applicability. Measures that could be important for Mexico include:

Energy-Efficient Fluorescent Lamps. Energy-efficient fluorescent lamps are one alternative that offers savings of about 15 percent. These lamps are constructed using a mixture of rare elements and gases to provide energy savings with minimal reduction in light output. These lamps are currently manufactured in Mexico.

Compact Fluorescents. Compact fluorescents are low-wattage, self-ballasted fluorescent lamps that can screw into incandescent lamp sockets on a one-to-one replacement basis. On average, these lamps draw 25 percent of the electricity of a standard lamp while providing the same lumen output. Like energy-efficient fluorescents, these lamps are currently manufactured in Mexico.

Electronic and Hybrid Electro-Mechanical Ballasts. Electronic ballasts use electronic circuitry to regulate the energy coming into lamps, reducing energy consumption and extending the life of the lamp. Energy savings are in the 15 to 20 percent range. Hybrid electro-mechanical ballasts represent a less expensive lighting alternative, also achieving savings with the use of electronic circuitry. Their savings are slightly less than those from fully electronic ballasts.

T-8 Lighting Systems. T-8 lighting systems save energy with an improved lamp design and materials. These features help reduce energy requirements 35 percent compared to standard T-12 systems. However, since T-8s operate only on T-8 electronic ballasts, their technical potential is somewhat limited and their high cost typically limits their achievable potential to new construction/remodeling applications.

Reflectors. Fluorescent fixture reflectors are polished or reflectorized metal inserts that are placed in an existing fixture. The reflectors direct more light out of the fixture toward the work surface, sometimes permitting half of the lamps in a room to be removed, thereby achieving substantial energy savings often at a competitive cost.

High-Intensity Discharge (HID) Lamps. HID lamps represent a family of lighting technologies which offer enhanced output per watt of electricity. These lamps, in order of increasing efficiency, include mercury vapor, metal halide, and high-pressure sodium lamps. These lamps differ substantially in color rendition and cost.

Lighting Controls. Lighting controls are another important energy-efficient lighting technology which could have significant application in Mexico. These controls include special switches, occupancy sensors and photocells. In each case, the controls ensure that lights are turned off when not in use.

Air Conditioning

Air conditioning is a particularly important end-use application in some parts of Mexico. Energy-efficient models offer savings usually at relatively low marginal cost, making replacements of broken machines or first-time purchases an attractive proposition in many instances. Because air conditioning is seasonal, however, yearly operating hours are low, lengthening the payback of energy-efficient units and thereby discouraging retrofit applications in some cases.

Low/No Cost Measures

In the U.S., many utilities use energy audit programs to assist industrial customers in identifying and implementing low/no cost improvements that can be easily implemented and provide noticeable energy savings. Based on U.S. experience, it is estimated that such measures will result in an average energy savings of 15 percent, peak demand reductions of 10 percent, and off-peak demand savings of 17 percent in Mexico. Such measures include: air compressor adjustment and repairs (such as reducing compressor air pressure or repairing leaks in hoses), improving motor rewinding techniques, turning off lights when not in use, turning off equipment when not in use, repairing damaged or missing insulation, closing doors to air conditioned or refrigerated areas, adjusting thermostats, using natural lighting to replace indoor lighting, and properly lubricating equipment.

Innovative Rates

Innovative rates offer customers incentives to change their patterns of consumption. These incentives can take the form of either lower or higher rates during peak periods.

The most common such rates in the U.S. are time-of-use (TOU) rates and interruptible and curtailable (I&C) rates; each is offered primarily to industrial customers. With TOU rates, higher demand and energy charges penalize customers for consumption during periods of peak demand. With I&C rates, customers pay reduced demand and energy charges for peak consumption in exchange for allowing the utility to cut or curtail power during selected days of maximum system demand. In Mexico, CFE has already begun to apply TOU tariffs for large customers and has estimated that these rates have reduced the system peak an estimated 500 MW with this program.

BENEFIT/COST ANALYSIS OF INDUSTRIAL DSM IN MEXICO

DSM analyses typically present savings in terms of demand (avoided peak capacity requirements given in megawatts or MW) and energy (consumption in gigawatt hours or GWh). The results of this study suggest that Mexico could realistically reduce its expected growth in electricity demand by over 5,000 MW by 2010 (about 18 percent of projected industrial demand and 10 percent of additional generating capacity investments for the period), with an aggressive, targeted DSM program to reduce or better manage industrial loads for various end-uses. Savings in energy by 2010 could reach over 17,000 GWh which is 11 percent of projected industrial consumption. Most of these investments have very attractive benefit/cost ratios.

Table 1 shows the industrial DSM potential which the analysis indicates can be realistically achieved in Mexico in 2010.

In terms of peak demand savings, low/no cost measures alone could reduce the peak demand requirement by 925 MW in 2010; innovative rates such as time-of-use and interruptible rates could together reduce the industrial sector's peak capacity requirement by over 3,100 MW.

Roughly 40 percent of the energy savings could be achieved at low or no cost (i.e., they could result from measures implemented as the result of energy audits). Another 40 percent could come from the successful implementation of a program designed to promote high-efficiency motors, drives and belts. Roughly 8 percent in energy savings could be achieved through DSM lighting programs.

None of the savings pertaining to any of the end-use sectors would require a sacrifice in comfort or economic growth in Mexico. All of the measures are designed to increase efficiency, not to reduce the number of end-use applications or decrease levels of comfort or industrial productivity. Additionally, we estimate that Mexico would avoid 19,247,895 tons of CO_2 emissions in 2010 if the penetration rates for energy-efficiency technologies in the U.S. are realized in Mexico. Cumulative avoided CO_2 emissions would be much higher.

Figure 1 presents the conserved energy supply curves for industrial DSM in Mexico, showing the levelized cost of each measure in U.S. cents/kWh. The figure compares the economic potential with the achievable potential. The economic potential curve shows the savings per kWh that would result if all the programs captured 100 percent of the eligible population in 2010, based solely on technology costs. The achievable potential curve maps the savings considering program costs and likely DSM penetration rates. Likewise, Figure 2 presents the cost of conserved demand supply curves for industrial DSM in Mexico, showing the levelized cost of measures in U.S. $/kW. The economic and achievable potentials are also contrasted. Finally, Table 2 summarizes the DSManager benefit/cost analysis results.

CONCLUSION

In this paper, the authors have reported the results of their work so far in assessing the potential for DSM in Mexico. We are now preparing to participate in the design and implementation of an industrial DSM pilot program to demonstrate how these potential DSM savings can be accomplished in Mexico. As we anticipate this effort, we note how different the implementation of industrial DSM will be in Mexico compared to the U.S. and Canada. For example, the utility is parastatal (meaning government owned and operated with no independent regulation), the utility and consumers are overcoming a lack of trust in the past, end-use data are hard to come by or don't exist, there is a lack of DSM awareness and technical capabilities, and the economy is rapidly evolving. And yet, there is tremendous pressure on Mexican industry to modernize and improve its competitiveness with the pending North American Free Trade Act. There is also money potentially available from the World Bank and other multilateral financial institutions if DSM can be proven to be a prudent investment. We hope that our efforts to document the cost-effective potential of industrial DSM in Mexico, including its environmental benefits, have started a process to show that such investments in DSM in Mexico are indeed prudent.

ACKNOWLEDGEMENTS

This paper is based on work performed by Ignacio Rodriguez and David Wolcott of RCG/Hagler, Bailly, Inc. under the Energy Efficiency Project sponsored by the U.S. Agency for International Development, Bureau for Research and Development, Office of Energy and

Infrastructure. The Fideicomiso de Apoyo al Programa de Ahorro de Energia del Sector Electrico (FIDE) and the Programa de Ahorro de Energia del Sector Electrico (PAESE) were the two Mexican organizations which served as counterparts for this study. Ing. Rogelio Magaña Castro and Ing. Salvador Herrera González of FIDE and Ing. Jesus Sada Gamiz of PAESE supported and contributed to this project.

Table 1

Industrial DSM Potential in Mexico

	Savings	**Percent Industrial Demand or Consumption**
Total Demand Savings	5,003 MW	17.9
-- Demand Savings from Conservation	1,894 MW	6.8
-- Demand Savings from Load Management	3,109 MW	11.1
Total Energy Savings	17,169 GWh	10.8
-- Energy Savings from Conservation	15,192 GWh	9.6
-- Energy Savings from Load Management	1,977 GWh	1.2

Figure 1
Cost of Conserved Energy Vs. GWh Savings
(Mexico Industrial Sector)

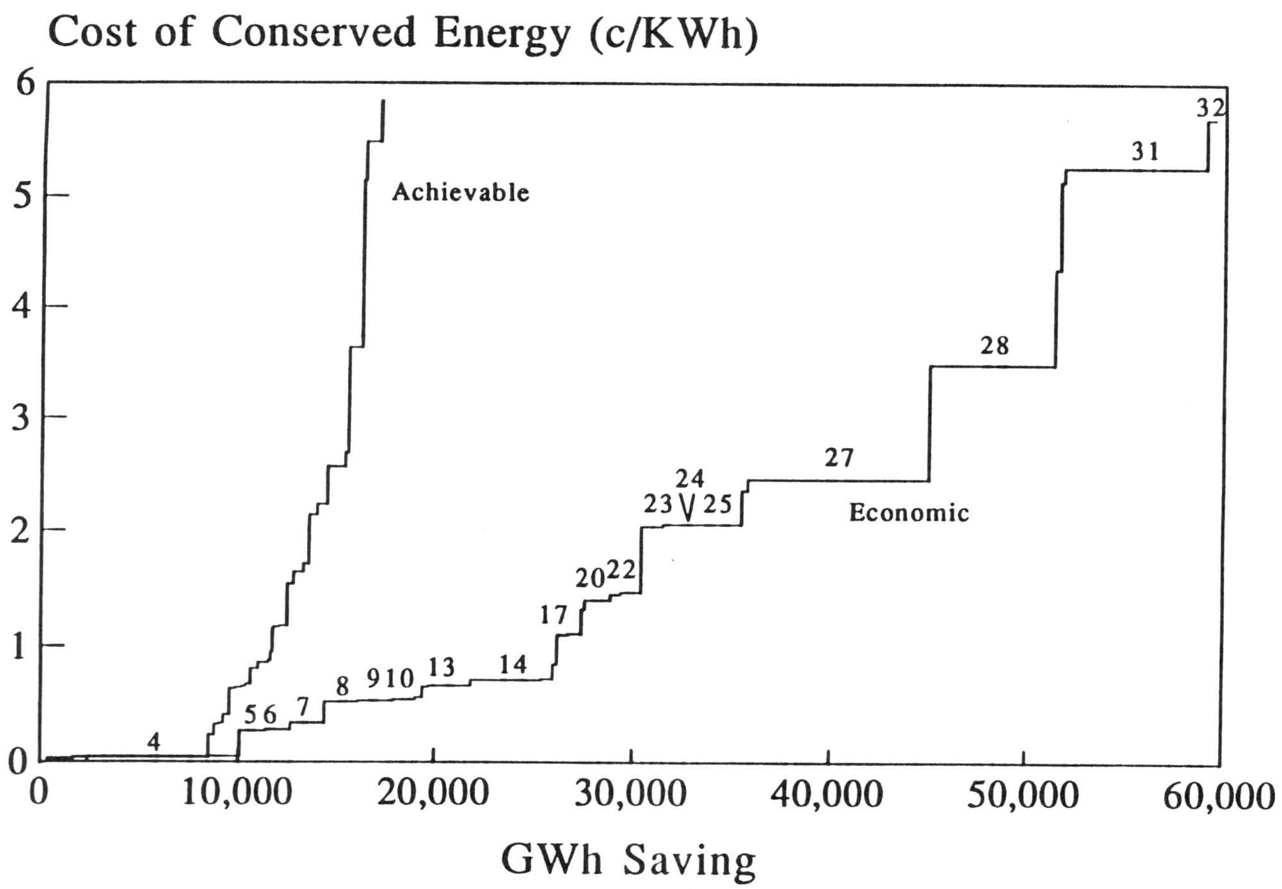

DESCRIPTION

1 Ind. Innovative Rate – Interruptible H–M
2 Ind. Innovative Rate – Interruptible H–S
3 Ind. Innovative Rate – Interruptible H–T
4 Ind. Misc – Low No Cost Measures
5 H.E. Motor 80.6 hp, Replacement
6 H.E. Motor 25.9 hp, Replacement
7 H.E. Motor 8.61 hp, Replacement
8 High Eff. Motor 25.9hp, Retrofit
9 High Eff. Motor 80.6hp, Retrofit
10 H.E. Motor 1.34 hp, Replacement
11 H.E. Motor 195 hp, Replacement
12 Ind Motor – Synch Belt, 195 hp
13 High Eff. Motor 8.61hp, Rerofit
14 Ind Motor – Downsize Std 10 hp to 5 hp
15 High Eff. Motor 195 hp, Retrofit
16 Ind Light – 2 F40W Reflector & Std. Ball
17 Ind Motor – Synch Belt, 25.9 hp
18 Ind Motor – Cogged V–Belt, 1.34 hp
19 Ind Motor – Synch Belt, 80.6 hp
20 Ind Air Con – 7.5 ton 10 SEER Air Con.
21 High Eff. Motor 1.34hp, Retrofit
22 Ind Motor – Synch Belt, 8.61 hp
23 Ind Light – 4 x F34 W & Mag. Ballasts
24 Ind Light – T–8 Lamps & Elec Ballasts
25 Ind Motor – ASD 195 hp
26 Ind Light – 4 x F40W & Electronic Ballas
27 Ind Motor – ASD 80.6 hp
28 Ind Motor – ASD 25.9 hp
29 Ind. 250W Mercury Vapor to 135W LPS
30 Ind Light – 4 x F40W & E.E.Mag. Ballasts
31 Ind Motor – ASD 8.61 hp
32 Ind. 250W Mercury Vapor to 150W HPS

Figure 2
Cost of Conserved Demand Vs. MW Reduction
(Mexico Industrial Sector)

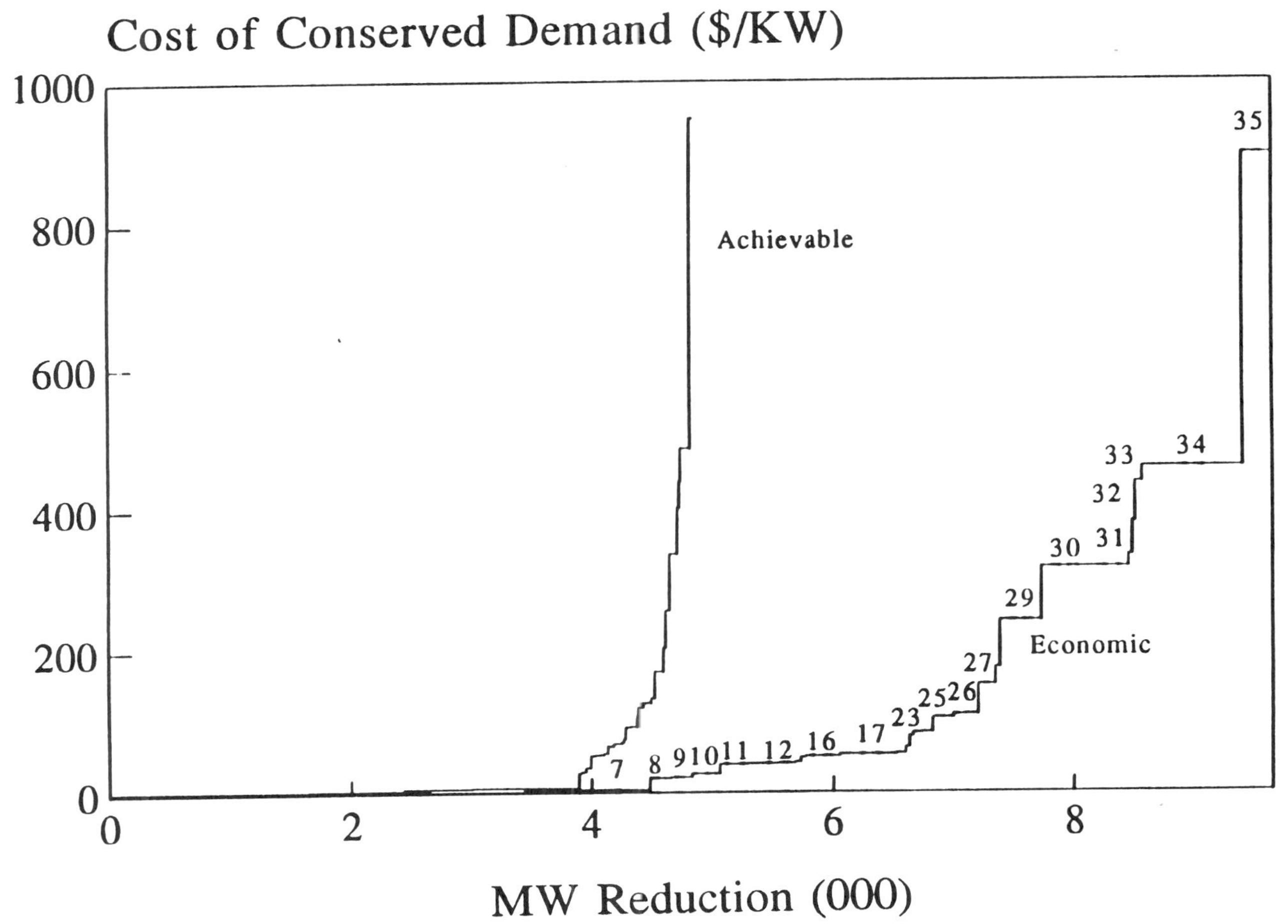

DESCRIPTION
1 Ind. Innovative Rate – TOU H–S
2 Ind. Innovative Rate – Interruptible H–T
3 Ind. Innovative Rate – TOU H–T
4 Ind. Innovative Rate – TOU H–M
5 Ind. Innovative Rate – Interruptible H–S
6 Ind. Innovative Rate – Interruptible H–M
7 Ind. Misc – Low No Cost Measures
8 H.E. Motor 80.6 hp, Replacement
9 H.E. Motor 25.9 hp, Replacement
10 H.E. Motor 8.61 hp, Replacement
11 High Eff. Motor 25.9hp, Retrofit
12 High Eff. Motor 80.6hp, Retrofit
13 H.E. Motor 1.34 hp, Replacement
14 H.E. Motor 195 hp, Replacement
15 Ind Motor – Synch Belt, 195 hp
16 High Eff. Motor 8.61hp, Retrofit
17 Ind Motor – Downsize Std 10 hp to 5 hp
18 High Eff. Motor 195 hp, Retrofit
19 Ind Light – 2 F40W Reflector & Std. Ball
20 Ind Air Con – 7.5 ton 10 SEER Air Con.
21 Ind Motor – Cogged V–Belt, 1.34 hp
22 Ind Motor – Synch Belt, 25.9 hp
23 Ind Motor – Synch Belt, 80.6 hp
24 High Eff. Motor 1.34hp, Retrofit
25 Ind Motor – Synch Belt, 8.61 hp
26 Ind Light – 4 x F34 W & Mag. Ballasts
27 Ind Light – T–8 Lamps & Elec Ballasts
28 Ind Light – 4 x F40W & Electronic Ballas
29 Ind Motor – ASD 195 hp
30 Ind Motor – ASD 80.6 hp
31 Ind. 250W Mercury Vapor to 135W LPS
32 Ind Light – 4 x F40W & E.E.Mag. Ballasts
33 Ind 250W Mercury Vapor to 150W HPS
34 Ind Motor – ASD 8.61 hp
35 Ind Motor – ASD 25.9 hp

Table 2: Summary of Economic and Achievable DSM Potential in Mexico's Industrial Sector

Economic Potential

	Energy Use	Energy Savings			Demand	Demand Reduction		
	2010 MWh	2010 MWh	Percentage of Total Saving	Percentage of Total Energy Use	2010 MW	2010 MW	Percentage of Total Reduction	Percentage of Total Demand
Industrial Sector:	159,445,348				28015.77			
Motors & Drives	92,478,302				16501.52			
Energy-Efficient Motors		13,356,706	22.69%	8.38%		1790.89	18.72%	6.39%
Adjustable Speed Drives		26,893,190	45.68%	16.87%		2145.62	22.43%	7.66%
Cogged V-Belts		70,090	0.12%	0.04%		9.40	0.10%	0.03%
Synchronous Belts		1,962,668	3.33%	1.23%		263.16	2.75%	0.94%
Motor Downsize Std 10 hp to 5 hp		3,509,547	5.96%	2.20%		470.57	4.92%	1.68%
Motor Program Sub Total:		45,792,200	77.78%	28.72%		4679.64	48.92%	16.70%
Lighting	9,327,553				1665.94			
Interior Lighting		2,604,820	4.42%	1.63%		352.37	3.68%	1.26%
Exterior Lighting		241,129	0.41%	0.15%		31.46	0.33%	0.11%
Lighting Program Sub Total:		2,845,949	4.83%	1.78%		383.83	4.01%	1.37%
Industrial A/C:	1,036,394				185.88			
Energy-Efficient Air Con		172,177	0.29%	0.11%		29.19	0.31%	0.10%
Low/No Cost Measures		7,663,128	13.02%	4.81%		1027.40	10.74%	3.67%
Innovative Rates		2,398,316	4.07%	1.50%		3445.90	36.02%	12.30%
TOTAL SAVINGS		58,871,771	100.00%	36.92%		9565.96	100.00%	34.14%

Achievable Potential

	Energy Use	Energy Saving			Demand	Demand Reduction		
	2010 MWh	2010 MWh	Percentage of Total Saving	Percentage of Total Energy Use	2010 MW	2010 MW	Percentage of Total Reduction	Percentage of Total Demand
Industrial Sector:	159,445,348				28015.77			
Motors & Drives	92,478,302				16501.52			
Energy-Efficient Motors		2,337,435	13.61%	1.47%		313.41	6.26%	1.12%
Adjustable Speed Drives		2,689,266	15.66%	1.69%		214.56	4.29%	0.77%
Cogged V-Belts		63,081	0.37%	0.04%		8.46	0.17%	0.03%
Synchronous Belts		1,198,513	6.98%	0.75%		160.70	3.21%	0.57%
Motor Downsize Std 10 hp to 5 hp		526,426	3.07%	0.33%		70.58	1.41%	0.25%
Motor Program Sub Total:		6,814,721	39.69%	4.27%		767.71	15.34%	2.74%
Lighting	9,327,553				1665.94			
Interior Lighting		1,302,411	7.59%	0.82%		176.19	3.52%	0.63%
Exterior Lighting		120,571	0.70%	0.08%		15.73	0.31%	0.06%
Lighting Program Sub Total:		1,422,982	8.29%	0.89%		191.92	3.84%	0.69%
Industrial A/C:	1,036,394				185.88			
Energy-Efficient Air Con		57,441	0.33%	0.04%		9.74	0.19%	0.03%
Low/No Cost Measures		6,896,816	40.17%	4.33%		924.70	18.48%	3.30%
Innovative Rates		1,977,076	11.52%	1.24%		3109.00	62.14%	11.10%
TOTAL SAVINGS		17,169,035	100.00%	10.77%		5003.07	100.00%	17.86%

Chapter 42

Managing the Risks of Performance Contracting

C. Kane

INTRODUCTION

Energy and environmental retrofit projects will receive much more attention in coming years. Efforts to save energy in the federal government are dictated by the Energy Policy Act of 1992. Furthermore, compliance deadlines under the Clean Air Act Amendments of 1990 will increase the value of energy conservation. Trends in using environmental externalities in the electrical generation planning process will also foster energy savings efforts. The use of performance contracting should play an increasing role in meeting all these challenges.

The definition of a performance contract is one in which the Owner contracts with a single entity which has all the responsibility for design, equipment procurement, installation, operations and maintenance (See Figure 1). The main advantages of this arrangement are as follows; 1) the single source of responsibility; 2) guaranteed performance of the project; and 3) payment based on performance through the energy savings. Design-build or "turnkey" contracting is often implemented in performance contracting. As such, all design and construction responsibilities are placed with a single entity. The engineer and builder typically enter into a joint venture or subcontract arrangement and the resulting single entity contracts directly with the Owner.

For the purpose of this paper, Energy Service Company or ESCO means any entity that undertakes a performance contract with a building Owner and is used interchangeably with a Contractor or design-builder.

As with any boom industry, the new opportunities will attract many unqualified and uninformed entrants. Owners must be aware of this fact in evaluating, selecting and contracting with the ESCO. This paper addresses some of the legal and contractual aspects of managing the risks associated with performance contracting. Although the principles should apply to any size project, the discussion assumes a relatively large project encompassing a mix of conservation/efficiency or environmental improvement efforts.

RESPONSIBILITY FOR DESIGN AND PERFORMANCE

In the traditional approach to capital improvement projects, project Owners have attempted to define their needs with the assistance of an engineering firm. The needs are spelled out in detailed plans and specifications and the work is then bid and awarded on the basis of low bid price and lump sum contract documents. This process in the public construction arena was thought to save money. Too often however, these savings are illusory. When applied to an area such as energy conservation retrofits, this traditional approach is ineffective.

Where the Owner produces the design, there is very little room for innovation and input on the part of the Contractor. Rather, the Contractor is left with a road map that, if followed, takes the Contractor off the hook, whether the project performs or not. The performance contract does just the opposite. The performance contract requires the Contractor to achieve certain results and leaves the route he chooses (the design) up to him. The other key aspect of performance contracting is that it puts the financial burden on the Contractor by providing his payments through the energy savings over a period of years.

One of the best known cases in construction law is the 1918 landmark decision of U.S. v. Spearin. In Spearin, the U.S. Supreme Court held that a project Owner extends an implied warranty to the Contractor that the plans and specifications are accurate and complete and if followed faithfully, will produce the desired result. Although many Contractors question whether Owners and engineers are still being held to this standard, Spearin remains good law, where the Owner maintains responsibility for design.

FIGURE 1 - PERFORMANCE CONTRACTING

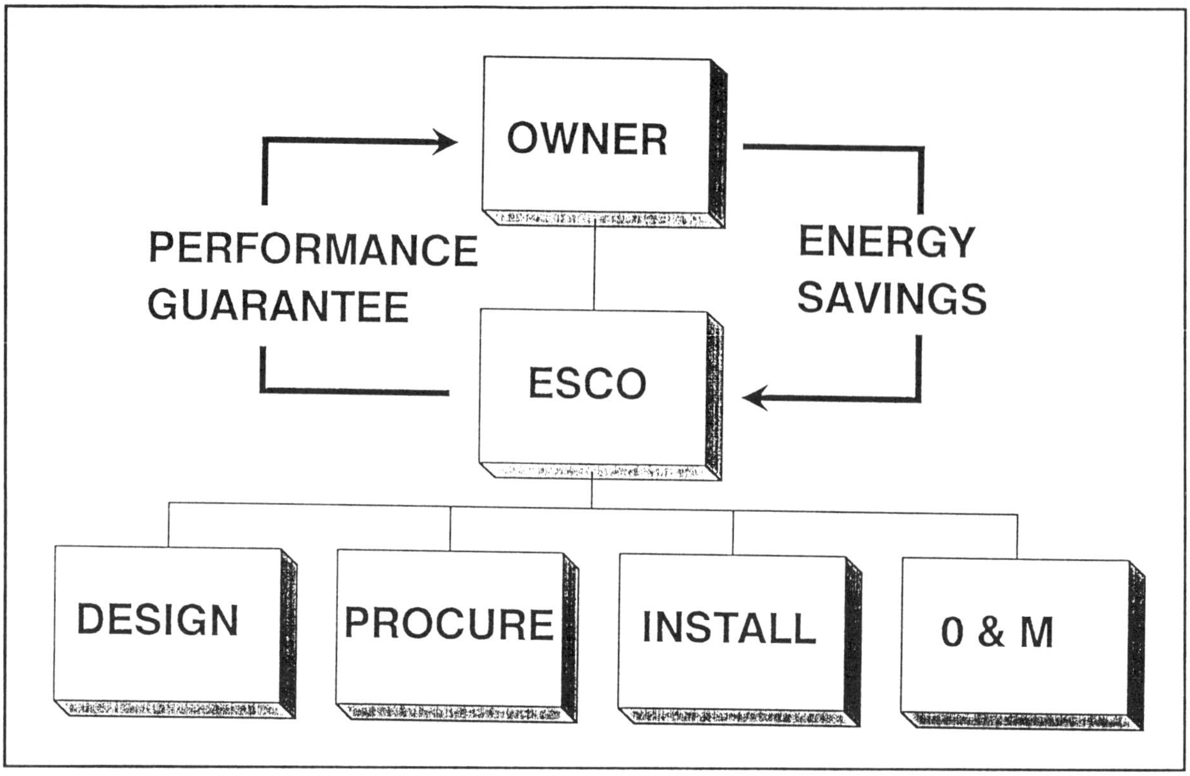

In applying this principle, there is a fundamental distinction in government contract law between projects based on "design specifications" and those based on "performance specifications". This distinction needs to be considered in outlining the requirements of performance contracting. A "performance specification" with defined parameters would typically be involved in the agreement between the Owner and ESCO. More detailed "design specifications" would normally be developed for the subcontract agreements between the ESCO and its subcontractors and suppliers.

This warranty of the Owner applies only to "design specifications". In general, "design specifications" provide in precise detail the material to be employed and the manner in which the work is to be performed. The Contractor consequently is not free to deviate from the specifications. In contrast, "performance specifications" merely set forth an objective and call upon the Contractor to use its own ingenuity to achieve the desired end. By definition, design-build contracting uses "performance specifications".

Design-Build Contracts

In design-build, the Spearin principles are reversed. The Owner outlines project requirements and the Contractor warrants that what he designs and builds will satisfy those requirements. Therefore, all of the Owner's requirements, standards of quality, and life cycle costing must be incorporated into its solicitation or first phase of design development and then negotiated into the final performance contract.

Design-build puts the maximum risk on the Contractor in terms of project costs and guarantees of performance, reliability and availability. Accordingly, the Owner has less control over the product. Qualified Contractors, with a competent team including design, construction and equipment elements, can control their risks in addition to obtaining a higher margin on the work. On the other hand, the Owner risks higher initial contract costs but reduces downstream risk over the course of the project. The trade off is the Owner's loss of control over the design details of its project.

One way to minimize risks to the Owner and the Contractor is to involve the design-build Contractor earlier and set up a two phase process. The first phase involves communication of the Owner's requirements for details and quality standards while allowing the Contractor to develop a better basis for determining his price. The Contractor is paid for this first phase of design development/preconstruction services before going on to the second phase of guaranteed pricing and construction. In this way, the price and the Contractor's risk can be reduced in addition to giving the Owner more influence over the design.

Owners sometimes attempt to shift certain design responsibilities to the Contractor by using "catch all" scope of work language even when using design specifications. Contractors are thus, required to pay the costs for missing design details which they probably were not able to price in their bids. In the design-build context, a different problem sometimes occurs when the Owner attempts to retain too much control over the design during construction. Either situation is likely to result in increased disputes and therefore, whoever has control over the design should bear the risk of performance.

REALISTIC RISK ALLOCATION AND MANAGEMENT

All performance contracts involve risks: risks in the building design, time, cost, quality and performance of the project. Over the years, the parties to the design and construction process have focused increasingly on "risk shifting" and "risk avoidance". This aversion for taking responsibility is both a product of and contributing factor to the current litigious nature of the construction industry. To avoid litigation and disputes, the focus must be on "risk sharing". Risk sharing involves allocating risks to the party who is in the best position to control that risk. Once allocated, risks must be managed throughout the performance contract. A summary of performance contract risk management is included in Figure 2.

Recent construction industry studies indicate that contracts which attempt to shift to Contractor, risks over which they have little or no control, are not cost effective. These risk shifting contracts are ineffective because they; 1) reduce Contractor competition, 2) increase prices due to increased Contractor contingencies, and 3) increase costs and reduce efficiency due to increased project disputes. These studies have concluded that imposition on parties of risks which they cannot manage and control is a primary <u>cause</u> of disputes.

Among other things, risk shifting clauses tend to create an adversarial relationship from the very start of a project. Walls are built rather than bridges and the odds for legal conflict increase greatly. In contrast, when risks are shared equitably, the need to operate defensively is eliminated and the odds for conflict are greatly reduced. Working relationships become more cooperative and less adversarial.

Performance contracting must include such realistic risk allocation, which involves the identification and allocation of risks in three key areas;

1) <u>Hardware Related Risks</u> (including construction costs, equipment performance, engineering/technical risks and operating and maintenance risks);

2) <u>Building Functional Risks</u> (including occupancy risks, establishing base year data, and building control risks); and

3) <u>Energy Price Risks</u> (including fuel options, total cost of fuel considerations, energy savings, and energy escalation risks).

Hardware Risks

In performance contracting, hardware related risks are invariably within the control of the ESCO. Therefore, these risks should be allocated to the ESCO. The ESCO also has the ability to pass these risks down to its subcontractors and suppliers by drafting clear and comprehensive subcontract agreements.

The ESCO through its design responsibility is also in control of the performance of the project. Thus, his payment is based on performance by passing on to him energy savings. The ESCO would be well advised to be conservative on these performance guarantees when setting up the performance contract. Recent studies have indicated efficiency projects frequently fall short of stated savings projections. Therefore, some experience or contingency factors should always be applied.

Building Functional Risks

These risks, to some extent, involve shared control by the ESCO and Owner. The ESCO will, in all likelihood, have participated in the development of the base year data and also will be involved in building control risks. Most ESCO's will remain involved with operations and maintenance and will have input into building control requirements. The Owner typically will have control over occupancy and therefore, should bear the risks of occupancy to the extent it negatively impacts energy savings. On the other hand, if one floor of the building becomes unexpectedly vacated, the ESCO should not be permitted a windfall for the positive effect on savings. Appropriate language and formulas should be included in the contract to account for these contingencies and address how they will be dealt with.

Every performance contract should contain schedules of requirements that address the following functional risk areas;

- Standards Of Comfort
- Owner/ESCO, O&M Responsibilities
- Facilities Changes Checklist
- Baseline Energy Consumption

Energy Price Risk

Energy price risk is one that neither party has control over. Therefore, it is appropriate to share this risk between the parties. One way to do this is to have the ESCO assume the risk of energy price, down to a certain minimum figure over the life of the contract. The Owner then would share the risk of energy prices escalating up, with a cap on the up side risk. Thus, the Owner receives some of the benefit in savings from energy price increases on the up side and the ESCO is protected at a certain minimum level on the low side.

Selecting An ESCO

The Owner's biggest risk is in the competence and financial stability of the ESCO it is dealing with. If the Owner is looking for some level of competitive prices in selecting its project, it must have a process that insures it ends up contracting with a reliable entity. This process involves a series of safety nets to maximize the chances for a successful project with little or no adverse impact on the Owner's business or activities. Some recommended aspects of the process are Ahighlighted below.

Prequalification - There must be an effort to narrow the field of proposers to eliminate the riff raff and only allow entities to propose who would be worthy of contracting. At a minimum, this process should require a detailed prequalification application, listing the following information;

- Years in Business
- Relevant Project Experience
- Projected and Achieved Savings
- Current Projects in Process
- Design, Equipment and Installation Resources
- Financial References/Resources
- Business References
- History of Litigation/Claims
- Credit Report
- Number and Experience Level of Employees
- Insurance and Bonding Information

Once submitted, these prequalification forms should be evaluated and scored in order to select the three to five proposers that should be allowed to proceed in the process. It is very important in the prequalification process to actually call references and explore the results of past projects. In the case of a government entity, prequalifing bidders may not permit a significant narrowing of the field and therefore, a two phase procurement process can be used.

Multi-Phase Procurement - In soliciting performance contract proposals, the Owner may have a specific retrofit already identified or may want to explore all options in order to maximize savings. In either case, some limited form of energy audit in the form of a site visit must be permitted by the potential proposers. Based on the solicitation and the energy audit, a technical proposal should be required as the first phase of the procurement process.

The technical proposal should list the selected energy conservation measures along with a range of potential savings. In addition, the technical proposal should identify with specificity, the make up of the project team describing design, equipment, installation, and O&M components. In particular, the key staffing for each function should be identified along with resumes. At this stage also, if a prequalification has not already been conducted, all prequalification information should be provided as discussed above.

Again, the Owner should develop a weighted point evaluation system to provide a ranking of the proposers based on the technical proposal. Based on this evaluation, there may be one obvious choice to proceed with or there may be two or three. If time permits, any proposers from the short list should be permitted to perform a further evaluation/energy audit prior to the final phase of price negotiation. From this a selected candidate should be identified for final contract negotiations.

The final price negotiation should be done with as much specificity of the equipment, scheduling, interface and coordination requirements as possible since all these factors will effect price. Even though this is a performance contract, the more specificity that can be established prior to contracting, the better for the Owner. Changes to a performance contract by the Owner can be very costly and jeopardize the benefit of its bargain. Scheduling requirements, in particular, need to be laid out especially if the Owner has continuing use of the facility.

MANAGING PERFORMANCE CONTRACT RISKS

PHASE I SELECTION OF ESCO

- PRELIMINARY PROJECT PLANNING
- PREQUALIFY BIDDERS
- PREPARE DETAILED RFP
- SITE VISITS BY BIDDERS
- EVALUATE AND SCORE PROPOSALS

PHASE II ENERGY AUDIT

- CONTRACT FOR ENERGY AUDIT
- ENERGY AUDIT AND PROPOSAL BY ESCO
- REALISTIC RISK ALLOCATION

PHASE III PERFORMANCE CONTRACT

- NEGOTIATE PERFORMANCE CONTRACT
- PROJECT ADMINISTRATION
- MONITORING

FIGURE 2 - MANAGING RISKS

Contract Negotiation And Administration

Both parties need to feel invested in the final contract agreement. There should be a discussion and clear understanding of the terms and conditions by all the participants before the contract is executed. Contract negotiations should not be performed in the mail. Face to face, extended discussions pay dividends in the long run in building trust and understanding. To the extent clarifications and changes to the contract can be made, they should be made in the final negotiations. Contracts are defined as a "bargained for exchange" and to be a viable contract, that bargaining must take place.

It is also important in managing risk that both parties make a commitment to the administration of the contract. An individual needs to be identified by the Owner to insure the Owner meets its obligations under the contract, in particular, with regard to providing information, access, and scheduling which will be required for any project to be successful. Scheduling of the work needs to be closely coordinated between the Owner and the ESCO and the ESCO's subcontractors. In like manner, the ESCO needs to identify a single project manager to be the contact in all matters with the Owner.

KEY CONTRACT PROVISIONS

In addition to identification, allocation and management of risk, certain key contract provisions are essential in preparing effective performance contracts. A few of these clauses are discussed below.

Payment

This is the most critical and distinct element of a performance contract. It is also a provision that ESCO's consider very proprietary. Therefore, not much is publicized on the variations of payment schemes and financing mechanisms for performance contracts. Each project team will participate differently in the financing developed for a given project. Some team members will be willing to take payment strictly from energy savings and some will not. Some large equipment suppliers may be inclined to take on financial and performance responsibilities to get an advantage over their competitors. The financing entities will also have various requirements geared toward the specific contract.

Owners will also differ in how much up front funding they may be willing to contribute. At a minimum, Owners should take responsibility for payment of the detailed energy audit, if the project does not go forward. Beyond that, Owners should look to the ESCO for creative payment and financial mechanisms since this is a principal advantage to the Owner in using performance contracting.

Warranties

Even though the ESCO is assuming the risk of performance, it is only for a fixed period of time. The contract must give the Owner a guarantee that the equipment will perform at a given level and be as specified or shall be repaired or replaced by the ESCO. This warranty must be for at least 12 months from completion of the project and should also include a second level of warranty throughout the period the ESCO is to operate and maintain the equipment. The Owner should make certain that these warranty obligations are passed down to all subcontractors, suppliers and designers and that the warranties run to the Owner.

Other representations should be included in the text of the contract. One in particular is that the project will not require the Owner to hire additional personnel or purchase any additional equipment which is not specified in the contract. For example, if a project included a new sophisticated computer automation and monitoring system requiring a full time operator, operator costs beyond the ESCO's period of performance should be considered in the contract price.

Scheduling And Delays

Most energy conservation and environmental retrofit projects are going to take place in existing facilities which may be occupied while the work is being performed. HVAC system modifications will have to be carefully coordinated in order to minimize disruption to operations. The scheduling and planning effort must be performed by the facility's management and the ESCO and written into the contract. Most performance contracts will have a "time of the essence" clause, which means that delays are a significant concern to the parties. Thus, the contract should have milestones and a specific end-date. The Owner should make certain to incorporate some flexibility or float in the schedule so that some unanticipated problems can be absorbed without delaying the project.

Time is money from the ESCO's point of view. Delays in installation will delay utility savings and will ultimately delay his compensation. The use of liquidated damages is one way to incentivize the parties to perform. Liquidated damages are based on a reasonable forecast of what a party's damages might be if it is delayed. Those amounts are then written into the contract as the sole and exclusive remedy for unexcusable delays.

There are normally three categories of delays in construction contracts; 1) excusable, non-compensable, 2) excusable, compensable and 3) non-excusable. The scheduling clause should define the delay categories which normally would be as follows:

1. Excusable, Non-Compensable. These include such things as acts of God, war, trade embargo and similar delays which neither party can control. Therefore, a party would get an extension of time in which to perform its contract. The delayed party would have no liability for liquidated damages but also no entitlement to additional costs which it may have incurred because of the delay.

2. Excusable, Compensable. These are delays to one party caused by factors within the control of the other party, such as the Owner denying the ESCO access to equipment or work space when it was scheduled, thereby delaying the ESCO's performance. To the extent the delay has caused the ESCO additional cost, he may be entitled to those costs (or if liquidated damages are provided, entitled to liquidated damages).

3. Non-Excusable. These are delays within the control of the party being delayed. For an ESCO, such things as lack of skilled or sufficient workman, equipment delivery delays or simply slow performance, would be included. With such delays, the ESCO must accelerate and still complete on time or be liable to the Owner for liquidated or actual damages.

Insurance And Bonding

In this age of lawsuits and business failures, minimum insurance coverages should be contractual requirements in every contract. Owners should require that the ESCO's and their subcontractors carry general liability insurance, workmen's compensation and errors & omissions insurance if engineering design is involved. A cost-benefit analysis should be performed in order to justify the types and amounts of insurance coverage. Whatever is required, should be spelled out in the contract. The Owner should also require that it be named as an additional insured on the policies and that it be provided copies of certificates of insurance before the ESCO starts work.

Bonding is a requirement on federal projects above a certain dollar amount. Normally, payment bonds are provided for the benefit of subcontractors and suppliers. Performance bonds are provided for the benefit of the Owner to insure that if the ESCO does not perform, the surety will step in. Even in private work, Owners should consider requiring bonds from the ESCO, especially for larger performance contracts. Requiring bonds will cost the Contractor a little more but he can easily roll the cost into his financing. ESCO's who are bondable, typically will be more substantial and financially viable then those who are not. Furthermore, payment bonds can eliminate some of the Owner's potential exposure to mechanic's liens by unpaid subcontractors and suppliers as discussed below.

Mechanic's Lien Waivers

In every state, mechanic's lien statutes have been enacted for the benefit of those who furnish labor and material improvements to property (other than state and federal). Every state is slightly different in terms of the procedural requirements and the extent of this right. In a nutshell, if someone provides labor or material to a project and is not paid, he can lien the property and force the Owner to pay off its lien or face potential foreclosure and sale of the property to satisfy its lien.

Owners, to the maximum extent permitted by law, need to include waiver of these mechanic's lien rights in their performance contracts with the ESCO's and contract that the ESCO will require its subcontractors and suppliers to similarly, waive those rights. Under performance contracts, the risk of payments to subs and suppliers should lay with the ESCO, since the Owner never controls the financing. If mechanic's lien rights are not waived and a payment bond is not provided by the ESCO, an unpaid lighting supplier, for example, could lien the job when the failure of payment was strictly the ESCO's.

Legal counsel should be consulted in drafting any such provision. In some states, it is void as against public policy to waive mechanic's lien rights in advance of payment.

Indemnification

These clauses are always of great interest to the lawyers and insurance agents when battling over contract terms. Normally, the Owner attempts to maximize the Contractor's duties to indemnify while the Contractor will try to minimize those duties. Studies indicate that 90% of the disputes between Contractors and Owners under indemnity clauses involve suits by Contractor employees against the Owner. Most worker's compensation laws bar suits against the employer (Contractor) and therefore, the Owner becomes a convenient "deep pocket".

The state of the law is confused on these provisions. If the parties can recognize the problem as a work place injury problem and deal with it as such, the situation becomes simplified. The Contractor should be required to give the Owner complete protection against suits by Contractor employees who are barred by the worker's compensation laws from suing the Contractor. This risk is insurable and therefore, the Contractor should be willing to assume the full risk of the Owner's liability at least up to the limits of the insurance without regard to the Owner's partial negligence.

Most courts will enforce this type of indemnity provision if clearly limited to work place injuries. However, the same result can be achieved even more effectively by requiring the Contractor's insurance to cover the Owner for this limited risk. This framework places the risk on the one most capable of controlling that risk. The Contractor is not only in the best position to prevent injuries in the work place, but also to handle resulting litigation.

Termination

A termination provision should describe the circumstances for which the contract can be cancelled, and how payments will be made upon cancellation. All government contracts will typically have a termination clause based on 1) the Contractor's default or 2) the convenience of the Owner. Most public contracts will include a termination for convenience clause which allows cancellation for virtually any reason, but also provides for payment of costs incurred up to the point of cancellation plus a reasonable amount of overhead and profit. A provision should also be included in any termination for convenience clause, specifically excluding any liability for loss of anticipated profits.

A termination for default provision is necessary for every performance contract. The problem is usually agreeing to what will constitute default. A typical listing of default events might include the following;

1. Failure to maintain specified standards of comfort, after notice and a failure to correct the situation;

2. Any representation or warranty made by the ESCO in the agreement which is false or misleading;

3. Failure to furnish and install equipment within times specified;

4. Any lien or encumbrance filed by a subcontractor or supplier; and

5. Failure to pay any subcontractor or supplier in a timely manner, without reasonable justification.

Upon a termination for default payment by the Owner for work performed should be severely restricted. Payments should not be required until the Owner has been permitted to complete the work and also has assessed all its damages with respect to the default. Delay costs, interruption of business, cost to complete the work, attorney's fees and other related costs are costs the Owner is entitled to upon default.

The default provision should make it clear that any work performed prior to the default automatically becomes the property of the Owner. Problems may arise in performance contracting if the ESCO financing entity has a security interest in the equipment provided. At least, if the proposed contract contains a properly written default provision, such issues will be identified and resolved during the contract negotiation process.

DISPUTE AVOIDANCE AND RESOLUTION

The dispute resolution procedures of the past relied too heavily on the "adversarial" processes such as litigation or binding arbitration. To be successful on projects in the future, greater emphasis must be given to multiple stages and more "collaborative" processes for resolving disputes (See Figure 3). The collaborative process relies on the parties working out the solution to their problems, sometimes with outside assistance. What follows is a brief discussion of dispute resolution methods that should be considered.

Teambuilding

Teambuilding on projects creates mutual trust and respect for the various roles necessary for the implementation of performance contracts. One teambuilding concept currently being promoted and achieving positive results is called "Partnering". While the contract establishes "legal" relationships, the Partnering process attempts to establish "working" relationships through a mutually-developed, formal strategy of commitment and communication. Partnering creates an atmosphere that avoids disputes.

A Partnering workshop is conducted in the early stages of the contract with the purpose of establishing and implementing the key elements of Partnering. These key elements include; 1) commitment from top management, 2) a sense of equity by developing win/win thinking, 3) trust among the parties and 4) the development of mutual goals and objectives. Although the elements appear self evident, unless these concepts are specifically addressed at the outset of and during a project, the adversarial and punitive ways of the past will creep back into the relationship.

The Partnering process continues throughout the project addressing problems head on and early on. One major point on Partnering is that it <u>does not</u> throw the contract or the specifications out the window.

Standing Neutral

One of the most effective techniques for resolving disputes that may arise on complicated projects such as energy conservation or environmental retrofit would be to use a

<u>DISPUTE RESOLUTION STAGES</u>

REALISTIC RISK ALLOCATION

CONTRACT NEGOTIATIONS

TEAMBUILDING

STANDING NEUTRAL

MEDIATION

ARBITRATION & LITIGATION

FIGURE 3 - DISPUTE RESOLUTION

standing neutral. The contract can be written to require the parties to mutually agree on a third party expert to be available throughout the project and act as an arbitrator of disputes. As disputes arise, the parties are required to present the dispute to the neutral who can quickly render a nonbinding decision concerning the dispute. The decision can be accepted by the parties or used as a basis for negotiation. This process allows for real time, third-party expert opinions on issues in order to facilitate timely resolutions.

Mediation

Described as the "sleeping giant" of alternative dispute resolution in the early 1980's, mediation has awakened and is quickly increasing in acceptance. The American Arbitration Association reports that of all cases referred to it for mediation, at least 80% settle. Within the last two years, mediation has been recommended in certain standard form agreements by the American Institute of Architects, the Associated General Contractors and the Engineer's Joint Contract Documents Committee.

Mediation is an extremely flexible settlement tool. The parties to a dispute agree to bring in a neutral third party to assist in finding a mutually acceptable resolution. The mediator's only role is to guide the parties towards settlement. No authority is granted to the mediator to render a binding or a nonbinding decision on the merits. Rather, the mediator serves to schedule and structure negotiations, acts as a catalyst between the parties, and serves as an assessor - but not a judge - of the positions taken by the parties during the course of negotiations.

With the parties' consent, the mediator may take on additional functions such as proposing solutions to the problem. Nevertheless, as in traditional negotiation, the parties retain the power to resolve the issues through an informal, voluntary process, in order to reach a mutually acceptable agreement. Having agreed to a mediated settlement, parties can then make the results binding.

Arbitration And Litigation

Every contract should have a binding final dispute resolution process. In private contracts, binding arbitration and litigation are still the most prevalent avenues. Each remedy has its proponents and its critics. However, if you incorporate some of the dispute avoidance and resolution mechanisms outlined above, there should be little need for the more costly and time consuming adversarial processes. The threat of either arbitration or litigation can provide a powerful incentive for the parties to negotiate and retain control.

CONCLUSION

Performance contracting has the potential to be a "win-win" solution on many fronts - efficiency, environmental improvement, capital investment, employment, innovation, financing and energy conservation. However, the performance contract is a relatively new and complicated commercial agreement which requires care, attention and knowledge in its preparation.

The parties to the performance agreement must invest the time, energy and resources necessary to prepare, negotiate and implement these agreements or the problems will quickly out number the successes. With proper care and up-front planning, there is no reason that performance contracting cannot become a major component in our arsenal for tackling the difficult energy, environmental and economic problems that we are facing this decade.

REFERENCES

1. Performance Contracting for Energy and Environmental Systems, By Shirley J. Hansen, Ph.D., 1992.

2. "Mitigating Construction Contract Disputes", Public Utilities Fortnightly, By Christopher Kane, July 1992.

3. "Energy Performance Contracting for Public and Indian Housing - A Guide for Participants", U.S. Department of Energy and U.S. Department of Housing and Urban Development, February 1992.

4. "Teambuilding", Independent Energy, By Christopher Kane, April 1993.

5. "Improved Construction Contracting for the 1990's", Power Engineering, By V. Frederic Lyon and Christopher Kane, June 1991.

Chapter 43

Protecting the "Host" in Performance Contracting for DSM

S.J. Hansen

If performance contracting is to be effective in satisfying DSM needs, the principal parties, the utility and the ESCO, must benefit. But that is not enough! It's very easy to lapse into thinking of the host as a passive third party -- like the shell a hermit crab borrows as long as it is useful to him. But the host site is where the savings happen -- or don't. If the host does not gain from the deal, the savings are less likely to happen. Then, the utility may not satisfy the IRP and very likely will have customer relations problems. The ESCO is apt to make less money, and may even end up paying the utility a penalty.

It is in the utility's interests and the ESCO's interests to make sure the host's needs are met -- throughout the life of the contract, even beyond. Otherwise, we have a shell game ... and crab soup.

Performance contracting from the host's perspective must start with an analysis of the host's:

- ✔ energy savings opportunities;
- ✔ facility and operating needs;
- ✔ financial needs; and
- ✔ management concerns.

This assessment should drive the initial project criteria, any proposal/response/evaluation procedure as well as contract negotiations and project implementation. Such a statement seems like a "given." But to the uninitiated, any effective performance contract, at first glance, can look like so much smoke and mirrors.

The term "effective" performance contract is stressed because some utility/ESCO agreements have been built on a bid/spec type of an arrangement, which to the utility may look more substantive than smoke and mirrors. This may be reassuring to the utility, but it is seldom in the ESCO's best interests -- and never in the host's best interests. And ultimately not in the utility's best interests either.

A Lose/Lose "Opportunity"

The more spec, the more bid, the more risk there is up front. One theme undergirds the relationship among all three parties: *money follows risk.* The greater the risk; the greater the share of the investment that must go into covering that risk. A greater portion of the investment, therefore, is devoted to higher interest rates and/or more insurance coverage. Consequently, a smaller portion of the investment goes into energy efficiency measures, which mean there is less savings. And the bottom line? There is less benefits the project can deliver. And, one way or another, its the host that picks up the tab! While the utility will get less verified savings and the ESCO ends up with less of a profit margin, the host gets less equipment and less energy savings for the same level of investment. Everybody loses, but particularly the host organization.

Building in all the specifications on the front end also narrows the host's options. Specifications dictate

---

1

Much of the material presented in this paper is drawn from the book, Performance Contracting for Energy and Environmental Systems by Shirley J. Hansen. The material is used with the permission of the publisher: The Fairmont Press, Lilburn, Georgia.

parameters. The innovative thinking from all the proposers is lost. All the savvy, all the expertise from ESCOs, whose business it is to know what works, is constrained by one set of specifications. In performance contracting, there is my "very old" adage: ***The more you ask for; the less you get.***

Perhaps even worse, the bid/spec then turns on dollar figures. ESCOs must be competitive. The price is forced down. This has the effect of funding just the short payback items. While accusations of "cream skimming" fly between the utilities and ESCOs, it's the poor host that is left with only a piece of its energy saving pie.

Bid/spec is reassuring to procurement people, in utilities and host organizations. If we are to protect the host's interests -- and assure greater benefits to the utility and ESCO -- these people must become convinced that the typical construction bid/spec is not to their advantage. This means convincing them that the uncertainty that surrounds a "smoke and mirrors" performance contract will give them the best deal. Not an easy task!

To protect the host, it is also important to resist the utilities' penchant to use "shared savings." The performance contracting business started as *shared savings*, an arrangement to split the energy cost savings at some pre-arranged percentage. This was an adequate procedure as long as energy prices were stable or going up. But as some defunct ESCOs (And the financiers that supported them!) can tell you, it doesn't work at all well when prices drop, as they did in 1986. After the shocks reverberated through the industry, *shared savings* as a financing mechanism was viewed as risky business by the financial houses. To repeat: "money follows risk." So in the late 1980s, interest rates for *shared savings* climbed.

When compared to guaranteed savings, *shared savings* requires that as much as one-third more goes into the cost of money; a major piece of the investment that does **not** go to buy equipment or generate savings. To make matters worse, *shared savings* deals may require the ESCO to maintain ownership of the equipment. While the ESCO can get some benefit from depreciating the equipment, tax-exempt institutions lose the advantages of tax-exempt financing and the not-for-profit host comes out on the short end of the stick -- again.

The worst part of this shared savings procedure, however, may be that many utilities ask ESCOs to guarantee cost savings for many years in the future. Talk about smoke and mirrors! In an area as volatile as guaranteed future energy prices with clear uncertainties in the Middle East and a myriad of other unknowns, the only way ESCOs can make such guarantees is to hedge their bets. Hedging bets means building in savings cushions. Being extra cautious!! Freely translated, that means less equipment investment, less benefit to the host -- often for the same dollar investment.

So if we go back to where we started and look at an "effective" performance contract, what is best for the host and indirectly the utility and the ESCO, we will rule out

~~bid/spec~~ and *~~shared savings~~.*

From the host's perspective, guaranteed savings performance contracts are more apt to meet its needs. The utility will also get greater verified savings and more likely better customer relations. The ESCO can offer a better package, save more energy and incur less risk for the same dollar figure.

The host is best protected if it first decides what it needs -- what results it wants. Then, get ESCOs best qualified to meet those needs.

It's Qualifications that Count

Absolutely key to protecting the host is recognizing that effective performance contracting turns on *QUALIFICATIONS*. Not engineering specs. Not bottom line bidding. Not shared savings.

Once this is internalized in the process, the rest of the process gets easier. To make it work, the host may need to go to outside technical, financial, legal and performance contracting support.

Step 1 for the host is to assess its internal expertise and to decide what outside services it needs to consider. A successfully implemented project requires:

Technical expertise - an energy engineer to evaluate proposers' technical capabilities. Not to write specs, but to evaluate the proposers and their test or sample audits.

Financial expertise - someone who can weigh financial options and understands net present value, discount rates and all that stuff. The financial person, or the performance contracting consultant, ought to be able to back out the numbers and tell the host organization exactly what it is buying -- what the host really receives for its avoided utility cost expenditures -- and the best financial structure for buying it.

Legal support - indispensable and hardest to come by -- someone who can walk the host organization through the legal intricacies and help negotiate the contract. Most attorneys are unfamiliar with performance contracting and reluctant to render an opinion. They become foot draggers because they hate to admit to the client that they don't know or understand what performance contracting is all about. If the host doesn't have an attorney with performance contracting expertise at hand, the next best thing is to groom one. Bring in an attorney who does know the business (chances are the host's attorney will only take another attorney's word for it) and let them talk legalese until a resident expert is born.

Performance contracting - a consultant can help the host avoid costly mistakes. A performance contracting consultant can even bring in the technical, financial and legal consultants the host may need. At the very least, an impartial third party consultant can help the host's managers overcome any internal performance contracting reluctance. This reluctance slows the sales cycle and can cut benefits for everyone.

Simply put, if the host organization does not have the built-in expertise, then it should go out and get it. Like the oil filter commercial, hosts can "pay now, or pay later." Later is more expensive. Fixing messes are costly. Horror stories abound. And horror stories travel like wild fire -- spreading fear that burns up future utility and ESCO opportunities.

A host that gets help early to decide what it needs and the results it wants and, then, obtains the needed help along the way to get the best firm to deliver those results, is a happy host. And happy hosts make projects work to the benefit of the utility and the ESCO.

HOST CONTRACT NEGOTIATIONS

So far we have looked at those aspects of performance contracting where the utility, the host and ESCO interact. The host must also look out for its own interests in negotiating a contract with an ESCO. Hosts will find some tough decisions must be made.

Energy service agreements **are negotiable**. Too many potential host administrators, when offered a printed contract, will view it as cast in bronze. With the wonderful age of word processors, any thing is changeable at the flick of a wrist. Owners and managers of host organizations need not, in fact should not, lose control of any facet of their operation that is key to comfort, productivity, safety or health. They can, and should, first determine the key elements the organization must have in the energy service and the energy financing contracts and the latitude within which they are willing to negotiate.

Understandably, an energy service provider has to have some assurance that it can protect its investment and can reasonably guarantee the savings. Contrary to fears engendered by some, this can be achieved without any negative impact on the internal environment. In fact the contract conditions can, and should, enhance the work environment.

Any firm that comes in with a "take-it-or-leave-it" contract and attitude is not a firm that will work with the host management to achieve the best results.

HOST DECISIONS

Criteria

First, the host needs to establish the criteria; then, prepare the request for qualifications (RFQ) or request for proposals (RFP) that reflects that criteria. The purpose of the RFQ/RFP procedure is to: (1) indicate the owner's interest in securing the services; (2) state the results the organization wants; (3) set forth the criteria for evaluating submitted proposals; and (4) provide sufficient facility, utility and finance information so an ESCO can assess its opportunities.

Neither the solicitation document nor the proposal should be considered all inclusive or binding. Items not in the solicitation or proposal can be placed on the table for discussion during negotiations. There is one caution, however, these items should not change the basic intent of the procedures as expressed in the RFP to the extent that ESCOs, which decided not to propose, feel they were mislead. This can lead to nasty accusations and lawsuits.

As an added protection, the solicitation might state that the host reserves the right to make the proposal part of the contract. This protects the host from getting overblown proposals. Interjecting this reality also helps set the stage for negotiations. All basic conditions, however, upon which the proposal was based should remain constant or the ESCO should not be expected to comply with this provision.

Modifications in what the organization asks for, or the firm proposes to do, are commonplace. Should the proposal, or parts of it by reference, become part of the contract, a statement should be included in the contract indicating that, in case of conflicting provisions, the contract prevails.

Contract Components

Topics generally addressed in energy service agreements are:

- ✍ equipment/building modifications and services to be provided;
- ✍ financial terms and conditions;
- ✍ user and ESCO responsibilities;
- ✍ savings guarantees offered; and
- ✍ the construction contract provisions.

These items may all be covered in one contract. Frequently, an energy service agreement, covering the general conditions, serves as a master agreement with separate schedules pertaining to work in a specific building, or a cluster of buildings, negotiated in phases of work.

Increasingly, however, two contracts are being used. A non-profit organization may enter directly into a installment/purchase agreement with the financier, so it can use its tax exempt status to obtain the equipment at a lower interest rate. In such cases, a parallel agreement with the ESCO can then be entered into to audit the facilities, install and maintain the equipment, provide other services and guarantee that the projected savings will cover required payments. The two contract procedure is frequently used today even when the host is not a non-profit.

Negotiation

A contract offered by a firm is designed to ensure that the ESCO's interest are protected. As in all contract negotiations, it is up to the customer to make sure <u>its</u> interests are protected as well. As previously stated, the host management should never get to the negotiations phase without first considering the implications of the various key components and the latitudes within which an item can be negotiated. In other words, decide what is not negotiable, what conditions can serve as "trading stock" and in what priority. Some key considerations are discussed below.

☞ <u>Equipment selection and installation</u>. The host should reserve approval rights on selected equipment; provided approval is not "unreasonably" withheld. ESCOs must retain some rights if they are to guarantee the savings. Under some bid procedures, the energy service company may take on the role of a construction manager; writing specs, monitoring bid procedures, and overseeing installation. These may be services the host organization needs, but they also serve to protect the ESCO's position and guarantees.

Contractual conditions the host organization would use in any construction project; i.e., liabilities, OSHA compliance, clean up, performance bonds, etc., should apply.

☞ <u>Equipment ownership</u>. The financing scheme used and the point at which the organization takes ownership can affect the organization's net financial benefit.

The useful life of the proposed equipment is a key factor in post-contract benefits. Statements from the manufacturer or some third party documentation is desirable. ESCO assertions are not enough. Without outside documentation, useful life provisions may need to be addressed in the contract.

ESCOs and/or their financiers usually insist on a first security interest in the installed equipment or collateral of equivalent value. Typically, the same principle as buying a car applies; you get the title when it's paid for. In performance contracting, however, many innovative procedures are being used.

In the case of buy-out provisions, termination and default values, the capitalized equipment cost should be established in the original contract. Terms, such as fair market value, need to be carefully defined.

Management must weigh the alternatives, make the calculations and comparisons and decide how much latitude the organization has for each option.

☞ <u>Maintenance</u>. The level of maintenance provided is key to savings persistence. A U.S. DOE study revealed that up to 80% of the savings in an effective energy program can be attributed to energy efficient operations and maintenance (O&M) practices. It is in everyone's interest for the ESCO to maintain the installed equipment. In fact, it will probably be desirable to have the ESCO maintain related energy consuming equipment as well. Or, at the very least, provide the host's O&M personnel training in energy efficient practices.

☞ **Malfunction and emergencies**. If the firm selected is not a local one, provisions for immediate, and back-up, service need to be spelled out. Local distributors for the selected equipment frequently serve this function with further back-up provided by the ESCO.

Maximum downtime needs to be considered. The allowable emergency response time will vary with the equipment installed and how essential it is to the host organization's operation.

☞ **Firm actions, damage**. Contracts proffered by ESCOS will discuss actions the host organization might take that could have a negative impact on savings, and remedies to protect the ESCO's interests. The host management needs to determine if these conditions are reasonable, and to similarly decide to what extent the organization should have the same type of protection. Should potential actions taken by the ESCO have a possible negative impact, redresses available to the host need to be incorporated in the contract to be sure ample protection is available.

☞ **Provisions for early termination**. From the host organization's point of view, contract language regarding termination should include removal provisions, including length of time required and a provision for restoration of the facility to its operable condition (at the time of contract origination or better).

ESCOs incur major exposure early in the contract, for they incur the major expenses at this time and must depend on eventual savings to cover these costs. Buyout provisions must provide for ESCO recovery of costs incurred and a proportionate profit.

For further protection, ESCOs (or their financiers) frequently specify that the organization cannot replace the equipment with equivalent equipment within a specified timeframe. If the contract carries non-appropriations language, this replacement provision usually becomes essential to obtaining the necessary capital.

☞ **Conditions beyond the control of the firm**. Usual contract language absolves the ESCO of certain contract responsibilities under specified conditions, which are beyond its control. These conditions should be examined to be sure they are reasonable and crystal clear. The merits of similar provisions for the organization should also be weighed.

☞ **Default language**. Language frequently limits the conditions of default for the ESCO, but may leave it wide open for the host organization. Since the financial burden is carried by the ESCO, this is not necessarily inappropriate. Similar language for the organization should be considered, however, especially if the host organization holds a separate debt service contract on the equipment.

☞ **Performance and payment bonds**. Depending on the assignment of risks, the host organization may need performance or payment bonds. The ESCO, for example, might conceivably become mired in circumstances where the host could become responsible for payments to subcontractors. In such cases, payment bonds should be part of the package.

If there is any question of the ESCO remaining solvent enough to cover any savings shortfalls, then a performance bond could be used to cover this contingency. Hosts, however, need to remember that costs to cover these risks come out of the project; meaning they ultimately come out of the savings; i.e., at some point, out of the host's pocket.

☞ **Indemnification**. Both the ESCO and the host organization should be indemnified from the actions of the other party and its agents.

☞ **Assignment**. Host management should insist on prior approval for any assignment changes of service responsibility or key personnel. Prior approval of subcontractors may also be desirable. The host should also expect to provide the ESCO some assurance of protection should the host organization change ownership.

☞ **Applicable law**. The ESCO typically presents a printed contract as the basis for agreement. This contract typically specifies the applicable laws of the state in which it is incorporated. Should court action be necessary, this places an additional burden on the host organization if it is located in a different state.

Since applicable law provisions may just as easily specify the host's state, this provision may become "trading stock" in the negotiations process.

☞ **Calculating the baseline**. A baseline that is the average of several years consumption will generally cover typical variations in usage. Provision for calculating baseline should consider particularly mild or severe weather in recent years. Recent changes in the structure, building function, occupancy, etc. should also be factored into establishing a baseline. Recent O&M work to reduce consumption or any recent renovation which affects energy consumption should also be considered. Re-open language should provide for some adjustment beyond an agreed upon variations in climate, occupancy, etc. This protects both parties.

☞ **Savings calculations formulas**. The reduction in units of fuel and electricity multiplied by the current cost of energy by unit is the standard procedure for calculating cost of saved energy. Attribution of demand charge savings also need to be negotiated and included.

Contract language that address formulas is frequently made far more complex than it needs to be. One contract had nearly thirty pages devoted to the treatment of degree days! Weather or occupancy changes, added computers, etc., can affect savings; however, extensive contract language trying to anticipate every contingency only benefits the legal profession. The simplest way is to have a broad-based baseline and then agree to re-open, or re-negotiate, changes of greater than ± "X" percent.

Changes in energy prices need to be considered. How is the burden of falling prices or the benefit of rising prices to be shared? If the ESCO insists on a price floor, then the host should enjoy the benefits of increased prices.

The share of the savings will vary with the length of payback, the services delivered, the length of the contract, financing scheme selected and the risks assigned to the ESCO, and the like. The interrelationship of these factors needs to be considered in negotiating the host's share of the savings.

☞ **Comfort standards**. The greatest fear employees, and often management, associate with performance contracting is the loss of control of the work environment. The "freezing in the dark syndrome" of the 1970s is still with us. The supposition that an energy service company will control the building operation is simply not warranted -- unless management abrogates its responsibility and gives the control away.

The organization can, and should, establish contractually acceptable comfort conditions; e.g., temperature parameters, lighting levels, air exchange, degree of building level control needed (and override required) to assure a quality environment and a productive workplace. The degree or latitude of control held by the organization can, of course, reduce savings. In such instances, the ESCO's risk becomes greater; so the host organization should expect some trade-offs -- the more control it holds, the smaller its share of the savings.

☞ **Projected compensation and guarantees**. The most attractive part of performance contracting is the idea that there is a third party out there who will make sure the host organization will have new capital equipment that works, and that the savings or ESCO will pay for it. Furthermore, the host can usually enjoy a positive cash flow without any capital cost to the organization.

The manner in which the energy savings are guaranteed to cover debt service payments is a key component of a contract and deserves careful consideration. Since the quality of maintenance on energy consuming equipment affects savings, most ESCOs require related maintenance contracts. They may, however, not guarantee that energy savings will cover the required maintenance fee. If the host organization regularly contracts for maintenance and the ESCO's fee is not greater than the existing maintenance contract fee, this may not pose a problem.

A major reason for a contract is to identify and assign risks and provide appropriate recompense. The *guarantees* are the bottom line in making sure a contract works to the host organization's favor. However, the greater the guarantees (and the risks shed); the lower the savings benefits will be.

The essence of performance contracting rests on the guarantees provided -- the risks assigned and accepted. Many of the host's most difficult decisions will rest on how much it is willing to pay for shedding those risks.

It is in the host's best interests -- and certainly the utility's and ESCO's -- that the respective risks, responsibilities and guarantees are fully understood by all parties at project inception. And that continued communication maintains those understandings. Such efforts on the part of the utility and the ESCO make for happy hosts. And happy hosts save energy, save money and save the environment. They reduce demand-side energy needs and make the SO_x , NO_x world a happier place for utilities -- and the rest of us.

Chapter 44

Energy Efficiency in Industry and Agriculture: A North Carolina Experience

R.N. Elliott

ABSTRACT

This paper looks at how energy is used in agriculture and industry in North Carolina based on fourteen years experience working on energy efficiency with both groups. Specific examples of energy saving technologies are presented along with a discussion of how these opportunities were identified, implemented and what benefits resulted. Suggestions are made on how to structure utility and government energy-efficiency programs to overcome market barriers.

INTRODUCTION

Objective

The objective of this paper is to share the lessons learned from working with North Carolina industry and agriculture on energy issues for fourteen years. This paper will attempt to provide a perspective of how energy is used by industry and agriculture in North Carolina, the opportunities for providing efficiency improvements together with other benefits, the barriers to realizing these improvements, and the strategies that will most effectively overcome these barriers.

Demography of North Carolina

North Carolina is a diverse state, geographically and economically. While high technology hubs, like the Research Triangle, exist as international centers for pharmaceutical and semiconductor research and manufacturing, agriculture still undergirds the state's economy. Energy use in the state is broken down by sector as shown in Table I.

Table I
1989 END USE ENERGY CONSUMPTION BY SECTOR IN N.C.

	Electric (Quads)	Fossil (Quads)	Total (Quads)
Residential	0.130	0.109	0.239
Commercial	0.076	0.058	0.134
Industrial	0.101	0.325	0.426
Transport	0.0	0.551	0.551
Total	0.307	1.043	1.350

Source: North Carolina Energy Policy Council, 1992

Agriculture is not broken out separately, but is included in all of the above categories. On-farm agriculture does not directly consume large amounts of energy. Transportation fuel represents the largest component. Most on-farm electricity is not separated out, but is billed as residential energy use except for a few large farms that are billed as commercial customers. Agricultural electric consumption is about 2 percent of the state total.[1] Farms do use many products with high embedded energy content such as water, feed, fertilizers and agricultural chemicals. Unique opportunities exist for changing the energy use patterns on the farm with dramatic co-benefits such as improved product quality and reduced losses, along with significant energy savings through better utilization of high energy content products.

Table II
1983 N.C. ENERGY USE BY MAJOR INDUSTRIAL GROUPS

SIC	Groups	Mfg. Jobs (%)	Electric (GWh)	Fossil (Quads)	Total (Quads)	% Total
20	Food	5	813	.011	0.014	4.8
21	Tobacco	3	805	.007	0.010	3.4
22	Textiles	31	7933	.050	0.077	26.5
24	Wood Prod.	4	397	.003	0.004	1.4
25	Furniture	10	721	.004	0.007	2.4
26	Paper	2	2622	.055	0.064	22.0
28	Chemicals	4	3259	.041	0.052	17.9
32	Stone, Clay, Glass	2	927	.028	0.031	10.7
	Others	38	4614	.016	0.032	11.0
	All Industry		22150	.216	0.291	

Source: Hinkle, 1983.

The food processing industries (SIC 20) are significant energy users, being the fifth largest industrial consumer as shown in Table II. This table is based on 1983 data, which is the most recent breakdown of energy use by SIC code. Tobacco processing, which is the second most important agricultural commodity in the state after poultry, accounts for an additional 3.4 percent of the industrial energy use.

The textile industry (SIC 22) is the leading manufacturing industry in North Carolina accounting for almost a third of the manufacturing jobs and over a quarter of the energy consumption as shown in Table II. To appreciate the true importance of the textile industry to the state, it needs to be combined with the allied synthetic fibers (SIC 2823 and 2824) and apparel industries (SIC 23). Combined, these industries account for over a third of the state's industrial energy use. The paper, chemical, and stone, clay and glass industries together account for only 8 percent of manufacturing jobs in N.C., but these industries are very energy intensive, accounting for over half of the total industrial energy use. Most of the energy consumed by the paper industry is at a few very large pulp mills that generate a significant portion of their energy from process waste.

OPPORTUNITIES FOR IMPROVED ENERGY EFFICIENCY

Energy use in industry and agriculture can be visualized as two groupings: "utility" energy and "process" energy. The utility energy uses, such as motors, lighting, compressed air and steam, can be changed without directly affecting processes and can yield significant energy savings. The Congressional Office of Technology Assessment describes "utility" energy efficiency measures as "house keeping, maintenance and equipment changes." Utility energy-efficiency measures are often very cost effective. Replacing a motor at failure with a high efficiency motor can have a payback of less than a year on energy savings alone, and with proper planning can provide additional savings from reduced down time as will be discussed in a later example.

Many sources agree that the greatest potential for energy savings lies in process modifications.[2,3,4] As an example, the energy savings potential of converting from a standard and high efficiency motors is on the order of 5 percent[5], while the savings potential from a change in the process, as a later example demonstrates, is on the order of 50 percent. Process modifications, however, confront the significant barrier of requiring fundamental changes to the primary activity of the facility: production. Thus it may be much easier to implement energy efficiency improvements in the utility area as compared to the process area.

Utility Uses

Utility energy represents generic forms of energy with well-defined parameters. Table III presents many utility energy forms with their relevant parameters. As long as these parameters are matched, different sources for the energy can be readily substituted without affecting the process.

The list in Table III is subdivided into two categories: energy carriers, and intermediate and end uses. Energy carriers are media that transport energy to a process, like electricity. They have well defined and straightforward parameters that are generally well understood. For example, chilled water is characterized by its temperature and pressure. It can be produced by various types of vapor compression chillers, absorption chillers, or from the melting ice. In some facilities, there may be several kinds of chillers operating together, as well as perhaps an ice storage system, all linked to a common distribution system. The water becomes an energy carrier much as the electrons in the electric grid.

Table III
Forms of Utility Energy

Utility Energy Forms	Parameters
Energy Carriers	
Electricity	Voltage, frequency and number of phases
Steam	Pressure and quality/superheat
Compressed air	Pressure, temperature and cleanliness
Chilled water	Temperature and pressure
Intermediate and End Uses	
Shaft power	Power, torque, speed and load-speed curve
Illumination	Intensity, distribution, color temperature and color rendering

The two remaining energy uses present more complex parameters that are more difficult to match, thus making substitutions more complex. In addition, many users may not be familiar with all the parameters and how changes in these parameters can adversely effect the production process.

Shaft power might at first glance appear a straightforward matter of matching motor size and speed. Other important considerations can include the actual load/speed and torque/speed curves, which are continuous functions rather than single values. Depending upon the application, these values may not be too critical, so any drive source can be used. The sources of shaft power can vary from different types of electric motors, to turbines, to internal combustion engines. It is common to see turbines and electric motors on the same shaft in critical applications at boiler plants. The turbines can be used to maintain boiler operations in the event of an electric failure. Some textile plants in North Carolina still use central power shafts that were originally driven by water turbines, now powered by electric motors.

For speed-sensitive applications like conveyers, fans and pumps, the substitution can become more complex. In the case of centrifugal loads, like fans, pumps and chillers, the magnitude of the load is very sensitive to motor speed since the load varies as the cube of the speed.[6] If we add the ability to vary the shaft speed, we then move the drive application into the area of a process energy use since changes in speed can have profound impacts on process, as will be discussed in an example in the next section.

Increasing the energy efficiency of lighting in agriculture and industry may appear to be a simple matter of substituting higher efficacy lighting systems for lower efficacy systems (i.e., combinations of lamps, ballasts and fixtures). But many applications require an experienced lighting designer to achieve the best results from both visual performance and energy efficiency standpoints. The more complex parameters of color temperature, color rendering, and restrike characteristics of the different light sources need to be considered.

In applications where the color of light is critical, the selection of different light sources can prove complex. In these cases, the first test is that the illumination is at least as good. If the color requirements are well understood, the wide variety of lighting sources available today can offer the potential for significant improvements in energy efficiency along with improved defect detection in manufacturing operations or improved productivity in agricultural production operations like greenhouses. Sometimes the energy savings come not from more efficient lamps, but from a lighting design that provides appropriate illumination of the task without wasting light on illuminating spaces like the tops of machines.

Task Lighting in Textile Spinning

An example of appropriate illumination is the lighting of textile, open-end spinning frames. These are long, narrow machines that are located in rows with a narrow isle between them for worker access. Traditionally, spinning rooms are lit with eight-foot fluorescent strips on the ceilings. An automated doffing rail, at the top of the machine, shadows the visually critical area of the spinnerets that is located under the rail. By placing a light source over the isles that directs light into the spinneret area, the total wattage of lamps in the room can be reduced. This approach has been taken by a South Carolina lighting firm, Mor-Lite, which uses a custom-formed fluorescent reflector fixture mounted over the isle to direct light onto the task. At one mill, two-lamp hooded high-output fluorescent fixtures were replaced with custom single lamp reflector fixtures. Lighting level on the task increased from 40 to over 70 foot-candles, while reducing the lighting energy consumption by 38 percent.[7]

Poultry Lighting

Another example comes from the poultry industry, where artificial lighting has been used in its chicken houses beginning in the 1950s. The N.C. Alternative Energy Corporation (AEC) undertook a project, in partnership with the electric utilities and the cooperative extension service, to promote the use of high efficiency lighting technologies for lighting poultry houses. Previously farmers used incandescent lamps because they were the only readily available source of artificial lighting. Conventional fluorescent lamps did not withstand the harsh environment (i.e., heat and corrosive gases) found in poultry houses. In the mid-1980s, compact fluorescent and high intensity discharge (HID) lamps became available. These lamps offered improved efficiency and lower operating costs, and could be obtained with sealed ballasts which could withstand the high temperatures and corrosive environment. The payback from energy savings for the conversion was very short as shown in Table IV with additional labor savings coming from the longer lamp life. In addition, a range of light colors were available.

While it is easy to substitute these higher efficiency lamps in the houses and dramatically reduce power consumption for an equivalent light level, the mixture of light spectra can have profound effects on how plants and animals behave. In the poultry industry these behaviors include aggressiveness, weight gain, egg production, egg quality and fertility. No information was available about what mixture of light spectra is optimal for the different sectors of poultry production: table egg production, fertile egg production (breeding), and meat production. Initial field trials, conducted jointly by AEC, the electric utilities and N.C. State University in cooperation with major poultry companies, showed that 2700K compact fluorescent lamps, which have a warm color with an appearance close to incandescent lamps, performed at least as well as incandescent lamps for table egg and meat production, but not for breeding. Parallel tests with 4100K lamps, which have a bluer color showed reduced rates of feed conversion and egg production making them inappropriate for poultry use.

Table IV
COMPACT FLUORESCENT POULTRY HOUSE CONVERSION PAYBACK (YRS)

Cost/ Fixture	Electrical Cost ($/kWh)			
($)	$0.04	$0.06	$0.08	$0.10
8	0.4	0.3	0.3	0.2
12	0.6	0.5	0.4	0.3
16	0.8	0.6	0.5	0.5
20	1.0	0.8	0.7	0.6
24	1.3	1.0	0.8	0.7

Source: AEC, 1987.

More basic research was required to look at light requirements for breeding. Initial results of research at N.C. State University indicate that potential exists for productivity, as well as energy, improvements by switching from incandescent to high pressure sodium lamps in breeder operations.[8]

Providing adequate technical information on all aspects of measures with potential for improved energy efficiency is a necessary but not sufficient condition for changing real world practices. Other factors such as interest in energy costs, financial incentives, promotion and market timing affect the degree to which viable efficiency measures get adopted. The initial trial results concerning the viability of converting to compact fluorescent lamps became available at an opportune time for the poultry industry. The industry was undergoing a major expansion with a large volume of new housing being constructed. The vertical structure of the industry was such that the poultry company, referred to as the integrator, was responsible for the chick, feed, processing and technical support costs, leaving only labor, energy and the capital cost of housing to the farmer. Since energy was about half of the farmer's operating costs, and lighting costs represented a third of his energy bills, the impact of improving energy efficiency on the farmer's bottom line was significant.

The extension service with support from AEC, the state's electric utilities and lighting vendors, undertook an aggressive promotional effort involving meetings with farmers arranged by the poultry companies. The integrators, who could block efforts, supported the program by publicizing the meeting to their growers, providing technical support, and in some cases arranging group purchases of lamps to reduce the farmer's costs. Compact fluorescent lamps made a 40 percent penetration into the table egg and chicken meat production sectors of the market within three years, before an industry wide recession slowed market implementation. The technology, however, had achieved market acceptance with much of the new construction installing compact fluorescent lamps.[9]

Process Energy

Process modifications can be approached at several levels. At the most superficial level, the process modification may only involve replacing one energy source with another as in replacing gas-fired infrared drying with electric infrared drying. This approach tends to provide only limited process improvement and energy savings.

If a technology substitution is considered along with an analysis of the actual process energy requirements, more substantial amounts of energy can often be saved. For example, several textile plants in the southeast that have considered installing industrial heat pumps have analyzed their process heat requirements only to discover that they can simply lower steam pressure or water temperature and realize significant energy savings without modifying the process.[10]

A more rigorous approach would be to analyze the particular process, or in many instances the system of processes, with respect to how best the production goals can be realized. Through reengineering the process, taking advantages of the latest technological innovations, the potential frequently exists for dramatic improvements in productivity, product quality, and energy intensity. Providing multiple benefits is a critical strategy for interesting a plant's management in modifying a functioning process. On the other hand, these potential benefits come at the cost of significant amounts of time, money and increased risks from unforeseen impacts of what may initially appear to be only minor changes in the production process.

Process modifications can be grouped by technology area: thermal, chemical and mechanical. In many cases with more significant modifications of the process, there may be at least a partial substitution of one type of energy process for another.

Radio Frequency Drying of Textiles

Thermal processes are used extensively in many industries. All three forms of heat transfer - conduction, convection, and radiation - are used, in many cases in combination. A thorough assessment and understanding of an individual process or systems of processes will often reveal opportunities to reduce or eliminate thermal inputs to the process. By selecting a process temperature as close to ambient as possible to reduce heat loss, an initial level of improvement can be achieved. This change will reduce the driving force of the heat transfer and increase the benefit of any insulation.

Convective heat transfer processes are inherently inefficient because they require the heating of a transfer medium that in turn heats the process material. This problem is particularly evident in forced air drying processes.

Drying is the largest energy end use in textile dyeing and finishing operations. Conventional dryers operate by blowing heated air through the wet product to evaporate the moisture and transport it away. To accomplish this, both the moisture and the product must be heated. This wastes the energy required to heat the product, which must be dissipated when the drying is completed, and results in thermal degradation of the fibers due to the heating of the already dry surface in order to dry the interior of the product. Radio frequency (RF) drying (a form of dielectric heating) selectively heats the water molecules by exciting them in an oscillating electromagnetic field in the same manner as a microwave oven. As a result, the drying is accomplished faster, more uniformly and with less heating of the fibers. In addition, the energy cost per pound is less than for the most efficient conventional dryers, even though the RF dryer uses more expensive electric energy compared to the conventional dryer which uses steam.[11]

Industry's motivation for investing in these dryers is not necessarily the energy savings but the other previously stated benefits. A dramatic example is a North Carolina firm that installed an RF unit to dry cashmere.[12] Cashmere is a very expensive fiber costing as much as $35 per pound in its raw state and over $70 when processed. Over-dying of cashmere during processing damages some fiber resulting in a loss of the product. By using an RF dryer the temperature of the fibers never exceeds 140°F, and the plant is able to increase their yield by 2 percent resulting in a savings of $1.40 per pound processed. This compares to an average energy savings of 4.8¢ per pound due to the switch from conventional to RF drying. The firm in this case justified the purchase on product quality improvements alone.

Ultrasound Enhanced Textile Dyeing

Sometimes the task performed by the thermal energy can be accomplished by the substitution of other energy, such as chemical or mechanical. An example is the application of ultrasound energy to the textile dyeing process. The N.C. State University College of Textiles is developing a new process in which ultrasound energy (sound waves) replaces some of the thermal energy (in the form of steam, hot water and direct fired gas) and some of the chemical energy (in the form of catalysts and surfactants) used to wet-out fiber, apply dye stuffs, fix reactive dyes, and wash-off excess dye stuffs.[13] Not only does this process promise to use less energy with reductions in:

- hot water volumes,
- the temperature of the process, and
- the volume of chemicals required by the process,

but also offers several co-benefits including:

- reduction of both the volume and temperature of the waste water stream,
- reduction the cost of chemicals required by the process,
- reduction of processing time,
- improvement in product quality, and
- allowing for more accurate shade control.

These co-benefits, as with RF drying, offer greater economic benefits than the energy savings alone.

Variable Air Flow in Lumber Dry Kilns

In N.C. industries, mechanical energy is used predominantly for manufacturing machines, such as loom, saws and cigarette makers, and for moving fluids using pumps, compressors or fans. Other than the option of changing the motor, machines are very difficult to modify because of the complexity of the operation. In the case of fluid flow opportunities do exist to save energy by accurately matching flow to process requirements. The advent of reliable and affordable variable speed drives allows many of these opportunities to be realized. The difficulty exists in gaining an understanding of what the process requirements are, how to sense an indicator of the requirement, and how to control the flow.

The AEC, in cooperation with the furniture industry and the state's electric utilities, undertook a project to demonstrate the potential of controlling airflow in hardwood lumber dry kilns. The first step was to understand the physics of the lumber drying process. The process involves three distinct phases: release of the mechanical or chemical bonds that hold the water in the wood; transport of the water to the surface; and transport of the water away from the surface by air flow. In furniture grade lumber, especially some hardwoods like oak, it is important that the moisture gradients within the wood be limited to minimize drying stresses that result in degrading of lumber quality. Since the first two phases of the drying process are driven by the wood temperature and the relative humidity at the surface, the amount of water requiring transport away from the surface will vary during the drying process. In conventional operations, the maximum air flow required is estimated and the requisite fan capacity is installed (usually with some excess for good measure). The drying rate is controlled by regulating the kiln temperature and recirculating air to control relative humidity. This means that the fan capacity installed is required for only a short period for the drying cycle.

An AEC study indicated that the required rate of air flow could be determined by measuring the wet bulb depression in the air immediately after it had passed through the lumber stack. Using this reading as an indicator, a control strategy was developed by AEC to vary the fan speed. Field trials with two industrial cooperators confirmed that varying the air flow resulted in at least as good a lumber quality with no effect on production rates. Though tests were not conclusive, the operators felt that lumber quality was better with the variable air, particularly on difficult woods to dry like oak. The total energy required for a kiln charge was reduced by about half as shown in Figure 1. Paybacks for the fan control systems vary from 2.4 to 8.7 years depending upon motor size and kiln configuration.[14]

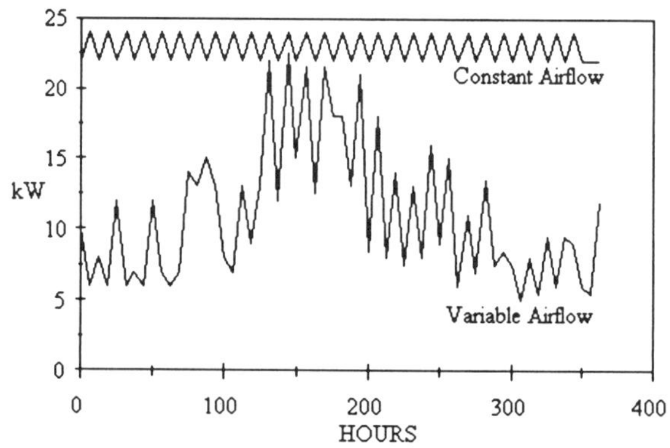

Figure 1. Fan Energy Comparison for Constant and Variable Speed Dry Kiln Trials.
Source: IEL, June 1992

The AEC, with its partners including the electric utilities and the Southern Dry Kiln Club, is now letting the hardwood lumber industry know about this technology. One of the demonstration partners is considering implementation of the technology company wide. A side benefit of the educational effort has been the opportunity to inform the hardwood lumber industry about electric motor selection and management.

THE DECISION MAKING PROCESS

Much has been made of the barriers to energy efficiency in industry. While competition for capital is frequently a problem,[15] some improvements can be realized with little if any additional capital expenditure. Often, it is competition for the attention and time of the work force that limits the improvements carried out. Industry often operates in a crisis mode, with whatever has the most immediate impact on the operations of a plant receiving staff attention. Anything that affects the ability of the facility to produce and meet customer requirements receives first priority. Next come issues that affect whether the plant will stay open. These lately have involved compliance with environmental regulations. After these issues are addressed, if any time remains, most plants look to preventing operating problems from occurring through preventive maintenance. It is frequently suggested that energy-efficiency measures can be implemented during annually scheduled plant shutdowns to prevent disruptions in the manufacturing process. During this period, the competition for time becomes even more intense since the engineering and maintenance staff usually are working overtime to install new equipment and do maintenance that cannot be done during normal operations. Discretionary tasks, like energy-efficiency improvements, are usually deferred.

Plant engineering and maintenance staff have generally been supportive of energy-efficiency improvements. But changes with both energy-efficiency and productivity benefits can be stymied by a purchasing system in many companies that operates on a lowest first-cost basis. With a focus on short-term results, life-cycle operating costs are frequently ignored. Carolina Power & Light's audit program has showed that opportunities with greater than a two-year payback are unlikely to be carried out unless there is a unique situation where the change is necessary for the customer's survival.[16]

One cause is the complexity of the energy decision-making process. Energy decisions are not made independently, but rather as a part of the operation of the plant system as a whole. Changes requiring process modifications may have to pass many reviews before receiving final approval. Many larger corporations have an established hierarchy for decisions starting at the plant level and then moving up to the division and corporate level. In many plants, facility operations are separated from the process operations. Depending upon the company, these parallel tracks may extend to the corporate level.

During the 1970s it was common to find an individual or entire department devoted to managing energy use and coordinating between departments. The decline in energy prices and the financial problems most manufacturing companies experienced during the 1980s caused many of these individuals or departments to disappear. The energy-efficiency advocacy role, when it remains, has been assumed by the engineering and maintenance staffs. A case in point in North Carolina is Burlington Industries, which used to have a department devoted to energy conservation and management. It served as an intermediary between corporate engineering and purchasing, and provided support to the plants on energy issues. This department shrank in size as the 1980s progressed, eventually disappearing by the end of the decade, with its role being assumed in part by the engineering department.[17]

Along with the loss of energy professionals, the research and development resources at many companies have shrunk or disappeared. For example, Burlington Industries closed their major research facility, losing their ability to develop and evaluate production modifications. R&D departments provide the technical expertise and dedicated staff necessary for the implementation of many process improvements. They identify the energy savings available through process modifications, work with plant personnel to design the process modifications, and provide the support to the plant necessary to carry out these changes. Of course the ramifications of these cuts extend beyond energy efficiency to other process advancements.

At the plant level, cuts have also been made in engineering and maintenance staff. Sometimes, the plant staff is so overworked that they can't even identify and implement utility energy improvements. Since many process decisions are not made at the plant level, plant staffs have limited input to improving the efficiency of the process, even if they have the technical expertise to make such recommendations. With the cutbacks in corporate staff, a void now exists that is only partially being filled by external groups such as the university extension services and utility technical staff.

While some utility and governmental energy-efficiency programs directed toward industrial customers have been successful[18], it is not easy for outside parties to play a useful role in influencing energy decisions within industry. These external parties often lack familiarity with the particular plant operations and are unable to stay with a modification from identification through design and implementation. Few, if any, changes go as anticipated. There is a shakedown period in which the modifications are fine-tuned to work within the plant system. It is difficult for a utility, industrial extension service or state agency to stay involved through all stages of this process.

OVERCOMING BARRIERS TO IMPROVED ENERGY EFFICIENCY

Appealing to the Bottom Line

An important concept that strikes a resonant cord with many industrial managers is the idea that "wasted" energy is lost profit, and that to make up for the loss, additional product must be sold. As the textile industry has discovered over the last fifteen years, often there is no market for that additional product. Opportunities for "utility" energy savings are frequently appealing in this regard. The savings are frequently easily documented, and the risks are low. Lighting in warehouses and outside buildings are particularly good candidates since the savings can be substantial and the paybacks are short. Motor projects can be more difficult because the improvements are small and may not be readily detectable.

Some inventive plant personnel have worked around the system to carry out energy-efficiency measures. At one textile plant, the maintenance supervisor wanted to purchase new, high-efficiency motors to replace existing standard motors when they failed, rather than repair them. Often the cost of overtime motor repairs equaled the cost of the new motors, but purchasing regulations prohibit the purchase and require a lengthy approval process for the new motor, while a standing purchase order exists with the rewinder. The maintenance supervisor started listing motors as replacement parts by catalog number and calling them "a drive assembly" rather than going through the normal motor procurement process. In this way he circumvented the lengthy procurement process and got the high-efficiency motors he wanted.

Securing Corporate Commitment

While plant staff may be interested, even enthusiastic about energy efficiency, practical demands and the conditions outlined above often don't allow most of them to advocate changes or work around the system. For those who do advocate change, the reality is that several years may be required to realize the fruits of their labors.

If corporate management becomes committed at the highest levels to improved energy efficiency, hurdles to energy efficiency can be eliminated. At one textile plant, corporate management was convinced that motors were important. They empowered staff to work with their motor suppliers to arrange to have high-efficiency replacement motors to be delivered within an hour, twenty-four hours a day. In contrast, an overtime rewind required a minimum of one day. The savings in reduced down time more than made up for the cost differential of buying a new, high-efficiency motor that was also more reliable and energy efficient than a rewound motor.[19]

Demonstrating Multiple Benefits

As mentioned in the previous section, energy decisions are not made independently but in the context of overall plant operations. To receive approval of a major process modification, it is important that it also offer benefits with regard to at least one priority issue for the company. For example, a consideration in the installation of the wastewater, heat-recovery heat pump installed at American and Efird's dyeing and finishing facility in Mount Holly, NC was the reduction in wastewater discharge temperature allowing compliance with municipal ordinances.[20]

Similarly, industry's initial interest in technologies like RF drying and ultra-sound enhance dyeing discussed earlier come not from the potential energy savings but from process enhancements. While the example of textile machinery lighting was in large part justified by the energy savings, a less easily quantified benefit of improved worker visual performance was present. Even in the case of poultry house lighting, the industry was very aware of the potential for controlling bird performance through the use of a more flexible light source.

Providing Technical Assistance

Since one barrier to the implementation of energy efficiency is lack of staff time, making technical assistance available may help address this issue. This approach can be particularly effective when the technical expertise being offered is not readily available in the marketplace. North Carolina has been a national leader in this activity. One early example was the Wood Assistance Team effort that started in 1978. The goal of the program was to promote the use of wood fuels in non-wood products industries. A need was identified for three areas of technical assistance: wood fuel combustion and handling, wood resource procurement, and environmental permitting. N.C. State University, the N.C. Forest Service and the N.C. Division of Environmental Management each contributed expertise to a team that worked with potential candidates to assist them in evaluating and implementing wood fueled systems.[21] The program was very successful in convincing 35 percent of the firms for whom case studies were prepared to install wood boilers. Even after the fuel price advantages of wood were erased in the early 1980s by drops in the price of coal and gas, the program continued a reduced level until 1986. The Extension Wood Products Department at N.C. State University continues to provide limited support today and continues to hear from companies who converted and are still happily using wood fuels.[22]

AEC has used the same team model for many of its technology dissemination projects. The poultry lighting project mentioned earlier brought together expertise in poultry science, lighting, and engineering to provide the assistance and support necessary for farmers to convert to fluorescent lamps. The project was delivered in cooperation with the Cooperative Extension Service who offered the most extensive network of contact with and were viewed as a credible source of information by poultry farmers.

The Industrial Electrotechnology Laboratory (IEL) is AEC's most ambitious technical assistance effort. Its goal to establish a physical location where technical assistance is available to industry in the application of energy-efficient technologies such as RF drying and high-efficiency electric motors. AEC joined with N.C. State University's College of Textiles to make state-of-the-art, production class facilities available to assist industry with initial investigation, through process development on through startup trouble shooting. IEL pooled research expertise from the university, with energy expertise from AEC and electric utilities, and hired staff from industry familiar with the needs manufacturing companies. A partnership was formed with the electric utilities. The utility customer service representatives served as the contact between the customer and IEL. No other group in the state has more frequent or substantive contacts with manufacturing plants than the utility staff. In addition, the utility staffs are technically oriented with many being registered engineers already familiar with energy issues.[23]

The IEL opened in April 1991 with a industrial lighting demonstration, RF dryer, vacuum microwave dryer and infrared dryer. Since then additional lighting demonstrations and RF dryers, and a motor test facility have been added. As of April 1993, over 200 trials had been run for customers in North Carolina, South Carolina and Virginia.[24]

One common characteristic of all these projects is an effort to develop an understanding of the particular issues facing the target audience and an attempt to identify benefits of the technologies, in addition to energy, which are important to the industry. Technical assistance is likely to be more successful if it makes use of an intermediary who is already in regular contact and has credibility with the target audience.

CONCLUSIONS

Experience in N.C. demonstrates that while the impacts that energy issues have on economic viability, security and environmental compliance are important, energy by itself is not important. Energy savings by itself will not usually motivate industry or agriculture to change its energy use patterns. Projects that produce economic, productivity or environmental co-benefits are more likely to capture industries' or farmers' attention, especially if the project addresses a current, pressing need.

If one is to influence the energy use of a group, it is important to understand how the industry functions, their markets, their diversity and the critical issues affecting their future. Expertise with time to devote to developing and implementing energy-efficiency projects is a commodity in short supply in industry. If industry sees a state or utility program as offering greater benefits to the company than the cost of participation (usually measured in time) then the program will likely receive a fair hearing.

The greatest potential for savings exists with changes in the process. These modifications frequently offer co-benefits whose value can exceed the energy savings. Projects involving process changes can be complex, costly and carry a high risk of adverse affects on the product. As a result, these project can frequently involve a lengthy and difficult corporate approval process.

While the elimination of waste and substitution of more efficient technologies in utility energy uses yields only incremental energy savings, these projects are usually low risk and frequently very cost effective. These projects can also be easier to implement since the approval process can be less complex.

It is important to understand the difference between utility and process changes. While changes to utility uses have little effect on a company's product, process changes can adversely affect a firm's reason for existence. To appreciate this difference requires an understanding of the company's process. Energy-efficiency recommendations made without this understanding can encounter unanticipated barriers and affect credibility.

In view of the how energy decisions are made in industry, a successful state or utility program to increase energy efficiency in industry should embody the following characteristics:

- consider working through groups who already have established credibility and channels of contact with the target audience
- build commitment to energy efficiency at all levels in a company. Plant staff are frequently receptive, and with support from top management, can become enthusiastic partners
- program staff should understand the product, processes, structure and current condition of the target audience
- make initial recommendations for utility energy changes since these projects can be cost effective and can more easily win approval than process changes
- focus on projects that offers benefits in addition to increased energy efficiency
- provide technical expertise and support throughout the decision-making process, from opportunity identification to final shakedown
- programs should persist for several years because of the long lead times often required to gain approval of energy-efficiency projects in industry.

REFERENCES

1. Gregory, Joe, personal communications, Carolina Power & Light Company, Raleigh, NC, 1986-90.
2. Ross, Marc H. and Daniel Steinmeyer, "Energy for Industry", *Scientific American*, Vol. 263, No. 3, pages 89-98, September 1990.
3. Office of Technology Assessment, *Industrial Energy Use*, OTA-E-198, U.S. Congress, Washington, DC, June 1983.
4. Office of Technology Assessment, *Industrial Energy Efficiency*, U.S. Congress, Washington, DC, April 1993.

5. Anon., *ENERGY SAVERS: Energy-Efficient Motors*, N.C. Alternative Energy Corporation, RTP, NC 27709, November 1990.

6. Anon., *Industrial Motor Selection Database*, pp ii-iii, Industrial Electrotechnology Laboratory, Raleigh, NC 27695-8301, 1992.

7. Hubner, John, personnel communication, Mor-Lite, Inc. Greenville, SC, 1993.

8. Siops, Thomas, *Spectral Effects of Light on Turkey Reproduction*, AEC Final Report, N.C. Alternative Energy Corp., RTP, NC, 1992.

9. anon., Renew America award application, N.C. Alternative Energy Corp., RTP, NC, 1989.

10. Puciano, Frank, personal communication, Georgia Power Company, 1991.

11. Dawson, C. H., "Economic Aspects of Yarn Package Drying," *Global Trend in Dyeing Science and Technology*, AATAC Conference Book of Papers, pp. 69-74, June 7-8, 1989.

12. Cato, M. J., et.al., *Handbook of Radio Frequency Drying of Textiles*, N.C. Alt. Energy Corp., RTP, NC, June 1, 1991.

13. Smith, Brent and Timothy Clapp, personal communications, College of Textiles, N.C. State Univ., Raleigh, NC 27695, 1989-1992.

14. Anon., *ENERGY SAVERS: Variable Airflow Control to Cut Lumber Drying Costs*, Industrial Electrotechnology Laboratory, Raleigh, NC 27695-8301, June 1992.

15. The Alliance to Save Energy, *Industrial Investment in Energy Efficiency: Opportunities, Management Practices, and Tax Incentives*, Washington DC, 1983.

16. Jordan, J. A. and S. M. Nadel, *Industrial Demand-Side Management Programs: What's Happened, What Works, What's Needed*, American Council for an Energy-Efficient Economy, Washington, DC, March 1993.

17. Lohr, A.C., personal communications, High Point, N.C., 1988-1992.

18. Jordan, J. A. and S. M. Nadel, March 1993.

19. Morgan, Charles, personal communication, Stowe-Pharr Mills, Gastonia, NC, 1990.

20. Batson, Barry, personnel communications, Duke Power Company, Charlotte, NC, 1989.

21. Elliott, R. N., *Wood Assistance Team: Final Report*, Industrial Extension Service, N.C. State Univ., Raleigh, 1981.

22. Jahn, Larry, personal communication, Extension Wood Products, N.C. State University, Raleigh, NC, 1993.

23. anon., *Annual Report*, Industrial Electrotechnology Laboratory, Raleigh, NC 27695-8301, Jan. 1992.

24. Dominic Dirisio, *1992 IEL Annual Report to Extension and Applied Research Area*, Industrial Electrotechnology Laboratory, N.C. State University, Raleigh, NC, April 1993.

Chapter 45

A Facility Monitoring System: The Single Most Valuable and Cost-Effective Tool Available to an Energy Manager

W.A. Holmes

ABSTRACT

Energy engineering and management combines engineering problem-solving and financial management techniques to reduce utility costs. At present, substantial amounts of time and money are being spent on audits, studies, portable instrumentation, calculations and modeling in order to attempt to quantify energy consumption and costs and define opportunities for savings, normally involving capital expenditures. Unfortunately, accurate verification of results is often overlooked. Advances in technology during the last few years have made the installation of a permanent, PC-based monitoring system possible for any facility, often for no more than the cost of a detailed study. By investing initially in a monitoring system rather than audits or studies, the actual consumption and cost data will be available on a continuing basis and can be used to produce immediate operational savings, more accurately analyze opportunities requiring capital investments, and to verify actual savings resulting from changes.

A permanent monitoring system, installed as the first step in a utility cost reduction effort, to identify where and how energy is used in a facility on a dynamic and real-time basis, can provide the most valuable and cost-effective tool available to an energy manager. The resulting data allows energy consumption patterns and utility costs to be understood and managed in the same manner as all other costs within a facility. It allows energy decisions to be information-based rather than product-driven and savings from investments in energy improvements to be verified and maintained. Holmes Energy Services has been using this approach since 1979 and has produced savings ranging from 15 to more than 50%, often with no capital investment by the owner. This presentation will use actual monitored data from case study projects to demonstrate the approach and results.

INTRODUCTION

Holmes Energy Services has provided energy engineering and management services to facilities ranging from schools to shopping malls to meat packing plants to foundries. Working for the first several years under shared-savings contracts with no up-front investment by the owner, an approach was evolved to produce savings as quickly as possible with the lowest initial cost on our part, maintain them at the highest practicable level, and provide verification and documentation on an on-going basis.

The key element and the first step in every project, is the installation of a PC-based monitoring system. Because the system was initially installed, operated and maintained at our expense, we designed and built a low cost, reliable and easy to use system which utilized off-the-shelf components wherever possble. Advances in hardware and software continue to increase capabilities while decreasing costs.

This approach has allowed us to produce exceptional savings in existing buildings with existing equipment. It has also allowed us to accurately assess opportunities for energy projects requiring capital expenditures and document the actual savings. The following description of our methods is provided.

FACILITY SELECTION

- Meet with CEO, determine interest in reducing utility costs.

- Review utility history, determine patterns, profiles and total annual utility costs.

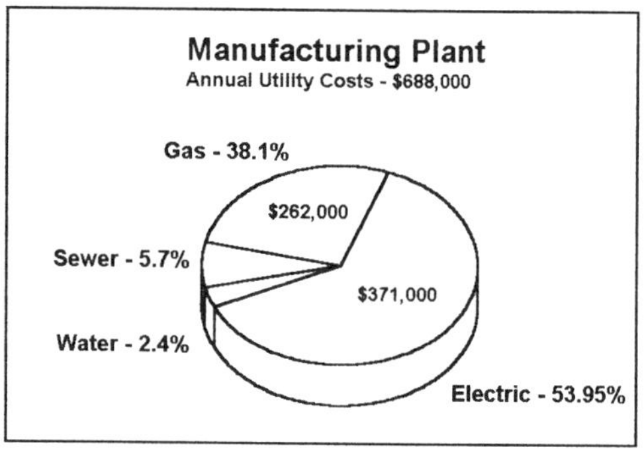

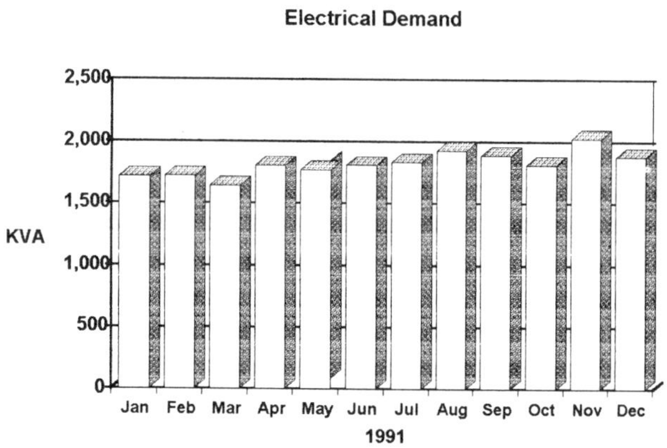

- Inspect facility, observe operation, talk to key people.

- Find out if any systems or processes are considered to be off-limits and cannot be changed to save energy.

- Find out if a monitoring system exists and is being used effectively.

The facility is a good candidate if:

1. Top management is serious about managing utility costs.

2. The annual utility costs are such that a 10-15% reduction will fund the cost of the monitoring system within one year.

3. There is no comprehensive monitoring system or effort in-place.

It may offer exceptional opportunities if there are areas with significant energy requirements that have been treated as off-limits in the past; this normally means that no one has ever looked objectively at the consumption patterns and costs.

INSTALL MONITORING SYSTEM

- Spend time becoming familiar with the facility, its energy consuming equipment, and operation.

- Prepare an inventory of significant energy consuming equipment.

- Draw preliminary one-line energy flow diagrams of the entire facility and major systems.

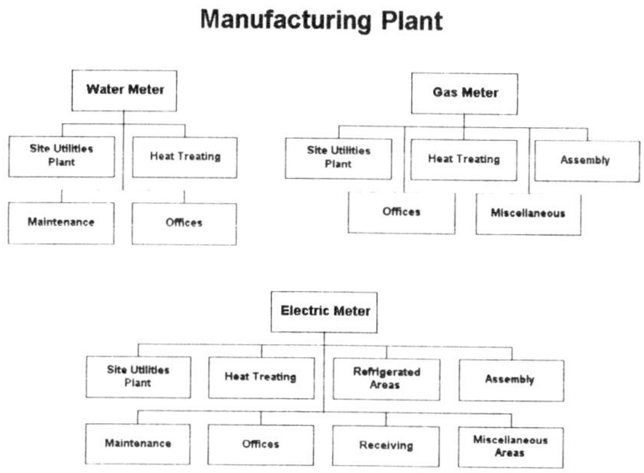

- Identify critical points for monitoring, prioritize and estimate costs.

- Install monitoring system.

QUANTIFY ENERGY USAGE AND UTILITY COSTS BY:

- Utility Meter.

- Time of day and day of week.

- System, process or department.

- Production volume and product mix.

- Weather and other external factors.

Daily Meter Report

Time	Electric	Refrigeration Compressors	Air Compressore	Gas
3:00	974.8	544.7	38.6	119.0
6:00	1,124.0	567.7	84.1	80.7
9:00	1,162.7	468.3	106.2	160.0
12:00	1,467.6	587.2	135.2	212.0
15:00	1,567.3	854.7	132.0	259.9
18:00	1,407.5	733.3	117.4	1860
21:00	1,417.8	850.6	111.7	51.6
24:00	1,274.3	762.4	40.8	48.0
Peak	1,583.9	898.8	136.8	262.0
Total	30,781.3	16,263.4	2,331.4	3,246.3

GENERATE IMMEDIATE COST SAVINGS

- Identify and quantify individual contributors to hourly consumption, demand and costs.

Electrical Demand Profile

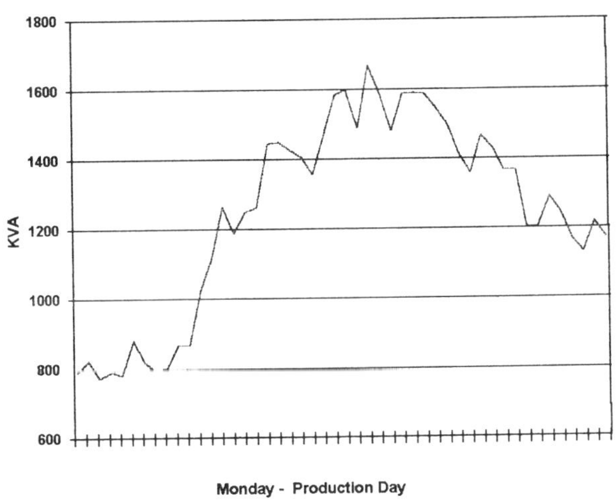

- Assign individual responsibility for producing savings, provide goals and incentives. Manage utility costs like all other costs.

- Coordinate equipment operation to reduce demand peaks and take advantage of special utility rates such as off-peak or interruptible.

- Determine optimum equipment operation and consumption. Eliminate waste. Set daily goals and monitor performance continuously.

- Add system alarms to notify personnel of potentially costly problems.

ANALYZE EFFICIENCY OF ENERGY USAGE

- Finalize one-line energy flow diagrams of the entire facility and major systems.

Site Utilities Plant
Electric Loads

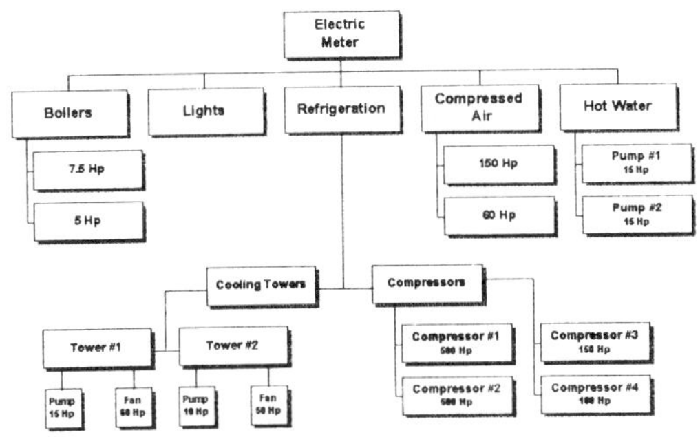

- Draw one-line energy balance diagrams for the entire facility, departments, processes, equipment and systems.

Refrigeration System Energy Balance

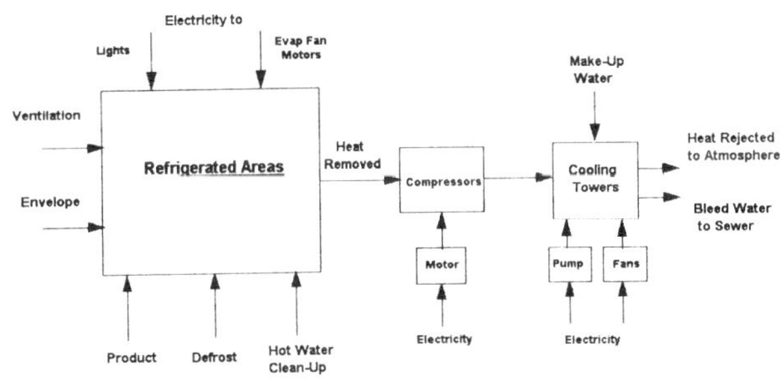

- Construct spreadsheet models and correlate them with monitored data and utility records to verify existing consumption and demand by department, process, equipment and system.

- Use the spreadsheet models to determine the theoretical consumption and demand by department, process, equipment and system.

- Compare theoretical costs with actual costs to determine opportunities for savings.

Refrigeration Compressors

Heat Load	Monday 27-Apr-92	Thursday 20-May-92	Saturday 13-Jun-92	Year 1992
Envelope	174	277	321	321
Ventilation	96	245	307	307
Product	1,205	1,205	0	0
Cleanup	126	126	0	0
Defrost	20	20	5	5
Total	1,622	1,874	633	633
Calculated KWH	2,433	2,811	949	831,800
Actual KWH	10,701	17,695	13,947	5,215,800

- Compare equipment capacities with actual usage to determine opportunities for downsizing and improving operating efficiencies.

SELECT AND IMPLEMENT COST EFFECTIVE ENERGY PROJECTS

- Identify specific opportunities for savings.
- Use spreadsheet model for "what/if" analysis of opportunities.
- Estimate cost, savings and payback.
- Select projects for implementation.
- Use monitored data to qualify for utility incentives.

TRACK AND MAINTAIN ENERGY & COST SAVINGS

- Generate daily, monthly and annual utility cost reports for the entire facility, departments, processes, equipment and systems.
- Integrate daily, monthly and annual utility cost reports into the existing facility financial management system.
- Assign costs to specific individuals or departments. Include utility costs in production reports, manager evaluations and incentives.
- Compare monitored data with utility records for verification.
- Compare actual performance of energy projects with projections.

OTHER FACTORS & SAVINGS

Through continuous monitoring combined with system alarms, significant additional savings can be realized by early detection and correction of:

- Excessive utility consumption due to operator error or equipment malfunction.
- Degradation of equipment performance prior to failure.
- Equipment failure, prior to noticeable effects on occupants or processes.
- Problems that could result in loss of production or revenue.
- Potential hazards.

DISCUSSION

The first step in any financial management process is setting up a system for determining and tracking individual expenditures on a continuing basis. Other than utility costs, a manager would never approve a major expenditure without a detailed accounting of where the money is being spent. The key to the success of any utility cost reduction effort, therefore, lies in convincing top management that utility costs must be treated and managed the same as all other costs within the facility. They represent opportunities, often significant and largely untapped, to improve the bottom line and must be recognized as such. The tremendous opportunity for creating utility savings through good management is an opportunity that will always be available in all facilities.

Portable instrumentation used to evaluate opportunities for energy savings provides a snapshot of energy use during the time that it is installed. Energy usage in a facility is complex and dynamic. It must be continuously monitored, measured, evaluated and managed. The use of portable instrumentation implies that technology is the primary means of reducing utility costs and that after improvements are made, the predicted savings will occur and no further detailed monitoring is required. Technology is only one part of energy efficiency, however. Many of the problems to be overcome are not technical problems, they are people problems; people who lack the understanding, training, time, interest or incentive to get the most from buildings and systems. We have encountered many energy improvements in the field that were saving nothing even though all of the projections, estimates and spot measurements said that they were.

There are so many vendors claiming that their products will produce big savings that many people are totally confused. Bad experiences in the past have turned many people off on energy management. A monitoring system puts the owner back in control by providing actual data to identify and quantify costs and verify actual savings after an improvement is made.

SUMMARY

The following summary of our approach as presented in this paper is provided:

- Work directly with the owner, president or plant manager and demonstrate how utility costs can be quantified, tracked and managed the same as all other costs associated with a facility. Once top management understands and accepts the concept and supports the effort, one of the most difficult hurdles has been overcome.

- Rather than investing in studies, audits and temporary monitoring equipment, apply that money toward the installation of a permanent monitoring system as the first step in reducing utility costs. A permanent monitoring system will produce more comprehensive and accurate data for identification of opportunities and can then be used to produce and document costs and savings on an on-going basis. The system can initially be tied to a few key points and, once the PC and software are installed, be continually expanded for only the cost of sensors and wiring.

- Apply the fundamental concepts of engineering problem-solving and financial management on an on-going basis.

  - Determine the actual utility consumption and cost profile of the total facility and significant areas and systems.

  - Determine opportunities for immediate savings through no cost or low cost changes in operating and maintenance practices.

  - Assign specific responsibilities and implement changes.

  - Use the monitoring system to provide feedback and track utility costs on a continuous basis. Integrate with the facility's financial management system.

- Develop a theoretical profile of the facility and use the differences between the actual and theoretical to determine opportunities for reducing costs that require capital expenditures.

Chapter 46

Improving Energy & Loss Efficiency: The Total Quality Way

F.J. Larranaga, V. Venkatesan

ABSTRACT

In the operating cost of a petroleum refinery, the cost of energy & the handling losses account for a major share; up to 45 % in some refineries. Any effort to cut down this cost is highly rewarding with additional spin-offs like reduced emissions to the environment.

This paper highlights the efforts being taken at a major petroleum refinery complex in the caribbean island to improve its energy & loss (E&L) efficiency. In a structured way, the paper identifies the problem, the methodology of approach, the planning of activities, progress assessment, steering activities and the involvement & commitment of all personnel. Since people play the most significant role in the whole improvement effort, the management of the refinery has adopted the Total Quality approach to achieve its mission and vision. Significant focus is given on the Total Quality's seven step method in identifying and solving the energy wastes problem. Special emphasize is given on the efforts undertaken during 1992, the problems faced, the progress achieved so far along with the specific action plan for 1993. The paper also outlines the awareness improvement programs presently undertaken . It is a long process towards achieving the best E&L efficiency standards and the paper highlights the case where a beginning has been already made and a clear vision established. The paper reflects the actual experience of a committed management and the natural working teams in improving the E&L efficiency of the refinery.

INTRODUCTION

Today the consumption of petroleum products is a barometer of the economy of a country. Petroleum refineries form part of a chain in the oil industry. The oil production facilities followed by transfer to oil refineries at distant places is a common feature today. Crude oil processed by these refineries to obtain marketable high value products are sold to the consumers through a network of distribution facilities.

For several decades, until the oil shock of the seventies, oil industry was booming. However, the competitive nature of the world economy forced the oil industry to look into each major part of this chain as a profit center. Also, the shift in the demand of product quality has increased the refining cgwts, whereas the market price remains relatively inelastic. These factors are exerting an increased pressure on the refineries to maximize the refining margin, which is mainly dependent upon the product yields and the operating costs. The major chunk of the operating costs is the energy & handling losses. In our refinery, the cost of energy & losses totals over 10 percent of the crude intake every year and has a great impact on the Gross Refinery Margin.

Efforts are being undertaken to improve the E&L efficiency as highlighted in the following paragraphs.

THE REFINERY

Refineria Isla is a 330 MBD medium to heavy crude processing refinery situated in the Caribbean island of Curacao north of the Venezuelan coast. It consists of 3 crude distillers, 5 high vacuum units, a Catalytic cracker, 2 Thermal crackers, Platformer, Alkylation and Polymerization units besides its Naphthenic and paraffinic lube units. A vast utility complex caters to the needs of steam, electricity, fresh water, compressed air and cooling water of the refinery. Since the island does not have any fresh water source, sea water is desalinated to meet all the fresh water requirements.

ENERGY SYSTEMS IN THE REFINERY

The utility units and the various energy systems of the refinery are briefly outlined below:

Fuels System

The refinery consumes about 2700 tons of various fuels every day, which include refinery fuelgas, residual asphalt, coke from catalyst regeneration, No 6 oil and gasoil. Among these fuels No 6 fueloil and gasoil are marketable products, with reasonable value, whereas the other fuels are not readily marketable.

Steam System

Steam generation forms the major activity of utility units as shown in figure 1. The refinery processes need about 350 t/hr of 18 barg superheated steam at a temperature of 370 deg C. The utility systems needs an additional of about 150 -170 t/hr of steam for the power generation, sea water distillers, various utility drives, soot blowing, deaerators and fuel atomization. The refinery processes generate about 130 t/hr of

steam through waste heat recovery. Turbine exhausts are routed to low pressure(LP) steam headers and supplied to deaerators and sea water distillers. LP steam demand/excess is managed by let down valves or dump condensers. A simplified schematic diagram of the steam system in the refinery is given in figure 2.

Electricity System

The refinery operations need about 50-55 MW electricity, depending upon the level of operation. All these power requirements are met through two back pressure steam turbines, three condensing turbines and three gas turbines. The three gas turbines are provided with waste heat boilers.

Fresh Water, Air & Cooling Water Systems

The refinery need about 850 t/hr of fresh water, 470 NM^3/min of compressed air and about 50000 t/hr cooling sea water. Sea water cooling is a once through system. Fresh water is distilled using 4 desalination units and further treated for boiler feed water requirements. The various condensate collections add to 250 t/hr.

PROBLEMS OF E&L EFFICIENCY AT ISLA

Historical trend of the refinery's energy performance shown in figure 3, indicates that the refinery accounts about 10 % of its crude intake for fuel consumption & hydrocarbon losses. However, refineries of similar complexity elsewhere account only 6-7 % of crude intake as their fuel consumption & losses.

A Macrowaste analysis done on the refinery margin during 1991 identified that improving the E&L efficiency alone can increase its gross refining margin by 24 cents/bbl. Bar chart from the macrowaste is given in figure 5.

The other problem observed is the non optimum consumption of fuels, meaning the saleable versus non saleable fuels. The pie chart in figure 4 highlights this problem. The unutilized refinery fuelgas was being flared and the heat from the coke burning was being rejected to atmosphere while consuming the expensive saleable fueloil & gasoil .

The refinery management took a serious look at the E&L patterns and a policy was formulated to improve this.

TOTAL QUALITY APPROACH

The refinery management is committed to the 'Total Quality Approach' in solving the problems, to achieve sustained results and continuous improvement in all areas. The E&L efficiency is one such problem, where the efforts to improve have to be taken by many individuals involved in the refinery operation. Although, the energy management efforts have produced immediate results, sustained improvement has not been achieved. This may be a result of the lack of involvement from all personnel. To get a total commitment from every one involved and for their well motivated efforts to achieve continuous improvement, the refinery management has decided to adopt Dr. Deming's 'Total Quality' approach (TQ).

THE SEVEN STEP METHOD

The principles of Dr Deming's TQ approach to solve any problem by a team is through seven steps which are outlined in Table 1.

As a first step, a preliminary plan outlining the approach was prepared and approved by the management. The players at strategic positions were grouped in to the E&L Steering Committee to guide the overall improvement efforts. The E&L Master Plan flow chart is enclosed in Figure 6.

The problem originally identified were again discussed by the steering committee and after the brainstorming sessions, pinpointed the following three specific aspects:

1. A higher energy consumption & losses compared to the international standards.
2. The desired performance levels of E&L efficiency of the refinery not well defined.
3. A lack of optimum fuel consumption pattern (increased use of cheaper non-saleable fuels).

At a macro level the cause and effects of the problem are summarized in the form of a fishbone chart as shown in Figure 7.

The following 'Mission' and 'Objective' of the E&L Steering Committee(ELSC) were defined and communicated to all operating personnel;

MISSION:
To provide the leadership in promoting the improvements on Energy & Loss efficiency at ISLA, and in establishing the policy guidelines to sustain a continuous improvement.

OBJECTIVE:
To reduce the refining cost by improving the Energy & Loss efficiency.

CURRENT SITUATION ASSESSMENT

To understand and monitor the current energy performance of the refinery, besides the 'percentage on Intake', three other options were identified, namely, the SHELL's corrected energy and loss(CEL) index system, BP's percentage of guideline(POG) system and Solomon associates's energy intensity index(EII).

The three systems mentioned above compare the actual energy consumption in each unit with the energy consumption of an ideal unit of similar capability. The ideal energy consumption is defined in terms of standard units like standard refinery fuel(SRF) or fuel oil equivalent (FOE) in each of these systems.

The ELSC members decided on the use of the SHELL's corrected energy and loss(CEL) index system for measuring and monitoring the E&L performance.

E&L performance monitoring through CEL:
The CEL system divides the refinery into a list of 'standard' process units and off-plot activities and determines the actual energy consumption in each activity. The system also allocates an appropriate theoretical energy and loss allowance to each of these activities and evaluates the performance in the form of indices for each activity. The performance indices are defined as;

$$\text{Index} = \frac{\text{Actual energy}}{\text{Theoretical energy}} * 100$$

The theoretical allowances are designed to take into account important operational factors that may have a significant influence on the energy consumption of that particular activity.

The various types of fuels used in the refinery are converted into equivalent quantities of standard refinery fuel (SRF), which is defined as a fuel with a lower heating value of 9673 kCal/kG (17411 BTU/Lb). Electricity and Steam at various pressures are also converted to SRF equivalents.

The CEL indices are evaluated and reviewed on a monthly basis, besides monitoring the saleable fuel consumption in the form of run charts. A set of monthly E&L performance charts are enclosed in Figure 8.

SPREADING THE ENERGY AWARENESS

After establishing the objective, the starting point and the E&L Master Plan, the steering committee formed five sub-committees in the operating divisions of the refinery to spread the energy awareness downwards and to effect the positive reactions from the natural work teams. The message of excessive energy consumption was conveyed to all the operating and support personnel through a 'Management Letter' issued by the refinery manager. Subsequently E&L workshops were held with the natural teams. These E&L contact workshops were repeated four to six times in each division covering all the operating personnel. Problems in energy consumption and improvements achieved elsewhere were explained by experienced engineers. At the end of each workshop the problems leading to high energy consumption were further brainstormed resulting in suggestions & contributions from the whole cross section of the refinery.

STRATEGY TOWARDS IMPROVEMENT

The following two step strategy has been conveyed through the E&L contact workshops to all operating personnel;

1. Short term strategy to cut down all obvious energy wastes, with minimum investments.

The major steps being undertaken by all divisions are the following:

Minimizing the consumption of saleable and high cost fuels like Gasoil & No 6 fueloil through proper utilization of fuels like fuelgas & asphalt.

Optimizing the combustion efficiency in all the boilers and fired heaters.

Better utilization and recovery of low pressure steam and condensate.

Better housekeeping for steam, water and compressed air systems.

2. Medium & Long term strategy to cut down energy wastes to resolve structural defects.

Eight decades of existence has made this refinery as a complex crude processing unit. Abnormal growth during World War II and prior to the oil shock has deviated it structurally from an efficiently designed system of late 80's and 90's. Most of the furnaces are designed for a stack temperature of 350 to 450 degC. Wide deviations from the PINCH concepts are quite common in most of the preheat trains in the refinery units. Hence, the medium and long term strategy is being developed at present aiming to bring a structural change in cutting down the wide deviation from the efficiency standards. The medium and long term strategy will also improve the measurement, control systems and process optimization.

PRESENT STATUS IN TQ APPROACH

Along the seven step method, the first two steps have been completed and the third step has commenced.

First step:
The problem has been defined and conveyed to all the concerned personnel.

Second step:
Current situation is understood by all the personnel.

Third step:
Causes of excess usage of energy at each unit are being analyzed. Some obvious wastes (short term strategy) in some units have been identified and the modification are being done. Further analysis on the medium & long term projects are being studied.

While subsequent steps have been taken on certain items of concern, on the whole, the exercise has reached its third step. Crossing this step is expected to take a considerable length of time. With an all round commitment this time may be reduced. A chart (Figure 9) reporting the status is sent to the management on a quarterly basis.

ACTIVITIES AHEAD

The following activities have been planned by the ELSC for completion along with its third step of TQ approach.

1. Additional E&L workshops to cover all the remaining operating personnel, spreading the awareness on E&L efficiency.

2. Improving the E&L data base in terms of reliability and accuracy.

3. Reinforcing the sub-committees with simple E&L audit kits, like portable pocket size furnace efficiency indicators, portable flow meters, portable power meters, portable digital thermometers.

4. Encouraging the sub-committees with necessary assistance from E&L Steering Committee, to make energy balance at micro levels in each of their areas and to make cause & effect diagrams for the identified E&L wastes in their area.

5. Establishing contact with at least one affiliate refinery to exchange E&L related information.

TOWARDS CONTINUOUS IMPROVEMENT

Continuous improvement through Total Quality is the objective of this whole exercise. During this short period of one year we have achieved some positive trends. Following are some of the highlights;

1. The refinery energy performance as in figure 10 is being monitored by the top management.

2. The awareness of the operating personnel on energy costs has increased.

3. The fuel consumption pattern is improving as indicated in figure 11, and steps towards further optimization are being undertaken.

4. The E&L efficiency monitoring system is being debated with a wider spectrum of operating personnel than ever before and time to time refinements are being incorporated.

5. The exercise of identifying and monitoring the obvious & structural energy wastes at unit levels are in progress.

6. Enthusiasm and participation of operating personnel has increased as indicated by the number of suggestions being made in the sub committee meetings.

- o O o -

Figures & Tables enclosed;

Figures
1. Steam system scheme
2. E&L % graph on intake
3. Macrowaste 1991 pareto
4. Fuel consumption pattern 1991 pie chart
5. E&L master plan
6. Mission & Objective of ELSC
7. E&L monitoring charts
8. Status report chart
9. CEL index chart (yearly)
10. 1991 & 1992 piechart
11. Fuel consumption pattern 1991/92

Table
1. Seven step method

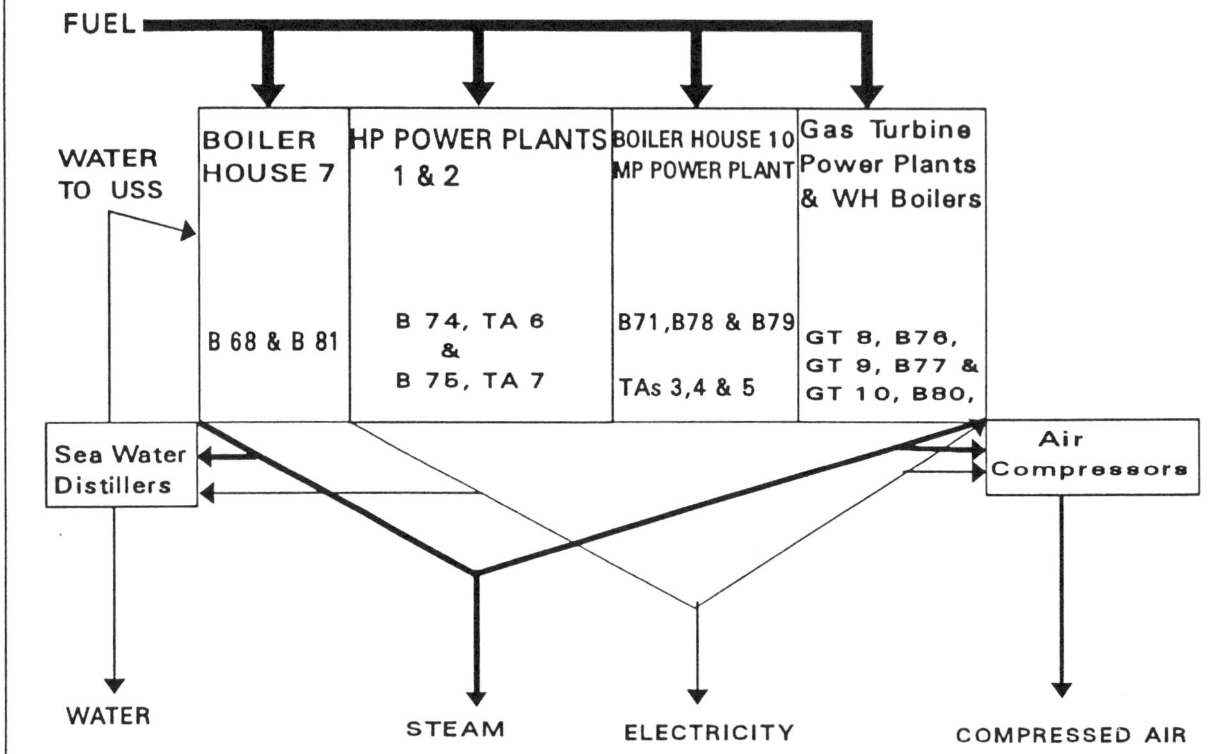

Fig 1. UNITS OF USS

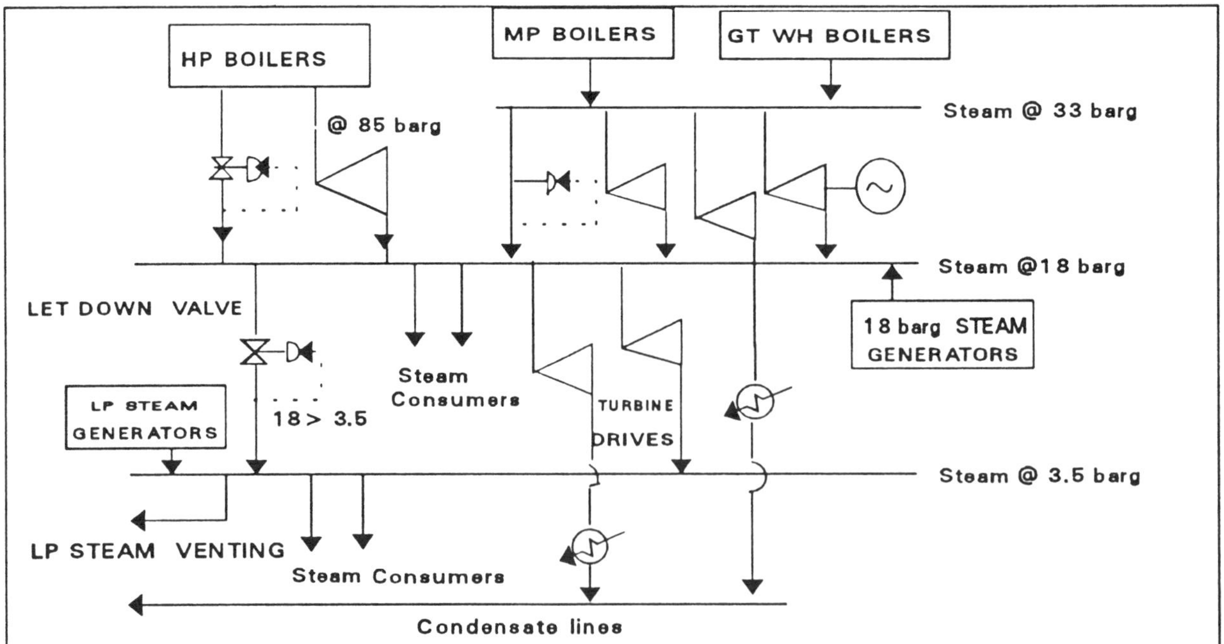

FIG 2. REFINERY STEAM SYSTEM

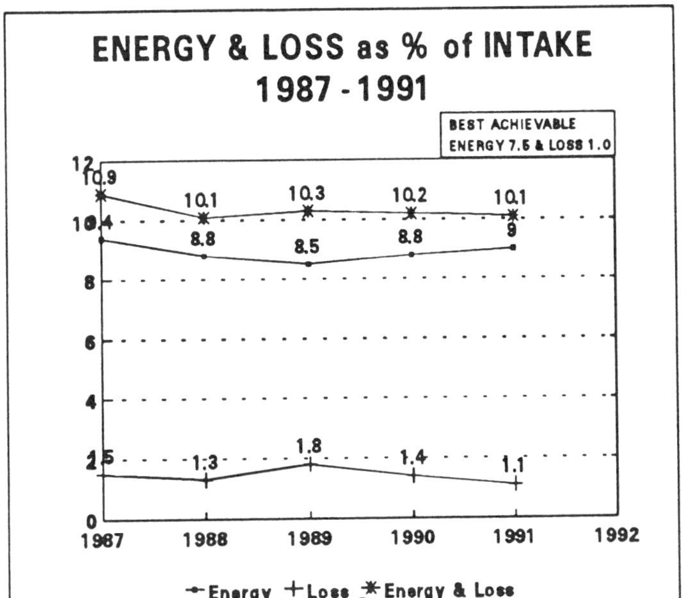

Fig 3

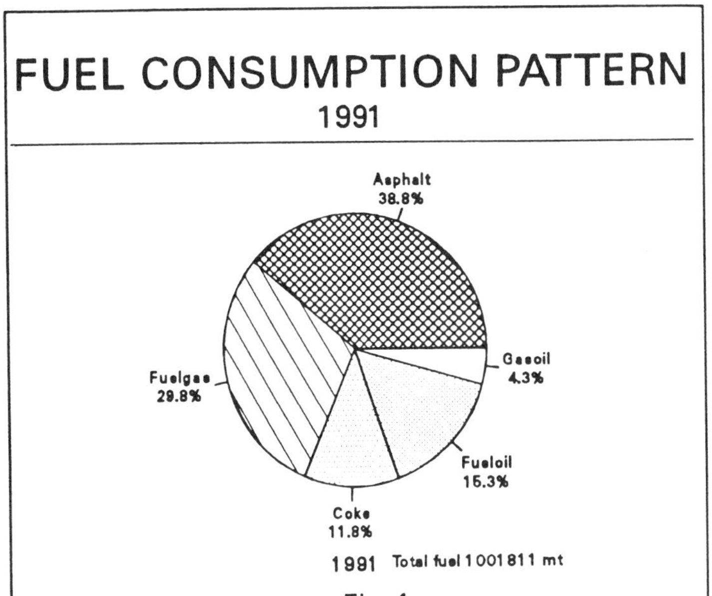

Fig 4

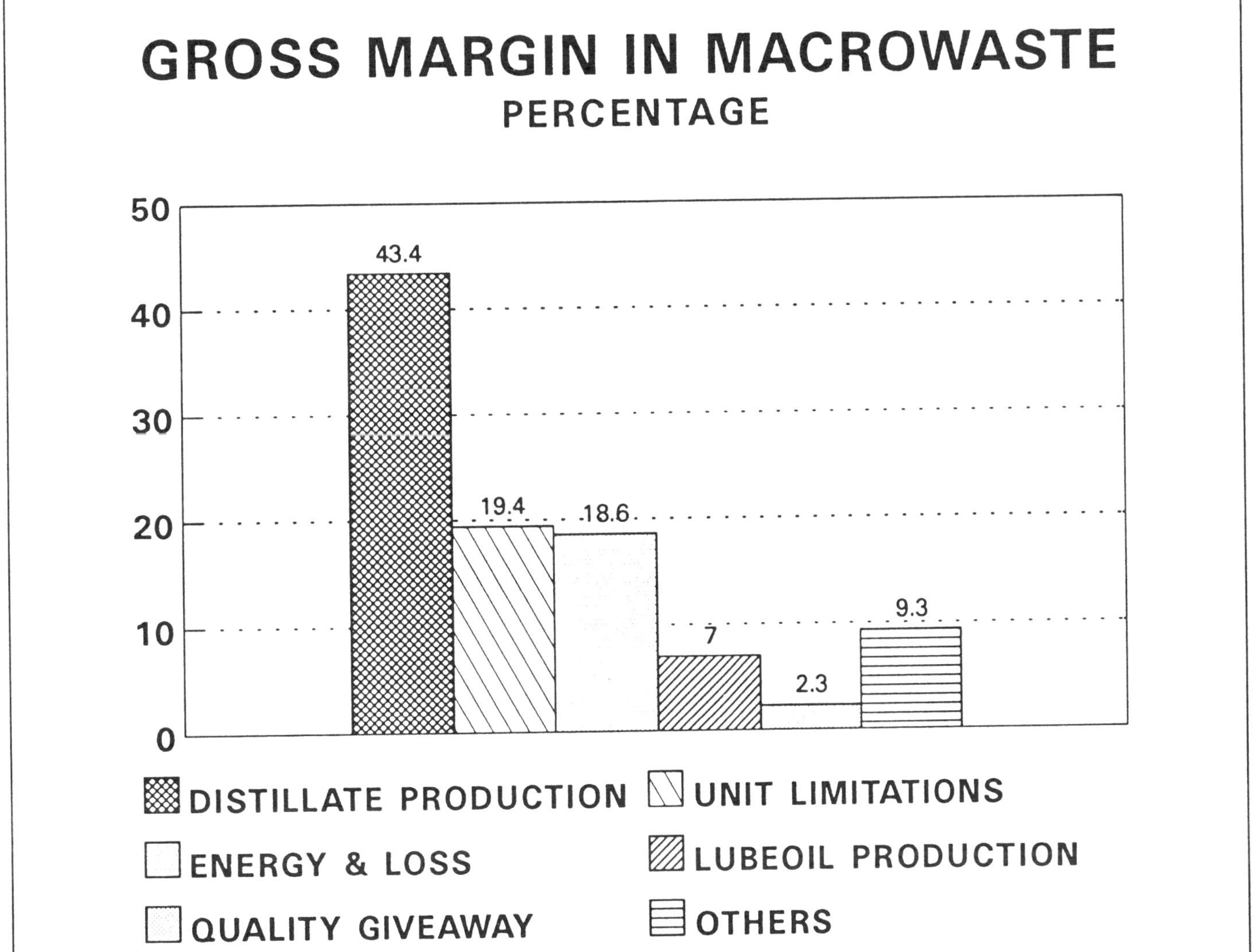

Fig 5

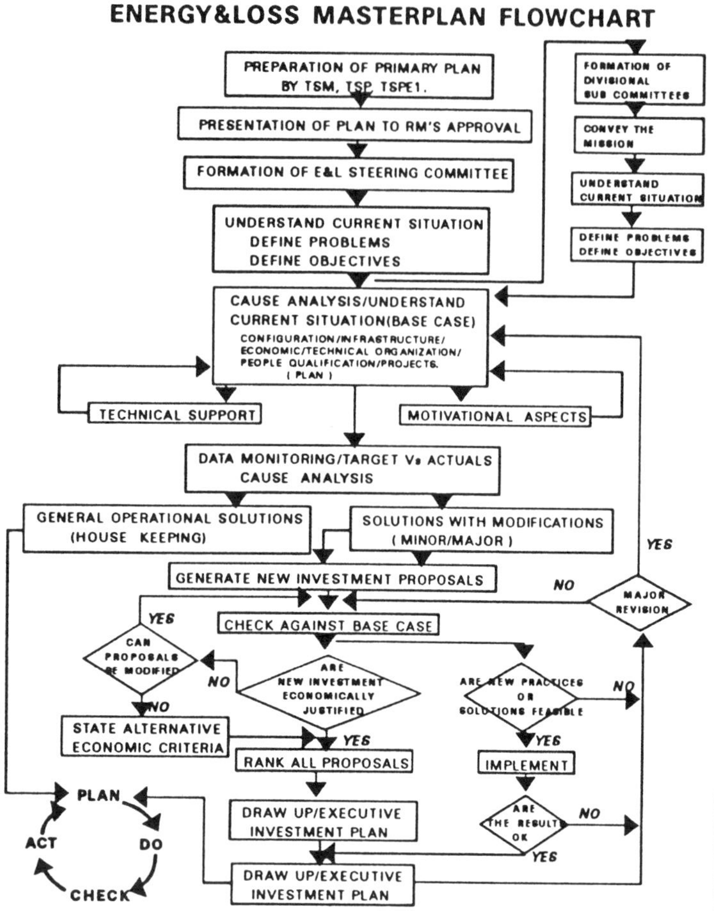

Fig 6

Table 1

SEVEN STEPS METHOD

STEP 1: DEFINE THE PROJECT
Define the Objective(s)
Define the team players

STEP 2: STUDY THE CURRENT SITUATION
Develop baseline data to verify solutions

STEP 3: ANALYZE THE POTENTIAL CAUSES
Find the root causes of the problem rather than the symptoms.

STEP 4: IMPLEMENT A SOLUTION
Select the best feasible solution and implement it.

STEP 5: CHECK THE RESULTS
Determine whether the decided solution is effective. Otherwise repeat earlier steps.

STEP 6: STANDARDIZE THE IMPROVEMENT
Ensure the problem stays fixed.

STEP 7: RESTART THE LOOP FOR CONTINUOUS IMPROVEMENT
Promote the practice of continuous improvement in terms of organizational and team effectiveness.

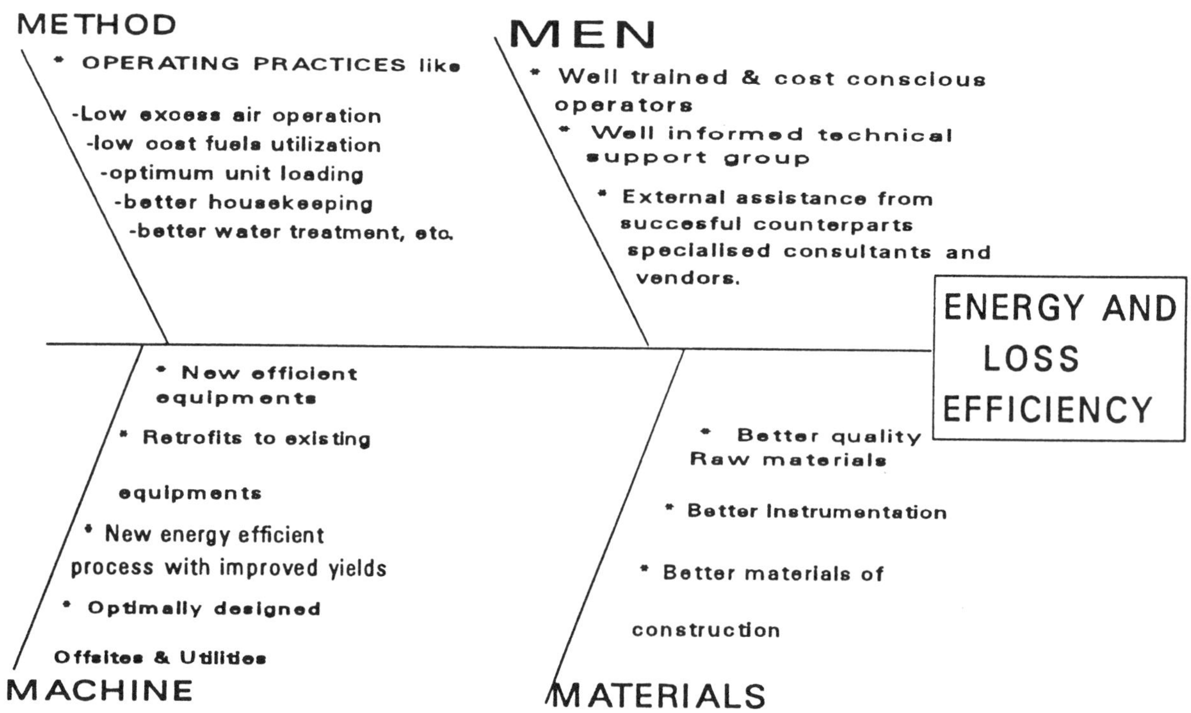

CAUSE & EFFECT DIAGRAM

Fig 7

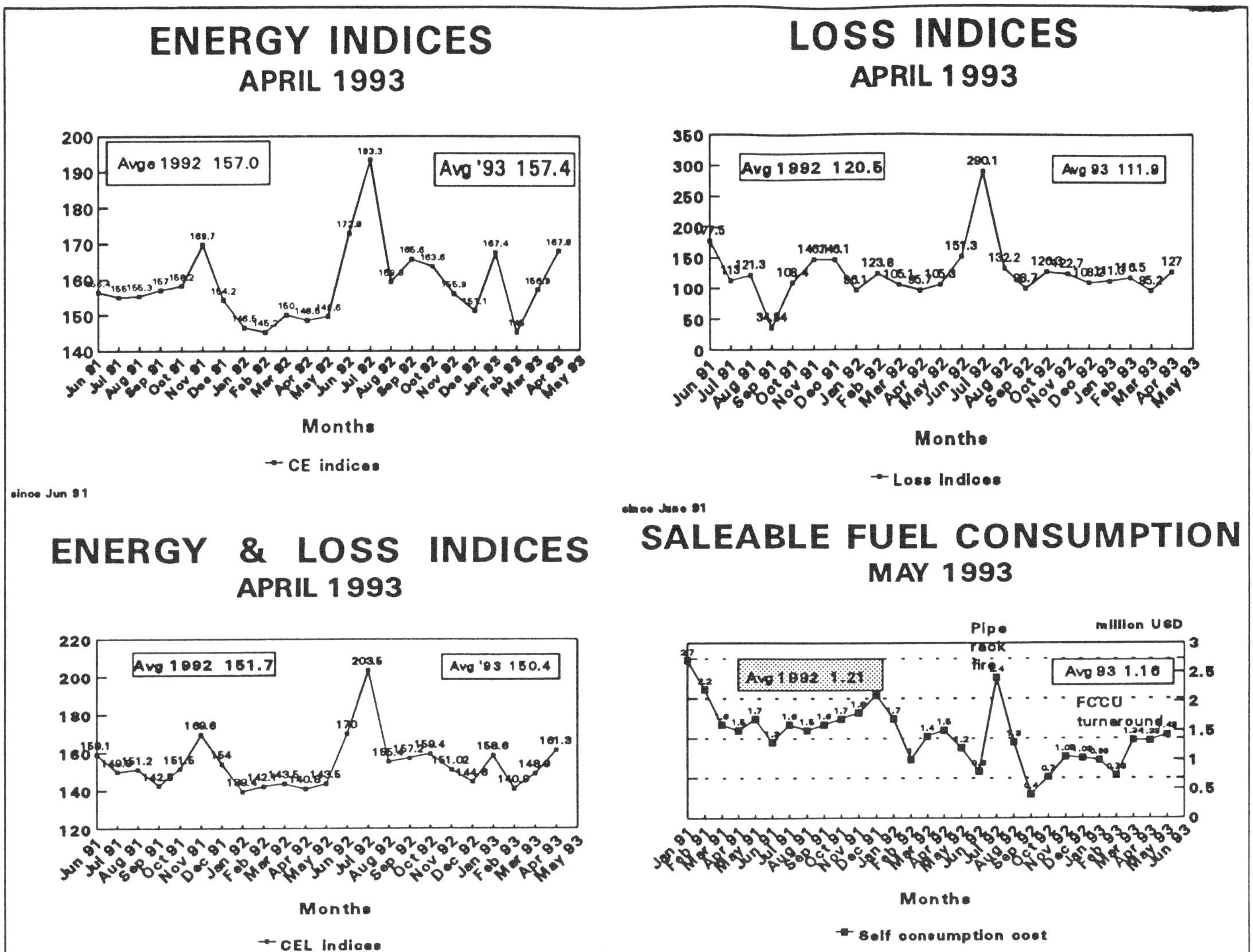

ENERGY INDICES
APRIL 1993
Avge 1992 157.0
Avg '93 157.4
Months
CE indices
since Jun 91
LOSS INDICES
APRIL 1993
Avg 1992 120.5
Avg 93 111.9
Months
Loss Indices
since June 91
ENERGY & LOSS INDICES
APRIL 1993
Avg 1992 151.7
Avg '93 150.4
Months
CEL indices
since Jun 91
SALEABLE FUEL CONSUMPTION
MAY 1993
Pipe rack fire
million USD
Avg 1992 1.21
Avg 93 1.16
FCCU turnaround
Months
Self consumption cost
since Jun '91

Fig 8

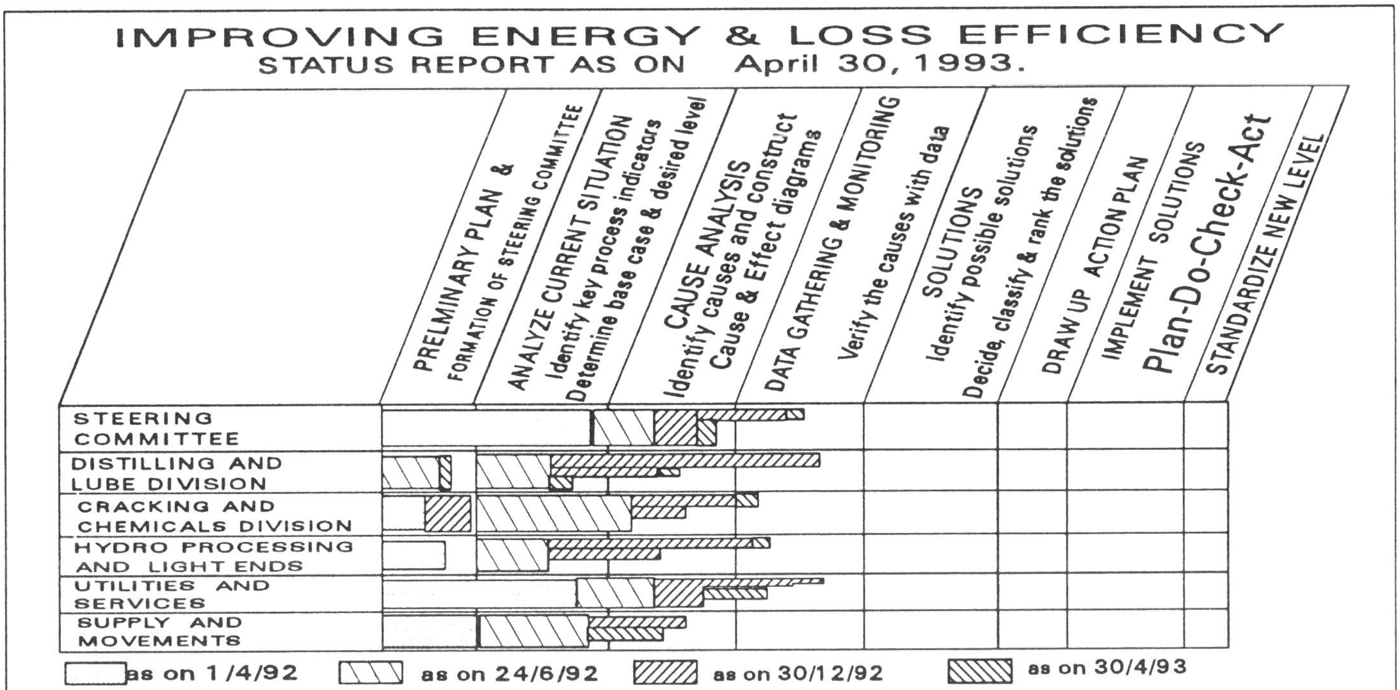

IMPROVING ENERGY & LOSS EFFICIENCY
STATUS REPORT AS ON April 30, 1993.
PRELMINARY PLAN & FORMATION OF STEERING COMMITTEE
ANALYZE CURRENT SITUATION
Identify key process indicators
Determine base case & desired level
CAUSE ANALYSIS
Identify causes and construct Cause & Effect diagrams
DATA GATHERING & MONITORING
Verify the causes with data
SOLUTIONS
Identify possible solutions
Decide, classify & rank the solutions
DRAW UP ACTION PLAN
IMPLEMENT SOLUTIONS
Plan-Do-Check-Act
STANDARDIZE NEW LEVEL
STEERING COMMITTEE
DISTILLING AND LUBE DIVISION
CRACKING AND CHEMICALS DIVISION
HYDRO PROCESSING AND LIGHT ENDS
UTILITIES AND SERVICES
SUPPLY AND MOVEMENTS
as on 1/4/92
as on 24/6/92
as on 30/12/92
as on 30/4/93

Fig 9

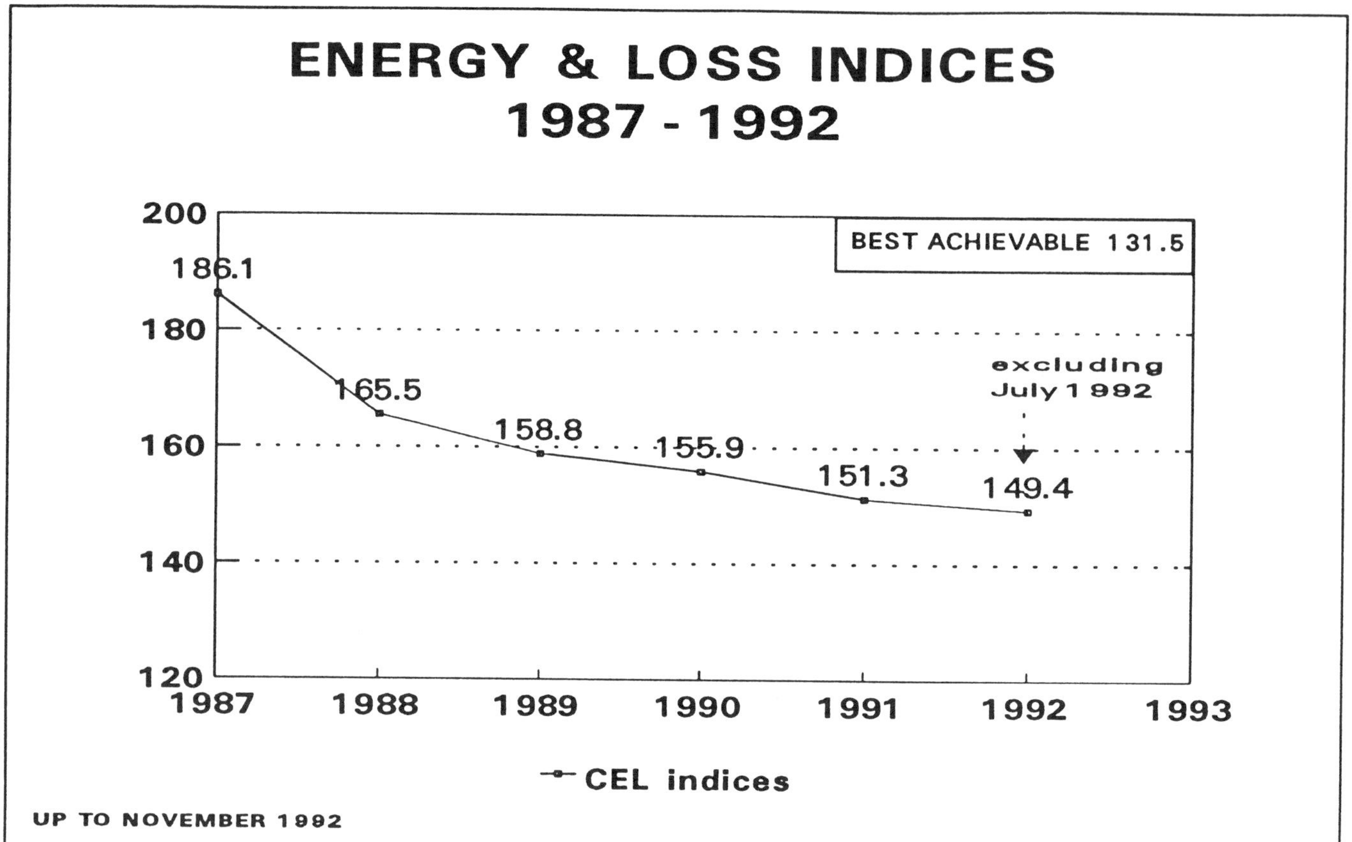
ENERGY & LOSS INDICES
1987 - 1992
200
180
160
140
120
186.1
165.5
158.8
155.9
151.3
149.4
BEST ACHIEVABLE 131.5
excluding
July 1992
1987
1988
1989
1990
1991
1992
1993
CEL indices
UP TO NOVEMBER 1992

Fig 10

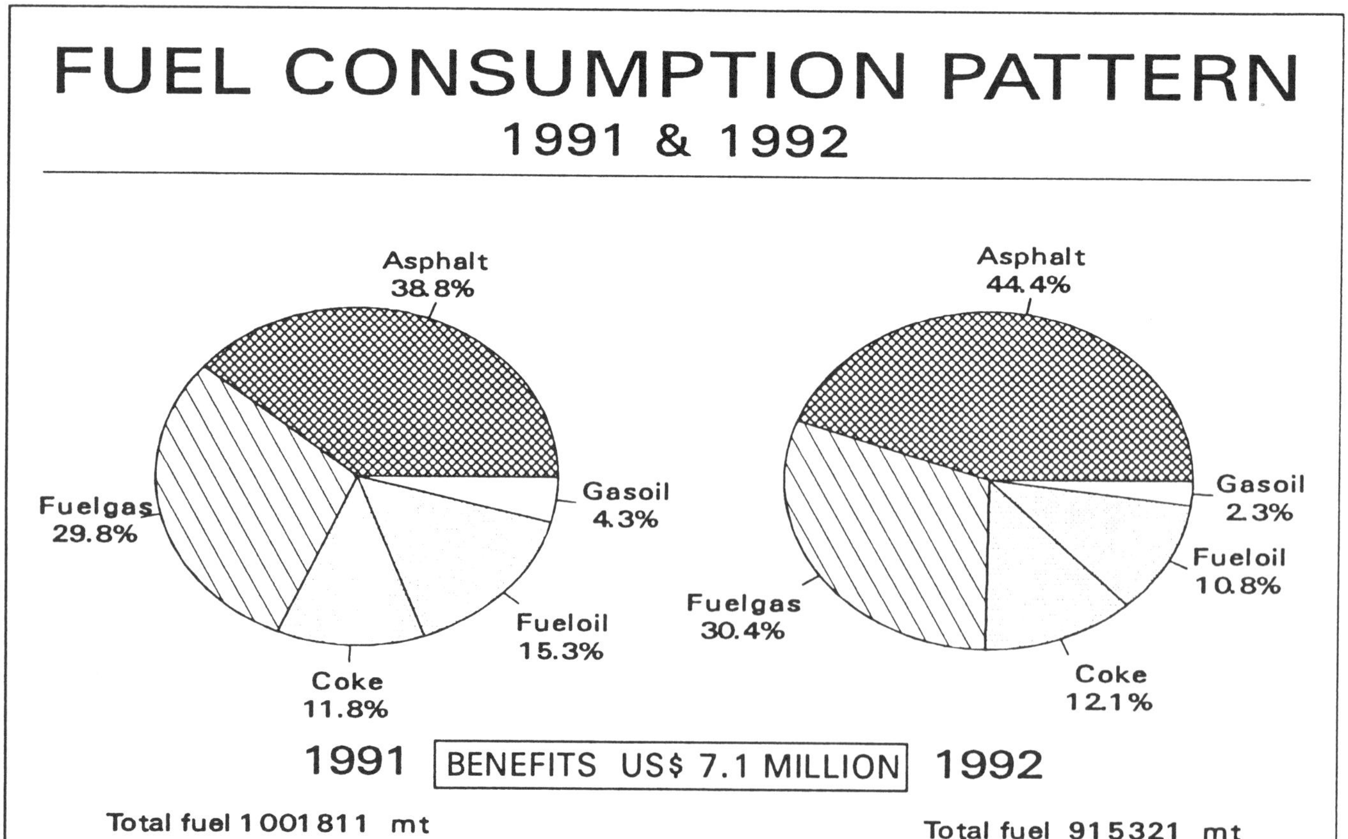
FUEL CONSUMPTION PATTERN
1991 & 1992
Asphalt
38.8%
Gasoil
4.3%
Fueloil
15.3%
Coke
11.8%
Fuelgas
29.8%
Asphalt
44.4%
Gasoil
2.3%
Fueloil
10.8%
Coke
12.1%
Fuelgas
30.4%
1991
BENEFITS US$ 7.1 MILLION
1992
Total fuel 1001811 mt
Total fuel 915321 mt

Fig 11

Chapter 47

The Development of Strategic Energy Management Planning at New Jersey Transit

R.E. Owan

ABSTRACT

The purpose of this paper is to present a planning approach for energy management and cost control in a transportation environment. The focus will be on how one entity - NJ TRANSIT is responding to the operational challenges of limited resources and environmental awareness.

There are certain transportation energy realities that are worth noting:

• Energy for transportation tends to move in the same direction with the level of service.

• Fixed guideway rolling stock is not easily abandoned once acquired.

• Fuel substitution is often difficult to attain due to transit stock of technology that is fixed in the short run.

• Transportation service providers are relatively unresponsive to changes in the price of energy.

• Price signals designed to encourage off-peak usage of electricity tend to be ineffective for propulsion needs.

• According to micro-economic thinking, the weak price response capacity (inelastic demand) means that the utility company can best increase total revenues by increases in the price of energy.

In view of the constrained energy environment, it is vital to develop and implement strategies that would mitigate the adverse effects of energy cost which in turn can maximize operational efficiency.

However, energy conservation in public transportation is not a glamorous topic as compared to funding and capital programs. The high profile of propulsion power or motor fuel consumption may also dwarf other areas of significant power usage. Therefore, it is indispensable for the transit operator to be aware of the overall energy picture.

The concept of systematic energy management in the context of strategic thinking and action will be advanced. The overall presentation will take the form of a case study. This paper is limited in that activities of electrification, electric traction, and traction power dispatch will not be covered.

INTRODUCTION

NJ TRANSIT (NJT) was created as an instrumentality of the state in 1979. It provides bus, rail and light rail subway services to about 273,000 passengers a day. Annual passenger trips of 161 million are provided to destinations in New Jersey, New York and Philadelphia. NJT covers an area of 5,325 square miles. It is the nation's third largest and only provider of statewide bus and commuter rail services.

The 1991 Section 15 farebox recovery ratio of 56 percent paid by the ridership is one of the highest in the nation. A list of selected data is shown below:

BUS OPERATIONS

Routes Operated	152
Annual Bus Miles	72.9 Million
Buses Operated	1,843
Bus Maintenance Shops	17
Total Shop floor sq. footage	3.42 Million

RAIL OPERATIONS

Rail Lines	12
Route Miles	426.7
Rail Cars in Service	687
Daily Revenue Trains	579
Stations With Service	158
Rail Maintenance shops	14
Horse Power of 115 locomotives	542,300
Weight of 115 locomotives	23 Mil. lbs.
Fuel tank capacity - 92 Loco.	250,200 gal

Above compiled from: (1) (2) and (3).

AN ENERGY EFFICIENT MANDATE

NJT is concerned about the energy resources it uses and the cost incurred. The responsibility for energy efficiency is also mandated in five ways.

1. In its 1979 formative declaration, the State Legislature listed first among five agency goals that: "The provision of efficient, coordinated, safe and responsive public transportation is an essential public purpose which promotes mobility, serves the needs of the transit dependent, fosters commerce, conserves limited energy resources, protects the environment and promotes sound land use and the revitalization of our urban centers". Emphasis supplied (4).

2. The company's mission statement also is imbued with the same sense of purpose regarding cost effective transit service..." (See 2). Energy cost is a transportation variable that is controllable.

3. A third of the state energy consumption is attributable to transportation usage. Thus, the State 1991 Energy Master Plan and the 1993 Implementation Sections stressed the need for energy, fuel savings, emissions control, and improved air quality. The plan calls upon NJT to contribute to the attainment of such goals. See (5) and (6).

4. The cost containment effort of the firm also necessitates paying closer attention to regions where savings are achievable. Energy is one such area.

5. Federal environmental emissions control and pollution abatement standards are regulatory constraints faced by the company.

NJT is being environmentally responsive in ways that include reducing congestion, using cleaner fuels, encouraging efficient usage, liquified natural gas bus R&D and the promotion of mass transit use among other things.

ENERGY ORGANIZATIONAL DESIGN

The recent amalgamation of Corporate, Rail, and Bus Divisions led to the energy management function being centralized in 1992.

The Management Engineering and Analysis group (ME&A) has overall energy management and tariff responsibilities. ME&A also performs classical industrial engineering functions and internal consulting services. It is organized within the Engineering, Development and Construction Department (EDC).

EDC is under the direction of Mr. Rick Richmond, Assistant Executive Director. Mr. Victor Strom is the Director of the ME&A Department. The author handles energy and tariff issues. The rest of the team is comprised of a manager, a Senior Industrial Engineer, and a Secretary.

A broad based summary of the ME&A energy management activities include:

1. Energy Accounting and Scorekeeping

2. Utility Rate Case intervention

3. DSM Planning and Implementation

4. Propulsion Energy Analysis & Monitoring

5. Energy Technology Assessment Coordination

6. Utility Tariff and Energy Contracts Analysis
7. Future Energy Education.

ENERGY ACCOUNTING and SCOREKEEPING

The aim of energy accounting is to quantify the end uses and cost pattern of energy consumption. This is an essential tool for basic decision making.

Scorekeeping deals with energy performance appraisal. A good illustration can be found in Sher (10).

UTILITY RATE CASE INTERVENTION
The objective is to maintain close watch on the electric and gas rate design petitions before the State Board of Regulatory Commissioners (BRC). Active intervention is advised based on the facts. In other instances, defensive participation may be appropriate in order to monitor the activities of others.

DSM PLANNING & IMPLEMENTATION
The sorting and ranking of conservation projects requires careful planning and setting of priorities. The model devised for evaluation of energy projects is as follows:

- Identification of energy problems, or issues
- Project ranking and selection

- Alternative considerations
- Economic/engineering analysis

- "Stakeholder Consultations"
- Organizational marketing of ideas

- Site visits to review installations
- Demonstration project

- Facility audit
- Bid and selection process

- Award & installation
- Project management

- Monitoring & verification

- Rebate filing or investment recovery
- Maintenance program

PROPULSION ENERGY ANALYSIS
The power to pull the trains constitutes over 75 percent of the utility bill. Therefore, special scrutiny on the consumption and billing is maintained.

ENERGY EFFICIENCY ASSESSMENT
Effort is made to interface with various disciplines, and personnel to ensure that energy efficiency aspects are incorporated in new or rehabilitation work. Coordination is vital in order to avoid duplication of effort. Emerging or existing energy technologies are also explored for their potential and applicability in the transit environment.

UTILITY TARIFF ANALYSIS
Utility structures can be very complicated. Effort is maintained on the analysis of rates and how they are applied to NJT. The results of previous rate work have been encouraging.

HISTORICAL BILL AUDITING
It is impractical to immediately examine each energy bill at the time of receipt. Therefore, it is beneficial to check the invoices for accuracy after the fact. Past experience has shown that mistakes do occur in the application of rates, estimated readings, and the billing process, just to name a few.

ENERGY EDUCATION
As a long term view, the performance of energy measures will depend on their upkeep. Therefore, maintenance personnel very often require training. Education will be used to increase energy awareness.

THE ENERGY FRAMEWORK
Power Consumption
Energy cost is a significant part of the operating budget. The annualized budget for propulsion and non-propulsion energy is approximately $30 million.

The 1991 Section 15 UMTA/FTA Report showed that NJT electric commuter rail and street car services alone consumed approximately 14.5 million kilowatt hours of electricity. The diesel fuel consumed for locomotives and motor bus was approximately 103 million gallons. NJT non-propulsion power was not included (7).

Power Purchases
NJT has approximately 1000 separately metered electric and gas accounts served by over a dozen utilities including municipalities in New Jersey, New York and Pennsylvania. Each utility has a different set of rate structures. Over 85 percent of the power is supplied by Public Service Electric & Gas (PSE&G) and Jersey Central Power & Light companies (JCP&L). Locomotive diesel fuel is supplied through contracts. Automotive gasoline and other energy sources such as propane round up NJT's energy demand. A sample of power suppliers includes: PSE&G, JCP&L, Atlantic Electric, Elizabethtown Gas, New Jersey Natural

Gas, South Jersey Gas, and Philadelphia Electric Company via Amtrak.

Power Supply Network
NJT operates two wholly owned electrified lines in the Northern and Southern parts of the state. It also operates on the North East Corridor NEC (New York-Trenton) under Amtrak agreements.

JCP&L feeds four substations at 230kV and one at 34.5kV high tension transmission voltages under the General Transmission GT tariff. PSE&G energizes a 230kV substation under the High Tension Service HTS tariff for traction power, and other 34.5kV and smaller supply points for large facilities and the subway. NJT owns, operates and maintains its own substations except for the subway. The GT and HTS rates have special explicit provisions for traction power users. Amtrak is a third party provider of NEC power to NJT. Philadelphia Electric Company in Pennsylvania feeds power to Amtrak.

Energy will continue to be a strategic issue due to the size of its facilities and new projects planned. The sources for a challenging energy future include:

The extensive overhead catenary system

The over 930 fixed guideway directional route miles.

The over 3.4 million square feet of existing or planned bus maintenance facilities

A proposed 20.5 mile light rail system to be developed along the Hudson River

New and proposed mobility enhancing connections to the fixed guideway systems.

The Newark-Elizabeth Rail link. See (2), (8) and (9).

The new projects also present opportunities to try new ideas and the application of state-of-the-art technology.

THE STRATEGIC PLANNING BASIS

The exploration of strategic planning and thinking is an attempt to minimize the number of surprises that can occur in dealing with complex energy issues and competing goals. The process has not yet been formalized. However, the ethos of strategic thinking and action are used for problem solving and dealing with stakeholders from within and without.

The planning process being followed is analogous to that advanced by Bryson (11). Further discussions on strategic planning were offered by Rupnow (12), and Kopfle (13). Bryson's approach specifically addressed strategic planning in public and non-profit firms. The steps can be summarized as follows:

1 Setting broad policy and strategic direction

2 Identifying mandates

3 Mission and values (with stakeholders)

4 External opportunities and threats

5 Internal strengths and weaknesses

6 Strategic issues

7 Formulating strategies

8 Future vision

Broad Direction
ME&A is charged with developing energy management strategies to improve consumption at a lower cost. It also ensures that NJT is charged no more than it costs to serve it by promoting rate design to reflect it's electric traction load diversity.

Mandates and Mission
The legislative mandates are specific as previously stated. The clarification of mission is vital to staying close to the demands and expectations of stakeholders. Through collaboration, it has been possible to get things done much faster.

The External Environment
Attention to the external envelope is based on the concept of managing PESTs - political, economic, social, technological forces and trends. The issues range from the current debate on energy taxation, the phase out of T12 lamps, to the developments in competitive markets for power. For open power markets, see Candlish (14) and Jaskow (15).

The Internal Environment
Internally, management endorsement of the energy conservation opportunities facilitates the planning and implementation of goals and objectives.

Strategic Issues
Three issues were identified.

1 Follow-up on the 1988 Rail Energy Study.

2 Should NJT intervene in JCP&L and PSE&G tariff proceedings?

3 Determination of NJT lighting DSM potential

Formulation of Strategies
Developing strategies for each issue relied upon certain aspects of SWOT analysis - strength, weakness, opportunities, and threats.

Future Vision
The future vision is to remain focused on the regulatory environment and intercede where possible. Another target is to increase awareness and encourage energy efficient behavior.

STRATEGIC ISSUES DISCUSSION
Energy Study Implementation
NJT commissioned Carnegie Mellon Rail Systems Center CMU to investigate the nature of rail energy consumption of the Morris and Essex Lines in 1988. Identified in part where opportunities in rate case participation, and the use of braking energy. The study's advise on rate case participation was endorsed. However, immediate participation was deferred because there were no active rate proceedings by utilities serving NJT. Certain aspects of regenerative braking power for on-board auxiliaries were already planned for the Arrow III car rehabilitation which is now underway. On the average, the Arrow III cars have logged 1.5 million miles each since 1977 (16).

RATE CASE INTERVENTION - JCP&L
In 1991 JCP&L petitioned the NJ BRC for an increase in electricity rates. Initially, NJT was not an intervenor party. However, in June of 1992, ME&A initiated an in-house quantitative analysis of the petitioners request using scenario construction. It was determined that NJT could potentially lose a great deal if it did not act. Therefore, the recommendation to intercede was made and endorsed by the company.

The principal reason for intervention was to ensure that the rate NJT pays for electric propulsion power did not exceed the cost to serve it. The following summarizes the steps taken:

1. The firm of R. J. Rudden Associates provided expert witness and cost-based technical positions. A mini cost of service study showed that it cost JCP&L less to serve NJT for traction power than other members of the GT class.

2. The public policy considerations for rate design affecting NJT were advanced by the Assistant Executive Director for Planning, Mr. Alfred H. Harf.

3. Internal litigation support was provided by Barbara S. Goldsmith, Esq. Deputy Attorney General.

4. NJT sought and was granted:

   An $0.88 discount for 230kV bulk transmission power. It asked for $1.25 relief.

   JCP&L was ordered to study the separation of NJT from the GT class in its next rate case because of it's load diversity.

   The treatment of energy consumption during certain Federal holidays as off-peak power.

   Conjunctive billing of contiguous substations that are abutted electrically.

   The use of hourly demand rather than the current 15 minute interval measurement for traction power was rejected. Instead, an engineering study of the issue was ordered.

The net benefit for the approved items was over $140,000 a year. In granting conjunctive billing and exception for federal holidays, the billable on-peak period was also modified as an offset. JCP&L was granted a rate increase on February 26, 1993 (17).

RATE CASE INTERVENTION - PSE&G
PSE&G filed for electric and gas rate increases in 1991. An examination of the petitioners request showed that the special provisions for traction power purchases were unaffected. However, a minimum level of intervention to monitor the activities of others was adopted. The defensive stance was also based on consultations with another

"stakeholder" the energy and rate group of the Port Authority of NY/NJ for PATH power. PSE&G rates were increased on January 1, 1993 (18).

DSM - LIGHTING PROGRAM

The decision to embark on lighting retrofit was based on three factors:

> The opportunity for a lasting reduction on lighting energy.
>
> Approved Utility rebate incentives which enhanced the return on investment or accelerated "negawatt" "no cost" installation offers.
>
> Advances in lamps, ballast, and reflector technologies.

The model for DSM Planning previously discussed was used. Some key milestones are as following:

1 Walk-through audit of major facilities.

2 Consultation with utility representatives, Energy Service Companies (ESCOs), and manufacturers.

3 Visits to sites with completed lighting projects.

4 Two varying lighting retrofit demonstrations were installed to ascertain the illumination efficacy and project feasibility. The results met with unqualified approval by both users at the test sites and senior management.

5 A site was selected for the first full scale project. An independent auditor provided room by room inventory of fixtures in readiness for the bid package.

The decision and action process are continuous activities. Under consideration are occupancy sensors, energy management systems, timers, and photo cells. Collaborative discussions on enhancing the color rendition of certain high intensity discharge (HID) fixtures and the phaseout of mercury vapor lamps is also covered. The finance options include "no cost," shared savings, and turnkey. The first project is expected to be underway in the Fall of 1993.

SUMMARY

It has been shown that because of the concern for the cost of energy, legislative mandates, and environmental responsiveness, NJT has stepped up it's effort in energy management. In response to the broad based goals, the Management Engineering and Analysis unit developed an energy decision making approach that is systematic and based upon the ideas of strategic thinking and action. Once the strategic issues were clearly defined and the consequences understood, it was relatively easy to act.

The internal endorsement and stakeholder mediation was found to be instrumental in the acceleration of action. NJT intervened in the tariff design process to ensure that the rates it pays for traction power were appropriate and just. The rate case participation proved beneficial. The adage that you have to be in it to win it was proven.

The test installations also proved invaluable in confirming the veracity of saving claims and helped with the organizational marketing of the DSM concept. Unsolicited testimonials received during a tour of the test sites assisted in securing management support.

REFERENCES

1 NJ Transit, "Quick Facts."

2 NJ Transit 1992 Annual Report, pp. 13-17.

3 US DOT/FTA, Transit Profiles: the Thirty Largest Agencies For the 1991 Section 15 Report Year, Govt. Printer, pp. 38-39.

4 New Jersey Statutes Annotated, 1992 West Publishing Co. St. Paul, Minn., pp. 230-257.

5 1991 New Jersey Energy Master Plan, State of New Jersey, Dept. of Environmental Protection and Energy, 1991 Trenton, NJ.

6 1991 New Jersey Energy Master Plan Implementation Section, State of New Jersey, Dept. of Environmental Protection and Energy, 1993 Trenton, NJ.

7 US DOT, Data Tables For 1991 Section 15 Report, December 1992. Govt. Printer.

8 Vantuono, William C., "NJ Transit: On A Growth Curve," Railway Age, May 1993, pp. 55-57.

9 The Star Ledger, "Hudson Light Rail Line Wins Federal Approval," June 8, 1993, pg. 20.

10 Sher, B. A. " Energy Accounting of Commercial Buildings," Strategic Planning For Cogeneration and Energy Management, 1986 The Fairmont Press, Atlanta, Ga. pp. 219-221.

11 Bryson, John M. "An Effective Strategic Planning Approach for Public and Nonprofit Organizations," in Strategic Planning for Public and Nonprofit Organization Jossey-Bass Publishing, 1989 pp.46-70.

12 Rupnow, R. F. and M. Bosch, " Strategic Planning: It's Merits and History," Strategic Planning For Cogeneration and Energy Management, 1986 The Fairmont Press, Atlanta, Ga. pp. 345-347.

13 Kopfle, J.T. "Southwire's Strategic Energy Plan," Strategic Planning For Cogeneration and Energy Management, 1986 The Fairmont Press, Atlanta, Ga. pp. 337-342.

14 Candlish, James " A Market Based Solution To The Rising Cost of Electricity," Massachusetts Bay Transportation Authority, Boston, Mass. Presented at the 1993 APTA Conference.

15 Joskow, Paul L. and Richard Schmalensee, Markets For Power: An Analysis Of Electricity Utility Deregulation, 1988 MIT Press

16 NJT Energy Study of The Morris and Essex Lines, by Carnegie Mellon Rail Systems Center, Pitts. PA. 1988.

17 Jersey Central Power and Light Rate Petition: BRC Docket# ER91121820J, OAL Docket No. PUC661-92.

18 Public Service Electric and Gas Rate Petition: BRC Docket# ER9111698J, OAL Docket# PUC111058-91N.

Chapter 48
An Integrated Approach to Providing Solutions to HVAC Systems

E.C. Newcomb

No three issues today command as much public attention, political activity, and facility management head-scratching as: CFCs, Indoor Air Quality, and Energy Conservation.

Through taxation, fines, incentives, and litigation, the powers that be will set a new course for how mechanical systems are designed, installed, operated and maintained based on these three issues. The complacency that was so characteristic of the HVAC community for decades will have to be replaced with proactive approaches to evaluation of designs for new facilities as well as for evaluation of existing facilities. Evidence that this has already taken place is in the rise in attendance at environmental forums, in the explosion in the publishing media, in the number of new businesses that have erupted from environmental/energy contracting not to mention manufacturers' commercialization of new technologies making use of environmental research and development.

Due to the increase in activity, there are a lot of players in this arena - all with associated pains and opportunities.

Government

The federal government is under numerous mandates to:

- reduce dependency on foreign energy supplies,
- reduce the deficit,
- put more people back to work,
- prevent unnecessary economic hardship from being inflicted on the business community,
- protect the environment,
- protect the health and safety of constituents, and
- decide which agencies are responsible for what regulation, verification, and enforcement.

Utilities

Utilities also have a complicated agenda. They are painfully trying to:

- prevent conversions from one energy source to another,
- effectively reduce energy consumption without reducing profits,
- develop a politically correct IRP with associated DSM plans that can be implemented (i.e., easy to understand and fair to manufacturers).

Facility Managers
Today owners/operators have to absorb more responsibility than ever before. Therefore, facility managers must be more:

- involved in the design and construction phases of the projects in that the ultimate liabilities that exist fall directly on the shoulders of the owner;
- knowledgeable of environmental issues due to the regulatory and absolute liability ramifications;
- attentive to health and safety issues such as Radon, Asbestos Abatement, ADA, Backflow, emfs, and how to interact with EHS managers;
- capable of working within reduced operation budgets ensuring the owner's competitiveness.

Manufacturers
Manufacturers are facing similar pains:

- needing to understand the pains of all of the above,
- keeping aware of all the new technologies,
- qualifying their use in the development of products that make sense,
- tying these products all together in effective and reliable systems,
- providing design and analysis tools to aid in application decisions,
- controlling costs in a globally competitive market, and at the same time,
- making sure that they are operating in an environmentally correct manner.

While all of these issues will cause the individual entities to make fundamental changes in the way they do business, the pain can be offset by the benefits that can be derived by approaching these changes as opportunities to improve the life style of constituents, relationships with customers, employment, and return on investment.

For instance, global government activity will create environmental policies that will reduce ozone depletion, prevent further global warming and unnecessary reductions in the natural resource supplies at the same time as providing jobs due to the new industries that arise. Consistent information dissemination regarding indoor air quality will directly improve the life styles of those people who are currently disabled from being productive due to chronic respiratory illnesses created by airborne contaminants.

The opportunity for the utilities is in the motivation to re-engineer the way they do business. Because they are forced to create their internal resource plans, keeping in mind environmental impact, new economics, new relationships with customers and new processes will naturally become the positive by-products of their efforts.

Facility Managers, either by incentive or by mandate, will have opportunities to upgrade existing systems or create new systems based on high efficiency, low emissions, and improved maintenance criteria. Because environmental issues have such a high public profile, management attention, and hopefully budgets, will be increased for mechanical systems. Special grants - Green Lights, Smart - Buildings - and rebates from utilities will provide even better returns on investments.

This increased attention has provided opportunity for manufacturers. Today more than ever before, we have the chance to develop equipment and quality system designs that make sense. All kinds of new customers from the government, utilities and private sectors are looking to us for issue updates, systems training, economic analysis tools, rebate opportunities, creative financing, turnkey management, as well as integrated system approaches. They are also looking for leadership in how to convert existing facilities into fully functional, environmentally correct systems.

With all of these players and the opportunities they have to offer, the owner must be careful to analyze his property keeping in mind the total picture including all of the requirements that are attendant to the new regulations, taxations, liabilities, and return on investment opportunities. Specifically, if mechanical systems are "fixed" for energy, they may not be "fixed" for CFCs or they might not meet the comfort and indoor air quality requirements of the occupants.

THE BIG PICTURE PROBLEMS

1. **CFCs Used in Chilled Water Systems**

How to prevent refrigerant emissions into the atmosphere.

Today, all HVAC manufacturers provide equipment that ships only with "alternative" refrigerants. "Alternative" according to US EPA, and building code officials, includes HCFC-22, HCFC-123, and HFC 134a. Therefore, if it is a new facility, there are numerous choices as to what chilled water equipment can be used.

However, if it is an existing facility, and replacement of the chillers is not possible due to access, the age of the chiller, sbestos abatement issues, budgetary constraints, the next choice may be to convert. It is critical that there is an understanding of the effects that conversion has on the capacity and efficiency. Before conversion is implemented, these factors must be included in the analysis.

Finally, if neither replacement or conversion is possible, it is the owner's responsibility to effectively contain the refrigerant inside the chiller. Because EPA allows only de minimus emissions during servicing, state of the art recovery equipment used by the servicer is required. Also, because EPA will not allow excessive leakage of equipment, high efficiency purges, pressurization equipment and effective monitoring equipment are necessary.

In all cases, following Guideline 3 which describes pressures, temperatures, and other points to be monitored insures protection of the environment. Also, in all cases, ASHRAE Standard 15, the safety standard, must be followed which requires proper refrigerant monitoring, ventilation, SCBA equipment, and purge rupture disk piping installation.

2. **Indoor Air Quality**

How to ensure good indoor air quality in a building.

In new systems, it is the system designer;s responsibility to design to ASHRAE 62 Standards; but, it is the owner's responsibility to make sure the construction manager/contractor follows those criteria and that the system is properly maintained and operated. No court in the world will let him "off the hook" if people get sick due to poor indoor air quality. In addition to meeting ASHRAE 62 is the market requirement of the owner to keep people comfortable and safe from air borne contaminants. Uncomfortable people are not productive; they are often times disgruntled; and they often leave facilities. Some are replaceable, some are not. Comfort can only be provided through a knowledge of all the system components, their opportunities to fail, and all the system requirements (varying codes, outside conditions and environmental restrictions) - therefore, ensuring clean air at the right flows and temperatures.

For existing facilities, again it becomes more complicated in that there may already be an existing problem. Only by proper appraisal and remediation can the facility be brought up to acceptable standards. The large amount of national publicity that "sick buildings" have created has heightened the senstivity of the American public to the point that EPA determined in a public survey that indoor air pollution in the work place ranked at the top of the list of major environmental concerns.

3. **Energy**

How to reduce the amount of energy used by the largest energy consumer in the building - the mechanical system.

Marriott saving 2% of energy per year means 23.7 million kwh ($1.7 million), 45 million pounds of carbon dioxide, 195,000 pounds of sulphur dioxide and 225,000 pounds of nitrogen oxide. There can be no argument as to whether Energy is an environmental issue.

Again, for new building designs, new technologies for improving the building envelope, glazing, polystyrene building blocks, insulation, will today reduce the building load. The luxury of creating intelligent chilled water system designs in a new structure (not always possible with existing building redesigns) affords the owner not only the opportunity to have a highly efficient but also a highly reliable system.

Hydraulic Design. The root cause of essentially all of the difficultities with parallel chiller control is the fixed relationship between chiller and system flow rates. If you hydraulically decouple the production piping from the distribution piping, it is possible to control them separately and the fixed relationships are broken apart.

This system, commonly called a decoupled system or primary/-secondary pumping arrangement, requires separate pumps dedicated to production and to distribution and a bypass line so that while the same water is pumped twice, there is no duplication of pumping energy. The individual chiller production pump - which only operates when that individual chiller is running - handles only the pressure differential from the return water tee to the supply water tee. The distribution pumps then take the water from the supply water tee and push it on to the return water tee.

A wide variety of distribution system arrangements is possible:

- A single, large, diversified pumping station which can appear as a single or multiplexed pumps sequenced by several methods;

- Alternately, each of several secondary distribution systems can be piped in parallel which has the advantage of providing an easy way to implement plant expansion.

In any case, significant energy reduction can be realized as part load pumping power approaches the theoretical cubic relationship to flow.

Experience with this system has proven that there are three advantages to this design:

- The quantity of water that flows is diversified in that not all chilled water loads peak simultaneously providing the opportunity to reduce pump and pipe sizes.
- With variable flow requirements, there is a reduced need for pumping energy.
- Decoupled systems provide higher return water temperatures since all water that is sent out for distribution is used to handle real loads and not to be bypassed through a 3-way valve. Therefore, all water has accomplished some cooling and the return water temperature will attempt to be at least as high as full load. Consistently warm return water provides other system design opportunities in that it permits preferential loading which is most useful in base loading the most efficient chillers as well as in heat recovery, ice storage and free cooling applications.

Further decoupling can be implemented on very large systems with tertiary pumping at the loads. However, the loads must be controlled so that only the water needed to perform cooling is taken from the distribution loop and must not be allowed to be returned until it has sustained a temperature rise.

Effective Controls and Efficient Equipment Design. Implicit in a design such as the decoupled system is the need for appropriate chiller plant control of the various components. Thorough understanding is necessary of:

- How the system works;
- How the equipment, collectively, operate the most efficiently and reliably;
- What system temperatures, pressures and flows must be sensed; and
- Where they should be sensed in order to determine if, when, and which chillers, pumps, and tower fans should be turned on and off.

Otherwise, all the pump energy savings that could have been saved in the decoupled system will be lost on starting and stopping the wrong chiller at the wrong time.

Also, appropriate system controls for special, energy cost improvement applications are necessary to ensure that the system works and that the savings are realized. These system designs require proper monitoring, scheduling and failure recovery control strategies.

- **Ice storage** implemented to reduce demand charges, reduce chiller size, or provide alternative cooling source. When this system is designed in combination with colder (i.e., less that 42 degrees Fahrenheit) leaving leaving water temperature is attractive for its humidity control potential and smaller distribution pipe size.

- **Free cooling** implemented to prevent chiller starts or to pre-cool return water by taking advantage of cold ambient temperatures.
- **Variable speed drives** implemented to reduce speed and energy consumption when building loads are reduced below full load design.
- **Two small chillers piped in series vs. one large chiller** By examining the development of centrifugal chillers and the economics that have driven that development, you will realize that R & D efforts have been primarily dedicated to smaller chillers. The old designs for large chillers (above 1200 tons) have not kept up with improvements that have been incorporated in smaller chillers. Especially with the influx of changes that have been implemented since alternative refrigerants have been introduced.
- **Multiple energy source chiller combinations (or hybrid systems)** implemented to reduce demand charges. Currently a popular design can accomplish this goal by utilizing a direct-fired absorption chiller piped in series with a centrifugal in combination with an ice storage system.

During off-peak hours, the system is isolated and the centrifugal has a 26 degree leaving glycol temperature and the tanks are charged. When mid-peak hours are reached, the ice tanks are dormant and one or both chillers will be loaded as required. Next, peak operation is enabled by totally eliminating operation of the centrifugal. The direct-fired takes as much of the load as it can and any remaining load requirements are passed onto the ice tanks. In this way, the benefits of a full storage system are realized without having to buy a full storage system, lowering the first cost requirement for ice and cutting in half the space requirement.

In addition, opportunities for savings on the air side of the system does exist with such technologies as air handling unit optimization using enthalpy economizers, variable speed drives, and colder air temperature designs.

High efficiency lighting and intelligent building automation controls will further help to reduce the internal load in the building. Working the rebate and grant programs available can even help reduce pay backs while ensuring good life cycle cost analysis.

Big Picture Solutions

In the past, EMS systems were "jumpered out" for several reasons. They were too complicated for operators to understand, they didn't work properly because the controls did not know how to control the equipment, and they sacrificed comfort for energy savings.

Also, it was not unusual for outside air dampers to be sealed shut because it created too large of a load on an underdesigned system or when capacity was lost due to equipment failure.

Mechanical construction designs and operating budgets were cut because of other priorities such as marble in the foyer.

These scenarios can be avoided if the justifications for design and redesign include opportunities that can be realized in solving CFC, IAQ, and energy problems simultaneously.

For example, an owner decides to implement a lighting retrofit, the economic feasibility based on the Green Lights Program - not an unusual scenario. Now, by virtue of the fact that the load is reduced, his VAV system will react instinctively by raising the temperature of the air and reducing the air flow. The result of this control activity would be that the space will become more humid and occupants will have the perception that the indoor atmosphere is "stifling."

At the same time, an evaluation is being conducted as to how to convert the chilled water system to remove CFCs.

Now, if the designer is thinking in terms of systems, he will conceptualize all of these events as an opportunity to solve all three problems: removing CFCs, improving IAQ, and reducing energy. First, by reducing the air temperature he not only provides better humidity control, he also can reduce pumping costs. (In new construction pump, pipe, and coil size reduction are also possible.) The chiller, on the other hand, must be properly designed to handle the lower water temperature. An ideal time to consider this redesign is during conversion or replacement so that energy optimization can be realized simultaneously.

In this example, as in many real life scenarios, the impact of IAQ and CFC issues can have minimal impact on energy usage if the system parameters are thoroughly explored.

Consultants and manufacturers should be able to provide tools to aid in this analysis and to provide accurate and comprehensive solutions.

CONCLUSION

The challenges are great. Plans must include evaluation on systems that not only are reliable and are economically feasible but are also environmentally correct and at the very least do not impair the health and safety of occupants and operators.

The opportunities are greater. Today, more than at any time before, those of us who design, create, provide energy to, and own HVAC systems have a clear opportunity to take a leadership position and to do the right things that positively impact our environment inside and outside the buildings that we occupy.

Chapter 49

Energy Savings in Computer/Data Centers

P.F. Hutchins, J.W. Wade, G. Sparts

ABSTRACT
As our society becomes more "computerized" and our world "shrinks" with advances in communications technologies, more and more computer facilities have been built. These facilities are the most energy-intensive of all commercial buildings, often operating 24 hours per day, 365 days per year with energy use three to four times that of typical office buildings. This paper addresses both operational and capital expenditure energy savings opportunities at these facilities. The observations and ideas presented here are a result of an energy survey of over 50 commercial buildings of major data processing corporations.

High energy use at data centers is exacerbated by practices established in their early development. Most centers are grossly over-lighted, over-cooled, and use far too restrictive tolerances for variations in temperature and relative humidity levels. Earlier electronic equipment required a higher degree of space temperature/humidity control and lower temperatures. Although newer equipment can operate over a wider range of temperatures and humidities, data center managers are very cautious of their equipment and would rather adhere to practices that worked in the past.

Since most computer room air conditioning units (CRACs) provide energy-intensive reheat coils and humidifiers, tight or low humidity settings coupled with low temperature setpoints can result in excessive energy use. Where units are in close proximity to one another, we have observed units "fighting" one another--one humidifying and the other dehumidifying. This paper will develop the costs associated with these and other energy wasting practices.

Areas of savings requiring capital expenditures are available in closely addressing ventilation requirements and providing space cooling via water-side economizer "free cooling" techniques. Since space cooling is a year-round requirement in data centers, there are many hours of energy savings available when the outside air temperatures drop below 55°F. How this is implemented varies with the type of cooling system used--central, packaged, air-cooled, water-cooled, etc. An evaluation of a variety of capital expenditures energy-saving projects are presented here, along with economic analysis results showing the simple payback that can be achieved.

Introduction

Computer room/data center HVAC (Heating, Ventilating and Air Conditioning) systems are as important as any system that exists. The small CPUs and their peripherals are prone to failure and errors when exposed to temperatures outside their operating range, while the larger machines have internal thermal switches that shut the equipment off when the high temperature limit is exceeded. Humidity also must be controlled to aid in keeping paper handling operations operating smoothly and to reduce static electricity problems. Where water-cooled mainframes are used, the quality, temperature and flow rates of the cooling water must be maintained within close limits or a thermal shutdown can occur. A thermal shutdown on these larger units can be lengthy and costly, as internal components may be damaged by the high temperature and required replacement.

However, because of the special space temperature and relative humidity requirements that are placed on computer rooms, they are intensive users of energy. Consequently, they are prime targets for energy savings. The following discusses Energy Conservation Opportunities (ECOs) in operations and retrofit projects.

Savings in Operations

ECO #1--Computer Equipment Power Down
Often, computer center equipment remains energized 8,760 hours per year, regardless of the use requirements. This not only uses energy directly as each piece of equipment operates, but requires space temperature and humidity control. Powering down tape drives, printers, some mainframe computers and other devices on a regular basis can show large savings. An example is described below.

A large computer facility in the Midwest uses 328 tape drives, 12 printers and several mainframe computers which run continuously, 24 hours a day, seven days a week throughout the year. The tape drives are used regularly during prime time (0700 to 1800 hours) and occasionally on weekends when saving secured files. About 50 percent of the tape drives and printers are not needed eight to 12 hours each night. Some of the mainframes are not scheduled for any user activities, especially on weekends.

The basic technology recommended is the on/off switch. Through a detailed analysis of the operations schedule and regular observations, optimum on/off schedules were determined. The objective was stressed to minimize power consumption with no impact to the computer facility users.

The results of turning off tape drives, printers, mainframes and lights were significant. In 1990, the savings were $86,300; in 1991, the amount was $168,000 and through the first quarter of 1992, $38,700. The total savings over the two-and-one-quarter-year period was $293,000.

ECO #2--Computer Room Air Conditioner Setpoints
Computer room air conditioning is typically provided through packaged, down-flow air handling units. These units may include stand-alone refrigeration systems (air-cooled, water-cooled, or glycol-cooled) or may require connection to a chiller. Computer room air conditioners (CRACs) are complete with blowers, coils, filters, humidifiers, reheat coils, and controls.

Generally, CRACs are capable of providing temperature and humidity control within much tighter tolerances than the computer hardware actually requires. As the individual controls are asked to maintain tighter tolerances, units will begin to "fight" each other. A unit may be running with reheat (dehumidification) while an adjacent unit operates with the humidifier on. This wastes tremendous amounts of energy with no benefit to the owner.

This condition can be avoided by judicious selection of setpoints and tolerances for temperature and humidity. For instance, consider the following example. A nominal 15-ton, air-cooled, DX split system CRAC has controls and tolerances as follows:

Temperature setpoint	=	72°F
Temperature tolerance	=	±2°F
Humidity setpoint	=	45%
Humidity tolerance	=	±2%

According to the vendor data, when the space temperature and humidity are at the setpoint, the unit can provide 14.5 tons of sensible cooling and zero tons of latent cooling (no dehumidification capacity, see Figure 1). If the space humidity rises two percent, then the reheat coil (25 kW) will energize. Even with the reheat on the unit has no latent capacity, so no dehumidification can occur. Not until the space humidity rises to 50 percent does the unit begin to dehumidify (about ten lbs/hr of water), but the reheat coil continues to operate anyway. If the space humidity drops two percent, then the humidifier (six kW) will energize, even though most hardware can withstand humidities down to 30 percent.

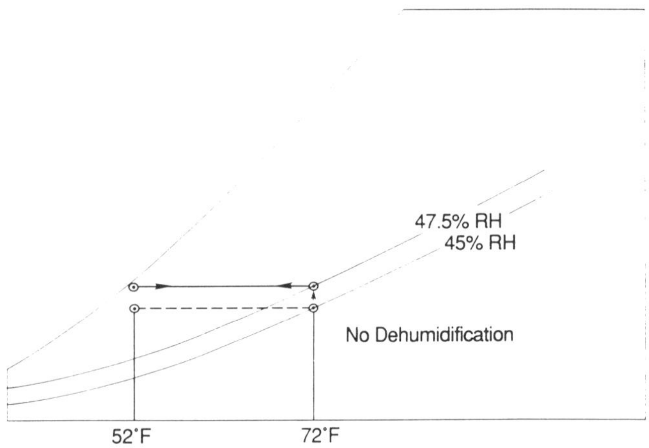

Figure 1. Effects of Small Relative Humidity Tolerances

Now consider the same unit with the following control setpoints and tolerances:

Temperature setpoint	=	75°F
Temperature tolerance	=	±2°F
Humidity setpoint	=	45%
Humidity tolerance	=	±7%

If the space temperature and humidity are at the setpoint, the unit will provide 15 tons of sensible cooling and zero tons of latent cooling (no dehumidification). If the space humidity rises two percent, then reheat coil stays off. If the space humidity rises five percent, then the unit begins to dehumidify at a rate of 14.4 lbs/hr of water--still without reheat. This is shown in Figure 2. Only if the space humidity rises to 52 percent does the 25 kW reheat coil energize and by then, the latent capacity of the unit will allow it to dehumidify at a rate of nearly 20 lbs/hour of water. In other words, the reheat coil will probably never energize. If the humidity in the space begins to drop, then the unit's latent capacity goes to zero at design conditions, so that the CRACs are not causing unwanted dehumidification. The humidifier does not energize until the space humidity reaches 38 percent. Since the most common cause of low humidity is cold ambient temperatures, this would limit the use of the humidifier to times when outdoor temperatures fall below 35°F.

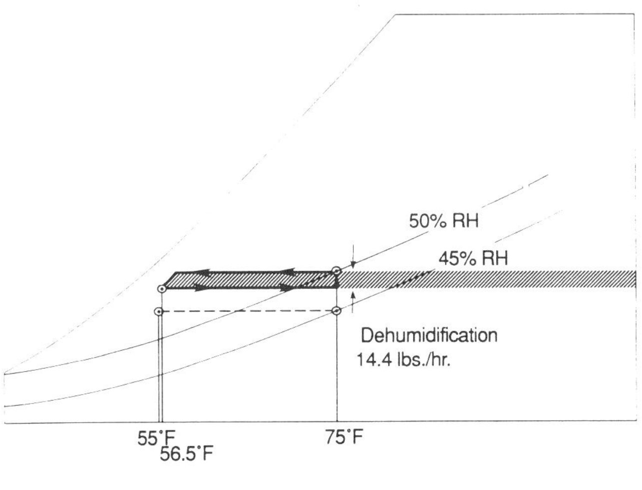

Figure 2. Effects of Larger Relative Humidity Tolerances

A review of a major computer manufacturer's literature for equipment temperature and humidity ranges shows that setpoints recommended above are not unreasonable:

Temperature and Relative Humidity Requirements
Examples

	Temp (°F)	RH (%)
Paper Tape Reader	50-110	8-80
Optical Reader	60-90	20-80
Printer	60-90	20-80
Magnetic Character Reader	65-80	20-65
Display Unit	50-90	8-80
Drum Storage	60-90	20-80
Fixed Head Storage	60-90	20-80
Disk Storage Drive	60-90	20-80
Direct Access Storage	60-90	20-80
Magnetic Tape Unit	60-90	20-80
Processor	60-90	20-80
Mainframe	50-90	8-80
Printer	60-100	8-80
Direct Access Storage	50-105	8-80
Mass Storage Facility	60-85	20-80
Document Processor (w/microfilming feature)	65-80	20-65

Estimated Savings @ 5¢/kWh are:

Annual Electricity Savings (kWh)	16,000/unit
Annual Cost Savings	$800/unit
Swings in Retrofit Projects	

ECO #3--Water-Side Economizer for Computer Room Air Conditioners

The typical computer room HVAC design calls for individual Computer Room Air Conditioners (CRACs) in a down-flow configuration providing cooling to the various pieces of equipment beneath the raised floor. The standard CRACs are air-cooled, direct expansion (DX) units.

Because of the high internal heat loads generated by lights, people and computer equipment, computer rooms have cooling loads all year round. Standard CRACs have compressors and air-cooled condensers, therefore, must operate even when outside air temperatures are cold enough to handle the heat load. This could be thousands of hours per year, especially in colder climates.

An energy-saving retrofit is to fit the CRACs with a chilled water coil pipe and connect same to an exterior dry cooler. During times of the year when the outside temperature is less than approximately 40°F, 45°F water can be supplied to the CRAC. Not only will this allow the computer room units to provide cooling in the winter without operating their compressors and air-cooled condensers, but it will also provide a backup cooling system for the computer room.

It should be noted that reduction in ambient temperature is usually accompanied by a reduction need for dehumidification. Therefore, higher temperatures in water circuits can normally be tolerated during the free cooling season. In many cases, typical 45°F of chilled-water temperature is allowed to rise to 55°F or higher in the free cooling mode of operation. This maximizes tower usage and minimizes system energy consumption.

Preliminary project evaluations showed paybacks were typically less than five years for electricity rates averaging 5¢/kWh.

ECO #4--Central System, Dewpoint Control

As mentioned earlier, a common problem occurs in computer rooms when individual floor-mounted computer room air conditioning units, each with the ability to dehumidify and humidify, begin to "fight" each other. Adjacent units can be found operating in reheat mode and humidifying modes, respectively.

For new construction, a central system with dewpoint control can provide an energy-efficient answer. Michael Cannon and Louis Hannauer in their ASHRAE Journal article, April 1988, discussed the advantages of this system. Since nearly all of the latent loads in data centers can be attributed to outside air ventilation, why not use the central system for dehumidification with virtually no reheat?

The basic concept is to allow the central system to perform humidification and dehumidification functions while the CRACs provide sensible cooling only. The chilled water coil provides a constant leaving air temperature using a modulating valve, resulting in a constant dewpoint temperature while in the dehumidification mode. A spray pump is used to provide humidity when needed.

The stability of this system is enhanced by providing a well-designed vapor barrier between the computer room and adjacent areas, and by allowing the minimum amount of outside air. The central unit could be out of service for days without humidity problems as long as the outside air is shut off. The authors noted that once a dewpoint-controlled central unit was down for six days during the winter and a drop of only seven percent in the room relative humidity was experienced.

Summary

Computer rooms are prime targets for energy savings. These savings can come through changes in operation or capital projects.

In data center design, it is important to allow for economizer use, especially in colder climates. The low first cost air-cooled DX CRACs can be the most expensive system when you include its high energy use.

On the operational side, something as simple as carefully selected temperature and relative humidity setpoints and ranges can create substantial savings.

Knowing the requirements of your equipment and working with the users can result in savings without impacting productivity.

Chapter 50

Changing Utility Markets for Renewable Energy

K. DeGroat, C. Shay

ABSTRACT

At their inception more than a decade ago, research, development, and commercialization strategies for renewable energy technologies (PV, solar thermal, geothermal, biomass, hydro and wind) used in electric generation focused on competitive central station systems as their goal. This objective was widely accepted and promoted by utilities, researchers, manufacturers, and developers. Remote and small-scale applications close to end-use operations were seen as intermediate steps to the final goal of large-scale, competitive power plants. However, since the mid 1970s when renewable energy development began in earnest, the utility industry has changed dramatically, moving toward a system that is less focused on central station generation and more on DSM and other energy service options. At the same time renewable energy technologies have matured, and as they have advanced it has become increasingly clear that many of the best uses for renewable energy technologies are in the distribution system and in end-use applications where electricity has a higher value to the customer and the utility. For some of the renewable technologies, competitive central station applications are becoming a less important goal as commercialization efforts focus on emerging utility interest in value-added energy service markets in DSM, dispersed generation and off-grid "service without a wire" applications.

This broadened definition of renewable energy's potential in utility applications mirrors a shift in utility markets and planning approaches that emphasizes integrated resource planning and demand side management, increasing investment in transmission and distribution assets and concern with their effective utilization, and greater competition from non-utility sources in all phases of utility operations. Central station generating resources which used to be the primary focus of utility planning and investment have become just one element in a much more complex planning and operating environment. As more sophisticated integrated resource planning approaches have developed, renewable energy technologies -- from remote PV to geothermal heat pumps and biomass cogeneration -- are increasingly recognized for their potential role in generation, transmission, distribution and end-use. These changes are only the leading edge of continuing shifts in the definition and operation of utility markets, as the industry struggles with transmission access issues, broader competition in wholesale generating markets, new environmental restrictions, global environmental challenges, and new overseas marketing opportunities.

This paper examines the changes occurring in the utility business as they relate to the renewable energy industry and its potential.

BACKGROUND

With the exception of hydropower, renewable resources did not expand into electricity production until the 1980s. After 1980 commercial development occurred primarily through qualifying facilities (QFs) selling power and capacity to utilities under the Public Utility Regulatory Policies Act (PURPA).

Geothermal, wind, solar, and biomass generating capacity doubled from 1970 to 1980, from 1.4 GW to 2.8 GW. From 1980 to 1990 capacity from these renewable resources quadrupled, to 12.3 GW of capacity, approximately 2% of total U.S. capacity. From 1970 to 1980 conventional hydropower increased by 7.6 GW, and by 3.8 GW from 1980 to 1990. [1]

Although much of this capacity is based on central-station geothermal, biomass, solar thermal and wind energy facilities, it also includes a significant proportion of biomass cogeneration and waste-to-energy that are as much industrial plants and waste disposal facilities as they are electric generating stations. These figures exclude the 137 MW of PV shipments since 1980, which have been deployed in a variety of applications both nationally and internationally, and the significant number of wind turbines deployed in hybrid or stand-alone remote applications.[2]

RENEWABLE ENERGY MARKET EVOLUTION

The underlying factors that have driven renewable energy growth the last 15 years are:

- rising, volatile fossil fuel prices in late 1970s and early 1980s;
- electricity rate shocks and increases in costs associated with the tail end of the utility construction boom of the 1970s;
- state and federal programs to encourage renewable energy use, including PURPA and investment tax credits; and
- significant improvements in technology performance and cost as a result of federal and private renewable energy research efforts.

Fuel Prices

The oil price spikes of 1973-74 and 1979-80 were the original impetus for concerted efforts to improve renewable energy technology and introduce it into utility markets. Projections of $50/barrel oil and gas shortages fueled early investment in renewable energy companies and provided a boost to the industry. Although oil prices collapsed in 1986 and the gas bust turned to a gas surplus, which also caused a major shakeout in the renewable energy industry, the industry was still well-established enough to weather the changed situation.

Electricity Costs

At the same time electricity prices were escalating and capacity surpluses were emerging in the utility industry as a result of the last wave of large nuclear and coal plant projects coming on line. The prudence reviews and disallowances of the early 1980s seriously disrupted the established pattern of utility construction and planning, which focused on large, utility-owned central stations. Just as both regulators and utilities were struggling to find a new way of doing business, PURPA began to take hold and open the industry to competition from non-utility generators for the first time.

PURPA and Tax Credits

The passage of the PURPA provided a major boost to the renewable energy industry. PURPA requires utilities to purchase electricity from qualifying non-utility generators that use renewable energy or cogeneration. The price for the energy and capacity were set at the utilities' avoided cost of production, which varied widely from utility to utility depending on whether they were in a capacity surplus situation and how state regulators chose to determine avoided costs. Since PURPA's inception over 33 GW of qualified non-utility generating capacity has come on line, including 16 GW of renewable energy capacity.[3]

PURPA created the non-utility generating industry. It also created a forum for renewable energy developers and cogenerators to make themselves heard in the utility industry and spur utilities to consider new technologies and new ways of doing business. Even in states with low avoided costs and very little renewable energy or cogeneration development, the PURPA process at least created a mechanism for these new technologies and their advocates to make their case to utilities and utility commissions.

Although PURPA was critical to renewable energy development, it had an even greater effect on cogeneration. Where PURPA opened the door a crack for renewable energy development, for cogeneration the door was flung wide open. And although PURPA benefits made small power production possible, it also imposed arbitrary size restrictions on renewable energy and limits on the use of fossil fuels that in retrospect were counterproductive. These restrictions basically required renewable energy to remain small and detached from conventional fossil fuel approaches to generation, foregoing opportunities for economies of scale and hybrid operations that could have benefitted some technologies. Both of these restrictions were significant factors in the demise of Luz, which developed 354 MW of solar trough generating plants that are still in operation in California under new ownership. It also affected the design and structure of individual wind energy and geothermal projects. These limitations are still in effect.

The avoided cost concept and the idea of purchasing generation from non-utility generators also led to the innovation of bidding, which was the first harbinger of open competition in the wholesale generation and DSM markets. Bidding started in Maine in 1984 and has now spread to over 28 states that have either adopted bidding, have permitted jurisdictional utilities to conduct bidding, are developing bidding programs, or are reviewing the use of bidding.[4] Competition and the success of the NUG market led to utility interest in competing in the wholesale generation market on an equal footing. This eventually led to the key electricity provisions of Title VII of the National Energy Policy Act (EPACT) which reformed the Public Utility Holding Company Act to allow utilities and regulated utility holding companies to compete in the wholesale generating market as electric wholesale generators (EWGs).

Federal tax credits for renewable energy development were also instituted in the early 1980s and played a major role in spurring industry development. The 10% business investment tax credit for renewable energy development was the most important incentive for utility applications of renewable energy, especially in conjunction with matching tax credits in key states such as California. Some of the major problems with the tax credits -- the constant uncertainty surrounding extensions and abuses involving a few projects and developers -- have been largely remedied by the EPACT which provides for a permanent extension of the solar and geothermal tax credits. EPACT also created a 1.5¢/kWh production tax credit for wind and close-looped biomass systems, and for renewable energy production by publicly owned utilities.

Technology Improvement

Finally, without major improvements in the performance and cost of renewable technologies as a result of intense research and development, the market opportunities provided by PURPA and tax credits would have had little effect. Technical advances have been the foundation for commercial applications, for without the advances in PV efficiency, solar thermal designs, wind turbine reliability and output, biomass combustion efficiency and pollution control, and new geothermal technologies, there wouldn't be any competitive applications of renewable energy technology. PV generating costs dropped from $19/kWh in the 1970s to roughly 30¢/kWh today, and in the process opened up cost-effective applications in remote power supply and consumer electronics that couldn't have existed before. Solar trough technology advanced enormously in efficiency, reliability and operations and maintenance from the original SEGS I designs to SEGS VIII. From a high of 60¢/kWh in 1980 solar thermal energy has now dropped to approximately 10¢/kWh.[1]

Wind energy systems are approaching 95% availability at the best wind sites, in contrast to the shaky performance of early turbines. Today the best turbines and wind resources in California are producing energy at less than 5¢/kWh, down from an average of 40¢/kWh in the 1970s. The best geothermal plants in California are producing power at 4.5¢ to 6¢/kWh. Waste to energy and biomass facilities are very sensitive to fuel costs and resource quality, but given the right situation are cost-competitive with utility generation at around 5¢/kWh.[1] These gains are the result of a combination of fundamental breakthroughs in key areas, and incremental engineering improvements and operating experience that has created the technologies and the capabilities that exist today.

Value Added Applications

Gross market penetration estimates and performance/cost curves only provide a rough indication of the potential of renewable energy because they obscure the how and why of renewable energy's inroads into utility markets. For example, much of the progress made in renewable energy electric technologies has been in end-use applications like biomass cogeneration and remote PV and wind systems where system characteristics have closely matched unique end-user requirements for low maintenance, DC power, waste disposal services, or small-scale modular designs.

Over 40 utilities have projects involving solar electric power generation ranging from airport lighting to providing backup power for uninteruptible power supplies. Representative utility-sponsored remote power projects include Public Service Company of New Hampshire (microwave relays), American Samoa Power Authority (island power), Arkansas Power and Light (irrigation systems), Florida Power and Light (irrigation systems), Austin Electric (water-flow meters), Kansas City Electric (stock water pumping), Arizona Public Service Company (communications, warning systems), Pacific Gas and Electric (warning systems, remote monitoring), Bonneville Power Administration (tower warning beacons), Florida Power Corporation (navigational aids, cathodic protection), and Southern California Edison (lighting).[5] While these applications are unlikely to ever constitute a major portion of a utility's generating capacity, they represent applications whose value to the customer or the utility go far beyond the cost per kWh of the energy used in the applications. They perform functions that in many cases would be impractical to provide in any other manner.

Recognition of the potential for high-value applications of renewable energy has also led to research at PG&E, the Department of Energy and EPRI on dispersed generation opportunities for renewable energy. PG&E's study of the benefits of a PV system at the Kerman substation is covering new ground in terms of looking at the benefits of dispersed generation for kVar support, reducing transmission losses, deferring transmission and distribution upgrades, increasing asset utilization in the transmission and distribution system, reducing environmental emissions, mitigating risk by deferring new capacity purchases, and enhancing reliability through diversification of energy sources.[6]

More and more renewable energy generation is being integrated into end-user appliances and processes -- two Japanese companies now offer air conditioners with integrated PV systems. The Department of Energy recently announced five finalists in its first round of awards under the PV:Bonus program to integrate PV into building envelopes. They include:

- Delmarva Power and Light for a dispatchable PV/battery system
- Energy Conversion Devices for PV module/roofing shingles
- Solar Design Associates, for PV modules with integral AC adapters
- Fully Independent Residential Solar Technology, Inc. for manufactured homes with integral PV panels.[7]

The paper industry has turned its waste materials into an energy source and an integrated chemical recovery process for recycling the chemicals in Kraft Black Liquors. Geothermal energy is expanding into end-use markets with ground-source heat pumps. Idaho Power has instituted a pilot program to offer customers in remote locations the option of a utility-installed remote PV system designed, installed and maintained by Idaho Power. Rather than paying for the energy output of the array, customers pay a fixed monthly fee for the electricity service.

The common theme behind these applications is a fundamental shift in the definition of utility services. While PV navigation buoys may seem well outside a utility's definition of its business, the line between utility and customer activities is blurring. More utilities are investing in service without a wire, like La Plata Electric Cooperative which offers customers PV water pumping and other off-grid systems in lieu of line extensions. Both utilities and utility commissions are increasingly looking to energy services as their products, and to the traditional generation, transmission, and distribution system as just one approach to delivering these products. More than 24 states allow utilities to capitalize or rate base DSM expenditures.[3] DSM programs are drawing utilities more and more into the details of their customers' operations and energy decisions.

The changes in renewable energy commercialization patterns are part of a larger shift in the utility planning environment that emphasizes demand side options and integrated resource planning methods. The spread of these new planning approaches has increased the number of markets available for renewables and the competition existing in these markets. While the new focus on demand side management has opened a market for ground source heat pumps and remote PV, it also puts them in competition with traditional conservation and load management techniques. The benefits of dispersed generation could just as easily be provided by a fuel cell or a PV system located at a substation.

Recognizing the importance of these regulatory and planning changes, the Department of Energy, the utility industry and the renewable energy industry are devoting more resources to analyzing and trying to influence utility planning procedures to insure that new technologies are given equal treatment and their benefits are fully considered. DOE's Integrated Resource Planning program is working on distributed utility issues and IRP planning paradigms that impact renewable energy and conservation. They are also investing in efforts like the Regulatory Assistance Project (RAP), which is designed to help regulators apply state of the art IRP planning methods that recognize the value of renewable energy and conservation.

DOE has also supported the Utility Photovoltaic Group (UPVG) and the PV for Utilities (PV4U) consortium. UPVG is a coalition of over 60 utilities and utility organizations that have joined together to develop strategies for encouraging the application and improvement of PV technologies in utility markets. They are currently working on an effort to fund a 50 MW joint purchase of PV equipment in order to help the industry improve product performance through enhanced manufacturing and economies of scale. The PV4U group is a coalition of utility regulators, environmentalists, consumer advocates and state energy officials who are working to identify and eliminate regulatory and institutional barriers to increased use of PV at the state and local level.

INFLUENCES ON FUTURE MARKETS

While both the utility and renewable energy industry have changed dramatically in the last decade, there is no sign that the pace of change is slowing. The two most current examples of continuing change, the Energy Policy Act of 1992 (EPACT) and the Clean Air Act Amendments (CAAA), create new market opportunities, different rules for competition and new competitors in the utility industry that didn't exist before, offering both fresh challenges and fresh opportunities for renewable energy.

PUHCA Reform

The markets PURPA opened have now given way to PUHCA reform and electric wholesale generators, which may largely eclipse the advantages of PUHCA exemptions under PURPA. PUHCA exemptions under PURPA once provided QFs with protection from competing with utilities in the wholesale market. With these restrictions removed, the only remaining effect of PURPA is the requirement to purchase power from QFs. This advantage has already been tempered by the spread of bidding and other new approaches to avoided cost determination and procurement that emphasize competition among all potential sources. The end result is going to be a much tougher competitive situation for all developers of power sources, whether they are EWGs or IPPs or QFs. While gas and DSM are still likely to dominate the generation strategy of most developers, there is an opportunity for the renewable energy industry and for utilities diversifying into the wholesale generating market to take advantage of renewables and renewable set-asides. One strategy is to begin differentiating their product more effectively by emphasizing the benefits of renewable energy in DSM and dispersed applications in comparison to traditional conservation or generation options. For instance, in contrast with many DSM programs the output of renewable energy technologies integrated into customer operations can be easily metered and valued. And unlike load management techniques like radio-controlled load interruptions, using renewable energy in DSM applications involves no real or perceived reduction in service.

Integrated Resource Planning

The continuing evolution of integrated resource planning in the context of PUHCA reform and increased transmission access will have a large influence on how PUHCA reform and the advent of EWGs affect the utility market. As the sources of generation and DSM programs have expanded to include more players, state regulators are moving toward more elaborate utility planning procedures that integrate supply and demand-side options with demand forecasting to identify the best portfolio of options to satisfy multiple objectives.

The methods and outcomes of IRP vary widely from state to state and even utility to utility. But there is a common thread of commission involvement and intervention in the planning process to achieve state and public goals that will temper trends toward wide-open competition in wholesale generation and DSM. Increasingly set-asides for renewables, DSM or other environmental options are specified during planning and then procured in bidding or other competitive processes from the lowest cost sources. It is also in the planning and procurement process that environmental benefits of renewable energy are incorporated into utility decisions, as evidenced by the increasing number of states that include some form of environmental valuation in their planning processes. As more utilities expand their affiliated operations into the wholesale market through EWGs, which are subject to FERC rather than state regulation, while playing more of a broker role in their own service territories, regulators are likely to rely more and more on planning requirements to shape utility decisions.

The Environment

The environmental benefits of renewable energy technologies have always been appealing, but in practice it has been difficult to translate environmental benefits into direct competitive advantage. As choices between generation and DSM have become more integrated, renewable technologies are increasingly competing with DSM's environmental benefits, which are roughly equivalent.

Until recently environmental regulation was based on a command-and-control approach that prescribed technology fixes for existing fossil-based resources, which largely ignored the alternative of encouraging approaches to avoid pollution altogether. The first major breaches in the command-and-control approach came in the form of the emission offsets markets in California. The offset program focused attention on overall pollution reductions and limits rather than on technical specifications for control, and in the process created a market for technology options that could completely avoid or radically reduce emissions through alternative processes.

The CAAA of 1990 brought the market-based approach to bear on sulfur emission control with the system of sulfur allowances and sulfur allowance trading. To date the net effect has been to encourage the installation of scrubbers, resulting in a sulfur allowance market that has been far less active and less expensive than originally predicted. But the fact that there is an explicit cost/value to avoiding emissions in the first place sets a strong precedent for more explicitly valuing the environmental attributes of renewable energy and DSM that is likely to continue. The award of the renewable energy and conservation bonus allowances will be an interesting experiment in recognizing and rewarding the efforts of utilities to reduce sulfur emissions through renewable energy and DSM.

The big uncertainty in future environmental controls is global warming and carbon dioxide. If there are major initiatives to control production of CO2 and other greenhouse gases, renewable energy has the advantage of being the only generating option that produces no greenhouse gas emissions. DSM and tree planting for offsets to fossil fuel emissions will be effective competitors, but they are limited to making inroads in reducing emission loadings that are on an upward trend worldwide. Particularly in developing countries, there is relatively little energy production to conserve, and limited interest in forest preserves when compared to new generation that can fuel economic development.

Globalization

The greenhouse gas issue is also representative of another major trend that will affect the future development of the renewable energy industry; the global implications of utility activities and global competition. The trend toward PURPA-style competition in generation and IRP approaches to planning are being adopted by many foreign utilities, especially in the developing countries, which will open their markets to competition from non-utility generators that might include U.S. utilities. The North American Free Trade Agreement (NAFTA) includes provisions that would allow U.S. utilities to compete on an even footing in Mexican wholesale generating markets. In terms of exports, the provisions of EPACT that lift restrictions on holding company involvement in foreign utilities may have much broader implications than anyone has envisioned so far.

Energy needs in the developing countries, Eastern Europe and the former Soviet Union are enormous, and represent a multi-billion dollar investment opportunity for the utility business. Renewable energy is already very competitive in these markets, particularly in the developing countries where off-grid applications are at an advantage, and where high-quality biomass and geothermal resources are available. The question is whether renewables will gradually lose market share in developing countries as centralized grids develop and utilities move to export their expertise in large-scale, centralized systems, or whether the two types of systems will expand together.

Ideally renewable energy technologies could become one of the areas of expertise offered by U.S. utilities moving into export markets, but to date there has been little indication that this will happen. A recent review of overseas activities by U.S. utilities identified 58 different utility efforts by 12 different companies, ranging from coal plants to transmission and distribution projects. Out of all the 58 efforts, only one involved renewable energy.[8]

However, renewables may be worth considering as part of a portfolio of energy services because they are becoming a higher priority for the World Bank and other international lending institutions. They also have a strong appeal in developing countries because they use indigenous resources and can bring the benefits of electricity to rural populations through phased investments in modular, small-scale systems rather than huge investments in central stations and extensive grids. The recent controversy over emissions from an SCE coal plant in Mexico is a strong indicator that environmental concerns are likely to follow utilities overseas, which will affect their investment strategies.

CONCLUSION

Renewable energy has made significant progress in the utility marketplace, although not along the path that was originally envisioned in the mid 1970s. Instead of the smooth progression to traditional, utility-owned and operated central station systems, renewable energy has evolved into a non-utility generating option that has found its most promising markets in applications where it is able to compete on broader terms than levelized cost of energy. While PURPA, tax credits and a successful research and development effort have resulted in many successful central station projects, it has also encouraged DSM applications and laid the foundation for dispersed generation in the transmission and distribution grid and for utility involvement in off-grid applications. The shift in market focus has paralleled a shift in the utility business toward energy services as the principle product of utilities, rather than raw kWhs. Through IRP and their own diversification strategies more utilities are moving into DSM and even off-grid generation as a key element of their product mix, with potentially as much value as new investments in central station generation.

While both the utility and renewable energy industry have undergone dramatic transformations in the last 15 years, the pace of change shows no signs of slowing. The future outlook for renewable energy and the utility sector will be shaped by the increased competition and transmission access instituted by EPACT, by market-based approaches to pollution control like the CAAA, and by new environmental challenges like global warming.

Globalization of environmental concerns and of the utility business will also widen the range of options and opportunities for both utilities and renewables. No less than the changes of the last 15 years, these new trends will impact renewable energy development in the utility sector. EPACT will create much broader competition between renewable energy, conventional generation and DSM options. It will erode some of the market advantages of PURPA, which were already being downplayed by competitive bidding and IRP processes that emphasized all-source competition for new energy needs.

As environmental concerns like global warming continue to grow, renewable energy's environmental advantages will become more valuable, although they will have to compete with DSM and other emerging non-polluting energy options like fuel cells. Finally, globalization of the utility business and environmental concerns will broaden the scope of competition and markets for both utilities and renewable energy with U.S. utilities playing the role of both potential allies and potential competitors, in the same way they are preparing to enter domestic wholesale generating markets as EWGs.

REFERENCES

[1] United States, Energy Information Administration, *Renewable Resources in the U.S. Electricity Supply,* February 1993.

[2] Maycock, Paul D., "1992 World PV Module Survey," *PV News,* February 1993.

[3] Cambridge Energy Research Associates, *Electric Power Trends 1992,* 1992.

[4] National Independent Energy Producers, *Competing for Power,* July 1991.

[5] Solar Energy Industries Association, 1990 and *Solar Industries Journal, 1993.*

[6] National Renewable Energy Laboratory, *Photovoltaics, New Opportunities for Utilities,* 1991.

[7] Stafford, B. and Taylor, R., "PV:Bonus -- Developing a U.S. Market for PV in Buildings, in *NREL Reprints for the 23rd IEEE Photovoltaic Specialists Conference,* National Renewable Energy Laboratory, 1993, and *The Solar Letter,* Volume 3, Number 9, April 1993.

[8] "U.S. Utility Overseas Activity," *Public Utilities Fortnightly,* February 1, 1993.

Chapter 51

Solar Integrated Pumping Applications

S.C. Turner

ABSTRACT

In January of 1993, the Florida Energy Office made several grant awards to promote the commercialization of Solar applications in Florida by Florida firms. GeoSolar Energy Systems Inc. received an award to commercialize photovoltaic (PV) powered water pumping systems for various applications in the Florida citrus industry.

There are over 700,000 acres in citrus in Florida. Irrigation is vital to good production. Citrus can tolerate neither drought nor prolonged wetting of root systems. Many growers in freeze-prone areas rely on water for freeze protection. Increasingly, "fertigation" and "chemigation" are augmenting traditional distribution methods.

Most pumping in groves requires power for rotating equipment in remote areas. Traditionally, these sites have been handled via grid extension for electric pumps or diesel engines with on-site tanks.

PV pumping can meet growers needs in several different situations. In the opening phase of this project, GeoSolar Energy Systems identified application criteria for PV powered pumps in citrus groves. Then, GeoSolar developed a design for each of six sites. Every site is representative of at least one criterion for PV applications. The Solar Energy Laboratory (SEL) at Florida Atlantic University then confirmed array requirements and simulated performance using computer modelling.

Final installations are proceeding over the summer of 1993. Field monitoring will yield data to be presented at the October 1993 World Energy Engineering Congress.

This paper discusses irrigation techniques in the citrus industry, characteristics of PV pumping systems, and key features of one system, with an overview of the other five systems. Detailed simulation results predicting system performance are included.

CITRUS GROWING AREAS IN FLORIDA

Over 700,000 acres are in citrus in Florida. Two major land types dominate. In the center of the peninsula, a spine-like raised area called the Ridge contains many groves. Along the east coast the famous Indian River Citrus grows in the Flatwoods. These two growing areas differ in many ways including soil types, weather, and availability of water.

The Ridge

The Ridge groves are characterized by deeper root zones, colder weather, and a paucity of surface water. The deeper root zones are a result of a thicker layer of top soil than is found nearer the coasts in Florida. Although the elevation of the Ridge compared to surrounding areas provides some frost protection against calm freezes in which cooler air rolls down off the Ridge, recent decades have seen this area periodically devastated by cold and windy winter weather systems. Ground water is typically pumped from deep wells. In some low areas, local topography permits use of horizontal de-watering wells.

The Flatwoods

Nearer the east coast and further south in general, citrus groves are planted where dry sandy scrub or slash pine with saw palmetto used to dominate. The topsoils here are thin to non-existent, resulting in an extremely thin root zone. The southerly location and proximity to the ocean greatly reduce frost problems except in the northern end of the area. Ground water is available from surface canals.

IRRIGATION TECHNIQUES

On the ridge, commonly used watering techniques are overhead sprinkler and low volume irrigation (LVI), including microsprinklering and drip. In the Flatwoods, techniques are volume gun, LVI, and seeper/flood.

Overhead sprinklers distribute high volume flow rates from above the tree canopy. Many groves with these systems experienced extensive freeze damage in the brutal freezes of the eighties.

LVI systems are usually microsprinkler emitters under each tree operating at flow rates of from twenty to ten gallons per hour (GPH) per tree. These systems afford freeze protection, preserving the vital growth base of the trees if not the fruit itself. The latent heat of fusion absorbed by the water as it turns to ice can maintain tree temperatures above the critical temperature below which irreversible freeze damage occurs.

Many growers are using lower flow rate LVI systems including either drippers or the Israeli developed tubing which incorporates self-regulating emitters. Flow rates in these systems range down to a few GPH per emitter, but each tree may have more than one emitter.

Volume guns are self-propelled high volume devices resembling fire-fighting nozzles with clappers which travel along a canal edge.

Seeper/flood systems are used to flood an entire block of a grove, which is then drained. The remaining water is afterwards used to flood the adjacent block.

PV PUMPING SYSTEMS FOR CITRUS

Costs for PV systems are currently comparable to traditional systems only when the following conditions apply:

1) Grid power is not on-site.

2) Automatic start/stop is required for diesel.

3) The block to be irrigated is less than ten acres.

4) LVI is used.

5) Water supply is at or near the surface.

6) Distribution piping minimizes pressure drop due to friction head and no excessive topography requires high discharge head.

The key components of a PV powered pumping system are the array and the pump. If used, the system may also include trackers, batteries, the charge controller, the pump controller, and the inverter.

This project uses fifty three watt panels in arrays with parallel/series wiring and quantities as determined by the pump power and voltage requirements.

Pump sizes maximize flow for the required TDH at each site. Since irrigation is needed only every second or third day during the dry season, and less frequently at other times of year, each pump can service multiple zones. Watering requirements coincide with times of high solar insolation, so arrays need not be oversized, and batteries or water storage are not required. DC pumps do not deteriorate with under-voltage or under-power conditions, and are thus well-suited for direct-to-array wiring.

Much heavy equipment operates in groves, so protecting arrays from physical contact is important. A shed or shelter roof is preferable except where arrays can be located out of the path of grove traffic. For similar reasons, tracking devices were not incorporated in any of the systems, despite their great benefit to array output.

Batteries and charge controllers are not required except in freeze protection applications, where night-time operation is mandatory. With DC pumps, system control requirements are limited to zone control, for which several good self-powered micro-processor based units are commercially available. Since inverters inherently require batteries, AC pumps were avoided.

The focus in system development has been on good performance, high reliability, and extreme simplicity of configuration. Once grower acceptance has developed, an increasingly high level of control and automation will eventually distinguish PV systems by virtue of their intelligence.

The critical limitations of PV systems at this time are:

1) Array cost.

2) Artificially low utility power costs.

3) Grower awareness.

4) Power limitations relative to array size.

5) Absence of environmental restrictions on portable diesel tank locations.

6) Low level of restrictions on water use by growers.

Water Availability

Consumptive Use Permits (CUPS) are issued by the local Water Management Districts to the grower. Although these permits have been readily available and volume of consumption has been estimated and not strictly monitored or regulated, CUPS are now issued for a limited term and renewals typically require the installation of metering devices to measure actual usage. This suggests a general trend towards volume restrictions and/or volume based costing.

Traditional Fuels

Power for irrigation systems has been grid electricity or diesel, with some gasoline used. Fuel costs for operation have been estimated at five to ten dollars per acre-inch. Typical horsepowers for irrigation pumps are in the ten to one hundred horsepower range.

CHARACTERISTICS OF PV SYSTEMS

Typical PV systems ranging in the one- half to one and one-half horsepower range due to limitations imposed by array costs and sizes. The need to identify niche applications for PV with limited volume and total dynamic head (TDH) requirements is compelling. PV pumping system development has focused on low volume, high lift applications such as remote home and village water supplies from deep wells and livestock watering in arid and semi-arid zones. In these applications, little or no energy is expended in distributing the water above the surface; pump energy requirements are dictated by the depth to the water source below the ground. By contrast, agricultural applications require extensive distribution, and sufficient pressure at the emitters to achieve requisite distribution patterns. Furthermore, the vast distribution areas in groves lead to substantial pressure drops in surface piping. Thus the primary requisite for suitable PV sites in citrus is a water source at or near the surface. Hence the Flatwoods, with their extensive canal systems, hold forth great promise for PV applications. However, many sites on the Ridge with nearby lakes or the potential for shallow horizontal de-watering wells are good candidates for PV.

Pressure and flow requirements at the distribution device also dominate the determination of eligibility of sites. Obviously, overhead sprinklers and volume guns have requirements not attainable with PV. However, the profligate use and inherent disadvantages of these systems have led to discontinuation of installations of these types in new groves.

Seeper/flood systems have low TDH's, but many of the systems have volume flow rate needs out of the range of PV systems. Some low lying areas in the Flatwoods require almost constant, medium volume, very low TDH drainage pumping to prevent excessive soil moisture, the "wet feet" situation which will harm root systems.

LVI groves with surface water available are ideal candidates for PV.

As these factors normalize, PV will become part of the mainstream of Florida agriculture.

SITES AND APPLICATIONS

Six sites were selected throughout the Florida citrus growing areas. Each was selected due to a combination of the eligibility factors enumerated above. These systems are currently at various stages of construction and testing. Further results and field measurements are available from the author.

Site Number One

Located in the Flatwoods, this site is on a large grove which utilizes the seeper/flood techniques described above. Each block of grove is surrounded by berms and service roads. Growing space is at an ever-increasing premium due to rising land values and expenses, and tough global competition in the citrus markets. This grower has thus been motivated to put narrow strips of land around the flood irrigated groves into production. However, these raised areas receive no water from the flooding. A Solar Integrated Pumping System (SIPA) provides water to several hundred trees in several rows around an existing block of trees. Water source is a canal; the pump and panels are located near a wing wall on the canal. A twelve by twenty foot shed roof without walls was constructed to provide secure mounting for the PV array out of the way of heavy equipment used in the grove, especially at harvest time. It also shelters the pump and appurtenances.

Suction piping and discharge mains are three inch PVC. Piping eventually down-sizes to three-quarter inch irrigation tubing. Emitters rated for ten gallons per hour (GPH) at fifteen pounds per square inch (PSIG) are fed by one-quarter inch tubing taps from the three-quarter inch tubing. Each tree is served by one emitter. The pump system serves three zones; one zone is over a thousand feet from the pump. Generous piping sizes and long sweep ells reduce friction losses.

The pump is an eighty-four volt D.C. direct drive centrifugal pump; it is nominally rated at 1300 watts. The brush type D.C. motor with permanent magnets operates at 3500 R.P.M. at rated voltage, but spins at lower speeds at lower voltages without harming the windings.

The simple centrifugal pump with brass impeller has low starting torque; it is self-priming after the initial prime.

The array consists of thirty each fifty-three watt panels; the nominal array output is 1590 watts. Six parallel sets of five series-wired panels provide a maximum open circuit voltage of eighty-five volts.

Working in the Solar Energy Laboratory at Florida Atlantic University, the author produced an hour by hour, year-round simulation of the array output, using PVFChart software from F-Chart Software, copyrighted by S.A. Klein and W.A. Beckman. The results are tabulated below. Data used is as follows:

1) NREL weather data for West Palm Beach, Florida.
2) Manufacturer's array efficiency and power temperature coefficient information.
3) Winter array slope of 40 degrees, summer slope 10 degrees.

January

Time	Solar kWh	Solar kW	Eff %	Array Output Watts	Estimated Pump GPM
7- 8	55.7	1.80	11.2	201	-
8- 9	113.6	3.66	12.0	440	-
9-10	171.0	5.52	12.2	673	12
10-11	217.0	7.00	12.2	854	16
11-12	242.7	7.83	12.2	955	18
12-13	242.7	7.83	12.2	955	18
13-14	217.0	7.00	12.2	854	16
14-15	171.0	5.52	12.1	667	12
15-16	113.6	3.66	11.9	436	-
16-17	55.7	1.80	11.0	198	-
Month	1600.1	5.16	12.1	625	

Daily: 10 hours power; 6 hours pumping; 5520 gallons pumped.

February

Time	Solar kWh	Solar kW	Eff %	Array Output Watts	Estimated Pump GPM
6- 7	12.7	0.45	6.9	31	-
7- 8	60.6	2.16	11.1	240	-
8- 9	116.2	4.15	12.0	498	-
9-10	170.5	6.09	12.2	743	12
10-11	213.9	7.64	12.2	932	16
11-12	237.9	8.50	12.2	1037	18
12-13	237.9	8.50	12.2	1037	18
13-14	213.9	7.64	12.2	932	16
14-15	170.5	6.09	12.1	737	12
15-16	116.2	4.15	11.9	494	-
16-17	60.6	2.16	11.0	238	-
17-18	12.7	0.45	6.8	31	-
Month	1623.6	4.83	12.0	580	

Daily: 12 hours power; 6 hours pumping; 5520 gallons pumped.

March

Time	Solar kWh	Solar kW	Eff %	Array Output Watts	Estimated Pump GPM
6- 7	22.8	0.74	8.0	59	-
7- 8	80.5	2.60	10.9	283	-
8- 9	145.7	4.70	11.8	555	10
9-10	208.5	6.73	12.1	814	14
10-11	258.2	8.33	12.1	1008	18
11-12	285.6	9.21	12.1	1114	20
12-13	285.6	9.21	12.1	1114	20
13-14	258.2	8.33	12.0	1000	18
14-15	208.5	6.73	12.0	808	14
15-16	145.7	4.70	11.7	550	10
16-17	80.5	2.60	10.7	278	-
17-18	22.8	0.74	7.9	58	-
Month	2002.7	5.38	11.8	635	

Daily: 12 hours power; 8 hours pumping; 7440 gallons pumped.

April

Time	Solar kWh	Solar kW	Eff %	Array Output Watts	Estimated Pump GPM
6- 7	37.3	1.24	9.0	112	-
7- 8	97.1	3.24	10.9	353	-
8- 9	162.0	5.40	11.7	632	10
9-10	223.1	7.44	11.9	870	16
10-11	270.6	9.02	12.0	1082	20
11-12	296.6	9.89	12.0	1187	22
12-13	296.6	9.89	11.9	1177	22
13-14	270.6	9.02	11.9	1073	20
14-15	223.1	7.44	11.8	878	16
15-16	162.0	5.40	11.6	626	10
16-17	97.1	3.24	10.8	350	-
17-18	37.3	1.24	8.9	110	-
Month	2173.2	6.04	11.7	707	

Daily: 12 hours power; 8 hours pumping; 8160 gallons pumped.

May

Time	Solar kWh	Solar kW	Eff %	Array Output Watts	Estimated Pump GPM
5- 6	3.9	0.13	11.5	15	-
6- 7	49.7	1.60	9.5	152	-
7- 8	107.0	3.45	11.1	383	-
8- 9	167.8	5.41	11.7	633	10
9-10	224.1	7.23	11.8	853	16
10-11	267.6	8.63	11.9	1027	18
11-12	291.2	9.39	11.8	1108	20
12-13	291.2	9.39	11.8	1108	20
13-14	267.6	8.63	11.8	1018	18
14-15	224.1	7.23	11.7	846	14
15-16	167.8	5.41	11.6	628	10
16-17	107.0	3.45	10.9	376	-
17-18	49.7	1.60	9.4	150	-
18-19	3.9	0.13	11.4	15	-
Month	2222.7	5.12	11.6	594	

Daily: 14 hours power; 8 hours pumping; 7680 gallons pumped.

June

Time	Solar kWh	Solar kW	Eff %	Array Output Watts	Estimated Pump GPM
5- 6	7.0	0.23	11.4	26	-
6- 7	48.4	1.61	9.8	158	-
7- 8	97.7	3.26	11.1	362	-
8- 9	149.5	4.98	11.6	578	10
9-10	197.2	6.57	11.7	769	14
10-11	233.9	7.80	11.8	920	16
11-12	253.8	8.46	11.7	990	18
12-13	253.8	8.46	11.7	990	18
13-14	233.9	7.80	11.7	913	16
14-15	197.2	6.57	11.6	762	14
15-16	149.5	4.98	11.5	573	10
16-17	97.7	3.26	11.0	359	-
17-18	48.4	1.61	9.7	156	-
18-19	7.0	0.23	11.3	26	-
Month	1975.1	4.70	11.5	541	

Daily: 14 hours power; 8 hours pumping; 6960 gallons pumped.

July

Time	Solar kWh	Solar kW	Eff %	Array Output Watts	Estimated Pump Output
5- 6	5.6	0.18	11.4	21	-
6- 7	50.0	1.61	9.6	155	-
7- 8	104.2	3.36	11.0	370	-
8- 9	161.5	5.21	11.6	604	10
9-10	214.3	6.91	11.7	808	14
10-11	255.0	8.23	11.7	963	18
11-12	277.2	8.94	11.7	1046	18
12-13	277.2	8.94	11.7	1046	18
13-14	255.0	8.23	11.7	963	18
14-15	214.3	6.91	11.6	802	14
15-16	161.5	5.21	11.5	599	10
16-17	104.2	3.36	10.9	366	-
17-18	50.0	1.61	9.5	153	-
18-19	5.6	0.18	11.3	20	-
Month	2135.8	4.92	11.5	566	

Daily: 14 hours power; 8 hours pumping; 7200 gallons pumped.

August

Time	Solar kWh	Solar kW	Eff %	Array Output Watts	Estimated Pump GPM
6- 7	39.5	1.27	9.3	118	-
7- 8	93.9	3.03	10.9	330	-
8- 9	152.6	4.92	11.5	566	10
9-10	207.3	6.69	11.7	783	14
10-11	249.8	8.06	11.7	943	16
11-12	273.0	8.81	11.7	1031	18
12-13	273.0	8.81	11.7	1031	18
13-14	249.8	8.06	11.7	943	16
14-15	207.3	6.69	11.6	776	14
15-16	152.6	4.92	11.4	561	10
16-17	93.9	3.03	10.8	327	-
17-18	39.5	1.27	9.2	117	-
Month	2031.9	5.46	11.5	628	

Daily: 12 hours power; 8 hours pumping; 6960 gallons pumped.

September

Time	Solar kWh	Solar kW	Eff %	Array Output Watts	Estimated Pump GPM
6- 7	23.0	0.77	8.8	68	-
7- 8	72.5	2.42	10.7	259	-
8- 9	127.1	4.24	11.5	488	-
9-10	179.0	5.97	11.7	698	12
10-11	219.6	7.32	11.7	856	16
11-12	241.9	8.06	11.7	943	16
12-13	241.9	8.06	11.7	943	16
13-14	219.6	7.32	11.7	856	16
14-15	179.0	5.97	11.6	693	12
15-16	127.1	4.24	11.4	483	-
16-17	72.5	2.42	10.6	257	-
17-18	23.0	0.77	8.7	67	-
Month	1726.1	4.80	11.5	552	

Daily: 12 hours power; 6 hours pumping; 5280 gallons pumped.

October

Time	Solar kWh	Solar kW	Eff %	Array Output Watts	Estimated Pump GPM
6- 7	14.5	0.47	7.3	34	-
7- 8	64.5	2.08	10.8	225	-
8- 9	122.1	3.94	11.7	461	-
9-10	178.1	5.75	11.9	684	12
10-11	222.7	7.18	11.9	855	16
11-12	247.4	7.98	11.9	950	16
12-13	247.4	7.98	11.9	950	16
13-14	222.7	7.18	11.9	855	16
14-15	178.1	5.75	11.8	679	12
15-16	122.1	3.94	11.6	457	-
16-17	64.5	2.08	10.7	223	-
17-18	14.5	0.47	7.2	34	-
Month	1698.5	4.57	11.7	534	

Daily: 12 hours power; 6 hours pumping; 5280 gallons pumped.

November

Time	Solar kWh	Solar kW	Eff %	Array Output Watts	Estimated Pump GPM
7- 8	56.9	1.90	11.0	209	-
8- 9	114.0	3.80	11.9	452	-
9-10	170.4	5.68	12.1	687	12
10-11	215.6	7.19	12.1	870	16
11-12	240.8	8.03	12.1	971	18
12-13	240.8	8.03	12.1	971	18
13-14	215.6	7.19	12.0	863	14
14-15	170.4	5.68	12.0	682	12
15-16	114.0	3.80	11.8	448	-
16-17	56.9	1.90	10.9	207	-
Month	1595.5	5.32	12.0	638	

Daily: 12 hours power; 6 hours pumping; 5520 gallons pumped.

December

Time	Solar kWh	Solar kW	Eff %	Array Output Watts	Estimated Pump GPM
7- 8	54.1	1.75	11.1	194	-
8- 9	112.2	3.62	12.0	434	-
9-10	169.9	5.48	12.2	669	12
10-11	216.4	6.98	12.2	852	16
11-12	242.3	7.82	12.2	954	16
12-13	242.3	7.82	12.2	954	16
13-14	216.4	6.98	12.2	852	16
14-15	169.9	5.48	12.1	663	12
15-16	112.2	3.62	11.9	431	-
16-17	54.1	1.75	11.0	193	-
Month	1589.7	5.13	12.1	621	

Daily: 12 hours power; 6 hours pumping; 5280 gallons pumped.

DATA SUMMARY

The key conclusion from data analysis is that during the driest months in Florida - March, April, and May - the pumping day extends to eight hours and the highest flow rates are attained. Thus the pumping system produces the most water distribution when it is needed the most. Pumping production is high enough at these times to water more than one zone per day, so that the number of total zones can be as high as four in this application, permitting future expansion of the number of trees to be irrigated.

Summary of Other Applications

The other five applications include:

* Low head constant drainage pumping in low lying areas.

* Irrigation with a horizontal de-watering well as source.

* Booster pumping in line with existing diesel pump to correct inadequate pressure for proper operation of emitters.

* New grove installation with young trees and surface water as source.

* Irrigation and freeze protection incorporating battery storage for nighttime operation.

Grower response has been tremendous, with much enthusiasm expressed by growers with all sizes and types of groves.

SOLAR PUMPING IN THE CITRUS INDUSTRY: ITS FUTURE IN FLORIDA

As array costs continue to plummet, and utility costs rise, the day is within sight when PV will compete at non-remote sites. Currently, substantial competing first costs as itemized above must exist before PV is cost effective.

There are possible benefits to the coincidence between solar pumping and tree water requirements which may improve tree growth, yields, and fruit quality.

More restrictive environmental regulations will only hasten the demise of the popularity of diesel pumping. Already, stationary tanks require masonry safety enclosures to prevent spills. Tightening water restrictions will mandate minimized flow rates and consumption; this too will make PV a natural choice. And the inherent promotional value of using sunshine to produce Florida oranges is not to be ignored.

Agriculture in Florida is hard pressed by an ever increasing burden of regulatory compliance and encroaching development. As it moves into this new environment, and moves into new growing areas, PV pumping can play an important role in reducing operating costs, improving production, and putting more land into production.

This project is providing invaluable experience and data for the ever-widening role of renewable energy sources in agriculture. It is also introducing growers, regulators, and the irrigation contracting industry to the simplicity and ease of modular solar integrated pumping applications.

Chapter 52

Recent Advances in Microprocessor Control and Variable Speed Drive Strategies for Small Wood-Fired Boilers: A Case Study

R. Clunie, R. Guibord, R. Gregoire

INTRODUCTION:

Located since the 1900's in the town of Westbrook, Maine, Saunders Brothers' wood products company has watched the steady growth of its neighboring community in Westbrook to the point where neighborhood housing is now within a few feet of the plant property. The Saunders Brothers' boiler plant consists of two HRT boilers of 4,500 LB/HR (150 HP) and 6,500 LB/HR (200 HP) respectively, which burn only bark and wood sawmill wastes, and shavings generated by the manufacturing plant.

As in the case of many older boiler installations, the boiler plant has been "grandfathered" and must comply with minimal stack emissions standards. Accordingly, up until the present time there has been no emission control device installed on either boiler, or in the stack.

However, in the last several years, as the surrounding population growth has intensified, emissions from the boilers had also become an increasing problem--both visible particulate ash and fly ash which, under certain wind and weather conditions, would spread around the neighborhood. Particularly troublesome was a very light fly ash resembling a "curly-cue" which was partially pyrolized birch bark carried over from the combustion process and then transported up the stack.

Although technically in compliance with State of Maine environmental laws, Saunders Brothers was concerned with the visible emissions and fly ash emanating from the plant site. Wishing to maintain its status as a good neighbor, in 1989 Saunders Brothers' boiler plant was extensively renovated. Part of these modifications consisted of changing over to automatic microprocessor combustion controls and installing AC variable frequency drives on the combustion air fans and fuel feed screws. Recent advances in single microprocessor control technology, allowing cross-limiting and other control functions not available previously, and the proliferation of reliable AC variable frequency drive technology has resulted in the ability to completely and effectively automate small and medium sized solid fueled boiler plants at a modest cost.

The purpose of adopting the microprocessor-based combustion control approach to solve the air pollution problem at Saunders Brothers was two-fold. The control package, as designed by Applied Engineering, Inc., of Wiscasset, Maine, takes complete control of the fuel feed and combustion processes. By accurately metering the fuel supply, as well as metering the air supply to the firebox, usual large air volumes and resultant velocites are reduced and strictly controlled. At the same time all particulate emissions are greatly reduced, and all visible emissions are eliminated. A side benefit of installing the new combustion control system was an expected fuel savings of approximately 20% to 25%. All this without resorting to the traditional approach of adding air pollution control equipment to the boiler flue gas stream--hence, air pollution control with a return on investment!

Saunders Brothers has not been disappointed. Since start-up, and with a three year operating experience, it can safely be said that the boiler plant renovation project has delivered what was promised. There have been no complaints from the neighboring community about particulate ash, and all visible stack emissions have been virtually eliminated. Fuel usage reductions have exceeded expectations with an approximately 33% savings--from 7,500 TPY average usage prior to installation, reduced to about 5,000 TPY each year. One additional benefit not figured into the original project R.O.I., was a full rebate for the installed variable frequency drive equipment which qualified under an electrical usage reduction rebate program sponsored by the local utility.

THE ORIGINAL BOILER PLANT AND OPERATING PARAMETERS:

The Saunders Brothers boiler plant consists of two HRT boilers: Boiler #1 with maximum capacity of 4,500 PPH steam at 150 PSIG; and Boiler #2 with maximum capacity of 6,500 PPH steam at 150 PSIG. The boilers provide heating and humidifying steam for 12 kilns operating at random, and general plant space heating during the winter months. Because of the varying steam load requirements throughout the year, the boiler plant has evolved into two modes of operation--wintertime and summertime, which is further complicated by two distinct interim transition periods.

During wintertime operation, both boilers have been on-line, each with separate ON/OFF operation according to steam pressure demand. Summertime operation has been characterized primarily by operating the smaller of the two boilers (Boiler #1). One transition period occurs in the Fall, between September 15 and October 15, when No. 2 Boiler is started-up after the summer layoff, while No. 1 Boiler is operated at reduced capacity or at low fire. The second transition period occurs in the Spring of the year from April 15 to May 15, where again No. 1 Boiler is cut back to low fire, and No. 2 Boiler has taken most of the required steam load.

The on-line, off-line operation of the two boilers, coupled with steam demand pressure control, is further complicated at Saunders Brothers by the inconsistent moisture content of the fuel. Fuel supplied by the saw mill operations consists of sawdust, bark, and hogged fuel with moisture contents ranging from 35% to 45% depending on the time of year. The other fuel component is 7% moisture kiln dry shaving wastes from wood turning and dowel manufacturing. Storage of the fuel wastes has been in two bins, essentially without segregating the fuels.

Fuel is transported from either bin directly to the boiler feeders by blowers. Thus, each boiler had been receiving inconsistent amounts of fuels with varying moisture contents. To compensate for these conditions, the boilers have stationary, refractory grates which were fitted with overfire and underfire air systems. A 25 HP induced draft fan is located on the stack to maintain the furnace drafts. All existing fans were equipped with manually adjusted dampers.

ELECTRONIC MICROPROCESSOR BASED CONTROL SYSTEM INSTALLED:

In the summer of 1989 Saunders Brothers installed state-of-the-art electronic controls and instrumentation, and variable frequency drives on both boilers for continuous, automatic, and optimal efficiency operation. This was Phase 1 of a three phase program to completely upgrade the boiler plant operation. Phases 2 and 3 will consist of streamlining the waste wood collection system for fuel, and installing electrical cogeneration for in-house consumption.

Automatic fuel blending:

As part of the overall boiler plant operation strategy, fuel blending capability has been incorporated into the new controls/instrumentation package. Fuel is now segregated into "dry" and "wet" storage bins. The purpose of the fuel blending system is to allow complete flexibility of the desired moisture content from 7% to 45%, and of the feed rates of the available fuels on hand to feed to the boilers. This system allows the operators to "dial in" a desired moisture content, and the fuel delivery system will automatically proportion the different fuels for feeding to the boilers.

Automatic combustion control:

Using control algorithms designed specifically for wood fired boiler applications, a fuel feed and combustion control strategy has been incorporated utilizing variable frequency drives on combustion air fans and fuel feed screws. All manually controlled dampers on the combustion fans have been eliminated. The combustion strategy uses the controller's configurable functional generators to provide ramping of various outputs to control boiler response. The result of this process is more controlled, consistent combustion and operation at all the various boiler loads and operating conditions.

The new system provides modulating control of the boilers based upon a primary steam pressure signal with adjustable setpoint. An additional steam flow signal is provided to anticipate load changes and adjusts the output accordingly. This is similar to an automobile cruise control which not only controls the car's speed, but will also anticipate approaching road grade changes (eg. such as a steep hill), and begins to make the necessary changes before the car actually reaches the grade change.

Independent, automatic, cross limiting of the fuel and air to the boilers is the secondary control function ("cross limiting" is the ability of the controller to allow the air to lead the fuel under increasing boiler loading, while conversely allowing the fuel to lead the air on decreasing boiler loading). Opacity and oxygen trim are also provided in the basic primary control package, and automatic fuel blending is provided as a separate, parallel control function.

A plant master designates which of the boilers is on line at any given time, and determines which of the boilers is in the lead/lag configuration, and thereby controls the boiler masters for the dual control system. Individual boiler furnace draft is controlled independently, as is the overall stack draft. A permanent record is continually provided for steam flow, opacity, and efficiency through a three pen recorder.

Final results:

Since the new control system has been operating this past year, consistent boiler efficiencies ranging from 78% to 83% have been the norm, while oxygen content readings in the flue gas have ranged from 4% to 5%. The automatic and continuous optimization of the fuel feed and combustion processes has resulted in fuel savings to 33%, and has eliminated visible stack emissions.

Another related benefit of the new system, is the virtual elimination of the 25 HP induced draft fan which previously had run full time at full throttle. The boiler plant electrical load has also been reduced from a constant average of 125 KW to 15 KW primarily due to the continually automated control of fan and fuel feed variable frequency drives which only operate as needed, and are then speed controlled 0% to 100% as required for prevailing combustion conditions. Saunders Brothers has also been able to recoup the cost of the variable frequency drives from Central Maine Power, the local utility, from the utility's electrical demand reduction rebate program.

In regards to the surrounding neighborhood, no complaints of stack emissions have been received since the new system has been installed and started-up. With the fuel savings inherent with the increase in boiler efficiencies, coupled with much of the installation work done by in-house labor, the project is predicted to have a straight payback in about one year. Saunders Brothers feels they have successfully implemented air pollution control with an attractive payback.

Chapter 53

Automatic Processing of Photometric Data

C.E. Gracias, L.L. Smith, E.C. Noble

Abstract

The process of data collection and manually transforming of hard copy data into usable computer format in manual mode is tedious and time-consuming. The New York State Energy Office (NYSEO) has developed procedures for the automatic processing of data. Of particular importance is the processing of IES photometric data which is the subject of the present paper. The methodology involves the use of several macros based on Lotus language which transforms the photometric data into useful listings and provides the means for establishing standards for defining TLE (Total Luminaire Efficiency) and VDT (Video Display Terminal) reflectance data, for example. The subsequent analysis of the processed data has suggested immediate applications such as the comparison of the various lighting products based on TLE and VDT values defined on a common set of criteria. It has also established reliable means for defining consistent and realistic levels of performance to support the standards used in the New York State Energy Code.

The availability of methods for determining such performance standards makes it possible for designers and specifiers of lighting systems to reliably carry out the selection and specification of lighting products with energy efficiency and photometric performance defined on a common set of standards.

Prologue

The most recent amendments to the New York State Energy Conservation Construction Code (The Energy Code) became effective March 1991. This set of code amendments updated standards for commercial building envelope, HVAC equipment and lighting equipment. The changes in lighting equipment standards include new efficiency requirements for lighting fixtures (luminaires) to be used for general lighting purposes. These requirements are expressed in terms of a Total Luminaire Efficiency (TLE) value which characterizes each type of fixture.

The Energy Code requirements apply to all new building construction and to existing buildings undergoing substantial renovation.

It should be noted that until recently there were no references upon which standards could be defined to gauge the efficiency of commercial lighting fixtures. Methodology existed for testing and calculating illumination levels[1] but lack of a common base made difficult the comparison of the lighting performance between the same type of luminaires from different manufacturers.

In order to establish the required lighting standards to be used in the Energy Code, an in-depth analysis was performed and a methodology was developed to determine and evaluate Total Luminaire Efficiency (TLE) levels for a broad range of fixture types. TLE is the ratio, expressed as a percentage, of the lumen output of the fixture (or luminaire) to the sum of lumens produced by the lamp/ballast combination(s) used in the fixture.

The starting point was to build a database of photometric data representative of the lighting fixtures from the manufacturer data contained in IES magnetic floppy diskettes. Subsequently, an extensive analysis was carried out to both sort the fixtures by categories containing similar characteristics and to determine the TLE values for each individual fixture represented in an IES file. Thus, the different fixture types were categorized by lamp type (e.g., fluorescent, compact fluorescent and HID). The rather broad category of fluorescent fixtures was further divided into subsets, according to certain specific fixture characteristics, such as: large-cell louver, small-cell louver, wrap-around diffuser/lens, flat diffuser/lens and open type. In some cases, some fixture types were further categorized on the basis of the efficiency differences associated with the lumen (illuminance) distribution features of the luminaire.

Concomitantly, the TLE levels determined for each fixture within the defined fixture category permitted the establishment of distinct levels of performance that were used in setting the efficiency requirements in the New York State Energy Code.

Subsequent to the work related to the amendments of the Energy Code, the database was used in the development of a Directory of Commercial Lighting Fixtures. The Directory was created as part of a joint project involving the seven investor-owned Utilities and the New York State Energy Office (NYSEO). A significant outcome of this joint project was the establishment of consensus levels beyond Code requirements which the participating Utilities used to define the criteria for granting rebates on energy efficient fixtures in Utilities Incentives Programs. In order to establish the basis for the definition of such consensus levels, NYSEO used the market compression analysis method to fully exploit the near-term potential for energy savings from financial incentives by concentrating on high-performance equipment currently in the marketplace that could be made readily available in response to incentive-induced consumer demand, but which was not generally purchased in this State. For further details, please refer to a concurrent paper by Denis Nadas, also of this Office.[2]

This work was initially performed by Eric Nobel, Lee Smith, Brian Dewey and John S. Edwards, Jr.. Further refinement of the Lotus 1-2-3 macros used to analyze and process the IES Photometric Data was carried out by Cecilio Gracias with the invaluable assistance of Lee Smith.

Introduction

Lighting is the indispensable element in the creation of an appropriate working environment. The amount of light to be provided to a specific area of the work space is a very important factor in lighting systems design. A simplified approach assumes initially the existence of light producing devices as point sources. The ideal point source of light provides a spherically symmetric illumination, in the sense that all points equidistant from the point source would have the same illuminance. However, for specific applications, the light distribution should be modified so that different levels of illuminance would be available to provide either an aesthetic effect or an effective illumination at the work area. Such modification of the original light distribution is achieved in practice by the use of devices called *luminaires* (also commonly known as fixtures).

From both the manufacturer and from the designer viewpoint, a given luminaire (fixture) is designed to provide a given light distribution so that the luminance at different areas of the working space would be controlled to certain levels in order to meet different work environments or to provide preestablished ambience effects. For this reason, photometric measurements are taken at different viewpoint angles from the luminaire, which for purposes of evaluation is assumed to be a point source. Of course variance from this assumption is determined by the shape and design of the luminaire and would be translated into different light intensities at the different angles. In other words, each luminaire could be characterized by photometric measurements taken at predetermined viewpoint angles from a given distance from the light source. As a result of the inverse square law, these measurements should be repeated for the same set of angles for different distances of the light source. These measurements would then provide the designer with known illuminances at different task levels. The designer would then select the luminaire with the most appropriate light distribution for the desired effects in the work environment.

The IES Standard Electronic File Format

Before personal computers became ubiquitously available, data from the IES Test Reports were reported in 'longform' (see Ref. 1). Since they were in ordinary text format, the calculation of illumination levels from photometric data required the handling of huge amount of calculations in a very cumbersome fashion.

To facilitate such manipulations and make easier the transfer of data, the Illuminating Engineering Society (IES) has developed a Standard File Format for Electronic Transfer of Photometric Data contained in magnetic media (Standard IES LM-63-1989). Most manufacturers have since adopted this format with the result that information of the fixtures they make is generally available in floppy disks.

Two main categories of information may be found in the IES File Format. They are: the 'Description' section and the 'Photometric Data' section. The 'Description' section contains in turn two groups of lines. The first set of lines in the IES File Format is reserved for *'labels'*, used to identify the model number, the test number, and the general description of the fixture. The second set contains basic characteristics such as tilt, number of lamps, rated lumen output, ballast factor, and dimensions and number of vertical and horizontal angles.

The 'Photometric' section lists both the vertical (θ) and the horizontal angles (ϕ), and candela values (R) for all vertical angles at each of the horizontal angles. The data can be envisaged in terms of the spherical coordinates, R, θ, and ϕ, and may be plotted using the appropriate graphic software.

Table 1 illustrates a typical IES Format for a fixture with two lamps, no shielding, no tilt, 3200 lumen/lamp, multiplier 0.258, 37 vertical angles, 5 horizontal angles and dimensions in feet (Width x Length x Height). The photometric data are displayed for each horizontal angle, the candela values being listed for the corresponding vertical angles.

Processing Photometric Data

The use of magnetic (floppy) disks for storage of photometric data is very convenient since such disks can pack an enormous amount of data in a small space. However, access to the data requires that the information contained in those floppy disks be readily available in usable form, either for display on a computer screen, or in a suitable printout. A method was developed to provide such printout. However, printouts contain photometric data in a format not readily usable for analysis (Table 1).

Methodology has been devised to process the photometric data into organized tabulations. Furthermore, the data for the various fixtures from a given manufacturer could also be processed to form a consistent database.

Using the terminology of a previous section, it is seen that the 'Description' part of the IES Format usually occupies up to ten lines at the top of the printout. The 'Data' part can be readily recognized from the listing of the angles, vertical and horizontal.

Properly processed, the original IES data, will yield important graphical information on the spatial distribution of luminance in terms of numerical data on the zonal distribution of light at specific vertical angles (see Table 2).

Methodology

Macros for Processing Photometric Data

Once the photometric data is organized in standard database format it is a relatively simple procedure to calculate the **Total Luminaire Efficiency** (TLE) and the glare-defining **VDT factors** at various significant angles. The latter are defined as the maximum average brightness of the fixture measured at 65 degrees from the vertical (altitude), and at 0, 45, and 90 degrees horizontal (azimuth), measured in terms of candela per cosine-corrected unit area, relative to 850 candela per square meter.

Lotus macros were designed to yield specific information on luminaire performance and to provide elements for a common database aggregating in the same format the multitude of information provided by the various manufacturers. The advantage of such a common base is that the performance of a given luminaire can be compared with that of any other luminaire regardless of type or manufacturer.

The application of the aforementioned macros yielded a universal Database containing data, specifications, and calculated performance indexes for all the luminaires in the original manufacturer listings (see Table 3). Among other information, the total luminance output (lumens) for each luminaire is of particular significance since it permits the determination of TLE values. The latter may be estimated by dividing the calculated total output value by the total lumens produced by the lamps in the fixture.

Similarly, the VDT factor is a calculation based on a recommendation made by the IES, contained in its publication IES RP-24-1989. VDT factors are provided for vertical angles of 0, 45 and 90 degree.

The Database also provides needed information for the calculation louver cell data in louvered fixtures and of the VDT factors.

Graphical Representation

Table 2 is the result of applying a subroutine of one of the macros the purpose of which was unwrap the photometric data into a form more amenable to mathematical analysis. It shows the measured candelas displayed in terms of both the vertical angles θ (altitudes or colatitudes) and horizontal angles ϕ (azimuth) as shown in the IES Data.
The corresponding spherical coordinates are:

$$Z = I*\cos(\theta)$$
$$X = I*\cos(\phi)$$
$$Y = I*\sin(\phi)$$

Because photometric data represent spatial distributions of light intensity (candelas), they are more readily understood if displayed in polar form. A simple methodology was created to provide such graphic displays. The data shown on Table 2 can be readily identified as the elements of a matrix $M(\theta,\phi)$, as illustrated in Table 4. The following relations may be established in spherical coordinates:

$$X_{i,j} = M_{i,j}*\sin(\phi_i)*\cos(\theta_j)$$
$$Y_{i,j} = M_{i,j}*\sin(\phi_i)*\sin(\theta_j)$$
$$Z_{i,j} = M_{i,j}*\cos(\phi_i)$$

Two types of graphical display may be derived from the matrix M: (1) a two-dimensional display on selected planes - such as X-Y and Y-Z (Figure 1); or (2) a three-dimensional display (X-Y-Z) of the values calculated from the matrix M (Figure 2). The curves shown in Figure 1 are similar to the ones often found in the technical specification sheets provided by a fixture manufacturer with the difference that the latter are displayed in polar form (i.e, polar radius vs. angle). Figure 2 is a surface plot of the three coordinates X, Y, and Z using MathCad vs. 4. In the particular example illustrated in Figure 2, the matrix $M_{i,j}$ is defined by elements in Table 4 in which the matrix elements are the actual photometric values in candelas, extracted from tabulations such as the one shown in Table 2.

The two sets of angles are defined as arrays of parameters m_i and n_j in terms of the indices **i** and **j**, respectively. These indices are labels used to identify the position of the elements in matrix M(i,j) and correspond to the two sets of angles which define the set of photometric data, namely, the horizontal angles (ϕ) and the vertical angles (θ):

$$i = 1,2,...,17$$
$$m_i = 0,15,...,360$$

$$j = 1,2,...,37$$
$$n_j = 0,22.5,45,67.5,90$$

with:

$$\phi_i = 2*\pi*m_i/360$$
$$\theta_j = 2*\pi*n_j/360$$

In this representation, the origin of the coordinates is the idealized light point-source. The vertical angle (θ)is measured from the 'nadir' (bottom of the XY-plane), the zenith being then at 180 degrees from the nadir, by definition. The vertical planes are then the X-Z plane and the Y-Z plane, in which the Z-axis is pointing outward of the plane of the paper and the X,Y-axes are lying on the said plane.

Figure 1 illustrates X-Y plots derived from the data contained in tables such as those contained in Table 2 and Table 4 for various types of luminaires.

Glare and VDT Factors

Glare[3] is defined as "the result of any excessively bright source of light in the field of view that causes a loss of visibility (*disability glare*) and/or discomfort (*discomfort glare*)". The effect of glare varies with the type of task, depending upon whether one is dealing with horizontal tasks or vertical tasks.

HORIZONTAL TASKS

Depending on how the glare-causing light impinges upon the field of view, glare may be **direct** or **indirect**. Indirect glare may in turn be divided into two other categories - reflected glare and veiling reflections. Direct glare occurs in the performance of head-up tasks with light produced in the zone from 45 degree to 90 degree and is measured empirically in terms of the *visual comfort probability*.[4]

Indirect glare is associated with light produced in the zone from zero degree to 45 degree and is related to heads-down viewing for horizontal tasks. The effect of indirect glare is to cause reduction in task contrast resulting in discomfort, loss in visibility and decrease of visual performance.

Reflected glare is caused by reflections from specular or glossy surfaces of light from bright sources and may result in disability glare or loss in visibility. On the other hand, veiling reflections stem from light reflected off matte or diffuse surfaces and may cause reduction in task level contrast. The effects of veiling reflections are not as apparent as those of reflected glare, yet they may result in considerable amount of eye fatigue and loss of visual performance.

VERTICAL TASKS

Modern office practice has led to an increased use of microcomputers and has thus caused an increase in the performance of vertical tasks. The latter corresponds to tasks requiring the viewing of material displayed on the screen of monitors. The acronym VDT (video display terminal) is currently being used to describe all devices which project a document or a picture on a screen located at the projection end of a cathode-ray tube (CRT). Because of the negative publicity of the real or potential effects of VDT on visual performance, many countries and some states in the USA have passed stringent laws regulating the VDT operations, in particular to provide "occupational safeguards for operators of video display terminals".

Indeed, the effects on the user of VDT operation in a given working environment is a combination of complex interactions between the luminous environment and the physical, psychological and emotional well-being of the user. For this reason, the design or retrofit of the lighting environment for vertical tasks involve considerations beyond the strict consideration of lighting levels and its distribution in the task area: it involves additional understand of the ergonomics of the VDT operator, as well as the ergonomics of the CRT screen and the VDT setup. The latter includes the microcomputer, the furniture supporting the hardware, the VDT user's chair, the source document holder (i.e., the monitor) and the CRT screen proper. Conditions for effective VDT lighting have been examined by the Illuminating Engineering Society of America in its publication IES RP-24-1989.[5]

IES RP-24-1989 also suggests luminance levels for the various types of illumination. The Energy Office has used some of its recommendations in calculating the VDT Factors for the azimuth (horizontal angle) values of 0, 45, and 90 degree for light sources at 65 degree elevation (vertical angle), using as reference the luminance level of 850 candelas per square meter at 65 degree. Converted to British units, 850 candelas/meter^2 < > 79 candelas/foot^2

To calculate the value of the VDT Factor for a given fixture, the peak value of the zonal lumen summary values M_ϕ at 65 degree vertical for azimuth angles ϕ = 0, 45, and 90 degree, is divided into the converted reference value of 79 candelas/foot^2. At VDT level, the luminous source (luminaire) appears as a rectangle whose cross-section S_{65} is the area of the opening of the luminaire, cosine-corrected for the vertical angle of 65 degree. Hence, the ratio M_ϕ /(79) - multiplied by this cross-section S_{65} and by an assumed relative light output of 0.85 - is then taken as the VDT Factor for the selected azimuth values.[6]
Values of VDT Factor < 1 indicate that the fixture meets or exceeds the IES recommendations. On the hand, values of VDT Factor greater than unity suggest that the fixture fails to meet the IES recommendations.

It should be noted here that the IES document emphasizes the importance of luminaire efficiency (as defined by the TLE value) since it directly determines the amount of energy used to provide a given illuminance level. The document further suggests that, in order to provide vertical surface illumination and modeling light for work station surfaces and people, it is advantageous to use high efficiency luminaires with a broad distribution of light within the confines of the required cut-off angle. Adherence to the cut-off levels recommendation will avoid objectionable reflections in the VDT screen for the most common ranges of operator viewing angles (from 5 to 20 degree) and tilt (0 to 15 degree).

Summary and Conclusions

The processed photometric data allow the definition of the Total Luminaire Efficiency (TLE) for each fixture. The value of TLE measures the effectiveness of light transmittal and distribution for the intended purpose. Moreover, its availability permitted the State of New York to established levels of efficiency for definition of standards in the Energy Code.

Once a tabulation of TLE values for different luminaires and for different manufacturers has become available, a tool was created for the designer for the selection of luminaires of given TLE range of values. Additionally, for a given TLE, the spatial distribution of illuminance levels permits the designer to chose among luminaires with equal TLE value that particular one that would provide the desired light distribution.

Glare from reflected VDT incident light causes eye fatigue and reduces screen visibility. The availability of VDT factors at critical angles allows the designer to select those luminaires that would provide minimal glare. Current SEO tabulations use the reference value of 850 candelas per square (or 78 candelas/foot^2) at a lighting angle of 65 degree to establish the listed values of VDT Factors. As mentioned, those fixtures for which the VDT factor is less than unity are deemed as satisfying the IES recommendations.

Because the spatial distribution of illuminance and luminance have an obvious effect on the level of glare observed at task surfaces, the processed photometric data readily permit the visual representation of the illuminance distribution in the form of polar plots, in both 2-dimensional and 3-dimensional display. This knowledge of the spatial distribution of light will permit the designer to determine the conditions for minimal glare and thus contribute to the reduction of operator discomfort and fatigue, with concomitant improvement in visual performance.

The most important practical consequence of developing a common Database for all luminaires from every participating manufacturer is that the Database provides the means for comparing the performance of different luminaires using the same format. Thus, because performance indexes such as TLE and VDT Factors are calculated in uniform fashion for all luminaires listed in the Database it is possible to effect the direct comparison of the performance of luminaires from different manufacturers. Additionally, it is possible to use the Database to categorize the performance of the various luminaires according to TLE (see Table 5), VDT Factor and/or other performance-related criteria.

In conclusion, the existence of an organized set of data on commercial lighting fixtures in the form of a Database, has made it possible for NYSEO to define criteria on the performance levels of such fixtures for both the Energy Code and Utility Incentive Programs. As mentioned earlier, these levels were established from market compression analysis using information from the Database.

Notes

1. IES LM-41-85, "Approved Method for Photometric Testing of Indoor Fluorescent Luminaires"; and IES LM-46-85, "Photometric Testing of Indoor Luminaires Using High Intensity Discharge or Incandescent Lamps"
2. Ref. 11, p.2.
3. Ref. 4, p.220; and Ref. 5,p.106
4. For more details, Ref.4, p.221; Ref. 5,p.155,176,188.
5. See Ref. 2.
6. It should be noted that the cosine corrected value of M_e represents the luminous flux impinging upon the VDT screen.

Appendix: Source List for Tables/Figures

Table 1: IES Photometric Data File - Automatic Printout
Table 2: 'UNWRAP' Printout
Table 3: Structure of Database
Table 4: Matrix of Photometric Data
Table 5: Energy Aid (Rebates) Levels
Figure 1 - Polar Representation of Luminance Distribution for a Luminaire
Figure 2 - Spherical Representational of Luminance Distribution for a Luminaire

Bibliographic References

1. 'IES Recommended Standard File Format for Electronic Transfer of Photometric Data', The IES Computer Committee, IES LM-63-86
2. 'VDT Lighting - IES Recommended Practice for Lighting Offices Containing Computer Visual Display Terminals', IES RP-24-1989. Also: 'IES Lighting Handbook', 1981, p.3-12
3. 'Applied Illumination Engineering' J. L. Lindsey, The Fairmont Press, 1991
4. 'Lighting for Energy-Efficient Luminous Environments "R. N. Helms and M. Clay Belcher, Prentice Hall, 1991
5. 'Architectural Lighting - Selected Papers' M. S. Rea, editor, SPIE, 1992
6. New York State Energy Code, as amended March, 1991.
7. New York State Directory of Commercial Lighting Fixtures, New York State Energy Office, 1993
8. 'Structure of a Database for Commercial Lighting Fixtures' Unpublished Notes by Cecilio Gracias, 1993
9. 'Mathematical Model for Representation of Photometric Data' Unpublished Research by Cecilio Gracias, 1993
10. 'Library of Lotus 1-2-3 Macros' Staff, New York State Energy Office, 1991-1993.
11. 'Equipment Efficiency Levels for Utility Rebates - What Are The Right Ones', by Denis Nadas, to be presented at the XVI WEEC, October 26,1993.

Table 1

File <> COL78024.PRN

Description	TEST #8024 <> T6−NL−421
	2−LAMP 5−3/4" "SUPERTUBE" LUMINAIRE WITH NO SHIELDING
	BALLAST: RQM−2S40, BALLAST FACTOR: .95
	TILT=NONE
	2 3200 0.258169 37 5 1 1 0.37 3.96 0.00
	1.0 1.0 92.0
Colatitude Angles	0.0 2.5 5.0 7.5 10.0 12.5 15.0 17.5 20.0 22.5 25.0 27.5
	30.0 32.5 35.0 37.5 40.0 42.5 45.0 47.5 50.0 52.5 55.0 57.5
	60.0 62.5 65.0 67.5 70.0 72.5 75.0 77.5 80.0 82.5 85.0 87.5 90.0
Azimuth Angles	0.0 22.5 45.0 67.5 90.0
Photometric Data	6025 6020 6005 5970 5920 5875 5800 5715 5620 5520
	5410 5285 5135 4990 4830 4660 4480 4284 4078 3867
	3647 3416 3170 2929 2668 2407 2152 1866 1590 1319
	1046 783 536 311 116 21 0
	6025 6025 6010 5980 5925 5870 5790 5700 5605 5495
	5365 5220 5065 4895 4715 4520 4315 4089 3858 3617
	3361 3110 2849 2578 2322 2044 1776 1513 1255 1010
	780 534 331 181 84 14 0
	6025 6045 6025 5990 5925 5845 5755 5650 5525 5390
	5230 5055 4865 4665 4460 4240 4005 3769 3523 3272
	3016 2755 2489 2228 1982 1728 1491 1199 926 688
	482 323 215 109 32 10 0
	6025 6020 6000 5945 5875 5795 5695 5570 5425 5260
	5080 4895 4700 4485 4270 4040 3800 3559 3298 3047
	2787 2531 2290 2042 1774 1476 1202 951 712 520
	373 266 163 61 31 5 0
	6025 5985 5960 5905 5830 5750 5635 5505 5355 5185
	5005 4815 4620 4400 4180 3950 3710 3459 3208 2953
	2697 2446 2206 1975 1660 1374 1113 865 648 472
	348 251 140 53 31 5 0

Table 2

REPORT NO. xxx <> CHART801
CATALOG NO. P345−123S
PERIM.STRIP WALL SLOT W/WHT.INTERIOR & SEMI−SPEC 20 BLADE LOUVER
LAMP F40T12CW LAMP LUMENS/MH=1.9
EFFICIENCY 29.0%   BALLAST UNIVERSAL 446−SLH
TILT=NONE   2 3150 1 37 17 1 1 .66 4 0   .95 1 84.1

AZIMUTH	ALTITUDE (COLATITUDE) 0	2.5	5	7.5	10	12.5	15	17.5	20	22.5	25	27.5	30	32.5	35	37.5	40	42.5
0	711	686	703	694	674	660	660	654	615	588	582	562	519	498	484	441	406	386
22.5	711	752	817	841	844	872	889	909	897	881	869	861	858	834	815	795	767	724
45	711	808	872	952	992	1021	1049	1085	1121	1193	1235	1231	1197	1142	1082	1006	931	839
67.5	711	831	921	997	1053	1116	1193	1309	1401	1416	1389	1354	1302	1242	1147	1043	975	921
90	711	831	939	1009	1070	1153	1272	1401	1447	1445	1414	1382	1335	1252	1156	1109	1076	1007
112.5	711	831	921	997	1053	1116	1193	1309	1401	1416	1389	1354	1302	1242	1147	1043	975	921
135	711	808	872	952	992	1021	1049	1085	1121	1193	1235	1231	1197	1142	1082	1006	931	839
157.5	711	752	817	841	844	872	889	909	897	881	869	861	858	834	815	795	767	724
180	711	686	703	694	674	660	660	654	615	588	582	562	519	498	484	441	406	386
202.5	711	664	636	616	568	527	492	463	435	379	344	327	315	272	248	247	250	227
225	711	631	584	539	487	420	359	311	283	271	290	297	286	290	294	302	325	330
247.5	711	619	552	480	414	359	319	286	294	302	314	330	343	355	358	358	358	358
270	711	608	535	455	376	311	266	250	270	294	317	337	357	370	378	390	384	387
292.5	711	619	552	480	414	359	319	286	294	302	314	330	343	355	358	358	358	358
315	711	631	584	539	487	420	359	311	283	271	290	297	286	290	294	302	325	330
337.5	711	664	636	616	568	527	492	463	435	379	344	327	315	272	248	247	250	227
360	711	686	703	694	674	660	660	654	615	588	582	562	519	498	484	441	406	386

AZIMUTH	ALTITUDE (COLATITUDE) 45	47.5	50	52.5	55	57.5	60	62.5	65	67.5	70	72.5	75	77.5	80	82.5	85	87.5	90
0	334	298	248	205	157	118	92	72	56	43	35	25	19	11	6	2	0	0	0
22.5	649	569	470	372	273	188	140	107	86	66	50	38	25	20	13	6	4	2	0
45	735	647	565	474	370	277	184	125	87	63	45	33	21	13	8	1	0	0	0
67.5	836	735	612	487	382	296	212	135	92	70	46	30	21	13	8	1	0	0	0
90	904	778	641	524	426	315	202	134	118	92	67	43	25	20	9	1	0	0	0
112.5	836	735	612	487	382	296	212	135	92	70	46	30	21	13	8	1	0	0	0
135	735	647	565	474	370	277	184	125	87	63	45	33	21	13	8	1	0	0	0
157.5	649	569	470	372	273	188	140	107	86	66	50	38	25	20	13	6	4	2	0
180	334	298	248	205	157	118	92	72	56	43	35	25	19	11	6	2	0	0	0
202.5	218	218	202	179	155	127	100	79	58	49	38	30	21	17	9	4	0	0	0
225	319	296	269	251	235	212	177	140	104	79	62	45	33	25	18	9	1	1	0
247.5	363	359	359	347	324	297	264	232	186	144	105	71	46	30	18	9	1	1	0
270	387	390	402	396	391	368	345	302	258	210	161	113	72	46	25	9	1	1	0
292.5	363	359	359	347	324	297	264	232	186	144	105	71	46	30	18	9	1	1	0
315	319	296	269	251	235	212	177	140	104	79	62	45	33	25	18	9	1	1	0
337.5	218	218	202	179	155	127	100	79	58	49	38	30	21	17	9	4	0	0	0
360	334	298	248	205	157	118	92	72	56	43	35	25	19	11	6	2	0	0	0

Table 3

DATABASE – Colibri Lighting

DESCRIPTION:

TEST #10009	2X4 TWO LAMP LUMINAIRE WITHSA–3 TWO BY SIX CELL BAFFLE.
TEST #10024	20" X 4 FT. THREE LAMP RECESS LUMINAIRE WITH 3X6 CELL SA/3 LOUVER
TEST #3853	2 X 2 RECESSED LUMINAIRE WITH 2 EA. FB40WW/6" LAMPS AND 4 X 4 CELL SA–3 ANODIZED PARABOLIC LOUVER

FIELD	FIXTURE #1	FIXTURE #2	FIXTURE #3
REPORT_NUM	TEST #10009	TEST #10024	TEST #3853
MANUFACT	Colibri Lighting	Colibri Lighting	Colibri Lighting
MODEL	4542G–43–242	4543G–43–043–20"	4544–43–222U
DESCRIPTION	(see above)	(see above)	(see above)
FIXTTYPE	FLL	FLL	FLL
LAMPTYPE	FL	FL	FL
LAMPSHAPE			
LAMPINFO			
LOUVX	0.870	0.470	0.438
LOUVY	0.623	0.623	0.438
NUMCELLS	12	18	16
NUMLAMPS	2	3	2
LUM/LMP	3200	3200	3025
MULTIPLIER	0.196	0.197	1
NUMVERT	37	19	37
NUMHOR	5	5	3
PHOTOMETRC	1	1	1
UNITS_TYPE	1	1	1
WIDTH	1.74	1.41	1.75
LENGTH	3.74	3.74	1.75
HEIGHT	0	0	0
RLO	1	1	1
BALLLPF	1	1	1
INPWATT	92	144	92
ZN20	2938.6	4856.1	613.3
ZN30	6422.7	10667.4	1315.3
ZN40	11154.9	17391.2	2129.2
ZN60	20832.5	26253.8	3295.1
ZN65	21852.4	27270.2	3372.3
ZN70	22266.9	27726.4	3389.2
ZN75	22414.5	27907.0	3399.9
ZN80	22477.6	27997.8	3405.9
ZN85	22510.8	28052.7	3409.0
ZN90	22522.6	28082.4	3409.7
ZN180	22522.6	28082.4	3409.7
TOTAL	22522.6	28082.4	3409.7

CALCULATION OF LOUVER CELL DIMENSIONS

Louver Description	SA–3 TWO BY SIX CELL BAFFL	3X6 CELL SA/3 LOUVER	SA–3 ANODIZED PARABOLI(
Item #	TEST #10009	TEST #10024	TEST #3853
Louver	2 x 6	3 x 6	4 x 4
N	12	18	16
Width	1.74	1.41	1.75
Length	3.74	3.74	1.75
L/W	2	3	1
Louver X	0.8700	0.4700	0.4375
Louver Y	0.6233	0.6233	0.4375
X	2	3	4
Y	6	6	4
Y/X	3	2	1
X*Y	12	18	16

Table 4

MATRIX FOR ANALYSIS OF IES PHOTOMETRIC TEST DATA

Source <> COLTIBSD.WKS

	COLUMN	1	2	3	4	5	6	7	8	9	10	11	12	13	14	15	16	17	18	19	20	21	22	23	24	25	26	27	28	29	30	31	32	33	34	35	36	37
	H\V	0	2.5	5	7.5	10	12.5	15	17.5	20	22.5	25	27.5	30	32.5	35	37.5	40	42.5	45	47.5	50	52.5	55	57.5	60	62.5	65	67.5	70	72.5	75	77.5	80	82.5	85	87.5	90
1	0	6025	6020	6005	5970	5920	5875	5800	5715	5620	5520	5410	5285	5135	4990	4830	4660	4480	4284	4078	3867	3647	3416	3170	2929	2668	2407	2152	1866	1590	1319	1046	783	536	311	116	21	0
2	22.5	6025	6025	6010	5980	5925	5870	5790	5700	5605	5495	5365	5220	5065	4895	4715	4520	4315	4089	3858	3617	3361	3110	2849	2578	2322	2044	1776	1513	1255	1010	780	534	331	181	84	14	0
3	45	6025	6045	6025	5990	5925	5845	5755	5650	5525	5390	5230	5055	4865	4665	4460	4240	4005	3769	3523	3272	3016	2755	2489	2228	1982	1728	1491	1199	926	688	482	323	215	109	32	10	0
4	67.5	6025	6020	6000	5945	5875	5795	5695	5570	5425	5260	5080	4895	4700	4485	4270	4040	3800	3559	3298	3047	2787	2531	2290	2042	1774	1476	1202	951	712	520	373	266	163	61	31	5	0
5	90	6025	5985	5960	5905	5830	5750	5635	5505	5355	5185	5005	4815	4620	4400	4180	3950	3710	3459	3208	2953	2697	2446	2206	1975	1660	1374	1113	865	648	472	348	251	140	53	31	5	0
6	112.5	6025	6025	6010	5980	5925	5870	5790	5700	5605	5495	5365	5220	5065	4895	4715	4520	4315	4089	3858	3617	3361	3110	2849	2578	2322	2044	1776	1513	1255	1010	780	534	331	181	84	14	0
7	135	6025	6045	6025	5990	5925	5845	5755	5650	5525	5390	5230	5055	4865	4665	4460	4240	4005	3769	3523	3272	3016	2755	2489	2228	1982	1728	1491	1199	926	688	482	323	215	109	32	10	0
8	157.5	6025	6020	6000	5945	5875	5795	5695	5570	5425	5260	5080	4895	4700	4485	4270	4040	3800	3559	3298	3047	2787	2531	2290	2042	1774	1476	1202	951	712	520	373	266	163	61	31	5	0
9	180	6025	5985	5960	5905	5830	5750	5635	5505	5355	5185	5005	4815	4620	4400	4180	3950	3710	3459	3208	2953	2697	2446	2206	1975	1660	1374	1113	865	648	472	348	251	140	53	31	5	0
10	202.5	6025	6025	6010	5980	5925	5870	5790	5700	5605	5495	5365	5220	5065	4895	4715	4520	4315	4089	3858	3617	3361	3110	2849	2578	2322	2044	1776	1513	1255	1010	780	534	331	181	84	14	0
11	225	6025	6045	6025	5990	5925	5845	5755	5650	5525	5390	5230	5055	4865	4665	4460	4240	4005	3769	3523	3272	3016	2755	2489	2228	1982	1728	1491	1199	926	688	482	323	215	109	32	10	0
12	247.5	6025	6020	6000	5945	5875	5795	5695	5570	5425	5260	5080	4895	4700	4485	4270	4040	3800	3559	3298	3047	2787	2531	2290	2042	1774	1476	1202	951	712	520	373	266	163	61	31	5	0
13	270	6025	5985	5960	5905	5830	5750	5635	5505	5355	5185	5005	4815	4620	4400	4180	3950	3710	3459	3208	2953	2697	2446	2206	1975	1660	1374	1113	865	648	472	348	251	140	53	31	5	0
14	292.5	6025	6025	6010	5980	5925	5870	5790	5700	5605	5495	5365	5220	5065	4895	4715	4520	4315	4089	3858	3617	3361	3110	2849	2578	2322	2044	1776	1513	1255	1010	780	534	331	181	84	14	0
15	315	6025	6045	6025	5990	5925	5845	5755	5650	5525	5390	5230	5055	4865	4665	4460	4240	4005	3769	3523	3272	3016	2755	2489	2228	1982	1728	1491	1199	926	688	482	323	215	109	32	10	0
16	337.5	6025	6020	6000	5945	5875	5795	5695	5570	5425	5260	5080	4895	4700	4485	4270	4040	3800	3559	3298	3047	2787	2531	2290	2042	1774	1476	1202	951	712	520	373	266	163	61	31	5	0
17	360	6025	5985	5960	5905	5830	5750	5635	5505	5355	5185	5005	4815	4620	4400	4180	3950	3710	3459	3208	2953	2697	2446	2206	1975	1660	1374	1113	865	648	472	348	251	140	53	31	5	0

TABLE 5

Commercial Lighting Fixtures

Category	Type	Energy Aid Level
Large Fluorescent	Small Cell Louver	narrow[1], TLE≥ 45% moderate, TLE≥ 57% wide, TLE≥ 50% [2]
	Large Cell Louver–Direct	narrow, TLE≥ 60% moderate, TLE≥ 67% wide, TLE≥ 70%
	Large Cell Louver–Direct/Indirect	narrow, TLE≥ 60% moderate, TLE≥ 67% wide, TLE≥ 70%
	Wrap–Around Diffuser/Lens	all, TLE≥ 79%
	Flat Diffuser/Lens–Direct	all, TLE≥ 70%
	Flat Diffuser/Lens–Direct/Indirect	all, TLE≥ 70%
	Open	narrow, TLE≥ 67% moderate, TLE≥ 74% wide, TLE≥ 77%
	Indirect	all, TLE≥ 80%
High Intensity Discharge	Round–No Refractor	TLE ≥70%
	Round–Refractor	TLE ≥88%
	Square	TLE ≥75%
	Indirect	TLE ≥73%

(Please reference Table 4–16, Section 7813.53 of the Energy Code)

1 – Lumen distribution characteristics are used to catagorize luminaires by the fraction of lumens delivered in the 0 to 40 degree segment divided by the lumens delivered in the 0 to 90 degree segment. Catagories are defined as follows: Narrow, > 0.55; Moderate, 0.55–0.45; wide, < 0.45.

2 – Current Energy Code requiremnet, insufficient product to identify a level above the Energy Code.

Figure 1

Files <> COL7IESD.WK3/PRUDMHPX.WK3

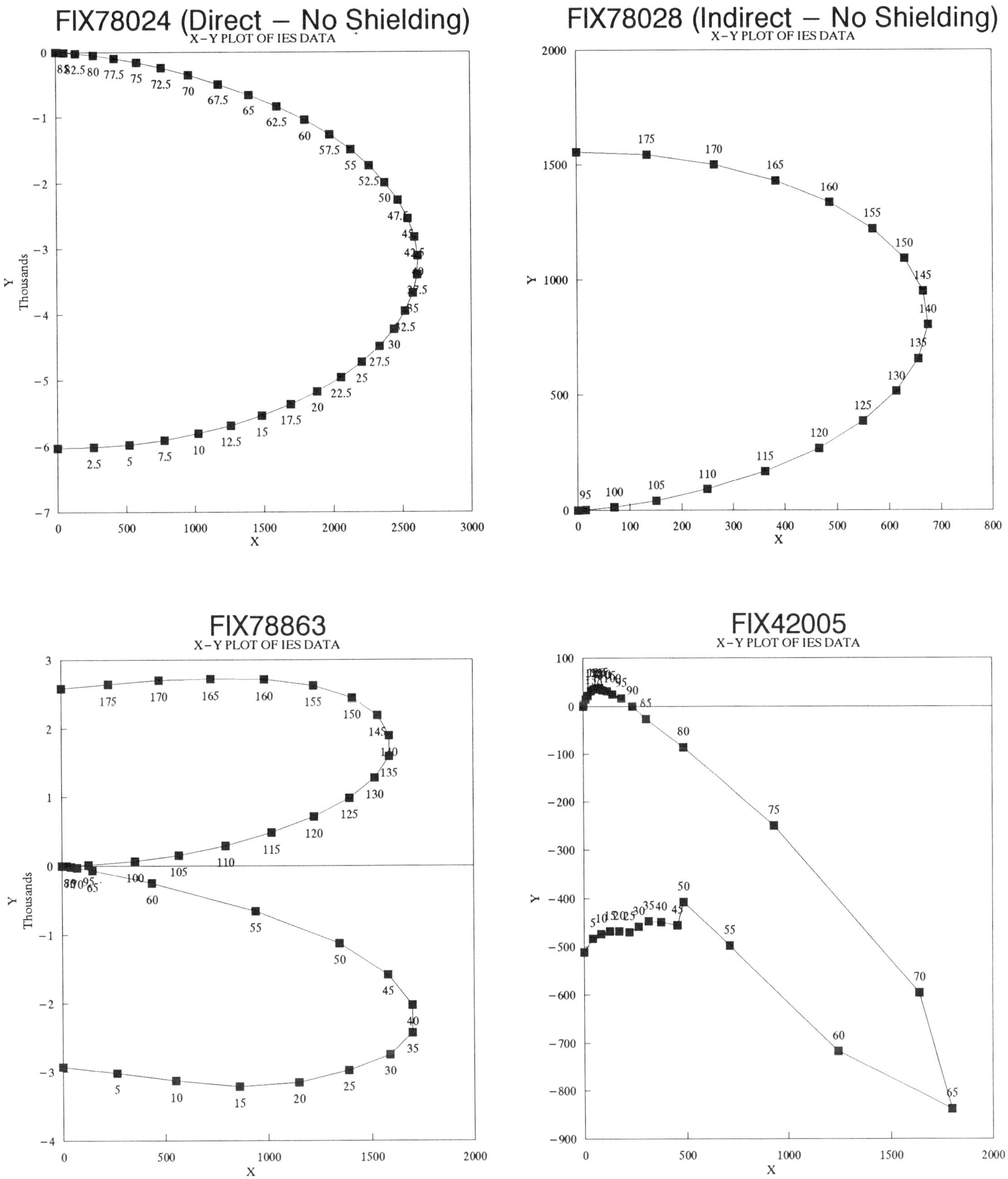

Figure 2

Surface Plot of IES Photometric Test Data

Tilt = 90 deg; Rotation = 0 deg

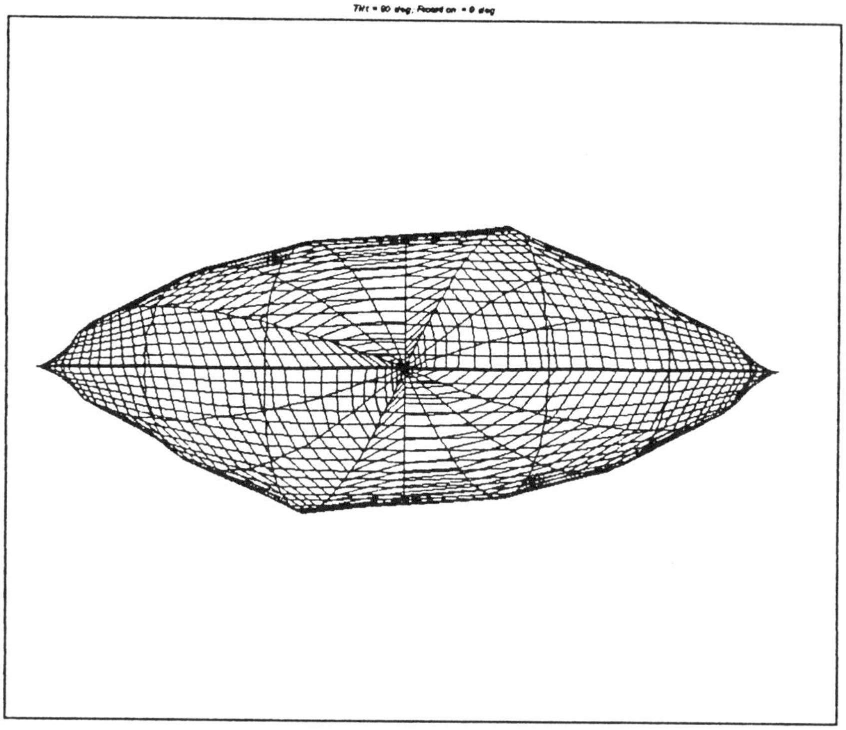

Z, Y, X

Tilt = 0 deg; Rotation = 0 deg

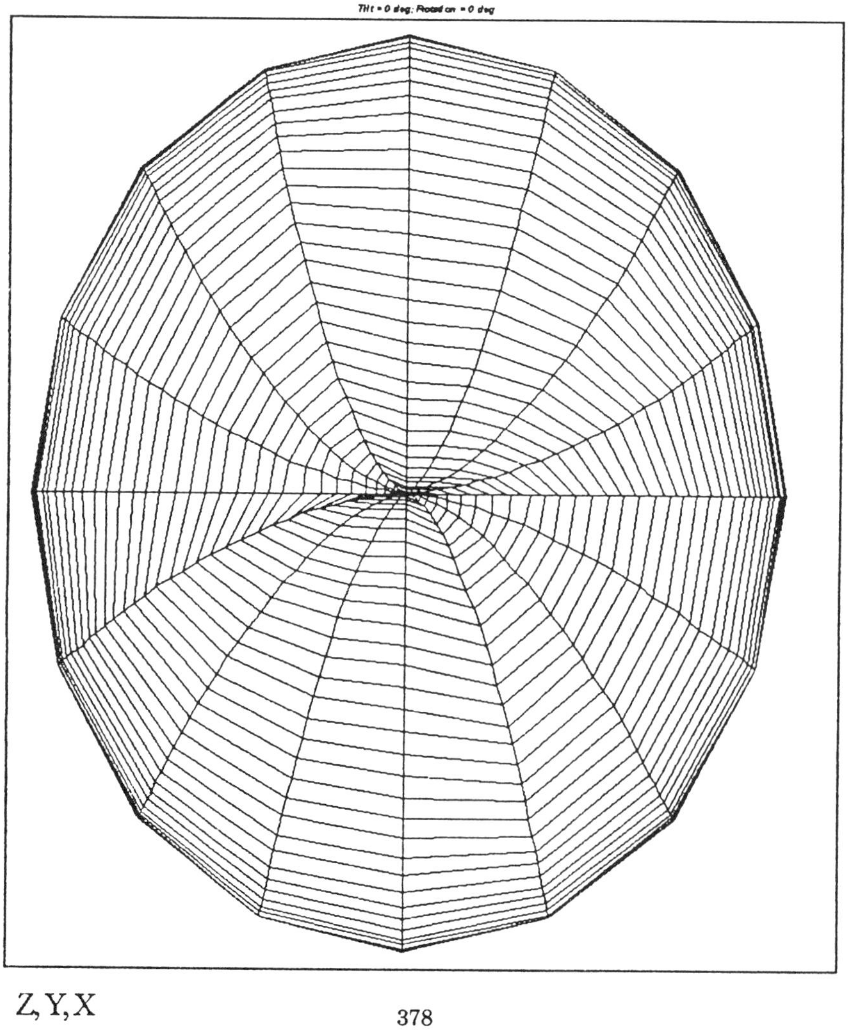

Z, Y, X

SECTION 5
ADVANCES IN LIGHTING EFFICIENCY & APPLICATIONS

Chapter 54
Relighting Program—Furman University

P. Simpson

It was in 1985 that Furman University, Greenville, S.C. became interested in improved lighting with a primary desire to reduce the electrical requirements. The highly touted claim was made at that time that through the use of highly polished reflectors in fluorescent luminaires, half the tubes could be removed while maintaining the same light level. Through extensive testing, we learned that that is not true - that is a false claim. However, during the same testing, we learned a very valuable lesson: reflectors can be a very valuable component in a relighting program. Reflectors are the primary component in the relighting program on the Furman Campus that has resulted in a 50% reduction in energy usage while improving the light level and quality.

First, the claim to reduce tubes by 50% with the addition of a reflector was based on selling maintenance. By adding a new tube, a reflector and cleaning the luminaire and lense, approximately the same light could be obtained. But you are not comparing apples with apples. With good lighting maintenance and care, the light level cannot be maintained with the removal of 1/2 the tubes and addition of reflectors. But do not throw away the baby with the wash water - as the old saying goes! Reflectors have tremendous value: reflectors are the basis of Furman's relighting program that has reduced electrical requirements by approximately 50% while improving light quality.

As we consider relighting, look at a few basics:

1. the value of energy
   -778 Ft. lbs in one B.T.U. (one wooden kitchen match)
   -500# bale of cotton lifted 1 1/2 F.
   -100 watt - equivalent to 4 slaves lifting 100# weights 3 ft. at rate of 4-min.
2. Lumen per watt ratio - Efficiency of light sources
3. Light Depreciation curve - old and new lamps or tubes
4. Inverse square light law.

One other item - get a good light meter so you know what you are doing.

With no other knowledge of lighting, Furman began to seriously consider lighting improvement. The fact that fluorescent is approximately 4 times as efficient as incandescent - indicated the replacement of all incandescent - if at all possible. And a close look at the typical 2x4 modular troffer, fluorescent fixture provided sufficient motive to become involved in relighting studies. I remember the day I stood on my desk, removed the lense from the 2x4 troffer over my desk and drew a sketch of the luminaire. It is obvious that much of the light produced is never able to get out of the box. Later I was to learn that the average luminaire efficiency of this design is approximately 60%. A little consideration of this and similar fixture showed the potential value of reflectors.

With that idea and the highly touted claim about the value of reflectors, we sent a typical luminaire to California to have reflectors properly designed for that fixture and we ordered sufficient reflectors to retrofit one complete classroom. Upon receipt, these were installed. But we were determined to compare apples with apples. Before installation, the fixtures were cleaned and 4 new lamps were put in each fixture. We laid out a grid and then took light level readings at 42 different places. With a watt meter, we also read the energy requirements. Then we retrofitted each fixture with 2 reflectors and two of the four lamps that had been removed. Again we took light readings at the 42 points and the power requirements. Results: power reduced by 50% but the light level was only 67% the original. The claim to be able to hold the original light level was false. We removed the fixtures and repeated the test in a different room: Results were exactly the same - only 67% of the original light. But it did represent an improvement and we did not want to fail to take advantage of the gain - if possible.

We were at very low light levels in the classrooms and no more reduction was acceptable. On one hand we wanted to avail ourselves of the improved efficiency of the reflectors but, on the other hand, the only available retrofit would result in an overall reduction in light level. For some time we were blocked in our efforts to develop a retrofit program.

But in 1988-1990 some very interesting technological improvements began to make their appearance on the market - high intensity fluorescent lamps. The Watt miser lamp - the work horse of our present lighting - had a rating of 2750 lumens and lost 30% of this light within their life span of 20,000 hours. New lamps were becoming available with a rating of 3400 lumen output and light depreciation of only 7% within their life span. This put new life into our stalled testing program. We quickly found that we could completely strip a 2x4 troffer, then install one ballast, 2 sets of tube holders and a set of reflectors and 2 new high intensity lamps and have the same light level as before when using 4 tubes. We were now realizing positive results with our relighting studies.

Now that we had reached our goal of maximizing the luminaire efficiency through the use of polished reflectors, we concentrated on energy efficiency. If there was anything else to be done while we were retrofitting the luminaire for improved lighting efficiency we wanted to incorporate that into the retrofit. We did not want to perform a retrofit only to find that there was something else we could have done.

Reminds me of the cat that got his tail broken in the screen and needed it cut off. Be sure and cut all the tail off at one time - don't do it in steps. The old cat will not stand still after the first time. Economics probably will not justify cutting the tail off the second or third time.

Now that we had obtained the maximum luminaire efficiency, we wanted to be sure we had the maximum system efficiency before doing the final economic analysis and proceeding with the retrofit. We know that electronic ballasts were available and reliable, but their cost was approximately 3 times that of iron core ballast. A test was improvised to determine their value and effectiveness.

This was to be a test of electrical efficiency only - as we were satisfied with light efficiency. A standard typical 2x4 fluorescent troffer was placed on a work bench with a watt meter connected in the supply liner. From many different suppliers, we had obtained a few of all the hi output fluorescent tubes and from the ballast suppliers, we had obtained energy efficient ballasts - both electronic and hybrid.

The test consisted of installing a trial ballast and then installing 2 tubes to be tested. When the input power had stabilized, we would record the watts, current and power factor. We would remove those tubes, install two more and repeat the test. Then we would continue until all the tubes had been tested.

Then, we removed that ballast, installed another test ballast and repeated the test; certain fluorescent tubes driven by particular ballast produced the same light but at approximately 25% less electrical power. This seemed to be a very strange phenomena and being unable to get much help from local reps, I, then, called the respective factories and talked with the product engineers. I was assured that our info was correct, and if certain tubes were matched with electronic ballasts, this degree of efficiency was available. Also, when properly matched, that company would guarantee the product line.

It seemed that we had finally reached the maximum lighting efficiency available today through the use of reflectors and the maximum electrical efficiency through a combination of hi output, tri phosphorous coated lamps and electronic ballasts. We could see nothing else to enhance the project, we proceeded to evaluate the economics. Only one assurance further was needed - as the economics looked good: the assurance that this combination would not introduce harmonics to the electrical circuits. We have a 'awful lot of "boob tubes" on campus!

A typical room was selected and before retrofitting we installed a watt meter and a wave form analyzer and read the light levels through the room. The results:

1) light level and quality - slightly improved
2) harmonic interference - slightly reduced
3) energy reduced to 45% original with nothing detrimental

Remember - approximately 86% of lighting cost is for the electricity.

Consider 4 reasons for relighting and all energy conservation practices:

1. Good economics - probably 2.5 year payback with utility rebates
2. Reduce load on power plants by 10-15%
3. Reduce atmospheric pollutants:
   Approximately 1.5# CO_2
   Approximately 2.5# NO_x
   Approximately 5.8 grams SO_2
   Uses approximately .85# COAL
4. 100 Watt is equivalent to 4 Slaves

Final

Chapter 55

Lighting Maintenance Practices That Save Energy

R.A. Tucker

INTRODUCTION

The most aesthetic, sophisticated, energy efficient lighting system a designer can devise cannot perform as intended if post occupancy maintenance is poor. Designers can ensure the upkeep of their energy saving designs by considering what happens to lamps and luminaires as they age and by recommending to new building owners various options for lamp and luminaire maintenance.

The accumulation of dirt, lumen depreciation, and discoloration of lenses, for example, can decrease light levels and the overall effectiveness of an energy efficient lighting system. Planned maintenance is the only way to ensure the upkeep of lighting systems so as to continue to save energy. It is important, therefore, to recognize factors that contribute to the overall loss of illumination, thereby increasing energy cost.

Lamp Burnout

Lamp burnout is the most obvious contributor to loss of illumination from a lighting system. The average rated life that a manufacturer publishes for a lamp is the point in time of operation when approximately 50 percent of the lamps in a large group have burned out. As lamps burn out, illumination levels decrease proportionately. In the case of a rapid-start fluorescent system, when one lamp burns out in a 2- or 4-lamp fixture, the companion lamp operating on the same ballast also goes out.

Lumen Depreciation

The light output from a lamp decreases as it ages. This light loss is caused by internal deposits from vaporized filaments, cathodes, or electrodes. Lamp lumen depreciation is an inherent characteristic of all lamps.

Luminaire Dirt Depreciation

The light loss caused by deposits of dust or dirt on the exterior of the light source, the fixture reflector, or the lens is referred to as luminaire dirt depreciation. This factor leads to a significant loss of light especially in some manufacturing plant environments. Dirt accumulation on the surfaces can be reduced if the reflector surface is sealed from the air, as in a dust-tight fixture. Ventilated lighting units tend to collect less dirt than fixtures with closed tops.

Fixture Selection and Accessibility
Designers should consider how the luminaire they choose for a specific location affects lamp replacement. Some lamps, for example, must be ordered from a specific supplier; others are readily available at local electrical distributors. Some fixtures are difficult to disassemble for lamp replacement. Others may be accessible only by scaffolding. Such problems may discourage maintenance workers from making prompt replacements of burned-out lamps.

Other Considerations
Other factors also can affect illumination levels. Materials used in luminaire construction differ in their resistance to dirt accumulation, discoloration, and deterioration. Dirt accumulation on room surfaces tends to reduce the amount of reflected light in the area. Voltage variation for incandescent lamps, and ambient temperature in the case of fluorescent lamps, can cause variations in daily light output.

Poor lighting maintenance habits can significantly affect illumination levels in the workplace. Typical light losses due to poor maintenance can range from 15 to 40 percent.

An effective lighting maintenance program must meet both short-term and long-term needs as well as budget requirements. For example, it is not practical to institute a planned lighting maintenance program that replaces every component of the system (lamps, sockets, ballasts, wiring, reflectors, etc.) before it malfunctions or burns out. Neither is it practical nor economical to totally neglect the routine maintenance of the lighting system until burnouts or breakdowns occur.

Designer's Maintenance Manual
Designers should prepare a maintenance manual for a lighting system where a number of different lamp types are used or there are unique features. Such a manual includes a set of electrical plans, manufacturers' information on each type of luminaire used on the project, and relamping information. If special lamps have been used in a specific location for a specific purpose, a good maintenance manual notes the locations, the types of lamps used, and supplier information.

Whenever possible, the designer of the lighting system should meet with maintenance personnel to acquaint them with the intricacies of the system. Most maintenance people want to keep a building in top condition and avoid tenant complaints.

Few Advantages to Spot Relamping
Spot relamping has a few perceived advantages: no planning is required, no cost is obvious, and no coordination of time or personnel is necessary. These advantages, however, are outweighed by many disadvantages. With spot relamping, light levels can fluctuate significantly, the benefits of cleaning are not realized and unit cost to replace one lamp at a time is high. Also, spot maintenance can disturb building users, safety hazards and electrical problems can go undetected and the aesthetics of an area can be affected. Other possible problems include color consistency which may not be

maintained because lamps of different colors may be used, ballast life which can be shortened by burned-out or flickering lamps, and finally, spot replacements which can interrupt maintenance personnel or take them away from other duties.

Evaluations usually indicate that a program incorporating planned maintenance is best. An analysis of the lighting installation will help in determining the most economical method based on replacement costs, desired illumination level, and method of replacement of burnouts that occur before a scheduled relamping.

SOFTWARE FOR EVALUATING AND PLANNING LIGHTING MAINTENANCE

A software package is available and can be used by the lighting community to play "what if" scenarios with different lighting systems. This enables buildings to be modeled with the latest energy efficient lighting systems, and helps users calculate performance and maintenance costs.

Starting Up

The software package is user friendly and operates on an IBM compatible personal computer with 512KB RAM. It consists of four 5 1/4 inch diskettes (360KB) or two 3 1/2 inch diskettes (720KB). Single or dual disk drive of 360KB or higher density with or without hard disk drive systems will also run the program. The monitor can be monochrome, color or liquid crystal display (LCD). Customizing or configuring the software is simply a matter of moving through the functions with diskettes #1 through #4 and following the instructions that appear on the screen or the prompts at the bottom of the screen. The configuration or customizing feature will identify the printer, set the style of print and set default to customer, client or user name.

USING THE PROGRAM

There are four functions from which to choose. The first is the INSTALL function. It is used to install the program on the hard drive for the first time. By following the instructions on the screen, you can install the program on the hard drive. You also can UPDATE the existing program or DELETE the program if it is already on the hard disk drive. The fourth function allows you to EXIT the program.

Login

The first screen you will see when running the program is the user identification window. Here you will be able to list the customer, company, client and the user for record keeping and identification. Once you have completed the Login you can move to the command menu which appears across the top of the screen by pressing END.

F1 Function Key

At any time while moving through the program you can press the "F1" function key and receive HELP. This key gives you a pop-up window on the screen displaying information about the screen or field that you are in and how the fields are calculated. Use the direction arrows on the keyboard to move around the screen and from field to field.

Command Menu

While in the command menu, using the direction arrow keys you highlight "File", press ENTER and a drop down menu will appear below "File". Again using the direction arrow keys you select

"New", if you wish to start a new application or begin an analysis for the first time; or select "Open" if you wish to use an existing case you've saved, then press ENTER.

A model window will appear and you choose the model you wish to use with the up and down direction arrow keys and press ENTER. Your selection can be from a life cycle cost model, a lighting maintenance program, a relighting energy analysis model, a return on investment analysis model or an economic analysis program. Select the lighting maintenance (Group relamp) model.

This model makes it possible to plan lighting system maintenance by analyzing and comparing the costs of the following three replacement methods:

- Spot relamping
- Group relamping
- Group relamping with spot replacement of early burnouts

The program considers the annual lamp and labor costs for each system based on both a maintained light level and the maximum savings that can be derived. Analysis results will provide the necessary information to plan and budget for lighting maintenance needs.

Data Entry

After you've selected the lighting maintenance model, two screens for data entry are made available. On Screen 1, lamp data for the existing lighting system can be provided. To help with selecting lamp information, you can press the "F2" function key. A pop-up window will appear allowing you to enter information.

For example, each lamp that appears in the lamp manufacturers catalog has a five digit Vendor ID number. The database will respond if you can provide the five digit number and press the "F2" function key again. A new pop-up window will appear allowing you to view the details of the lamp you have selected. After viewing the information, accepting is simply a matter of pressing ENTER and filling in the blanks under lamp data for the existing system. You will have to provide the net cost per lamp, the number of lamps per luminaire, and the number of luminaires for the system.

If you do not have the five digit number but you are aware of some descriptive information on the lamp you are interested in evaluating, then type in that description and proceed as before. For example, if the lamp you desire is a warm white (WW) fluorescent, just type in WW under the description field then press ENTER and type F under the category field and press the "F2" function key. Every fluorescent lamp listed in the database with a WW in the first two letters of lamp color description will be listed in a pop-up window on the screen.

Use the up and down direction arrows to select your lamp and press ENTER. Then follow the prompts at the bottom of the screen to complete the data entry screen.

Enter the luminaire dirt condition code taken from the IES Lighting Ready Reference (RR-85) which best describes the environment where the lighting equipment is to be used. Choices range from a very clean space (1) to a very dirty space (5).

Operating and maintenance data is to be inserted on Screen 2. The annual operating hours, the cycle time in hours per day for use of the lighting system and the total area lighted is to be inputted here.

The ballast factor (BF) is the ratio of lumens produced on a commercial ballast to the lumens produced on an American National Standards Institute (ANSI) defined reference ballast in a controlled test environment. This represents how much light will be produced in the real world. This factor is to be indicated on this screen. Enter the fixture coefficient of utilization (CU) from the luminaires manufacturer's data to provide the percentage of light that the fixture allows into the space.

Next enter the class of luminaire (1-6) again taken from the IES RR-85 handbook represented by the equivalent Roman Numeral (I-VI).

The percentage of the initial lighting level in footcandles (fc) that represents the lowest acceptable level you want to maintain with your lighting system may be one set of determining criteria for the analysis results. All calculations will be based on the point in time when that lowest fc level occurs or the lamp life is reached. A second option will be the point in time when group relamping with spot replacement represents the lowest additional cost per fixture per year and maximizes the savings in lighting maintenance.

Analysis Screens

Analysis results are provided on the next three screens. From the information provided, this model will calculate the lowest fc level using the Zonal Cavity Method, the relamping period in hours of use and in years, the annual number of lamps to buy for spot relamping, the annual labor cost for relamping and the total relighting cost per year for each replacement method. Then it summarizes the information to provide a total annual relamping savings. Finally, this model will calculate the relamping cost per fixture per year and the annual group relamping savings per fixture when compared against the spot replacement method. By comparing the different lighting maintenance strategies, the program will calculate the costs associated with each and will determine the one most economical.

Pressing the F1 function key will now allow you to examine the various equations used to calculate lamp replacement cost per fixture per year.

Either new lamps or working lamps from the previous relamping can be used to replace burnouts. The spot replacement lamp cost for early burnouts can therefore run from the cost of a new lamp to zero. Typically, spot replacement is done with used lamps and the cost is nothing.

AN EXAMPLE OF THE SOFTWARE CAPABILITIES

The following example, using estimated lamp and labor cost figures and estimated times for spot and group relamping, analyzes the lighting maintenance methods for a small industrial concern. The lighted area is approximately 1200 feet by 1200 feet and involves 22,600 two lamp, 8 foot industrial reflecter fixtures with a coefficient of utilization of .65 and ballast factor of .9. An energy saving, slimline instant start lamp is being used to provide almost 100 footcandles initially, but 65% of that initial level is the lowest acceptable maintained lighting level. The analysis results appear in Printout 1.

Results of Analysis

In this hypothetical analysis, group relamping with early burnout replacement is the most economical method. If it is important for aesthetic reasons to maintain uniform lighting between group relampings (in retail store operations, for example), the above calculations indicate a minimal cost requirement.

In general, building management should consider group replacement when the labor cost per lamp for group relamping is significantly less than the labor cost per lamp for spot relamping. Records or work studies on lamp replacement may help management to determine the time and labor cost involved in either spot or group replacement. The computer program is available to determine the most economical relamping schedule by considering lamp and labor costs for a specific installation.

Building management and maintenance staff will ultimately determine lighting maintenance choices. Designers, however, can create lighting systems that are more likely to be well maintained.

This software package is designed to provide an analysis and comparison of different lighting systems and to determine which is the most economical. It allows the user to quickly and accurately analyze lighting systems in the convenience of their office, at a job site or wherever the personal computer is located. The "user friendly" program comes with easy-to-follow, system prompted instructions.

Now anyone who wants to maximize a lighting investment in terms of operating and maintaining the most efficient lighting system while achieving the required light levels has access to a software package that will help accomplish this goal.

DETERMINING ENERGY SAVINGS

Lighting systems are designed for maintained light levels. Without scheduled routine maintenance, light output after a few years may drop to less than 65% of the initial light output. This was indicated in Printout 1 for the Spot Replacement Method. A situation like this can be averted only by overdesigning the lighting system by at least 650 fixtures, as indicated in Printout 2 or by providing a viable routine lighting maintenance program. Good lighting maintenance minimizes the overdesign of lighting systems and reduces fixture costs, installation costs, energy costs and cooling costs.

Although the hypothetical building is a new structure, the same thinking can apply to existing buildings with an initial cost for disconnecting fixtures, not just removing lamps, and for possibly relocating some of the fixtures to correct light distribution problems which may occur. This might effect the payback for the first year but should be readily justified.

To determine power saved, energy savings and the effect of this lighting maintenance on the environment, the equations in TABLE 1 were used.

PRINTOUT 1

Date: 06/08/93 SYLVIA Group Relamp Analysis - Light Level GRP

Customer: EXAMPLE OF SOFTWARE CAPABILITIES
Rep: ARNOLD TUCKER
Comments:

Note: The calculations are believed appropriate for most lighting installations.

Common System Data			
Lamp Type Code	29815		
Description	F96T12/CW/SS		
Rated Lamp Life	12,000		
Initial Lumens	5,400		
Number of Luminaires	22,600		
Total Number of Lamps	45,200		
Total Area Sq. Ft.	1,440,000		
Area per Luminaire	63		
Initial Footcandle	99.16		
Acceptable Footcandle	64.46		
Replacement Method	**Spot**	**Group**	**Group w/Spot**
Lowest Footcandle Level	62.68	64.50	64.48
Operating Data			
Burning Hours per Year	4,720	4,720	4,720
Group Relamp Period (Hrs)		10,000	15,100
Group Relamp Period (Yrs)		2.11	3.19
Aver # Lamps Spot Replace	10,665		2,456
Cost Data			
Lamp Cost per Year	$61,003.80	$122,035.31	$80,817.73
Relamp Labor Cost per Year	$119,803.49	$47,932.28	$59,324.87
	=============	=============	=============
Total Relamp Cost per Year	$180,807.29	$169,967.58	$140,142.60
Total Relamp Savings/Year		$10,839.71	$40,664.69
Fixture Data			
Relamp Cost/Fixture/Yr	$8.00	$7.52	$6.20
Grp Relamp Sav/Fixture/Yr		$0.48	$1.80

Form: SYL2R21 GTE Sylvania - Danvers, MA 01923

PRINTOUT 2

Date: 06/09/93 SYLVIA Group Relamp Analysis - Maximum Savings GRP

Customer: EXAMPLE OF SOFTWARE CAPABILITIES
Rep: ARNOLD TUCKER
Comments:

Note: The calculations are believed appropriate for most lighting installations.

Common System Data	
Lamp Type Code	29815
Description	F96T12/CW/SS
Rated Lamp Life	12,000
Initial Lumens	5,400
Number of Luminaires	23,250
Total Number of Lamps	46,500
Total Area Sq. Ft.	1,440,000
Area per Luminaire	61
Initial Footcandle	102.01
Acceptable Footcandle	66.31

Replacement Method	Spot	Group	Group w/Spot
Lowest Footcandle Level	64.49		
Operating Data			
Burning Hours per Year	4,720		
Group Relamp Period (Hrs)			
Group Relamp Period (Yrs)			
Aver # Lamps Spot Replace	10,972		
Cost Data			
Lamp Cost per Year	$62,759.84		
Relamp Labor Cost per Year	$123,252.13		
Total Relamp Cost per Year	$186,011.97		
Total Relamp Savings/Year			
Fixture Data			
Relamp Cost/Fixture/Yr	$8.00		
Grp Relamp Sav/Fixture/Yr			

Form: SYL2R22 GTE Sylvania - Danvers, MA 01923

TABLE 1

EQUATIONS FOR ENERGY SAVINGS:

$$\text{Power Saved} = \text{No. of Fixs.} \times \frac{\text{Watts/Fix}}{\text{1000 Watts/KW}}$$

$$= 650 \times 140/1000 = 91\text{KW}$$

Energy Saved = Power Saved x Hours Use/Year

= 91KW x 4720 Hours/Year = 429,520 KWH/Year

Cost Savings = Energy Saved x Average Cost/KWH

= 429,520 x $.08 = $34,361.60

Emission Reductions = Energy Saved x Weighted Avg. Emmissions/KWH

for CO_2 = 429,520KWH X 1.6 Lbs/KWH x 687,232 pounds

for SO_2 = 429,520 KWH x 5.3 grams/KWH = 2,276,456 grams

for NO = 429,520 KWH x 2.8 grams/KWH = 1,202,656 grams

TABLE 2

TOTAL COST OF LIGHTING:

Maintenance Method is Spot Replacement

Energy Cost = $23250 \times \frac{140}{1000} \times 4720 \times \$.08$	=	$1,229,088.00
Relamping (Lamp & Labor) Cost = 23250 x $8	=	$ 186,000.00
TOTAL		$1,415,088.00

Group Replacement:

Energy Cost = $22600 \times \frac{140}{1000} \times 4720 \times \$.08$	=	$1,194,726.40
Relamping (Lamp & Labor) Cost = 22600 x $7.52	=	$ 169,952.00
TOTAL		$1,364,678.40

Group Replacement with Spot Relamping:

Energy Cost = $22600 \times \frac{140}{1000} \times 4720 \text{ X } \$.08$		$1,194,726.40
Relamping (Lamp & Labor) Cost = 22600 X $6.20	=	$ 140,120.00
TOTAL		$1,334.846.40

Power Saved = 91 KW
Energy Saved = 429,520KWH/Year
Cost Savings = $34,361.60
Emission Reductions
in CO_2 = 687,232 Pounds
in SO_2 = 2,276,456 grams
in NO = 1,202,656 grams

SUMMARY

Therefore, if all of the building fixtures are cleaned and relamped as indicated in Table 2 after 10,000 hours of operation (approximately 2.11 years), the reduction of 650 fixtures results in an annual savings of $50,409.60. If all the fixtures are cleaned and relamped after 15,100 hours of operation (approximately 3.19 years), and the lamp failures before the scheduled group relamping date are spot replaced, the 650 fixture reduction will result in an annual savings of $80,241.60. Thus, continued good maintenance of the lighting system can ensure a maintained light level and can result in a lower overall total cost of lighting by requiring fewer fixtures, reducing maintenance cost and reducing energy cost.

Chapter 56

Upgrading "Modern" Lighting Systems

D.J. Burke

During an inspection tour of our newly upgraded Corporate Headquarters Lighting System, a visitor marveled at its beautiful appearance. When informed that it also provided a 40% reduction in lighting costs, the immediate response was"

"Then why doesn't everyone do it?"

The purpose of this paper is to help others install an attractive, energy efficient and money saving lighting system upgrade. It is focused particularly on occupants of recently constructed office buildings who are laboring under the false assumption that meaningful cost deductions with their "Modern" lighting systems are not feasible.

BACKGROUND

Our firm occupies 21 floors at Two World Trade Center, New York. Occupancy began in June 1986 and continued, in stages, through December 1991 as we matched lease expirations at other locations with vacancies in the Trade Center. Every floor was completely gutted and rebuilt from slab to slab prior to move-in day. Included in the renovation was a new, thoroughly 'Modern state-of-the-art' lighting system. And modern it was - by 1986 standards!

Unfortunately, our lighting system designer was lavish in the application of recessed ceiling fixtures with incandescent bulbs. Excessive fixture and ceiling temperatures doomed these incandescent bulbs to a 2 to 4 week life expectancy. For those of us in the Facility Department this was a maintenance nightmare. We searched for a solution. After much experimentation, we found the perfect answer: dual bi-axial fluorescent bulbs with a 10,000 hour rated life, housed in a shallow fixture which produced superior illumination and a 60% reduction in energy. Replacement of the fixtures, however, required extensive revisions to our ceiling support grid. The net result: an 8 year payback to replace fixtures and modify ceiling grid! Not a cost effective solution even though it offered such attractive energy and maintenance savings.

Late in 1987, the Local Utility provided a needed stimulus - a menu of cash incentive rebates for substitution of energy saving bulbs and ballasts. Here was an opportunity to combine other less costly renovations with our incandescent replacements to create a cost effective package. Every aspect of our 'Modern state-of-the-art' lighting system was scrutinized for potential energy savings and utility rebates.

Prime targets for Upgrade were our 2 x 4, 2 x 2, and 1 x 1 fluorescent ceiling fixtures. Months of research, seminar participation, interrogation of suppliers/salesmen and manufacturers plus visits to other retrofitted/upgraded buildings led to several trial installations utilizing the new technology and products coming on the market. Experiments proved the superiority of these new products:

- ♦ New Electronic Ballasts
- ♦ New T-8 Bulbs
- ♦ Custom Designed Specular Reflectors

[See Table 1]

Together with utility rebates, this technology provided a cost effective, energy efficient, environmentally protective upgrade. Test installations demonstrated illumination levels equal to or superior to the original design. Most important, outstanding employee acceptance eliminated any fears of reduction in comfort or productivity. We were now ready to begin our upgrade. Or so we thought!

THE OPPOSITION

The upgrade of any lighting system requires up-front money, undoubtedly the greatest obstacle to any undertaking. To obtain funds, our firm, a Financial Services organization, demands that management be presented with a sound business plan detailing how funds are to be utilized and the proposed Return on Investment (ROI). Preparation of this plan is critical. Technical details and accurate cost estimates require cooperation of all interested parties, i.e., contractors, equipment suppliers, labor unions, affected department heads for work scheduling, architect approval for compatibility and, in tenant occupied buildings, the landlord. A formidable task! Consider:

The Architect

Having orchestrated a professional team of Architects/Engineers/Consultants/Contractors plus appeasing various executives with ever-changing design criteria plus scheduling and re-scheduling move-in dates plus appeasing the wounded sensitivities of new occupants for over-sights and/or inconsistencies, (both real and imagined) plus managing a staff engaged full time in a continuing design and construction, the architect was now confronted with a new challenge: Facilities wanted to alter his 'Modern state-of-the-art' design! Imagine, if you can, what an affront this must have presented to professional pride and ego.

When approaching an Architect, employee comfort and productivity take primary concern. You must be prepared to discuss, in detail, Photometric Test Reports, Color Rendering Index (CRI), Visual Comfort Probability (VCP), glare and PL work station impact, as well as illumination measurements. Energy savings, maintenance reductions and environmental concerns must be of secondary importance!

The most controversial item in our proposal was the widespread use of Customized Specular Reflectors, not only a critical component but the keystone of our upgrade. The Architect, in refusing to endorse this product, had the solid backing of all our Engineering Consultants as well as Professional Societies (including AEE). They remembered, too well, the early published reports of initial design failures. All parties urged caution! Gradually, trade publications detailed a succession of successful installations as metal fabrication techniques improved and new energy saving products reached the marketplace. Our successful test areas finally overcame the resistance and, grudgingly, approval was obtained.

Other Departments

Competition for money is always intense. Other highly visible departments such as Computers, Communications and Marketing are skilled in techniques for obtaining funds. They pursue their objectives with vigor and determination and they do no welcome additional competitors. One typical challenge: "What's this I hear about you trying to change our lighting system?"
Here the response was a shamelessly blunt appeal to environmental protection. We translated kilowatt savings into tons of carbon dioxide, nitrous oxide and acid rain which would no longer be spewed into the atmosphere and we aligned ourselves with the angels!

[See Table 2]

Since our Headquarters is located in the World Trade Center this approach was very personal. From these lofty heights one can look out the window and see the pollution! Then, almost as an afterthought: "Oh, by the way, if this upgrade is accepted your departments lighting budget will be dramatically reduced"! It is amazing how zealously protective of the environment people can be when it puts dollars in their pockets.

The Landlord

For multi-tenant office buildings, as well as other rental spaces, the crafting of lease agreements is both science and an art. Landlords compete for tenants by advertising rental cost on a dollars per square foot basis. What is not normally advertised are the ancillary charges such as rubbish removal, cost for after hours services, cleaning services and the cost for electrical energy usage all of which are in addition to the base rental rate.

The Landlord establishes his charge for electrical energy by computing the utility demand and usage costs plus provisions for his routing maintenance, repairs plus the amortization of existing equipment and its eventual replacement. With today's competitive marketplace many tenants have negotiated very favorable rental and ancillary charges.

Consider now the dilemma of the Landlord when a tenant announces his intention to upgrade his lighting system to reduce <u>his</u> cost and qualify <u>his</u> firm for a utility incentive cash payment. Remember it is the Landlord, not the tenant, who pays the utility and it is the Landlord who is listed as the utilities' customer. (The Landlord, anxious to please his tenant, would now be required to reduce his electrical charges to the tenant based on the merits of an upgrade and still protect his amortization, maintenance repair and replacement projections). The Landlord now has a bona-fide concern: This tenant upgrade might indeed cost him revenue. No wonder Landlords may not encourage lighting upgrades, especially if his current electrical charges are favorably calculated and, in effect, provide a profit center.

The ideal solution would be for the Landlord to undertake the Lighting Upgrade, reduce his costs and pocket the utility rebate. With current vacancy rates and cash-flow problems the upgrading of Lighting Systems may not be financially feasible.

Our building renovation included electronic sub-metering. Whatever we did would not have an impact on our lease agreement. Still, we suspect, our Landlord's organization deliberately stalled all proceedings. His concern was legitimate, how to arrange for other tenant renovations which would surely follow when the news of our upgrade leaked out, as indeed it has. More important to all parties was the legal procedures for "passing through" utility incentive rebate payments to a tenant. This involved the cooperation of three sets of lawyers; the Landlord's, the utility's and the tenant's.

Without the enlightened and aggressive actions by one member of the Landlord's staff, our upgrade might have withered on the vine from Lawyer over-kill. It was his aggressive pressure on all three organizations that finally produced the breakthrough. His rationale was straight-forward: The Landlord <u>does indeed</u> profit from tenant upgrades by:

- Reduction in Base Building Air Conditioning Load
- Freeing of Electrical capacity to attract high tech firms requiring 10 watts per square foot.
- Postponing of Electrical System Upgrades
- The modernization and upgrading of the Landlord's property.

Not only did he ramrod this proposal through his organization, he proved a most valuable ally in pressing our own Corporate Executives.

Enlisting the cooperation and support of the Landlord <u>must be a high priority</u> for a tenant contemplating a lighting upgrade in a multi-tenant building. ABSOLUTELY!

The **Unions**

Our ROI was predicated on employing "A" Grade electricians for incandescent Replacement and "M" Maintenance grade for Fluorescent upgrades. There is a 2:1 differential in "A" vs. "M" pay scales. In addition, considerable work would have to be scheduled after-hours at a premium surcharge.

To enhance our ROI, we approached the Electrical Union Shop Steward and requested concessions. This was done, we thought, with confidence knowing the scarcity of work and number of electricians unemployed. No concessions were granted! In fact, the Union Shop Steward raised the ante: "A job of this magnitude will be considered new construction, not renovation, therefore only "A" class electricians will be allowed on the job". Ironically, he indicated that had the union enjoyed full employment he could make concessions but with so many men "on the bench" he was upgrading all the work he could find to put men back on the payroll.

Shop stewards are notoriously hard bargainers and tough negotiators. Our Shop Steward was no exception, but he was not stupid! When convinced that his proposal jeopardized the entire project he relented and agreed to our original request. Remember he was simply doing what was expected of a Shop Steward.

Once on the job, the electricians proved our most valuable asset. They offered many suggestions that saved time and money and eliminated potential code violations. We had their full cooperation and valued it. On one occasion, an essential truck delivery was delayed and arrived after the electricians had left for the day. Against all Union rules and agreements, we unloaded this cargo with non-union porters and handymen. At 8:00AM the next morning, we explained our actions to the electrical foremen and offered to pay full compensation including overtime premium, to his men who should have performed the work. What we didn't know: by 8:00AM the foreman already knew who unloaded the truck, how long it had taken, the elevator they used, where the equipment was stored and had personally inspected and counted the number of pallets. The foreman's reply: "Forget it! We'll let you make up for it some other time". They never did. Simply by demonstrating our respect for their Union rules and regulations and being completely honest we had earned their trust. We never violated it!

An observation: Pride in workmanship is alive and well at the World Trade Center. We suspect it is thriving in your location as well. Union technicians are skilled and competent. Utilize them. Request their input. solicit their ideas, their suggestions, their advice and listen! They know their trade and can be of inestimable value. Maintain their confidence and cooperation. You will never regret it!

General Contractors

To keep costs down and enhance our ROI, we acted as the General contractor. Not recommended, unless you have many empty hours in your daily work fare to track shipments, schedule work, co-ordinate trades, resolve installation conflicts, review job progress and maintain copious records. Better to prepare detailed and clear-cut specifications and plans, then solicit bids and let the successful contractor perform the General contractor duties. He will do a more professional job. Believe it - a General Contractor earns his money!

Manufacturers

It's hard to visualize a manufacturer as an obstacle to an upgrade but that is precisely what we encountered. Or more correctly stated, some manufacturers advanced their products as the ultimate solution to all energy problems. Repeatedly, alternate proposals were advanced to our Architectural Department which they claimed would "do the same thing" as our proposal, Examples: Use 34 watt bulbs; use programmable controls; install occupancy sensors; purchase watt reducers, etc.

These suggestions proved an advantage. It permitted rebuttal by demonstrating the superiority of our approach. In addition, many of these proposals were already incorporated in our future plans. To have incorporated them at this stage would have jeopardized our basic plan by escalating costs to over one million dollars. At that level, Management approval becomes even more competitive and vicious.

Top Management

Ms. Virginia Hines, Editor-in-Chief of Energy User News in an article in Strategic Planning for Energy and Environment, listed the 3 big "E's" for the 1990's as "Energy, the Environment and Economics".

Our analysis say Virginia has it wrong. The 3 big "E's" are: Economics, Economics and Economics. Take care of Economics and Energy and Environment will follow.

Certainly when requesting money ($820,000 in our case) you must clearly document your ROI. We have found that a simple Texas Instrument Business Calculator can quickly and objectively establish your ROI with four entries as follows:

Cost $ = PV
Savings $ = PMT
# = NUMBER OF YEARS INVOLVED
CPT = COMPUTE ROI

[See Table 3]

This data must accompany any request. More important to management is how you propose to quantify savings. In our case, it was simple: Our electric system was already metered therefore, a before vs. after comparison for both kilowatts and dollars would permit excellent verification.

Management has been repeatedly exposed to "Magic Black Boxes" and inflated savings projections. Treat them as skeptics who lack conviction. If your area is not already separated metered, propose the installation of same as a separate cost for savings verification. This will demonstrate your confidence and give credence to your projections. Always, always be conservative in your calculations. One overstatement could prejudice an entire proposal.

ADDITIONAL SOURCES OF INFORMATION

If you are contemplating a lighting retrofit or upgrade you have a great many sources of information and assistance:

seminars: Certainly your attendance at professional sessions can only be beneficial. Don't forget your co-attendees. They, too, are in the same market. Question them. They might have your answer. My personal experience attests that AEE, ASHRAE and BOMA seminars attract an audience of professionals who are after knowledge, hard facts and that most valuable of all commodities: **IDEAS!**

PUBLICATIONS

The Energy User News Monthly provides names, locations and detailed information on what is new, what has worked and the equipment employed. Don't overlook the many books published by professional organizations. You need them to expand your horizons, clarify your thinking and formulate plans.

SUCCESSFUL INSTALLATIONS

Call and/or visit others who have accomplished a successful upgrade. They are most willing to share their problems, their solutions, their mistakes and will offer advice and encouragement. Utilize them. Their assistance proved invaluable to us.

Here are our results from just one floor!

[See Figure 1]

Time frame to accomplish the above, 3 1/2 years! We are now in Phase II: Occupancy sensors and programmable controls.

PHILOSOPHY

To plan and execute a successful Lighting Upgrade, you must be firmly convinces of its importance and its contribution to your organization's goals. **DO YOUR HOMEWORK**. Be careful with your numbers. Outline how you plan to quantify your savings. Be it's severest critic. And above all don't expect the world to love you or your plan. Be assured, your conviction will be tested.

Always keep in mind that you are an advocate of change. Never mind that what you propose is economically sound, saves energy and protects the environment. Some people oppose change, any change! Here is how Woodrow Wilson, our 28th President, a Doctor of Philosophy and an educator expressed this fact:

"If you want to make enemies, try to change something."

It's been said that today, in our nation's capitol, this view has been superseded by a quote attributed to Clair Booth Luce, Author, Lecturer and Politician:

"No good deed goes unpunished."

Lest you become too cynical, I offer as an anti-dote to those pessimists, the motto of an American Army World War II General, by name, Joseph Stillwell. (A clue to his personality was his nickname, "Vinegar Joe".) It states, in pseudo-Latin:

"Illegetimis non Carborundum"

The translation:

"Don't let the bastards grind you down"

I have adopted it and found its message appropriate.

POST-MORTEM

Our firm like many others is constantly re-structuring, revising and rearranging personnel, departments and objectives in an effort to be more efficient and profitable. Just a little over two years ago, our Facilities Department was absorbed by, alas, the Architecture Department! There have been subsequent changes. As the sole survivor of these revisions, let me assure you that change evokes entirely different emotions when you are the changee, not the changer! Always be professional in your comments and charitable in your appraisal of others. You may end up working for them - like me!

Not only has our successful upgrade proven itself, but we are well into Phase II: Occupancy Sensors and Programmed Lighting.

We have proved our mettle. We have credibility and, more importantly, two colleagues have gone on to bigger and better positions of responsibility at different buildings where they will carry-on the message.

SUMMARY

If you are contemplating an upgrading of your "Modern" Lighting System:

- ♦ Do your homework.
- ♦ Start with test areas.
- ♦ Quantify your results.
- ♦ Remember: Economics, Economics, Economics.
- ♦ Believe!

And most important:

- ♦ Don't let <u>them</u> grind you down!

GOOD LUCK!

TABLE 1

OUR PLAN

2' x 4' FIXTURES

FROM:	3 EACH - 40W T-12 W/BALLAST	109 W/FIX
TO:	2 EACH 32 W T-8 ELECTRONIC BALLAST CUSTOMIZED REFLECTOR	58 W/FIX
	SAVINGS	51 W/FIX
QUANTITY:	6177 FIXTURES	
	TOTAL ENERGY SAVED	315.0 KW

2' X 2' FIXTURES

FROM:	2 EACH 40W T-12U W/BALLAST	72 W/FIX
TO:	2 EACH 17W T-8 ELECTRONIC BALLAST CUSTOMIZED REFLECTOR	33 W/FIX
	SAVINGS	39 W/FIX
QUANTITY:	550 FIXTURES	
	TOTAL ENERGY SAVED	21.45 KW

1' X 1' FIXTURES

FROM:	3 EACH PL13	45 W/FIX
TO:	1 EACH PL13 CUSTOMIZED REFLECTOR	15 W/FIX
	SAVINGS	30 W/FIX
QUANTITY:	621 FIXTURES	
	TOTAL ENERGY SAVED	18.63KW

REPLACEMENTS: INCANDESCENTS

FROM:	75 WATTS/INCANDESCENT	75W
TO:	2 EACH PL 13 FLUORESCENT	30W
	SAVED	45W
QUANTITY:	842 FIX X 45 W/FIX	37.89KW
ELIMINATED:	264 FIX X 75 W/FIX	19.8 KW
	TOTAL SAVINGS	412.77KW

TABLE 2

ENVIRONMENT PROTECTION

ENERGY REDUCTION - 412.77KW

ANNUAL REDUCTION IN ATMOSPHERIC EMISSIONS

CARBON DIOXIDE (1) - 286.8 TONS

NITROUS OXIDE (1) - 77.0 TONS

"ACID RAIN" (SO) (1) - 1.34 TONS

FUEL OIL CONSUMPTION (2) - 550.0 BARRELS

REFERENCE:

(1) U.S. EP "GREEN LIGHTS" BROCHURE - MARCH 1991

(2) CON EDISON "ENLIGHTENED ENERGY" BOOKLET, QUESTIONS AND ANSWERS QUIZ - MARCH 1991

TABLE 3

RETURN ON INVESTMENT

COSTS:		
	$820,000.00	MATERIAL, LABOR & CLEAN UP
	-175,000.00	UTILITY REBATE
	$645,000.00	TOTAL COST

SAVINGS:

ENERGY: $412.77\text{KW} \times .10\frac{\$}{\text{KWH}} \times 3650 \frac{\text{HR}}{\text{YR}}$

$= \$150{,}661/\text{YR}$

MAINTENANCE: $35,000 LABOR & MATERIALS
$185,661 TOTAL SAVINGS

PROJECT LIFE: (MINIMUM)

10 YEARS

TEXAS INSTRUMENT BA.35 BUSINESS CALCULATOR

ENTER:

$645,000 = PV (PRESENT VALUE)

$185,661 = PMT (ANNUAL SAVINGS)

10 = n (NUMBER OF YEARS)

CPT (COMPUTE) = %

READ: 25.91%

SIMPLE PAYBACK

$\frac{645{,}000 \text{ COST}}{185{,}661 \text{ SAVINGS}} = 3.47 \text{ YEARS}$

LIGHTING RETROFIT FLOORS

Meter#	Date Retrofit Done	Meter Type	1991 Jan	Feb	Mar	Apr	May	Jun	Jul	Aug	Sep	Oct	Nov	1992 Jan	Feb	Mar	Apr	May	Jun	Jul	Aug	Sep	Oct	Nov	63 Floor
1,134	33,739	LP63E	6,401	7,439	7,226	7,290	7,669	7,121	7,441	6,569	6,883	6,787	7,538	7,439	7,501	6,785	6,698	5,427	3,905	3,718	3,677	3,990	3,828	4,357	
2,134	5/15/92	LP63W	8,548	9,662	8,416	8,937	8,932	8,011	8,009	7,735	8,089	8,246	9,500	8,264	8,260	7,582	7,967	6,859	5,313	5,168	5,211	5,070	5,090	5,991	
63rd Floor Total:			14,949	17,101	15,642	16,227	16,601	15,132	15,450	14,304	14,972	15,033	17,038	15,703	15,761	14,367	14,665	12,286	9,218	8,886	8,888	9,060	8,918	10,348	6,102 $689.51

KWH ELECTRICAL CONSUMPTION
Jan-Nov 1991 vs. Jan-Nov 1992

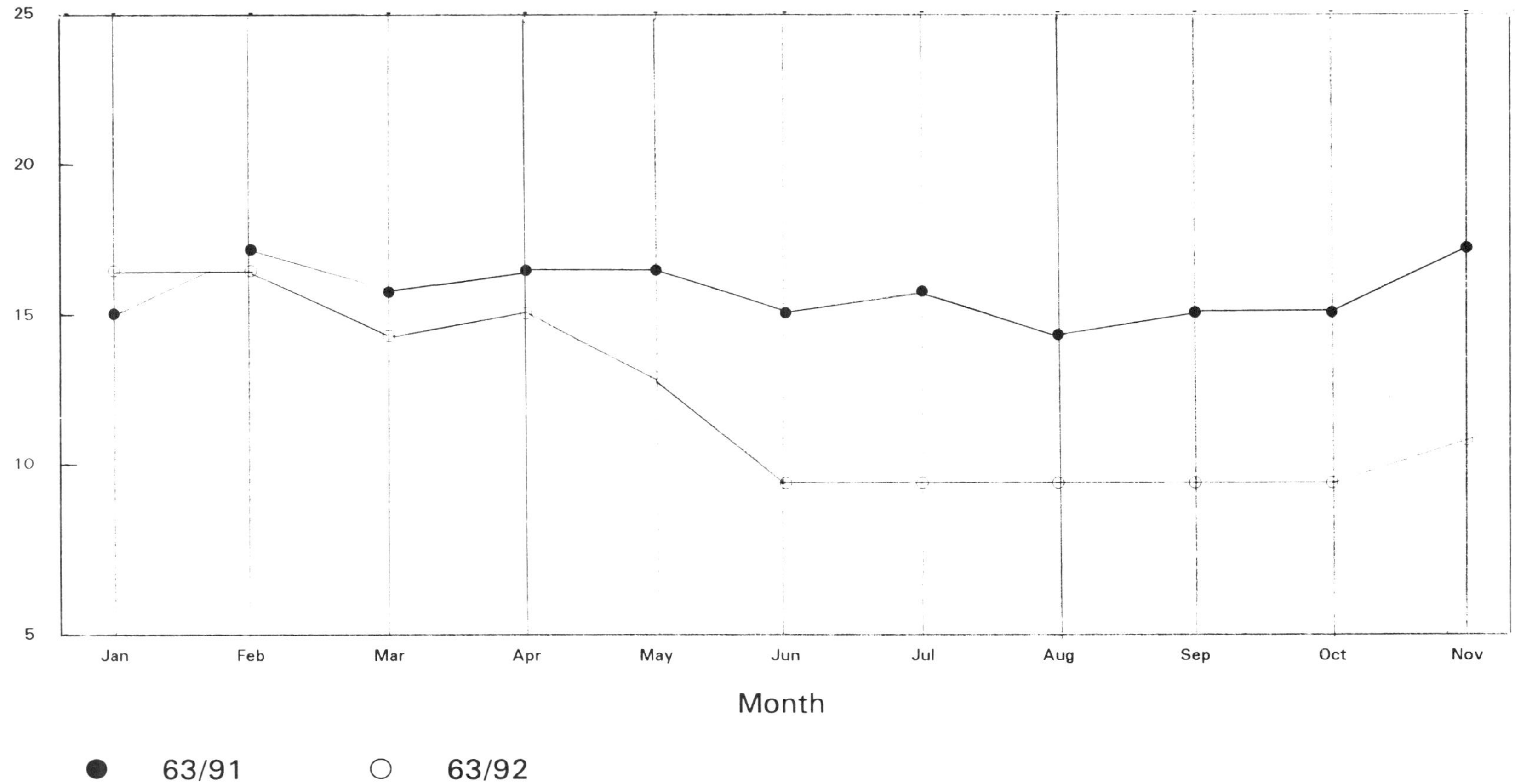

Note: 1991 and 1992 graph lines for a given floor have, respectively, solid and hollow icons of the same shape.

FIGURE 1

Chapter 57

Lighting Systems That Promise Energy Efficiency: Fluorescent System Technology C 1993

G.V. Preston

I. INTRODUCTION

The area of energy efficient lighting continues to increase in popularity and accelerate in application due to a wide variety of forces. One such force is the great success that energy professionals have had in demand reduction and energy savings as described at the 14th and 15th World Energy Engineering Congress [1,2]. This success contributes to broader trial and further successes.

Third party intervention, such as active utility DSM programs, which now extend through 23 states, has had a major impact in overcoming initial cash outlay barriers. According to an industry survey, this contribution may grow to $500 million per year by the mid 1990s. Another third party intervention is the EPA's Green Lights program, which had 752 participants through January 1993 who had already avoided 95 million KWH annually - which equates to $9.4 million in energy cost avoidance.[4] The most recent and most far reaching external force is the National Energy Policy Act of 1992 (NEPA) which has also focused nationwide attention on energy efficient lighting. According to one estimate the impact of NEPA alone will account for savings of 49-123 TWH/year by the year 2030[5]. The lighting industry is also doing its part by developing innovative, energy efficient light sources to keep pace with market demands. It is this new technology in light sources and systems that will be reviewed in this paper. The criteria for inclusion were 1.) innovations that have potential for major energy savings and 2.) sources that did not exist this time last year.

These new light sources have widened the applicability of energy efficienct lighting from the standard 4 foot and 8 foot fluorescent systems into the realm of 2'x2' square luminaires and recessed downlights. The latest system alternatives can provide 40% energy savings in retrofit and a 79% savings in new construction. Let's look at each of these in detail.

II. 2'x 2' FIXTURES

A. Opportunity

One of the fastest growing segments of commercial space lighting is the two foot square fluorescent luminaire. This segment has been growing at a rate near 20% annually in what is an otherwise flat market. The primary light source that is used in this luminaire is the U-shaped 1½ inch diameter fluorescent lamp with a 6" leg spacing. Although this source exhibits the generally high, 60 LPW efficacy of fluorescent lighting it has not benefited from any suitable replacement innovation in the last 12 years. Further, this standard lamp is on the endangered species list of the National Energy Policy Act of 1992 and is slated for extinction in 1995.

There exists an installed base of over 35 million fixtures in this 2'x 2' catgeory. Assuming a weighted average power of 85 watts per fixture, this represents an installed load of 3 Gigawatts! Saving just 30% of this it would mean savings of over $288 million per year for the private sector.

B. Technology

Several source options have existed with some recent attempts being made to develop alternatives. Until December 1992, however, no good retrofit option was available. Last December, however, a new lamp was introduced that combined the exciting benefits of the 1 inch (T8) diameter tube with high-efficiency, high color rendering (CRI) phosphors, as presented here in past years[6], with the premium efficiency of high frequency ballasts and a convenient 6" leg spacing. The resulting system provides an exciting energy saving retrofit opportunity as shown here:

LAMP BALLAST	STD U/T12/CW STD MAGNETIC	ENERGY SVG U/T12/CW STD MAGNETIC	NEW U/T8/TRI HF ELECTRONIC
NOMINAL LAMP LUMENS	2,650	2,400	2,800
RATED BALLAST FACTOR	.95	.88	.88
SYSTEM LUMENS	2,518	2,112	2,464
SYSTEM POWER (WATTS)	96	82	58
SYSTEM EFFICACY (LPW)	26.2	25.8	42.5
CRI	62	62	85

TABLE 1

Results:

For application then, the direct retrofit of the new 6" T8 U-lamp system can provide energy savings, based on 4,000 operating hours per year and $0.08/KWH, of

POWER SAVINGS = (96-58) = 38 WATTS
OR 40%
ANNUAL ENERGY SAVINGS
= 38 WATTS * 4000 HOURS * $0.8/KWH * 1KW/1000H
= $12.16

Thus 40% energy savings is realized, with no change in light distribution, and an improvement in light quality (CRI) with the combination of a lamp and ballast upgrade.

III. COMMERCIAL RECESSED DOWNLIGHTS

A. Opportunity

One of the remaining commercial lighting applications that clings to the older incandescent technology is the realm of downlights. In each of the years 1990 and 1991 over ***30*** *million recessed downlights were sold. Assuming that on average each downlight represents 100 Watt of power then the total load is* ***3 Gigawatts*** *of new demand being installed each year! It is clear that this represents an enormous opportunity for us as energy professionals.*

B. Technology

There are several alternatives for light sources to fill this specific application but the most promising is a new compact fluorescent source known as a Triple. This Triple is a non-integrated source, which means that it is designed as a replaceable lamp element. This lamp pairs with a separate electronic ballast to produce an energy saving system which can provide 79% savings over incandescent sources with comparable illuminance.

A Triple is a narrow diameter (10mm) compact fluorescent that utilizes a combination of state-of-the-art technologies to reach this performance level.

These technologies include:

- *Trichromatic, Rare Earth phosphors which are used to provide high efficiency, superior color quality (CRI >80), and good maintenance over lamp life.*
- *A combination of bend and bridge technologies which are employed to bring high light output into a very compact package (32W, MOL= 140mm).*
- *High frequency, constant current, electronic ballasts which are used to provide enhanced system efficacies.*
- *Amalgam mercury pressure control that is employed*

SYSTEM COMPARISON
6" APERATURE RECESSED DOWNLIGHT

	100W INCANDESCENT	26W QUAD CFL	26W TRIPLE CFL
Lamp Lumens	1480	1800	1800
Fixture Efficiency	.75	.50	.65
Effective Lumens	1069	900	1170
		-16%	+9%
System Power (W)	99	34	28
Fix. Efficacy (LPW)	10.8	26.5	41.8
Ann. Energy Savings*	-	$20.80	$22.72

*Assume 4K Hours/year, $.08/KWH

TABLE 2

to provide optimal light output over a wide temperature range that is often found in real applications.

- *The lamp:ballast system constant-current co-design to provide lamp interchangeability for the flexibility that previously was available only from incandescent sources.*

For example: *Suppose we were illuminating a lobby with 6" recessed downlights in which you were analyzing your different alternatives to 100W incandescent sources. The prime qualities you might compare are shown in Table 2. Please note both the higher light output and lower energy consumption of the Triple system.*

Why does the new Triple exhibit such improved efficiency? The key is that new techniques are employed to solve old problems. First consider contributors to Fixture Efficiency:

$$FE = OE * TF$$

where

FE = fixture efficiency
OE = optical efficiency
TF = thermal factor

In quad type compact fluorescent lamps, the lamp length (M.O.L.) is long enough that much of the lamp is positioned in the luminaire's neck, or outside of the reflector in order to avoid projecting out of the recessed ceiling fixture. Therefore, the optical efficiency of the luminaire suffers. A common optical efficiency is 50%-60% with a vertically mounted lamp and 30% with a horizontally mounted lamp. Moreover compact fluorescents, as all "uncorrected" fluorescent lamps, are temperature dependent. As Figure 1 shows, a typical quad lamp will operate at about 85% of peak output at the ambient temperatures found in an open, commercial grade, recessed downlight. The new Triple lamp employs an amalgam, or mercury pump, which mitigates against this temperature effect. The results are a flatter temperature response curve and a 98% thermal factor.

As a result in a 6" Aperature:

	OE	TF	FE
QUAD 26W	.58	.85	.50
TRIPLE 26W	.66	.98	.65

TABLE 3

Therefore a 30-50% improvement in efficiency is obtained.

C. Results:

What does all this esoteric light bulb technology mean to us as energy professionals?

Previously, a very common lighting scheme was 150W incandescent lamps in a 7" aperature downlight as the general lighting source for commercial spaces.
Table 4 provides the data for this set of fixtures. For

Luminous Flux

as function of Temperature

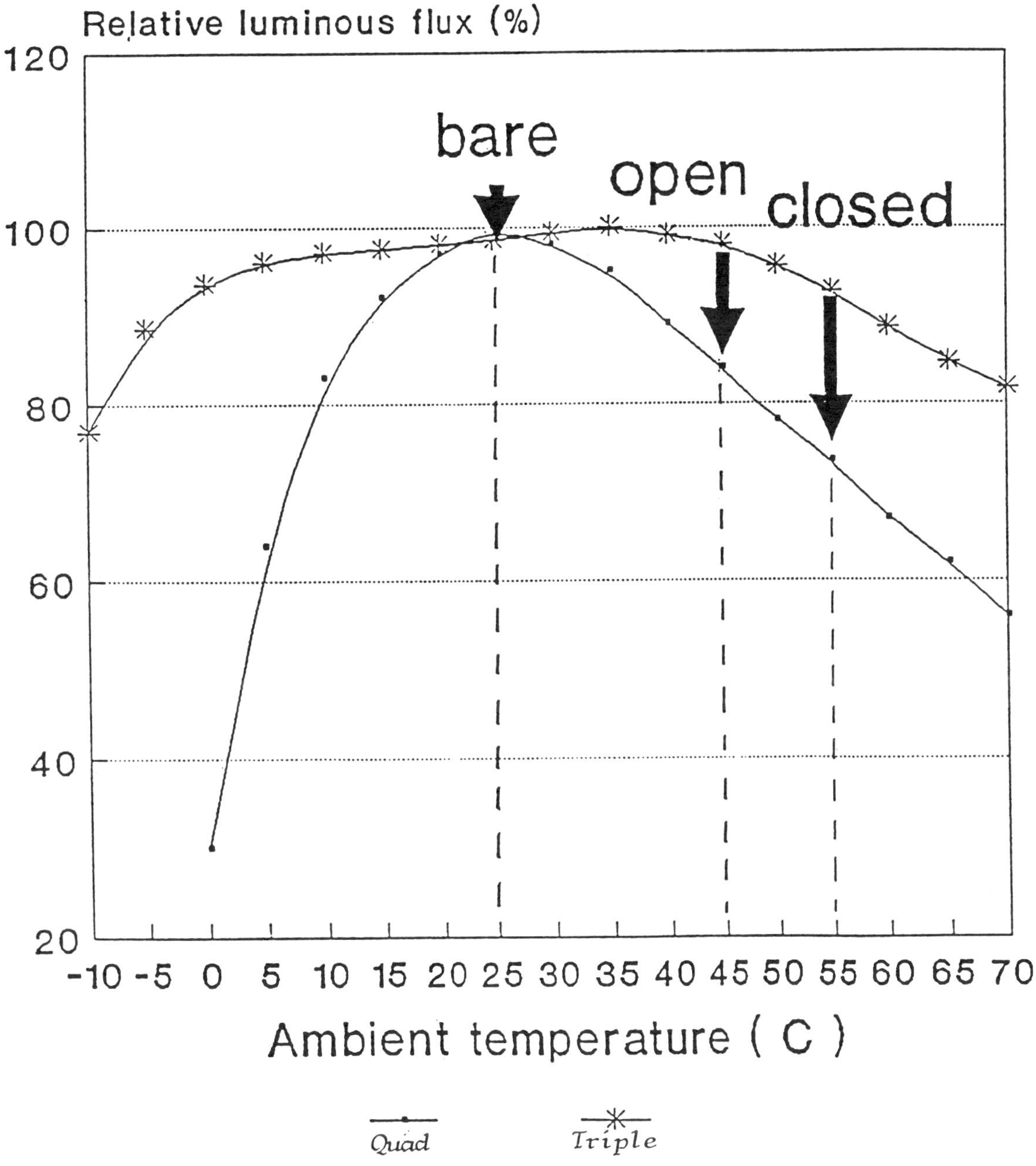

Figure 1

running cost we'll use the cost of energy plus the cost of and replacement lamps and will calculate the savings of the alternative systems using $.08/KWH and 4,000 operating hours per year. All costs will be converted to present values utilizing a 7% interest rate over 5 years. Then the difference in running costs provides the present value of the cost savings of using the Triple compact fluorescent technology.

For new systems with equal light level vs. 150W incandescent:

LAMP	#FIXTURE (%)	POWER (%)	PRESENT VALUE COST SAVINGS ($)
QUAD 26W	+57	-63	-
TRIPLE 26W	+34	-75	-
TRIPLE 32W	-11	-79	$158

TABLE 5

Plus the additional savings of fewer fixtures both initial and running costs are $158/.89= $177.50 which is the incremental amount that it is worth to you to invest in the newer Triple compact fluorescent sources.

IV. CONCLUSIONS

These newest sources offer energy professionals powerful weapons to add to their energy efficient lighting arsenals. The T8, 6 inch, U lamp complements the collection of T8 straight lamps. T8 lamps are by a factor of almost 2 to 1 the most commonly installed lighting upgrade by the EPA's Green Lights partners. The T8 U lamp broadens their application arena to include 2'x 2' square luminaires while providing energy savings of up to 40% with a simple system retrofit.

The compact fluorescent, non-integrated, Triple contributes shorter, more powerful light sources to the growing family of compact fluorescent lamps. This lamp provides both high lamp efficacy and, by overcoming optical and thermal

RECESSED DOWNLIGHT SYSTEMS COMPARISON

7 INCH APERATURE

	INCANDESCENT 150 WATT	FLUORESCENT QUAD 26 WATT	FLUORESCENT TRIPLE 26 WATT	FLUORESCENT TRIPLE 32 WATT
FIXTURE POWER (WATTS)	150	35	28	36
LAMP LUMENS	2,400	1,800	1,800	2,400
FIXTURE EFFICIENCY	66%	52%	67%	73%
FIXTURE LUMENS	1,584	936	1,206	1,752
COMPARISON		59.1%	76.1%	110.6%
FIXTURE EFFICACY (LPW)	10.6	26.7	43.1	48.7

TABLE 4

barriers, high real-world fixture efficiencies. This combination of features can save up to 79% of the lighting load while providing comparable illumination and longer lamp life than traditional incandescent sources.

These major innovations were introduced within the last year. They represent the commmitment that the lighting industry has made to continuous improvement in energy efficient systems. A systems emphasis ensures that the components of the lighting system, lamp, ballast and luminaire are not sub-optimized but are integrated to provide application benefits. The result is systems that will provide ease of installation, cost savings for the user, and reduced demand for the utilities.

Footnotes

1 Bleeker, N.C., Benefits of Energy Efficient Lighting Part 1, October 1991.

2 Bleeker, N.C., Benefits of Energy Efficient Lighting, Part 2, October 1992.

3 Survey and Forecast of Marketplace Supply and Demand for Energy-Efficient Lighting Products, Lighting Research Institute and Plexus Research Inc., October 1991.

4 Green Lights The Second Year, United States Environmental Protection Agency, March 1993.

5 Atkinson, B.A. etal, Analysis of Federal Policy Options for Improving v.s. Lighting Energy Efficiency:Commercial and Residential Buildings, 1992, Lawrence Berkely Laboratories.

6 Krsyzckowski, A.M. and Bleeker, N.C., Fluorescent System Performance & Economic Cost Comparison for New and Existing Installations, W.E.E.C., October 1988.

Chapter 58

Bright Ideas: A Comprehensive Lighting Retrofit in an Academic Setting

E.H. Selmon, D.G. Groves

ABSTRACT

A $100,000 lighting retrofit project in the Terman Engineering Center (TEC) at Stanford University has improved the quality of light and will earn a return on investment triple that of the University's endowment. Initiated by student concern for energy management and environmental degradation, the proposal culminated in a comprehensive project involving students, faculty, staff, and outside experts. After extensive time-of-use surveys, occupant surveys, lighting measurements, and economic analyses, a design was chosen which incorporated the latest in lighting technology. Various control technologies, included occupancy sensors, timers, and dual-switching, as well as advanced components such as electronic ballasts and T-8 fluorescent lamps were used. The project involved many settings, including offices, classrooms, a library, an auditorium/TV studio, public areas, and labs. On completion, a comprehensive follow up study was done, evaluating changes in lighting use, lighting quality, energy use, the effectiveness of the individual technologies, and user satisfaction. Metering was done before and after the project in order to verify energy and dollar savings. This project not only serves as a model for other Stanford buildings, but also for other universities attempting to improve lighting quality, energy efficiency, and user comfort while saving money.

INTRODUCTION

Stanford University, located in Stanford, California, spent approximately $26 million on total energy costs in 1992. By Fiscal 1993, this spending is projected to rise to $32 million due to new construction and increased load from computers and other equipment. Stanford, not insulated from the trend of budget cuts and fiscal shortages in the university system, is in the process of reducing costs in every department. As energy technologies continue to improve, Stanford, as well as other institutions, is looking to reduce costs through energy efficiency.

The TEC is a large multi-use structure consisting of 222 offices, classrooms, labs, a multimedia recording auditorium, and a library. The TEC used $114,000 of electricity 1991, and based on surveys and observation, it is estimated that 50% of all electricity was used for lighting. Although the TEC was recognized for its innovative energy saving design when built in 1978, changing technology has made it a prime candidate for a test energy management retrofit project. This paper discusses a $100,000 lighting retrofit project for the TEC. It has a projected saving of more than $27,000 a year. This project was developed by Stanford undergraduate students with the guidance of faculty, staff, and outside experts. We began fall of 1992 and will be completed in September 1993.

This retrofit project focused on saving energy and increasing the quality of light in all areas of the TEC. This was accomplished by updating older lighting technology, replacing wasteful and nonessential lights with more efficient systems, and using control technology to reduce the operation of lights in unoccupied rooms.

In most of the areas, the main lighting before the retrofit, was provided by 1x4 fixtures with two F34 T-12 fluorescent lamps and one electromagnetic ballast for each fixture. We retrofitted all these fixtures, replacing the T-12 lamps with F32 T-8 fluorescent lamps and using the minimum number of electronic ballasts possible (each ballast can power two 2-lamp fixtures). Not only was this a large energy saver, but it also increased lighting levels by 5%. We replaced all incandescent floodlights with the appropriate compact fluorescent equivalent. We also used the following control technologies: dual-switches for areas that received sunlight, occupancy sensors for areas that are used periodically, timers for the hallways, and switches for the library stacks.

We estimate that Stanford will reduce electricity consumption due to lighting by more than 317,000 kWh per year in the TEC, or 46%. This will translate into over $27,000 of savings per year. The pay-back period will be under four years. In addition, this retrofit improved lighting in many areas, especially the open areas favored by many students for studying.

PROJECT SETTING

The project was initiated, researched, and developed by undergraduates in the research atmosphere of Stanford University. Throughout the entire course of the project we worked with the faculty, staff, and administration. We had to understand the way that the University's energy was distributed, learn how the bureaucracy functioned, and develop a project which was of interest to assisting faculty and staff members.

Administrative and Energy Environment

Stanford University incorporates not only traditional university facilities, but also includes a large hospital and other non-traditional, high energy use buildings. To power all these facilities, Stanford buys electricity from an on-campus co-generation plant, and to a minor extent from the Palo Alto Utility and Pacific Gas and Electric (PG&E). All of the academic buildings receive electricity from Cardinal Co-gen. Cardinal Co-Gen also provides steam for Stanford and sells excess electricity to PG&E. Stanford University in the past few years has recognized that with its physical growth and rising energy budget, energy conservation could be an effective way to control costs and save money. At first look, the energy independence that Cardinal Co-gen provides to Stanford appears to be a great benefit. However, Stanford is no longer eligible for the full DSM rebates offered by PG&E. Stanford's rebates are prorated based on the amount of electricity purchased directly PG&E, now around 8-10%. This has forced Stanford to be very careful about any retrofits it does, because they are financed entirely by Stanford itself. Stanford's Energy Management Group is in charge of finding and/or developing energy management projects. In the past, each retrofit's return on investment in relation to that of Stanford's endowment (currently 11.4%) has been the criteria which has been used for determining project feasibility.

Stanford is divided into zones for most facilities activities. Retrofits are initially financed out of the appropriate zone's University facilities budget. The zone is then reimbursed by the Utilities Department which has a general fund for retrofit projects. In the past, the retrofits have been mostly initiated by the Utilities Department. However, as the University moves towards making each department fiscally responsible for its energy use, more and more projects are expected to be initiated by the departments themselves.

Academic Environment

Stanford University has a large and heavily research-oriented School of Engineering. Even though the Stanford administration encourages undergraduate student to take advantage of these research resources, it is difficult for some of the students to find areas of research that are mutually interesting to themselves and to the faculty. We found, however, that lighting energy management is an area where the students can take the initiative and devise their own project, while learning and producing worthwhile results.

For students interested in energy management research, Stanford has few official classes. However, in spite of this modest number of energy related classes our group was able to find an interested faculty member to give us guidance throughout our project. Professor Gil Masters of the Civil Engineering department after being approached and presented with our idea, served as a very essential liaison between the often intimidating school administrative bureaucracy and ourselves.

HISTORY AND PROCEDURE

This project began in the fall of the 1991-1992 academic year. We started as a group of six students called the STanford Energy Action Movement (STEAM), with an interest in energy issues and environmental degradation. Under the direction of Professor Masters, we formed a special research class in which we were able to learn about energy use at Stanford and apply this knowledge towards energy management projects. As we began to realize the possibilities for energy savings at Stanford, we directed our attention to the TEC to study the energy use patterns more closely. This individual project was to be part of a longer term project to change the way energy-related decisions were made at Stanford.

Direct experience was the emphasis of this research class. We began our study with a thorough walk-through of the building to identify the types of spaces and their lighting as well as the number of hours in use. Total energy use and demand were also estimated by hourly readings of meters in the electric rooms.

After gaining a general understanding of the building's lighting plan by the walk-throughs, and a broad picture of consumption patterns by reading the meters, surveys were sent to all of the office occupants to model their energy use in more detail. Questions were asked about the amount of time they used their offices, the other types of office equipment they used, as well as their opinions of various energy saving proposals like occupancy sensors. This became an important part of our research, as our goal was to learn about energy management both economically and socially.

By circulating questionnaires, we attempted to determine if the occupants were willing to reduce their light levels to achieve significantly greater savings in energy. While some of the occupants were amenable, it was concluded that in order to maintain overall support for a possible retrofit, we could not lower any of the lighting levels. In addition, recognizing the inevitable objections to any changes in lighting by some occupants of the building, we decided that we should focus on improving lighting quality where deficient. This focus led our group to do extensive light level measurements and it helped us formulate a retrofit strategy which would account for its potential impact on the occupants' working environment.

Throughout the formulation of our proposal, people in energy related positions at the University spoke to our group about Stanford's energy use. These talks included the current process of energy management and possible schemes to improve it. They also addressed the decision-making process for energy management proposals. This gave us a better understanding of the type of retrofits Stanford University would not only approve of, but also fund. In addition, it established a possible framework for the project plans of action that could be taken to get the University to make changes.

After all the basic research was done, we worked through many possible retrofit strategies. We were interested in formulating a strategy that would be commendable not only on the grounds of its benefit to the environment, but also by its financial feasibility. With this in mind, we devised a general strategy and did the economic analysis necessary to prove that the retrofit made financial sense.

All of these parts culminated in a paper recommending changes in the lighting at the TEC. This was sent to the president of Stanford, the dean of the School of Engineering, the Stanford utilities manager, and the energy engineers. By doing this, the top administrative and academic positions related to energy decisions were presented with a proposal by the normally underrepresented students.

When school resumed in the fall of 1992, we learned that the facilities department was interested in our proposal. Two members of the original group decided to improve our retrofit by making it more detailed and accurate so that it would be funded and eventually implemented. The original study had been more of an educational tool, describing a possible retrofit, rather than specifications for an actual retrofit. With the knowledge accumulated over the course of the previous year's study, we reassessed the original proposal. We determined a method of presenting our ideas in an accurate and plausible way, focusing more on actual implementation.

In order for our proposal to be presented for funding and official approval, it was necessary to expand on our previous study. We first determined the actual change in lighting levels that would occur in the offices after the retrofit, through an experiment in one of the offices. We were interested in not only how the lighting levels would change by adding the new T-8s and electronic ballasts, but also simply by cleaning the fixture. We found that cleaning the fixture increased the light levels by 5% and replacing the older T-12s with new ones increased light levels an additional 20%. Changing to T-8s then increased the light levels by only another 5%. From these results we determined the net increase of lighting levels that the occupants would notice after the retrofit. We also interviewed the occupant the next day to determine her reactions. She explained that the quality of lighting improved so much that the neighboring professor was actually "covetous of her lights." The purpose of this experiment was not only to learn about the actual retrofitting of lighting through firsthand experience, but also to make the occupants of the building aware of the possibilities of the forthcoming retrofit. The building's occupants were very receptive to these and other similar experiments.

To put together a complete retrofit recommendation we needed to conduct a thorough walk-through of the entire building, which had not been done with the original STEAM report. We recorded every lamp in the building, noted emergency circuits, measured light-levels, and gained a better understanding of the deficiencies of the building. By using this new and more detailed information, we were able to work with Professor Masters, Scott Gould of the Stanford Energy Management Group and Matt Tracy of T. Marshall Associates, a local lighting energy management firm, in formulating a retrofit which would satisfy the occupants and save enough money to pay for itself in less than four years. From this information we produced a single spreadsheet which clearly presented our suggested changes on a room/area by room/area basis. We also included the projected energy use change, cost of the retrofit, and projected pay-back period. The final analysis and decisions on the lighting retrofit were put into a final proposal which was presented to the Stanford Energy Management Group for approval. After it was approved for funding, we wrote up the specifications and it went to bid in June, 1993.

THE RETROFIT

This retrofit was designed to make the most cost-effective changes in each area. In planning the retrofit, the appropriate technologies were determined based on the layout and pattern of use of the area. The final design consisted primarily of replacing all of the fluorescent tubes and ballasts with T-8s and electronic ballasts. Additional changes were also made that were specific to each area.

There are eight types of areas in the TEC, the most numerous being the offices. The offices are the most sensitive to user satisfaction because they are where most of the occupants of the building will directly experience the changes. There are 222 offices in the TEC and 21 spaces with the same layout as two or three offices combined. All of the offices had four 1x4 fixtures laid out either in a strip of four or two pairs. Initially we planned to remove one lamp from each fixture and install reflectors and occupancy sensors. We chose not to install reflectors because it would result in a 20% drop in light levels and from the survey results we decided to maintain at least the same light levels as before the retrofit in all areas. We decided against occupancy sensors in the offices because, from the earlier surveys, we estimated that the offices were used only 6.5 hours-per-day. At that point it was no longer cost effective to use sensors because the lighting savings would probably be no more than one to two hours-per-day. The final plan for most offices consisted of replacing all of T-12s with T-8s and replacing the magnetic ballasts with half the number of electronic ballasts.

For some of the offices, however, there was another option: daylighting. Nearly 50 of the offices have south facing windows and receive enough sun during long periods of the day to light the offices. In these areas we decided only to use dual switching because the savings did not justify the installation of photocells and dimmable ballasts. The dual switching would allow the occupant to turn on one lamp in each fixture, the other lamp in each fixture, or both. We felt this would be fairly effective if accompanied with some education on design and purpose of the retrofit. It also gives the occupants more flexibility to become an active part of the system.

The next area we examined, and nearly as sensitive, was the library. The library occupies a part of three floors of the TEC. It is an important study and research area for the students. The lighting consisted primarily of 1x4 fixtures; strips of two, three, or four T-12s in the stacks; incandescent down lighting; and incandescent track lighting. Savings were much greater in the library than the offices because the library is used fourteen hours-per-day. This meant that a simple change from T-12s and electromagnetic ballasts to T-8s and electronic ballasts yielded major savings. It also allowed more flexibility in designing the retrofit. In the stacks, the major change, in addition to T-8s and electronic ballasts, was to install a switch on each stack to allow users to turn off the lights as they leave the stacks. We estimate that this will result in a three to four hour use reduction each day. The incandescent down lighting, consisting of 75W reflector lamps with some 50W lamps, was changed to 13W modular compact fluorescents (the ballast can be used with new lamps when the original burns out) because the fixtures had built-in reflectors. The track lighting was replaced with integral 15W compact fluorescents with reflectors.

In the copy room, an occupancy sensor was included because the room's lights are on fourteen hours-per-day, yet the room is only used three or four hours-per-day. In the other area of the library it did not make sense to use occupancy sensors because the areas were either used consistently throughout the day, or, like the stacks, had a layout that made sensors difficult to use.

In some areas, like the large classrooms scattered around the TEC, occupancy sensors were also very cost effective. The six classrooms in the TEC range in capacity from forty students to a few hundred. Some are also used for televised lectures, restricting our options in designing the retrofit. In rooms used for televised lectures there were incandescents on dimming circuits as well as the usual 1x4 fixtures. Due to the sensitivity of video cameras to various kinds of light, we chose to use diodeless quartz-halogen lamps (quartz-halogen lamps with diodes can flicker when dimmed). This was not expected to cause any color shift in the broadcasts and would allow some savings in energy. The classrooms are occupied at scheduled and unscheduled times throughout the day with significant blocks of unoccupied time. The lights are on for 8 to 10 hours a day, including periods when they are unoccupied. We planned for occupancy sensors in all of the classrooms, estimating a 25% reduction in lighting use.

Another significant area of the TEC that allowed use of occupancy sensors were the various lab and non-office areas. These include various workshops, computer clusters, and other rooms that did not fit into any of the categories above. All of these rooms used the basic 1x4 fixture. Hours of use varied from 8 to 16 hours-per-day. In some, such as the computer clusters, occupancy sensors would not be appropriate because the rooms were in use at all times. However, many rooms, such as the labs, were unoccupied for long periods of time while the lights were on. In the labs which are used to study hazardous waste, extra care was taken in the placement of sensors to prevent the lights from turning off at wrong times.

The next major areas we examined were the open areas, the lobbies and hallways. These areas had a variety of lighting, all of which were on 24 hours-per-day. In the lobbies and hallways, down lights with 15W compact fluorescents were being used. In most hallways there were a number of single T-12s and in one area there were 1x4 fixtures. In the lobbies, the light provided by these fixtures was not adequate for the many students who studied there. After measurements with a number of types of compact fluorescents, we determined that a higher wattage of lamps was not needed. The problem was the diffuser on the lamp blocked 50% of the light. Using a compact fluorescent with no diffuser doubled the light levels. Using a 15W Osram Dulux with reflector raised the light levels even more, but we decided to use a modular compact fluorescent because the long hours of use made the replaceable lamps more cost effective.

In the rest of the open area fixtures, T-12s and electromagnetic ballasts were replaced with T-8s and electronic ballasts. The long hours of use made this very cost effective, even in the single lamp fixtures. However, the savings were improved because the hallways are only unlocked 10 hours-per-day. Therefore, all of the fixtures on non-emergency circuits (about 2/3 of them) could be attached to a timer to reduce the hours of use from 24 to 10.

Another category of major energy savings was the exit signs. Before, the exit signs used two 20W incandescent lamps. We considered various options including retrofitting the signs with compact fluorescents and replacing the old signs with new LED signs. In the end we decided to do the latter, and replace the signs with LEDs. This cost slightly more than retrofitting the signs with compact fluorescents, but the LED signs draw extremely low amounts of power (1 watt) and require no maintenance for at least 25 years (the length of the warranty). These differences made up for the premium.

The final two areas, bathrooms and stairways, primarily involved replacing T-12s with T-8s. The other major technology which was in use was 30W circular fluorescent lighting which we decided not to change due to the minimal savings and relatively high costs of the other options. In the bathrooms there were also 150W incandescents in use. We decided to remove them because they had little impact on the light levels due to their placement and the condition of their fixtures. We decided that the light levels did not need to be improved. We also proposed to use occupancy sensors in the bathrooms which we found to be cost-effective.

We estimate that the total retrofit will reduce energy use of the lighting by 46%. This will reduce the total energy use of the TEC by 23%. A detailed look at all of the areas and their users allowed us to make changes that will improved light levels where necessary and saved energy by giving control over the systems to individual users wherever possible.

FIGURE 1: ENERGY USE AT THE TEC

Area	Before (kWh)	After (kWh)
Offices	201,000	118,900
Library	150,950	83,500
Classrooms	55,800	28,800
Labs & Non-Office	121,000	69,300
Open Areas	108,000	63,400
Bathrooms	15,800	3,500
Exit Signs	32,200	1,600
Stairways	5,700	4,700

CONCLUSION

At the time that this paper was written, we did not have the results of the retrofit, but the steps we followed would be a good guideline for anyone doing their first lighting retrofit project. More specifically, this paper has a number of ideas that will be of special use to people working on energy management in an academic environment, particularly students.

We found that the fact that students were the main proponents and designers of the project was a key selling point to the administrators. Also, doing all of the advance planning and footwork made it very easy for the zone manager to approve the project because little extra work would be required beyond the actual overseeing of the contractor. Finally, the use of actual small-scale projects to judge specific retrofit ideas proved invaluable. They not only allowed us to judge the effects on light levels and occupants, but they also provided excellent opportunities to give advance warning of the forthcoming retrofit and to increase public awareness of the potential benefits of the project. We anticipate results that will provide enlightening conclusions on the various lighting technologies and their effectiveness in different settings.

ACKNOWLEDGMENTS

We would like to thank the members of the original STEAM group: Thomas Butler, Van Hiemke, Amy Richard, and Rafe Sagarin. We are also grateful for all the help and support from Professor Gil Masters, Scott Gould, and Matt Tracy.

Chapter 59
Lighting Retrofits: The Good, The Bad, and The Ugly

L. Czeschin, D. Doubek

Energy usage has become one of the nation's most publicized concerns. It influences everything from the refrigerators we buy, to the cars we drive, and the light we read by. And though we are more aware of energy usage today, consumption continues to increase. Power companies have realized that unless this trend changes they will have to deal with the tremendous cost of building new power plants as well as the ever present threat of depleting our natural resources and polluting our environment. This incentive has led to massive utility rebate programs to encourage a reduction in power consumption through the use of efficient products. In addition, state and federal legislation has mandated certain minimum efficiency reguirements for not only electrical products, but also building electrical design in terms of watts per square foot. With these programs come many devices designed to save energy and directly replace existing technology - retrofits. The lighting field has been inundated with these due to the universality and accessibility of lighting systems and, while retrofits look good at first glance, they may have long term detrimental effects. In the wrong application they can cause losses of illumination and/or introduce glare which can ultimately lead to reductions in productivity and possible health problems. For these reasons it is important to not only keep abreast of all new technology, but to also be aware of how each is incorporated in a lighting system, including any cautions which may be associated with them. It is the combination of these aspects that makes for an effective retrofit and maximizes the energy savings without sacrificing the integrity of the system. The following will address these lighting retrofit issues with a combined emphasis on: The Good - the advancements in technology; The Bad - the cautions associated with these advancements, and; The Ugly - the results of their improper application.

Improvements in light sources continue to be the most aggresive of all changes in lighting technology. Each of the source types - incandescent, fluorescent, and high intensity discharge - have had major advancements and continue to improve in not only efficacy (lumens per watt - see FIGURE 1) but also color and size. The following itemizes the changes that are available in each source type, as well as improvements in the other components of a complete lighting sustem - ballasts, transformers, and fixtures.

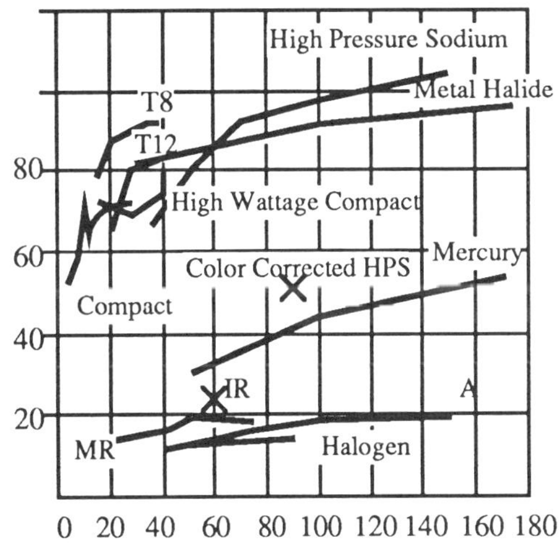

FIGURE 1. LAMP EFFICACY COMPARISON

Incandescent

Changes in the incandescent family are predominantly geared toward an increase in efficacies. This is demonstrated most significantly with the introduction of the halogen cycle. The halogen cycle is a process in which halide gases are used as a catalyst to extend the life of a standard tungsten filament and increase its efficacy. The filament is surrounded by the halide gases (bromine or iodine) and enclosed in an envelope. Here, through the normal process of incandescence, the gases are heated and pressurized to a point where, as the tungsten atoms evaporate off the filament, it is chemically mixed with the gases and taken back to the hottest element in the envelope - the filament. The gas is then released to repeat the cycle. This prevents the tungsten from being deposited on the envelope providing longer life (2000-3500 hrs vs. 750-2000 hrs). In addition, the increased pressure and temperatures required to generate this process improve the efficacy, producing 16-18 lm/W versus the standard 13-16 lm/W. For these reasons, and the added benefit of a whiter light, halogen lamps have become an increasingly popular retrofit. However there are inherent differences associated with the design of these lamps that the user should be aware of. The pressure and, more importantly, the 1200°F temperature of the halogen process are so much higher than that produced by standard incandescents that the standard lamp materials will not withstand them. Quartz glass, in lieu of soda lime, is most commonly used as the envelope to contain the halogen gases. Although quartz is appropriate for this purpose it creates a few concerns. First, it is sensitive to dirt and oil. When left exposed to pollutants in the air or the salt, or sodium, present in the oil of a person's skin, the molecular structure of the quartz can be weakened. If the quartz is heated, pressure increases, causing potential failure at the weakened points. In lamp technology this becomes hazardous. Quartz filament burners may fail - nonpassively. For this reason, it is mandatory that all quartz be contained in an enclosure that will withstand this type of failure as well as protect it from contaminants. This enclosure is usually as a 1/8" thick piece of glass. This could be borosilicate such as that used in standard PAR lamps or a small circular soda lime cover glass. Standard thin wall A and R type lamps formed from soda lime glass, are not approved for halogen containment and cannot be used with this technology. Another small, but none the less important, concern with the use of quartz is that it does transmit ultraviolet radiation (energy with wavelengths between 200nm and 400nm) which the Food and Drug Administration has deemed a possible cause of skin cancer. Of the entire power distribution of an incandescent source, approximately 0.2% is ultraviolet radiation. Soda lime or borosilicate glass absorbs this small amount of UV from the spectrum so that any risk of UV exposure from a standard incandescent lamp is elliminated. Quartz does not. For this reason it is required that all quartz filament burners be properly shielded with a glass lens, again such as that used in standard PAR lamps. Both lamp and fixture manufacturers are required to abide by these safety precautions associated with the use of halogen lamps, however it is equally important that the user be aware of the potential concerns in order to make educated decisions when using energy saving retrofits. Both PAR and, most recently, thick glass A type incandescent lamps which meet both the containment and UV radiation criteria are now available in wattages from 20-150W. These are available most readily as direct replacement options for 25-300W incandescent lamps providing a wide range of approved energy saving possibilities.

One of the more popular halogen replacement packages is the halogen PAR30. This lamp has the same diameter as the standard incandescent R30 with a significantly higher efficacy making for an easily justifiable replacement. However, until recently, the only PAR30 available was 1 3/4" shorter in maximum overall length than the R30 causing problems in retrofit situations. Placing a lamp in a reflector which is designed for a longer lamp causes changes in the light distributioin characteristics of the fixture and/or, worse, traps light and heat within the fixture housing causing the initiation of the thermal protection and a cycling on and off of the fixture. A possible, but not permanent, solution is the use of socket extenders to add the 1 3/4" to the position of the lamp. A preferred solution is the introduction of long-neck PAR30 lamps which match the R30 maximum overall length for direct retrofits. The user should be advised however, that new fixtures are designed specifically around the dimensions of either the standard or long-neck versions of the PAR30. In general, care should be taken when replacing any lamp. Retrofit or not, it is important that the fixture requirements match the specifications of the lamp.

A more recent addition to the incandescent family incorporates the halogen process with a high temperature heat reflective film to produce even more efficacious results. Thin layers of interference material, are applied to the quartz filament burner to form a filter for parts of the electromagnetic energy spectrum. These layers are tuned to separate the infrared energy (heat), which is approximately 87% of the power distribution of an incandescent lamp, from the visible energy (light) - approximately 7% of the power. It is then possible to redirect the infrared portion back onto the tungsten filament raising the temperature of the tungsten which in turn produces more light. Efficacy is increased to 19-22 lm/W allowing for retrofits of maintained lumen output with reductions in wattage of as much as 60%. The cautions associated with this technology are the same as those referred to with quartz halogen - containment and UV radiation- and at this time dichroic infrared halogen is only available in PAR and linear T3 lamp types.

Another technology available in PAR packages, which provides similar lumen output at reduced wattages, incorporates a diode. A diode will reduce the voltage at the lamp allowing for the use of a smaller filament. This provides for improved lamp optics and increased efficacies. A drawback, however is that the diode only conducts half of the wave which can result in visible lamp flicker as well as cause adverse effects on dimming and power quality. Depending on the application, these concerns may override the energy savings gained by the use of these lamps. Regardless, diodes have become another means of achieving higher incandescent efficacies.

An ever increasing area of technological advancement within the incandescent family is that of low voltage. Low voltage, meaning anything less than 120v, but most commonly between 5.5v and 12v, provides many advantages. First, incandescent undergoes a phenomenon at lower voltages in which, for the same wattage, more light is produced. Combining this with the halogen process makes low voltage lamps very efficacious light sources. It also makes possible a physically smaller filament which improves lamp optics and allows for smaller reflectors. This in turn means smaller fixtures. The small fixtures and controlled beam patterns, as well as high efficacies, are the factors that make this a popular source for accent lighting. In addition, the most common low voltage lamp, the MR16, has, on most types, incorporated a dichroic coating on the reflector directing approximately two thirds of the heat out through the back of the lamp. This makes it ideal for highlighting objects which are temperature sensitive while maintaining energy efficiency. A caution with this lamp, however, is that it requires the use of a fixture designed to properly dissipate the heat which is being concentrated in the back of the fixture. If it is placed in an improperly vented fixture it can cause the temperatures to exceed the UL rating of the fixture and at the same time exceed recommended temperatures for the lamp, causing short lamp life. Again, as cautioned, in any situation the fixture requirements must match the specifications of the lamp. Another possibly detrimental characteristic associated with the dichroic MR16 is that any degredation of the thin dichroic coatings, such as that caused by contaminants in the atmosphere, will cause a change in the color of the light emmitted by the lamp. This can produce noticable color variations between lamps which, in some cases, may create an unacceptable "rainbow" affect on a wall or display. In response to this problem, lamp manufacturers now offer both an improved dichroic reflector and an aluminized reflector MR16. Both reflector types provide color stability. However, the aluminized reflector does not maintain the cool beam in affect of the dichroic type and the majority of the heat is now directed out the front of the lamp. This lamp should not be used to illuminate objects which are heat sensitive. Other options and improvements to the low voltage family include new, more efficient reflector shapes, lensed lamps rated for halogen containment, and most recently the introduction of new twist-type pin and socket configurations for ease of maintenance and improved electrical connections.

Low voltage is the only incandescent lamp type that requires an auxilliary component for its use - a transformer. This equipment is required to perform the function of reducing the line voltage to the voltage used by the lamp. Until recently, all transformers were of a simple magnetic type which are considerably large, both in dimension and weight, and produce fairly high amounts of heat. These factors must be taken into consideration in the design of the fixture, resulting in rather large housings for this small lamp type. Magnetic transformers also consume affective amounts of power in addition to that of the lamp. An optional solid state or electronic transformer is now available. Its power consumption is less, as well as the amount of heat produced and its physical size. (See FIGURE 2) Smaller, more efficient low voltage fixture packages are now available for retrofit as well as new construction. One caution, however, associated with the use of either the magnetic or solid state transformers, occurs when they are subjected to dimming. More than one solid state transformer on a standard incandescent dimmer can produce interference which causes noisy operation, visible lamp flicker, and possible short lamp life. These same problems can also occur in the opposite combination, magnetic transformers on solid state dimmers. The solution is to always match the dimmer type to the transformer type for the most efficient and effective low voltage application.

MR16 Lamp Wattage	Wattage Difference Between Magnetic and Solid State
20	5 W
42	7 W
50	8 W
65	13 W
75	14 W

FIGURE 2 TRANSFORMER WATTAGE LOSS

Each of these advancements in incandescent technology provide numerous energy saving retrofit options and each has specific inherent operating concerns as noted. However, another caution concerning all incandescent reflector type lamps that are being retrofitted (R, PAR and MR types) lies in their distribution classifications. Not all beam spread designations are the same. For instance, flood can be used to designate beam spreads ranging from 58 degrees to 130 degrees. A spot can be anywhere from 10 degrees to 50 degrees. This can be a problem when a layout of fixtures designed for a specific lamp and beam spread is retrofitted with lamps of an entirely different distribution. Spotty or nonuniform illumination is the resulting affect and may be unacceptable from not only an aesthetic standpoint but functionally as well. When retrofitting, it is important to compare actual beam angles to determine if the change will be visually correct in addition to providing an energy savings advantage. See FIGURE 3.

	Rated Life Hours	Beam Spread [1]	Average Max. Candlepower[2]
150 PAR 38/FL (Common lamp)	2000	36°	3900
120 PAR 38/FL/WM (Redesigned reflector)	2000	36°	3400
90 PAR 2/FL/H (Halogen (Quartz)	2000	32°	3400
60 PAR 38/FL/HIR (Halogen and infrared)	2500	32°	3200
Q 50 MR 16/NFL (Halogen low voltage)	4000	32°	2450

[1]Beam spread is to 50% cd.

[2]Average of cd in 10° cove

FIGURE 3. LAMP COMPARISON

Fluorescent

The advancements in fluorescent technology exceed that of any other source. This is due, in part, to the fact that it is inherently a long life and efficacious lamp type. Its use is versatile making its volume of usage the highest of all of the lamp families. Unlike the advancements in efficacy in the incandescent line however, fluorescent technology has been consumed mainly with improvements in color rendition.

Triphosphor is a term describing the most recent changes to the phosphor coatings used on the inside of a fluorescent tube. These newer phosphors include a mixture of three rare earth materials that when activated, or made to fluoresce, release light in wavelengths of 450nm (blue), 540nm (green) and 610nm (red). These colors match those that the sensors in the human eye (cones) use to produce all other colors. Because of this relation to our sense of sight, the color rendition of a fluorecent lamp with triphosphors is significantly higher than that of a standard lamp. As an added plus, the lumen output and lumen maintenance are also improved with as much as a 7% increase in output and 90% lumen maintenance at end of life. T12 triphosphor lamps are available as direct retrofits with no major concerns other than a higher initial cost. An option at a lower initial cost is the T8 lamp. The T8 lamp, a smaller diameter lamp, was developed with the use of triphosphors. Because of the composition of the triphosphors, it is possible to push them closer to the arcstream of the lamp producing similar lumen output with less phosphors and at a reduced wattage. A 32W T8 lamp provides within 9% of the lumen output of a 40W T12 - an energy savings of 20%. In addition, less of the rare earth triphosphors are required meaning a lower initial cost. Since the introduction of the T8, many more small diameter fluorescent lamps have been developed, each through the use of the high color rendering triphosphors. As the diameters have shrunk so have the overall sizes of the lamps. Fluorescent lamps are now available in sizes as small as some incandescent A lamps with approximately the same lumen output. Good color, long life, and high efficacies are all factors which make these fluorescents the leading options for retrofits. However, as with everything discussed thus far, there are a few concerns.

The main concern with any retrofit is, again, how the fixture requirements match the lamp specifications. This is more important with fluorescent than with any other source mainly because fluorescent is usually used as a retrofit for incandescent. There are important differences between these two sources. First, fluorescent is a line source - light is produced across its entire length. No matter how compact the fluorescent lamp is, it will still produce light in this linear fashion. Incandescent, on the other hand, distributes light from a point - the small tungsten filament. An incandescent fixture is designed to distribute the light emitted from this point source and when a linear source is used in its place, much of the light is either trapped in the fixture or not controlled at all. This change in distribution and delivery results in losses in illumination, even though the lumen outputs of the lamps may be the same, and an introduction of unwanted glare. Both situations are detrimental and defeat the purpose of the retrofit. In response, some fixture manufacturers now provide their own retrofit kits, complete with lamps, which provide for the proper positioning of the lamp within the reflector. This reduces glare and maximizes efficiencies. Lamp manufacturers are continuing to reduce the maximum overall lengths of compact fluorescents while still attempting to increase lumen output. Again, however, fluorescent will always be a diffuse line source and, where retrofits call for any type of directional control from the fixture, it is best, both aesthetically and economically, to use an incandescent. Additionally, compact fluorescents can not be interchanged when more or less light is needed. For lamps between 5W to 26W, there are 9 different base configurations. These lamps must be specified carefully.

Another important characteristic of fluorescent which may affect the performance of a fixture is its sensitivity to temperature. Fluorescent lamps are designed to produce 100% lumen output at a bulb wall temperature of 100°F. Subsequently, if the temperature surrounding the lamp varies such that the bulb wall temperature is not 100°F, the lumen output will drop. This affect can be seen in situations where, for example, a fluorescent lamp is used outdoors in a cool climate. The light produced by this lamp is reduced dramatically. In actuality, for every degree below the 100°F mark, 1% of lumen output is lost. On the other end, for every degree above 100°F, 1/2% of lumen output is lost. This fact has previously been overlooked and a few fixtures, especially those being retrofitted with compact fluorescents, fail to provide proper lamp temperature. In some cases, as much as 40% of the light output is lost due to excessive temperatures. Retrofits with only one lamp are usually not at risk for this type of light loss, however those that offer the promise of more light with multiple lamps should be reviewed.

Most manufacturers now carry lines of fixtures specifically designed for compact fluorescent taking into consideration this temperature issue. Recent introductions of almagum filled compact fluorescent lamps designed to reach 100% lumen output at higher temperatures above the normal temperatures. The alrojun acts as a sponge, capturing and releasing mercury to keep the pressure constant. Again, base configurations vary and must be specified accordingly, and the fixture must still provide proper lamp cavity temperatures. In addition, these lamps must undergo a warm-up time before reaching full light output. (See FIGURE 4)

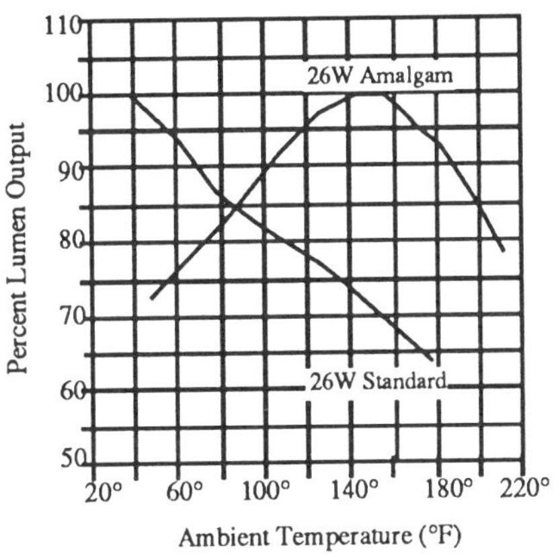

FIGURE 4. LUMEN OUTPUT RELATIVE TO TEMPERATURE

An aesthetic concern related to fluorescent, other than the glare from a poor retrofit, is tied to triphosphors. Although this technology has paved the way for the introduction of countless new lamps, when combined with certain control media its use becomes, in some cases, unacceptable. Anodized aluminum is a material that is used in fixtures as a reflective surface for directional control. It is used specifically for parabolic louvers in large fluorescent fixtures as well as the reflector portions of recessed downlights. This aluminum is coated with a thin, clear, hard aluminum oxide film which varies in consistency depending on the grade of aluminum. When light strikes this surface it is either transmitted or reflected depending on its wavelength. Due to the reactions with the anodic coating, wavelengths of the colors violet, blue and green are enhanced while red, yellow, and orange are cancelled, and, depending on the angle at which the light stikes, some of the enhanced colors can become visible on the surface of the aluminum. This results in a "rainbow" affect called iridescence. Because triphosphor lamps have a higher amount of blue and green in their light ouput, iridescence is enhanced with their use. This affect can be irritating and intolerable. Most fixture manufacturers now offer a low iridescent option for fixtures using triphosphor lamps. Reduced iridescence is achieved by using a more expensive, higher grade aluminum with a more consistent thickness to its anodized coating. Another solution is to use a painted finish which eliminates any possibility of iridescence. In recessed downlights, this is becoming popular. White painted baffles or cones, for example, are now standard options on several fixture types and are the standard for compact fluorescent retrofit kits. Other solutions include the use of highly specular silver film which produced by vaccum deposition, acts as a mirror and avoids any irredescents, or any type of lenses. Be aware, however, that a retrofit involving only the installation of a fluorescent triphosphor lamp in a recessed downlight or parabolic fixture will most likely have iridescence.

Advancements in fluorescent technology also include some major accomplishments in the auxilliary components, specifically ballasts. A ballast is required for the operation of every type of fluorescent lamp. Its purpose is to provide the voltage necessary to start the lamp as well as limit the current. Depending on the lamp type, ballasts perform these functions in different ways. A preheat ballast and lamp combination requires a separate starter to warm the cathodes of the lamp before the starting voltage is applied. This is the most common type of system used with compact fluorescents because the ballast is small and can be placed in retrofit adaptors with a minimum use of space. A significant caution concerning the use of these ballasts is that the standard package has a normal power factor, or a power factor of less than 90%. This means that the current is not being used efficiently, and between fixtures of the same wattage, the amperage will be much higher for a fixture with this type of ballast. This is especially critical in a retrofit situation. If, for example, a corridor of incandescent fixtures is retrofitted by compact fluorescents with normal power factor ballasts, even though the wattage may be the same or less, the current draw may be enough to overload the circuit. This can be a problem in new construction as well, forcing the use of more circuits which in turn means more materials and more labor. This hazard is avoidable with the use of high power factor preheat ballasts - ballasts with 90% power factor or higher. Though they are available, again, in most cases, they are not standard and must be designated specifically at an increased cost.

The most common type of fluorescent system is the rapid start. This system incorporates a ballast which contains windings which provide for continuous warming of the cathodes and a faster start. Rapid start ballasts, which are regulated to meet high power factor standards, are available in many different forms. The most common is the magnetic energy saving type. It is commonly available in one or two-lamp versions, each for a specific lamp type. The matching of ballast to lamp type is critical. An example of this is in the retrofitting of a T12 installation with a more efficient T8 system. The lamp sockets are the same, however, the ballasts are not. The T12 is a 430ma lamp while the T8 is 265ma meaning that, on the existing ballast, the T8 will be operated under extreme conditions that will result in detrimental effects on lamp life. The ballast must be compatable with the lamp type and designated as such. This is also the case with an increasingly popular rapid start ballast type - the solid state or electronic ballast. This ballast operates at a much higher frequency than magnetic (25000-50000 Hz versus 60 Hz) making possible several advantages. First, the higher frequencies produce more light from the fluorescence process of the lamp.

Less time between cycles in the lamp generates an overlap of the pulses in lumen output of the phosphors, thus increasing the amount of light produced. This higher light output is taken as an energy savings. The wattage is reduced to maintain the standard lumen output as set by the magnetic ballast system. In addition, lamp flicker which is visually detected by some people at 60 Hz is eliminated at the higher frequencies. Another advantage is that the solid state ballast runs cooler meaning less wear on the internal components and, in some fixtures, an improved lamp lumen output due to a reduction in the bulb wall temperature. It is also usually quieter and much lighter in weight than the magnetic. Along with these advantages, solid state ballasts are available in various styles - single, double, three, and four-lamp versions for most lamp types, offering many options for energy savings and controllability. In addition, it is now possible to dim certain electronic ballasts, adding tremendously to the flexibility and energy saving capabilities of fluorescent systems. Early versions of the electronic ballast were plagued by problems of reliability, high total harmonics distortion, and radio frequency interference. Today each of these concerns have been improved upon and more and more electronic ballasts are being used, both in new construction and in retrofit situations. (See FIGURE 5) They are even being introduced for compact fluorescents, providing the same advantages to this growing family of lamps.

Lamp	Ballast Type	System Watts	Percentage Base System Wattage
(2)F40T12	Standard	95	114%
(2)F40T12	E.S.	83	100%
(2)F40T12	Electronic	74	89%
(2)F40T12/WM	E.S.	72	87%
(2)F40T12/WM	Electronic	62	75%
(2)F32T8	E.S.	71	86%
(2)F32T8	Electronic	62	75%

FIGURE 5. BALLAST COMPARISON

High Intensity Discharge

The high intensity discharge sources, specifically metal halide and high pressure sodium, are the most efficacious of all of the lamp types. Unfortunately, they are also associated with some of the poorest performances of color rendering. This is one of the main reasons their usage is limited in many applications and is the largest area of technological advancements in H.I.D. Improvements in color as well as the addition of reduced wattage packages and improved restrike times in both metal halide and high pressure sodium have now made each of them possible solutions to an ever growing number of applications. This is happening at the same time that the oldest H.I.D. type, mercury vapor is becoming obsolete. Mercury vapor, although it still exists, has the poorest efficacy of any H.I.D. source, very poor color, and the worst lumen maintenance of all sources combined. However, there are many existing mercury installations, all of which would benefit by a retrofit. In most cases, the optimum solution is to replace the entire system of fixtures with metal halide or high pressure sodium fixtures. The savings in energy will quickly offset the cost of most new systems. Another option, although not recommended on a long term basis, is to retrofit only the lamps. For example, some higher wattage metal halide lamps will operate on certain mercury vapor ballasts. This is only feasible for a short term application however, because, although they will operate, they will no longer perform to the lamp manufacturers' specifications. Lumen output and lamp life will be reduced. With advancements in internal circuitry, other lamp retrofits of this type are becoming possible, however, in general, for long term applications it is always recommended to match the H.I.D. lamp type to a specific ballast.

In conclusion, there are major advancements taking place in all areas of lighting - the sources, the auxillary components, the fixture designs, and the materials they are composed of. With each of these changes comes the opportunity for retrofits and with each retrofit comes the concern for proper application. This trend will continue and, most likely, at an elevated pace. Federal legislation has ensured this through the passage of the National Energy Policy Act. Minimums have been established for acceptable efficacies in both incandescent and fluorescent sources with provisions for scheduled reevaluations of the minimums and the affected lamp types. For these reasons, it will become even more critical to remain educated on all aspects of retrofit technology and energy efficient design. It is only through this knowledge that we can optimize the Good, control the Bad, and eliminate the Ugly, making the right choices today for an energy efficient tomorrow.

Chapter 60
Environmentally Responsible Disposal of Lighting Equipment

R.A. Penny

INTRODUCTION

More and more people are interested in knowing how to properly dispose of fluorescent ballasts, mostly spurred to action by a fear of potential liability for environmental noncompliance. This issue is becoming more of a concern as environmental regulations become more strict. The trend in public opinion is to become more environmentally cautious, and utilities are responding by formalizing the environmental restrictions for their rebate programs. It would, therefore, be prudent for all those involved with lighting retrofits to take a proactive approach to the lighting equipment disposal issue by becoming informed and taking decisive action to avoid future liability.

Environmental regulations can be difficult to understand, and most people involved with lighting retrofits just want to know what they have to do without having to interpret complex regulations. Following these guidelines does not relieve you from consulting with the proper authorities about the more detailed regulations. It is intended to give you a good overview, provide the contacts and phone numbers of the authorities involved, and prepare you to ask good questions, understand the answers, and make more informed decisions.

PCBs: AN ENVIRONMENTAL CONCERN

With some ballast disposal methods, polychlorinated biphenyls (PCBs) contained in ballast capacitors can find their way into the environment. PCBs can leak through municipal landfills and sometimes even hazardous waste landfills, and end up in ground water. PCBs can also get into the air from a municipal incinerator, and then end up in surface water. Since potting compound (a asphalt compound used in ballasts to dissipate heat) may be made of waste oil, there is some chance that it contains PCBs as well.

Since the PCBs in ballasts are very highly concentrated, one gallon of PCBs can contaminate a one-million gallon water supply. PCBs are chemically stable, and do not readily break down or degrade into other compounds. When introduced into the environment, they tend to enter the food chain. PCBs can then work their way up the food chain, eventually affecting people.

Although the exact effect of minor exposure to PCBs is still disputed, the U.S. Environmental Protection Agency (EPA) has taken a conservative approach to the handling and disposal of products containing PCBs. With longer exposures, PCBs can accumulate in organisms and can cause cancer in some animals. Workers exposed to PCBs have developed skin disorders, nausea, dizziness, bronchitis, liver injury, and have experienced adverse reproductive effects.

Since 1978, other dielectric materials have been used in place of PCBs. Their degree of toxicity is of concern, since it may contain substances such as trichlorobenzene that may be regulated.

MERCURY: AN ENVIRONMENTAL CONCERN

Mercury can find its way into the environment in the same ways that PCBs do. Mercury can also vaporize as broken lamps are dumped into a landfill or crushed in shredders. Mercury is also highly concentrated, and it takes only a few teaspoons to poison a lake for centuries.

Exposures to high concentrations of mercury vapors for brief periods can cause acute symptoms such as pneumonitis, chest pains, shortness of breath, coughing, stomatitus, gingivitis, and salivation. Chronic exposure may cause tremors and neurological problems, and may also cause redness and irritation as a result of contact with skin or eyes. These dangers would generally only concern workers that are regularly exposed to mercury. However, it is reported that mercury is starting to find its way into the human food chain.

As far as where mercury comes from, spent lamps account for approximately 5 percent of the mercury in landfills (batteries account for approximately 88 percent). It takes over 5,000 4-foot fluorescent lamps to yield one pound of mercury. Fluorescent tubes contain 35-75 milligrams (mg) of mercury, while compact fluorescent lamps contain 5-10 mg. The lighting industry is working hard to reduce the amount of mercury used in lamp production. The amount of mercury in a standard fluorescent lamp was reduced about 14 percent from 1985 and 1990, and another 35 to 50 percent decrease is expected by 1995. However, no replacement for mercury have been found yet. Manufacturers of batteries are making similar efforts to reduce the mercury content of their products.

An important question to consider is whether the hazardous waste produced by using energy-efficient lamps would be worse for the environment than the energy savings they generate. Four-foot fluorescent lamps, which account for the great majority of the lamps in the U.S., contain mercury, glass, aluminum, and phosphors, mercury usually being the only toxic material. Lamps made before 1988 also contain cadmium. The amount of mercury emitted by power plants to produce the additional energy needed to power less efficient lamps has been estimated to be 2-10 times greater than the amount of mercury in the lamps. The ratio of mercury levels would vary with the source of power generation in a region. Coal-fired plants produce the most mercury, while hydropower plants produce none.

In addition to mercury, a compact fluorescent lamp, contains glass, plastic, metal, phosphors, and in some cases, radioactive isotopes, all of which could end up in landfills which could leach toxics into our water supply. However, by replacing an incandescent lamp, each 28-watt lamp over its lifetime will help utilities avoid producing the by-products of power generation. If that power is produced of a coal-fired plant, in addition to mercury, this could be roughly 1,020 pounds of carbon dioxide, 3 pounds of nitrous oxide, 0.5 pounds of sulfur dioxide, 4.75 grams of lead, 2 grams of arsenic, 375 micro curies of radioactive gases, and 145 micro curies of radioactive waste, in addition to beryllium, cadmium, chromium, copper, manganese, nickel, formaldehyde, and other pollutants.

All this seems to indicate that even if lamps end up in the landfill, they have less effect on the environment than the production of the additional energy needed by less efficient incandescent lamps.

BALLAST IDENTIFICATION

Exactly how ballasts are to be properly disposed depends on how the ballasts in question were made. Ballasts manufactured prior to 1979 have PCBs in them. In 1976, EPA banned the manufacture of PCBs after 1979. Ballasts manufactured since that 1979 which do not contain PCBs should be marked by the manufacturer with the statement "No PCBs". For those manufactured prior to that time and those which contain no statements regarding PCB content, you should assume that they do contain PCBs. Ballasts can operate for 20 years or more, resulting in 25 million pounds of PCBs in one billion ballasts currently installed in U.S. buildings, roughly half of which contain PCBs.

Fluorescent ballasts are generally composed of a transformer to regulate the incoming voltage and current, a small capacitor (which may contain PCBs), and possibly a thermal cut-off switch. Sometimes the capacitor may leak, allowing the PCB oil (clear or yellow in color) to leak out of the fixture. If this happens, contact EPA or the Clearinghouse for more detail instructions. If the potting compound (an asphalt substance that meant to help dissipate heat from electrical components) leaks out, as it may when heated, this could also be a problem because the potting material may contain PCBs.

THE LIABILITY OF INACTION

Because proper disposal of ballasts is confusing and costly, some facility managers may elect to leave the ballasts in the ceiling, even after they are no longer operating. This is not recommended. Ballast may leak at any time. It may be some time before the leak is detected, and at that time other building materials may be contaminated, requiring a costly cleanup.

If the building were mortgaged or sold, auditors or financiers may require the PCB ballasts to be removed, and that would be more costly than removing the used ballasts as part of a planned lighting upgrade.

Building owners that choose to leave old ballasts in place to avoid disposal costs risk the greater costs of ballast leak cleanup and PCB contamination in a fire. If ballasts are left in place and there is a building fire, the temperature will not be enough to destroy the ballast PCBs, but can be enough to make the PCBs volatilize and contaminate the building, including the floors, ceilings, fixtures, and furniture, costing a building owner millions of dollars to decontaminate.

BALLAST HANDLING GUIDELINES

Removal

When removing ballasts, wear chemically-resistant gloves. When removing leaking PCB ballasts, it may be advisable to also wear a chemically-resistant suit and face mask. Always dispose of personal protective gear along with the ballasts. Remember that if PCBs remains in contact with your skin, it could lead to chloracne, a painful skin illness. The Occupational Safety and Health Agency (OSHA) can provide assistance in this area.

Be careful not to drop non-leaking ballasts because that could cause them to leak. If more than one pound of PCBs (the amount contained in 12-16 ballasts) is spilled, it must be reported to the EPA National Response Center (800-424-8802). It must also be cleaned up immediately according to EPA procedures. Note than all materials that come in contact with PCBs leaking from a ballast must be disposed of as hazardous waste. It is best to have the handling and disposal of leaking ballasts performed by contractors that are specifically trained to do so. You can consult the Yellow Pages *(Waste Disposal--Hazardous)* or contact you state environmental agency for assistance.

Storage

Put all leaking ballasts along with any other materials the PCB oil came in contact with into a double plastic bag inside a steel drum. If there has been a fire, shut off the power, ventilate the room well, and be especially watchful for PCB oil and potting compound leaking from ballasts.

When moving non-leaking ballasts within a building, cardboard boxes can be used, but don't load too many in a box, as twelve ballasts from four-foot fixtures weigh 40-50 pounds. Don't leave them in the boxes too long because you may have not noticed a leaking ballast and you don't want to contaminate your floor; a small spot can end up being very expensive to deal with. Clip off wires to reduce ballast weight.

A typical 55-gallon drum will hold 150-300 ballasts from 4-foot fixtures or 60-100 ballasts from 8-foot fixtures, depending on how tightly they are packed. A full drum will weigh 600-1000 pounds. Some disposal facilities limit the weight for each drum to 600 pounds. If you plan to send the ballasts to a landfill, use 6" to 12" of non-biodegradable absorbent material (such as kitty litter) on the bottom of the drum under the ballasts and more in the interstitial spaces between the ballasts. Don't seal the drums until the transporter has had an opportunity to inspect the contents. Pack leaking ballasts separately from non-leaking ballasts. For more exact packing instructions, contact the disposal facility that you plan to ship the ballasts to.

To buy drums, check your local yellow pages under "drums" or "barrels". Be sure to specify 17C or 17H, 55-gallon, open-head steel drums for transportation of hazardous materials. It is permissible to purchase used drums if they have been properly cleaned, tested, and labeled as per the Department of Transportation regulations. These usually cost $30-$40 each.

Drums of leaking ballasts should bear the yellow "Caution: Contains PCBs" label. All drums should be clearly labeled with the name and address of the generator, weight, the date the ballasts were first removed, a description of the material (discarded fluorescent lighting ballasts), a DOT shipping description (i.e. RQ Polychlorinated Biphenyls, 9, UN2315, PGII), and a tracking number unique from any other hazardous waste container ever shipped from the same building.

It is recommended that non-leaking ballasts not be stored on site for over three months because leaks may develop over time and because the incinerator or chemical waste landfill must dispose of leaking ballasts within one year of the date that the ballast was removed. Leaking ballasts should not be stored on site for more than 30 days. If more than 70 cubic feet of regulated PCB material is stored on site, EPA approval as a Commercial Storer is required. For more information about ballast handling and storage, call the TSCA Hotline, listed at the end of this report.

Transportation

Use the Federal Uniform Hazardous Waste Manifest or a specially designed hazardous waste manifest for ballasts. Some states have their own hazardous waste manifest which may also be used. The manifests would be signed by the waste generator (the building owner) and the transporter, telling what it is and where it's going. Less than one pound of PCBs (12-16 ballasts), is not a reportable quantity, so a manifest may not be required.

PCB ballasts should be transported by a licensed hazardous waste transporters. Lists of such transporters may be available in each state, from the EPA Regional offices, and from consulting the Yellow Pages *(Waste Disposal--Hazardous).* Under Federal Law, the transporter must file a Notification of PCB Activity with EPA. Some transporters will not accept leaking ballasts. Some states allow a limited number of non-leaking ballasts containing PCBs to be dropped off at specified hazardous waste collection centers.

The cost of transportation is calculated as cents per pound per mile, and varies according to the number of drums to be shipped and the distance to be shipped. For more information on the transportation of PCB ballasts, call the TSCA Hotline.

LAMP HANDLING GUIDELINES

EPA recommends minimizing mercury content by using smaller lamps (a T-12, 1 1/2" fluorescent lamp contains more mercury than a T-8, 1" lamp) and to utilize group relamping practices to allow better management of spent lamps by encouraging coordination among maintenance and facilities crews for collection and recycling efforts. They also recommend arranging for your electrical or lighting contractor to provide disposal services as part of their contract, being careful to ask for certification, licenses, and references as well as documentation of transportation and disposal of the lamps.

Lamp recyclers generally require that the lamps arrive unbroken. This can usually be accomplished by inserting old lamps into the boxes that the new lamp came in. If the new lamps are not replacing old lamps, or if the new lamps are smaller than the old lamps (a T-8 lamps replacing a T-12 lamp), boxes are often available from lamps suppliers and distributors.

If a lamp breaks, ventilate the area well and clean up by vacuuming or other method that avoids dust generation. Also use safety glasses, goggles, or face shields as well as NIOSH- approved respirators. Lamp disposal workers must be provided with special education about hazardous waste handling, and an materials data safety sheet (MSDS) must be posted at the site. Consult OSHA for further regulations.

Some facilities use a "lamp eater". This is a machine that sits on top of a drum and crushes fluorescent tubes as they are fed in. Some have better filters and mercury monitoring equipment than others, and they cost up to \$5,000. OSHA allows 0.1 milligrams per cubic meter of exposure over an eight-hour day; state regulations may differ. You can purchase passive badges to measure this for approximately \$50. Consult OSHA and your appropriate state agency for more information on this. One drawback to this method is that it creates what EPA calls D009 waste. Such waste needs to go to a mercury recycler. The cost to process a drum of lamps (roughly 500 F40 lamps) can be as high as \$2,000. This waste should not go to a municipal landfill, and some states prohibit the use of lamp eaters.

BALLASTS DISPOSAL OPTIONS

Select a disposal/recycling vendor very carefully. If a vendor, contractor, transporter, or anyone else contaminates the environment with PCBs from ballasts, the owner of the building the ballasts come from will be liable as well as others. In selecting a service, make sure that they know all the regulations, have the necessary state and local permits, and has no past violations on file with EPA. Make sure that they have a good procedure for dealing with contaminated potting compound and are capable of testing for PCB contamination. They should have sufficient pollution liability insurance or environmental impairment liability coverage, and should demonstrate good worker training and quality assurance. Recycling firms should have signed contracts with a PCB incinerator as well as with foundries or mills.

Facilities accepting hazardous waste generally require a profile submitted and approved ahead of time describing just what sort of waste and how much waste will be shipped. They will then provide any specific requirements they may have regarding transportation containers, labeling, and pricing. It is wise to have a signed document mailed to you by the disposal facility after your ballasts have actually been disposed of, closing the loop of liability.

Since ballasts are very heavy, the total cost of disposal will vary with the proximity of the building being retrofitted to the disposal facility chosen. The cost of disposal will also include a fee for documenting the hazardous waste, called profile fee. This will range from \$100-\$300. This fee may be waived if a certain volume or frequency of deliveries is assured. Ballast disposal options are as follows:

Municipal Waste Landfill

These are also known as a sanitary landfill.

Some facility managers may choose this option because of the low cost (about 10 to 20 cents per ballast), but this is not recommended. Sanitary landfills are not designed to handle liquid PCBs in steel cans which will eventually corrode or break, leading to future Superfund liability for the ballast disposers. Such disposal of any large quantity of non-leaking ballasts violates the interpretation of federal regulations by the Region 10 EPA office in Seattle.

Disposal of leaking PCB ballasts in a municipal landfill is against TSCA regulations. In many parts of the country you would be hard pressed to find a municipal landfill that would knowingly accept a shipment of ballasts anyway unless a spot check reveals that all the ballasts are clearly marked "NO PCBs". This would even apply to industrial landfills, which have a liner, but are still not set up to handle hazardous waste.

It is important to realize that EPA considers the disposal of more than 12-16 PCB ballasts, leaking or non-leaking, within a twenty-four hour period, into a municipal waste landfill a reportable event under CERCLA regulations. The National Response Center should be contacted.

Hazardous Waste Landfill
Also known as a chemical waste landfill. This is acceptable for non-leaking ballasts, and is relatively inexpensive ; roughly \$1-\$2 per F40 ballast. There is some concern that hazardous waste landfills are not designed to handle liquid PCBs or PCB concentrations over 500 ppm (ballast PCB oil is liquid and is 700,000 to 1,000,000 ppm). Leaking ballasts cannot be brought here, but rather must be incinerated. Additionally, if the PCBs have absorbed enough lead from ballast solder to fail the TCLP test, RCRA regulations would prohibit even non-leaking ballasts to be brought to any landfill.

Even though hazardous waste landfills listed are approved by EPA to dispose of non-leaking PCB ballasts, this does not preclude the landfill from becoming a Superfund site in the future, so there is still some liability.

Municipal Waste Incinerator
This is about the same cost as putting the ballasts in a municipal waste landfill. It is not recommended because these incinerators lack the combustion conditions and monitoring equipment to thoroughly destroy all the PCBs.

Hazardous Waste Incinerator
Also called a high temperature, PCB, or TSCA (Toxic Substance Control Act) incinerator. This method will destroy the PCBs completely, permanently removing them from the waste stream, so Superfund liability is minimal. In a PCB incinerator, the ash and air emissions are carefully tested to insure 99.9999% destruction of the PCBs. The primary drawback of using this method for whole ballasts is cost; \$6-\$8 per ballast. This method also generates ash and air pollution and uses a great deal of energy because the ballast is 80% metal and is difficult to burn. It also wastes valuable metals which can be recycled. Recyclers can send only the PCB contaminated ballast components to an incinerator as part of an effective solution.

Ballast Recycler
These facilities remove the 4-ounce capacitor from the 4-pound ballast, incinerating the PCB-contaminated material (including the asphalt potting materials), and recycling the remaining metals. Over 80% of the ballast by weight is recoverable. Some of these firms X-ray various brands and models of ballasts so that they know exactly how to cut them open safely. Others put ballasts into a deep freeze, then pry the ballast apart such that the potting compound crumbles out in pieces. This option is 60% lower in cost than whole ballast incineration (about \$3-\$5 per F40 ballast), and it generates less air pollution and ash, uses less fuel, and reclaims valuable materials. It only costs about \$2-\$3 per F40 ballast to recycle a non-PCB ballast. This process is very similar to that used for PCB ballasts. There is some concern about the risk of having great volumes of hazardous waste at one site.

Disassemble Ballasts In-house
Another choice is to saw or guillotine open the ballast, removed the PCB capacitor, and then send the capacitor to a hazardous waste incinerator or hazardous waste landfill and the rest to a sanitary landfill. This is a very dangerous practice because of possible spills and does not deal with the problem of PCB contaminated asphalt.

Treatment, Storage, and Disposal Facility (TSDF)
This category includes hazardous landfills and incinerators as well as hazardous waste brokers. Such brokers can simplify the whole process of packing, manifesting, transporting, and disposing of ballasts. The firms listed in this report have interim status within EPA Region 10 as PCB commercial storers. They can often provide a turn-key operation including all of these services. Not surprisingly, they add on a fee for this, but for small waste generators this may be well worth it. Disposal costs are typically \$300 per drum after a \$200 profile fee. Remember that the ballasts must be destroyed within a year of removal, and that no matter who handles them, if they came from your building you will be liable if they are disposed of incorrectly.

Lighting Contractor
Like using TSDFs, this is a way to skip over many of the complications involved in ballasts disposal. Simply add a portion to your contract with a lighting maintenance firm

or lighting contractor which stipulates that they take care of the proper disposal of ballasts. It is important, however, that you insist on specifying hazardous waste incinerator or hazardous waste landfill as your method of disposal and on receiving some documentation explaining exactly how the ballasts were packed and transported, and where they were shipped, and remain informed enough yourself to make sure that they are complying with all regulations. Otherwise liability may find its way back to you.

Community Hazardous Waste Collections
Some communities have annual or semi-annual collections of hazardous waste. Some communities have drop-off sites when homeowners can drop off up to five ballasts anytime.

BALLAST DISPOSAL REGULATION

Once it has been determined that your ballasts contain PCBs, you need to determine who regulates their disposal. The laws are complex. There are two federal laws which are related to PCB ballast disposal. The Toxic Substances Control Act (TSCA) and the Comprehensive Environmental Response, Compensation, and Liability Act (CERCLA or "Superfund"). There are also state regulations. The best approach to take is to follow the strictest regulations for each aspect of PCB removal, storage, transport, and disposal.

Toxic Substances Control Act (TSCA)
This act was passed in 1976 to control chemicals being produced and used in commerce that present an unreasonable risk of injury to human health and the environment. The act specifically mentions PCBs by name, banning their manufacture and distribution and regulating their disposal, marking, and storage.

Old fluorescent lighting ballasts have a high enough level of PCB concentration (over 50 parts per million, or 50 ppm) to fall under TSCA regulations. However, TSCA has a small capacitor exemption that covers non-leaking capacitors with less than three pounds of dielectric fluid, under Federal Register 40 CFR 761.60(b)(2)(ii). However, EPA strongly encourages the disposal of large quantities of ballasts to follow TSCA guidelines voluntarily in the 1979 preamble to 40 CFR 761, and reiterates this in the final ruling in 1982 (47 FR 37342).

EPA is currently investigating the typical leakage of PCB oil from ballasts capacitors into potting compound. If this is common, the exemption could be removed. A working group is considering a modification of existing TSCA regulations. This would further encourage recycling and would identify the maximum number of ballasts that could be disposed of under the TSCA exclusion. This number would likely be between 16 and 50.

The EPA Region 10 office in Seattle has interpreted "large quantities" as being five or more ballasts per project (no, NOT per day). Therefore, virtually any commercial lighting retrofit project should consider disposing of all old ballasts as PCB ballasts as hazardous waste. If the exact weight of the dielectric fluid is not known, assumptions can be made based on ballast size and weight. If the ballast is smaller than 100 cubic inches, it's considered small. If it's over 200 cubic inches, it's considered large. Between these two volumes, if it weighs less than 9 pounds it's considered small.

TSCA requires the PCB-contaminated parts of ballasts to be disposed of in a high temperature incinerator or a chemical waste landfill. TSCA also requires leaking ballasts and ballasts with PCB-contaminated potting compound to be labelled, manifested, transported, and disposed of in a hazardous waster incinerator. Leaking ballasts can not be stored on site over 30 days unless the building owner meets special PCB storage requirements and files a form. If more than ten drums of regulated PCB material are stored on site, EPA approval as a Commercial Storer is required.

Comprehensive Environmental Response, Compensation, and Liability Act (CERCLA or Superfund)
This act provides for extensive liability for improper PCB disposal practices rather than regulating disposal practices directly. Superfund Law covers the disposal of one pound of PCBs (12-16 ballasts) or more within a 24-hour period. Anyone disposing of more than one pound of ballasts in a landfill is a potential Primary Responsible Party (PRP) in any subsequent Superfund cleanup of the landfill, and is required to call the National Response Center when they do this. Both the waste generator and transporter that could be liable. The owner of the building the ballasts were removed from is the generator. Compliance with federal and state law does not preclude later Superfund liability. Disposing of old ballasts at a hazardous waste incinerator either directly or through a recycler is the best protection against future liability.

Resource Conservation and Recovery Act (RCRA)
RCRA 40CFR, Section 761, exempts PCBs from RCRA regulation, but RCRA would regulate any other hazardous waste that they may contain. Whether they contain enough

lead for this is unclear but the criteria would be 5 milligrams of lead per liter of landfill leachate, using the Toxic Characteristic Leachate Procedure (TCLP). This test is discussed further in the following section lamp disposal regulations. Non-PCB ballasts may be regulated under RCRA because the dielectric fluid they use may contain regulated substances such as trichlorobenzene.

The disposal options under RCRA depend on how many pounds of hazardous materials you generate per month, including the PCBs and any other hazardous materials. A RCRA Orientation Manual (order #055-000-00364-5) is available for $13 from the U.S. Printing office that outlines regulations more clearly than section 40CFR, parts 260-299, of the formal regulations (order #869-013-001-478).

Occupational Health and Safety Administration (OSHA)
OSHA's Office of Health Compliance Assistance follows the Hazardous Communication Standard (29CFR 1910.1200). This, as well as a directive (CPL2-2.38C) which explains the standard in more understandable language, can be obtained from the U.S. Publications Office.

States
Most states essentially follow the federal regulations for ballast disposal. Some have adopted the EPA Region 10 interpretation for quantity allowance.

LAMP DISPOSAL REGULATIONS

Federal
The U.S. Environmental Protection Agency (EPA) passed the Materials Disposal Restriction Regulations (known as the "land ban") that went into full effect in May of 1992. They require anyone wanted to dispose of lamps containing mercury to determine if they are toxic or not. They can do this by obtaining documentation from the manufacturer, but unfortunately manufacturers are not providing such documentation that clearly establishes lamp toxicity.

The alternative is either to assume the lamps are toxic or to take some sample lamps to a laboratory and have a Toxic Characteristic Leachate Procedure run on them. The cost to have a sampling of ten lamps tested would be roughly $300. If the concentration of mercury is over 0.2 parts per million (PPM), they are considered toxic, and cannot be disposed of in a landfill. Such waste is called D009 waste.

Some trade associations found that performing the TCLP on fluorescent lamps dis not always yield consistent results. EPA commissioned a study to investigate this issue, and concluded that the laboratories with varied results were performing the test incorrectly and that with some clarification this can be corrected.

EPA is currently considering altering the requirements of the lamp disposal regulations in terms of transportation and disposal requirements. There is also some chance that an exemption will be granted for fluorescent lamps. EPA may further encourage the recycling of lamps.

States
State regulations vary, most deferring to federal regulations.

- California: Lamp disposal is regulated by the California Environmental Protection Agency Department of Toxic Substances Contol. Waste containing over 20 mg of mercury per kilogram (about 2.2 pounds) of waste must be delivered to a hazardous waste facility or to a recycler. Four-foot lamps and compact fluorescent lamps would included in this. A maximum of 25 lamps per day per company (not per building) may be disposed of in a municipal landfill. Spent lamps must be packaged to avoid breakage. Three lamp disposal facilities operate in the state, but it is estimated by one recycler that only 10 percent of the spent lamps are recycled currently.

- Minnesota: The Minnesota Pollution Control Agency (MPCA) requires fluorescent lamps from a commercial, industrial, or multifamily facility to be shipped to a hazardous waste facility or a recycler. Any lamps shipped out of state must be manifested. All lamps sold to such facilities in the state must include a statement about the mercury content of the lamp.

- Other States: Wisconsin now has lamp disposal regulations similar to those of Minnesota. Connecticut and Florida are considering lamp disposal regulations. Other states follow RCRA guidelines, some adding their own, slightly stricter regulations.

LAMP RECYCLING
The lamp recycling facilities currently in operation take accept unbroken lamps, usually shipped via common carrier in the original lamp manufacturer's boxes. They crush them, and separate the components. They charge 7

to 12 cents per foot for this services, depending on transportation and quantity. In California, where three or four recyclers handle over 13 million lamps per year and state law requires recycling of all lamps disposed more than 25 at a time, only about 10 percent of spent lamps ever reach the recyclers. In Europe, it's only a little better than that. The glass recovered can be made into non-food containers or used as filler in cement, the aluminum end caps can be made into anything, the phosphor can be used in fertilizer or concrete unless contaminated by cadmium, and the mercury can be sent to a refinery for use in thermometers and switches.

The following is a partial list of lamp recyclers. Mercury Technologies was not reachable as this paper was published, but may still be in business.

- Lighting Resources, Inc. in CA and AZ (soon to include IN and NJ) charges roughly $2 per ballast for non-PCB ballast or per PCB if the capacitor goes to hazardous waste landfill. They charge $3 per PCB ballast if capacitor goes to hazardous waste incinerator. They also recycle lamps for 7 to 10 cents per foot.

- Mercury Recovery Services charges roughly 10 cents per foot of lamp and 50 cents per U-tube. They provide pick and boxes with packing in CA. From out of state, the fee is roughly 6 cents per foot plus shipping.

- Mercury Technologies has a retorting process to separate the phosphor and mercury themselves. They charge 7-10 cents per foot depending on quantity. They do not have to accept broken lamps. They can process over 10,000 per eight-hour shift, 250,000 lamps per month, and will soon double that capacity. They are making a good profit at the operating level, but have spent heavily for research and expansion, so have been in the red for six years. They were recently acquired by Advanced Environmental Technology Corporation.

- Recyclights charges 59 cents per lamp up to four feet, 83 cents for longer lamps.

The mercury phospher compound need to go to a refiner. Currently, refiners include Bethlehem Apparatus in Bethlehem, PA, and Mercury Refining Company (MERECO) in Albany, NY.

The National Electrical Manufacturers Association (NEMA) is concerned that there have not been any reviews of the uses of the end products of recycling and the environmental, safety, and health risks associated with the storage or processing of the phosphor/mercury mixture.

NALMCO encourages a national recycling policy or a national lamp recycling fund paid for by charging a deposit on lamps. They also encourage EPA to promote development of a recycling infrastructure and educational programs on the subject.

LAMP DISPOSAL ECONOMICS

Some proponents of energy conservation are concerned that regulations which can discourage large-scale relamping by restricting or prohibiting lamp disposal may jeopardize the success of these programs.

Looking at the effect quantitatively, if a recycler charges ten cents per foot of lamps, and a utility charges ten cents per KWH, recycling a four-foot, T-12 (1 1/2") lamp would extend the payback of an investment to replace that lamp with a 32-watt, T-8 (1") lamp by 500 operating hours (about 2 1/2 months), or about 2 percent of its rated life. Looking at the effect qualitatively, lamp disposal restrictions may if fact discourage some good candidates for energy conservation projects.

NEMA estimates the cost to dispose of lamps in a landfill as 2 to 3 cents per lamp. If the lamps need to be handled as a separate waste stream, this cost would rise to 8 to 10 cents per lamp. Chemical Waste Management charges $77 per 55-gallon drum.

The National Association of Lighting Management Companies (NALMCO) feels that if EPA believes that lamps are hazardous waste, they need to warn prospective participants in the Green Lights energy conservation program of the economic and environmental consequences of this. NEMA estimates the costs of recycling, including collection, transportation, and the net value of any recovered materials, to be 51-56 cents per lamp, or $275-300 million for all of U.S. production. They estimate that an initial investment of about $115 million would be required to build recycling centers with sufficient capacity.

Although the cost of recycling lamps is expensive, that cost could be considered "peanuts" for large-scale users worried about future contamination lawsuits.

Some utilities such as Arizona Public Utility and San Diego Gas and Electric Company have modified their rebate programs to include an allowance for proper disposal of lamps. This is a trend that is expected to continue due to utility's fear of implication into improper disposal of hazardous waste. The Bonneville Power Administration has scheduled a meeting to discuss their response to this issue.

CONCLUSION

Ballasts

PCBs in old ballast capacitors are hazardous to the environment and to people. Leaving them in a building involves a risk of an expensive cleanup from a leak and a contaminated building in the event of a fire, and doesn't take advantage of new, energy-efficient ballasts.

Even though federal environmental regulations permit disposal of non-leaking PCB ballasts in municipal landfills, this should be avoided for several reasons. It is impossible to determine whether the capacitor within a ballast is leaking or not, and there may be PCBs in the ballast's potting compound, so it is best to treat all old ballasts without the "NO PCBs" label as hazardous waste. EPA strongly encourages that a significant quantity of ballasts be treated as hazardous waste. The interpretation of "significant quantity" by the regional office of U.S. EPA in Seattle is five or more ballasts. The last reason is that approximately 85% of the materials in old ballasts can be recycled, saving landfill space, energy, and resources.

There are a number of methods of PCB ballast disposal, each with a different cost and liability level. The least expensive and highest risk of liability is disposal in a municipal landfill, if you can find one that will accept them. A more expensive method with the lowest liability level involves a ballasts recycler who will send the PCB capacitor inside to a hazardous waste incinerator and recycle almost all the materials left over.

Lamps

The latest research indicates that all old fluorescent lamps will fail the TCLP test for mercury. Building owners can assume that this is the case or have the test run on a sample of their lamps. If it passes, they can dispose of the lamps as they wish. If if fails, they need to comply with state or federal regulations, whichever is more strict.

General

Building Owners should keep lamps and PCB ballasts out of municipal landfills to avoid future liability. Ideally, they will ship the lamps to a recycling facility since the almost all the materials in a lamp are recyclable. This saves the energy and mining involved with the manufacturing of new materials, saves the pollution or incineration or valuable landfill space. Relamping makes the disposal process easier, as does including disposal as part of a lighting maintenance contract or utilizing a hazardous waste broker.

Although proper disposal of fluorescent ballasts will increase the cost of a lighting retrofit, this will not significantly alter the project economics, and allows building owners to avoid the substantial economic liability that could result from a having sent PCB ballasts to a municipal landfill that becomes a Superfund cleanup site.

Regulations covering ballast removal, labeling, manifesting, transporting, and disposal can be complex, difficult to interpret, and state-specific. You are, therefore, strongly urged to check with the your state to make sure you are complying with their requirements. Keep your own record of ballasts removed, and insist on copies of documentation of transportation and destruction or landfilling of your ballasts. Remember that if your old lighting equipment is disposed of incorrectly for any reason, you can be liable.

MORE INFORMATION

Federal Contacts

U.S. EPA TSCA Hotline (202) 554-1404
U.S. EPA RCRA Hotline (800) 424-9346
U.S. EPA Green Lights Information (202) 775-6650
U.S. EPA National Response Center (800) 424-8802
OSHA (Occupancy, Safety, and Health Administration) (202) 219-8036
U.S. Printing Office (202) 219-4667

State Contacts

CA EPA Dept. of Toxic Substances Control (916) 322-3670
Minnesota Pollution Control Agency (612) 296-6300
Wisconsin Hazardous Waste Management Program

Hazardous Waste Landfills

CECOS International in OH (513) 720-6114
CECOS International in NY (716) 282-2676
Chemical Waste Management Alabama, Inc. (205) 652-9721
Chemical Waste Management in CA (209) 386-9711
Chemical Waste Management of the Northwest in OR (800) 962-4987
CWM Chem. Services Control, Inc. (716) 754-8231
Envirosafe Services of Idaho (208) 384-1500
U.S. Ecology in NV (702) 553-2203
U.S. Pollution Control, Inc. in UT (801) 534-0054

Hazardous Waste Incinerators

Aptus Environ. Services in KS & UT (800) 292-2558
Chemical Waste Management in TX (409) 736-2821

Rollins Environmental Services in TX (713) 930-2334

Ballast Recyclers

Lighting Resources in CA (John Chilcott): (714) 622-0881

FulCircle Ballast Recyclers (Bob Hatcher) (310) 921-1857 in CA, (617-876-2229 in MA, (212) 328-4667 in NY

Salesco USA in AZ (Frank Sales) (800) 368-9095 or (602) 233-2955

Environmental Energy Group in TX (817) 383-3632

Ensquare, pInc. in MA (617) 969-9238

Eastern Environmental Technologies in CT (203) 856-2014

Nine West Technologies in NH (201) 623-0007

Environmental Energy Group in TX (817) 383-3632

Lamp Recyclers

Marpan Supply in FL (904-224-9353)

Mercury Recovery Services in CA (818-301-1372)

Mercury Technologies in CA MN

Lighting Resources in CA (714) 622-0881

Midwest Recycling and Recovery Services, Inc. in IW (319) 556-8037

Recyclites in MN (800) 831-2852

Mercury Recycling

Bethlehem Apparatus in PA (215) 838-7034

Mercury Refining Company in NY (800) 833-3505

REFERENCES:

- "The Practical Guide to Ballast Disposal" by FulCircle Ballast Recyclers, June 1992.
- Home Energy Magazine, May-June, 1992
- Energy User News, October, 1992
- Building Operator Magazine, October, 1992
- Lighting Management and Maintenance, October, 1992
- Conversation with Charles Occhino, President of the National Association of Lighting Management Companies
- Material data safety sheets from Sylvania Lighting
- "The Management of Spent Electric Lamps Containing Mercury" by the National Electrical Manufacturers Association, October 1992.

Chapter 61

Turn Out Those Lights! Here Comes the Energy Patrol! A Step by Step Guide

W.B. Hardin

As a result of the concern of Springs Industries, the Sandhills Chapter of the Association of Energy Engineers (AEE), and myself, I have developed the following presentation as a definite energy conservation method for schools. A similar program was initiated in DeVargas School near San Jose, California, resulting in at least $1,000 per month savings on their energy bills for eight consecutive months. This prompted me to design the following program for local and state benefits.

For the cost of four t-shirts, jackets or hats, your school could save as much as $1,000 a month. How? By beginning a student "ENERGY PATROL". Energy patrols are one of the most popular and effective energy action programs being implemented today. Students on energy patrols monitor classrooms to ensure that lights are turned off when rooms are vacant, reducing school electrical costs considerably. Lighting is a good target for any school energy program. Lighting is typically the largest energy user consuming from 38% to 54% of total classroom energy. The patrol will also check for efficient energy usage by monitoring the use of windows blinds, open or closed doors, and the reporting of any water leaks throughout the school. The patrol gives students an opportunity to practice leadership skills by taking responsibility for their school. By getting kids actively involved, the program makes the school a more exciting learning environment, and at the same time can reduce the school's energy costs by roughly 1/3 or possibly up to $1,000 per month. This money can be used for other school programs. It is important not to omit special recognition for the students. That is where t-shirts, jackets or hats come in. Special jackets with the school name and energy patrol slogan would spark student interest and give them the recognition they deserve. Tracking your success is also important to see what percentage of the school's energy bills your patrol is saving.

The following is a packet of materials, including check sheets and procedures to help you get started in your energy savings program. It's well worth the effort!

THE ENERGY PATROL

The Energy Patrol is made up of twenty students who check each classroom, the office areas, portables and store rooms at recess, lunch and after school. When lights are found on unnecessarily, they are quickly turned off. Periodically, thermostats are also checked to see that they stay at an energy saving 70 degrees in the winter and 78 degrees in the summer. All windows and doors will be observed for efficient solar energy usage and all water leaks will be reported.

PATROL REQUIRES LITTLE TEACHER INVOLVEMENT

There needs to be something special about the group to attract and motivate students. For approximately $100, which can be requested from donations, the school district, school business partners or local utilities, four jackets printed with "THE ENERGY PATROL" on the front and the school symbol with a lightning bolt on the back can be purchased. Photo identification name tags for each Energy Patrol member might also be considered. Members need to wear these when they are on duty.

Teachers are asked to recommend good, responsible students that can be trusted and who would need little supervision. There is no other teacher involvement other than enthusiastic support of the program. Administrative involvement after initial start-up is minimal.

SIMPLE DAILY ROUTINE SAVES DOLLARS

Four students are on duty each day. One team of two students check half the rooms and a second team checks the other half. After a short while, it should only take about 5-10 minutes. The members go to the office where the jackets, name tags and clipboards are kept in a designated area. They get into uniform, do their rounds, keep records of whose lights are on or off, put up reminder notices on light switches, adjust window blinds, note any water leaks, bring back their materials and go back to their school schedules. It's easy and quick!

STUDENTS CREATE LIGHT SWITCH REMINDERS

There are other suggestions for prompting teachers and fellow students to be energy conscious. Members can make bulletin boards on conservation, put "SAVE ENERGY" signs by every light switch, visit their local utility to hear more suggestions for conservation, or perform an air flow study and send results to the school district office. The results of an air flow study in the individual school could lead to definite action and school district reaction. One recommended activity that can most directly save energy is creating a reminder that says, *"OOPS! YOU FORGOT TO TURN OFF YOUR LIGHTS!"* A 3" x 3" piece of paper with a picture of an unhappy energy man can be taped over the light switches that are left on in unused rooms. They can be copied and then cut apart. Eight will fit on one page. It will increase the awareness of students and staff.

Nobody wants an unhappy mascot to remind them that they are not helping in the conservation efforts. The conservation reminders are made the same way, but they can be colored very brightly and laminated.

KIDS CAN MAKE A DIFFERENCE

Awards can be presented at regularly scheduled school assemblies to classes that have done a good job of conserving energy. The Energy Patrol checklist will make this easy to judge. However, the basic activity of getting lights turned off at recess, lunch and after school and encouraging other students and teachers to SAVE ENERGY is still their job and all that is necessary for success.

Can one school save $100 a month? That and more! Kids CAN make a difference!

NOTE: Light switch reminders can be adapted to fit on door handles. Your school might consider a competition to design reminders so that the entire school can be involved. The program is adaptable. It's common after several months of operating an energy patrol to find that lights are almost never left on. If this happens, patrols can change their schedules so that surprise "spot checks" are made once or twice a week or month. Patrols can also invent new areas to check to see that shades are opened and closed at the proper times or checking thermostat settings to see that they stay at an energy saving 70 degree in the winter and 78 degree in the summer.

A complimentary energy patrol start-up kit is available upon request.

Contact William B. Hardin
TEL # (919) 462-3928 or
(919) 276-7066

ENERGY PATROL START-UP PROCEDURES

ACTIVITY	RESPONSIBILITY
1. Obtain support from your District Office	*Principal*
2. Hold staff meeting to present Energy Patrol Ideas a) Rough overview of Patrol activities. 1. Entering classrooms unsupervised. 2. Record keeping. 3. Reporting to students, etc. b) Generate staff support & enthusiasm.	*Principal*
3. Develop list of possible Patrol members	*Teachers*
4. Evaluate list and make final choices	*Patrol Advisor*
5. Meet with students and explain the general procedures for Energy Patrol	*Patrol Advisor*
6. Develop procedures and materials a) Rules, regulations and agreements. b) Check sheets for each room. c) Location of supplies. d) Schedule of inspections. e) Draw and copy light switch reminders.	*Student/Advisor*
7. Have students sign agreement to serve as Energy Patrol members. Keep agreements on file	*Student/Advisor*
8. Walk-through with all members noting locations of light switches & thermostats	*Students/Advisor*
9. Publicize starting date and purpose a) Students. b) Community. c) Staff.	*Principal*
10. Go and save energy!	*Students*

ENERGY PATROL AGREEMENT

STUDENT'S NAME: ______________________

GRADE LEVEL: ______________________

HOMEROOM TEACHER: ______________________

I agree to assume the responsibilities of the Energy Patrol. I will follow all Energy Patrol procedures and will do the best job possible.

STUDENT'S SIGNATURE: ______________________

DATE: ______________________

ENERGY PATROL GUIDELINES

1. Mature and responsible behavior is expected at all times.

2. Be prompt, pick up supplies, complete rounds, replace supplies and return to class.

3. Be attentive only to your task.

4. Follow your assigned routes. Record information on the checklist.
   * Enter the principal's office only if the door is open.
   * Limited lighting is needed during teachers' planning periods.

5. Failure to follow guidelines will result in dismissal from the Energy Patrol.

ENERGY PATROL PROCEDURES

1. Get jacket, clipboard, identification tag, reminder notices and record sheet from the storage room. Check for notices on Energy Patrol meetings.

2. Inspect your area and record information neatly on a checklist.

3. Return supplies to their proper area. If you run out of reminder notices, leave a note so more can be copied.

REMEMBER:

- *ONLY* Energy Patrol members may enter designated area. Do not bring your friends along or allow desperate students who need snacks, jackets, pencils, books, etc. to enter the classrooms!

- Members must *ALWAYS* wear their jackets and I.D.'s when on duty.

- Work quickly, quietly and politely in rooms where people are working.

ENERGY PATROL CHECKLIST

BUILDING: ______________ MONTH: ______________

LOCATION/DATE	1	2	3	4	5	6	7	8	9	10	11
ROOM # 3	/	/	/	/	/	/					
ROOM # 4	O	O	O	O	X	X					
ROOM # 5	O	O	X	X	X	X					
ROOM # 6	X	X	O	X	O	O					
ROOM #											
ROOM #											
ROOM #											

RECESS ______

LUNCH ______

AFTER SCHOOL ______

INSPECTORS: MONDAY ______
TUESDAY ______
WEDNESDAY ______
THURSDAY ______
FRIDAY ______

KEY: / - Area In Use.
O - Energy In Use/No People.
X - No Energy In Use.

ENERGY CONSERVATION INCENTIVE PROGRAM

JANUARY - DECEMBER 1993 ENERGY SAVINGS

MONTH				
JANUARY				
FEBRUARY				
MARCH				
APRIL				
MAY				
JUNE				
JULY				
AUGUST				
SEPTEMBER				
OCTOBER				
NOVEMBER				
DECEMBER				
TOTALS				

* Savings calculated at $0.? per Kwh, the average rate paid for electricity during 1993.

NOTE: Savings may vary due to energy consumption other than electricity. Yearly weather differences are not reflected.

SECTION 6
HVAC CONTROLS

Chapter 62

Energy-Efficient Office Equipment: An Overview

L.A. Smith, M. Ledbetter

ABSTRACT

Office equipment uses over 5 billion kWh/year, and is the fastest growing end use of electricity in the commercial sector. In this paper, the authors give a brief overview of the energy use characteristics of personal computers, computer printers, copiers, and facsimile machines. Various technologies are discussed within each equipment category, as well as the energy impacts of computer and printer networking and retrofit power management devices for computers and peripherals. New and emerging technologies and their potential for energy savings are outlined. The authors also discuss indirect energy effects (i.e., HVAC loads, backup power costs, etc.) and the impact of various imaging technologies on typical office paper consumption. An example is presented, comparing the direct and indirect energy and dollar impacts of setting up a new office with inefficient vs. efficient equipment. The paper concludes with a summary of current government, utility, and private sector efforts to promote the acceptance of new energy-efficient office equipment, as well as education of end users on the energy-efficient operation of office equipment.

DETERMINING OFFICE EQUIPMENT ENERGY USE

It is relatively difficult to differentiate between various types of equipment based on energy performance. Except for copiers (for which test methods already exist), there are no standardized tests, labels, or other simple means of determining office equipment's potential energy or power use. The wattage or amperage ratings found on nameplates indicate maximum draw, and have been found to be typically three to four times higher than actual power consumption. Test procedures to determine typical energy consumption are being developed by the American Society for Testing and Materials (ASTM) as well as the newly formed Council on Office Products Energy Efficiency (COPEE) in response to the Energy Policy Act of 1992 (EPAct). However, a handful of researchers have monitored equipment energy consumption and use patterns; for the purposes of this paper, we will cite their data.

PERSONAL COMPUTERS

There are two distinct areas of focus when considering PC energy use: machine power draw (during operating and idle periods) and use patterns. The average PC with a CRT monitor draws about 135 Watts during active and inactive periods (Piette et. al., 1991). Figure 1 shows the

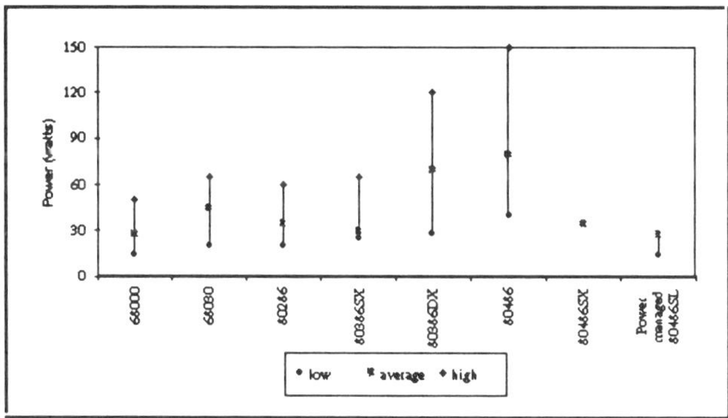

FIGURE 1: MICROCOMPUTER POWER DRAW, BY PROCESSOR TYPE, PER UNIT

ranges of measured power levels for various microcomputer systems. Note that there is not a direct linear relationship between increased computing power/speed and energy consumption; 386-based PCs average 70 Watts, while 486-based machines draw 80 W (not including the monitor) (Norford, 1993).

However, laptop power management technologies have shown that it is possible to significantly reduce PC energy consumption during inactive periods. While low energy consumption has not been a selling point traditionally with desktop machines (large power supplies have been seen by many users as advantageous), manufacturers are now beginning to incorporate low-energy laptop technologies into desktops. The new IBM "Green PC" has a 3.3V 486 SLC2 microprocessor and 10.4 inch active matrix color LCD display. The processor consumes only 16 W during full operation and 8 W during standby (Nadel, 1993).

The most comprehensive study to date on patterns of PC use, done by researchers at the Canadian National Research Council, showed that PCs are actively in use (measured by keyboard or mouse activity) between 30 and 40% of on-time, and that about 20% of machines were left on all night (Newsham and Tiller, 1992). Thus, there is significant opportunity for using power management to reduce total energy consumption by shutting off nonessential system components, or perhaps the entire system, during idle periods.

A handful of manufacturers have recently developed retrofit devices for conventional desktop machines that turns the system off after a user-specified period of inactivity; they are available with a wide range of sophistication, including some that can recognize active files or 'wake-up' with a modem ring, etc. These devices range from $100 to $200, yielding a marginal return (Stickney and Lovins, 1992). Additionally, their market potential is questionable given user concern about compromised productivity and data loss. Therefore, it appears that desktop PC energy use will be most affected by the proliferation of new machines designed with energy-efficiency in mind.

DISPLAYS

There are two potential areas of reducing energy use by displays: better control of CRTs and increasing the usability of flat panel displays. Consumer demands for larger, color, high resolution displays have resulted in increased display energy use. For example, a 12" monochrome CRT uses approximately 65 kWh/year, while a 17" SVGA monitor uses about 235 kWh annually. The scaling of power consumption with size and resolution is shown in Figure 2. However, given the profile of typical office computer system usage discussed above, better control of CRTs (to put them to sleep or turn them off) during idle periods can more than offset the increase in power consumption during active periods.

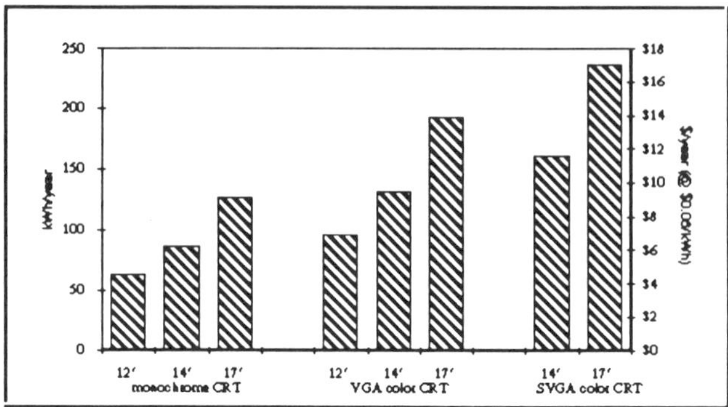

FIGURE 2: ANNUAL ENERGY CONSUMPTION, BY DISPLAY SIZE

As with computers, low-energy display technologies have developed primarily for use in laptop systems. Passive matrix LCDs, gas plasma displays, and electroluminescent displays are all currently available, but provide inferior viewing as compared to CRTs. Active matrix color LCDs, however, compete fairly well with CRTs and a fraction of the energy, as shown in Figure 3. However, their cost is currently four to five times higher than CRTs and thus have been used exclusively in laptop systems where their low-energy characteristics and light weight are critical. Display experts anticipate that LCD prices may decrease with higher production volumes. and thus may emerge for desktop systems within the decade.

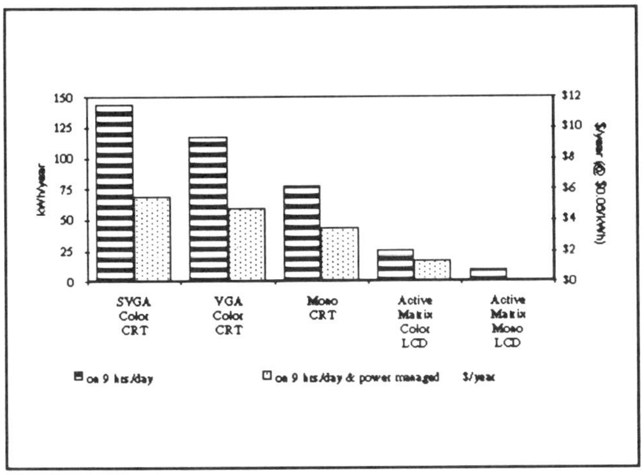

FIGURE 3: ANNUAL ENERGY CONSUMPTION, PER UNIT, FOR VARIOUS 14" MONITORS

COMPUTER PRINTERS

Because most computer printers in commercial offices are actively printing for 10% or less of on-time, energy consumption during standby periods is most critical when considering total energy use. The two most common nonimpact printing technologies, laser and ink jet, have markedly different energy consumption as shown in Table 1.

TABLE 1: TYPICAL PRINTER ENERGY RATINGS

Printer Type	Printing Speed (pages per minute)	Idling (Watts)	Printing (Watts)
ink jet	3	8-12	15-23
laser	4	40-55	130-220
laser	8	75-100	250-300

However, manufacturers are increasingly designing laser printers with integrated power management to reduce their consumption during standby periods (primarily by turning off the energy-intensive fuser rollers). Currently available laser printers with identical features and printing speeds can vary in terms of energy consumption as much as 50%.

Ink jet technology, which does not employ a heat fusing mechanism, is inherently energy efficient. Most current ink jet machines provide somewhat inferior printing quality as compared to lasers, and are also limited in terms of printing speed. However, ink jet manufacturers are continually improving print quality and are developing faster printing capabilities; one manufacturer reportedly plans to release the first 7 ppm ink jet printer this year.

Because standby energy consumption is the dominant portion of the total energy consumption of a printer, the fewer printers on-line, the lower office energy use (as well as up-front and maintenance costs) will be. For example, the energy use and cost of operating individual versus networked laser and ink jet printers for 10 microcomputers is displayed in Figure 4. Despite longer periods of printing, networked printers use substantially less energy than the sum of several individual printers.

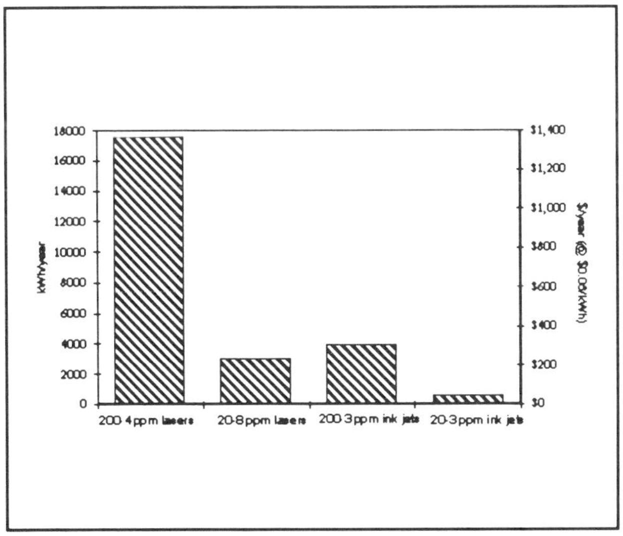

FIGURE 4: ANNUAL ENERGY CONSUMPTION OF INDIVIDUAL VS. NETWORKED PRINTERS

COPIERS

As with printers, the dominant portion of copier energy use is standby consumption. The majority of copiers are equipped with an "energy-saver mode," but this feature does not always result in a significant reduction in power draw; in fact, some machines simply turn off panel lights during energy saver operation, saving only a few watts. Additionally, energy-saver features are frequently not activated (machines are generally shipped with the feature disabled, requiring installer or user activation). The development of more aggressive power management for heat and pressure fusing copiers, as well as increased user awareness of the need for its activation and use, can significantly reduce conventional copier energy consumption.

Alternative copier technologies may create a low-energy opportunity for consumers in the future. Low-heat or cold pressure fusing technologies are currently used in large volume applications and, if successfully scaled down to smaller machines, could provide significantly lower energy use. A few ink jet copiers are currently available and should become more common in the coming years. Liquid ink copiers, currently recommended for lower quality, low-volume applications may also be improved. The most promising current development is in digital duplicator technology, used for high-volume applications at a small fraction of the energy use of a high-volume conventional copier. Again, if this technology could successfully be scaled down to medium- and low-volume machines, it would afford a significant opportunity for energy savings.

FACSIMILE MACHINES

There are a variety of facsimile machine technologies currently available, including laser, direct thermal, thermal transfer, ink jet, front end to printers, and fax cards for PCs. As with other imaging equipment, the energy use during idle periods accounts for the majority of the total consumption. This is of particular importance when considering the efficiency of fax front ends or fax cards; if a printer or PC are left on just to receive faxes, these devices are poor energy performers (unless they are power-managed). However, if properly used, they can eliminate the need for a separate fax engine in the office.

Of the four main direct imaging fax technologies (laser, direct thermal, thermal transfer, and ink jet), ink jet machines generally consume the least energy, while laser (or LED) machines use the most energy in all modes of operation. Ink jet machines can also utilize paper that has previously been printed on to image on the unused side, while lasers cannot. Direct thermal and thermal transfer fall somewhere between lasers and ink jets in terms of energy use.

Because fax machines must be left on 24 hours per day, there is little opportunity for improving user control. However, some laser and LED faxes are now being designed with energy-saver features, reducing their power draw during prolonged periods of inactivity to lower levels than during normal idle periods.

INDIRECT ENERGY EFFECTS

Besides direct energy use costs, office equipment use results in other indirect energy use and associated costs. Heat dissipation from equipment increases cooling loads, and both direct equipment and air conditioning equipment use also results in electrical demand charges in most commercial buildings. With demand charges of $10 per kW and higher in some areas, this portion of total electric costs can be significant. Using energy-efficient office equipment can reduce these costs and decrease the burden upon older buildings' electrical supplies.

INDIRECT NON-ENERGY EFFECTS

There are other non-energy related, typically one-time, costs associated with office equipment. The costs of installed air conditioning capacity required to cool office equipment and additional wiring and transformers needed to power the equipment can be significantly reduced if use of energy-efficient office equipment is considered when designing a building. Of concern to many businesses are uninterruptible power supply system costs. Use of low-energy equipment, or equipment with built-in battery backup such as laptop PCs can reduce these costs substantially. Manufacturers have also found that consumers often favor energy-efficient equipment for non-energy reasons; for example, the new high efficiency desktop computer mentioned above does not require a fan to cool the machine, resulting in a quieter machine. Electromagnetic field emissions, an ongoing concern for CRT users, are significantly lower with energy-efficient flat panel displays.

AN OFFICE-WIDE EXAMPLE

Suppose a company decides to lease a space for two hundred employees, all of which will have a microcomputer and need access to a printer, copier, and fax machine. The type, quantity, and interconnection of equipment purchased to serve this company's needs can result in dramatically different energy use related to office equipment. The table below shows two possible combinations of equipment that could be purchased to meet the company's needs and the resultant direct and indirect energy, paper, and one time costs associated with the chosen equipments' use.

TABLE 2: ALTERNATIVE OFFICE-WIDE EQUIPMENT SCENARIOS

	High-Energy-Use Scenario	*Energy-Efficient Scenario*
Equipment Purchased	▪ 200 conventional desktop PCs ▪ 100 laser printers ▪ 20 laser fax machines ▪ 20 conventional copiers	▪ 150 desktop PCs, 50 laptops ▪ 10 laser printers, 20 ink jet printers (networked) ▪ 20 ink jet fax machines ▪ 20 copiers with power-management
Annual Direct Energy Cost	$33,536	$5,500
Annual Indirect Energy Cost	$10,515	$2,050
Annual Paper Costs	$13,750	$8,870
Total Annual Costs	***$57,750***	***$16,400***
Total One-Time Costs (for A/C capacity and UPS)	***$59,250***	***$37,400***

Significant savings can thus be achieved by carefully evaluating office equipment purchases in light of their energy costs without compromising quality or convenience. Emerging technologies for aggressively power-managed desktop microcomputers, combination peripheral equipment, and other equipment will afford consumers the opportunity to reduce office equipment-related energy use and costs even more dramatically.

PROGRAMS TO PROMOTE ENERGY-EFFICIENT OFFICE EQUIPMENT

The federal government is undertaking several efforts to promote the development and use of energy-efficient office technologies. The Environmental Protection Agency's Energy Star program allows manufacturers who make PCs and monitors capable of being automatically powered down to 30 W or less during periods of inactivity sign a voluntary agreement with the U.S. EPA. This agreement entitles them to use a special Energy Star logo in labels and promotional materials for these products. There are also guidelines for Energy Star printers, which vary based upon printing speed and color capabilities. There are strong indications that the program will have a very important effect on the kind of equipment that will be offered in the market, and on the ability of buyers to differentiate equipment based on energy use. Additionally, the General Services Agency has begun developing guidelines for federal agencies regarding office equipment purchasing and use.

Sweden, Canada, and several countries also have begun programs for office equipment efficiency within their own government agencies and for their citizens. States have also begun similar efforts for their government purchases; for example, California requires copier manufacturers to submit information on life-cycling costs for machines, including operating energy costs, when responding to public bids.

Private companies, including office equipment manufacturers, are also developing programs. The Council on Office Products Energy-Efficiency was recently formed in response to EPAct, which requires that manufacturers and other interested parties develop voluntary test standards for office equipment by 1995. Many private companies, most notably computer manufacturers, have campaigns to encourage employees to turn off unused equipment in their own offices.

Electric utilities are also becoming increasingly interested in promoting office equipment efficiency. The Energy-Efficient Office Technologies Consortium is a voluntary group of electric utility, public interest group, government, academic, and manufacturer representatives that promote the development and dissemination of information on energy-efficient office products. They sponsored the writing of ACEEE's *Guide to Energy-Efficient Office Equipment*, as well as research at MIT on future technology possibilities in this area.

CONCLUSION

There are significant opportunities for energy savings through the procurement and proper use of less energy-intensive office products, and significant potential for technological developments in equipment design to decrease energy use by equipment itself as well as indirect effects.

ACKNOWLEDGEMENTS

The authors thank Les Norford and Cyane Dandridge of MIT who provided the bulk of technical information for this paper. We would also like to thank the members of the Office Technology Consortium who have sponsored ACEEE's research in this area.

REFERENCES

Nadel, Brian, *The Green Machine*, PC Magazine, Vol. 12 No. 10, May 25, 1993.

Newsham, Guy R. and Dale K. Tiller, *Energy Consumption and Desktop Computers*, National Research Council of Canada, Ottawa, Ontario, May 1992.

Norford, Leslie K. and Cyane B. Dandridge, *Near-Term Technology Review of Electronic Office Equipment* (forthcoming), Massachusetts Institute of Technology, Cambridge, MA, June 1993.

Piette, M.A., J.H. Eto, and J.P. Harris, *Office Equipment Energy Use and Trends*, LBL-31308, Lawrence Berkeley Laboratory, Berkeley, CA, September 1991.

Stickney, Bristol L. and Amory B. Lovins, *Tech Update, September 1992, TU-92-2*, E-Source, Boulder, CO, September 1992.

Chapter 63

Case History: The Retrofit of a Run-Around Loop for Economical Humidity Control in a Surgery Center

W. Beckwith

ABSTRACT

Recently, the World Health Organization has identified microbiological contamination in buildings as number five of the top five health threats to humans[1]. Because suitable growth substates and adequate nutrients are generally available in properly maintained modern buildings, the most practical way to control microorganisms is to maintain reduced indoor humidity[2]. In fact, the ASHRAE IAQ Standard 62-1989; Section 5.12 states that the relative humidity in all ductwork should be kept below 70%. This condition is easy to meet using the technologies described in this paper. Control of indoor fungal growth can prevent a wide variety of serious health impacts for building occupants[3].

This paper will introduce the dehumidifying Run-around loop technology via a utility company monitored case history. Run-around loops (RAL's) passively transfer heat from the return air to the supply air in an HVAC system. When used to replace reheat systems, RAL's provide a cooling load reduction, and virtually "free" reheat which is derived from heat loads which increase return air temperatures.

In this project, the RAL's also allowed an economical increase in fresh air ventilation to provide a positive pressurization of the facility. Furthermore, drier air, rather than saturated air, was introduced into the supply ductwork, thus reducing the likelihood of fungal growth in the supply air ducts. Finally, the RAL's maintained a relative humidity in the conditioned space between 40 and 50%, which is the range in which fungal growth is reduced[4, 5]. On the economic side, energy savings realized after installation of the system provided a return on investment of 80% with a payback of 1.2 years.

INTRODUCTION

High humidity can have a detrimental effect on many industrial processes, equipment, and products. Building materials, books, and records deteriorate more rapidly in humidity levels above 60% due to bio-degradation. High humidity increases electrical costs by reducing the temperature required for occupant comfort. Fungal growth in humidities above 70% can have many detrimental health effects, depending on the particular species encountered[5, 6]. In fact, high humidity is often an underlying cause of many air quality disorders that result in indoor air quality (IAQ) problems.

Moist environments (above 70% rh) allow the proliferation of microbiological contamination that may cause lethargy, allergic reactions, and disease in people[3, 4, 6]. Airborne pathogens may survive freely in air which has a high relative humidity [5]. The effectiveness of air filters is reduced in high humidity. Finally, the offgassing of chemicals such as formaldehyde, from urea-formaldehyde products, and the perception of odors all increase in high humidity [5].

Prior to the mid-1970s, humidity was controlled by using a reheat system. In a reheat system, air is overcooled with an air conditioner in order to condense excess water vapor and is then reheated to a temperature suitable for introduction into the occupied space with supplemental heat such as resistance heaters or hot water. In effect, the reheat adds an artificial sensible heat load to the space and achieves humidity control with higher capital and operating costs.

COST FACTORS

There are two operating cost factors involved with reheat. First, is the obvious electrical power consumption by the electric heat strips, or fuel consumption for a hot water system. Less obvious is the additional power consumption required by the air conditioner to remove the reheat from the return air. Besides increased annual operating costs, capital costs are higher. Overcooling the moist air and handling the supplemental heat load requires an air conditioner which is two to five times larger than a unit that does not overcool.

HOW A RUN-AROUND LOOPS WORK

When a RAL is positioned around the cooling coil of an air conditioner, it transfers heat out of the warm air before it reaches the cooling coil. This heat is transferred around the coil and then added to the cold air which leaves the cooling coil. The return air is precooled by the RAL typically 5 to 15°F (depending on design) before it reaches the cooling coil. Now the cooling coil can overcool the air to remove excess moisture. Next the overcooled air is warmed (reheated) by the RAL to a comfortable temperature for delivery to the room. Rather than saturated air entering the supply duct, the RAL produces drier air. Moisture removal is typically increased 50 to 100%. RAL's work equally well in direct expansion or chilled water systems of any size.

PROJECT SITE DESCRIPTION

The St. Petersburg Surgery Center (SPSC) is located in St. Petersburg, Florida, adjacent to the Gulf of Mexico. The facility is a medical clinic specializing in outpatient surgery. The building is single story, with 6,200 sq ft of floor space constructed of concrete and stucco in 1988. The facility contains a large waiting room, three operating suites, a recovery room, a break room, and several offices. The HVAC system operates approximately 4,100 hours per year.

The original HVAC system was comprised of a 25-ton and a 7.5-ton direct expansion, rooftop package system. The units are constant air volume with terminal electric reheat. The connected electrical load for the 10 reheat units totalled 66 kW.

PROJECT DESCRIPTION

The indoor environmental conditions had been a continuing problem. The relative humidity level was typically 60%. The doctors were concerned because of rusting of delicate medical instruments and the growth of fungi in the pathology microscope optics.

As is often the case, the first measure taken by the HVAC service contractor to control humidity was to close the dampers in the outside air intakes. This measure resulted in the indoor environmental conditions stated above. These relative humidity conditions are assumed to be an improvement over those which occurred with the outside air dampers open.

Closing the outside air dampers is not recommended. Without outside air make-up, the exhaust fans create a negative air pressure in the building. This allows the introduction of unconditioned outside air at many uncontrolled locations in the building fabric, increasing relative humidity at points of intrusion.

At this point, Mr. Beckwith arranged for SPSC to become one of the field verification projects used in the development of Florida Power Corporation's (FPC's) heat-pipe rebate program. The purpose of this program is to reduce electrical demand by replacing electric reheat with heat-pipe heat exchanger reheat.

Preliminary monitoring of the project by FPC began in March 1991. The parameters monitored were: kilowatt demand, kilowatthour consumption, indoor relative humidity, and indoor dry bulb temperature. The heat-pipe heat exchangers were installed on May 25 and 26, 1991. Monitoring continued through November 1991.

In addition to installing the Ral's, portions of the terminal electric reheat were disabled and the outside air dampers were re-opened to provide 2,500 CFM of outside air to a total supply air flow of 10,000 CFM. This outside air flow provides sufficient make-up air to satisfy the exhausts and provide a positive pressurization of the building interior.

UTILITY MONITORING RESULTS

Figure 1 is a graph which shows the relative humidity throughout the day in the Recovery Room prior to the installation of the Ral's. During this time the outside air dampers were closed. As the graph indicates, relative humidity was maintained at approximately 60% during March, April, and May.

Figure 2 is a graph of the daily relative humidity in the Recovery Room after the installation of the Ral's with the outside air dampers open. June and July represent the "building dry-out" period. This is the period of time during which the equilibrium moisture content of the building materials and furnishing is declining due to the reduced relative humidity in the space. In other words, the furnishings and materials were giving up moisture which was removed by the HVAC system. The amount of moisture entering the occupied space in this fashion can be very significant. For this facility, the quantity of moisture removed during dry-out was estimated to be near 1,000 pounds. During dry-out, this moisture also elevates the relative humidity of the air in the facility. In contrast, during the period of August through November, the relative humidity in the Recovery Room averaged 45%. This reflects the design conditions after dry-out.

Figure 3 is a bar chart which depicts the kilowatt demand of the electric heat strips used for reheat. For the nine week period prior to the installation of the Ral'ss, kilowatt demand averaged 34.62 kW. After the installation of the Ral's, demand averaged 21.27 kW. This represents a decrease in demand of 13.35 kW, or 39%. At this point, the reader is reminded that this decrease in demand was achieved at the same time the outside air ventilation rate was increased 2,500 CFM. This increased the heat load on the cooling system.

Figure 4 is a bar chart of the daily electrical consumption for the electric reheat before and after the installation of the Ral's. During the months of April and May, prior to the Ralo's installation, daily electrical consumption averaged 675 kWh. After the installation of Ral's, for the period June through November, daily electrical consumption averaged 433 kWh. This is a reduction of 242 kWh per day, or 36%.

FPC projected annual operating cost savings for reheat energy using the decrease in electrical demand and the operating hours of the mechanical system as follows:

13.35 kW X 4,108 hrs/yr = 54,842 kWh
54,842 kWh X $.06/kWh = $3,291.

To calculate the annual cooling load savings provided by the Ral's precooling, FPC used the kW of reheat saved converted to Btuh and applied to the HVAC system operating at an Energy Efficiency Rating (EER) of 8. The following formulas were used:

13.35 kW X 3,413 Btu/kW = 45,564 Btu ÷ 8,000 = 5.695 kW cooling
5.695 kW X 4,108 hrs/yr = 23,397 kWh
23,397 kWh X $.06/kWh = $1,404.

TOTAL SAVINGS WITH IMPROVED IAQ

Because the utility's monitoring data does not account for the fact that outside air ventilation was increased 2,500 CFM when the Ral'ss were installed, calculations will be performed to account for the outside air heat load. Table 1 shows the engineering weather data for nearby MacDill Air Force Base in Tampa, Florida, from the National Oceanic and Atmospheric Administration. The hours of occurrence for the BIN temperatures (temperature ranges occurring during the year) are adjusted to reflect the operating hours of the facility. The enthalpy of each BIN temperature is multiplied by the hours of occurrence. These products are totaled and this total is divided by the total number of hours of operation to arrive at a weighted mean average enthalpy for the outside air. The enthalpy of the air in the conditioned space (70°F @ 45% rh) is subtracted from the weighted mean outside air enthalpy. This difference is used in the following formulas to calculate the outside air load:

2,500 CFM X 4.45 X 9.93 = 110,471 Btuh
110,471 Btuh @ EER 8 = 13.8 kW
13.8 kW X 3,773 hrs/yr = 52,067 kWh
52,067 kWh X $.06/kWh = $3,124.

If the outside air dampers had been open during the pre-Ral's monitoring period, the HVAC system would have expended additional energy to cool this load. As is known, the outside air raises the entering mixed temperature of the return air. Because of a constant temperature leaving the cooling coil, the elevated mixed temperature creates a greater temperature differential across the cooling coil. This temperature differential is the primary driving force behind the heat transfer of the Ral. This causes the Ral to transfer more heat from the return-side, thus creating a greater cooling load reduction and increased reheat to the units supply-side.

Another area of savings must be discussed. This is the fact that the conditions of temperature and relative humidity in the facility were improved after the installation of the Ral's. To calculate the additional energy cost that would have been expended by the system before the Ral's to achieve the same humidity level as with Ral's, the amount of moisture removed from the return air and the energy efficiency of the HVAC system are used in the following formulas:

Conditions before Ral's: 68°F; 60% rh; 60 grains/lb
Conditions after Ral's: 70°F; 45% rh; 49 grains/lb
Change in water vapor ratio: 11 grains/lb

7,860 CFM X 4.45 X 11 grains/lb ÷ 7,000 grains/lb = 55 lbs/hr
55 lbs/hr X 1,070 btu/lb = 58,850 Btuh
58,850 Btuh ÷ EER 8 = 7.36 kW
7.36 kW X 4,108 hrs/yr = 30,235 kWh
30,235 kWh X $.06 US/kWh = $1,814.

Totaling the utility projected savings, the outside air load savings, and the improved indoor conditions, the annual operating cost savings is $9,633/year. Finally, the annual energy savings result in significant outdoor environmental benefits. Savings in coal are 69 tons per year. The atmospheric loading of carbon dioxide is reduced by 110 tons per year.

CONCLUSIONS

This paper clearly shows that dehumidifying Run-around loops can economically improve indoor air quality. In this project, the Ral's allowed an economical increase in fresh air ventilation to provide a positive pressurization of the facility. Furthermore, drier, rather that saturated, air was introduced into the supply ductwork. Finally, the Ral's maintained the relative humidity in the conditioned space between 40 and 50%, in which range microorganism growth is reduced. On the economic side, the cost of this project was $15,000. Under the FPC rebate program, SPSC received a rebate of $3,040 based on electrical demand reduction. The net project cost was $11,960. The annual operating cost savings of $9,633 yields a return on investment of 80% with a payback of 1.2 years.

ACKNOWLEDGEMENTS

A very special thanks is extended to Mr. Leonard Zyki, P.E., of Florida Power Corporation for the considerable extra effort he expended on site and in his acquisition, reduction, and presentation of data. Finally, appreciation is extended to Dr. David Harlos, Chief Scientist, Environmental Science and Engineering, Incorporated, for his thoughtful review of the text.

REFERENCES

1. Morey, P. et al., "Biological Contaminants in Indoor Environments", ASTM STP 1071 Baltimore, MD 1990

2. Toth, Cynthia. "Microbials in the Overall Context of Indoor Air Quality Investigation", Proceedings of First Annual IAQ Conference and Exposition, National Coalition of Indoor Air Quality, Tampa, Florida. 1992 pp 255-259

3. Kay, J.G., et al. "Biological Pollutants in the Indoor Environment", Indoor Air Pollution. Vol 78, pp 13-24, 1991

4. Samet, J.M., et al. "Health Effects and Sources of Indoor Air Pollution.", Part II. Am Rev Respir Dis, 137:221-242, 1988

5. Arundel, A.V., et al., "Indirect Health Effects of Relative Humidity in Indoor Environments", Environmental Health Perspectives. Vol. 65, 1986, pp 351-361

6. Buge, H. et al. "Sick Building Syndrome: A Study of 4373 Workers"' Ann. Occup. Hygiene 31:493, 1987

Table 1
ST. PETERSBURG SURGERY CENTER
WEIGHTED AVERAGE ENTHALPY OF OUTSIDE AIR

OUTSIDE AIR BIN TEMP.	ADJUSTED HOURS DURING YEAR (*)	ENTHALPY (Btu/lb)	PRODUCT HOURS X ENTHALPY (Btu/lb)
95/99	4	41.57	166.28
90/94	186	40.59	7,549.74
85/89	676	38.65	26,127.40
80/84	826	37.74	31,173.24
75/79	777	35.04	27,226.08
70/74	566	31.70	17,942.20
65/69	420	27.92	11,726.40
60/64	318	25.18	8,007.24
TOTAL	3,773		129,918.58

Weighted Mean:	34.43 Btu/lb
Conditioned Air:	70°Fdb/57°Fwb
Enthalpy of Conditioned Air:	24.50
Enthalpy Change:	9.93

(*) Hours within each BIN are adjusted to reflect the operating hours of the mechanical system.

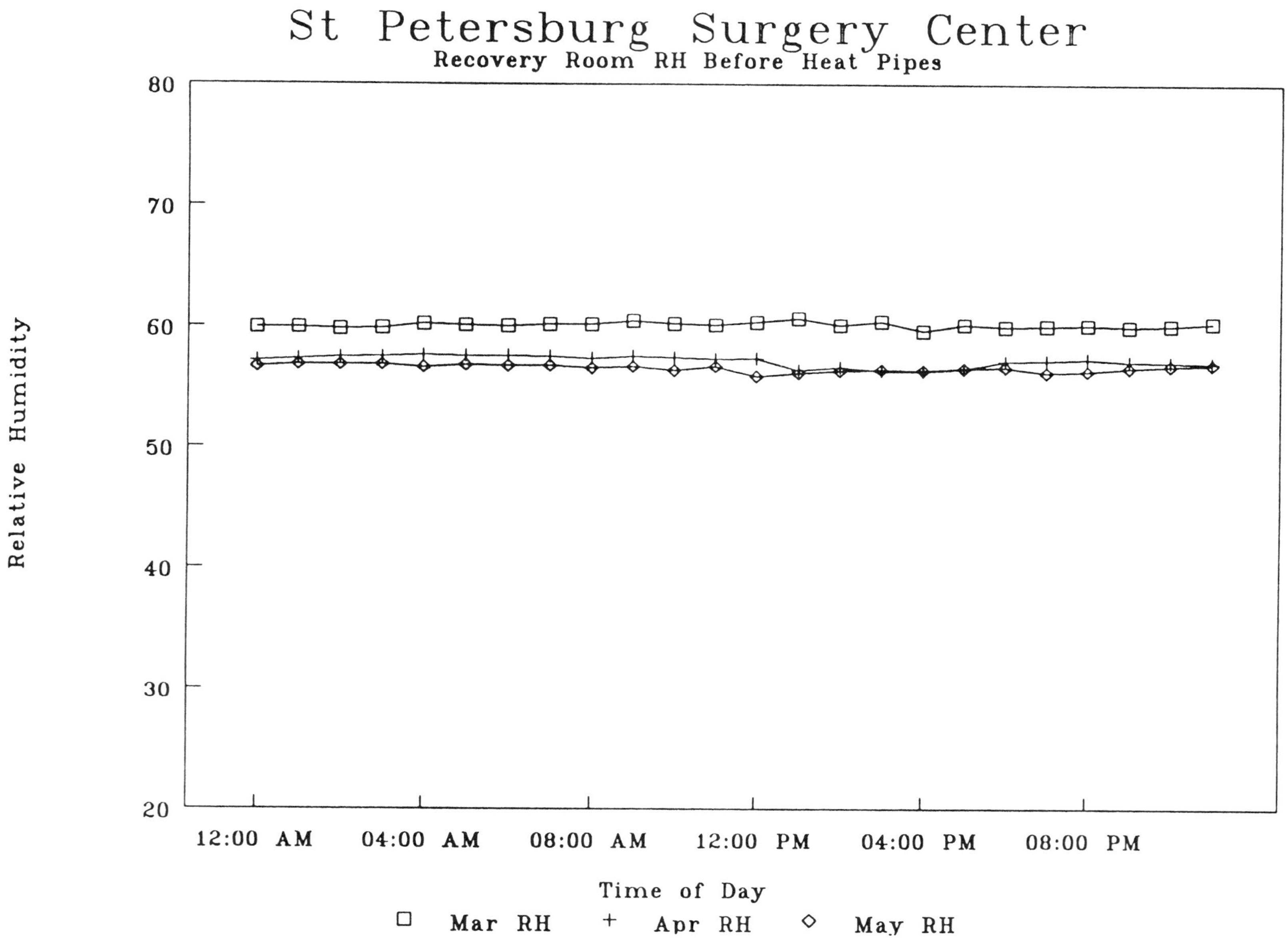
St Petersburg Surgery Center
Recovery Room RH Before Heat Pipes
Relative Humidity
80
70
60
50
40
30
20
12:00 AM
04:00 AM
08:00 AM
12:00 PM
04:00 PM
08:00 PM
Time of Day
Mar RH
Apr RH
May RH

FIGURE 1.

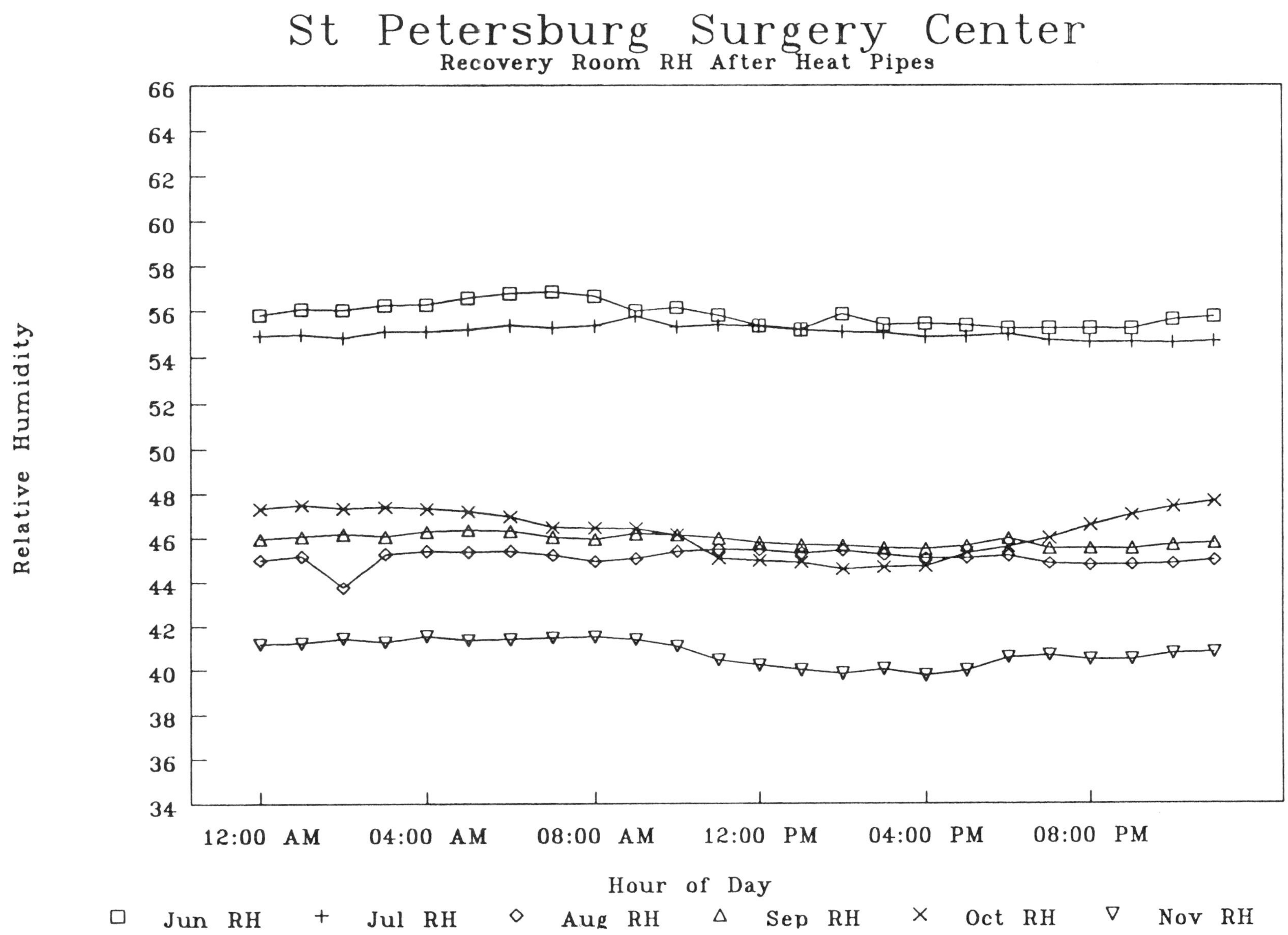
St Petersburg Surgery Center
Recovery Room RH After Heat Pipes
Relative Humidity
66
64
62
60
58
56
54
52
50
48
46
44
42
40
38
36
34
12:00 AM
04:00 AM
08:00 AM
12:00 PM
04:00 PM
08:00 PM
Hour of Day
Jun RH
Jul RH
Aug RH
Sep RH
Oct RH
Nov RH

FIGURE 2.

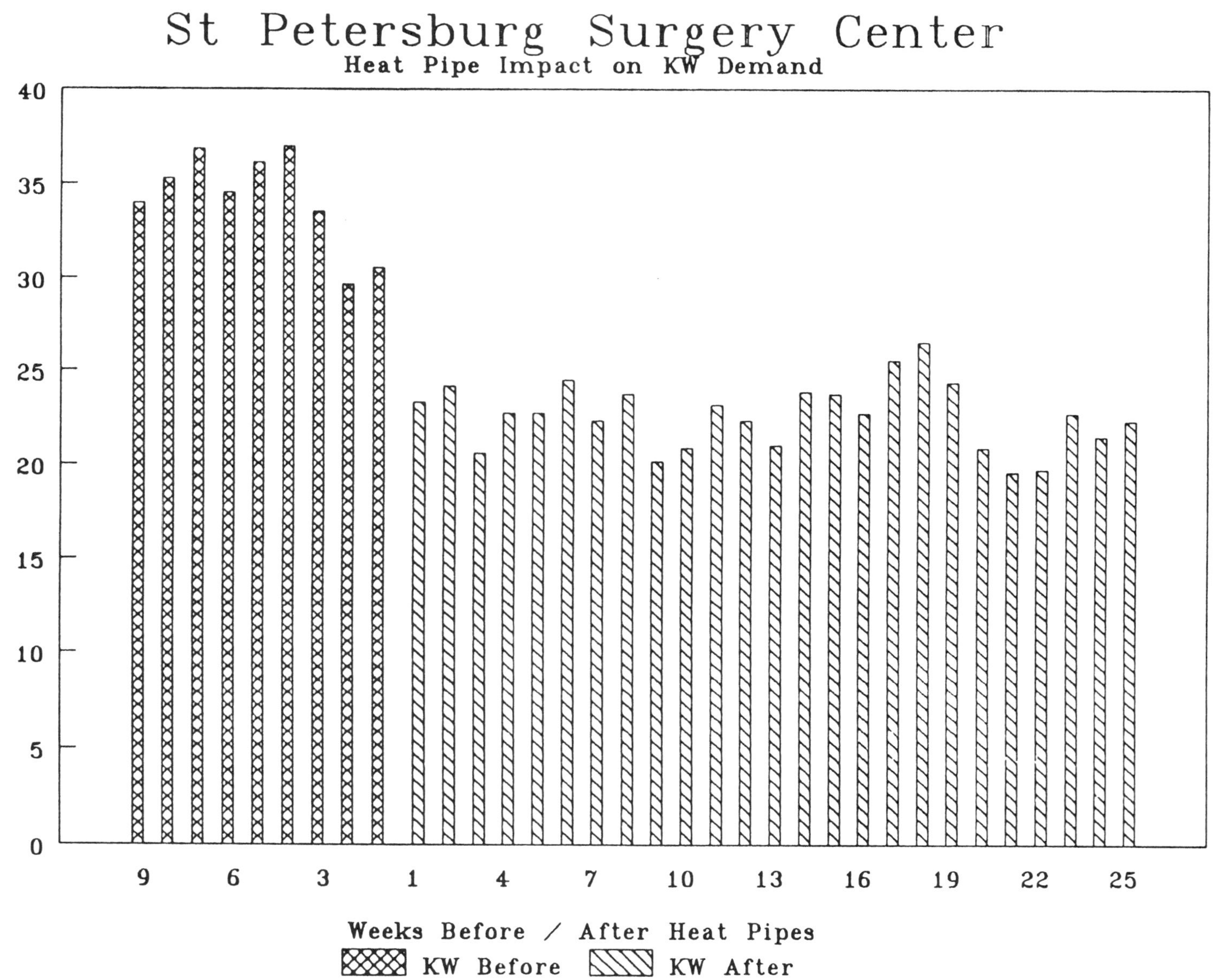
St Petersburg Surgery Center
Heat Pipe Impact on KW Demand
40
35
30
25
20
15
10
5
0
9
6
3
1
4
7
10
13
16
19
22
25
Weeks Before / After Heat Pipes
KW Before
KW After

FIGURE 3.

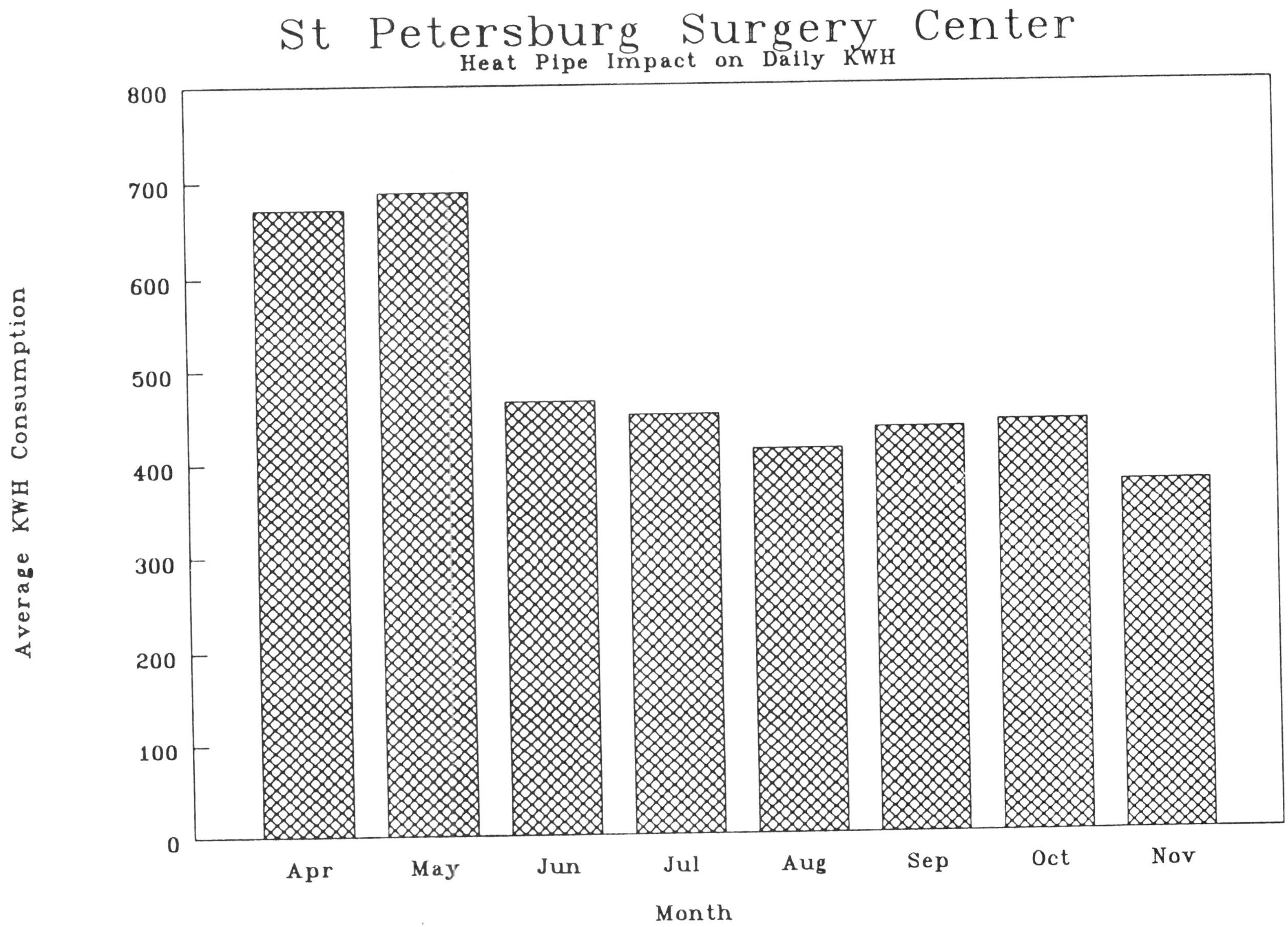
St Petersburg Surgery Center
Heat Pipe Impact on Daily KWH
Average KWH Consumption
800
700
600
500
400
300
200
100
0
Apr
May
Jun
Jul
Aug
Sep
Oct
Nov
Month

FIGURE 4.

Chapter 64

Ground-Source Heat Pumps: A Feasibility Study

H.G. Nezhad

ABSTRACT

This paper summarizes a detailed study of ground-source heat pumps conducted by the Energy Research Institute at Moorhead State University.

The paper provides technical as well as financial data on several ground-source heat pumps installed in this region. The results of a marketing survey of 243 heat-pump users is presented. Finally, total energy use, environmental impacts and environmental costs to society for ground-source heat pumps is compared to air-source heat pumps and four other conventional systems.

INTRODUCTION

There are many factors to be considered when determining the feasibility of the ground-source heat pumps. First, and perhaps the most important factor is the customer acceptance. Customer acceptance itself is a function of many other factors such as initial cost of a heating/cooling system, operating and maintenance costs, reliability, comfort and many other subjective factors.

Second, one must consider the benefits to those utility companies which are promoting expanded use of heat pumps. Since heat pumps use electricity, they provide a new market for the electric utility, particularly with those customers using sources other than electricity for heating. Also, with the combination of another fuel or thermal storage, the utility can offer their customers lower interruptible rates which would benefit both the customer and the utility.

Finally, one must consider the benefits to society at large. Because heat pumps are highly energy efficient, they extend the lives of our exhaustible energy resources and have less adverse environmental impacts.

1. CUSTOMER ACCEPTANCE

A telephone survey of 163 heat pump users in the North Central region of the United States was conducted.The collected data were then compiled and evaluated using StatPac Gold statistical analysis software. Here is a summary of the results.

Ninety-eight percent of the respondents owned their homes, and one fourth of the sample have lived in their homes for 16-20 years, and 63% resided in the survey region for 21 years or more.

Seventy percent of heat pump systems were installed initially in new home construction of which 65% were superinsulated. Eighty-two percent of the respondents made the decision to install the heat pump. Among those surveyed, 96% stated that the heat pump adequately heated and cooled their home.

Gas and propane heating systems accounted for 57% of the backup systems used.

Fifty-six percent of the survey respondents had not required heat pump repairs in the last two years. When questioned about their electricity bills, 83% found them to be lower than expected.

All the survey respondents living in their homes less than two years liked the heat pump better than other types of heating systems they used before. They all said that they would recommend ground-source heat pumps to others.

Ninety-eight percent of the respondents indicated that their electricity bills were either lower than they expected or about the same as they expected, while 5% said they were higher. However, one hundred percent of the respondents said that they were satisfied with their heat pump.

2. ENERGY EFFICIENCY ANALYSIS

The energy efficiency of a heat pump is measured by the coefficient of performance (COP). COP is a dimensionless measure, therefore the heat provided by the system must be in the same units as the energy consumed by the system. The inputs and outputs are usually measured in British Thermal Units (BTU) or in kilowatt hours (KWH).

We measured energy inputs and outputs of the three units under study both during heating and cooling operations. The results were very impressive. The average COP with outside temperature of 20 degrees and water temperature of 50 degrees fahrenheit was 3.21 and the average cooling efficiency with outside temperature of 70 degrees fahrenheit was 2.2 (or 220%).

3. FINANCIAL ANALYSIS

The most commonly used methods for determining financial feasibility of a project is the simple payback method. However, simple payback does not take into account the time value of money. Another common method is Life Cycle Cost Analysis (LCCA). LCCA method considers all the significant costs of ownership during the service life of the system under consideration. Furthermore, LCCA considers the time value of money.

The LCCA method is a very effective way of comparing financial costs and benefits of two or more projects under consideration. In this method all the costs and savings over the life of the project(s) are discounted to determine their present value. The difference between the present value of savings and costs is called the Net Present Value (NPV) of the investment. Obviously, the higher the NPV the more attractive the project would be. Projects with negative NPVs are unacceptable.

In addition to using NPV, one should also determine the rate of return on the investment. The most commonly used method for measuring the return on investment for a project with projected cash flows is to calculate the Internal Rate of Return (IRR). The IRR is defined as "the discount rate which equates the present value of a project's expected cash inflows to the present value of the project's expected cost." When NPV is zero the IRR will be equal to the discount rate used to calculate the NPV.

In our heat pump project, the projected cash inflows are the annual savings produced by installing a ground-source heat pump. When the IRR is greater than the discount rate, the homeowner is better off investing in the heat pump rather than at the discount rate. In the following sections two common cases of deciding on a heating/cooling system are analyzed.

A. Replacing an Older Heating and Cooling System with a Ground-Source Heat Pump

In this case we will use an energy efficiency of 99% for the electric resistance, and 50% for an old propane, fuel oil, and natural gas systems. Furthermore, we assume the annual increase in electric rate to be 0.5%, while for propane, fuel oil, and natural gas to be 2%. The heating units being compared to the ground-source heat pump have a capacity of 100,000 BTU/h, and the ground-source heat pump has a capacity of 120,000 BTU/h. Two five ton (1 ton=12,000 BTU) ground-source heat pumps will be installed at a total cost of $20,000 ($3,500 for each heat pump system and $13,000 to drill the wells and install a backup system).

The results of our analysis indicate that although all the systems being compared have a payback period within the service life of the ground-source heat pump, only one system (propane) had a positive NPV. As a result, this is the only case where the IRR is higher than the discount rate used to calculate the NPV (see Figure 1).

B. Installing a Ground-Source Heat Pump in a New Building

Given an annual heating load of 200 million BTU for a house in our region, one can calculate the annual operating cost of different systems. Based on the data available in our region, costs and benefits of ground-source heat pump is compared to five other systems: natural gas, propane, fuel oil, electric resistance, and air-source heat pump. The cost of a conventional air conditioner is added to those systems lacking air conditioning capability.

Based on our assumptions, ground-source heat pump has a negative NPV compared to other five systems. Although the payback periods for electric resistance, propane, and fuel oil are between 11.5 to 16.5 years, there is no payback period within the twenty-year service life of a ground-source heat pump when compared to a natural gas system (see Figure 2).

4. SOCIETAL COSTS OF DIFFERENT HEATING AND COOLING SYSTEMS

Societal costs include, among other things, environmental costs, national security costs, as well as impacts on employment. Placing a dollar value on the societal costs of a heating/cooling system is very difficult and controversial. However, it is obvious that one system would have more harmful effects on the environment than another. The environmental impacts of a heating and cooling system can be measured by the amounts of carbon dioxide (CO^2), sulfur oxides (SO_x), and nitrogen oxides (NO_x) that are produced by burning a fossil fuel to generate the needed energy. Table 1 shows an estimated amount of pollutants generated by various energy sources.

TABLE 1 EMISSIONS FOR DIFFERENT ENERGY SOURCES (LBS/KWH)

Energy Source	CO_2	NO_x	SO_x
Fuel Oil	2.0	0.004	0.012
Natural Gas	1.3	0.005	0.000
Electricity (from coal)	1.9	0.005	0.009

Source: EPA, June 1992

Propane(% of weight)	11.1%	0.0%	0.0%

Source: National Propane Gas Association

Note: Propane has 0.126 lb/ft^3.

By associating a cost per unit of pollution, one would be able to calculate the cost of producing that pollutant. Table 2 lists the data used in our analysis.

TABLE 2. POLLUTION COSTS

SO_2	$0.400/lb
NO_x	$2.920/lb

Source: Tellus Institute

CO_2	$0.011/lb

Source: New England Electric System

To compare the societal costs of different systems to ground-source heat pump we consider a house in our region with the annual heating load of 208 million BTU and the annual cooling load of 12.2 million BTU. Total pollution costs to society over the twenty-year life of a ground-source heat pump are compared in Figure 3 to five other heating systems.

CONCLUSION

Although ground-source heat pump is more than three times as efficient as any other heating system available today, the initial cost is very high for a homeowner. If the homeowner's decision is solely based on the monetary costs and benefits of a heating/cooling system, ground-source heat pump would not be a logical choice without any financial incentives from the utilities or government agencies. But as our marketing survey indicated, people's decisions are not based only on the dollar value attached to a system. Other factors, such as the environmental benefits and the pride of extracting some of their heating and cooling needs out of their backyard and being in the forefront of a new technology, among others, affect their decisions.

From the utility's perspective, ground-source heat pump would provide growth opportunities through Demand-Side Management. At the same time, utilities will be contributing to the global efforts in cleaning up our environment. Thus, from the societal point of view, ground-source heat pump is a wise choice.

ACKNOWLEDGEMENT

The author would like to thank the Otter Tail Power Company for providing the financial support for this project. Also, the following Energy Management students assisted in collecting and analyzing the data: Wesley Leier, Nick Ethen, and Douglas Rapske.

FIGURE 1. LIFE CYCLE COSTS ANALYSIS FOR GROUND-SOURCE HEAT PUMP V.S. PROPANE

	(A)	(B)	(C)	(D)	(E)	(F)	(G)	(H)	(I)	(J)	(K)
		G-Source Heat Pump 0.50%		Alternative Heating 2.00%		Alternative Cooling 0.50%					
Yr.	Invest-ment	Annual Energy Price Increase	Annual Energy Cost	Annual Energy Price Increase	Annual Energy Cost	Annual Energy Price Increase	Annual Energy Cost	Gross Savings	Maint.	Net Savings	(Cash Flow) Simple Payback
0	($20,000)									($20,000)	($20,000)
1		$9.67	$692	$6.62	$2,648	$9.67	$88	$2,044	$0	$2,044	($17,956)
2		$9.72	$696	$6.75	$2,701	$9.72	$88	$2,094	$0	$2,094	($15,863)
3		$9.77	$699	$6.89	$2,755	$9.77	$89	$2,145	$0	$2,145	($13,718)
4		$9.82	$703	$7.03	$2,810	$9.82	$89	$2,197	$0	$2,197	($11,522)
5		$9.86	$706	$7.17	$2,866	$9.86	$90	$2,250	$0	$2,250	($9,272)
6		$9.91	$710	$7.31	$2,924	$9.91	$90	$2,304	$0	$2,304	($6,968)
7		$9.96	$713	$7.46	$2,982	$9.96	$91	$2,359	$0	$2,359	($4,609)
8		$10.01	$717	$7.60	$3,042	$10.01	$91	$2,416	$0	$2,416	($2,193)
9		$10.06	$720	$7.76	$3,103	$10.06	$91	$2,474	$0	$2,474	$281
10		$10.11	$724	$7.91	$3,165	$10.11	$92	$2,532	$0	$2,532	$2,813
11		$10.16	$728	$8.07	$3,228	$10.16	$92	$2,593	$0	$2,593	$5,406
12		$10.22	$731	$8.23	$3,292	$10.22	$93	$2,654	$0	$2,654	$8,060
13		$10.27	$735	$8.40	$3,358	$10.27	$93	$2,717	$0	$2,717	$10,777
14		$10.32	$739	$8.56	$3,425	$10.32	$94	$2,781	$0	$2,781	$13,557
15		$10.37	$742	$8.73	$3,494	$10.37	$94	$2,846	$0	$2,846	$16,403
16		$10.42	$746	$8.91	$3,564	$10.42	$95	$2,913	$0	$2,913	$19,316
17		$10.47	$750	$9.09	$3,635	$10.47	$95	$2,981	$0	$2,981	$22,296
18		$10.53	$754	$9.27	$3,708	$10.53	$96	$3,050	$0	$3,050	$25,346
19		$10.58	$757	$9.45	$3,782	$10.58	$96	$3,121	$0	$3,121	$28,467
20		$10.63	$761	$9.64	$3,858	$10.63	$97	$3,193	$0	$3,193	$31,660
								Net Present Value		$3,625	
								Internal Rate of Return		10.32%	

Conditions:

A1)	Discount rate	8.00%
A2)	Amount of energy required for cooling per year in million BTU	20.00
A3)	Amount of energy required for heating per year in million BTU	200.00
A4)	Cost to install the ground-source heat pump system	$20,000
A5)	Estimated additional maintenance costs/year for the ground-source system	$0.00
A6)	Energy cost for heat pump system in $s per million BTU	$9.67
A7)	Heating efficiency of the ground-source heat pump system	320.00%
A8)	Cooling efficiency of the ground-source heat pump system	220.00%
A8)	Annual rate increase for the energy to operate the heat pump system	0.50%
A9)	Energy cost to operate propane heating in $s per million BTU	$6.62
A10)	Heating efficiency of a propane system	50.00%
A11)	Annual rate increase for the energy to operate propane heating	2.00%
A12)	Energy cost to operate alternative cooling system in $s per million BTU	$9.67
A13)	Cooling efficiency of the alternative cooling system	220.00%
A14)	Annual rate increase for the energy to operate the alternative cooling system	0.50%

Assumptions:

1) Initial investment cost does not include the duct work that is necessary for the system to operate, it is assumed that the duct work will be required for all forced air heating and cooling system.
2) Estimated economic life of ground-source heat pump is 20 years.
3) Discount rate of 8.0%

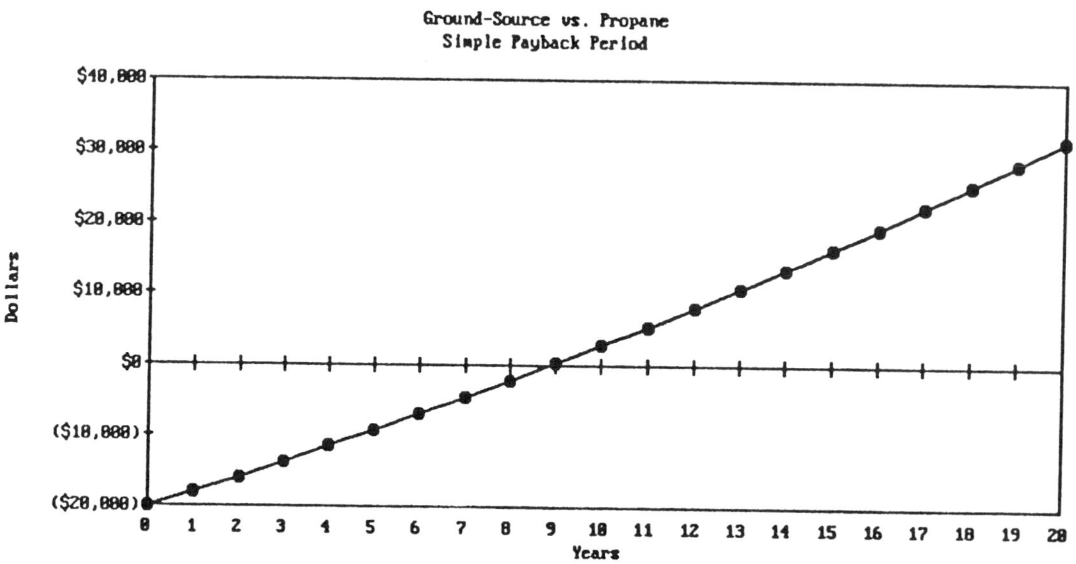

FIGURE 2. LIFE CYCLE COSTS ANALYSIS FOR GROUND-SOURCE HEAT PUMP V.S. NATURAL GAS

	(A)	(B)	(C)	(D)	(E)	(F)	(G)	(H)	(I)	(J)	(K)
		G-Source Heat Pump 0.50%		Alternative Heating 2.00%		Alternative Cooling 0.50%					
Yr.	Invest-ment	Annual Energy Price Increase	Annual Energy Cost	Annual Energy Price Increase	Annual Energy Cost	Annual Energy Price Increase	Annual Energy Cost	Gross Savings	Maint.	Net Savings	(Cash Flow) Simple Payback
0	($20,000)									($20,000)	($20,000)
1		$9.67	$692	$6.62	$2,648	$9.67	$88	$2,044	$0	$2,044	($17,956)
2		$9.72	$696	$6.75	$2,701	$9.72	$88	$2,094	$0	$2,094	($15,863)
3		$9.77	$699	$6.89	$2,755	$9.77	$89	$2,145	$0	$2,145	($13,718)
4		$9.82	$703	$7.03	$2,810	$9.82	$89	$2,197	$0	$2,197	($11,522)
5		$9.86	$706	$7.17	$2,866	$9.86	$90	$2,250	$0	$2,250	($9,272)
6		$9.91	$710	$7.31	$2,924	$9.91	$90	$2,304	$0	$2,304	($6,968)
7		$9.96	$713	$7.46	$2,982	$9.96	$91	$2,359	$0	$2,359	($4,609)
8		$10.01	$717	$7.60	$3,042	$10.01	$91	$2,416	$0	$2,416	($2,193)
9		$10.06	$720	$7.76	$3,103	$10.06	$91	$2,474	$0	$2,474	$281
10		$10.11	$724	$7.91	$3,165	$10.11	$92	$2,532	$0	$2,532	$2,813
11		$10.16	$728	$8.07	$3,228	$10.16	$92	$2,593	$0	$2,593	$5,406
12		$10.22	$731	$8.23	$3,292	$10.22	$93	$2,654	$0	$2,654	$8,060
13		$10.27	$735	$8.40	$3,358	$10.27	$93	$2,717	$0	$2,717	$10,777
14		$10.32	$739	$8.56	$3,425	$10.32	$94	$2,781	$0	$2,781	$13,557
15		$10.37	$742	$8.73	$3,494	$10.37	$94	$2,846	$0	$2,846	$16,403
16		$10.42	$746	$8.91	$3,564	$10.42	$95	$2,913	$0	$2,913	$19,316
17		$10.47	$750	$9.09	$3,635	$10.47	$95	$2,981	$0	$2,981	$22,296
18		$10.53	$754	$9.27	$3,708	$10.53	$96	$3,050	$0	$3,050	$25,346
19		$10.58	$757	$9.45	$3,782	$10.58	$96	$3,121	$0	$3,121	$28,467
20		$10.63	$761	$9.64	$3,858	$10.63	$97	$3,193	$0	$3,193	$31,660
								Net Present Value		$3,625	
								Internal Rate of Return		10.32%	

Conditions:

A1)	Discount rate	8.00%
A2)	Amount of energy required for cooling per year in million BTU	20.00
A3)	Amount of energy required for heating per year in million BTU	200.00
A4)	Cost to install the ground-source heat pump system	$20,000
A5)	Estimated additional maintenance costs/year for the ground-source system	$0.00
A6)	Energy cost for heat pump system in $s per million BTU	$9.67
A7)	Heating efficiency of the ground-source heat pump system	320.00%
A8)	Cooling efficiency of the ground-source heat pump system	220.00%
A8)	Annual rate increase for the energy to operate the heat pump system	0.50%
A9)	Energy cost to operate propane heating in $s per million BTU	$6.62
A10)	Heating efficiency of a propane system	50.00%
A11)	Annual rate increase for the energy to operate propane heating	2.00%
A12)	Energy cost to operate alternative cooling system in $s per million BTU	$9.67
A13)	Cooling efficiency of the alternative cooling system	220.00%
A14)	Annual rate increase for the energy to operate the alternative cooling system	0.50%

Assumptions:
1) Initial investment cost does not include the duct work that is necessary for the system to operate, it is assumed that the duct work will be required for all forced air heating and cooling system.
2) Estimated economic life of ground-source heat pump is 20 years.
3) Discount rate of 8.0%

Ground-Source vs. Propane
Simple Payback Period

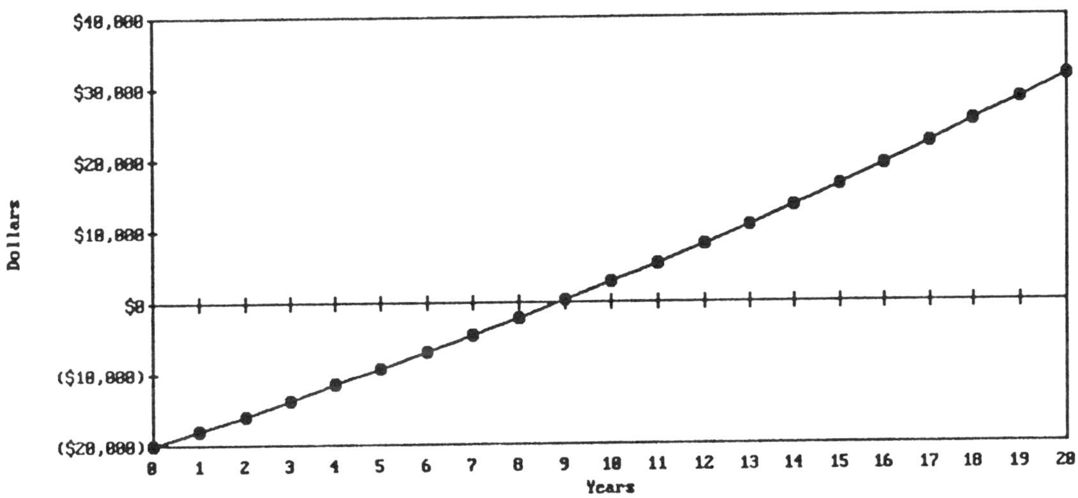

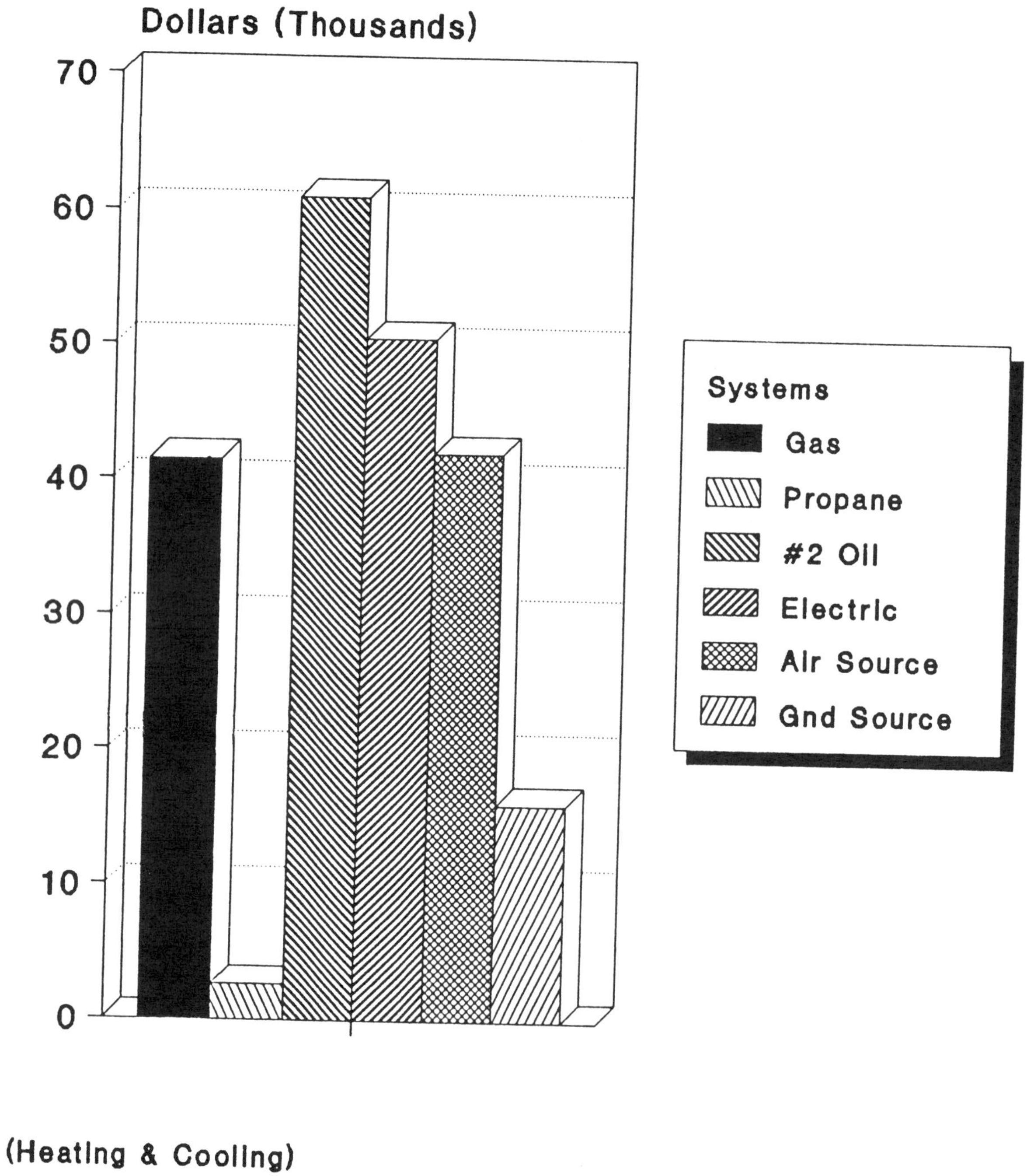

FIGURE 3. POLLUTION COSTS TO SOCIETY OVER TWENTY YEARS

Chapter 65

Marketing Opportunities With Thermal Storage

F.N. Broberg

TERMINOLOGY

Thermal storage is a principal that has been used by man over the ages to provide comfortable living conditions. For example, the Native Americans in the southwest built thick adobe cliff dwellings facing south. The adobe would store the winter sun's heat during the day to keep their homes warm at night. Thermal storage is a general term referring to storing thermal energy for heating and cooling applications. It is often used today to mean thermal energy storage for shifting operation of air conditioning equipment to off-peak electrical demand hours.

The general term was used in the title for familiarity, but a more accurate description: Off-Peak Air Conditioning (OPAC) will be used for the remainder of this paper to include the entire cooling system and the Thermal Energy Storage (TES) system .

BACKGROUND

Building owners today are increasingly concerned about environmental issues such as CFC's and Indoor Air Quality, as well as efficiency and reliability. Unfortunately, funds to make improvements to existing HVAC systems and additional funds for more efficient systems in new construction are limited especially in the present economy.

Electric utilities are also caught between the need to provide greater environmental controls and the decreasing industrial base. Over the last decade the country has changed to a predominantly white collar economy from one that was previously industrial. There has been a resulting shift to more office buildings which require most of their electrical power during the day, but little at night. Air conditioning has also become standard in buildings where it was seldom used 10 or 20 years ago. For these reasons, the electric utilities are offering incentives to shift air conditioning loads to off-peak periods for a more uniform load profile.

Off-Peak Air Conditioning (OPAC) is not new. Seventy years ago when electricity and refrigeration equipment were relatively expensive, many theaters and places of assembly were cooled with OPAC. Present electrical rates have once again made OPAC an attractive alternative. There have been numerous installations of ice storage and chilled water storage over the past ten years. Some of those installations also use low temperature air distribution to minimize duct size and reduce fan power requirements [6].

It is estimated that in 1992 $30 million worth of thermal storage tanks were sold. The manufacturers of ice storage equipment have had rapid growth in the past few years. For example, Calmac Corporation, who has about 45% of the ice storage market, grew about 50% in 1991 and about 10% in 1992, this is good considering the economy. The thermal storage industry only makes the tanks and not the piping, chillers, cooling towers, insulation, wiring, controls, and many other things necessary to install the system. If these were included, the total would be about $120 million.

TES Systems

The two major classifications of TES systems for OPAC applications are chilled water storage and ice storage. The disadvantages of chilled water storage are that it can take about four times the volume as an ice system, and the weight of the water can become a predominate factor. Where there is ample space outside the building, standalone tanks can be installed. To place a water storage tank within a building requires geotechnical investigation for the loading of the tank (which can be over 1,000 pounds per sq. ft.) and then the additional loading that is placed on structures and foundations. There is also the concern of leakage from a tank that is built in the field. There are, however, many advantages of chilled water storage. The major one is the higher efficiency of the system. The chillers can operate at higher temperatures and thus higher KW per ton and the need for heat exchangers are often eliminated. Both of these will reduce the operating cost of the system. There are many good references on chilled water storage and there have been many installations within the past 10 years. In general, chilled water storage should be considered on systems of 5,000 ton-hours and above. Over 20,000 ton-hours, chilled water storage would almost always be more cost effective than ice storage [16].

Ice Storage

There are three major types of ice storage equipment. One makes the ice on the evaporator and stores it on the evaporator. This is typically referred to as an ice builder. BAC has systems like this with large tanks filled with water and evaporator coils running within them. A second type makes ice on the evaporator and stores it remotely. This is called a shucker or harvester [17]. Mueller and Turbo are examples of these systems. The third type makes ice remotely and stores it remotely. This would be similar to a Calmac or Fafco system where brine or glycol solution is circulated through small coils in a tank filled with water and ice is built on the coils. These coils are commonly packaged into prefabricated round or rectangular tanks or may be placed in field built tanks.

Full Storage

A full storage system is one in which the TES system is designed to provide all of the cooling load during the utility company's on-peak period. This is cost effective if there is a large differential rate between on-peak and off-peak and the on-peak period is relatively short.

Partial Storage

A partial storage system requires that the primary air conditioning equipment (chillers, cooling towers and associated pumps) run during the on-peak period. The equipment may be sized at less than the full design load and use the TES to supply the balance of the load during on-peak periods. The cost and space requirements for TES often make partial storage the most cost effective.

I will not attempt to go into the details of all of the advantages and disadvantages of these systems. The Engineer who performs the study to determine if OPAC is a viable alternative should be familiar with these systems to be able to choose the system which best suits the application.

APPLICATIONS

The following is a partial list of applications for OPAC systems and a brief discussion of their relative merits and limitations:

Office buildings
Places of assembly
Educational facilities
Hospital operating rooms
Computer rooms
Industrial process
Roof top unit retrofit
Supermarkets
Insufficient electrical service
Increased cooling loads
CFC chiller replacement
Cooling tower replacement
District chilled water systems
Energy management

Office Buildings

The occupancy of office buildings is well suited to OPAC systems. The people, lighting and equipment loads consistently occur between 8:00 a.m. and 5:00 p.m. weekdays with little use during unoccupied hours. Unless the utility company offers a short on-peak period, a partial storage system will generally prove more cost effective than full storage.

Places of Assembly

Some places of assembly, such as churches, have very predictable loads and might use OPAC to downsize primary equipment and reduce the demand charge. Other places of assembly may have unpredictable schedules, for example, a convention center which is occasionally occupied all day may not be as cost effective.

Educational Facilities

Like office buildings, the occupancy is normally predictable and well suited for OPAC systems. Caution should be exercised for auditoriums and gymnasiums to determine what (and when) the maximum summertime occupancy might be.

Hospital Operating Rooms

In existing facilities, the combination of increasing heat

gains within operating rooms and heavier clothing to limit the spread of infections diseases has required additional cooling requirements. A partial OPAC system may allow the existing primary equipment to serve the increased load.

Procedures such as open heart surgery require much cooler temperatures. An ice TES can supply colder chilled water than conventional systems and easily satisfy the requirements.

The other consideration is reliability. Should there be a power failure or primary equipment failure, emergency generator capacity would only be required for the pumps and fans, not the chillers and cooling towers. The TES would provide the cooling for a predictable period.

Computer Rooms

Emergency cooling capacity is available just as with operating rooms on a loss of power or primary cooling equipment. The size of the TES can be designed for the expected downtime or the time it would take to shut down the facility in an orderly manner.

Industrial Process

Where there are predictable cooling requirements, OPAC can be used to keep from adding the HVAC demand to the process demand. For example, a gas turbine generator peaking plant uses ice TES to cool the intake air. This increases the efficiency of the plant, but does not decrease the plant capacity by running the chillers when the plant is in operation.

Rooftop Unit Retrofit

OPAC is normally thought to be applicable only to large chiller plants. Techniques have been developed to convert existing DX coils in rooftop units to glycol water coils and economically convert to OPAC [8].

Supermarkets

The humidity control is critical around refrigerated display cases to prevent condensation. OPAC and low temperature air distribution can help to maintain the low humidity required [4].

Insufficient Electrical Service

On an existing building with growing daytime electrical requirements, the use of OPAC will allow reduction of HVAC/electrical daytime loads. This will make available additional capacity for other loads which cannot be shifted to off-peak use.

Increased Cooling Loads

As electrical loads increase within a building, the air conditioning requirements also increase. In existing buildings, it is normally difficult to increase the size of the existing chilled water piping or duct risers. Without increasing the size of the risers, it would take larger pump and fan motors and use more energy to circulate the additional water and air required to increase the HVAC system capacity.

An alternate solution would be to circulate colder water and cooler air to increase system capacity without increasing the pump and fan motors or the distribution energy use. OPAC systems which use ice TES can easily supply 34°F. to 40°F. chilled water and cool 44° F. to 51°F. supply air [6, 13]. Special consideration should be given to the insulation system used on cold temperature piping and duct systems [9].

It has been calculated that for new construction, the use of cold chilled water and cool air distribution may actually decrease the initial construction cost of a building as well as save operating costs [13] for a comparison of conventional versus OPAC systems in three different localities).

CFC Chiller Replacement

It is now becoming increasingly apparent that existing CFC chillers will need to be either upgraded to non-CFC refrigerants or replaced with new chillers design for non-CFC refrigerants. Upgrading of existing chillers to non-CFC refrigerants often will cost 30% to 60% of the value of a new chiller. When the chiller is more than 10 years old, it is often not cost effective to upgrade the chiller. New chillers will normally have higher efficiencies and service will be more readily available. For any CFC upgrade, a study should be done which should evaluate not only the initial cost of the upgrade but the total operating and owning cost over the remaining life of the chiller versus the life of a new chiller. As more and more chillers are being replaced for CFC upgrades, there will be an increasing demand on the manufacturers of chillers, taxing their production capacity. Availability and lead time should be considered in any such evaluation.

If it is determined that a new chiller should be installed or if a new compressor is being installed on an existing chiller, there is a good opportunity to design the new chilled water system for OPAC. Chillers designed for ice making need to have a much higher lift than chillers designed for production of chilled water at normal temperature. This design should be optimized for the conditions which are expected during ice making. The chillers should also be efficient in operation when connected directly to the load (at conventional chilled water temperatures). Chillers designed for chilled water storage, on the other hand, operate very close to the operating conditions of a conventional chiller (about 38° F. to 40°F. minimum). Where chillers are already scheduled to be replaced, the cost of the chiller replacement may not need to be included in the cost of the OPAC conversion, only the cost for the TES system.

Cooling Tower Replacement
On an OPAC retrofit job the chillers are often downsized and the existing cooling towers would remain oversized. If the cooling tower is in good operating condition, it may remain as is or may need to have the distribution nozzles adjusted to the revised flow rates through the new chillers. In either case, the capacity of the tower will allow a closer approach in higher operating efficiencies for the chiller. If the cooling tower is not in good operating condition and needs replacement, then a smaller tower may be used.

If the only work in a particular building is replacement of the cooling tower, the Owner should be asked if any additional work, such as chiller replacement, is contemplated in the near future. Typically, the cooling towers and chillers were installed together and may need replacement at the same time. If both are to be replaced in the future, this would be a good time to recommend to the Owner that a study be done to see if OPAC should be considered for his building. It would not be prudent, for example, to replace the existing 2,000 ton cooling tower and two years later downsize it to 1,200 tons.

District Chilled Water Systems
Systems serving office buildings will have the same advantages for converting to OPAC as an individual building described above. The large TES system will generally be chilled water stored in a free standing or underground tank.

Energy Management
In the past "energy management systems" were little more than central time clocks, turning equipment on and off. Some equipment was even "duty cycled" on 10 minutes, off 20 minutes, etc. This increased the frequency of bearing and belt failures as well as compromising the environmental conditions in areas served by the cycled equipment.

With our present awareness of indoor air quality, the standards for ventilation air are being increased. The increased cooling loads can be provided with existing or new equipment augmented by OPAC systems. If the TES is sized adequately, the building management system can truly manage the energy use. Strategies can be programmed into the software to optimize when the ice is made and what chiller capacity to use during on-peak hours to stay below a pre-determined electrical demand. OPAC now allows the plant operator the opportunity to manage and control energy use without compromising building operation.

THE OPPORTUNITY
Before recommending OPAC for any of the above applications, it is important to consider the following:

Utility Company Incentives
Design Professionals' Experience
Local Experience
Owner's Expectations

Utility Company
In areas which have experienced a large growth of daytime peaking customers, utility companies have a dilemma. It sometimes takes 20 years to plan and build a new power plant with the need to obtain necessary permits from various environmental, agency, and public utility commissions. New transmission lines are also under more public scrutiny then ever before. The erosion of the steady industrial base load and replacement with office buildings having a high daytime peak demand overloads the existing distribution system each day, but leaves unused capacity available at night.

Utilities in this situation have responded by increasing their demand charge, the charge for the greatest 15 minute use during the day. Often this is ratcheted so that this maximum use (one month) will be reflected on the electric bills for the next year. This is significant when the demand (KW) portion of the monthly bill exceeds the use (KWH) portion.

Another method is to offer time of day and seasonal rates. Daytime (on-peak) demand and/or use rates might be significantly higher than nighttime (off-peak) rates. There may even be shoulder rates between the day and night rates. Rates might be different for the summer and winter seasons depending upon the utility company's needs.

More recently, utility companies have started offering cash incentives to engineers for OPAC Feasibility Studies and for the additional engineering services required. They are also offering cash payments to the Owner for a percentage of the additional costs of OPAC systems over a conventional system.

Some neighboring utility companies have such different customers and sources of power that their incentive programs are quite different. For example, in Connecticut, Northeast Utilities serves much of New England's industrial and residential customers and has a fairly steady load year round. The southern shore of Connecticut is an area characterized by rapid growth of office buildings with a high summer daytime peak, this is served by United Illuminating Company which has had heavy demands placed on their coal and oil-fired power plants. Each utility has designed their incentives around their needs. As an Engineer, I recommended energy conservation measures to reduce electrical use in Northeast Utilities areas and OPAC to reduce peak demand in United Illuminating Company areas, each one was consistent with the best interest of the utility company program and the client.

Design Professional

The design professional needs to be proficient in design of OPAC systems in order to conduct a comprehensive feasibility study. There are many sources of information available (see references). The designer must have a thorough understanding of standard heating, ventilating, and air conditioning (HVAC) systems and then also become familiar with various TES systems as well as control strategies.

Proper sizing of an OPAC system requires a higher level of detail and accuracy than sizing of a conventional system. This is because a conventional system usually runs about 8 to 12 hours each day and if it is slightly undersized, it can simply start up a little earlier each day. An OPAC system is sized much closer to its actual requirements and some partial systems might run 24 hours on a design day; there is no room for error. Even small internal gains that are present all night can have a large impact on an OPAC system.

The other important role of the design professional is to appraise the client of the advantages and disadvantages of the proposed OPAC system before the design is "locked in". This should include not only construction cost and operating cost information, but also personnel, maintenance and training requirements. All of these should be compared to a conventional system with which the client is familiar.

Local Experience

No matter how well a system is designed, it is up to the Contractor to install the system so that it performs as intended. The Engineer should check with local contractors who are bidding the project if there is any doubt about their ability to perform the work.

For example, if there is no local representative for a specified product, the Contractor may substitute with something that is not adequate. One such substitution of fiberglass insulation for the 22°F. glycol piping resulted in condensation in the insulation and replacement of all of the insulation within one year.

If there have been any OPAC systems operating satisfactorily in the area, others will be more easily accepted, but if one has ever had problems, few will be willing to "try" another one. This might be true even if totally different OPAC concepts are used.

Owner

The Owner is referred to in this paper as the one who engages the Engineer (direct or through an Architect) to design the HVAC system for a specific project. He is also responsible for completing the project within the allocated construction cost and for operation and maintenance after the project is complete.

Some Owners are motivated largely by first cost and any operating and maintenance savings would need to have a simple payback of less than three years to be considered. Other Owners have more difficulty obtaining operating funds, but would prefer to finance the additional construction cost; they might accept a simple payback up to 10 years for energy conservation measures.

If there are few or no OPAC systems near the proposed project, a conservative Owner may be reluctant to be the first one to "try" it. An innovative Owner may, on the other hand, use the OPAC system in marketing to environmentally or technologically conscientious prospective tenants for the building.

The major marketing opportunity consideration must be the utility incentives, without them there will be little reason for OPAC. The Engineer should then work directly with the Owner to determine if an OPAC system would meet his needs and expectations. If there is enough potential, the Engineer should request that a feasibility study be done to quantify the opportunity.

Moorhead Federal Building

This is an example of a 27-story office building built in Pittsburgh in 1960. It has two 1,000 ton chillers with an efficiency of about .9 KW per ton and one 2,000 ton cooling tower on the roof. Our office was requested to consider using variable frequency drives to run the two chillers. Our study noted that even though the efficiency could be improved, a greater increase in performance would be achieved by complete replacement of the 32 year old chillers with new high-efficiency non-CFC chillers. Upgrading of the existing chillers for use with non-CFC refrigerant would not have improved the efficiency. The construction cost for the new variable frequency drives would have been $300,000 and the cost for new chillers would have been about $600,000. After consideration by the Owner, it was decided that the chiller replacement was the better choice.

A second study was conducted to determine the optimum choice for the chiller replacement. It was determined that two 900 ton high efficiency chillers would save about $110,000 per year in operating costs. In addition, the study pointed out that conversion of the chilled water system to primary/secondary, with the secondary pumps having variable frequency control, would save more electrical energy.

It was during the study that we learned that the local electric utility company had a proposed rate pending which offered large incentives for OPAC systems which would shift all HVAC/electrical loads outside the 12 noon to 4 p.m. on-peak period. Chiller plants would be metered separately and only be charged the $14/KW demand charge during the on-peak period. There would be no demand charge outside of that period. Lower usage rates also applied during off-peak hours. This proposed rate was used with a preliminary OPAC design to simulate the potential energy use and savings. Taking into consideration only the additional cost of the TES system and controls, there would be an additional $100,000 per year savings if the new thermal storage rate was approved by the PUC. Even if the rate were never approved, the existing rate, with its $14/KW demand charge would allow a savings of about $40,000 per year. Based on the preliminary information, the Owner authorized us to proceed with design of an OPAC system. Additional studies utilizing the Comtech 3.0 software projected peak demand savings of about 1,000 KW and $88,000 per year for a full storage system. Projected savings for a 50% partial storage system were about 180 KW demand and $25,000 per year.

Studies were done to determine and optimize the best TES system to use. Chilled water was considered and after the geotechnical investigation to determine if the floor loadings were acceptable, a 5,000 ton-hour chilled water storage system was tentatively designed and priced at about $500,000 construction cost. Ice harvesters were considered, but the need for additional height ruled them out. Incremental ice storage systems such as Calmac, Fafco, and BAC were considered for a projected construction cost of about $400,000. The operating efficiency of the chilled water system would have been higher than that for the ice making systems, but the Owner selected the incremental ice TES because of savings in floor space and the ability to put in a partial system now and add onto it easily in the future. This would fit his budget limitations better than one large chilled water storage tank.

SUMMARY

The Engineer now has available software that can accurately simulate operation of various HVAC, OPAC and TES systems. Manufacturers of TES equipment and chillers have also developed software which can predict chilled water and ice making operations. This software should be not only used for the feasibility studies and life cycle cost analysis, but also for the final design of OPAC systems [15].

When an Owner and Engineer determine through a feasibility study that OPAC benefits everyone, the marketing or promoting of the concept becomes easier. The utility company is able to reduce the peak demand and avoid the cost of building new plants. This not only keeps the cost of generating electricity lower but also avoids the environmental impact of siting new plants. The user benefits by saving operating costs on his chiller plant and in the case of a new building, may also save on construction costs and floor space allocated to chases and mechanical rooms. This is truly a win-win situation, where properly applied.

Once an Engineer has demonstrated his experience in this field, it will help to distinguish him from other Engineers being considered for similar work. It is essential that a feasibility study be done before proceeding with design of an OPAC system, but Owners are sometimes reluctant to invest in a study when the results are uncertain. The local utility companies often assist by partially funding the study. This will allow the Engineer to thoroughly review the alternatives with the Owner prior to making a final selection of systems. After the scope of work is established, the Engineer can sometimes negotiate additional fees which would be outside the normal scope of work for engineering services; for example, the level of detail involved in design of the project and the additional field services required to insure that the contractor fully understands the differences between OPAC systems and conventional HVAC systems. There may also be additional commissioning and training requirements to assure that the system will function properly after the project is turned over to the Owner.

CONCLUSION

OPAC systems are not only feasible but are a proven alternative to conventional HVAC systems. Over 1,000 systems are currently operating with more being installed each year.

It is up to design professionals to lead the way by becoming proficient in OPAC systems so that they can recognize where they are suitable and to be able to explain their advantages and limitations to the Owner.

References

1. EPRI, EU.3024, Commercial Cool Storage
2. EPRI, BR-020252, Thermal Energy Storage
3. EPRI, EM-3981, Commercial Cool Storage Design Guide
4. EPRI, CU.3031.9.90, Supermarket Air Conditioning and Dehumidification
5. EPRI, EM-5545, Water Treatment Technologies for Thermal Storage Systems
6. EPRI, EM-5730, Cold Air Distribution Design Guide
7. EPRI, TR100945, Cool Storage Ethylene Glycol Design Guide
8. EPRI, TR-101038, Ice Storage Rooftop Retrofit Performance
9. ASHRAE Journal, May 1993, Cold Air Distribution and Concerns about Condensation, John J. Marmon, P.E., Hsig Chug Yu, P.E.

The following references were taken from *Thermal Energy Storage Systems,* Dated November 17-18, 1992

10. Load Calculations - Don Gatley
11. Cool Storage Strategies - Jim Denkmann
12. Economics - Chuck Dorgan
13. Cold Air Distribution - Chuck Dorgan
14. Energy Efficiency of TES - Chuck Dorgan
15. Analysis Tools - Jim Denkmann
16. Chilled Water Storage - Ian Mackie
17. Ice Harvesters - Jim Denkmann
18. Ice Module Selection - Don Gatley
19. Example Sizing of Static Brine TES - Don Gatley
20. Ethylene Glycol Chiller - Don Gatley
21. Modeling of Ice Storage Systems - Don Gatley
22. Application of Ice Builders - Jim Denkmann

About the Author

Frederick N. Broberg, P.E. is Project Engineer for H.F. Lenz Company in Johnstown, Pennsylvania. He has been the Project Engineer for numerous thermal storage systems as well as cool storage systems over the past 20 years. He is presently Project Engineer for design of a 5,000 ton-hour cool storage system in a 27-story office building in Pittsburgh, Pennsylvania.

Chapter 66

Thermal Energy Storage—The Natural Way

K. Scheiss

ABSTRACT

This paper introduces Thermal Energy Storage (TES), also known as Off-Peak Air Conditioning (OPAC) to the novice and the engineer. It draw the attention of designers or users to the many challenges (read problems) that some TES installations have encountered when "natural" occurrences have been ignored. Almost all TES systems are addressed and some recommendations are made for prospective TES users.

INTRODUCTION

These days it is in vogue to do things the "natural" way. How about TES the natural way? Let's accept the laws of nature and design systems accordingly. We know that hot air rises or water expands when it freezes, so let us not fight nature. If we keep that in mind, we would have less challenges (read "problems") in commissioning TES systems.

BASIC CONCEPTS

We need to be very specific about the two major concepts of TES:

1. The total storage capacity (Ton-Hours).

2. The rate of discharging and charging. The storage tank must perform as a chiller, this means it must satisfy cooling on a per time basis. The system must provide a certain cooling rate right at the end of its discharging cycle and it must also charge at a given rate.

We can have enough storage capacity (ton-hours), but if we cannot get the cooling out fast enough and also put it back in fast enough, then we have some unnatural challenges or "OS" problems. "OS" problems? Well, we have all seen that poster which shows the left rail track being connected to the right track and the two word caption under it "Oh, s...".

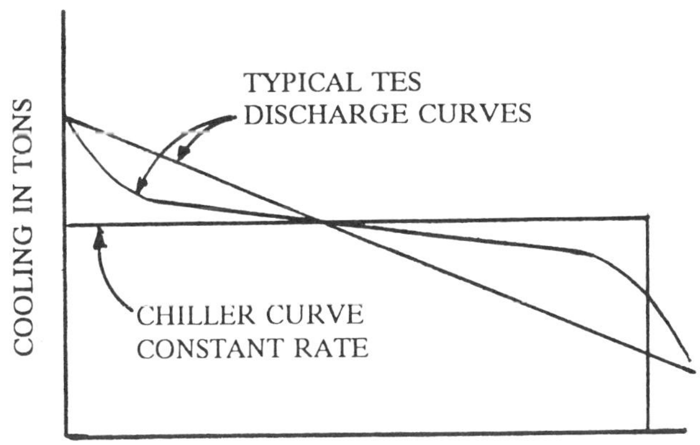

CHILLED WATER STORAGE

When storing chilled water, we attempt to separate water at different temperatures, preferably with a constant temperature difference and the larger the better. Usually air conditioning considerations limit this temperature difference (delta-T°) to about 20°F.

The challenge nature gives us is to separate the two bodies of water at different temperatures. We have gone through labyrinths, cascading tanks, empty tank system, separating membrane, but the trend now is to use the natural method of stratification. If certain conditions are adhered to, then the thermocline (layer of blending) can be kept to an acceptable depth. But what are these conditions?

To achieve proper stratification, we should have a constant delta-T^{O} between the in-coming and out-going water. If we allow the return temperature to vary, we get the natural phenomena of blending which will destroy the thermocline and thus reduce total storage capacity . Also during the charging cycle, the chiller will not "see" full charging delta-T^{O} and will not be able to load fully. To achieve a constant delta-T^{O}, a constant return water temperature is required, which leads to the use of variable flow pumping at least during the discharging cycle with 2-way valve controls and balancing to ensure a common constant return temperature.

Hydraulic Aspect

An open tank (open to atmosphere) constitutes an open surface and if located below the highest point in the distribution system it is "naturally" just a plain hole in the system. Worse there are two holes, one on the supply and one on the return. The supply is usually plugged by the check valve associated with the pump which is required to supply the system pressure in addition to any system resistance to be overcome. The return line also needs to be plugged otherwise the return water will drain very fast and we do get the pressure back which the pressurizing pump required to deliver water into the system. A pressure sustaining valve is required to keep the system pressure. But a pressure sustaining valve is nothing else but a "natural" way of creating resistance or pressure drop.

Considerable pumping energy is required (read wasted) during charging and discharging if the head is more than 30 feet. Atmospheric pressure allows us to accept vacuum conditions, but in practice it should be avoided.

Often this extra energy loss due to having to pump the water "up to the roof" is neglected. We all want TES to succeed but the feasibility study must provide realistic information. In high rise buildings this energy loss can be considerable and bite into the projected savings.

The "natural" way is to separate the open tank from the closed system with a flat plate heat exchanger. This works well if a large delta-T^{O} is available (ice systems). With chilled water storage systems however it cuts deeply into the efficiency of the tank. Even if we take an expensive flat plate heat exchanger and allow only a 2^{O} drop, the drop occurs twice, so we lose 4°. The 4° in 20° represents a 20% loss.

Energy Recovery

A quick word about recovering energy on the return with a turbine or a pump running in reverse. I have read articles of a manufacturer who claims a recovery of as close as 5% below the efficiency running as a pump. This may be true for one tiny point on a pump iso-efficiency curve. The chance to hit that point in practice is minute, impossible with a variable pumping system. The real recovery is in the 25% region.

In general, chilled water systems have a very good chance for favorable payback periods if the water volume can be combined with some other usage, such as using it as a fire reservoir.

EUTECTIC SALT STORAGE SYSTEMS

The open tank principle applies. When used as a new installation, we can specify the chiller for the charging conditions normally cooling water from eutectic salt freezing point of 47° down to 40°.

It is important to realize that this is a delta-T^{O} of 7° and not the 10°F - 18°F usually specified for chillers. For a retrofit application this fact must be kept in mind to estimate charging periods. If we push the design water quantity through the evaporator with a delta-T^{O} of 7°F, the chiller will run on part load and the "natural" charging period may take longer than one has planned. Also a chiller operating at part load of 7°/18°F to charge the tank is not such a good idea.

ICE-ON-COIL DX TYPE

Various ice-on-coil system are available. The ice-on-coil with direct expansion (DX) seems to be fading out because of the problems with measuring ice thickness. Also, the melting and refreezing occurs at a point farthest away from the freezing medium. To counter the efficiency loss due to higher energy consumption required by lower suction temperatures, computerized control sequences are proposed to predict to freeze only as much ice as expected for the next day. How do you know how much to freeze the night before. There will always be a weather forecast factor input to predict loads ahead. A margin for error is always evident.

The problems with uneven burning off and uneven freezing makes it difficult to tell the chiller when to stop. To counter this effect air is bubbled through the tank. Increasing the oxygen content in the water is an undesirable "natural" ingredient within a distribution system containing corrodible surfaces.

Also note that usually the ice-on-coil systems are contained in an open tank.

ICE-ON-COIL GLYCOL TYPE

If a mixture of glycol and water is used as the cooling medium inside the coil, various "natural" occurrences improve matters. The water to be frozen (storage medium) is stored in an open tank but with the cooling media (glycol mixture) inside the coils which can be pressurized. Efficiency is improved as the water will freeze and melt first closest to the coil transporting the glycol. The glycol inside the tube also affords the benefit of allowing a certain pressurization up to the limits of the plastic material used. So direct connection to the distribution system is possible.

Tanks that can be frozen solid without the danger of bursting are preferable to those where the controls have to protect the structural integrity of the tank. Controls do have a habit of failing.

Other "natural" occurrences with glycol systems: Glycol is used at temperatures as low as 20°F. Extra insulation may be required on piping. Valves may have to be heated to avoid freezing of condensate on the outside, which may physically hamper valve action.

Similarly, if it is possible that glycol below a temperature of 32°F may get through a flat plate heat exchanger, water may freeze on the other circuit and cause damage. This can occur towards the end of the charging cycle with very low load being supplied to the system.

ICE-IN- CONTAINERS

There are systems that encapsulate the water in containers usually made of plastic. Two important "natural" factors need to be considered:

1. Water expands considerably (approx. 8%) when freezing.

2. Flowing water chooses the path of least resistance.

The containers are usually placed into pressurized tanks. If the glycol flows in the vertical over the containers, equal flow patterns and even distribution seems assured if the containers expand or should move in relation to each other. However, if the flow is in the horizontal, the buoyancy of the ice affects the charging and discharging rate of the tank. The total ton-hour capacity is not affected, but as mentioned at the beginning, a TES system is also a chiller which has to deliver and absorb cooling energy at a certain rate. Also the containers should make allowance for expansion within the container, otherwise the plastic can be subjected to considerable cycling internal pressures.

ICE HARVESTERS

The ice harvester is a very simple method of storing energy. Ice is produced to fall into a tank and the heat transfer is large. The ice manufacturing industry quickly saw the opportunity to enter the TES market. They have been making ice for a long time, but there are subtle differences. The ice produced was removed to be used elsewhere. With TES we recirculate water, possibly introducing impurities from the distribution system. A heat exchanger reduces this problem and will also avoid problems associated with the open-tank principle.

Another of those "natural" effects that may cause problems is the fusion effect of ice under pressure. We all know how pressure makes a snowball. Some method of keeping the ice in the original small pieces should be provided. Incomplete burning of ice during extended periods of low cooling load requirements can lead to conditions where the fusion effect makes a solid lump of ice. "Channeling" occurs and the heat exchange surface is then reduced considerably. This is another potential "OS" problem where we have the storage capacity but we cannot get it out at the required rate.

SLUSH SYSTEMS

It is probably due to the problems with this fusion effect that has caused the slush system to disappear from the market. The idea of slush is great as we could pump slush, which is water with increased heat content due to the latent energy of the ice particles.

CHILLED WATER DISTRIBUTION SYSTEMS

Chilled water systems are very forgiving systems. In spite of that or perhaps because of it, chilled water systems are often unbalanced and not tuned for maximum efficiency. This occurs not only from the point of view to friction but also the balance between chiller capacity and water quantities being pumped around the system. If a chilled water system is unbalanced, the chances are very small that the TES system will solve balance problems at the same time. So, beware when adding a TES system to an existing system. One must do a lot of homework to check out the existing system. The best designed TES system will from that moment on be blamed for any problem that existed before.

CONCLUSION - NATURAL CAUSES

Why are we experiencing problems or challenges with TES systems? A lot of those challenges occur because we neglect these natural occurrences. There is a learning curve as well for designing TES systems, even though it may look easy at the beginning. It is those natural ingredients that make TES so spicy.

CONCLUSIONS - HUMAN CAUSES

Humans are also a natural ingredient to TES. Often there is a natural tendency of making things difficult. Some of us engineers have trouble making designs and operational sequences simple. Control sequences must be simple. The person who operates the system must understand it. Automatic controls have a tendency to fail or go out of calibration. Then the operation of the system reverts to the level of the competence of the operator. Another slogan that can affect TES performance: Tempoary measures tend to become permanent.

Often problems are caused by a lack of a single authority in the line of responsibility to get the system to function as designed. Where is the final authority to ensure the system will perform as expected by the client: Consultant, general contractor, storage manufacturer, chiller manufacturer, automatic controls or the Operator? It is our opinion that the design engineer should be in charge of the all the various trades until the system is commissioned.

It is our experience that TES systems must be designed by an engineer experienced in TES design. Too often capable engineers under-estimate the natural ingredients of TES. Support of manufacturers is given up to the point where the equipment is sold. It is then always expected that the automatic controls, (the famous black magic box) will sort out everything. The control people speak "controls", it still takes the experienced engineer to speak "TES" and instruct what and how the black control box is supposed to control. Control sequences must be simple.

It is our recommendation that the TES engineer commissions the system and even provides or thoroughly checks the "as-builts" drawings with the final set points and values achieved at the take-over clearly marked. We have repeatedly encountered projects where the operators had to struggle through years of trials and errors before reaching satisfactory results. Operators do change jobs. Clear directions with actual values must be available to the owner at a latger stage stating exactly how the system is supposed to operate.

Prospective owners can save themselves a lot of expense if they are prepared to employ an TES engineer for design and specification inclusive contract management and commissioning of the TES system on a direct basis.

Chapter 67

Procedure for Analyzing Performance of a Thermal Energy Storage System Using an Energy and Demand Balance Approach

R. Troncoso, S. Reddy

ABSTRACT

This paper presents the methodology developed to analyze the performance of a thermal energy storage system retrofit for a community college in Houston, Texas, under the Institutional Conservation Program (ICP). The study uses a system of energy and demand balance calculations to accurately determine energy use and demand profiles over all representative periods in a year. The advantage of this approach is that it reduces the uncertainty in TES analyses that use 'typical' load profiles, by providing a convenient and simple tool for developing load profiles unique to a building, using utility bills, demand data, site observations and one-time equipment power draw measurements. Since time and resources are always a constraint for most facilities undergoing a retrofit, this approach offers an inexpensive alternative to setting up an elaborate energy and demand monitoring system. Furthermore, the procedure demands minimal computer resources and expertise as all the required calculations can be easily performed on any standard spreadsheet program.

The following sections discuss the procedure in detail through an illustrative case study, involving the analysis of a chilled water storage system retrofit for San Jacinto College-North Campus, Houston, Texas. All representative tables used for analyses, a discussion of Houston Lighting & Power's (HL&P) rate schedule as it applies to a thermal storage retrofit, and calculations to determine savings that would accrue as a result of TES retrofit are also presented.

INTRODUCTION

Due to the lack of sufficient information about building load profiles, performance analysis of a TES system in retrofit applications involves considerable uncertainty. The uncertainty stems from the fact that, apart from the utility bills and demand profiles for a few days, for most facilities which undergo a TES retrofit, monitored end-use demand and energy use profiles are not available. The inadequate information causes engineers to rely on judgment and use of 'typical' patterns in determining hourly load profiles. Although load profiles based on aggregated energy use values and typical building load patterns provide a good first approximation, the hourly variation of loads, unique to a facility, are often masked by such an approximation, and can lead to an inaccurate analysis. Furthermore, computer programs that aid in TES analysis require extensive load profile information, spanning all daily, weekly, and seasonal variations.

The objective of this paper is to bridge this gap by providing an effective methodology for accurately determining building load profiles. This paper presents, through an illustrative case study, a detailed procedure for developing load profiles using a process of energy and demand balance calculations. The salient feature is the systematic approach that was used for establishing total energy use and peak demand values as the sum of individual energy use components. HVAC equipment operation was simulated as a function of outdoor temperature and building operation intensity, while non-HVAC loads were treated as a function of building operational intensity alone. These individual load components were then aggregated to generate a set of representative 24-hour load profiles, and calibrated with metered demand data. The energy and demand calibration process was fine-tuned until a satisfactory match was obtained. The technique was to isolate and calibrate equipment energy use values with the greatest uncertainty through a process of elimination using site data. Once a satisfactory energy and demand balance was achieved, TES system was simulated as an additional plant load during its charge cycle, and as a negative plant load during its discharge cycle over all representative 24-hour periods. A match between the sum of hourly charge and discharge rates indicates satisfactory performance. The model was then be used to analyze the economics of various TES operational strategies for different utility rate schedules.

DESCRIPTION OF CASE STUDY BUILDING

For the purpose of illustrating this approach, the case study of a community college recently analyzed for a cool storage (chilled water) retrofit is presented. The college is a 309,537 ft^2 facility located in Houston, Texas, consisting of five buildings for academic and administrative activities, and a central physical plant that serves the entire campus HVAC needs. The primary HVAC equipment consists of two chillers (750 ton and 500 ton) for cooling, and two steam boilers (250 HP and 300 HP) that produce hot water via a heat exchanger to meet campus heating requirements. Chilled water and hot water are distributed to individual buildings through a central piping and pumping network. Individual buildings have secondary pumps to boost water from the primary loop to feed heating and cooling coils in air handler units.

As part of the Technical Analysis for an ICP grant, several ECMs were analyzed and ECM dependencies were accounted for by analyzing measures affecting loads first, followed by systems, control, and lastly central plant. Since the TES comes towards the end in this sequence, it was analyzed by assuming that all other ECMs have already been implemented. It should however be noted that these dependencies have no effect on our discussion, and that it merely helps to explain the difference in energy use values between Tables 1 to 5 and Tables 6 through 10.

METHODOLOGY

Energy and demand balance is a systematic process of determining total campus energy use and demand as a sum of individual energy using components, and calibrating it with metered energy use obtained from utility bills and 15-minute demand profiles provided by the utility. This process serves to distinguish the two types of loads required for TES analysis:

1. Plant Load (the central plant energy profile includes chiller and accessories, primary chilled water pumps, and primary hot water pumps)
2. Base Load (all loads that are not plant loads; lighting, equipment, secondary HVAC systems etc. constitute the base load)

The following sections describe the process of demand and energy balance in detail. All calculations were performed using data from the most recent twelve month period, starting from September 1991 to August 1992. Houston bin temperature data was used to determine cooling and heating system operating hours.

Base Year Energy Use

Since the twelve month period from September 1991 to August 1992 was chosen as the base year, a brief discussion of energy use patterns during this period is beneficial in understanding the general trends in energy use over a year. Figure 1 shows On-Peak kW (as the sum of plant and base kW) and Figure 2 shows the electric energy consumption for all twelve months. It is evident from the high electric energy consumption during summer months compared to winter months, that cooling loads constitute a major portion of energy use. It is very interesting, however, to note that the peak demand and energy use for the year were reached in September and October respectively, and not in July or August which typically experience warmer weather. This phenomenon is attributed to the unusually high base loads (non weather dependent loads like lights, equipment etc.) at the beginning of each academic year due to new enrollments and admission/other associated activities. The overall pattern can be summed up as: a peak load and high energy use at the beginning of each academic year, followed by reduced loads from November through January due to both lower attendance (student population) as the fall session approaches its end, and cooler weather (since heating is on gas, no significant HVAC electrical loads are experienced during winter); and a steady increase from February on as spring semester starts, and weather begins to get warmer.

Demand Balance

Although demand profiles for four typical weeks spread over different weather conditions, provided by Houston Light & Power were useful in gaining an insight into the overall demand profile of the campus, very little was understood in terms of individual equipment load profile. Also, since the TES retrofit affects primary HVAC equipment alone, it was important to disaggregate plant loads from base loads.

Table 1 shows the procedure used to develop a 24-hour load profile for the design day (annual peak demand for the reference year 1991-92 was recorded on September 08, 1991). As shown in the table, campus cooling load is determined as a function of time of day and outside-air temperature (provided by HL&P for design day), and since there is no TES tank the entire load is passed on to the plant. Depending on the magnitude of load either the 500 ton chiller or the 750 ton chiller along with the respective condenser water pump and cooling tower fan are brought on line. Again, depending on the load, either one or two primary chilled water pumps are brought into operation. Column R in Table 1 shows plant load as the sum of chiller, cooling tower and pumps kW, and base load profile for design day is determined as the difference between metered kW and plant kW for the 24 hour period. Base load calculated for design day was used as the base load for establishing all other representative day profiles. However, the validity of this assumption was verified by measuring power of all primary and secondary equipment, lighting, major equipment loads and miscellaneous equipment loads as explained in the following section.

Once the base load profile was established, twelve 24 hour load profiles for each month were developed using hourly outside-air temperature from Test Meteorological Year (TMY) weather tape for Houston, Texas.

Energy Balance

The primary purpose of developing an energy balance chart was to arrive at an accurate energy consumption profile of the whole facility and its components. Further, the sophisticated rate structure employed by HL&P to calculate the institution's energy bill, makes energy use predictions based solely on demand profiles very questionable. Therefore, an energy balance was deemed necessary to provide more credibility to the analysis. Energy balance would help, not only in consolidating the energy use profile obtained using the demand profiles, but also in desegregating energy use by various components/buildings. Such a procedure where components energy consumption are determined individually and then aggregated to perform a global check with actual metered energy consumption for each month and a full year, ensure a high degree of accuracy and confidence in all assumptions and calculations employed in the analysis.

Plant operating and HVAC equipment operating schedules were obtained through consultations with facility maintenance staff and by inspecting load profiles. Lighting and miscellaneous equipment operating hours were determined based on information provided by college staff and observations made during our surveys. Energy use calculations of primary and secondary HVAC equipment were based on field measurements. An exhaustive inventory of all lighting and equipment installed was compiled, and name plate specifications combined with suitable diversity factors were used to calculate lighting and equipment energy consumption.

Energy balance calculations were made for each month of a year, and were divided into the following categories based on their functions:

1. Primary and Secondary HVAC Equipment

Primary equipment consists of two chillers (750 and 500 ton), primary chilled water pumps (two 100 hp), primary hot water pump (one 50 hp), condenser pumps (30 hp and 40 hp) and cooling tower fans (two 40 hp). Secondary HVAC equipment consists of secondary chilled and hot water pumps, air handler units, fan coil units, exhaust and ventilation fans.

A two step procedure was used to determine total energy use. First, primary equipment energy use was determined using Table 2 and then values calculated in these tables were plugged into Table 3 to calculate the sum of energy use by all primary and secondary HVAC equipment. Tables were set up on a spreadsheet to perform repetitive calculations for each month and the whole year. A brief explanation of the calculations performed for the whole year is shown, with a similar procedure used for individual months.

2. Lights

During our several surveys, a thorough listing of all lighting fixtures and their wattage was compiled. This information was processed using spreadsheets to calculate the energy consumption based on lighting wattage and operating hours. Calculations were performed for each wing/building for all twelve months, and the values aggregated to determine whole building lighting energy use for each month and a whole year.

3. Equipment

An exhaustive inventory of all computers, and other office equipment such as copiers, printers etc., and other major equipment such as vending machines, hair dryers etc. was compiled by our audit team. Despite our laborious efforts, a complete inventory could not be obtained. Therefore, based on our on site observations, past experience and recommendations of researchers, the following equipment wattage were assumed for the buildings/wings:

Main Building: Fine Arts wing-0.3 W/ft^2, Administration, Brightwell, and Wheeler wings-0.7 W/ft^2, Learning Resource Center-0.5 W/ft^2, Physical Education Building-0.15 W/ft^2. Equipment was assumed to operate on class room schedule, but a diversity factor of 25% (only 75% operating at any given time) was applied since some of the equipment may be turned off at any given time.

Having arrived at this value, our next step was aimed at improving, further, the credibility of our equipment energy use calculations by accounting for all known equipment compiled during our site surveys. To accomplish this task, audited equipment was subdivided into two categories for analysis:

Computers: Power draw by PCs was estimated at 150 W per PC (reference 3), printers were estimated to draw 25W [reference 1] a piece, and copiers at 2000 W (reference 1) a piece. Based on our audit observations, only 50% of all computers in student labs were assumed to be on at any given time. A diversity of 50% was applied to copiers since they are not in continuous use. Operating hours were estimated at 2200 hrs per year based on a 9 hr. week- day and 4 hr Saturday operation.

General Equipment: Power draw for vending machines and hair dryers was obtained from name plate specifications. Standard ASHRAE recommended values were used for cash registers power draw. Vending machines were assumed to operate all round the year, but a diversity factor of 20% (80% usage) was estimated for soft drink vending machines and ice cream machines to account for thermostat control of their refrigeration systems.

Miscellaneous Equipment: The remaining equipment energy use unaccounted by the audited equipment is termed as miscellaneous equipment energy use. It amounts to about 5.6% of the total campus energy use, and includes kitchen equipment, water fountains, fax machines, lab equipment, audio-visual teaching aids etc. Due to the diverse nature of miscellaneous equipment and their operation, a different approach was taken in distributing the energy use over the year. As discussed earlier, the monthly demand and energy use patterns of the whole campus have a unique bi-annual feature; where at the start of each semester, fall and spring, there are higher loads that gradually ebb as the level of activity decreases. It is assumed that much of the miscellaneous loads occur during these months (regular semesters), and an activity factor based on monthly energy use values was estimated for such months. The energy use was then divided among the participating months, proportional to their respective activity factors.

4. Other Buildings and Miscellaneous Campus Loads
Miscellaneous campus loads include energy use of other buildings on campus such as the day care center, natatorium (swimming pool), workshops and central plant, and other energy users such as exterior lighting, swimming pool pump etc. These loads were estimated to make up the difference between the reference year energy consumption and the calculated energy use for all other buildings. Total miscellaneous energy use was estimated at 363,976 kWh per year. An approach similar to that used for miscellaneous equipment was used to distribute miscellaneous campus loads over all twelve months, accounting for higher use of exterior lighting during winter months compared to summer months.

Table 4 shows a summary of lighting and equipment energy use by different buildings/wings for each month.

Comparison of calculated monthly energy use and energy use from monthly utility bills (Figure 2) indicate an excellent match between the two values, and lends credibility to all further analysis.

SIMULATION OF TES MODEL

Once the 24 hour representative load profiles are accurately determined, the next step is to incorporate the thermal storage tank into the campus energy model, and simulate it to project monthly campus peak demand values subsequent to the installation of TES. The TES system is simulated by incorporating an additional column for the tank in the campus energy profile model, and is modeled as an additional plant load during the charge period, and as a negative load during the discharge period. The process is discussed for two scenarios:

Scenario 1: Table 5 shows campus profile for the design day after all ECMs (upto and including high eff. chiller but excluding TES tank) have been incorporated.

Col. H and Col. I simulate the proposed 650 ton chiller.

Col. L and Col. M reflect central plant piping and pumping modifications (Central Plant - ECM1).

Col. R shows the new plant kW

Col. S shows the new base kW. The new base kW is derived from base kW by subtracting kW reduction due to all previous ECMs from design base kW during all operating hours (hours 7 through 22).

Note that storage tank is not simulated, and hence Col.E is inactive.

On-Peak hours window: 8 a.m. to 10 p.m.

Peak demand: 1,327 kW at Hour 12 (Col.T)

Scenario 2: Table 6 shows campus load profile for the design day after all ECMs and with TES. With the TES option, utility On-Peak window now changes to 1 p.m. to 8 p.m.; therefore the storage tank is simulated in its discharge cycle during these hours (negative number, hence reduces plant load) and as an additional plant load during Off-Peak hours to simulate tank charging. Since this ECM's objective is shifting chilled water plant demand entirely from On-Peak hours to Off-Peak hours, the tank meets all campus cooling loads during the On-Peak window, effectively reducing all plant loads except secondary chilled water pumps to zero. A match between the positive and negative numbers (charge and discharge tons) in Col.E indicates satisfactory performance of TES. The model assumes tank losses to be negligible.

On-Peak hours window: 1 p.m. to 8 p.m.

Peak demand: 751 kW at Hour 20 (Col.T)

Table 6 was then extended to all typical days for each month of a year to determine peak demand for a typical twelve month period, as required to perform an HL&P cost savings analysis. Table 7 shows energy and demand savings due to the TES retrofit (includes replacement of old 500 ton chiller with high efficiency 650 ton chiller). Note that energy savings are due to the chiller alone, while demand savings are due to both TES and the chiller.

UTILITY RATE ANALYSIS

A spreadsheet was developed to determine energy costs according to HL&P Large General Service (LGS) Rate Schedule. In the pre-TES scenario, a ratchet clause defines primary kVA as the highest of monthly On-Peak kVA or Off-Peak kVA or 85% of annual On-Peak kVA or 600 kVA. Table 8 calculates energy costs under this scenario. With a TES option, primary kVA (billed kVA) is the peak 15 minute demand (or 600 kVA whichever is higher) recorded between 1 p.m. and 8 p.m. for each month. Table 9 calculates energy costs under this scenario.

Total Cost Savings = \$337,455/yr - \$244,595/yr
= \$92,860/yr
Implementation Costs = \$754,637 [reference 2]
HL&P Rebate = \$210,000
(for approx. 700 kW peak demand reduction)
Payback = (\$754,637 - \$210,000)/\$92,860
= 5.9 years

CONCLUSIONS AND RECOMMENDATIONS

Energy and demand balance offers significant advantage over normally used typical load profiles by aiding in a quick and more accurate TES analysis. The approach offers an inherent advantage when analyzing a large campus with many buildings, compared to using a modeling and calibrating tool such as DOE-2 or other energy analysis software which require considerable time and expertise to obtain similar results. Also, unlike commercially available software which are applicable either to hot or cold storage alone, the generic nature of energy and demand balance approach, allows it to be used effectively for both hot and cold storage system performance analysis.

Furthermore, field measurements and other input/data required for this approach are very easily and inexpensively obtained. However, if time and resources permit, continuous monitoring of major equipment continuously over a 24 hour period could offer significant advantages over the one time measurements used in this analysis.

REFERENCES

1. ASHRAE, *1989 Handbook of Fundamentals*, American Society of Heating, Refrigeration, and Air-Conditioning Engineers, Atlanta, Georgia, 1989.
2. ACR Engineering, Inc., *Technical Assistance Report, San Jacinto College North, Physical Plant Building*, ACR Engineering, Inc., report to the State of Texas, Governor's Energy Office, March 1993.
3. Norford, L.K., Rabl, A., Harris, J., and Roturier, J., "The Sum of Megabytes Equals Gigawatts: Energy Consumption and Efficiency of Office PCs and Related Equipment, "*Proceedings of the 1988 ACEE Summer Study*, Vol. 3, American Council for an Energy Efficient Economy, Washington, DC, August, 1988.

TABLE 1
DAILY PLANT KW PROFILE IN EXISTING CASE AND WITHOUT TES

MO/DAY September 08, 1992 - Peak-Day

A	B	C	D	E	F	G	H	I	J	K	L	M	N	O	P	Q	R	S	T	U
TIME OF DAY	OAT	Pct. Full Load	Total Campus Tons	Storage Tank Tons	Load on Plant Tons	Pct Full Cap.	YORK Chiller-1 500 Tons	YORK Chiller-1 kW	YORK Chiller-2 750 Tons	YORK Chiller-2 kW	CHWP-1 kW	CHWP-2 kW	CWP-1 kW	CWP-2 kW	CTFAN-1 kW	CTFAN-2 kW	Plant kW	Sep-8 Base kW	Total kW	Sep-8 Metered kW
1	78	0.00	0	0	0	0	0	0	0	0	0	0	0	0	0	0	0	280	280	280
2	77	0.00	0	0	0	0	0	0	0	0	0	0	0	0	0	0	0	279	279	279
3	76	71.58	501	0	501	72	0	0	501	407	70	70	0	26	0	25	597	305	902	902
4	76	71.58	501	0	501	72	0	0	501	407	70	70	0	26	0	25	597	323	920	920
5	76	71.58	501	0	501	72	0	0	501	407	70	70	0	26	0	25	597	428	1,025	1025
6	76	71.58	501	0	501	72	0	0	501	407	70	70	0	26	0	25	597	412	1,009	1009
7	76	71.58	501	0	501	72	0	0	501	407	70	70	0	26	0	25	597	743	1,340	1340
8	77	73.04	511	0	511	73	0	0	511	416	70	70	0	26	0	25	606	1,003	1,609	1609
9	82	79.87	559	0	559	80	0	0	559	461	70	70	0	26	0	25	651	993	1,644	1644
10	86	84.82	594	0	594	85	0	0	594	495	70	70	0	26	0	25	685	1,040	1,725	1725
11	89	88.25	618	0	618	88	0	0	618	520	70	70	0	26	0	25	710	1,004	1,714	1714
12	90	89.34	625	0	625	89	0	0	625	528	70	70	0	26	0	25	718	1,051	1,769	1769
13	89	88.25	618	0	618	88	0	0	618	520	70	70	0	26	0	25	710	976	1,686	1686
								0												
14	88	87.14	610	0	610	87	0	0	610	512	70	70	0	26	0	25	702	895	1,597	1597
15	78	74.46	521	0	521	74	0	0	521	425	70	70	0	26	0	25	615	845	1,460	1460
16	78	74.46	521	0	521	74	0	0	521	425	70	70	0	26	0	25	615	820	1,435	1435
17	82	79.87	559	0	559	80	0	0	559	461	70	70	0	26	0	25	651	758	1,409	1409
18	85	83.63	585	0	585	84	0	0	585	487	70	70	0	26	0	25	677	716	1,393	1393
19	84	82.40	577	0	577	82	0	0	577	478	70	70	0	26	0	25	668	858	1,526	1526
								0												
20	81	78.56	550	0	550	79	0	0	550	452	70	70	0	26	0	25	642	913	1,555	1555
21	79	75.86	531	0	531	76	0	0	531	434	70	70	0	26	0	25	624	923	1,547	1547
22	77	0.00	0	0	0	0	0	0	0	0	0	0	0	0	0	0	0	575	575	575
23	77	0.00	0	0	0	0	0	0	0	0	0	0	0	0	0	0	0	340	340	340
24	77	0.00	0	0	0	0	0	0	0	0	0	0	0	0	0	0	0	262	262	262

TABLE 2
ANNUAL CENTRAL PLANT ENERGY CONSUMPTION - EXISTING CASE

A	B	C	D	E	F	G	H	I	J	K	L	M	N	O	P	Q	R
Temp. Bin	Plant HRS /Year	Systems HRS /Year	Pct. Full Load	Total Tons	YORK Chiller-1 500 Tons	YORK Chiller-1 kW	YORK Chiller-2 750 Tons	YORK Chiller-2 kW	CHWP-1 kW	CHWP-2 kW	CWP-1 kW	CWP-2 kW	CTFAN-1 kW	CTFAN-2 kW	HWP-4 kW	Total kW	Total kWh
107	0.00	0.00	0.00	0	0	0	0	0	0	0	0	0	0	0	0	0	0
102	0.82	0.82	100.00	700	0	0	700	612	70	70	0	26	0	25	0	803	657
97	32.49	32.84	96.00	672	0	0	672	580	70	70	0	26	0	25	0	771	25,091
92	236.36	240.38	92.00	644	0	0	644	548	70	70	0	26	0	25	0	739	175,349
87	457.57	470.57	88.00	616	0	0	616	518	70	70	0	26	0	25	0	709	326,415
82	544.92	574.36	80.00	560	0	0	560	462	70	70	0	26	0	25	0	653	359,737
77	709.48	757.51	72.00	504	0	0	504	409	70	70	0	26	0	25	0	600	432,577
72	570.37	607.17	64.00	448	448	378	0	0	70	0	21	0	25	0	0	494	284,298
67	436.64	465.07	56.00	392	392	319	0	0	70	0	21	0	25	0	0	435	191,944
62	348.74	372.16	48.00	336	336	267	0	0	70	0	21	0	25	0	0	383	135,123
57	291.62	311.32	41.00	287	287	226	0	0	70	0	21	0	25	0	0	342	101,150
52	245.33	262.36	27.00	189	189	158	0	0	70	0	21	0	25	0	0	274	68,422
47	181.29	194.24	14.00	98	98	109	0	0	70	0	21	0	25	0	35.1	260	48,508
42	120.14	127.84	0.00	0	0	0	0	0	0	0	0	0	0	0	35.1	35	4,487
37	65.22	69.49	0.00	0	0	0	0	0	0	0	0	0	0	0	35.1	35	2,439
32	26.93	28.31	0.00	0	0	0	0	0	0	0	0	0	0	0	35.1	35	994
27	7.66	7.99	0.00	0	0	0	0	0	0	0	0	0	0	0	35.1	35	280
22	2.49	2.66	0.00	0	0	0	0	0	0	0	0	0	0	0	35.1	35	93
17	0.40	0.40	0.00	0	0	0	0	0	0	0	0	0	0	0	35.1	35	14
12	0.00	0.00	0.00	0	0	0	0	0	0	0	0	0	0	0	35.1	35	0
	4,279	4,526	---														2,157,579

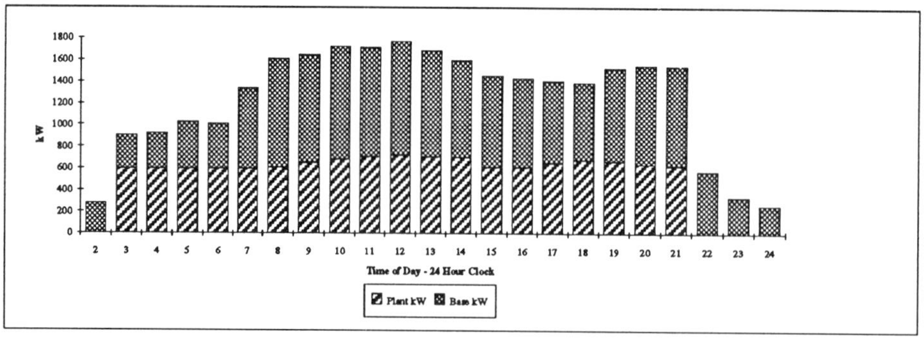

FIGURE 1: DESIGN DAY LOAD PROFILE

TABLE 3
ANNUAL PRIMARY AND SECONDARY HVAC SYSTEMS ENERGY CONSUMPTION

A	B	C	D	E	F	G	H	I	J	K	L	M	N	O	P	Q	R	S
	Plant	Systems	Primary Equipment			Secondary Pumps/AHUs			Fan Coil	Exhaust	Total	Total	Total	Cooling	Heating	AHUs/Fans	Total	Dist.Sys.
Temp. Bin	HRS /Year	HRS /Year	Chill. Plant kW	CHWPump kW	HWPump kW	AHUs kW	CHWPump kW	HWPump kW	kW	kW	Cooling kW	Heating kW	AHUs/Fans kW	Energy kWh	Energy kWh	Energy kWh	Dist.Sys kW	Energy kWh
107	0.00	0.00	0	0	0	181.5	74.9	0	24.1	9	74.9	0	214.6	0	0	0	290	0
102	0.82	0.82	663	140	0	181.5	74.9	0	24.1	9	878.1	0	214.6	719	0	164	1,093	883
97	32.49	32.84	631	140	0	181.5	74.9	0	24.1	9	845.7	0	214.6	27,550	0	6,551	1,060	34,101
92	236.36	240.38	599	140	0	181.5	74.9	0	24.1	9	814.4	0	214.6	193,354	0	47,950	1,029	241,304
87	457.57	470.57	569	140	0	181.5	74.9	0	24.1	9	784.3	0	214.6	361,660	0	93,867	999	455,527
82	544.92	574.36	513	140	0	181.5	74.9	0	24.1	9	727.5	0	214.6	402,756	0	114,570	942	517,326
77	709.48	757.51	460	140	0	181.5	74.9	0	24.1	9	675.1	0	214.6	489,314	0	151,104	890	640,418
72	570.37	607.17	424	70	0	181.5	74.9	0	24.1	9	568.8	0	214.6	329,774	0	121,115	783	450,890
67	436.64	465.07	365	70	0	181.5	74.9	0	24.1	9	509.9	0	214.6	226,778	0	92,770	725	319,548
62	348.74	372.16	313	70	0	181.5	74.9	0	24.1	9	457.7	0	214.6	162,998	0	74,237	672	237,234
57	291.62	311.32	272	70	0	181.5	74.9	0	24.1	9	417.0	0	214.6	124,468	0	62,101	632	186,569
52	245.33	262.36	204	70	0	181.5	74.9	0	24.1	9	348.9	0	214.6	88,073	0	52,334	564	140,407
47	181.29	194.24	155	70	35.1	181.5	0	35.4	24.1	9	225.0	70.5	214.6	41,691	13,694	38,746	510	94,130
42	120.14	127.84	0	0	35.1	181.5	0	35.4	24.1	9	0.0	70.5	214.6	0	9,013	25,501	285	34,514
37	65.22	69.49	0	0	35.1	181.5	0	35.4	24.1	9	0.0	70.5	214.6	0	4,899	13,862	285	18,761
32	26.93	28.31	0	0	35.1	181.5	0	35.4	24.1	9	0.0	70.5	214.6	0	1,996	5,647	285	7,643
27	7.66	7.99	0	0	35.1	181.5	0	35.4	24.1	9	0.0	70.5	214.6	0	563	1,594	285	2,157
22	2.49	2.66	0	0	35.1	181.5	0	35.4	24.1	9	0.0	70.5	214.6	0	188	531	285	718
17	0.40	0.40	0	0	35.1	181.5	0	35.4	24.1	9	0.0	70.5	214.6	0	28	80	285	108
12	0.00	0.00	0	0	35.1	181.5	0	35.4	24.1	9	0.0	70.5	214.6	0	0	0	285	0
Year	4,278	4,525	1,696,882	445,570	15,126	752,928	306,683	15,255	109,064	40,729				2,449,134	30,381	902,722	1,093	3,382,237

Campus HVAC Electric Energy Use = 3,382,237 kWh

Primary Equipment		**Secondary Equipment**	
Plant	1,696,882	AHU	752,928
CHWP	445,570	CHWPump	306,683
HWP	15,126	HWPump	15,255
		Fan Coil	109,064
		Exhaust	40,729
Totals	2,157,578		1,224,660

TABLE 4

LIGHTS

	FINE ARTS	ADMIN	BRIGHTWELL	WHEELER	LRC	GYM
Jan	27,037	26,824	26,974	29,239	22,480	15,180
Feb	25,137	25,271	25,835	27,519	21,644	14,154
Mar	28,231	27,963	28,062	30,485	23,366	15,845
Apr	28,231	27,963	28,062	30,485	23,366	15,845
May	26,332	26,410	26,923	28,765	22,531	14,819
Jun	28,231	27,963	28,062	30,485	23,366	15,845
Jul	28,937	28,377	28,112	30,960	23,316	16,207
Aug	27,526	27,550	28,010	30,010	23,418	15,484
Sep	27,037	26,824	26,974	29,239	22,480	15,180
Oct	27,526	27,550	28,010	30,010	23,418	15,484
Nov	23,454	23,406	23,713	25,503	19,820	13,185
Dec	21,771	21,542	21,589	23,486	17,996	12,217
TOTAL	319,453	317,646	320,321	346,187	267,201	179,446

EQUIPMENT

HRS	FINE ARTS	ADMIN	BRIGHTWEL	WHEELER	LRC
184	1,924	2,067	1,555	1,141	4,081
172	1,917	1,977	1,457	1,066	4,052
192	1,929	2,127	1,621	1,190	4,100
192	1,929	2,127	1,621	1,190	4,100
185	1,924	2,074	1,563	1,147	4,083
192	1,929	2,127	1,621	1,190	4,100
196	1,931	2,157	1,654	1,215	4,109
188	1,926	2,097	1,588	1,166	4,090
184	1,924	2,067	1,555	1,141	4,081
188	1,926	2,097	1,588	1,166	4,090
164	1,912	1,917	1,391	1,017	4,033
148	1,902	1,797	1,260	918	3,994
TOTAL	23,071	24,631	18,475	13,547	48,912

ALL TOTALS

LIGHTS TOTAL	EQUIP. TOTAL	HVAC TOTAL	ALL TOTALS
147,735	11,906	200,789	387,396
139,560	11,607	201,464	444,681
153,953	12,105	255,098	473,837
153,953	12,105	301,075	506,857
145,780	11,930	311,191	456,971
153,953	12,105	355,926	535,793
155,909	12,204	371,266	553,089
151,998	12,005	356,805	521,760
147,735	11,906	331,347	569,783
151,998	12,005	301,804	583,374
129,082	11,408	219,594	439,377
118,601	11,009	175,683	365,164
1,750,254	142,294	3,382,242	5,858,100

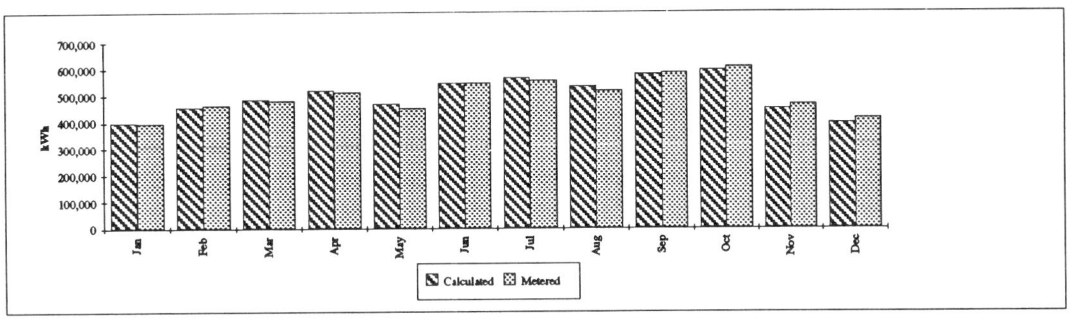

FIGURE 2: CALCULATED VS METERED BASE YEAR ENERGY USE

TABLE 5
DESIGN DAY CAMPUS LOAD PROFILE WITHOUT TES

MO/DAY: September 08, 1991 - Peak-Day / Design Day

A	B	C	D	E	F	G	H	I	J	K	L	M	N	O	P	Q	R	S	T
TIME OF DAY	OAT	Pct. Full Load	Total Campus Tons	Storage Tank Tons	Load on Plant Tons	Pct Full Cap.	NEW Chiller-1 650 Tons	NEW Chiller-1 kW	YORK Chiller-2 750 Tons	YORK Chiller-2 kW	Primary CHWP kW	Sec'dary CHWP kW	CWP-1 kW	CWP-2 kW	CTFAN-1 kW	CTFAN-2 kW	NEW Plant kW	NEW Base kW	NEW Total kW
1	78	0.00	0	0	0	0	0	0	0	0	0	0	0	0	0	0	0	280	280
2	77	0.00	0	0	0	0	0	0	0	0	0	0	0	0	0	0	0	279	279
3	76	0.00	0	0	0	0	0	0	0	0	0	0	0	0	0	0	0	902	902
4	76	0.00	0	0	0	0	0	0	0	0	0	0	0	0	0	0	0	920	920
5	76	71.58	501	0	501	72	501	277	0	0	22	19	21	0	25	0	365	428	793
6	76	71.58	501	0	501	72	501	277	0	0	22	19	21	0	25	0	365	412	777
7	76	71.58	459	0	459	66	459	250	0	0	22	19	21	0	25	0	338	581	918
8	77	73.04	469	0	469	67	469	256	0	0	22	20	21	0	25	0	345	841	1,186
9	82	79.87	517	0	517	74	517	288	0	0	22	25	21	0	25	0	381	831	1,212
10	86	84.82	552	0	552	79	552	312	0	0	22	30	21	0	25	0	411	877	1,288
11	89	88.25	576	0	576	82	576	330	0	0	22	34	21	0	25	0	432	841	1,273
12	90	89.34	583	0	583	83	583	336	0	0	22	35	21	0	25	0	439	888	1,327
13	89	88.25	576	0	576	82	576	330	0	0	22	34	21	0	25	0	432	813	1,245
14	88	87.14	568	0	568	81	568	324	0	0	22	33	21	0	25	0	425	732	1,157
15	78	74.46	479	0	479	68	479	263	0	0	22	21	21	0	25	0	352	683	1,035
16	78	74.46	479	0	479	68	479	263	0	0	22	21	21	0	25	0	352	658	1,010
17	82	79.87	517	0	517	74	517	288	0	0	22	25	21	0	25	0	381	596	977
18	85	83.63	543	0	543	78	543	306	0	0	22	29	21	0	25	0	403	554	957
19	84	82.40	535	0	535	76	535	300	0	0	22	28	21	0	25	0	396	695	1,091
20	81	78.56	508	0	508	73	508	282	0	0	22	24	21	0	25	0	374	751	1,125
21	79	75.86	489	0	489	70	489	269	0	0	22	22	21	0	25	0	360	761	1,120
22	77	73.04	0	0	0	0	0	0	0	0	0	0	0	0	0	0	0	413	413
23	77	73.04	0	0	0	0	0	0	0	0	0	0	0	0	0	0	0	340	340
24	77	0.00	0	0	0	0	0	0	0	0	0	0	0	0	0	0	0	262	262

TABLE 6
DESIGN DAY CAMPUS LOAD PROFILE WITH TES

MO/DAY: September 08, 1991 - Peak-Day / Design Day

A	B	C	D	E	F	G	H	I	J	K	L	M	N	O	P	Q	R	S	T
TIME OF DAY	OAT	Pct. Full Load	Total Campus Tons	Storage Tank Tons	Load on Plant Tons	Pct Full Cap.	NEW Chiller-1 650 Tons	NEW Chiller-1 kW	YORK Chiller-2 750 Tons	YORK Chiller-2 kW	Primary CHWP kW	Sec'dary CHWP kW	CWP-1 kW	CWP-2 kW	CTFAN-1 kW	CTFAN-2 kW	NEW Plant kW	NEW Base kW	NEW Total kW
1	78	0.00	0	650	650	93	650	389	0	0	22	0	21	0	25	0	457	280	737
2	77	0.00	0	650	650	93	650	389	0	0	22	0	21	0	25	0	457	279	736
3	76	71.58	501	149	650	93	650	389	0	0	22	19	21	0	25	0	477	902	1,379
4	76	71.58	501	149	650	93	650	389	0	0	22	19	21	0	25	0	477	920	1,397
5	76	71.58	501	82	583	83	583	335	0	0	22	19	21	0	25	0	423	428	851
6	76	71.58	501	0	501	72	501	277	0	0	22	19	21	0	25	0	365	412	777
7	76	71.58	459		459	66	459	250	0	0	22	19	21	0	25	0	338	581	918
8	77	73.04	469	0	469	67	469	256	0	0	22	20	21	0	25	0	345	841	1,186
9	82	79.87	517	0	517	74	517	288	0	0	22	25	21	0	25	0	381	831	1,212
10	86	84.82	552	0	552	79	552	312	0	0	22	30	21	0	25	0	411	877	1,288
11	89	88.25	576	0	576	82	576	330	0	0	22	34	21	0	25	0	432	841	1,273
12	90	89.34	583	0	583	83	583	336	0	0	22	35	21	0	25	0	439	888	1,327
13	89	88.25	576	0	576	82	576	330	0	0	22	34	21	0	25	0	432	813	1,245
14	88	87.14	568	-568	0	0	0	0	0	0	0	33	0	0	0	0	33	732	765
15	78	74.46	479	-479	0	0	0	0	0	0	0	21	0	0	0	0	21	683	704
16	78	74.46	479	-479	0	0	0	0	0	0	0	21	0	0	0	0	21	658	679
17	82	79.87	517	-517	0	0	0	0	0	0	0	25	0	0	0	0	21	658	679
18	85	83.63	543	-543	0	0	0	0	0	0	0	29	0	0	0	0	25	596	621
19	84	82.40	535	-535	0	0	0	0	0	0	0	28	0	0	0	0	29	554	583
20	81	78.56	508	-508	0	0	0	0	0	0	0	24	0	0	0	0	28	695	723
21	79	75.86	489	0	489	70	489	269	0	0	22	22	21	0	25	0	24	751	775
22	77	73.04	0	650	650	93	650	389	0	0	22	0	21	0	25	0	360	761	1,120
23	77	73.04	0	650	650	93	650	389	0	0	22	0	21	0	25	0	457	413	870
24	77	0.00	0	650	650	93	650	389	0	0	22	0	21	0	6	0	457	340	797
																	438	262	700

TABLE 8
Total Savings Due Thermal Energy Storage

Month	Plant Energy Prior To TES, kWh	Plant Energy After TES, kWh	Primary Deman Pre TES, kVA	Primary Deman Post TES, kVA	TES Retrofit Savings Energy-kWh	Demand-kVA
January	78,529	64,700	1,522	895	13,829	627
February	82,358	67,453	1,425	869	14,905	556
March	115,135	90,570	1,375	856	24,565	519
April	152,214	113,874	1,384	836	38,340	549
May	166,343	122,820	1,375	826	43,523	549
June	198,779	146,489	1,375	836	52,290	539
July	210,086	154,655	1,375	850	55,431	525
August	201,546	148,431	1,375	880	53,115	495
September	181,720	133,912	1,618	830	47,808	788
October	154,570	115,773	1,581	781	38,797	800
November	101,520	78,984	1,617	781	22,536	837
December	73,171	59,488	1,588	788	13,683	800
Annual	1,715,971	1,297,149	1,618	895	418,822	722

TABLE 9
CAMPUS ENERGY COSTS PRIOR TO TES RETROFIT

Energy Charge for First 295 kWh / Primary kVA = $ 0.025734
Energy Charge for Additional kWh = $ 0.00754
Primary Demand Charge / kVA = $ 6.82
Secondary Demand Charge / kVA = $ 2.4
Annual On-Peak kVA = 1,618

Customer Charge/month = $530
Avg. Fuel Charge / kWh =$ 0.019639
Avg. PCRF Charge/kWh=$ 0.000128

Month	kWh	On-Peak kVA	Off-Peak kVA	Primary kVA	Secondary kVA	Primary kWh	PCRF	kWh Charge	kVA Charge	Total Cost
Sep-91	468,442	1,618	1,425	1,618	0	477,197	0.000128	$21,314.56	$11,032.14	$32,714.94
Oct-91	489,248	1,530	1,581	1,581	0	466,422	0.000128	$21,845.97	$10,783.05	$33,000.91
Nov-91	374,323	0	1,617	1,617	0	477,147	0.000128	$17,032.05	$11,031.00	$28,431.30
Dec-91	338,508	0	1,588	1,588	0	468,389	0.000128	$15,402.43	$10,828.51	$26,602.17
Jan-92	305,279	0	1,522	1,522	0	448,987	0.000128	$13,890.48	$10,379.96	$24,648.24
Feb-92	369,734	0	1,425	1,425	0	420,304	0.000128	$16,823.25	$9,716.85	$26,927.63
Mar-92	369,677	0	1,342	1,375	0	405,617	0.000128	$16,820.65	$9,377.32	$26,590.48
Apr-92	391,723	0	1,384	1,384	0	408,332	0.000128	$17,823.77	$9,440.08	$27,655.44
May-92	338,190	941	1,232	1,375	0	405,617	0.000128	$15,387.96	$9,377.32	$25,157.79
Jun-92	425,730	1,306	1,119	1,375	0	405,617	0.000128	$19,005.20	$9,377.32	$28,775.02
Jul-92	434,913	1,339	1,217	1,375	0	405,617	0.000128	$19,255.96	$9,377.32	$29,025.78
Aug-92	399,005	1,277	1,198	1,375	0	405,617	0.000128	$18,155.11	$9,377.32	$27,924.93
	4,704,767							$212,757.39	$120,098.21	$337,454.63

TABLE 10
CAMPUS ENERGY COSTS AFTER TES RETROFIT

Annual On-Peak kVA = 895

Month	Current kWh	Saved kWh	New kWh	New Primary kVA	New Primary kWh	New kWh Charge	New kVA Charge	New Total Cost
Sep-91	468,442	47,808	420,634	895.3	264,128	$16,292	$6,106	$22,839
Oct-91	489,248	38,797	450,451	869.0	256,345	$16,964	$5,926	$23,334
Nov-91	374,323	22,536	351,787	856.3	252,615	$14,202	$5,840	$20,487
Dec-91	338,508	13,683	324,825	835.6	246,511	$13,355	$5,699	$19,500
Jan-92	305,279	13,829	291,450	826.1	243,710	$12,393	$5,634	$18,474
Feb-92	369,734	14,905	354,829	835.6	246,511	$14,174	$5,699	$20,320
Mar-92	369,677	24,565	345,112	850.0	250,750	$13,986	$5,797	$20,228
Apr-92	391,723	38,340	353,383	880.0	259,600	$14,373	$6,002	$20,817
May-92	338,190	43,523	294,667	829.5	244,705	$12,499	$5,657	$18,603
Jun-92	425,730	52,290	373,440	780.8	230,338	$14,388	$5,325	$20,165
Jul-92	434,913	55,431	379,482	780.8	230,338	$14,553	$5,325	$20,330
Aug-92	399,005	53,115	345,890	787.9	232,424	$13,674	$5,373	$19,498
	4,704,767	418,822	4,285,945			$170,854	$68,384	$244,595

Chapter 68

Integration Enables Coordination of Diverse Building Systems

J.D. Petze

ABSTRACT

Advancing microprocessor technology has made it possible for manufacturers to include intelligent controllers on all types of mechanical equipment: chillers, boilers, packaged HVAC units, heat pumps and variable air volume (VAV) boxes. The rapid proliferation of microprocesser controlled devices, however, has presented the building owner/manager with a new challenge - how to achieve an integrated control system that consists of a variety of different, intelligent devices. The author believes that the need for integrated systems will dominate the building automation industry by the year 2000.

The industry is responding with a variety of potential solutions to allow different systems to interoperate. These include: proposals for a standard protocol for building automation systems, formal inter-manufacturer relationships, and "open protocols", meaning that manufacturers provide documentation to allow third parties to write interface software to their microprocessor devices.

While the primary focus of industry activity has been on addressing the need for different manufacturers building automation systems (BAS) to be able to communicate with each other, the rapid and somewhat unexpected appearance of microprocessors on all types of equipment **(including non-HVAC equipment)** has created the need for an immediate and practical solution to the problem of systems integration.

This paper will discuss the issue of open systems integration, present a historical basis for interoperability between different manufacturers control products, briefly review the activities being undertaken by various organizations within the industry, and present some examples of unique, non-HVAC related applications of open systems integration. These include interfaces to: particle counters used in monitoring of clean rooms, variable speed drives, broadcasts of information over "walkie talkie" systems, data exchange with security systems, chiller control panels, and industrial I/O modules.

UNDERSTANDING THE NEED FOR INTEGRATED SYSTEMS

Building operators want a single, coordinated user interface, through which they can view and control all of the equipment systems in a building. This capability is a necessity to enable operators to effectively manage today's sophisticated buildings with their complex mechanical and electrical systems, and to control operating costs.

Due to the low cost, increased reliability and widespread acceptance of microprocessor technology, more and more of the equipment systems found in buildings have built-in microprocessor controllers. In the HVAC industry this is referred to as "factory packaged, digital controls". This trend presents a new challenge: creating an **"integrated building automation environment"** for the operator, from a diverse range of intelligent devices provided by different manufacturers. To be effective, this environment must create the appearance of a single system for the operator.

In order to meet this challenge, the building automation system must become more than just a computer-based control system, it must act as an **integrator** of these various microprocessor based systems, enabling them to act as one system. This capability is referred to as **"interoperability"**.

The Trend Towards Factory Packaged Controls Creates New Challenges In The Application Of Building Automation Systems

In the early and mid 1980's, building automation systems were typically applied to a facility as a "retrofit". They were not part of the original equipment design and installation. As these systems proved their value to owners and engineers by lowering energy costs, improving occupant comfort, and providing easier, more efficient building operation, they began to become specified as a standard part of most new buildings. Even at this point, however, these systems were applied on top of HVAC equipment which typically had no controls, or in some cases had conventional electromechanical controls. Interfacing between the building automation system and these conventional control devices was relatively easy, and well understood. It required knowledge of "instrumentation-type" electrical principles and a basic understanding of electronics.

With the advent of low cost microprocessors, HVAC equipment manufacurers themselves began to include factory packaged digital controls as a standard part of the equipment package in order to make their products "intelligent" and gain some of the benefits of this new technology. This trend, which is still in its early stages, creates a very different situation for the application of a building automation system. It affects the capability required in the BAS (communications, programmability, processing power), the design and installation process, and the capability of the organization performing the installation. In short, it requires a new level of expertise and knowledge.

Examples of HVAC Equipment That Include Built-in Microprocessor Controls

Examples of HVAC equipment that now often include digital controls include:

- Chillers
- Boilers
- Fan Coil Units
- Variable Air Volume (VAV) Boxes
- Heat Pumps
- Variable Speed Drives

A HISTORICAL BASIS FOR INTEROPERABILITY

The operator's desire for an integrated system, where products from different manufacturers can "talk" with each other, is not really new. In a way it is a response to something that the control industry has inadvertently taken away from the building operator. In the pneumatic control systems of the past for example, it was common to be able to utilize an actuator from one manufacturer with a receiver/controller from another, and a temperature sensor from yet another. This was possible because there was a common, standard, open protocol which all manufacturers products could communicate with. This "protocol" was **3 - 15 psi air** pressure.

As control systems advanced and embraced microprocessor technology, this seemingly simple feature was lost, at the same time that the need to integrate a wide variety of devices increased.

This desire by operators to have a building automation system that provides interoperability with various computer-based systems has led to a number of activities by various manufacturers and industry societies.

OPEN PROTOCOL, STANDARD PROTOCOL AND CURRENT INDUSTRY ACTIVITIES RELATED TO INTEROPERABILITY

The concept of having different systems work together is often referred to by the general term "open protocol". As is often the case with new concepts, confusion results and a single term becomes a buzzword for a much larger issue. The term open protocol has come to mean the ability for building automation products from different manufacturers to communicate with each other to provide coordinated control. This concept is more accurately referred to as interoperability as mentioned earlier.

The terms **open protocol** and **standard protocol** which are often used when discussing this issue actually represent two very different methods of achieving interoperability, and deserve some clarification. Open protocol means that the communications protocol used to "talk" to a microprocessor-based device is openly available to the general public, at no more than a nominal charge. The existence of an open protocol allows a controls manufacturer or third party to develop an interface to another manufacturer's digital devices. This interface is accomplished through the development of a hardware or software "translator".

A standard protocol on the other hand refers to a single protocol that all manufacturers would use in their products, thereby allowing products from different manufacturers to talk to each other without the need for a translator. The development of a standard protocol is the subject of ASHRAE Committee 135P which has produced the BACNet Protocol Standard. The first public review draft of the BACNet communication protocol standard was released in August of 1991. The Council on Open

Protocols of the Intelligent Buildings Institute (IBI) is also working on "adapting, adopting or building a standard or open building systems protocol." IBI is an international trade organization formed in 1986 to promote the concept of intelligent buildings.

OPPORTUNITIES FOR INTEGRATION NOW GO BEYOND HVAC EQUIPMENT

As we consider the possibilities for, and benefits of a standard protocol for BAS an important fact must be recognized: the types of systems an operator may need to interface to is **no longer limited only to HVAC equipment.** Because of this a standard protocol for HVAC products will not necessarily provide all of the interoperability solutions required by users. This adds a "new twist" to the problem and causes the need for interfaces to continue to grow with every new product released to the market. For example, will all variable speed drive manufacturers conform to a communication standard for building automation systems? What about manufacturers of laboratory monitoring equipment? Is there any reason to believe that they will consider it important? What if they have standards of their own which they must conform to for their primary market?

The key point is that the range of products which need to be interfaced to has become so diverse that a standard protocol for HVAC systems will not be a complete solution. This should not be taken to mean that a standard protocol for HVAC and building automation products is not needed and would not be good for the industry. It simply means that it will not negate the need for open systems integration. The existence of a standard(s) should make open systems integration easier in the future, and will most likely result in even more open systems integration activity.

Examples of "Non-HVAC" Intelligent Microprocessor-Based Systems Found In Buildings

Opportunites for integration are presented not only by intelligent HVAC equipment, but also by other computer based systems that are now found in today's buildings. Examples include:

- Security/access control systems.
- Fire/life safety systems.
- Guest management systems in hotels.
- Scientific and laboratory monitoring equipment, such as particle counters used in clean rooms.

By integrating these systems with the building automation system, the operator is given the ability to manage his facility as a whole, instead of as a group of separate islands of data.

In the author's opinion, the fact that there will continue to be a growing variety of microprocessor-based devices used in buildings, means that **"open systems integration"** will be a growing trend in the building automation marketplace. It is the fundamental method of accomplishing interoperability between different intelligent systems.

Understanding Open Systems Integration

Open systems integration means to take a microprocessor-based product from one manufacturer and make it communicate with another manufacturers product so that they can operate in a coordinated fashion.

The key word in the phrase open systems integration is the word **open.** In order to accomplish an interface to a microprocessor-based system, the protocol for communication to that system must be openly available as described previously. With an open protocol, and a flexible, programmable building automation system, it is possible for an application programmer to create custom interfaces to virtually any system.

THE SUPPORT ISSUE

A number of control system manufacturers and HVAC equipment supplier have already made their protocols "open". Some, however, continue to resist. They have arguments to support their belief that open protocol is not a good idea. The most common argument is that "control installations require a single-source supplier, and that combining products from a variety of manufacturers will cause problems." Competent support of an automation system has been, and will continue to be, a significant consideration. The real issue, however, is not single-source supply of the components, but **single-source responsibility** for the installation as a system.

The mechanical system of a building is by definition combination of products from a variety of manufacturers (chillers, boilers, cooling towers, air handlers, valves, dampers, controls, etc.). A building owner looks to the mechanical engineer to design a **system** from these diverse components, and to the mechanical contractor to install these components and commission them as a system. Each party has a different area of expertise and a different responsibility. Together they make a complex building system work. The same model applies to control systems. The control system installer must be able to design, install, program, commission and support the various devices that they bring together to create the control system (controllers, actuators, sensors, etc.).

After over a decade of Direct Digital Control and Energy Management System experience, a whole new group of specialists has emerged in the industry: organizations that have hands-on experience working with products from a variety of manufacturers, and the knowledge to understand the compexities of computer-based control systems. These organizations are best referred to as **"systems integrators."** They provide the single-source responsibility that is required. Their experience gives them the ability to evaluate the feasibility of installing a hybrid system, design it, install it, program it, commission it and provide warranty and service for it. These firms cannot offer their customers the benefits of a hybrid system, however, if the communication protocols are not available.

Microprocessor-based products from outside of the HVAC controls market almost always offer open protocols. The author believes that these companies understand the need for integration in order to allow users to exploit the full advantages of these intelligent devices.

LEVELS OF INTEGRATION

There are various "levels" of integration which can be acheived between different systems. By this we mean the sophistication of the interface, and the amount and type of data that can be shared by the two systems. For example, we may only need our building automation system to respond to an alarm message produced by an access control system so that it can turn on the lights in all areas. This could be accomplished by an interface that supports only one-way communications, and a software program that is only designed to read an ASCII data string and interpret it in a predefined way. We would consider this to be a limited integration of these two systems. Similarly, we may only need the building automation system to send out an ASCII alarm message to a hand-held transceiver ("walkie talkie") so that all building operations personnel are notified of the problem immediately. Again, this would be considered a low level of integration between the systems. The key point is to match the level of integration needed to the needs of the user.

Contrast this with an interface that allows a specific manufacturer's building automation system to communicate fully and completely with a network of VAV controllers from a major HVAC equipment supplier. The interface will give the building operator access to all of the data and features of the VAV controller directly through the workstation of the building automation system. This is an example of **"complete integration"** of these two systems. Here again the level of integration must meet the needs of the building operator.

PRACTICAL EXAMPLES OF INTEGRATION

The capability for implementing the open systems integration applications described in this paper exists today. It has proven to be cost effective and reliable. Some building automation systems available today offer an **open systems communications port** capability which allows the application programmer to write custom communication interfaces to other microprocessor-based devices. The following examples of existing integration projects demonstrate the capabilities that exist.

- Critical alarms generated by a building automation system are sent in ASCII form to a Motorola Verbal Response Annunciator to send verbal commands over the security/engineering communication system (walkie talkies).

- A building automation system sends an ASCII message via modem to a security system using DS6500 protocol.

- A building automation system monitors building door contacts and sends ASCII alarm messages to the operator workstation, a security printer, and during after hours a special security transmitter is activated.

- A manufacturer of medical ultrasonic devices operates class 100 clean rooms. They utilize a microprocessor based particle counter to monitor conditions (number and size of particles) in the clean room. Before the installation of a building automation system, personnel had been monitoring the data from this device manually. The particle counter contained an RS232 port for communications, however, and an interface was developed to allow the particle counter to be polled by the building automation system. The data from the particle counter was then be presented to the operator in the same manner as all other environmental data, and extended data logging and analysis was provided.

CONCLUSION

The widespread use of microprocessor-based devices on building equipment has created a new challenge for the building controls industry - interfacing these various products to create a simple, integrated system for the operator. This can be accomplished via open systems integration. The need for open systems integration will continue to grow along with the continued use of factory packaged microprocessor controls on equipment of all types. The need for integration is so great, and the types of systems to be interfaced to so diverse, that open systems integration will be a major area of activity in the building automation industry, even once a standard protocol is developed for HVAC and control system products.

The capability to interface different microprocessor-based systems together to create a coordinated system exists today, and is a viable solution for a variety of building control needs.

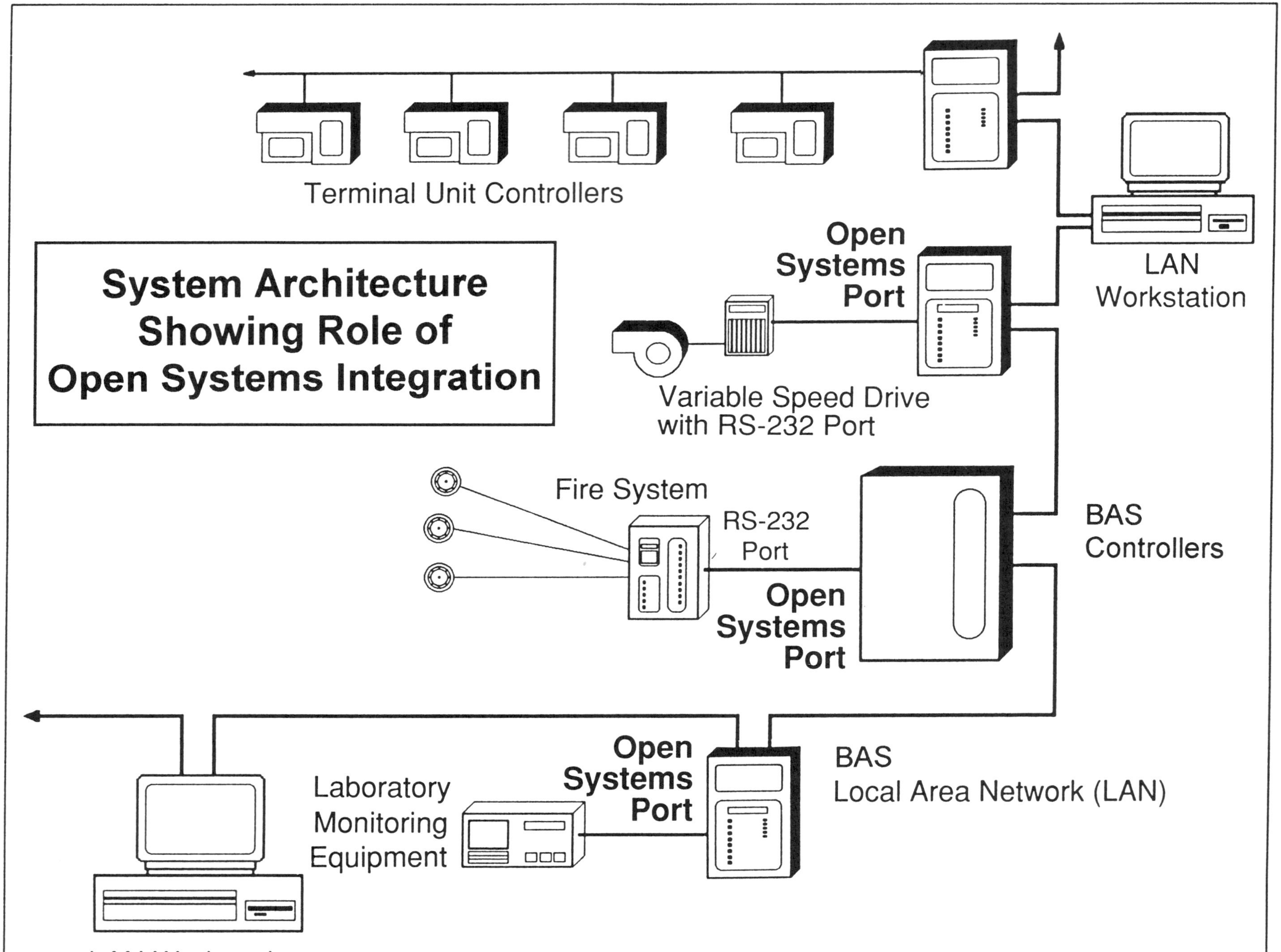

Terminal Unit Controllers
System Architecture
Showing Role of
Open Systems Integration
Open
Systems
Port
LAN
Workstation
Variable Speed Drive
with RS-232 Port
Fire System
RS-232
Port
Open
Systems
Port
BAS
Controllers
Open
Systems
Port
BAS
Local Area Network (LAN)
Laboratory
Monitoring
Equipment
LAN Workstation

FIGURE 1

GLOSSARY

Building operator - A person responsible for maintaining the operation of all equipment systems in a building.

Building automation system - A microprocessor-based system designed and implemented to monitor and control the HVAC and lighting systems in a building. A building automation system monitors environmental conditions in a building and optimizes the operation of these equipment systems to minimize energy consumption while providing desired comfort levels.

Complete integration - An interface between two microprocessor-based systems that allows all of the data and features of one device to be accessed via the other system.

Factory packaged digital controls - Microprocessor-based controls which are installed as part of a piece of equipment by the original equipment manufacturer.

Integrated building automation environment - A coordinated control system which ties together direct digital control of HVAC equipment with other intelligent microprocessor based systems found in the building, to provide the operator with a single, comprehensive operator interface through which the building operator can view and control all of the equipment systems in their buildings.

Open Systems Integration - Interfacing a microprocessor-based product from one manufacturer with another manufacturer's product to enable data communication between the two devices so that they can operate in a coordinated fashion.

Open Systems Port - A communications port on a building automation system which can be programmed to communicate with another system.

Systems Integrators - Contractors that specialize in the design, installation and programming of electronic and microprocessor-based systems to perform automation of building environments and equipment systems.

BIBLIOGRAPHY

Teletrol Systems Inc. "Open Systems Port Interfacing Guidelines." Teletrol Systems Inc. October 1992.

Petze, J. P. "Open Protocol for Terminal Untit Controllers," Heating Piping and Air Conditioning, September 1989, by Penton Publishing Inc., Cleveland, Ohio, USA 44114.

Parsonage, H. N. "Serial Interface Library," Gamma Infinity, Pleasant Hill CA, USA 94523.

"Proceedings of the Second Open Protocols Symposium," Intelligent Buildings Institute, Washington, D. C. USA 20037

"BACNet: A Communication Protocol for Building Automation and Control Networks", American Society of Heating, Refrigeration, Air-Conditioning Engineers, Inc. Atlanta, Georgia, USA, 1991.

Chapter 69

Are You Certain You Understand the Economics for Applying ASD Systems to Centrifugal Loads?

W.L. Stebbins

ABSTRACT

The advantages gained by using ac adjustable-speed drives (hereafter referred to as ASDs) on fans, pumps, compressors and other equipment have been widely publicized in recent years and their use in industry has been on a steady increase.

In HVAC systems, the use of variable-air-volume approaches (known as VAV systems) have gained wide acceptance since they first gained popularity during the energy crisis of the early 1970s. Outlet dampers, inlet guide vanes, and ASD systems are the usual choices for providing VAV control.

Using an outlet damper system to reduce airflow to 60% reduces input power requirements to approximately 88%. For a system using inlet guide vanes, a similar reduction in air volume reduces input power to 61%. Using the well known "cube rule fan law," a reduction in air volume to 60% theoretically reduces the input power requirement to approximately 22%.

Now for the rest of the story. It is very important to realize that this theoretical relationship between fan speed and horsepower does not take into account the effect of total system interactions, and thus does not include the effect of static back pressure in the system duct work (or the effect of static head in a centrifugal pump application). Despite what the ASD vendor's sales brochures claim, the magnitude of the static back pressure will have an adverse effect on the potential power and energy savings.

Experience over many years has indicated that the actual system curve using ASD equipment in centrifugal fan and pump systems will approximate a square function instead of the theoretical cube function. The prudent approach is to take this into consideration when developing potential savings using ASD equipment.

INTRODUCTION

Details presented include a layman's guide to fan laws, calculation of potential savings based on the three choices of VAV control, and evaluation of claims made by vendors. All dollar values presented are approximate, and are based on 1994 energy and equipment costs in the Southeast. Brand names are included to present a cross-section of equipment available and the listing is not intended to be all inclusive.

There are a variety of firms with the capabilities to supply satisfactory equipment, hardware, and software. Application of this equipment, hardware, and software for these specific requirements should not be construed as a general endorsement by either Hoechst Celanese or by the Author. It is important to note that a variety of brands and suppliers should be evaluated by anyone considering a similar application.

Values given for flows, temperatures, pressures, and unit energy costs approximate actual numerical quantities and are used to illustrate the methodology employed.

BACKGROUND

Since 1973, the Celanese plants of Hoechst Celanese Textile Fibers have had an extensive energy management process in place. Results of this process have yielded, within the Celanese plants, a reduction of over 30% in BTU/NGP.

Discussions with many people involved in energy management at a variety of facilities indicates that the majority still use the cube rule law for calculating potential savings when considering the use of ASD systems. The intent of the following discussion is to clarify that issue, allowing for a more realistic estimate of savings and greatly increasing the credibility of the estimate and of the engineer involved in the ASD application.

METERING SYSTEMS

Energy cannot be well managed unless there is a way to identify the major users and then easily trend the consumption patterns to identify potential reductions in energy per unit of production or per service performed. In the words of one old sage, "If you can't meter it, you can't manage it."

Adequate metering is a key ingredient to allow before and after economic evaluation of ASD system installations for VAV applications. These metering systems can be complex and sophisticated [1] or simple systems. [2] Applying trend analysis techniques provides the energy manager with the data necessary to make intelligent decisions regarding proper control of VAV systems. [3]

SIX REASONS TO METER:

1. CHARGE OUT ENERGY TO INDIVIDUAL DEPARTMENTS

This is the most basic reason to meter. Each month the total energy bill is proportioned to the various departments. This data is used to compare costs against the department's budget and thus develop a variance-dollar value. Also included is the use of meters for revenue billing when energy is sold to a third party such as in a cogeneration arrangement.

2. ACCOUNTABILITY FOR ENERGY USED

Trending of energy consumption per unit of production or per service performed is the basis for initial analysis and resulting corrective actions.

3. EFFICIENCY OF UTILITY EQUIPMENT AND SYSTEMS

The experience HCC has gained from their Utility Test Process [4] provides the following guidelines values:

1) Centrifugal air compressors at 125 psig:
   3.2-4.0 kWh/1,000 standard cubic feet (scf)

2) Centrifugal chiller drives producing 45 degree F. water:
   0.6-1.0 kW/ton refrigeration or
   0.6-1.0 kWh/ton-hour refrigeration

3) Refrigeration delivered to the conditioned space:
   1.0-2.0 kW/ton refrigeration or
   1.0-2.0 kWh/ton-hour refrigeration

4) Steam boilers at 250 psig:
   Number 6 fuel oil = 7.1-9.0 gal/1,000 lb steam
   Natural gas = 1,000-1,300 scf/1,000 lb steam
   Pulverized coal = 100-120 lb/1,000 lb steam
   (Fuel Consumption values will be 25 percent higher at 600 psig.)

4. PROVIDE INFORMATION FOR AUDITS OF ENERGY PROJECTS

With funding becoming increasingly difficult to obtain, audits of cost reduction energy management projects have been required more frequently over the past five years.

5. MAINTENANCE WORK, IDENTIFY PERFORMANCE PROBLEMS, FEEDBACK TO MANAGERS

The collection of energy-consumption data in support of maintenance work is seen as a viable tool, greatly aiding in the identification of equipment performance problems. As a side issue, performance problems associated with the people operating the equipment are also identified, allowing managers to take any necessary corrective action.

6. IDENTIFY POTENTIAL FUTURE ADDITIONAL ENERGY SAVINGS

With the long-term goal of management being one of continual improvement, metering and trending systems provide data on which to base resource allocation decisions.

THE APPLICATION OF ADJUSTABLE SPEED DRIVES

In recent years, energy managers have increasingly turned to adjustable-speed drives as technology has improved and product cost has come down. That trend apparently will continue. According to projections by the Electric Power Research Institute and Power Electronics Application Center, as much as 60% of all electricity in the United States will flow through power electronic systems – including adjustable-speed drives – by the year 2000. [5]

The reason for the trend is a simple question of efficiency. HVAC system designs are based on an area's weather extremes. Thus, a system designed to handle an extreme temperature of 95°F requires only a portion of its full capacity on a typical day. In fact, actual air-volume requirements for a typical HVAC system tend to follow a fairly normal distribution curve. Approximately 87% of the time, the system requires less than 70% of its maximum airflow. If adjustable-speed drives are used on fans and pumps in the HVAC system to reduce their speed (and volume) accordingly, the energy savings can be significant.

ENERGY SAVINGS WITH ADJUSTABLE-SPEED DRIVES

Induction Motors [6]

The AC induction motor is the mainstay for energy conversion in the U.S. It is found in industry, in commerce, and in the home, and is the major converter of electrical energy into another useable form. For this purpose, about two-thirds of the electrical energy produced is fed to motors, which then provide mechanical power to a wide variety of loads.

Much of the power that is consumed by AC motors goes into the operation of fans, blowers and pumps. It has been estimated that approximately 50% of the motors in use are for these types of loads. These types of loads are particularly attractive to look at for energy savings.

Basically, fans and pumps are designed to be capable of meeting the maximum demand of the system in which they are installed. However, quite often the actual demand could vary and be much less than the designed capacity. These conditions are accommodated by adding outlet dampers to fans or throttling valves to pumps. These are effective and simple controls, but severely affect the efficiency of the system.

There is another form of control which adapts fans and pumps to varying demands and does not decrease the efficiency of the system as much. This is adjustable-speed control of the fan or pump, a method providing a more efficient means of flow control than the existing methods.

Fans

The most common fan is the centrifugal fan that imparts energy in to air by centrifugal force. This results in an increase in pressure and produces air flow at the outlet of the fan. These types of fans are used in a wide variety of applications from the simplest residential heating systems to the largest commercial and industrial systems.

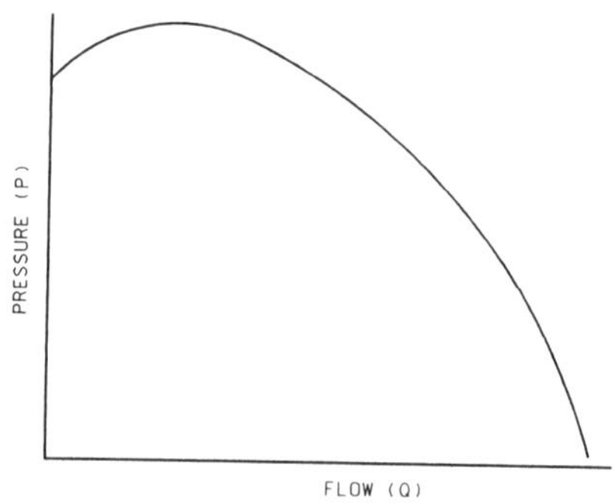

FIGURE 1. TYPICAL FAN CURVE

Figure 1 is an example of what a typical centrifugal fan can produce at its outlet at a given speed. The curve is a plot of outlet pressure (in static inches of water) versus the flow of air (in cubic feet per minute). Standard fan curves will usually show a number of curves for different fan speeds and includes fan efficiencies and horsepower requirements. These are all useful for selecting the optimum fan for any application. They also are needed to predict fan operation and other parameters when the fan operation is changed.

Many systems however require operation at a wide variety of points. Figure 3 shows a profile of the typical variations in flow experienced in a typical system. How these variations are produced and controlled will have a direct effect on the energy saved.

There are several methods used to modulate or vary the flow (or CFM) of a system to achieve the optimum points. Some of these include:

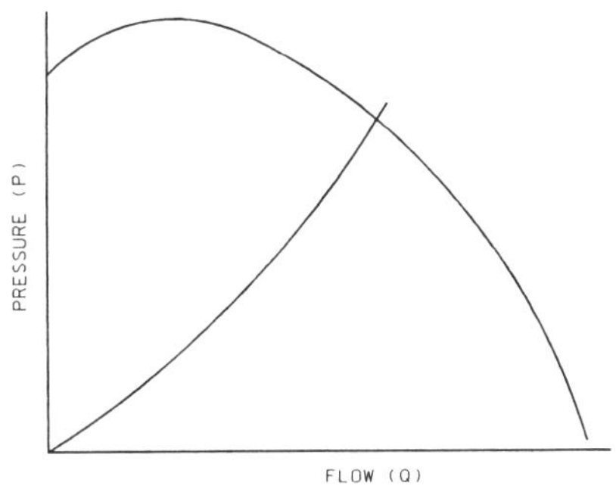

FIGURE 2. TYPICAL SYSTEM CURVE

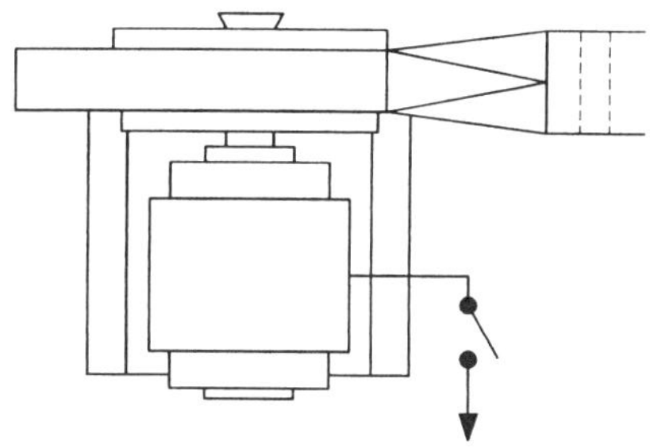

FIGURE 4. CYCLING: TYPICAL FOR HOME HEATING SYSTEMS. BECAUSE OF THE ERRATIC AIR FLOW, THIS APPROACH IS UNACCEPTABLE FOR COMMERCIAL OR INDUSTRIAL USES.

Figure 2 shows a system curve added to a typical fan curve.

The system curve is not a product of the fan, but a curve showing the requirements of the vent system that the fan is used on. It shows how much pressure is required from the fan to overcome system losses and produce an air flow. The fan curve is a plot of fan capability independent of a system. The system curve is a plot of "load" requirement independent of the fan. The intersection of these two curves is the natural operating point. It is the actual pressure and flow that will occur at the fan outlet when this system is operated. Without external influences, the fan will operate only at this point.

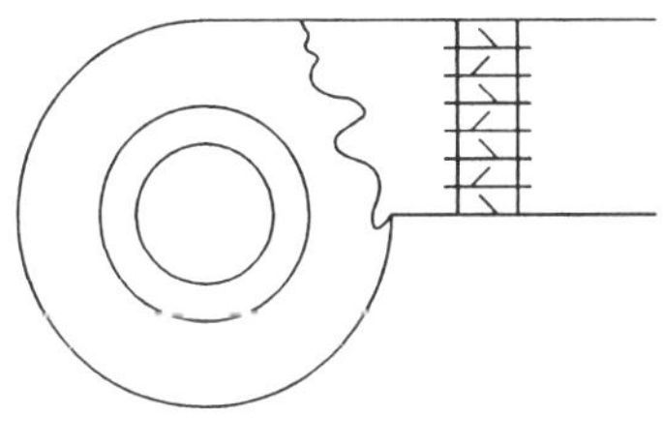

FIGURE 5. OUTLET DAMPERS: CONTROL LOUVERS OR DAMPERS ARE INSTALLED AT THE OUTLET OF THE FAN. TO CONTROL AIR FLOW, THEY ARE TURNED TO RESTRICT THE OUTLET WHICH REDUCES THE AIR FLOW.

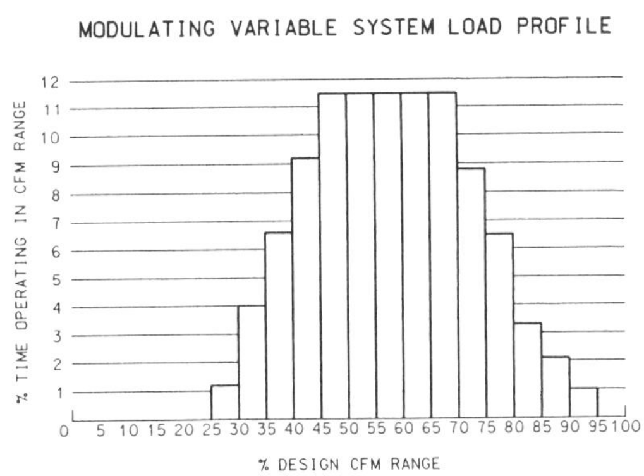

FIGURE 3. EXAMPLE OF FAN OPERATION

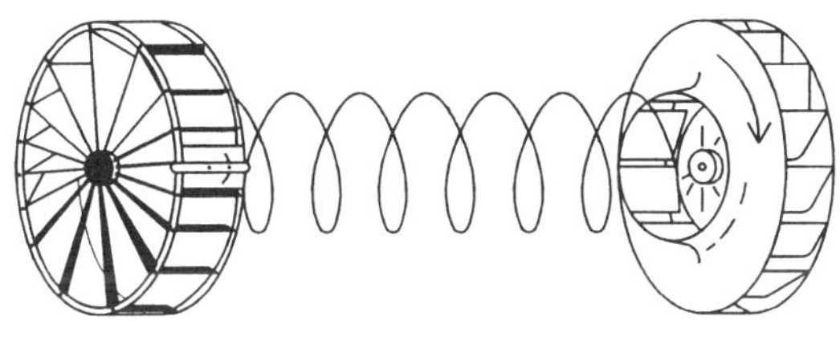

FIGURE 6. VARIABLE INLET VANES: BY MODIFYING THE PHYSICAL CHARACTERISTICS OF THE AIR INLET, THE FAN OPERATING CURVE IS MODIFIED WHICH CHANGES THE AIR FLOW.

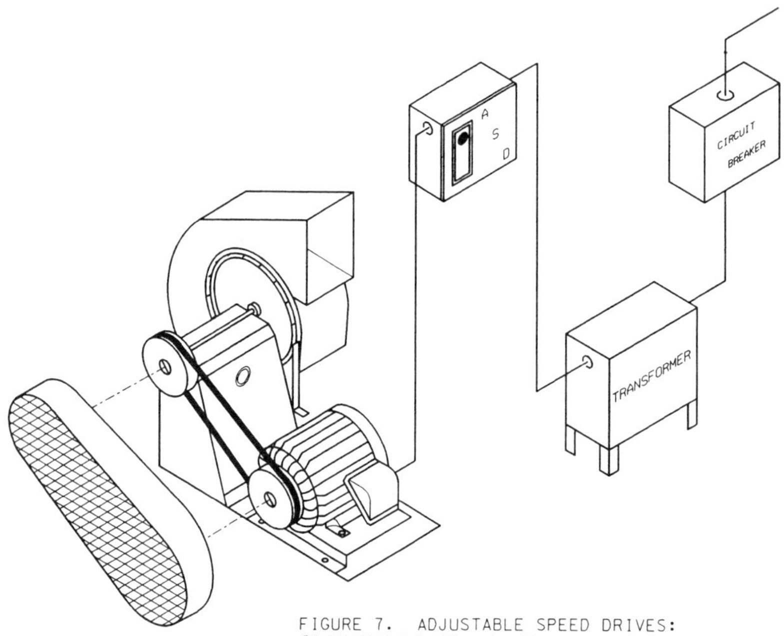

FIGURE 7. ADJUSTABLE SPEED DRIVES:
BY CHANGING THE ACTUAL FAN RPM, THE PERFORMANCE OF THE FAN CHANGES PRODUCING A DIFFERENT AIR FLOW.

Each of the three latter methods affect either the system curve or the fan curve to produce a different natural operating point. In so doing, they also may change the fan's efficiency and the horsepower requirements. Below are examples of each function.

Outlet Dampers

The outlet dampers affect the system curve by increasing the resistance to air flow. The system curve is a simple function that can be stated as **P = K x (CFM)²**. **P** is the pressure required to produce a given flow in the system. **K** is a function of the system and represents the friction to air flow. **CFM** is the air flow desired. The outlet vanes affect the **K** portion of the formula. Figure 8 shows several different system curves indicating different outlet damper positions.

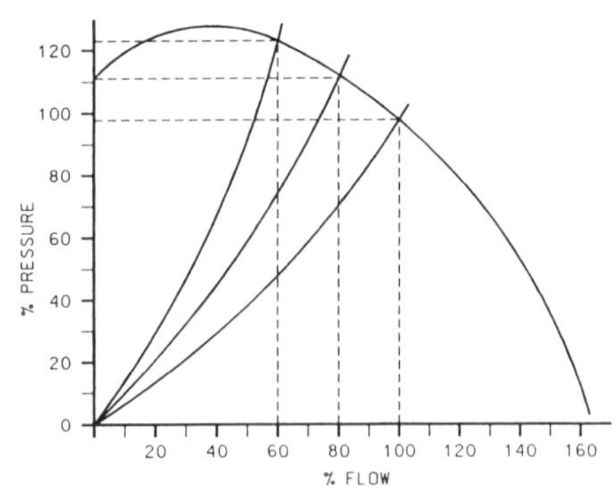

FIGURE 8. OUTLET DAMPER-STYLIZED SYSTEM CURVES

Figure 9 is a curve of the horsepower requirement for this type of operation. From this curve, it can be seen that the horsepower decreases gradually as the flow is decreased.

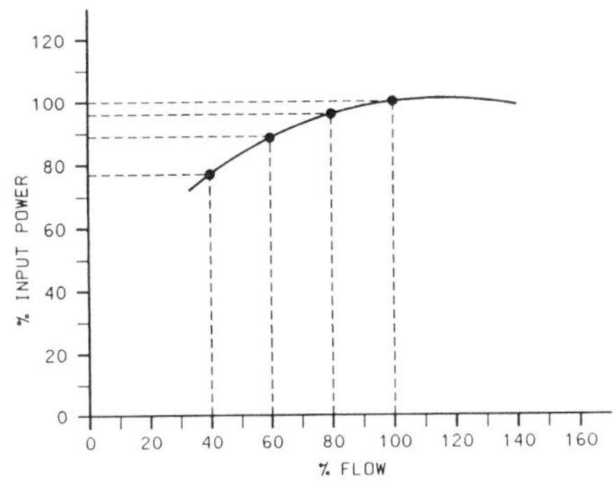

FIGURE 9. OUTLET DAMPER-STYLIZED HORSEPOWER REQUIREMENTS

Variable Inlet Vanes

This method modifies the fan curve so that it intersects the system curve at a different point. Figure 10 is a representation of the changes in the fan curve for different inlet vane settings.

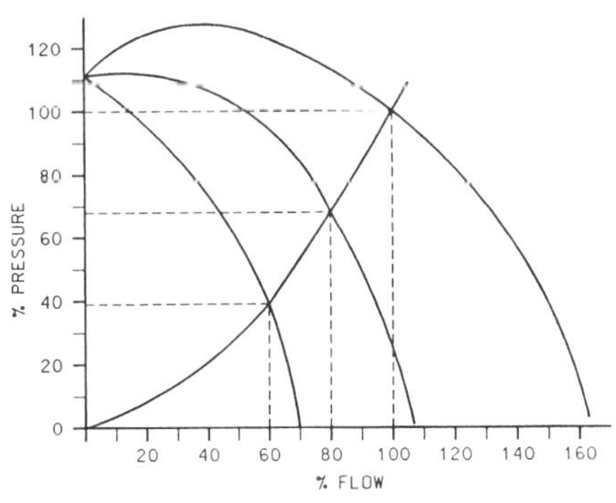

FIGURE 10. VARIABLE INLET VANE STYLIZED SETTINGS

Figure 11 illustrates that the horsepower required by this method drops and to a greater extent than it did in the outlet damper method.

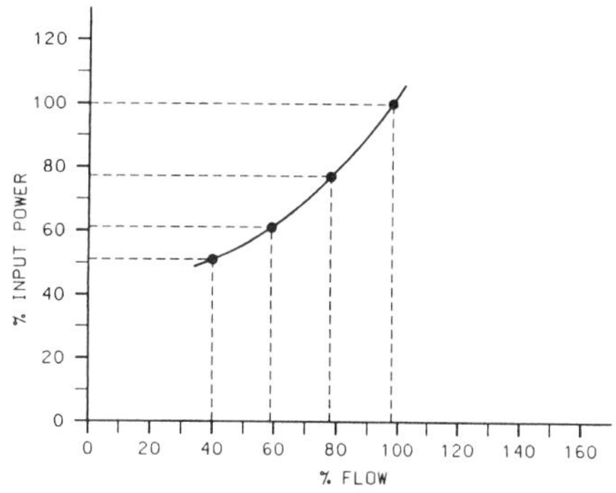

FIGURE 11. VARIABLE INLET VANE-STYLIZED HORSEPOWER REQUIREMENTS

Adjustable-Speed Drives

This method takes advantage of the change in the fan curve that occurs when the speed of the fan is changed. These changes are quantified in a set of formulas called the affinity laws. These laws are as follows:

AFFINITY LAWS

$$\frac{Q_2}{Q_1} = \frac{N_2}{N_1} \qquad \frac{P_2}{P_1} = \frac{(N_2)^2}{(N_1)^2} \qquad \frac{HP_2}{HP_1} = \frac{(N_2)^3}{(N_1)^3}$$

WHERE **N** = Fan Speed
Q = Flow (CFM)
P = Pressure (Static Inches of Water)
HP = Horsepower

Note that when the flow and pressure laws are combined, that the result is a formula that matches the system curve formula: [SYSTEM CURVE **P** = **K x (CFM)²**]. Thus the fan will follow the system curve when its speed is changed.

Figure 12 is a representation of the adjustable-speed method. Note the effect with a 30% back pressure in the duct.

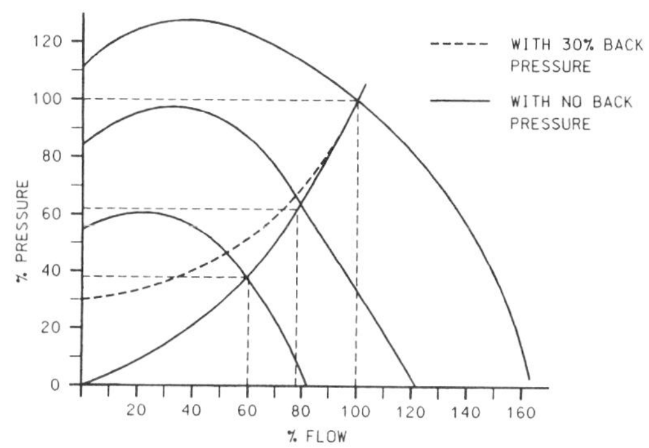

FIGURE 12. ADJUSTABLE SPEED METHOD-STYLIZED CURVES

Figure 13 illustrates the significant reduction in horsepower achieved by this method. The theoretical horsepower formula from the affinity laws shows the effect of the cube functions. Experience has shown that the actual horsepower requirement will approximate a square function when static back pressure is present. Only fans with little or no static pressure, such as cooling tower fans and roof vent fans with domes, will approximate a cube function.

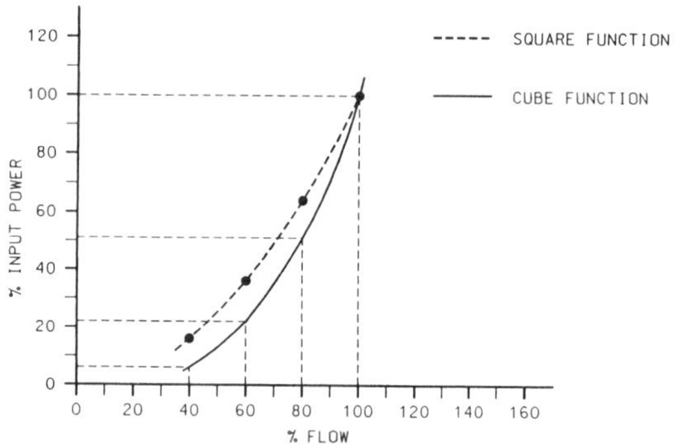

FIGURE 13. ADJUSTABLE SPEED-
STYLIZED HORSEPOWER REQUIREMENTS

The adjustable-speed method achieves flow control in a way that closely matches system or load curve. This allows the fan to produce the desired results with the minimum of input power. The other two methods modify some system parameter which generally results in the reduction of efficiency of the fan. This is why the horsepower demand is greater than the adjustable-speed method.

CALCULATION OF ENERGY SAVINGS

It is now appropriate to estimate the fan power consumption if no VAV action is taken, and then compare the power consumption for each of the various methods to determine potential savings.

The fan selected for this example is a unit operating at 300 RPM producing a 100% flow rating of 100,000 CFM, with a motor shaft power requirement of 35HP.

With an assumed downtime of approximately 9% for cleaning and maintenance, the fan will operate 8,000 hours per year. With no VAV action available and at an energy cost of \$0.08 per kWh, the annual electricity cost will be:

$$(35\ \text{HP})\left(0.746\ \frac{\text{kW}}{\text{HP}}\right)(8{,}000\ \text{HRS})\left(\frac{\$0.08}{\text{kWh}}\right) = \$16{,}710$$

To determine the potential savings from any VAV method, a load profile must be developed similar to that shown in Figure 3. This is no small task and a great deal of system analysis is likely to be required using weather data, variations in the occupation or activity in the conditioned space and other pertinent factors.

For this example, the following load profile is determined:

CFM	DUTY CYCLE (% of time)
100%	10%
80%	40%
60%	40%
40%	10%

Required Horsepower

For each operating point, a required horsepower from the fan curve can be obtained. This horsepower is multiplied by the percent of time (divided by 100%), that the fan operates at this point. These calculations are then summed to produce a "weighted horsepower" that represents the average energy consumption of the fan.

From Figure 9, the weighted horsepower calculations for the outlet damper method are as follows:

% CFM	% DUTY CYCLE	HP REQUIRED	WEIGHTED HP
100	10	(35)(1.0) = 35.0	3.50
80	40	(35)(0.96) = 33.6	13.44
60	40	(35)(0.88) = 30.8	12.32
40	10	(35)(0.77) = 26.9	2.69
		TOTAL HP	31.95

From Figure 11, the weighted horsepower calculations for the variable inlet vane method are as follows:

% CFM	% DUTY CYCLE	HP REQUIRED	WEIGHTED HP
100	10	(35)(1.0) = 35.0	3.50
80	40	(35)(0.78) = 27.3	10.92
60	40	(35)(0.61) = 21.4	8.56
40	10	(35)(0.51) = 17.8	1.78
		TOTAL HP	24.76

From Figure 13, the weighted horsepower calculations for the adjustable speed drive method can be developed for both the square function and the cube function. It is of interest to do both and then compare the results:

Using the Square Function

% CFM	% DUTY CYCLE	HP REQUIRED	WEIGHTED HP
100	10	(35)(1.0) = 35.0	3.50
80	40	(35)(0.64) = 22.4	8.96
60	40	(35)(0.36) = 12.6	5.04
40	10	(35)(0.16) = 5.6	0.56
		TOTAL HP	18.06

Using the Cube Function

% CFM	% DUTY CYCLE	HP REQUIRED	WEIGHTED HP
100	10	(35)(1.0) = 35.0	3.50
80	40	(35)(0.51) = 17.9	7.20
60	40	(35)(0.22) = 7.7	3.08
40	10	(35)(0.06) = 2.1	0.21
		TOTAL HP	13.99

COMPARISON OF SAVINGS

The data in Table 1 on the following page illustrate the operating cost for each system, and the savings compared to no VAV action.

TABLE I

	NO VAV	Outlet Damper	Inlet Guide Vanes	ASD@N² (Realistic)	ASD@N³ (Theoretical)
Weighted HP	35	31.95	24.76	18.06	13.99
x 0.746 kW/HP	0.746	0.746	0.746	0.746	0.746
x 8000 HR/YR	8000	8000	8000	8000	8000
= kWh/YR	208,880	190,678	147,768	107,782	83,492
x $0.08/kWh	$0.08	$0.08	$0.08	$0.08	$0.08
= $/YR	$16,710	$15,254	$11,821	$8,623	$6,679
Annual Savings Over No VAV Action	0	$1,456	$4,889	$8,087	$10,031
Estimated Cost to Implement	0	$3,210	$7,620	$8,145	$8,145
Simple Payback in Years	--	2.20	1.56	1.01	0.81

Cost Basis

The costs to implement the above types of VAV control are based on historical experience. Engineering costs are assumed to be similar in all three cases.

Discharge Damper

Damper Assembly	$2,500.00
Pneumatic Actuator complete with linkage for damper control	350.00
12 MH Labor @ $30/MH	360.00
TOTAL	$3,210.00

Inlet Vanes

Inlet Vane Assembly	$6,500.00
Pneumatic Actuator Complete with linkage. Floor Mounted Assembly	640.00
16 MH Labor @ $30/MH	480.00
TOTAL	$7,620.00

ASD Inverter

40 HP Inverter ($8,192) x (0.60 multiplier)	$4,915.00
65 kVA Isolation Transformer	1,500.00
25 MH Labor at $30/MH	750.00
Materials (conduit, wire, fittings, brackets)	250.00
Spare Parts (boards, controls, power devices)	250.00
4 MH Training for 4 Men (16 MH at $30/MH)	480.00
TOTAL	$8,145.00

What is the simple payback if a VAV control scheme already exists?

In the case where an outlet dampen control scheme already exists, the simple payback by installing an ASD system is:

SYSTEM	$/YR
Annual Cost for Outlet Damper	$15,254
Annual Cost for ASD Invertor	-8,623
Annual Savings	$6,631

$$\text{Simple Payback} = \frac{\$\ \text{Cost}}{\$\ \text{Savings}} = \frac{\$8{,}145}{\$6{,}631} \approx 1.23 \text{ years}$$

In the case where an adjustable inlet guide vane control scheme already exists, the simple payback of installing an ASD system is:

SYSTEM	$/YR
Annual Cost for Inlet Damper Vanes	$11,821
Annual Cost for ASD Inverter	-8,623
Annual Savings	$3,198

$$\text{Simple Payback} = \frac{\$\ \text{Cost}}{\$\ \text{Savings}} = \frac{\$8{,}145}{\$3{,}198} \approx 2.55 \text{ years}$$

OTHER CONSIDERATIONS

No cost has been assigned for an instrumentation system which will sense the duct pressure and produce an output signal to maintain the desired CFM flows for the example load profile. This is because the same instrumentation system can be used for any of the VAV methods. Including this cost will of course lengthen the simple payback period compared to no VAV action.

Typical costs for a single stand alone instrumentation system can range from $500 to $2,500. If part of a large multi-point DDC system, the cost for any single control point may be as low as $100.

Different costs per kWh can be used to reflect energy rates in various parts of the country by a simple ratio to the $0.08 value used in the example.

This example is also somewhat basic in that it only takes into account the operation at the fan. In practice, the motor efficiency and the drive efficiency should also be taken into account. It is impractical to list or chart the motor and drive efficiencies for all possible load conditions that could occur. However, since the same motor would be used in all of the examples shown here, the difference in motor efficiencies would be minimized and not significantly affect the results shown. The possible exception would be if the motor efficiency decreased to a great degree as motor speed was reduced.

A more detailed calculation including motor and drive efficiencies can be done by the ASD vendor if there is a specific application to be analyzed. For that analysis to be meaningful, the load profile, system curve and fan curve must be available.

Consideration should be given to addressing the potential problems associated with harmonics produced when applying ASD systems. Details on the types of problems and various solutions are beyond the scope of this paper. The reader is directed to a variety of recent publications which cover the subject in greater detail. [7][8][9]

CONCLUSION

Adjustable-speed drives applied to pumps and fans in a variable-air-volume HVAC system can reduce the system's energy consumption significantly. Improved technology has brought the price of drives down significantly. This, coupled with often substantial utility rebates designed to encourage energy conservation, has helped make adjustable-speed drives economically viable in many commercial, industrial and institutional applications.

It is important to keep in mind that the prudent approach is to use the square function when developing the potential savings resulting from installing ASD equipment.

REFERENCES

[1] W.L. Stebbins, "Metering Revisited – Innovative Concepts For Electrical Monitoring and Reporting Systems." In *Proceedings Of The 15th World Energy Engineering Congress,* The Association of Energy Engineers, Atlanta, GA, October 1992.

[2] W.L. Stebbins, "Utility Monitoring Systems: Key To Successful Energy Management." In *Proceedings Of The 6th World Energy Engineering Congress,* The Association of Energy Engineers, Atlanta, GA, November, 1983.

[3] W.L. Stebbins, "Highly Efficient Energy Metering And Trend Analysis Techniques For Maximum Control." In *Proceedings Of The 14th World Energy Engineering Congress*, The Association of Energy Engineers, Atlanta, GA, October 1991.

[4] W.L. Stebbins, "Implementing An Effective Utility Testing Process: A Keystone For Successful Energy Management." In *Proceedings Of The 13th World Energy Engineering Congress*, The Association of Energy Engineering, Atlanta, GA, October, 1990.

[5] Adapted from "Adjustable-Frequency Drives Reduce HVAC Costs." Paul E. Beck, *Consulting – Specifying Engineer*, June 1992.

[6] Adapted from "Energy Savings With Adjustable Frequency Drives." James Van Zile, in *Allen-Bradley Publication DGI-2.1*, December 1984.

[7] Adjustable Speed Drives, Applications Guide. EPRI Distribution Center, 207 Coggins Drive, P.O. Box 23205, Pleasant Hill, CA 29523. (There is no charge for reports requested by EPRI member utilities and affiliates.)

[8] IEEE Std 1100-1992 (Emerald Book), "Recommended Practice for Powering and Grounding Sensitive Electronic Equipment." IEEE Service Center, P.O. Box 1331, Piscataway, NJ 08855-1331.

[9] "IEEE Recommended Practices and Requirements for Harmonic Control in Electric Power Systems," (Project IEEE-519), IEEE Service Center, P.O. Box 1331, Piscataway, NJ 08855-1331.

Chapter 70

Simple Solutions to VSD Pumping Measures

R.R. Vaillencourt

INTRODUCTION

In this day and age, pumping energy conservation means variable speed drives (VSD). A good engineer will recognize those situations where it will be more cost effective to downsize the pump or change the operating strategy first. But these situations will be highlighted during a VSD evaluation.

Some energy evaluations are quite easy. Some others are extremely difficult, but mostly unnecessary for the first stage of an evaluation. The engineers who write the energy auditing handbooks seldom take the trouble to show you how easy the easy evaluations are. They seem to desire to impress you with their understanding of the difficult ones by going into great detail whether you want to or not.

By developing a standard approach that strikes a balance between the need for accuracy and the amount of time available, you can quickly zero in on projects that are worth the effort. The purpose of this approach is to provide a decision making tool to help you to decide whether to invest real time and money into developing the details of a project.

Most inexperienced energy engineers discover the Affinity Laws and go crazy. The potential for energy savings is almost unbelievable. In the simplest situations, the savings are absolutely true. The true test is to be able to recognize when things are not simple and the evaluation needs to be modified.

Rest assured, the modifications all follow the Affinity Laws, the calculations are simple, and the savings are usually considerable. But the savings potential will be less, and in some cases too small to consider when certain easily recognized situations exist.

DEFINITIONS

Affinity Law

I had to look up the word "affinity" in the dictionary to get an idea why it was chosen for the formulas defining the relationships between speed ratios, pressure ratios, and horsepower ratios for centrifugal pump and fan applications. The word "affinity" means "relationship". It is my feeling that a deliberate effort is made every now and then to name useful and clever formulas with words that tend to hide the connection to their use in the real world. Possibly it is just a result of engineers taking a cue from doctors and lawyers.

Anyway, the ratio equations are straight forward and simple to manipulate to determine useful information with a few known facts:

$$\frac{N_1}{N_2}=\frac{F_1}{F_2}; \quad \left(\frac{N_1}{N_2}\right)^2=\frac{H_1}{H_2}; \quad \left(\frac{N_1}{N_2}\right)^3=\frac{HP_1}{HP_2}$$

$N = rpm$; $F = flow$; $H = head$; $HP = horsepower$

Head

With respect to fluid flow, this term is synonymous with the word "Pressure". By far it most often refers to pressure given in units of Feet of Water Column, other times it will be Pounds per Square Inch.

Static Head

The static head is the pressure at the pump caused totally by the elevation of the water above the pump (which is why the units often are in feet). In a closed loop the static head at the outlet of the pump must be the same as the static head at the inlet to the pump. Therefore the static head that the pump must overcome to move water is zero.

The static head is also the minimum pressure required by the system header piping to provide the necessary flow through all connected devices. If, for example, you are supplying cooling water to several parallel fan coils designed for operation with a 30 psi inlet pressure, the static head that the pump must deliver into the piping system is the 30 psi (approximately 70 feet).

Friction Head
The friction head is the pressure loss due solely to the friction of the water moving against the pipe and fittings. (If you have started with units of feet, remember to use the same units in all your calculations.)

Pressure Drop (PD)
The pressure drop is the change in pressure between the inlet of something and the outlet of something. In the simplest system of a closed loop without any equipment, the pressure drop from the beginning to the end is the friction head. In the next simplest system of a fluid transfer arrangement, like from a sump to the top of a cooling tower, the pressure drop from the beginning to the end is the friction head plus the static head.

All pieces of equipment that require cooling or heating water will specify what the pressure drop through the equipment is expected to be when you operate it at its rated heat exchange level. This is the friction head that needs to be overcome in order to push water through the equipment at the flow rate for which it is designed.

If equipment is in parallel, the equipment with the highest PD is the controlling factor. If the equipment is in series, the sum of each PD is the controlling factor.

Total Dynamic Head (TDH)
The TDH is the highest pressure that the piping and equipment system will impose on the pump. This is simply a sum of the static head, the friction head and the controlling pressure drop of the various pieces of equipment.

WHAT TO LOOK FOR

Centrifugal pumps used to circulate cooling (or heating) water to HVAC or process applications offer major savings potential. Look for throttling valves. Not just at the outlet from the pump, but at the various end use equipment. Many process cooling water loops are supplied by a pump that runs constantly, pressurizing a plantwide header with various machine taps ending in a valve (solenoid or manual) at each machine.

Each of those machine valves varies the flow requirement from the pump. I usually recommend a variable speed drive (VSD) controlled by a pressure sensor in the header. As a valve opens, the flow increases and the pressure drops. This will speed up the pump to maintain the pressure setpoint. As a valve closes, the flow will decrease and the pressure will rise, slowing the pump down.

This measure should also be evaluated for HVAC chilled water loops with two-way, modulating valves at each coil. If the existing system uses three-way modulating bypass valves, the savings will be optimized if the valves are replaced with two-way valves or modified with the bypass isolated and capped.

Cooling towers transfer heat from the inside of a facility to the outside. When used with refrigeration equipment the heat is transferred from the condenser side of refrigeration machines to the outside air. When used with process equipment the heat is transferred from a process heat exchanger or cooling jacket to the outside air. Installing a VSD controlled by return water temperature will ensure that the gpm (mass flow rate) is matched to the Btu rejection requirements. (**Caution!** Check with the equipment manufacturer to determine the minimum flow rate required to maintain turbulent flow through the condenser and evaporator tubes.)

Another common application is a boiler feed water pump that runs continuously while the actual inlet flow to the boiler is controlled by a modulating float valve. The calculations must take into account the fact that the pump must provide a minimum pressure higher than the boiler pressure. This means that there is a minimum horsepower requirement even if the flow is negligible.

Therein lies the difficulty. There are many applications where the minimum speed will be dictated by the minimum TDH. The trick is to recognize that there is a minimum TDH and then determine what it is.

How To Find the Minimum Speed
Any pump is selected to operate at a certain maximum design flow rate, F_D and maximum design total dynamic head; H_D (D=Design) (Fig. 1). H_D is the total of static and friction head. For each system there is a specific minimum static head requirement, H_{min} (unless the water is just flowing in a pipe loop without any devices in the way). The friction head, however, varies as the square of the flow rate. Since the static head will remain constant for all flows, the system curve does not start at zero H because of the static head requirement (Fig. 2). To use a VSD means that you are developing the "family" of pump curves, i.e., a curve for each RPM. The minimum RPM curve (RPM 4) is the one closest to the minimum static head (H_{min}) at zero flow (Fig. 3).

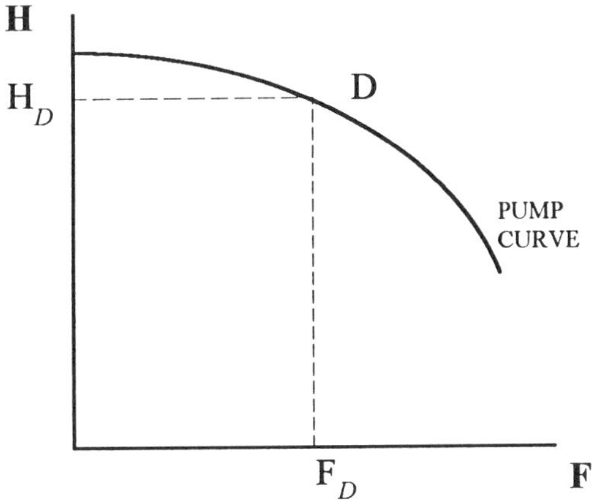

FIG. 1

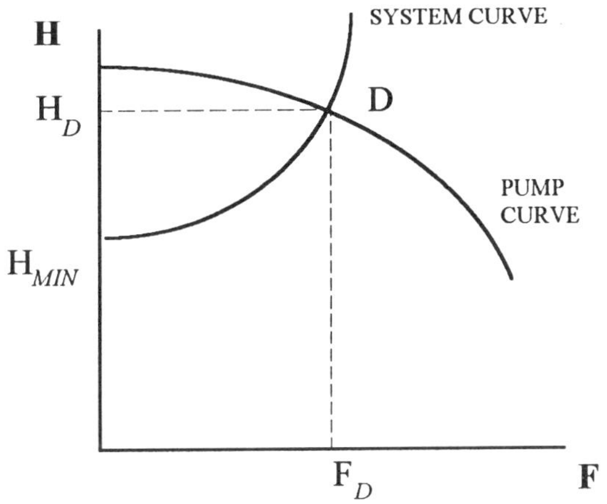

FIG. 2

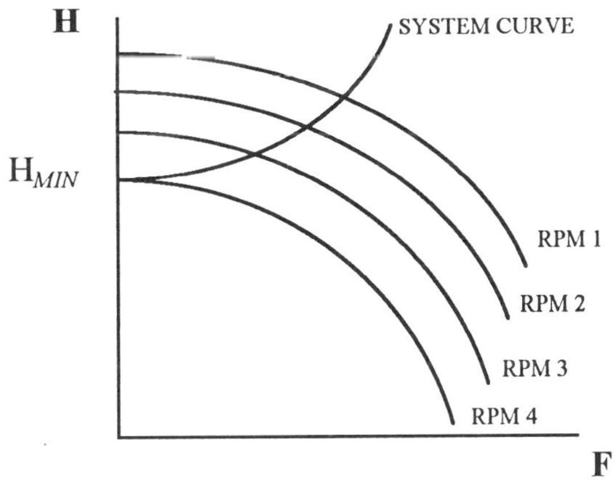

FIG. 3

That minimum RPM curve can be found by the Affinity Laws. The same laws identify the minimum HP requirement. All you need is simple algebra skills. Potential energy savings will be available between (H_{min}) and H_D

$$\frac{N_{min}}{N_D} = \frac{F_{min}}{F_D}; \quad \left(\frac{N_{min}}{N_D}\right)^2 = \frac{H_{min}}{H_D}; \quad \left(\frac{N_{min}}{N_D}\right)^3 = \frac{HP_{min}}{HP_D}$$

$$\left(\frac{N_{MIN}}{N_D}\right)^2 = (\%SPEED)^2_{MIN} = \frac{H_{MIN}}{H_D}$$

$$(\%SPEED)_{MIN} = \sqrt{\frac{H_{MIN}}{H_D}}$$

Therefore:

$$HP_{MIN} = HP_D * \left(\sqrt{\frac{H_{MIN}}{H_D}}\right)^3$$

In the real world, the amount of pump HP savings is determined usually by the amount of friction head variation caused by the variation in flow rates. The greater the friction losses at high flow as compared to low flow the greater the HP savings available. The best situation is one that will also have a reduced pressure requirement. However; this is also a situation that should be evaluated for downsizing the pump first. The application of a VSD to a smaller pump results in and optimization of cost and savings.

THINGS TO LOOK OUT FOR

Constant high flow requirements

If there is no variation in flow <u>and</u> the pump is maxed out, there is no place to save energy. If, however, there is no variation in flow <u>and</u> the pump is grossly under loaded, first try to change the pump to the right size. If operating personnel can't part with their existing equipment (see Pump Retrofit Rule #2), then the installation of a VSD will save a lot of money.

Non-variable fluid transfer systems

In many process pump applications the goal is to move a quantity of fluid from one place to another as fast as possible. The size of the pump and motor are determined by first cost economics. If they could afford to buy a 100 hp pump and reduce the transfer time by 50% they probably would. There is no potential for savings because the process requires the short burst of consumption and then the pump is shut off (see Pump Retrofit Rule #1).

Positive displacement pumps

The use of a VSD on a positive displacement pump is not a problem. The only reason to be aware of such pumps is that they do not necessarily follow the affinity laws. The affinity laws apply to centrifugal loads. Centrifugal loads exhibit the characteristic of requiring variable torque with the variable flow. Positive displacement loads require constant torque over the total range of flows. The horsepower to speed relationship can be a straight line at worst or a square law relationship. Either way saves energy, but the calculated savings will be less than with a centrifugal load that follows the cubed relationship.

Typical examples of positive displacement loads are piston type pumps for highly viscous fluids and mechanical aerators (paddle wheels) for aeration lagoons.

CALCULATIONS

The calculations for pump retrofit applications with variable speed drives may, at first, seem anything but simple. Indeed, the concepts behind the equations and their derivations can be intimidating. However, you don't have to derive the formulas every time you use them. And, while it is always better to understand the finer points, you don't need to be a mechanic to drive a car. Certainly driving the car without knowing the rules of the road can be very dangerous, and just driving around without knowing the route can land you someplace that you don't want to be. But going slow at first in familiar territory by yourself can be very productive and give you valuable experience before you have to deal with a super highway in rush hour.

Remember, you have two goals:

1. Correctly define the existing conditions in terms of total energy consumption

2. Estimate the total energy consumption that will occur after the project is implemented.

The first goal is simply the HP of the pump motor times the hours it runs.

The second goal is simply the HP of the pump motor at reduced speeds (flows) times the hours that it runs at those reduced speeds.

To find the horsepower at any flow you simply have to fill in the unknowns in the following equation:

$$(1)\ Horsepower = (GPM*TDH) \div (3960*Pump\ Efficiency)$$

Determining the various flow rates (gpm) that are required to meet the system's functional needs is what you get paid to do. If you didn't think that there would be any variations then you wouldn't be evaluating a VSD. To determine the TDH you can use the next two equations:

$$(2)\ TDH = Static\ Head\ (ft.) + Friction\ Head\ (ft.)$$

$$Friction\ Head\ (ft.) = (3)\ \left(0.001246*GPM^2*Equivalent\ Lineal\ Feet\ of\ Pipe\right) \div (Pipe\ Diameter\ in\ inches)^5$$

(Note: the constant 0.001246 is the factor for galvanized or black iron (rough) pipe. For brass, copper, or plastic (smooth) pipe, the factor is 0.000623.)

The following information must be gathered or guessed in order to arrive at an answer to the question of TDH:

1. Static Head

This parameter should be the easiest to determine. If there are no gages, the facility personnel should know the pressure requirements and/or the physical elevation changes of the systems involved. If available, use the original construction plans and specifications. If all else fails, research equipment specifications for similar equipment. You can at least base your assumptions on realistic values from someone's published data.

2. Equivalent lineal feet of pipe

This is an undeniable wild guess. If the system is a plant wide cooling loop, use the perimeter of the area served. In any case, don't forget to figure on a round trip for a circulating system. A good estimation of the effect of valves and fittings is to take the estimated distance and multiply it by 2.5 to get the equivalent lineal feet. Get a feel for the complexity of the run. At a city water supply facility it was more appropriate to use a 1.5 multiplier because the pumps served several miles of relatively straight pipe.

This parameter is, however, a value that can, and should, be "tweaked" to establish the estimated system curve. Once all the other data is satisfied, making realistic changes in this value can bring the horsepower calculated at your peak flow gpm to match the installed motor.

3. Average pipe diameter

Always start off with the header diameter at the pump outlet. This will be most accurate for short runs such as boiler feed water systems. If the resulting peak horsepower in the spreadsheet is unrealistically low, adjust this value downward to a realistic effective average diameter that will reflect the effects of longer branch systems with smaller diameter pipes in distant branches.

If you followed closely, you will note that you have solved the two steps. The energy before was the pump running at a constant load all the time. The energy after is the pump running at various gpm for various periods. Just run the horsepower equation for the various flow rates and you will know how much energy it will take to run the system with a VSD. Subtract the before energy from the after energy and you know how much energy you will save.

I told you it was simple.

The following information must be gathered or guessed in order to arrive at an answer to the questions of economics:

1. Active pump horsepower.
The key word is "active". The pump has to be running to save energy. This may sound obvious to some, but if you find a duplex pumping system that is alternated, both pumps need to be controlled but only one uses energy at a time.

2. Motor efficiency.
If the efficiency isn't stamped on the nameplate then make a logical assumption (guess).

3. Assumed load factor.
No motor is loaded to 100%. If a design engineer calculates the horsepower required to be 4.5 hp he won't select a 5 hp motor (if he's smart). Why? Because he knows that the chances of his design information being that correct are slim. He should select a 7.5 hp motor. If you come across a motor that is fully loaded, you have also come across a situation where there is no potential for conservation.

If you assume the existing condition utilizes the full nameplate horsepower you will be making the all too common error of overstating the savings potential. If I don't have anything else to base my assumptions on, I use a load factor of 80%.

4. Pump run time.

5. Assumed normal "OFF" time.
These may sound like a redundant way to say the same thing. What I am addressing is the situation where the controls are automatic. All the facility personnel will be able to tell you is that the pumps are "enabled" during certain time periods. You will have to develop some level of understanding about the control strategy so that you can arrive at an assumed percentage of the enabled ON time that the pumps will be automatically turned OFF.

6. Maximum gpm.
Most of the time this will be given. If you find that you have to guess, you can start with the formula:

$$GPM = 15 * D^2 \qquad D = \text{Pipe diameter; inches}$$

Again, if available, use the original construction plans and specifications.

7. Pump efficiency.
For gpm ranges less than 1,000 gpm use 70% to 75%, for 1,000 to 1,500 gpm use 75% to 80%; for large pumps: 80 to 85%.

8. Operating scenario; percent time at percent flow.
I usually use four steps: 100%, 85%, 75%, & 65% for circulating pumps that are not used in comfort heating or cooling. Try to get the facility personnel to make these assumptions for you. They should have a better feel for what's right and wrong. (This may sound like it contradicts Retrofit Rule #3. It does. But unless you have better data than they do, let them make the guess.) Don't forget to check the minimum speed requirement for the system parameters. Even if the facility personnel give you a speed percentage that is lower this minimum, only use your calculated minimum value for your evaluations. You don't want to look like you don't know what you're doing. Especially if you don't.

PUTTING IT TOGETHER

Cooling and Heating Pumps

For pumps that are used in comfort cooling applications, utilize the local area temperature history through its bin temperature observation data. If you assume that the highest temperature represents 100% load (the system is designed correctly), the lower temperatures represent a lower percentage load and thus a corresponding lower flow requirement.

For pumps that are used in comfort heating applications, utilize the local area bin temperature winter observation data. If you assume that the lowest temperature represents 100% load (the system is designed correctly), the higher temperatures represent a lower percentage load and thus a corresponding lower flow requirement.

Sample Cooling Load Spreadsheet
MAY 1 TO SEPTEMBER 30
24 HOURS PER DAY

DESIGN IAT: 72

BIN TEMP	BIN HOURS	ACCUM. BIN HOURS	% LOAD	VSD KWH
97.5	7	7	100%	70
92.5	34	41	93%	276
87.5	122	163	86%	788
82.5	249	412	79%	1,254
77.5	396	808	73%	1,521

Boiler Feed Water Pumps

The only application for a VSD on a boiler feed water pump exists when the pump runs continuously against a modulating valve that controls the flow. But keep in mind the minimum speed to overcome the boiler pressure is usually higher than 90%.

So where are the savings? Even 95% cubed results in a 14% savings. Since these are typically long run time situations (all winter for heating or all year for process applications) the savings are often sufficient to justify the implementation. In addition, very often the pumps are so oversized that the flow control valve is wasting a lot of energy by throttling, and the output pressure of the pump is very much higher than needed.

There are reasons for over sizing boiler feed water pumps. First, having each pump sized to serve all the boilers in a multiple boiler installation adds redundancy. Second, if the pump is sized only to handle the equivalent gpm of the steam flow it will never be able to raise the boiler level at peak usage. Finally, if it can only just keep up, it will not be able to handle any emergency requirement to raise the water level in the boiler in a hurry. This is one application where I don't recommend downsizing. If you don't change the ability of the pump to deliver whatever peak flow it was designed to produce, you can't be sued for screwing things up. (Successfully sued, that is. You can always be sued.)

Typically when you apply the water horsepower formula (as I do in my spreadsheet) you will be amazed when you calculate a required horsepower 1/2 to 1/3 of the connected horsepower. This is the horsepower that will be delivered by a VSD.

Circulating Pumps

Process cooling, process heating, and process licquer circulation (such as dyeing, tanning, pharmaceutical processes, etc.) usually have variations that are significant but stable for long durations. This is where I use the four-step approach.

The potential for savings is usually the highest because there are long run times and the load is primarily friction loss.

Sample Boiler Feed and Circulating Pumps Load Tables

OPERATING ASSUMPTIONS:

25%	HRS @	100%	FLOW
25%	HRS @	75%	FLOW
25%	HRS @	50%	FLOW
25%	HRS @	25%	FLOW

WATER PUMP LOAD DATA:

GPM	FT. TDH	% SPEED	MOTOR HP	PSI TDH	MOTOR KW
3.4	288.69	97%	4.76	125	4.1
6.7	288.90	97%	4.76	125	4.1
10.1	289.26	97%	4.77	125	4.1
13.4	289.76	98%	4.78	125	4.1
16.8	290.41	98%	4.80	126	4.1
20.1	291.20	98%	4.82	126	4.1
23.5	292.13	98%	4.84	127	4.2
26.8	293.20	98%	4.87	127	4.2
30.2	294.42	98%	4.90	128	4.2
33.5	295.78	99%	4.93	128	4.2
36.9	297.29	99%	4.97	129	4.3
40.2	298.94	99%	5.01	129	4.3
43.6	300.73	99%	5.06	130	4.3
46.9	302.67	100%	5.11	131	4.4
50.3	304.74	100%	5.16	132	4.4

ANECDOTES

Richard's Retrofit Rules

Pumping Rule #1: "You can never save any more energy than shutting it off."

Pumping Rule #2: "You will never get a production manager to accept downsizing."

Pumping Rule #3: "No one in production knows the flow requirements."

The very first VSD installation I designed involved a 50 hp pump in a textile dyeing application. This pump was attached to a 6" diameter pipe with two short elbows immediately connected to the outlet that fed into a pneumatic butterfly valve. The operator made sure that the valve was closed during start up, then he would gradually open the valve until the flow in the open tank was visually satisfying to him. The pump would then run for two to four hours at a time with fifteen minute breaks while the quality was checked. The total daily run time was close to 22 hours.

Because I could visually see that the butterfly valve was only open a fraction I knew that there was potential here. I installed pressure gages on both sides of the valve. The pressure on the inlet side of the valve was 60 psig and the pressure on the outlet side of the valve was only 5 psig.

This confirmed that there was a major energy loss across that valve. In order to quantify the savings I needed to find out what the flow rate was. See rule #3. Not even the dye house Chemist who is responsible for designing the chemical formulas knew how much water was going through the fabric. Fortunately, the equipment manufacturer was willing to make a guess at 750 gpm. The chemist thought that sounded OK, so that's what we used.

Running through the standard pump formulas, I calculated that the required pump brake horsepower for that system without a valve was 5 hp! My first reaction was that I made some stupid mistake. Since I didn't want anyone to know that I was stupid I spent the next day studying the application and pump formulas in general. I couldn't get away from the 5 hp answer.

Finally, I called the equipment manufacturer and asked them where I was going wrong. Here I was saying that there is only a 5 hp load and there they were using 50 hp on all their equipment. They were very cooperative and reassuring. They agreed that 5 hp probably was all that was needed in our application of their equipment, but they sell their equipment to all kinds of dyeing operations. Some of the other operations will need that 50 hp. Since you are paying for it, they keep their design and inventory standardized at 50 hp.

Armed with new self-confidence, I recommended that the 50 hp motors be removed and replaced with 10 hp motors. (Why go too far out on a limb when you don't have to?) Then we can install the VSD equipment at the much cheaper 10 hp level.

Not a chance.

The production supervisors got sweaty palms and the production manager broke out in a rash. "But we may need that 50 hp some day!"

Although my conversations with the equipment manufacturer outlining the different dyeing operations that would require 50 hp had led me to conclude that we would never, never, never need 50 hp, I explained that the motors could easily be reinstalled when the time called for it. And that the difference in the cost to install a 10 hp VSD vs. a 50 hp VSD was about $6,000 for each one of the five pumps.

Money was no object when it came to production reliability, translated: their peace of mind. The 50 hp units were installed. They still saved a lot of money. They even improved quality with better control of the flow and pressure. But the cost to implement could have been much cheaper.

Chapter 71

Industrial District Cooling With Stratified Chilled Water Storage

D.P. Fiorino

ABSTRACT

Stratified chilled water storage is a strategic energy management concept that has been successfully applied in retrofit by Texas Instruments (TI) to reduce energy costs at two of its large defense electronics manufacturing facilities in North Texas. The first application, at TI's 1.1 million sq. ft. Forest Lane defense electronics manufacturing facility in Dallas, won AEE's 1991 "Energy Project of the Year Award." Now, stratified chilled water storage is being applied in retrofit to a very large district cooling system at TI's world headquarters and manufacturing complex in Dallas. This paper will discuss the 175,500 ton-hour "MegaStorage" project in some detail, with particular emphasis on project planning and design.

ENERGY COST ENVIRONMENT

The impetus for stratified chilled water storage in North Texas has been the large demand charge increases caused by the Comanche Peak Nuclear Generating Station. Comanche Peak is now complete – more than eleven years behind schedule and more than ten times as expensive as originally estimated. Its two 1,150 MW generating units began commercial operation separately – the first in 1990 and the second in 1993. Rates increased by $442 million/yr (10.2%) in August 1990 to recover the costs of the first generating unit and by $760 million/yr (15.3%) in July 1993 to recover the costs of the second generating unit.

Characteristic of nuclear rate cases, large percentage increases in the demand charge were necessary for the utility to recover its $13 billion total investment. In fact, the high voltage demand charge increased by 168% from $5.19/kW before Comanche Peak Unit #1 to $13.93/kW after Comanche Peak Unit #2 (See Figure 1). Furthermore, due in part to the financial burden of Comanche Peak, the electric utility deferred all of its previously planned capacity additions indefinitely and applied additional emphasis to demand-side management. As a result, the 1993 thermal energy storage incentive scale increased to $250/kW for the first 1,000 kW shifted <u>plus</u> $125/kW for all additional kW shifted.

These circumstances make stratified chilled water storage attractive to industrial electricity users in North Texas. Consider that water chilling typically accounts for 33% of peak electrical demand and 20% of annual electricity usage at an air conditioned electronics manufacturing facility equipped with an all-electric water chilling plant Once retrofitted with a "full-shift" stratified chilled water storage reservoir at a gross cost of approximately $60/ton-hour, such a facility can shift 100% of its cooling electricity demand to off-peak periods when the demand charge is only $3.48/kW. Furthermore, stratified chilled water storage enables such facilities to take advantage of indirect evaporative chilling, a.k.a., water-side "free" cooling, to reduce cooling electricity usage during the wintertime.

EXISTING COOLING LOADS

TI's Expressway Site in Dallas is a 4.2 million sq. ft. campus-style complex of multi-story buildings having continuous manufacturing operations and year-around cooling loads (See Figure 2). The buildings' ages vary from four years old to thirty years old. Cooling loads consist of: (1) space air cooling and dehumidifying; (2) fresh air cooling and dehumidifying (to make-up for exhausted space air); and (3) heat removal from water-cooled manufacturing equipment. Air handling equipment consists of: (1) built-up units for fresh air cooling and dehumidifying; (2) multi-zone units for cooling and dehumidifying most spaces; (3) dry-coil units for sensible cooling in clean rooms; and (4) single-zone units for clean room dehumidifying and pressurization. Equipment cooling is generally accomplished by chilled water heat exchangers and non-condensing cooling water for dry-coil units is generally produced by chilled water blending stations. Control valves are mostly two-way pneumatic valves with low spring closing force. The site's peak composite cooling load is approximately 22,500 tons (5.4 tons/1,000 sq. ft.) and peak cooling electrical demand currently accounts for approximately 25% of site peak electrical demand.

EXISTING CHILLERS

Regarding chilled water generation, the Expressway Site has 18 water chillers totalling 30,600 tons. Of this total, 17,200 tons are located in the Central Utility Plant, 6,250 tons are located in the East Building Equipment Room, 4,150 tons are located in the Mechanical Equipment Building, one 1,500 ton unit is located in the Research Building, and one 1,500 ton unit is located in the Semiconductor Building. Like the ages of the buildings, the ages of the water chillers vary from four years old to thirty years old. Five of the water chillers are field-erected 3,000 ton centrifugals - four having 3,245 hp, 4,140 rpm steam turbine drives consuming 14 lb./ton-hour of 230 psig saturated steam (0.6 COP) and one having a 4,160 volt, 1,800 rpm synchronous electric motor (4.6 COP).

Of these, four use medium pressure CFC-12 refrigerant and one uses high pressure HCFC-22 refrigerant. Another field-erected chiller is a 1,000 ton low pressure centrifugal driven by a 1,000 hp natural gas combustion turbine, having a high temperature exhaust gas-to-boiler feedwater heat exchanger, and using CFC-11 refrigerant. The remaining twelve water chillers are factory-packaged, low pressure centrifugals using CFC-11 refrigerant and ranging in size from 625 to 1,500 tons. Their efficiencies range from 0.60 to 0.90 kW/ton, depending on age, and several are equipped with auxiliary heat recovery condenser bundles. Due to their low efficiency, high evaporator pressure losses, and high condenser heat rejection, the 3,000 ton steam drive chillers are only operated during peak summertime cooling conditions, and the electric drive chillers and gas drive chiller are "base-load" operated year-around.

EXISTING PIPING

Regarding chilled water distribution, the Expressway Site is served by a 30 year old underground piping network radiating outward from the Central Utility Plant. The distribution system is a closed, all-primary system with 500 hp constant speed pumps located in the main return header in the basement of the Central Utility Plant. Pressurization of 35 psig is applied at that point and a single expansion tank serves the entire site. The chilled water supply temperature is maintained at 41°F in the summertime and the pressure differential between the supply and return headers in the Central Utility Plant is maintained at 25 psid year-around. The water chillers located in other buildings are connected in parallel to the water chillers located in the Central Utility Plant and all returns are direct-connected. Chilled water flow rates vary as equipment is staged in response to changes in the cooling load, but the chilled water temperature rise is often less than 12°F.

PERFORMANCE ANALYSIS

One major discontinuity affecting the performance of the existing chilled water system is that the evaporators of the 3,000 ton water chillers are sized for a 15°F temperature differential and the highest temperature differential provided by the cooling load is only 12°F. This imbalance is due to inadequate flow control and reduces chiller capacity by 20% at design flow rates. Alternatively, if evaporator flow is increased by 25% to fully load the 3,000 ton chillers, the evaporator pressure losses increase by 60%. The latter course of action is required in the summertime, along with much higher pumping energy

The above condition is interrelated with the all-primary chilled water distribution system and inadequate flow control at many of the cooling terminal devices. The single site supply temperature and single site pressure differential must be adequate to satisfy the "worst case." Unfortunately, many of the control valves are unable to close-off against 25 psig, resulting in unnecessary cooling and dehumidifying, as well as below-design return temperatures. Also, low temperature differentials necessitate high flow rates and additional friction losses, causing differential pressure to deteriorate as distance from the Central Utility Plant increases. Finally, the constant speed secondary hot water distribution pumps in the buildings, in conjunction with inadequate flow control at the heating terminal devices, caused false loading of the cooling system under most load conditions.

The conditions described above prompted the installation of additional water chillers nearer to the cooling loads during the 1980's. These water chillers have evaporators sized for 12°F temperature differentials - consistent with the 12°F peak temperature rise provided by the cooling load. Now, with the expenditure of additional energy, 41°F supply temperature and 25 psid pressure differential can be maintained.

ENERGY MANAGEMENT OBJECTIVES

After consideration of the energy cost environment in North Texas and existing cooling loads, chillers, and piping at the Expressway Site, several definitive energy management objectives were established:

First, rapidly increasing cooling energy costs needed to be slashed by 33% or more in order to reduce operating expenses and improve competitiveness and profitability.

Second, 50% or more of existing CFC refrigerants needed to be replaced with alternative refrigerants, then recycled to keep remaining CFC chillers in operation.

Third, a 3,300 ton increase in the peak cooling load needed to be satisfied beginning in June, 1994, without the installation of an additional water chilling plant.

Fourth, cooling combustion emissions needed to be slashed by 90% or more.

Fifth, low differential pressures and temperatures associated with inadequate flow control needed to be cured - not accomodated with additional energy and equipment.

Sixth, any retrofit project undertaken to accomplish the first five objectives needed to earn an attractive after-tax rate-of-return with little systematic risk.

Seventh, any retrofit project undertaken to accomplish the first five objective needed to be time-phased in order to conserve capital and take advantage of the time value of money.

Eighth, any retrofit project undertaken to accomplish the first five objectives could have no adverse environmental impact.

Ninth, any retrofit project undertaken to accomplish the first five objectives could not interrupt site manufacturing activities.

DISTRICT COOLING WITH STORAGE

A number of alternative energy management concepts were evaluated for technical feasibility, economic attractiveness, and conformance to the nine energy management objectives outlined above. These alternatives included, but were not limited to: (1) baseload cogeneration; (2) stratified chilled water storage; (3) applied ice storage; (4) chiller refrigerant conversions; (5) chiller driveline conversions; (6) replacement electric chillers; (7) replacement absorption chiller/heaters; and (8) combinations of the above.

The alternative satisfying all evaluation criteria was a combination of chiller refrigerant and driveline conversions, along with stratified chilled water storage. The deciding factors were: (1) the low cost of off-peak electricity relative to either natural gas or on-peak electricity; (2) the high COP's of fully-loaded electric chillers and auxiliary equipment at nighttime; (3) the low unit cost ($/ton-hour) of stratified chilled water storage; (4) and large utility incentive payments for thermal energy storage. Furthermore, the excellent performance, reliability, and profitability of stratified chilled water storage had already been proven in two large applications totaling 5.8 million gallons and 66,000 ton-hours.

PHASE I

Construction of Phase I of the MegaStorage project began in May, 1993, and completion is scheduled for May, 1994. Phase I incorporates a storage reservoir, a low pressure transfer piping/pumping system, an intermediate pressure primary piping/pumping system, a high pressure secondary piping/pumping system, and numerous flow control improvements and load modification measures inside the South Building.

The main feature of Phase I will be a 5.2 million gallon stratified chilled water storage reservoir - referred to as the South Tank (See Figure 2). The South Tank will have an integrated cooling capacity of 58,500 ton-hours and its discharge rate is designed to be 7,300 tons (10,000 gpm at 40°F supply and 58°F return) from 12:00 noon to 8:00 p.m. daily. Until additional stratified chilled water storage is installed during Phase III, the site will operate as a partial storage cooling district with storage priority. Load modifications will reduce the South Building's peak cooling load by 1,100 tons (13%) and the South Tank will satisfy the remaining 7,300 tons. Thus, 8,400 tons (8.4 MW) of existing electric water chillers located in the Central Utility Plant and the Mechanical Equipment Building will no longer be operated on-peak. And, the remaining 14,100 tons of site on-peak cooling loads, plus the new 3,300 ton on-peak cooling load beginning in June, 1994, will be satisfied by the remaining 22,200 tons of steam, natural gas, and electric water chillers, with 4,800 tons (27.6%) of reserve capacity remaining.

The South Tank will be a precast, prestressed cylindrical concrete reservoir designed and constructed according to ANSI/AWWA Standard D110-86 (Type III). It will have a column-supported flat roof and be fully-buried underneath an existing parking lot adjacent to the South Building. Its diameter will be 140 ft. and its water level will be 45 ft. deep, yielding a near-optimum height-to-diameter ratio of 0.32. Four concentric octagons constructed of 12" diameter PVC pipe having 3/8" wide by 90° lateral slot-shaped openings will provide stratification with a common inlet velocity of 0.45 fps during charging and 0.90 fps during discharging. Inlet Reynolds numbers (Re_i) during charging range from 526 for the innermost octagon to 397 for the outermost octagon.

The South Tank will be interconnected with a below-grade transfer pump station by a pair of 250 ft. long, 24" diameter, low pressure underground pre-insulated PVC transfer lines. The transfer pumps will overcome dynamic losses in the diffuser and transfer piping as well as boost/reduce pressure to/from the intermediate pressure underground primary distribution piping. The three transfer pumps each consist of: (1) a 5,000 gpm forward-rotation booster pump; (2) a 5,000 gpm reverse-rotation recovery turbine; and (3) a variable speed 125 hp motor with shaft extensions at both ends. One of the transfer pumps is designated as a dedicated stand-by throughout the entire cycle of operation.

During the discharge cycle, the shaft rotation is clockwise and the cold pump boosts the pressure of the 40°F cold water withdrawn from the bottom of the South Tank from 0 psig to 15 psig. Simultaneously, the warm pump reduces the pressure of the 58°F warm water returning from the South Building from 35 psig to 0 psig. The energy recovered by the reverse-rotation turbine unloads the motor by approximately 35 bhp (30%) while, at the same time, system sustaining pressure is automatically maintained at 35 psig.

During the charge cycle, the shaft rotation is counter-clockwise and the warm pump boosts the pressure of the 58°F warm water withdrawn from the top of the South Tank from 0 psig to 35 psig. Simultaneously, the cold pump reduces the pressure of the 39.5°F cold water generated at the Central Utility Plant and the Mechanical Equipment Building from 55 psig to 0 psig.

During cycle switch-over, the 125 hp motors are electronically braked, reversed, and accelerated by their variable speed drives, one-at-a-time, with no hydraulic shocks or loss of system sustaining pressure.

The transfer pump station will be interconnected with three at-grade secondary pumps stations, the Central Utility Plant, and the Mechanical Equipment Building by intermediate pressure underground pre-insulated PVC primary lines. The secondary pump stations, each consisting of three variable speed pumps, with one of each set designated as a dedicated stand-by, will maintain differential pressure inside the South Building.

During the discharge cycle, the secondary pumps will boost the pressure of the 40°F primary chilled water from 15 psig to 65 psig. Primary return pressure will be maintained at 35 psig, so the differential pressure inside the South Building will increase by 5 psid (20%) to 30 psid.

Operation of the secondary pumps will be identical during the charge cycle, except that the primary chilled water pressure will increase from 15 psig to 55 psig and the primary chilled water temperature will decrease from 40°F to 39.5°F due to operation of the water chillers and primary pumps in the Central Utility Plant and the Mechanical Equipment Building.

Because the Mechanical Equipment Building's electric water chillers will only serve the the South Tank and South Building, their primary pumps will be downsized from 125 hp to 60 hp. The primary pumps in the Central Utility Plant must continue to serve the entire site and cannot be downsized until Phase III. In the meantime, pressure-reducing valves will be employed to coordinate the pressures of the site's existing high pressure primary supply piping and the new intermediate pressure primary supply piping serving the South Building.

Inside the South Building, all four of the existing 75 hp constant speed secondary hot water pumps will be retrofitted with variable speed drives in order to keep secondary hot water pressure from rising at partial heating loads. New digital controls will also be added to provide more accurate control of secondary hot water temperatures. Together, these measures will eliminate false loading of the chilled water system.

A total of 198 existing chilled water control valves on multi-zone unit cold decks will be replaced with new valves rated for 35 psid operating pressure differentials and 100 psig close-off pressures. New digital controls will also provide more accurate control of cold deck temperatures.

Eight 70,000 cfm built-up fresh air units will be retrofitted with 15 hp pumps and 400 gpm "run-around" coils to pre-cool 102°F outdoor air to 86°F upstream of the existing cooling/dehumidifying coils, then reheat 52°F cold air to 68°F downstream of the existing cooling/dehumidifying coils.

The 1-1/2" bypass ports in 31 existing three-way chilled water control valves in two large computer rooms in the South Building will be plugged. Also, self-regulating temperature control valves will be installed in the return lines of five unthrottled air-to-water heat exchangers in the compressed air plant in the Mechanical Equipment Building.

Taken together, these flow control improvements and load modification measures will increase the South Building's composite water temperature rise from 12°F to 18°F (50%) at peak summertime cooling conditions and, at the same time, reduce the peak cooling load from 8,400 tons to 7,300 tons (13%).

PHASE II

Phase II of the MegaStorage project is currently being designed and construction is expected to begin early in 1994. Phase II incorporates refrigerant conversions for four 3,000 ton CFC-12 water chillers and one 3,000 ton HCFC-22 water chiller, driveline conversions for four 3,000 ton steam turbine water chillers, high pressure secondary piping/pumping systems for the remainder of the site (less the East Building), and numerous flow control improvements and load modification measures for the remainder of the site (less the East Building).

Four 3,000 ton water chillers currently charged with 12,000 lbs. of CFC-12 refrigerant apiece and one 3,000 water chiller currently charged with 6,000 lbs. of HCFC-22 refrigerant - all located in the Central Utility Plant (See Figure 2) - will be converted to HFC-134a refrigerant. In conjunction with refrigerant conversion, the mineral-based oils currently used to lubricate the compressors will be flushed from the systems and replaced with ester-based oils.

New two-stage compressors will be furnished for the HCFC-22 machine and two of the CFC-12 machines, and the rotors in the existing two-stage compressors on the other two CFC-12 chillers will be modified to operate with HFC-134a. The conversion will require compressor speeds to be increased from 4,000 rpm to 4,360 rpm (9.0%) for optimum performance.

Because the compressors will operate at the most efficient point on their "map" in this particular conversion, shaft horsepower is reduced from 2,936 bhp to 2,897 bhp (1.3%). Also, the combination of HFC-134a refrigerant, higher evaporator entering water temperatures (due to flow control improvements), and lower condenser entering water temperatures (due to removal of steam condensing load from the cooling towers, as well as nighttime-only operation) results in a capacity increase from 3,000 to 3,220 tons (7.3%).

The optimized compressors will be driven by 3,000 hp, 4,160 volt, 1,800 rpm synchronous electric motors having 0.8 leading power factors and 0.985 efficiencies. These exceptionally high motor efficiencies will be offset by 3% losses associated with the 1,800/4,360 rpm speed reducers required between the motors and compressors. Nonetheless, the retrofitted water chillers will have a demand of 2,260 kW at 3,220 tons, resulting in an overall efficiency of 0.70 kW/ton (5.0 COP).

The driveline conversion will also eliminate 740,000 MCF/year of cooling natural gas combustion, with associated reductions of 74.0 tons/year of NO_x emissions and 7.4 tons/year of CO emissions.

The 48,000 lbs. of CFC-12 refrigerant recovered during Phase II will be exchanged for CFC-11 refrigerant, allowing the 13 remaining CFC-11 water chillers to be operated until their future scheduled conversion/replacement dates.

PHASE III

Phase III of the MegaStorage project is currently being designed and construction is expected to begin in mid-1994. Phase III incorporates a second storage reservoir, a second low pressure transfer piping/pumping system, completion of intermediate pressure piping/pumping system site-wide, a high pressure secondary piping/pumping system for the East Building, and numerous flow control improvements and load modification measures inside the East Building.

The main feature of Phase III will be a 10.4 million gallon stratified chilled water storage reservoir – referred to as the North Tank (See Figure 2). The North Tank will have an integrated cooling capacity of 117,000 ton-hours and its discharge rate is designed to be 14,600 tons (20,000 gpm at 40°F supply and 58°F return) from 12:00 noon to 8:00 p.m. daily.

Once the North Tank is installed, the site will operate as a full storage cooling district, with the sole exception being the 1,000 ton low pressure water chiller driven by a 1,000 hp natural gas combustion turbine. Load modifications will have reduced the site's peak composite cooling load by an additional 1,800 tons (to 22,900 tons) during Phases II and III in order to permit this.

Once the intermediate pressure primary distribution piping system is in service site-wide, the primary pumps in the Central Utility Plant and the East Building will be downsized and operation of the pressure-reducing valves installed during Phases I and II will be discontinued.

SUMMARY

The MegaStorage project is a bold undertaking. It achieves nine challenging energy management objectives in three phases via its broad scope and high degree of integration. Although its design features and operating strategies are at the leading edge of thermal energy storage practice, the MegaStorage project employs relatively simple, inexpensive stratified chilled water storage and builds on a proven track record. Furthermore, the project's underlying analysis is thorough and its planning and design are sound. A series of future papers will discuss the progress and results of the MegaStorage project.

REFERENCES

Fiorino, D.P., "Thermal Energy Storage Program for the 1990's," in Energy Engineering, Vol. 89, No. 4, 1992, pp. 23-33.

ABOUT THE AUTHOR

Donald P. Fiorino is a Facility Engineer and Member of the Group Technical Staff at Texas Instruments in Dallas, Texas. He received his M.S. in Industrial Engineering from the University of Texas – Arlington in 1986 and his B.S. in Applied Science and Engineering from the United States Military Academy at West Point, New York in 1977. Fiorino serves on the National Advisory Council for the Thermal Storage Applications Research Center at the University of Wisconsin – Madison.

DALLAS EXPRESSWAY SITE
HIGH VOLTAGE DEMAND CHARGE

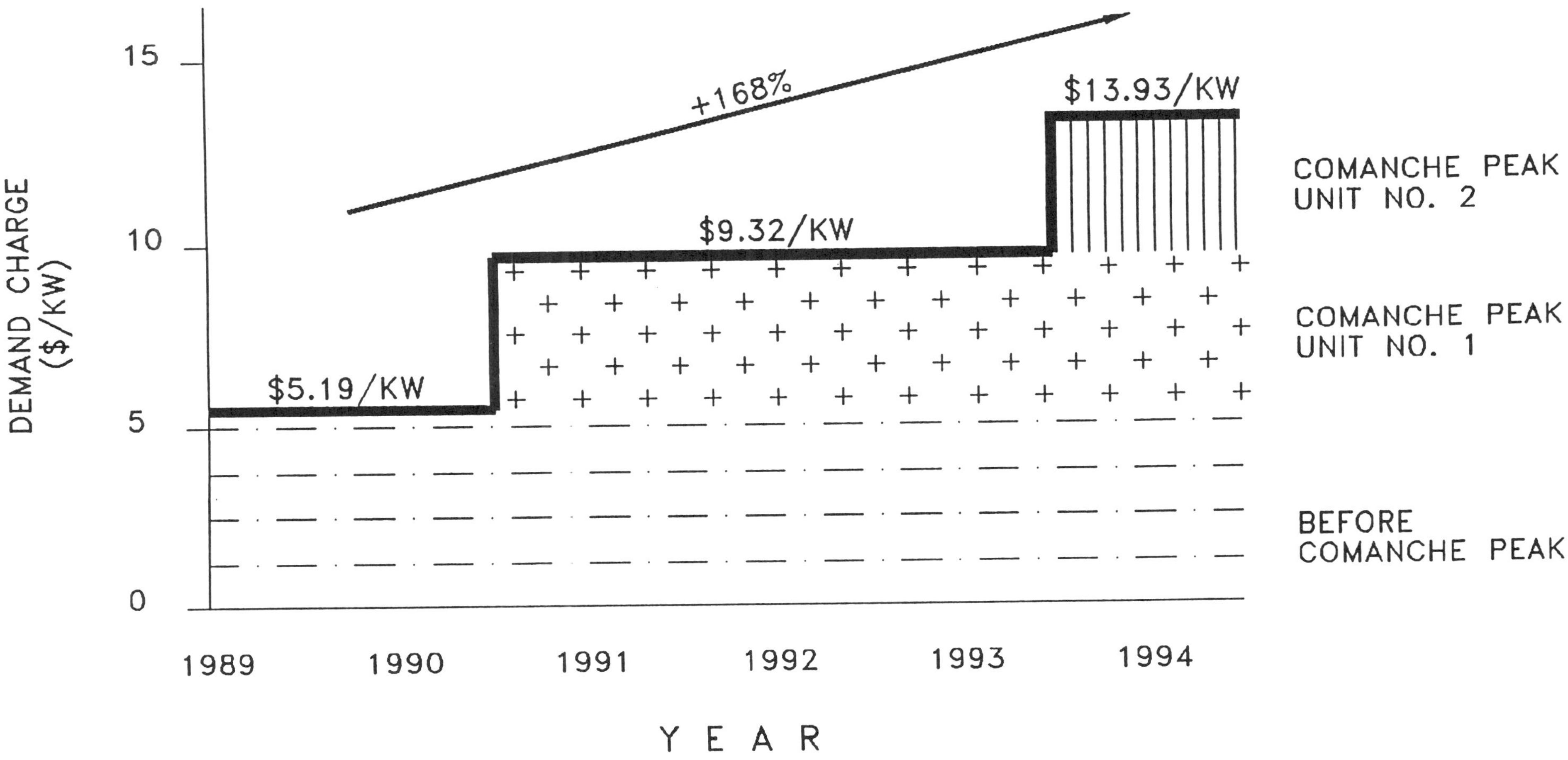

Figure 1

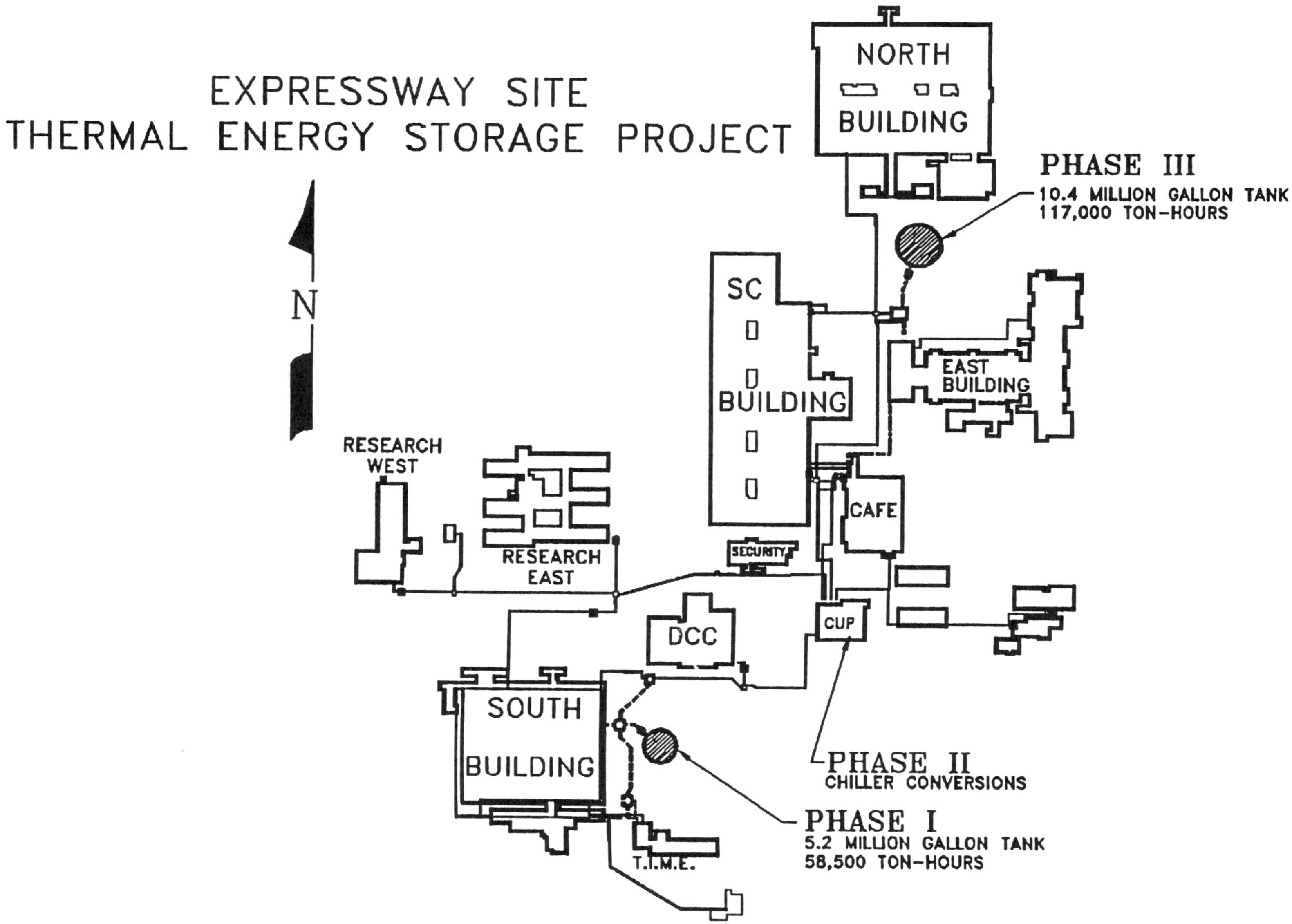
EXPRESSWAY SITE
THERMAL ENERGY STORAGE PROJECT
N
NORTH BUILDING
PHASE III
10.4 MILLION GALLON TANK
117,000 TON-HOURS
SC BUILDING
EAST BUILDING
CAFE
SECURITY
RESEARCH WEST
RESEARCH EAST
DCC
CUP
SOUTH BUILDING
T.I.M.E.
PHASE II
CHILLER CONVERSIONS
PHASE I
5.2 MILLION GALLON TANK
58,500 TON-HOURS

Figure 2

Chapter 72

Evaluation of the Technical and Economic Performance of Energy-Efficient Engine Driven Chillers (EDC)

D.J. Anderson

INTRODUCTION

The use of internal combustion, reciprocating engine drives, was extremely popular about 30 years ago to power air conditioning and refrigeration compressors. By far the most popular fuel of choice for these drives was natural gas, supplied by the local gas utility. After demonstrating its capabilities on the job, the reciprocating engine drive earned industry acceptance as one of the "conventional" drives.

Because of the increasing number of electric generating plants being constructed during that period, the availability of electrical power grew, and the rates for electric energy became lower. This caused the operation of many of the engine driven air conditioning and refrigeration systems to become uneconomical. However, there were some systems installed during that period that remain operational today.

Once again, the use of internal combustion reciprocating engine drives has become acceptable and considered one of the "conventional" drives. The use of Engine Driven Chillers (EDC) and the associated new technology and controls is being endorsed by the American Gas Association (AGA), American Gas Cooling Center (AGCC), Gas Research Institute (GRI) and local gas utilities as one of the most economical and efficient chilling methods available. Of course, there remain areas of the United States and the World, where abundant and low priced electricity continue to be the most economical means of powering air conditioning and refrigeration compressors.

In many areas of the country, the electric utilities are looking at or incorporating gas chillers as part of their programs for Demand Side Management (DSM). According to the Electric Power Research Institute (EPRI), comfort cooling in the commercial sector places the highest summertime on peak demand and the second highest annual consumption demand on electric utilities. Lighting is the highest annual consumer of electricity. The demand associated with cooling, left unchecked, without DSM programs would increase almost 45 % by the year 2010 and consumption would increase nearly 35%. To relieve the large demand and consumption associated with chilling, electric utilities may offer some incentives to encourage their customers to purchase and use natural gas engine driven chillers as an alternative.

It is the intent of this presentation to provide material and guidance useful to all who have an interest in the application of Engine Driven Chillers and Refrigeration Systems. This information can be used to prepare or develop a preliminary feasibility or economic study but is not intended for system sizing or design. Determination of the actual size and characteristics of a cooling system for a given facility is the responsibility of a professional engineer.

The term "air conditioning" encompasses year round control of not only temperature, but humidity and the quality of air in the building. This usually requires cooling in the summer and heating in the winter. Since conditioning during the summer calls mainly for a cooling operation, all air conditioning systems incorporate a refrigeration circuit to perform this function. It is this component that consumes the majority of the energy or power required to operate the system. It also increases the facility demand for power and could even place a burden on the electric utility's capacity.

Cold or cool refrigeration compression is even more cost intensive because of the additional horsepower per ton required for the lower temperatures developed. Because economic sense in areas of the country that have high demand and electricity costs.

Internal combustion reciprocating engine drives usually offer a very attractive operating economy, high C.O.P., as their most outstanding feature. They also offer an opportunity for the user of such systems to quickly recover the price differential, if any, between electric drives and the EDC. Over time and operation of the system, the user can recover all the costs associated with the system. Cost recovery will never occur with the use of an all electric system. C.O.P.'s become higher and the recovery time is shortened when the facility captures and uses the normally wasted heat from the engine jacket water and exhaust.

TECHNICAL PERFORMANCE OF THE EDC

Performance has many definitions; however, in the equipment industry it may be defined simply as an evaluation of how well a piece of equipment performs the job for which it was designed. As applied to engine driven chiller or refrigeration systems, performance is primarily an evaluation of the behavior of the system with respect to the following items:

A. Energy Input, Horsepower per Ton

B. Fuel Economy, BTU per Horsepower Hour

C. Operating Range, RPM

D. Operating Range, Load

E. System Controls and Interface

F. Reliability and Availability

G. Torsional Compatibility of Components

H. Heat Recoverable and Usable

Energy Input Per Ton

Since the energy input per unit of refrigeration, like the power requirements per unit of capacity, for centrifugal and rotary screw compressors is so much a function of system design, it is not considered practical to attempt to apply average values, except for comfort cooling. For an accurate analyses of low temperature systems, the energy requirement per ton (or applicable unit) of refrigeration should be secured from the compressor manufacturer or the packager of Engine Driven Chiller Systems for each and every installation being considered.

The energy (power consumption in horsepower-hour) per ton-hour of chilling will, under most conditions, be less than one HP-HR. It is a characteristic of positive displacement compressors for the efficiency to improve as the load is reduced from maximum to approximately 50 to 60 % of rated capacity. Also, since compressors in air conditioning service very seldom operate at maximum output for more than a few hours per season, power consumption, and thus, fuel consumption calculations for preliminary economic studies will be safe when based on the ratio of one HP-HR per Ton-Hour of chilling. In low temperature refrigeration applications, the required HP-HR per Ton-Hour increases based on the needed temperature and could even exceed a 1.5 to 1.0 ratio.

There will be a horsepower loss in the drive train of between two to four percent, depending on the type of drive system and whether a speed increaser is utilized. However, the ratio of approximate engine output per Ton-Hour of refrigeration is adequate to compensate for all losses in the drive train when completing a rough economic analysis. If a more accurate analysis is needed, use the "Annual Cost Method" form in the Economic Performance section of this presentation.

Fuel Economy

The fuel economy of any EDC System will be largely contingent upon five major factors:

1. Energy input required per ton of refrigeration. (Horsepower-Hour per Ton-Hour)

2. Annual hours of operation and at what load.

3. Load factor. The actual load in tons divided by the rated capacity of the EDC System, assuming the engine is properly matched to the compressor.

4. Revolution Per Minute (RPM) of the engine and whether it is constant or variable speed design.

5. Brake Specific Fuel Consumption (BSFC) of the engine and the related part load fuel consumption.

Of all the factors listed, only the load factor is likely to be difficult to determine. In air conditioning applications, the load factor can usually be estimated within acceptable accuracy for fuel consumption calculations. Low temperature or refrigeration applications will be even easier to estimate load factor because of the steadier state load and the longer annual hours of operation. In cooler climates, such as the northern United States or other areas with similar conditions, the average seasonal load factor for properly sized and applied comfort cooling load and the longer annual hours of operation. In cooler climates, such as the northern United States or other areas with similar conditions, the average seasonal load factor for properly sized and applied comfort cooling systems will range from 50% to 65%. Fuel consumption estimates based on a 60 % load factor conform quite closely to actual recorded data from metered plants in that geographic region. In warmer climates, such as the southern portion of the United States or similar climates, the average seasonal load factor will usually be somewhat higher. The normal range will be between 70% to 85%, for again, properly sized and applied systems.

Factor number 1, energy input per ton of refrigeration is a variable quantity which is affected by a.) load factor b.) temperature to be maintained c.) condenser water temperature and d.) efficiency of the compressor. However, proper application of the data and suggestions offered in the preceding paragraphs will lead to quite reliable estimates of the average Brake Specific Fuel Consumption (BSFC) or average consumption per ton of refrigeration.

Factors number 2, annual hours of operation and number 4. Revolutions Per Minute (RPM), are easily determined by the facility profile, by the type of engine used, and by the site loads to be served. This will, in part, establish either a constant or variable speed design.

Factor number 5, Brake Specific Fuel Consumption characteristics of the engine portion of the system should present no problem as "Part Load" fuel consumption curves or charts are normally available from the chiller system manufacturer for the different models and configurations being offered. Because EDC Systems and their applications do not favor the prime mover or engine with optimum load conditions for long periods of time, it is necessary to know the fuel consumption habits of the engine when operating at part load and with some systems at various speeds. While engine manufacturers or dealers may be able to supply approximate data, they may not be aware of the drive being used or losses associated with the drive or any parasitic loads of the EDC.

Operating Range, RPM

Manufacturers of EDC systems using a constant speed design or facilities that have a constant load condition need not be concerned with the operating range of the engine being used. The engine will be sized to operate the system based on the HP per Ton requirement of the compressor with associated losses. Most centrifugal and rotary screw compressors have a means to modulate the unit capacity from 100% to 10% of the designed load. This is normally accomplished with vanes or valves that are part of the compressor unit.

Some manufacturers offer both constant and variable speed designs. The variable speed systems are controlled based on the actual site load at a given time and the temperature required.

The operating or "working" speed range of the variable speed EDC systems may be limited by either the compressor or the engine, assuming all torsionals are eliminated. Where speed modulation is to be used as a method of capacity control, the speed range acceptable to the compressor must be provided by the manufacturer. The working RPM range of the reciprocating engine is the range between rated operating RPM and the RPM at which maximum torque occurs. This usually includes a range equivalent to approximately 33% of the rated RPM. That is, the "working" range will be between 67% and 100% of the rated RPM. In actual practice , however, when applications are made, each engine must _______to applied according to is particular torque characteristics.

With both constant speed or variable speed designs, it is recommended that, when possible, a fluid coupling is used in the drive train to assure "soft starts" and eliminate potential torsional problems and generally cushion the system. Without the use of a fluid coupling, a complete torsional analysis must be performed at all operating speeds.

Operating Range, Load

With centrifugal and rotary screw systems designed and operated at a constant speed, the load range will vary from 100% all the way down to 10% of the rating. Load control is accomplished with valves or vanes in the compressor and there is no concern about surging or torsional problems at lower loads.

Centrifugal compressors that are designed to accept speed modulation as part of load control will usually accommodate only about a 20% working range, or deviation from the rated RPM, and still remain safely out of the surging area. This speed modulation range will normally provide a capacity control down to around 50% of the compressor rating. Load control below the engine peak torque is accomplished through the compressor down loading.

Most rotary screw compressors that normally operate at higher speeds, 3600 RPM, can be coupled with the medium or slow engines through a speed increaser or derated and run at 1800 RPM. Screw compressors designed to operate at variable speeds will run in the 3600 RPM to 1800 RPM operating range as part of the load control. Again, the engine peak torque will determine where the load control is accomplished by the compressor down loading.

System Controls and Interface

Automatic controls for the Engine Driven Chiller Systems are supplied as a common practice by most suppliers. These Micro Computer Control Centers (MCCC) will start, operate, control, supervise and stop the systems on signal from either the compressor or the engine. This can be accomplished locally or by remote control through a personal type computer. Part of the MCCC's job is to control the engine as to emulate an electric motor drive to the compressor.

As part of the overall system control, it is critical to equip the engine with a governor that will properly respond to whichever type of system it is to operate. The governor will either control the engine RPM at any speed within the working range of the compressor, or maintain a constant speed on that type of system. Although speed modulation control may reduce the cycling frequency or "on/off" operations, some constant speed designs react faster and more smoothly to load change by using the engine manifold pressure to operate the sliding vanes or valves on the compressor.

Integral controls for any system should be specifically designed and built to work in conjunction with the particular EDC design and should not be general type controls or safeties. Make sure you ask the manufacturer or his representative to provide any additional data on their MCCC supplied with the system. Chances are good that the same control will be able to handle additional site control requirements and/or may be expandable to meet your needs.

Reliability and Availability

Reliability and availability of mechanical equipment are quite simply the products of quality and durability of the equipment, coupled with the preventative maintenance service provided.

Maintenance and operating records of currently installed and operating industrial grade EDC systems confirm that their record of availability has been, and is today, either equal to or better than the electrically powered systems in the same area or application. Redundant units are sometimes employed based upon the facility cooling curve, seasonal variations on cooling loads, or critical environments where a standby system would normally be required anyway.

Where large seasonal variations occur, the EDC systems can operate without penalty of electrical demand charges during the light to moderate load months when chilling may only be required but a few hours per month. The systems can also be employed as part of your internal peak shaving program or as part of your electric utilities Demand Side Management (DSM) program. In some cases, there are incentives offered to reduce electrical demand on the utility grid.

Torsional Compatibility of Components

The total absence of serious torsionals throughout the operating speed range is a must for any combination of "prime mover" and driven equipment. For all Engine Driven chiller Systems, this quality can be assured by requiring the manufacturer to provide a torsional analysis throughout its operating speed range on any hard coupled systems. This must include all components that make up the drive train.

Both constant and variable speed designed systems can and should be equipped with some type of "soft" coupling which will assure the absence of critical torsionals in the system.

Heat Recoverable and Usable

Direct fired "prime movers", engines, covert only a part of the fuel energy input to mechanical energy, and the balance is rejected in the form of heat energy, part of which can be recovered and used. Recovery systems on reciprocating engines can provide hot water and low pressure steam for use in the installed facility.

The total recoverable heat is equal to approximately 40% of the input fuel, in BTU's. Heat recovered from the engine in the EDC system can also be used in an absorption chiller to provide additional capacity. This type of system is referred to as a "piggy back" system.

Technical Performance Conclusion

There is nothing secret, mystical or magic regarding the application of a reciprocating engine to drive a cooling compressor. The engine is once again accepted as one of the "conventional" drives.

Some may want to construct their own system from components that they select and purchase. This presentation, I am sure, will be an excellent guide to proper design and application. For those that plan to simply purchase a system from a manufacturer or packager of EDC systems, this will act as a guide regarding the heavy duty design and performance that you can expect and demand from the supplier that you may select. Be sure to ask the questions that concern you.

ECONOMIC PERFORMANCE OF THE EDC

Economics as used in this presentation, refers simply to the effect of a given course of action upon the profit or fiscal position of a respective consumer of purchaser of an Engine Driven Chiller (EDC) System. An analysis is a measure of whether this course of action is unfavorable or favorable, depending on the prevailing conditions, and to what degree. Since the purchase or lease of any type of EDC system entails the investment of capital, as well as the expenditure of money for operating and maintenance costs, the economics associated with the venture are usually best expressed in language or terms which relate to the ultimate gain to the capital investment. It is not sufficient to simply determine that an investment will make or save money, but at what rate will it make or save money. What Rate Of Return (ROR) may be expected from the money invested in an EDC system? The analysis in this presentation will permit a fair comparison of the prospective benefits from the investment.

There are no simple short cuts for making a complete and somewhat accurate appraisal of any proposed business venture, but there are no particular mysteries about the procedure either. The purchase or lease of an EDC system can be viewed as an investment, made for the purpose of yielding an attractive return. EDC systems do yield a return, where the standard electric motor driven systems do not, but are a liability.

The emphasis on the capital investment results from the assumption that the use of a reciprocating engine drive always entails an add initial cost as compared to electric drives of absorption systems. EDC systems have become somewhat competitive to the other forms of cooling, however, we will assume that there will be some additional incremental or initial investment required.

For those cases where using the EDC system does not require an additional investment or the incremental investment is rather small, the most practical method of making an economic appraisal involve nothing more than a simple comparison of the projected annual costs of the system being considered. The comparison should include both owning and operating costs.

Economic Analysis Factors

There are several easily identifiable factors which collectively determine the economics of the EDC system. In addition to the items that affect the mechanical performance, as discussed earlier, we must complete an "Annual Cooling Cost" (attachment A) method of determining savings. In order to do that and before we analyze the cash flow and Rate Of Return (ROR), we need the following information:

1. Tonnage of existing and new chiller system.
2. Kilowatt and horsepower per ton of capacity.
3. Cost of energy for current system.
4. Maintenance costs. (incremental/actual)
5. Annual operating hours of chiller system.
6. Load factor.
7. Cost of EDC fuel.
8. BSFC of the engine drive.
9. Usable heat, if any.
10. If heat is used, boiler efficiency.

With the above information, we can complete the rough cut "Annual Cooling Cost" (attachment A) method of solving for potential savings.

Once we have determined our potential annual savings, and run the cash flow analysis, and determine our ROR, we will need the following:

1. Incremental or initial capital investment.
2. Number of capital recovery years.
3. Corporate tax rate.

Any savings on the operational cost of a facility is obviously attractive! But how attractive is it when considering the amount of capital investment required to accomplish these savings? Capital investment can be either the incremental investment or the total spent to implement the EDC system. This total must deduct any cost savings for deferred equipment investment or retro fit, salvage value of any removed equipment and any incentive money paid by a utility to the owner. The following straight line method example illustrates just one approach for reducing the economic evaluation to more meaningful comparative terms.

To determine Cash Flow, follow these steps:

1. $\dfrac{\text{System Cost}}{\text{Capital Recovery Years}}$ = Annual Capital Recovery

2. * Annual Savings - Capital Recovery = Savings Subject to Tax

3. ** Tax Rate X Taxable Savings = After Tax Savings

4. After Tax Savings + Capital Recovery = Cash Flow

*Use "Annual Cooling Cost" method savings.
**Example: 34% tax rate, use .66 multiplier.

In this instance, Cash Flow is the amount of money to be retained by the owner of the EDC system after all expenses, except capital costs are paid. To determine the rate of return on the capital invested, it is only necessary to assume that the entire Cash Flow will be applied to defray capital costs or interest plus capital recovery.

This example uses one method of solving for the interest rate by using the data Table # 1 when the amount of Cash Flow and the Capital Recovery Years are known.

This type of an analysis fulfills the objective to reduce the economic gains to a useful common denominator. It should be emphasized that this is an economic analysis only, and is not necessarily any indication of how a specific accountant may record the economic gain or performance as seen by the owner/user of an EDC system. Each company or corporation will have its own way of calculating the benefits and gains associated with the use of the EDC. This is known as the Internal Rate Of Return and could vary greatly from the above example.

Table 1

Capital Costs per Year, per $ 1000.00
(for capital recovery years as indicated)

Int. Rate	5 Yrs.	10 Yrs.	12 Yrs.	14 Yrs.	15Yrs.	16 Yrs.	18 Yrs.	20 Yrs.
3 %	218.35	177.23	100.46	88.43	83.77	79.61	72.71	67.22
4 %	224.63	123.29	106.55	94.67	89.94	85.82	78.99	73.58
5 %	230.97	129.50	112.83	101.02	96.34	92.27	85.55	80.24
6 %	237.40	135.87	119.28	107.58	102.96	98.95	92.36	87.18
7 %	243.89	142.38	125.90	114.34	109.79	105.86	99.41	94.39
8 %	250.46	149.03	132.70	121.30	116.83	112.98	106.70	101.85
10 %	263.80	162.75	146.76	135.75	131.47	127.82	121.93	117.46
12 %	277.41	176.98	131.44	150.87	146.82	143.39	137.94	133.88
15 %	298.32	199.25	184.48	174.69	171.02	167.95	163.19	159.76
20 %	334.38	238.52	225.26	216.89	213.88	211.44	207.44	205.36
25 %	371.85	280.07	268.45	261.50	259.12	257.24	254.59	252.92
30 %	410.58	323.46	313.46	307.82	305.98	304.58	302.69	301.59

$$\frac{\text{Cash Flow}}{\text{Investment (thousand \$)}} = \text{\$ return/year/thousand dollars invested}$$

The dollar return, per year per thousand dollars invested, can be applied to Table # 1 under the recovery years, to determine the approximate or actual Rate Of Return. If the ROR dollars fall between two percentages and you want to know the actual ROR, then you can use the following:

Use the percentage that applies to the next lowest dollar amount (up on the chart) and to that percentage add the sum of the calculation below:

$$\frac{\text{\$ return/year/thousand - lower amount, table 1}}{\text{higher amount, table 1 - lower amount, table 1}} \times .03$$

This will give you the actual Rate Or Return.

Also, there may be other methods of depreciation available that could materially increase the Cash Flow and improve the owners tax position.

Tax Exempt Facilities
Should the EDC system be installed at a non-profit hospital, city, county, federal, or any other quasi-governmental tax exempt facility, the economics would be considerably more attractive. In that case, the Gross Savings per Year and the Cash Flow are one and the same, thus, the dollar return/year/thousand dollars invested increase significantly.

Heat Recovery Economics
Like any venture which requires the investment of capital, the economics of heat recovery will be attractive only when the gains justify the investment. Heat recovered from an EDC system will be worthwhile where

there is a sizable need for domestic or process hot water during the cooling season.

For those installations where the recovered heat can be utilized to produce additional chilling without increasing the total plant cost, the economics will certainly be attractive. This condition prevails when the capacity of an existing absorption plant is being increased by the addition of an EDC system.

For new plants, the economics of using heat recovery to produce additional chilling becomes attractive when the size of the plant reaches about 500 tons or more. At this point, the additional capital outlay required for the combination EDC and absorption system, as compared to the EDC system only, becomes either non-existent or is reduced to a small amount.

Economic Conclusion

A simple economic analysis is recommended in all cases, when contemplating the application of an Engine Driven Chiller System . Once the installed costs have been determined or estimated for the system being considered, either the Annual Cooling Cost method and/or the ROR method will serve to place the economics of the system in the proper perspective for making an intelligent, learned and profitable decision.

To assist any companies interested in the possible application and use of EDC systems, my company offers to do a rather through financial evaluation for you. We offer the seven page preliminary analysis at no charge to the prospective customer or for our friends with the utilities. You may contact us to obtain the questionnaire form that must be completed by you and returned to us for computer analyzation.

Overall EDC System Conclusion

With all the changes in chiller system technology and the different systems that are available to you, it is important to know what your options are regarding a system that is right for your facility and for your needs. Hopefully, this presentation will assist you in the review of your options and help you to make a learned decision on what is best for your facility.

One are we have not addressed is the effect of the "Montreal Protocol", the international agreement protecting the stratospheric ozone layer, which speeds up the phase out of Chlorofluorocarbons (CFC's) from the year 2000 to January 1, 1993. The available alternatives for the short run are CFC retention, compressor retrofit or system replacement with non CFC refrigerants. The replacement HCFC of HFC refrigerants are being tested and evaluated to determine the best substitute. Various compressor manufacturers are promoting various replacements based on their equipment and their opinion as to the best refrigerant.

At best, there is an extreme amount of confusion on the end user as to the best refrigerant to use.

The number of systems that need to be addressed or replaced vary from around 80,000 to as high as 170,000, depending on the size and configuration of the chiller package and who is estimating. The larger number is from the Environmental Protection Agency (EPA) and does not address cool or cold storage systems currently in operation.

Each and every facility that operates chiller or refrigeration systems must establish a plan for retainment, retrofit or replacement or their system now. Time is short and the alternatives are few.

As part of your plan, you will now be able to even better evaluate the alternative systems being offered and judge them first on their technical and performance merits. This will assure that your decision will incoude heavy duty, twenty to thirty year life components and systems that will give you dependable and economical performance.

Once you have selected the system type and design that best fits your requirements and needs, then you can perform the economic analysis. This will tell which system costs or saves you money and to what degree. Look at the CASH FLOW and see what you can anticipate regarding savings. Determine the Rate Of Return (ROR) and then, and only then, will you be able to finalize your decision and if need be sell your project to those who manage your company or those who control the capital expenditures.

Remember, the decision you make today will affect your operation for years to come. So take the time and thoroughly investigate and examine all your alternatives.

Attachment A

ANNUAL COOLING COSTS

Electric Plant

ENERGY: (current system)

Chiller rating in tons ___________ X _____._______ KW/Ton X $___._______/KWH (1)

(1) Average KWH cost Energy Cost per Operating Hour $___._______

MAINTENANCE:

Chiller rating in tons ___________(2) X $0.0075 /ton/hr

(2) Use full rated tonnage Maintenance Cost per Operating Hour $___._______

Current Cost per Operating Hour $ _________

Annual Operating Hours X _________

(3) **Load Factor % X ___.______**

(3) Average annual load versus system capacity

CURRENT ANNUAL OPERATING COST = $ _________

Natural Gas Engine Driven Chiller

(no heat recovery)

ENERGY: (proposed system)

Chiller rating in tons _______ (2) X ________Horsepower per Ton = ________ Engine Horsepower ** X ._____________Therms/ HP(4) X $._____________ Per Therm=

Energy Cost per Operating Hour $_____._______

(4) Brake Specific Fuel Consuption (BSFC) per HP-HR divided by 100,000

MAINTENANCE: (EDC system)

Engine full load horsepower ____________ X $0.014 per HP-HR =

Maintenance Cost per Operating Hour $ ___._______

EDC Cost per Operating Hour $ ___.______

Annual Operating Hours X _________

(3) Load Factor % X ___.______

EDC ANNUAL OPERATING COST = $ _________

CURRENT ANNUAL OPERATING COST $ _________

WITH EDC ANNUAL OPERATING COST $ _________(minus)

ANNUAL SAVINGS $ _________

HEAT RECOVERY CREDIT

Value of Recovered Heat:

Engine full load horsepower ** __________ X 0.039 (5) Therm per HP ÷ .____ Boiler Efficiency X $ ____._______ per Therm

Heat Value per Operating Hour $ ___.______

(5) Average Available Heat per HP-HR

Annual Operating Hours X _________

EDC (3) Load Factor % X ___.______

Heat Use Load Factor % X ___.______

Recovered Heat Value = $ _________

CHILLING SAVINGS PER YEAR $ _________

HEATING SAVINGS PER YEAR $ _________

TOTAL SAVINGS PER YEAR $ _________

EXAMPLE

In this EXAMPLE, we are considering the replacement of an older and expensive to operate centrifugal chiller rated at 500 tons. We are investigating the use of a natural gas fueled Engine Driven Chiller (EDC) System with a low horsepower per ton requirement and a high Coefficient Of Performance (C.O.P.).

The conditions are as follows:

1.) The chilling portion of the air conditioning system operates around 1715 hours per year and requires .85 Kilowatt/ton. The average electrical cost, including demand charge is \$ 0.08/kwh.

2.) The annual load factor on the system is 65%.

3.) The cost for natural gas is \$ 0.45 per therm.

4.) The engine horsepower required for the EDC is .78 BHP/Ton and the Brake Specific Fuel Consumption (BSFC) is 7200 BTU/HP-HR.

5.) We do have a need for hot water during the cooling season so we wish to recover the heat to supplement the 80% efficient boiler at our facility. Our heat load factor is 60%.

<u>For determining cash flow and Rate Of Return (ROR) we need the following:</u>

1.) The incremental cost difference between the electric driven and the EDC system is \$ 120,000.

2.) Capital recovery period is ten (10) years.

3.) Corporate tax rate is 34 %.

Calculation Steps

Step #1. Calculate the approximate potential savings using the EDC system with heat recovery using the "Annual Cooling Cost" method per Attachment A.

Step #2. Calculate the CASH FLOW from use of the EDC.

Step #3. Finally, calculate the Rate Of Return based on the installation of the EDC.

Attachment A

ANNUAL COOLING COSTS

Electric Plant

ENERGY: (current system)

Chiller rating in tons 500 X .85 KW/Ton X $ 0.08 /KWH (1)

(1) Average KWH cost Energy Cost per Operating Hour $34.00

MAINTENANCE:

Chiller rating in tons 500 (2) X $0.0075 /ton/hr

(2) Use full rated tonnage Maintenance Cost per Operating Hour $ 3.75

Current Cost per Operating Hour $ 37.75

Annual Operating Hours X 1715

(3) **Load Factor %** X .65

(3) Average annual load versus system capacity

CURRENT ANNUAL OPERATING COST = $ 42,082

Natural Gas Engine Driven Chiller

(no heat recovery)

ENERGY: (proposed system)

Chiller rating in tons 500 (2) X .78 Horsepower per Ton = 390 Engine Horsepower ** X .072 Therms/ HP(4) X $.45 Per Therm=

Energy Cost per Operating Hour $ 12.64

(4) Brake Specific Fuel Consuption (BSFC) per HP-HR divided by 100,000

MAINTENANCE: (EDC system)

Engine full load horsepower 390 X $ 0.014 per HP-HR =

Maintenance Cost per Operating Hour $ 5.46

EDC Cost per Operating Hour $ 18 10

Annual Operating Hours X 1715

(3) **Load Factor %** X .65

EDC ANNUAL OPERATING COST = $ 20,177

CURRENT ANNUAL OPERATING COST $ 42,082

WITH EDC ANNUAL OPERATING COST $ 20,177 **(minus)**

ANNUAL SAVINGS $ 21,905

HEAT RECOVERY CREDIT

Value of Recovered Heat:

Engine full load horsepower ** 390 X 0.039 (5) Therm per HP + .80 Boiler Efficiency X $ .45 per Therm

Heat Value per Operating Hour $ 8.55

(5) Average Available Heat per HP-HR

Annual Operating Hours X 1715

EDC (3) Load Factor % X .65

Heat Use Load Factor % X .60

Recovered Heat Value = $ 5,718

CHILLING SAVINGS PER YEAR $ 21,905

HEATING SAVINGS PER YEAR $ 5,718

TOTAL SAVINGS PER YEAR $ 27,623

EXAMPLE

To Solve for CASH FLOW:

$$\frac{\$120{,}000 \text{ Incremental Cost}}{10 \text{ Years}} = \$12{,}000.00 \text{ Annual Recovery}$$

\$ 27,623 Savinge - (minus) \$ 12,000 Annual Recovery = \$ 15,623 Taxable Savings

34% tax rate (.66) X \$ 15,623 Taxable Savings = \$ 10,311 After Tax Savings

\$ 10,311 After Tax Savings + \$ 12,000 Capital Recovery = **\$ 22,311 CASH FLOW**

To Solve for RATE OF RETURN (ROR):

$$\frac{\$ 22{,}311 \text{ CASH FLOW}}{\$ 120 \text{ (thousand)}} = \$ 185.92 \text{ \$ Return/Year/Thousand \$ Invested}$$

ROR Table: \$ 185.92 falls between 12% and 15% under the ten (10) year recovery period

To Solve for actual ROR:

$$\frac{\$ 185.92 \text{ from above} - \$ 176.98 \text{ from Table}}{\$ 199.25 \text{ from Table} - \$ 176.84 \text{ from Table}} \times .03 + 12\% = \textbf{13.20\% ACTUAL ROR}$$

Chapter 73
Electric Motor Do's and Don'ts

R.K. Hoshide

ABSTRACT
This paper provides important practical information for those considering replacing standard electric motors with energy-efficient (EE) ones. Electric motors consume about 67% of the electricity produced in the United States[2] (Figure 1). Reduced manufacturer's costs, electric utility incentives and rebates, increased selection and availability, along with increased electric rates have made installing EE motors increasingly popular. Typical gains of 2 to 10 points in efficiency are normally expected. Larger gains can be obtained with proper engineering analysis and application.

Changing from standard to EE motors can sometimes result in only minuscule or no energy savings. Correct engineering analysis and technical monitoring should be conducted to assure that motors are properly selected, installed, operated, and maintained.

EFFICIENCY STANDARDS

Electric motor manufacturers have been improving their product efficiencies since 1974. There are various motor

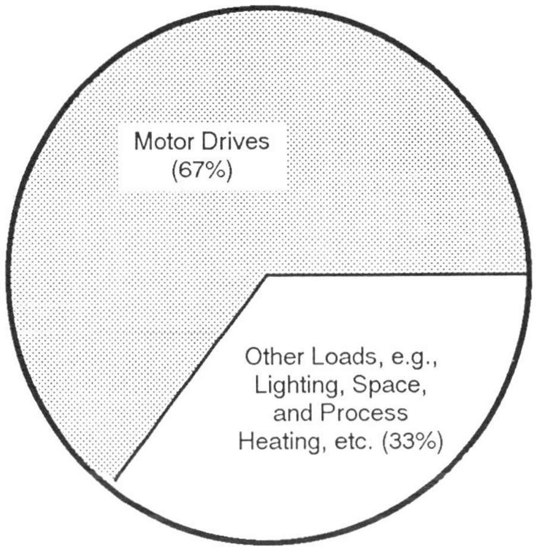

FIGURE 1. ELECTRICITY USE BY APPLICATION (IN THE U.S.)

TABLE 1. EFFICIENCY RATING STANDARDS

Standard	Full Load Efficiency	
	7.5 hp	20 hp
U.S. (NEMA MG-1)	80.3%	86.9%
Canadian (CSA C390)	80.3%	86.9%
Japan* (JEC-37)	85.0% (+4.7)	90.4% (+3.5)
British* (BS-269)	82.3% (+2.0)	89.3% (+2.4)
International* (IEC34-2)	82.3% (+2.0)	89.3% (+2.4)

*Calculates efficiency.

testing standards (Table 1), but it is best to use either the United States (NEMA MG-1) or the Canadian (CSA C390) efficiency rating standard. The others are less accurate since they assume a fixed stray loss or ignore it altogether.

Numerous nominal and minimum efficiency standards have been proposed over the years including the National Electrical Manufacturers Association (NEMA), American Society of Heating Refrigeration and Air-Conditioning Engineers (ASHRAE), U.S. DOE, Canadian Utilities, etc. The National Energy Policy Act of 1992[3] has also specified nominal efficiency standards (Table 2). Typical existing motor efficiencies for standard and EE motors are listed in Table 3.

DATABASE

The Washington State Energy Office (WSEO) has generated an electric motor database for more than 9,000 three-phase motors ranging in size from 1 to 600 hp. This database, called "Motormaster," is on computer disc, and it is available from the WSEO.[4] Table 4 shows typical data obtained from this database for 20 hp, 1,800 rpm (4 pole), open drip type motors. These data are useful in the preliminary selection of candidate motors where efficiency, full load speed, power factor, cost and other information are available. Because no one manufacturer has the most efficient motor in every size or type, users should comparison shop. Manufacturers also change their designs from time to time, so it is very important to obtain up-to-date efficiency, full load speed, cost, etc., information from the manufacturers.

TABLE 2. NOMINAL FULL-LOAD EFFICIENCIES FROM THE NATIONAL ENERGY POLICY ACT OF 1992

Motor (hp)	Number of Poles					
	Open Motors			Closed Motors		
	2	4	6	2	4	6
1	—	82.5	80.0	75.5	82.5	80.0
1.5	82.5	84.0	84.0	82.5	84.0	85.5
2	84.0	84.0	85.5	84.0	84.0	86.5
3	84.0	86.5	86.5	85.5	87.5	87.5
5	85.5	87.5	87.5	87.5	87.5	87.5
7.5	87.5	88.5	88.5	88.5	89.5	89.5
10	88.5	89.5	90.2	89.5	89.5	89.5
15	89.5	91.0	90.2	90.2	91.0	90.2
20	90.2	91.0	91.0	90.2	91.0	90.2
25	91.0	91.7	91.7	91.0	92.4	91.7
30	91.0	92.4	92.4	91.0	92.4	91.7
40	91.7	93.0	93.0	91.7	93.0	93.0
50	92.4	93.0	93.0	92.4	93.0	93.0
60	93.0	93.6	93.6	93.0	93.6	93.6
75	93.0	94.1	93.6	93.0	94.1	93.6
100	93.0	94.1	94.1	93.6	94.5	94.1
125	93.6	94.5	94.1	94.5	94.5	94.1
150	93.6	95.0	94.5	94.5	95.0	95.0
200	94.5	95.0	94.5	95.0	95.0	95.0

Notes: 1) Manufactured by October 25, 1997
2) Manufactured by October 25, 1999 for motors requiring certification from a nationally recognized safety testing laboratory.

ENERGY SAVING CALCULATIONS

Typically, engineers choose EE motors using simple and straightforward calculations. Common equations for determining motor replacement savings are shown below.

Demand savings:

$$kW_{saved} = hp \times load \times 0.746 \frac{kW}{hp}\left(\frac{1}{\eta_{standard}} - \frac{1}{\eta_{energy\ efficient}}\right)$$

where: $\eta_{standard}$ = standard motor efficiency
$\eta_{energy\ efficient}$ = EE motor efficiency

Energy savings:

$$\frac{kWh}{yr} saved = kW_{saved} \times \text{hours operation per year}$$

Cost savings:

$$\frac{dollars}{yr} = \frac{kWh}{yr} saved \times \frac{\$}{kWh} + \frac{\$}{kWyr} \times kW_{saved}$$

The two areas of concern with these calculations are:

(1) Engineers often assume that the motor is fully loaded.
(2) The change (increase) in speed of the EE motor is usually not considered nor accounted for.

Motor Load

About 1/4 of the electric motors operate up to a maximum load of 30%. Another 1/4 operate at a maximum load of over 60%, and the remaining 1/2 operate between 30 and 60%[5] (Figure 2).

TABLE 3. TYPICAL EFFICIENCIES OF ELECTRIC MOTORS

Typical Standard Motors			Typical Energy-Efficient Motors		
Full Load Efficiencies of NEMA Design B Induction Motors			Full Load Efficiencies of Three Phase, Four Pole Motors		
hp	Nominal Efficiency Range	Average Nominal Efficiency	hp	Nominal Efficiency Range	Average Nominal Efficiency
1.0	68.0–78.0	73.0	1.0	80.0–84.0	83.0
1.5	68.0–80.0	75.0	1.5	81.0–84.0	83.0
2.0	72.0–81.0	77.0	2.0	81.0–84.0	83.0
3.0	74.0–83.0	80.0	3.0	83.5–89.5	86.5
5.0	78.0–85.0	82.0	5.0	85.0–90.2	87.6
7.5	80.0–87.0	84.0	7.5	86.0–91.0	88.5
10.0	81.0–88.0	85.0	10.0	87.5–91.7	89.6
15.0	83.0–89.0	86.0	15.0	89.5–92.4	91.0
20.0	84.0–89.0	87.5	20.0	90.0–93.2	91.6
25.0	85.0–90.0	88.0	25.0	91.0–94.1	92.6
30.0	86.0–90.5	88.5	30.0	91.0–94.5	92.8
40.0	87.0–91.5	89.5	40.0	91.5–94.5	93.0
50.0	88.0–92.0	90.0	50.0	91.5–95.0	93.2
60.0	88.5–92.0	90.5	60.0	91.7–95.0	93.3
75.0	89.5–92.5	91.0	75.0	92.0–95.0	93.5
100.0	90.0–93.0	91.5	100.0	93.0–95.0	94.0
125.0	90.5–93.0	92.0	125.0	93.0–95.4	94.2
150.0	91.0–93.5	92.5	150.0	93.5–95.8	94.6
200.0	91.5–94.0	93.0	200.0	94.4–96.2	95.3

A quick method for checking motor load is to take advantage of the nearly linear relationship between motor slip and load (Figure 3). It shows that slip is proportional to load from about 8% to 110% load. Current, on the other hand, is not linear over the lower range of motor load (see Figure 3). The

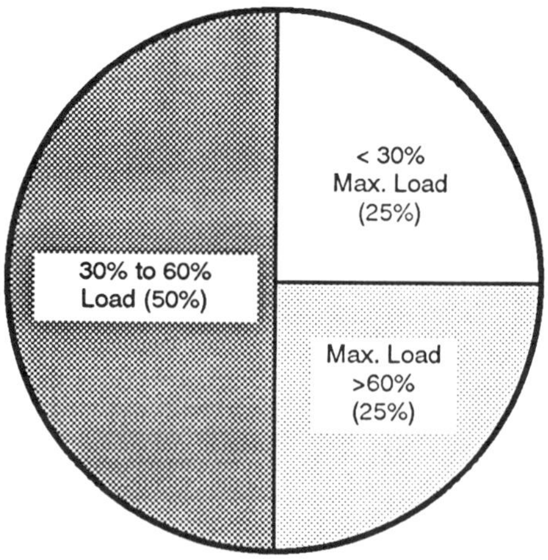

FIGURE 2. U. S. MOTOR OPERATION

TABLE 4. TYPICAL DATABASE FOR 20 HORSE-POWER, 1800 RPM, OPEN DRIP TYPE

Manufacturer	Model	Full Load Speed (rpm)	Full Load Efficiency (%)	Motor Cost
Toshiba	E.Q.P./Premium Eff.	1773	93.6	$1,055
G.E.	Energy Saver	1770	93.6	1,055
Reliance	XE Premium Eff.	1762	92.4	1,055
MagneTek	E-Plus III	1765	92.4	1,055
Baldor	High Eff.	1780	92.4	1,150
Sterling	K Series	1770	91.0	780
Reliance	E-2000 Energy Eff.	1757	91.0	933
Marathon	Series E/High Eff.	1760	91.0	1,055
---	**NEMA Standard**	---	91.0	---
US Motors	Premium Eff.	1760	90.9	1,055
Leeson	Standard	1750	90.7	771
Toshiba	Standard Eff.	1759	90.4	843
G.E.	Standard Eff.	1760	90.2	922
Dayton	Standard	1750	90.2	828
TECO	Standard Eff.	1750	89.5	843
Sterling	J Series	1750	89.5	780
Lincoln	Signature Series	1750	89.5	577
Marathon	Standard Eff.	1755	88.5	843
US Motors	Normal Eff.	1760	87.5	843
Reliance	Standard Eff.	1753	87.5	843
MagneTek	Standard Eff.	1750	87.5	843
Siemens	Standard Eff.	1750	86.5	843
Delco	T Line	1750	86.5	843
Baldor	Standard Eff.	1760	86.0	850

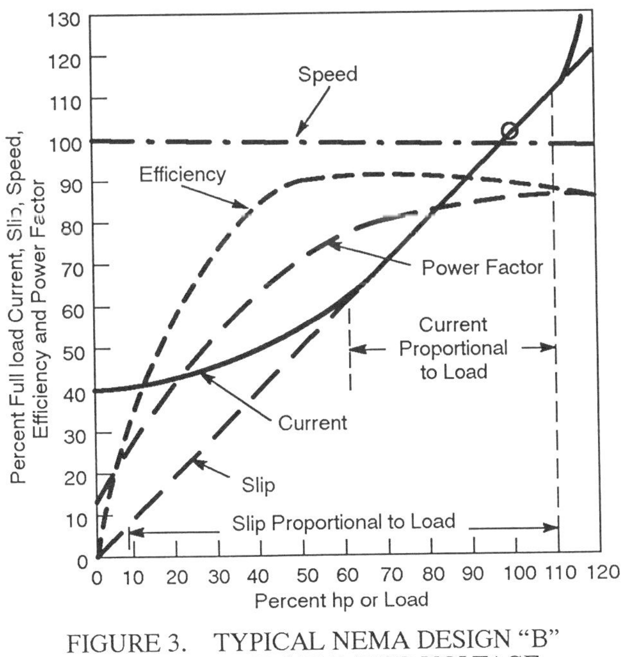

FIGURE 3. TYPICAL NEMA DESIGN "B" MOTOR CURVE (AT RATED VOLTAGE AND FREQUENCY)

percentage slip change varies as the inverse of the voltage variation squared. Frequency variations of ±5% have practically no change in slip.

TABLE 5. INDUCTION MOTOR DATA AND CALCULATIONS

- Nameplate Data:
  - (1) Rated hp
  - (2) Full load speed, rpm
  - Phase, Φ
  - Frequency, Hz
- Measured Data:
  - (3) Operating speed
  - (4) Average kW
- Calculations:
  - (5) Full load slip = Synchronous - (2)
  - (6) Running slip = Synchronous - (3)
  - (7) Load on motor = (6)/(5)
  - (8) Hp output = (1) x (7)
  - (9) kW output = (8) x 0.746
  - (10) Efficiency, % = ((9)/(4)) x 100

TABLE 6. EXAMPLE OF INDUCTION MOTOR DATA AND CALCULATIONS

- Nameplate Data:
  - (1) Rated hp = 7.5 hp
  - (2) Full load speed = 1,740 rpm
  - Phase, Φ = 3
  - Frequency = 60 Hz
- Measured Data:
  - (3) Operating speed = 1,768 rpm
  - (4) Average kW = 3.6 kW
- Calculations:
  - (5) Full load slip = Sync - (2) = 1,800 - 1,740 = 60 rpm
  - (6) Running slip = Sync - (3) = 1,800 - 1,768 = 32 rpm
  - (7) Load on motor = (6)/(5) = 32/60 = 0.533
  - (8) Hp output = (1) x (7) = 7.5 x 0.533 = 4 hp
  - (9) kW output = (8) x 0.746 = 4 x 0.746 = 2.98 kW
  - (10) Efficiency, % = ((9)/(4)) x 100 = (2.98/3.6) x 100 = 82.8%

Running slip can be obtained by measuring motor speed using a digital strobe type tachometer. These are very accurate, safe, and easy to use and can be obtained for about $225. Select units that do not require installing a reflecting tape on the rotating shaft. Contact type tachometers are unsafe, difficult to use, inaccurate, and are not recommended.

Running slip should be obtained when the motor's driven equipment is operating at maximum load. This may require simultaneous operation of all pumps in a multi-loop parallel system. Chilled water and condenser water pumps may fall into this category. Fans should be checked at maximum power operation. Two-speed motors should be checked at the high-speed operating condition.

Once running slip is obtained, the motor load can easily be calculated. Table 5 lists typical nameplate and measured data and the calculations for obtaining motor load. Also listed is the measurement of average kW using a kW meter for obtaining motor input power. With motor load and input power, one can easily calculate efficiency (Table 5). Table 6 shows an example of obtaining motor load and efficiency using the slip method.

If motor performance data or curves are available, you should check on how close your efficiency value comes to the manufacturer's data. If motors have been exposed to severe operating conditions, poorly maintained, etc., their efficiency will be lower. Very old motors usually do not have full load efficiency specified on their nameplate or performance data available.

Three very important pieces of motor information derived when following the above procedure are:

(1) Operating load
(2) Operating efficiency
(3) Full load speed (nameplate).

Motor Downsizing
From the slip method, motor load can be estimated. Motors operating at 50% load or less are strong candidates for downsizing. By installing smaller motors, their load will increase, which increases efficiency and power factor. Capital investment cost is also decreased. Typically, engineers replace standard motors with EE motors on a one-to-one basis (a 10 hp standard with a 10 hp energy efficient one).

Table 7 shows a typical electric motor replacement procedure. For this example, full load operation is assumed and only the improvement in motor efficiency is used.

Table 8 shows operating data obtained from these 11 motors. The operating speed along with input kilowatts were obtained. Motor load and motor efficiency were then calculated. From this table, it is easy to see that most of these motors were oversized. Efficiencies were very low for most of these motors.

Table 9 shows the first 9 recommended downsized EE motors. Same size EE motors are recommended for the last two motors. The recommended motors estimated loads are reasonable, motor efficiencies and power factors are high, and savings increased dramatically (from 12.54 hp saved to 66.5 hp saved). The capital investment is also reduced (from about $10,000 to $7,261).

It should be noted that costs for the downsized motor installation will increase some, since both the mounting system and the shaft may be a different size. This may include using an adapter plate for mounting the motor and changing

TABLE 7. TYPICAL ELECTRIC MOTOR REPLACEMENT PROCEDURE

Motor No.	Motor I.D. No.	Nameplate hp (FL)	Existing ST Motor Estimated Efficiency (%)	Replacement EE Motor Estimated Efficiency (%)	Power Saved		Replacement Motor Cost[6] EEM TEFC ($)
					(hp)	(%)	
1	65	10	85	89.6	0.60	5.4	471
2	94	15	86	91.0	0.96	5.8	646
3	91	15	86	91.0	0.96	5.8	646
4	97	15	86	91.0	0.96	5.8	646
5	34	15	86	91.0	0.96	5.8	646
6	61	20	87.5	91.6	1.02	4.7	787
7	69	20	87.5	91.6	1.02	4.7	787
8	63	20	87.5	91.6	1.02	4.7	787
9	71	40	89.5	93.0	1.68	3.9	1,500
10	32	40	89.5	93.0	1.68	3.9	1,500
11	33	40	89.5	93.0	1.68	3.9	1,500
11	-	250	-	-	12.54	-	9,916

TABLE 8. ACTUAL MOTOR LOAD DATA

Motor No.	Nameplate hp (FL)	Full Load Speed (rpm)	Measured Speed (rpm)	Calculated Load (%)	Output hp	Input hp	Motor Efficiency (%)
1	10	1,160	1,179	53	5.3	7.7	69
2	15	1,750	1,782	36	5.4	12.9	42
3	15	1,750	1,784	32	4.8	7.9	61
4	15	1,750	1,783	34	5.1	10.2	50
5	15	1,750	1,777	46	6.9	12.5	55
6	20	1,750	1,785	30	6.0	14.6	41
7	20	1,750	1,779	42	8.4	20.0	42
8	20	1,755	1,777	51	10.2	17.0	60
9	40	1,150	1,192	16	6.4	21.3	30
10	40	1,770	1,780	66.5	26.6	32.0	83
11	40	1,770	1,780	66.5	26.6	30.9	86
11	250	--	--	--	111.7	187.0	--

TABLE 9. RECOMMENDED ENERGY-EFFICIENT MOTORS

Motor No.	Recommended hp	Estimated Load (%)	Estimated Motor Efficiency (%)	Input hp	Power Saved		Replacement Motor Cost[6] EEM TEFC ($)
					hp	%	
1	7 1/2	71	90	5.9	1.8	23.4	389
2	7 1/2	72	91	5.9	7.0	54.3	389
3	7 1/2	64	90	5.3	2.6	32.9	389
4	7 1/2	68	91	5.6	4.6	45.1	389
5	10	69	91	7.6	4.9	39.2	471
6	10	60	92	6.5	8.1	55.5	471
7	15	56	92	9.1	10.9	54.3	646
8	15	68	92	11.1	5.9	34.7	646
9	10	64	93	6.9	14.4	67.6	471
10	40	66.5	94	28.3	3.7	11.6	1,500
11	40	66.5	94	28.3	2.6	8.4	1,500
11	170	--	--	120.5	66.5	--	7,361

pulleys or putting on a sleeve for attaching the existing pulleys to the new shaft.

Since EE motors have lower starting torque (due to lower slip), it should also be checked.

The benefits for selecting correctly sized and properly operated EE motors include:

(1) Saves energy (higher and flatter η curve)
(2) Reduces demand
(3) Saves energy in the cables and transformers that feed the motor
(4) Increases power factor
(5) Lasts longer
(6) Runs cooler
(7) Less maintenance.

Speed Matching
Electric motor manufacturers have improved motor efficiency by reducing electrical, magnetic (core), mechanical (windage and friction), and stray losses. High efficiency usually means less slip for a given load. By reviewing full load slip of the motors listed in Table 4, it shows that most of the high efficiency motors have higher full load speeds compared to the standard efficiency models.

Speed becomes very important when motors being replaced drive centrifugal fans and pumps. One of the affinity laws state that the power required by a fan or pump increases with the cube of its speed. Replacing standard motors with EE ones without considering speed changes has resulted in much lower energy savings than expected.

Table 10 shows an example of what can happen when the increase in speed of an EE motor is not accounted for. The expected savings of 5% in power amounted to miniscule saving. Figure 4 shows the increase in power of the EE motor as a function of standard and EE motor speed differences. The example from Table 10 is also shown.

It is normally hard to match speeds of EE motors to standard motors. Do not compromise motor selection by trying to do this.

TABLE 10. EXAMPLE OF THE IMPORTANCE OF SPEED MATCHING

- Typical 20 hp, 1,800 rpm open drip-type at 100% load
  - Old motor: Standard Efficiency
    Efficiency = 87.5%
    Speed = 1,750 rpm
    Input Power = 1.00 x 20 hp/0.875
    = 22.86 hp
  - New motor: High Efficiency
    Efficiency = 92.3%
    Speed = 1,780 rpm
    Input Power = 1.00 x 20 hp/0.923
    = 21.67 hp
- Apparent saving = 22.86 - 21.67 = 1.19 hp (5.21%)
- Increased load on new motor because of speed increase:
  - $\left(\frac{1780}{1750}\right)^3 = 1.0523$ (fan/pump affinity laws)
  - Or a 5.23% increase in load on the new motor from rotating equipment
- Actual saving = 22.86 - 21.67 x 1.0523 = 0 hp

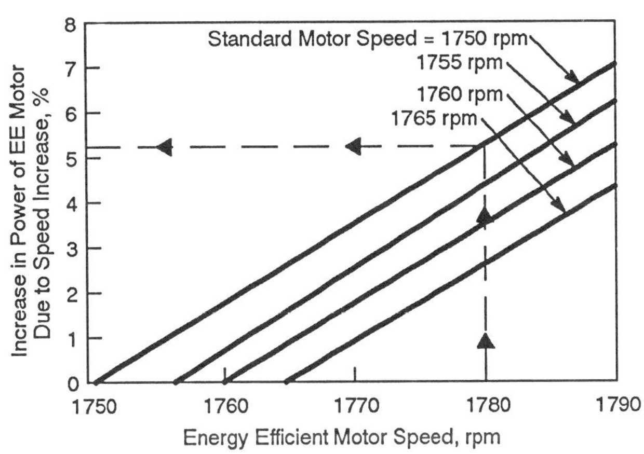

FIGURE 4. INCREASE IN POWER OF EE MOTOR DUE TO SPEED INCREASE (TYPICAL 20 HP, 1800 RPM OPEN DRIP-TYPE ELECTRIC MOTOR)

Ways to maintain the same fan speed include adjusting the sheaves or changing its pulleys. Pumps should be modified or changed out so that their output head and flow rate remain the same. Ways to keep the same head and flow of an existing pump, operating at a higher speed, include changing to a smaller diameter impeller or trimming the existing impeller.

DO'S AND DON'TS

Recommended do's and don'ts are divided into four steps which include: (1) Pre-Replacement, (2) Motor Selection, (3) Motor Installation, and (4) Post-Installation.

Pre-Replacement Recommendations

Do:

(1) Inventory motors.

- Record motor function, nameplate data, condition, and estimate operating hours.

(2) Measure motor and driven equipment speeds.

- Operate at maximum load if possible.
- Monitor absorbed power under different operating conditions using data logger or chart recorder.
- Calculate motor load, hp, and kW output.
- Check starting torque requirements.

(3) Measure average input kW.

- Check voltages for unbalance.
- Calculate efficiency.

(4) Select candidate motors for changeout.

- Motors loaded 50% or less should be checked for downsizing.

Don't:

(1) Wait until motor fails.

- No longer possible to check load or determine pump or fan operating speed.
- Usually replaced by like or larger motor.
- EE motor may not be available.
- No downsizing ever contemplated or incorporated.
- Purchase is piecemeal instead of volume with discount.

Motor Selection Recommendations

Do:

(1) Consider EE motors for:

- Existing and all new facilities.
- Replacing oversized or undersized ones.
- Replacing failed motors and spares.
- New equipment package purchases.

(2) Obtain up-to-date performance data of specific motors under consideration from the manufacturers.

(3) Specify NEMA Standard MG-1-12.53 or CSA C390 efficiency rating.
(4) Select motors that exceed NEMA and UL requirements.
(5) Use same voltage, frequency, and phase.
(6) Have recommended capacitor.
(7) Check utility for financial incentives and manufacturers for discounts (cost multiplier).

Don't:

(1) Use indoor (open) motors outdoors.
(2) Substitute ball with sleeve-bearing motor (and vice versa).
(3) Use sleeve-bearing motor for adjustable frequency drive operation below 500 rpm.

Motor Installation Recommendations

Do:

(1) Maintain the same fan speed (or close to) after EE motor is installed.

- Adjust sheaves or change pulleys.

(2) Maintain the same output head and flow rate of pumps.

- Change or trim impeller.
- Change out pump.

(3) Change out old heaters (if used)
(4) Correct loose connections, bad contacts, power leads, etc.
(5) Ensure proper alignment of motor pulleys and fan.
(6) See that pump motors are properly aligned.

Don't:

(1) Bypass thermal protector.
(2) Replace manually protected motor with auto-reset.
(3) Remove capacitor without shorting its terminals.

Post-Installation Recommendations

Do:

(1) Check driven equipment rotation.
(2) Eliminate excessive vibration.
(3) Check motor load and efficiency.
(4) Grease bearings as specified.
(5) Check for overheating and provide adequate ventilation.

- Use your sense of smell.

(6) Keep motors clean by regularly cleaning fans and protective grill.
(7) Maintain proper belt tension.
(8) Monitor motor efficiency and voltage regularly.

- Voltage level of $\pm 10\%$ is okay.
- Voltage unbalance >1% should be remedied.

(9) Maintain drivetrain equipment.
(10) Draw up an effective preventive maintenance schedule and follow it.

Don't:

(1) Repaint motors.

- Extra paint makes motors run hotter which reduces life and efficiency.

(2) Overlubricate motors.

- Follow motor manufacturer's recommendations.

SUMMARY

Information is provided on how motors are rated and the National Energy Policy Act of 1992 motor efficiency standards. The importance of determining electric motor operating load for possible downsizing where energy consumption, energy cost, and capital investment can be substantially reduced is discussed. The importance of speed matching of fans and maintaining the same pump output (head and flowrate), when motors are replaced is described.

Estimated savings can vanish without proper engineering analysis. The do's and don'ts recommendations are listed by methodology (pre-replacement, selection, installation, and post-installation).

REFERENCES

(1) By providing technical assistance during the past 14 years for the U.S. Department of Energy (DOE) and to 27 State Energy Offices (SEOs), the Energy Technology Engineering Center (ETEC) has obtained practical experience in energy efficiency, energy management, and technical engineering information. This experience includes technical and economic reviews of over 28,000 energy-efficiency projects valued at over $630 million for the Institutional Conservation Program (ICP). Technical auditing and technical monitoring have also been conducted for the ICP and the Federal Energy Management Program.

(2) Electric Power Research Institute Journal, June 1992, 3412 Hillview Avenue, P.O. Box 10412, Palo Alto, CA 94303, Phone (415) 855-2411.

(3) Energy Policy Act of 1992, Conference Report to Accompany H.R. 776, October 5, 1992 - Ordered to be printed, U.S. Government Printing Office, Washington, DC 20402.

(4) MotorMaster Electric Motor Selection Software, Washington State Energy Office, 809 Legion Way, SE, P.O. Box 43165, Olympia, Washington 98504-3165, Phone (206) 586-8588.

(5) Air Conditioning, Heating, and Refrigeration News, July 15, 1991, Business News Publishing Co., P.O. Box 2600, Troy, MI 48007, Phone (313) 362-3700.

(6) *Energy-Efficient Motor Systems: A Handbook on Technology, Programs, and Policy Opportunities*, American Council for an Energy-Efficient Economy, 1991, 1001 Connecticut Avenue, NW, Suite 535, Washington, DC 20036.

D030-0040-I5

Chapter 74

Improving the Performance of Small Type Air Conditioners

R.K. Alnajjar, S.A. Said, J.M. Saleem

ABSTRACT

Many experiments have been conducted to enhance the performance of the vapor compression air-conditioners (small type), and to study the effect of the following parameters: outdoor air temperature, indoor air temperature and the volumetric air flow rate (fan speed) before and after carrying out the modification. the modification involves adding a secondary condenser by fixing it in front of the evaporator. The additional condenser is then cooled by the room recirculated cold air.

Results show an enhancement in the coefficient of performance (COP) ranged from 31.32 to 51.50% and a reduction in the electric power consumption (KW per unit ton of refrigeration) ranged from 23.85 to 35.27% within the range of the experimental conditions.

INTRODUCTION

Since the power consumed by the small type air-conditioner mainly depends upon the condenser pressure, any reduction in this pressure will lead to a power reduction and enhancement of the COP. The condenser pressure is directly proportional to the saturated condensing temperature, therefore any additional sub cooling will yield a satisfactory result.

In the present work, the following steps were considered:

↳ Study the effect of carrying out the modification by adding a secondary condenser in series with the main condenser and fixed in front of the evaporator as shown in Figure (1).

↳ Study the effect of the following parameters; before and after carrying out the modification:

- ➲ Outdoor air temperature.
- ➲ Indoor air temperature.
- ➲ Volumetric air flow rate (fan speed).

↳ A comparison between the performance of the air-conditioner before and after carrying out the modification.

The enhancement of the COP of the small type air-conditioners has been explored by many investigators. Guinn and Novell carried out an experimental procedure to study the effect of using a commercial condenser supplied with a water sprayer added to a small (3 tons of refrigeration) air-conditioner. Results show a decrease in the power consumption ranged from 5 to 9%, and an increase in the COP ranged from 11.6 to 18.9% when 51.1 l/hr of spraying water is used continuously [1].

Zarkrzowski performed experimental measurements on a single stage refrigeration cycle with and without heat exchange between the condensed refrigerant and the vapor leaving the evaporator. Results showed an increase in COP [2].

EXPERIMENTAL APPARATUS

Modifications were carried out to facilitate controlling and measuring of the necessary parameters. These modifications and the experimental device are shown in Figure (2). Here the evaporator and the secondary condenser are kept outside the container, while other components of the air-conditioner are kept inside the container.

The present modifications to the experimental apparatus proved satisfactory in controlling the measured parameters under study.

Controlling the outdoor air temperature (ranging from 40°C to 50°C) was done by varying the rear air duct attached to the box container. To control the volumetric air flow rate, a fan speed selector switch was used (990, 1120 and 1225 rpm). The indoor air temperature ranged from 20°C to 26°C [3].

THEORETICAL ANALYSIS

Figure (3) shows the simple P-h diagram for the vapor compression refrigeration cycle before and after carrying out the modifications.

The main thermodynamic equations used in the present work are:

$$CE = M_{ae} * (H_{ie} - H_{oe}) / 3600 \quad (1)$$

$$COP = CE / W \quad (2)$$

$$P = W / CE * 3.5168 \quad (3)$$

$$U_a = \sqrt{2g / \rho_a} * [\sqrt{hw_1} + \sqrt{hw_2} + .. + \sqrt{hw_{24}}] / 24 \quad (4)$$

$$Q_{ae} = A_e * U_a \quad (5)$$

$$M_{ae} = \rho_a * Q_{ae} \quad (6)$$

RESULTS AND CONCLUSION

The effect of the modification carried out in the present work on the vapor compression refrigeration cycle (P-h diagram) is shown in Figure (3). Since the heat rejected by the condenser (Hc) is equal to the heat absorbed by the evaporator (CE) in addition to the power consumed by the compressor (W). Therefore any increase in Hc will lead to a decrease in the condenser pressure or temperature, the latter affects a reduction of power consumption. This results in an increase in CE at the same time [4]. Figures (4, 5, 6 and 7) show the effect of carrying this modification on CE, W, COP and P respectively.

Results show that:

❶ An increase in the CE by adding a secondary condenser ranged from 17.68 to 29.89%, Figure (4).

❷ A reduction in W ranged from 9.36 to 17.42%, Figure (5).

❸ An enhancement in the COP ranged from 31.32 to 51.5%, Figure (6).

❹ A reduction in electric power consumption P ranged from 23.85 to 35.27%, Figure (7).

within the range of the experimental conditions. Therefore carrying out such a modification is visible after taking into account the initial cost.

REFERENCES

[1] Guinn, G.R. and Novell, J.B., "Operating performance of a water spray on an air-type condensing unit", ASHRAE TRANS., Part 2, 1981, pp. 373-381.

[2] Zarkrzewski, B., "The effect of heat regeneration in refrigeration cycles with Chlorofluorocarbon refrigeration", Int. J. of Refrigeration, Vol. 6, Number 2, March 1983, pp. 100-102.

[3] Alnajjar, R.K., "Improving the performance of small type air-conditioners", M.Sc. Thesis, University of Technology, Mechanical Engineering Department, August 1987.

[4] Tanaka, N., Ikeuchi, M. and Yanamaka, G., "Experimental study on the dynamic characteristics of a heat pump", ASHRAE TRANS., Part 2, 1982, pp.323-331.

NOMENCLATURE

A	Air duct cross sectional area, m^2.
CE	Cooling effect, KW.
COP	Coefficient of performance.
g	Gravitational acceleration, m/sec^2.
H	Enthalpy, KJ/Kg.
Hc	Heat rejected by the condenser, KW.
hw	Head, mm H_2O.
M	Averaged mass flow rate, Kg/sec.
P	Electric power consumed per unit ton of refr., W/ton of refrigeration.
Q	Averaged volumetric flow rate, m^3/sec.
U	Averaged air speed, m/sec.
W	Compressor work, KW.
ρ	Density, Kg/m^3.

SUBSCRIPTS

a	air
e	evaporator
i	inlet
o	outlet

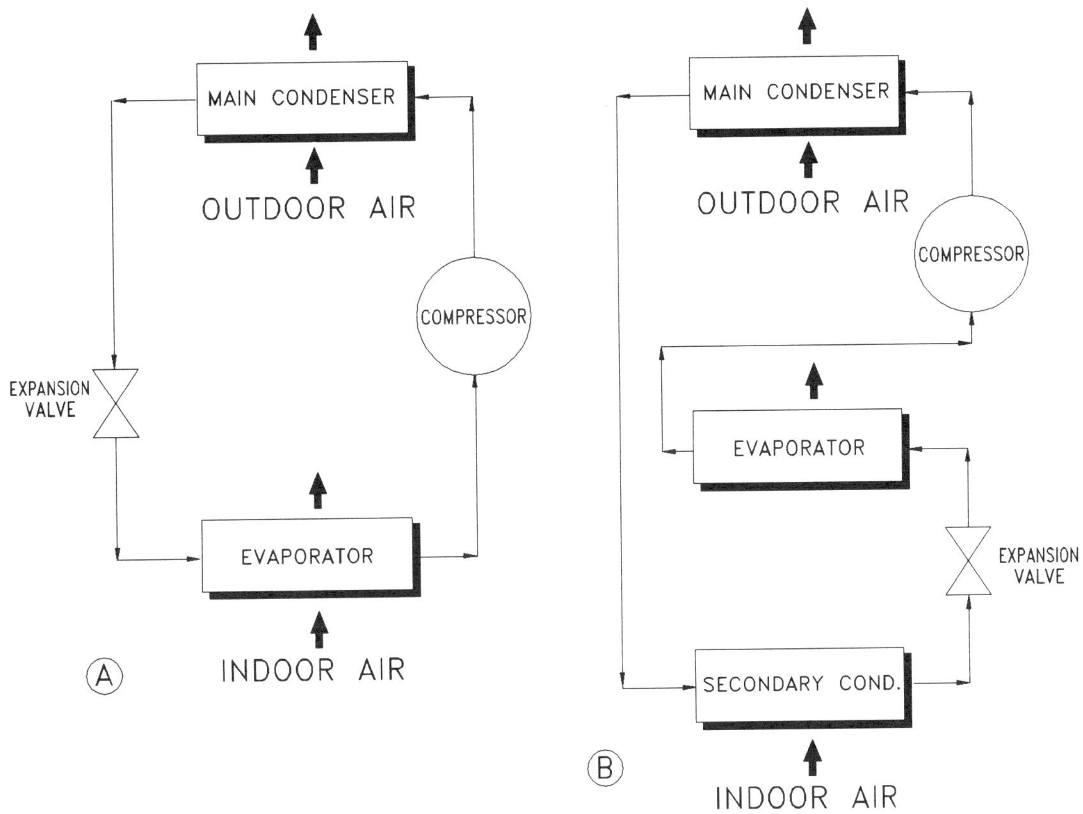

FIG. (1): COMPONENTS OF AIR-CONDITIONER
(A) BEFORE MODIFICATION (B) AFTER MODIFICATION

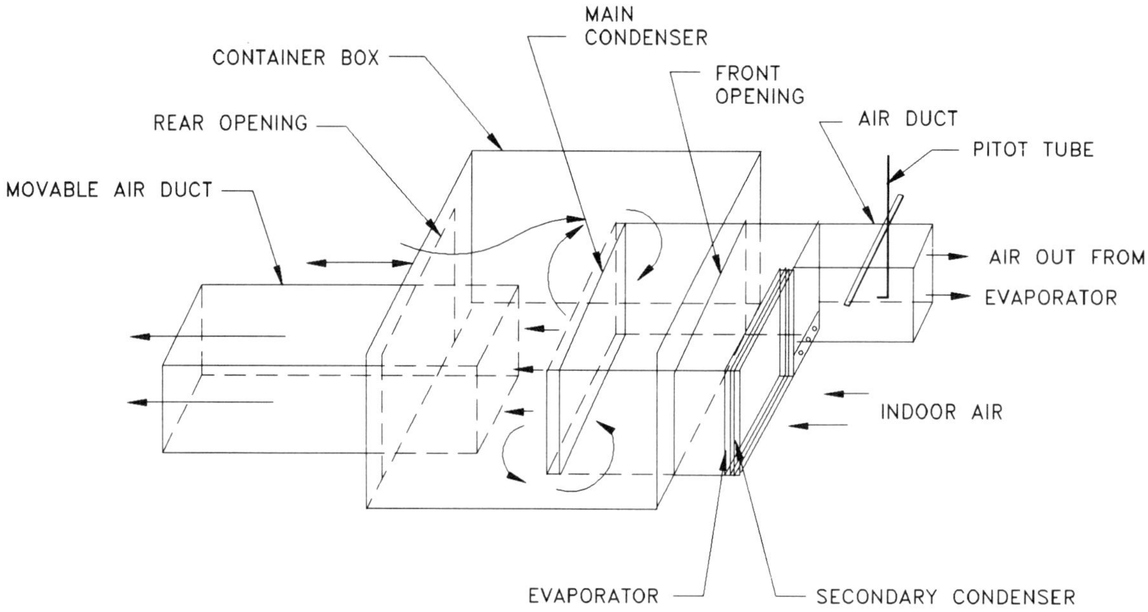

FIG. (2): EXPERIMENTAL DEVICE

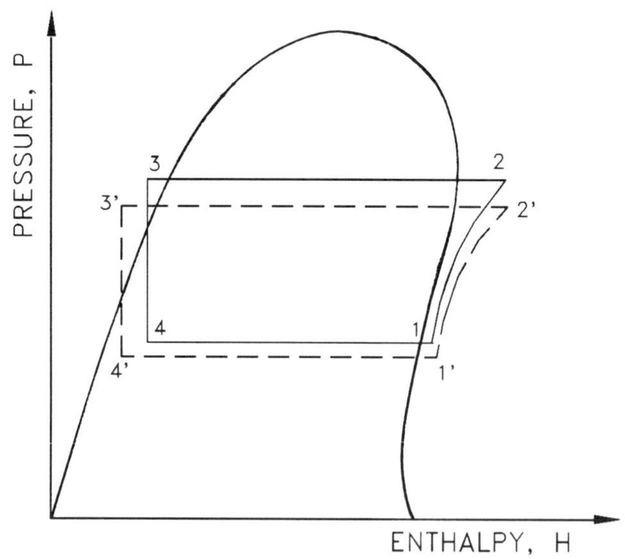

FIG. (3): VAPOR COMPRESSION REFRIGERATION CYCLE, WHERE:
1–2–3–4–1 BEFORE MODIFICATION
1'–2'–3'–4'–1' AFTER MODIFICATION

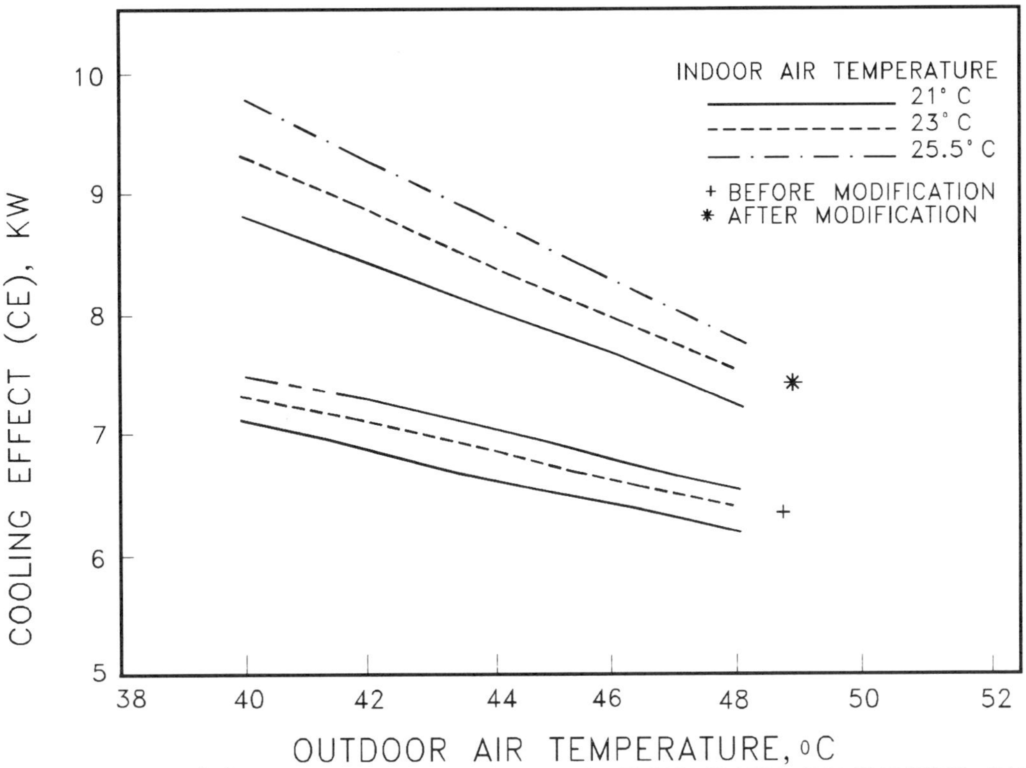

FIG. (4): THE EFFECT OF USING SECONDARY CONDENSER ON CE AT FAN SPEED 1120 RPM.

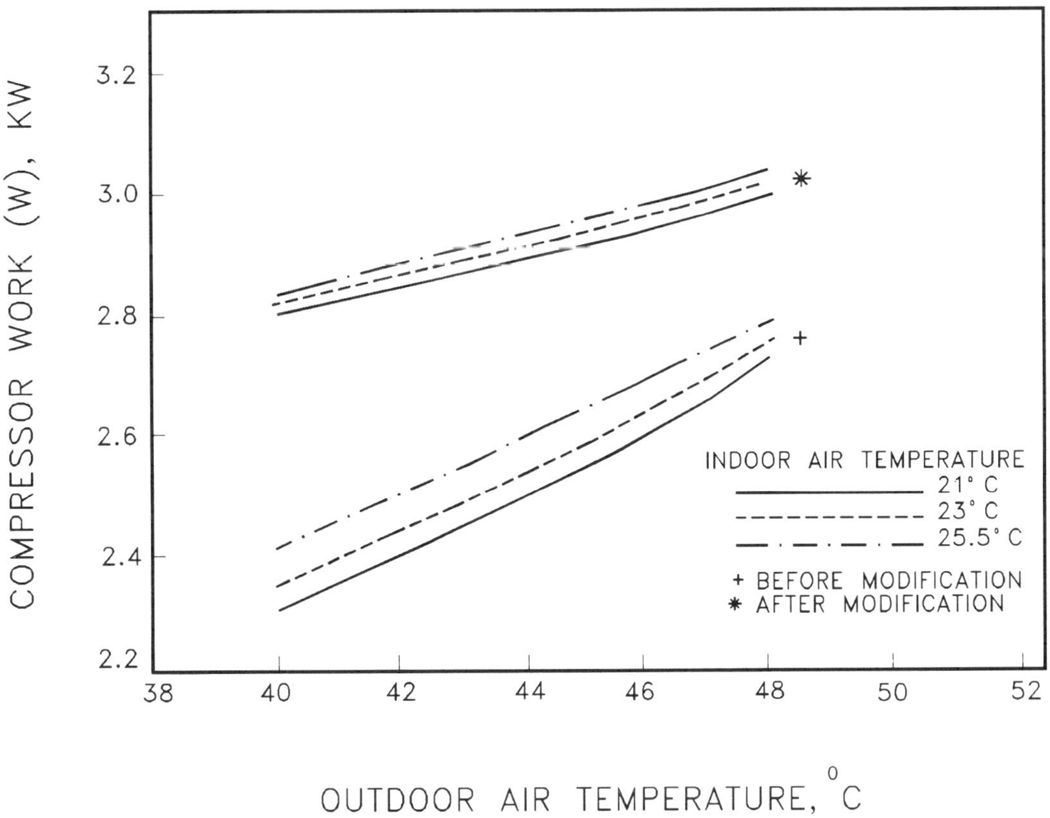

FIG. (5): THE EFFECT OF USING SECONDARY CONDENSER ON W AT FAN SPEED 1120 RPM.

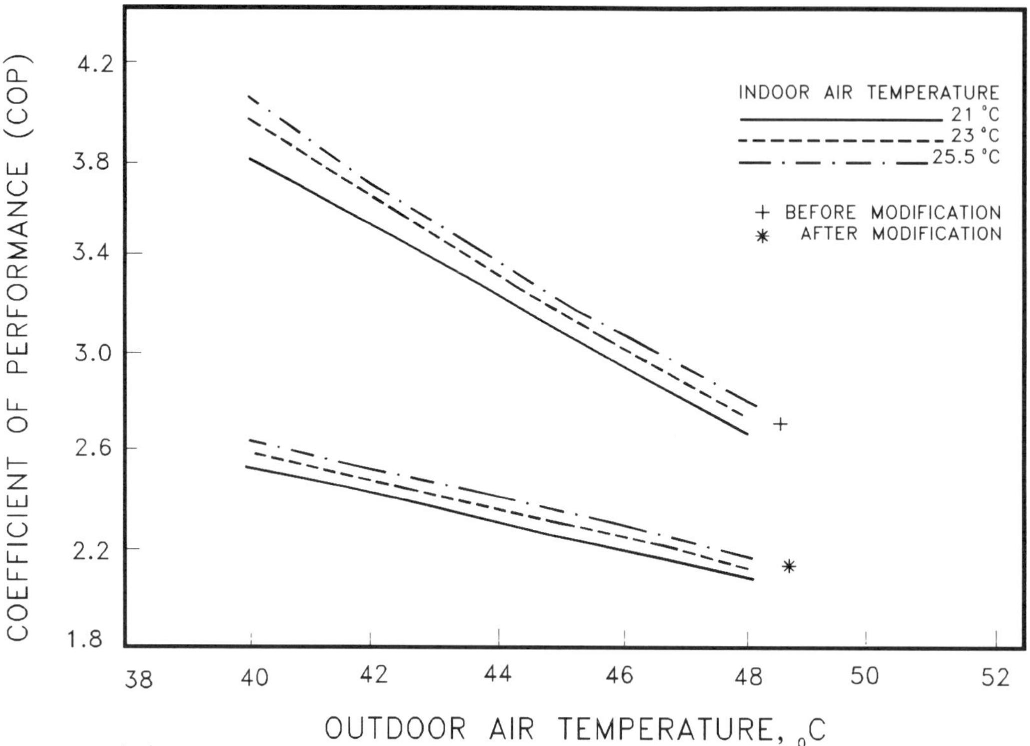

FIG. (6): THE EFFECT OF USING SECONDARY CONDENSER ON COP AT FAN SPEED 1120 RPM.

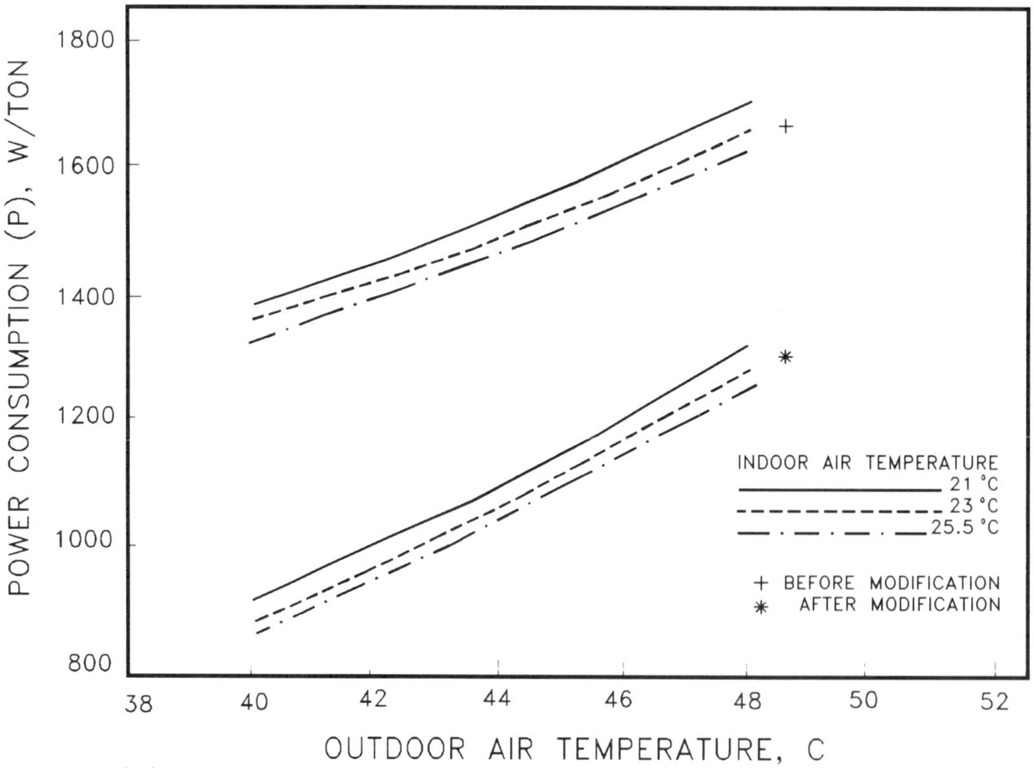

FIG. (7): THE EFFECT OF USING SECONDARY CONDENSER ON P AT FAN SPEED 1120 RPM.

Chapter 75
Process Cooling Primary/Secondary Renovation

K.W. Gerken, C.D. Svard

ABSTRACT

This technical paper presents an overview of a process cooling renovation project implemented in a 700,000 square foot magnetic disk/tape storage manufacturing facility in Louisville, Colorado. The project consisted of the evaluation, design, and implementation of several modifications to the process chilled water cooling system located in the manufacturing plant. The primary objective of the project was to provide additional source and distribution capacity to serve the increasing process chilled water load. Engineering objectives also included both optimizing the efficiency of the system and maximizing its future flexibility. The modifications included expansion of the cooling system capacity, conversion of the distribution configuration to a primary/secondary piping system, conversion of the distribution pumping to variable volume, conversion of the chiller controls to direct digital controls, and implementation of new chiller control sequences of operation. Significant operational cost savings resulted from reduced energy consumption, maintenance costs, and minimized system downtime.

BACKGROUND

The Corporation

Storage Technology Corporation is a manufacturer of computer data storage devices. Headquartered in Louisville, Colorado, the company employs approximately 4,500 people. Activities at this site include research, engineering, manufacturing, testing, product support, warehousing, and administration.

The computer peripheral industry is a very dynamic one. The use of space within any given building may be changed to accommodate process requirements several times in a single year. For this reason, the utilities serving that space must allow for easy retrofit.

The Building

The Louisville site consists of ten buildings totaling approximately 1.8 million square feet. These buildings are grouped together creating five logical building complexes. Each building complex has its own heating and cooling systems, independent of the other building complexes. This project involved the conversion of a chiller plant and associated distribution system to a primary/secondary piping system in the building 4 and 5 complex.

The area of buildings 4 and 5 totals 700,000 square feet: 128,000 square feet is manufacturing area; 69,000 square feet is engineering and research lab space; 57,000 square feet is computer room raised floor; and 429,000 square feet is general office space. The remainder is support services space.

Chilled Water System Description

The chilled water system for the building complex consisted of four electric-drive centrifugal chillers located in building 5 [FIGURE 1]. Since its construction in 1980, three Trane model CVHA-032FH chillers, each with a design capacity of 295 tons, provided chilled water for process cooling to the building complex. In 1989, a York model YTK6K6F1 chiller with a design capacity of 800 tons was added. Chilled water produced in the plant was supplied to buildings 4 and 5 through four chilled water distribution zones. Two zones served the loads in building 4, and two zones served the loads in building 5.

Approximately 90 air handling units serve both process and environmental loads throughout the building complex. Environmental loads are served by 18 of the units while the remaining 72 serve process loads, including computer room raised floor. Prior to the conversion, nearly every unit used a three-way control valve and booster pump assembly [FIGURE 2]. The valves were installed to direct water through the cooling coil of the air handler when cooling was required, and to bypass the cooling coil and circulate the water back into the return piping when no cooling was required. The typical coil arrangement had a booster pump in series with the chiller pumps.

Chilled Water System Problems

In 1989 a study was commissioned to evaluate and identify deficiencies in the performance of the chiller plant and the chilled water distribution system. Operational problems were present; in some sections of the distribution system, the return water pressure was higher than the supply water pressure, and the temperature differential through the evaporator of the chiller was as low as 3.5 degrees F. Also, some air handling units were no longer able to cool the space adequately.

The study identified an excessive chilled water flow rate through the chiller plant and distribution system [FIGURE 1]. The design flow rate for the chiller plant was 2.2 gpm/ton. Under the conditions present at the time of the study, although the flow through the York was less than 2.0 gpm/ton, the flow rate through the Trane chillers was over 6.0 gpm/ton. This excessive flow resulted in the low temperature differential and high water pressure drop through the evaporator section of the chiller. The average temperature differential was measured at 4.1 degrees F, compared to a design temperature differential of 11.0 degrees F. The drop in water pressure through the Trane chiller evaporators was measured from 60 feet of head to 90 feet of head at the time of the study. This drop is far above the 21 feet of head loss for which the system was designed.

The chilled water flow for the chiller plant and the distribution system was provided by three identical base-mounted, end-suction, centrifugal pumps. These pumps were operating in a range near the limits of the pump and motor capacity. At this extreme operating condition, the net positive suction head requirements of each pump was in the range of 35 feet to 48 feet. With all three pumps operating, the available suction pressure was measured at only 14 to 18 feet.

The excessive chilled water flow rates measured in the chiller plant were also found to be present throughout the distribution piping system. Typical design water flow velocities are in the range of five to eight feet per second (fps) for chilled water flow through steel pipe. Measurements in three of the four chilled water distribution zones found flow rates exceeding 10.0 fps.

These problems in the chilled water system were attributed to the use of booster pumps throughout the piping distribution system. The typical coil installation had a booster pump in series with the main chiller pumps. Because the booster pump was placed upstream of the three-way control valve, the system pressure available to the main chilled water pumps was lowered. The lowered head resulted in increased flow.

The multiple problems which surfaced during the study were serious enough to warrant action. Potentially, the high flows could cause premature wearing of the evaporator tubes; they also increased the possibility of piping failure. The low suction pressure at the chiller pumps made cavitation and pump failure likely. In addition to the increased risk of system failure, the series pumping arrangement was costing the corporation energy dollars.

PROJECT DESIGN

Design Requirements

A project to resolve the problems in the chilled water distribution system in buildings 4 and 5 was budgeted for 1991. The three main considerations were: solve the flow problems in the system; provide for future expansion; and minimize chilled water system down time.

While the primary consideration of the design was to rectify the flow problems identified by the 1989 study, the design also needed to address flexibility for future modifications. Without flexibility, the system would eventually become imbalanced again.

Many individual processes in the building required 168 hours per week operation. Because the building 5 chiller system was the primary source of process cooling for the building complex, it was necessary to limit system down time to a minimum during installation. However, there were opportunities for shutting off chilled water to localized areas for short periods of time without adversely impacting the business.

Another design consideration was the availability of an in-house crew of highly skilled pipe fitters for system installation. While the crew was not large enough to accomplish the entire project, it was available to assist in the construction as appropriate.

Final Design

The solution chosen to correct the flow problems was to convert the existing constant volume piping system to a primary/secondary piping system with variable flow in the secondary loops [FIGURE 3]. There were four main aspects of this conversion. First, it was necessary to establish secondary zoning and variable volume pumping requirements. Second, planning was required for a constant volume primary chilled water loop, with considerations for zone bridal loop take-offs. Third, all of the cooling coil control valves had to be converted from three-way operation to two-way operation and the booster pumps removed. Finally, a control scheme had to be implemented which would allow efficient operation of the chillers and distribution pumps.

The secondary zones were divided using the previously established distribution piping. This created four main secondary zones, all of which converged outside the chiller plant. The majority of the load was projected for building 4; loads in building 5 increased only slightly. The zone pumps were sized based upon projected future zone cooling requirements. The design engineer investigated the possibility of locating all four zone pumps in the chiller room. However, the existing chiller plant inside building 5 did not have sufficient space to house all four new secondary pumps. The two pumps feeding the secondary zones in building 5 were located inside the chiller room, and a second pump room was designed to house the two larger pumps feeding the building 4 secondary zones [FIGURE 3].

Establishing the primary loop with the zone take-offs was the most challenging aspect of the piping design. The four chillers in the plant fed supply water into two main headers; those headers fed into the corridor outside the chiller plant. At this point, the headers were piped into an air separator used as a distribution manifold for the separate chilled water zones. This configuration resulted in extremely congested piping in this area. Because all of the zone piping, as well as the chiller mains, were located in the general area, this corridor was chosen for the installation of the bridal loops. A significant engineering effort was required to measure the existing piping and design a series of bridal loops that could tie into the distribution mains. The biggest difficulty was fitting this piping within the tight confines of the corridor.

Four new dedicated chiller pumps were specified for the primary chilled water system. These pumps were intended to replace the main chiller pumps and to accommodate the lower head losses of the primary loop. The pumps were positioned in the existing chiller room in the same location of the old pumps. Locating the new pumps in this manner allowed the accessories and piping from the old pumps to be re-used.

For the coil/valve conversion, each coil and valve installation was surveyed, then a detail of the required modifications was prepared. Where booster pumps existed, they were specified to be removed and replaced with a spool of pipe. In order to save money, the design indicated that most valves be left in place and the bypass piping be removed. The bypass ports on the three-way valves were to be plugged, allowing only two-way operation from the valves [FIGURE 4]. After consulting with the valve manufacturers, it was confirmed that this valve conversion would allow proper valve operation in most applications. In the cases where valve operation was questionable, the decision was made to go forward with the modification and monitor the valve after the conversion. If, in the future, the valve was found to provide poor modulation or chatter it would be replaced at that time.

Storage Technology Corporation has an ongoing program to install direct digital controls (DDC) whenever possible. The use of DDC enables more complex control sequences to be implemented and allows maintenance and engineering remote, real-time access to equipment and system operating information. In keeping with the corporate effort, the decision was made to control the secondary pumping using DDC. Existing DDC for the chiller plant were not compatible with the new controls network on the site. These controls were therefore scheduled to be upgraded to the new standard at the time of the primary/secondary conversion; in this way, the entire chiller and distribution system would be kept on a single control system. The secondary system was designed to be controlled on system pressure. To facilitate this, a differential pressure transmitter (DPT) was located in the secondary loop. Each secondary pump was equipped with a variable frequency drive (VFD) to control the motor speed using the feedback from the DPT to maintain a constant differential pressure between the supply and return water piping headers [FIGURE 5]. The chillers were designed to be cycled on or off based on the percent of full load of the operating chillers as determined by the chiller motor amperage.

In addition to the DDC control of the chillers and the distribution pumps, several monitoring points were included in the design of the DDC system. Supply and return water temperature on each secondary loop was to be monitored, while a flow meter installed in each secondary loop would measure flow. This combination of monitoring points made it simple to calculate and trend the load on each zone.

Because the building complex housed some manufacturing and support functions that could not be shut down for long periods of time, it was important to minimize any shutdowns. In addition, any system shutdown needed to be scheduled well in advance of the construction event. The shutdown for this project was scheduled for 72 hours over the Memorial Day weekend of 1992. The negotiations with the building occupants for this shutdown started in early February as the project design was beginning to take shape. The fact that the shutdown was scheduled prior to design completion allowed for timing constraints to be dealt with during the design phase.

PROJECT CONSTRUCTION

Construction Prior to Shutdown

Contractor selection was based on negotiations using the 65 percent drawings. Because of the complexity of the construction and tight scheduling, a contractor with considerable experience on the site was chosen. Both the engineering team and the construction contractor benefitted from the ability to address all anticipated issues before the shutdown. Several suggestions by the contractor for saving time and money were incorporated into the design. Also, the team was able to determine that the proposed work could in fact be accomplished during the scheduled 72 hour shutdown.

Prior to the Memorial Day shutdown, all advance work which could be done while the system was in operation was completed. New pumps were set in place and the accessories were installed prior to the shutdown. New mains were put in place and spool pieces were fabricated to connect them into the system. Insulation was removed from the existing chilled water piping where new connections were to be made over the shutdown. In addition, the contractors had made arrangements with their suppliers in case unanticipated materials were needed during the holiday weekend.

Three-Way Valve Conversion

The conversion of the coil valves from three-way to two-way was done by the in-house personnel. This allowed a greater flexibility in scheduling the coil conversions. Because the system was operating with such high flows, it was possible to complete several of the conversions while the main pumping system was still constant flow without adversely impacting the system operation.

In isolated areas, the plumbing crew was often able to negotiate with occupants of a space to schedule work which included shutting the cooling coil down for a short period of time to perform the valve conversion. The plumbers would prefabricate piping spools and valve end plugs whenever possible in order to minimize the length of the cooling shutdown. If pump removal was required at the same time, the plumbers would coordinate the electrical demolition with the electrical contractor. By working closely with the occupants in this manner, the crew was able to complete all valve conversions in noncritical areas without a shutdown of the entire system. The remainder of the coil conversions were done during the major system shutdown.

Construction During Shutdown

Crews were scheduled by the contractor for the entire 72 hour weekend shutdown. Draining the chilled water system was the first step in the construction. As soon as it was feasible, demolition of the old main distribution header outside the chilled water plant was begun. This demolition was necessary to provide additional room for the installation of the new primary loop, bridal loops, and secondary zone take-off. The construction in this area progressed slowly because of the height of the piping and because a large amount of new piping was required in such a tight space. In addition to the primary loop installation, crews were working throughout the building complex to install new mains, remove booster pumps, convert three-way valves, perform final connections on secondary pumps, and install control sensors.

System Startup

All of the piping modifications were completed several hours ahead of schedule. The piping was tested and filled with water prior to startup. After the air was bled from the distribution piping and air handler coils, the system pressure was set at the expansion tank. All seven primary and secondary pumps were checked for alignment and rotation. After the pumps had been allowed to run for a period of time, the construction debris was removed from the strainers. The flows in each primary pump were adjusted using the triple duty valves. Next, the flows were balanced through each of the bridal loops to ensure that all zones had adequate water flow available. The final task of the weekend was to adjust the pressure in each of the four secondary zones manually, using the VFD's that had been installed on the pump motors.

During the week following the shutdown, details of the piping were completed. Leaks were identified and corrected, insulation was installed on the new chilled water piping, and construction clutter was cleaned up.

Controls

In order to simplify the shutdown construction, controls were not switched to the new DDC system at that time. After the shutdown, the chiller plant was still on the old control system, and the secondary pumps required manual adjustment. The controls contractor switched each secondary pump to automatic individually; engineering and maintenance personnel were consulted to ensure proper operation of the system. The chillers were also switched over to the new DDC system one at a time. Each chiller was tested with the new DDC system during a nonpeak shift to minimize the exposure of a chiller failure. When all the chillers were tested, the chiller plant was switched over to the new DDC system. Finally, a formal "point to point" check was performed to verify that all chiller and pump controls would function during various normal and emergency modes of operation.

Balancing

Maintenance personnel had been closely involved in the design and construction of this project. Because they had offered to assist during the balancing of the system, it was decided to instruct the maintenance personnel on the balancing of the system instead of hiring a balancing contractor. An ultrasonic flow meter was purchased to assist in this task. During the balancing, the design engineer was available to assist with technical issues; however, the maintenance personnel performed all work. When the balancing was complete, the engineer compiled the data into a test and balance report.

CONCLUSIONS

Common Sense Approach

The primary/secondary piping concept has been available since the 1950's and it is still a viable solution to some piping system problems today. The use of DDC allows the engineer to optimize the sequence of operation to fit the design criteria. The retrofit of the building 4 and 5 chilled water system has resolved past flow problems and established an expandable system to ensure that the problems will not recur. The primary/secondary piping system allows for loads to be added or removed from a zone with minimal impact on the rest of the system.

The old, constant volume system required the pumps to route water through the distribution system even when it was not required. The two-way valve, variable flow secondary loops prevent excessive pump flows since only the chilled water required to satisfy the load is pumped to the valve. This reduction in flow equates to reduced pump energy usage.

Benefits

The conversion to a primary/secondary piping system enabled the chilled water plant to function at an increased load handling capability with a decrease in operating horsepower. With the incorporation of the primary/secondary piping scheme, the distribution was optimized to permit full utilization of the chiller plant capacity. This utilization potential allows for maximum flexibility when changing cooling loads in the future. The removal of 33 booster pumps reduced the preventive maintenance requirements for the building complex. Additionally, the energy costs for the delivery of the chilled water to the loads were decreased by approximately 36 percent, resulting from the removal of 60 horsepower of unnecessary pumps.

The primary equipment was brought back to design conditions. The new piping configuration ensures that the flow rates through the evaporator section of the chiller will remain at design conditions. This will reduce the amount of wear on the equipment.

Lessons Learned

Working with the strategic planning organization to determine the future loads was a very time-consuming process. However, without this effort the end result would not have met the needs of the corporation. The information gained by engineering during this process led to a 67 percent increase in the secondary pump sizing for future load requirements. The total design flow for all secondary zones started at 2,885 gpm, while the final design flow was 4,830 gpm.

For each of the four bridal loops, a balancing valve was specified and installed. This feature proved to be critical during the system startup. Initially, most of the water flowed through the bridal loop closest to the chiller plant. The balancing valves on the bridal loops were then adjusted to ensure that each bridal loop had adequate flow to accommodate the loads on the appropriate zone.

"Partnering" with the contractor on this project was instrumental in its successful completion. Because of the brevity of the shutdown, there was no room for error. The additional communication gained by hiring the contractor early in the process allowed the optimal use of engineering and construction resources.

The maintenance personnel actively contributed ideas and input from the start of the project, thereby also contributing to the project's success. These personnel were invaluable in the data collection, design, construction, and commissioning of the chilled water system renovation project. Now that the project is complete and operational, their ownership and intimate knowledge of the system help them keep the plant functioning at optimum performance.

SUMMARY

Following are some significant actions to be considered to guarantee the success of a project of this type:

Design

- Evaluate system problems fully and define goals prior to initiating the design.
- Coordinate with maintenance, engineering, and planning.
- Design installation to meet user scheduling constraints.
- Design to avoid the past system problems.
- Design the system to include monitoring and troubleshooting "tools".
- Keep it simple.

Installation

- Utilize all available resources, in-house and outside.
- Bring contractor in early in the process.
- Keep engineering involved and working with the contractor and maintenance.
- Coordinate with the user regarding shutdowns, outages, and installations.

Startup/Balancing

- Get maintenance involved.
- Get engineering involved with system startup.
- Use the project as a vehicle to acquire "tools" for the system.

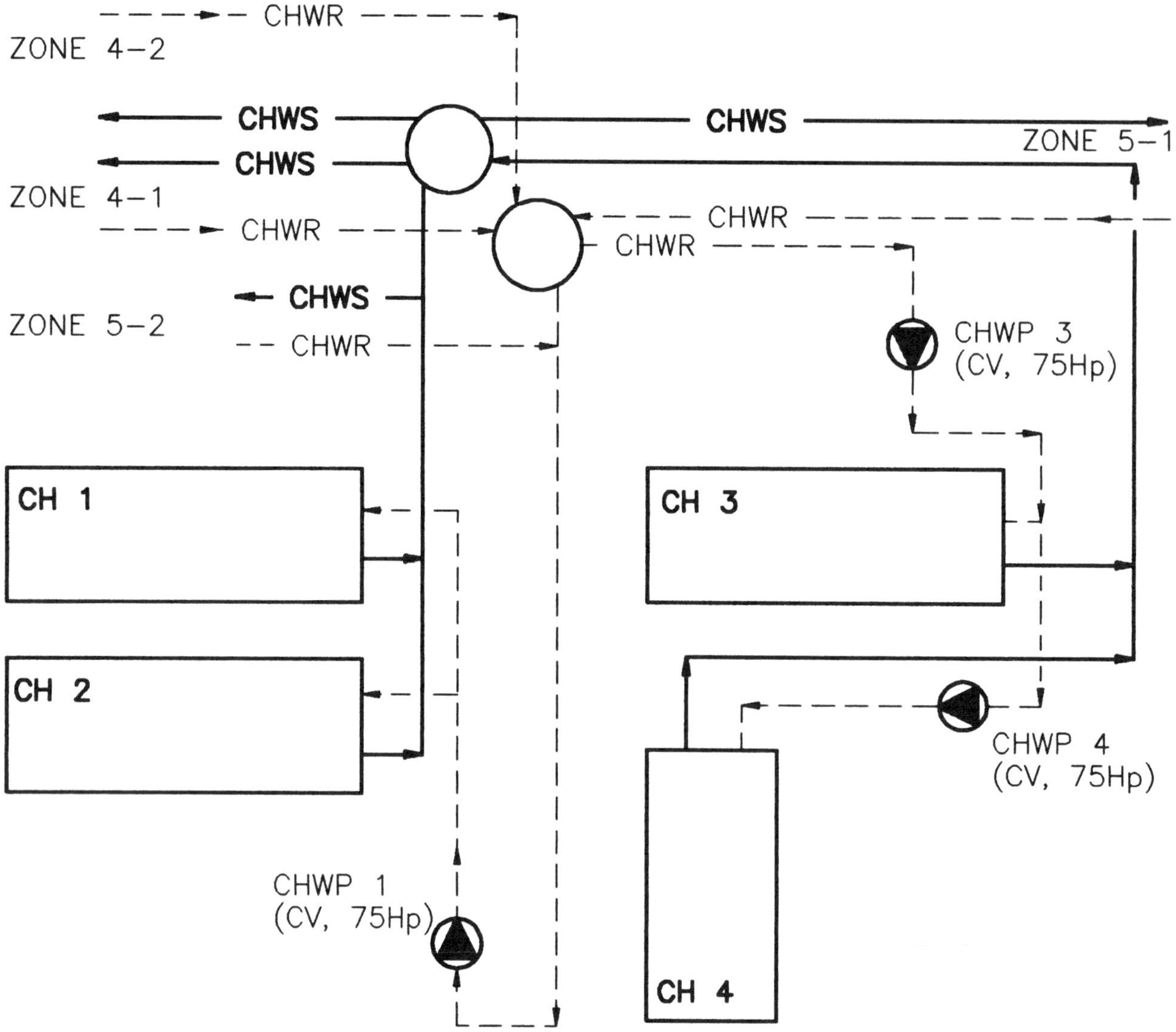

CHILLER PERFORMANCE

DESIGN:	CH1	CH2	CH3	CH4
FLOW (GPM)	643	643	643	1920
AP (FT)	21	21	21	31
GPM/TON	2.2	2.2	2.2	2.4

OPERATING:	CH1	CH2	CH3	CH4
FLOW (GPM)	1300	1350	1500	1400
AP (FT)	6.2	75	90	20
GPM/TON	4.4	6.9	6.8	1.8

FIGURE 1
CHILLED WATER SYSTEM (PRIOR TO UPGRADE)

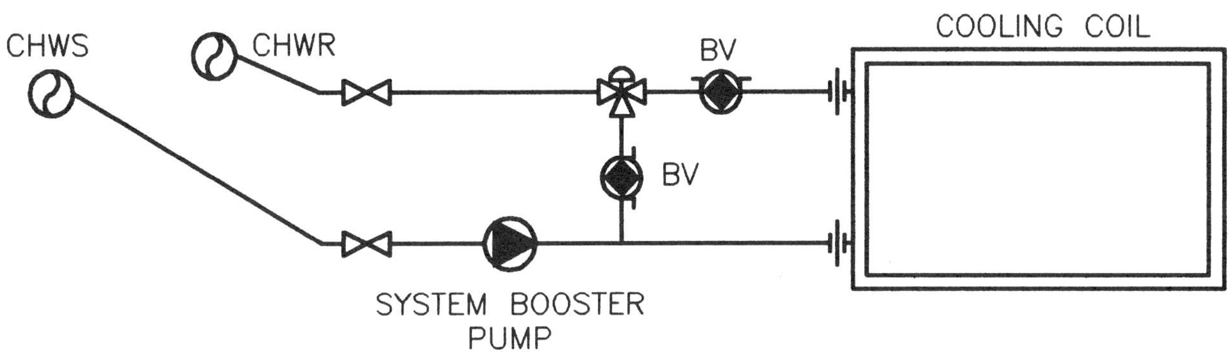

FIGURE 2
BOOSTER PUMP/COIL PIPING DETAIL

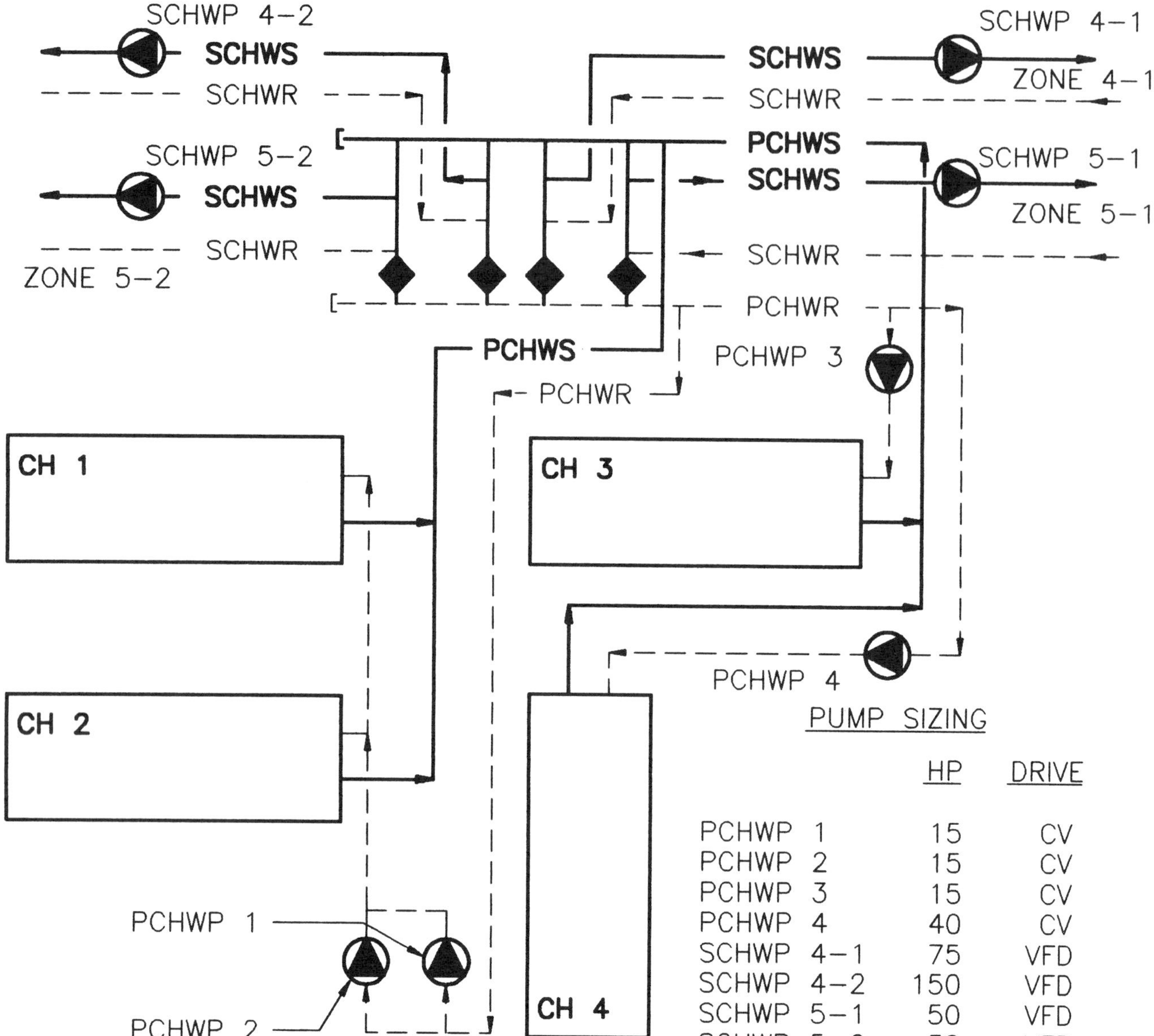

PUMP SIZING

	HP	DRIVE
PCHWP 1	15	CV
PCHWP 2	15	CV
PCHWP 3	15	CV
PCHWP 4	40	CV
SCHWP 4-1	75	VFD
SCHWP 4-2	150	VFD
SCHWP 5-1	50	VFD
SCHWP 5-2	30	VFD

CHILLER PERFORMANCE

DESIGN:	CH1	CH2	CH3	CH4
FLOW	645	645	645	1920
AP (FT)	21	21	21	31
GPM/TON	2.2	2.2	2.2	2.4

OPERATING:	CH1	CH2	CH3	CH4
FLOW	640	640	640	1900
AP (FT)	21	21	21	30
GPM/TON	2.2	2.2	2.2	2.4

FIGURE 3
CHILLED WATER SYSTEM (AFTER UPGRADE)

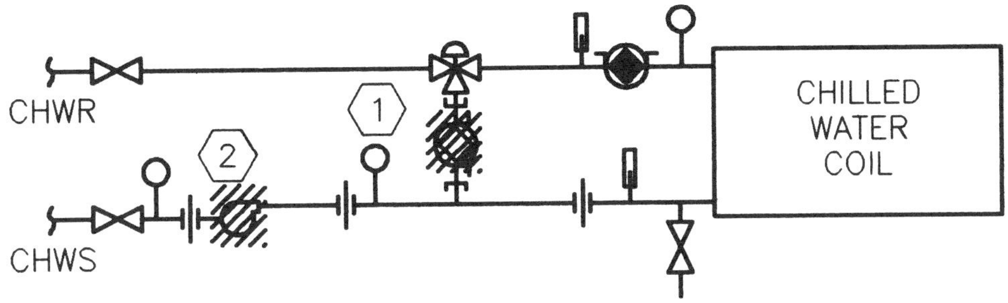

⟨1⟩ REMOVE BYPASS PIPING AND BALANCE VALVE, CAP BYPASS PORT ON VALVE, AND CAP T ON CHWS PIPING.

⟨2⟩ REMOVE CHWS BOOSTER PUMP. REPLACE WITH LINE SIZE PIPING.

FIGURE 4
CHW COIL 3–WAY TO 2–WAY CONVERSION

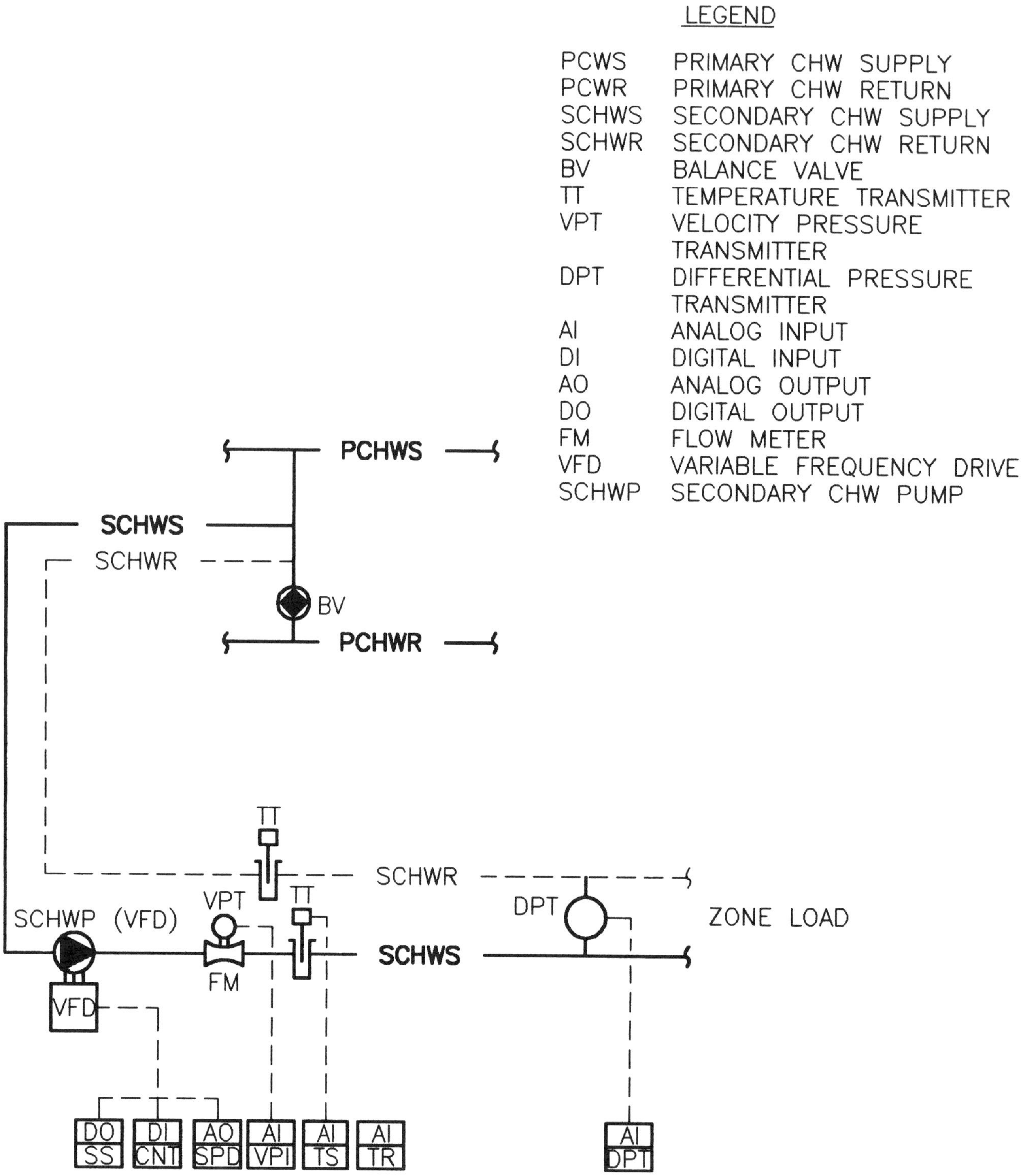

FIGURE 5
VFD SECONDARY PUMPING PID (TYPICAL OF 4 ZONES)

Chapter 76

Thermodynamic Analysis of a Combined Vapor Compression/Absorption Refrigeration System

D.R. Swalla, S.A. Sherif

ABSTRACT

This paper reports on results of a thermodynamic analysis of a combined vapor compression/absorption refrigeration system. The system employs R-134a as the working fluid for the vapor compression system (VCS) and a lithium bromide/water solution as the working fluid for the absorption system (ARS). The performance of the individual systems were only considered in relation to their effects on the performance of the combined system.

INTRODUCTION

In many large capacity chilling applications a steam turbine drives the compressor for a vapor compressor system (VCS). The turbine exhaust is then used as a heat input into an absorption refrigeration system (ARS) generator, which eliminates the need for an additional heat source. Because the absorption system makes use of the turbine exhaust, the utilization of a combined system is more advantageous for chilling applications than using either system independently.

Many studies have been performed on the use of absorption refrigeration systems that use lithium bromide/water as the working fluid. A computer program for simulating absorption heat transformer operation was developed by Jiexiu et al. [1] based on thermodynamic data of LiBr-ZnC12-CaBr2-/H20. They concluded that LiBr-ZnC12-CaBr2-/H20 is a promising working pair for absorption heat transformers. A few studies document the performance of vapor compression systems using refrigerant 134a. Bansal [2] concluded that the use of refrigerant 134a was advantageous for use in heat pump installations for low to medium temperature applications. It was also concluded in a separate study by Bansal [3] that refrigerant 134a was a suitable replacement for refrigerant 12.

It appears that there is very little information regarding the use of refrigerant 134a in vapor compression systems, and little or no existing information regarding the performance of a combined system. This paper reports on performance results of a combined vapor compression/absorption refrigeration system for a host of conditions.

ANALYSIS

Figure 1 is a schematic of the combined vapor compression/absorption refrigeration system. Superheated steam enters the turbine, which generates the power required to drive the compressor. Saturated steam leaves the turbine and is used as a heat source for the absorption system generator. The conditions of the steam entering and leaving the turbine are assumed constant throughout the analysis. Also, the steam flow rate through the turbine is assumed equal to the steam flow rate through the generator.

In order to express the performance of the combined system, two methods of analysis were devised. The Variable Condensate Enthalpy (VCE) method assumes that the heat input to the generator is constant given set conditions in the absorption system. This assumption necessarily implies that a change in the steam flow rate would cause a corresponding change in the generator condensate enthalpy. Thus, for a specific set of absorption system conditions, the generator condensate enthalpy becomes a measure of the performance of the combined system. The Constant Condensate Enthalpy (CCE) method, on the other hand, assumes that the generator condensate enthalpy is kept constant at a base value, while the steam flow rate is allowed to change. This necessarily forces the heat input from the turbine exhaust to change, which in turn causes the condenser temperature and heat rejection rate to change. Thus, when the vapor compression system conditions are changed, the generator refrigerant enthalpy can become a measure of the combined system performance.

The solution enthalpies and all flow rates in the absorption system were kept constant for all calculations. These parameters are functions of the concentrations in the absorber and generator, respectively, and of the temperatures in the absorption system condenser and evaporator. To allow these values to change (instead of the generator refrigerant enthalpy or the generator condensate enthalpy) would greatly increase the complexity of the analysis.

Use of the VCE and CCE methods produces two performance extremes for the combined system. The performance extreme associated with the VCE method is observed by the inability of the absorption system to respond to any change in the heat input, which is manifested by a change in the generator condensate enthalpy. The performance extreme associated with the CCE method is observed in the fact that the absorption system must respond to all changes in the heat input, which is manifested by a change in the generator refrigerant enthalpy. For a realistic combined system, it is unlikely that either condition will exist. However, use of the two performance extremes provides boundaries within which to operate the combined system for a given set of conditions.

RESULTS AND DISCUSSION

The performance of the combined system will be presented here for a few cases employing both the VCE and CCE methods. Results of the base case calculations are based on the following conditions for each system:

Turbine
HT_I = 3080 kJ/Kg
HT_o = 2676 kJ/Kg
Psat = 101.33 kPa
Tsat = 100°C

Vapor Compression System (VCS)
Refrigerant: R134a
Refrigerant Flow = 1.5 kg/s
Te_{VCS} = -10°C
Tc_{VCS} = 40°C
Qc_{VCS} = 253.1 kW
Qe_{VCS} = 202.7 kW
$\dot{W}_c$ = 50.4 kW
COP = 4.02

Absorption System (ARS)
Absorbent: Lithium Bromide
Refrigerant: Water
$\dot{m}_{1A}$ = .35 kg/s
$\dot{m}_{2A}$ = .264 kg/s
$\dot{m}_{3A}$ = .086 kg/s
Te_{ARS} = 10°C
Tc_{ARS} = 40°C
Tg_{ARS} = 100°C
Ta_{ARS} = 31°C
GCONC = 66.4%
ACONC = 50%

VCE Method
HC = 517.4 kJ/kg
H_{3A} = 2574.4 kJ/kg
COP_{ARS} = .755
COP_{COMB} = 1.271
Qe_{ARS} = 203.4 kW
Qg_{ARS} = 269.23 kW
Qc_{ARS} = 208.1 kW
Qa_{ARS} = 264.5 kW

CCE Method
HC = 419 kJ/kg
H_{3A} = 2716.4 kJ/kg
COP_{ARS} = .722
COP_{COMB} = 1.223
Qe_{ARS} = 203.4 kW
Qg_{ARS} = 281.5 kW
Qc_{ARS} = 220.3 kW
Qa_{ARS} = 264.5 kW

Results based on the VCE method are given in Figures 2 and 3. Figure 2 displays the absorption system generator condensate enthalpy as a function of the vapor compression system evaporator temperature for several vapor compression system condenser temperatures. As can be seen, the generator condensate enthalpy increases with an increase in the condensing temperature or a decrease in the evaporating temperature. This is expected since an increase in the condensing temperature or a decrease in the evaporating temperature would cause an increase in compressor power and a subsequent increase in the steam flow rate. This in turn causes the condensate enthalpy to increase.

The base case value for the condensate enthalpy is 419 kJ/kg. For the VCE method, all condensate enthalpies along a curve that fall above the base case value represent an inability on the part of the absorption system to accept additional heat input from the turbine exhaust. Values below the base case value represent a higher heat input requirement for the absorption system than the turbine exhaust can supply for a given set of absorption system conditions. As an example, for the case when the condensing temperature is equal to 40°C, the point at which the heat input requirement for the absorption system is met by the turbine is approached when the evaporating temperature is approximately equal to -8°C. If the latter is decreased to -10°C, for instance, then either the condensing temperature would have to be lowered, or the conditions in the absorption system would have to change in order to make best use of the available heat energy from the turbine.

Figure 3 is a plot that illustrates the effect of changing the absorption system conditions for constant vapor compression system conditions. For the vapor compression system conditions listed in Figure 3, the base case value for the generator condensate enthalpy is 517 kJ/kg. As the absorption system condenser temperature is decreased and/or the absorption system evaporator temperature is increased, the heat input from the turbine is increased, and the condensate enthalpy is increased. This may be explained by noting that a decrease in the condenser temperature is associated with an increase in the heat rejection rate (for the absorption system condenser). Also, as the evaporating temperature increases, the cooling capacity of the absorption system would increase. The net effect on the absorption system is a decrease of the generator input energy requirements. The optimum working range for the absorption system given a specific set of vapor compression system conditions can be found by determining where each curve intersects the base value of the condensate enthalpy. For instance, the optimum absorption system evaporator temperature for the case when the absorption system condenser temperature is equal to 45°C is approximately 14°C.

Results associated with the use of the CCE method may be seen in Figures 4 and 5. Figure 4 corresponds to the data presented in Figure 2. However, for this method the generator condensate enthalpy was kept constant at 419 kJ/kg, while the generator refrigerant enthalpy was allowed to vary. This means that the absorption system is forced to adapt to any fluctuations in the heat input from the turbine exhaust. The base case value for the generator refrigerant enthalpy was calculated as 2574.4 kJ/kg, at which point the refrigerant is saturated vapor at the same temperature as the condenser (listed on Figure 4). As expected, the generator refrigerant enthalpy increases as the vapor compression system condenser temperature increases and/or the vapor compression system evaporator temperature decreases. For the case when the condenser temperature is equal to 45°C, the optimum vapor compression system evaporator temperature for this plot is -4°C. If the latter is allowed to drop, then the generator refrigerant enthalpy would be superheated, or at a different condenser temperature than specified. This would force the absorption system heat rejection rate to decrease.

In Figure 5, the coefficient of performance of the absorption system is plotted versus the vapor compression system evaporator temperature for several vapor compression system condenser temperatures. In this case, the absorption system coefficient of performance is mainly a function of the heat input from the turbine exhaust because the absorption system cooling capacity is kept constant for all cases. The base case for the absorption system coefficient of performance is 0.755. As the vapor compression system condenser temperature decreases or its evaporator temperature increases the generator energy requirements decrease, which causes the coefficient of performance to increase.

CONCLUSIONS

This paper presented a summary of performance results of a combined vapor compression/absorption refrigeration system employing two methods. The first method (VCE) imposes the restriction that the absorption system cannot accept any changes in the heat input from the turbine exhaust. The second method (CCE), on the other hand, forces the absorption system to accept all changes in the amount of heat exhausted from the turbine. The VCE and CCE methods act as boundaries within which a real system would work. The application of these two methods give considerable insight into the behavior of the combined system, without having to conduct a more complex iterative or numerical procedure.

LIST OF SYMBOLS

ARS - Absorption refrigeration system
CCE - Constant condensate enthalpy
COP - Coefficient of performance
$\dot{W}_c$ - Compressor power for vapor compression system (kW)
HC - Generator condensate enthalpy (kJ/kg)
HT_I - Enthalpy of fluid entering steam turbine (kJ/kg)
HT_O - Enthalpy of fluid exiting steam turbine (kJ/kg)
$\dot{m}_s$ - Steam flow rate (kg/s)
$\dot{m}_v$ - Vapor compression system refrigerant flow rate (kg/s)
Qa - Absorber heat rejection rate (kW)
Qc - Condenser heat rejection rate (kW)
Qe - Cooling capacity (kW)
Qg - Generator heat input (kW)
$\dot{W}_T$ - Steam turbine power
VCE - Variable condensate enthalpy
VCS - Vapor compression system
$\dot{m}_{1A}$ - Absorption system absorber solution flow rate (kg/s)
$\dot{m}_{2A}$ - Absorption system generator solution flow rate (kg/s)
$\dot{m}_{3A}$ - Absorption system refrigerant flow rate (kg/s)

ACKNOWLEDGEMENTS

The authors would like to acknowledge the support of the Department of Mechanical Engineering at the University of Florida. The authors are indebted to the help received from Drs. R.S. Basu and A. Zaltash. The second author would like to acknowledge the help of Mrs. Alice Jempson of the University of Florida.

REFERENCES

1. Jiexiu, D., Minduan, T., and Jingchang, Y., "Computer Prediction and Experimental Study of Operation Behavior for Absorption Heat Transformer," AES-Vol. 27/HTD-Vol. 228, ASME, New York, pp. 149-152, 1992.

2. Bansal, P.K., Dutto, T., and Hivet, B., "Performance Evaluation of Environmentally Benign Refrigerants in Heat Pumps 1: A Simulation Study," *International Journal of Refrigeration*, Vol. 15, No. 6, pp. 340-348, 1992.

3. Bansal, P.K., Dutto, T., and Hivet, B., "Performance Evaluation of Environmentally Benign Refrigerants in Heat Pumps 2: An Experimental Study With HFC134a," *International Journal of Refrigeration*, Vol. 15, No. 6, pp. 349-356, 1992.

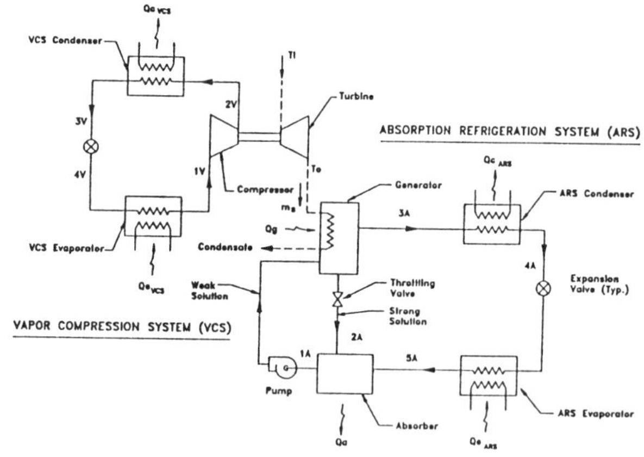

FIGURE 1. SCHEMATIC OF COMBINED VAPOR COMPRESSION/ABSORPTION REFRIGERATION SYSTEM

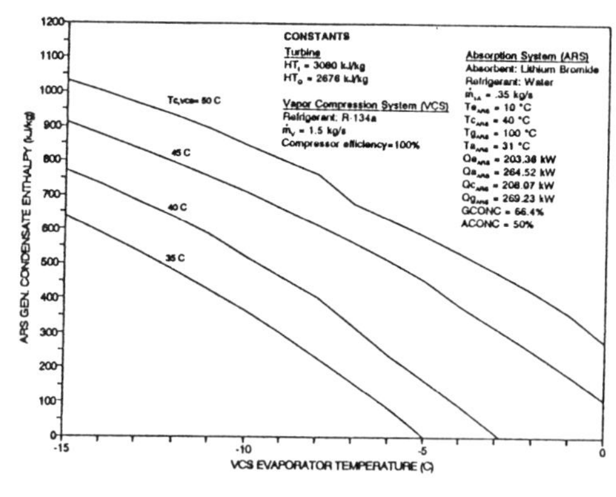

FIGURE 2. ABSORPTION SYSTEM (ARS) GENERATOR CONDENSATE ENTHALPY VERSUS VAPOR COMPRESSION SYSTEM (VRS) EVAPORATOR TEMPERATURE

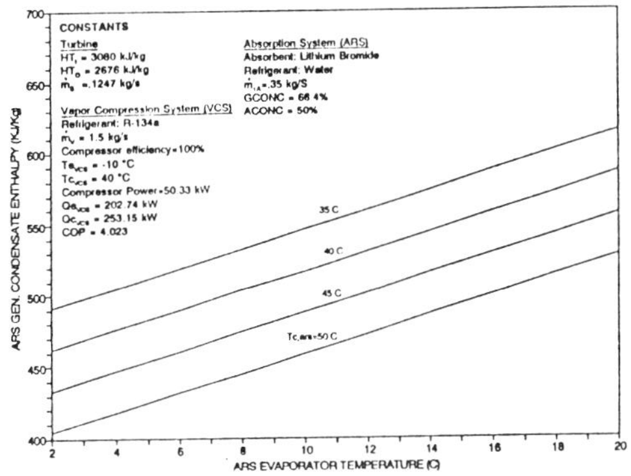

FIGURE 3. ABSORPTION SYSTEM (ARS) GENERATOR CONDENSATE ENTHALPY VERSUS ARS EVAPORATOR TEMPERATURE

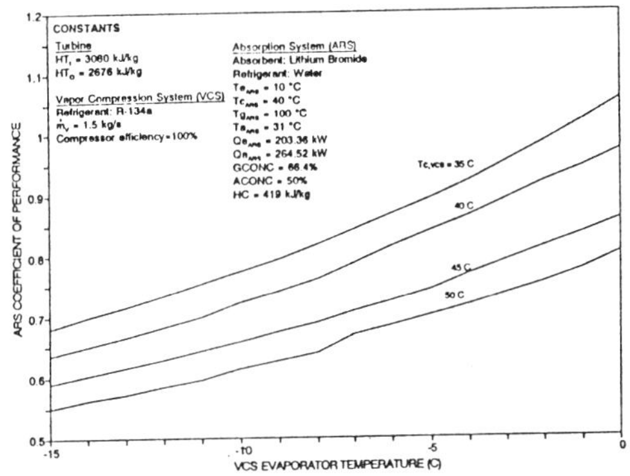

FIGURE 5. ABSORPTION SYSTEM (ARS) GENERATOR REFRIGERANT ENTHALPY VERSUS VAPOR COMPRESSOR SYSTEM (VCS) EVAPORATOR TEMPERATURE

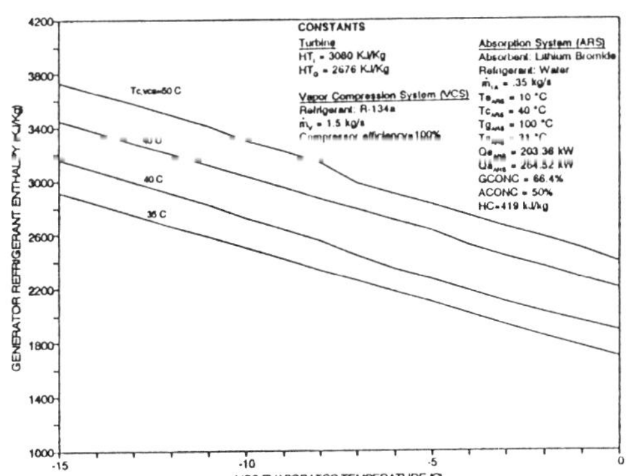

FIGURE 4. ABSORPTION SYSTEM (ARS) GENERATOR CONDENSATE ENTHALPY VERSUS VAPOR COMPRESSION SYSTEM (VCS) EVAPORATOR TEMPERATURE

Chapter 77

Frosting and Defrosting Effects in Refrigeration Coils

S.A. Sherif

ABSTRACT

The process of frost formation on cold surfaces is important in a variety of refrigeration, air conditioning, and cryogenic applications. This paper is one in a series reporting on frosting and defrosting effects of evaporator surfaces held at subfreezing temperatures. An overview of the nature of the problem and the available remedies will be examined.

INTRODUCTION

The process of simultaneous heat and mass transfer under frosting conditions is important in a variety of refrigeration and air conditioning applications. This process is extremely complicated at the fundamental level because of the following reasons: (a) Frost properties vary continuously during the development of the frost layer both spatially and temporally. (b) The thermal resistance of the frost layer changes, thus causing a change in the air-frost interface temperature. This, in turn, causes a change in the partial pressure of water vapor in the vicinity of the frost surface which affects the frost deposition and heat transfer rates. (c) The upper layers of the frost surface are likely to experience repeated cycles of melting and refreezing in high humidity and temperature environments. This causes structural changes to the frost layer, with high density ice collected near the coil surface and lower density frost near the air-frost interface.

Frost will generally form on a surface if two conditions are met. First, the surface temperature has to be equal to or less than the dew-point of water vapor in air. Second, the surface temperature has to be below the freezing point. Investigators have argued that the initial frost deposits may indeed be desirable since they act as fins, thus temporarily improving the evaporator heat transfer performance. Additional deposits, however, are more likely to degrade the performance because of the dominance of the insulating effect of the frost over other effects. Also, over time, the frost layer may experience density stratification, whereby the lower sublayers tend to gradually transform into a thick solid ice layer, thus increasing the frost insulating effect. This phenomenon is primarily attributed to the occurrence of repeated and rapid cycles of melting and refreezing near the frost of surface. In this process, the upper frost layers melt and the melt seeps into the pores of the frost layer. As the melt migrates inward it freezes into ice because of the lower temperatures found in the middle of the frost layer. Over long deposition times, the whole frost layer is bound to transform into solid ice, thus blocking any further inward migration of melt.

Given the aforementioned facts, engineers and researchers spent a great deal of time and effort investigating means to get rid of the frost. Traditional defrost methods include hot-gas bypass or reverse cycle, electric, hot-water, and time defrost. In all of these methods, undesirable defrosting heat is passed on to the refrigerant during the defrosting cycle. This represents an additional cooling load that needs to be removed during the refrigeration portion of the cycle.

Previous work on frost formation is numerous. A fairly complete review can be found in Padki et al. [1], O'Neal and Tree [2], Kondepudi and O'Neal [3], and Sherif [4]. Defrosting work is far less extensive and more qualitative in nature. Furthermore, most of the defrosting studies dealt with hot-gas bypass or reverse cycle methods as opposed to hot-water, electric, or time defrosting. For a more comprehensive review of these methods, the reader is advised to consult Sherif and Hertz [5].

FUNDAMENTAL ISSUES

It appears that there are a number of basic issues that need to be addressed for both frosting and defrosting before a complete understanding of the different processes can be achieved. For the process of formation, a knowledge of how the frost-air interface temperature change with both time and position appears to be crucial. Previous investigators have repeatedly demonstrated that many of the frost formation parameters depend in whole or in part on the frost-air interface temperature. Some even derived empirical relationships between the interface temperature on the one hand and both the frost density and thermal conductivity on the other. Current analytical methods for predicting the interface temperature depend on a quasi-steady state conduction/convection analysis of the frost layer/boundary layer combination. Experimental methods rely on infrared temperature measurement techniques. Sherif [4] reported that both the rates of frost formation and heat transfer on a cold surface are heavily dependent on the air-frost interface temperature.

Another important issue in the formation process is that of deposition in a high humidity/high temperature environment. As explained in the Introduction, density stratification occurs because of the occurrence of rapid cycles of melting and refreezing. A quantitative description of the variation of the frost density with time from the onset of melting is badly lacking. However, Padki et al. [1] suggested a density formulation in terms of an experimental constant describing the nature of melting and refreezing.

A third issue related to the formation process is how much of the frost formed actually deposits on the upper layers as opposed to diffusing inward. The latter frost tends to increase the density of the lower frost sublayers without an appreciable increase in the frost thickness, whereas the former tends to cause the opposite effect. White and Cremers [6] contend that approximately half the water vapor transfer to the frost layer diffuses inward, condenses, and increases the density of the frost layer.

Other important issues include the nature of the deposited frost at freezer temperatures. Some investigators have argued that at low temperatures, the cooling process line on the psychrometric chart may invade the supersaturated zone and hence influence the nature of frost/snow/ice crystals that may form. A fundamental study to address some of these issues is currently being planned for ASHRAE funding.

Issues associated with coil defrosting are also equally significant. One of these issues continue to be coil defrosting loads and their nature (sensible or latent). At present, ASHRAE is funding a project to experimentally determine defrosting loads under different environmental conditions for a host of coil geometries. Other significant defrosting issues include determining the optimum defrosting frequency and duration, deciding on the criteria for initiating defrosting, and evaluating the defrosting efficiency.

CONCLUSIONS

This paper presented an overview of some of the problems and remedies of frosting and defrosting air conditioning and freezer coils. The paper attempted to identify some of the significant research problems that investigators need to examine in order to increase the level of understanding of the problem physics.

ACKNOWLEDGEMENTS

The author would like to acknowledge the help of Mrs. Alice Jempson of the University of Florida. Support from the Department of Mechanical Engineering at the University of Florida is also gratefully acknowledged.

REFERENCES

1. Padki, M.M., Sherif, S.A., and Nelson, R.M., "A Simple Method for Modeling the Frost Formation Phenomenon in Different Geometries," *ASHRAE Transactions*, Vol. 95, Part 2, pp. 1127-1137, June 1989.

2. O'Neal, D.L. and Tree, D.R., "A Review of Frost Formation in Simple Geometries," *ASHRAE Transactions*, Vol. 91, Part 2, June 1985.

3. Kondepudi, S.N. and O'Neal, D.L., "The Effects of Frost Growth on Extended Surface Heat Exchanger Performance," *ASHRAE Transactions*, Vol. 93, Part 2, Paper No. 3070, pp. 258-274, June 1987.

4. Sherif, S.A., "Theory of Frost Formation," *Proceedings of the 32nd Heat Transfer and Fluid Mechanics Institute*, California State University at Sacramento, F.H. Reardon and N.D. Thinh (Editors), Sacramento, California, pp. 239-252, June 6-7, 1991.

5. Sherif, S.A. and Hertz, M.G., "Electric Defrosting Analysis of a Cylindrical Coil Cooler," *Proceedings of the 33rd Heat Transfer and Fluid Mechanics Institute*, California State University at Sacramento, F.H. Reardon (Editor), Sacramento, California, pp. 199-209, June 3-4, 1993.

6. White, J.E. and Cremers, C.J., "Predictions of Growth Parameters of Frost Deposits in Forced Convection," AIAA Paper 74/746, 1974.

SECTION 7
COMPETITIVE POWER TECHNOLOGIES

Chapter 78
State PURPA Proceedings: Effects of Competition

R.D. Feldman

PURPA Revisited: The Legal Setting
Fundamental modification of PURPA may be initiated by Section 712 of the Energy Policy Act of 1992. The result could be harm to the competitiveness of independent power options, regardless of ownership sources. It is imperative to know why that could be the case; what the issues will be; and what will be necessary to address them effectively.

The Act amends PURPA to effectively require that within one year from its enactment each state regulatory authority must conduct a proceeding for "general evaluation" of certain specified issues, (also referred to in the section as "standards") and determination of how the standards will be implemented. As a result of such evaluation, and "notwithstanding any other provision of federal law", broad remedial action may be taken by state commissions, including action with respect to the "allowable capital structure of exempt wholesale generators", when it is determined in the public interest to do so. State proceedings must be conducted in the same manner as the first round of PURPA proceedings. Proceedings commenced prior to statutory enactment do not fulfill the statutory requirement.

Two of the "standards" are interlinked:

- How utility purchases of long term wholesale power (from whatever supply source) - rather than construction of new generation facilities - will affect utility cost of capital, and with what resultant retail rate impacts;

- Whether utility reliability will be threatened as a consequence of their supply by EWGs, or whether utilities will be subject to an "unfair advantage" in competition with EWGs, since EWGs use proportionally more debt in their "capital structure" than do utilities.

Two additional "standards" require state regulatory commission scrutiny of two finance-related operational aspects of wholesale power supply (again, from whatever supply source):

- Whether there should be advance approval of specific power supply purchases;

- Whether approval of power purchase should be conditioned on "reasonable" assurances of fuel supply "adequacy".

PURPA Revisited: The Possible Consequences
Obviously, the future of EWG/QF (herein, "IPP") power suppliers could be directly affected by these proceedings. Some of the potential adverse results which do not seem beyond include the following:

- EWG capital structure could be subjected to a mandatory debt ceiling cap. (This was originally proposed in the Senate.)

- Rules imputing higher capital cost to competitive bid proposals from an IPP agency could be established as a result of the implicit impact on utility cost of capital. (This would follow along the lines of the approaches already taken by the rating agencies and propounded by utilities in proceedings.)

- Fuel supply and transport contracts coterminous with power sales agreements could become a prerequisite for acceptability of competitive bids from IPPs. (This has already begun to occur in some jurisdictions.)

- Prior regulatory commission approval of power sales agreements is traditionally important from the standpoint of mitigating the risk of denial of utility passthrough to its customers of the cost of purchased power. Prior approval may involve more frequent evaluation of the comparison of power purchases with the then-annual costs to the utility if it were the pendency of the proceedings in any state could have a chilling effect on the ability to close the financing for projects currently in development and may also affect the ability to complete competitive bids and power contract negotiations. Issues posed by the imprecise language of the Act include:

- Vulnerability of QFs to state modification of PURPA avoided cost, power contracting, and competitive bid rules in connection with state evaluation of the standards.

- Grandfathering of existing power contracts, bid awards, and financing and transactional arrangements;

- Adequacy and enforceability of existing power procurements;

- Feasibility of final FERC approval of rates for sales by an EWG pending finalization of capital structure and fuel supply rules by state regulatory authorities.

All sectors of the electric power industry must be concerned with these possibilities. Supply growth is jeopardized. Utility affiliate development strategies may be retarded. The feasibility of launching an EWG-based development strategy may be set back. Effective integrated resource planning through comparison of conservation and supply options will be delayed.

To avoid state Commission gridlock, there must be submission of objective and detailed analysis, which rises above rhetorical posturing. Commissions must be assisted in sifting through what will inevitably be a plethora of submissions. The outlines of the issues to be considered already are discernible from testimony, articles and rating agency activity which has already occurred. The conflicting arguments are summarized briefly below, as a roadmap to where data and analysis must be conducted. The potential difficulties summarized above can be forestalled if precedent for constructive consensus can be established.

The Cost of Capital/Rate Impact Debate

The objective predicate for determining whether to constrain EWG capital structure (standard #2) is examination of the issue of whether increased purchases of power from IPPs will raise utility cost of capital, with resultant rate increases. The essence of the argument is two-fold. First, attention focuses on whether the IPP-utility power sales agreement represents a capacity cost recovery and plans performance risk to the host utility that is higher than if the utility were to build the needed capacity. Second, concern is raised whenever the capacity charge payment obligation increases a utility's fixed cost obligations. While the resultant payments may be treated for financial accounting purposes as operating costs, they are argued as having the same effect on cash flow as fixed charges. If the utility had constructed the facility itself, the increased "debt" and associated risk would have been offset in part by new equity, which is not true in the purchased power case.

Standard & Poors has begun to impute a certain percentage of the future present value of capacity payments of power sales agreements as debt, and to compute an adjusted leverage ratio which treats the related debt obligation as part of the total debt. The resultant deterioration of a utility's debt/equity ratio over time can ultimately lead to reduction of the utility's bond rating, and increase in its cost of debt funds. The down-rating of several utilities during the past year has been attributed to excessive reliance on purchased power. Some utilities have sought limits on the amount of purchased power, or have claimed a premium value for purchased power, out of stated concern with this possible result.

Another adverse impact of purchased power arguably may be to reduce the return on and increase the cost of utility equity. Rate base erosion resulting from purchased power reliance can reduce rate based generated revenue and earnings growth. It is also argued that ---- base erosion may leave a utility with less invested capital over which to spread risk, including the risk said to be shifted by the IPP through the project finance process. In this frame of reference, the only certain way to offset risk is to increase equity, with attendant cost to the utility. In turn, this type of analysis has led to the recommendation, which has not yet accepted outright in any state proceeding, that utilities need to adjust IPP bids to reflect the effect of purchase on their cost of capital.

IPPs challenge the characterization of risk shifting from IPPs to utilities, and also contest the and the resulting characterization of power contract risks as capitalizable debt risks. The use of the perceived leveraged IPP structure power demand risk can be mitigated through power purchase acquisition strategies, and efficient use of competitive bidding. Construction risk, is in fact shifted from utilities to third parties via turnkey contract arrangements. The current IPP operating record is such that it cannot be said to be a major risk. Regulatory risk falls upon utilities as well as IPPs. Utilities have a history of being plagued by prudence proceedings. They have not experienced any disallowances of purchase power, and the competitive framework in which IPP power is procured reduces the likelihood regulatory disallowance. From a utility operational standpoint, reliance on IPP power should not affect utility cost of capital.

Analysis of the actual structure of power sales arrangements argues against their characterization as a financial risk as well. As utility flexibility has increased, this is increasingly true. The vast majority of contracts are take and pay, and not take or pay are subject to cancellation for non-performance or significant capacity derating; and frequently are dispatchable or curtailable.

Most fundamentally, what would the utility do in the absence of purchased power? The effect on utility financial condition of the additional financial burden on that would be imposed in the absence of purchased power is the key to this issue. Purchases from wholesale suppliers, with pre-approved passthrough, may be preferable to utility construction, with frequent prudence proceeding, from the standpoint of preserving----

Finally, overall, its has not been demonstrated that, taken by itself, debt equivalence arising from power sales arguments, and prudently regulated power purchases, are themselves key factors, near term, affecting utility rates. As rate base erosion, and related fundamental structuring issues become more important, this scenario may change.

Effects of EWG Leverage

The historical record shows that highly leveraged IPPs can be counted upon to maintain the same continuing quality of service as utility-owned facilities. Project finance loan covenants have the further effect of mitigating the key components of reliability risk, finance, fuel and environment.

For example, lenders not only review construction agreements to assure they will be kept whole in the event of a deficiency, but arrange for developer commitment to monitor construction standards and obtain liquidated damages. Operating risk is particularly covered by conditions precedent in loan agreements regarding conversion from construction to permanent loan, covenants against operating risks and operator replacement rights. Operating shortfalls are addressed by cash infusion obligations to deal with unexpected cash shortfalls and scheduled periodic expenses. Special provisions are also included to prevent siphoning off of funds from the project via the mechanism of a tightly defined and controlled "waterfall", permitted investment limitations and restriction on cash withdrawal from the project.

Concern with adequate operating coverage is dealt with through debt service reserves and explicit maintenance covenants. Utility indentures likely will not be as plant specific as IPP project financing arrangements.

Regarding the unfair competitive advantage issue, it is pertinent to note that while project finance uses proportionately less equity and more (relatively lower cost) debt, each element has, however, a higher cost in a project financing. That reflects the maturity of that debt and the need to fund debt service reserves.

While IPPs initial debt/equity ratio is significantly higher than that of utilities, it runs off faster, with the result that, even taking IPP refinancing into account, the average debt level of an IOU is higher than that of an IPP, and substantially less than that of a public power agency. Second, the weighted after tax cost of capital of IPPs only appears to be less than utilities if there is a failure to acknowledge the impact on IPP debt of its shorter term and the cost of carrying reserve funds. There does not appear to be any generic advantage for either form of finance, when embedded cost, weighted average analysis is utilized. Indeed, project finance has a capital charge disadvantage compared to utility debt, because the markets have imposed a more rapid amortization schedule on it. This may be even truer if a marginal analysis is conducted, since during major multi-year construction cycles, debt financing is sometimes used to reduce equity. The alleged unfair advantage is not present.

Conclusions

Section 712 of the Act can have very adverse effect on IPPs if state regulatory commissions interpret it as the basis for restrictive standards. Such restrictive concerns are not necessary to provide desired consumer protection. The key factors contributing to this conclusion are:

- The improved risk sharing profile which utilities face as a result of IPP purchases

- The numerous and fine-tuned covenants lenders impose on IPPs

- The absence of the need to treat as debt power contract capacity credits

- The absence of an objective weighted cost of capital gap between IPPs and utilities, when IPP capital costs (including those arising from short debt term) are taken into account.

These are not self evident arguments, and ones that are likely to be disputed. It is critical for these matters to be resolved expeditiously in state commission proceedings, because their continuing status as matters for analysis may delay or defer project development and utility purchases.

\09900\021\10ARTRDF.012

Chapter 79

Wholesale Ratemaking for EWG's

E.L. Selgrade

I. OVERVIEW

Over the last fifteen (15) years, there has been spreading recognition that the bulk power market is an emerging competitive market in which enough buyers and sellers may exist to allow the price of bulk power to be determined by market forces. In light of the risks associated with the production of bulk power, competitors in this emerging industry have conventionally chosen to structure their generation investments on a "free-standing" basis supported by the power sales agreements pursuant to which bulk power from individual power facilities is sold to buyers. As a result of their power sales, such competitors were potentially exposed to regulation under two significant federal laws, the Federal Power Act and the Public Utility Holding Company Act of 1935. In 1978, Congress passed legislation which allowed certain qualified competitors to escape regulation under both federal laws. In 1992, Congress addressed the issue of regulation a second time in response to concerns that certain competitors who did not qualify under the 1978 law could not, and should not be forced to, comply with the Public Utility Holding Company Act. Regulation under the 1935 Act severely restricted the competitive efforts of all entities which could not qualify under the 1978 law. The 1992 law exempts most competitors from the 1935 Act, but leaves competitors subject to the Federal Power Act. This was not considered to be a significant impediment to competition in light of recent rulings enforcing the Federal Power Act which have allowed buyers and sellers to set the prices they are willing to pay in freely negotiated, competitive sales agreements.

II. DEFINITIONS, FILING REQUIREMENTS AND EWG STATUS DECISIONS

The Energy Policy Act of 1992 ("EPA") created a new category of "exempt wholesale generators" ("EWG's") under the Public Utility Holding Company Act of 1935 ("PUHCA"). The purpose of the new definition is to free EWG's from the requirements of PUHCA, which have proven to be a significant constraint on the development of competition in the wholesale power market. New Section 32(a)(1) of PUHCA defines an EWG as:

"any person determined by the [FERC] to be engaged directly, or indirectly through one or more affiliates, . . . and exclusively in the business of owning or operating, or both owning and operating, all or a part of one or more eligible facilities and selling electric energy at wholesale."

"Eligible facilities" are defined as facilities that are used for the generation of electric energy exclusively for sale at wholesale, although retail sales are permitted for purely foreign facilities. PUHCA, Section 32(a) and (b).

Section 32(a)(1) allows utilities or non-utilities to form EWG's without triggering the corporate or geographic restrictions of PUHCA. In the world of energy acronyms, EWG's equal the sum of IPP's and APP's, where IPP's are the commonly known "Independent Power Producers" and APP's are the slightly less well known "Affiliated Power Producers". APP's earn the "affiliated" label because of their relationship to traditional rate-base utilities. NUG's, or "Non-utility Generators", can then be equated with the sum of EWG's and QF's, or "Qualified Facilities" under the Public Utility Regulatory Policies Act of 1978 ("PURPA"). Not wishing to revisit an old QF problem, Congress has allowed EWG's to own facilities and goods, such as fuel and related transmission, storage and handling facilities, that are necessary for the operation of their businesses. Section 32(a)(1).

Sections 32(c) and (d) allow existing rate-based facilities and hybrid facilities to qualify as EWG's if certain state approvals or other requirements are obtained or met. Existing powerplants can be "spun-off" from a traditional utility, leaving that utility's rate base, if each state public utility commission ("PUC") having jurisdiction over the retail rates of the utility in question makes a specific determination that the transaction (i) will benefit consumers; (ii) is in the public interest; and (iii) does not violate state law. EWG's under Section 32(d) may own part of "hybrid" facilities in which the remaining parts of the facility are owned by utilities, as long as no such utility is an affiliate of the EWG's which are part owners.

To guard against affiliated abuses which may exist between an EWG and its affiliated utility, the EPA created a specific role for state PUC's which is in addition to the ratemaking review that will be required by the Federal Energy Regulatory Commission ("FERC") under the Federal Power Act ("FPA"). The EPA added a new Section 32(k)(1)(A) to PUHCA pursuant to which an electric utility may not enter into a contract to purchase power from an affiliated or associated EWG unless each state PUC with jurisdiction over the buying utility's retail rates makes the following determinations prior to the sale:

(A) that the PUC has sufficient regulatory authority, resources, and access to books and records to exercise its duties to oversee affiliate transactions (including enforcement provisions and protection of trade secrets); and

(B) that an affiliate sale will benefit consumers, will not violate state law, will not provide an EWG with a competitive advantage, and is in the public interest.

Under the EPA, EWG's must be certified as such by the FERC and, on February 10, 1993, the FERC issued its final rule establishing the mechanism under which it will determine EWG status, 62 FERC Para. 61,127. The FERC must rule within 60 days of the filing of an application for determination of EWG status or the application will be deemed to have been granted. Each application must contain (1) a sworn statement attesting to all facts or representations demonstrating eligibility, including the representation of compliance with the EWG definition; (2) a description of the facility(ies) and any affiliated or associated utility; (3) copies of any specific determinations from state PUC's; and (4) a draft Federal Register notice. 18 C.F.R. Sec. 365.3 (new). The FERC will not hold hearings or issue deficiency letters, although denials are without prejudice to a subsequent refiling with additional information.

On November 25, 1992, the FERC issued a determination in Commonwealth Atlantic Limited Partnership that the partnership was an EWG under Section 32 of PUHCA. On February 18, 1993, in Richmond Power Enterprise, L.P., the FERC granted EWG status to the limited partnership, as well as the limited and general partners, which own the 240 MW Richmond Power facility. The FERC also addressed two significant issues related to EWG status, whether the sale of steam by an EWG precluded EWG status under the definitional requirement of being engaged in the "exclusive" business of generating at wholesale and whether EWG's can also own QF's. The FERC held that the sale of steam did not preclude EWG status and that EWG's could also own QF's. An EWG may simultaneously own a QF and an "eligible facility", as long as the applicable requirements of PURPA and PUHCA are satisfied. On March 11, 1993, in Louis Dreyfus Electric Power Inc., 62 FERC Para. 61, 254, the FERC denied EWG status to an electric power marketer which did not own or operate any generation, transmission or distribution facilities. Power contracts were not deemed for these purposes to be "eligible facilities".

III. WHOLESALE RATEMAKING

Except in Hawaii, Alaska, and part of Texas, ratemaking for EWG's located in the U.S. will be, by definition, wholesale ratemaking by the FERC under the FPA. The EPA did not change the jurisdiction of the FERC and, subject to the limited jurisdictional exceptions for municipal and co-op EWG's, the rates charged for power at wholesale by EWG's will be wholesale rates subject to the jurisdiction of the FERC under the FPA. As such, the rates for power sold by EWG's will either be cost of service based or meet the FERC's evolving standards for market based rates.

Historically, the FERC has for sometime interpreted the "just and reasonable" standard in Section 205 of the FPA as requiring that rates for sales of wholesale power in interstate commence be designed to recover from the buyer the seller's costs of production, including a reasonable rate of return set by the FERC on equity invested in the power producing facility(ies). Since 1978, when Congress enacted the Public Utility Regulatory Policies Act ("PURPA"), many in the power industry have come to recognize that the supply of electric power in bulk lacks natural monopoly characteristics and that the generation industry might conduct business more efficiently if the price for bulk power were set by competition and not by regulators based on the costs incurred by sellers. More recently then, the FERC has allowed power sellers to sell power at rates set in competitive procurements or in negotiations which buyers conduct with one or more competing sellers. Rates set by competitive forces are referred to as "market based" rates.

In a line of cases over the last three to five years, the FERC has stated its standards for the approval of market based rates. In general, where an EWG is an IPP, unaffiliated with a traditional utility, the FERC can be expected to accept market based rates for the sale of power by the EWG if the EWG demonstrates that it lacks market power over identified buyer(s), or that it has adequately mitigated its market power, because neither the EWG or any of its affiliates:

(1) is a dominant firm in the sale of generation in the relevant market;

(2) owns or controls transmission facilities through which the buyers could reach alternative sellers; and

(3) can erect or control any other barrier to market entry.

Enron Power Enterprise Corporation, 52 FERC Para. 61,193 (1990); Commonwealth Atlantic Limited Partnership, 51 FERC Para. 61,368 (1990); Doswell Limited Partnership, 50 FERC Para. 61,251 (1990); Dartmouth Power Associates Limited Partnership, 53 FERC Para. 61,119 (1990). Hartwell Energy Limited Partnership, 60 FERC Para. 61,143 (1992).

Expanding the market power analysis to a potential class of buyers in a particular geographic market, the FERC has granted blanket approval of future power sales at negotiated market based rates where the identity of the buyer(s) was not known and where a specific rate or price formula had not yet been negotiated. Louis Dreyfus Electric Power Inc., 61 FERC Para. 61,303 (1992); National Electric Associates Limited Partnership, 50 FERC Para. 61,378 (1990); Citizens Power & Light Corporation, 48 FERC Para. 61,210 (1989); PSI Energy, Inc., 51 FERC Para. 61,367 (1990); Entergy Services, Inc., 58 FERC Para. 61,234 (1992) ("Entergy I"). In such cases, the applicant must demonstrate a lack of market power in the relevant geographic and product markets by performing market power analyses and submitting the results to the FERC. If the applicant is a stand-alone IPP that is engaged exclusively in the business of making bulk power sales from its single powerplant, it is not likely to own or control transmission facilities, generation sites or other essential inputs to generation in the relevant geographic market. For such an IPP, an abbreviated market power analysis should be sufficient.

For the short-run firm bulk power market, the seller's ownership share of installed capacity and the seller's ownership share of excess or uncommitted capacity are the two indicators used by the FERC to measure the seller's potential ability to dominate the market, Entergy I, 58 FERC at 61,756. As found in Entergy I, Id. at 61,755, this market is characterized as a market in which delivery of bulk power is needed relatively soon and can be made only from existing or nearly completed capacity. The FERC recently confirmed that a "firm's share of the uncommitted capacity available in a market is a general indication of its ability to dominate firm sales in the short-term market." Louisville Gas & Electric Company, 62 FERC Para. 61,146 (1993) ("Louisville"). In Louisville, Id. at 61,146, the FERC held that a market share in uncommitted capacity of 14% or less, and a market share in installed capacity of 20% or less, is sufficiently small to indicate that Louisville did not have market power. For a stand-alone IPP, shares below these numbers can be readily expected.

For the non-firm energy market in a particular geographic market, an IPP would be called upon to demonstrate how its energy available for non-firm sales compared to non-firm sales volumes in the region in question. Because of the volume of "economy flows" in many regions, it is likely that a stand-alone IPP will have the ability to comprise only a small percentage of the non-firm sales volumes in question.

For the long-run firm bulk power market, the FERC assesses the following factors to determine market power: (1) the potential for entry by other suppliers into the relevant market; (2) the ability of the applicant to use its control over transmission facilities to restrict competition; and (3) the ability of the applicant to restrict market entry by potential competitors by controlling essential inputs to the construction or operation of generation facilities (i.e., through control over scarce plant sites, fuel resources, etc.). Entergy I, 59 FERC at 61,759. For an IPP which does not own or otherwise control transmission facilities other than those related to its facility and which is not affiliated with entities which are in a position to erect barriers, or otherwise control market entry, it can be relatively straightforward to show a lack of market power in the long-run firm bulk power market. The potential for entry by other suppliers is an important consideration in assessing how market power may change over the long run. Entergy I, 58 FERC at 61,759. Recent power supply solicitations may provide proof that many suppliers are willing to enter the generation market in question if given sufficient time to develop and construct new facilities. Such evidence may be sufficient for an IPP to complete its required showings that it lacks market power in the long-run firm bulk power market.

For APP's, the FERC has engaged in a much more probing inquiry due to the potential for abuse of the affiliated relationship between the APP and its related utility. The FERC has shown great concern that a utility with a monopoly franchise may have an economic incentive to exercise market power through its affiliate dealings. "The potential abuses include such practices as affiliates selling products to a franchised utility at excessive prices or a franchised utility providing inputs to an affiliate at preferentially-low prices -- both of which are examples of market power that is exercised to the disadvantage of captive customers and other potential non-affiliated power suppliers." Boston Edison Company, Re: Edgar Electric Energy Company, 55 FERC Para. 61,382 (1991) at Footnote 56; TECO, 52 FERC Para. 61,191 (1990) at 61,697. In short, the concerns of the FERC originate with the affiliate relationship itself between the APP and its franchised utility and extend to APP sales to unaffiliated, as well as affiliated, buyers.

Before the APP demonstrates that it lacks, or has mitigated, market power in the same type of analysis required of an IPP, the FERC first requires that there be a clear showing of no potential abuse of self-dealing or reciprocal dealing.

"In these circumstances, as in prior cases involving affiliate transactions (e.g., Ocean State [44 FERC Para. 61,983 (1988)], Terra Comfort [52 FERC Para. 61,241, (1990)] and TECO [53 FERC Para. 61,191 (1990)]), the critical first step of our analysis is to ensure lack of abuse of self-dealing." Edgar Electric Energy Company, 55 FERC Para. 61,382 (1991).

Even when APP's sell to unaffiliated buyers, the FERC is concerned how self-dealing or reciprocal dealing can result in the APP exercising market power over buyers. Self-dealing abuses can involve cross subsidies between the affiliated utility and the APP in question or exclusionary behavior by the affiliated utility which serves to limit the alternatives that unaffiliated buyers have access to when negotiating with the APP. Cross subsidies and exclusionary behavior provide the potential for the APP to earn greater than competitive profits in "market based" sales to unaffiliated buyers. Reciprocal dealing provides the same possibility and can result in higher than market prices being paid by the ratepayers of the affiliated utility which engages in one side of the reciprocal deal while the APP engages in the other side of such a deal. Ratepayers of both utilities involved in reciprocal dealing can be exposed to this risk.

In a line of cases, the FERC has rejected market based pricing in what would appear to be "arms-length" transactions between power sellers and power buyers. As indicated, the potential for exclusionary behavior and cross-subsidiaries were the motivating factors. For example, when the affiliated utility transfers power or provides transmission services to its APP at preferentially low rates which the affiliated utility would not offer to unaffiliated entities, the APP acquires market power because it can offer attractive rates to unaffiliated buyers which still represent greater than competitive profits to the APP because of its subsidized power or transmission costs. In effect, the affiliated utility has abused its franchised control over its captive ratepayers who may unknowingly allow profits to be transferred from the utility's books (where they would be credited by the utility's regulators to the benefit of the captive ratepayers) to the books of the APP (where they will be credited to the utility's shareholders.) An APP also possesses market power by virtue of its affiliation if, for example, the affiliated utility uses its control over transmission to restrict access by unaffiliated buyers to suppliers competing with the APP. With its competition restricted, the APP can earn supra-competitive profits even if its costs are unsubsidized.

On the other hand, APP's which appear to have little chance or option for cross-subsidy or exclusionary preferences have fared well before the FERC. When APP's seek market based pricing for sales to unaffiliated buyers geographically distant from the affiliated utility's transmission network, the APP cannot use its affiliate's transmission network to exclude competing sellers and the transaction looks more like an IPP sale. The geographic distance also seems to lessen concerns about cross-subsidies since the major inputs which can be subsidized, such as wholesale power, transmission or fuel, are unlikely to be subsidized over the geographic separations at work. Commonwealth Atlantic (1990); Wallkill Generating (1991).

However, whenever APP's seek market based pricing for sales to unaffiliated buyers which are geographically adjacent, concerns about exclusionary behavior and cross-subsidies have resulted in disapprovals. Nevada Sun Peak Limited Partnership ("Sun Peak"), 54 FERC Para. 61,264 (1991), reh'g and rates accepted on other grounds, 55 FERC Para. 61,058 (1991); Teco Power Services Corp. ("TECO"), 52 FERC Para. 61,191 (1990), reh'g denied and rates accepted on other grounds, 53 FERC Para. 61,202 (1990); Terra Comfort Corp., 52 FERC Para. 61,241 (1990). In rejecting the TECO application, the FERC established a "market value" test for judging potential cross-subsidies in the transfer pricing of inputs such as power, transmission, fuel or other services which the APP receives from its affiliated utility. An external standard of value is sought. The APP can submit "bid" evidence of other offers the APP receives for the same inputs or "benchmark" evidence of other sales by the affiliated utility of the same inputs to unaffiliated buyers or other sales of comparable inputs between unaffiliated sellers and either the APP or other unaffiliated buyers. Where "bid" or "benchmark" evidence is absent, the FERC has been unable to conclude that the transfer pricing was not preferentially low. If the input transaction is tainted, the FERC will not approve market based rates for the APP's sale to unaffiliated buyers.

Finally, in Edgar Electric Energy Company, an affiliated utility buyer was involved in the classic example of self-dealing. The APP in question was selling power to its affiliated utility. The FERC was not only concerned about cross-subsidies and exclusionary behavior, see: Footnote 67, 55 FERC Para. 61,382, but also was required to address directly the possibility that the APP was preferred to its competing power sellers without justification. "[t]he Commission must ensure that the buyer has chosen the lowest cost supplier from among the options presented, taking into account both price and non-price terms (i.e., that it has not preferred its affiliate without justification)." To eliminate concerns about preferential pricing, the Commission relied upon its market test which applies a "bid" or "benchmark" standard to determine external market value.

Three non-exclusive examples of ways to demonstrate a lack of affiliate abuse under the market test standard were suggested:

(A) evidence of direct head-to-head competition between the seller and competing unaffiliated suppliers in either a formal solicitation or in an informal negotiation process;

(B) benchmark evidence of the prices that nonaffiliated buyers were willing to pay for similar services from the seller; and

(C) benchmark evidence of the market value, based on both price and non-price terms and conditions, of contemporaneous sales made by non-affiliated sellers for similar services in the relevant market.

In applying the "bid" or "benchmark" standard for market value to the sales price for power sold by the APP back to its affiliated buyer, the FERC in Edgar Electric Energy Company was applying to the "downstream" transaction between the APP and its affiliated utility the same market value standard it had applied to the "upstream" transactions between an affiliated utility and its APP when the utility transfers factor inputs to the APP. In the "upstream" transactions, the FERC is concerned that transfer prices, due to affiliate abuses, could be preferentially low. In the "downstream" transactions, the FERC is concerned that transfer prices, due to affiliate abuses, could be preferentially high. In both cases, the potential for higher than competitive profits exist because the abuse of the affiliated relationship allows the APP to exercise market power to the disadvantage of ratepayers and competing power sellers.

After examining the benchmark evidence, the FERC in Edgar Electric Energy Company determined that the possibility of self-dealing abuse had not been effectively eliminated. "The evidence does not support a finding that Edgar's price is similar to the price at which nonaffiliates sell comparable power." Because there was no clear showing of lack of potential affiliate abuse, the Boston Edison application for market based prices for its APP sale of power to itself was rejected.

A year after the Edgar Electric Energy Company decision, the FERC accepted the benchmark evidence relating to an APP's sale to its affiliated utilities which was presented in Ocean State II, 59 FERC Para. 61,360 (1992). The Commission required and favorably reviewed evidence of power sales transactions that were contemporaneous with, and involved sales that were comparable to, the APP sale of power to three affiliate utilities. Of 29 projects offered into evidence, the FERC focused its analysis on 10 contemporaneous projects of comparable size and technology offering long term sales similar to the APP in question. Pricing formulae were projected over the life of the contracts and uniform assumptions were applied for the escalation rates for fuel and other expenses, availabilities and capacity factors. Sensitivities were run on O&M costs, availabilities and returns on equity (in the APP's pricing formula). Nine non-price factors were also analyzed. A final inquiry was made to determine whether the APP or its affiliated utilities exercised market power over the benchmark buyers' supply choices. The FERC found no evidence in Ocean State II that market power had been exercised to skew the benchmark. There was no evidence that the APP or any of its affiliated utilities dominated the generation supply choices in the relevant market. In addition, there was no evidence that the APP or any of its affiliates exercised market power over transmission or created any other barrier to entry so as to render the benchmark unreliable.

An APP's ability to obtain market based pricing, both for sales to affiliated buyers and for sale to unaffiliated buyers in service territories nearby its affiliated utility's transmission network, is dependent, in many cases, on the

APP's success in demonstrating that its potential to exercise market power through exclusionary affiliate abuse has been mitigated by open access transmission, available to the APP's competing suppliers on terms comparable to those available to the APP. Because "[t]he most likely route to market power in today's electric utility industry lies in ownership or control of transmission facilities", Citizens Power & Light Corp., 48 FERC Para. 61,210, open access transmission deserves special attention in EWG ratemaking.

IV. MITIGATING POTENTIAL APP MARKET POWER THROUGH OPEN ACCESS TRANSMISSION

Market power, and the potential for self-dealing abuse through exclusionary behavior, are likely to be found to exist whenever an APP attempts to market power on a competitive basis to nearby unaffiliated utilities which are dependent, at least to some significant extent, on the affiliated utility's transmission network in order to gain access to the full market of competing power suppliers. The FERC is prepared to allow such an APP to enter into power sales with market based pricing if the affiliated utility is prepared to mitigate its transmission-related market power.

In addressing the request of a traditional utility for pricing flexibility for its future power sales to a class of unidentified potential buyers, the FERC in Public Service Company of Indiana, Inc. ("PSI Energy, Inc."), 51 FERC Para. 61,367, reh'g denied, 52 FERC Para. 61,260, order on clarification, 53 FERC Para. 61,131 (1990), dismissed, No. 90-1528 (D.C. Cir. January 21, 1992), approved market based pricing on the condition that the seller mitigate its market power by offering cost based firm and non-firm transmission service to any utility, IPP or QF which wanted to reach, and compete for, the potential buyers. Importantly, the terms and conditions on which transmission was to be provided were comparable to the terms and conditions of the service the seller provided itself. The commitment also included an obligation to expand the transmission network at the wheeling customer's expense.

In 1992, the FERC approved two market based tariffs, one for service from a traditional utility and another for service from its APP. Entergy Services, Inc. ("Entergy I"), 58 FERC Para. 61,234 (March 3, 1992), order on reh'g, 60 FERC Para. 61,168 (August 7, 1992). The Commission set forth the basic features that it regards as necessary to include in a transmission tariff in order to mitigate transmission related market power. These features are the following:

(i) open access transmission service must be offered to any requesting utility or non-utility power supplier (restrictions on retail wheeling, new distribution utilities and QF's sales rights were allowed);

(ii) the utility must commit to expand transmission facilities if necessary to accommodate wheeling requests (wheeling customers would be required to finance the new facilities to the extent they did not contribute to the system as a whole);

(iii) both the utility and the APP would be required to obtain transmission under the same open access transmission tariff in order to make sales under the market based tariffs approved by the Commission (i.e., the market based sellers would enjoy no transmission related advantage over their competing suppliers);

(iv) transmission priority for native load customers should be on a par with priority for firm transmission customers if the pricing of firm service protects native load customers; and

(v) a wheeling customer should be permitted to reassign transmission capacity provided that the assignee is an eligible customer.

In Louisville Gas and Electric Company, 62 FERC Para. 61,016 (January 14, 1993), the FERC again approved a market based tariff for a traditional utility which had agreed to mitigate its transmission related market power by providing open access transmission similar to that provided in Entergy I .

V. PROTECTING CAPTIVE RATEPAYERS WITH COST OF SERVICE APP PRICING

In two cases, sales made by an APP to an unaffiliated buyer which were rejected by the FERC on a market basis were subsequently approved when justified on a cost of service basis. TECO, reh'g, 53 FERC Para. 61,202 (1990); Sun Peak, reh'g, 55 FERC Para. 61,058 (1991). In an earlier case, the FERC had suggested such a solution to the issue of the potential transfer of profits from the captive ratepayers of the affiliated utility to the shareholders of the APP. Portland General Exchange, Inc., 51 FERC Para. 61,108 at 61,246-247 (1990). Cost of service power sales are judged on distinctly different standards and applicants are relieved of the burden of showing a lack of market power and the absence of affiliate abuses of self dealing or reciprocal dealing. For example, neither "bid" or "benchmark" evidence is required with respect to the "upstream" transfer of inputs from the affiliated utility to its APP.

Captive ratepayers are protected in the sense that profits are not shifted from the affiliated utility to the APP since the APP is recovering no more than the costs that it incurs in receiving inputs from its affiliated utility. Whatever level of profit exists in the transfer price is passed on without mark-up to the non-affiliated power buyer. Regulators which review the rates and sales of the affiliated utility are able to determine whether the profit level implicit in the transfer price was prudent or not (the transfer would be imprudent if the price and profit were below the level that the affiliated utility could have obtained in the open market) and can always assure the captive ratepayers that whatever that level of profit was, it will be credited to their benefit.

While captive ratepayers may be protected when an APP sells to non-affiliated buyers on a cost of service basis, competing suppliers may not be protected. This is because the affiliated utility may possess market power over the non-affiliated buyer and could exercise that power to restrict the ability of the buyer to reach or contract with suppliers who wish to compete with the APP. For example, the APP may be sold power, at cost, from the affiliated utility's capacity that is over-priced in comparison to current market prices. The affiliated utility may wish to relieve its captive ratepayers of the burden of the over-priced plant. If the affiliated utility controls transmission facilities through which the non-affiliated buyer has access to competing suppliers, the APP may be able to complete the sale on a cost of service basis because the buyer cannot economically reach lower priced market based or cost of service suppliers.

It is not clear at this point in time whether the FERC will re-visit the issue whether APP's should be allowed to complete cost of service sales without review of the question whether such rates reflect the existence and exercise of market power.

VI. UTILITY MARKET BASED RATES WITHOUT BID OR BENCHMARK REVIEW OF INPUT COSTS

The FERC appears to be neutral whether a utility chooses to set rates for power sales to the open market with cost of service or market based pricing. In this instance, no APP is involved and the FERC makes no inquiry concerning the "bid" or "benchmark" value of the inputs to the utility's power sales. If market based pricing is chosen, the FERC in a clearly defined line of cases does require a showing that the seller lacks market power, or has mitigated its market power, as for example, with an open access transmission tariff, in the product and geographic markets in question. Pacific Gas & Electric Company, 44 FERC Para. 61,010; PSI Energy Inc., 51 FERC Para. 61,367 (1990); Entergy I, 58 FERC Para. 61,234 (1992); Louisville, 62 FERC Para. 61,016 (1993).

From a practical point of view, there appears to be little reason to attempt to limit a utility with a system of plants to cost of service pricing. Historically, utilities with excess capacity of various vintages, with differing fuel sources and differing unamortized capital costs, have been able to package different combinations of plants into different sales agreements with overall costs which could fall almost anywhere in a broad range of values. In short, even with cost of service pricing, utilities have been able to price their excess capacity to market conditions.

Apart from requiring that the power sales price recovers at least the incremental cost of producing the power in question, TECO, Entergy I, the FERC, when it reviews applications for market based pricing from utilities, engages in no investigation as to how the power sales revenues may compare to the input costs allocated to the sale or how the allocated input costs compare to market benchmarks. The allocation of utility costs to the sale in question is subject to the review by the regulators which set rates for the utility. That regulatory control can address the question whether the sales price was prudent in comparison to the prices which could have been obtained in the market, as well as assuring that revenues are credited to the captive ratepayers who are presumably supporting in rate base the facilities from which the sales are made.

In the absence of an APP with its potential for profit shifting, the FERC appears unconcerned how a utility's market based prices compare to those of the IPP's which may want to compete with the utility in question. The FERC will assure that the utility will not use its control over transmission to prevent competition from occurring. However, that competition may be very uneven to begin with if the utility has excess supply supported by captive ratepayers. Utilities with such captive ratepayers are then in a significantly better competitive position than IPP's which lack a customer base which supports the capital costs of new facilities that the IPP's may want to build in order to compete in the geographic market with the "long" utilities. Whether the capital costs of excess supply are supported by ratepayers or not, the FERC may reason that the costs are "sunk" and the power seller, whether it is a utility or an IPP which overbuilt, is better off pricing its power at any level over incremental costs than not making a sale at all. If this is the reasoning involved, the FERC appears to be willing to let the market in generation operate as an unregulated market in which suppliers with excess supply will drive the price of such excess down to marginal costs until demand or loss of supply eliminates the excess supply. IPP's which wish to construct new facilities will get the message of the unregulated market and, like other competitors in other unregulated industries, will seek geographic markets where demand and supply are more in balance.

[June 24, 1993]

Chapter 80

Tracking the Bulk Power Market: Opportunity Among Municipal & Cooperative Systems

K.S. Knutson

The utility industry faces a transitional period as it begins to sort out the advantages and disadvantages of moving from a regulated to a competitive environment. However, the industry is not on the verge of change; it has already begun. Evidence of this is documented by the tremendous rise in activity in the nation's bulk power markets.

According to the Department of Energy (DOE), during 1991, more than 1,827,000 gigawatt hours (GWh) of electricity were either purchased or received as interchange and wheeling transactions.

As Chart 1 demonstrates, that volume represents more than sixty percent of all the power generated and made available for sale in the United States. Just five years ago, only fifty-five percent was transacted in the nation's bulk power markets. Among all types of utilities, from the largest investor owned to the smallest publicly held company, bulk power purchases have risen by more than 40.6 % since 1986.

In the investor-owned sector alone, bulk power purchases have more than doubled since 1980. As Chart 2 displays, the largest proportion of the growth in bulk purchases has come from the non-utility sector. During the past decade, purchases of non-utility generated bulk power by investor-owned utilities have risen by over 130,000 GWh. In addition, today bulk power purchases account for nearly a quarter of all electric operating and maintenance expenses among investor-owned utilities — in 1980, bulk power accounted for less than 14% of the total investor-owned operating and maintenance expense.

Along with an ever-expanding wholesale market, every company, small or large, is feeling the impact of a changing regulatory environment. Several states already incorporate

Chart 1

Electric Energy Receipts & Available Energy

(Contiguous U.S. -- 000s of Gigawatt Hours)

Energy Sources	1986	1987	1988	1989	1990	1991
Available Energy						
Generation	2,553	2,632	2,731	2,811	2,821	2,835
NUGs	39	50	68	90	116	138
Int'l Imports	39	50	37	24	21	26
Total Available	2,631	2,732	2,836	2,925	2,958	2,999
Total Energy Receipts						
Purchases of Power	898	918	987	1,046	1,111	1,263
Exchange Received	407	399	382	421	427	241
Wheeling Received	195	194	269	299	322	323
Total Receipts	1,500	1,511	1,638	1,766	1,860	1,827
% Receipts of Available Energy	57.0	55.3	57.8	60.4	62.9	60.9

Note: Energy receipts include the sum of purchased power, exchange received and wheeling received.
NUGs refers to non-utility generator sales of energy to regulated utilities.
Source: Department of Energy -- Form EIA-861, "Annual Electric Utility Report".

Resource Data International, Inc., 1993.

Chart 2

Wholesale Power Purchases by Investor-Owned Utilities

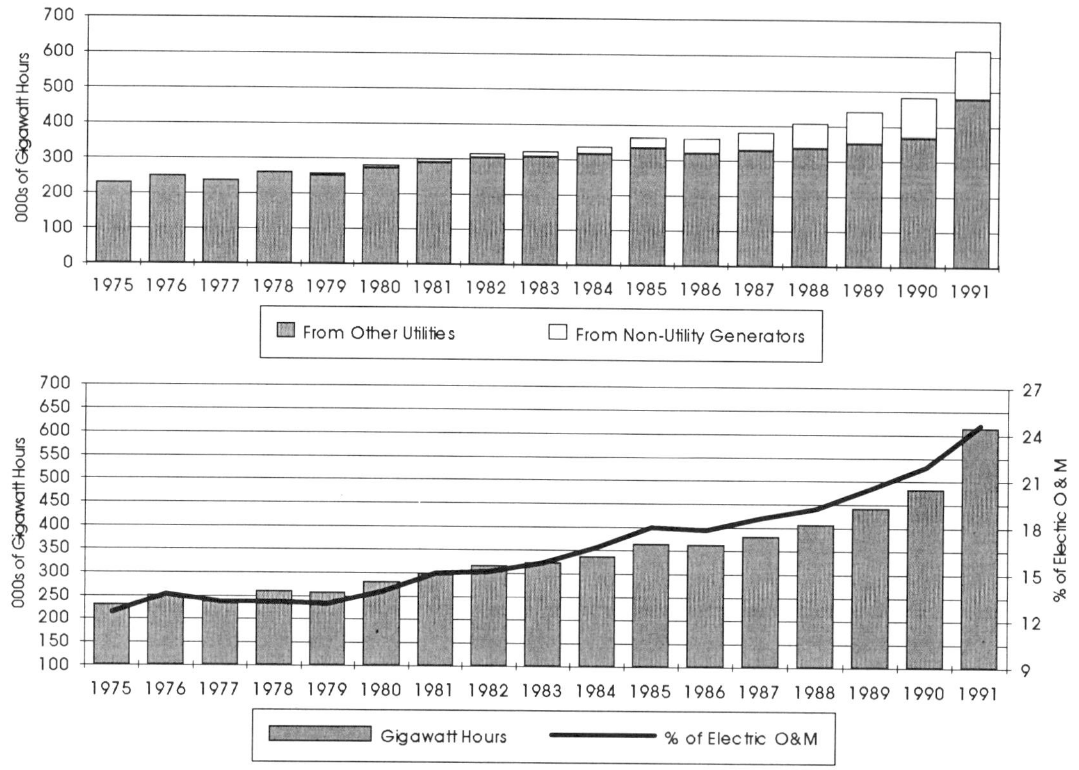

Note: Transactions compiled from FERC Form 1 annual reports. Due to a change in required reporting, 1991 figures include some wholesale purchase transactions formerly recorded as interchange agreements.

Resource Data International, Inc., 1993.

some level of Integrated Resource Planning which mandates cost-effective methods for meeting load. With the enactment late last year of the Energy Policy Act that promotes open access to transmission and investment opportunity in generation, wholesale power markets are expected to be even more expansive and competitive in the years to come.

Considering these events, it is clear that today's utility industry will become increasingly competitive in the future. Even though many argue that competition and mandatory access may jeopardize reliability and cause the loss of native load, others believe that wholesale power buyers, including municipalities and large industrials, will soon be able to shop for power in broader markets. As was quoted in a recent issue of Electric Week, "The era of competition is upon the electric utility industry. Utilities that want to be winners rather than losers must cut costs, pamper customers, and revolutionize their corporate cultures."

The wholesale power market will be lively with numerous opportunities as well as risks. In order to keep current bulk power customers, generating companies must design creative and cost competitive rates. Often, that will mean cutting profits on wholesale transactions. On the other hand, for those aggressive systems with efficient generating and transmission resources, new opportunities in wholesale markets abound. In particular, due to rising rates and thinning margins, small electric companies represent a significant target.

Opportunity Among Small Electric Companies

For large companies with satisfied, stable customer bases and surplus low cost capacity, future opportunities in wholesale power markets look very promising. Though other large companies with shrinking reserve margins represent a significant market for them, there is an equal if not larger market among small distribution systems. With the advent of open access to transmission, municipals and cooperatives will become prime targets for bulk power sales during the coming years.

As a group, municipals and distribution cooperatives represent by far the largest single class of wholesale power buyers in the country — more than 2,700 nationwide. Generally, these companies purchase power on a long-term requirements basis from investor-owned utilities, federal administrations, and generation and transmission companies. As Chart 3 displays, altogether, municipals and cooperatives purchase in excess of 565,000 GWh of wholesale power per year with expenditures topping $22.7 billion — well over 60% of their operating expenses annually. Together, they account for nearly half of all expenditures for wholesale power nationwide.

Municipals and cooperatives represent not only wholesale opportunity, but a significant acquisition opportunity as well. Findings from a recent study prepared by RDI, titled *U.S. Electric Utility Industry Mergers & Acquisitions,* indicate a wave of consolidation among small electric systems may be poised to happen during the next several years. The reasons include high and increasing rates, financial instability, declining profits, high cost power contracts, a declining customer base, and the inefficient operation of small generating plants. Chart 4 provides the distribution of companies by average residential rates. Considering only those systems with double digit rates (>10.0 ¢/kWh) during 1991, the range in number of companies across company types is fairly even. Due to a number of factors, mainly large debt accumulation during the early 1980s, a number of investor-owned utilities nationwide are supported by relatively high rate bases. However, considering the range of 8 to 10 ¢/kWh, the number of companies from municipals (320) and distribution cooperatives (242) increase dramatically — a formidable market for bulk power considering the companies relatively high average rates alone. So what could be done to improve efficiency in operations and rates for these companies?

Take, for example, a medium-sized distribution cooperative in the Midwest. The system has 55 employees on the payroll and serves just over 16,000 customers. During 1991, the electric company purchased 130,000 MWh for which it paid just over $8.1 million. During the past 6 years, rates have increased from 5.88 ¢/kWh ($730/customer per year) to 6.24 ¢/kWh ($809/customer per year). During 1991, electric operating and maintenance expenses topped 8.34 cents per kilowatt hour of retail sales. The average rate charged retail customers was 11.25 ¢/kWh. The difference between operation and maintenance expenses and sales revenues totals is significant. While any single variable can be scrutinized in detail and the findings may point to a just and necessary cost, it is clear in most instances that there is a great deal of room for improvement. The margin between current O&M expenses (which can be improved) and revenues is nearly 3 ¢/kWh of sales. If expenses could be shaved by 2 ¢/kWh and rates reduced by 1 ¢/kWh, a significant gain in profits could be generated for the overall system while at the same time improving customer satisfaction.

Another finding from the study indicates that a large number of small companies, particularly distribution cooperatives, are netting less income today than three years before. Chart 5 identifies two groups: first, those companies reporting negative net income during 1991; second, those companies reporting lower net income in 1991 than three years before. Within the first group, 28 municipal systems and 21 distribution cooperatives reported negative net income in 1991. One

Chart 3

Electric Utility Wholesale Transactions: 1991 Summary

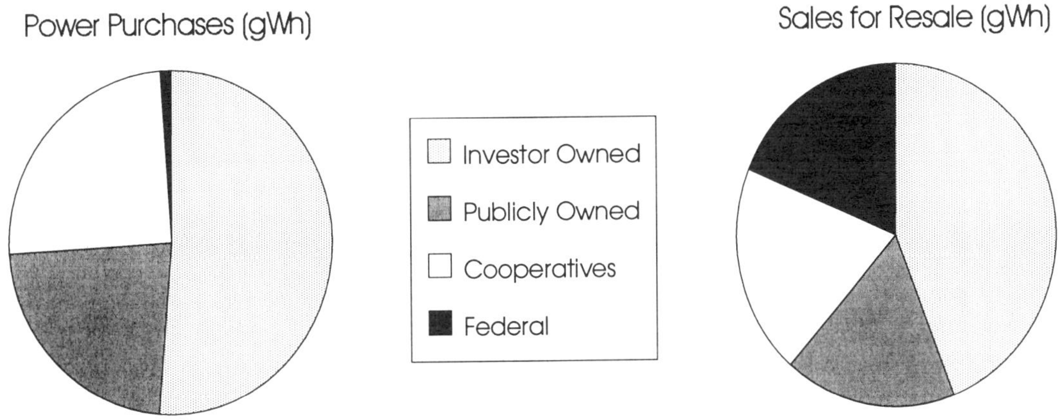

	Purchases		Sales for Resale	
	Energy 000s of gWh	Expense (Millions $)	Energy 000s of gWh	Revenues (Millions $)
Investor Owned	615	24,169	489	17,168
Publicly Owned	270	10,571	192	7,022
Cooperatives	295	12,175	219	9,171
Federal	13	314	204	6,342
Total:	1,193	47,229	1,104	39,703

Resource Data International, Inc., 1993.

Chart 4

Distribution of Companies by Average 1991 Residential Rates

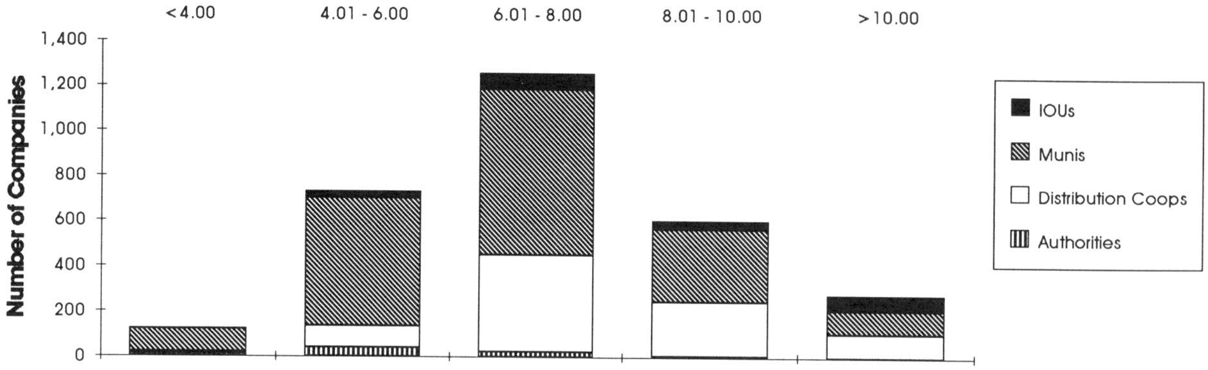

Cents / kWh Range in Residential Rates

	<4.00	4.01 - 6.00	6.01 - 8.00	8.01 - 10.00	>10.00	Total
Authorities	12	38	22	5	0	77
Distribution Coops	8	99	429	242	104	882
Munis	102	563	733	320	106	1,824
IOUs	1	31	72	36	68	208
Total	123	731	1,256	603	278	

Resource Data International, Inc., 1993.

important consideration is that the sample for municipals included only those operations with at least 100,000 MWh of sales, generation or purchases. That group would only account for about one-third of the total number of municipals nationwide. The coverage among other company types in the analysis was complete. With that consideration, it is clear that there is a large group of small companies losing money today. Within the second group, considering those who have recorded declining net incomes during the past three years, the distribution cooperatives stand out with 383 of the estimated 882 companies reporting less income today.

Chart 5

Net Income Profile

Distribution of Companies Reporting . . .

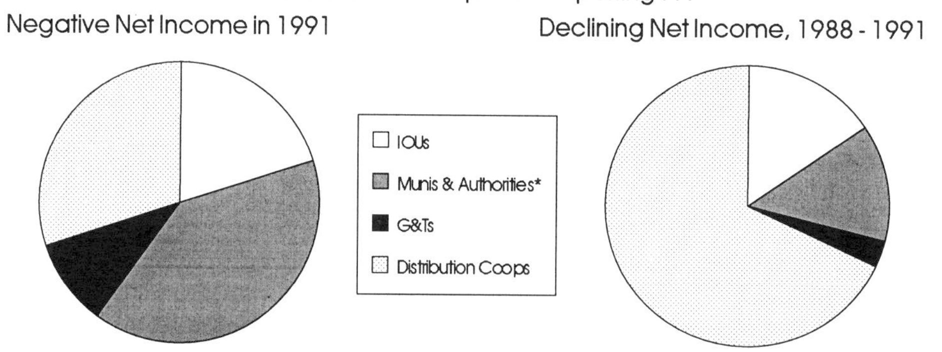

	Negative Net Income in 1991	Declining Net Income, 1988 - 1991
IOUs	14	87
*Munis & Authorities**	28	76
G&Ts	7	18
Distribution Coops	21	383

* 'Munis and Authorities' includes only those companies reporting at least 100,000 MWhs in sales, purchases or exchanges, as required for filing operations data with the EIA.

Resource Data International, Inc., 1993.

Along with the statistics on changes in residential rates, the figures on changes in net margins indicate a great deal of inefficiency among small companies nationwide. Those inefficiencies will be met either with creative new wholesale power contracts or by acquisition. The redesign of power contracts will necessitate competitive pricing.

Small Generating Plants: A Benefit or a Hindrance?

Another important factor to consider when analyzing market opportunity is the portfolio of small generating plants operated by small electric companies. Are they cost effective? Theory has it that operating a small generating facility provides leverage for a company in contract negotiations and provides more flexibility within the grid. However, an analysis of small companies operating generating facilities indicates a great deal of inefficiency in today's market.

Chart 6 provides a profile of small electric plants operated by either publicly-owned companies or by distribution cooperatives. Only those companies where the total operating capacity was less than 50 MWs were examined. The sample consisted of 475 companies operating 817 plants. The number of units operating at these plants totaled 3,120 — or approximately 30.3% of the total number of units operating nationwide. However, the total nameplate capacity of these units totaled 15,775 MWs — only 2.15% of the nations' total operating capacity. Ironically, these units generated less than one-tenth of one percent of the energy output from electric plants in 1992. It is clear that most of these plants are used only for peaking power and sparingly even then. The largest number of these units are fueled by oil (1,663) and natural gas (845); yet, the total energy output from these fuels was less then 1,000 GWh during 1992.

Most of the systems identified in the chart represent wholesale power sales opportunity since they all require a staff of engineers and maintenance personnel, and each represents a certain degree of inefficiency due to high fuel cost and other operations expenses.

To better illustrate what cost savings could be achieved, consider a small municipal system in the Midwest. The company's annual energy load is approximately 13,600 MWh. The utility operates five internal combustion units totaling 8.5 MWs. The units are fueled by oil and vary in age from 22 to 45 years. Last year the plant generated 2,712 MWh. In addition to its generation resource, the company purchased wholesale 12,445 MWh at an average price of 2.66 ¢/kWh. The company's power sales netted 7.98 ¢/kWh. A large gap between the retail rate and wholesale purchase price exists. The question is whether the operation of an aging oil-based plant is necessary in today's open bulk power markets. If a creative cost-saving full-requirements contract could be agreed upon, the savings to customers could be significant while at the same time providing an expanded rate base for the wholesaling company.

Chart 6

Profile of Small Electric Power Plants

Publicly Owned Companies and Distribution Cooperatives
With Under 50 MWs of Operable Generating Capacity

Primary Fuel	# of Units *	Operable Capacity MWs	1992 Generation MWhs
Coal	137	2,342	609,550
Oil	1,663	4,153	252,546
Gas	845	5,958	717,260
Hydro	452	3,027	1,110,865
Other	23	295	50
Total	**3,120**	**15,775**	**2,690,271**
% of U.S.	**30.30%**	**2.15%**	**0.10%**

* The analysis consisted of 475 companies reporting operating statistics on 817 electric plants.

Resource Data International, Inc. 1993.

Federal Power Allocations Shrinking

Another important factor to consider when targeting wholesale opportunity, particularly in the West, is the availability of low-cost federal power. The large federal power administrations like Bonneville (BPA) and Western (WAPA) currently dominate bulk power markets in the West. These two hydro-based federal utilities deliver the most competitively priced wholesale power in the region with combined deliveries far exceeding any other producer — 88,121 GWh in 1991. Many remote western rural electric cooperatives and municipals rely on these administrations as a means of keeping overall costs low. Since neither BPA or WAPA has plans to add significant blocks of capacity in the future, more and more distribution companies will be forced to turn to non-federal sources to meet growing demand. That means opportunity for surplus low-cost systems. In addition to a thinning of the overall allocations due to growth in energy demand, energy output from the federal hydro projects will be dampened further due to environmental restrictions on discharge at certain times of the year and the day.

Summary

Greater access to transmission, combined with mandatory wheeling at competitive rates, will bring bulk power markets to a new level of activity during the next several years. The opportunity to sell bulk power to small electric companies looks promising for a number of reasons including; inefficiency in operations, financial instability, and rising customer rates. The era of competition is here. With legislation mandating open access to transmission, hence promoting competitive markets, look to the smaller distribution companies to begin weighing their alternatives. A large number of municipal and cooperative power systems provide the underpinnings for significant future market opportunity. The potential to provide this market with bulk power lies in their diminishing current resources and in the seller's ability to competitively price wholesale power.

Chapter 81

Analysis of Cogeneration Systems

C.E. Gracias, L.S. Hoffstatter

Abstract

The New York State Energy Office (NYSEO) manages several Energy Conservation Programs, which are funded in conjunction with federal funding sources. Applicants for funding under the aegis of such programs are Institutional and Not-for-Profit Organizations which often submit cogeneration proposals. These proposals are reviewed by NYSEO staff - under tight timeframes - for technical soundness, feasibility and acceptable paybacks. Due to the relative complexity of technical calculations submitted in support of these cogeneration proposals, a need has been felt for a reliable, yet quick method to corroborate such calculations. The authors have developed over a period of years a template based on Lotus language that has proven to meet the requirements for speed and reliability in the analysis of cogeneration projects.

The methodology used in the NYSEO template is very sensitive to fluctuations in fuel consumption and actually shows variations between results based on annualized fuel consumption and those based on actual monthly consumption data. It also detects variations in performance of the cogeneration system as the size of the system varies over a wide range of capacities. The method accommodates multiple-fuel cogeneration systems and takes into account the various means used in these systems to recover heat developed in the cogeneration process.

The NYSEO template has been tested against numerous other cogeneration calculation methods. Where the data are correctly inputted and the technical analysis is correctly established, a very close agreement has been observed between the results of the NYSEO template analysis and that conducted by consultants in support of the funding application.

Introduction

Cogeneration is the process whereby the Energy User generates its own electricity to satisfy in whole or in part the facility's electrical power needs and simultaneously recovers the otherwise wasted heat of combustion in an attempt to meet at least part of the facility's thermal load.

Cogeneration is only cost effective if there is a thermal load which fairly matches or exceeds the heat output of the cogenerator. Ideally, the cogeneration thermal output should be able to satisfy all of the existing thermal load, but the condition is not necessary as long as there is enough usable recovery to yield an acceptable payback.

The prime mover in a cogenerating unit is either a fossil-fuel-fired engine, or a gas turbine or a steam turbine powered with steam from a primary or a waste heat boiler (viz. Combined Cycle Systems). For most facilities, an engine powered by direct combustion of gas or oil is the most common solution.

The general requisites for **Cogeneration** are: (1) a high electric demand coupled with heavy electric energy consumption; (2) a suitably large thermal load (for heating or process or both); (3) high demand and/or energy unit costs. **Cost effectiveness** results from a suitable combination of these three factors. Yet, a very high demand lessens the effect of a high unit cost of demand when cogeneration is implemented. The following points should be considered when evaluating cogeneration:

1. Cogeneration to meet demand primarily;
2. Cogeneration to meet electrical consumption;
3. Cogeneration to meet thermal load primarily.

Cogenerating systems designed to meet peak demand alone are not cost effective. Viable alternatives would be: a) to consider meeting average demand, with occasional excess capacity; or b) to accept a lower cogenerating capacity to just satisfy energy requirements, paying the possible penalties of frequent utility backup and dependency on existing boilers to satisfy residual thermal load.

There is a further consideration which may be the overriding factor in deciding the final cogenerating configuration - that of availability of the appropriate cogeneration engine fuel. Once the main objective for installing cogeneration is established, the selection of the best operating strategy should be based on site load duration curves.

Following are potentially effective operating strategies:

a. "Base Load" thermally and electrically (continuing to buy power and make steam);
b. "Follow" thermal load, displace part of purchased power (idle boilers, purchase supplemental power as required);
c. "Follow" electrical load, displace a portion of the thermal load (minimum purchase: standby only);
d. "Follow" electrical load, idle all boilers and supplement steam demand with extraction condensing combined cycle;
e. "PURPA ENGINE' - generate power for sale and buy steam.

Once the cogeneration system is installed, use the load duration curves to match generation and heat recovery for optimum utilization of cogeneration capacity.

Proposed Cogeneration Systems

Through the DOE ICP program, potential cogeneration situations are first analyzed in the form of Technical Assistance Studies (TAS) and may be submitted concurrently for funding in a concurrent or subsequent Energy Conservation Measures (ECM) Application. The following is a partial list of the Cogeneration Proposals that have been filed with the SEO under the ICP program under Cycles XIII, XIV, and XV:

1. S.U.N.Y.[1] Stony Brook, NY (Mar/81) - 2x16.3MW[2] PB = 6.3 yrs
2. S.U.N.Y. Oneonta, NY (Apr/81) - 3400 KW PB = 7.1 yrs
3. S.U.N.Y., Oswego, NY (Apr/81) - 4300 KW PB = 7.2 yrs
4. S.U.N.Y., Brockport, NY (Apr/81) - 6000 KW PB = 7.2 yrs
5. Creedmore Psychiatric Center, Queens Village, NY (Oct/81) - 1410 KW PB = 3.0 yrs
6. Yeshiva Flatbush High (Mar/89) - 60 KW PB = 10.4 yrs
7. Long Island College Hospital(Feb/89) - 1200 KW PB = 2.7 yrs
8. Lutheran Medical Center, Brooklyn, NY - 2x860 KW PB = 2.5 yrs
9. Albany Memorial Hospital, Albany, NY - 525 KW PB = 3.8 yrs
10. Montefiore Hospital(Feb/89) - 3200 KW PB = 2.8 yrs
11. St. Johns' Riverside Hospital, Yonkers, NY (Apr/89) - 1260 KW PB = 5.4 yrs
12. St.Joseph's Med. Ctr. (Oct/88) - 350K PB = 11.4 yrs
13. Waverly CSD (Nov/89) - 300 KW PB = N/A
14. St. Barnabas Hospital, Bronx, NY (Dec/89) - 1000 KW PB = 2.4 yrs
15. Bayley Seton Hospital (Apr/90) - 548 KW PB = 3.9 YRS
16. Phelps Memorial Hospital (Apr/90) - 2x350 KW PB = 3.7 yrs
17. Calvary Hospital (1993) - 450 KW PB = 4.3 yrs
18. Catholic Med. Ctr. St. Johns Hosp. (1993) - 600 KW PB = 3.9 yrs
19. Saratoga Hospital (1993) - 515 KW PB = 4.3 yrs
20. Carthage High School (1993) - 72 KW PB = 5.5 yrs

Energy Balance in Cogeneration

Consider a cogeneration system consisting of the prime-mover connected to an electrical generator. The heat engine may be a gas turbine, a steam-operated turbine or a reciprocating engine. Thermodynamically, the energy balance in the cogeneration system can be defined in terms of the following relationships:

Available Heat = Heat of Combustion - Exhaust Losses
Available Power = Available Heat - Engine Losses
Shaft Power = Available Power - Derating Losses
Cogenerated Electric Power = Shaft Power - Generator Losses

All of the above quantities can be calculated, but unfortunately input data are not readily available. Instead, the designer uses the performance curves provided by the manufacturer.

Methodology

The Cogeneration Process

Before going into the details of the spreadsheet construction, a review of the cogeneration process is in order. By definition, ***Cogeneration*** is the process by which electricity is generated on site by the user to displace electrical energy purchased from the Utility. Cogeneration is only economic when part of the existing thermal load can be replaced by the recovery of the heat contained in the effluent gas. The following quantities are then essential in characterizing the performance of the proposed cogeneration system:

P_c = cogenerating capacity (KW)
Q_c = rate of fuel combustion = Q_f*R_f (BTUH)
Q_a = heat rate (BTUH/KW) or available fuel heat/KW
Q_r = exhaust heat recovery rate (BTU/KWH)
H_c = number of cogeneration hours (H/Y)
K_e = 3142 BTUH/KW
Q_w = unrecovered exhaust heat (BTUH/KW)
Q_f = BTU per unit fuel
R_f = units of fuel per hour (BTU/Gal or Therm)

In a practical application, Q_a for a given cogeneration unit is taken from the manufacturer's data sheet for a desired value of P_c the cogeneration capacity of the unit. Otherwise, Q_a would have to be calculated from Q_c, using the fuel heating value Q_f and its consumption rate R_f. Considerations of energy balance would then lead to the following relationships:

For the thermal process: $Q_a = Q_r + Q_w = Q_r/\epsilon_h$

For the generation process: $Q_r = P_c*H_c*K_e/\epsilon_e$

For the overall process:

$$Q_r = P_c*H_c*K_e/(\epsilon_e*\epsilon_h) = P_c*H_c*K_e/\epsilon c$$

$$\epsilon_c = \text{cogeneration efficiency} = \epsilon_e*\epsilon_h$$

where ϵ_e and ϵ_h are the efficiencies for the electric and thermal conversion processes, respectively.

The Cogeneration Template

In its present state, the SEO Template allows a quick evaluation of cogeneration potential in a wide variety of situations or any given capacity of Cogenerator, it will analyze cost effectiveness of full or part time cogeneration with mixed fuels for both Boiler and Cogenerator based on an average or period fuel costs.[3] The attached Sample (Table 1) shows a blank Cogeneration Spreadsheet.

The Template consists of three main Sections:

1. The Data Display Section, situated at the leftmost area of the spreadsheet; this section comprises three tables which respectively contain: the Fuel Data; the Cogeneration Data; and the Calculated Parameters.

2. The Fuel Consumption Section, which occupies the top of the spreadsheet. This section contains all the Utilities data, namely quantity of fuel used and respective costs for the indicated periods (usually monthly). Because this section represents a reading table for the computational section, care must be exercised to input the data in the allocated slots.

3. The Computational Section, which contains all the formulae needed to calculate the various intermediates quantities required to estimate net savings from cogeneration and the simple payback over the period selected for analysis. The computational section contains the following: the calculated energy data; the estimated costs of fuel; the projected energy and energy cost savings; and a summary table, which among other quantities, shows the simple payback for the period.

The usual period for analysis is that of twelve consecutive months of operation. The last column on the Fuel Consumption Section shows the total over this period. From this column the annualized consumption data and fuel unit costs may be calculated, as shown in the Calculated Parameters section. The latter also includes the calculated values for the various efficiencies which may be defined for the cogeneration system (i.e., thermal, electrical and overall). These calculated efficiencies usually show a close agreement with values reported elsewhere (e.g., TA Studies).

Although for NYSEO purposes the use of average cost is sufficient for the purposes of the ICP program, period-based costs are also included in the section below the Summary of Results line. The latter are more useful for assessing cogeneration needs for specific periods (e.g., for seasonal variations) and later on to control operation of the installed cogeneration system.

Input Data

These refer to the information contained in the Input Table and consists of design data needed to define the proposed cogeneration system(s). They are: the contemplated ***capacity*** (KW), the cogeneration fuel, and the ***operating schedule*** (# hours/year), with indication of fuel types used. Also included is the information on the heating system whose output will be, at least in part, replaced by the cogenerating system thermal output. In the case of dual fuel systems, an estimate or assumption of the relative consumption is made by assigning consumption fractions (listed in the Fuel Data Table).

Besides cogenerating capacity, the following additional information is needed: the ***heat rate*** (BTUH/KW or BTU/KWH), which describes the thermal energy released by combustion of the cogeneration fuel; and the ***exhaust heat recovery rate*** (BTU/KWH), which expresses the amount of heat available for displacement of current heating load.

Consumption Data

These refer to the information derived from the Utilities bills and constitute the Fuel Consumption Table. As mentioned, the Fuel Data Table allows for the often encountered situation in which two fuels are alternately used to accommodate temporary scarcity or to compensate for fluctuations in fuel allocation (as is often found in natural gas supply). Once the two Input Tables are filled with the appropriate information the Template quickly calculates the following quantities for a variety of fuel utilization ratios:

1. The source-related savings (MBTU/yr);
2. The cost per KW cogenerated; and
3. The project simple payback period.
4. The cogeneration thermal output efficiency,
5. Cogeneration electric output efficiency,
6. Cogeneration overall efficiency.

Also calculated are MBTU costs of boiler fuels, cogenerator fuels and billed electricity, as well as engine fuel costs and thermal and electric cost savings.

Analysis of Cogeneration Systems

An application, using a single fuel scenario, is illustrated by Table 2. This example corresponds to the following hypothetical situation: A large building complex consumes electricity for lighting, refrigeration and air conditioning. Peak demand is 1944 KW (January) and the low occurs at 1224 KW (April), encompassing a span of 720 KW. Space heating requirements are satisfied by several boilers burning #4 oil, occasionally supplemented by LPG (propane).

The utility bills are:

	ELECTRICITY	#4 F. O.
QUANTITY	7,369,200 KWH	288,253 Gal.
COST	$ 523,772	$194,128
UNIT COST	$ 0.0711/KWH	$0.673/Gal.

"What If" Analysis
Once the Consumption and Input Data are typed into the Template, the values of the remaining cells will be quickly and automatically calculated. Because of this particular feature of the Template, **what if analyses** may be carried out by varying the Cogeneration parameters, the Utilization factors[4] and any other input parameters such as daily operating hours, fuel unit costs, etc. This type of analysis is the most valuable feature of the template application to cogeneration analysis as discussed in the next section.

Cogeneration Cost Analysis
To determine the most effective cogeneration engine size, the Template was used to analyze the costs of cogeneration for two different fuels - oil and natural gas - over a capacity ranging from 100 KW to 1500 KW. Lack of space does not permit the displaying of the cost parameters versus size. However, analysis of this type of plot reveals that the optimum size should occur, when the following condition is verified:

$EngFuel = $NetBoilerFuel

However, If one were to draw a plot of savings vs. size, the curve will show that maximum savings occurs right near the point where above condition is satisfied. Table 2 contains the data necessary for the discussion that follows. The optimum size is 900 KW for oil-fired engines, and 1000KW for gas-fired engines. However, much larger savings are attained in the latter case. It should be noted that payback rises steeply beyond this savings maximum, as expected.

For fuel oil fired engines[5], the 900 KW engine appears to be the better solution: Net savings from cogeneration indeed reach a maximum for engine size = 900 KW. Any engine larger than 1100 KW is wasteful. At all times, the electrical energy requirements are fully covered up to 1100 KW (without excess availability), but engine fuel cost is higher for the larger engine $683,250/Y as compared to $ 599,023/Y for the 900 KW engine. On the other hand, net savings from cogeneration for the 1100 KW engine is $406,145, while this quantity peaks at $437,773 for the 900 KW engine. Note, that although the net boiler fuel cost decreases with engine size, there is increased thermal dumping for larger engines.

The <u>total avoided cost of electricity</u> is what contributes to the cost effectiveness of cogeneration. As mentioned, the 900 KW engine has the highest net savings figure in the range 100KW to 1500 KW covered by this analysis.

A closer analysis of the performance of cogeneration systems may be achieved by means of suitably defined parameters as discussed in the next sections.

Performance Indexes
The following **Performance Indexes** are defined:

1. <u>Cogeneration (Avoided) Cost Index (CCI)</u>:

CCI = {$NetSavCogen/$TotCost}

$NetSavCogen = $DisplKWH + $AvoidedFuel
$SavCogen = Avoided fuel cost (displaced thermal load)
$TotCost = $CogenFuel + $NetBoilerFuel

Because $NetSavCogen effectively defines the net cost of avoided energy, CCI increases with savings resulting from the cogeneration process.

2. <u>Fuel Utilization Ratio (FUR)</u>:

FUR = {BTU eq. KWH displaced + Avoided fuel}/(Engine Fuel)

By definition, FUR << 1 due to losses in the cogeneration process. It measures how effectively the engine fuel is used to generate electricity while satisfying part of the existing thermal load. Because FUR is directly dependent on the system's ability to replace purchased energy, engines designed to meet demand should have high FUR values.

3. <u>Thermal Load Matching Factor (TLM)</u>:

TLM = 1 - {(Boiler fuel - Avoided Fuel)/(Boiler fuel)}

This index indicates how well the cogeneration thermal output matches existing thermal load. Boiler fuel refers to fuel consumed by a boiler used to satisfy current thermal loads.

(1) When TLM = 1: Avoided fuel = Boiler fuel, a perfect match.

(2) When TLM < 1: Avoided fuel < Boiler fuel, with less than ideal thermal match.

(3) When TLM >>1: Avoided fuel > Boiler fuel, thermal dumping.

Thus, engines designed to meet peak demand will meet the condition that TLM >> 1, with resulting dumping of excess cogeneration thermal output. The term **avoided fuel** applies to the portion of the cogeneration thermal output which *effectively* meets current thermal load on the premises.

Parametric Analysis of Cogeneration

The preceding discussion indicates that the selection of engine size for the cogeneration system will depend on the scope of cogeneration. As is generally known, unless the thermal load matches the useful thermal output of the engine designed to satisfy the electrical demand requirements, there usually will be occasional shortages of cogenerated thermal output. However, since cogeneration is primarily designed to displace KWHs purchased from Utility, the main objective in sizing the engine is to select a capacity that will satisfy the energy requirements, leaving the Utility the role of peak shaving in instances of peak demand or where demand exceeds the capacity of the cogenerating engine.

The Performance indexes - **CCI**, **FUR** and **TLM** - are useful to measure the extent to which the cogeneration criteria are met and to guide the designer in the selection of a suitable engine capacity. Thus, CCI measures how well cogeneration contributes to reduce the cost of purchased energy. The index **FUR** would indicate how effectively the cogenerating unit satisfies its double function of displacing Utility purchased KWH and of providing a suitable thermal output to meet existing thermal loads. On the other hand, the index **TLM** would indicate how closely the engine thermal output matches existing thermal loads. As defined, **CCI** gauges the net amount of avoided cost through cogeneration. Thus, it should be able to measure the cost effectiveness of cogeneration more accurately than simple payback. Payback is simply an indication of how fast the initial investment is recovered: it is an accounting entity and it is rather insensitive to short term load variations.

The indexes defined above may be used to analyze the performance of the cogeneration engine on a periodic basis, as shown in Table 2. This table shows how the various parameters CCI, FUR and TLM vary with engine size. The table was constructed with data derived from the analysis of individual spreadsheets for each engine size. It includes results for both the fuel alternatives (fuel oil and natural gas) discussed previously. The indexes listed on Table 2 are the average of monthly figures for a 12-month period for each engine size used. They are also displayed in graphic form in Figure 1.

At all times, the index FUR for natural gas cogeneration exceeds that for fuel-oil fired engines indicating better fuel utilization for gas-fired engines[6]. As expected, the value of TLM increases continuously with engine size: however, beyond 400 KW, TLM > 1 thus indicating considerable thermal dumping. Sometimes, thermal dumping of excess thermal cogeneration output is inevitable as long as the cost of purchased energy is reduced. For this situation, the index CCI becomes rather useful. In both fuel types, the value of CCI attains a **maximum** around 850 KW. However, the values of CCI are substantially larger for gas-fired engines than for oil-fired engines as would be expected, since net savings from cogeneration are higher in the former case (gas).

It is interesting to note that payback (PB) begins to increase as the fuel utilization ration (FUR) drops for increasing engine size. The deflection occurs as the value of CCI approaches a maximum (beyond ca. 700 KW).

Environmental Considerations

The main purpose of cogeneration is to reduce the cost of purchased energy. This objective is further reinforced by the fact that part of the existing thermal load can be replaced by the heat recovered in the cogeneration process. Additionally, from the utility view point, expectations arising from demand-side management programs (DSM) have until recently led some utilities to provide incentives for cogeneration projects. However, a few utilities are now facing the prospect of excess generating capacity, thus reducing or in some case nullifying the appeal of the cost avoidance theme which has so far propelled the utilities to promulgate and promote their DSM programs.

Aside from these and related economic problems, cogeneration may contribute to an ecological problem due to the further deterioration of the environment by emissions from cogeneration systems. This potential for pollution would become particularly significant when poorly managed cogeneration systems operate under far from ideal conditions thus contributing to the contamination of the environment with toxic effluents.

As is well known, the combustion of fossil fuels when properly done yields CO_2 and, in the case of hydrocarbon containing fuels, water vapor as well. Thus, even under ideal conditions, burning fossil fuels always raises the CO_2 content in the atmosphere, with an increase in the 'greenhouse' effect. There is also a direct thermal pollution effect due to the heat carried by the hot exhaust gases. Under worse conditions, sulfur-containing fuels may add sulfur dioxide to the environment as an air pollutant, which when deposited on the soil increases the acidity of the medium, thus altering the pH balance of the soil and contributing to the corrosion of pH sensitive materials.

One of the authors[7] has examined the inherent contamination potential of the ideal combustion of fossil fuels. Based on this study, a **CO_2 emission index** is defined as follows:

$$\delta = 1000*(3.664*c)/HHV$$ [8]

where:

c = Carbon content in the fuel, lbmC/lbmFuel
HHV = Higher heating value, BTU/lbmFuel

Values of δ have been calculated for various types of fuel. Based on this Emission Index, we can classify the various types of fuel under three broad categories:

Fuel Category	Type	Emission Index[9]
Low emission	Molecular fuels	$\delta < 0.160$
Med emission	Fuel Oils	$0.16 < \delta < 0.180$
High emission	Coals	$\delta > 0.180$

As expected, this tabulation suggests that CO_2 emission in a given fuel increases with its carbon content. Thus, coals would be the worst offenders. Therefore, the index $\boldsymbol{\delta}$ may be used as a measure of the potential environmental effect of any given fuel.

The assumption of ideal combustion conditions is unattainable under practical operating conditions of any engine, including those used as prime-movers in cogeneration systems. Without embarking into a complex environmental analysis of the kinetics of fossil fuel combustion, one can restrict the analysis to two major contaminants that would arise directly from incomplete combustion of fossil fuels:

Formation of carbon monoxide: $C + \frac{1}{2}O_2 \rightarrow CO$

Water gas effect: $H_2 + CO_2 \rightarrow CO + H_2O$

The water gas effect is more pronounced in natural gas because of its higher hydrogen content relative to its molecular weight as compared to other hydrocarbon-based fuels. The phenomenon of incomplete combustion is rather complicated and is beyond the scope of this paper. A suggestion is made here that one of the reasons for the incomplete combustion of methane may be ascribed to the higher diffusivity of both CH_4 and hydrogen relative to that of the other molecules (such as the components of air) in the combustion mixtures. Hence, these two molecules tend to diffuse ahead of oxygen in such mixtures, thus favoring a secondary reaction with the combustion products, viz. CO_2, to yield among other products, water gas.

Therefore, attention must be paid to the proper operation of the cogeneration engine, particularly in the case of gas-fired engines in order to reduce adverse environmental effects due to the cogeneration process. Of course, the presence of sulfur would tend to aggravate the situation by the introduction of a highly corrosive component to the combustion effluent gases.

Because cogeneration systems operate at relatively high temperatures, other noxious pollutants, such as NO_x and organic peroxides and hydroperoxides, are produced which will add to further contamination of the environment. This fact is taken into consideration in the ranking system used by NYSEO in the ICP Program which penalizes those cogeneration proposals that present the prospects of a higher environmental impact.

The other environmental aspect, which is easily overlooked, is that of thermal pollution which stems from the exhaust heat carried into the atmosphere by the effluent gases. To reduce this effect, properly designed heat sinks should be included in the cogeneration system at the time of its installation. A beneficial effect of such heat sinks would be the partial recovery of low-grade heat which can be used to preheat combustion air or other process fluids. The combined effect of cogeneration heat recovery with the heat sinks can significantly reduce the environmental impact of the process of cogeneration in comparison with the power generating systems it is intended to displace.

Summary and Conclusions

The analysis presented in the preceding sections suggests that there are certain performance parameters that may be defined from data calculated by the Cogeneration Template that are very useful in evaluating the performance of the various alternatives to a given Cogeneration Scenario. These are:

CCI = cogeneration cost index =
= {$NetSavCogen/$TotCost}

FUR = fuel utilization ratio =
= {BTU equiv. of KWH displaced + Avoided fuel}/(Engine Fuel)

TLM = thermal load match =
= 1 - {(Boiler fuel - Avoided Fuel)/(Boiler fuel)}

In particular, the variations in CCI values for various engine sizes yield better indication of the performance of the different alternatives, particularly when coupled with an analysis of the other parameters, FUR and TLM. Altogether, the analysis based on such performance indexes provides a more solid basis for the selecting of the most cost effective cogeneration system.

For the illustration used in this paper, application of this methodology leads to selection of the appropriate engines for either fuel-oil-fired or gas-fired cogeneration engines. Selected data are shown below:

COGENERATION ALTERNATIVES

Alternative 1: Fuel = Natural Gas

	ENGINE SIZE		
	950 KW	**900 KW**	**850 KW**
KWH/Y	7274400	7196400	7044000
NetSavings	$675,639	$669,809	$656,576
Boilerfuel	$368,132	$389,558	$413,099
Payback	2.81	2.69	2.59
FUR	0.548	0.562	0.575

Alternative 2: Fuel = #4 Oil

	ENGINE SIZE		
	950 KW	900 KW	850 KW
KWH/Y	7274400	7196400	7044000
NetSavings	$433,255	$437,773	$437,003
Boilerfuel	$422,097	$443,092	$464,087
Payback	2.81	2.69	2.59
FUR	0.548	0.562	0.575

As expected, Alternative 1 is the most cost effective. This table also indicates that net savings from cogeneration peak around 900 KW with a payback around 3.5 years (4.11 years for oil and 2.7 years for gas). Alternative 2 was included in the event of unavailability of natural gas. This situation actually occurs for the complex building for which the current analysis is a simplified adaptation. Because two types of fuels were used for thermal loading, the analysis was carried out on a boiler fuel cost which is a composite of the unit costs for fuel oil #4 and LPG. The economic feasibility of this alternative predicates upon the possibility of buying the fuel at a bulk price, ideally lower than the current lot price.

Since attaining maximum savings is the economic criterion, it is concluded that the smaller engine for either type of cogeneration fuel that would yield the highest savings (about 900 KW) would appear to be the most cost effective. Furthermore, the 900 KW value is the threshold beyond which payback rises rapidly. The analysis presented in this paper suggests that payback is not very effective in the evaluation of cogeneration systems. Certainly it is not to be used as the sole criterion for selecting cogeneration engine size.

The methodology used so far in the analysis of cogeneration systems may soon have to be revised to include savings resulting from demand reduction, as per recent mandate from DOE as expressed in Document 10 CFR Part 455 (Docket No. CE-RM-91-130) issued by the Office of Conservation and Renewable Energy, US Department of Energy. Conformance with the new guidelines may foster the need for an extensive overhaul of the SEO technical review process used in the evaluation of cogeneration applications submitted to the ICP Program which is supported by federal funds. For instance, demand savings and cost avoidance projects may henceforth be favored even though their cost effectiveness as expressed by the simple payback criteria may be deemed marginal under current policy. New ICP Regulations and Procedures are under study by New York State Energy Office but as of this date it is not possible to predict how deeply the criteria for evaluation of cogeneration will be affected by the forthcoming changes.

Concerns with environmental contamination stemming from the combustion of fuels used in powering cogeneration system brings further emphasis on the proper management of the operation of such systems. Thus environmental considerations will reinforce the need for proper handling of the cogeneration process, beyond those of purely economic nature (such as cost avoidance and DSM incentives). However, concerns for environmental purity may not necessarily increase the cost of cogeneration, since an improvement in combustion conditions and the adoption of heat sinks (waste heat recovery) may contribute to an increase in overall efficiency of the process.

Acknowledgements
The authors wish to thank the NYSEO management for allowing publication of this Paper. The Cogeneration Template is the result of the joint collaboration of both Authors. The Parametric Analysis concept was developed as part of private research carried out by the senior author (CEG) and was initially applied to the evaluation of a cogeneration system for a large complex building situated in one of the all-year round resorts in New York state. The authors also wish to acknowledge the assistance provided by Mr. Tom Eapen, also of NYSEO, who stressed the importance of environmental effects of cogeneration and provided updated information on changes in the ICP Program guidelines arising from DOE's final ruling on the ICP Grants as expressed in Document 10 CFR Part 455 (Docket No. CE-RM-91-130) issued by the Office of Conservation and Renewable Energy, USDOE.

Notes

1. State University of New York.
2. Proposed cogeneration capacity: MW = megawatt; KW = Kilowatt
3. NYSEO uses as cost of electricity the average cost calculated by dividing the total energy consumption by the combined billed cost of energy and demand.
4. Discussed in the Analysis section.
5. The combined fuel is the composite of the following:
   Fuel Oil #4 261,803Gal/Yr $194,126
   Liq. Propane 135,233Gal/Yr $116,253
   The LPG was converted to Equivalent Fuel-Oil by using the ratio of their HHV values (95457/145000=0.658).
6. Unfortunately the scale used does not indicate the differences as clearly as the values shown on Table 2.
7. C.E. Gracias, unpublished study, "CO_2 Emissions and the Heating Power of Fuels - A Brief Analysis", Sep/89.
8. Similar indexes may be defined for other combustion products, such as SO_2 (sulfur dioxide) and CO (carbon monoxide) by adjusting the parameters accordingly.
9. For those who are chemically oriented, the trend observed for the δ values can be readily explained by comparing values with the corresponding carbon:hydrogen ratio for a given fuel.

Appendix - List of Tables and Figures

Bibliographic References

"New York State Energy Office's Experience with Retrofit Cogeneration Under the Institutional Conservation Program," R.P. Stewart and Naresh Ghiya, XI WEEC Proceedings, 1988, p. 359.

"A Cogeneration Template" - Status Summary, C.E. Gracias and L.S. Hoffstatter, NYSEO, Apr. 1988, "Small-Scale Cogeneration: NYSEO Seminar Notes, 1992," L. Hoffstatter.

"Guide to Natural Gas Cogeneration," ed. N. Hay, AGA, AEE Energy Books, Lilburn, GA 30226.

"Gas Turbine Cogeneration - Design, Evaluation and Installation," Robin Mackay, The Garret Corporation, Los Angeles, Feb. 1983.

"Cogeneration: Regulation, Economics and Capacity," Resource Dynamics Corporation, McLean, VA 22101, DOE/NBB-0031, Apr. 1983.

"Making Cogen Practical for Various Commercial Buildings," G. Meckler, Cons. Spec. Eng., Jan. 1987, p.73.

"Industrial and Commercial Cogeneration Case Studies," Synergic Resources Corporation, Bala Cynwyd, PA 19004, EPRI EM-5083, Mar. 1987.

"Proceedings: 1986 EPRI Cogeneration Symposium," Synergic Resources Corporation, Bala Cynwyd, PA 19004, EPRI EM-5285, Jun. 1987.

"Cogeneration And Utilities: Status and Prospects," Synergic Resources Corporation, Bala Cynwyd, PA 19004, EPRI EM-6096, Nov. 1988.

"Cogeneration - Generating Power and Profits"Engine Heat Balance, Cooper-Bessemer Reciprocating, Grove City, PA 16127.

"Handbook of High-Efficiency Electric Equipment and Cogeneration System Options for Commercial Buildings," N. Richard Friedman, Resource Dynamics Corporation, and M. H. Blatt, EPRI, EPRI CU-6661, Dec. 1989.

"All Fired Up About Cogeneration", Newsfront, Chem. Eng., Jan. 1992, p. 39.

"Cogeneration - A Technology That Has Come of Age," R.L. Oliverson, Cons.Spec.Eng., Jul. 1987, Editorial.

"Cogeneration Now Feasible For Small Buildings, W. Endicott, Cons. Spec. Eng., Jul. 1987, p. 54.

"Packaged Cogeneration - A Maturing Technology," L.J. Kostrzewa and K.G. Davidson, ASHRAE Journal, Feb. 1988, p. 22.

"The Challenge of Packaged Cogeneration" J. Douglas, EPRI Journal, Sept. 1988, p. 29

"Packaged Cogeneration for Small Buildings", J.A. Orlando, HPAC, Dec. 1988, p. 49

"Requirements for Cogenerating in Parallel With The Utility," J.Daley, EC&M, Apr. 1988, p. 63.

"Small Cogeneration System Costs and Performance," Science Applications International Corporation, EPRI EM-5954, Aug. 1988.

"Cogeneration - An Up and Coming Technology," Milton Meckler, Cons. Spec. Eng., Mar. 1989, p. 82.

"An Engineer's Guide to Cogeneration," Editors, HPAC, Jul. 1990, p. 84.

"Cogeneration Plant Slashes College's Electricity Costs," ITT Bell & Cossett, HPAC, Dec. 1990, p. 21.

"The Cogeneration Connection," J. Sinclair, HPAC, Dec. 1990, p. 43.

"Wastewater Treatment Plant Benefits from Cogeneration," B.L. Steiger, HPAC, Dec. 1990, p. 55.

"Cogeneration Revisited," Letter, P.O. Marron, Marron Associates, Eng. Systems, Jan./Feb. 1992, p. 21.

Document 10 CFR Part 455 (Docket No. CE-RM-91-130) issued by the Office of Conservation and Renewable Energy, US Department of Energy.

"Btu Tax Revisions Boost Cogen/IPP Industry," A. Donnelly and A. Wetz, World Cogeneration, May/June 1993, p. 1.

Table 1

COGENERATION TEMPLATE

Worktable

COGENERATION ANALYSIS

				Jan	Feb	Mar	Apr	May	Jun	Jul	Aug	Sep	Oct	Nov	Dec	Total
C.Gracias & L.Hoffstatt28/Sep/89e(		1	Month --->	Jan	Feb	Mar	Apr	May	Jun	Jul	Aug	Sep	Oct	Nov	Dec	Total
File: DulFuel9.wk3	FINAL	2	***Utility Billings***													
	EDITION	3	No. days 0	0	0	0	0	0	0	0	0	0	0	0	0	0
FUEL DATA		4	KW demand	0	0	0	0	0	0	0	0	0	0	0	0	0
BOILER FUEL:	Fraction	5	KWH billed	0	0	0	0	0	0	0	0	0	0	0	0	0
Natural Gas	0.00	6	$Cost billed	0	0	0	0	0	0	0	0	0	0	0	0	$0
Other Fuel	1.00	7	No. days	0	0	0	0	0	0	0	0	0	0	0	0	0
COGEN FUEL:		8	#therms billed	0	0	0	0	0	0	0	0	0	0	0	0	0
Natural Gas	1.00	9	$Cost billed	0	0	0	0	0	0	0	0	0	0	0	0	$0
Other Fuel	0.00	10	#Gal billed	0	0	0	0	0	0	0	0	0	0	0	0	0
		11	$Cost billed	0	0	0	0	0	0	0	0	0	0	0	0	$0
		12	***Energy Data***													Total
		13	BoilerFuel,MBTU	0	0	0	0	0	0	0	0	0	0	0	0	0
		14	EngFuel, MBTU	0	0	0	0	0	0	0	0	0	0	0	0	0
		15	Tot.Ht.avail, MBTU	0	0	0	0	0	0	0	0	0	0	0	0	0
INPUT DATA		16	Avoided fuel, MBTU	0	0	0	0	0	0	0	0	0	0	0	0	0
		17	NetBoil.fuel, MBTU	0	0	0	0	0	0	0	0	0	0	0	0	0
KW rating	0	18	Cog+BoilFuel, MBTU	0	0	0	0	0	0	0	0	0	0	0	0	0
Heat rate, BTU/KWH	0	19	Max Cogen KWH	0	0	0	0	0	0	0	0	0	0	0	0	0
#HRS/Day cogen.	0	20	KWH displ.Cogen.	0	0	0	0	0	0	0	0	0	0	0	0	0
Exh.recov.,BTU/KWH	0	21	***Costs***													
BTU/KWH, Source	11600	22	$Cost Eng. Fuel	$0	$0	$0	$0	$0	$0	$0	$0	$0	$0	$0	$0	$0
		23	$Cost NetBoil.Fuel	$0	$0	$0	$0	$0	$0	$0	$0	$0	$0	$0	$0	$0
Boiler, Efficiency	0	24	$TotCost w/Cogen.	$0	$0	$0	$0	$0	$0	$0	$0	$0	$0	$0	$0	$0
BTU/UnitBoilNGas	0	25	***Savings***													
BTU/UnitBoilFOil	0	26	TotElec. Savings	$0	$0	$0	$0	$0	$0	$0	$0	$0	$0	$0	$0	$0
		27	Thermal Savings	$0	$0	$0	$0	$0	$0	$0	$0	$0	$0	$0	$0	$0
BTU/UnitCogNGas	0	28	NetSav from Cogen	$0	$0	$0	$0	$0	$0	$0	$0	$0	$0	$0	$0	$0
$Cost/UnitCogNGas	0	29	NetMBTUSav,Source	0	0	0	0	0	0	0	0	0	0	0	0	0
BTU/UnitCogFOil	0	30														
$Cost/UnitCogFOil	0	31	***Summary:***	SourceRelSavings,MBTU =			0		$InstCost/KWCog =		$0			Payback =	0.00Yrs	
CogenInst. Cost	0	32														
		33	PERIOD-BASED COSTS:													
		34	$Cost/KWH	$0	$0	$0	$0	$0	$0	$0	$0	$0	$0	$0	$0	$0
		35	$/UnitBoilGas	$0	$0	$0	$0	$0	$0	$0	$0	$0	$0	$0	$0	$0
		36	$/UnitBoilFOil	$0	$0	$0	$0	$0	$0	$0	$0	$0	$0	$0	$0	$0
		37	$CostBoilFuel	$0	$0	$0	$0	$0	$0	$0	$0	$0	$0	$0	$0	$0
CALCULATED PARAMETERS		38	$CostEng.Fuel	$0	$0	$0	$0	$0	$0	$0	$0	$0	$0	$0	$0	$0
		39	$CostAvoidedFuel	$0	$0	$0	$0	$0	$0	$0	$0	$0	$0	$0	$0	$0
El. Cost, $/KWH	$0.0000	40	$CostNetBoilFuel	$0	$0	$0	$0	$0	$0	$0	$0	$0	$0	$0	$0	$0
		41	$TotCost w/Cogen	$0	$0	$0	$0	$0	$0	$0	$0	$0	$0	$0	$0	$0
$Cost/MBTUBoilNGas	$0.000	42	Boil.FuelSavings	$0	$0	$0	$0	$0	$0	$0	$0	$0	$0	$0	$0	$0
$Cost/MBtuBoilFOil	$0.000	43	TotElecSavings	$0	$0	$0	$0	$0	$0	$0	$0	$0	$0	$0	$0	$0
		44	NetSavfrom Cogen	$0	$0	$0	$0	$0	$0	$0	$0	$0	$0	$0	$0	$0
$Cost/MBTUCogNGas	$0.000	45												Payback =	0.00 Yrs	
$Cost/MBTUCogFOil	$0.000	46														
		47														
CogThermEff.	0.000	48														
CogElecEff.	0.000	49														
TotCogEff.	0.000	50														
		51														
$Cost/KWCog	$0	52														

Table 2–1

COGENERATION ANALYSIS: Summary

File: Cogt2elt.wk3

Fuel = Fuel Oil

EngineSize: KW	**100**	**200**	**300**	**400**	**500**	**600**	**700**	**800**	**900**	**1000**	**1100**	**1200**	**1300**	**1400**	**1500**
EngFuel,MBTU	12080	24160	36240	48320	60400	72480	84560	96640	108720	120800	132880	144960	157041	169121	181201
Ht.avail,MBTU	3066	6132	9198	12264	15330	18396	21462	24528	27594	30660	33726	36792	39858	42924	45990
Max Cogen KWH	876000	1752000	2628000	3504000	4380000	5256000	6132000	7008000	7884000	8760000	9636000	10512000	11388000	12264000	13140000
KWHdisplCog	876000	1752000	2628000	3504000	4380000	5256000	6062400	6762000	7196400	7323600	7369200	7369200	7369200	7369200	7369200
% KWHdispl	12.13	24.26	36.39	48.53	60.66	72.79	83.62	92.69	98.08	99.49	100.00	100.00	100.00	100.00	100.00
$EngFuel(G)	62114	124227	186341	248455	310568	372682	434796	496909	559023	621137	683250	745364	807478	869592	931705
$NetBoil(FO/LPG)	871844	815290	758735	702181	645627	589072	534148	487200	443092	401102	362775	325279	293022	265687	242446
$TotCostFuel	933957.6	939517	945076.3	950635.7	956195	961754.4	968943.6	984109.4	1002115	1022239	1046026	1070643	1100499	1135279	1174151
$NetSavCoge	56703	113406	170109	226812	283515	340218	390345	424904	437773	426690	406145	381527	351671	316891	278019
10*CCI	0.607127	1.207069	1.799952	2.385901	2.965037	3.537477	4.028561	4.317646	4.368492	4.174072	3.882741	3.563534	3.195557	2.791309	2.367832
CCI–Avg	0.060713	0.120707	0.179995	0.23859	0.296504	0.353748	0.402856	0.431765	0.436849	0.417407	0.388274	0.356353	0.319556	0.279131	0.236783
FUR – Avg	0.5648	0.5648	0.5648	0.5648	0.5648	0.5648	0.5607	0.5482	0.5286	0.5031	0.4782	0.4560	0.4349	0.4148	0.3959
TLM – Avg	0.2492	0.4983	0.7475	0.9966	1.2458	1.4949	1.7441	1.9932	2.2424	2.4915	2.7407	2.9898	3.2390	3.4881	3.7373
Payback	3.53	3.53	3.53	3.53	3.53	3.53	3.59	3.77	4.11	4.69	5.42	6.29	7.39	8.84	10.79
10*FUR	5.648	5.648	5.648	5.648	5.648	5.648	5.607	5.482	5.286	5.031	4.782	4.560	4.349	4.148	3.959
$EngFu/KWH	0.0709	0.0709	0.0709	0.0709	0.0709	0.0709	0.0717	0.0735	0.0777	0.0848	0.0927	0.1011	0.1096	0.1180	0.1264
$TotFuel/KWH	13.1244	11.8313	10.7815	9.9123	9.1807	8.5566	8.0239	7.5901	7.2226	6.9079	6.6429	6.4109	6.2211	6.0669	5.9411
EngineSize: KW	100	200	300	400	500	600	700	800	900	1000	1100	1200	1300	1400	1500

Table 2-2

COGENERATION ANALYSIS: Summary

File: Cogt2alt.wk3

Fuel = Natural Gas

EngineSize: KW	X	100	200	300	400	500	600	700	800	900	1000	1100	1200	1300	1400	1500
EngFuel,MBTU		12080	24160	36240	48320	60400	72480	84560	96640	108720	120800	132880	144960	157041	169121	181201
Ht.avail,MBTU		3066	6132	9198	12264	15330	18396	21462	24528	27594	30660	33726	36792	39858	42924	45990
Max Cogen KWH		876000	1752000	2628000	3504000	4380000	5256000	6132000	7008000	7884000	8760000	9636000	10512000	11388000	12264000	13140000
KWHdisplCog		876000	1752000	2628000	3504000	4380000	5256000	6062400	6762000	7196400	7323600	7369200	7369200	7369200	7369200	7369200
% KWHdispl		12.13	24.26	36.39	48.53	60.66	72.79	83.62	92.69	98.08	99.49	100.00	100.00	100.00	100.00	100.00
$EngFuel(G)	A	42280	84560	126840	169121	211401	253681	295961	338241	380521	422801	465082	507362	549642	591922	634202
$NetBoil(FO/LPG)	B	863765	799131	734498	669864	605231	541307	487200	437094	389558	346705	305174	272768	245766	219204	193541
$TotCostFuel		906044.9	883691.5	861338.1	838984.8	316631.4	794987.8	783161	775334.7	770079.4	769506.7	770255.1	780129.9	795407.8	811126.3	827742.7
$NetSavCogen	C	84616	169232	253847	338463	423079	506985	576127	633678	669809	679423	681915	672040	656762	641044	624428
10*CCI	a	0.933903	1.915053	2.947128	4.0342	5.180783	6.377268	7.356437	8.172966	8.69792	8.829326	8.853109	8.614467	8.256927	7.903135	7.543739
CCI-Avg		0.09339	0.191505	0.294713	0.40342	0.518078	0.637727	0.735644	0.817297	0.869792	0.882933	0.885311	0.861447	0.825693	0.790313	0.754374
FUR - Avg		0.6102	0.6102	0.6102	0.6102	0.6102	0.6095	0.5984	0.5836	0.5623	0.5338	0.5078	0.4806	0.4554	0.4337	0.4145
TLM - Avg	c	0.2492	0.4983	0.7475	0.9966	1.2458	1.4949	1.7441	1.9932	2.2424	2.4915	2.7407	2.9898	3.2390	3.4881	3.7373
Payback	d	2.36	2.36	2.36	2.36	2.36	2.37	2.43	2.52	2.69	2.94	3.23	3.57	3.96	4.37	4.80
10*FUR	b	6.102	6.102	6.102	6.102	6.102	6.095	5.984	5.836	5.623	5.338	5.078	4.806	4.554	4.337	4.145
$EngFu/KWH		0.0483	0.0483	0.0483	0.0483	0.0483	0.0483	0.0488	0.0500	0.0529	0.0577	0.0631	0.0688	0.0746	0.0803	0.0861
$TotFuel/KWH		1.0343	0.5044	0.3278	0.2394	0.1864	0.1513	0.1292	0.1147	0.1070	0.1051	0.1045	0.1059	0.1079	0.1101	0.1123
EngineSize: KW		100	200	300	400	500	600	700	800	900	1000	1100	1200	1300	1400	1500

Figure 1

PARAMETRIC ANALYSIS

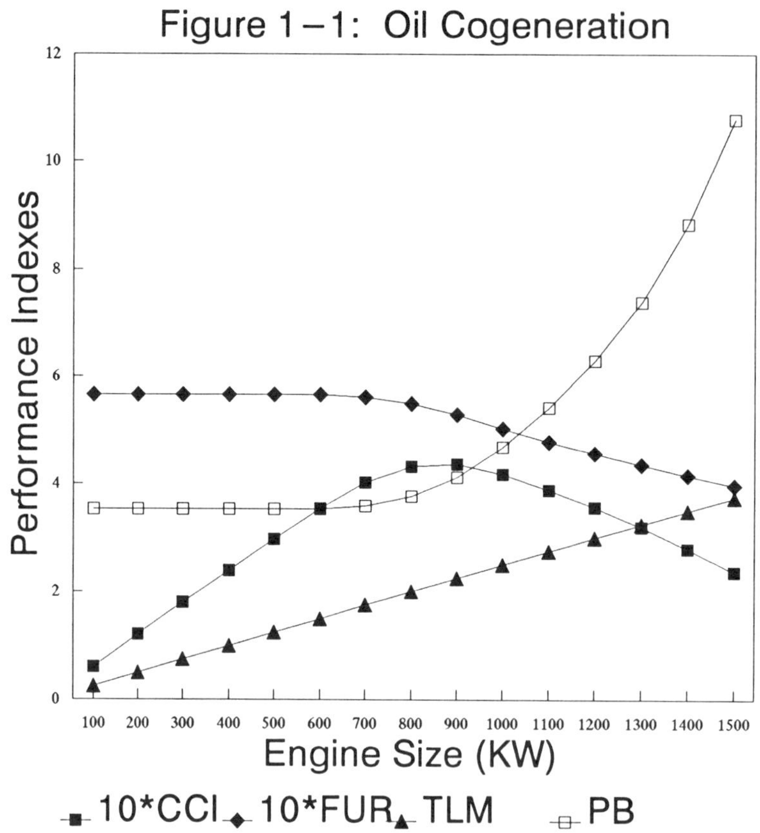

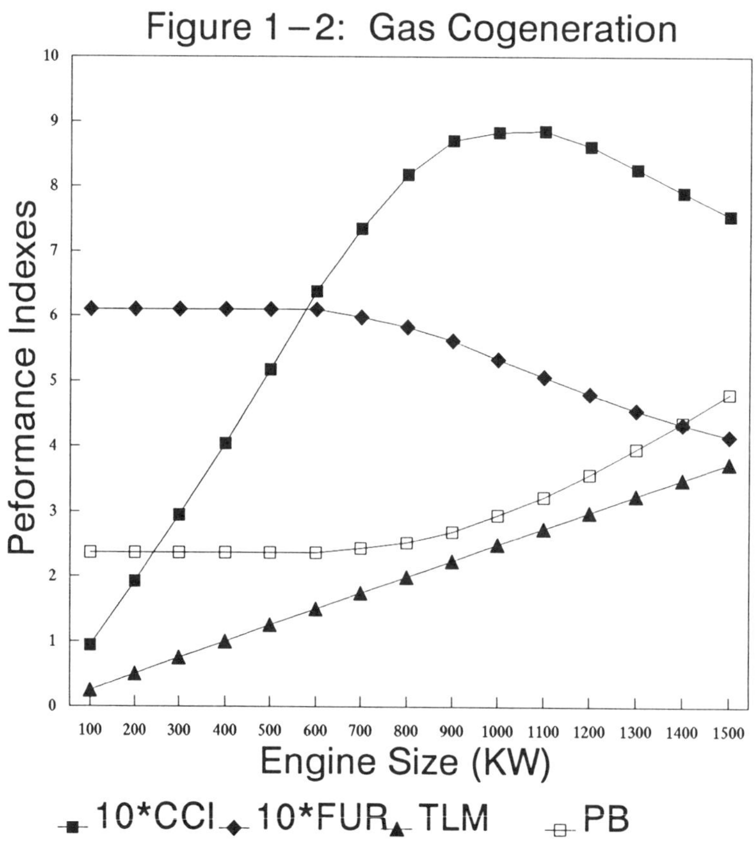

Chapter 82

Cogeneration Emissions Breakthrough

R.H. Warren

ABSTRACT

Failure to achieve air emissions levels stipulated in environmental permits, if uncorrected, can spell financial disaster for a cogeneration project, especially in an EPA non-attainment area.

Recently, a natural gas-fueled internal combustion cogeneration plant was able to overcome its failure to meet emissions criteria through what was, at the time, experimental combustion technology. The effort was a success however, and critical emissions dropped dramatically below permitted levels.

This paper will review the history of extreme site operating conditions that caused one compound (carbon monoxide) to fall marginally short of requirements in the most strictly enforced air quality district in the world, California's South Coast Air Quality Management District. It will also describe in detail the actions the project participants did to solve the problem.

Although under limits in the other critical pollutant, NOx, regulations then in effect prohibited any trade-offs between NOx and CO. An over-limit reading in one could not be waived for an under-limit reading in the other. Accordingly, an operating permit would not be issued. The project seemed to be faced with either a permanent shutdown or installation of extremely expensive catalytic reduction equipment that would have rendered it economically unviable at the time.

However, a third alternative, though not "tried and proven", fortuitously appeared. The system supplier reported that recently completed controlled environment tests of a design modification showed promising results. After reviewing the data yielded by laboratory and limited field tests, the project owner opted to have the design modification installed. After a week of down time and a month of carefully monitored operation, the cogenerator was ready for another emissions test.

Results of the test were significantly better than expected: carbon monoxide went from eight percent over limits to 16 percent under, an improvement for the system of 22 percent, while other emissions remained well below permit requirements.

Consensus is that, since laboratory and field test data has been validated through duplication in severe industrial operating conditions, it is likely that this technology will soon become BACT (Best Available Control Technology) for field applications.

INTRODUCTION

Cogeneration is part of a growing collection of the Works of Man that are at once good for the environment and good for the economy. Contrasted with the electric utility plants they supplement, cogeneration plants are far more efficient and far less polluting.

Even with the use of decades-old technology, energy generated by cogeneration plants today can be thermodynamically twice as efficient, economically two-thirds as expensive, and half as polluting as utility-generated energy.

But, in some areas of the country even these achievements are not enough, especially with regard to air emissions. In EPA "non-attainment" areas such as the Los Angeles (geological) basin, stationary emissions sources are about to go through another iteration of EPA-mandated reductions under new regulations promulgated by South Coast Air Quality Management District (SCAQMD).

In an attempt to reach compliance with EPA requirements, local regulators, focusing on the 20 percent of the air pollution problem created by stationary sources, are imposing even tighter controls on those industries that remain in the Los Angeles Basin. Figure 1 shows the decreasing progression of air emissions limits during the past 10 years, and the new limits set out by SCAQMD for the next 10 years, to 2003.

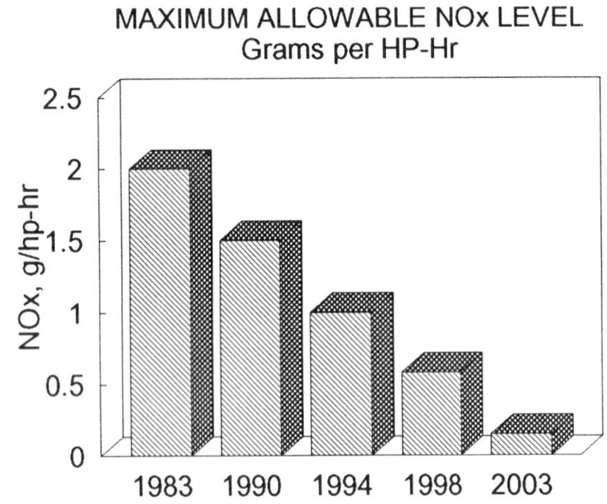

(FIGURE 1)

PAST, PRESENT AND FUTURE LIMITS

Cogeneration system manufacturers have significantly improved the emissions chemistry of their products through on-going research and development. As new plateaus have been reached, old engine designs have been discontinued and updated versions brought to market. In addition, various combinations of add-ons, such as catalytic converters and ammonia injection, have been offered to meet the challenge of tighter emissions control. Figure 2 shows the emissions reduction progression of one manufacturer, Caterpillar.

The results of this emissions reduction effort are distinctly impressive. Note that in 1980, available control technology produced 20.0 grams of NOx per horsepower-hour (g/hp-hr). By the end of 1992, this had decreased to 1.2 g/hp-hr.

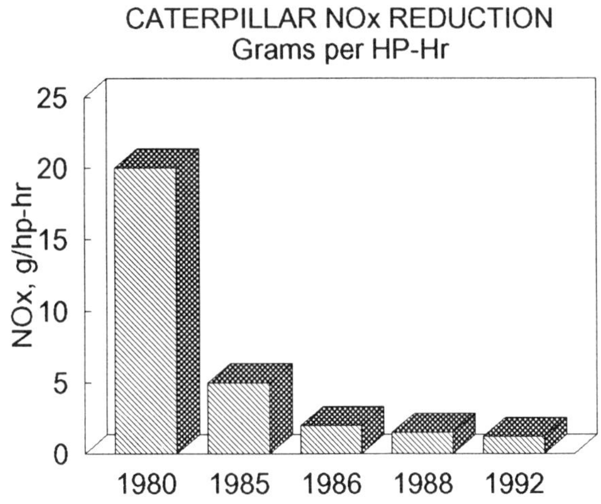

(FIGURE 2)

Until 1993, an actual field installation utilizing a medium size, natural gas-fired cogeneration plant typically could meet emissions requirements either through a "lean-burn" design (a configuration utilizing a very low fuel-air ratio), or by addition of an automotive-type catalytic converter to an older "rich-burn" design.

But, this will change in 1994 as SCAQMD implements its "RECLAIM" (Regional Clean Air Incentives Market) program that will, as shown in Figure 1, reduce the allowable emissions to 10 percent of 1992 values. To meet these allowable limits, RECLAIM will force a cogeneration operator to buy emissions credits on the open market if it cannot meet its limits through actual reductions. It has been estimated that such credits may cost between $1,000 and $10,000, per ton-year.

One probable effect of RECLAIM will be to stimulate further advances, perhaps sequels to the one described herein, in emissions reduction technology. It is also likely that the cogeneration industry will be stimulated, since getting dual use of energy output from a single fuel source will obviously reduce emissions, as illustrated below.

ENVIRONMENTAL DIMENSIONS

In replacing thermal energy developed by a given heat source such as a duct burner, a natural gas-fired lean-burn cogenerator reduces, by about 30 percent on an equivalent BTU basis, the NOx and CO otherwise produced by the other heating source.

On the electrical side the same cogenerator, considering for purposes of analysis, a thermodynamic system comprised of the cogenerator itself and that portion of the utility whose electrical output the cogenerator replaces, will eliminate burning of more than six cubic feet of natural gas per kWh generated. As shown in Appendix A, this adds up to 30 MMCF annually for a 600 kW unit.

In addition, by elimination of extra heat sources and the perhaps surprising economies of down scaling, an in-plant dedicated cogeneration unit can reduce the net cost of electricity by as much as 33 percent, compared to power furnished by the utility.

Taken together, these two "E"s - Environment and Economy - are mutually enhanced by a properly sized cogeneration system. And, contrary to most environment-versus-economy relationships, this situation is getting better, not worse. The synergistic influences of market place competition and the aforementioned regulatory pressures have combined to bring about still another breakthrough in cogeneration design.

CASE HISTORY

A typical environmental permit for a cogeneration system issued in the Los Angeles area in the late 1980's specified limits for NOx and CO as follows:

NOx	1.50 g/hp-hr
CO	1.60 g/hp-hr

Although the breakthrough described herein relates primarily to CO emissions reduction, it must be viewed in association with NOx because, due to the inverse relationship between these two compounds in natural gas engines (described below), a decrease in one causes an increase in the other.

Existing rich-burn engines with catalytic converters and lean-burn engine designs met the above permit criteria under normal conditions. Severe field conditions could, however, push the engine design to its limits and in some cases, beyond.

One such installation was in a ceramics industry plant in an area in which normal summer temperatures reached 115 F (db) and 75 F (wb). The system consisted of a Caterpillar Series 3512 SITA engine, coupled to an 600 kW Kato synchronous generator. Manufacturer's test cell data indicated that the cogeneration system would be able to meet the air quality criteria at the site. But when an emissions test was performed in actual field conditions, the results were mixed, as follows:

NOx	1.10 g/hp-hr	(Pass)
CO	1.72 g/hp-hr	(Fail)

The NOx emission, normally the more villainous of the two, was more than 25 percent below limits, but the CO was over by eight percent, or 12 g/hp-hr.

Emission levels for both CO and NOx are determined by adjusting the air - fuel ratio of the engine carburetion system. In this case, the ratio had been adjusted prior to the test, but due to varying ambient conditions, the optimum mixture was not achieved during the test.

Because of the inverse relationship between NOx and CO, as illustrated by Figure 3, a trade-off with SCAQMD was sought.

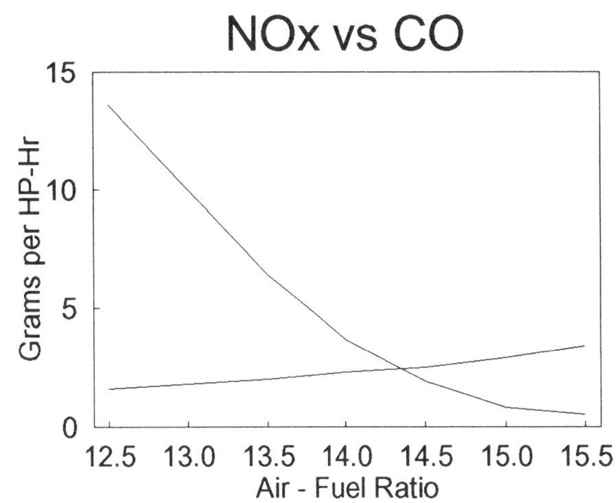

(FIGURE 3)

However, regulations then in effect prohibited the SCAQMD from approving a trade-off, and since test results were above permitted limits of CO, the installation was denied an operating permit.

The manufacturer's R&D group was apprised of the problem and a crash development program was initiated. Several ideas were tried in the laboratory before one with a high potential for success surfaced.

PISTON REDESIGN

A considerable amount of research had been performed earlier with regard to reducing particulate emissions in diesel engine applications. Particulate emissions from a diesel engine are primarily carbon. One significant finding was that the carbon emission level could be reduced by decreasing the volume of the piston-cylinder crevice. Physically, the "crevice" is defined vertically by the piston and cylinder walls, and horizontally by the top of the piston and the first compression ring. Tests showed this was a region in which carbon collected and remained un-burned, thus creating the particulate matter. By reducing the volume of the crevice, they could produce a decrease in the quantity of particulates discharged from the engine.

In a technical leap of faith, researchers reasoned that a reduction in CO emission might be similarly achieved in natural gas applications.

The underlying principle, they found, was that as the combustion process begins, the lighter-end hydrocarbons, those with more hydrogen and fewer carbon atoms, tend to burn in the combustion chamber while the heavier ends, with more carbon atoms, settle and concentrate in the crevice. Thus, since

1) The mass in the crevice is exhausted un-burned,

2) This mass is comprised mostly of carbon atoms, and

3) The air-fuel ratio presents an excess of oxygen, which is also exhausted,

a chemical reaction takes place, and carbon monoxide is formed in accordance with the following (simplified) formula.

$$CH_4 + O_2 \text{ -> } CO + 2H_20$$

It was further reasoned that, by relocation of the compression ring further toward the top of the piston to reduce the volume of the crevice, as was done to solve the diesel particulate problem, the CO quantity could be reduced enough to meet limits.

Calculations indicated that a decrease in crevice depth of 9 mm would be sufficient to reduce the manufacturing of CO to acceptable limits.

A set of redesigned pistons, with the first compression ring moved from 25 mm to 16 mm below the top of the piston, was manufactured and tested in the lab. Also, the piston material was changed to a stronger alloy to preserve structural integrity.

CO reduction was significant and the researchers decided to run a full test cell program (1,000 hours). Again, results were encouraging: enough so to try in the real world conditions of a high temperature industrial environment.

FIELD EXPERIMENT

Three months after initial identification of the problem, a set of the new piston assemblies, comprised of what was now called the Low Crevice Volume (LCV) piston, was delivered to the site and installed by the engine supplier, Hawthorne Power Systems of San Diego.

After startup, the engine was run-in for approximately 750 hours to seat the new components. Hawthorne technicians sampled the emissions and found that the CO content of exhaust gas had fallen significantly. Moreover, they found that NOx had not increased at all: the troublesome inverse relationship had been defeated. A re-test was scheduled and performed with a representative from SCAQMD in attendance.

RE-TEST RESULTS

To assure consistency of technique, the same independent testing laboratory, equipment and team were engaged to perform the re-testing of the rebuilt cogeneration system.

Test protocol was such that definitive results were not immediately available. Instead, raw data obtained in the field had to be returned to the office and processed before a determination of pass-fail could be made.

However, after the data was put into the computer model and the calculations made, the final results were as follows:

NOx	1.10 g/hp-hr	(Pass)
CO	1.34 g/hp-hr	(Pass)

NOx was 0.4 g/hp-hr below permit limits, and CO was 0.16 g/hp-hr below. Thus, the engine successfully passed the test, and an operating permit was at last issued.

CONCLUSION

This dramatic reduction in CO (22 percent below previous results), without an accompanying increase in NOx, serves as the genesis of a new class of low emission engines.

Earlier research and development efforts by the manufacturer had already reduced NOx by orders of magnitude - 20 g/hp-hr to 1.2 g/hp-hr in 12 years. Now, with this CO breakthrough, CO levels are similarly at and below permissible limits in non-attainment areas. New cogeneration projects will be more readily accepted by both the air quality districts and the general public since they are indeed so much cleaner burning.

As regulatory pressures continue to increase, it seems likely that market forces will continue to combine with engine manufacturers' R&D to produce even more spectacular emissions control achievements. The eventual result should be reduction of SMOG from NOx, and human respiratory problems due from CO, to a level near extinction.

APPENDIX A

NATURAL GAS CONSUMPTION - BEFORE VS AFTER COGEN

Determine the annual difference in total gas consumption before and after installation of cogeneration for a system ("System") comprised of (a) Utility's electrical generation to supply electricity to Host, (b) Host natural gas consumption and (c) the cogenerator.

BEFORE				
Host total power consumed		5.576	MMkWh	(1)
Host power purchased from Utility =	100.0%	5.576	MMkWh	
Assumed Utility line losses @	4.0%	0.223	MMkWh	
Utility gross power generated for Host		5.799	MMkWh	
Utility annual average heat rate		9763	BTU/kWh	(2)
Utility natural gas burned for Host		566,180	TH	
Host natural gas consumption		3,130,000	TH	(1)
Cogenerator gas consumption		0	TH	
Total gas consumed in System		3,696,180	TH	

AFTER				
Host total power consumed		5.576	MMkWh	
Host power supplied by cogen =	80.1%	4.465	MMkWh	(3)
Host power purchased from Utility =	19.9%	1.111	MMkWh	
Assumed Utility line losses @	4.0%	0.044	MMkWh	
Utility gross power generated for Host		1.156	MMkWh	
Cogenerator energy furnished to Utility		0.199	MMkWh	(4)
Net Utility power generated for System		0.957	MMkWh	
Utility annual average heat rate		9763	BTU/kWh	
Utility natural gas burned for Host		93,470	TH	
Host natural gas consumption		3,130,000	TH	
Cogenerator gas consumption		550,732	TH	
Less nat gas displaced by cogenerator		(377,975)	TH	
Total gas consumed in System		3,396,227	TH	
Reduction in gas consumed in System		299,953	TH	
Assumed heating value of natural gas		1000	BTU/SCF	
Reduction in gas consumed in System		29,995	MSCF	
Reduction in gas consumed in System, per kWh		6.4	SCF/kWh	

Notes:
(1) Actual Host Consumption
(2) Utility published annual average heat rate.
(3) Calculated cogeneration output consumed by Host.
(4) Calculated surplus power to be exported to Utility grid.

BIOGRAPHICAL SKETCH

ROBERT H. WARREN, PE, CEM

Mr. Warren is Senior Vice President of Commercial Energy Systems Management Corporation, San Diego, California. During a thirty-five year professional engineering career, he has managed execution of more than $1 Billion in capital projects. Included have been numerous small power, combined cycle and cogeneration plants, and thermal energy applications on manufacturing and process plant projects.

Prior to joining CES, he was Group Vice President with Ultrasystems, an energy project developer specializing in alternative fuel power plants and cogeneration. Earlier, he was associated with Fluor Daniel where he managed engineering and construction of major projects that included power, chemical, synthetic fibers, mining, paper, and industrial projects.

Mr. Warren earned degrees at Rensselaer Polytechnic Institute and Washington and Lee University. He maintains professional engineering and contractor registrations in several states and has been accredited as a Certified Energy Manager and as a Certified Cogeneration Professional by the Association of Energy Engineers. He is author of several articles and a recently published text, *Motivation and Productivity in the Construction Industry* (Van Nostrand Reinhold Company).

Chapter 83

State of Florida Capitol Center Cogeneration Plant

J.R. Geers, C.D. Svard

INTRODUCTION

The basic purpose of this study was to determine the feasibility of cogeneration located in Tallahassee, Florida. Tallahassee is the capitol of the State of Florida with a large number of State buildings downtown, referred to as the Capitol Center, and two large Universities.

The scope of the study included the following:

Cogeneration for the Capitol Center alone, as well as adding Florida State University (FSU) and Florida A&M University (FAMU) thermal loads to the plant were also studied. Both of these Universities are within 5,000-7,000 feet of the new plant site. The consultants were to provide a general concept, recommendations and back-up analysis in a final report. This has been Phase I of what is intended to be a four-phase project. Since Cogeneration is feasible, the State intends to approve a Phase II effort to determine the financing option, participants, and a final concept design.

The State of Florida procured these A-E services through Sunbelt Engineering, Inc. (Sunbelt) and assigned Mr. Robert W. Mohrfeld of the Department of Management Services as the Project Director. Sunbelt sub-contracted to PLM Technologies, Inc. (PLM) to lead the study phase while Integrated Planning & Engineering, Inc. (IPE) assisted PLM throughout the Study.

A. Statement of Services - Phase I

The following summary is included for quick overview.

1. Collected and projected thermal and electrical load data for all three locations.
2. Assessed electrical, heating, and cooling load data.
3. Performed field survey of mechanical rooms, existing chilled water and steam plants, potential plant sites, and routing of distribution lines.
4. Assessed fuel options for Cogeneration Plant.
5. Developed plant alternatives and selected equipment types.
6. Selected absorption and centrifugal chillers and steam turbines.
7. Developed preliminary O&M and Capital budgets.
8. Performed economical analysis and Power Plant Modelling.
9. Prepared a preliminary report.
10. Presented preliminary report results.
11. Submitted a final study report.

B. Status Of Project

The Preliminary Report was submitted on February 1, 1993, and presented to various State agencies and organizations on February 11. The presentation was followed by individual meetings with FSU, FAMU, and the City of Tallahassee, as well as a joint meeting with all parties during the week of March 1-5. The Final Report was submitted on June 1, 1993. The starting date of the Phase II effort has not been determined. In addition, the City has retained the services of two outside engineering firms to review and comment on the Preliminary and Final Reports. The State and City also intend to procure an independent third party to review and comment on the report.

Depending upon the participants, the Cogeneration Plant could serve the Capitol Center only, Capitol Center and FSU or FAMU or, as recommended, all three sites. The Phase II effort will take the project through full concept design, education of all participants, development of four working sub-groups, overall concept for the cogeneration agreement with the City, and other major project tasks.

Phase II will last six to eight months. The Phase III effort would include the financing effort as well as full design of the plant and distribution systems. Phase III is estimated to take twelve months. Phase IV would be the testing and commissioning of the Plant. This final phase would take two months and be completed by early 1996.

COGENERATION PLANT ALTERNATIVES

A. Description of Alternatives

1. **General Overview**
   The selection of the central plant alternatives was complicated by the fact that the load data for the Capitol Center and FAMU was not very specific. However, the alternatives were developed using the steam and cooling base loads at each site. The new 33 MMBH boilers in the existing Heating Plant were included in all alternatives as backup boilers to the new plant.

   There are several potential cogeneration plant alternatives which could serve the thermal and electrical loads of the Capitol Center, FSU, and FAMU. These alternatives vary in the type and quantity of equipment and level of thermal and electrical output. However, all provided a significant and adequate supply of thermal and electrical requirements at each site. Plant configurations include variations of the following types of equipment:

   a. Gas-turbine generators with heat recovery boilers.
   b. Supplemental firing for the heat recovery boilers.
   c. Condensing steam turbine generators.
   d. Electric driven centrifugal chillers.
   e. Steam driven absorption chillers.

2. **Basecase - Capitol Center - Replacing existing chillers in-kind**
   This case basically includes the replacement of all building chillers over a thirty year period. They were grouped into five year replacement cycles based upon age and efficiency. The total chiller tonnage of this alternative was 11,430 tons.
3. **Alternative A - Capitol Center - Central Cooling Plant Only**
   This alternative was evaluated, but the annual savings of installing a new chilled water distribution system and chiller plant could not offset the large capital investment. The payback far exceeds the other alternatives. This concept works well when the project involves all new buildings and the distribution system can be a part of the central plant complex.
4. **Alternative B - Capitol Center - Central Cogeneration Plant**
   There were three sub-alternatives selected and each included four 1500 Ton electric chillers and one 3750 kW gas turbine. Each sub-alternative also included combinations of the following equipment: a 1000 Ton electric chiller, a 1000 Ton absorption chiller, a 1200 or 3000 kW steam turbine and supplemental firing on the gas turbine. This alternative is feasible.
5. **Alternative C - Capitol Center/FSU - Central Cogeneration Plant**
   There were three sub-alternatives which included either four or five 1500 Ton electric chillers; one 1200 Ton absorption chiller (single effect); two or three 1200 Ton absorption chillers (double effect) at FSU; two gas turbines with supplemental firing and two 1200 kW steam turbines. The maximum steam to FSU was set at either 50,000 PPH or 75,000 PPH. This allowed the equipment to supply the maximum thermal requirement at FSU, which dropped the payback of these alternatives. This alternative is also feasible.
6. **Alternative D - Capitol Center/FSU/FAMU - Central Cogeneration Plant - Preferred Alternative.**
   There were three sub-alternatives for this option, which involved serving all three sites. The equipment for the preferred sub-alternative (D-0) included: four 1500 Ton electric chillers, and one 1200 Ton double effect absorption chiller at the new cogeneration plant; two 1200 Ton double effect absorption chillers at FSU; and one 1200 Ton double effect absorption chillers at FAMU. There would be four 3750 kW gas turbines. The maximum steam to FSU plus FAMU was set at 75,000 PPH. This allowed the equipment to supply the maximum thermal requirements which again reduced the economic payback. This alternative is also feasible and was selected as the preferred alternative.

B. Alternative Considered - Base Case vs. Preferred Alternative Only.

1. **General**
   All alternatives for the new cogeneration plant were specified for a 30-year economic life. All replacement costs were determined based on recommended major overhauls or complete replacements.

2. **Base Case: Replace Existing Chillers In-Kind**
The base case scenario assumes a "business as usual" situation. Each of the Capitol Center buildings would continue to be served by individual electric chillers and gas/oil fired boilers. The central steam system from the Heating Plant would continue to serve the steam/hot water requirements of eight buildings. The electrical requirements for the Capitol Center include buildings served by the State owned distribution system plus buildings served directly by the City of Tallahassee.

3. **Alternative D: Capitol Center Central Cooling Plant with Cogeneration and Steam Generation for FSU and FAMU**
The selected alternative embodies the following components:

a. **A New Central Cooling Plant.**
The central cooling plant would contain four to five electric-drive, high-efficiency, centrifugal chillers. A new direct-buried primary/secondary chilled water distribution system would be installed to provide chilled water to all of the Capitol Center buildings, including the "Distant Buildings": the FSU Law School, Law Library, Center for Professional Development, and the Tallahassee-Leon County Civic Center.

b. **A Central Cogeneration Plant.**
The Capitol Center cogeneration plant would consist of four 3,750 kW gas-fired turbine generators with waste heat boilers for the production of steam. A steam-absorption chiller would be utilized in addition to four electrical chillers to generate chilled water for use in the Capitol Center through the new distribution system outlined above. During Phase II additional absorption chillers would be modelled for an optimum solution.

c. **Steam Turbines and Supplemental Firing.**
In sub-alternative D-1, one of the gas turbines is equipped with supplemental firing to help drive two 1,200 kW steam turbines, permitting all-electrical chillers at Capitol Center. Life Cycle Analysis showed this additional equipment actually resulted in a longer payback than the preferred D-0 alternative.

d. **Steam Distribution to FSU & FAMU.**
A new steam supply and condensate return system would be installed, in Alternative D, from the cogeneration plant to the central boiler/chiller plants on FSU and FAMU campuses.

The generators in this alternative are configured to supply, from Waste Heat Recovery alone, the pre-determined load of 75,000 PPH of steam for FSU+FAMU, plus Capitol Center's heating requirements, plus one absorption chiller at the Capitol center.

This results in a surplus of electrical energy available to the State for use at FSU and FAMU.

C. **Selection Analysis**

1. **Energy Analysis**
The various alternatives studied involved the following datasets (combinations of individual load profiles), which were used in the cogeneration modeling program called PC-CUBE as synthetic profiles.

a. **Electrical**
The study assumes that the only Electrical demand directly served by the cogeneration plant is that of the Capitol Complex, not including the Distant Buildings.

The assumption is also made that excess electrical energy generated by the plant is fed into the City electrical grid to FSU/FAMU for credit. The credit is equivalent to the applicable energy rate ONLY, with no demand component. If the City rejects this approach, the State would install electrical feeders to FSU and FAMU.

b. **Chilled Water**
The modeled chilled water load for the Central Plant includes the composite chilled water load from the 21 Capitol Center buildings, plus the composite cooling load from the distant building load.

The modeled Chilled Water loads for the FSU and FAMU central plants were converted to steam equivalents (at 12 lbs of 150 psig steam per Ton of cooling), enabling these loads to be combined with the Hot Water/Steam loads.

c. **Hot Water/Steam**
For Hot Water/Steam, the Capitol Center is always modeled using the buildings which are to be served by the Central Boiler Plant. It is assumed that the other buildings in the Capitol Center will continue with their current individual heating configurations.

The HW/Steam loads of FSU and FAMU were combined with the steam-equivalent Chilled Water loads, to give an overall steam demand for those sites. It was then possible to model a steady supply of steam that met the cooling and heating needs of all sites.

2. **Equipment Sizing Philosophy**
Specific equipment choices were selected for modeling in PC-CUBE.

a. **Electrical**
All electrical generation is supplied by multiples of the SOLAR Centaur "H" Gas Turbine, at times with supplemental firing of the waste heat resource.

b. **Hot Water/Steam Backup**
The pair of existing boilers at the Central Boiler Plant is available in all alternatives as part of the backup to the new plant.

c. **Chilled Water**
Two types of chillers were modeled: electrical driven centrifugal, and steam driven absorption. All chillers were modeled with built-in electrical auxiliaries for chilled water distribution requirements.

3. **Cogeneration Plant Modeling and Analysis**
This section summarizes the configuration for the selected "D" alternative.

a. **Alternative D-0:**
(4) Electrical Chillers + (1) Absorber at Capitol Center, 75,000 PPH Steam to FSU+FAMU.

In this alternative, four Gas Turbines supply electricity to meet all Capitol Center needs (including four 1500T electrical chillers) and generate Waste Heat Steam to power one additional 1,200T steam absorption chiller at Capitol Center, and a pre-determined steam flow of 50,000 PPH to FSU, and 25,000 PPH to FAMU.

FUELS

The types of fuels available for use at the site were analyzed. Natural gas and fuel oil were considered as the primary energy source of the plant. It appears that the best fuel for use in the plant would be natural gas. It is available within the Capitol Center; it has the lowest cost, and is readily available from the City. Its' clean burning characteristics reduce the maintenance and environmental permitting problems. Interruptible gas service offers additional cost savings. No. 2 Fuel Oil should be considered as the backup energy source. No. 2 Fuel Oil is a relatively clean fuel, is readily available, and is easy to store. The greater cost of this fuel is offset by the low expected usage rate due to the demonstrated (10 year) reliability of the natural gas service in the Tallahassee area. We have assumed that a 10 day fuel oil storage supply will be installed.

ENERGY LOADS - SUMMARY

Summaries of peak and minimum load demands for each of the energy quantities for each site are presented below in Table A.

TABLE A

ENERGY DEMANDS	CAPITOL CENTER	FSU	FAMU
ELECTRICAL DEMANDS			
MAXIMUM	13,471 kW	–	–
MINIMUM	4,450 kW	–	–
HOT WATER/STEAM DEMAND			
MAXIMUM	21,700 #/hr	72,000 #/hr	30,000 #/hr
MINIMUM	900 #/hr	10,100 #/hr	12,300 #/hr
CHILLED WATER DEMAND			
MAXIMUM	6,200 Tons	4,800 Tons	2,000 Tons
MINIMUM	670 Tons	1,000 Tons	500 Tons

CAPITAL COST ESTIMATES

The cost estimating approach was straightforward. The estimates were developed using Means Cost Data, historical figures, recent bids within the Capitol Center for a similar gas turbine, and using Sunbelt's knowledge of the labor and material costs for northern Florida. These cost estimates are only "study" quality. Once a specific concept is selected then a plus or minus 15% cost estimate will be developed during Phase II. The total capital cost varied from $14,250,000 to $51,750,000. The gas turbine cost was within plus or minus 5% since Sunbelt had just bid a similar gas turbine for Baptist Medical Center in Jacksonville.

The contingency for each alternative varied between 15% to 25%. The lower percentage was used on the large capital investment while the higher percentage was used on the lower investments. The contingencies were varied during the analysis, but did not significantly increase or decrease the payback for the alternatives.

<u>It was assumed that the State would finance this plant as a normal budget appropriation.</u> Financing cost and interest during construction were not included.

LIFE CYCLE COST ANALYSIS

A Life Cycle Costing program developed by PLM was used throughout the analysis. Significant effort was expended in performing a sensitivity analysis of the key economic parameters to determine the impact on the economic viability of each alternative.

Escalation rates for the energy categories follow the average values used for these factors. For electrical energy 3%, 4%, and 5.7% escalation rates were used in order to see their effect on the payback. The conservative value of 4% was selected for the summary figures.

The payback varied from just over 3 years to more than 30 years, using the annualized method, and from 6.5 years to over 30 years using the actual cash flow method. The district cooling plant concept had the longest payback and was eliminated from further consideration. The cogeneration alternatives for the Capitol Center have acceptable paybacks and, if the Universities elect not to participate, then this option is still an economical choice. However, the FSU and FSU/FAMU alternatives all have very similar paybacks and either one or both could participate without significantly impacting the payback of the project. The shorter payback for these alternatives basically results from offsetting higher-cost electrical energy with lower-cost natural gas, and sizing the plant to supply almost all of the base energy load for the Capitol Center, FSU, and FAMU.

The State has established 15 years as the maximum payback for this type of project. All of the cogeneration alternatives meet this criteria. The State should use both the annualized and actual payback methods when it is making a decision on this project. Tables B and C below provide a summary of parameters and paybacks.

TABLE B
LIFE CYCLE COST ANALYSIS
Economic Parameters Summary

Yearly Discount Rate (%)	8.2%
Number of Alternatives Analyzed (including Base Cases):	15
Economic Life (years):	30

Energy Source	Current Cost ($/MMBtu)	Yearly Rate
Electricity	$20.22	4.00%
EL Credit	$16.69	4.00%
Natural Gas	$ 2.75	3.00%

TABLE C
LIFE CYCLE COST ANALYSIS
Summary of Payback - Preferred Alternative

Payback Options	EL%	D-0
Annualized	3.0 4.0 5.7	5.9 4.6 3.1
Cash Flow (Actual)	3.0 4.0 5.7	7.7 6.9 6.5
Special Cases (all on annualized basis)		
No Demand Side Rebate	--	4.8
No EL Credit for excess back to utility	--	8.9
Worst Case Scenario (see box)	--	13.4

DISTRIBUTION SYSTEMS

The distribution systems evaluated for in this study are a high capital cost. However, if designed and installed by a qualified contractor, these systems will provide a long life. Proper maintenance is also required.

The Central Heating Plant distribution system located within the Capitol Center will continue to be a part of all alternatives. The Plant will provide backup steam service whenever the new plant needs the steam for base load or under planned maintenance or emergency shutdowns. This system was not expanded for purposes of this study. All of the cogeneration alternatives include a new direct buried chilled water distribution system for the Capitol Center. The direct buried approach is recommended and will reduce the initial capital cost. All of the mechanical rooms in the Capitol Center buildings would be modified for a centralized distribution system. These costs were included in the life cycle costing analysis. The alternatives which included cogeneration also included a new lift-top tunnel system from the plant site locations to the main steam line outside the Heating Plant. This connection will allow the new plant to pickup the steam load which was being served by the new natural gas-fired boilers.

The alternatives which include FSU and FAMU have new steam lines in lift-top tunnels from the plant site to the central plant at each campus. These lines will provide the base steam load at each site and additional steam which will be used for the production of chilled water via new absorption chillers at each campus. The new tunnels will be similar to the tunnels recently installed in the Capitol Center.

SITING

Although there were several sites considered during the initial stages of the study only four remain based on the evaluation to date. The effort to date can be classified as preliminary at best. Three of the locations are on the extreme south boundary of the area referred to as the Capitol Center while the fourth is on the property just east of the City's substation number 6.

There may be other locations, but the limitations of this study restricted the effort to basic evaluations. Some proposed sites were eliminated due to soil contamination while others were within restricted areas based on the recently-published State/City Master Plan. A hard siting effort will occur during the initial stages of Phase II. The basic building varies from 17,000 square feet to 35,000 square feet, depending upon the alternative.

PURCHASED UTILITIES

Currently the City does not have any cogeneration plants connected to its system. Therefore, extended negotiations will be necessary with the City. The City is actively pursuing new baseload generation capacity for the 1996 thru 2010 timeframe. This new plant could be on-line by 1996 if the decision to proceed were given during fourth quarter of 1993.

The new cogeneration plant will significantly change the mixture of energy purchases from the City. The plant will be a base load plant which will use large quantities of natural gas but minimum amounts of electrical energy will be purchased for the Capitol Center. In addition, most of the alternatives produce excess electrical energy which can be used to offset purchases at FSU and FAMU. The rates for the energy produced at the plant will be determined as part of the Phase II effort. There are several methodologies which can be used.

Assuming that the State owns/operates this plant, new electric, natural gas, and cogeneration agreements must be negotiated with the City. If the State elects to have the City own and operate the new plant, then the savings will be lower because the State would not have to operate this plant. The economical, political, and other factors needed to make this decision must be carefully considered by the State.

Presently, the City's position is unknown. They have evaluated the plant from a "Utility Perception" and found it uneconomical. However, they are currently evaluating the economics based on a State/City partnership.

OPERATIONS AND MAINTENANCE

1. **General**
   Cogeneration Plants require a qualified staff and an annual operating budget to operate and maintain the equipment and building. The magnitude of each will vary depending on the type and size of the plant. Considering the size of this proposed plant and the facilities which it will serve, it is assumed that an unmanned plant is not desirable due to the importance of steam and chilled water throughout the sites, and the need for a rapid response in case of failure of any system. Only continuously manned plants have been considered. The total O&M Budget for the district cooling/steam plant referred to as Alternative A is approximately $600,000 per year. The preferred alternative budget is $1,500,000.

   It is assumed that major repairs will be accomplished by an annual maintenance contract, or by contracting out the work on an as needed basis. The recommended staff was for operation of the plant on a daily basis and performance of the necessary routine maintenance.
2. **O&M Budget**
   The operating and maintenance costs have been estimated based on a similar plant in the Jacksonville, Florida area. Staff salaries and benefits were estimated along with the maintenance costs.
3. **Service Contract (Gas Turbine Alternatives Only)**
   The utilization of a service contract arrangement in connection with the new Cogeneration Plant is recommended if the State operates the Plant. A five year service contract with a five year option (if possible) will provide the ability for the State to ensure a reliable operation and provide time to train and license their own operators. If the training of in-house personnel does not work out, the State can exercise the five year option and ensure continued operation. The intent was to provide for the safe operation of a multiple gas turbine Cogeneration Plant which must be staffed with highly qualified personnel.

ENVIRONMENTAL PERMITTING

1. **Overview**
   The following discussion addresses the types of

environmental permits required and provides some insight into the specifics of what may be required, based upon conversations with regulatory agency representatives. This discussion is presented in two subsections: analysis of permitting requirements and anticipated permit schedule.

2. **Analysis of Permitting Requirements**

a. **General**

Agencies that could be involved in the permitting process, include the Public Service Commission (PSC), Florida Department of Environmental Regulation (FDER), Florida Department of Transportation (FDOT), and local County/City permitting agencies, including County/City Road Departments (where off-site utility construction in public rights-of-way is undertaken). PSC involvement would be restricted to the provision of electric and gas service to the new Cogeneration plant. The PSC would also be involved should the State go into a program of Cogeneration and make excess power available beyond its own uses.

FDOT involvement would occur for any gas transmission facilities. However, if the supplier of the gas is responsible for installation of the gas transmission lines to the site, then they would assume all responsibility for appropriate permits and regulatory requirements. Similarly, work within FDOT, County or City rights-of-way will require road use/utility permits and would apply to off-site work for piping distribution systems.

b. **Plant Siting:**

Whatever site is selected, it will trigger appropriate water and sewer main connections, relocations and/or removals as needed. It will be very important to conduct adequate site planning and field verification of all utility sizes, locations and depths early in the design to minimize the necessity of relocations, and to plan for maintaining utility services during construction. Plant siting will also require review by a number of governmental agencies, including the FDER, U.S. Environmental Protection Agency, and the City of Tallahassee.

3. **Anticipated Permit Schedule**

a. The timetables for preparation of the permit application forms and acquiring the approved permit vary substantially from a matter of weeks to an estimated six (6) months or more for each permit.

b. **Summary**

In summary, the anticipated permit schedule is estimated at 12 months for the new Plant. As pointed out in the discussion, most, if not all, of the 12 months would be occupied by normal design and construction activities. Forthcoming changes in the permitting process do not appear to significantly alter the time factors discussed in the above.

EXECUTIVE SUMMARY

A. Summary

Observations

1. The concept of having a single cogeneration or energy plant serving multiple complexes is not new.
2. The State should maintain the option of owning and operating this plant regardless of whether the City or a third party would be the owner/operator of the plant.
3. The various State agencies, as well as FSU and FAMU, have significant "joint venture" issues to resolve once a decision to proceed has been made.
4. The plant can be a showcase/training facility for both elementary, middle school, and senior high school, as well as for the FSU/FAMU engineering school.
5. The quality of the load data at each facility was not as good as was desired, but adequate for a "feasibility" study effort.
6. FSU and FAMU have a base steam load which can be served by the new plant. They also have a base cooling load which could be served by new absorption chillers located on each campus.
7. The City is seeking electrical generation capacity during the late 1990's and early 2000's. A portion of that capacity could come from this new plant.
8. A major issue of the State concerns the involvement of the City. Will the City choose to participate as owner/operator, operator, or not at all?

Conclusions

1. Cogeneration is feasible for the Capitol Center with or without the FSU and/or FAMU loads.
2. The new plant should serve all three facility "base" loads. However, the total load at each site should be factored into the Cogeneration Plant service to optimize the economics of the project.
3. The majority of the cooling loads for all sites occurs during a 10 hour weekday period (50 hours per week), not a 24 hour weekday period or throughout the weekend.
4. As the result of a variable cooling load and the high cost of the Capitol Center chilled water distribution system, a new cooling-only Capitol Center central plant is not economical. However,

a new chilled water distribution system can be economically justified when combined with the cogeneration plant concept.

5. The new cogeneration plant steam distribution system would tie into the existing Capitol Center steam system.
6. The Capitol Center steam distribution system would not be extended under the Cogeneration project.
7. Thermal energy to FSU and FAMU campuses should be supplied in the form of steam. Chilled water production would occur at each site via absorption chillers in the respective plants.
8. Electrical energy generation should be maximized by adding thermal loads, due to the relatively high cost of electrical energy purchases. The significant operating cost savings from the new cogeneration plant result from the use of the cogeneration concept plus utilizing natural gas at $2.75/MMbtu versus electrical energy at over $20.22/MMbtu.
9. The concepts in this report minimize the offsite electrical purchases and maximize the on-site generation by picking up the maximum of thermal load at the three locations. There is excess electrical energy in most all of the alternatives.
10. The general siting constraints restrict the plant sites to the southern portion of the Capitol Center area, but outside the main campus area.
11. A new steam distribution system should be installed to FSU and FAMU using a lift-top tunnel concept.
12. The chilled water distribution system should be direct buried and not be placed in common steam tunnels.
13. The Tallahassee Civic Center load is very minimal except for the large events which occur several times per month. The new plant can serve this load, but it was not a major factor in sizing the plant since it is an off-peak load for most workdays.
14. The new plant should be developed with multiple units in lieu of a single large gas turbine which would present serious reliability and redundancy problems when considering the critical nature of the Capitol Center loads.

Recommendations

1. The State should proceed with all Phases of this Project.
2. The State should determine the City's plans for participation as soon as possible.
3. The Universities should participate.
4. A task force should be developed with teams for policy, technical, marketing, and utility contracts. A project director should be appointed with a single State agency taking the lead.
5. The State should begin negotiations with the City so that a general cogeneration agreement can be developed.
6. The Public Service Commission should be contacted and a brief overview of the project provided. This agency will be involved throughout the project.

B. Recommended Alternative

The bottom line of this feasibility study is that in all basic cogeneration alternatives, the paybacks are much lower than the maximum allowable paybacks of fifteen (15) years. Combining the Capitol Center with FSU and/or FAMU provides the paybacks. These paybacks result from the significant reduction in electrical purchases and from maximizing output on the gas turbines.

There is a significant difference in producing steam and electricity using a cogeneration process with $2.75/MMbtu fuel vs. using very costly electrical energy at $20.22/MMbtu. This difference provides the very basis of the large annual savings, *and its magnitude is the reason the other cost variables have a secondary impact on the viability of the project.*

<u>The study shows that the new cogeneration plant should serve all three campuses and thus maximize the savings.</u> Allocation of these savings will need to be determined; but, in any case, all parties will receive tremendous operational savings. The specific plant equipment selection requires further refinement which will be accomplished during the preliminary concept design effort.

The rationale for selecting the preferred alternative (Alternative D) included the following:

a. Provides "optimal" choice for the State.
b. Lowest overall life cycle cost with reliable operation.
c. Lowest payback, whether measured by annualized or actual methods.
d. Provides for easy maintenance of major components
e. Provides generating capacity to the City.
f. Meets "Demand Side Management' objectives of the City.
g. Meets the State Energy Executive Order.

h. Avoids the replacement of individual building chillers in-kind.
i. Meets the intent of the State/City joint agreement for a cooperative effort in cogeneration projects.
j. Increases energy efficiency.
k. Enhances State control over energy prices.
l. Provides major benefits to the State and City.

Obstacles

The major obstacles in implementing a partnership in cogeneration include:

- Obtaining the support of the City Commission/FSU/FAMU.
- Coordinating facility planning.
- Negotiating with the City.
- Securing the necessary permits and energy agreements with the Regulatory Agencies.
- Appointing a full-time Project Manager.

Chapter 84

Solar Thermal Electric Technologies for Power Generation

D. Kumar, G.D. Burch

ABSTRACT

Significant progress has been made in the last two decades toward generating economically competitive solar thermal energy for electrical applications. Research and development (R&D) efforts in the U.S. during the past decade have improved the performance and reliability of solar thermal systems and reduced both the capital installed costs ($/kW) and energy costs ($/kWh) of solar thermal systems to one-fifth the cost of early systems. This large reduction is a result of lower cost components and increased system efficiency. However, further cost reductions in the 1990s will be required to make solar thermal energy cost-competitive with energy generated by fossil fuels.

The fundamental motivation for developing solar thermal electric technology is to bring this renewable energy source to the marketplace as an economically competitive option for generating electric power. Current efforts in the U.S. represent a collaborative approach among industry, the research and development community, and potential users toward achieving this goal. Solar thermal electric systems are beginning to show potential in the utility marketplace just as the demand for modular, environmentally clean, fossil/fuel/independent power generation is beginning to surge; and they are poised to become dynamic competitors in both on- and off-grid power markets. Today's solar thermal electric systems, making use of parabolic dish/engine, power tower, and parabolic trough technologies, are proving that they can produce substantial amounts of electricity in an environmentally acceptable way.

INTRODUCTION

During the 1970s and early part of the 1980s, few people thought electricity could be produced economically from sunlight. Today, scores of utilities in the U.S. are looking closely at the performance of commercial solar thermal electric power systems, which convert the energy of the sun to heat energy and use this heat energy to generate electricity.

Versatility is a key attribute of solar thermal electric (STE) technology. STE systems are well suited to both grid-connected and off-grid applications, in both national and international markets. The basic requirements for a successful STE application are a sufficient solar resource, available and inexpensive land area for an installation, and a demand for electric power. Hybrid STE systems using another renewable or conventional fuel for use during sunless periods can supply continuous power at attractive life-cycle costs. Solar thermal electric technology has a number of attractive attributes that make it a very desirable energy supply option: it is an unlimited energy source; it has low environmental impact; and it requires short construction time and can be built in modules to respond to demand growth.

BASIC SOLAR THERMAL TECHNOLOGY CONCEPTS

Solar thermal electric power-generation systems convert the energy in sunlight to heat by using concentrators such as mirrors to focus sunlight onto a receiver, where heat is absorbed in a fluid. In larger plants, standard power-generating equipment such as a conventional steam turbine extracts the heat energy form the fluid to generate electricity. In smaller systems, a stand-alone engine/generator produces the electricity.

The main difference between a solar thermal electric power plant and a conventional power plant is that a solar energy collection system is used instead of a conventional fuel-supply system. The two types of systems are so similar, in fact, that hybrid designs have been constructed that can use heat energy input from either sunlight or fossil fuels. Not so similar, however is the manner in which the plants are operated. Solar power plants, due to the fluctuating nature of solar insolation, operate on a far greater cyclical basis than most conventionally fueled systems. Special attention must therefore be given to choosing components and operation strategies that minimize the negative effects of thermal cycling.

Solar thermal power systems are usually characterized by the type of concentrator they use (Figure 1). Concentrators focus sunlight onto a receiver in one of three architectures:

- Line-focus systems concentrate sunlight onto tubes running along a focal line (in a parabolically-shaped reflective trough). Troughs concentrate the sun up to 100 times onto a fluid-filled receiver tube positioned along the line of focus in the trough. Heat can be produced efficiently up to 400°C (750°F) and used as heat or to generate electricity.

- Large point-focus power-tower systems use fields of heliostats -- very large sun-tracking reflectors, each with an area of about 100 m^2 -- to concentrate sunlight at a focal point on top of a tower. The sun heats a fluid in the receiver typically to temperatures up to 650°C (1,200°F). The heated fluid is converted to steam that drives a turbine to produce electric power.

- Small point-focus systems use parabolically-shaped dishes that reflect light onto an individual receiver mounted at the focal point of that dish. The solar energy ultimately heats a fluid powering a small engine/generator mounted at the focal point of the dish (the dish/engine concept). Operating at about 800°C (1500°F), a single dish module can generate up to 50 kilowatts of electric power. Like trough systems, many dishes can be grouped together to produce more power.

The type of concentrator/receiver employed not only has a great effect on the system's performance, but it also affects the size of the system. Line concentrators predominate in today's commercial plants; each is about 80 MW_e in size. However, point concentrator systems, dish concentrators and power towers, can achieve higher energy conversion efficiencies than can line concentrators because they can operate at higher temperatures. When commercial-scale power tower systems are built (probably by the year 2000), these modular solar thermal power systems will likely reach sizes of 100 to 200 MW_e.

By contrast, dish/engine systems are much smaller, typically about 5 to 25 kW_e. At this size, they are ideal for remote, stand-alone applications. The size of individual dish/engine

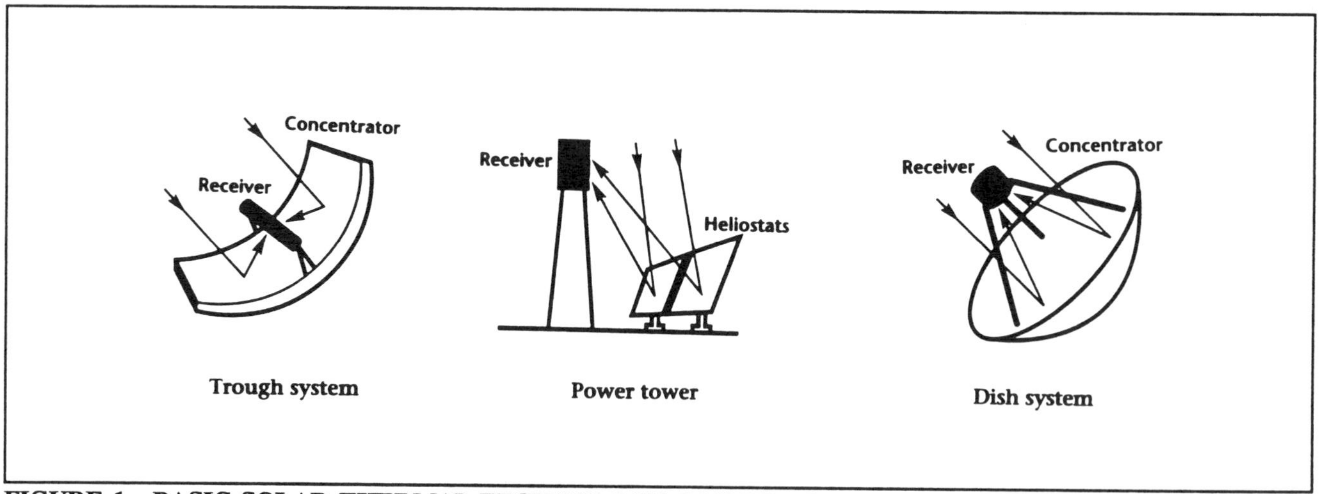

FIGURE 1. BASIC SOLAR THERMAL TECHNOLOGY CONCEPTS

systems is limited by the amount of energy that one dish concentrator can collect. Commercial parabolic dish/Stirling engine units, currently under development through a joint venture between DOE and the Cummins Engine Company, are expected to have extremely low maintenance requirements and are ideal to operate unattended in remote settings. The 7.5 kW_e units now being tested are intended to compete with the diesel genset market. Pre-production versions of the 7.5 kW_e units are being tested in water pumping and DSM applications. Hybrid power systems with diesels, wind, small hydro, PV, or other complementary power source are also anticipated. A 25 kW_e dish/engine joint venture project is under negotiation by DOE. These units are expected to find applications which range form remote power for isolated sites to bulk power serving an integrated utility grid. To meet larger power requirements, several dish/engine systems can be connected together electrically to form power systems in sizes ranging from 100 kW_e to 100 MW_e.

PROGRESS TO-DATE

Over the past decade, the combined efforts of the Federal STE Program, private industry, and research organizations have proven the technical viability of solar thermal systems for electric power generation. Thus far, the nine parabolic trough solar electric generating systems (SEGS) installed in California by the private sector represent over 90 percent of the world's installed solar electric capacity. The cost of electricity from line-concentrator solar thermal power plants decreased by 83% in less than a decade. This manufacturers taking advantage of economies of scale in production.

With over 350 MW_e of installed capacity, parabolic trough power plants are a proven technology for producing reliable electric power. Installed costs of the most recent parabolic trough plants are around \$3,100/kW_e for 80 MW_e solar/gas hybrid plants. With an annual capacity factor of 35%, the levelized energy cost (LEC) of current generation trough electric plants is around 13 cents/kWh (Figure 2).

For summer-peaking utilities, these generation costs can be attractive today even without factoring in the environmental benefits of solar power. Further commercial development of the technology combined with larger plant sizes could eventually reduce the capital cost of the technology to around \$2,200/kW_e and the LEC to 10 cents/kWh.

Power tower systems have been successfully demonstrated by the 10 MW_e Solar One pilot plant located near Barstow, California. The STE Program is continuing development of power tower technology by retrofitting the Solar One Plant. This retrofitted plant, known as Solar Two, will also be rated at 10 MW_e, replacing the water/steam system with a molten salt heat transfer fluid. The molten salt system, which includes thermal storage, will increase efficiency by permitting higher temperature operation (1050°F). It will also permit operation for up to four hours after sundown, leading to capacity factors of greater than 60%.

Power tower technology is currently not as commercially developed as trough technology, but is expected to eventually be available as the technology of choice for utility electricity production. Industry design studies have projected that the first commercial power tower plants, which should be ready near the end of the decade, will have a capital cost of

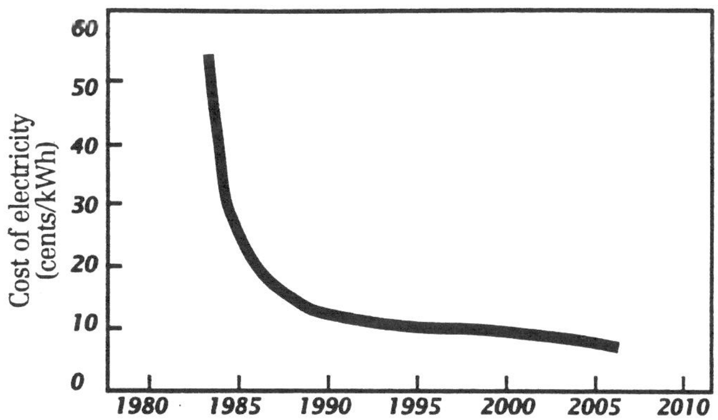

Source: The Potential of Renewable Energy: An Interlaboratory White Paper, 1990.

FIGURE 2. SOLAR THERMAL COST OF ENERGY

\$3,000/kW$_e$ and a capacity factor of 40%. These plants, which will use energy storage rather than natural gas hybridization, will have an LEC of about 11 cents/kWh. Costs for power tower systems should drop substantially over the first few plants as design and manufacturing experience builds, allowing scale-up of plant sizes to 200 MW$_e$ and reduction of the costs of solar components. Industry estimates are that power tower plants built early in the next century can expect to have capital costs around \$3,000/kW$_e$ for plants with a 63% capacity factor and LEC of 7 cents/kWh.

Dish/Stirling systems will enter the market differently than did either troughs or power towers. With their small module size (5-25 kW$_e$) and potential for unattended operation, the initial markets for dish/Stirling systems are likely to be small, remote applications where the value of power is quite high. The STE Program, in cooperation with industry, is presently focusing on the development of a 7.5 kW$_e$ dish/Stirling system over the next two to three years, and a 25 kW$_e$ dish/engine system over the next five years. The early commercial systems that will be available before the end of the decade are likely to have capital costs near \$3,000/kW$_e$ initially. The first dish/Stirling systems will have capacity factors around 23% in nonhybrid designs and an LEC of around 18 cents/kWh. Increasing the module size to 25 kW$_e$ and designing distributed utility (1-10 MW$_e$) installations are expected to allow further reductions in cost. Coupled with efficiency improvements in the receiver and engine, grid-connected dish/Stirling systems are expected to eventually achieve capital costs of \$1,500/kW$_e$ for systems with capacity factors augmented by the ability to operate on both solar and fossil fuels. These long-term systems will achieve an LEC of 6 cents/kWh.

FEDERAL SOLAR THERMAL ELECTRIC PROGRAM

The role of the Federal Solar Thermal Electric Program is centered on the development of improved cost effectiveness and reliability of solar thermal electric components and development of additional energy markets with high strategic or economic value to U.S. industry. The program is supporting the development of solar thermal electric commercial applications to increase acceptance of this technology for generating cost-competitive power by utilities, both in the U.S. and abroad. By aggressively supporting development of the technology's industrial base through cost-shared, joint-venture projects with the private sector, the program seeks to help solar thermal electric technology penetrate energy applications and markets, creating new jobs and business opportunities.

Components that are being investigated include stretched-membrane heliostats with features such as a front-surface or sol-gel reflector. Stretched-membrane technology is also being developed for parabolic dishes. The solar concentrator is the most expensive component in any solar thermal system, with costs ranging rom 40% to 50% of the total system cost. The reliability of the concentrator has a strong impact on overall performance of the system. Inexpensive and readily maintained solar concentrators are essential for the successful commercialization of solar thermal technologies, and are an integral part of the STE Program.

The program is also pursuing the development of power conversion systems for both dish/engine systems and power tower systems. Power tower receiver development is focusing on advanced salt-in-tube receivers, molten salt film receivers, and volumetric air receivers. Dish receiver development, particularly of the reflux type, is critical to the long-life reliable operation of parabolic dish/Stirling engine systems. The heart of a solar thermal dish/engine system is the subsystem that converts thermal energy into electricity: the engine generator. Power conversion development and technology validation efforts synchronize R&D activities with the needs of users, expanding the availability of resource data and improving system performance.

SOLAR THERMAL ELECTRIC PROGRAM THRUST FOR THE 1990S.

From a market perspective, from an environmental viewpoint, and considering the status of the technology development, solar thermal electric systems are poised to become dynamic competitors in both on- and off-grid power markets. The technology is entering the marketplace just as the national and international demand for modular, environmentally clean, fossil/fuel/independent power generation is beginning to surge.

The decade of the 1990s will see progressive advances of U.S.-designed and -built solar thermal electric systems moving into the marketplace. Building on past successes in the areas of parabolic troughs, power towers, and parabolic dishes, and viewing past problems as hard-won lessons, the technology will continue moving forward:

- Power tower systems will shift from the demonstration stage to modular bulk power production. The Solar Two molten salt power tower project led by Southern California Edison will provide the technical base for Solar 100, the first 100-MW$_e$ utility-scale power tower module, which is due for installation by the end of the decade.

- Trough systems, which have already made an aggressive market entry with more than 350 MW$_e$ connected to the southern California grid, will continue their validation of solar thermal electric technology. The operations and maintenance cost reduction study for parabolic trough plants will be completed by 1995, thereby providing for lower levelized energy costs for power tower and dish/engine solar systems as well as trough plants.

- Dish/electric systems will make their market debut in both remote and grid-connected applications. The Cummins Engine Company 7.5 kW_e dish/Stirling system will be commercially available by 1996 for remote and grid-connected applications. Contracts will be awarded under the Utility-Scale Joint-Venture Program for 25 kW_e dish/engine systems, with the last phase of this program resulting in at least one megawatt of dish/engine system capacity installed by utilities by the late 1990s.

SOLAR THERMAL ELECTRIC INDUSTRY THRUST

The solar thermal electric industry will coalesce into a cohesive, competitive market force with a product mix ranging in capacity from a few kilowatts to hundreds of megawatts. The national and international market for solar thermal electric technology includes utilities, independent power producers, and off-grid applications.

The modern solar thermal electric industry began to be distinguishable from the broader solar industry in the mid-1970s. Even so, the industry is still in the formative stages; and in many ways, it is still evolving. As the technology moves further into the marketplace, expansion of the industry is expected to continue both in membership and in the level of commitment of its members to the technology.

By the early part of the coming decade, individual solar thermal electric power tower systems could be competitive in utility-scale applications in the range of 100 to 200 megawatts. Individual parabolic dish/Stirling engine units (5 to 25 kW_e range) could be available on a widespread basis for smaller-scale applications, either in place of utility line extensions, in remote environments, or in clusters for larger power requirements. Multi-megawatt parabolic trough solar electric systems such as the SEGS units will be operating on a commercial scale in a utility environment in the western part of the U.S.

Of the 100 gigawatts of capacity projected to be required by the U.S. over the next decade, approximately 20 gigawatts are not yet committed for construction. Construction and permitting lead times for large fossil and nuclear power plants almost eliminate them from the list of candidates for filling this capacity gap. Environmental pressures worldwide will also favor the use of renewable energy technologies. Among various renewable energy power generation options, solar thermal electric technology is an excellent candidate for filling this capacity gap because it features the following attributes:

- Short Construction Lead Times—Installation times of 12 to 18 months have been projected for solar thermal electric systems; the most recent SEGS plant was installed in less than 8 months. Individual dish systems can be installed in less time still, just in a matter of weeks.

- Less Onerous Regulatory Hurdles—The environmental effects resulting from the installation and operation of solar thermal electric systems are small, especially in comparison to conventional nuclear and fossil-fueled plants.

- Modularity—The economical size of solar thermal electric systems ranges from 5 kilowatts to 100 megawatts or greater. Systems will be available in economical sizes that can be combined to serve virtually any power requirement—from remote villages to grid-connected high-capacity factor applications.

- Known Cost of Energy—At any given site, costs for energy produced by solar thermal electric systems are a function of capital costs required for initial installation as well as the costs of operation and maintenance personnel and equipment. For solar-only systems, energy costs are virtually independent of the sometimes volatile costs of fossil fuels and can be predicted with accuracy throughout the life of the plant.

- Potential for Dispatchability—Utilities and other users value the ability to bring a power plant on-line with minimal delay when it is needed. The dispatchability of co-fired solar thermal electric plants has been clearly demonstrated by the SEGS plants. All other types of solar thermal systems can also be hybridized. Thermal storage can accomplish the same effect without the use of fossil fuels or the creation of emissions.

Efforts in the late 1990s will continue to focus on dissemination of available evaluation tools that accurately reflect the true strengths of solar thermal electric technology. The basis for this effort stems from the unique characteristics of solar thermal electric technology when compared to conventional power generating options, as shown in Figure 3. First, it has a higher up-front capital cost per unit than conventional options, but a much lower operating cost because it uses a free-fuel source. Thus, solar thermal electric and other solar technologies are not subject to fuel price increases

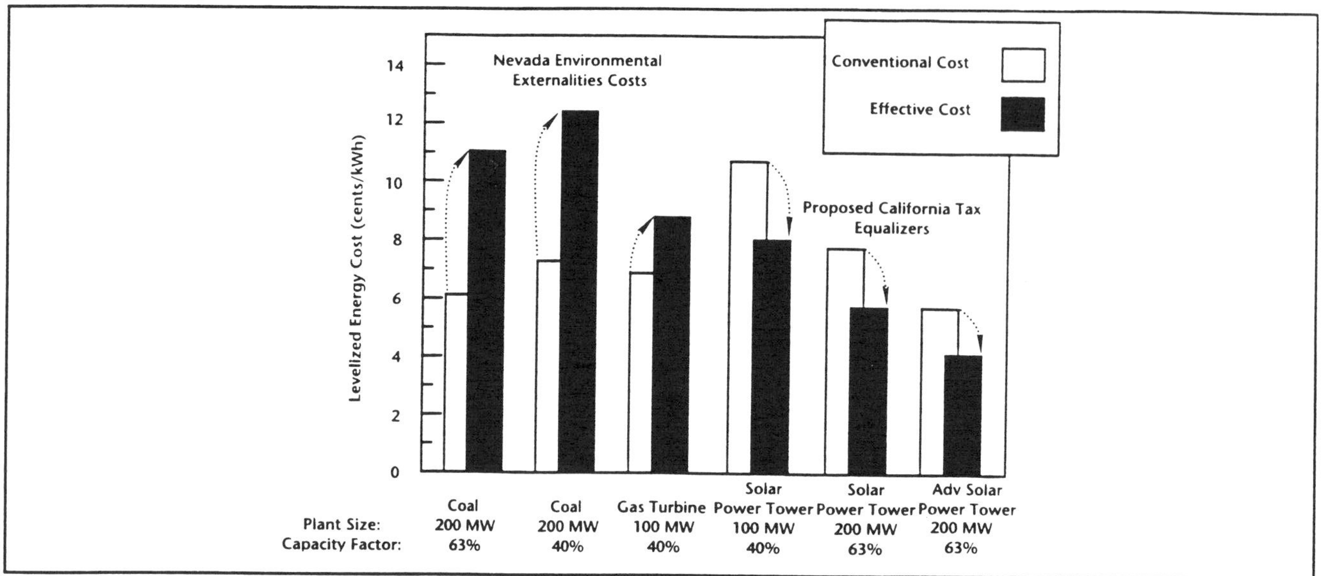

FIGURE 3. FOSSIL VS. SOLAR ENERGY COSTS: THE IMPACT ON ENVIRONMENTAL AND TAX EQUALIZERS

and escalation rates as are fossil-based options. Second, STE systems are modular in nature, and thus can be brought on-line in small increments by directly meeting customer needs, while reducing construction costs, lead times, and regulatory hurdles. Lastly, the technologies do not produce harmful emissions, and thus provide a solution for decision makers seeking ways to meet power requirements in an environmentally sound manner.

SOLAR THERMAL ELECTRIC SUMMARY

Solar thermal electric technology is poised to enter the commercial power generation marketplace. Some systems already are available to meet the needs of a broad range of users from remote power applications of just a few kilowatts to utility-scale applications of hundreds of megawatts. The market potential for this technology is substantial. The combined new capacity requirements for all domestic and international power markets is estimated at 600 gigawatts during the 1990s.

Current realities in the power market, such as increasingly stringent environmental regulations, smaller incremental capacity requirements, short planning horizons, and uncertainty in costs of conventional energy due to fluctuating fuel costs, have created favorable conditions for the rapid deployment of solar thermal electric technology. Remote power users, who have been straining under the steadily increasing cost of diesel fuel, are ready for a clean, renewable alterative. While not a panacea for all of the nation's energy problems, solar thermal electric technology has a formidable potential for filling a critical need for electric capacity requirements in the 1990s.

REFERENCES

1. Department of Energy, *Solar Thermal Electric Five-Year (FY 1993-97) Program Plan*, April 1993.

2. Department of Energy, *DOE's Solar Thermal Electric Program* brochure, April 1993.

3. Department of Energy, Energy Information Agency (EIA), *Renewable Resources in the U.S. Electricity Supply*, February 1993.

4. Department of Energy, Energy Information Agency (EIA), *Annual Energy Outlook 1993 with Projections to 2010*, January 1993.

5. Department of Energy, *Solar 2000: A Collaborative Strategy, Volume I and II*, February 1992.

6. Department of Energy, *Solar Thermal Electric and Biomass Power Program Overviews (Fiscal Year 1990-91)*, June 1992.

Chapter 85

Biomass Energy Generation—Trends and Outlook in California

J.H. Green

ABSTRACT

Among renewable energy resources, biomass is one of the most promising, with the potential for providing electricity and gaseous and liquid fuels that can compete with conventional energy sources in large-scale applications. The production of biomass for energy purposes can also offer environmental benefits. The most notable is the potential for providing energy with little or no net buildup of carbon dioxide in the atmosphere, if the biomass is produced renewably. Biomass also has the potential to help revitalize the rural sector of the economy. A domestic natural resource, biomass can be grown and harvested. This requires labor. The biomass power industry can create jobs in harvesting and transporting biomass and in the related industries of fertilizers, pesticides, and agricultural equipment. In the future, biomass facilities will be larger and more efficient and as such an important alternative for energy producers.

Due to the availability of forestry and agricultural residues in California, the active encouragement of the California Energy Commission, and past favorable power purchase contracts, a substantial number of biomass projects have been implemented in California. This paper summarizes the key factors related to the use of biomass as a fuel source and lessons learned from past projects, as well as examining the trends and outlook for biomass energy generation in California.

INTRODUCTION

Through the end of the 1990s, California is expecting to add, depending on the marketing projections selected, anywhere from 6,000 to 11,000 MWe of power. A large percentage of this will be new co-generation/small power or repowering opportunities. This market for new power will be emissions-driven. Any new power development must therefore contend with lower emissions profiles than presently in place. This thinking must be factored into the economics and development strategy of any new California energy sources, including the availability of offsets. Technology is in place now, however, to address the future emissions-driven market.

Among the viable contenders for new generation sources is biomass energy. Biomass to energy already accounts for 6 GWe of utility generation and 3 GWe of private power generation in the United States. By comparison, coal-fired electric utility units in the U.S. accounted for 297 GWe capacity, approximately 43% of the total U.S. generating capacity. The resources for fueling biomass power stations could evolve from current waste sources to dedicated high-productivity energy crops specifically developed for fuel applications by the mid-1990s. The end result would be a new, widely available, low-sulfur renewable fuel for utility and independent power production, and a potentially profitable new crop for U.S. agriculture. Biomass fuels contain very little sulfur and, on a production-to-power generation basis, generate no net carbon dioxide (CO_2). With emissions of sulfur dioxide (SO_2) now strictly controlled by law and carbon dioxide being monitored for possible future control, biomass fuels provide an important option for energy producers.

BACKGROUND

There has been a recent and rapid growth of biomass combustion industry in California in direct response to the Public Utilities Regulatory Policies Act (PURPA) of 1978 which guaranteed small electricity producers that utilities would purchase their surplus electricity at a price based on the utilities' avoided cost of producing electricity. This encouraged competition in electricity production and in addition encouraged the use of renewable energy sources.

California has a biomass potential of 47 million bone dry tons (bdt). This is comprised of lumber and mill waste, fruit and nut crops, vegetable and nursery crops, food processing wastes, forest slash, energy crops, urban wood and yard waste, livestock and chaparral. Of this only 7 million bdt (15%) is used for electricity generation. Approximately 8 million bdt (17%) is used for non-energy purposes such as manure for fertilization, plywood production, and field and seed crop residues for animal feeds.

Therefore, the gross biomass resource available for energy production is 32 million bdt, which has an equivalent energy of 489 trillion BTUs or nearly 0.5 quad of electricity (approx. 7% of the total yearly energy used in California).

Currently, direct combustion, and gasification are the technologies of choice. Direct combustion is used for 75% of the biomass resources, but anaerobic digestion amounts to only 0.1%.

Of the electrical energy generation facilities in California, biomass produces 805 MW, landfill gas 182 MW, and municipal solid waste 85 MW, for a total of 1,072 MW. [1]

ELECTRICITY

As of June 1991, the California Energy Commission estimates that there were a total of 66 biomass combustion facilities operating. [1] Of these, 19 produced steam only and the rest produced electricity. Out of the 47 electricity producers, 24 of the facilities are cogenerators with the rest producing electricity only.

The total gross electrical output from the 66 direct combustion biomass facilities is 805 MW with an average output of 17.1 MW per electricity producing plant. By 1993, California will have added three new biomass direct combustion facilities to come on line with an additional scheduled combined gross output of 77 MW.

ENVIRONMENT

California is an emissions-driven energy market and as such places a great deal of restriction on the choice of energy technologies utilized. In 1992, the majority of the Air Quality Districts in California identify a non-attainment status in at least one of the major pollutant constituents. The most prevalent are carbon monoxide, nitrogen oxides (as NO_2) and sulfur dioxide. As a consequence, most energy applications of any significant size, are subject to Prevention of Significant Deterioration (PSD) regulations. These regulations include review for application of Best Available Control Technology (BACT) emissions controls and without offset, limit major pollutants from a major stationary source. In addition, certain districts also require hydrocarbons (VOCs) to be maintained below 13-28 tons per year (70-150 lb/day).

The emissions limits on plants were therefore tied to an allowable maximum of any one pollutant. Following the protocol for management of air emissions, BACT must first be applied to determine the emissions levels which results. If these are above the PSD triggers, and modeling does not allow relief, then offsets must be applied at certain ratios to reduce the pollutant to net zero or less (i.e., pollutant reduction for the offset area within the district).

As offsets are expensive to buy and in most places difficult to find and take time to implement, California plants must be targeted to generate whatever power/cogeneration level is possible given the pollutant constraints. Biomass energy will be one of the technologies used to meet the growing energy demand and environmental constraints.

ADDITIONAL BIOMASS RESOURCES

In California, municipal solid waste, tires, ethanol and methanol are also being actively considered and to some degree utilized as an alternative energy sources.

<u>Municipal Solid Waste (MSW)</u>. Californians generated more than 50 million tons of waste

in 1990. Of these 50 million tons, approximately 1% consisted of rubber and tire waste, 12.4% yard waste, 26.8% paper, 5.2% plastics, 3.7% glass, 6.0% metals, and 44.9% other waste. Other waste is comprised of disposable diapers, food waste, wood waste, textiles and leather, manure, and agricultural crop residues.

As of mid-1991, there were four waste-to-energy (WTE) facilities operating in California: [2]

- A 380 tons per day (tpd) mass burn plant in Commerce produces 10 MW of electricity (gross), and uses a dry scrubber and spray dryer systems with a fabric filter (DSCR/FF). This system was designed by Foster Wheeler and has been on line, owned and operated since 1986 by L.A. Sanitation.
- Another mass burn in Long Beach (SERRF) operates at 1,380 tpd, producing 33 MW (gross) of electricity. This DSCR/FF with a thermal DeNOX was designed and operated by Montenay.
- A 800 tpd mass burn in Stanislaus produces 24 MW (gross, 20 MW net) of electricity. This DSCR/FF and thermal DeNOX plant monitors emissions on a continuous basis. This system was designed by Ogden Martin, who also owns and operate it.

In the planning stage is a 1,600 tpd refuse derived fuel (RDF) plant in North San Diego County. This system is designed by Thermo-Electron and should be on line sometime this year. It will produce 41 MW (gross) of electricity

A large proportion of the future U.S. MSW power potential is located in California. However, due to strict air quality regulations and environmental concerns, there has not been any significant development of MSW combustion capacity to date. One advantage of the air quality regulations is that this could encourage the production of MSW ethanol. California regulations also require the sale and use of alternative fueled vehicles, such as ethanol vehicles, that could operate on dedicated ethanol fuel supplies.

Tires. California has more registered vehicles than any other state, which means it has the largest inventory of used tires in the nation. Today, about 33 million waste tires exist and added to this inventory is the fact that every year another 28 million used tires will be discarded. [3] Of the 28 million tires discarded in 1992, only 6 million were used as a fuel substitute or in energy production. Nearly 60% were landfilled, stockpiled or dumped along the side of the road or near riverbeds.

Tires are made from, among other things, oil. An 18 pound passenger car scrap tire contains energy equivalent to about two gallons of gasoline. [4] Scrap tires have, pound for pound, an energy content greater than coal. [5]

Plants that use tires as a fuel can do so by either utilizing whole tires or tire-derived fuel (TDF), depending on the type of combustion facility and the auxiliary equipment developed for loading the tires or TDF into the facility. Tire-derived fuel is processed (tires are shredded, ground, or chipped) for supplemental fuel. Burning whole tires can be more economical in the long-term because it is more expensive to process tires. The U.S. EPA estimates that it costs between 17.9 cents and 24.4 cents per tire to shred tires. [6] To burn 1 million tires at 20 cents a tire, shredding would cost a company $200,000 a year. At this volume it would be cost-effective to burn whole tires even though special equipment must be designed and installed to accept whole tires as fuel. This equipment can cost between $0.5 million and $1.0 million dollars. [6]

Presently, there is only one tires-to-energy plant in California. At the Oxford Energy facility in Westley, five million tires are combusted annually to produce 14.4 MW of electricity with excess capacity sold to Pacific Gas and Electric. A report produced for the EPA calculated that the payback period for this energy plant is 6.7 years which is within the industry guidelines of seven years for similar power plants. [7] It can be concluded from these and other calculations that this use of whole scrap tires can be economically feasible, especially in areas of the country where fuel cost and tire disposal fees are high.

The Oxford Energy Company (OEC) recycles other byproducts from its scrap tires and burning process: recaptured zinc oxide can be sold for zinc production or as a component in fertilizer; and gypsum generated by the plant's scrubber is sold as a soil conditioner to local farmers. [7] However, the OEC has difficulty finding markets for the bottom ash which contains oxidized steel.

Two other plants use tires as a fuel supplement:

- The Calaveras Cement Company in Reading uses whole tires as a fuel in its kiln at a 20%

mix (20% tires, 80% coal). About 1.5 million tires are consumed annually.

- Southdown, Inc. in Victorville, which used 0.5 million tires in 1992 This facility has the potential to burn 2.6 million whole tires per year.

A waste-to-energy plant in Modesto also consumed an additional 0.3 million tires from a large stockpile in 1992. Recently, RMC Lonestar Cement in Davenport completed testing tires for supplemental fuel use.

The California Integrated Waste Management Board (CIWMB) has issued a contract to assess the current state of waste tire liquefaction, pyrolysis and gasification technologies worldwide. Texaco is also in the developmental stage of producing oil, which can be used as a fuel, through a tire liquefaction process.

In addition, California has the Waste Tire Act (PRC 42800 et seq.) which directs the CIWMB with other state agencies to develop a permit program for waste tire facilities and to report to the Legislature on the feasibility of using waste tires as a fuel supplement for cement kilns, lumber operations, and other industrial processes.

The current R&D efforts in addition to the Waste Tire Act (PRC 42800 et seq.) may lead to the increased utilization of tires as a fuel supplement and the development of more tires-to-energy plants.

Biofuels for Transportation Sector. In response to air quality concerns, California has mandated that by 2003, ten percent of the vehicles sold in the state must have zero emissions. In order to meet this goal, ethanol and methanol have been studied as a means of reducing emissions.

Ethanol is currently used in very small amounts in California. The primary feedstock for this is corn. At present there is one station in La Habra dispensing E85 (85% ethanol and 15% gasoline). However, it is felt that ethanol production is very energy intensive and because of this methanol is a much more popular biofuel.

Currently, there are 45 stations dispensing M85 (85% methanol and 15% gasoline). The cost varies from $0.85 to 0.95/gallon which compares favorably to ethanol which cost approximately $1.80/gallon. [8] Methanol has a higher octane rating (102) than ethanol therefore giving better performance but the energy content of both ethanol and methanol is lower than gasoline. All the methanol in California uses natural gas as a feedstock due to the lower cost; however, it is hoped that more methanol will be generated from biomass.

There are approximately 3,000 flexible fuel vehicles presently operating on either M85 or gasoline. At this time, there are only 15 vehicles operating on E85. Another 3,000 flexible fuel vehicles will be added in 1993.

Ethyl tertiary butyl ether (ETBE) and methyl tertiary butyl ether (MTBE), which are both fuel oxygenates, are being used to reduce CO_2 emissions as this is an EPA requirement in certain non-attainment areas. [8]

Energy Plantations. Studies have been performed to assess the viability of energy farms in California. Biomass crops are being studied for compatibility in different regions of the state. Sacramento Municipal Utility District seems to be leading in this field. The University of California at Davis, which has a strong agricultural program, has also been performing studies in this field. However, no new project of this type has gotten past the stage of paper analysis.

BARRIERS TO BIOMASS ENERGY

The following are some of the factors affecting the dissemination of biomass energy.

Biomass

- Competition for resources between the food industry, paper and furniture industry, as well as for biomass usage as a fertilizer and animal feed. Feedstock producer risks of planting and harvesting a new crop (for a new market) with a 2-8 year growth/harvest cycle.
- Reliability of supply and the added energy input which increases the cost of harvesting, collecting and transporting biomass. This includes the lack of an established infrastructure for conveying crops to power producers.
- Diversity of fuels and the need to establish standardized systems to accommodate such diversity. Potential for scale-up problems and hidden environmental cost.
- Fuel handling and storage considerations. Conventional fuel handling systems generally rely on gravity for transporting fuels into the boiler. However, the lower density of some biomass fuels can create difficulties.

- The need to increase plant efficiency and power output of current technology. Engineering and construction contractor risks for guaranteeing plant construction schedule, availability, and heat rate.
- Investor risks in financing power production systems dependent on an entirely new fuel supply system and fluctuating tax incentives.
- California has many non-attainment air districts, making the siting of a biomass facility very difficult. Several facilities in these areas have experienced difficulty in obtaining offset fuels. Stiffer permitting regulations in non-attainment districts typically raise the power plant's capital, operational, and maintenance costs.

Municipal Solid Waste

- There is often an unwillingness of municipalities to make long-term MSW management decisions in favor of short-term fixes.
- Resource recovery is generally perceived as complex, expensive and risky. Communities are more concerned with the economics of the waste-to-energy facility than solid waste management.
- The necessity of developing fair and equitable purchase power agreements.
- Component screening of the solid waste stream.
- Public concern on WTE emissions of dioxins and mercury and on ash residues.

Tires

- Scrap tire supplies may be unreliable; it is difficult to supply fuel (tires) to remote locations.
- Whole tires will have to be stockpiled nearby, which may cause a fire or health hazard unless properly contained and handled.
- Environmental permits have to be modified and new emissions standards have to be met.
- Public perception is often negative concerning the storing and burning of scrap tires.
- There is only a marginal cost advantage of using tires over competing fuels; fees make using tires cost-effective.

OUTLOOK FOR BIOMASS IN CALIFORNIA

The most likely scenario for the next decade is the continued exploitation of existing sources of wood wastes and agricultural residues. Fuel handling and preparation systems will continuously improve and net plant efficiencies well over the typical 20% reported by PG&E for California biomass power plants should become the norm. Average plant capacities should double in size and improvements in steam turbine technology will have the greatest impact on plants in the 50 to 80 MWe range. By the end of the next decade, gas turbine combined cycle or steam injected technology is expected to dominate the market.

Methanol demand and supply will continue to grow due to the stringent emissions mandate in the transportation sector. As electric vehicles struggle to overcome the limiting factor of their batteries, gasoline substitutes and additives will have to play a stronger role in lowering current emissions as these alcohol based fuels are cleaner than fossil fuels.

As the decade progresses, the WTE numbers will only increase. Under California state standards, facilities must control pollutants. For instance, EPA's final New Source Performance Standards (NSPS), promulgated in January 1991, require all new WTE facilities with units that have a burning capacity of 250 tons of MSW to install acid scrubbers that will reduce metals emissions and organic compounds by 99% and acid gases by 90 to 95%. Existing facilities burning over 1,100 tpd must meet Emissions Guidelines, which require retrofitting with scrubbers to reduce acid gases emissions.

EPA regulations are undergoing refinement due to the Clean Air Act Amendments of 1990 (CAAA) Maximum Achievable Control Technology (MACT) concept. MACT applies to existing WTE plants, all of which are now expected to achieve the same level of control as that reached by 12% of existing facilities. However, many existing units in California have scrubbers/baghouse emission controls in place that already meet fairly stringent restrictions.

Currently, environmental, economic, and technical barriers exist for tires-to-energy facilities, and there is some public opposition to this technology. In 1994, with special exceptions, all waste tire facilities in California will be required to have a permit from the California Integrated Waste Management Board. These regulations will provide procedures for obtaining a waste tire facility permit, and set technical and operational standards for indoor and outdoor storage and disposal of waste tires at these facilities and at solid-waste sites. Although initially this may involve added cost for tires-to-energy facilities, these requirements will do much to remove public concerns relating to the environmental consequences of tires-to-energy generation as well as set technical standards which could further promote the dissemination of this technology.

Biomass-based power, with proper management of biomass production, offers major environmental benefits compared with fossil-fuel based power, and it is likely that biomass power will be competitive with power from conventional sources in a wide range of circumstances. Given the current political and social climate, the use of biomass for energy generation, in all its forms, will continue and no doubt increase.

REFERENCES

1. California Energy Commission, Biomass Resource Assessment & Facilities in California, presented to Indian Study Tour, Indian Renewable Resource Delegation, June 9, 1992.
2. Kiser, Jonathan, A Comprehensive Report on the Status of Municipal Waste Combustion, Waste Age, November 1990.
3. California Integrated Waste Management Board, Market Status Report, Tires, Draft Copy, February 1993.
4. California Integrated Waste Management Board, Tires as a Fuel Supplement: Feasibility Study, January 1992, p. xv.
5. Arthur Little, Recycling Handbook for Selected Materials, prepared for the New Jersey Department of Environmental Protection, March 1989, p.96.
6. U.S. Environmental Protection Agency, Markets for Scrap Tires, October 1991.
7. Franklin Associates Ltd. and Robert L. Hershey, Market Development Study for Tires (Draft), prepared for the Office of Solid Waste, U.S. Environmental Protection Agency, p. 2-7.
8. Telephone conversation with Bill Blackburn of the California Energy Commission, May 14, 1993.
9. Draft Copy of Sacramento Municipal Utility District Development Plan, 1993.

BIOGRAPHY

Dr. Green has worked for the Deutsche Gesellschaft fur Technische Zusammenarbeit (GTZ) GmbH, Eschborn, Germany as an independent energy expert and has been Visiting Fellow–Energy at the Asian Pacific Development Centre, Kuala Lumpur, Malaysia. She received her PhD in Electrical Engineering at the University of Edinburgh, Scotland and currently works for Bechtel Group, Inc. in the Renewable Energy Research and Development Division. Her professional interests include energy policy and planning for the Asia-Pacific region.

California's Total Electric Capacity

(in MW's)

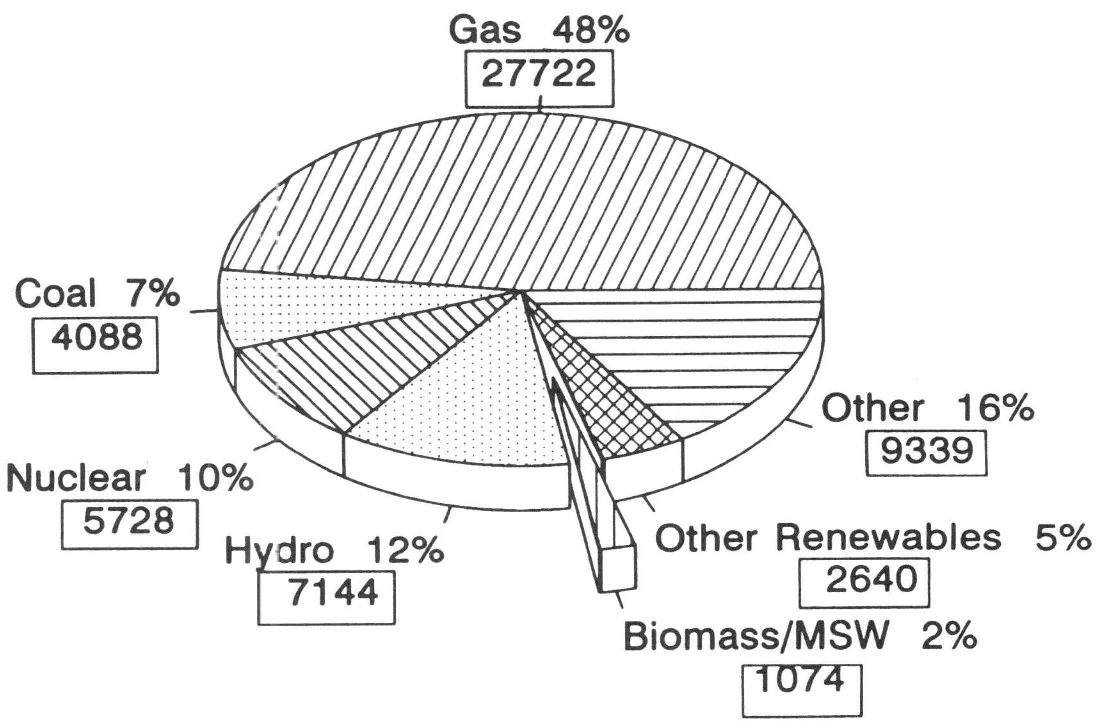

Source: California Energy Comission
Other - includes out of state power and pumped hydro.
Other renewables - includes geothermal, wind, and solar.

California's Generation Facilities

• Biomass	805 MW
• Landfill Gas	182 MW
• Municipal Solid Waste	85 MW
• Total	1,072 MW

Source: California Energy Commission

California's Biomass and MSW Gross Electrical Capacity

(in MW's)

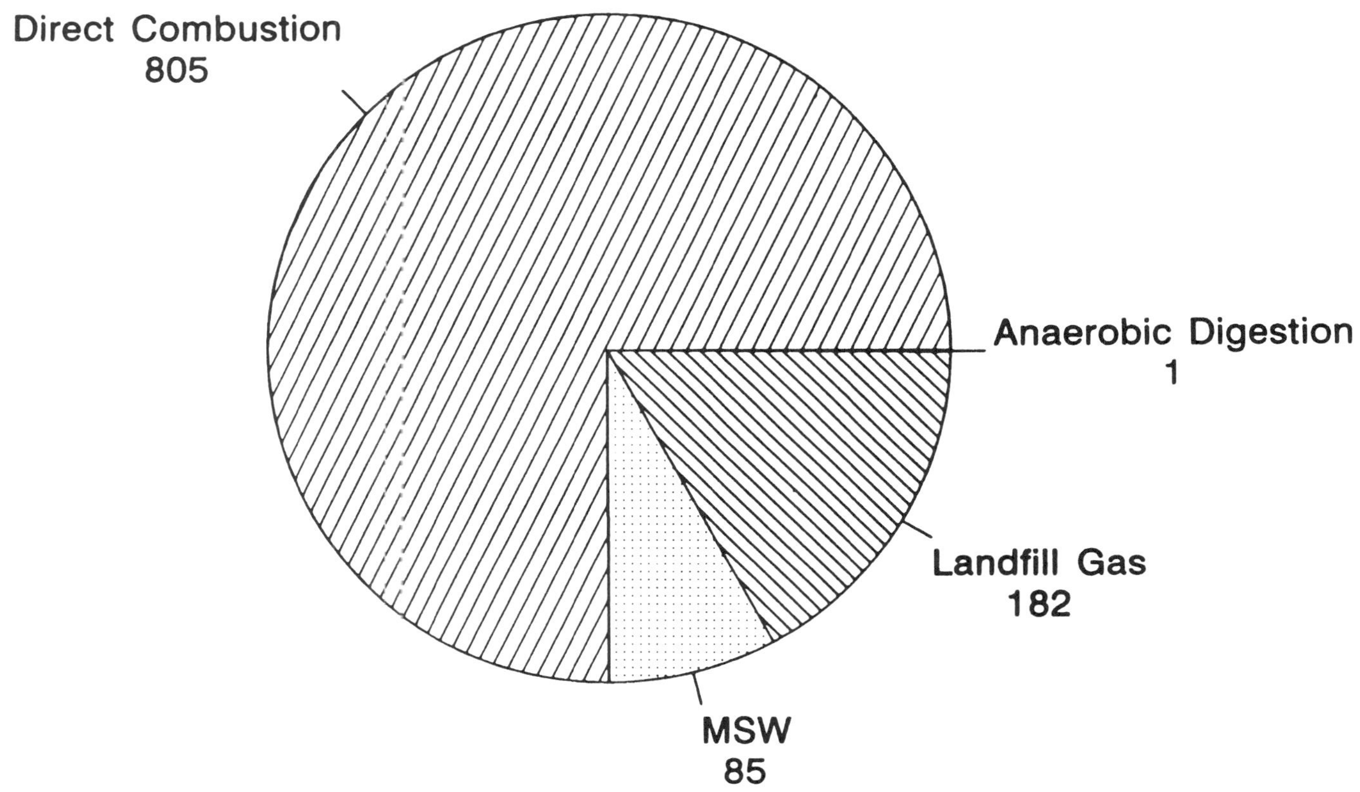

Source: California Energy Commission

Biomass Fuel Usage in California

units are in thousands of bone dry tons (kbdt)

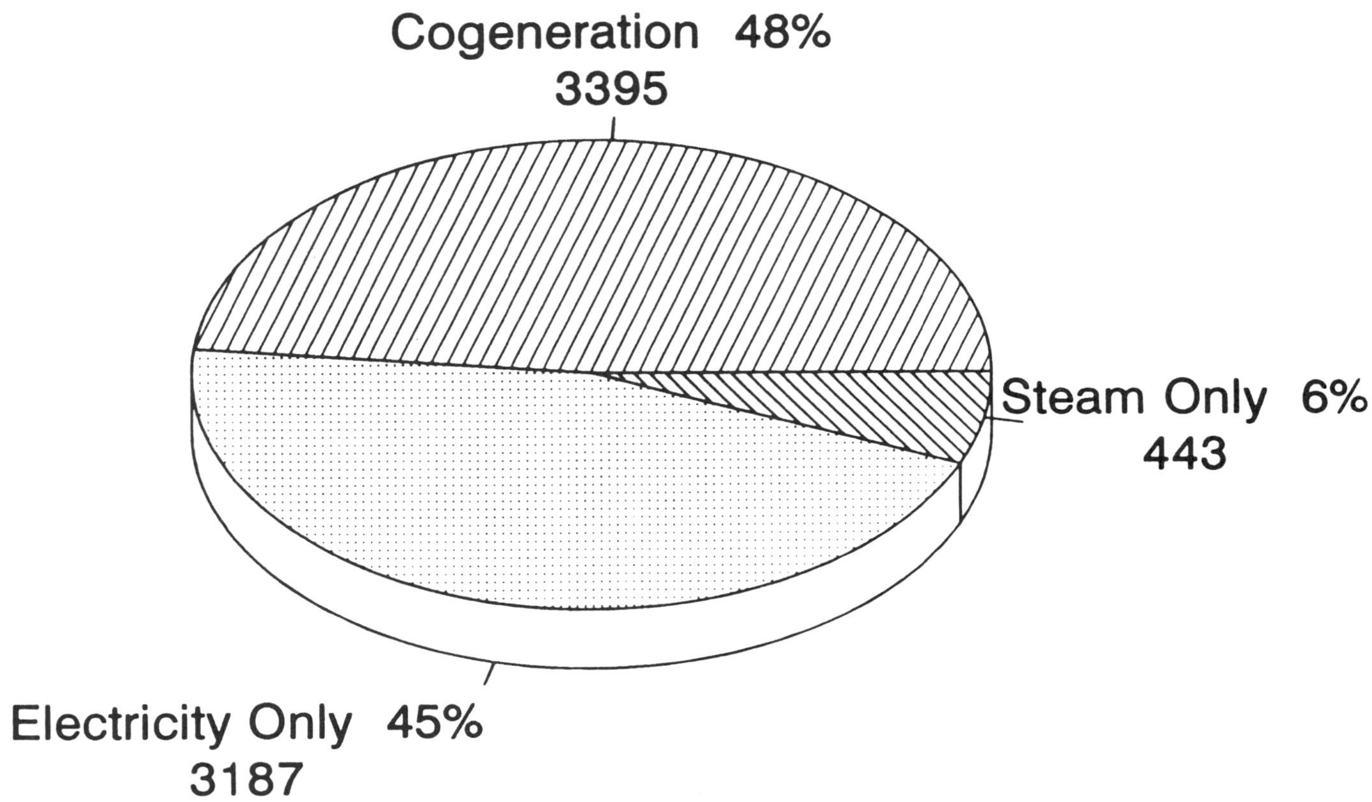

Source: California Energy Commission

Chapter 86

AT&T's 138 KV Substation Improves Power Quality and Reduces Costs

J.L. Fetters, W.M. Lewis

INTRODUCTION

An isolated source of electrical power has been installed for the Columbus (Ohio) Works of AT&T. The new state-of-the-art high-voltage substation now provides all of the power for the Works and has been in operation since July 1992. This paper will discuss the reasons that owning a primary substation makes sense, how power quality has been significantly improved, and how the technical objectives were achieved by a sound engineering design.

Columbus Works

The Columbus Works is a 2.2 million square-foot manufacturing and software development center located on the east side of Columbus, Ohio. Originally built as a Western Electric telephone crossbar manufacturing plant, it is now the leader in providing highly technical communications products and services to the world. These services are provided by advanced computer systems, linked by complex and intricate networks. When power is lost, even for a short time, AT&T loses the capability to serve many of its global customers.

Power Quality Problems

It has been recognized for several years that the quality of power provided to the Columbus Works did not meet high-tech computer power requirements. With the Works on the same utility substation supplying other major users, any voltage dip caused by normal power protection operations, such as circuit breaker reclosures, resulted in computer systems malfunctions. The isolation obtained from a dedicated high-voltage substation was expected to solve these problems. Power quality was therefore one of the major incentives to providing a dedicated substation for the Columbus Works.

Power Quality History

A history file of power "events" has been kept since 1987. An event is defined as any abnormal power condition that results in an observable effect or complaint within the facility. Power sags and dips were found to be the most numerous, culminating in a record number in 1990—with an average of one event every ten days. Since the substation has been on line, voltage regulation has been excellent and only two events have been reported, compared to the 37 events recorded in 1990.

Substation Development

The local AT&T power engineer communicated the need for better power quality to the electric utility and they suggested that the idea for a separate substation be explored. They further suggested that the substation be located adjacent to the present source of AT&T power at the Broad Street substation. In addition, the utility had requested a major rate increase that was to go into effect soon. The utility was subsequently granted a large rate increase to be implemented over a three-year period. The first increment of the increase would go into effect before a substation could be completed. This made the outlook for a short-term payback on a costly investment even brighter. By accepting power at 138 kV, a lower cost transmission rate would provide a projected payback in approximately two years. But the power quality and reliability improvements that could be realized by feeding this new substation from two separate and independent transmission lines, and from a fully redundant/fully duplicated system was the real attraction for proceeding with the project. When it was learned that the utility rate increase would take effect early in 1992, the normal one-year installation was fast-tracked to seven months.

OBJECTIVE

The primary objective in constructing a new substation was to provide the power quality and reliability needed by AT&T. An ancillary objective was to reduce the Works' operating costs by reducing the cost of electricity. As discussed below both objectives were accomplished.

Power Quality

The prior means of supplying electrical power to AT&T was by two 13.2 kV parallel, underground circuits originating at the utility company's substation. The 13.2 kV bus also supplied the utility's area distribution system which was mostly of overhead construction and was utilized by numerous other customers in the immediate area, including several other industrial facilities. These two circuits appeared to be reliable in the sense that no extended outages had occurred in recent years. However, the 13.2 kV system had a recent history of many events with the potential to cause memory loss or disruption in on-line terminal work-in-progress. It was to be expected that the disturbances and voltage depressions inherent in a multi-user distribution circuit also appeared on the circuits to AT&T. These types of events are particularly troublesome to computers and their peripherals. In addition, while there was no confirmed evidence of significant harmonic content in this service, there had been ballast failures at AT&T which may indicate harmonic presence. Since the 13.2 kV system was known to serve variable frequency drives, there was, at a minimum, a distinct possibility of odd integer harmonic currents and voltages being impressed upon the 13.2 kV system and thus into AT&T's service.

Since a major part of the Columbus Works' efforts is in the area of computer utilization, it seemed prudent to consider a move to a direct feed from the utility's 138 kV transmission system. Such a move would reasonably ensure AT&T that any harmonic voltage problems were of AT&T origin and could be remedied internally. The utility's 13.8 kV substation is adjacent to and radially fed from one of the utility's transmission substations which is a connection point for five 138 kV transmission lines. This arrangement is illustrated in Figure 1.

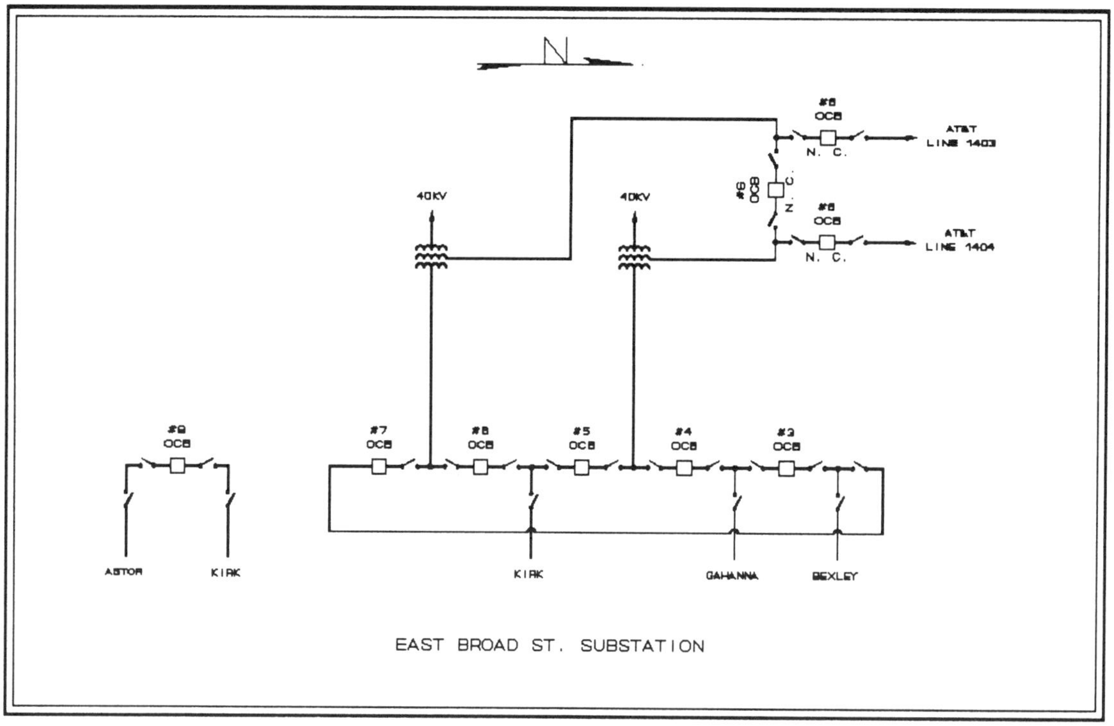

FIGURE 1

Two of these 138 kV lines were not connected to the substation's ring bus and were available to feed the new AT&T substation. These lines occupied a separate bay of the substation and had experienced only one outage of either line in the last ten years. Therefore, the increased reliability of a 138 kV supply was overwhelming!

Cost of Electricity

AT&T was being billed for its electrical use under a distribution voltage rate schedule. This schedule had been in effect for some eight years. The utility had requested major increases in most of its electric rates, including this schedule. If these requests were granted, it would have resulted in an increased cost to AT&T of approximately $1.3 million over a twelve-month period. This represents an increase of over 50 percent to AT&T in annual electric costs. However, if AT&T was to be supplied at transmission voltage, then it would be eligible for a more economical rate schedule. Based on 1990-91 usage, studies indicated that the annual savings to AT&T for service under the transmission rate in effect prior to the proposed increase, rather than the corresponding distribution rate, would be approximately $657,000. The savings under the proposed rate increase were approximately $1.3 million. These savings were essentially the same as the amount of increase facing AT&T if the service remained at the distribution rate. Therefore, by constructing its own 138 kV substation and receiving service at transmission voltages, AT&T could avoid the large increases in electric charges that were proposed by the utility. These potential savings, and other savings available, were evaluated versus the initial cost of such a substation, the cost effects of annual maintenance requirements, and reliability considerations.

On the transmission rate, additional savings were available by increasing the power factor of the AT&T load. Under the distribution rate, a penalty is assessed if the average monthly power factor is less than approximately 87 percent. However, no benefit or reduction in charges flowed to AT&T if the facility increased its power factor toward 100 percent. Demand billings under the transmission rate are calculated on the basis of kVA rather than kW. Billing kVA is determined by multiplying the peak 15-minute monthly kW by the reciprocal of the power factor. Therefore, as the power factor approaches 100 percent, the reciprocal approaches 1.00, and the resulting demand charge is minimized to the amount of monthly peak kW times the unit charge.

AT&T's power factor could be made to approach 100 percent by adding capacitors to the Works' supply circuits. Studies indicated that the inclusion of a variable bank of capacitors to the proposed 138 kV substation would result in annual savings of approximately $73,000 at 1991 rates and about $119,000 at the proposed increased rates. In addition, the capacitors would add the ancillary benefits of voltage stabilization and the dampening of harmonics in the Works' power circuits.

APPROACH

The above-described improvements in power quality, reliability enhancements, and economic gains are all dependent upon the cost and reliability of the proposed 138 kV-13.8 kV substation. The proposed substation, which is described in more detail later, is premised upon a cost-effective design which incorporates redundant circuit features and real-time monitoring for reliability and ease of maintenance. By the use of dual power transformers, each of sufficient capacity to serve the present and projected loads of AT&T, a failure of either transformer will not result in a substation outage or loss of supply to AT&T. Nor will periodic preventive maintenance or testing require an interruption or curtailment of the electric supply to the Works. This is of utmost importance as AT&T normally had no maintenance outages of the Works' electrical system. The incorporation of real-time monitoring of major equipment and system parameters contributes to an otherwise unattended and low maintenance installation. A single-line diagram of the new substation is illustrated as Figure 2.

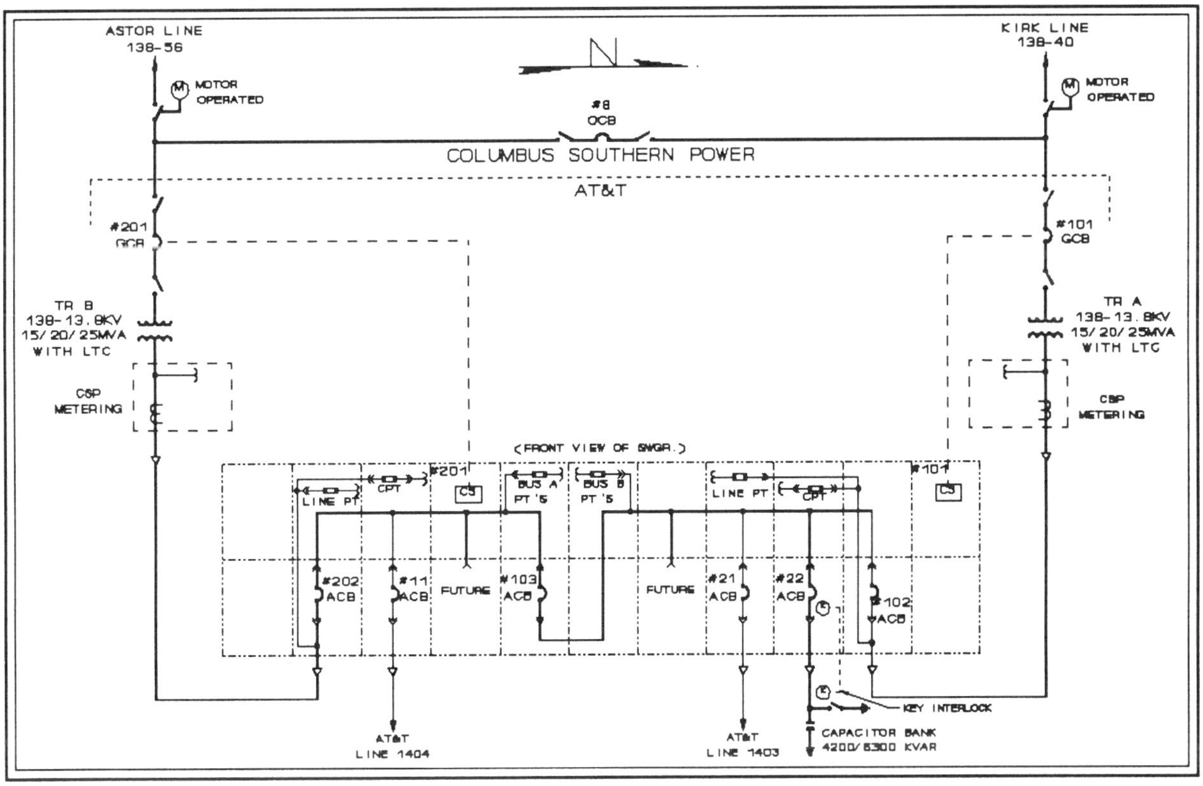

FIGURE 2

DESIGN AND CONSTRUCTION

Fast-track Construction

In July of 1991, AT&T's electrical power consultant submitted the results of its feasibility study for the new 138 kV supply. The study detailed the improved power quality and potential savings, and proposed a preliminary design for the new substation. The study was the result of several months of work and many discussions with the utility.

Later in July during a meeting among representatives of AT&T, the consultant, and the utility, the utility was informed that AT&T was proceeding with the new substation and that the operational date would be November 1992. This date had been primarily determined by the utility and was based on a need for installing new relaying on the transmission lines and the installation of motor operators on the transmission line disconnect switches.

In August and September of 1991, an electric rate increase was announced and scheduled to take effect in early 1992, thus the savings as a result of the new substation became more apparent. AT&T wanted to move the inservice date forward. In an October 1991 meeting, the consultant proposed an April 30, 1992 inservice date to the utility. The utility did not think that this date was attainable but would investigate the possibility of improving the November 1992 date.

In November of 1991, AT&T decided to accept additional costs for an accelerated schedule and instructed the consultant to expedite construction of the substation. Specifications for the power transformers, regulators, 138 kV switches, and 138 kV circuit breakers were prepared. On November 27, 1991, a meeting of prospective bidders was held to discuss the project and distribute the specifications. The local AT&T power engineer explained to the prospective bidders that delivery time would be a major consideration in addition to the price.

When the bids were received on December 13, 1991, the 138 kV circuit breakers were the longest delivery items (mid-June 1992). The power transformers were quoted for an April 15, 1992 delivery, if a purchase order was issued by December 31, 1991. To meet this request the local AT&T power engineer and the consultant traveled to the transformer manufacturer's plant on December 20, 1991, to finalize various technical and contractual matters.

A purchase order was issued on December 30, 1991, for the 138 kV breakers establishing a June 1992 delivery. The regulators were not required because the transformers purchased were of a Load Tap Changer (LTC) design. The delivery of the 138 kV switches was not a problem for the April 30 inservice date.

In late December 1991 and January 1992 negotiations for the 15 kV shelter aisle switchgear were ongoing. Three proposals were finally received, with only one being able to meet the April 30 inservice date.

In most cases for the 138 kV circuit breakers, the switches, and the 15 kV switchgear, the lowest bidder also had the quickest delivery. The one exception was that a combination of non-LTC transformers and three-phase regulators were approximately $9,000 less than the LTC transformers. Because of a long delivery on the regulators, the LTC transformers were purchased. Thus, there was not a large premium in equipment cost for the improved inservice date. However, to obtain these dates, the consultant agreed to approve drawings "on the boards" at the switchgear manufacturer's plant.

With the June delivery for the 138 kV circuit breakers, June 30, 1992, was the new target date. With the promised delivery dates, renewed requests were made by AT&T to the utility to accelerate its construction. It was proposed that AT&T should be placed on the transmission rate when its substation was complete even though the utility was not able to serve the new substation. After some rescheduling the utility was able to improve its schedule to a mid-July inservice date.

With a completion target date of mid-July, the construction schedule was revised and the project accelerated. The site grading contractor started work in mid-April. Due to interference with another contractor who was relocating a storm sewer and with unusually wet weather, the completion of the initial site work was delayed until May 20, 1992. This delay was recovered by paralleling later tasks.

The electrical construction contractor started work on May 20, 1992. At all times when work was being done at the site, a representative of the consultant was present, not only to check the quality of the work, but to expedite the construction by solving problems quickly and to ensure that the contractor was performing the work in the most expeditious manner.

On July 22, 1992, the power transformers were energized, and on Sunday morning, July 26, 1992, the AT&T load was served by the new substation.

The fast-track construction program resulted in moving the inservice date from November 1, 1992 to July 26, 1992, and also resulted in a reduction of the electric cost of more than $270,000.

SUBSTATION EQUIPMENT

The new substation was designed to be located adjacent to the utility's East Broad Station on land which was owned by the utility. The land for the substation and access road, and an easement for the underground 13.8 kV feeders were purchased by AT&T from the utility.

The utility's station is located to the south of the Columbus Works across East Broad Street, a four-lane divided street. It was originally served from this station by two dedicated, underground, 13.8 kV, 1,200-ampere feeders. The feeders were routed from their respective circuit breakers to a vault located just inside the station's fence, then 150 feet across East Broad Street to another vault on the AT&T property, and finally another 1,080 feet through two more vaults to the

Works' main switchgear.

The conductors on one of the two feeders had been replaced in 1985. This feeder consisted of three conductors per phase of 15 kV, 500 MCM copper EPR cable. These conductors were owned by AT&T. For this feeder it was decided to install new conductors from the new AT&T substation (through a new duct run) to the first vault and to splice the new conductors to the existing conductors.

Records indicated that the second feeder, which was owned by the utility, was more than 20 years old and each phase consisted of two 15 kV, 500 MCM copper paper-insulated lead covered (PILC) and one 15 kV, 750 MCM aluminum EPR. It was decided to completely replace this feeder. A new feeder was installed from the new substation to the main switchgear through the new duct run to the first vault and then through the existing ducts and vaults to the switchgear.

In order to minimize AT&T's downtime to the scheduled July 26 shutdown, a plan for the sequence of cable replacement and splicing was developed. The first replacement was made by de-energizing the feeder cable that would remain and splicing in the new underground cable from the new substation. This was done just prior to the outage for commissioning the new substation. Service to AT&T was maintained on the second feeder from the utility during splicing and testing of the new cable. On the morning of commissioning, the older cable was de-energized at the utility's breaker and placed in standby.

Immediately following the inservice of the substation, the second new cable, which was a full run from the substation to AT&T's switchgear, was prepared for installation. The installation consisted of removing the existing lead-covered cable which was in a standby mode followed by the pulling of the new cable. The contractor was instructed to test and cut the new cable into pulling lengths and to have all required splices and connectors on hand before any removal of the standby cable could commence. This was stipulated to minimize the duration of service on only one feeder.

The above procedure was followed and the second cable was installed to complete the redundant service. The original lead-covered cable was removed and cut into specified lengths for the reclamation of the lead by AT&T.

The site for the new substation had an existing storm sewer with one catch basin running diagonally across it. A single span of 138 kV transmission line was located directly across the proposed site. It was decided to reroute the storm sewer to just outside the new substation and install a new pole in the transmission line so that the conductors would not be directly over the new power transformers and switchgear. The utility was responsible for the engineering and work to accomplish both relocations. The cost for these relocations was paid by AT&T.

There are two 138 kV feeds to the new AT&T substation. These terminate at the 138 kV SF_6 gas circuit breakers (GCBs). The utility provided motor operators for the transmission line air break switches. If there is a fault on one of the transmission lines, relaying will trip the utility's oil circuit breaker (OCB) and one of AT&T's 138 kV GCBs. A reclosing relay will attempt to reclose the line by closing the utility's OCB; if this is successful, the AT&T 138 kV GCB will automatically reclose one second after voltage has been established on the line. If the reclosing is unsuccessful, the motor-operated transmission line air break switch will open, isolating the line, and then the utility's OCB is reclosed, re-energizing the feed to AT&T. One second after voltage is established, the AT&T GCB will automatically reclose. Service to AT&T is maintained during the above by the second 138 kV feed and GCB.

Initially, the fault duty on the 138 kV feeds to AT&T is 3,500 MVA three phase and 2,570 MVA single line to ground. If the two 138 kV lines feeding AT&T are connected into a ring bus at a later date, the fault duty will increase to 6,000-7,000 MVA. Therefore, the 138 kV portion of the substation was designed for a fault duty level of 10,000 MVA.

The main 138 kV bus is three-inch aluminum bus at eight-foot phase spacing. Because of the eight-foot phase spacing, vertical break switches, as opposed to horizontal, are used to isolate the 138 kV GCBs. The switches are three- insulator vertical break rated at 145 kV and 600 ampere continuous. The 138 kV GCBs are SF_6 insulated and rated at 145 kV maximum voltage and 1,200 ampere continuous. The GCBs have a 40,000-ampere maximum symmetrical interrupting rating and 64,000-ampere close and latch capability. During design, it was decided to use the SF_6 GCB design rather than oil-type. This was based upon the better reliability of the GCB, ease of maintenance, a smaller profile, and reduced oil containment.

On September of 1990, AT&T had a peak demand of 17,140 kVA at an 89.9 percent power factor. The peak monthly kVA demands for 1990 ranged from a low of 14,327 to a high of 17,140. The monthly load factors varied between 71 and 85 percent. In order to provide for at least 33 percent near-term load growth and to allow that one of the two power transformers should be able to carry the complete load, it was decided to purchase two 15/20/25 MVA transformers. This resulting reduction in capacity utilization on the transformers provides an additional reliability enhancement.

Initial plans were to purchase non-LTC transformers and a three-phase regulator to operate with each transformer. This was thought to be the most reliable configuration. During bidding, it was learned that only one vendor would bid on the three-phase regulators; therefore, an addendum to the specifications was issued, asking that the transformer manufacturer bid both LTC and non-LTC transformers. In this substation design where there is 100 percent redundancy in the transformer capacity, it was decided that the use of LTC transformers would be acceptable. The deletion of the regulators would also result in additional space being available in the substation. A criterion for bid evaluation was developed to assess the value of the alternative designs.

The lowest bid, as submitted, was for a combination of non-LTC transformers and three-phase regulators, but a quoted delivery time of 40 weeks was not acceptable. The second lowest bid (approximately $9,000 higher) was for LTC transformers with a delivery of 15 weeks. The LTC transformers were purchased.

A loss evaluation formula and dollar value for the losses were provided in the specifications. The transformer price evaluation was made taking into account the losses. The LTC transformers purchased had the lowest initial price and the lowest loss evaluated price. Of the six LTC transformer bids, the purchased transformers had the second lowest losses.

The three-phase oil-filled power transformers are 15/20/25 MVA, OA/FA/FA, at 65°C rise and rated 138,000-volt Delta to 13,800-volt Wye. They are provided with two ± 2-1/2 percent primary de-energized taps and a ± 10 percent LTC in thirty-two 5/8 percent steps. The construction is of disc-wound copper primary coils and helical-wound copper secondary coils. The transformers are equipped with 84 kV MCOV station class arresters on the primary and 8.4 kV MCOV station class arresters on the 13.8 kV secondary.

A 15 kV metal-clad, sheltered-aisle switchgear was purchased for the 13.8 kV switching. A total of six 15 kV circuit breakers was provided as:

- 2 - 2,000 ampere transformer low side breakers
- 1 - 1,200 ampere bus tie breaker
- 2 - 1,200 ampere AT&T feeder breakers
- 1 - 1,200 ampere capacitor bank breaker

In addition, space was provided for two future breakers.

The substation design precluded the typical separate control building by locating the following relaying and control equipment in the switchgear enclosure:

138 kV Lines
- One-shot reclosing relays for AT&T's 138 kV breakers supervised by 13.8 kV bus and 138 kV line potential. Line protection relaying is provided by the utility.
- Breaker control switches

Power Transformers
- Transformer differential relays
- Phase overcurrent relays
- Neutral overcurrent relay
- Ground overcurrent relay
- Reverse power relays
- Sudden pressure relay

13.8 kV Bus
- Bus differential relays
- Control switches

13.8 kV Feeders
- HCB pilot-wire differential relay
- Phase overcurrent relays
- Neutral overcurrent relay
- Control switches

Capacitor Bank Feeder
- Phase overcurrent relays
- Neutral overcurrent relay
- Control switch

Control Power
- 2 - 25 kVA control power transformers with an automatic transfer switch for the control power
- Control power distribution panel
- 48 V. sealed battery charger and batteries
- 48 V. distribution panel

For the power factor correction, a metal-enclosed automatic power factor control system was specified and procured. The system consists of a bank of 4,200 kVAR of fixed capacitors and three switched banks of 300 kVAR, 600 kVAR, and 1,200 kVAR. The switching provides for seven steps of 300 kVA and a total of 6,300 kVA. The bank has been sized to maintain a 99.7 percent power factor at the September 1990 peak demand of 17,145 kVA and a unity power factor for lower demands.

This capacitor bank is supplied from a breaker in the switchgear through a 15 kV, 500 MCM, underground cable. A ground switch in the capacitor enclosure is key interlocked with the 15 kV breaker and with the doors on the capacitor enclosure to prevent access to the capacitors while they are energized.

A ground grid of bare 4/0 copper stranded wire was installed. The consultant's ground grid design is based on ANSI/IEEE Standard 80-1986 and utilized a computer-aided design. The ground grid consists of 2,200 feet of bare 4/0 copper conductors and 14 eight-foot ground rods. The ground grid was connected to the existing East Broad Station ground grid at three points.

System Surveillance

A supervisory control and data acquisition (SCADA) system was provided to transmit various status data and system parameters to a computer terminal at AT&T so that AT&T personnel can monitor the substation's performance and status. Examples of data included are breaker positions and power transformer tap position, temperature, and oil level. System parameters monitored and recorded include bus voltages, kW demand, energy, RkW demand, and feeder loading. The SCADA system will allow early detection of abnormalities and prevent major problems. An auto-dialer system is included in the SCADA system to inform AT&T personnel and the maintenance contractor of abnormal conditions.

Oil Drainage System

AT&T's corporate policy is one of concern for environmental impact and of strict observance of prudent environmentally sensitive design and construction. As the substation includes some 6,000 gallons of oil in each power transformer, the substation design criterion was to provide containment even though this oil is mineral-based. Therefore, a system of site confinement and recovery of oil loss was designed and implemented to retain the oil in the case of a transformer tank rupture. A trench was placed around each transformer in the clay substrata and a perforated PVC pipe was installed. The trench was backfilled with large stone. Under normal conditions the trench serves as a storm drain for the substation. If there is an oil spill from the transformers, the oil will be collected by the perforated pipes and drained to a 6,000-gallon oil-water separator tank. A hand hole in the top of the tank will allow the oil to be pumped into a tank truck for recovery and proper disposal.

Construction Problems

With the substation's being placed into service in just three months after the start of site grading, there were no major problems that caused long delays. There were several control wiring errors in the switchgear and in the power transformers, which caused problems in placing the substation into service and in paralleling the transformers. The consultant remained at the substation through August 1992 solving these problems one at a time as they arose. They included the calibration of the reclosing relay system, pilot-wire relay adjustments, and adjustments to the paralleling circuitry. The design of the substation allowed these wiring corrections to be made without loss of service.

Cost Savings and Cost of Construction

As stated previously, the new substation allowed AT&T to be served under a more favorable rate. The utility has received regulatory approval of annual increases in rates over a three-year period. The initial increase was effective in August 1992 with two subsequent increases effective on May 21 of 1993 and 1994. The actual nine-month cost savings to AT&T from September 1992 through May 1993 have been approximately $763,000. The projected savings for the initial 12 months of operation—through July 1993—are nearly $1 million. The projected 12-month cost savings in 1994-1995, at the full roll-in of the three year increase, are over $1 million. These savings are depicted below in graphical form:

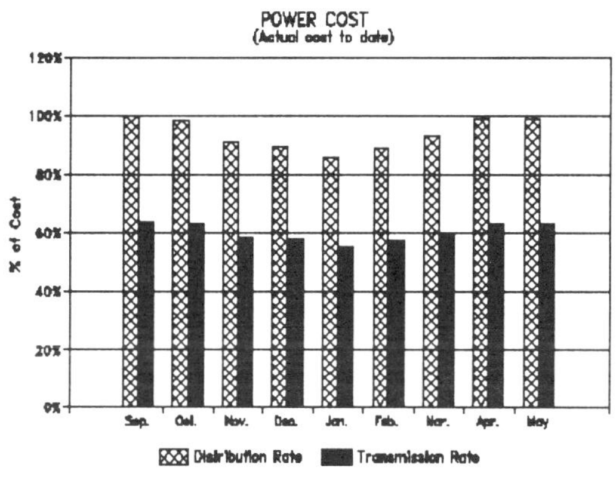

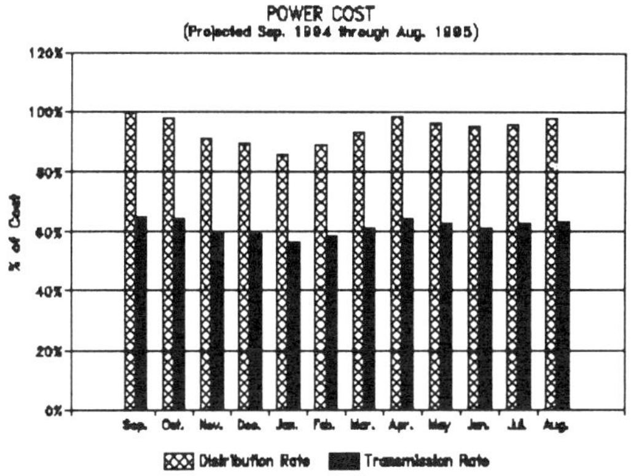

At the projected initial 12 months of savings, the substation has an approximate two-year payback.

SUMMARY

The installation of a dedicated 138 kV electrical substation by AT&T Network Systems shown in Figure 3 has resulted in significant improvements in reliability of service and power quality. Power abnormalities, with related voltage fluctuations, have been reduced from a previous frequency of some 30 events annually to a total of two events to date since substation commissioning. Neither of these two events was of a severity to cause disruptions to work or equipment at AT&T. Reliability has been further enhanced by the redundant power transformers and the feeder design. This allows normal maintenance and repairs to proceed with no corresponding outage or curtailment of electrical supply to the Columbus Works. The relay reclosing and circuit breaker design incorporated in the substation allow the service to be maintained during fault clearing and network disturbances. The addition of oil spill containment features and a low-profile design has resulted in an environmentally sound and aesthetic installation.

As an additional benefit, the installation of the substation has resulted in major savings in electric service costs. These demonstrated savings are projected to result in a less than two-year payback of the approximate $2 million investment. Additional cost savings will continue to flow to AT&T as the increased reliability and enhanced quality of service result in increases in productivity and avoided losses of work-in-progress. This stems from the fact that the avoidance of voltage excursions and short laps in voltage profile have been shown to result in disc errors, loss of RAM, and other complications to on-line terminals.

The 2 x 25,000 kVA, 13.8 kV substation was placed into service in eight months from the start of preparation of specifications for equipment, seven months from the start of procurement. Site preparation, installation of equipment, and construction were accomplished in approximately three months. This resulted in advancing the commissioning of the substation by some 3.5 months which resulted in additional savings to AT&T. There were only minor extensions of the planned four ten-hour work days per week. While normally encountered problems during construction and installation were magnified by the fast-track schedule, all were handled on-site as they occurred without schedule impact. As the substation approaches its initial annual preventive maintenance, there have been no operational abnormalities or equipment failures.

The installation of a dedicated 13.8 kV substation by AT&T's Columbus Works has resulted in service directly from the utility's 138 kV transmission grid. This design has effected significant improvements in reliability and power quality. The control of power outages for utility maintenance is now under control of AT&T and AT&T is now isolated from the effects of neighboring facilities on the electrical distribution system. When these benefits are considered, in addition to the significant cost savings flowing from the change in service, the installation of the substation can be seen as a multi-faceted advantage to AT&T.

FIGURE 3

SECTION 8
FEDERAL ENERGY MANAGEMENT PROGRAMS

Chapter 87

Energy Conservation Technology Selection: An Agent-Based Approach

L.P. DeBaille, R.J. Nemeth, M.P. Case

ABSTRACT

Organizations desiring to invest in energy conservation are faced with the challenge of determining which energy conservation technologies are optimal for their situation. This process is often characterized by a body of data from which ranking criteria must be derived. The required amount of data can quickly become overwhelming and difficult to acquire, as is the case with the United States Department of Defense (DOD) considering such improvements on a national scale.

This paper presents an automation model designed to manage the required data and the calculation of technology ranking criteria such as investment payback, energy savings, and pollution abatement on a national level. The model is under development at the United States Army Construction Engineering Research Laboratories (USACERL) in MicroSoft Windows 3.1 on a personal computer using Designer Software; an agent-based and object-oriented modeling environment. Small rule-based systems called *agents* evaluate each technology based upon information contained in a shared database called a *blackboard* The blackboard contains data such as installation fuel prices, building data, and weather information. Each agent represents a particular energy conservation technology and evaluates the potential of that technology based upon this shared data. The model is constructed in a modular fashion to facilitate the modification and addition of new technologies by adding or modifying agents. A rule-based system applies each agent's heuristics (rules-of-thumb) and algorithms to produce analysis results. A spreadsheet data link has been implemented to input external data or to output analysis results.

The proposed approach provides *variable levels of analysis*; i.e., evaluations can be conducted that are applicable to a single installation, region, or to the national level. For example, a user at an installation can view and change assumptions to conduct local parametric investigations, while a national policy analyst could estimate next year's energy conservation expenditures. Since many of the heuristics introduce relatively gross assumptions, analysis is more accurate at a national level. A new graphical logic tree called a *checklist* used to direct users through the required evaluation steps and options. *Combinations of analyses* such as many technologies applied to one location or one technology applied to many locations are also addressed by the application.

INTRODUCTION

The Problem

Executive Order 12759 [1] defined new energy efficiency requirements and goals for the Federal government. For the DOD to comply, new techniques are required to assess the current facility infrastructure and determine how best to invest funds to meet mandated goals. This becomes a complex problem when considering the distribution of installations throughout various climatic regions, the diversity of building types and construction methods, and varying building vintages and states of repair. Energy conservation tends to be approached in a piecemeal fashion rather than as a nationally coordinated program. Analysis at a national level affords a more global perspective of the problem and provides a single dollar figure to support requests for funding sent to the congress. A further problem with past energy studies is that older studies applied technologies that are currently obsolete or standard fare. Technological advancements in the last five years alone demand a reassessment of many energy conservation opportunities.

The Renewables and Energy Efficiency Planning (REEP) model was developed to support budget requests to the congress to be allocated towards energy conservation and renewable technologies. This model will help to conserve energy, reduce dependence on foreign oil, reduce levels of harmful emissions, and maintain responsible stewardship of federal facilities.

Assessing energy savings potential can be approached in a number of ways. Very detailed studies can be conducted that require large amounts of specific information and apply to very particular situations. Another approach is to generalize an Energy Conservation Technology (ECO) and apply it to a minimum amount of specific information to assess its potential across a much larger population of buildings and infrastructure. While the generalized approach does not afford the accuracy of the detailed approach, applying a detailed approach across all DOD installations would be virtually impossible due to the quantity, quality, and breadth of information that would be required. Therefore, the REEP program uses a generalized approach for studying the potential of various ECO's across many Army installations. The approach allows for the rapid analysis of numerous ECO's using a minimum amount of data from each installation. A more detailed approach would increase the amount of installation data required, which is often difficult or impossible to obtain, and bury the analyst in data. The generalized approach provides estimates at a level of detail suitable for planning purposes.

Following is a brief description of the analytic methodology used for the REEP program. A software-based agent is developed for each ECO using appropriate engineering heuristics and algorithms. The agent is kept as simple as possible to minimized data requirements. Agents use heuristics triggered by installation data to determine their applicability to each installation. If applicable, energy savings, economics, and pollution abated are estimated. A supervisory agent analyzes the case studies generated by each of the ECO agents to select conservation opportunities with the best payback. The results of the case studies are presented to the user or may be exported to a spreadsheet using a Dynamic Data Exchange link for further analysis.

APPROACH

Technology Specific Analysis

The goal of this research was to determine the potential for energy savings at fifty different Army installations across the nation that meet certain economic criteria, rather than to determine exact projects at each installation. The agents were kept fairly simple so that they could be used in different climates, and, in certain instances, be applied to more than one building type. Furthermore, the heuristics and algorithms needed to be kept simple due to the lack of detailed data at each installation, and the desire to keep the model from becoming overly complex.

The agents calculate how each ECO would perform at each location based on installation data. In most cases, the algorithms make conservative assumptions about the existing conditions of the mechanical system or building envelope for a given building type and area, and then calculate the differential between an energy baseline and a retrofit situation. The determination of which building type(s) an ECO applies to was straightforward. Some ECO's applied to a single facility type, while others applied to two or three building types. Again, assumptions had to be made about characteristics of certain building types and whether an ECO would apply.

Of the fifty ECO's evaluated, there were a few instances where ECO's were mutually exclusive. In a preliminary (spreadsheet) model, a hierarchy of mutually exclusive technologies was determined, based on the highest efficiency technology being the first to be considered, and then progressively less efficient technologies being evaluated. Once an ECO passed the payback criteria at an installation, further mutually exclusive technologies were not considered. The synergistic effects between numerous ECO's being applied to the same building are currently not considered. The ECO's are evaluated as if they applied to an unaltered building. Naturally, if numerous ECO's were applied to the same building, the cumulative savings would not be as great as the sum of each being evaluated independently.

Determine analysis requirements

Determining energy savings potential of ECO's constitutes just one of three phases of the ECO analysis. The other two phases are an economic and pollution abatement analysis.

The economic analysis portion of the program follows a recommended simplified economic analysis summary format developed for the U.S. Army's Energy Conservation Investment Program (ECIP). The economic analysis calculates simple payback, savings to investment ratio, and internal rate of return. These calculations follow analysis methodologies as specified by the National Institute of Standards and Technology [2]. ECIP criteria specify that for a project to be funded it must have a Savings to Investment Ratio of greater than 1.25, and have a discounted payback of ten years or less. The evaluation of each ECO involves calculating the energy savings and then converting the energy savings into dollar savings based on each installation's energy costs. Each fuel type saved is then discounted over the life of the ECO using NIST discount factors to derive a total discounted savings. The discounted savings is then divided by the total investment cost to arrive at the savings to investment ratio. The total investment cost also includes project overhead, management, design, salvage value, and utility fees or rebate credits.

A measure of pollution abatement is also estimated for each ECO at each installation. Following the energy savings calculations, an approximation for pollution abatement due to the reduced energy consumption can be calculated. The reduction in pollutants is regionalized by state to account for different methods of electricity production. Because of this, similar levels of electrical energy cost savings can result in very different pollution reduction values.

Installation Data

ECO analyses require information specific to each installation. All of the installation specific data is contained in one file, and each installation consists of one row of data. Data in this file were obtained from numerous sources, however the majority of the information came from an annual summary of operations published by the Department of the Army [3]. This source provides data on annual utility usage, capacity and consumption of heating and cooling plants, total square footage of facility types, and other miscellaneous data. Except for global radiation, climatic information for each installation was obtained from an Army Technical Manual [4]. This manual contains latitude and longitude, heating and cooling degree days, summer and winter design temperatures, bin data, and other weather specific data used to determine if installations qualify for air-conditioning. Global radiation was obtained from a Solar Energy Research Institute (SERI) publication [5]. Fuel oil, gas, and coal costs for each installation were obtained from the Defense Energy Information System (DEIS). Since DEIS only provided an overall cost for electricity with no breakout for demand charges, electrical cost and demand charges were obtained from utility contracts. In order to properly account for the differences in how electrical energy is produced by state, regionalized energy production characteristics were obtained from the Energy Information Administration [6]. This information was used to assess pollution abated due to electrical savings in different regions of the country. ECO costs were also required to conduct the analysis. Most ECO costs were obtained from Means cost estimating books [7]. Costs for ECO technologies not found in Means were obtained from research reports from various sources.

Requirements for Automation

A concern in developing software is producing computer code that is easy to modify at a later date. If the algorithms and data structures are created in a rigid format, the application will have little chance to evolve with time, and hence become obsolete. Another concern is allowing flexibility in the combinations of analyses that may be performed. It is often difficult to predict the exact needs of future software users. In light of this, it is desirable to keep user options as flexible as possible. For example, a user at a particular site may wish to evaluate many technologies at that location, while a user at the national level may wish to evaluate a single technology over many locales. In addition, when performing an analytic study, it is often desirable to alter the controlling parameters to observe the effects on the results. The relative importance of particular parameters can then be observed, or a series of "what if" scenarios can be executed. Another desirable feature is the ability to question the origin of a result; i.e., *what* algorithms using *what* data led to this result? Results can then be questioned if they appear suspicious, rather than being isolated from their origins by a "black-box" type of analysis. A meaningful analysis of a technology must also include the influences of other technologies that have been selected. This coupling of analyses leads to optimization schemes where the best combination of technologies, interactions considered, is selected. An additional requirement of the application is the presence of an easy to use, centralized user interface. The interface should pull together all of the various facets of the project while allowing the user as much freedom as possible in choice of analyses, or the level of detail in viewing results.

ANALYSIS OVERVIEW

To gather an overall view of the REEP software application, a brief example of the process is presented. Details regarding the various steps will be covered in later sections.

The first step in running the REEP software application is to open a user-interface in Designer Software called a *checklist*. From this checklist the REEP agent is loaded. The REEP agent is the overall body of code coordinating the REEP application, similar to the main program in traditional algorithms. Next, the data for installations of interest are loaded from a choice of several data formats. ECO's are then selected to be evaluated and are allowed to "act" upon the installation data and evaluate their effectiveness at each installation. The results of this analysis can be displayed at a variety of levels; from generalized graphical displays to a detailed spreadsheet format. Parametric investigations can also be performed by altering basic ECO assumptions, the effects of which are automatically propagated to the results and available for display.

IMPLEMENTATION

The REEP application was developed in Windows on the PC using Designer Software; a software design platform developed at USACERL. Designer Software is built upon a *blackboard system*. The blackboard is a shared database where global information is stored and accessed. The data stored in this central location are structured in object-oriented fashion, a programming method conducive to later modifications. Object-oriented data structures also possess the ability of *inheritance*, a way to reuse pre-defined attributes and behavior without re-definition. Besides data, the blackboard also contains *rules*. Rules are statements of cause-and-effect logic and take the form of if-then statements. Rules manipulate data. Links are preserved by the system so that the validity of a result can be questioned by tracing back the rules that lead to its creation. These links also maintain the consistency of the system if data changes. Finally, Designer Software is designed to be *distributed*, meaning it has the potential to allow several users to interact on the same project via a network.

Computer code that interacts with the blackboard data is split up into discreet units, each with a particular expertise. These units are called *agents,* and are programmed with data and rules to carry out a particular task. For the REEP application, each ECO technology is represented as an agent, while the blackboard contains the installation data (see Figure 1).

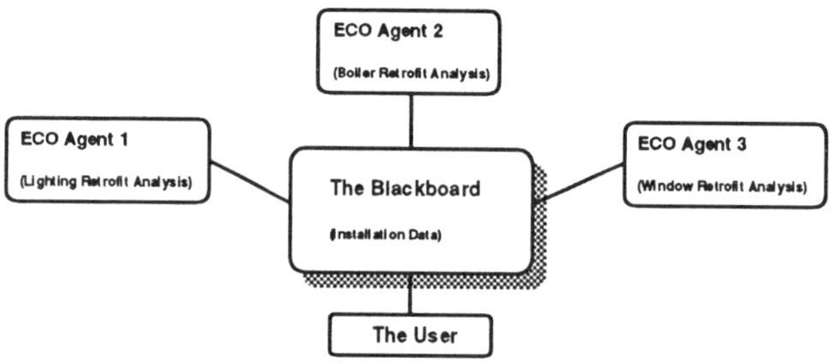

FIGURE 1: THE BLACKBOARD SYSTEM AND AGENTS

The ECO agents provide a way to organize the approach to the project. They are modular and can be connected to or removed from the blackboard, therefore allowing for the addition of new ECO's or the removal of those obsolete. In addition, when ECO's are connected to the blackboard, they can be turned "on" or "off" to limit participation. Each ECO agent contains data and rules specific to that ECO's task only. Because the data and rules are ECO-specific, ECO's can be removed or disabled without affecting other ECO's. Like the blackboard, the data in each ECO is implemented in object-oriented fashion to facilitate modification. Each ECO consists of two files, the *frame* file and the *rule* file. The frame file contains frames that store data that are specific to that agent's task. The rule file contains rules that manipulate both agent and blackboard data. The ECO rules encode the REEP heuristics and algorithms as if-then statements. These heuristics and algorithms were first encoded into spreadsheets by energy engineers. The spreadsheets were then given to software developers to translate into rules and frames for use in Designer Software.

A user interface has been developed that centralizes the REEP analysis options into a step-by-step format. Figure 2 displays the Designer Software utility called a *checklist*. The checklist is a custom user menu for running applications such as REEP. The checklist can be constructed to steer the user through the required evaluation steps or to present user options. The options are presented in logical "branches" of and/or logic. The branches can be set up to force the user to perform the checklist items in a required order, or flexibility may be allowed. The checklist items are activated by selection with a mouse and change color when completed. Several important steps can be seen in the checklist for the REEP project; namely, loading the REEP agent, loading and saving installation data, selecting ECO's to evaluate, starting the analysis, and displaying the results graphically or in spreadsheet form. If a checklist item requires further information, pop-up windows are displayed to query and guide the user.

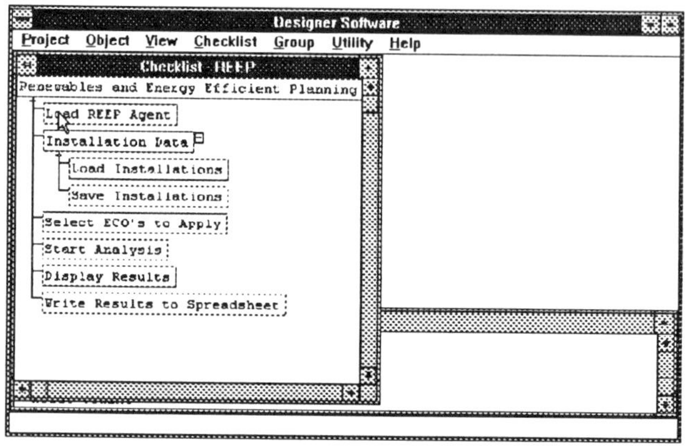

FIGURE 2: USER INTERFACE - THE REEP CHECKLIST

A spreadsheet interface has been developed to input installation data or to output analysis results. The installation data were collected from many different sources and were centralized in a spreadsheet. A spreadsheet was chosen to be the common repository of input data based upon its relative ease-of-use and the general familiarity many users have with it. The spreadsheet installation data is imported into REEP by selecting the *Load Installations* item from the checklist. The input data is not restricted to spreadsheet format only, but can also be read in comma-separated or binary format. These last two formats are created by saving data initially input by the spreadsheet under a new format using the *Save Installation Data* checklist item. If the user selects the spreadsheet as the input file, he/she is prompted for the location of the installation data file. The user is also prompted for the location of the spreadsheet application if the system cannot find it. In reverse fashion, the results of different analyses can be output to the spreadsheet with the *Write Results to Spreadsheet* item within the checklist. Here, the user is prompted for the particular ECO technology results that are of interest. Each output spreadsheet file is limited to the results of a single technology's evaluation at any number of locations. Before writing to the spreadsheet, the user is prompted for the location of an output template file. This is a spreadsheet file that is copied for use as a standard output file template. The format consists of pre-made headers, borders, and spreadsheet formulas designed to summarize the results of installations that pass federal criteria.

ANALYSIS SESSION

The following is a description of an analysis session in REEP.

The first step is to open the checklist as shown in Figure 2. The installation data must then be loaded from the desired file format. Figure 3 displays the installation data spreadsheet as it links with the REEP application. At present, data for fifty installations are available.

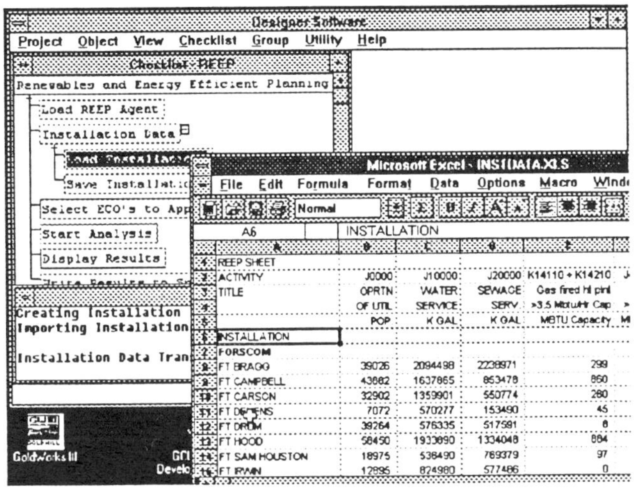

FIGURE 3: LOADING INSTALLATION DATA

Next, the technologies that are to be evaluated are selected by enabling their respective agents as shown in Figure 4. Currently, forty-five separate technologies are available for analysis.

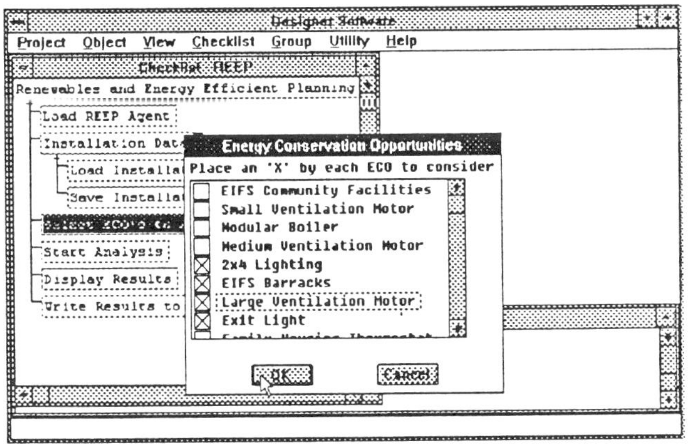

FIGURE 4: SELECTING TECHNOLOGIES TO EVALUATE.

The analyses are initiated by selecting the *Start Analysis* checklist item. ECO's being evaluated are displayed in the message window as shown in Figure 5.

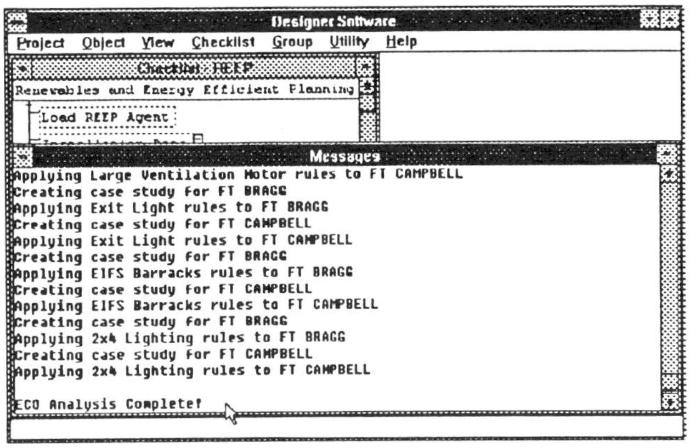

FIGURE 5: CONDUCTING ECO ANALYSES.

The results of the analyses can be viewed in several ways, one of which is the simple graphical display seen in Figure 6.

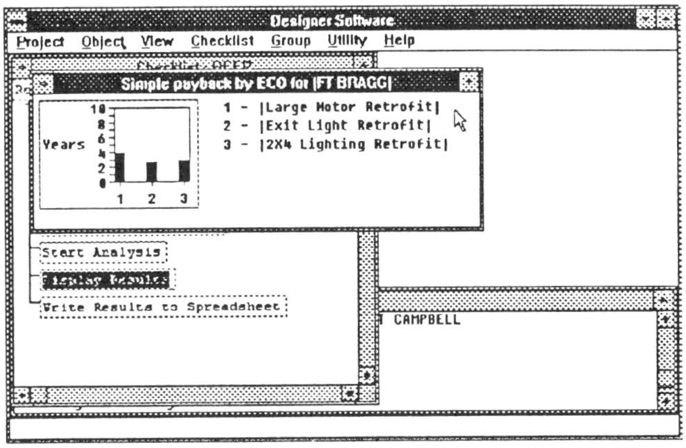

FIGURE 6: DISPLAYING RESULTS GRAPHICALLY.

The results and any numerical assumptions can be also displayed through the Designer Software *View* utility as shown is Figure 7. The data are displayed next to buttons labeled with their print-names. Selecting the button with a mouse evokes a pop-up window displaying three user options. *Help* provides explanation text describing the value and providing units, etc. *Why* displays the rules and data leading to the value so the user can question its derivation. *Override* is a way for the user to manually override the rule system to assert a constant value. The *View* utility also allows the user to manually change any value. The system then re-calculates any dependent values to allow for parametric investigation.

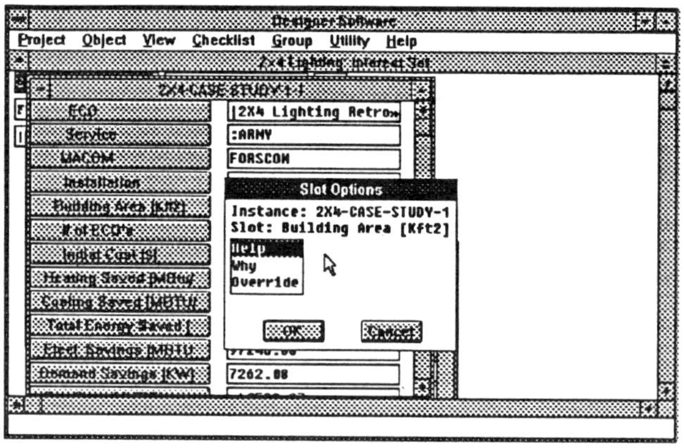

FIGURE 7: THE DESIGNER SOFTWARE *VIEW* UTILITY

Finally, the results of the analysis can be written to a template spreadsheet file as shown in Figure 8.

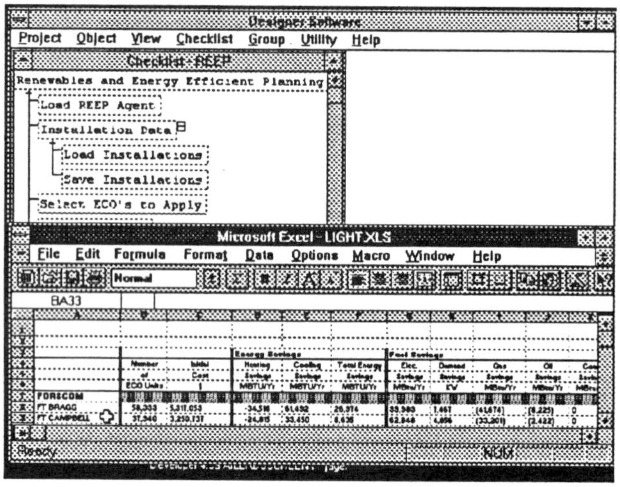

FIGURE 8: WRITING RESULTS TO SPREADSHEET

RESULTS

The result of the REEP effort is a flexible software analysis package that provides a rapid analysis of many technology alternatives over many locations. The checklist user-interface has proven itself to be readily understandable to new users, while the ability to perform parametric analysis and rationale tracking has appealed to technical users. The spreadsheet interface has been a surprisingly useful feature that allows a variety of output formats while remaining a relatively familiar software package. The object-oriented implementation eased a major data structure modification occurring in mid-development and continues to allow the easy addition of new ECO technologies. The REEP model has been generated interest in several other fields with similar problems. Some examples are: air compliance strategies at U.S. DOD facilities, family housing health-hazard mitigation, integrated utility resource planning, and overall environmental compliance.

ADDITIONAL RESEARCH

Development of the REEP application continues with new investigations. Presently, it is not considered that selection of a single technology may exclude evaluation of another. Also, the *evaluation* of a particular technology is often dependent upon the presence or absence of other technologies. These phenomena can become quite complex when attempting to determine not only adequate technologies, but optimal *combinations* of the technologies. A proposed approach is a "bidding" process where agents compete for minimum payback, maximum savings, or some other desirable parameter. The winning technology is then assumed to be "installed" and the process continues under the influences of the installed technology. The process ends when criteria are met such as depletion of investment capital or achievement of an energy-reduction or effluent-reduction goal. In keeping with the intent of maximum flexibility, it is desirable to leave the optimization and "end analysis" criteria as selectable user options.

Another promising area of research lies in the multi-user potential of Designer Software. Currently, REEP is a single-user application. It is conceivable, however, that users at different locations and different levels of analysis would benefit from working with the same data and sharing an environment that would coordinate their efforts and objectives. This scenario implies the use of a centralized installation energy database that all levels of users could access. This would be an improvement upon the current situation where energy-relevant data resides in scattered sources in a variety of formats.

REFERENCES

[1] Executive Order No. 12758, "Addition to Level IV of the Executive Schedule" 56 F.R. 14631, April 5, 1991.

[2] National Institute of Standards and Technology, Handbook 135, "Life Cycle Cost Manual for the Federal Energy Management Program."

[3] Facilities Engineering and Housing Annual Summary of Operations, Vol. III - Installations Performance, Department of the Army.

[4] Facility Design and Planning Engineering Weather Data, Tri-Service Manual, Army TM 5-5785 July, 1978.

[5] Insolation Data Manual, Solar Energy Research Institute, October 1980.

[6] Electric Power Monthly, Energy Information Administration, DOE/IEA-0226(93/02), February 1993

[7] Means Cost Estimating Guides, R.S. Means Co., 16th Annual Edition, 1993.

Chapter 88

Establishing a Money-Saving Energy Management Program at a DOD Research Facility

G.R. Kelly, M.E. Agudo, M.W. Delurey

ABSTRACT

The Naval Surface Warfare Center, Dahlgren Division (NSWCDD) is a DOD research facility composed of 569 buildings situated on approximately 4,321 acres. The total occupied area of buildings is approximately 2,304,000 square feet. NSWCDD has long strived for the goal of reduced operating costs in its facilities through energy conservation.

In 1990 the Utilities Management Group decided that while conservation efforts are certainly worthwhile, attaining the goal of reduced costs could only be accomplished by the adoption of an energy management philosophy which considers energy economics, maintenance, and occupant needs in parallel with energy conservation.

With this in mind, a comprehensive energy management plan was developed and implemented that takes a more holistic approach to NSWCDD's energy needs and attempts to provide the highest quality service to the building occupants, while reducing the operating costs as much as practical.

The plan uses a two level approach to manage the energy usage at NSWCDD. The first level considers the macro-level electrical usage of the Center as a whole, and takes the steps necessary to monitor and control the demand. The second level takes the monitor and control functions down to the individual building at NSWCDD.

At the heart of this plan is a merger of the maintenance staff and the energy management staff through the use of a Computer Based Facility Management System. The system is currently serving approximately 700,000 square feet of laboratory / office space, and is supported by a 24 hour roving watch crew.

In the space of two and a half years, NSWCDD has gone from a limited energy management program to a highly successful, expanding program that is providing a net savings in excess of $100,000 per year.

INTRODUCTION

Over the past 15 years, many engineering evaluations and studies have been performed documenting potential savings to be derived from energy conservation measures. While there is no question as to the validity of these proposed savings, very few have been implemented and fewer still have been successful. The reason for this is that NSWCDD had no clearly defined process for implementing energy-related projects.

In 1990, the Utilities Management Group decided that, while conservation efforts are certainly worthwhile, attaining the goal of reduced costs could only be accomplished by the adoption of an energy management philosophy which considers energy economics, maintenance, and occupant needs in parallel with energy conservation.

With this in mind, a comprehensive energy management plan was developed and is currently being implemented that takes a more holistic approach to NSWCDD's energy needs and attempts to provide the highest quality service to the building occupants, while reducing the operating costs as much as practical.

ENERGY USAGE AT NSWCDD

Historically, most energy plans at NSWCDD have emphasized only conservation. The result has been an unfortunate tendency to equate energy conservation with energy deprivation and the associated reduction in an activity's ability to carry out its function. At a facility whose primary function is Research and Development, it is particularly difficult to convince Program Managers that the few kilowatt-hours saved are worth the perceived disruption and inconvenience. Historic energy consumption data is presented in Tables 1, 2 and 3. The data presented is for fiscal years 85 through 92.

TABLE 1 - TOTAL ANNUAL ENERGY DATA SUMMARY FOR NSWCDD, FY 85 THROUGH FY 92

Year	Total MBtu Consumed	Percent Change	Yearly Cost	Cost In FY85 $	% Change In FY 85 $	$/KSF
FY85	207,783		$2,258,994	$2,258,994		1,673
FY86	218,703	5.26%	$2,262,045	$2,185,551	-3.25%	1,537
FY87	245,547	12.27%	$2,417,683	$2,256,933	3.27%	1,631
FY88	264,443	7.70%	$2,559,003	$2,308,074	2.27%	1,727
FY89	273,274	3.34%	$2,696,543	$2,349,881	1.81%	1,653
FY90	298,619	9.27%	$2,924,981	$2,462,756	4.80%	1,677
FY91	358,665	20.11%	$3,808,471	$3,098,194	25.80%	2,078
FY92	327,986	-8.55%	$3,268,161	$2,568,745	-17.09%	1,752

TABLE 2 - ANNUAL ELECTRICAL ENERGY DATA SUMMARY FOR NSWCDD, FY 85 THROUGH FY 92

Year	Total MBtu Consumed	Percent Change	Yearly Cost	Cost In FY85 $	% Change In FY 85 $
FY85	138,097		$1,795,582	$1,795,582	
FY86	150,776	9.18%	$1,872,144	$1,808,835	0.74%
FY87	159,466	5.76%	$1,965,758	$1,835,056	1.45%
FY88	168,647	5.76%	$2,123,132	$1,914,943	4.35%
FY89	189,773	12.53%	$2,316,612	$2,018,794	5.42%
FY90	205,063	8.06%	$2,550,676	$2,147,601	6.38%
FY91	268,095	30.74%	$3,155,466	$2,566,974	19.53%
FY92	227,494	-15.14%	$2,782,924	$2,187,353	-14.79%

TABLE 3 - ANNUAL FUEL OIL DATA SUMMARY FOR NSWCDD, FY 85 THROUGH FY 92

Year	Total MBtu Consumed	Percent Change	Yearly Cost	Cost In FY85 $	% Change In FY 85 $
FY85	69,686		$463,412	$463,412	
FY86	67,927	-2.52%	$389,901	$376,716	-18.71%
FY87	86,081	26.73%	$451,925	$421,877	11.99%
FY88	95,796	11.29%	$435,872	$393,132	-6.81%
FY89	83,501	-12.83%	$379,931	$331,088	-15.78%
FY90	93,556	12.04%	$374,305	$315,155	-4.81%
FY91	90,570	-3.19%	$653,005	$531,220	68.56%
FY92	100,492	10.96%	$485,237	$381,392	-28.20%

Tables 1, 2 and 3 also present summaries of the energy costs from fiscal years 85 through 92. For comparison, these costs have also been calculated in constant 1985 dollars. In addition, Table 1 indicates the annual energy cost per thousand square feet of occupied buildings at NSWCDD.

Tables 1 and 2 indicate there is clearly a trend toward both increased usage and increased costs at NSWCDD (with the exception of FY 92 which represents electric usage after the implementation of the energy management plan). In terms of total cost of energy, the marked increase in electrical energy has been partially offset by the fluctuating cost of fuel oil; however, the overall trend is one of increasing costs.

OVERVIEW OF THE NSWCDD ENERGY MANAGEMENT PLAN

The starting point for the plan was a differentiation between energy conservation and energy management. Webster defines conservation as, "*A careful preservation and protection of something.*" By contrast, management is defined as, "*The judicious use of means to accomplish an end.*"

The plan was developed around the simple concept that energy is a resource, and as such is clearly subject to the management process. Instead of attempting to reduce costs by applying only standard conservation methods, we would develop a methodology that accomplished mission objectives at the lowest possible cost. In effect, we set out to manage the utility usage to minimize the utility bill by using both conservation and energy management techniques.

Target Areas and Approaches

There are two primary sources of energy used at the Naval Surface Warfare Center Dahlgren Division: electricity and fuel oil. In developing the energy management plan, it was necessary to assess these primary sources and decide which should be targeted first. The assessment was made on the basis of unit energy costs.

To make a meaningful comparison of energy costs, it was necessary to convert the conventional costs (i.e. $/kWH and $/gallon) to a common base of MBTU. Using the proper conversion factors, the 1991 unit cost of electrical energy was $12.45/MBTU, while fuel oil was running at $3.99/MBTU. More important from the cost point of view was the "ratchet clause" contained in the service contract with Virginia Power, which placed the demand charge at

$10.78/kW. Any effort to reduce operating costs should clearly be directed at the electrical energy first.

The approach decided upon was to manage the energy usage at two levels. The first level involves considering the macro-level electrical usage of NSWCDD and taking the steps necessary to monitor and control the demand. The second level takes the monitor and control functions down to the individual buildings at NSWCDD.

ELECTRIC DEMAND MANAGEMENT

As is true for any large facility, demand management is best accomplished as a Center-wide endeavor. This means a process of coordinated efforts orchestrated by the Public Works Department. The objective is to reduce the monthly peak during the summer months, and thus the charge incurred for demand. The primary elements of the load management plan are described below.

Energy Management Monitoring System.

In order to efficiently implement the electric demand management strategies, it was determined that remote metering of large electrical loads and critical points in the NSWCDD's electric distribution system would be required. An Energy Management Monitoring System (EMMS) using Dranetz® meters was designed to monitor the total energy use and demand. In addition, the EMMS was intended to provide more accurate electric bills for tenant facilities located at NSWCDD .

The heart of the EMMS is a PC terminal driven by Dranetz® communication software. The monitored points such as substations and building loads are equipped with Dranetz Remote Energy/Demand Analyzers®. As shown in Figure 1, the remote analyzers can be interrogated manually or automatically at preset time periods via the local telephone network. The monitored data includes real power, reactive power, apparent power, power factor and energy use. Currently, there are two permanently installed remote analyzers located at the two metered delivery points for commercial power, and a number of portable units. Installation of additional remote analyzers is on-going, with seven distribution substations, one additional utility delivery point and nine buildings slated to be monitored in the coming fiscal year.

For smaller buildings (typically 400 KW of demand or less) E-MON® meters are used to monitor their energy use (kWH) and demand (kW). One benefit of this system is to quickly identify buildings whose usage exceeds the norm. These meters are interfaced with the FMS via telephone lines. There are approximately 50 small buildings on schedule for E-MON® metering.

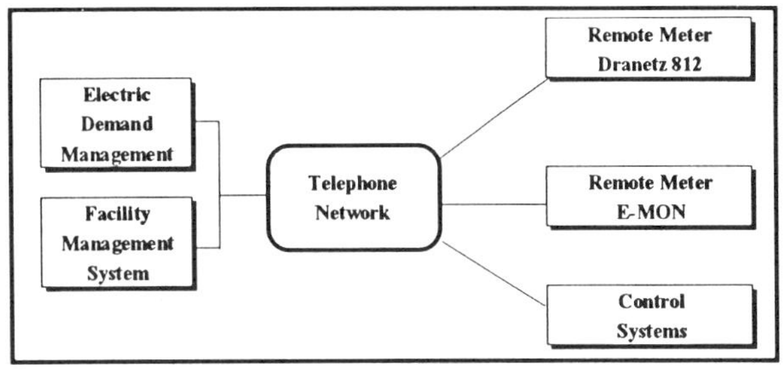

FIGURE 1 - BLOCK DIAGRAM OF THE ENERGY MANAGEMENT AND MONITORING SYSTEM

Electric Demand Management

The first demand management strategy that was implemented at NSWCDD was the Peak Shaving Program. "Shaving off" of monthly electric demand peaks during the summer season was considered the most appropriate demand management strategy because of the Center's existing power plant. The power plant, which consists of two diesel-driven 750-kW generating units, was primarily used for emergency standby. With minor modification in the control circuitry and operational procedure, the generating units were converted into "peak shaving" units in 1991.

Figures 2 and 3 depict the electric demand profile of NSWCDD from 1987 to 1993. Since 1987, the demand charge has accounted for more than 50 percent NSWCDD's electric bill, or about $4.8 million. Virginia Power has imposed a "ratchet clause" that establishes the minimum electric demand charge during a given year to 90 percent of the peak demand of the preceding year. The horizontal line shown in the Figures represents 90% of the peak demand of the preceding summer season. Notice that as the demand peak increases, the 90% peak demand line also increases and stays at this level until another summer demand peak is recorded. The horizontal line indicates how much NSWCDD is charged for the peak demand, even during winter billing months. Taking the billing month of November 1988 as an example, the actual electric demand was about 8800 KW, yet NSWCDD was billed for about 10,200 KW in demand charge (i.e. 90% of the peak set that summer). With a rate of $10.19 per KW of peak demand (the rate in 1988), NSWCDD paid approximately $14,200 in November and continued to pay this every month until a new peak was established.

As indicated in Figures 2 and 3, the Center has experienced generally increasing growth in electric demand until October 1991. The first drop is a result of a change in energy accounting procedures between Virginia Power and NSWCDD. The second drop is attributable to the energy management plan.

Until 1991, NSWCDD was supplied by Virginia Power through a single service point. In October 1991, a new service point was added and some of the Center's loads were transferred to the new service point. As a result, the electric demand at the main service point drastically dropped. The 90% ratchet was adjusted to 9.4 MW.

In the Summer of 1992, the electric demand management program was implemented. The generating units were run in parallel with Virginia Power during peak hours. As a result, the summer peak demand of 1992 was drastically reduced, causing the new 90% ratchet to drop from 9.4 MW to 8.5 MW, as shown in Figure 3. The estimated net annual savings from the reduction of the 1992 summer peak demand is $30,000.

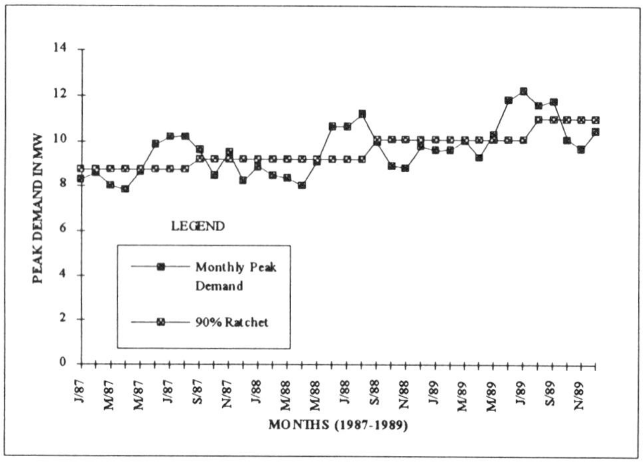

FIGURE 2 - MONTHLY PEAK DEMANDS FOR NSWCDD 1987 - 1989

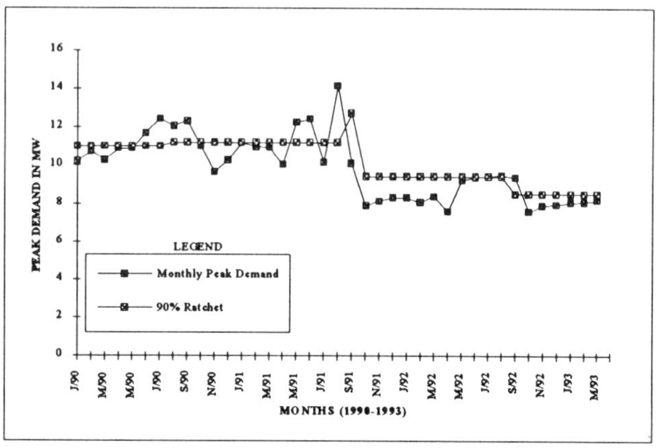

FIGURE 3 - MONTHLY PEAK DEMANDS FOR NSWCDD 1990 - 1993

Other Demand Management Efforts.
NSWCDD has entered into a contract with Virginia Power to participate in its Standby Generator Program for reducing peak demand. This program will begin when Virginia Power has completed installing its metering devices. Due to the available standby generators, which have total capacity of 2,650 KW, the Center can transfer several loads to their respective standby generators during peak hours upon notification by Virginia Power. Not only will the summer peak demand will be reduced, but substantial financial credit is anticipated to offset the demand charge.

FACILITY-LEVEL ENERGY MANAGEMENT

This portion of the plan is intended to reduce total operating cost through effective maintenance, control and monitoring of major mechanical and electrical systems at the individual facilities through the Center.

To attain the objective of reduced operating cost at the facility level, the Facility-Level Energy Management portion of the plan stresses two areas:

1. Measurement and Control
2. Maintenance

Measurement and Control is an attempt to optimize the daily functioning of building and systems in terms of energy usage. Clearly, the first step is to measure the energy usage of the facility and assess its relative efficiency. This effort is intended to develop bench marks by which to make comparisons, target key opportunities, and measure performance related to building operation. This is accomplished by sub-metering key facilities at the Center using E-MON® meters, as discussed earlier. The control function is the traditional optimization of energy usage.

In 1990, a navy Fleet support building was constructed that included the first major Direct Digital Control (DDC) system at the Center, and included a Computer Based Facility Management System (FMS). The results have been so favorable that DDC systems are being installed wherever practical. However, to avoid multiple host systems, cut hardware costs and support a uniform controls network, all new buildings and all major retrofits use DDC systems that can interface with the existing FMS.

In 1991, Mr. Jeff Johnson, Head of the Utilities and Maintenance Division, initiated an aggressive campaign to expand the number of buildings served by the FMS. As indicated in Table 4, since the initial building in FY

91, fourteen more buildings have been added representing over 700,000 square feet of laboratory / office space. There are also firm budget plans for an additional thirty buildings to be added in the next fiscal; this represents an additional 600,000 square feet of monitored space.

TABLE 4 - LISTING OF BUILDINGS SERVED BY THE FMS (ACTUAL AND PLANNED)

Building Number	Year Installed	Year Planned		Building Number	Year Installed	Year Planned
1560	FY 91			199		FY 94
1520	FY 92			193		FY 94
1500	FY 93			218A		FY 94
1281	FY 93			219		FY 94
115	FY 93			198		FY 94
182	FY 93			197		FY 94
180	FY 93			220		FY 94
106	FY 93			150		FY 94
112	FY 93			TL-11		FY 94
1580	FY 93			119		FY 94
470	FY 93			214		FY 94
959	FY 93			212		FY 94
963	FY 93			190		FY 94
1200	FY 93			216		FY 94
1510	FY 93			217		FY 94
960		FY 93		192		FY 94
997		FY 93		2007		FY 94
1450		FY 93		101		FY 94
221		FY 93		152		FY 94
962A		FY 93		404		FY 94
202		FY 93		194		FY 94
1700		FY 93		189		FY 94
200		FY 94		13 Additional Bldgs		FY 94

In addition to buildings, the Energy and Water Management Branch has been tying other Center-wide infrastructure networks into the FMS. To date, the Center's four water towers and pump houses have been tied in, as well as the two metered delivery points for commercial power. Plans are also in place to monitor the sewage treatment plant and critical sewage lift stations. The ability to remotely monitor these sites has enabled the maintenance and utilities staff to use their time much more effectively.

The critical role maintenance plays in a successful energy management program cannot be overstated. Therefore, at the heart of this plan is a merger of the maintenance staff and the energy management staff through the use of the FMS. This change has removed the organizational barrier between the people responsible for maintaining the systems and those charged with managing the energy usage. The result has been an improvement in the overall quality of service provided, with a simultaneous reduction of costs.

The FMS has also been used to improve the scheduled maintenance function. The FMS can generate preventive maintenance reminders based upon actual run time or by preset schedules. This information can then be integrated into the Public Works Preventive Maintenance Program to improve maintenance of facilities and equipment.

Key Benefits of the System

The following list summarizes the key benefits derived by NSWCDD from the implementation of this portion of the plan.

Remote Monitoring - The FMS allows monitoring of the mechanical and electrical systems from a remote point, or by the building occupants. If an abnormal condition occurs, an alarm is automatically sent via telephone lines to an area within the Energy and Water Management (EWM) Office which is manned 24 hours a day. This benefit provides constant monitoring, while reducing the manpower expenses associated with this level of attendance.

Remote Controlling - The FMS also allows remote operation of critical equipment to maximize efficient system operation. This is especially important in buildings with extensive security requirements. Crucial time can be lost obtaining access to the building in the event of a mechanical system failure during non-business hours. With remote controlling ability, this function can often be performed from the EWM Office without the need to enter the building.

System Reliability - The FMS is based upon a distributed intelligence system. This means that the "intelligence" driving the controls is distributed throughout the building and placed near the piece of equipment to be controlled. The system is controlled by the units in the field, and not by a central system. The primary benefit of this approach is that a single failure of an individual controller will not cause the entire systems to crash.

Accuracy - The DDC systems supported by the FMS are electronic and hold their calibration much better than pneumatic controls, which need to be recalibrated periodically. This benefit reduces required maintenance and energy consumption while increasing the comfort level within the facility.

Flexibility - The DDC systems can also be reprogrammed to accommodate a new sequence of operations, without the cost of additional equipment required for any changes in a comparable pneumatic panel.

Information Gathering - With the FMS, it is possible to gather and store information related to the mechanical and electrical systems within a building or any other controlled point. This information can be used to analyze trends, investigate abnormal conditions and alarms, and present results in graphic form. This feature has played a particularly important role in "selling" the system to upper management.

Increased Energy Efficiency - Naturally, the FMS has the capability of implementing various energy management strategies. A sampling of the strategies available for use by NSWCDD includes:

1. Timed Programmed Commands
2. Duty Cycling
3. Optimum Start/Stop
4. Chilled / Hot Water Temperature Reset
5. Enthalpy Optimization

Reliable Utilities Usage Information - The system has the capability of automatically monitoring water, electric and fuel oil usage to provide accurate Utilities Cost Analysis Reports (UCAR) information and billing information for tenants and service costs centers.

INTEGRATING THE TWO APPROACHES

The two approaches discussed in the preceding sections represent both the macro and micro levels of energy management. Yet, to be totally effective they must be managed in tandem. As indicated in Figure 4, this is done within the Energy and Water Management Branch using the Computer Based Facility Management System.

From a 486/PC host computer located in the EWM Office, all facets of energy management are conducted. There is a core staff of 9 highly trained people, but what really makes the system work is the empowerment of the maintenance staff with energy management responsibilities. Instead of assigning one maintenance person total responsibility for energy related matters, all staff members share responsibilities. One of the unanticipated benefits has been a steady flow of helpful suggestions for improvements from the people who really know the systems.

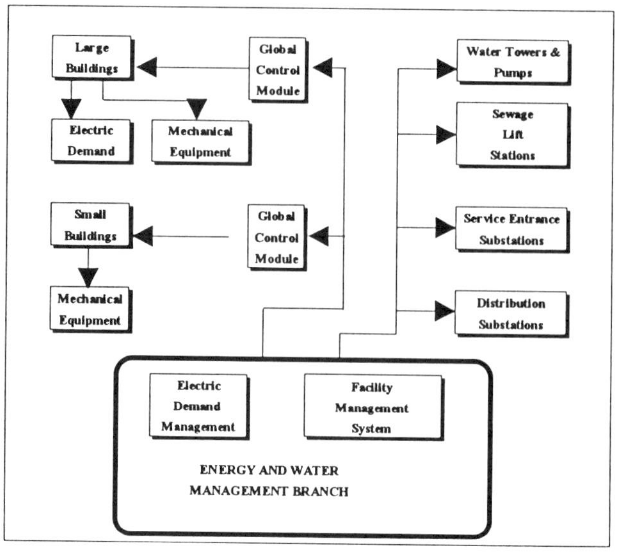

FIGURE 4 - DIAGRAM OF THE CENTRALIZED CONTROL FUNCTION OF THE FMS

CONCLUSIONS

The energy management plan for NSWCDD was developed to provide a consistent framework within which the energy management function could be carried out. Instead of attempting to reduce costs by standard conservation methods alone, we have develop a methodology which, in effect, manages the utility usage to minimize the utility bill by using both conservation and energy management techniques.

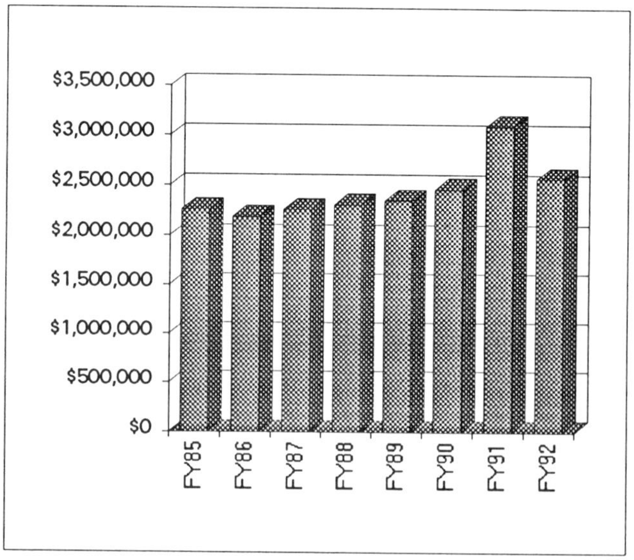

FIGURE 5 - TOTAL ANNUAL ENERGY COSTS FOR NSWCDD IN CONSTANT 1985 DOLLARS

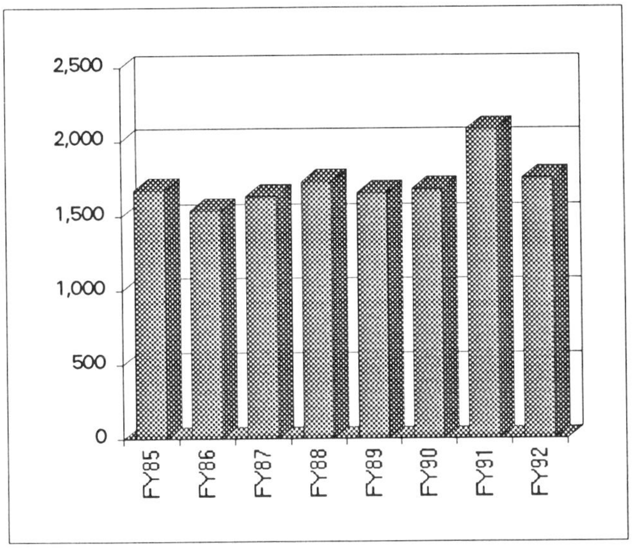

FIGURE 6 - ANNUAL TOTAL ENERGY COSTS PER THOUSAND SQUARE FEET FOR NSWCDD IN CONSTANT 1985 DOLLARS

To assess the effectiveness of the plan, we plan to rely on "Bottom-Line" techniques. That is, we will look at the data and ask, "Did we save money this year?"

Figure 5 shows the total annual energy costs for NSWCDD in constant 1985 dollars. Figure 6 presents the same data in unit measure of the cost per thousand square feet. While it is true that all the savings may not be directly attributable to the plan, we believe the data shows this program is having a marked impact on energy costs for NSWCDD.

With the recent Federal legislation requiring a substantial reductions in energy usage in federal buildings by 2000, the importance of this type of program has never been more acute.

Chapter 89

Automated Utilities Damage Assessment (AUDA) System

R. Salavani, G.C. Laventure, M.D. Smith

ABSTRACT

The United States Air Force (USAF) requires continuous availability of mission-critical utility systems for wartime, contingency, and peacetime operations. Utility systems include electrical power; petroleum, oil and lubricants (POL), potable and wastewater; heating, ventilation and air-conditioning (HVAC); and natural gas. Review of bomb damage information on facilities and utility systems from World War II, "Desert Storm", and USAF air base capability exercises; as well as natural disaster damage caused by hurricanes to USAF air bases revealed significant damage to utility systems. Widespread utility damage causes massive confusion, limits base damage assessment and repair capability, and delays the restoration of essential base operations. Utility systems management and damage assessment is a complex and slow effort involving the collection and processing of a large volume of data. Current methods of gathering this data by telephone, radio and onsite assessment are time-consuming and hazardous. Collection and processing of this critical utilities data can be expedited and simplified by the use of an Automated Utilities Damage Assessment (AUDA) system.

A Research and Development (R&D) prototype AUDA system was designed, installed, and successfully tested at Tyndall AFB, Florida in October 1992 by the Air Base Operability and Repair Section, Wright Laboratory. Wright Laboratory was assisted by Applied Research Associates, Inc, Centaurus Systems, Inc, Pate Electric and by the 325th Civil Engineering Squadron.

AUDA system testing included a variety of utility system's hardware and software. Hardware included approximately 1000 real sensor points and 250 simulated sensor points distributed throughout the various electrical, HVAC, water and waste systems on Tyndall AFB. POL and natural gas systems were not addressed during this testing because of time and dollar constraints. Monitoring and control of these sensor points were through Tyndall AFB's Energy Management and Control System (EMCS). Additionally, hardware development included modification of a 60 KW mobile generator, development of a universal Signal-Conditioner Cabinet (SCC) and Serial Interface Multiplexer (SIMUX) panel, and design of a hardware simulator. Software system development consisted of modifying EMCS, developing a new AUDA Graphic System (AGS), interfacing the EMCS and AGS software, and developing a SIMUX simulator program. Software development also involved using AutoCAD release 12, AutoCAD Development System (ADS), and Tyndall AFB's Base Comprehensive Plan (BCP) drawing files.

The AUDA system provides an enhanced graphical representation of the airbase, along with the location and status of all sensor points. It also provides enhanced reporting and limited diagnostic capabilities.

This paper will discuss the development of the AUDA system, operational capabilities, system prototype testing, conclusions and recommendations.

OBJECTIVES/BENEFITS

The main objective of this R&D effort was to develop and test a prototype AUDA system. An additional objective was to show the feasibility of using an AUDA system for wartime/contingency damage assessment and peacetime utilities management.

wartime/contingency damage assessment and peacetime utilities management.

The AUDA system is being developed to enhance Wing, Base and Civil Engineering Commander's ability to recover the base from wartime or contingency operations and to more effectively manage peacetime utility systems operations. During both wartime and contingency operations the AUDA system will be used to speed detection and isolation of utility failures, reduce exposure of base personnel to hazardous conditions, and speed up the base recovery process by reducing utility outages. During peacetime operations the AUDA system will supplement base EMCS or Supervisory Control and Data Acquisition (SCADA) systems and will provide increased status reporting and graphical capability. Eventually, the AUDA system will provide the capability for Civil Engineering (CE) Operations and Engineering personnel to access supply data and utility system operational characteristics and to perform network analysis. Use of the AUDA system will help CE maintenance personnel to identify, locate, and correct operational utility system deficiencies, resulting in increased system efficiency and reliability.

BACKGROUND

Review of bomb damage reports and photos of World War II, Desert Storm, and natural disaster damage caused by hurricanes to USAF airbases revealed significant damage to utility systems. Figure 1 shows representative bomb damage from Desert Storm to facility and utility systems. Figure 2 shows similar damage from Hurricane Andrew on Homestead AFB, FL. These photos highlight the types of damage that need to be assessed following an attack or natural disaster.

A literature review of existing capabilities in monitoring and controlling utility systems revealed that significant advances had been made in the last 6 years and that EMCS and SCADA system technology was equally applicable to wartime and peacetime operations. This review also revealed that previous USAF R&D efforts {1, 2} concentrated on paper studies and concept development of an Automated Postattack Utilities Damage Assessment System (APUDAS).

This effort {3} implemented recommendations from previous R&D studies and enhanced airbase EMCS capabilities. The AUDA system concentrated mostly on the Main Operating Base environment although some consideration was given to Bare Base operations. It used Tyndall's EMCS and BCP drawing files and combined them through modification of EMCS and development of new AGS software. Additionally, this effort developed simulation hardware (AUDA DEMO Box) and software (SIMUX Simulator) to provide proper training and testing of the AUDA concept. This effort concluded by conducting hardware and software functionality tests and by successfully field testing an integrated AUDA prototype system.

CONCEPT OF OPERATIONS

At the start of this project there was no defined Concept of Operations (CONOPS) for integration of an AUDA system. A brief review of EMCS and SCADA systems on USAF airbases showed that no systems existed for or Bare Bases, only very limited systems existed on OVERSEAS airbases, and that most continental United States airbases had EMCS or SCADA systems. Additionally, we found that EMCS equipment is purchased separately by each major command and varies considerably from base to base. We also found that existing USAF airbases use both centralized and decentralized EMCSs.

Review of wartime utilities damage assessment procedures revealed that damage assessment inputs are received by radio from CE Damage Assessment and Response Teams or by telephone from other base control centers or individuals. Damage assessment data are recorded on CE work orders and processed manually by CE Damage Control Center (DCC) personnel. In peacetime, utilities damage assessment is done by various shop personnel as situations arise, recorded on CE work orders, and processed through the normal CE job control process.

The following CONOPS was developed based on Wright Laboratory's experience and through coordination with Tyndall AFB shop personnel. During peacetime on airbases that use decentralized EMCS operations, (different shops have separate systems), an AUDA system would operate as shown in Figure 3. Each shop EMCS computer is connected to associated hardware and may be connected to an AGS post-processor. Additionally, a centralized EMCS computer and an AGS post-processor would be located at the main CE control center. This central EMCS computer would handle EMCS shop operations and have the capability to download other CE EMCS information. During wartime, contingency, or peacetime operations on airbases that use centralized EMCS operations (one system controls all shop systems), an AUDA system would operate as shown in Figure 4. The Master AUDA computer set

FIGURE 1 "Desert Storm" Bomb Damage

FIGURE 2 "Hurricane Andrew" Damage

would monitor and control shop operations from a centralized location in the EMCS shop or DCC. Up to 15 additional AGSs could be installed at the DCC via a local area network (LAN).

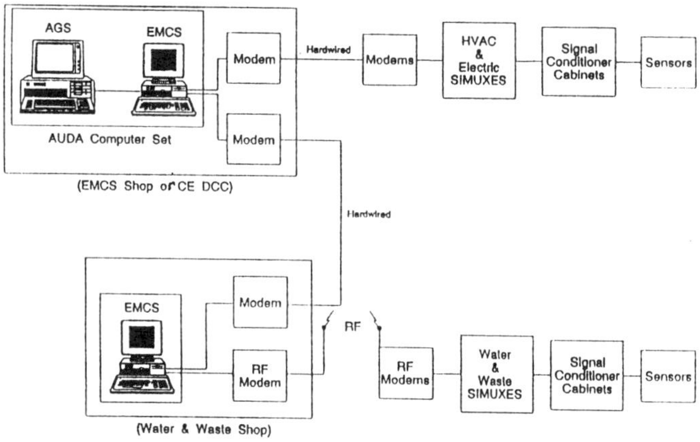

FIGURE 3 System Architecture (Decentralized EMCS)

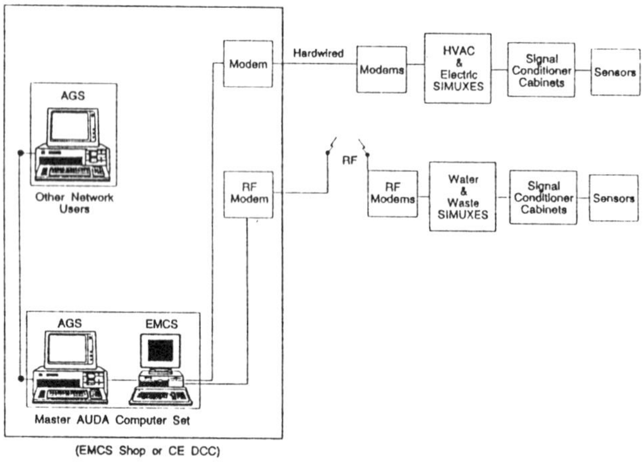

FIGURE 4 System Architecture (Centralized EMCS)

DESIGN CRITERIA

The following is the design criteria relative to this phase of AUDA development:

a) AUDA system development shall utilize existing EMCS capability, including off-the-shelf hardware and software, to the greatest extent possible. AUDA system development shall not duplicate EMCS capabilities but will enhance graphical and reporting capabilities.

b) Commonality among the AUDA system hardware and software for all tested utilities shall be maximized.

c) The AUDA system shall provide real-time damage assessment. Complete utility systems status shall be updated within 15 minutes following an attack.

d) The AUDA system shall identify failures at multiple damage locations over repeated attacks.

e) The AUDA system shall demonstrate the capability to manualy or automatically by-pass failed utilities and be capable of turning equipment on and off.

f) The AUDA system shall demonstrate redundant (hardwired and RF) data communication link capability.

g) AUDA system components shall have battery backup or generator power to support operations following loss of commercial power.

h) Any new software developed shall be coded in ADA in accordance with MIL- STD-1815A. When existing off-the-shelf software is used, conversion to ADA will be based on feasibility and cost.

i) The AUDA system shall use Tyndall AFB's BCP drawing files.

TEST ARCHITECTURE

The AUDA system test architecture is shown in Figure 5. Representative pressure, flow, current, voltage, and power sensors/meters were distributed throughout each selected utility systems. Each sensor/meter was connected to a SCC which conditioned the sensor signal. The SCC was connected to a SIMUX panel which digitized, formatted and transmitted the sensor data via hardwired or radio frequency (RF) communications equipment to the Master computer set. The EMCS computer was a 386 DX machine, running concurrent DOS, and the AGS computers were 486 DX, 33 MHz personal computers using DOS. The Master computer set processed and displayed sensor data and passed this information through a LAN to one additional network user. The SIMUX Simulator was incorporated into the test system to expand the number of test points and simulate outages. The AUDA Demonstration Box was used during testing to demonstrate AGS monitor and control functions and to simulate field handware.

The EMCS computer performed normal monitor and control functions and provided historical data files, sensor status reports, and limited graphical capability. The AGS computer performed as a post-processor with enhanced graphical and reporting capabilities. The AGS computers graphically displayed alarm/sensor data on the BCP drawings in a manner that allowed the operator to quickly determine a utilities operational status and locate the area where the utility was damaged.

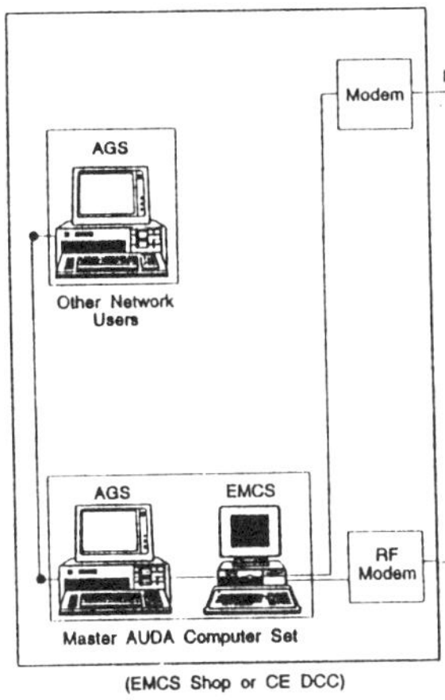

FIGURE 5 System Test Architecture

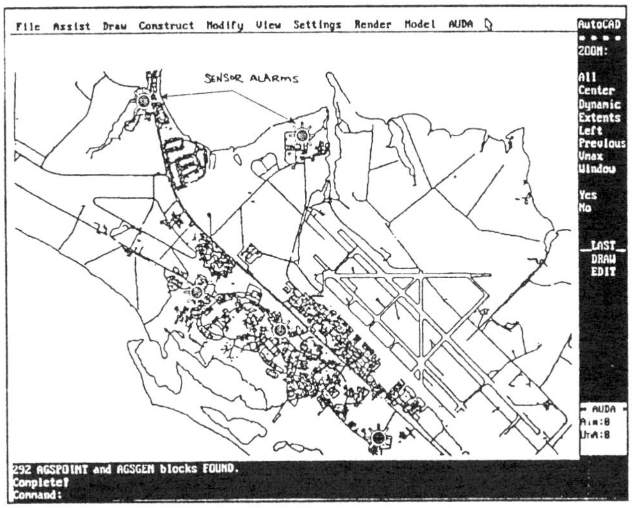

FIGURE 6 AGS Alarms Overlay

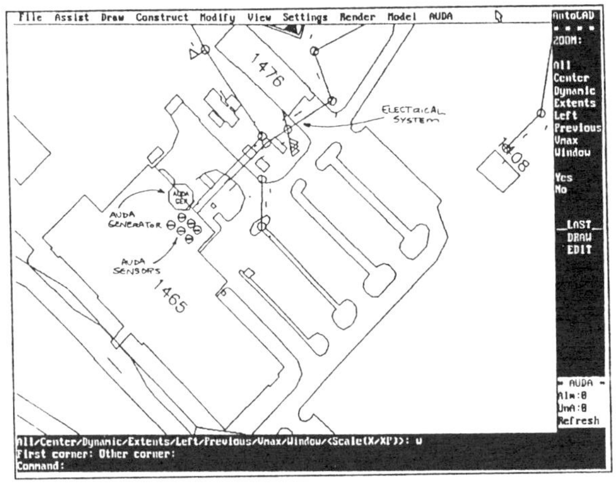

FIGURE 7 AGS Sensors Overlay

FIGURE 8 AGS Aerial Photo

OPERATIONAL CAPABILITIES

AUDA system hardware and software capabilities are discussed in detail in the AUDA technical report {3}. Some examples of AUDA system capabilities follow:

1) AGS Graphics: Figures 6 and 7 show typical AGS alarms and sensors overlaid on the Tyndall AFB BCP drawings. Figure 8 shows a scanned photo (aerial view of Tyndall AFB) and Figure 9 shows a typical building floor plan.

2) AGS Reports: Figures 10, 11 and 12 show specially designed AGS system, point and generator status summary reports.

3) AGS Diagnostics: Figure 12 was constructed based on a truth table of generator operating characteristics and Figure 13 includes diagnostic information that would be available for electrical and water outages on a fully developed AUDA system.

4) AUDA System Control Capabilities: The AUDA system controlled existing HVAC equipment, a building lighting circuit, several automatic battery test circuits, a mobile generator and a water tower altitude valve. Figure 14 shows the a photo of the mobile generator test setup.

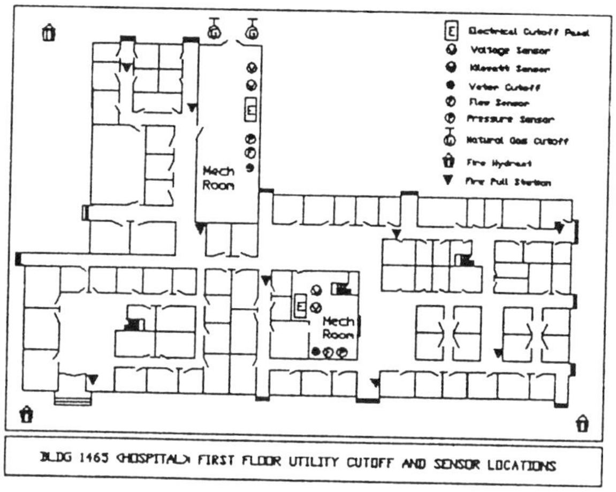

FIGURE 9 AGS Floor Plan

AGS System Status

System	Total	Exceed	Alarm
ELEC	68	0	14
HVAC	2	0	0
Water	56	0	22
Waste	20	0	1
POL	0	0	0
NGas	0	0	0
Generator	136	0	5
TOTAL	282	0	42

OK

FIGURE 10 AGS System Status Report

Index Top Bottom Update reMarks Report Location Detail eXit

Point Status Report Mon Mar 15 09:00:35

Priority	Location	Point Name	COS Date	COS Time
15	217	S217-CTRLTOW-POWER	930309	1421
2	219	S219-CMPST-G-FUEL-LV	930309	1421
2	219	S219-COMMPOST-POWER	930309	1421
2	219	S219-DP-FLOW-METER	930309	1602
2	219	S219-CP-PRESSURE	930309	1602
20	227	S227-PHA-VOLT	930309	1421
100	235	S235-DPS-FLOW-METER	930309	1602
100	235	S235-DPS-PRESSURE	930309	1602
100	235	S236-DST-TANK-LEVEL	930309	1421
48	2401	S2401-(MOB)LS7-POWER	930309	1421
48	2401	S2401-LS7-G-FUEL-LEV	930309	1421
44	249	S249-(MOB)CS-POWER	930309	1421
44	249	S249-CS-G-FUEL-LEV	930309	1421

Remarks :
Current Value : 1199.705688
Sensor Type :
System :GENERATOR

State: NORMAL

FIGURE 11 AGS Point Status Report

Index Top Bottom Update reMarks Report Location Sensors eXit

Generator Status Report Mon Mar 15 09:01:40

Priority	Location	Manufacturer	Type	Rating
1	485	CAT 2 EA		600KW
2	219	ONAN		150KW
4	216	DETROIT		175KW
5	32	ONAN		30KW
8	649	ONAN		180KW
9	909	NoData		125KW
11	1280	NoData		250KW
13	421	POW'R GARD		30KW

Remarks :
State :STANDBY

Summary

Generators On Stand-By : 26 Generators Inoperable : 0
Generators Running : 0 Generators Exceeding Limits: 0

Total Generators Available : 26

FIGURE 12 AGS Generator Status Report

AGS Fault Diagnostics

System Affected:
ELECTRIC

Condition:
Loss of base power on MISSISSIPPI feeder at substation (Bldg 1230).

Probable Cause:
Vacuum breaker trip-out or regulator failure.

Recommended Action:
On-site investigation.

OK

AGS Fault Diagnostics

System Affected:
WATER

Condition:
Excessive flow/pressure drop in main 16 inch water line at Bldg. 5002. Excessive flow/pressure drop and reverse direction in 10 inch water line at 4000 Area.

Probable Cause:
Break in water line between Bldg. 5002 and 4000 Area.

Recommended Action:
Perform street assessment between Bldg. 5002 and 4000 Area.

OK

FIGURE 13 AGS Diagnostic Information

FIGURE 14 AGS Mobile Generator Test Setup

FIELD TEST DESCRIPTION

Initial AUDA system testing involved the functional pretesting of existing EMCS hardware, new AUDA hardware, modified EMCS software, and the new AGS software. Hardware pretests included calibration, alignment and verification tests of individual sensor points and functional systems tests at each test location. Software pretests consisted of a series of functional tests to evaluate the operation of the modified EMCS and newly developed AGS software. No major problems occurred during pre-testing. The results from the functional pretests are contained in the AUDA technical report {3}.

Before the start of the integrated AUDA system field demonstration test, all major components were individually retested. The following were tested successfully:

- EMCS Hardware
- Mobile Generator
- Universal SCC and Universal SIMUX Panel
- AUDA Demonstrator Box
- Simux Simulator Program
- Master AUDA Computer Set

The following component failed to operate properly:

- Remote AGS/LAN

The Remote AGS/LAN had been tested successfully during functional software pretesting. The problem that arose during the Field Demonstration was that the Remote AGS appeared to lose or miss some network message packets being issued by the Master AGS. Initial functional software system testing did not reveal this problem because the high rate of network traffic on the "live" system was not adequately simulated during pretesting. This failure was due to a limitation of our LAN configuration and to an AGS prototype software limitation.

AUDA system testing concluded with an integrated hardware and software Field Demonstration at Tyndall AFB during the week of 5 through 9 October 1992. Testing was conducted in accordance with the AUDA Test Plan {4} which includes individual test objectives and details on operational tests performed to validate system capabilities. The Field Demonstration consisted of a series of individual tests on each utility system . And, sample electrical and water outages to demonstrate AGS software capabilities (Table 1).

TEST RESULTS

The results of the AUDA system field demonstration test (Table 1) were very successful. All field hardware tested successfully and all integrated software tested successfully with two exceptions. The first exception was the Remote AGS/LAN described earlier and the second was a communications problem between the EMCS and AGS computers. This second problem was an unforseen software problem that occurred when the Master AGS was not able to keep up with the high rate of change-of-states (COSs) that were being sent from the EMCS computer. This problem arose because the over 1000 test points were generating COSs at a rate of approximately two per second. This problem was overcome during testing by increasing the communication speed between the EMCS and AGS computers to 9600 baud from 4800 baud, by reducing the token pass interval down from 1/2 of a second to 1/4 of a second, and by locking out some SIMUX panels not necessary for the test.

Both of the software problems encountered during field testing could have be avoided if more integration time was available. These type of integration problems can be expected when software is developed at dispersed locations (Raleigh, NC and San Diego, CA) and a compressed testing schedule limits integration time.

CONCLUSIONS

1.) AUDA system development accomplished all R & D objectives. The prototype system was successfully tested at Tyndall AFB in October 1992. The developed prototype demonstrated the feasibility of using an AUDA system for wartime/contingency damage assessment and peacetime utilities management. The developed AUDA system capability would increase a base's ability to identify damaged/failed utility systems.

2.) Insufficient time was allocated for hardware and software integration testing. Hardware and software problems that occurred during pre-testing or during the Field Demonstration could have been avoided if

TABLE 1 AUDA System Field Demonstration Test Results

Test(s) Conducted	Successful/Unsuccessful		Comments
PEGASUS/EMCS Capability Tests			
a) PEGASUS/EMCS Setup	X		
b) PEGASUS/EMCS Graphical & Text Screens	X		
c) PEGASUS/EMCS Status Reports	X		
AUDA System Capability Tests (Electrical System)			
a) Electrical System Setup	X		
b) Electrical Substation, Building 1238	X		
c) Primary Distribution, Poles M202 & M141-1	X		
d) Building Loads, Secondary Input Voltage (530+1725)	X		
e) Building Loads - Energy Management (Monitoring)	X		
f) Building Loads - Energy Management (Control)	X		
AUDA System Capability Tests (Generator System)			
a) Generator System Setup	X		
b) Critical Fixed Facility (Building 1725)	X		
c) Fuel Tank Measurements (Bldgs 1446, 1640 + 1725)	X		
d) Mobile Generator (Monitoring)	X		
e) Mobile Generator (Control)	X		
AUDA System Capability Tests (Water System)			
a) Water System Setup	X		
b) Water Distribution Line, Building 5002	X		
c) Storage - Water Towers (Monitoring)	X		
d) Storage - Water Tower, Building 733 (Control)	X		
e) Building Loads - Energy Management, Building 1550	X		
AUDA System Capability Tests (Waste System)			
a) Waste System Setup	X		
b) Final Lift Station, Building 1722 & Pit	X		
c) Intermediate Lift Stations, Buildings 1446 & 1640	X		
d) Main Waste Treatment Plant, Building 1725	X		
AUDA System Capability Tests (HVAC System)	X		
a) HVAC System Setup	X		
b) HVAC Facility , Building 1550			
ADDITIONAL AUDA System Capability Tests			Note: The Remote AGS/LAN worked fine in pre-testing but failed in the Field Demonstration. The communications speed between the EMCS/AGS computers had to be adjusted during field testing due to an excessive amount of change-of-states (COS) overloading the AUDASCAN COS queue. (see text for discussion).
a) Remote AGS/Local Area Network Test		X	
b) Miscellaneous AUDA System Performance	X		
SAMPLE OUTAGES - **Electrical Feeder Outage With Backup Generator Operations**	X		
- **Commercial Water Outage**	X		

more system integration time had been available.

3.) The development of the Simux Simulator program and the AUDA Demonstrator Box extended AUDA system capabilities. However, care must be taken not to rely to much on simulation and to provide adequate "live" system testing.

RECOMMENDATIONS

1.) AGS development should continue to improve diagnostics and graphics capabilities. Use of utility system network analysis programs, expert systems, and automated utilities mapping programs should be considered.

2.) AGS system design should be modified to a "Server/Client" architecture that more evenly distributes the AGS processing load and allows for operation with or without an EMCS computer.

3.) AGS software system design should consider "porting" the AGS software to a multi-tasking environment, such as Windows 3.1 (and ultimately Windows NT) or going to workstation/UNIX environment.

4.) AUDA system development should investigate Reliability, Maintainability, Supportability, and Survivability of EMCS/SCADA systems and the communication paths that support these systems, to ensure that these systems and communication paths will support AUDA system operations.

1. Merkle, D.H., D'Arcy, G.P. and Rochefort M.A.," Automated Postattack Utilities Damage Assessment System (APUDAS) For Sewerage And Mission-Critical HVAC Systems," ESL-TR-90-18, HQ AFESC, Tyndall AFB, FL, October 1990.

2. Storm, J. H., Moats R.K. and Fertal M. J., " Automated Postattack Utilities Damage Assessment System (APUDAS)," ESL-TR-88-73, HQ AFESC, Tyndall AFB, FL, December 1990.

3. Laventure, G. C., Smith, M.D., Weston, D. J., and Caldwell, B., " Automated Utilities Damage Assessment (AUDA) System Validation And Field Demonstration," WL-TR-93-3508, Wright Laboratory, WL/FIVCO, Tyndall AFB, FL, June 1993.

4. Test Plan For Automated Utilities Damage Assessment (AUDA) System Validation And Field Demonstration, HQ AFCESA/RACO, Tyndall AFB, FL, 23 September 1992.

Chapter 90

Federal Energy Decision Screening (FEDS) Process at Fort Drum, New York

D.R. Dixon, K.K. Daellenbach, S.E. Rowley, A.H. Gillespie

ABSTRACT

The federal energy manager has been directed by the Comprehensive Energy Policy Act of 1992 (EPAct) to reduce energy consumption by 20% from 1985 levels by the year 2000. However, the tools and funding to capture this resource in a cost-effective manner have not been provided. In an effort to assist federal agencies in meeting EPAct requirements, the Pacific Northwest Laboratory (PNL)[(a)] has been tasked by the U.S. Army Forces Command (FORSCOM) to identify, evaluate, and acquire all cost-effective energy projects at selected federal facilities. PNL has developed and applied the Federal Energy Decision Screening (FEDS) methodology at the Fort Drum FORSCOM facility near Watertown, New York. The FEDS methodology is a systematic approach to evaluating energy opportunities that result in a roadmap to significantly reduce energy use in a planned, rational, cost justified fashion over a 5 to 10 year period. At Fort Drum, the net present value (NPV) of the installed cost of all cost-effective energy resource opportunities (EROs) is over \$16 million (1992 \$). The NPV of the savings associated with this investment is nearly \$47 million (1992 \$), for an overall NPV of approximately \$31 million. By implementing all the cost-effective EROs, Fort Drum will reduce annual energy use by over 230,000 MBtu, or 15%. Annual energy expenditures will decrease by over \$2.4 million, or a 20% reduction.

(a) Pacific Northwest Laboratory is operated for the U.S. Department of Energy by Battelle Memorial Institute under contract DE-AC06-76RLO 1830.

INTRODUCTION

The Comprehensive Energy Policy Act of 1992 (EPAct) [1] directs federal agencies to reduce energy consumption by 20% from 1985 levels by the year 2000. In an effort to assist federal agencies in meeting EPAct requirements, PNL has been tasked by the U.S. Army Forces Command (FORSCOM), as the lead laboratory supporting the U.S. Department of Energy (DOE) Federal Energy Management Program (FEMP), to identify, evaluate, and acquire all cost-effective energy projects at selected federal facilities. PNL has developed and applied the Federal Energy Decision Screening (FEDS) methodology at the Fort Drum FORSCOM facility near Watertown, New York. This paper presents a description of the FEDS process as applied to a case study of Fort Drum, the complete results of which are documented in PNL-8424 [2,3].

FEDERAL ENERGY MANAGEMENT

Some of the most difficult problems that a federal site has in reducing its energy consumption in a cost-effective manner revolve around understanding where the energy is being used, and what technologies could be employed to decrease the energy use. Many large federal sites have one or two meters to track electric energy use for several thousand buildings and numerous industrial processes. Even where meters are available on individual buildings or family housing units, the meters are not consistently read. When the federal energy manager has been able to identify high energy users, he or she may not have the background, training, or resources to determine the most cost-effective options for reducing this energy use. This can lead to selection of suboptimal choices that prevent the site from achieving the full energy savings potential.

The FEDS process is a systematic approach to evaluating energy opportunities that 1) identifies the building groups and end uses that use the most energy (not just have the greatest energy-use intensity), and 2) evaluates the numerous options for retrofit or installation of new technology that will result in the selection of the most cost-effective technologies. In essence, FEDS provides the federal energy manager with a roadmap to significantly reduce energy use in a planned, rational, cost-effective fashion that is not biased by the constraints of the typical funding sources available to federal sites. The results from FEDS can easily be turned into a five- to ten-year energy management plan that identifies where to start and how to proceed in order to reach the mandated energy consumption targets.

Another part of the FEDS process is to work with electric and fossil fuel utilities to develop a demand-side management (DSM) program that supports the site's energy management goals. Federal sites are authorized and encouraged to actively participate in DSM programs offered by regulated public utilities. The goal of FEDS is to negotiate a DSM program where the utility will:

- finance 100% of the installed cost of the cost-effective energy projects,

- enlist the services of an energy services contractor (ESCO) to design, install, maintain, and evaluate the equipment acquired in these projects, and

- cost share a substantial portion of the total cost installed cost.

FEDS: THE FIVE-STEP PROCESS

The FEDS process consists of five distinct steps: screening, analysis, negotiation, implementation, and verification. The process begins with the screening step in which the project's potential viability is determined. The analysis step identifies, evaluates, and prioritizes EROs at the site. The negotiation step includes a review of the integrated resource plan (IRP) developed in the analysis step, along with utility DSM programs and other federal agency funding sources, to determine a strategy for implementation. The implementation step involves enlisting the assistance of an ESCO who will conduct detailed audits and install technologies. The final step verifies energy savings through an established metering program.

Step 1: Screening

The purpose of the screening step is to provide information about the project's viability, from which the initial decision to proceed with a FEDS assessment will be made. A FEDS project manager is contacted by a prospective federal client with a need to improve energy efficiency at a site. The client will most likely be the facility energy manager and/or the major command. Federal client management supports the assessment with staff, data, and other resources. Following up this contact, the FEDS manager initiates conversations with the prospective federal client and the utility servicing the federal client. Interest is identified and groundwork layed for subsequent interaction.

Preliminary information gathered through phone calls, meetings, and a site visit assist in the initial identification and characterization of potential cost-effective energy projects. Information to be collected includes facility location and purpose; energy consumption and price by fuel type; and quantity, operation schedule, and age of facility buildings and energy-consuming equipment (e.g., boilers, furnaces, pumps). In addition, the utility is studied to understand how its resource mix and system energy balance will respond to changes in future load. This information includes resource generating capacity, system load characteristics (e.g., utilization factors for residential, commercial, industrial, and federal customers), and avoided energy costs. From this, a preliminary estimate of savings potential is made.

To complete this step, an evaluation of the project's potential viability is assessed by considering the energy savings potential for the federal customer and utility interest in developing a prototype DSM program for this customer. If the decision is made to continue beyond the screening phase, a schedule for assessment deliverables is established and the analysis step begins.

Step 2: Analysis

A detailed study is conducted in the analysis step to identify, evaluate, and prioritize energy resource opportunities (EROs) available to the site. First, energy consumption is characterized by fuel type, energy end use, and building type. This assessment of baseline energy use is more rigorous than the rudimentary analysis of the screening step, as it considers a more comprehensive set of data.

An initial data request is sent to and reviewed by the client. This list includes specific electronic and paper documents such as real property lists; scope and distribution of energy conservation projects; fuel consumption records; and inventories indicating model, capacity, age, and operating hours for energy-consuming equipment. Following up this data request is a site visit to collect additional information about typical buildings and equipment. This first-hand inspection allows one to become more familiar with the facility and the specific needs at each site. In addition, evaluation of metered data provides important information useful in determining electric consumption and energy-use intensities for various site facilities. Collecting this data is an iterative process which requires extensive collaboration and review of data with site personnel.

Second, EROs are identified and evaluated using the federally mandated life-cycle cost (LCC) methodology. Based on engineering experience with other energy assessments, discussion with site personnel and external expert energy consultants, promising EROs are selected from a master list for further LCC evaluation. This preliminary screening considers technology maturity; resource conservation potential; economic feasibility; utility incentives, rates, and cost sharing level; impact on mission-critical function; data availability; climate; condition of site equipment and buildings; and site preference and experience.

With new experience gained at each site, the master ERO list expands. ERO categories include: boilers, furnaces, domestic hot water (DHW), building envelope, fans, pumps, motors, compressors, air conditioning, lighting, transmission and distribution, process-related energy use, fuel-switching and alternative or renewable energy. The performance of each ERO is evaluated for energy saving based on thermodynamic, heat transfer, and other engineering calculations. Baseline energy use information and assumptions such as climate, operating hours, equipment condition, and age are inputs for these calculations. Installed cost calculations are determined using current product literature, published costs, and vendor quotes. Environmental impact of each ERO is also considered. Next, the LCC economic analysis employs a methodology detailed in and mandated by 10 CFR Part 436 [4] to determine the ERO's net present value (NPV).

Finally, EROs are prioritized in a fuel-neutral manner by NPV. It is this ranking which defines the optimal return on site and utility investment and from which negotiations can begin.

Step 3: Negotiation

The negotiation step strives to produce a contract that matches the utility's program offering with that proposed by the federal customer. It begins with an iterative review and refining process involving the federal client and servicing utility. Efficiency improvement calculations and estimates of individual ERO cost-effectiveness in the IRP are agreed upon and accepted as an accurate representation of potential savings. This review also examines the willingness of the utility to modify, if necessary, its DSM program offering to include the range of EROs proposed by the federal customer.

Broader input from federal and/or private organizations may be necessary. For example, Public Utility/Public Service Commissions (PUC/PSCs) may be called upon to encourage the utility to develop an implementation plan acceptable to the federal customer. In addition to utility negotiation of implementing specific EROs or groups of EROs, contracting requirements for ESCO services are discussed. Difficult technical and procedural issues are resolved. The result is a successful contract between the utility and federal customer which determines a strategy for implementation.

Step 4: Implementation

This step is closely associated with the preceding negotiation and results in installation of equipment. The negotiated contract is developed using a systematic, objective, and rational approach to install technologies in the most cost-effective manner. The selection and ordering of projects considers: NPVs of individual and groups of EROs, verification requirements of the utility and federal customer, expected energy savings, impact on mission-critical requirements, and available funding from different sources. Project implementation periods may range from five to ten years. From this contract and using the FEDS assessment as a guide, the contracted services of one or more ESCOs are enlisted in a competitive bidding process. The ESCO conducts detailed energy-use audits, and designs, procures, installs, and maintains the new and retrofit equipment. The federal customer approves and inspects all ESCO projects. The utility pays the ESCO for the installed cost of the project. The federal customer pays the utility so it may recover its portion of the project cost.

Step 5: Verification

This final step of the FEDS process relates back to the mandate of the EPAct. By evaluating energy consumption after successful completion of the implementation step and then comparing it with consumption prior to implementation (baselined in the analysis step), the federal customer will be able to verify how successfully the statutory energy savings requirements of the EPAct are being met.

Utilities will direct the verification process and will rely on fuel consumption, metering programs, and other accurate methods established with the assistance of the contracted ESCO. Initial assessments of energy savings upon implementation as well as detailed tracking of retained energy savings over time will be necessary to meet utility and DoD retained energy savings requirements. The actual difference between what would have been consumed by pre-implementation equipment and what was actually consumed by post-implementation equipment cannot be directly measured. One can only measure pre-implementation and post-implementation energy consumption and extrapolate pre-implementation consumption to the post-implementation period using weather, building occupancy, and other factors to normalize the separate periods.

FEDS AT FORT DRUM, NEW YORK

Fort Drum provides training and support to the 10th Mountain Division of the U.S. Army, and had over 10,000 active military personnel in 1991. Most of these military personnel and their dependents live in housing provided by the Army either on the Fort, or in leased housing off post. In addition, the Fort employed over 1,500 civilian personnel in 1991 that do not live in military housing. Including all the military dependents, the total Fort population can exceed 30,000 people during the working week.

The Fort was divided into six major categories for the FEDS analysis. These categories include old post commercial, new post commercial, on-post housing, on-post utility (including exterior lighting, pumping, and transmission and distribution losses), off-post housing, and the off-post commercial (one large office complex is located in Syracuse, New York). The old post commercial consists primarily of about 700 one- and two-story frame structures constructed in the 1940s. Many of these facilities have been renovated over the years with metal siding, wall and attic insulation, storm windows, fluorescent lighting, and fuel oil furnaces and boilers. These old post facilities are used for barracks, office space, dining, warehouse/storage, recreational purposes, medical services, etc. The new post commercial was constructed between 1985 and 1988, and consists of about 200 large, concrete block and brick buildings. Most of these facilities are used as barracks, office space, dining, motor vehicle maintenance, recreation, retail sales, motel, and data processing. These new post facilities are heated with high-temperature hot water (HTHW) from a central, coal-fired co-generation plant operated by a private party.

The off- and on-post housing consists of single family, 2- and 4- plexes, and multi-apartment complexes, with over 4,000 separate buildings. Most of these units are relatively new, and use natural gas or propane for heating and cooking. Some of the older units are all-electric.

Energy Use Baseline

The first step in evaluating energy savings opportunities at Fort Drum was to gather information on the energy usage across the Fort. On the electric side, there were four sources of electric metering and consumption data: utility whole post meter readings, off-post housing meters, on-post meters (some buildings, areas, other), and substation feeder readings made by PNL. The utility readings and housing meters were used to determine the total electric energy consumption by building category (i.e., old post commercial, etc.). The substation feeder readings were collected for over a year in 15-minute intervals, and gave insight into the time of day consumption at the Fort (see Figure 2). This feeder data, along with estimates of energy-use intensities (EUIs), were used to reconcile the energy consumption by building type and end-use technology (lighting, heating, cooling, etc.). The energy that could not be directly associated with an end-use was included in the other category (plug loads, process equipment, line losses, etc.). These data are shown in Figure 1.

Table 1 shows a summation of the typical yearly energy consumption for all facilities at Fort Drum. For each energy type, the yearly total is shown in units appropriate to the energy type and in a common unit as a basis for comparison. The total consumption values represent typical current yearly usage, from the best available data during 1990 and 1991. The yearly energy consumption was 1,600,903 MBtu.

Energy Resource Opportunities

After gaining a better understanding of the energy usage, the next step was to identify and evaluate the EROs at Fort Drum, provide the LCC analysis for each ERO, and prioritize the EROs based on their NPV. The NPV is defined as the difference between the LCC of the existing technology and that of the proposed alternative. Analysis results were presented in 14 common energy end-use categories (i.e., building lighting; streetlighting; HVAC system; refrigeration; boilers and furnaces; building envelope; high-temperature hot water (HTHW) systems; domestic hot water heating; fan, pump and compressor

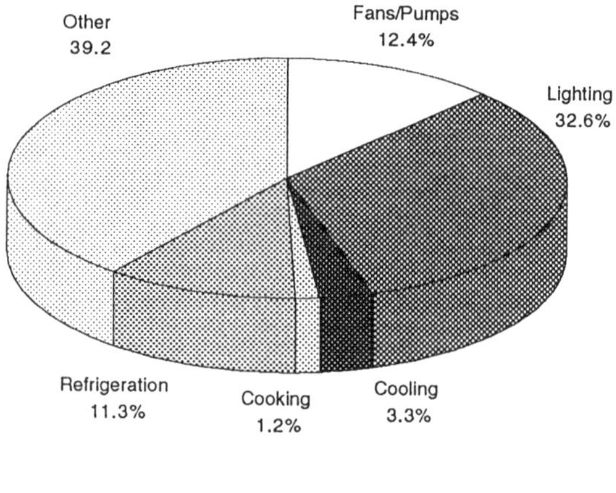

FIGURE 1. ENERGY USE BY END USE FOR ELECTRICITY (MBtu)

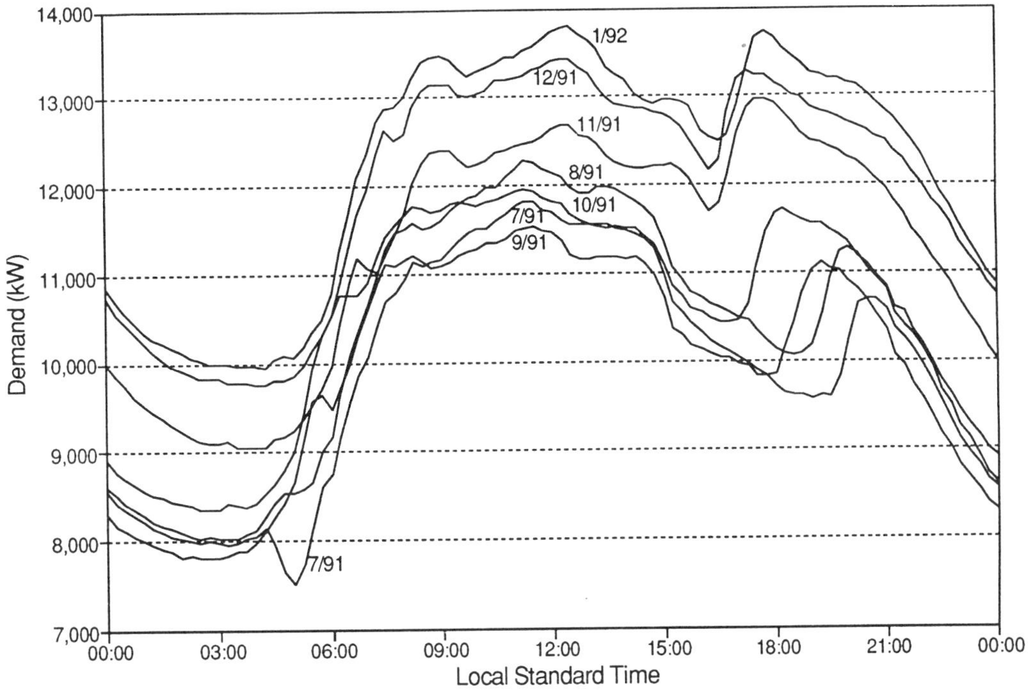

FIGURE 2. MONTHLY AVERAGE WEEKDAY DEMAND PROFILES AT FORT DRUM, NEW YORK (07/91 - 01/92)

TABLE 1. TYPICAL YEARLY ENERGY CONSUMPTION AT FORT DRUM

Energy Type	Yearly Total	Yearly Total [MBtu[a]]	Percent of Total Energy Consumption
Electricity	105,849 MWh	361,260[b]	22.6
Natural Gas	3,857 ktherm	385,720[c]	24.1
#2 Fuel Oil	2,407 kgallon	334,146[d]	20.9
Propane	652 kgallon	59,524[e]	3.7
HTHW	458,048 MBtu	458,048	28.6
Coal	98 Ton	2,205[f]	0.1
Totals:		1,600,903	100.0

(a) 1 MBtu = 1,000,000 Btu.
(b) 3,413 Btu/kWh.
(c) 100,000 Btu/therm; 1,050 Btu/ft^3 of natural gas.
(d) 0.1388 MBtu/gal of #2 fuel oil.
(e) 0.0913 MBtu/gal of propane.
(f) 22.5 MBtu/ton of coal.

motors; transmission and distribution; fuel-switching; electric demand reduction measures), and four building groups representative of Fort Drum: new post commercial, old post commercial, on-post family housing, and off-post family housing. Savings were indicated by fuel type: fuel oil #2, natural gas, propane, HTHW, and electricity.

After applying the analysis step described in the FEDS process section above, the following results were obtained (see Table 2). The present value of the installed cost of all cost-effective EROs at Fort Drum is over $16 million (1992 $). The present value of the savings associated with this investment is nearly $47 million (1992 $), for an overall NPV of approximately $31 million. By implementing all the cost-effective EROs, Fort Drum will reduce annual energy use by over 230,000 MBtu, or 15%. Annual energy expenditures will decrease by over $2.4 million, or a 20% reduction.

Table 3 provides a complete breakdown and summary of the electric EROs at Fort Drum.

The lighting ERO category has the greatest energy conservation potential at Fort Drum. Lighting EROs alone represent almost 70% of the potential reduction in electricity use identified in this study and include the replacement/retrofit of existing lighting equipment with T-8 bulbs, electronic ballasts, optical reflectors, high-pressure sodium and compact fluorescent fixtures, and occupancy sensors.

The use of energy-efficient and two-speed motors accounts for 12% of the kWh savings, and transmission and distribution EROs such as power factor correction and conservation voltage reduction account for slightly under 3% of total electrical savings.

Demand charges constitute approximately 10% of Fort Drum's monthly electric bill. These charges can be reduced by using existing emergency service diesel generators to carry part of the on-post electric load whenever demand approaches the monthly peak. The most reliable and effective way to accomplish this is to modify selected generators, so that they can be used as a power source that can feed into Fort Drum's electrical distribution system.

A summary of the savings potential categorized by fossil fuel EROs is presented in Table 4.

The old post commercial buildings, consisting predominantly of World War II-era buildings heated with fuel oil, offers opportunities for fuel-switching to natural gas and replacement or conversion of aging and inefficient

TABLE 2. TOTAL SAVINGS, COST, AND NPV (1992 $)

Total Present Value of Installed Cost	Total Present Value of Energy Savings	Total Present Value of Demand Savings	Total Present Value of O&M Savings	Total Present Value of All Savings	Total Net Present Value
$16,478,259	$34,018,474	$6,233,134	$6,621,546	$46,873,154	$30,644,466

TABLE 3. SUMMARY OF THE ELECTRIC EROS AT FORT DRUM

Energy Resource Opportunity Category	Present Value of Installed Cost[(a)] (1992 $)	Annual Energy Savings (kWh)	Value of Annual Energy Savings (1992 $)	Value of Annual Demand Savings (kW-mo)	Annual Demand Savings (1992 $)	Total Net Present Value (1992 $)
Air Conditioning	31,850	25,093	1,470	225	1,241	8,393
CO_2 Sensors	56,430	154,028	7,659	0	0	37,639
Lighting	7,940,028	18,078,203	1,011,494	46,501	256,218	16,578,499
Fluorescent	4,227,477	9,512,174	513,182	28,250	155,656	7,014,250
Incandescent	3,123,671	6,922,986	419,014	15,396	84,833	6,551,533
Exits	530,740	1,020,584	50,825	1,398	7,703	2,476,115
HIDs	58,140	622,458	28,473	1,457	8,026	536,600
Motors	477,618	3,387,177	172,059	-1,907	-11,635	1,549,979
Peak-Shave	194,000	4,000,000	234,360	31,200	171,912	1,362,873
Streetlights	83,784	1,073,557	49,735	2,596	14,304	938,318
Trans. & Dist.	8,522	756,954	38,291	1,332	7,340	883,150
Weatherization	63,850	197,066	9,767	583	3,212	118,860
Total	$8,856,082	27,672,078	$1,524,835	80,530	$442,592	$21,477,711

(a) Analysis assumes no utility cost-sharing.

TABLE 4. SUMMARY OF FOSSIL FUEL EROS AT FORT DRUM

Energy Resource Opportunity Category	Present Value of Installed Cost (1992 $)	Annual Energy Savings (MBtu)	Value of Annual Energy Savings (1992 $)	Total Net Present Value (1992 $)
Peak-Shave	194,000	-43,451	-216,000	1,362,873
CO_2 Sensors	56,430	2,221	9,539	37,639
Controls	66,090	6,790	31,112	458,647
Furn. & Boilers	2,953,444	51,658	362,185	2,752,818
Hot Water	710,524	8,679	70,730	448,266
Trans. & Dist.	70,130	52,750	226,561	3,749,684
Weatherization	3,821,989	60,513	318,136	1,757,340
Total	$7,872,607	139,161	$802,263	$10,567,266

(a) Analysis assumes no utility cost-sharing.

boilers, furnaces, and water heaters to new high-efficiency units.

Although most of the buildings on the old post have been renovated over the years and have storm windows or double-pane glass and some thermal insulation, significant opportunities still exist for weatherization measures such as weatherstripping, floor and additional ceiling insulation, and window quilts.

The heating systems in many of the old post buildings are currently regulated by a radio-control system that consists of a central computer and an FM transmitter, with a receiver and a day-thermostat and a night-thermostat in each building. Fort Drum has already purchased much of the hardware required to add most of remaining old post buildings to the FM control system. A very cost-effective opportunity exists by funding the installation of this equipment.

Large energy and dollar savings can also be realized through the repair or replacement of pipe insulation on the HTHW distribution system servicing the new post. In addition, several sections of the HTHW loop that currently do not provide service to buildings could be decommissioned, resulting in reduced pipeline heat loss and additional dollar savings.

On- and off-post housing units, although relatively well insulated, are prone to draftiness. A "weatherization package" consisting of attic hatch insulation, switchplate gaskets, caulking, and filling plumbing holes and other penetrations with insulation was found to be life-cycle cost-effective, as were window quilts, which provide insulation and reduce infiltration. Finally, the all-electric, on-post family housing units have insufficient ceiling insulation; increasing the insulation in these units up to an R-60 level will be life-cycle cost-effective.

Implementation Plan

The purpose of the FEDS process is to develop an analytical and rational approach to reducing the energy consumption (and energy cost) at Fort Drum. The implementation step of this process reviews the EROs, groups them into manageable projects, and develops a five-year energy management plan. When fully developed this plan would discuss the types of projects, timing, sources of funding, and other considerations for Fort Drum. In addition, a strategy for working with the electric and gas utility (Niagara Mohawk) would be part of this plan, in order to take full advantage of the utility incentives offered through demand-side management.

An implementation planning workshop was held with U.S. Army staff to develop a strategy for implementing the opportunities identified during the FEDS process. Resulting from the workshop were nearly 400 EROs divided into nine projects to be implemented over a three- to five-year period. Major projects to be initiated in the early years included:

- A fossil fuel project to provide natural gas for space heating and water heating on the old post; to enhance the FM control system on the old post; and to decommission, insulate, and provide package boilers for parts of the HTHW system on the new post. The space and water heating element converting fuel oil to natural gas use avoids emissions from fuel oil use and retrofit of aging storage facilities.

- An electric project to retrofit lighting on the new/old post, provide 2-speed motors for building ventilation and sewer lift stations, replace and reconfigure streetlights throughout the Fort, and improve the site's power factor by installing capacitors, which corrects for the load's reactive component.

- A transportation project to convert portions of the existing vehicle fleet to compressed natural gas (CNG)-powered vehicles. Niagara Mohawk may pay for the refueling station, and government funding could be used to convert the vehicles.

- A maintenance project to reduce the HTHW supply temperature from the central steam plant. This project involves working with the contractor that operates the central steam plant to lower the supply temperatures to reduce the thermal losses from the distribution pipeline.

Other projects to be implemented in later years were also identified.

CONCLUSIONS

The federal energy manager needs a systematic process by which energy reduction can be achieved in a cost-effective manner. The FEDS process is designed to:

- conduct a fuel-blind analysis to identify all potential electric and fossil fuel energy efficiency, fuel switching, demand management, self-generation, improved operations and maintenance, energy purchase, etc., energy projects,

- evaluate the cost effectiveness of these projects based on the federally mandated life-cycle cost analysis procedures, and

- acquire all cost-effective energy projects at selected federal facilities through working with local utilities to fund, finance, and manage these activities.

This process is attractive to utilities to reduce DSM administrative costs by working with one large federal customer and benefit from a low-cost source of energy resources by avoiding large capital investment in new generating capacity. The federal customer will reduce energy costs and be able to verify how successfully statutory energy savings requirements are being met. At Fort Drum, the FEDS process is being successfully applied and the desired objective—to significantly reduce energy consumption at the Fort—is well on the way to being realized.

ACKNOWLEDGMENTS

The authors gratefully acknowledge the effective and creative staff at Pacific Northwest Laboratory involved in developing the FEDS process, the U.S. Army Forces Command, and the significant and conscientious assistance provided by the Fort Drum staff.

REFERENCES

1. The Comprehensive Energy Policy Act of 1992 (EPAct). Public Law PL 102-486, Washington, D.C., October 24, 1992.

2. D. R. Dixon, P. R. Armstrong, J. R. Brodrick, K. K. Daellenbach, F. V. Di Massa, J. M. Keller, E. E. Richman, G. P. Sullivan, and R. R. Wahlstrom. December 1992. *Fort Drum Integrated Resource Assessment, Volume 2: Baseline Detail*. PNL-8424 Vol. 2, Pacific Northwest Laboratory, Richland, Washington.

3. D. R. Dixon, K. K. Daellenbach, J. E. Dagle, F. V. Di Massa, D. B. Elliott, J. M. Keller, E. E. Richman, S. A. Shankle, G. P. Sullivan, and R. R. Wahlstrom. December 1992. *Fort Drum Integrated Resource Assessment, Volume 3: Resource Assessment*. PNL-8424 Vol. 3, Pacific Northwest Laboratory, Richland, Washington.

4. 10 CFR Part 436. 1990. U.S. Department of Energy. "Federal Energy Management and Planning Program." (Revision as of January 1, 1990). *U.S. Code of Federal Regulations*.

BIBLIOGRAPHY

Defense Energy Program Policy Memorandum (DEPPM), 91-2, Washington, D.C. (March 19, 1991).

Executive Order of Federal Energy Management, E.O. 12759. *Federal Register*, Vol. 58, No. 76 (April 19, 1991).

G. B. Parker, T. J. Secrest, N. Flood, and A. Gillespie. "DSM/Federal Energy Efficiency Partnership Program at Fort Lewis, Washington." PNL-SA-21066, Pacific Northwest Laboratory, Richland, Washington. In *Proceedings of the 15th World Energy Engineering Congress*, October 27-30, 1992. pp. 177-184.

J. W. Currie, L. R. Harris, and G. B. Parker. February 1993. "Designing and Implementing DSM Programs for Federal Customers." PNL-SA-22045, Pacific Northwest Laboratory, Richland, Washington. In *Proceedings of the Integrated Resource Planning Conference*, March 18-19, 1993, Denver, Colorado.

J. W. Currie, R. W. Reilly, R. W. Brancato, and J. A. Dirks. 1991. "Modernizing Federal Energy Systems: The FEDS Development Approach." *Energy Engineering*, 88(4):6-19.

J. W. Currie. "Testimony by J. William Currie, Ph.D., Manager, Energy Systems Modernization Office, Pacific Northwest Laboratory, before the Senate Committee on Governmental Affairs, Washington, D.C., 18 February 1992." PNL-SA-20228, Pacific Northwest Laboratory, Richland, Washington.

Chapter 91

Implementing Energy Performance Standards in Commercial and Federal Buildings

C.S. Warren, P.F. Hutchins

ABSTRACT

On January 30, 1989, the U.S. Department of Energy (DOE) published 10 CFR Part 435 Subpart A, which established energy conservation voluntary performance standards for the design of new commercial and multi-family high-rise residential buildings. While the standards are strictly voluntary for the private sector, their use is mandatory for designing new Federal buildings.

Standards are set for building envelope, auxiliary systems and equipment, lighting, and energy management. There are several methods of achieving compliance with the standards once the minimum compliance requirements for equipment, systems, and the building envelope have been established:

1. Prescriptive compliance - used when the minimum amount of calculation and effort to achieve compliance is desired; allows few design trade-offs;

2. System performance compliance - allows more flexibility in building design; requires more calculations than the prescriptive method;

3. Building energy/cost compliance - allows more innovative design concepts; requires computer simulation of building energy use.

Each of the alternatives is examined and explained. Application to design of a commercial building is shown.

INTRODUCTION

The Energy Conservation Standard for New Buildings Act of 1976 was passed by Congress to establish standardized energy efficiency measures for the buildings industry. Under the terms of the Act, the U.S. Department of Energy (DOE) was to develop, issue, implement, and monitor energy performance standards for the design of new buildings.

As part of the implementation of the Act, in January 1989 DOE published interim energy conservation standards for the design of new commercial and multifamily buildings (10 CFR Part 435 Subpart A). These interim standards became mandatory for all federal agencies in July 1989. Since the interim standards are voluntary for the private sector, DOE has worked closely with the American Society of Heating, Refrigerating and Air-Conditioning Engineers, Inc. (ASHRAE) to define standards consistent with the federal interim standards. Late in 1989, ASHRAE issued a similar commercial building standard, ASHRAE/IES Standard 90.1-1989.

On October 24, 1992, the Energy Policy Act of 1992 was signed into law (PL 102-486). The Act amended Title III of the Energy Conservation and Production Act to include updating of state residential and commercial building codes. Under the 1992 Act's requirements, each state must certify by October 24, 1994, to the DOE Secretary, that its commercial building code meets or exceeds the requirements of ASHRAE/IES Standard 90.1-1989. The Secretary of DOE must also establish, by rule, Federal building energy standards within the same time period. The federal standards must become enacted within one year of publication.

In the meantime, the interim standards remain in effect, and any buildings designed for federal agencies must meet the standards.

APPLICATION OF THE INTERIM STANDARDS

The Interim Performance Standards establishes minimum performance standards for lighting, auxiliary systems and equipment, building envelope, electric power and distribution, HVAC systems and equipment, service hot water, and energy management. Flexibility of application of the standards is fulfilled by offering three alternative methods to follow; (1) Prescriptive/System Performance Compliance, (2) Building Energy Cost Compliance, and (3) Building Energy Compliance. The alternative methods of compliance are illustrated in Figure 1.

The Prescriptive/System Compliance involves two applications of the least complicated of the alternatives. A minimum amount of calculations are required under the Prescriptive method, and its requirements can easily be specified in construction documents. If more design flexibility is required, then the System Compliance method may be used. However, the System method involves more calculations.

Either of the two Building Energy methods may be used when more innovative building designs are preferred. These methods allow exchanges of energy consumption accountability between building systems as long as the total calculated design annual energy consumption does not exceed the limit prescribed. These applications are relatively more complicated than the above methods, and require computer simulation.

Minimum requirements for the components address the following:

1. Lighting - controls and ballast efficacy factors;

2. Auxiliary Systems - elevators/conveyors, freeze protection systems, and retail food and food service refrigeration;

3. Building Envelope - overall thermal transmittance, below grade thermal resistance, envelope thermal transmittance, gross area of envelope components, shading coefficients, wall heat capacity, air leakage and moisture migration, shell buildings, buildings located in climates with greater than 15,000 heating-degree-days (HDD), and daylight credits for skylights;

4. Electric Power and Distribution - electrical distribution system, transformers, and electric motors;

5. HVAC Systems - calculation procedures, system and equipment sizing, separate air distribution systems, temperature controls, off-hour controls, humidity control, materials and construction, and completion requirements.

6. HVAC Equipment - equipment efficiency, field assembled equipment and components, equipment controls, comfort heating equipment, and maintenance;

7. Service Hot Water - sizing of systems, equipment efficiency, piping insulation, controls, equipment and control requirements for hot water conservation, and swimming pools;

8. Energy Management - system metering, instrumentation, HVAC system controls, central monitoring and control systems, completion requirements, energy performance testing, and documentation requirements.

PRESCRIPTIVE/SYSTEM COMPLIANCE

Figure 2 illustrates the methodology for the application of the Prescriptive/System compliance procedures. The proposed building design must first meet the minimum standards listed in the above paragraphs. Two components, lighting and building envelope, may be satisfied through application of either prescriptive or performance standards. In addition, performance requirements for the HVAC systems and the service hot water systems must be met. In the cases of lighting and the building envelope, the performance standards application offers more flexibility in that certain credits may be taken for energy-conserving design, such as the use of skylighting. Both methods rely on energy budgets and require calculations to arrive at the final power requirement of the building.

BUILDING ENERGY COST COMPLIANCE ALTERNATIVE

Figure 3 shows the Building Energy Cost Compliance Alternative procedure. As in the case of compliance achieved by the Prescriptive/System method, the minimum requirements must first be met in the building design.

The purpose of this method is to allow flexibility in the energy efficient design of a building using an annual energy cost. Using unit costs rather than units of energy or power allows the energy use contribution of different fuel sources at different times to be added and compared. It is intended to allow for innovation in designs, materials, and equipment, such as daylighting, passive solar heating, heat recovery, better zonal temperature control, thermal storage, and other applications of off-peak electrical energy.

Energy Cost Budget

The Energy Cost Budget (ECB) is a numerical target for annual energy cost. It is intended to provide a target that is reasonably independent of building system choices and architectural design. The ECB for a given building size and type will vary only with climate, the number of stories, and the choice of simulation tool.

The ECB is determined either through the Reference Building Method or the Prototype Building Method. Both methods calculate the ECB by summing the 12 monthly ECBs. Each monthly ECB is determined by the product of the monthly Building Energy Consumption (BECON) of each type of energy and the monthly energy cost (ECOS) per unit of energy for each type of energy used. This is illustrated by the equation

$$ECB = \sum_{jan}^{dec} ECB_m$$

$$ECB_m = \sum_{i} BECON_{mi} * ECOS_{mi}$$

where

ECB_m = monthly Energy Cost Budget
$BECON_{mi}$ = monthly building energy consumption for the ith fuel
$ECOS_{mi}$ = monthly cost of the ith fuel

The monthly ECB is determined by using current rate schedules or contract prices available at the building site. These costs include demand charges, rate blocks, time of use rates, interruptable service rates, delivery charges, taxes, and all other applicable energy charges.

Prototype Building Method

Prototype building characteristics are listed in the Interim Standards, and cover orientation and shape, internal loads, building exterior envelope, HVAC systems and equipment, service water heating, and controls. The Prototype Building procedure must be used for the following building types:

1. Assembly;
2. Office (business);
3. Retail (mercantile);
4. Warehouse (storage);
5. School (educational);
6. Hotel/motel;
7. Restaurant;
8. Health/institutional; and
9. Multi-family.

Simulation of the energy cost is accomplished by using the same gross floor area and number of floors for the prototype building as in the proposed design.

Reference Building Method

If the proposed design of the building cannot be represented by one or a combination of the prototype buildings, or if the assumptions for the prototype building do not reasonably represent the proposed design, the Reference Building Method is to be used. The reference building must have each floor oriented in the same manner as in the proposed design; the form, gross and conditioned floor areas of each floor and the number of floors are also to be the same as the proposed design. All other characteristics and systems must meet the prescriptive/system requirements discussed above.

Design Energy Cost

The Design Energy Cost (DECOS) of the proposed design is calculated and modeled by similar methods as were used for determination of the ECB, according to the following equations:

$$DECOS = \sum_{jan}^{dec} DECOS_m$$

$$DECOS_m = \sum_{i} DECOS_{mi} * ECOS_{mi}$$

where

$DECOS_m$ = monthly design energy cost
$DECON_{mi}$ = monthly building energy consumption of the ith fuel
$ECOS_{mi}$ = monthly cost of the ith fuel

Assumptions for the orientation and shape, internal loads, building exterior envelope HVAC systems and equipment, service water heating, and controls must follow guidelines specified in the Standards.

If the Design Energy Cost is less than or equal to the ECB, and all of the minimum requirements are met, then the proposed design is within compliance.

BUILDING ENERGY COMPLIANCE ALTERNATIVE
Figure 4 illustrates the procedure for using the Building Energy Compliance method. Under this alternative, a detailed energy consumption analysis of the proposed design is required. A life-cycle cost (LCC) analysis is used to select the fuel source for the HVAC systems, the service hot water, and the process loads from available alternatives.

Once the LCC analysis has been completed, the procedures for compliance are similar to the energy cost alternative described above, using either a prototype or a reference building. The difference is that an Energy Use Budget (EUB) is calculated in Btu/sf/yr and the simulation must segregate the calculated energy consumption by fuel type producing an EUB for each fuel type. The total EUB is the calculated by:

$$EUB = \sum_i EUB_i * f_i$$

where

EUB_i = energy use targets for the ith fuel
f_i = fuel conversion factor (BTU/unit)

The Design Energy Use (DEU) is determined for the design building by:

$$DEU = \sum_i DEU_i * f_i$$

where

DEU_i = design energy target for the ith fuel

Compliance with the Standards is achieved if the DEU is less than or equal to the EUB.

EXAMPLE
In order to illustrate an application of the Interim Energy Conservation Standards, a recently constructed high-rise office building located in New York City was used. An energy cost simulation for the building, as designed and constructed, was calculated using DOE 2.1, and compared to a simulation with minimum Standards as input.

The existing building meets the minimum Standards with few exceptions:

1. HVAC supply fan control - the Standards require a variable-air-volume (VAV) system with an air-foil centrifugal fan and an AC frequency variable-speed drive motor; the existing building has variable inlet vanes on the fan and a constant-speed motor.

2. Cooling - the minimum Standards require the use of a dry-bulb economizer, and the supply air be reset by the zone with the greatest cooling demand; the existing building has neither.

Three utilities are purchased by the building owners - steam, chilled water, and electricity. In addition, supplementary chilled water is also made by the facility.

The results of the comparative simulations are shown in Figure 5. As can be seen, the existing design would be out of compliance with the Energy Conservation Standards, with a greater utilities cost of approximately \$347,000, or 6.7 percent. Were this a case where the minimum standards were required, certain redesigns would have to be accomplished.

REFERENCES
(1) Performance Standards for New Commercial and Multi - Family High Rise Residential Buildings, DOE/CE-0304T, January 30, 1989.

Figure 1
Alternative Methods of Achieving Compliance

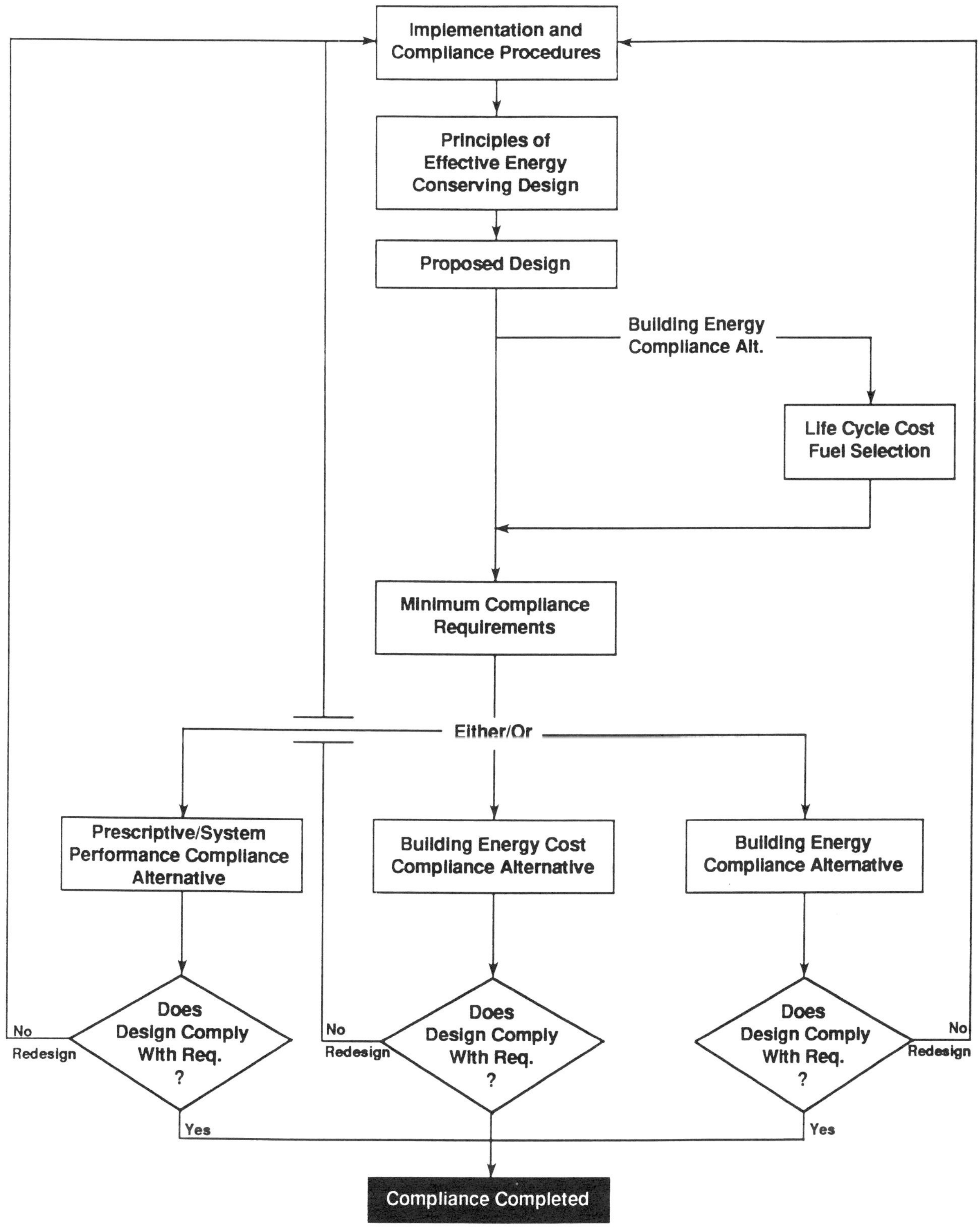

Figure 2

Prescriptive/System Performance Compliance Alternatives

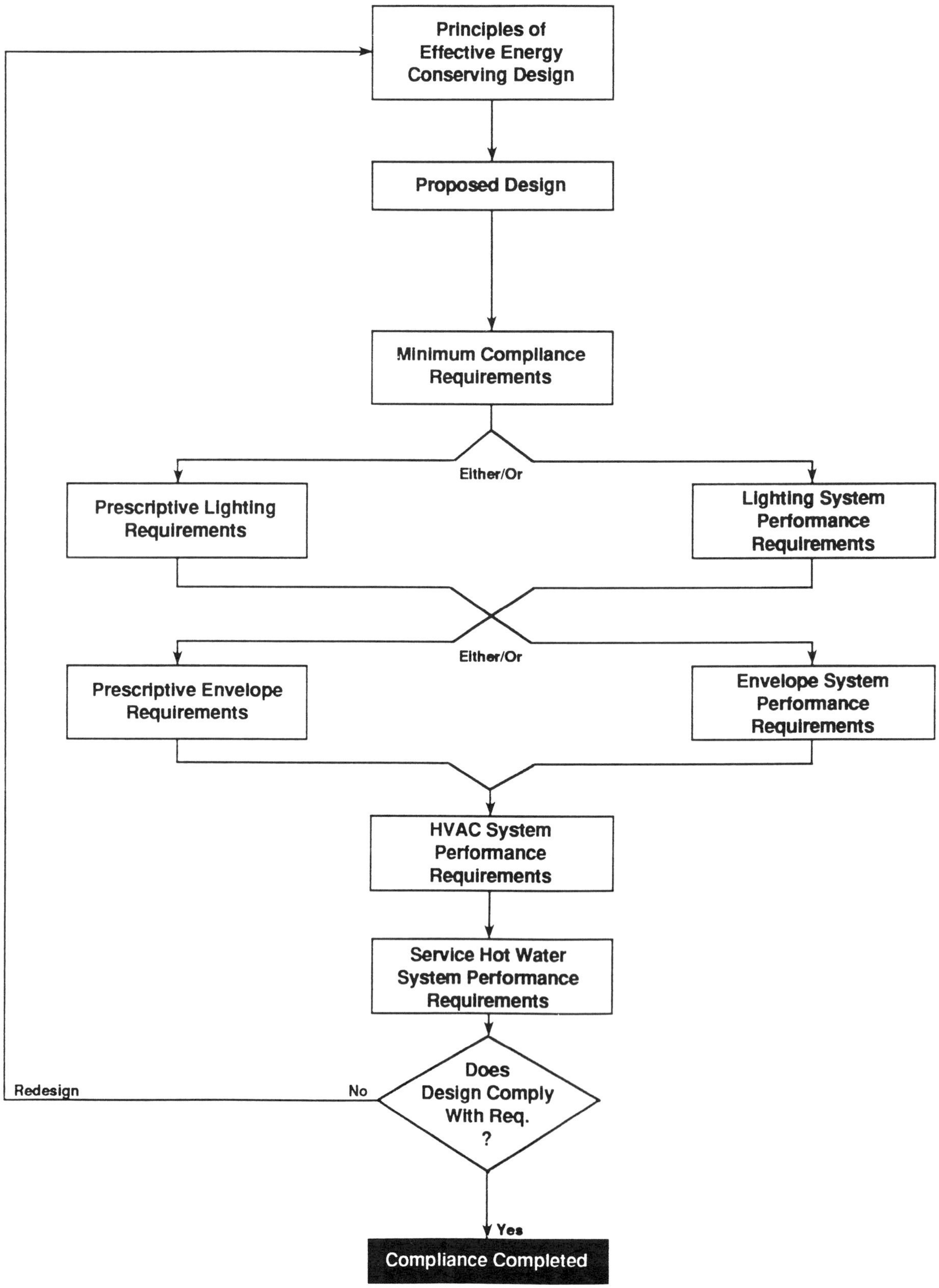

Figure 3
Building Energy Cost Compliance Alternative

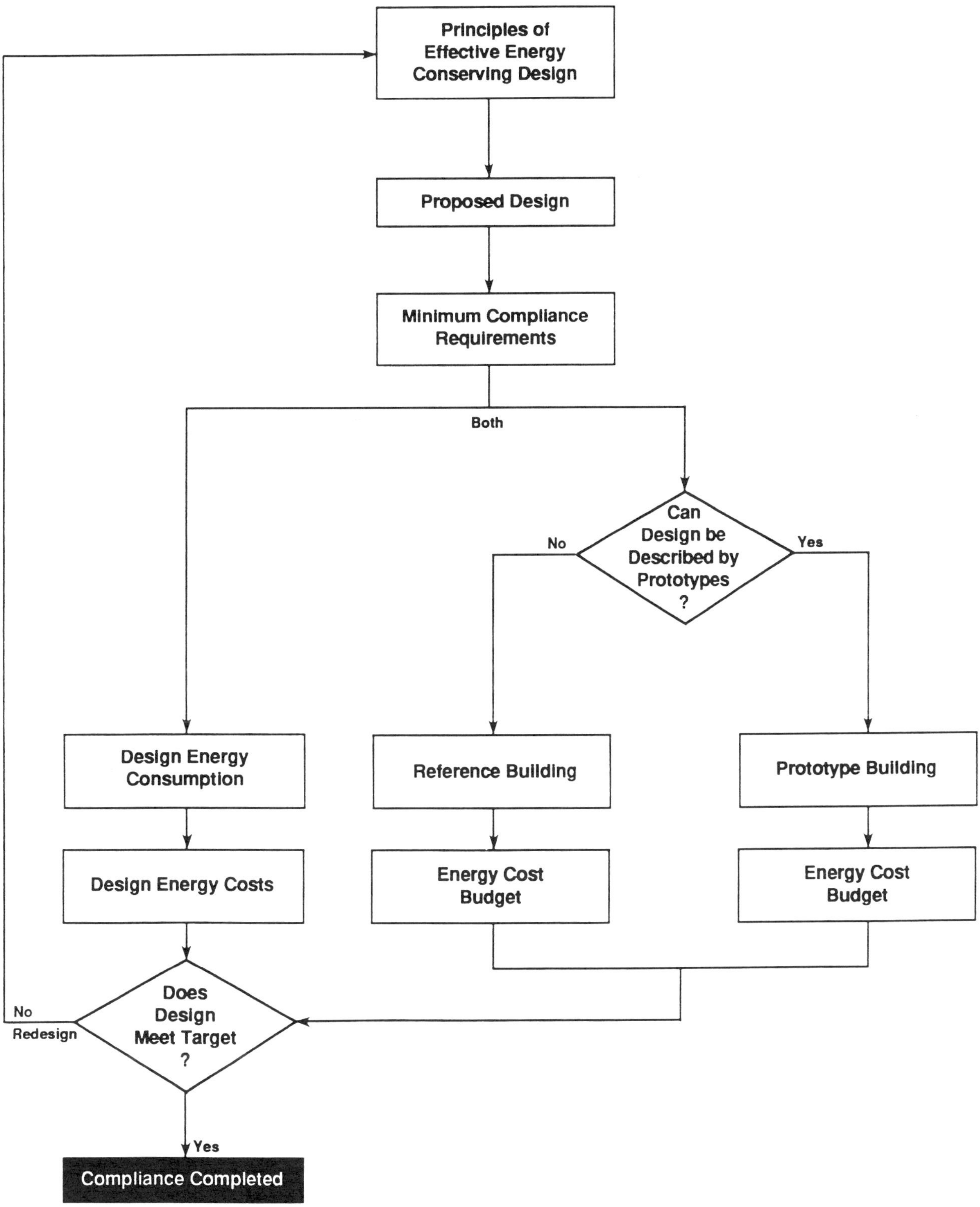

Figure 4
Building Energy Compliance Alternative

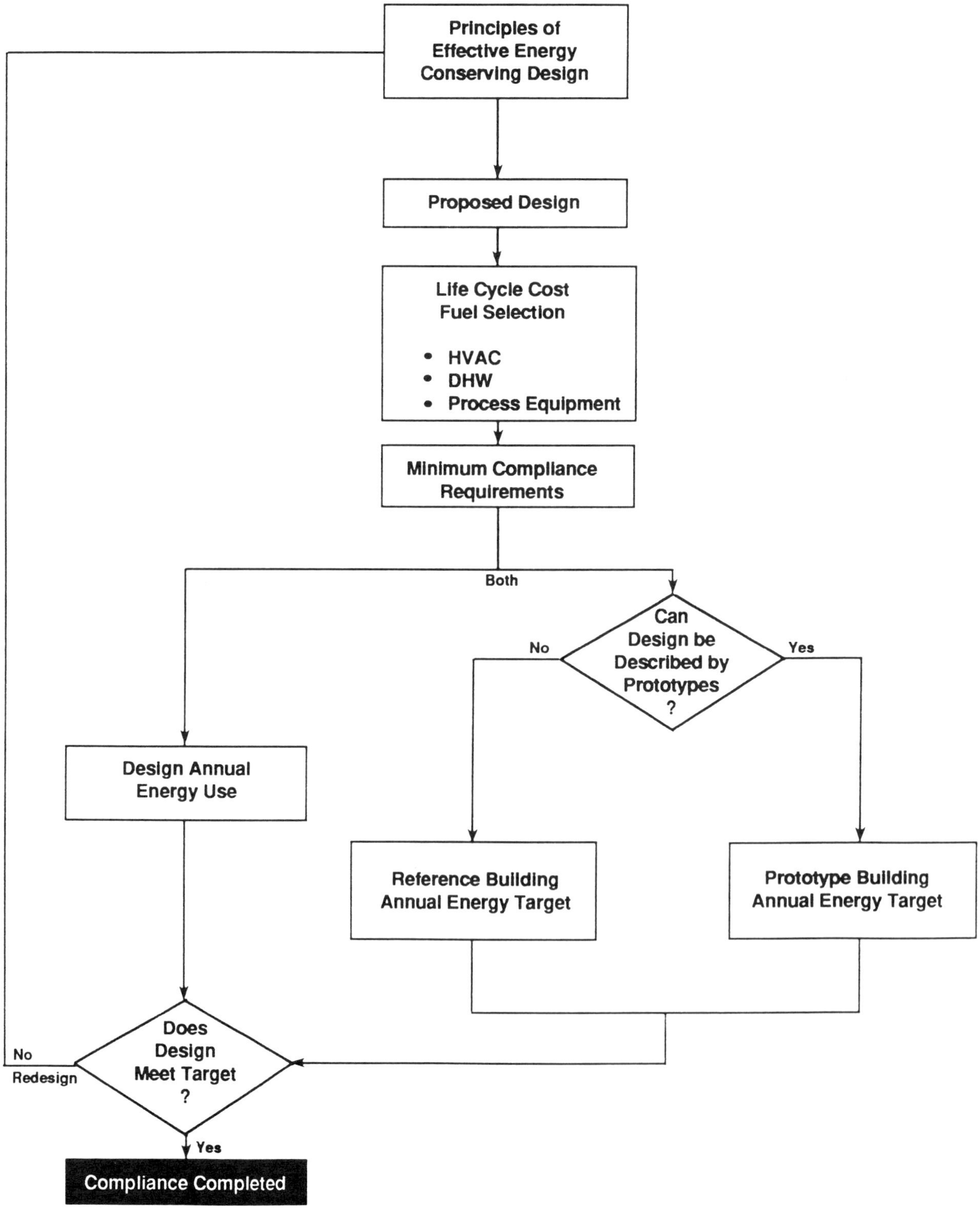

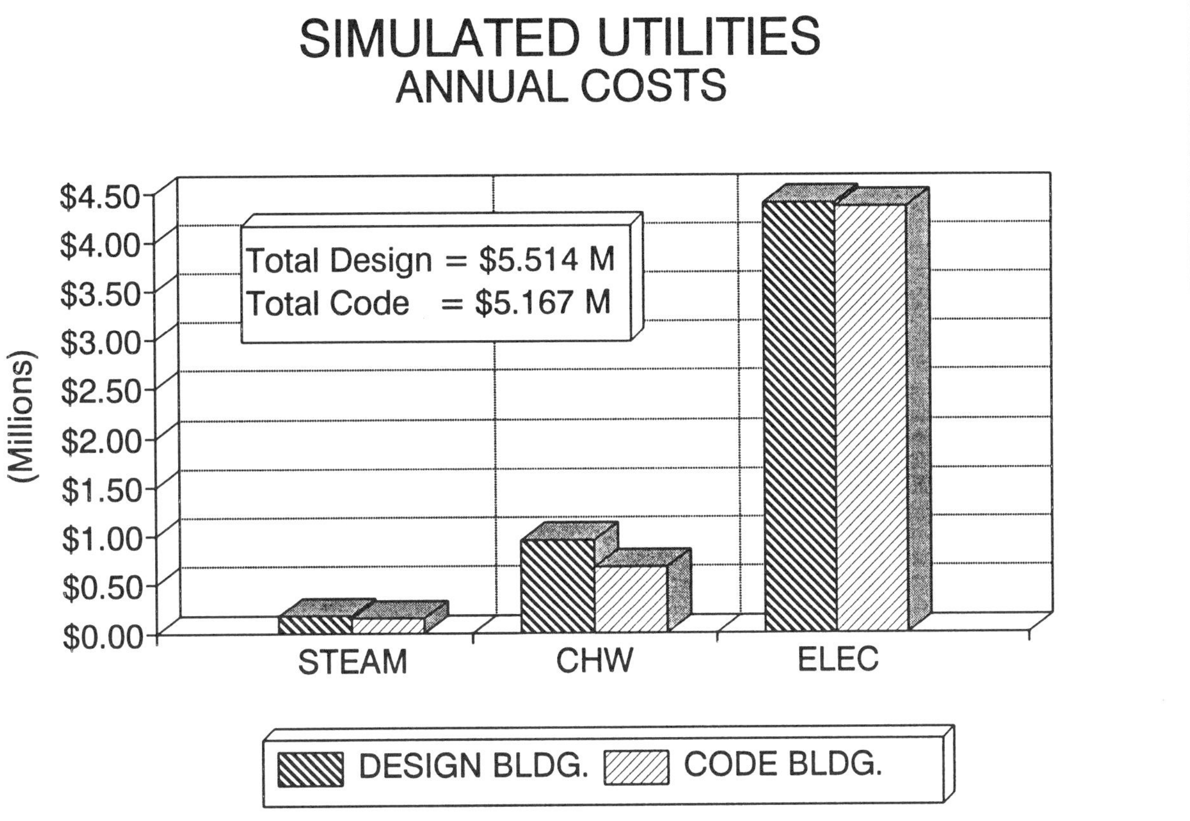
SIMULATED UTILITIES
ANNUAL COSTS
$4.50
$4.00
$3.50
$3.00
$2.50
$2.00
$1.50
$1.00
$0.50
$0.00
(Millions)
Total Design = $5.514 M
Total Code = $5.167 M
STEAM
CHW
ELEC
DESIGN BLDG.
CODE BLDG.

Figure 5

Chapter 92

Financing the Best Federal Projects

M. Ginsberg

The Federal government is about to embark on the last lap of one of the largest energy savings challenges in the world, to reduce Federal energy consumption by 20% by 2000, over a 1985 baseline. When achieved, we will have reduced demand the equivalent of four 500 megawatt power plants, achieved energy savings of $1 billion a year by 2000, produced significant environmental benefits and created new jobs while stimulating a new generation of energy efficient products.

To achieve those results, the Federal Energy Management Program has determined a need to invest $2-4 billion in energy efficiecy opportunities. We have also concluded there are $6-8 billion worth of cost-effective measures that would pay back in ten years or less and which would far exceed the 20% savings goal.

There are two essential steps in achieving energy efficiency goals of that size, whether in the private sector or in Federal facilities: identifying the best projects and then paying for them. This morning, I spoke about ways we in the Federal sector are identifying opportunities. Now I want to take it a step further and discuss how to fund them.

Once the determination is made that there are worthwhile projects to address, the issue of paying for them arises. Often the initial cost can be an impediment to achieving rapid energy savings. Now there is good news, with four ways to finance projects in the Federal sector so the savings can begin to accrue more quickly:

- o Appropriations
- o Utility-based rebates
- o Energy savings performance contracts
- o Federal Energy Efficiency Fund

APPROPRIATIONS

The President has laid out an investment plan that calls for a significant increase in the next four years for energy efficiency projects. It would triple the level of appropriations that has been invested over recent years. That proposed increase is above and beyond what agencies have directed from existing construction and operation and maintenance funds. These dollars would be directly targeted to energy efficiency improvements.

The Office of Management and Budget has responsibility to establish a specific line-item budget for energy investments and may specify a training budget also. The large increases will be directed principally at the four largest energy consuming departments: Defense, Energy, General Services Administration, and Veterans Affairs. Other agencies may see shifts from existing accounts into energy efficiency to make clear their commitments to using construction and O&M funds for energy savings.

UTILITY-BASED REBATES

The utilities in many parts of the country have found the need to encourage greater energy savings to avoid having to build expensive new power plants. They have been providing demand-side management (DSM) rebates and energy reduction strategies, such as design assistance and audits. The pace of rebates has accelerated in recent years with several very large examples in the Federal sector.

However, not all Federal sites are served by utilities that offer DSM rebates. Each utility must assess its own supplies in integrated resource plans and work with its public utility commission to determine if demand side management makes sense in their circumstances. For example, utilities in Virginia and throughout much of the midwest have determined that their plant capacity and load use patterns do not justify rebates or demand-side management incentives.

DSM programs are receiving added scrutiny as regulators become more insistent on documenting that energy savings are actually achieved and are direct results from the measures that were installed with rebate funding. There are increasing requirements for before and after metering and monitoring to accurately verify savings. The result has been a change in the pace and nature of utility incentives. Several utilities are now considering loans instead of cash rebates. Others are being much more selective about the projects they undertake. The net result is some uncertainty about the size and availability of utility rebates for the near term.

Where available, agencies should take full advantage of utility rebates, but it is important to establish strong relationships with utility partners to determine whether rebates can be counted on as new projects are initiated.

ENERGY SAVING PERFORMANCE CONTRACTS
Innovative financing has become common practice in many parts of the country. Of particular interest is shared energy savings, where private sector funds are invested in an energy savings project in return for a portion of the resulting energy cost savings. Shared Energy Savings legislation has been available in the Federal sector since the late-1980's. However, difficukt procurement has resulted in just 15 contracts worth some $45 million of investments. Critics cited cumbersome procurement practices in the Federal sector and a lack of will. Others acknowledge lack of adequarte training and information. Either way, the Congress recognized the slow pace and addressed it with new legislation.

The Energy Policy Act of 1992 (EPACT) established energy saving performance contracts for Federal agencies. These contracts have been used in the private sector and by states, local governments and schools. Energy service companies (ESCOs) have formed to finance and install energy savings measures. Through negotiations, they guarantee certain levels of performance, based on measured savings, and retain a negotiated part of the energy savings for their expenses and share the remaining savings with the agency.

EPACT required DOE to develop and issue a rule, with the concurrence of the Federal Acquisition Regulations (FAR) Council, on performance contracting. As this goes to press, the Department is working on it. It is a challenging rule with procurement requirements. Congress intended that this rule speed the pace and availability of ESP contracts. We are hopeful that with new training and procedures in place, the use of private sector funding can increase the number and size of energy efficiency projects in the Federal sector.

Among the most important provisions are the following:

o **Qualified Vendors List.** DOE will prepare a list of qualified vendors that will consider their experience and track record. This should simplify some of the procurement steps needed by agencies when they seek eligible contractors. Some advocates of this provision had hoped this would allow an agency to simply select from this list, but any time one is chosen, the procurement official must justify why the others were not selected. In Federal procurement language, that is tantamount to a complete procurement selection process.

o **Unsolicited Proposals.** An ESCO may submit an unsolicited, sole source proposal. At that point the agency must advertise it in the Commerce Business Daily to allow other potential bidders to compete. It is unclear how actively this provision will be used or whether it will increase the number of projects. Once the rule is in place, this process may be used.

o **A How-To Guide and Model Solicitation.** DOE is preparing these tools to help facility managers and the procurement specialist to implement ESP contracts. DOE intends to provide support to agencies undertaking these contracts and using the guide and model solicitation.

In addition, the EPACT requires the following:

- an annual audit must be conducted
- the contractor has responsibility for O&M, where appropriate
- payments cannot exceed utility payments
- savings must be guaranteed
- contracts cannot exceed 25 years
- projects with cancellation ceiling over $750,000 must be reported to Congress
- DOE will issue rules covering selection, monitoring and termination of contracts
- the program has a five year sunset
- the comptroller will conduct an annual report

FEDERAL ENERGY EFFICIENCY FUND
Congress determined that a fund was needed to stimulate projects. EPACT authorized $10 million in FY 1994, $50 million in FY 1995 and sums as needed in the following years. DOE issued guidelines to agencies this summer to explain how to submit proposals for funding if appropriations become available in FY 1994. As this goes to press, the FY 1994 budget contains a request for $6 million. Funds may go to all agencies but a priority will go to those agencies that have not received direct appropriations from Congress for projects.

DOE will conduct a competitive assessment based on cost-effectiveness, energy and cost savings, amount of agency funds, amount leveraged from non-Federal funds and other factors set by DOE. With limited funds, agencies should submit proposals quickly although they will be accepted at any time. Grants will be made as funds are available.

FEMP SHOWCASE PROJECTS
There are several outstanding examples of the use of several of these financing mechanisms.

o At Fort Lewis, Washington, the local utility, Tacoma Public Utilities, and the supplier, Bonneville Power Administration, have joined in a project with the Army base to install $12 million in lighting and motor upgrades. The utility will finance the entire project through a shared savings contract and will pay for 85% of the entire cost. By the time the project is complete, additional measures may bring the total investment to $25 million.

o At DOE's headquarters, the Forrestal building, a utility rebate from PEPCO and a shared energy savings contract will pay the entire cost of replacing over 30,000 lighting fixtures with high efficiency T-8 bulbs, electronic ballasts, new reflectors, and controls where needed. During the seven year life of the project, the Department will reduce lighting energy use by 63% and avoid spending over $1 million in utility charges at no capital cost to the Federal government.

o In New York, DOE, GSA and Consolidated Edison joined in an effort to improve 26 Federal Plaza, the second largest Federal building. GSA has received a check for $1.2 million for lighting improvements and anticipates another check for $1.6 million in motor efficiency improvements. With a DOE-2 simulation run by Con Ed, two new buildings were found to have efficiency improvements worth $2 million in rebates from Con Ed to GSA.

o In Denver, FEMP worked with the GSA staff in charge of the Denver Federal Center as the first Fast Track pilot program. With support from NREL and Environment Plus, a bid was prepared for Public Service of Colorado in response to a DSM bidding program. As this goes to press, the project looks very positive to produce a large retrofit with utility participation. The Denver Federal Center is also a possible site to rehabilitate an older solar system to bring it back to life and make it an efficient contributor to energy supplies there.

o Finally, the Federal Correctional Institution north of Phoenix gets all of its hot water from electricity. We are working with NREL and the Federal Bureau of Prisons to demonstrate shared energy savings as a way to pay for a solar water heating system at no capital or operating costs to the government with a guaranteed cost savings for hot water. We are exploring a Cooperative Research and Development Agreement (CRADA) as a way to conduct the demonstration.

CONCLUSION

With hard work and commitment to exceed the goals set, the Federal government can save billions of dollars, stimulate new technologies and create new jobs, protect natural resources and improve the environment. Perhaps more important, the Federal government can take leadership and set a proud example for others to follow. It won't be easy, but we at DOE are prepared to help the agencies and private sector to make this a success.

Enclosure

Support Office Contacts
Training Schedule

FEDERAL ENERGY COSTS IN BUILDINGS

(000's)

	1985	1992	Percent Change
Defense	2,793,005.4	2,383,327.5	-14.7%
Energy	179,360.1	186,796.5	4.1%
Veterans Affairs	210,364.7	204,562.2	-2.8%
Postal Service	242,143.7	306,995.0	26.8%
General Services Admin.	224,578.2	215,855.8	-3.9%
Other	294,341.6	357,290.1	21.4%
Total	3,943,793.7	3,654,827.1	-7.3%

12313.007

FEDERAL ENERGY CONSUMPTION IN BUILDINGS BY AGENCY, 1985 AND 1992

(BILLIONS OF BTUs)

	1985	1992	Percent Change
Defense	349,076.675	295,719.764	-15.3%
Energy	32,146.678	28,859.752	-10.2%
Veterans Affairs	24,551.974	24,619.971	0.3%
Postal Service	16,238.338	18,858,402	16.1%
General Services Admin.	16,887.150	13,061.415	-22.7%
Other	30,318.332	30,283.611	-0.1%
Total	469,219.147	411,402.915	-12.3%

12313.005

FEDERAL ENERGY CONSUMPTION IN BUILDINGS BY AGENCY, 1985 AND 1992

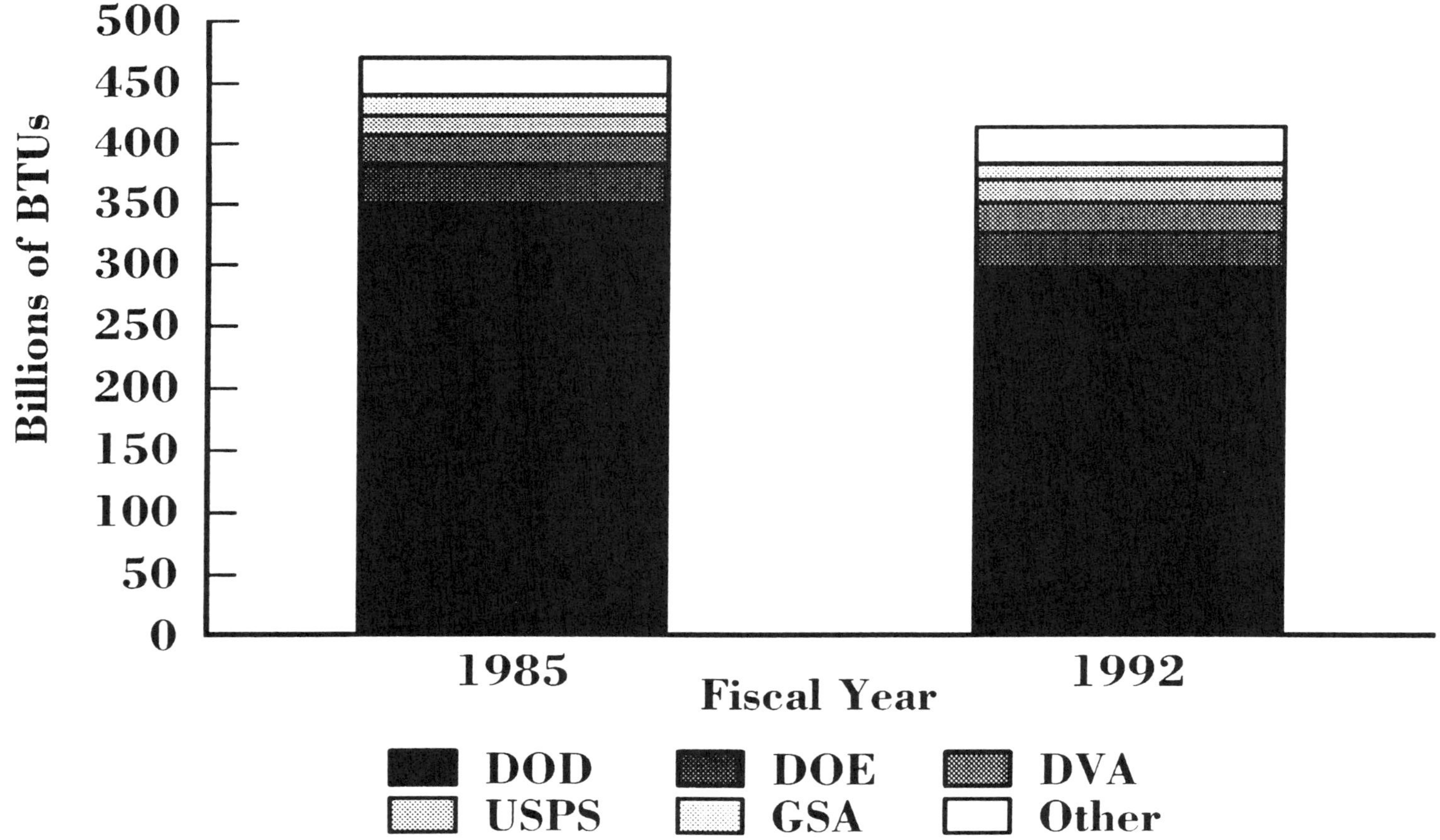

12313.006

FEDERAL GOVERNMENT PROGRESS TOWARD MANDATED ENERGY REDUCTION GOALS FOR BUILDINGS AND FACILITIES 1985 TO 1992

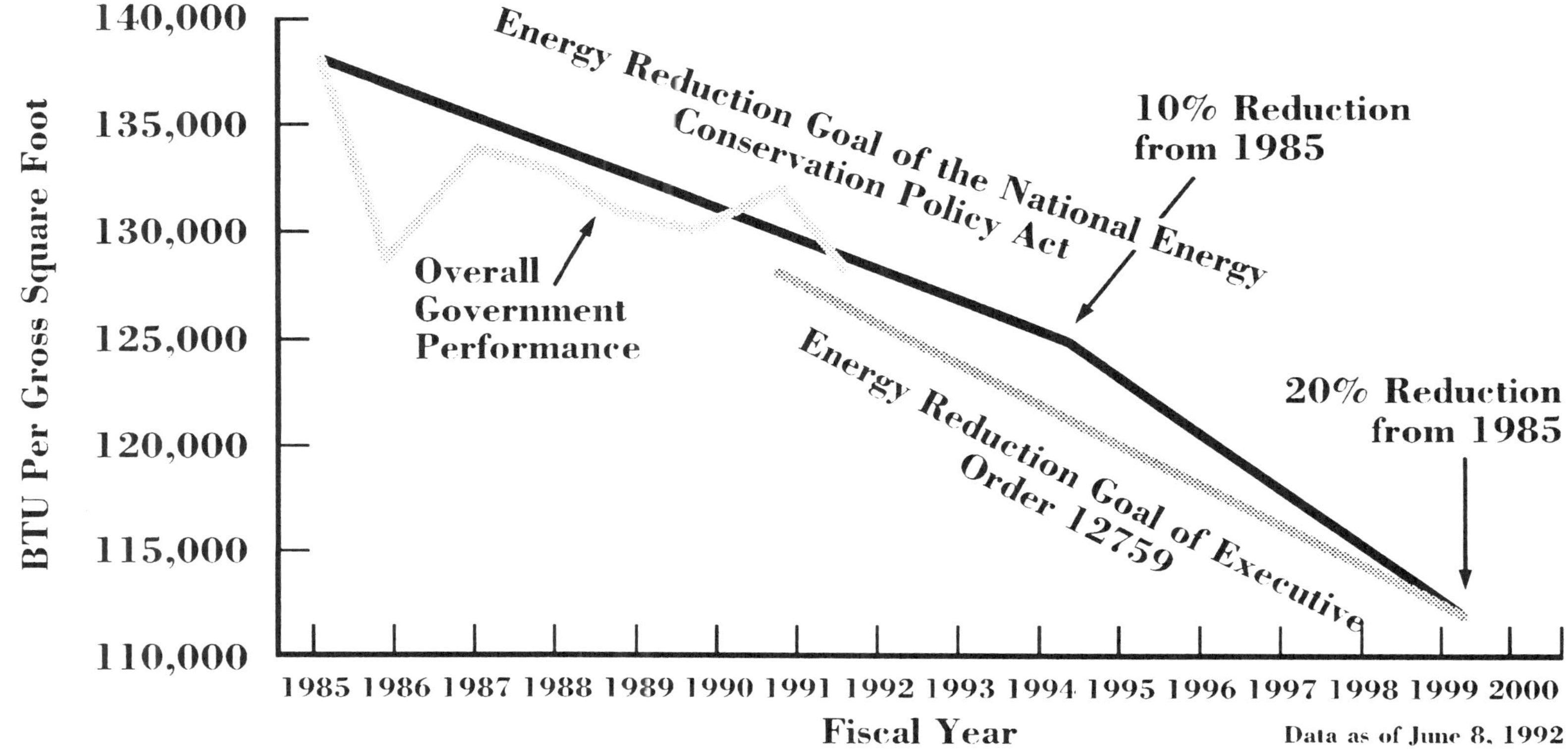

Office of Federal Energy Management Programs

12313.001

1993-1994 DOE Support Office List

U.S. DEPARTMENT OF ENERGY
SUPPORT OFFICE DIRECTORS AND FEMP CONTACTS

Mr. Hugh Saussy, Director
Boston Support Office
U.S. Department of Energy
1 Congress Street, Suite 1101
Boston, Massachusetts 02114
617/565-9700
FAX: 617/565-9723
Paul King, 617/565-9712

Mr. Charles Baxter, Director
New York Support Office
U.S. Department of Energy
26 Federal Plaza, Suite 3437
New York, New York 10278
212/264-6335
FAX: 212/264-9805 or
FAX: 212/264-2272
Wayne Belgrave, 212/264-0012

Mr. Christopher McGowan, Director
Philadelphia Support Office
U.S. Department of Energy
1880 JFK Boulevard, Suite 501
Philadelphia, Pennsylvania 19102
215/656-6950
FAX: 215/656-6981
Anthony Pontello, 215/656-6951

Mr. Buddy L. Jackson, Director
Atlanta Support Office
U.S. Department of Energy
730 Peachtree Street, N.E., #876
Atlanta, Georgia 30308
404/347-2837
FAX: 404/347-3098
Fred Singleton, 404/347-2380

Mr. William S. Becker, Director
Chicago Support Office
U.S. Department of Energy
9800 South Cass Avenue
Argonne, Illinois 60439
708/252-2220
FAX: 708/252-2071
Allie Mansker, 708/252-2119

Mr. Curtis E. Carlson, Jr., Director
Dallas Support Office
U.S. Department of Energy
1420 W. Mockingbird Lane, Suite 400
Dallas, Texas 75247
214/767-7248
FAX: 214/767-7231
Robert Gabour, 214/767-7232

Mr. Ronald Brown, Director
Kansas City Support Office
U.S. Department of Energy
911 Walnut Street, 14th Floor
Kansas City, Missouri 64106
816/426-5533
FAX: 816/426-6860
Chris Cockrill, 816/426-5182

Mr. Al Hymer, Director
Denver Support Office
U.S. Department of Energy
2801 Youngfield Street, Suite 380
Golden, Colorado 80401-2266
303/231-5750, ext. 133
FAX: 303/231-5757
Randy Jones, 303/231-5750, ext. 135

Ms. Martha Dixon, Director
San Francisco Support Office
U.S. Department of Energy
3101 Clay Street
Oakland, California 94612-5219
510/637-1941
FAX: 510/637-2017
Demitrius Fowler, 510/273-4463

Ms. Kathy M. Vega, Director
Seattle Support Office
U.S. Department of Energy
800 Fifth Avenue, Suite 3950
Seattle, Washington 98104
206/553-1132
FAX: 206/553-2200
Arun Jhaveri, 206/553-2152

Chapter 93

Identifying the Best Projects in the Federal Sector

M. Ginsberg

As the nation's largest energy consumer, the Federal government has a special responsibility to lead the way to greater energy efficiency. Not only will it reduce the governments's costs, but it will have benefits to the energy suppliers, natural resources, and the environment. Perhaps the two most significant benefits would come in the form of economic stimulus and leadership for others to follow.

It is particularly timely that the Association of Energy Engineers has convened this Congress and has included once again the Federal track. I applaud AEE for its willingness to include these valuable sessions here in Atlanta. I hope it is useful for the participants and that we can continue our relationship with the Association.

FEDERAL ENERGY PERSPECTIVE

The Federal government saw an increase in its energy consumption of over a billion dollars to $11.2 billion from 1990 to 1991, due in large part to fuel costs associated with Desert Storm. The good news is that there has been a drop to $8.6 (?) billion in 1992. From 1990 to 1992, there was a reduction of $400 million in the cost of operating the energy systems for the buildings in the Federal sector, from over $4 billion to about $3.6 (?) billion. With 500,000 buildings at some 8000 sites, using over 3.2 million square feet of space, the energy implications are clear. The opportunities for continued savings are also clear.

The challenge my colleagues in the Federal sector and I have is not only to identify but to pay for the best projects to increase energy efficiency, water conservation and the use of renewable energy. This morning, I will address the identification aspect and later today, in a second session, I will deal with the funding. I hope to give you some concrete ideas to take home and begin to implement quickly.

REQUIREMENTS IN THE FEDERAL SECTOR

Doing work in the Federal sector is a bit different from the private sector and even other public entities. That's not to say it's better or worse, just different. For example, there is a clear life-cycle costing requirement to see that we get the "biggest bang for the buck." It looks beyond the first cost to the entire cost that includes the monthly utility bills over time. It essentially optimizes energy savings in the context of available budget. Related to that is "fuel neutrality." That suggests that the life-cycle price of a replacement is more important than the technology or fuel source.

With the increased scrutiny of utility regulators and financial investors, it is becoming more important that we conduct before and after metering to document the energy savings that result from a specific retrofit. Engineering calculations are good for many proven technologies, but we are finding that the impact of the retrofit on <u>the entire building</u> may be different from the engineering calculations made for a specific technology.

There are also new requirements in the Federal sector for facility managers and agencies to report their energy use, funds invested in projects and facility energy manager training. I won't go into all of those but look for your agency's guidance documents or keep your eye on the FEMP Focus newsletter that comes out every two months or so. If you like, you can get your very own subscription by writing or calling us.

IDENTIFYING THE BEST PROJECTS

It is particularly important in the Federal sector to make good choices in selecting which projects to pursue. There are two important ways FEMP is addressing this essential step: training and on-site audit teams. In addition, I will discuss issues of solar and water conservation and the need for verification of savings.

Training. It is not always obvious how to select the best, most cost-effective technology for a particular site. For example, in some circumstances, lighting is a clear choice, but in others, it might be motors or the HVAC system. That is where an understanding of life-cycle cost (LCC) analysis comes in. With our colleagues at the National Institute of Standards and Technologies (NIST), we have developed a new half-day course in LCC that is being offered by the DOE regional Support Offices. I have included a list of those contacts so you may find out when they are offering the course in your region. With a knowledge of LCC, a facility energy manager can understand how to select from among several technologies and energy consuming systems. We often refer to this as whole building analysis. It is important to consider all the system in the building. Similarly, a facility manager must know how to select the best opportunities from among all the buildings at his or her site. Since some Federal facilities can include literally thousands of buildings, it is hard to know how or where to begin.

To address whole building and whole base or facility analysis, we have several tools and approaches. These are incorporated into our training programs. A schedule of our training is enclosed with this presentation and is updated regularly in the FEMP Focus. The following descriptions summarize the software developed and constantly improved to help Federal energy managers. We also believe there is strong transferability to other public sector organizations: state and local governments, schools and universities.

ASEAM. A Simplified Energy Analysis Method (ASEAM) has been around for several years. Developed for DOE, we include it in the three-day training course on Life-Cycle Costing as a practical example of applying LCC to whole building analysis.

FEDS. The Federal Energy Decision Screening (FEDS) software tools allow whole base analysis. With a two day training, facility managers can learn how to spend 2-4 hours of data entry to get the benefits of an analytical tool that imbeds 118 technologies for consideration. It is ideal for large bases and facilities and will serve as an umbrella for other FEMP software.

FLEX. As part of our effort to identify the likeliest candidates for action, we recognized that lighting represents about 25% of the energy used in a Federal building. We developed the Federal Lighting Expert (FLEX) system to help facility managers make choices in lighting technologies using an "expert" system that incorporates consensus decisions in a chain driven process. Developed with the Lighting Research Institute, there are training courses scheduled this fall and winter. FLEX will be incorporated into whole site and building analysis as part of FEDS in the near future.

One important new tool we are using to help Federal energy managers know about training opportunities is called the FEMP Training Locator. We encourage all public and private, university and non-profit organizations to submit the training they offer that might be of interest to Federal energy managers. We will make it available either in hard-copy or, perhaps, on a computer bulletin board. That way, those interested in training can learn what courses are offered, where, when and how much they will cost.

FEMP Audits. The Energy Policy Act of 1992 provided for Audit Teams to assist facility managers. DOE is developing a plan that, if funded, would provide on-site audits at facilities either selected by the agencies or based on request and availability. We hope to have specialized and general teams able to address a wide variety of situations.

What FEMP and the audit teams can do is bring the latest technologies to the Federal site and explore not only the most cost-effective approaches but learn the critical next step of how to fund the project. That is integral to our way of thinking about audits. We all know too many audits have been done that never were implemented. We want to take the FEMP Audit past the technical language and into implementation.

We anticipate there will be a variety of approaches with public sector and private sector audit programs. I know we will count on the ten regional DOE Support Offices to assist in providing training and letting Federal energy managers in their regions know about audit possibilities.

Utility, ESCO and Engineering Audits. Many utilities offer audit services and energy service companies can submit a sole source proposal to a facility. Often these services are free to the facility. DOE encourages the use of any free services such as these. In some cases, facilities may have their own Architect/Engineer contracts with companies who could perform these services for a fee. There are several important reminders to the service provider as well as the facility manager. Whether free or paid, it is important to understand life-cycle cost, fuel neutrality and Federal procurement requirements. Since we cannot "validate" or "certify" private sector software programs, we urge the service provider to become fully conversant with these requirements and incorporate them into their analyses. Facility managers should request that their private sector audit include these requirements. Later today I will also describe the requirements if a facility receives an unsolicited sole source proposal from an energy service company.

IDENTIFYING SOLAR AND WATER CONSERVATION OPPORTUNITIES

The Energy Policy Act of 1992 specifies that both solar and renewables and water conservation be increased in the Federal sector. Those of you who know I come from Arizona might think I had something to do with that since we old desert rats love solar and need water. Even though I had nothing to do with those requirements, I embrace them. Both solar and water conservation make sense in the Federal sector.

FEMP is working with the National Renewable Energy Lab and several agencies interested in solar. We are exploring ways to restore some older systems and are looking aggressively at new opportunities in both solar thermal and solar electric applications as well as passive solar design. We have already sponsored a Passive Solar

Design workshop with the Passive Solar Industries Council and participated in the SOLTEC annual solar industries meeting with a workshop on funding solar and energy efficiency projects in the Federal sector.

The Federal Bureau of Prisons has a need for hot water and in some cases gets all of it from electricity. Solar can compete favorably with electric hot water, so we are working on ways to install solar water heating systems at Federal prisons through shared energy savings concepts and other innovations. The Fish and Wildlife Service of Interior has several interesting projects. And the Department of Defense has already installed over 2000 solar devices in military applications through the exceptional work of Garyl Smith and his colleagues at China Lake Naval Weapons station in California. Garyl earned the distinction of being the winner of the Special Award for Solar Energy at the 1992 Awards Cenremony.

Water conservation offers large opportunities in Federal facilities. There are new water service companies forming, much like the energy service companies, willing to share in the savings from reduced water costs. New low flush toilets, low flow shower heads, xeriscaping and low water landscaping offer exciting opportunities to protect our water resources.

METERING, EVALUATION AND DIAGNOSTICS

These days, it's not enough to identify good projects. We must also be conscious of the real results, not just engineering calculations. There is increasing interest among utility regulators and the energy service industry to verify that the actions taken actually produced the results anticipated. Before and after metering has become very important in this process as companies and end users try to document savings. All of this is still in its infancy, but there are excellent examples of efforts to overcome this information barrier.

FEMP is exploring a cooperative project with several partners to verify savings at a few key Federal sites. We would work with the designated Federal site energy manager and the utilities, with possible help from the national energy labs and private sector. These pilots would help establish a protocol that would be used in the Federal sector.

The use of computer monitoring as a diagnostic tools is gaining speed. Several private sector firms have installed sensors and monitoring equipment to get real time feedback on building energy system operations. At the DOE Headquarters, Forrestal Building, Texas A&M installed monitoring equipment and provides monthly recommendations on ways to improve the efficiency of the building. Often this approach can more than pay for itself quickly as recommendations turn to energy savings.

CONCLUSION

With these approaches, Federal facility managers can begin to determine how to make the best investments in energy efficient projects. With training, on-site audit assistance, with software tools geared for the Federal sector, and with an eye to translating the needed cost-effective retrofits into action, we can make great progress in meeting and exceeding the energy savings goals in the Federal sector. In the process, we will not only save energy but reduce pollution, save billions of dollars and demonstrate national leadership. It would be my hope that the Federal government will be in a position to say: Do as I do, not just as I say.

Enclosures:

List of Support Office Contacts
Training Schedule

Chapter 94

Exploring Energy and Environmental Opportunities: How the CFC Issue Can Work to Your Advantage

E. Dunham

BACKGROUND

Since the passage of the Federal Energy Management Improvement Act of 1988 (FEMIA), I have directed the energy conservation efforts of the General Services Administration for Federally-owned and -leased buildings. This has been a challenging, fast-moving, educational opportunity for me and a small staff of dedicated professionals who work either in the Energy Branch of GSA's Public Buildings Service or as one of ten Regional Energy Coordinators. Collectively, and with the support of our buildings management and design staffs, GSA has reduced energy consumption by 43 percent in its owned Federal buildings between 1973 and 1985. Since the passage of the FEMIA, GSA has reduced energy use in both our owned facilities and our leased facilities (where we pay utility costs separately from lease payments), by about 8 percent. The buildings GSA leases use less than 87,000 BTUs per GSF and our owned buildings use less than 76,000 BTUs per GSF - about 12.5 percent less energy than in the buildings we lease. The fact that GSA's owned buildings average 40 years of age (almost 60 years of age in Washington, DC), and that the buildings we lease are considerably younger, attests to the long term commitment GSA has had to energy conservation . As a matter of fact according to the Department of Energy's 1992 report to Congress and the President, GSA uses less energy per gross square foot than any other major Federal property management agency, as reported in the Department of Energy's latest annual report, and about 40 percent less energy than all other civilian agencies.

The level of success in energy conservation achieved by GSA, to date, has affected GSA in several ways. First, from the positive standpoint, conservation efforts since 1973 have achieved an estimated cumulative $1.7 billion cost avoidance for building energy expenditures. This is only a small part of the efforts at GSA to reduce expenditure and provide better service to our clients. From a negative standpoint, however, new legislation and Executive Orders continue to require percentage reductions in energy consumption without regard to past performance. Considering the age of our building inventory, the state of conservation technologies, and the increasing difficulty in identifying viable energy conservation retrofit projects, we are beginning to wonder if we are approaching an optimum, if you will, BTU per GSF threshold that cannot be reduced without a major turnover in our building inventory or the development of new cost effective conservation technologies.

We exploited low cost/no cost technologies in the 1970's and early 1980's and have pursued relighting Federal buildings in the 1990's along with a host of other projects including everything from air handling to window replacements. Lighting projects that are cost effective, in buildings that do not contain asbestos, or in buildings not scheduled for a major renovation within the next five years, are beginning to dwindle.

At the same time, funding specifically earmarked for energy conservation retrofits began to increase. GSA began to set-aside funds specifically for energy conservation in 1990 and has set-aside $30 million or more each of the last three years. In 1994, we expect the Federal budget, for the first time, to specifically earmark $37 million funding for energy conservation. We expect that number to grow to $44 million in 1995 and maintain that level through 1998. Further, GSA has significantly expanded its participation in utility company demand side management programs and could collect an additional $6 million a year in rebates which could be utilized to fund additional energy projects in 1994 and beyond. Along with budgeted amounts, we could experience a 67 percent increase in available funding over the next two years. Given GSA's history in conserving energy, and aging building inventory, we are starting to become concerned about whether we can continue to identify good conservation opportunities that have reasonable payback periods.

GSA's approach to developing energy conservation projects has focused on single system projects, i.e., lighting, HVAC, windows, energy management and control systems, etc. There are several reasons for that approach. GSA has discretion over approximately $300 million in repair and alterations funds which is used to replace outdated and non-functional building equipment and make general repairs to our inventory of 1700 buildings and facilities. However, GSA cannot spend more than $1.6 million per building without the national budget containing a specific line-item approval for that specific project. While this may seem like a lot of money for a small office building, several million dollars does not go far in older buildings many of them larger than 500,000 square foot . Additionally, the processing of receiving line item approval for a major renovation project could take several years or more.

In fiscal year 1994, we anticipate a "generic" budget line item for energy conservation projects. This line item sets aside funding of energy conservation efforts but does not identify specific projects. If approved by Congress, and included in the final budget, GSA will have the flexibility to pursue some of these larger, multi-system projects which will help us make significant progress toward future energy conservation goals. However, these projects will be much more complex; require a well developed scope of work, be difficult to analyze and estimate conservation potential, and require extensive design work.

At the same time that energy conservation funding became an important part of the President's budget, eliminating the use of cloro-fluoro-carbons, or CFCs, also became an administration priority. GSA developed a program to replace thousands of tons of refrigeration equipment, currently using R-11 or R-12 refrigerants, with new equipment using approved refrigerants or other technologies. Again, a generic line item for the replacement of building chillers is in the President's budget request and is expected to be in the final budget.

Because of the age of GSA's building inventory, many of our buildings were originally constructed before air conditioning became a standard part of office building construction. Many of these buildings were air conditioned in the 1960s and the chiller plants in these buildings still include the original equipment. Though it is possible to retrofit chillers to be compatible with new refrigerants, in the majority of our aging chiller inventory, a complete system replacement was the prudent option. Because both energy conservation and environmental concerns have strong support with the Administration and Congress, we looked for opportunities to combine these efforts into both energy conserving and environmentally conscious building retrofits.

As we developed this idea, we found out that while research into fuel cells, improved energy storage, and renewable energy technologies are proceeding, little research is readily available to property managers about how to best use the combined technologies available today. As we looked at major office buildings, (both Government and private sector), we found no consistent pattern in chiller plant configuration, control systems, water treatment, cooling towers, or the use of other technologies which make chillers easy to maintain and efficient to operate. We also found that some design firms gave little consideration to opportunities to reduce heat load within a building and that routinely, engineering firms oversized chillers, towers, pumps, and other equipment to handle some expected additional but unspecified load resulting from yet to be developed office technologies or other factors. As an example, the energy conservation manager for a major hotel chain recently said in a seminar that during energy audits, they found chilled water pumps were routinely oversized by a factor of four!" We felt it would be critical to our energy conservation efforts, to identify what we could do to make a chiller plant efficient from a from a load reduction standpoint, an equipment standpoint, and a plant configuration standpoint.

LOAD REDUCTION

Lighting

As discussed previously, GSA has been heavily involved in relighting Federal buildings. Through FY-93, GSA has funded over $60 million of lighting retrofits or replacements. A lighting retrofit using lamp/ballast combinations which provide about 100 lumens per watt of electricity used will generally payback in less than five years; often in less than three years. The lighting in most office buildings today produces about 70 lumens per watt. When you put electricity into a fixture, you get both light and heat in return, the more electricity you turn into light, the less is dissipated as heat. Since building lighting may account for 25 percent or more of building energy, the heat load from lighting is significant.

Windows

Until recently, most window replacements in Federal buildings, were to maintain the integrity of the building structure. Air and water leaks, condensation, broken and rotting window sashes contributed to sill and wall damage. Many manufacturers today are marketing a line of low emissivity windows which, for windows, have marvelous insulating capacities. Paybacks for window projects using this technology are approaching 10 years. However, when performed in combination with lighting and other opportunities, the resulting project could net a payback of five to seven years.

Other HVAC work

An assessment of the buildings current air distribution system could reveal additional opportunities for load reduction. Some distribution systems leak almost half the conditioned air supplied by the air handler. While air distribution will never be 100 percent efficient, their are significant opportunities for improvement which could affect system sizing.

Office equipment

GSA has worked with the Department of Energy and others on efforts to improve the efficient use of office equipment; systems furniture and computers. In terms of system furniture, GSA has done some preliminary work on systems furniture design, lighting, color selection, and work station configuration, and GSA space planners were briefed on our efforts in June, 1993. This briefing stressed coordinating design considerations of overhead lighting, work station layout, glare reductions, partition heights, blockage of heating and cooling supplies and returns, natural lighting and other considerations.

New requirements for energy efficient task lighting will be incorporated into future system furniture procurements. While many manufacturers now market energy efficient task lighting, such lighting will become the standard rather than the exception.

Many systems furniture workstations incorporate task lighting which doubles the existing ceiling lighting load. Reducing the need for task lighting and using more efficient task lighting only when necessary can also assist in reducing head load.

GSA collaborated with DOE and others to improve the efficiency of personal computers. In June, 1993, an exposition was held in New York City where manufacturers unveiled these new energy efficient computers. These computers use less than half the electricity of standard computers and can "put them self to sleep" when not in use to save even more energy. By Executive Order on Earth Day, 1993, President Clinton mandated that all agencies will purchase these new personal computers in future ADP procurements. Personal computers use about twice the electricity is takes to adequately light a standard work station. A reduction by half in energy consumption will reduce heat load significantly.

Load Reduction Summary

The opportunities to reduce heat load in buildings can be significant. In combination, attention to design of systems furniture, the careful use of energy efficient ceiling and task lighting, energy efficient office equipment, windows and air distribution could, in combination, provide the basis for significantly downsizing chiller plants at the time of replacement. The practice of overdesigning cooling systems for some new, unexpected heat load that will come with the next wave of new office technology should be carefully reconsidered. Lighting, computers, printers, fax machines, and other technologies have become more efficient and improved efficiency innovations can be expected.

CHILLER PLANT CONFIGURATION

Many of the buildings in GSA's inventory were built prior to 1960 which was about the time when the practice of airconditioning new Federal buildings became common. In these new buildings, and as the existing buildings in GSA's inventory were air conditioned, the general practice was to install a chiller sized to handle the calculated heat load with a "safety factor" of 20 to 25 percent. A backup chiller of the same size was also installed. These chiller plants were reliable, but as energy conservation became a consideration with the advent of foreign oil embargoes, we found that these plants could not be operated efficiently, particularly under low load or partial load conditions. GSA pursued other operating efficiencies, specified lighting levels, controlled building temperatures, and implemented a number of low cost and no cost energy projects to meet energy reduction goals.

As new Federal buildings were built in the 1980's, a variety of different chiller configurations were used. Many buildings continued to be designed with the two chiller configuration discussed previously. However, some new building designs included multiple chiller plants and an energy management and control system which could be utilized to operate building equipment much more efficiently. Yet, there does not appear to be standard design practices within the engineering community --- either in Federal buildings or in those buildings built for the commercial sector. Variable speed drive units are being installed in some chiller plants in an attempt to operate large chiller efficiently under low load conditions. However, this technology is used much more heavily on large motors in air handlers and pumps than on chillers.

In the 1980's, office automation was beginning to affect Federal buildings. Computer centers added significantly to heat load. First electric typewriters and calculators, then copiers, then word processors, then personal computers, laser printers and fax machines. The Federal appetite for technologies that would enhance productivity and/or reduce the need for workers became as voracious in the public sector and it was in private sector. The equipment in early computer centers was temperamental at best, generated tremendous amounts of heat, and required humidity control and cooling 24 hours a day, 365 days a year. To gain efficiency, computer centers in Federal buildings were cooled with packaged air conditioning units so that the large main chillers would not need to be operated year around to cool the relatively modest portion of the building which housed a computer center. A look at the roof of any Internal Revenue Service Center would show hundreds of tons of additional cooling for any number of small computer centers spread throughout the building. Clearly, the standard two chiller configuration was not able to efficiently meet the cooling needs in many Federal buildings.

CHILLER FEATURES AND OTHER TECHNOLOGIES

Different combinations of features and equipment can be found on chiller plants in many Federal buildings. Many of the original chillers purchased by GSA included a marine box (a manifold, if you will.) Chilled water pipes are attached to the side of the box instead of directly into the front of the chiller. An access plate is provided on the end of the marine box allowing inspection and maintenance of tubes without removal of chilled water piping. In GSA's largest chillers, some as large as 3000 tons, a marine box could reduce the down time for maintenance significantly by saving the effort of removing and replacing hundreds, and in some cases, thousands of pounds of piping.

Tube cleaners were installed in some new and existing chillers during the 1980s with mixed results. Essentially, a brush-like device is forced through chiller tubes by air pressure and periodic intervals. The cleaner chiller tubes can be kept, the more efficient the heat transfer process. Tube cleaners are still installed and used occasionally in new and existing chiller plants.

A number of water treatment options are used in chiller plants in Federal buildings. Standard chemical treatment is the most common treatment for condenser water, however; some agencies, NASA in particular, utilizes ozone water treatment extensively in chilled water systems. In some countries, magnetic water treatment is used along with other methods of water treatment though only limited experiments with this technology have been conducted in the United States. Effective water treatment, regardless of the technologies used, can ensure effective heat transfer and reduce water consumption through reduced blow downs.

As mentioned previously, chilled water pumps are routinely, and often significantly, oversized. The old saying that "no engineer has ever been sued for overdesign" may be the key contributing factor for this practice.

THE OPPORTUNITY --- THE CHALLENGE

Mandates to save energy and eliminate CFC's provide a significant opportunity to coordinate energy and environmental efforts with improved service to building tenants. In GSA's aging building inventory, the success of this effort will significantly affect our ability to meet currently mandated energy conservation goals and posture us to meet future goals. Since 1973, Federal property management agencies have been pushed to conserve energy by national interests, Executive Orders and a variety of legislation. The most recent legislation, the Energy Policy Act of 1992 contains requirements for Federal agencies through the year 2005, and there is every reason to believe that additional legislation requiring even greater conservation will follow.

To successfully posture ourselves to meet current and future energy goals, we must change the way we have traditionally looked at energy conservation projects. Instead of system by system retrofits or replacements, we must look at the interactions between multiple building systems. We can ensure that designers understand both our environmental and energy conservation goals and that designs consistently incorporate the technologies that make plants efficient to operate, inexpensive and easy to maintain, and durable. We must ensure that in-house maintenance personnel and contract maintenance companies are trained to operate these more complicated systems efficiently and are trained to maintain them in a manner that will extend their useful lives.

We will also be challenged by financial limitations. We expect future funding to be severely limited and must strive to maximize every dollar spent. Because of the additional space and relocation costs associated with vacating a building so that major renovations can proceed, the scope of many projects may need to be scaled back in order to maximize conservation and minimize disruption or displacement of tenants.

SUMMARY

The purpose of this paper was neither to bring to light emerging technologies which will conserve significantly more energy nor to present new ideas, methods, or procedures which will help us to meet environmental goals. Everything discussed in this paper is a standard industry practice included in many thousands of chiller plants. Rarely, however, do we consider all, or even most of these ideas, technologies, or features in any one chiller plant. As energy managers, your challenge, and our challenge, is to form a partnership with the designers, the buildings managers, the installers, and safety and environmental managers to ensure that, to the maximum extent practicable, these new chiller plants provide efficient and effective operations over the long term and provide reliable productive space for tenants.

Within GSA, the planning process has started. A year from now we hope to be using our combined chiller replacement and energy conservation funding to make significant strides in improving overall building energy efficiency. As energy managers, <u>this is your greatest long term conservation opportunity</u>. You scratch and claw every day for funding to improve efficiency, but your influence over projects related to CFC reduction will affect conservation in your buildings for the next 30 years. The efficiency of these systems or lack thereof, will drive your future conservation efforts and, in large part, determine the magnitude of the savings you can achieve. Work with your design people. Know what their plans are for chiller replacements and retrofits. A replacement in kind approach will be devastating to your conservation efforts. Your commitment can change this liability into your greatest conservation asset. GOOD LUCK!

Chapter 95

Renewable Energy in the Department of Defense

G.D. Smith

ABSTRACT

With increasing public and administration focus on the efficient use of energy, the Department of Defense (DoD), as the largest consumer of energy in the United States, must play an important role in addressing U.S. energy issues. As indications of this effort, the DoD has been directed by Congress and Presidential Executive Order to reduce energy consumption and find alternative energy sources. Through the assistance of two investment programs for renewable technologies, the DoD has launched a determined effort to decrease its dependence on fossil fuels by integrating renewable energy technologies into military applications. Four renewable technologies -- geothermal, photovoltaics, solar thermal, and wind -- have already proven to be successful in, and show promise for, future military applications.

INTRODUCTION

The Department of Defense (DoD) is the single largest energy consumer in the United States. The Energy Information Administration estimates that the various branches of the military accounted for 1.24 quadrillion Btu or 88 percent of federal energy consumption in 1991. With ongoing budgetary cutbacks and environmental concerns, the DoD has come under increasing pressure to reduce energy (and specifically fossil fuel) consumption and to control energy costs wherever possible. Renewable energy resources are playing a significant role in accomplishing these goals for the DoD.[1]

The DoD has been directed by several federal mandates to utilize alternative and renewable technologies to offset fossil fuel production and consumption. In the House Defense Authorization Act (1990), Congress directs DoD to install 100 MW of renewable energy by 1996. Also, by Presidential Executive Order 12579, all federal agencies (including the DoD) are required to reduce energy consumption by 20% and to offset energy production using renewable energy where it is cost effective using life cycle cost analysis. Finally, the DoD must place emphasis on actions which achieve environmental compliance through renewable energy as a least cost approach to power production. Complying with these mandates will lower energy costs for the DoD and will enhance energy security, particularly in overseas bases vulnerable to energy supply disruptions.

The Department of Defense has established an aggressive program encouraging the use of renewable energy in a variety of military applications and technologies. Since 1991, two federal investment programs have been made available to the DoD for renewable technology utilization in military applications. The Energy Conservation and Investment Program (ECIP) was established to encourage the development of energy conservation techniques and programs. The Strategic Environmental Research and Development Program (SERDP) has been established for the development of technologies that address environmental issues, including clean energy programs.

The primary driving force in selecting a technology for a specific military application will be the technology's ability to reliably and cost-effectively enhance or support the its intended mission. The renewable technologies that are commercially available have proven to be successful in a variety of military situations. Among the renewable resources involved in DoD missions, geothermal, photovoltaics, solar thermal, and wind are the central focus for this paper. These four technologies have either already established a successful history in military utilization, or they show significant promise for future applications.

INVESTMENT PROGRAMS FOR RENEWABLE ENERGY

Previous to 1991, any development of renewable technologies for military uses was accomplished primarily through the efforts and resources within the DoD. Since 1991, the availability of two investment programs has greatly assisted the DoD to reach its goal of increasing the integration of these new technologies into various military operations.

The first mechanism is the Energy Conservation and Investment Program (ECIP) which is a part of the Military Construction Program. Historically, ECIP has been used exclusivelyto fund military energy conservation projects, but the criteria for qualifying for ECIP funds changed to include renewable energy projects. Renewable energy technologies that achieve similar reductions in the use of conventional fuels as conservation programs may now also be submitted under ECIP. Photovoltaics programs have received $6 million in FY 1992, and renewable and hybrid energy systems received $10 million in FY 1993. The DoD continues to encourage other renewable technologies to compete for an additional $50 million available through ECIP.[1]

The Strategic Environmental Research and Development Program (SERDP) is the second investment program available for renewable resources. This fund is a Congressionally mandated program designed for projects that address environmental restoration, waste minimization, hazardous waste substitution, research involvingglobal environmental change, clean energy, and technology transfer. Funds are allocated for various uses through a joint SERDP Council consisting of representatives from the DoD, the Department of Energy (DOE), and the Environmental Protection Agency (EPA). For FY 1992, $4 million was allocated for photovoltaics projects as well as an additional $2.8million for wind installations. In FY 1993, another matching $4 million has been tentatively awarded for photovoltaics.[1]

GEOTHERMAL

Energy generated from geothermal resources is produced using the naturally occurring heat in the earth. Depending on the properties of the available resources, geothermal energy can be used for generating electricity using steam from geysers, or it can be used for process space heating.

FIGURE 1: GEOTHERMAL WELL AT THE COSO FIELD AT CHINA LAKE. POWER IS USED AT THE FACILITY AND IS ALSO SOLD TO AN AREA UTILITY.

Geothermal has shown to have limited potential as an alternate energy source for the military simply because of the limited number of potentially economic resource sites. However, in the instances where a geothermal resource exists under DoD property, significant cost savings are possible.

The geothermal program within DoD works toward increasing geothermal's utilization in military applications. In achieving its goal, the program has three primary objectives: management of encroachment to protect facility missions, lessen the dependence on fossil fuels using geothermal technology, and provide financial benefits to project development. The program operates in an environment of joint ventures between the government and industry; therefore, its efforts are at no cost to the government.[6]

Guidance for the geothermal program in the areas of definition of new resources, the procurement process, and the implementation of the program's objectives is provided by the Geothermal Coordinating Committee (GCC). The GCC is an informal group consisting of representatives from each branch of the military as well as the Deputy Assistant Secretary of Defense - Office of Energy Policy and the Defense Logistics Agency.[6]

The Army and Air Force have concentrated their efforts on identifying possible geothermal resources. The Army has identified 35 potential sites, and is currently concluding a two-year joint effort with the Navy to investigate the geothermal potential of the National Training Center at Fort Irwin, California. Results of this effort indicate that there is only a low to moderate grade geothermal resource available to utilize.[6]

The Air Force has identified 15 potential sites, and it has drilled test wells on Ascencion Island. Engineering estimates at Ascencion Island indicate that the volume of fluid may not be sufficient for development.[6] Similar tests are being performed in cooperation with the Portuguese government on the Azores Islands Air Force Base.[4]

The most active geothermal program in the DoD has involved the Navy. This project developed the Coso geothermal resource at the Naval Air Weapons Station (NAWS), China Lake, California under a private venture. The project consists of two 80 MW power plants that were financed and constructed by a private entity. This private contracted party continues to operate and maintain the facility. This venture requires the private organization to rent the land from the Navy,pay the Navy a royalty, and supply the base with electricity while the Navy retains control of 51% of the field. In return, the operator gets an inexpensive source of energy for its own electricity generation.[5]

The project at China Lake actually pays the government money -- a concept unique in the federal government. The Coso unit consists of two phases -- Navy One and Navy Two. In the project's first year of operation, 1988, the Navy One unit accounted for a direct reduction in NAWS, China Lake's utility bill of almost 34%. The 1992 savings from this private venture amounted to 45%. For 1993, the U.S. taxpayers received almost $9 million in revenue from power sales from Navy One and Two. It has been estimated that the Coso geothermal facility will bring in several hundred million as a result of the sale of electricity to Southern California Edison during the 30-year contract.

The life expectancy of the Coso geothermal field is 1000 MW; far beyond the life of the contract. Finally, the NAWS, China Lake facility will realize a total energy cost savings of $47 million in the first 10 years of operation.[5]

A comprehensive plan for powering all DoD facilities in the eastern Mojave Desert region using geothermal technology is a long term goal for the DoD geothermal program. Facilities that would be served by this resource include: MCAGCC-Twentynine Palms, Fort Irwin, Edwards Air Force Base, NAWS China Lake, and Nellis Air Force Base.[6]

Although geothermal production is limited by resource availability and high development costs, it has proven itself to be a significant cost saving technology for the DoD. The Coso geothermal field developed on the China Lake Naval Base signifies a new possible avenue for renewable energy development for the DoD.

PHOTOVOLTAICS

Photovoltaic (PV) technology uses semiconducting materials to convert sunlight, or radiant, energy into electricity. It is the only technology that does not convert renewable energy into mechanical energy in order to generate electricity. According to DOE, photovoltaics have shown a dramatic price drop over the last 15 years -- from $3.00/kWh to 25-50 cents/kWh now. It has been projected that the price will drop to 10-20 cents/kWh in the 1995 - 2000 time period and eventually to 5-6 cents/kWh in the 2010 - 2030 time period.

FIGURE 2: THE COAST GUARD HAS CONVERTED OVER 10,000 OF ITS NAVIGATIONAL AIDS TO PHOTOVOLTAIC TECHNOLOGY.

The United States Coast Guard (USCG) has been particularly successful in implementing photovoltaic technology by converting navigational aids to PV. Its first PV conversion of navigational aids was completed in the mid-1980's. This was the beginning of a planned 5-year program to convert over 10,000aids to navigation to PV-powered systems. The USCG reports that the conversion program has resulted in increased reliability and decreased operations and maintenance costs. [2]

The use of photovoltaics is the focus of a significant level of activity in the DoD. Photovoltaic technology has been very successful in past military applications and possesses strong potential for future applications. However, realizing that certain barriers needed to be addressed before PV technology could reach its fullest level of utilization in the DoD, the Photovoltaics Review Committee (PVRC) was formed in 1985. Similar to the GCC, the PVRC identifies cost effective applications, increases awareness of PV advantages, and provides assistance and information to potential PV system users.

Through the efforts of the PVRC, thousands of potential PV military applications have been identified. These projects include both small remote applications such as lighting, instrumentation, and navigational aids and large applications such as diesel generator replacements. Additionally, grid-interactive systems are being considered for the future as PV technology continues to develop.

One of the main criterion that makes photovoltaic technology attractive is its cost-effectiveness for remote applications. For this reason, the present thrust of the PVRC is to concentrate on integrating PV technology into small, remote applications. Following this implementation, larger, remote systems will be a focus for the committee. Finally, as technology develops, grid-interactive systems will become a long-term goal for the PVRC.

Presently, the PVRC is concentrating on the development of several PV projects. These projects are direct results of the ECIP and SERDP mechanisms.

- The Navy will develop a hybrid PV system for their Range Electronic Warfare Simulator (REWS) on San Clemente Island in California. The remote island hybrid system will consist of 100 KW of PV array combined with diesel generators.[3]

- The Navy will develop another PV-diesel hybrid system at Junction Ranch site at NAWS, China Lake, California. This system is expected to have a PV array of 100 - 200 KW.[3]

- The Air Force is developing a down-range instrumentation site for the Eastern Test Range at Ascencion Island, Patrick Air Force Base, Florida. The Ascencion Island project is a PV-diesel generator hybrid with the PV array being less than 50 KW.[3]

- The Army is involved in developing a grid-interactive PV system at the Yuma Proving Grounds in Arizona. This endeavor will also be a part of PVUSA (a collaborative effort between the U.S. government and industry). The system is estimated to be as large as 200 KW.[3]

- The Army is also developing a PV system at Fort Carson, Colorado. This project consists of 39 PV powered water pumps. Each well will have a 200 - 500 watt PV system.[3]

Photovoltaic technology is a successful power option that can assist the DoD to achieve its renewable energy goals. The technology has made steady progress in efficiency, reliability, and cost competitiveness over the last 15 years. Today, photovoltaics is the technology of choice for many remote applications and is making inroads into application niches formerly held by grid power, primary batteries, and diesel generators. The utilization of photovoltaics for military applications is gaining significant importance as awareness of its advantages becomes more widespread through the efforts of the PVRC.

SOLAR THERMAL

Solar technology encompasses a group of reflective surfaces that use concentrated radiant energy to create steam for the production of electricity. Heat engines mounted to a central receiving tower convert solar heat to electricity. Solar thermal technology is categorized into two technologies. Active solar systems are those that use mechanical devices such as fans and/or pumps to collect, store, and transport heat. Passive solar systems, on the other hand, collect and distribute heat naturally by conduction, convection, and/or radiation.

Active Solar Thermal

The DoD has devoted a major effort to the application of all types of active solar energy systems. Experience indicates that the most feasible mass market for active solar thermal systems in the DoD is in the preheating of domestic and service hot water. A wide variety of innovative active solar thermal applications are being considered by the DoD as a result of the following basic research efforts by DOE and industry.

Salinity Gradient Solar Ponds (SGSP): Salt gradient solar ponds have been used for many applications internationally and in the private sector. In addition to their capability to collect and store available solar energy in the summer for use during the winter, they are also the cheapest form of solar thermal storage currently available. SGSPs can supply thermal energy directly for process heating or can be used to drive heat engines for peaking or desalination plants.[4]

Solar Thermal Electric Systems for Power Generation (STE): As of 1991, STE capacity in the United States exceeded 350 MW. This technology is rapidly growing and is becoming a cost-effective means to provide peak power in areas of large amounts of sunlight. STE technology is most appropriate for the DoD when shared with a utility since an STE system is most efficient in the 30 to 80 MW range and the peak power requirements for a military base is only 15 MW. Military installations use the power indirectly through a local utility.[4]

Industrial Process Solar Water Heating: Although this technology has been available for over ten years, recent advances in reliability and performance have raised interest in possible military applications.[4]

Ventilation Air Preheating with Transpired Solar Collectors: This technology captures incident sunlight on buildings that is used to preheat outdoor air being drawn into buildings for ventilation. These systems fit into the DoD requirements of low maintenance and high reliability as they do not need storage, have low capital costs, and are simple to operate and maintain.[4]

FIGURE 3: PICTURED IS THE SOLAR ONE CENTRAL RECEIVER NEAR BARSTOW, CA. THE CONSTRUCTION OF A SECOND SIMILAR TOWER IS BEING CONSIDERED.

Passive Solar Thermal

Some form of passive solar application is found in almost all new defense construction. Passive solar energy systems have been applied extensively in new family housing. The military has thousands of family housing units that are being built to utilize passive solar techniques to supplement their space heating requirements. Passive solar design features have included thermal mass window boxes, trombe walls, unique glazing, earth berms, louvers, and integrated and enhanced natural lighting.[4]

Solar thermal technology has a variety of application possibilities within the DoD. While passive technologies are currently more popular, active solar technologies will increase their presence in military applications as research continues.

WIND

The kinetic energy of wind is used to generate electricity as the wind turns a turbine rotor driving an electric generator. With today's technology, wind resources are sufficient on some military installations to be a viable alternative energy source. Activity in the Wind Energy Conversion Systems (WECS) area has concentrated on small, cost-effective, off-grid applications.

The Army has been primarily involved in wind power generation through Civil Works projects. In December 1983, two 25 KW turbines with 44-foot diameter rotors were installed at Wilson Lake, Kansas. In the first year of operation, the generators provided 140,000 kWh of electrical energy. A smaller wind generator has been operational for more than three years at the Saylorville Reservoir, Illinois. A turbine with a 26-foot diameter rotor has produced approximately 10,000 kWh of electrical energy in an area with wind speeds averaging 11 miles per hour. The Air Force has also been involved in wind energy production. It has two 25 KW turbines at Reese Air Force Base that have been operating since July 1983.[4]

The Navy has an active research and development WECS program as well. The Navy has pioneered the use of private party contracting to provide wind generated electricity. A contract has been formed for electricity at 10.5 cents per kWh at the Navy REWS facility on San Clemente Island, California. The electricity is to be generated from five 20 KW wind machines.[4]

The Navy has also been involved in hybrid wind/photovoltaic systems. The most notable of this type of system is at the Naval Air System Command's, Tactical Aircrew Combat Training Systems (TACTS). This system consists of two 16 KW wind machines which carry from zero to ninety percent of the electrical load on this remote military installation. When the force of the winds is not great enough for electricity generation, the photovoltaic system and battery storage systems provide the power.[4]

FIGURE 4: PHOTOVOLTAIC AND WIND GENERATOR ON TACTS, OFFSHORE NEAR CHARLESTON, SC.

Presently, wind turbines have been limited to these and other demonstration sites; however, as ECIP and SERDP funds become available, wind technologies could become increasingly utilized in military installations.

CONCLUSION

The Department of Defense is under increased pressure to address its energy use and consumption practices. As the largest energy consumer in the United States, the DoD is looking toward alternative energy resources as cost saving, reliable technologies to meet its federally mandated goals. In answer to these mandates, as well as to internal goals, the DoD has developed, and continues to develop, renewable energy technology applications for various military installations. The availability of investment mechanisms such as the Energy Conservation and Investment Program and the Strategic Environmental Research and Development Program have enhanced DoD's efforts and have made many projects a reality that previously may not have been possible. The involvement of private entities into renewable energy technology development has also proven to be very successful at the Coso Geothermal Facility, and this type of venture can become a significant avenue for future renewable energy development.

The Department of Defense is committed to accelerating renewable energy technologies by focusing on those energy systems that have the highest economic payback, proven reliability, and cost effectiveness in military applications. As research programs grow and demonstration projects prove to be successful, renewable energy technologies have the potential of becoming mainstream systems at many military installations.

REFERENCES

[1] "Department of Defense, Photovoltaics Review Committee Master Plan," U.S. Department of Defense Photovoltaic Review Committee, June 8, 1993, Draft.

[2] "Photovoltaics for Military Applications: A Decision Maker's Guide," U.S. Department of Defense, Photovoltaics Review Committee, Office of the Secretary of Defense, and Photovoltaic Energy Technology Division of the U.S. Department of Energy, January 1988.

[3] "Recent Advances in the Department of Defense Photovoltaics Program," Garyl Smith, SOLTECH Conference, April 26, 1993.

[4] "Department of Defense Renewable Energy Applications," Department of Defense, 1991.

[5] "Geothermal Exploration at Coso," Department of Defense, Naval Weapons Center, China Lake, CA, April 1990.

[6] "Department of Defense, Geothermal Coordinating Committee Master Plan," U.S. Department of Defense Geothermal Coordinating Committee, March 1993.

Chapter 96

120 KW Lightweight Mobile Generator

T.C. Hardy, D. Tormey

ABSTRACT

In the past, military interests endorsed the large, heavy mobile power production system because the design attitude was that they were more durable and reliable than lighter built systems. Currently, the Air Force uses 1960 technology to produce mission-essential electrical power. However, power production technology advancements have made it possible to drastically reduce system size and weight. This reduction is necessary to accomplish the Air Force's Global Reach and Global Power operational concept. The 1960 electric power generator (genset) system requires more manpower and a large level of logistics for mission support.

Wright Laboratory's Air Base Operability and Repair Section (WL/FIVCO) determined that the genset weight could be significantly reduced by using recent technological advancements in the prime mover, permanent magnets, power rectifiers, power switching devices, and by integrating the two major components. We also considered the need to improve power quality to maintain military mission control in peace and wartime for field operations.

Traditionally, the 100 kW genset uses a diesel engine which rotates to 3600 RPM. Its four output terminals provide three-phase, 60 Hz, 120-volt rms voltage from phase to neutral. Air Base Systems Branch (FIVC) contracted with EML Research, Inc., to develop a smaller volume-per-kilowatt output genset that would have an overall improved system efficiency. This was accomplished by eliminating the 3600 RPM speed barrier to increase generator power density. Instead of using the generator's output directly as with a typical genset, the generator output is rectified to make a DC voltage. An electronic inverter is then used to synthesize the output voltage from the DC. This AC to DC, back to AC conversion process eliminates the link between rotating speed and output electrical frequency, thus allowing the generator to operate at 6000 RPM to minimize size and weight. A modified Mazda 13B rotary multifuel engine with a supercharger attached was used. The supercharger replaced the turbocharger to provide a constant speed application. The supercharger, intake, and exhaust system were tuned for the 6000 RPM operation of the genset. The system electronics provided the power conditioning to the user. The Air Force is predicting a 3:1 weight reduction and a 30 to 40 percent thermodynamic efficiency improvement over current technology

INTRODUCTION

The current U.S. Air Force readiness philosophy of Global Reach has a slightly different logistical responsibility than past readiness concepts. The Air Force is interested in developing strategy in which an unoccupied site, composed of a few essential assets, can be converted into a working, operational location. This would include a usable runway, a taxiway, parking areas, and a source of water that can be made potable. When called upon for a mission, the site must be capable of rapid buildup to support assigned aircraft and to provide other mission-essential resources, including a logistical support and services infrastructure composed of people, facilities, equipment, and supplies. The operational location concept requires mobile facilities, utilities, and support equipment that can be rapidly deployed and installed to quickly transform into a functional operating location.

Global Reach calls for a rapid, mobile military force. This is becoming even more important as countries resist locating fixed facilities in or near regions of potential conflict. The Air Force generally locates mobile power systems at fixed facilities to ensure rapid resource movement to support and defend our allies.

To perform this task, combat units require hundreds of mobile, electrical power generator units between 50-750 kW to support the combat mission. These units are needed for operating locations--each similar to a small city--in a posture of readiness that requires large amounts of electric power to operate defense systems.

WL/FIVC, in coordination with U.S. Army Mobile Electric Power Program Office, has investigated emerging electric power production technologies to ensure most cost-effective power systems to meet Air Force mission requirements. Many energy technologies are approaching maturity, and others are under development.

Users and the Air Base Operability (ABO) community expressed a need for lightweight mobile generators that are more deployable and require less logistical support than the current bare-base gensets. In 1992, WL/FIVCO developed a lightweight generator program to reduce genset size below the weight and volume of the current electric power production system.

This project is an entirely new concept in electric power gensets, using a rotary engine as prime mover and axial field, permanent magnet alternator. The primary development goal was to show the feasibility of producing a genset that has a weight-to-power ratio in the 8 to 12 pounds per kW range.

PRIME MOVERS

Currently, the Air Force mobile electrical power systems are powered by gasoline, diesel, and gas turbine prime movers (not listed in the priority of importance). Diesel appears to be more popular than gasoline; some reasons being that diesel fuel has 13 percent greater energy content than gasoline, and diesel engines have an approximate 35 percent fuel economy advantage (gallons per hour) over gasoline engines. This would equate to approximately 20 percent savings on an energy basis. However, from an Air Force mission standpoint, a multifuel prime mover is desirable. The Mazda 13B rotary engine was selected because it is compact and easier to turbocharge or supercharge than piston engines (see Figure 1). A program is underway to convert the Mazda 13B to run on diesel fuel. Although a gasoline-fueled engine is used in the prototype genset, future units will use the multifuel engine.

FIGURE 1
120 KW Genset Skid
(Mazda 13B Engine Only)

Mazdatrix of Signal Hill, California, constructed and tested the modified Mazda 13B, built the genset frame, and mounted the engine and its auxiliaries to the frame. Mazdatrix is very familiar with the Mazda 13B's operating at speeds of up to 9000 RPM, mainly in automotive racing applications. The normally carbureted 13B engine is capable of 200 HP at 7000 RPM, but Mazdatrix performed some modification to the engine to guarantee 200 HP at the 6000 RPM-design speed. A supercharged unit was considered to achieve the necessary performance goal. The supercharger unit was selected because Mazdatrix has used it effectively in the past, and it is easy to install on the 13B.

ALTERNATOR

Currently, the most popular mobile alternator is the induction alternator. The three-phase alternator coils are arranged in a wye configuration, and the instantaneous EMF is displaced by 120 electrical degrees as illustrated in Figure 2.

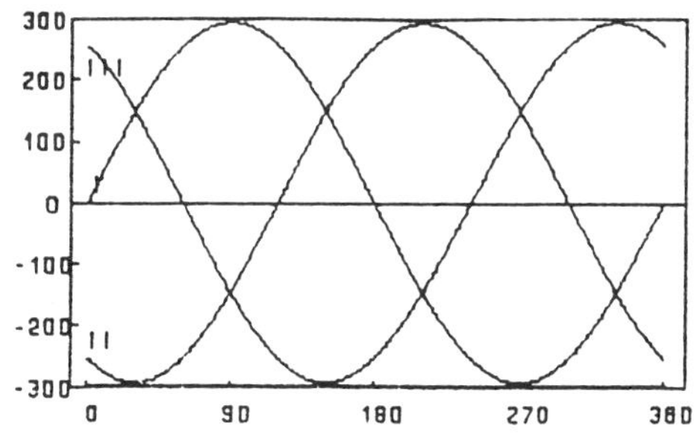

FIGURE 2
Three-Phase Voltage

The conventional alternator is large and very heavy because of the silicon steel rotor and the alternator's massive steel casing. A number of sources contribute to power loss in conventional alternators:

1. Friction and windage from bearings, brushes.

2. Core loss, which is the result of eddy currents and hysteresis in the iron core.

3. Resistance heating loss in the armature and field conductors.

4. Exciter losses.

5. Ventilation losses.

WL/FIVCO performed literature searches to decide which of the emerging generator technologies would offer lighter, easier deployable electric power generators. EML Research, Inc., was developing an axial-gap electric machine with substantial power capabilities. We found that EML was working on two programs that were indirectly related to FIVCO's Lightweight Power Generator program.

Since the start of this development program, EML has been involved in a number of related technical efforts. Although the applications range from ship proplusion to traction vehicle drives, the core technology is directly related to applications like the lightweight generator set. The most recent testing effort at EML demonstrated a complete drive system (motor, inverter, and control electronics) which delivered 430 HP at 3800 RPM. Many of the circuit board and mechanical subassemby designs from this drive system were used on the lightweight genset. EML's motor and generator designs range in speed from 120 to 20,000 RPM and range in power from less than 10 to 3000 HP. The lightweight genset program was able to take advantage of this wide range of design experience to ensure success.

The advance concept in this program is the imposition of a rectifier-inverter between the generator output and the load terminals. This frees everything upstream from the necessity to generate power at 60 Hz. The resulting design flexibility permits substantial reductions in generator weight and, by removing an upper limit of 3600 RPM on generator speed, makes it possible to achieve weight savings with a higher speed prime mover without paying any weight penalty for a gear box.

Until recently, high power solid-state switching elements were limited to switching frequencies below 10,000 Hz. The recent introduction of the Insulated Gate Bipolar Transistor (IGBT) switches now makes it possible to consider switching in circuits at the subject power level at rates in the 30,000 Hz range. There is an overall weight reduction with use of an inverter. The introduction of the inverter gives other potential benefits, resulting in a very attractive overall system configuration.

Figure 3 shows a conceptual cross section or point design of the generator. The estimated alternator weight is 125 pounds.

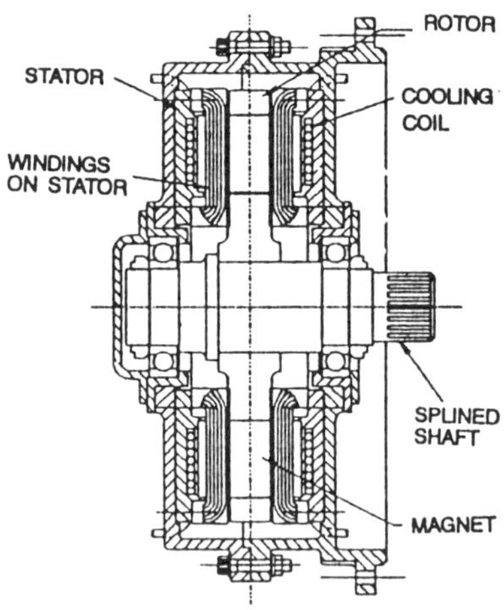

FIGURE 3
120 kW Generator Cutaway

EML Research, Inc., designed, built and tested several high-power-density permanent magnet disk motors and generators. All aspects of the motor and generator designs focus on making small lightweight machines, but three design features stand out as technical breakthroughs. These are listed below:

1. The brushless, axial field disk type topology

2. Neodymium-Iron-Boron (NdFeB) Permanent Magnets

3. Aggressive thermal management

The disk-type permanent magnet topology is attractive from a number of standpoints. In the disk topology, unlike the more conventional drum machine, the magnetic and centripetal forces on the permanent magnets are orthogonal to each other. As a result, the problem of securing magnets to the rotor is much more easily solved in an axial field machine. The magnetic and centripetal forces on the permanent magnets are orthogonal to each other. As a result, the problem of securing magnets to the rotor is much more easily solved in an axial field machine. The disk generator design uses two stators, one working each side of the rotor. This offers much more surface area for producing magnetic shear than does the drum machine, just as disk brakes have more braking surface than drum brakes. The two-stator approach also offers an inherent redundancy since the generator can operate (at reduced power) on only one stator.

The idea for disk type machines is not new and is not an EML invention. However, high-power permanent magnetic disk machines are not common because there are many obstacles to creating practical designs. EML Research has invested a great deal of effort to develop creative solutions for manufacturing stators and winding coils, securing permanent magnets, and reducing electrical losses. This investment has resulted in EML's ability to design practical, high-power axial field permanent magnet machines. The generator designed for the lightweight genset makes use of all three technology drivers identified previously. This unit is a liquid cooled disk machine which uses high-temperature grade NdFeB permanent magnets. Generator specifications are given in Table 1.

Output Power (@6000 RPM)	132 kW
Output Voltage(line-neutral @ 6000RPM)	258Vpeak
Line current (@6000 RPM)	61.5 Arms
Frequency (@6000 RPM)	800 Hz
Shaft Speed	6000 RPM
Weight	125 Lbs
Dimensions	15"(Diameter) X 6" (L)

Table 1
Lightweight Genset Specifications

The alternator thermal management is controlled as the heat generated in the windings is conducted through the impregnated epoxy to the core structure, then flows axially through the core structure, in parallel to the core losses, to cold plates on the end faces of the stators.

RECTIFIER-INVERTER

The rectifier/inverter section serves the basic function

of converting the high-frequency AC produced by the alternator into 60 Hz AC for the system load. Like the alternator, the rectifier/inverter is designed to minimize the weight of the complete genset. Other necessary system attributes, including short-circuit protection and good maintainability, are achieved in a manner consistent with the overall goal of the project.

An important link to the Lightweight Generator is the rectifier/inverter. A schematic for one phase is shown in Figure 4. Each phase is driven by two half bridges. The output of each half bridge is fed to the load through an inductor. A capacitor is connected between the load point and ground. The inductor and capacitor function as a single-stage filter, reducing high-frequency ripple to the load.

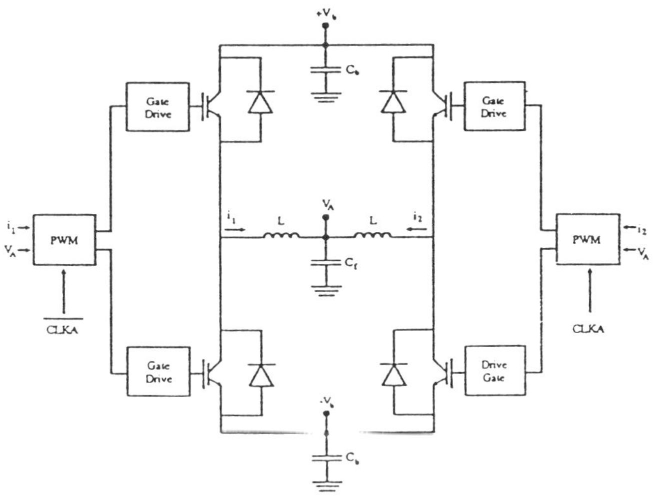

FIGURE 4
Phase Schematic

Both the half bridges in Figure 4 are PWM controlled at a frequency in excess of 60 Hz (normally 25,000 Hz). The phases of the PWM clocks for the two bridges are opposite to one another. One desirable result is an effective doubling of the PWM frequency seen by the output filter. The filter can therefore be made of relatively small components and still produce a very clean output. A second result of staggering the clock phases is that the ripple current draw is reduced on the DC supply.

The circuit in Figure 4 uses IGBTs as the basic switching elements. These newly evolving switches offer a previously unobtainable combination of speed and power-handling capability. The high-speed feature results in low switching losses, which permit use of a high PWM frequency. The high PWM frequency makes possible both the good waveform quality and the low weight of the filter components. The good power handling capability allows a design with only one device per leg.

The absence of parallel switches, combined with the small filter components, facilitates packaging the rectifier/inverter into a small number of discrete modules. The result is an easily maintainable design. Field repair can be limited to replacement of complete modules. The modules are the bolt-in type, requiring only a low-skill individual to troubleshoot and replace. The size and weight of individual modules are small enough so that carrying spares is not a major logistical problem.

SYSTEM INTEGRATION

EML and University of Colorado have worked closely in integrating the prime mover, generator, and rectifier/inverter subsystem on one skid. The genset skid is a 60 in. (L) by 40 in. (W) by 40 in. (H), welded assembly made with sections of 0.25 in. thick steel tubing. See Figure 5. The 60 in. tubes along the bottom on either side of the skid are open at both ends so that the genset can be easily moved with a forklift. The engine, the electronic enclosures, the radiator, and the engine muffler are all mounted to the skid rubber isolation mounts to minimize vibration transmission.

The engine system is packaged so that the engine and support equipment are completely contained at one end of the genset skid. The genset face at this end of the skid is almost entirely occupied by the radiator used for cooling the rotary engine. The radiator is larger than that used in automotive applications because the genset is stationary. A very high-volume DC-powered fan is used to blow air across the engine and out through the radiator. This radiator is used exclusively to cool the engine. A separate cooling system is used to cool the generator and electronics. A water/ethylene glycol mixture is used as a cooling fluid for both cooling systems.

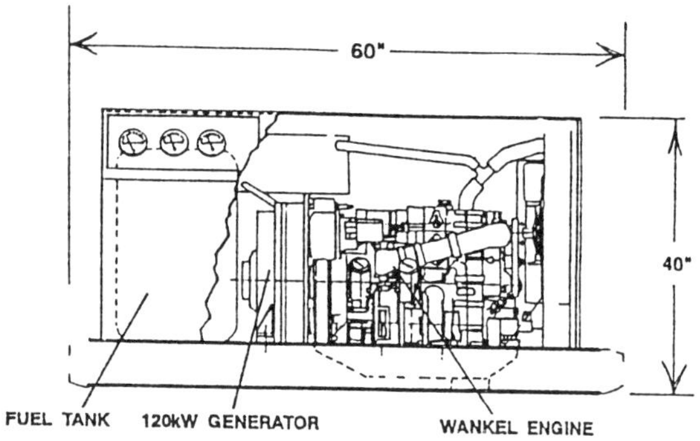

FIGURE 5
Lightweight Genset Prototype

There are three power takeoffs from the engine; two are used to generate 12-volt DC for the radiator fans and the electronics, and the third pumps cooling fluid through the generator/electronics cooling system. Two 12-volt, 60-amp alternators are required to meet the power demand of the DC power fans and the control electronics. The system battery is a standard automotive cell that is mounted to the skid near the engine. The third power takeoff drives an 1800 RPM roller pump to provide a coolant flow of 8 GPM to the generator and electronics cooling system.

There are several auxiliary items which have been designed into the system. There is an electronic overspeed controller which will shut the engine down if the engine speed exceeds the set-point. The alternator has an upper speed limit of 8000 RPM, but the set-point was set at 6500 RPM.

An array of engine diagnostic gauges have also been incorporated into the skid to provide information on engine speed, temperature, and oil pressure. The gauges will be mounted on the control panel at the opposite end of the skid so that all genset control and diagnostic information is provided in one location.

The generator shown in Figure 3 is designed to mount to the rotary engine via an EML-designed mounting/interface flange. The engine mount is located on the engine itself, and the alternator is cantilever mounted from the mount/interface flange. The engine and alternator shafts are connected through a torsional isolation coupling to dampen torque ripple and to take up any misalignment between the shafts.

GENSET TEST PLAN

The test began in July and continued through August. A detailed test plan is outlined below:

Alternator/Rectifier Feasibility Test: Initial testing concentrated on the alternator/rectifier to confirm operation for a 132 kW load at 6000 RPM shaft speed. Data confirmed that temperatures on the alternator, rectifier, and cooling system were in agreement with design goals. EML has a variable, high-power resistive load bank which was used as a load in this and subsequent tests. Before this test, the alternator rotor was balanced and spin-tested to verify mechanical integrity at 6000 RPM and above.

Genset Preliminary Test: After verifying the alternator and rectifier, the inverter and controls were connected. Before this test, the inverter and control electronics had undergone a series of tests to confirm basic functionality. Under no load or light load conditions, measurements were made on the genset output voltage and current waveforms. Specific focus was on the Total Harmonic Distortion (THD) and the frequency stability of the voltage waveform. Some adjustments to the control electronics were made during this test stage to optimize output waveform quality. The load was increased incrementally up to about 60 kW to confirm waveform stability and temperatures in the inverter and output filter components. Based on measurements at the 60 kW power level, peak genset power capability was projected to determine some of the set-points for the next set of tests.

Genset Full-Load Test: Loading was increased incrementally to verify the 120 kW output power capability. Testing was picked up at the 60 kW level, and component performance was measured carefully as the load was increased. The basic objective of this final test was to confirm the 120 kW output power capability of the genset.

CONCLUSION

The Air Force is investigating technologies that will help conserve resources and improve mission support systems. This is another of the Air Force development programs that we predict will have an excellent pay back. The 120 kW Lightweight Mobile Generator Set technology will be beneficial to the commercial market as well as to the military. The commercial benefits will be the 8 to 12 pounds per kW power density and also the 15 percent reduction in operating cost. The fine tuning of the subsystems that make up the genset promises to return all the theoretical total system savings. The commercial technology transition will be seen in smaller domestic and industrial portable type generators that are used in recreational vehicles, homes, and industrial back-up power systems.

References

1. Arbtin, A.; Atkinson, S.A.; Bowman, AL; Fridlund, P.S., Lopez, D.A.; Lussie, N.G.; Marts, DJ.; Stevenson, D.H.; Bare Base Power Supply Systems, ESL-TR-8725, June 1988.

2. EML Research, Inc., 120 kW Lightweight Generator Set Proposal, July 1991.

3. University of Colorado, Rotary Information Fax and Phone Communication, May 1993.

Chapter 97

The Centralized Purchase of Natural Gas for DOD: The Details and Dollar Benefits

M.K. Iden

NATURAL GAS INDUSTRY BACKGROUND

Until the mid-1980s, most large commercial, industrial and Department of Defense (DoD) natural gas end users were limited to acquiring natural gas supplies directly from their local distribution companies (LDCs). The LDCs generally provided a complete range of services for their end users under standard, regulated tariff sales rates monitored by the regulatory bodies. The range of services provided by the LDCs included the sale of natural gas supplies, inter/intra-state transportation, back-up service, storage, meter reading and billing services. In general, most end users had few responsibilities within this highly structured and regulated environment, other than to request gas service and pay the bills.

With the advent of deregulation of the natural gas industry in the mid-1980s, new means of acquiring natural gas supplies became available to many end users. Among the most significant changes within the industry, Federal Energy Regulatory Commission (FERC) Orders 436/500 provided for open access, nondiscriminatory transportation of natural gas by the interstate pipelines. Prior to this action, most natural gas supplies delivered by the interstate pipelines were sold either directly to the LDCs (for resale to end users in their service territory) or to a few large industrial customers having direct connections to the interstate pipeline system. FERC Orders 436/500 provided the opportunity for many classes of end users to acquire their natural gas supplies in a competitive environment, thereby offering the opportunity for these customers to achieve significant cost savings. As a result of FERC Orders 436/500, the role of the interstate pipelines has changed from one of a merchant (sales) and transportation function to one now consisting mainly of the transportation function.

DEPARTMENT OF DEFENSE INITIATIVES

Reacting to the broad acceptance of FERC Orders 436/500 by the various interstate pipelines, many large commercial and industrial natural gas end users moved quickly to implement revised methods of acquiring their natural gas supplies. Significant cost savings were achieved by this class of end users as a result of their timely and competitive acquisition efforts. The Military Departments, however, were not as quick to react to the cost savings opportunities presented by the deregulation of the natural gas industry. Beginning in 1983, the Military Departments (Air Force followed by Army and Navy) independently began the process of pursuing competitive natural gas acquisitions for their individual installations. As early as 1987, centralizing natural gas acquisition within the DoD had been suggested as a means to increase participation and the potential savings opportunities for Defense installations in the evolving natural gas industry. At the time, however, the DoD instead pursued a policy of individual Military Department responsibility for natural gas matters with interdepartmental cooperation on a voluntary basis. The lack of a formal structure to facilitate cooperation by and between the Military Departments was a major impediment to the DoD's efforts in acquiring natural gas in this rapidly evolving industry.

Although the evolving natural gas industry provided end users the flexibility to obtain their natural gas supplies from sources other than their LDC, the LDC still remained an integral part of the acquisition effort because of their role in distributing the natural gas to the various customers within their service territories. Figure 1 depicts the organizational structure of the natural gas supply and distribution systems. As the DoD natural gas acquisition program evolved, it became apparent that the ability to consolidate the natural gas requirements of all DoD installations located within an LDC's service territory would enhance the buying power for these installations - larger volume, more market power, therefore, better prices. In line with this concept and a subsequent 1989 Defense Acquisition Board Study, the Deputy Assistant Secretary of Defense (Logistics)

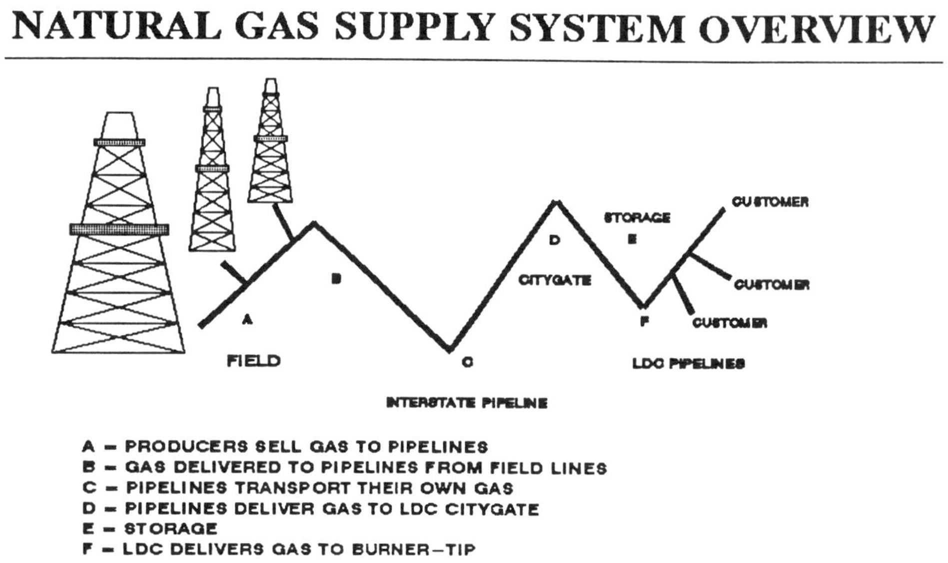

Figure 1. Natural gas supply and distribution systems.

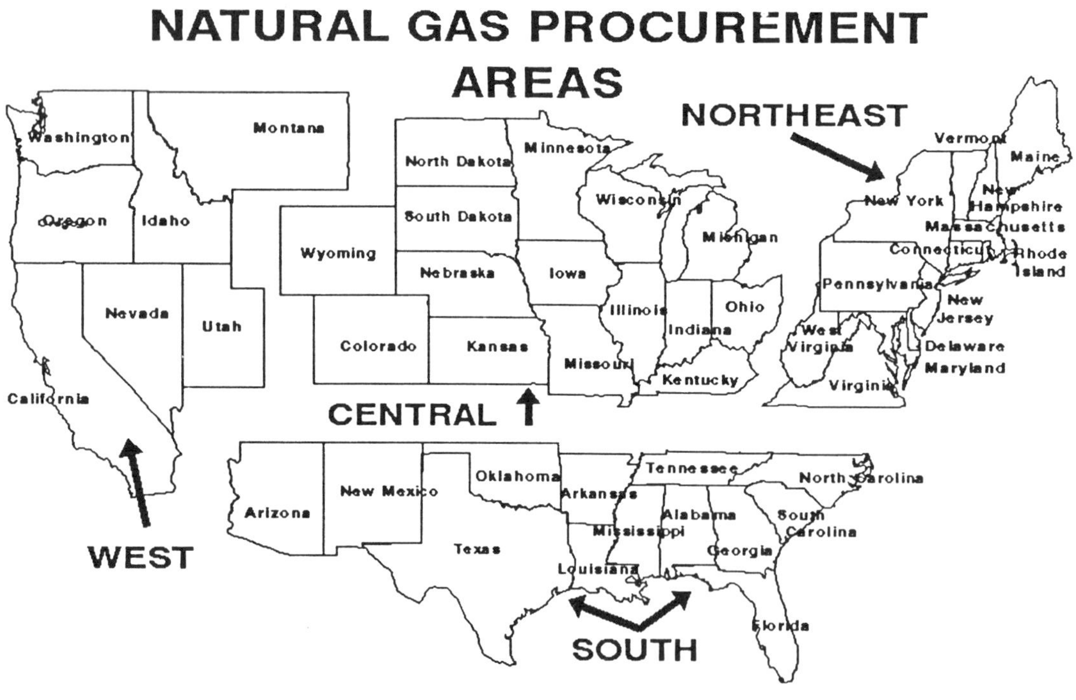

Figure 2. DFSC natural gas procurement areas.

directed the centralized procurement of direct supply natural gas for the DoD beginning in 1989. The Defense Logistics Agency (DLA) was given the sole responsibility within the DoD for competitive direct supply natural gas acquisition. The Defense Fuel Supply Center (DFSC) was designated the implementing agent for this effort.

DFSC NATURAL GAS PROGRAM

The DFSC centralized competitive natural gas acquisition program was initiated to obtain the most cost effective supply of natural gas for DoD installations without sacrificing supply reliability. In order to ensure the best price for the installation, the DFSC performs a series of economic analyses to determine whether to solicit the installation's natural gas requirements in a competitive environment (direct supply natural gas) or to utilize LDC "sales" gas under established tariff rates. To further complicate this matter, some LDCs allow an installation to switch between LDC "sales" gas and direct supply natural gas on a monthly basis, depending on which is the most cost effective option. Accordingly, the DFSC was challenged to develop an acquisition program which took into account many of the anomalies of the natural gas industry, but still complied with the regulatory and statutory guidelines mandated for Government contracting.

Beginning in 1989, the DFSC began the organizational and structural development necessary for their centralized natural gas acquisition program. Although centralizing the competitive acquisition of direct supply natural gas in one location (DFSC) was a key factor, a decentralized approach was still utilized in several aspects of the acquisition program. The reason for this dual approach was the fact that the LDCs could still remain a potential supplier of natural gas to the DoD installations under certain conditions. Under the decentralized approach, installation personnel remained responsible for issues involving their LDCs. In addition, installation personnel are also responsible for the ordering, balancing and payment actions associated with the DFSC natural gas acquisition program, with the exception of California where DFSC personnel retained these functions due to special circumstances.

Program Start-Up

As part of the organizational and structural development for their natural gas acquisition program, the DFSC originally identified seven geographical procurement areas within the Continental United States (CONUS), reflective of factors such as interstate pipeline distribution patterns, customer base, and LDC service territories. In subsequent years the procurement areas have been realigned into four regions which mirror the geographic boundaries of DFSC's Defense Fuel Regions (see figure 2). As a result of this realignment, the DFSC is now in position to award contracts for the four procurement areas on a six month staggered basis, each having two year delivery periods. In addition, the contracts are structured so as to start deliveries in either the months of April or October, thereby avoiding new contract start-ups during peak demand times.

The first DFSC solicitation, patterned after earlier USAF contracts, was developed for deliveries of natural gas to 17 DoD installations located in the Central United States was issued in May 1990. The solicitation developed for the natural gas program was similar to DFSC's other major program buys in that it was a fixed price requirements-type contract with monthly economic price adjustment (EPA) provisions. The monthly escalators were based on published market indicators which focused on the spot market price of natural gas. A unique aspect of the natural gas contracts was the provision that allows DFSC to evaluate the projected costs of the DFSC contractor's natural gas for the coming month versus the published LDC tariff rate for the same time frame. In situations where the LDC allows switching of "sales" gas and direct supply natural gas on a monthly basis, this provision ensured the installation would have the opportunity to access the most cost effective source of gas on a monthly basis.

The initial DFSC natural gas contracts required transportation that was classified as "interruptible". This classification indicates that in times of peak demand (i.e., winter months), the delivery of natural gas on the pipeline systems may be interrupted or curtailed due to lack of pipeline capacity to move all of the needed natural gas. Customers must be able to accept these interruptions in their natural gas service or have some type of back-up or alternate fuel capability in order to meet their requirements. In addition, the normal delivery point for the DFSC contracted natural gas was to the LDC "citygate", the entry point for the LDC local distribution network. For these natural gas contracts, DFSC considered supply and transportation costs to be passed through costs, so suppliers only offered bids on what is called the "adjustment factor". The adjustment factor therefore constitutes the offeror's overhead, margin and profit in relation to his overall bid price for the natural gas supplies.

Program Status

Since the initial DFSC natural gas contracts were awarded, the number of participants in the DFSC natural gas program throughout the country has grown to 135 installations by the second quarter of fiscal year 1993. These 135 installations reflect annual natural gas requirements of approximately 49 million dekatherms. Of the 80 participating DoD installations, 28 are Army installations, 29 are Navy, 22 are Air Force and 1 is a DLA activity. The remaining 55 activities are Federal Civilian Agency installations comprised mostly of Veterans Affairs Medical Centers and Department of Energy facilities (see figure 3). The number of participants in the DFSC program is expected to continue to rise as additional DoD and Federal Civilian Agency activities are incorporated into the program. Recent DoD policy guidelines now mandate that DoD installations must participate in the DFSC program as long as cost savings exist and supply reliability is not jeopardized. All of the Federal Civilian Agency installations are voluntarily participating in the DFSC program, a good indicator of the value and benefits of this acquisition program.

As mentioned earlier, a key aspect of the DFSC natural gas program has been the savings opportunities available as a result of the competitive acquisition process. The means of quantifying the savings potential as a result of this program has been to compare the monthly DFSC contract index prices versus the established LDC tariff prices during comparable time frames. This particular method has also been used on a monthly basis to determine whether to utilize LDC "sales" gas or switch to the DFSC contractor's direct supply natural gas. Figure 4 depicts the recent monthly price differentials between LDC "sales" gas and the DFSC contracted gas. Since the inception of the DFSC program in 1990, over $20 million in savings potential has been identified and realized by the program participants (see figure 5). The amount of savings is also expected to grow significantly as more installations participate in the DFSC program over a longer period of time.

Program Flexibility

In order to develop a meaningful, realistic and cost competitive acquisition program, the DFSC recognized that certain unique parameters of the natural gas industry would need to be incorporated into the new DFSC acquisition program. Such a unique contract provision that recognizes a part of the natural gas industry but that normally is not found in other supply contracts is one that deals with natural gas interruptions or curtailments to either the source of supply or the pipeline system. Since interruptions and curtailments are an inherent problem in the gas industry, the DFSC developed a contract provision that requires the DFSC contractor to use his "best efforts" to obtain an alternate pipeline route and supply source for delivery to the installation when a curtailment or interruption occurs. This "best efforts" alternate plan concept always came with a price higher than the DFSC contract price, but would only be accepted when it was lower than the price of a back-up or alternate fuel. The alternate plan looked good on paper but had a major flaw. Curtailments and interruptions are announced by the pipelines with very little notice; therefore, the DFSC had very little time to verify the contractor's proposed price increase. The end result was that at the end of the 1991/92 heating season, 14 of the 26 installations in the Northeast region had accumulated less cost savings and experienced less reliability than originally anticipated. Since the major premises of the DFSC natural gas program are cost savings and supply reliability, the DFSC had to reevaluate the curtailment issue and take actions to try and limit the number of curtailments and avoid the additional costs of the emergency/alternate plans.

In the Spring of 1992, the DFSC reviewed all the alternate plans submitted by those contractors who delivered after the major pipeline system was interrupted during the previous winter and found a common denominator. Alternate plans that were successful during times of heavy curtailment all transported gas from the same area of the country. Since the DFSC knew that interruptions and curtailments on the pipelines in the areas indexed during the winter were most likely to be curtailed again, the DFSC looked at the possibility of using a second index in the same areas that supplied gas to the DFSC contractors during the primary source interruptions. Even though the price of gas in the "alternate plan" would be higher, the DFSC would only pay the price of gas as stated by the contract's escalator, thereby keeping a lid on costs that were previously uncontrollable. In other words, the DFSC required the contractors to propose a secondary "built-in" alternate plan that could automatically be switched to if the primary pipeline was curtailed or interrupted. The secondary method of delivery is only activated when a curtailment or interruption occurs on the primary delivery method and is, therefore, in effect, an alternate plan with a price that is negotiated prior to the award of the basic contract.

The idea appeared sound and on the surface appeared capable of accomplishing two goals: more reliability for the Northeast installations and guaranteed fair and reasonable prices under the alternate plan concept. Since the DFSC anticipated different offered prices for each delivery method, a special evaluation procedure was factored into this approach. Provided in the solicitation schedule was a Government estimate of that part of the

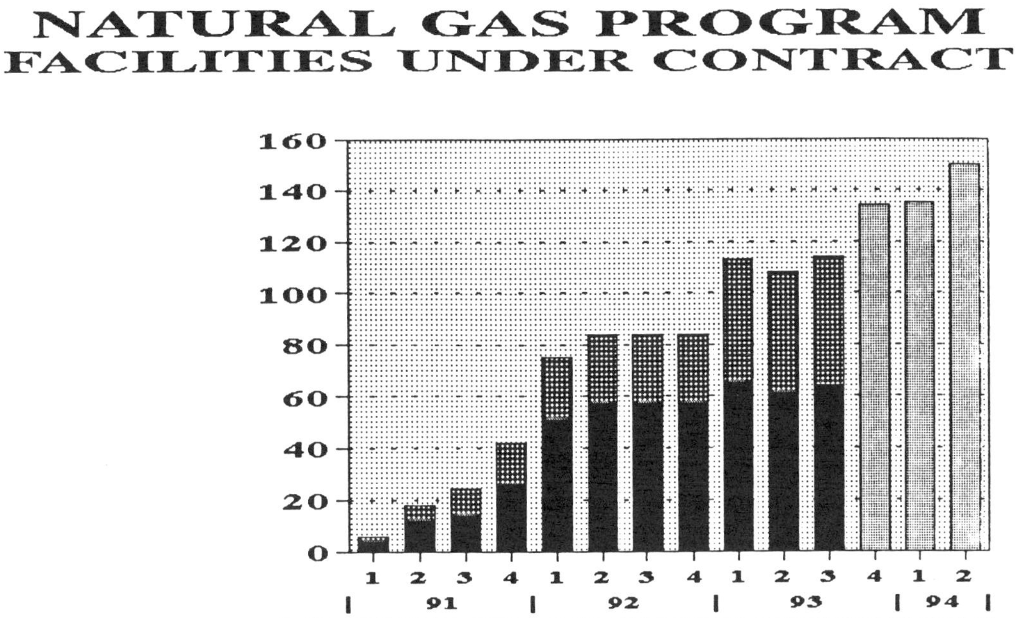

	91-1	91-2	91-3	91-4	92-1	92-2	92-3	92-4	93-1	93-2	93-3	93-4	94-1	94-2
PROJECTED ON LINE												134	135	150
CIVILIAN	2	6	10	16	24	27	27	27	48	47	50			
MILITARY	4	12	14	26	51	57	57	57	65	61	64			
ACTUAL ON LINE	6	18	24	42	75	84	84	84	113	108	114			

Figure 3. DFSC natural gas program participants.

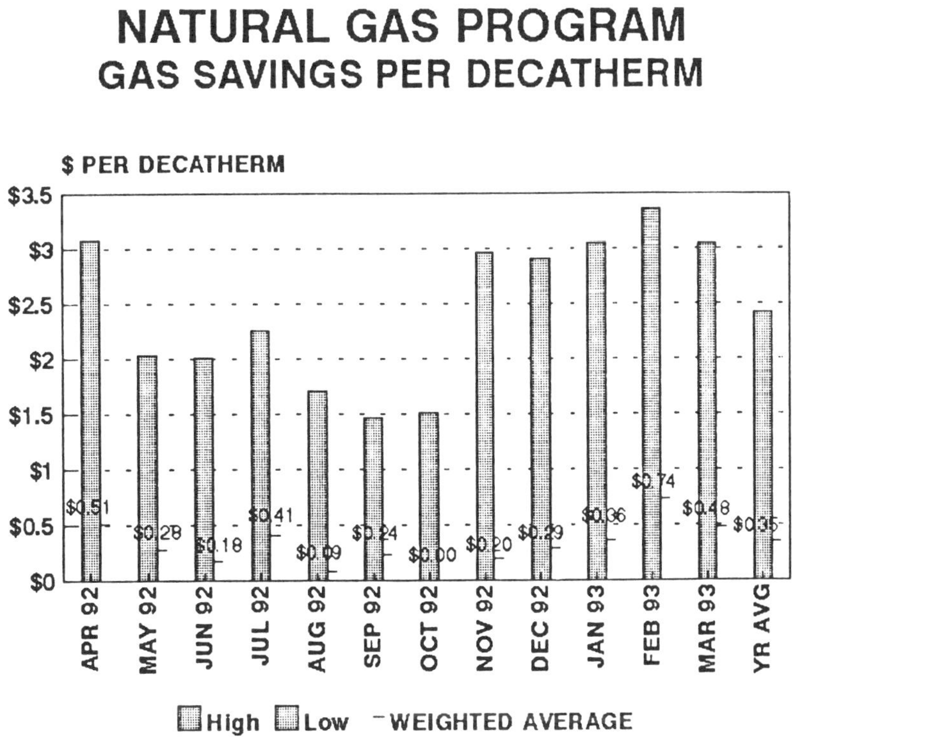

Figure 4. Cost differentials between LDC gas and DFSC-contracted gas.

NATURAL GAS SAVINGS
BY SERVICE

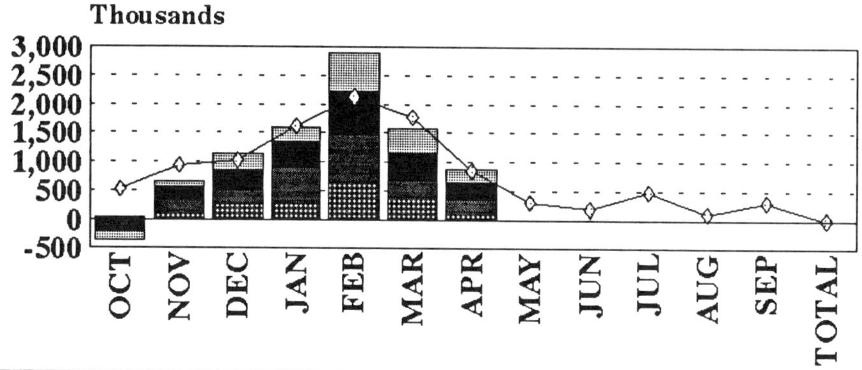

	OCT	NOV	DEC	JAN	FEB	MAR	APR	MAY	JUN	JUL	AUG	SEP	TOTAL
FY 92	516	943	1,026	1,620	2,133	1,776	841	298	180	473	105	293	10,204
FY 93	-325	643	1,137	1,594	2,894	1,566	861						8,369
NON-DOD	-141	75	286	248	649	415	218						1,750
AIR FORCE	26	250	347	451	771	473	306						2.623
NAVY	-191	166	251	565	830	284	194						2,099
ARMY	-19	151	254	330	643	395	143						1,898

POINT: CUMULATIVE PROGRAM SAVINGS $22.6 MIL. (1 0CT 90)

Figure 5. Potential savings realized by DFSC natural gas program.

total estimated contract quantity that would be delivered under the primary approach, and an estimate of the quantity of gas that would require delivery under the secondary proposal. Both estimates were based on 1991/92 winter actual deliveries, and each supplier's offer was evaluated utilizing a weighted average. This meant that a supplier could not bid low and "buy in" on the primary, and then "get well" on his secondary price. Simply stated, it placed all offerors on a level playing field.

At time of best and final offers, the DFSC realized that this approach just might work. The average difference in the low offerors' primary and secondary adjustment factors was less than one cent. The contractors who would receive awards under the resolicitation had guaranteed supply sources for their secondary "alternate plan" proposals and also were the shippers of record on the applicable pipelines. The DFSC determined that terminating the original contracts and putting in place contracts with the prepriced alternate delivery plan concept and the more definitive technical specifications were in the best interest of the Government. The DFSC awarded new contracts by September 30, 1992, and gas under the new contracts flowed on November 1, 1992.

CHANGING CONDITIONS - NEW DIRECTIONS

FERC Order 636 represents the latest in a series of changes being incorporated into the evolving natural gas industry. Specifically, FERC Order 636 acts to complete the restructuring of the interstate natural gas pipeline business by "unbundling" the services that the interstate pipelines perform. The merchant or sales function of natural gas by the pipelines is now a separate service from the transportation function. In addition, pipeline and storage capacity is now available to end users on an equal basis, although existing customers still retain first rights of refusal to their original capacity. The effects of these changes will result in a restructured and more competitive environment for all participants in the natural gas industry. End users will need to adapt their acquisition strategies in order to ensure they obtain a reliable supply of natural gas at the most economical rates possible.

The DFSC continues to track the ongoing changes in the natural gas industry and retains the flexibility in their natural gas acquisition program to provide the most economic and reliable supply of natural gas to their customers. The primary/secondary delivery plan approach to curtailments in the Northeast United States was a unique and innovative manner of resolving a recurring problem for the DFSC customers. The DFSC has also instituted a policy whereby the contractor also must have the interstate pipeline transportation agreement in his own name, an action which further helps the issue of supply reliability and contractor accountability within this program. As the natural gas industry continues to evolve, the DFSC is now pursuing other alternatives such as "firm" transportation, "burner tip" deliveries, and a "bundled service" in order to ensure the most reliable and cost effective supply for installations participating in the DFSC natural gas program. The DFSC track record of reliable service and program flexibility, coupled with the significant cost savings achieved for the participating customers, reflects an extremely successful and dynamic acquisition program.

Chapter 98

Central Energy Supply Planning for DOD Installations

M.J. Savoie, R.E. Moshage

ABSTRACT

Development of improved planning methods and procedures for DOD Energy Supply Systems is a critical element in the long term economic viability of these installations. The average age of DOD central energy plants is about 30 years. This means that many of the plants are nearing the end of their useful life. As expected, many of these plants are experiencing poor combustion efficiency, excessive emissions, reduced reliability and increased maintenance costs. However, the actual causes of these problems are often overlooked due to equipment age and appearance. This has resulted in the tendency to replace equipment without a thorough evaluation of all viable options. Compounding this is the fact that fuel and electric costs have changed dramatically for many installations, making current energy supply methods less economic. In addition, Base Realignment and Closure (BRAC) activities will affect over 150 DOD installations. While some installations will be closing, most will gain or lose portions of their mission, resulting in significant changes in energy supply needs. Finally, the recent passage of the Clean Air Act Amendments of 1990 will affect the applicability and economics of certain fuel technologies.

For these reasons an Integrated Utility Resource Planning (IURP) approach is being promoted for DOD installations. Although the full IURP approach has not been utilized at any DOD installation to date, several central energy plant modernization studies have been completed by USACERL. Implementation of these studies has resulted in the identification of existing tools, as well as the development of new tools, that are necessary for an IURP analysis. This paper will discuss these tools, their application methodology and present case studies of their application at two DOD installations.

ARMY RESEARCH PROGRAM

The Central Energy Supply Planning research program, led by USACERL, is organized around several thrust areas: evaluation of energy supply needs, status quo evaluation, modernization opportunities evaluation, and economic analysis of alternatives. The economic analysis of central energy supply alternatives, whether it's maintaining the status quo, modernization, or a new plant, needs to be performed in a logical and consistent manner. USACERL has developed several computer tools and incorporated others into a method for analyzing energy supply alternatives. These computer tools are HEATLOAD, Central Heating Plant Economics (CHPECON), STATUS QUO Evaluation, and Life-Cycle Cost in Design (LCCID). The Steam Heat Distribution Program (SHDP), developed by the Navy, has proven useful in steam distribution system analysis. Finally, the HEATMAP program, recently developed by the Washington State Energy Office (WSEO), promises to significantly enhance the evaluation of district heating and cooling opportunities. The IURP method has been demonstrated, to various degrees, at several installations - including Picatinny Arsenal, the Defense Personnel Support Center, Fort Dix, and Aberdeen Proving Ground.

The analysis procedure consists primarily of (1) thermal energy supply evaluation and (2) energy supply alternative economic evaluation. Figure 1 is a simplified flowchart of the thermal energy supply evaluation that uses HEATLOAD and SHDP to develop a model of the

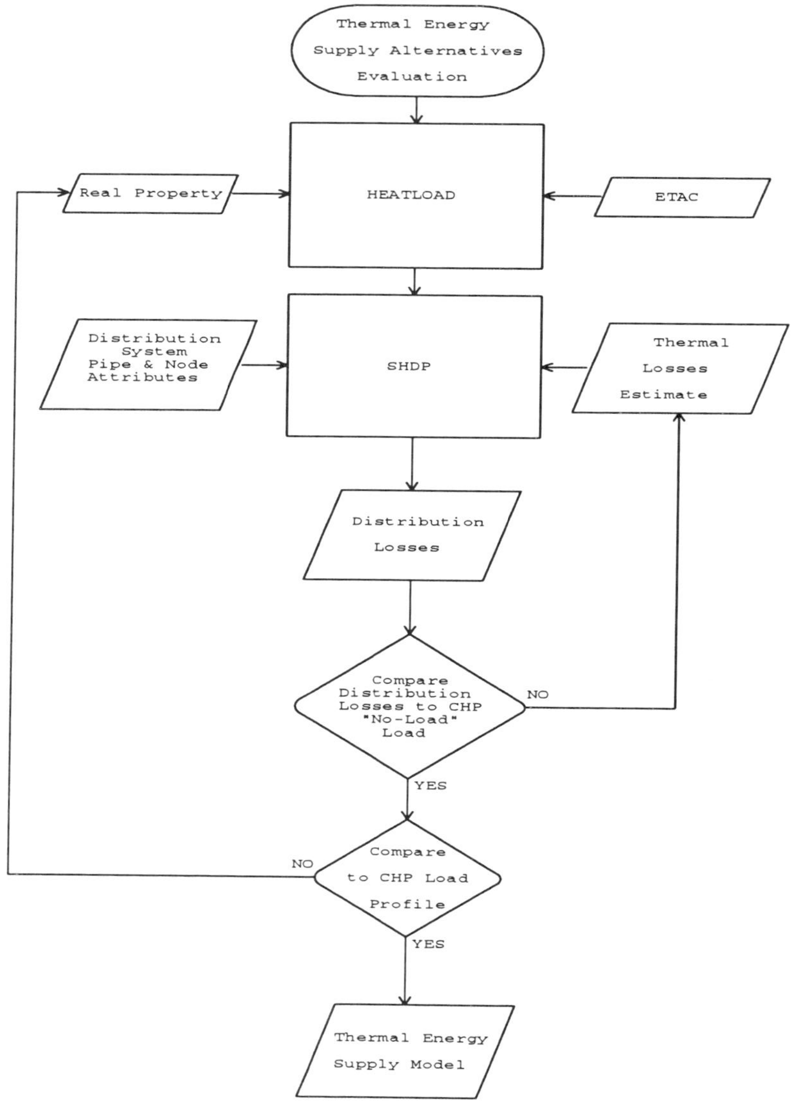

FIGURE 1. THERMAL ENERGY SUPPLY EVALUATION FLOW DIAGRAM

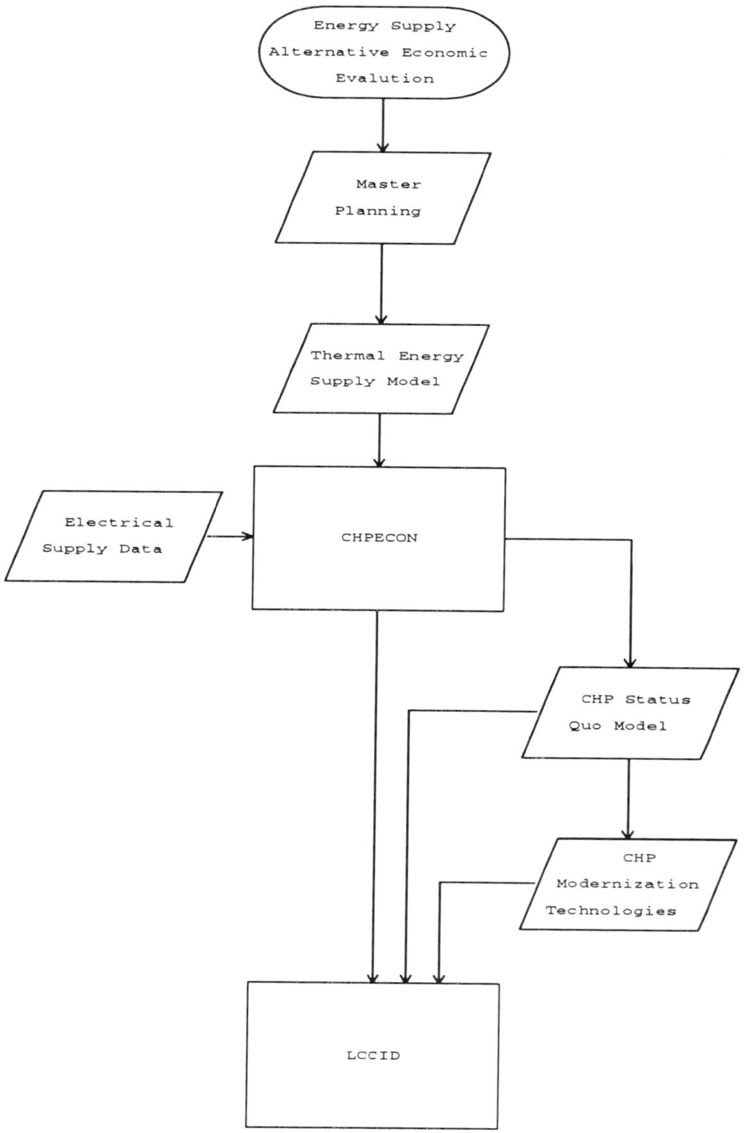

FIGURE 2. ENERGY SUPPLY ALTERNATIVE COST EVALUATION FLOW CHART

thermal energy use for an installation. Figure 2 is a simplified flowchart of the energy supply alternative economic evaluation, which uses the thermal model from Figure 1, CHPECON, STATUS QUO, and LCCID to select and perform the economic analysis of alternatives. The following is a brief description of how each program works:

HEATLOAD

HEATLOAD provides a simple method of calculating building heat requirements. Other computer programs such as Building Loads Analysis and System Thermodynamics (BLAST) and DOE2 can provide more accurate analyses, but require more information to develop a heat load estimate. Some installations may have unusual buildings that would warrant a BLAST evaluation; others may have had a BLAST analysis performed at the time the building was designed. If more accurate information is available, it should be used. Experience with HEATLOAD has shown it to be quite accurate for estimating installation-wide building heat requirements for central energy plant alternatives.

HEATLOAD is based on a series of linear regressions developed from heating use measurements at typical facilities on several Army installations. It calculates the peak hourly heating load, average monthly loads, maximum monthly loads, and total annual heating load.

STEAM HEAT DISTRIBUTION PROGRAM

The Steam Heat Distribution Program is a pressure-flow-thermal efficiency computer program for modeling steam district heating systems. In addition to modeling system capacity, the program can help evaluate benefits of various steam distribution system modifications including isolation of lines during the summer, renovation of manholes, line insulation, operating at lower pressures, and improved steam trap maintenance.

The entire installation distribution system is entered into SHDP along with the heat loads for each building (calculated from HEATLOAD). SHDP calculates the distribution system losses and the total heating requirement for the central heating plant. The distribution losses calculated by SHDP are compared to the "no-load" load from the plant records. The "no-load" load is the lowest load at temperatures above 65oF. This load is primarily distribution losses, however it may contain some process loads. For plants that shut down sections of the distribution system during the summer months or operate at lower pressure in the summer, there are two "no-load" loads, one for summer and one for winter.

The SHDP monthly steam load profile is then compared to the load profile developed from the central heating plant boiler log steam readings. The two profiles should fall within 10% of each other; if not, then the real property building information is reviewed for missing buildings or process loads and partially heated space. Once there is reasonable confidence in the estimates, the peak hourly steam load is calculated for the CHPECON program.

CENTRAL HEATING PLANT ECONOMICS

CHPECON development was funded by the Department of Defense (DOD), in response to the 1986 Appropriation Act (PL-99-190, Section 8110) that directed the DOD to increase coal use for facility energy supply. CHPECON is designed to identify candidates for new or retrofit coal thermal energy/power generation facilities. However, because DOD must still use the best life-cycle cost (LCC) alternative, CHPECON was developed with capabilities to analyze oil and natural gas fuel alternatives for comparison to coal fuel alternatives. This means CHPECON is a tool for estimating most new central energy plant (CEP) alternatives applicable to DOD installations.

CHPECON provides a consistent approach for evaluating competing technologies. The program applies screening criteria that are used in the preliminary stages of evaluation. After the preliminary screening, CHPECON provides detailed conceptual facility design information, budgetary facility costs, and economic measures of project acceptability, including total LCC and levelized cost of service, for the technology under evaluation. The program provides flexibility to vary critical design and operating parameters to determine project sensitivity and parametric evaluation.

CHPECON provides costs for the building, boilers, ancillary equipment, fuel, and operations and maintenance - all of which are required for a complete LCC analysis. The program creates a computer file that is read directly by the LCCID program to produce the LCC report. The model can analyze steam or high-temperature hot water plant sizes ranging from 60 to 500 MBtu/hr, with individual boiler sizes from 20 to 200 MBtu/hr. It can evaluate costs for most fuel technologies including natural gas, oil, stoker, bubbling fluidized bed, circulating fluidized bed, coal/water slurry and coal/oil slurry boilers. For smaller plants up to 150 MBtu/hr with boilers ranging from 5 to 50 MBtu/hr, another model called HPECON is available. This model was developed for U.S. Army Europe to assist in district heating alternative evaluations. It functions similarly to

CHPECON; however, it has fewer options because of the smaller plant sizes considered. It has no options for co-generation or fluidized bed boilers.

STATUS QUO MODEL

The STATUS QUO model is a microcomputer-based technique for determining the LCC of an existing CEP, also funded by the DOD coal use program. The "status quo" situation implies the continued operation of the CEP by performing routine maintenance and repair. The STATUS QUO model will provide a baseline alternative with which to compare new CEP plant alternatives. It will also provide a baseline for comparison of improvements through modernization which can, in turn, be compared to new plant alternatives.

The evaluation of the CEP's status quo is determined through a field survey of major plant components. Currently, the model is capable of estimating the LCC of coal-, oil-, and natural-gas-fired equipment for boilers in the 20 to 200 MBtu/hr range and a maximum plant capacity of 600 MBtu/hr. Electric generation and thermal distribution components are planned for future development. The plant component data input is quite simple, consisting primarily of equipment size (dimensions, capacity, power requirements, etc.) and year of installation. Costs are based on average industry prices and the replacement year is based on industry experience.

The program contains defaults for labor, maintenance, spare parts, and utility costs. Actual costs should be used to obtain an accurate economic analysis. Similar to CHPECON, the STATUS QUO model uses the LCCID program to perform the LCC analysis. The STATUS QUO program produces an LCCID input file containing all the plant components with their replacement year and cost and the plant labor, maintenance, spare parts, and utility costs. It then runs the LCCID program.

MODERNIZATION MODEL

Although the STATUS QUO model provides economic analysis for replacing existing equipment at the end of its expected life, it does not analyze the improvements from upgrading or modernizing plant equipment. There are many opportunities for achieving lower life cycle costs by applying new technologies or operating strategies that increase equipment efficiency and reduce operating costs. Modernization is particularly important for DOD coal plants because most are past their expected useful life. Modernization will be the key to maintaining DOD's existing coal capabilities.

USACERL plans to develop a Modernization Model based on the methodology used by the STATUS QUO Model. The model will contain a set of modernization opportunities for major subsystems: such as water treatment, feedwater system, fuel handling and storage, combustion controls, boilers, air pollution control devices and heat recovery equipment. These opportunities will be linked to a particular CEP by boiler size, fuel type, O&M condition and other status quo characteristics. The model will calculate reductions in energy use through improved efficiency. The model will optimize combinations of modernization opportunities to achieve the lowest life cycle cost modernization alternative.

LIFE CYCLE COST IN DESIGN

LCCID is an economic analysis computer program tailored to DOD's needs. LCCID provides the present-worth LCC, payback period, and savings/investment ratio. Each energy supply alternative can be compared and ranked according to each of these cost methods, providing a consistent method of evaluating any central heating system energy supply alternative.

HEATMAP

HEATMAP is a computerized system that provides a fast and reliable means for modeling the operation and economics of district heating and cooling (DHC) systems. Although primarily designed to model proposed DHC systems, HEATMAP functions equally well in modeling existing systems. HEATMAP will take accumulated information related to the study area, production plants and distribution network, and size the district heating and cooling system to meet thermal requirements. HEATMAP will optimize the mechanical facilities associated with DHC, (i.e. pipe size and plant size). In addition, HEATMAP has the ability to model building loads and to determine the environmental impact of various DHC options. HEATMAP uses actual information, where available, and provides engineering estimates elsewhere.

HEATMAP is a stand-alone program that interfaces with the proprietary software AutoCAD and LFLOW-2F. The AutoCAD program provides a means for graphically representing the DHC system. Within an AutoCAD "map" of an installation, the user can identify and locate production plants, consumers and existing or proposed distribution lines. The LFLOW-2F program then models and analyzes the distribution system operation based on the map developed with the AutoCAD program.

HEATMAP is being developed by the Washington State Energy Office in conjunction with several public and private organizations. USACERL is a partner with the WSEO in the continuing development and enhancement of the HEATMAP program. Although not shown in Figures 1 and 2, HEATMAP promises to enhance the evaluation of DHC options by combining several analysis tools into one package.

CASE STUDIES

Picatinny Arsenal

This study was completed in 1990 for the Directorate of Engineering and Housing, Picatinny Arsenal by USACERL. The objective of the study was to provide recommendations on the most life cycle cost effective techniques for producing both thermal and electrical energy to meet the current and future demands of Picatinny Arsenal.

The Picatinny Arsenal central power plant (CPP) consists of three oil fired boilers. Two boilers are 1942 Combustion Engineering pulverized coal boilers, each originally rated at 190,000 pounds of steam per hour (pph) at 430 psig and 750 ^{o}F. In 1971 these boilers were converted to No. 6 oil and derated to 160,000 pph. The third boiler is a 1971 Riley packaged oil-fired boiler rated at 50,000 pph at 430 psig and 750 ^{o}F. The boilers produce superheated steam, which can flow through three condensing steam turbine-generators (TG) to produce up to 7.5 MW of electricity. The low-pressure saturated steam, about 125 psig, provides heating and process steam via 18 miles of steam line to approximately 350 buildings spread over the 2000 plus acre installation.

The CPP had been plagued with operational problems that had caused the plant to be unreliable for producing both steam and electricity. Currently, all three turbine-generators are out of commission because of mechanical failures occurring over the last 10 years. The combustion controls could not be operated in automatic and so had to be operated manually.

The Picatinny study utilized the approach outlined in Figures 1 and 2, with the exception of the STATUS QUO model, which was not available at the time of the study. A major step in the study was the evaluation of the existing thermal and electrical supply systems. USACERL contracted with Schmidt Associates, Inc., Cleveland, OH, to evaluate the current condition and maintenance status of both steam and electrical generating equipment. The U.S. Army Engineering Housing and Support Center (EHSC) evaluated the electrical distribution system. HEATLOAD and SHDP were used to develop a model of the thermal supply system. The model was fine-tuned using climatological and boiler log data for one year. Figure 3 shows the model prediction and the actual steam output from the Picatinny CPP. The model also identified two choke points in the steam distribution system caused by undersized piping that was preventing adequate steam flow to users. To project future load demands, the model was updated using master planning information, adding future buildings and removing buildings scheduled to be demolished.

TABLE 1. TOP 15 ALTERNATIVES FOR PICATINNY

ALTERNATIVE	NPW $1000
NEW/GT (7.0 MW)/HRSG/NG	66,253
NEW/GT (7.0 MW)/HRSG/#2 OIL	66,764
REN/BPTG (4.5 MW)/NG	72,673
REN/BPTG (4.5 MW)/ RECIP (5 MW)/NG	74,330
REN/BPTG (4.5 MW)/#6 OIL	76,578
REN/GT-CHENG (4.77 MW)/NG	76,920
REN/GT (3.38 MW)/NG	77,162
REN/GT-CHENG (4.77 MW)/#6 OIL	77,807
REN/100% ELECT PURCH/NG	77,986
REN/GT (3.38 MW)/#6 OIL	78,047
REN/BPTG (4.5 MW)/ RECIP (5 MW)/#6 OIL/#2 OIL	78,236
REN/100% ELECT PURCH/#6 OIL	81,300
INDIVIDUAL BLDG BOILERS/#2 OIL	83,750
REN/BPTG (4.5 MW)/COAL	85,779
SATELLITE PLANTS/NG	89,443

REN - RENOVATE
GT - GAS TURBINE
BPTG - BACK PRESSURE TURBINE GENERATOR
HRSG - HEAT RECOVERY STEAM GENERATOR
CHENG - CHENG STEAM INJECTION CYCLE
RECIP - RECIPROCATING GENERATOR

This projection was entered into the CHPECON program, which calculated the number of boilers and sizes and ancillary equipment required for the appropriate combustion technology. CHPECON was used to calculate cost of all the new plant alternatives. The other alternatives were performed using manual methods. In all, 35 different alternatives were evaluated on the basis of cost, including initial capital, annual operation and recurring maintenance costs. Table 1 shows the top 15

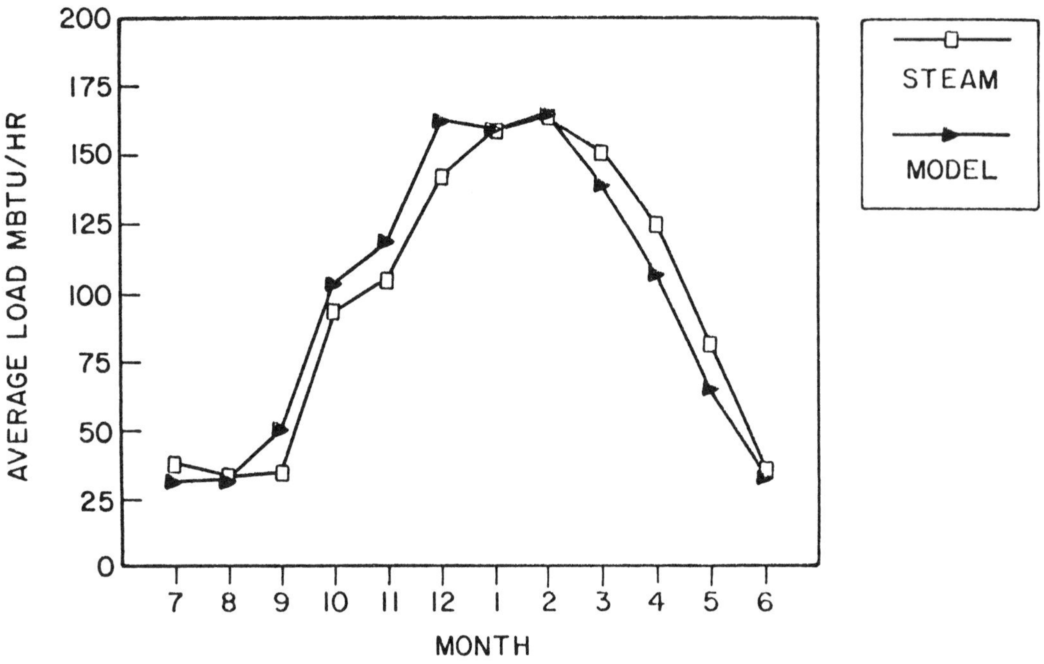

FIGURE 3. COMPARISON OF MODEL AND ACTUAL STEAM LOADS FOR PICATINNY ARSENAL

alternatives ranked by life cycle cost. Eight of the top ten alternatives involve renovation of existing facilities.

Defense Personnel Support Center

This study utilized the methods described earlier, including the STATUS QUO model, to develop alternatives to modernize the central heating plant at the Defense Personnel Support Center (DPSC), Philadelphia, PA. The central heating plant (CHP) contains five boilers; four are 50 years old and one is 14 years old. The age of this equipment and rising electrical costs warranted an investigation of alternatives for providing thermal and electrical energy to the installation.

The CHP provides steam for heating and process loads to 15 buildings via steam lines. The total steam distribution system measures about 6.3 miles. Figure 4 shows the main distribution system pipes (in bold). Because the lines run primarily through the building ceilings and utility tunnels, the heat losses were minimal. The lines were also well insulated with no apparent steam leaks at fittings and traps.

To provide additional confidence in the thermal model, end-user loads were estimated with both HEATLOAD and BLAST. These loads were entered into the SHDP program to develop thermal supply models. Both recorded steam and fuel were also used to check the accuracy of the model. A 75 percent fuel conversion efficiency was assumed to estimate the steam produced. The monthly load profiles are compared graphically in Figure 5. The HEATLOAD profile compares most favorably to the CHP steam load profiles. The peak load was projected using a linear regression based on HDD, shown in Figure 6.

Thirteen alternatives were developed for upgrading the existing system, installing new boilers, co-generation and absorption chilling. Limited space on the installation and ground water pollution problems (from non-DOD offsite sources) eliminated new CHP options. Air quality regulations also eliminated coal and heavy oil fuel technologies.

Based on LCC, new natural gas boilers and a natural gas turbine-generator with a heat recovery steam generator (HRSG) was the best alternative. The net present worth of this alternative was $64,868,000. On an annual basis, the

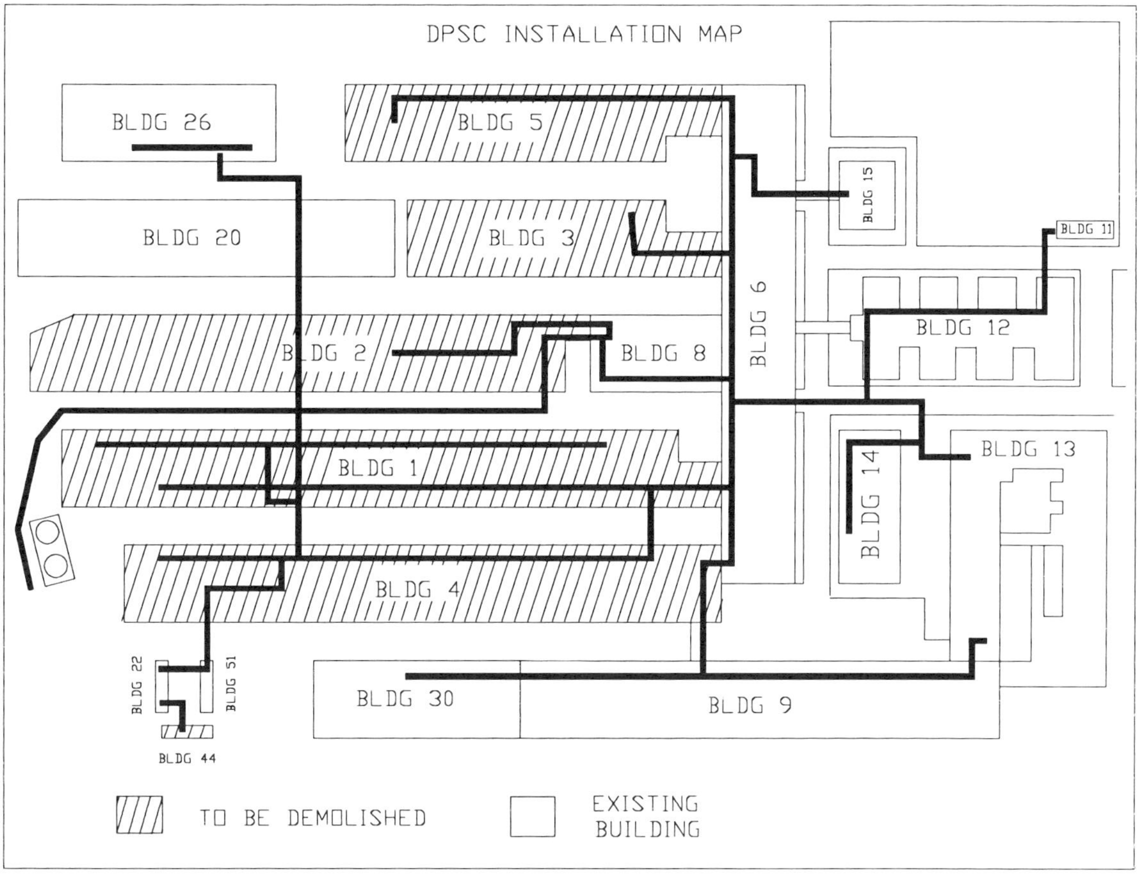

FIGURE 4. DPSC STEAM DISTRIBUTION SYSTEM LAYOUT

turbine-generator will produce about 75 percent of all the electricity needed and reduce the peak electrical demand by about 50 percent. The HRSG will produce about 109 million pounds per year or 47 percent of the required steam. The increase of almost $700,000 in fuel costs is offset by the $1,800,000 decrease in electrical costs for an estimated annual energy savings of approximately $1,100,000.

The initial cost for this alternative was about $7.1 million dollars. The economic analysis also showed a first year savings of $1,043,012, a total net discounted savings of $13,607,660, a discounted savings-to-investment ratio (SIR) of 1.98, and a simple payback of 6.59 years.

CONCLUSION

The decision making tools described in this paper are not the complete answer to energy supply problems for an installation. However, they do provide a solid foundation upon which to build. The energy supply analysis tools HEATLOAD, SHDP and HEATMAP, for example, can be used throughout the life of an installation, providing an analysis of the impact of every new facility added or removed from the distribution system. The energy supply alternative economic evaluation tools can be used to estimate the cost of individual plant components such as feedwater pumps, deaerators systems or air pollution control devices.

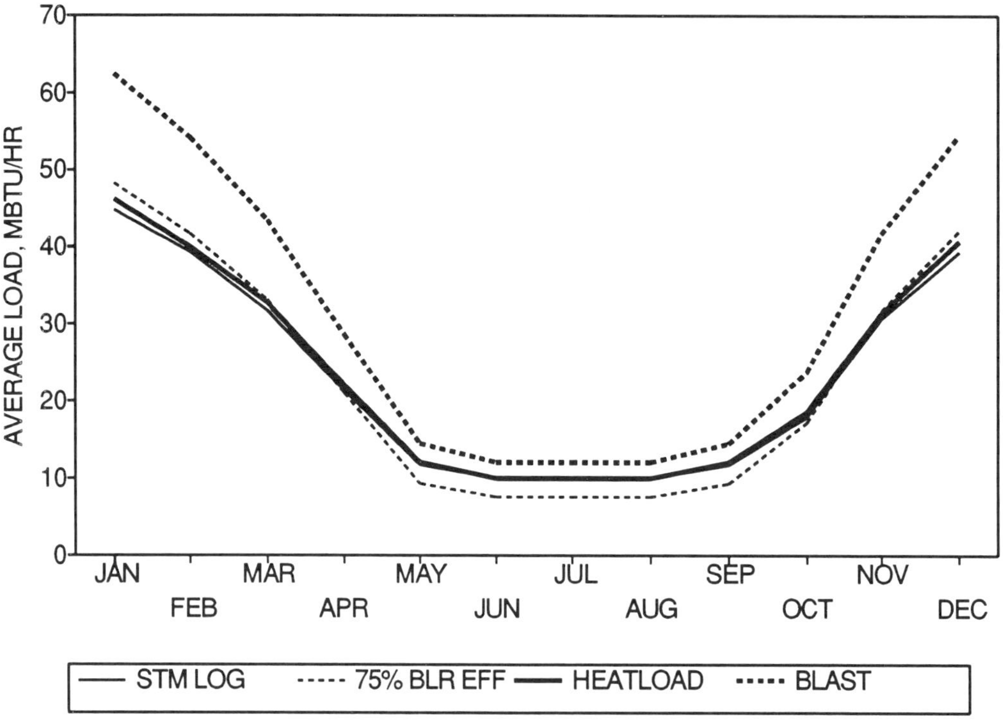

FIGURE 5. COMPARISON OF MODEL AND ACTUAL STEAM LOADS FOR DPSC

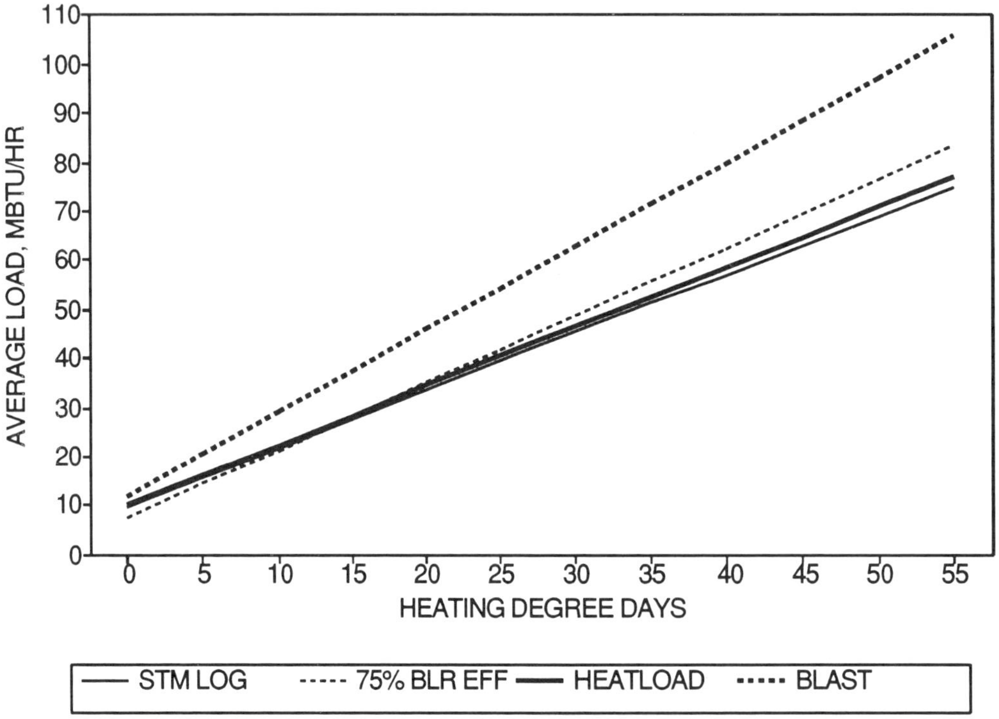

FIGURE 6. LINEAR REGRESSION OF STEAM LOAD AND HDD

The concept of IURP is important to the Corps of Engineers Districts because these agencies provide evaluation, design and construction services for installations requiring major renovation or replacement of central energy supply systems. In addition, because project requirements analysis begins at the installation level, this concept is equally important to the installation Directorate of Engineering and Housing (DEH). The IURP concept should be included in the decision-making process for any central energy supply project. Although some of these tools presented in this paper are still under development, most are available to DEH and USACE District engineers through USACERL.

REFERENCES:

HEATMAP Version 1.0 - A Computerized Tool for Analyzing District Heating and Cooling System Feasibility, Washington State Energy Office, WSEO #91-372, June 1992

Lawrie, L. K., et al., Life Cycle Cost In Design (LCCID) Economic Analysis Computer Program User's Manual, BLAST Support Office, Department of Mechanical and Industrial Engineering, University of Illinois, 1988.

Miller, J. A., et al., Steam Heat Distribution Program User's Manual, Navy TM M-73-89-01CR, August 1989.

Moshage, R. E., et al., Heating Plant Options Economic Analysis System (HPECON) User's Manual and Technical Reference, USACERL ADP Report E-91/03, 1991.

Savoie, M. J., et al., Energy Supply Alternatives for Picatinny Arsenal, USACERL Technical Report FE-92/04, 1992.

SECTION 9
DEMAND-SIDE MANAGEMENT

Chapter 99

Lessons from the Trenches—Niagara Mohawk's Commercial/Industrial Energy Survey Program

T.K. DePaull, T.C. Voigt

ABSTRACT

Comprehensive energy survey (audit) programs are frequently one of the first parts of a utility's overall DSM program to be implemented. Niagara Mohawk Power Corp. has recently completed just such a program for more than 5000 of its commercial and industrial customers. This paper examines the goals, performance and results, and provides some lessons for those about to embark on this path. The paper opens with a discussion of Niagara Mohawk's program goals, its customer base and how this base was segmented for the survey program. Next is an examination of the types of energy conserving measures (ECM's) recommended, the frequency of their recommendation and associated energy savings, and the relationship of concurrent incentive programs to the measures recommended and customer participation. These issues are correlated by customer segment. This is followed by a discussion of the energy survey program as a part of Niagara Mohawk's overall DSM efforts. Lastly, the authors draw upon their experience to highlight the issues that need to be addressed not only at the beginning of these programs but periodically throughout their life.

INTRODUCTION

Program Goals

The Niagara Mohawk Power Corp. initiated a large-scale energy audit program for its commercial and industrial customers in July of 1990. This program was designed as an "umbrella" program within Niagara Mohawk's overall DSM efforts and represented an extension of programs already in place. The audit program itself had no set goals in terms of demand or usage reduction; rather, its purpose was to serve as a lead generator for Niagara Mohawk's load reducing DSM programs.

Customer Base and Segmentation

The potential customer base for this program was Niagara Mohawk's commercial and industrial customers, representing approximately 48,000 accounts at the beginning of the program. These accounts cover the full spectrum of business, from small retail establishments to office buildings to food processors to paper mills. These facilities ranged upwards of 34,000 KW peak demand.

As originally planned, the program did not utilize any criteria for segmenting Niagara Mohawk's customer base in terms of the audit services to be provided. Rather, it was envisioned that a fairly uniform basket of services would be provided, with the cost to Niagara Mohawk directly related to the physical size of the facility being surveyed. However, during the RFP/Evaluation/Award process, Niagara Mohawk decided upon a segmentation scheme based upon customer peak demand. Table 1 shows the demand ranges for this segmentation and the audit services provided. It was assumed that customers within similar demand ranges would exhibit similar needs and capabilities with respect to energy conservation and efficiency, and that there would be common patterns of energy use. The results of the program bear out these assumptions.

PROGRAM IMPLEMENTATION

The program was initiated in July of 1990 with the first field audits performed in August of that year. The original schedule for the audit program called for approximately 450-500 audits to be performed in 1990 and 1900-2100 each in 1991 and 1992. The initial portion of the program was targeted at customers scheduled to participate in Niagara Mohawk's "Helmsman" energy tracking program. Initially, outside contractors would perform the bulk of the work with an emphasis on performing audits. As Niagara Mohawk strengthened their own internal capability, the emphasis would switch to follow up and implementation.

PROGRAM RESULTS

The results of Niagara Mohawk's program are presented in two parts. First, we will look at the overall results such as degree of market penetration, the types of energy conserving measures recommended and the overall impact in terms of potential demand and usage reduction. Next we will look at the projected savings on a segment by segment basis.

Overall Results

The penetration, or number of customers audited correlated strongly with peak demand, as one would suspect. This is due to the relatively smaller number of large demand customers versus small demand customers and the greater effectiveness of program marketing in these segments. Table 2 shows the breakdown by customer segment of the number of potentials, the number audited and the degree of penetration on both a customer segment basis and a system basis.

The energy conserving measures recommended to Niagara Mohawk's customers were divided into six broad categories corresponding to core DSM programs. As Table 3 illustrates, measures dealing with lighting and motors and drives accounted for the lion's share of recommendations, potential demand reduction, potential usage savings, and potential rebate dollars. Given that the audit program represented one of the first steps for Niagara Mohawk in this arena and that these two broad areas represent such a large portion of the energy consumption in commercial and industrial facilities, these results are not surprising. Rather, they serve to confirm the fact that Niagara Mohawk' system is fairly typical in its demand and usage characteristics.

Segment Results

When we break the above figures apart and look at them on a segment by segment basis, some interesting differences appear. Tables 4-8 illustrate these differences.

Specifically, as customer demand increases, the following trends become apparent: lighting accounts for a decreasing number of recommendations, a decreasings fraction of the total recommendations made and a decreasing portion of the potential energy savings as measured both by demand and usage. Contrast this with the area of motors and drives. Here we find that as the level of customer demand increases, the fraction of conservation opportunities uncovered rises as does the potential energy savings. Both of these trends follow overall nationwide averages.

Similarly, the segment results for HVAC, Building envelope and hot water production confirm our expectations. As the demand level increases, the fraction of recommendations made, and the potential energy savings hold at an essentially constant low level for HVAC, start at a low level and vanish for building envelope recommendations and are only of significance for the smallest customers for water heating.

CONCLUSIONS

Niagara Mohawk's commercial and industrial audit program accomplished the goals set for it. Penetration rates for audit services were high for larger customers and projected energy savings and demand reductions were significant on both an individual and system basis. An added benefit to Niagara Mohawk was the accumulation of a great amount of customer specific information as to the nature of electric loads.

During the course of a program of this nature, particular attention must be paid to the mix of the services offered. This mix must be adjusted over time as program awareness and customer sophistication increases.

TABLE 1 - PROGRAM SERVICES BY LEVEL

DEMAND LEVEL (KW)	AUDIT LEVEL	SERVICES PROVIDED		
		FACT SHEETS	COMPUTER ANALYSIS	FOLLOW-UP VISIT
0 - 99	1	X		
100 - 499	2A		X	X
500 - 999	2B		X	X
1000 - 4999	3A		X	X
>5000	3B		X	X

TABLE 2 - SEGMENT PENETRATION RATES

	LEVEL 1	LEVEL 2A	LEVEL 2B	LEVEL 3A	LEVEL 3B
POTENTIALS	44,151	5,742	853	725	83
NO. AUDITS	3,252	736	247	301	64
% PENETRATION	7.4	12.8	28.9	41.5	77.1
% TOTAL AUDITS	70.7%	16.0%	5.4%	6.5%	1.4%

TABLE 3 - SYSTEM WIDE RESULTS, ALL SEGMENTS

OVERALL PROGRAM						
	LIGHTING	MOTORS /DRIVES	HVAC	BUILD ENV	H2O HEATING	OTHER
# RECOMM	11783	1314	303	358	686	1212
% RECOMM	75.3%	8.4%	1.9%	2.3%	4.4%	7.7%
KW SAVED	53628	20922	6533	0.3	163.6	40234
% KW SAVED	44.2%	17.2%	5.4%	0.0%	0.1%	33.2%
KWH SAVED	242626401	219769821	33714921	1224230	1612533	215964280
% KWH SAVED	33.9%	30.7%	4.7%	0.2%	0.2%	30.2%
THERMS SAV	0	0	277790	735012	92885	3563735
% THERMS SAV	0	0	5.7%	15.1%	1.9%	73.0%
REBATE $	$25,339,028	$6,521,564	$892,813	$905	$2,722	$736,676
% REBATE $	75.7%	19.5%	2.7%	0.0%	0.0%	2.2%
REBATE $/KW	$472	$312	$137	$3,017	$17	$18
REBATE $/KWH	$0.104	$0.030	$0.026	$0.001	$0.002	$0.003

TABLE 4 - TOTAL DEMAND 0 - 99 KW

LEVEL 1 SEGMENT						
	LIGHTING	MOTORS /DRIVES	HVAC	BUILD ENV	H2O HEATING	OTHER
# RECOMM	8537	256	153	283	658	274
% RECOMM	84.0	2.5	1.5	2.8	6.5	2.7
KW SAVED	9046	321	517	0.3	3.6	838
% KW SAVED	84.3	3.0	4.8	0.0	0.0	7.8
KWH SAVED	32112882	597694	923114	85543	346453	2172221
% KWH SAVED	88.6	1.6	2.5	0.2	1.0	6.0
THERMS SAV	N/A	N/A	10418	294493	47732	44044
% THERMS SAV	N/A	N/A	2.6	72.4	11.7	10.8
REBATE $	3,992,289	49,456	49,531	905	622	19935
% REBATE $	97.1	1.2	1.2	0.0	0.0	0.5
REBATE $/KW	441.00	154.00	95.80	3,016.67	172.78	23.79

TABLE 5 - TOTAL DEMAND 100 - 499 KW

LEVEL 2A SEGMENT						
	LIGHTING	MOTORS /DRIVES	HVAC	BUILD ENV	H2O HEATING	OTHER
# RECOMM	1877	397	71	41	13	301
% RECOMM	69.5	14.7	2.6	1.5	0.5	11.1
KW SAVED	12394	2889	605	0	0	5612
% KW SAVED	57.6	NMF	2.8	0.0	0.0	26.1
KWH SAVED	41453120	11573363	1513752	147392	96397	14420157
% KWH SAVED	59.9	16.7	2.2	0.2	0.1	20.8
THERMS SAV	N/A	N/A	26824	258398	15016	451726
% THERMS SAV	N/A	N/A	3.6	34.4	2.0	60.1
REBATE $	5,326,295	614,838	118,609	0	2,100	124,447
% REBATE $	86.1	9.9	1.9	0.0	0.0	2.0
REBATE $/KW	430	213	196.05	NMF	NMF	22.18

TABLE 6 - TOTAL DEMAND 500 - 999 KW

LEVEL 2B SEGMENT						
	LIGHTING	MOTORS /DRIVES	HVAC	BUILD ENV	H2O HEATING	OTHER
# RECOMM	511	186	32	19	4	160
% RECOMM	56.0	20.4	3.5	2.1	0.4	17.5
KW SAVED	8162	3970	1018	0	160	5893
% KW SAVED	42.9	NMF	5.4	0.0	0.8	31.0
KWH SAVED	30325484	19262196	5085945	649990	140026	15155816
% KWH SAVED	42.9	27.3	7.2	0.9	0.2	21.5
THERMS SAV	N/A	N/A	30409	57970	10287	410768
% THERMS SAV	N/A	N/A	5.3	10.1	1.8	71.6
REBATE $	4,058,315	680,726	148,137	0	0	96,318
% REBATE $	81.4	13.7	3.0	0.0	0.0	1.9
REBATE $/KW	0.13	0.04	145.52	NMF	0.00	16.34

TABLE 7 - TOTAL DEMAND 1000 - 4999 KW

LEVEL 3A SEGMENT						
	LIGHTING	MOTORS /DRIVES	HVAC	BUILD ENV	H2O HEATING	OTHER
# RECOMM	767	404	40	13	9	364
% RECOMM	48.0	25.3	2.5	0.8	0.6	22.8
KW SAVED	16655	9619	3223	0	0	17740
% KW SAVED	35.3	NMF	6.8	0.0	0.0	37.6
KWH SAVED	85263266	108418628	15302550	329920	1029657	69362659
% KWH SAVED	30.5	38.8	5.5	0.1	0.4	24.8
THERMS SAV	N/A	N/A	86376	78635	19790	2035055
% THERMS SAV	N/A	N/A	3.7	3.4	0.8	86.9
REBATE $	6,678,886	3,901,351	97,136	0	0	58,301
% REBATE $	62.2	36.3	0.9	0.0	0.0	0.5
REBATE $/KW	0.08	0.04	30.14	NMF	NMF	3.29

TABLE 8 - TOTAL DEMAND GREATER THAN 5000 KW

LEVEL 3B SEGMENT						
	LIGHTING	MOTORS /DRIVES	HVAC	BUILD ENV	H2O HEATING	**OTHER**
# RECOMM	91	71	7	2	2	**113**
% RECOMM	31.8	24.8	2.4	0.7	0.7	**39.5**
KW SAVED	7371	4123	1170	0	0	**10151**
% KW SAVED	32.3	NMF	5.1	0.0	0.0	**44.5**
KWH SAVED	53471649	79917940	10889560	11385	0	**114853427**
% KWH SAVED	20.6	30.8	4.2	0.0	0.0	**44.3**
THERMS SAV	N/A	N/A	123763	45516	60	**622142**
% THERMS SAV	N/A	N/A	15.4	5.6	0.0	**77.2**
REBATE $	5,283,243	1,275,193	479,400	0	0	**437,675**
% REBATE $	70.7	17.1	6.4	0.0	0.0	**5.9**
REBATE $/KW	0.10	0.02	409.74	NMF	NMF	**43.12**

Chapter 100

Moving the Industrial Motor Market: Development and Performance of BG&E's Efficient Motor Rebate Program

C.P. Packard, B.K. Hedden, P. Parfomak

ABSTRACT

In 1991, Baltimore Gas & Electric Company (BG&E) entered into a collaborative process with its regulators, customers, and environmental groups, to design aggressive gas and electric conservation programs. One of the programs developed within the collaborative encourages the purchase of high-efficiency motors by providing financial incentives. Energy-efficient electric motor systems offer significant opportunities for efficiency improvements. The targeted market includes industrial and commercial customers, which account for nearly half of BG&E's electric load and revenues.

The design of the program required technology evaluation, market research and the development of incentive levels. The implementation of the program required overcoming market barriers, coordinating with various stakeholders and developing marketing strategies to address customers' needs and concerns. After ten months, program participation has been three times higher than expected and impacts have been twice as high as expected, although impacts per customer were one-third lower than expected.

BACKGROUND

Electric motors are the dominant electric end use in industrial and commercial buildings. According to the American Council for an Energy-Efficient Economy (ACEEE), motors account for 78% of industrial energy consumption and 43% of commercial energy consumption. The ACEEE estimates that 3.7% of this energy could be saved through the use of energy-efficient motors.[1] Due to the widespread use of motors and the new energy-efficient motor designs, energy-efficient motors represent perhaps the single greatest source of electricity conservation potential in nonresidential buildings. For this reason, the last several years have seen an ever increasing interest in capturing this potential through utility-sponsored demand-side management (DSM) programs.

Baltimore Gas & Electric Company

BG&E provides natural gas and electric service to central Maryland, serving approximately 1,000,000 residential customers, 100,000 commercial customers, and 3,000 industrial customers. The industrial customers in BG&E's service territory account for approximately 11% of the energy sold annually and nearly 10% of the peak demand. Commercial customers account for 50% of energy and 39% of peak demand.

In 1992, through a collaborative process involving regulatory agencies, intervenors, customers and environmental agencies, BG&E designed and implemented a number of DSM programs for its industrial customers. BG&E's DSM programs include time-of-use rates, and credits for reducing demand during critical periods. In 1992, BG&E also designed its Plant Expansion and Major Equipment Replacement Program, intended to provide incentives to promote energy-efficiency in manufacturing facilities, minimizing lost opportunities. One component of this program encourages customers to purchase energy-efficient motors. Although the program was developed primarily for industrial customers, eligibility was also extended to commercial customers who might be using motors in space conditioning, refrigeration, or other non-manufacturing applications.

DESIGNING BG&E's MOTOR PROGRAM

The design of BG&E's motors program involved three principal steps:

- Characterizing the market
- Developing a program approach
- Projecting program performance

Characterizing the Market

The first priority in designing BG&E's motors program was characterizing both its customer base, and the use of motors within that customer base. To this end, BG&E drew upon data from its load forecasting and customer billing databases to identify its industrial customers. BG&E also used secondary data from sources such as the Electric Power Research Institute (EPRI) and the American Council for an Energy-Efficient Economy (ACEEE) to estimate motor use among its particular mix of industrial customers. These estimates provided BG&E with a clearer context within which to develop its program.

Through focus groups, discussions with trade allies, a review of motor-related publications, and communication with motor experts, BG&E was able to learn more about motor applications, design and maintenance issues, and customer purchasing behavior. This research revealed a number of critical issues to be considered in program design.

- Energy-efficient motors were generally available in the BG&E area but often not stocked, nor widely purchased, due to their higher first costs.

- Customers were unlikely to replace functional motors just to upgrade their efficiency.

- Most motors above 5 hp were rewound (rebuilt) upon failure rather than replaced by a new motor because rewinding was less expensive.

- Customers often maintained a shelf stock of motors to minimize system shutdown in the event of an unexpected motor failures.

- Customers had strong ties with local motor vendors from whom they purchased most of their motors and repair services.

Developing a Program Approach

Having familiarized itself with the local motors market, BG&E sought to review other utilities' experiences with motors programs. The accumulation of this type of information, however, proved to be a difficult task. While a number of utilities had, indeed, offered efficient motor rebates, differences in program approach, gaps in reported data and variations in reporting formats limited the usefulness of this data in BG&E's program design.

One good source of participation and impact data for 13 energy-efficient motor rebate programs was an ACEEE study of commercial and industrial DSM program experience.[2] The report was very helpful in summarizing general approaches (e.g. customer rebates, vendor rebates, and analysis software) taken by other utilities in promoting energy-efficient motor applications. Unfortunately, variations in incentives, customer support and marketing approaches made it impossible to associate differences in participation and impacts with particular program elements. Consequently, it became difficult to identify a single best program approach based on these experiences.

Having reviewed this secondary utility information, BG&E decided to develop its own unique program approach incorporating many of the elements found in other utilities' programs with a critical modification. BG&E decided to offer rebates for the purchase of new energy-efficient motors as defined by the National Electrical Manufacturers' Association (NEMA) efficiency standards. For new motors and replacement motors smaller than 60 hp, incentives were to equal to the incremental cost of an energy-efficient motor over a standard-efficiency motor. For replacement motors 60 hp in size or larger, the incentives were to equal the difference between an energy-efficient motor and the cost of rewinding the motor minus the first two years of electric bill savings (a two-year payback buydown).

This incentive strategy was adopted to discourage rewinding of standard motors, which was considered more common than purchasing of new standard motors in sizes above 5 hp. The program also extended its incentives for the purchase of stock motors, rather than for replacement motors only, under the assumption that, once purchased, a motor would eventually enter service. Scrapping of old, standard-efficiency motors was required.

Projecting Participation and Impacts

Projecting program performance is the most difficult aspect of DSM program design. The lack of critical primary and secondary data, and limitations in data transferability, issues related to program delivery (e.g. trade ally cooperation, aggressiveness of marketing, scheduling, etc.) combine to introduce a great deal of uncertainty to performance projections. To minimize this uncertainty, BG&E opted to draw on all available sources of data to refine its program projections. This data included other utility data, data from industrial energy audits conducted by the DOE Energy Analysis and Diagnostics Center (EADC) program, input from collaborative participants, and Barakat & Chamberlin's estimates of the motor population and motor turnover in BG&E's territory. Based on this analysis, the program projected impacts of 6.2 kW per customer, with 18 customers participating in the first year of the program.

IMPLEMENTING THE PROGRAM

In addition to offering attractive financial incentives for efficient motors, BG&E realized that it needed an aggressive marketing plan to maximize participation in the program. The first priority was to identify trade barriers to efficient motors that could be influenced by program delivery. The second was to address these barriers in program implementation. As a result of this analysis, BG&E developed a two-pronged approach to program promotion involving both customer education and extensive trade ally cooperation.

Market Barriers

In addition to the cost-related issues noted above, BG&E identified a number of other market-related issues which needed to be addressed in implementation planning.

- Motors were often labeled as "high-efficiency" even though they did not meet minimum energy-efficiency criteria specified by NEMA or utility DSM programs. This labeling created confusion among customers as to whether particular motors were really energy-efficient.

- Customers were often uneducated about the advantages of energy-efficient motors over standard motors — energy savings, greater reliability and longer life. They were also often unaware about the energy economics associated with motor use in their facilities.

- Distributors often had a great deal of influence in customers' motor purchase decisions. Also, customers were often highly loyal to distributors with which they had a long-term relationship. DSM programs often failed to adequately motivate distributors to promote energy-efficient motors over standard motors or over-rewinds.

- Complicated rules and rebate applications often scared potential participants away from utility DSM programs.

- Plant staff might be reluctant to switch to different types of motors for technical reasons. For example, plants employing older U-frame motors might not wish to switch to efficient T-frame motors out of concern that the change in frame type would adversely affect motor performance.

- Smaller customers were often more concerned about motor first costs than operating costs. Their staff were also often overworked, with little time to spare on evaluating energy-efficient motors, comparing models and doing utility program paperwork.

- Customers were often unaware that utility DSM programs promoting energy-efficient motors were available to them.

Customer Education

BG&E realized that one of the principal barriers to energy-efficient motors in its territory was a lack of customer education. To overcome this lack of awareness, BG&E developed a comprehensive set of marketing materials designed to both educate customers about energy savings opportunities in motor applications, and to inform them about BG&E's program to promote motor efficiency. This set of materials included:

- A full color brochure describing BG&E's program, and program benefits to businesses and society as a whole.

- A package of straightforward, self-duplicating, program rebate forms providing customers and vendors with all necessary administrative documents at once, along with pre-addressed, post-paid return envelopes.

- Excerpts from an efficient motors handbook (reprinted with permission).

- Free copies of a customized version of DOE's "Motor Master" software, which uses BG&E rebate levels to calculate cost comparisons for 10,000 motors by brand, size and type.

Working with Trade Allies

Recognizing the close relationships between its industrial customers and local motor vendors, BG&E sought from the start to work closely with trade allies in promoting the motors program. To this end, BG&E held a meeting with 60 local motor trade allies (90% of local area vendors) to introduce them to the program and solicit their support. This meeting included an informational presentation, followed by a question and answer session and a luncheon, with BG&E staff assigned to each table to continue these discussions informally. After lunch, motor vendors were given an opportunity to try the "Motor Master" software and ask questions about its use. BG&E also assisted in hosting a national conference on Efficient Motor Systems held in Baltimore, which provided additional exposure for the program and interaction with the motor industry.

Administration

The motors program is supported primarily by professional staff in BG&E's Marketing and Energy Services Department. The role of these staff has been to facilitate the rebate process, clarifying terms of the program, coordinating inspections of motor installations, and tracking the flow of rebate applications. These staff received training on energy-efficient motors and BG&E's motors program in an in-house training session developed specifically in support of the program.

RESULTS OF THE PROGRAM

BG&E's motors program has achieved dramatic results. In the first 10 months, 71 customers had applied for rebates for 583 motors (7,338 total horsepower). Table 1 summarizes program projections and estimated performance after the first ten months of the program.

As Table 1 shows, program participation has been over three times higher than expected at the start of the program. This unexpected level of participation appears to be driven by several factors:

- Basing incentives on rewind cost yields substantially higher rebate levels than most other utility-sponsored motors programs. The higher rebates appear to make efficient motors much more attractive to customers.

- The program was designed to serve primarily the industrial sector, although non-manufacturing customers were also eligible. Because motor vendors serve both sectors, they appear to have marketed the program strongly to commercial customers, which outnumber industrial customers 30 to 1. As a result, 42% of program participants have been non-manufacturing customers.

Table 1.
BG&E Motors Program Performance After 10 months

Program Result	Projected	Actual	Actual/Projected
Participation (customers)	15	48	320%
Participation (hp)	2,295	7,338	319%
Participation (motors)	N/A	583	N/A
Avg. Horsepower	153	153	100%
Avg. Demand Savings (kW/customer)	6.2	4.2	68%
Total Demand Savings (kW)	93	201	216%
Avg. Energy Savings (kWh/customer)	35,000	23,837	68%
Total Energy Savings (kWh)	525,000	1,144,167	218%

- The general level of awareness among customers regarding energy-efficient motors has been increasing more rapidly than expected due to other utility programs, trade publications, and other outside influences.

OTHER ACTIVITIES AND FUTURE DIRECTIONS

Having implemented an effective energy-efficient motors incentive program, BG&E is considering additional ways to improve energy-efficiency in industrial motor systems. BG&E already offers incentives to improve the efficiency of compressed air production, a major motor end use in industrial plants. This year the company also filed a custom incentive program offering incentives for more specialized motor efficiency improvements such as adjustable-speed drives and improved mechanical power transmission systems. BG&E will also be offering on-site facility audits to its industrial customers to help them identify a full range of energy savings opportunities, including motor system efficiency opportunities. Finally, BG&E will be enhancing its marketing of the current motors program to non-manufacturing customers to take greater advantage of the interest they have shown in current program.

REFERENCES

1. Steven Nadel, Shepard, M., Greenberg, S., Katz, G., de Almeida, A. *Energy efficient Motor Systems.* American Council for an Energy-Efficient Economy (ACEEE). Washington, D.C. 1991.

2. Nadel, S., 1990. *Lessons Learned: A Review of Utility Experience with Conservation and Load Management Programs for Commercial and Industrial Customers.* New York State Energy Research and Development Authority (NYSERDA).

Chapter 101

Equipment Efficiency Levels for Utility Rebates—What Are the Right Ones?

D. Nadas

ABSTRACT

Market compression analysis, a statistical technique, was applied in a unique project linking seven investor-owned electric utilities in New York State with a public agency to establish consensus on energy-efficiency levels for equipment considered for utility Demand-Side Management (DSM) Incentive Programs. The analysis covered the entire efficiency range of manufactured equipment rather than only what is typically being sold in a particular region. These results could be used to support selection of higher efficiency levels than would be likely to result from surveying existing purchasing practices. These efficiency levels are not unique to each utility service area but are common factors to which other utility-specific DSM planning and program criteria can be applied. The efficiency levels were used to generate lists of specific models of equipment meeting these levels which were then incorporated into fifteen equipment-specific directories. These directories assist utility staff, consumers, building contractors, and design professionals in the selection of equipment; provide a uniform basis for manufacturers to identify energy-efficient equipment that would be potentially eligible for financial incentives and which they would target for shipment to the state; and increase consumer access to energy-efficient models of equipment not previously available within the state.

BACKGROUND

For several years, many electric utilities in the United States have run incentive programs as part of their Demand-Side Management (DSM) Programs, whose purpose is to increase efficiency in the use of electricity by their customers. While such DSM programs are often mandated by state public utility commissions, utilities have found that they are often the cheapest way to meet their customers' increasing demand for electricity and that they reduce the need for developing new generating capacity. Most of the utility incentives to date have been for lighting equipment (electronic ballasts, fluorescent lamps, lighting fixture reflectors, etc.) but incentive programs also exist for air-conditioning equipment, electric motors and adjustable speed drives. In addition, there are so-called "custom-rebate" programs for other demand-reducing and energy-conserving measures not contained in standardized incentive lists, such as for process applications.

DSM incentives are given for two different types of efficiency measures. One is the substitution of an inherently more efficient technology or system operation, such as replacing incandescent lights with fluorescent ones, or using adjustable speed drives on electric fan-motors to control air-flow instead of using outlet vanes. The other is the use of more efficient versions of existing devices, such as high efficiency motors and air-conditioners. Typically, incentive programs for the first type don't specify minimum levels of efficiency. For example, any fluorescent lamp is far more efficient than the incandescent lamp it replaces. However, such equipment frequently is available in a wide range of efficiencies, independent of characteristics which affect the serviceability of the equipment. State and federal laws may compress the market by limiting selection of some equipment to models that meet minimum energy-efficiency requirements. Incentive programs can compress the market further by rewarding the selection of models at levels of energy-efficiency beyond those levels that market forces alone would achieve. A low degree of market compression could contribute to widespread participation in an incentive program but result in low per-unit demand-reduction or energy savings, while a high degree of market compression could limit program participation but result in higher per-unit demand-reduction and energy savings. For example, an energy-efficiency level set at a 75% market compression would indicate that only 25% of all available models would meet or exceed this level.

INTRODUCTION

Before the Appliance Directory Project was initiated by the New York State Energy Office (NYSEO), utilities in New York State (and in other states) ran incentive programs with energy-efficiency levels set at widely different market compressions for different equipment and with no consensus on them among the utilities. As late as 1992, incentives were offered by some utilities in New York State for equipment that just met the energy-efficiency requirements of the New York State Energy Conservation Construction Code. This resulted in a very low market compression. In at least one case, because of a change in the Energy Code, incentives were offered for equipment that did not meet the code. Also, there have been examples of efficiency levels set so high that very few models of equipment, if any, qualified for the incentives. This situation reflected the approach that utilities took to set efficiency levels for the equipment in their incentive programs. The utilities collected information about current purchasing patterns through techniques such as customer focus groups, dealer surveys and reviews of census data, all of which tended to lead to very low market compressions. Sometimes they targeted future technological potential, which produced very high market compressions. These techniques led each utility to focus on the uniqueness of their customer base rather than the availability of equipment not currently used within their service territories.

In 1990, NYSEO began receiving many information requests from consultants assisting the seven investor-owned utilities with their DSM Program incentive plans (Central Hudson Gas and Electric, Consolidated Edison Company of New York, Long Island Lighting Company, New York State Electric and Gas Corporation, Niagara Mohawk Power Corporation, Orange and Rockland Utilities, and Rochester Gas and Electric Corporation). The utilities needed information concerning the performance of a wide range of residential and commercial appliances to support decisions about proposed financial incentives. Much of the information requested already existed at NYSEO as a result of previous research for the Energy Code amendments, but was neither complete nor consolidated into a format suitable for consumer use. NYSEO proposed, and the utilities accepted, the idea of a three-year cooperative effort to create a series of fifteen appliance directories and six updates. The project was carried out during 1991-1993. During that period, NYSEO collected and analyzed performance data and compiled listings of efficient equipment in support of the utilities' incentive programs. The equipment included in these directories are: fluorescent lamp ballasts, commercial lighting fixtures, commercial lamps, compact fluorescent lamps and ballasts, commercial and industrial motors, commercial central air conditioners, residential central air conditioners, commercial heat pumps, residential heat pumps, residential room air conditioners, residential refrigerator/freezers, residential freezers, residential water heaters, electric chillers, and fossil-fueled cooling equipment.

Instead of using the utilities' approach to setting equipment energy-efficiency levels as described above, NYSEO used the market compression analysis method to fully exploit the near-term potential for energy savings from financial incentives by concentrating on high-performance equipment currently in the marketplace that could be made readily available in response to incentive-induced consumer demand, but which was not generally purchased in the state. A natural result of this approach was the achievement of utility consensus on equipment energy-efficiency levels where previously none existed.

NYSEO identified two levels of energy-efficiency for each product category. Level 1 typically represented the most stringent energy-efficiency level at which a sufficient selection of models was available to operate a DSM incentive program. A less stringent Level 2 typically represented a threshold below which a utility was likely to experience a disproportionate number of participants who would have installed the equipment anyway, without an incentive. An Advisory Committee, comprised of a representative from each of the utilities, met to discuss and evaluate the identified energy-efficiency levels. Agreement was reached by the Advisory Committee on a consensus threshold (typically, Level 2). NYSEO then prepared lists of models, verified by the manufacturers, which met or exceeded the chosen efficiency level. The utilities typically combined the consensus energy-efficiency information with other utility-specific parameters, such as economic indicators, peak usage, and customer demographics to determine the feasibility of a financial incentive program. Such programs, where determined appropriate, are generally offered at energy-efficiency levels at or above identified consensus thresholds. The lists of equipment provided in the directories are used as references in identifying models that meet program criteria.

ANALYSIS

The method used to arrive at the recommended efficiency levels was an empirical one. Equipment performance data was collected from a variety of sources, depending on the type of equipment, and stored in a computerized data base using a data base program. For example, some cooling equipment performance data generated according to standards set up by the Air-Conditioning and Refrigeration Institute (ARI) was available from them and independently verified by the manufacturers. Performance data for domestic hot water heaters (electric, gas and oil) was available from the American Gas Association (AGA). However, performance data for other equipment, such as fluorescent lamps and ballasts, high-intensity discharge lamps, compact fluorescent lamps and ballasts, and fluorescent and high-intensity discharge lamp fixtures, while generally derived from standardized tests, was available only from the manufacturers. Prior to data collection from manufacturers, NYSEO staff held meetings with representatives of each group of manufacturers to explain the project, to discuss technical aspects of their equipment performance, and to develop a consensus on the kind of information requested and in what format. As a result of these meetings, NYSEO built relationships with the manufacturers and has become a clearinghouse for performance data for the equipment listed in the directories. (For a description of a portion of the work done for the Directory of Commercial Lighting Fixtures, see "Automatic Processing of Photometric Data," by NYSEO colleagues Cecilio Gracias, Lee Smith and Eric Noble, which was accepted for a presentation at a poster session at the 16th World Energy Engineering and 2nd Environmental Technology Congress on Wednesday, October 27, 1993.)

The first step in the analysis was to group the equipment into categories so that the equipment performance data could be analyzed. Categories were either established directly by analysis to identify logical groupings or by using existing categories established by the New York State Energy Code or the National Appliance Energy Conservation Act (NAECA). For example, Table 1 shows the NYS Energy Code requirements for heat pumps. Figure 1 from the 1992 Directory of Commercial Heat Pumps shows that for air-source single-package, single-phase, heat pumps under 65 MBtuh cooling capacity, about 95% of all models in the data base met or exceeded the 1992 New York State Energy Code requirement of 8.5 SEER. (Analysis by NYSEO has demonstrated that heating seasonal performance factors (HSPFs) are highly correlated with SEERs.) Figure 2 shows that most of these models had capacities between 18 and 48 MBtuh with fairly evenly distributed clusters. Consequently, the resulting efficiency listing of these products grouped the equipment into the following cooling capacity categories within which the products are listed from most efficient to the least efficient, above the consensus efficiency levels: less than 21 MBtuh, between 21 and 27 MBtuh, between 27 and 33 MBtuh, between 33 and 39 MBtuh, between 39 and 51 MBtuh and between 51 and 65 MBtuh.

In contrast, Figure 3 shows that the distribution of air-source single-package three-phase heat pumps under 65 MBtuh cooling capacity had most of the models between 33 and 48 MBtuh, with fairly evenly distributed clusters. However, Figure 4 shows that less than 5% of all these models met or exceeded the New York State Energy Code requirement of 10.0 SEER. Therefore, no energy-efficiency levels for this category was identified. In other words, in this category, the New York State Energy Code compressed the market so much that there were very few models available and it would not make sense for a utility to offer an incentive for the purchase of this catogory of equipment.

On January 1, 1993, the situation changed for the air-source, single package 1-phase heat pumps when the Energy Code requirements for this category of equipment rose from 8.5 to 9.7 SEER (Table 1, footnote 6). Figure 1 shows that this compressed the 1992 selection significantly.

Similar analyses for other categories of equipment have led to our identification of energy-efficiency levels and subsequent agreement by the utilities on consensus thresholds at or above which incentives would target sales of high-efficiency equipment.

Market compression analysis revealed some quirks in the distribution of models over the range of equipment efficiency. For example, figures 5, 6 and 7 were taken from the 1993 Directory of Commercial Central Air Conditioners. Figure 5 shows that for air-cooled, single-phase split systems under 65 MBtuh, all models in the data base meet the Energy Code SEER of 10.0, about 25% meet 11.5 SEER, and about 20% meet 12.0 SEER. Figure 6 shows that for the three-phase versions of the same equipment, about 90% of the models meet the Energy Code requirement of 10.0 SEER. Of the models which meet or exceed 10.0 SEER, Figure 7 shows that about 5% meet 11.5 SEER, and only about 2% meet 12 SEER. There is no technical reason for the disparity in the distributions of single-phase and three-phase high-efficiency models. It turned out that high-efficiency single-phase systems are sold primarily to homeowners in the South and Southwest where air conditioning use is high and high-efficiency air conditioning systems are important. Therefore, the competitive marketplace produced many high-efficiency models. On the other hand, three-phase systems are purchased primarily by small businesses (retail shops in strip malls, for example), where low first-cost is important and operating costs are deductible business expenses. Therefore, there is little demand for high-efficiency three-phase systems, so few such models are made. However, by providing incentives for three-phase systems with efficiencies at or above the same efficiency levels as for single-phase systems, utilities easily can increase the consumer demand and resulting supply of high-efficiency three-phase systems or encourage their replacement with the high-efficiency single-phase systems.

NYSEO has found that most manufacturers' distribution networks are such that products can be made available anywhere. A case in point is that manufacturers were shipping all of their most efficient refrigerators to California because of the existence of very favorable rebate programs there, while consumers and utilities in the Northeast didn't know that such models even existed. When NYSEO discovered that these high-efficiency models existed and recommended a high energy-efficiency level for refrigerator rebates for a program jointly administered by NYSEO and a New York utility, the manufacturers' shipping patterns changed and high-efficiency refrigerators started appearing in stores in the utility's service area. In this way, NYSEO has brokered changes in the distribution of products. An advantage for the manufacturers is that they can then sell more of their highly efficient top-of-the-line products by filling the demand created by utility DSM incentive programs.

In contrast with the experience with single-phase and three-phase central air conditioners and high-efficiency refrigerators is that of electric motors. Analysis of performance data of integral horsepower (1 to 600 hp), low voltage (208 to 575 v), Design B squirrelcage induction motors revealed that the efficiencies vary with nominal RPM (1200, 1800, and 3600, for example) and with enclosure type (open dripproof (ODP), or totally-enclosed fan-cooled (TEFC)). There are legitimate physical reasons for these differences. Therefore, the efficiency levels established for incentive programs for such motors should reflect these variations instead of the typical procrustean method of setting one efficiency level regardless of RPM and enclosure. An extreme example is that until the Directory of Commercial and Industrial Motors was produced, one New York State utility had required a full-load efficiency of 84% for a 1.0 hp motor to qualify for an incentive. This efficiency level was a good choice for 1800 RPM TEFC and ODP motors, caused very high market compressions for ODP 3600 RPM motors and TEFC 1200 RPM and 3600 RPM motors, and excluded ODP 1200 RPM motors from the incentive program because the most efficient one had an efficiency of only 82.5%.

Another example of the value of market compression analysis is in the area of equipment that is either unregulated or barely regulated. Lighting fixtures fall into this description because the New York State Energy Conservation Construction Code (Energy Code) is the only energy code in the country that regulates total luminaire efficiency (TLE). The first tier of these new code requirements, which took effect on March 1, 1991, is fairly lenient, with more stringent requirements scheduled to take effect on July 1, 1995. NYSEO managed to collect a large amount of photometric data from the fixture manufacturers. In many cases, the manufacturers were not very concerned with the TLEs of their products since, until recently, no utility had an incentive program aimed at getting high TLE fixtures installed. Analysis of the photometric fixture data revealed a tremendous range of TLEs for most fixtures. Specifically, the most common fixture type in the data base is a 1-foot by 4-foot (1 X 4) fluorescent fixture with a flat diffuser/lens. There are 593 fixture models of this type in the data base with TLEs ranging from 25% to 90%, while the Energy Code requires 53% TLE. This says that the most efficient fixture will put out 70% more light than the least efficient one permitted by the Energy Code, lamp lumens being equal. Another way of looking at it is that the most efficient fixture will require approximately only 60% as much power and energy to provide the same illumination as the code-minimum provides, all other things being equal.

Because lighting is such a large load in commercial buildings, the utilization of the information in the Directory of Commercial Lighting Fixtures should be of great interest to most utilities. As a result, utility incentives to encourage the installation of much more efficient fluorescent luminaires should become much more common. Compact fluorescent fixtures are not regulated by the Energy Code because when the code was developed, no compact fluorescent fixtures used more energy to produce the same amount of light produced by its incandescent counterpart. The code requires 46% TLE for general service incandescent fixtures. The most common compact fluorescent fixture in the NYSEO data base is the 1 X 1 fixture. The 47 models have TLEs ranging from 80% down to about 17%. Therefore, it would be quite possible to replace an incandescent down-light, for example, with a compact fluorescent fixture and end up with significantly less illumination than before. If the original illumination was marginal to begin with, the lighting project would have to be redone, leading to greater cost and/or lower energy savings, depending on the remedy.

CONCLUSIONS

The use of market compression analysis in the production of the Appliance Directories has been a very valuable tool for promoting the aims of the DSM programs of New York State's seven investor-owned utilities. Some of the most important results are the following:

1. Market compression analysis leads to the determination of appropriate energy-efficiency levels for equipment in utility incentive programs, as opposed to the use of alternative methods, such as customer focus groups and equipment dealer surveys.

2. Because market compression analysis focuses on the range of efficiencies of each type of manufactured equipment and not on local customer equipment-buying patterns nor on utility-specific DSM program considerations, utilities were able to come to consensus on equipment energy-efficiency levels for the first time in New York State.

3. Market compression analysis can reveal facts about equipment efficiencies and product distribution that otherwise may not be known, thereby acting as a catalyst in promoting the use of more efficient equipment and nudging the equipment industries to design, build and market more efficient products.

4. The New York State Energy Office has estimated that the Appliance Directories, which all used market compression analysis, increased the overall DSM energy savings for New York State in 1992 by approximately 5% or 51 million kwh in avoided electrical usage. The concomitant environmental benefits attributed in 1992 to the Appliance Directories were estimated as reductions of emissions in the amount of 240 tons of sulfur dioxide, 90 tons of nitrogen oxides, 36,500 tons of carbon dioxide, 0.5 tons of volatile organic compounds, and 5.8 tons of particulate matter.

5. Additional energy savings and environmental benefits will accrue due to the incorporation of the results of the Appliance Directory project into the many financial assistance programs offered by the New York State Energy Office. On average, the energy-efficiency requirements for equipment in these programs have increased by 10 to 15%.

ACKNOWLEDGMENTS

The author greatly appreciates the insightful contributions of Aimee McKane, the Appliance Directory Project Coordinator, and Eric Noble, the Appliance Directory Project Manager, without which this paper could not have been written.

TABLE 1

FROM NEW YORK STATE ENERGY CONSERVATION CONSTRUCTION CODE

Heat Pumps[1/]
Minimum Performance

Category	Air Source[2/]		Water[3/] Source	Groundwater[4/] Source
	Single Phase	Three Phase		
<65,000 Btu/hr				
Heating	2.6 COP (6.8 HSPF[5/])	2.6 COP (6.8 HSPF[5/])	3.3 COP (3.8 COP)	2.8 COP (3.0 COP)
Cooling	8.5 SEER (10.0 SEER[6/])	8.9 SEER (10.0 SEER)	9.0 EER (9.3 EER)	10.0 EER (11.0 EER)
≥65,000 Btu/hr ≤135,000 Bth/hr,				
Heating	2.8 COP (3.0 COP)	2.8 COP (3.0 COP)	3.4 COP (3.8 COP)	2.8 COP (3.0 COP)
Cooling	8.3 EER (8.9 EER)	8.3 EER (8.9 EER)	9.5 EER (10.5 EER)	10.0 EER (11.0 EER)
>135,000 Btu/hr				
Heating	2.8 COP (2.9 COP)	2.8 COP (2.9 COP)	3.0 COP (3.0 COP)	3.0 COP (3.0 COP)
Cooling	8.2 EER (8.5 EER)	8.2 EER (8.5 EER)	7.5 EER (7.5 EER)	7.5 EER (7.5 EER)

1/ Except as provided in footnotes 5 and 6 below, performance requirements in parenthesis applicable to equipment manufactured after January 1, 1992.

2/ COP @ 47° db

3/ COP @ 70° ent.

4/ COP @ 50° ent.

5/ Applicable to split systems manufactured after January 1, 1992; 6.6 HSPF applicable to single package systems manufactured after January 1, 1993.

6/ Applicable to split systems manufactured after January 1, 1992; 9.7 SEER applicable to single package systems manufactured after January 1, 1993.

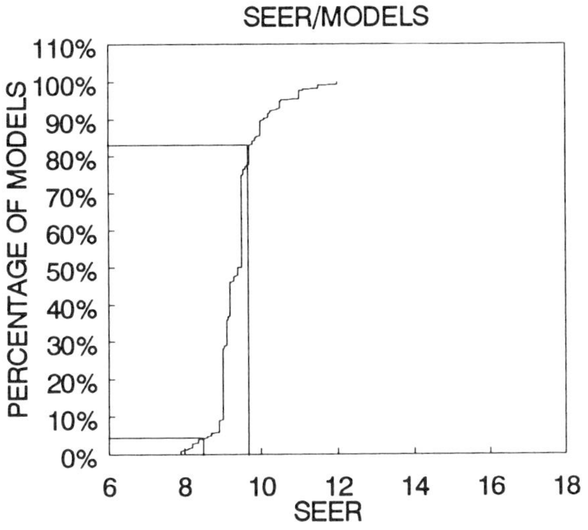

FIGURE 1

AIR–SOURCE, SINGLE PACKAGE,
1–PHASE HEAT PUMPS
< 65 MBTUH COOLING CAPACITY

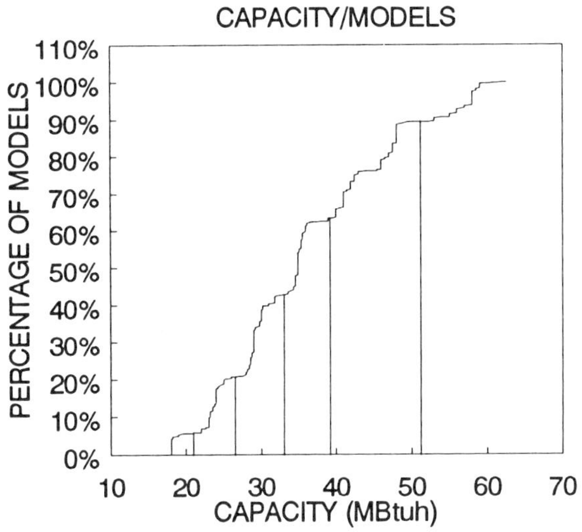

FIGURE 2

AIR–SOURCE, SINGLE PACKAGE,
1–PHASE HEAT PUMPS
< 65 MBTUH COOLING CAPACITY

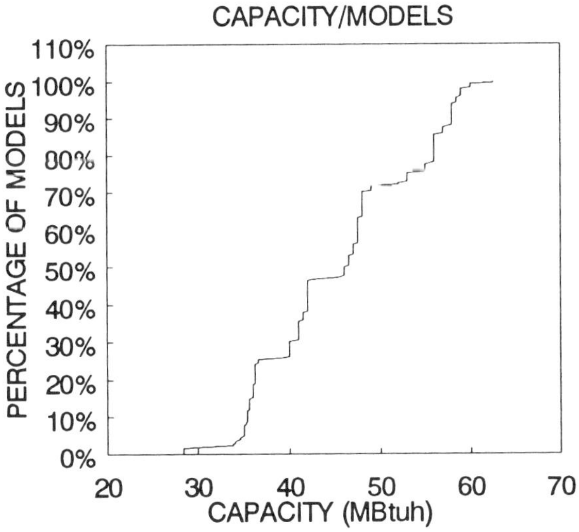

FIGURE 3

AIR–SOURCE, SINGLE PACKAGE,
3–PHASE HEAT PUMPS
< 65 MBTUH COOLING CAPACITY

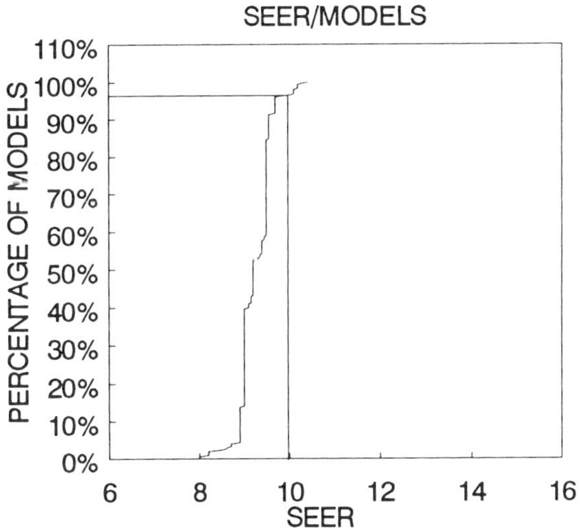

FIGURE 4

AIR–SOURCE, SINGLE PACKAGE,
3–PHASE HEAT PUMPS
< 65 MBTUH COOLING CAPACITY

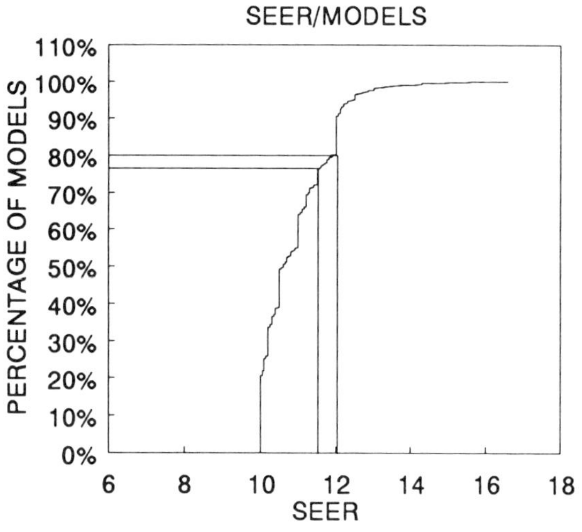

FIGURE 5

AIR–SOURCE, SPLIT, 1–PHASE
CENTRAL AIR CONDITIONERS
< 65 MBTUH CAPACITY

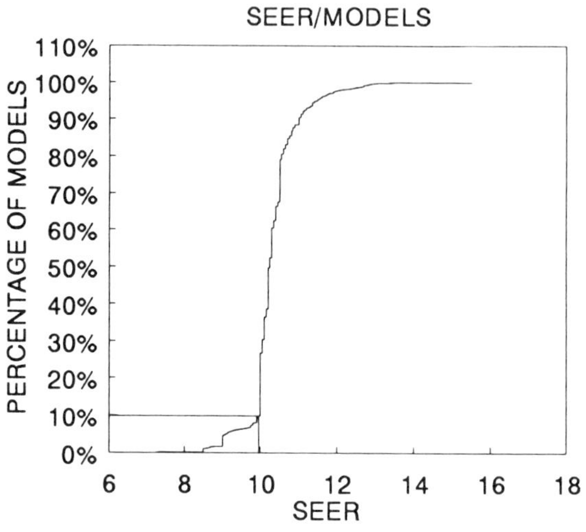

FIGURE 6

AIR–SOURCE, SPLIT, 3–PHASE
CENTRAL AIR CONDITIONERS
< 65 MBTUH CAPACITY

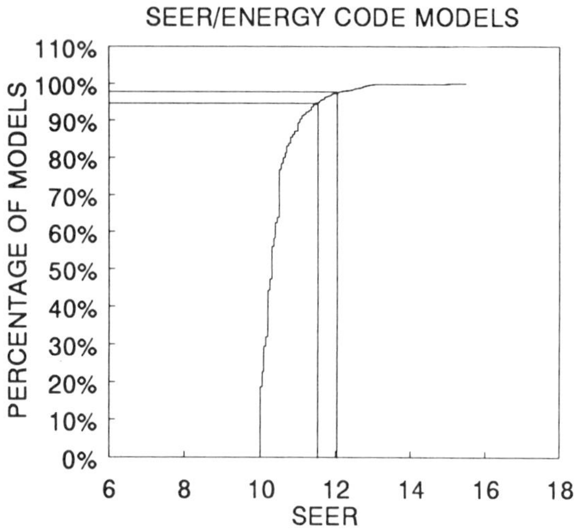

FIGURE 7

AIR–SOURCE, SPLIT, 3–PHASE
CENTRAL AIR CONDITIONERS
< 65 MBTUH CAPACITY

Chapter 102

Understanding DSM Technologies Through Electronic Information Transfer

R.P. Penney

THE INFORMATION CHALLENGE

Successful demand side management (DSM) program designers and operators know which energy conservation measures will work best in specific applications. They not only understand the theory behind energy conservation technologies but also know which conservation measures are most effective in the climate and building types found in their region.

When utilities discovered least cost planning and DSM programs began to proliferate, many program managers seemed to be focused on achieving the highest possible levels of participation. This resulted in a wide distribution of some energy-saving products early in their development that were misapplied and not always well received. While a compact fluorescent lamp may save 75% of lighting energy, the light quality of early models was harsh, they flickered as they came on, they didn't fit in many fixtures, and their power quality characteristics were terrible. Similar experiences may have occurred with low-flow shower heads, heat pump water heaters, and other products.

Since that time, these drawbacks have been corrected by manufacturers, but such experiences can result in low customer acceptance of new energy saving products; "Oh no, it's those energy people again; remember what happened last time" they may say. DSM programs are now more interested in accurate, honest assessments of product performance and application guidelines.

Acquiring up-to-date information on DSM technologies can pose a challenge for several reasons. One is the speed at which conservation technologies are developing. Many publications are outdated before they achieve an effective level of circulation. Expensive books, periodicals, codes, and standards must be purchased to stay current in the field. Training costs for staff continue to increase.

The various perspectives that sources of information have may present another difficulty. For example, building operators can get recommendations on whether or not to install an energy-saving device from their utility representative, a contractor, or a manufacturer's representative, but it may be more useful for them to hear from other building operators who have installed such a device and share their specific concerns and perspectives. What works well in a testing laboratory may not work well in a specific building. As energy-consuming equipment becomes more electronic, unforeseen incompatibilities among such devices occur more often due to an increased production of or sensitivity to poor power quality or due to conflicting control signals.

Imagine that the building operators mentioned above could instantaneously access the knowledge of other building operators 24 hours per day, from an office in Seattle, a cabin in Aspen, or a hotel room in Brazil.
Imagine that they could similarly access a database of a thousand training events; thousands of software programs and data files they could download and use; millions of people they could communicate with; directories of energy experts they could contact; late-breaking, energy-related news; descriptions of energy-related programs, products, and research; and much more. This world of information at your fingertips isn't in the future; it is feasible--today--by "telecomputing" with electronic bulletin board systems and networks.

Anyone with a computer and a modem can access thousands of electronic bulletin board systems. Electronic information transfer can save a great deal of time and money by reducing travel and postage costs.

Electronic information transfer has even reached The White House. Last June, President Clinton and Vice President Gore announced that they would be connected to a number of widely used networks, reaching many of the estimated 25 million electronic mail addresses now existing worldwide--twice the number that existed just two years ago. Regular mail is now referred to as "snail mail" at The White House.

Some people may associate any use of computers beyond specific software programs as "play". They resist electronic information transfer because they don't have time for "that sort of thing". While some telecomputing technologies can be credited to computer enthusiasts and some systems have recreational themes, the services described in this paper are not toys. Busy people in the energy industry don't have time to not utilize these resources. What was once considered a "gadget" used by university researchers is now becoming mainstream, and those not making use of it will be at a disadvantage.

AVAILABLE RESOURCES

Thousands of bulletin boards are available in the United States alone, and more abroad. Bulletin boards may be accessed directly, through communications services companies, or through larger information networks. Some bulletin boards only offer discussion groups, while others offer a full collection of information services. These bulletin boards cover every topic imaginable. Some networks include bulletin boards as well as many computers with directories of files only. For simplicity, this wide variety of systems will be referred to as "bulletin boards" within this paper.

A detailed primer on how to connect to bulletin boards, an explanation of common bulletin board features, and some helpful guidelines for using bulletin boards are provided at the end of this paper for those not already familiar with telecomputing terms such as modem and E-mail.

Some of the energy-related bulletin boards that are available for your use are described in the following text. Anyone considering starting up a bulletin board to serve a group of people in the energy industry would be well advised to consider making arrangements to customize an existing system rather than starting from scratch, as there is a significant learning curve involved. The bulletin boards described here may well have changed since this paper was written, especially in terms of access speed, interconnections, number of users, and access options. In addition to the information/assistance phone numbers listed for each bulletin board, you can send electronic mail to the System operator of each system for assistance.

*** The Architects Online Network--The BSA BBS**
Operated by:
The Boston Society of Architects (BSA).

Description:
This bulletin board covers a wide range of architectural issues, including some renewable energy and environmental topics related to architecture. This is an example of a small bulletin board with limited scope that is accessible nationally and may be useful. It is also an example of a more friendly bulletin board. All BSA committees are required to use the bulletin board

Program Information/Assistance:
(617) 951-1433

Number of Users:
450

Access Options:
Direct dial to (617) 737-8102

Parameters:
Up to 2400 baud, 8 data bits, 1 stop bit, no parity, full duplex (no echo).

Cost:
None

Services:
- E-mail with file attachments
- Discussion forums on architectural issues, CADD, solar energy
- Information bulletins
- BSA committees
- Calendar of events
- Libraries of files, with demo software and symbols

*** The ASME BBS**
Operated by:
The American Society of Mechanical Engineers (ASME).

Description:
This is for the benefit of ASME members, but is accessible to anyone. You are allowed about 1 1/4 hours for each day. This is an example of a large engineering bulletin board with energy as one of many topics.

Program Information/Assistance:
Send E-mail to Sysop

Number of Users:
5,000

Access Options:
Direct dial (608) 233-3378

Parameters:
Up to 14,400 baud, using a V.32bis modem. 8 data bits, 1 stop bit, no parity, full duplex (no echo).

Services:

- E-mail with scanning, text search and exchange with Internet, MCI, Compuserve, and others. They are considering a system that would Fax you your E-mail if you don't retrieve promptly.
- 350 groups with 15,000 files.
- 55 conferences you can join, including 6 local, one supporting Internet E-mail, 45 Usenet newsgroups on various mechanical engineering topics, only a portion of which deal directly with energy.

Cost:
One-time $15 sign up fee includes first hour, then $10 per month that includes up to one hour of access, then $5/hour off-peak, $10/hour on-peak, $3/hour for direct calls. Internet and public data access from abroad is $3/hour on top of local packet switching or local node charges, which vary.

Services:

- E-mail, with aliases, searchable directory, and gateways to telex, fax, and other international networks.
- Over 1,000 public conferences, including a read-only list of energy research papers, articles, fact sheets, and directories, discussions of implementation of energy efficiency, renewable energy, and transportation technologies and policies.
- Calendar of Events
- 6 News Services with keyword searches. Searching two of the services for "energy" came up with 70 articles.
- Legislative Alerts
- Bibliography and library catalogs

*** EcoNet**
Operated by:
Institute for Global Communications (IGC)

Description:
EcoNet is an international, computer-based communications system committed to serving organizations and individuals who are working for environmental preservation and sustainability. This is an example of a huge, international network of information.

Program Information/Assistance:
(415) 442-0220

Number of Users:
9,000 in U.S., 16,000 in 94 countries plus Internet users.

Access Options:
A local call in most U.S. cities, as well as from 10 other countries with partner networks, including Australia, Canada, Great Britain, Sweden, Germany, Uruguay, Russia, Brazil, Ecuador, and Nicaragua. Through Telnet on Internet (igc.apc.org) and through public data access systems (X25) in 94 countries.

Parameters:
For SprintNet, 7 data bits, 1 stop bit (2 at 300 baud), odd parity, Xon/Xoff flow control, full duplex (echo off), no line feeds/carriage returns. Call for other access mediums.

*** EEI Online**
Operated by:
The Edison Electric Institute (EEI)

Description:
EEI is a trade association of investor-owned electric utilities. This service is operated for the benefit of members and non-members who are interested in the electric utility industry, such as consultants.

Program Information/Assistance:
(202) 508-5599 or (202) 508-5005

Access Options:
Using EEI Online software, provided with your subscription, you can dial directly.

Parameters:
EEI software takes care of this automatically.

Cost:
Primary Service costs 53 cents per minute. Committee Talk, and ROAD are for EEI members only. Committee Talk costs $100 per year for individuals or $2000 per year for corporations, plus 53 cents per minute. Custom Services are $90 each, and ROAD is $105/month plus $1.30 per minute.

Services:

- E-mail, with file attachments.
- Search EEI bookstore for information and order publications.
- Pose questions to EEI staff and get a response within 24 hours.
- Committee Discussion Boards (forums) for Committee Talk users only. This helps EEI committee members communicate.
- ROAD (Regulatory Online Abstracts Database) includes electric utility rates, economic issues, and regulatory decisions.

CUSTOM SERVICES:

- Utility Newsline: Access to national and international newspapers, business magazines, AP wire service, Dow Jones News, and special updates and statistical information and newsletters provided by EEI. It also includes the Daily News which is an abstracting services which provides daily summaries of what the media is reporting about the electric and gas industry.
- Envirowatch: EMF news and alerts, legislative updates, and Greenwire for environmental issues.
- RISE: Materials inventory exchange service and communication tool for utility engineers. Available only to utilities.
- Utility Policy Watch: Daily tracking of all federal and selected state legislation and regulatory issues, including StateNet, EEI Washington Letter, and Congressional Quarterly Washington Alert, as well as other legislative papers.

*** Electric Ideas Clearinghouse BBS**

Operated by:
The Washington State Energy Office.

Description:
This bulletin board is sponsored by the Bonneville Power Administration for the support of commercial and industrial energy conservation programs in the Pacific Northwest. This steadily expanding system is used by a wide variety of people in the energy conservation field, and is generally easy to use. Special arrangements can be made to customize the system to better serve an organization, providing special privileges to their members if needed. This will soon be arranged with Western Area Power Administration, along with toll-free access expanded to serve 17 western states. E-mail and forums will soon be shared with Energy and Regulatory Matters bulletin board.

Program Information/Assistance:
(800) 872-3568 in WA,OR,ID,MT, or 206-586-8588

Number of Users:
1600

Access Options:
800-762-3319 in WA,OR,ID,MT, or 206-586-6854 (also through Internet, described in this section). Expecting to expand toll-free to 17 Western states and to be accessible by a number of communication services by August 1993.

Parameters:
Up to 9600 baud, 8 data bits, 1 stop bit, no parity, full duplex (no echo), and supports V.32bis, MNP-5, and V.42bis.

Cost:
None

Services:

- E-mail with off-line mail reader, file attachments, group mail, searchable registry, and an exchange of messages with Internet.
- 20 files directories including lighting/photometric, energy accounting, California Energy Commission databases, Northwest state energy codes, HVAC, maintenance, motors, small Hydro, simulation software, and computer utilities.
- Forums with keyword search, scanning, and off-line mail reader. Topics include CAD, building modeling, building envelope, codes, hospitals, schools, HVAC, lighting, motors and drives, renewable resources, environment, power quality, policy/resource acquisition, building commissioning, and water/waste water.
- Databases including 1050 training events and videos searchable by 250 technologies, date, sponsor, or event type as well as commercial and industrial referrals directories with 5000 entries, searchable by technology, name, company, and location.
- Searchable Text Files with keyword search, including jobs, resume, energy codes, energy and environmental programs, and users manual.
- Library Services, including card catalog search of the largest energy library in the Northwest, newsletters, articles, and magazine list.
- Other Services, including teleconferencing, chatting, surveys and polls, acronyms, and list of other energy and environment bulletin boards.

*** Energy and Regulatory Matters**

Operated by:
The Michigan Public Service Commission with support from National Association of Regulatory Utility Commissioners and the U.S. Department of Energy.

Description:
This national system was established to increase the frequency, quality, and speed of information transfer and communications transfer among state and federal agencies and individuals and businesses interested in energy and regulatory issues. Viewing the notes from some of the Public Service Commission regulatory hearings can be very informative for DSM planners. E-mail and forums will soon be shared with the Electric Ideas Clearinghouse.

Program Information/Assistance:
(517) 334-6240.

Number of Users:
800

Access Options:
Direct dial: (517) 882-1421

Parameters:
8 data bits, 1 stop bit, full duplex (no echo), 2400 baud with limited access to 9600 baud.

Cost:
None

Services:
- E-mail.
- 53 forums, mostly public, with file attachment capability.
- Files libraries of energy and environmental topics. Most of the public environmental files relate to acid rain legislation.
- Many databases, mostly of Michigan public utility commission hearings, notices, and filings.
- Press Releases
- Legislative Updates
- National integrated resource planning database

*** EPRINET**
Operated by:
The Electric Power Research Institute (EPRI).

Description:
EPRINET is operated for EPRI member utilities and others worldwide. Limited access is offered through 1993 to non-members for free to encourage wider use. You can have documents or E-mail downloaded to your computer or directly to your printer. All in all, it is quite a sophisticated system. Next year EPRINET-2 will be available. It will be linked to Internet and other networks, will send graphics, and will use a Windows (icon) format.

Program Information/Assistance:
(415) 855-1000 (very helpful)

Number of Users:
Approximately 10,000

Access Options:
With the EPRINET software package, users can dial directly, utilize a commercial service such as Tymnet or the IBM Information Network, or by using a VM-based host (a mainframe computer).

Parameters:
This is written into the scripts in the EPRI software.

Cost:
The cost of EPRI membership varies. Limited access is free.

Services:
- E-mail that allows you to word process messages and then send them through E-mail, to set up distribution lists, and to search a directory of users
- Special Interest Forums on such subjects as EMF, environmental issues, lighting, renewable resources, and software. Scanning can be configured.
- Demand-Side Information Service that includes a publications database, Tech Notes Database (descriptions of end-use technologies, products, applications, research, and software), DSM Survey Database, and Industrial Database.
- EPRI research plans
- Calendar of events searchable by category and keywords
- Resource Catalogs searchable by keywords
- Directory of experts at EPRI
- Products database
- Reports and articles
- 30 News services, some with more services nested inside them, that can be scanned.
- Publications ordering
- File Transfer Services
- Bulletin Board Services

*** Green Light Bulletin Board**
Operated by:
The U.S. EPA Global Change Division

Description:
This is operated for participants in EPA's Green Lights program. This program is aimed at reducing the amount of energy used in buildings, and indirectly the greenhouse gases produced by power generation facilities.

Program Information/Assistance:
(202) 775-6650

Number of Users:
10,000

Access Options:
Direct dial (202) 775-6671

Parameters:
8 data bits, 1 stop bit, full duplex (no echo), 2400 baud

Cost:
None

Services:
- E-mail, mostly to communicate with EPA
- National finance directory of DSM rebates
- Files library with software for analyzing energy conservation measures such as Quickcalc
- List of weekly participants used by vendors

*** Internet**
Description:
This is the grandaddy of all telecomputing systems. It was originally set up by the U.S. government for communications between government agencies and research centers, but is now used by a great variety of people worldwide. It includes 5-10 million users as well as 10,000 bulletin boards. It is not terribly user friendly. The trend seems to be for more bulletin boards to link to Internet to exchange E-mail and pull off the news services and forum information that its users may find interesting.

Program Information/Assistance:
None. A number of books are available, however, such as Zen and the Art of Internet and The Whole Internet.

Number of Users:
5 to 10 million, with 15 million addresses.

Access Options:
You must be connected through a Internet access provider.

Parameters:
Determined by access provider.

Cost:
Only the cost of subscribing to access provider.

Services:
The services on Internet are too numerous and complex to be concisely summarized in this paper. A paper devoted to accessing energy information through the Internet may be available at the WEEC. You can contact the author through Internet using the address robpen@wseo.wa.gov.

*** WATTSLINE**
Operated by:
The Association of Demand Side Management Professionals (ADSMP).

Description:
This service is operated for the benefit of ADSMP members, but non-members are permitted limited access to four databases for downloading only.

Program Information/Assistance:
(408) 297-4779

Number of Users:
2000

Access Options:
Direct Dial, (408) 984-3667

Parameters:
Up to 9600 baud, 8 data bits, 1 stop bit, no parity, full duplex (no echo).

Cost:
No cost for limited access, and no cost to members beyond telephone charges.

Services:
- Public Databases accessible by public for downloading:
  - Calendar of Events
  - Calls for Papers
  - Available Publications
  - Requests for Proposals
- Membership database, including areas of expertise
- DSM news clipping service
- Job postings

GETTING CONNECTED
Adapting to a major change in the basic way we communicate and access information day to day can be difficult. Many very intelligent people resist the idea of telecomputing; some of them are the same people that took some time to overcome "computer-phobia".

Other people have had negative experiences attempting to use telecomputing. Maybe they couldn't get their boss to authorize the purchase of a modem, or they never got their modem to work with their phone system, or they finally

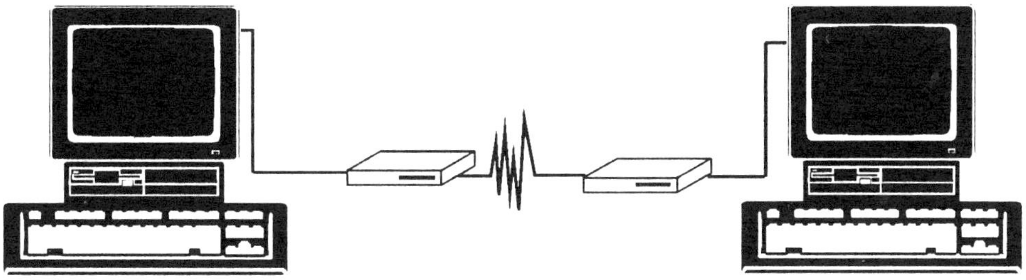

accessed a bulletin board but weren't able to extract useful information from it. Although these can be obstacles to making a successful initial connection, once the process is accomplished it is simple to repeat.

The following guidelines are to serve as a primer on telecomputing that will hopefully help you navigate through the experience successfully.

Once you have a computer, the next step is to add a modem to it. A modem is a box or board of electronics that allows two computers to communicate. A modem can be confusing at first, but once you install one and program the needed phone numbers into your communications software, this equipment becomes almost unnoticed.

Modems cost as little as $35 for an internal, 2400 baud model that installs inside your computer and transfers data at 2400 bits of information per second. Make sure that you have an available space inside your computer before you buy one. An external, 14,400 baud modem that plugs into the serial port on the back of your computer costs over $200. Most computers have a serial port--a long, D-shaped port with 9 or 25 pins sticking out. Check to see if it is currently being used by a mouse or something else; if so, you can often have another one installed. External modems can be used with a wide variety computers and have indicator lights that can be useful.

How fast a modem do you need? If you plan to transfer a lot of large files or may access a graphical information system, get a 9600 or 14,400 baud model. If you only want to exchange ideas, a 2400 baud modem may be enough, although you will be obsolete sooner. Your choice also depends on the speed that the bulletin boards you plan on accessing can handle. Many can now handle 9600 baud, but you can access them at that or any lower baud rate. Another consideration is the cost of the bulletin board access. If you are paying a fee on top of telephone charges each minute, a faster modem may pay for itself quickly. Modems faster than 2400 baud should be V.32, and preferably V.32bis to assure compatibility with many information systems. It is preferable to use software that makes use of V.42, V.42bis, or MNP-5 to increase data compression and error checking.

If you are using a multi-line phone system, you need to have a dedicated phone line run to your desk from a point before the phone line enters the multi-line system, as is done for FAX machines. If you have "call-waiting", you may want to turn it off while using your modem, because an incoming call will disconnect you; often typing "*70" on your phone will accomplish this. Contact your telecommunications coordinator for modem assistance.

The next step is to set up your communications software. This software tells the modem who to call, what language to speak, and how to transfer data. Modems generally come with communications software, but if you don't like what you get, other software is available. Some very good software is available as shareware such as ProComm from DataStorm. Shareware is software that you can get for free, but if you like it you're expected to send in a registration fee. In choosing a software package, look for simplicity, a variety of file transfer protocols to choose from) and powerful features that allow you to "capture" data scrolling across your screen into a text file.

Setting up your software involves a some terminology you may not be familiar with. However, you can take comfort in knowing that once it's set up you never have to do it again. The five primary parameters of data transfer that need to be set for each bulletin board you want to connect to are: *parity, data bits, stop bits, baud rate, and duplex/echo*. When you read about a new bulletin board, it should specify these parameters. If not, call the organization operating the bulletin board and ask.

Some bulletin boards can be accessed with a direct phone call through your modem. Others require the use of a communication service such as Compuserve or SprintNet. You connect your modem to such a service using a local or 800 number, and the service provides access to a number of bulletin boards. These services charge a monthly fee and allow a certain number of hours of use. In the near future, you will be able to make one call and access many bulletin boards through one bulletin board, using a process called "gateways".

FEATURES TO LOOK FOR

As you start to explore the many bulletin boards and networks available, useful attributes to look for include:

Simple

Some bulletin boards are so complex that even veteran computer users may get discouraged and frustrated. As a wider variety of people make use of bulletin boards, more simplicity is being demanded. Graphical icons or windows formats are friendly, but are not commonly found on existing bulletin board software and may exceed average user capability.

Powerful

Even if a bulletin board is simple to use, if you have to go through eight steps to accomplish a simple task, the process is inefficient. Powerful bulletin boards have features that allow you to very quickly get right to the information you need, using scanning, key-word searches, and databases to facilitate rapid data retrieval. In the 1990s, quick access to specialized information seems to be more lacking than quantity of information.

Popular

For many bulletin board services, the rule is "the more the merrier," especially if the users are well informed, articulate, and prolific. More good users means more discussions and perspectives; more entries on databases of training, referrals, jobs, and programs; and more files contributed. For energy and environmental professionals, a bulletin board would ideally be used by "key players" at research facilities that can make valuable contributions. If all the users are asking many more questions than they answer, progress is slow. Diversity of users is similarly important; if all the users are building owners, other perspectives won't be represented.

Fast

Only a few years ago, 1200 baud was the standard rate of information transfer. Now it is 9600, and rates as high as 56,000 baud are used. Busy people won't wait an hour for a file to transfer, and as software becomes more complex, files get larger and therefore slower to transfer.

Accessible

The more ways that a bulletin board can be accessed, the more people are going to find it and use it. Some bulletin boards are accessible by direct phone line, by communications services such as Compuserve, SprintNet, and ADP, and through other bulletin boards and networks. Connection to Internet or to X-25 public data access (also called packet switching) provides access to millions of users all over the world. Linking E-mail and forums among bulletin boards increases the value of all of them. Some bulletin boards are completely private, requiring a paid subscription or membership in an organization, and some allow "visitors" to access to parts of the bulletin board. A useful bulletin board should also be cheap enough to attract a great number of users. Some bulletin board services will reduce your access fee if you are a regular contributor of useful information.

Well Managed

Large bulletin boards offer a multitude of services and resources, and require the constant attention of a system operator (Sysop) to keep things running smoothly and answer questions promptly. Ideally, this Sysop would have sufficient technical knowledge to make judgments on the organization and usefulness of the information provided. Many bulletin boards make use of volunteers to help monitor specific discussion areas. Good documentation and on-line help screens are also helpful.

COMMON FEATURES

Some features are found on most bulletin boards. The exact terminology used by a particular bulletin board may vary somewhat, but the features will be similar.

E-Mail

Electronic mail, or E-mail, is used to communicate with other users of a bulletin board and sometimes with users of other bulletin boards. One benefit of using E-mail is that the receiving party doesn't need to be on-line when you send it; they can collect their accumulated E-mail when it is convenient. This can help you avoid the common game of "phone tag". Many systems can notify you when the receiver has read your message. Many allow you to attach files to an E-mail message, saving you the time and expense of mailing diskettes. They also let you copy and forward messages, organize messages in electronic folders, and send a message to a group of users. One wind energy researcher in Vermont sends a weekly file of wind energy news to 500 people in less than five minutes, and his only expense is a monthly fee for connection to access to Internet. If you attach a file to an E-mail message that is executable (not just text or a spreadsheet; usually having extensions .COM, .EXE, or .BAT), check for viruses first (data files cannot contain a virus). A number of adequate virus checking software packages are available. Users need not have a major concern about having a virus transfered to their computer from one of the bulletin boards listed in this paper; quality control is quite good.

The key to using E-mail effectively is knowing who else is on the system. Some systems have a user registry that can be searched in a variety of ways. Other systems have too many users for that (Internet has approximately 15 million E-mail addresses), so you may have get your associates to provide you with their bulletin board addresses. Some bulletin boards are linked so that E-mail is exchanged between them regularly, and some provide an off-line mail reader which allows you to download messages to a file to read later, which shortens the amount of time you are connected to the bulletin board. Using E-mail as opposed to U.S. mail or a telephone can save money if you pay a small fee or no fee for bulletin board use.

File Libraries

Sharing computer files through file "libraries" can be a very effective form of networking. If you have written a program or developed a spreadsheet and don't mind sharing, it may save someone else time. It is critical that all executable files be checked for viruses before they are made available to users. A letter from the author giving permission for distribution should be on file.

To be most useful, these files should be well organized and edited, to minimize searching time for what you want and to avoid "dead wood". Some bulletin boards have many thousands of files but a number may be of little value. If you upload a file with anything but ASCII text (an agreed upon standard of simplicity), be sure to clearly note that in the description so no one will try to read a spreadsheet with their word processing software. It is best to transfer files in a compressed format. The most popular utility is PKWARE, which is available as shareware. Such file compression software will reduce the size of many files by over 50%, greatly reducing transfer time.

To transfer files, you and the bulletin board need to agree upon a protocol. Choices include such strange names as Kermit, Xmodem, and Zmodem. You need to pick a protocol supported by both your communications software and the bulletin board. Some are faster or more reliable than others. Xmodem is the most common, but experiment and see what works best for you. You can usually select "list" as a protocol for text files. This will make the document scroll across your screen. If you have arranged for your communications software to "capture" your screen information into a text file, this can be painless way to get a file onto your computer.

Discussion Forums

Forums, sometimes called conferences or special interest groups, allow users to share information with others in a timely and efficient manner. Forum messages might include information learned at a seminar, a summary of an article from a technical journal, or an experience someone had installing occupancy sensors in a warehouse--basically what works (or doesn't work) where and why. Conversely, forum messages can be used to pose a question to other users, such as "who has installed those new 8' T-8 lamps, and how did the project work out?"

The most effective forums are those used by people with a wide variety of perspectives, including utilities staff, engineers, consultants, researchers, building owners and operators, architects, environmental professionals, and others. Some organizations use a forum to communicate between meetings and share minutes and notices.

As a general rule, summarize your forum communication, including additional details in an attached file for those who are interested. As with E-mail messages, check executable files for viruses. You might write a short message saying "I just attended LightFair in San Francisco. If you are interested, my notes are attached" or "I just finished a DOE2 model of an office building with thermal storage and a greenhouse; attached is my input."

Keep messages short, usually two screens or less. Think of a clear title so that people scanning the messages can easily judge the content of the message. If you have more than one subject to discuss, write separate messages. If you put the same message in a number of forums, include a note asking that all responses be sent to one forum. Use plain "vanilla" text, with no fancy word processing characters that could cause problems for people.

Some forums are set up for political or business purposes, and some restrict such uses. Most will provide an introduction that explains the purpose and limitations of the forum; respect these guidelines. Some systems have a profanity index built in to restrict more emotional discussions.

To make discussions easier to read, many systems allow you to follow the "thread" of a discussion from its initial statement to all those made in response to it. This process can be clearer than chronologically viewing messages from a number of simultaneous discussions.

Keyword searches are also quite useful. If a lighting forum has 200 messages in it but all you want to know about is occupancy sensors, you can search for the word "occupancy" and go right to the messages discussing them.

Scanning is a feature that allows you to view messages you haven't read yet (it keeps track of which message you've read) in the forums that you have identified as being within your field of interest. Out of 30 forums on a variety of energy subjects, if you only work with lighting, you could specify the lighting and power quality forums.

On some bulletin boards, each forum is moderated by someone very familiar with the topic of the forum. This person throws out some thoughts when dialog is slow, tries to make sure that those asking a question get an answer, mediates if a discussion starts to get out of hand, and decides how long a message will be kept on the forum. If a message is about a meeting in two weeks, it can be deleted soon after, while other messages may have value years later. One function that the system operator can perform is to coordinate messages between forums. A Sysop might write a message in a forum for hospital engineers saying "I see you have started a great discussion about CO_2 monitoring for indoor air quality. The HVAC forum has been discussing that for months, so I suggest that you participate in that discussion instead."

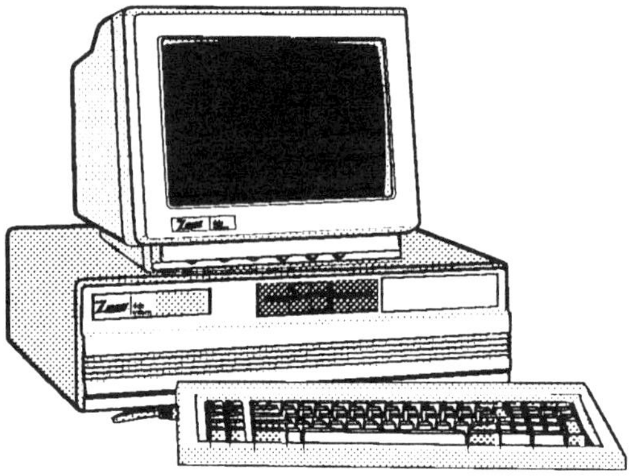

Databases
Databases of information can provide rapid access to a variety of information. Unlike file libraries, databases are used to look up information without having to transfer a large file to your computer. Databases are available for training events, calls for papers, energy experts, building products, and much more.

Searchable Text Files
For some information, text files (searchable using keywords) are more appropriate than databases, due to the freeform nature of text rather than the restricted, structured format of databases. Examples of text files are job opportunities, resumes, energy codes, lists of books, magazines, and research papers.

CONCLUSION
Everyone working with demand side management needs up-to-date, practical information about energy conservation measures, both theoretical background and application guidelines that will avoid customer dissatisfaction with DSM programs. Technology continues to advance at a dizzying pace and new applications of existing technologies are being explored. Therefore, the information transfer mechanisms used must be very flexible and widespread.

To be effective, they should also include a wide variety of people, including utilities, engineers, contractors, building owners and operators, architects, and consultants. To get all these people to participate in this information exchange, the mechanism used should be simple, powerful, inexpensive, and well managed.

An increasing number of energy-related electronic bulletin boards and networks are developing to meet this need. Anyone with a computer and a modem can have instant access to a tremendous amount of information 24 hours per day. More of these services are linking together, allowing a consultant with a laptop at a vacation cottage to communicate with millions of people around the world, including most major universities and research centers.

Telecomputing was once the domain of university researchers, but now is accessible to everyone for little or no cost. Although some technical and psychological hurdles are involved, the multitude of benefits are well worth it. The best way to approach this may be to simply relax and have fun with it. Remember that you can telecompute from anywhere, so next time you're in a hotel room with some time to spare, plug your laptop into a phone jack and explore.

REFERENCES
The author would like to credit the following helpful sources of information:

- The many system operators and program managers behind the bulletin boards and networks described in this paper.
- "User's Manual" version 2.5, by the Institute for Global Communication, Fall 1991, published by the Institute for Global Communications.
- Vicky Johnson at the Electric Power Research Institute.
- The technical and support staff of the Washington State Energy Office.

Chapter 103

The Three E Project—Integrating Environmental Sensitivity, Energy Efficiency & Electricity

P.E. Conlan

The *Three E* Project[1] is an initiative of America's Electric Utilities to promote energy efficiency and environmental sensitivity in both new and existing homes. Sponsors of the project include utility company members of the Edison Electric Institute (EEI) and the National Rural Electric Cooperative Association (NRECA). These utilities include electric and gas combination utilities and electric only companies. They serve more than half of the residential customers in the United States.

OBJECTIVES

Energy efficiency improvements can result in substantial environmental benefits. The *Three E* project seeks to go beyond traditional in-home energy efficiency measures to incorporate other home features which result in environmental benefit within the home, at the home site, and within the community and the world at large.

The objective for the environmental component of *Three E* is to develop criteria which can be used to:

- Evaluate and certify environmental components of existing utility home programs[2]
- Serve as a framework for the incorporation of environmental criteria within utility-sponsored home programs that currently have no environmental component.

Three E qualifying programs will:

1. Provide home buyers and owners with a healthy and environmentally sensitive home in which to live and from which to demonstrate ongoing environmental concern without sacrificing comfort and convenience.
2. Provide homeowners and prospective home buyers with packaged environmental options and information for:
   - Enhancing the environment of their new or existing home.
   - Contributing to the improvement of the global environment.
3. Advance residential sector current practice related to environmental sensitivity.
4. Provide utility residential sector trade allies (i.e., developers, builders, contractors, realtors and lenders) with added environmental value for the sale of homes.
5. Link energy efficiency and other environmental features at the home level.

THE *THREE E* SYSTEM

Market research and working contacts with environmental organizations, government agencies, builders, utilities and others was initiated to obtain input into the *Three E* environmental component. Related programs and pilot and demonstration projects were also studied.

Environmental features have been identified and organized into categories (or modules). They are:

- Indoor Air Quality
- Home "Commissioning"
- Home Waste Management
- Construction & HVAC Practices
- Water Quality & Conservation
- Transportation
- Home Site Evaluation
- Environmental Testing
- Custom

Additional modules have been added to represent the following areas addressed by *Three E*:

- Energy Efficiency
- Comfort, Safety & Convenience
- Renewable Energy
- Information
- Energy & Load Management

Taken together, these 14 modules are the framework of the draft *Three E system* by which utility-sponsored home programs will be evaluated for certification. Figure 1.0 presents a graphic overview of the system. The individual criteria listed are only representative of the criteria being considered.

Terminology

To help ensure understanding of the proposed environmental component and *Three E* system, it is important to define how we are using certain words within the *Three E* context.

Criteria. We are using the term *criteria* to indicate a specific environmental, energy, comfort, safety, or convenience guideline or option (e.g., incorporation of an in-home recycling center).

Module. The term *module* is being used to denote a category, or set, of related criteria (e.g., home waste management). Each module has multiple criteria, of which, some may be *basic* and some *optional*.

Basic. The term *basic* (or *basics*) is used to denote a program element that is required, or fundamental to *Three E* (not optional).

Basic Modules. Basic modules are required for utility program certification. Qualifying *Three E* utility programs must have criteria which meet or exceed those specified. Criteria within individual modules are of two types - *basic* and *optional*. Basic criteria must be met. Sometimes, alternative approaches will be offered for meeting required criteria. Utility programs may also incorporate optional criteria from basic modules.

Every qualifying *Three E* program must meet the minimum guidelines for each of the following seven basic modules:

- Energy Efficiency
- Renewable Energy
- Indoor Air Quality
- Home Waste Management
- Water Quality & Conservation
- Comfort, Safety and Convenience
- Information

Optional Modules. *Optional* modules (or *options*) may be chosen at a utility's discretion to provide more thorough environmental assurances, address areas of local environmental concern, and to customize the program. Optional modules, like basic modules, include basic and optional criteria (or require that a sufficient number of criteria be included in their program). The seven optional modules are:

- Energy & Load Management
- Construction Practices & HVAC
- Transportation
- Home Site Evaluation
- Environmental Testing
- Home "Commissioning"
- Custom Environmental

Three E System Interaction

It is anticipated that a point, or trade-off, system will be used to tie-together the various *Three E* modules. This will accomplish the following:

- Provide a system to ensure that a sufficient minimum range of environmental modules and criteria are included in addition to the energy requirements
- Provide weighting factors for the various modules and criteria to recognize their:
  - approximate environmental benefit
  - relative contribution to the project objectives
- Serve as the quantifying mechanism for certification of utility programs.

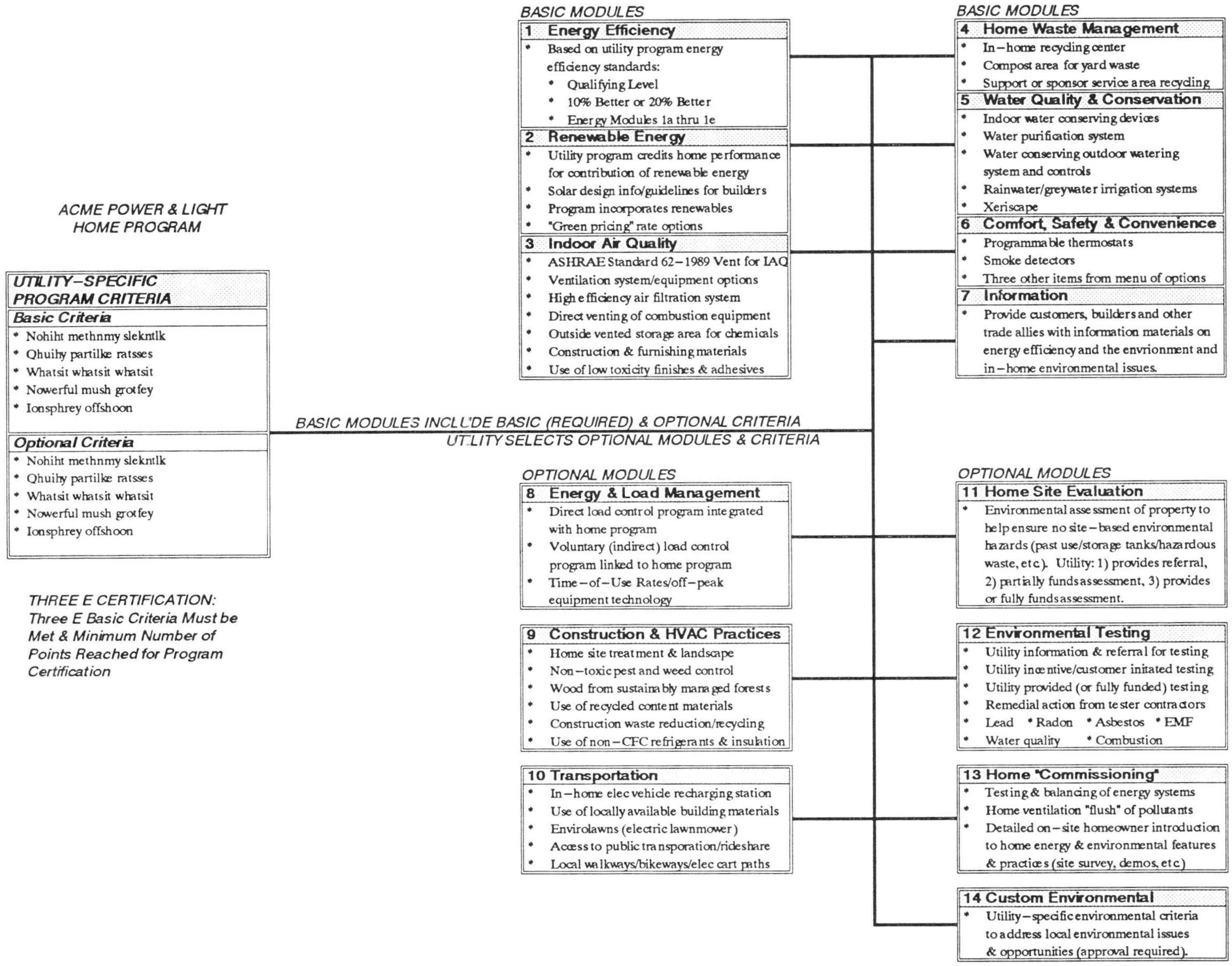

Figure 1.0 – Overview of Three E System

This system concept has been approved with the definition and division of basic and optional modules and criteria. The detailed criteria themselves and the point/weighting system are still under development. Input is being obtained from a wide range of technical advisors regarding design of the detailed criteria.

Basic Criteria Modules

The following sections are brief discussions of the seven basic modules.

Module 1 - Energy Efficiency. Though this paper focuses primarily on the non-energy environmental components of *Three E*, the energy efficiency component is at the heart of the system. It is intended that *Three E* certification require a significant degree of energy efficiency and that the energy criteria be linked to a nationally accepted energy standard. Based on research, it is currently recommended that the minimum *Three E* new home efficiency criteria include the following:

- Overall energy *performance* for heating, cooling and water heating set at a minimum of ten percent better than the combined requirements of the 1992 CABO Model Energy Code thermal envelope requirements and the NAECA efficiency standards for HVAC and water heating equipment.

- *Prescriptive* criteria will address air distribution system efficiency, lighting and appliances.

Energy criteria are also being established for existing homes.

With the primary environmental thrust of the *Three E* project, it is important to recognize the environmental benefits that accrue from utility home programs that go beyond the minimum *Three E* energy requirements. It is therefore proposed, when a utility program's level of minimum energy efficiency exceeds the *Three E* minimum, it be credited for this toward certification. In the example below, efficiency levels are presented in percentage improvements (i.e., higher efficiencies) over the minimum *Three E* levels for space heating and cooling, and water heating. More points will be available for programs that offer higher required efficiency levels.

An example of how this could be structured is as follows:

Level of Energy Efficiency	Points
Three E Minimums (MEC/NAECA + 10%)	Points
10% Better than *Three E* Minimums	Points +
20% Better than *Three E* Minimums	Points ++
30% Better than *Three E* Minimums	Points +++

This system will allow utilities with programs offering very high levels of energy efficiency to trade-off the incremental environmental benefit of these levels against other *Three E* criteria. This trade-off will help keep the cost of utility program homes from being excessive while the overall point (or weighting) systems will ensure that a sound mix of criteria are included. For example, under a possible trade-off scenario, one optional environmental module could be traded for a +10% increment of energy efficiency above the *Three E* minimums.

Module 2 - Renewable Energy. Utility home programs which provide credit for the use of passive or active renewable energy designs/systems will receive a set number of bonus points. A program could also receive these points if it has an active *green pricing* program for residential customers. At the home level, utilities must be able to evaluate renewable energy contributions or a customer's overall energy use from renewables through the green pricing mechanism.

Module 3 - Indoor Air Quality. Indoor air quality (IAQ) requirements will specify minimum ventilation rates and offer equipment/system options for realizing them. Other factors affecting IAQ such as selection of construction materials, adhesives and finishes, home furnishing materials, use and storage of toxic materials and direct venting for combustion equipment will also be addressed.

It is expected that the energy modules (1 and 2) will be weighted approximately 50 - 60 percent of the total number of points needed for program certification. Module 3, Indoor Air Quality, will also be rated higher that the other non-energy modules.

Module 4 - Home Waste Management. This module will address homeowner waste management opportunities. Included will be home features that facilitate in-home and yard waste recycling. Some credit may also be given for utilities that support or sponsor recycling programs in their service areas.

Module 5 - Water Quality & Conservation. Water conservation is of concern in many areas of the country.

Criteria will address water-efficient indoor devices, site treatment and outdoor watering devices and controls. Optional water criteria will be available for utilities in arid climates where more serious measures are appropriate. Also included will be an option for electronic water purification systems to help provide water quality.

<u>Module 6 - Comfort, Safety & Convenience.</u> In addition to the energy and environmental criteria, a number of other home features are being considered. Some of these are likely to be required (e.g., programmable thermostats, smoke detectors, etc.), while others can be selected from a menu of features related to homeowner comfort, safety or convenience.

<u>Module 7 - Information.</u> A comprehensive homeowner information package is required as part of a qualifying utility program. Information pieces should address home energy systems and their operation and maintenance as well as transportation related energy issues. Home environmental criteria and especially in-home practices should be covered. Topics can include air quality, asbestos, radon, water quality, lead, toxic materials storage and disposal, recycling, etc. Information should be provided to allow the homeowner to live in a more environmentally sensitive manner. Information pieces may be approved by EPA or other environmental organizations. Innovative information formats such as video tapes will be encouraged.

Builder and contractor information and/or training will be required for new construction programs.

Optional Criteria Modules

The following seven sections are brief discussions of the optional *Three E* modules.

<u>Module 8 - Energy & Load Management.</u> This module will recognize and credit utilities which have incorporated active load management strategies into their home programs. Examples include direct load control of air-conditioners and electric water heaters, time-of-use rates and residential thermal storage. In-home energy management systems will also be included as an option.

<u>Module 9 - Construction & HVAC Practices.</u> The construction practices module will include a significant number of environmentally sensitive options from which the builder/remodeler can choose. Specific criteria will address treatment of the home site, selection and use of construction materials, reduction or recycling of construction waste materials, etc. There may also be prohibitions on the use of some materials. The use of CFC refrigerant alternatives will also be an option.

Interested utilities can include optional criteria to encourage the use of *local* construction materials or products. This reduces product transportation energy and costs, and makes a strong statement on behalf of the local economy.

<u>Module 10 - Transportation.</u> This module includes criteria related to transportation and other factors which address clean air. Credit is offered for homes with electric vehicle recharging capabilities, and where electric lawn mowers (Envirolawns project) are used. Use of and/or access to public transportation and local community transportation alternatives are also recognized.

<u>Module 11 - Home Site Evaluation.</u> This optional environmental module addresses "home site evaluation." This is proposed since it makes little sense to construct an environmentally sensitive home on an environmentally hazardous site. Site hazards include old landfills, leaking storage tanks, industrial waste storage or disposal sites, accident/spill sites, local dry cleaners, paint plants, etc. Evidence of environmental hazards may not be readily apparent.

Database services are now available that can search for all recorded environmental hazards. These databases are compiled from federal and state databases and records and also include sites under investigation. A number of companies offer this type of service for a single-service cost of $75 per home site.

Utility programs that provide information *and* referral lists for qualifying services (or reference professional associations which provide lists of qualified services/companies) will receive one level of points. Utilities which actually require homeowners, builders or developers to provide a "clean bill of health" for the site will receive more points. Builders, developers and lenders have additional liability reasons for checking this information and we believe that an increasing number will be doing so in the future. This is already common practice for commercial properties.

<u>Module 12 - Environmental Testing.</u> Environmental testing includes home testing in the following potential areas:

- Asbestos
- Radon
- Water quality
- Lead (paint & water)
- Combustion safety

- Electro-magnetic fields

Utility-sponsored testing represents a departure from energy-related in-home services such as energy auditing and direct installation with which utilities are familiar. Additionally, environmental testing could open doors to liability that utilities rather avoid.

On the other hand, if testing is not incorporated, the chances are high that some *Three E* homes will have serious environmental/health hazards. A possible compromise would be for the utility to provide the homeowner (directly, or through the builder) information on these hazards and suggest that the homeowner initiate testing.

Utilities could provide listings of testing contractors in their service areas as they now do for energy trade allies. Home evaluation, reporting and recommendations for remedial action would then be the responsibility of the testing contractors. It may also be possible to work through professional associations representing environmental consultants and testing services.

Project research indicates that some utilities feel it is very important that *Three E* address electro-magnetic fields (EMF), while others felt just as strongly that it should not be addressed. Including EMF as an option under this module will accommodate utilities holding both views.

EMF informational literature can, at the utility's option, be provided customers under Module 7 - Information. Utilities that want to offer more could include additional criteria related to EMF-reducing electrical system design for new construction and remodel applications. In-home testing for EMF is a new field and presents some of the same concerns for utilities addressed in the discussion of Module 9 - Environmental Testing.

Module 13 - Home Commissioning. *Commissioning* is a building start-up procedure which is gaining popularity for commercial buildings. The same concept can be applied, with modification, to homes. Residential commissioning could include the following:

- Systems start-up, testing and balancing
- Review of equipment/systems maintenance and service requirements
- "Flush" period to provide high ventilation rates to help drive-out hazardous emissions from new materials and systems
- On-site homeowner demonstration or survey to help ensure their understanding of the home's energy systems and environmental features. Information materials (Module 7) could also be provided the customer at this time.

Module 14 - Custom Environmental Criteria. It is recommended that utilities be able to propose alternate, or additional, environmental criteria not addressed in the current modules. This will allow for flexibility in addressing very localized environmental issues and for new developments in environmental practice or products. A review process will be established to determine whether the proposed criteria is valid and how it should be weighted and incorporated into the *Three E* criteria for the utility.

CONCLUSION

As previously mentioned, the detailed criteria within each module is now under development together with the point/weighting system. Criteria will address both new construction and retrofit applications. These criteria will be organized into the same system of modules, but one set will address new construction and the other existing homes.

Additional information on the *Three E* initiative can be obtained from Mike McGrath or Tom Farkas at EEI (202/508-5557), or Paul Conlan at A&C Enercom (404/633-9099).

[1] *Three* E is a temporary place-holder name which will be replaced prior to market introduction. At the time of writing this paper, the *Three* E project is still underway. Finalized, and more detailed, information will be available at the time of the conference.

[2] It is important to understand that the *Three* E initiative is not a stand-alone utility program, but a certification system. Existing, or new, utility programs will be evaluated by the *Three* E criteria. Those which meet the criteria will be certified.

Chapter 104

From Demand-Side to Integrated Resource Planning: EPRI's DSManager and IRP-Manager Planning Models

M.D. Haakenstad, R.M. Running

The Electric Power Research Institute (EPRI) has introduced two state-of-the art products to enhance the productivity of utility planners: DSManager Version 2.0 and IRP-Manager Version 1.0. These new tools can help utility planners meet the challenge of increasing management and regulatory expectations in the face of diminishing human resources. Both models are supported by Electric Power Software, Inc. (EPS).

DSManager 2.0 enables planners to evaluate the costs and benefits of demand-side management (DSM) options, design specific DSM programs, and forecast impacts of DSM programs. DSManager is now in use at over 100 utilities, regulatory bodies, and consulting firms.

IRP-Manager 1.0 dynamically evaluates demand and supply resources to develop fully integrated resource plans. IRP Manager provides a versatile "Tool Box" of integrated resource planning capabilities, including chronological simulation of demands and resources, integrated resource optimization, and decision/risk analysis. The integrated nature and chronological algorithms used by these models allow unparalleled insights into the consequences of integrating demand-side and supply-side resources of all types.

Together, these two EPRI products permit planners to apply unprecedented rigor to both the demand-side and integrated resource planning processes while being freed up from the tedium of data transfer and compatibility issues. EPRI has gone beyond the concept of merely automating file transfer or linking model inputs and outputs. In a revolutionary approach to integrating the total process, EPRI has achieved a seamless union between DSManager and IRP-Manager. For example, IRP-Manager directly accesses the DSManager database for review and selection of DSM options, instead of relying on ASCII file transfer. Since both models share the same file handling system and graphical user interface, planners may prefer to view this process as requiring a single model.

The integrated planning process is inherently iterative as opposed to linear. This paper proceeds through demand-side analysis prior to addressing integrated analysis, but the reader should realize that the actual process is a continuum. For example, the starting point may actually include an IRP-Manager analysis for the purpose of defining or updating utility avoided costs, just as there may be a secondary assessment in DSManager following the integrated analysis (to see what impact, if any, the revised set of avoided costs has on DSM program design and ranking).

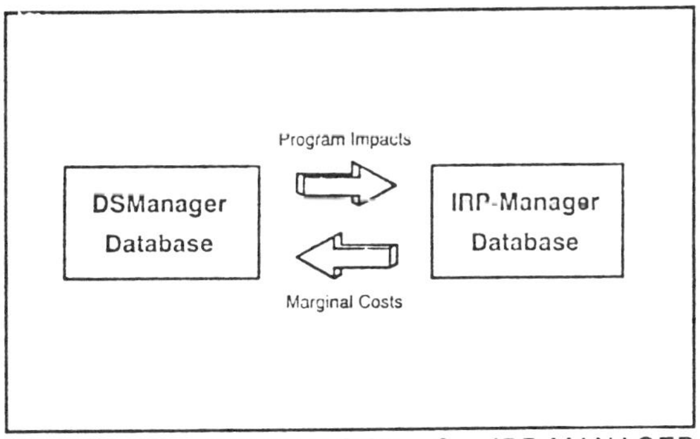

FIGURE 1 DSMANAGER & IRP-MANAGER INTERACTION

DEMAND-SIDE ANALYSIS WITH EPRI DSMANAGER

EPRI DSManager is a demand-side screening tool which allows users to analyze the economic and forecast impacts of demand-side management technologies and programs. DSManager has a Graphical User Interface which allows easy data entry and graphical representation of data. It is fully compatible with many planning and end use models such as

EPRI LMSTM, EPRI IRP-Manager, EPRI EGEAS, EPRI MIDAS, EPRI RELOAD, EPRI MarketTREK, as well as other models such as PROMOD™, PROSCREEN™, and UPLAN™.[1]

DSManager has a wide array of reporting capabilities providing the user with both summary and detailed reports from various perspectives including the utility, participant, ratepayer, total resource, and societal. DSManager is used extensively by more than 100 utilities, regulatory agencies, and consulting companies.

Key DSManager Features

* Program design is streamlined through DSManager's transparent structure.
* Cost effectiveness analysis includes benefit/cost, net present value, annual cash flow, and levelized calculations from multiple perspectives.
* Detailed forecasting capability projects individual or aggregate program impacts over a multi-year time horizon.
* Electric and gas analysis is ideal for evaluation fuel-switching and load building options, and wherever non-electric and electric measures should be compared on equal footing.
* Comprehensive database manager effectively and efficiently manages large amounts of data, including numerous data transfer functions.
* Documentation editor simplifies entry and tracking of program assumptions, and provides complete output reporting.
* Sensitivity analysis module allows users to assess uncertainty by using probability analysis. This helps identify key program variables, and allows efficient usage of limited resources.
* Participation module provides forecasts of market penetration by using both economic and non-economic factors to predict customer acceptance.
* Separate distribution system option allows users to split off the G and T utility from the distribution cooperative.

DSM Analysis Using EPRI DSManager

Step 1: Determine the Utility's Objectives for Demand-Side Programs

The first step is to determine the goals or objectives of the demand-side activities of the utility. Some utilities will provide general goals and guidelines to their demand-side planners and managers. One possible goal might be to find programs that provide a positive net benefit for the utility and its customers. Alternatively, these general guidelines might specify that the utility wants to delay a base load generating plant scheduled for a certain future year by two years, or reduce the cost to serve specific customer end uses by ten percent. Other utilities will provide direction to demand-side personnel in the form of an integrated planning process or a least-cost planning process. Still other utilities will establish more specific guidelines on desired results for end use programs, such as clipping peaks or increasing sales.

1 PROMOD and PROSCREEN are trademarks of Energy Management Associates, UPLAN is a trademark of LCG Consulting

Step 2: Identify Candidate Programs

Next, design specific programs that can achieve the goals defined in Step 1. The DSM program data definitions indicate the relevant energy use impacts and costs for the programs. Planners will often have insufficient information available to completely define all programs of interest, so the analysis team will frequently rely on "best estimates". Important assumptions can be refined later in the cycle.

Step 3: Preliminary Analysis of the DSM Programs

In this step DSManager is used to evaluate each of the programs. By examining the results, one can develop an understanding of the types of programs that offer the potential of achieving the goals defined in Step 1. For example, a goal to lower customers' energy costs can be achieved much more effectively using programs that sell additional energy and conserve energy at the end use than by using load management programs that shift electric usage to off-peak times. These insights can be developed by looking at the detailed reporting of the cash flows and energy use or the summary results of costs and benefits produced by DSManager. For example, the revenue from additional sales might greatly exceed the variable cost of production, or there may be several end uses that are particularly amenable to conservation programs.

Generally, it is possible to develop a list of the big winners among the candidate DSM programs. Plotting a stack graph of the net benefits to a party for all the programs with positive benefits reveals the few key programs that provide most of the benefits. Figure 2 shows such a graph for the net benefits to participating customers. Although there are ten programs that provide positive net benefits to customers, four of the programs provide about 85 percent of the total.

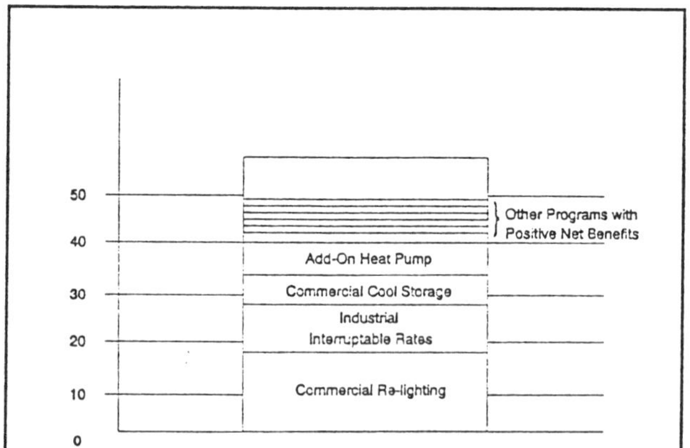

FIGURE 2 A FEW PROGRAMS WILL PROBABLY CONTRIBUTE MOST OF THE NET BENEFITS

Because of the state of the analysis and the preliminary nature of the data being used, the insights developed in this step are preliminary. However, they will form the basis for understanding the results obtained in future steps.

Step 4: Develop Participation Estimates

Forecasting utility demand side programs requires the ability to realistically predict consumer choices among fuels and technologies involved in the end use consumption of energy. DSManager incorporates a unique and highly flexible consumer choice module that:

* Is driven by customer economics;
* Integrates non-economic physical constraints, (e.g. feasibility);
* Considers non-economic customer attributes, (e.g. awareness);
* Allows simultaneous consideration of unlimited alternatives; and
* Is understandable and explainable.

The consumer choice module combines a technique known as microsimulation with a probabilistic approach developed by EPS. Microsimulation is analogous to how survey sampling uses a small sample to characterize a much larger population. A small number of representative consumers (households) with differing economic perspectives and energy using characteristics are used to explain the diversity of consumer fuel and efficiency choices for a service area. Applicability, feasibility and various other diffusion constraints are explicitly integrated with the microsimulation using a probabilistic framework.

The economic algorithm explicitly recognizes that consumers differ significantly regarding how far into the future they consider the costs and benefits of an alternative and that costs and benefits differ among consumers as well. The sample for microsimulation is then generated from a representative partitioning of the population. Within each consumer segment, the alternatives are sorted from least expensive to most expensive based on the time horizon and other energy using characteristics of the consumer segment. Lastly, each consumer segment chooses one or more technologies based on any feasibility and/or diffusion constraints applicable to the economically preferred technology, i.e., the consumer within each segment will select the least expensive alternative:

* That the consumer is aware of and willing to consider; and
* That it is technically feasible and applicable.

An overview of the methodology is illustrated in Figure 3. The process described in the figure is repeated twice. The first execution determines customer choices in the absence of any utility DSM program. The inputs which define customer awareness and willingness without a DSM program are used in the calculations. This provides free rider participation for each alternative.

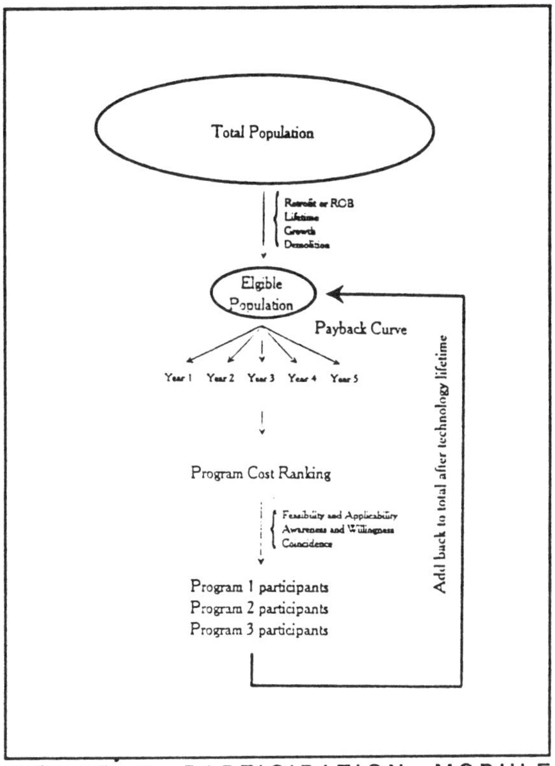

FIGURE 3 PARTICIPATION MODULE METHODOLOGY OVERVIEW

The second execution of the process determines customer choices when the utility DSM program exists. Customer paybacks are recalculated with the utility rebates, if any are offered as part of the program. The inputs which defined Customer Awareness and Willingness with a DSM program are used in the calculations. This provides the projection of total participation for each alternative.

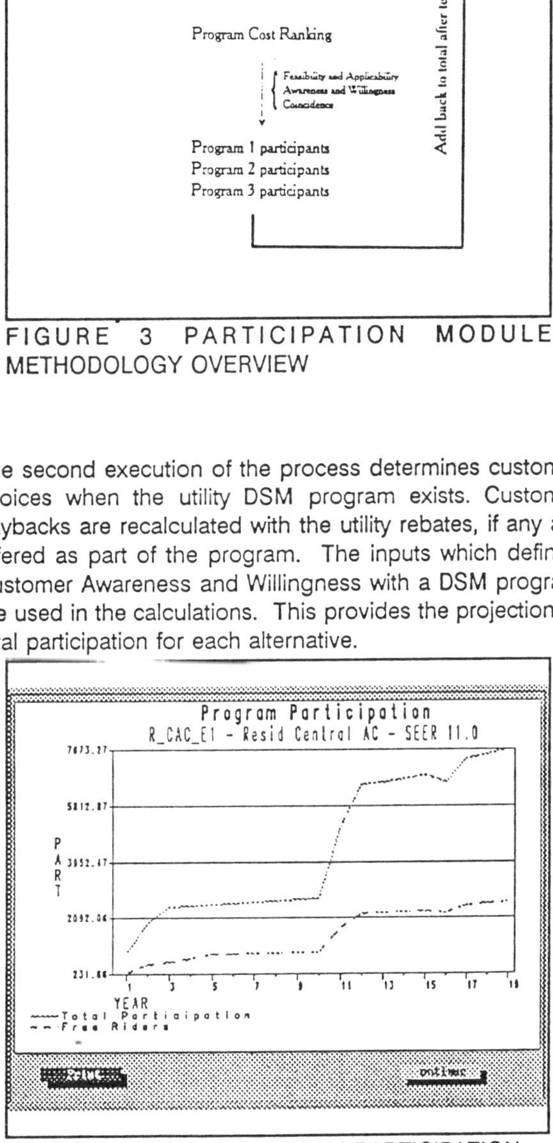

FIGURE 4 CURRENT PROGRAM PARTICIPATION

DSM programs may affect consumer choices in the model either directly as a function of program incentives or indirectly through the various diffusion constraints, e.g. and educational program to increase awareness and willingness, or both.

Step 5: Sensitivity Analysis on Preliminary Analysis of Programs

In this step, a sensitivity analysis is performed for the program(s) that have been designed. When evaluating a demand-side program DSManager provides a benefit/cost (b/c) ratio for each of the Cost Tests (Participant Test, Ratepayer Impact Measure Test, Utility Cost Test, Societal Test, Total Resource Cost Test). These b/c ratios, displayed in the Summary of Present Values reports, will have positive values such as 0.76, 1.4, or 10.2. These b/c ratios assist the analyst in determining whether or not the demand-side program should remain under consideration. A program with a b/c ratio less than 1 is typically considered less than acceptable; a b/c ratio greater than 1 is typically considered acceptable. The sensitivity analysis feature produces two graphs for each Cost Test perspective to assist the analyst in determining a comfort level of the data entered.

The first graph, Figure 5, of the Sensitivity Analysis feature can be used to determine the possible ranges of b/c's for each perspective (Participant, Ratepayer, Utility, Societal and Total Resource). A probability distribution or range of b/c tests is determined for each perspective. If the probability distribution is entirely above one or less than one, the user can be confident that the program is either acceptable or unacceptable given the inputs. Therefore, no further refinement is needed. However, if the probability distribution spans a b/c ratio of 1.0, refining the quality of key inputs will determine if the program b/c is acceptable or not. This information is extremely important to the analyst because it is suggesting that the acceptability of a program (based on the b/c test) is directly related to the quality of the data input into DSManager.

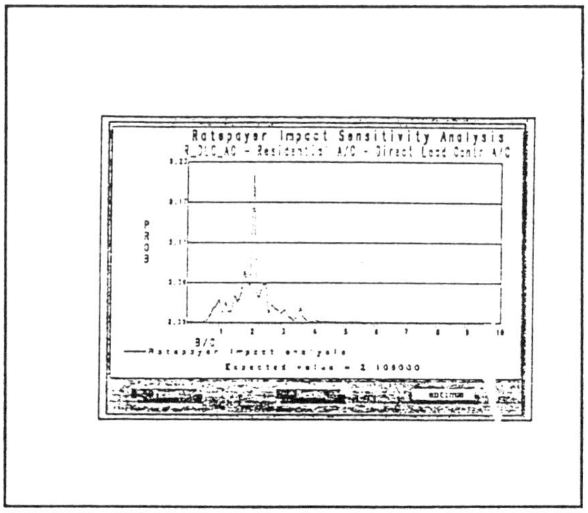

FIGURE 5 SENSITIVITY PROBABILITY DISTRIBUTION

The second graph, Figure 6, of the sensitivity analysis feature helps the analyst determine which variables contribute most to the uncertainty of the b/c results. From the graph, the analyst can determine which variables have a large impact on the b/c uncertainty. It would probably be beneficial to further research or refine these variables. On the other hand, those variables which have very little impact, require no further research.

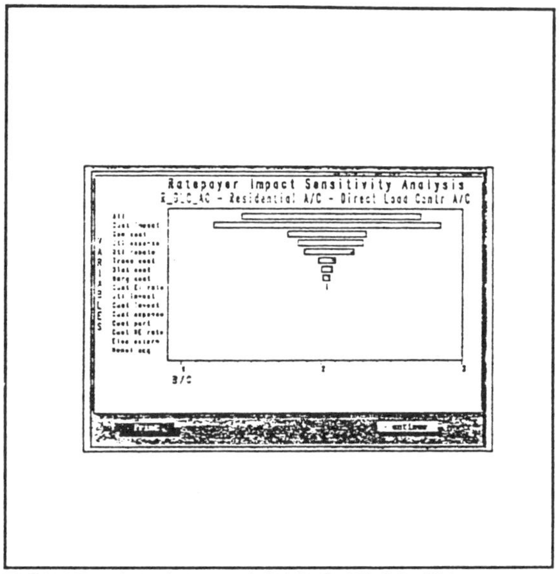

FIGURE 6 SENSITIVITY CYCLONE GRAPH

The sensitivity analysis assists the analyst in deciding which variables contribute most to the uncertainty of each b/c ratio. It assists them in determining areas where further research should be done to gather quality data to design acceptable demand-side management programs.

DSManager utilizes user defined confidence intervals (CI) when conducting sensitivity analysis. These confidence intervals assist DSManager in the quantification of the risks associated with uncertainty. The confidence intervals are defined when the user supplies the upper and lower multipliers for both the 70% and 99% confidence intervals. The 70% CI low and high values should be considered as the range where the user is relatively confident the value associated with that variable will fall. The 99% CI low and high values should be considered as values that define a range where the variable value will almost always fall.

By employing a variation of the Latin Hypercube Sampling Methodology and conducting between 1,000-6,000 statistical iterations for each Cost Test, DSManager calculates a demand-side program b/c probability distribution for each Cost Test. Probability distributions are used as a graphic medium allowing the analyst to summarize the uncertainty associated with the program evaluation.

DSManager also uses the 70% confidence interval to calculate the b/c range for each variable. The size of the b/c range corresponds to the impact that the particular variable has on the demand-side program's b/c ratio (i.e.

those variables with the largest b/c range have the greatest impact). DSManager calculates the b/c range for each variable by calculating low and high values corresponding to the 70% confidence interval while holding the other variables constant.

Step 6: Refine the Data

The results of the sensitivity analysis will help the analyst identify the critical areas where improved data will improve the quality of the analysis. One should consider the possibility of further analysis to improve the estimates for these data. Some data improvement is relatively easy. For example, one could contact other utilities to find out their success in increasing participation in DSM programs. Or, the vendors that provide cool storage equipment could be contacted to get better estimates of the cost and performance of such equipment. Other data are more difficult to obtain, for example data on appliance saturations in a specific utility's service territory. The extent of data improvement efforts will depend on the time available, budgets, the difficulty of obtaining the improved data, and the value of improved data for achieving the goals of the utility's demand-side activity.

Figure 7 summarizes the use of sensitivity analysis and data refinement in the analysis process. This figure summarizes the work that is performed in Steps 3, 4, 5 and 6.

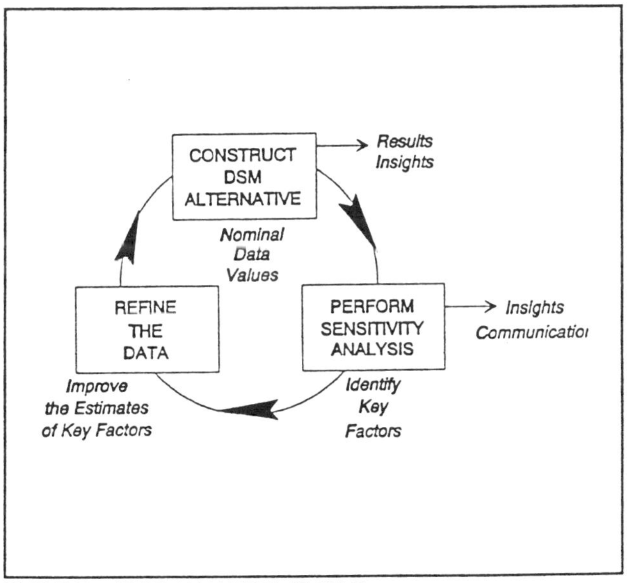

FIGURE 7 THE ANALYSIS CYCLE

Step 7: Final Analysis of DSM Programs

In this step, the evaluations of DSM programs are re-run to produce the final evaluations. These runs will probably form the basis for communicating the analysis team's work to management and the basis for decisions made concerning demand-side management programs.

The process described in these steps will probably be repeated on a regular basis. As time passes, changes will affect in the utility's loads, capacity, and fuel prices as well as the success of demand-side activities. The planner will also learn more about customers and their technical options. Thus, the analysis process not only provides insight into the effects of many factors on the utility system; it also challenges the planner to develop better information.

At this point the planner now has a set of DSM programs that have been designed to match (or consider) utility objectives. These programs must now be compared to other resource options using an integrated model. IRP-Manager is the ideal choice for comparing demand-side options such as efficient lighting or interruptible rates against supply-side options such as combustion turbines or base-load coal units. The following section describes how IRP-Manager facilitates this process.

INTEGRATED RESOURCE PLANNING WITH EPRI IRP-MANAGER

Integrated resource planning (IRP) has evolved to the point that quality analysis is most challenged by the ability of planners to manage information and applications. Planners must concurrently consider large numbers of resource alternatives, many scenarios of the future, and several independent perspectives ranging from ratepayer and shareholder to environmental. No single planning tool or evaluation technique provides all the necessary capabilities. And the variety of separate planning tools, do not adequately interface or share data to meet industry planning needs. In response EPRI has developed EPRI IRP-Manager, an integrated planning model offering unprecedented analytical and data management capabilities.

As illustrated below, IRP-Manager's graphical user interface and database management system envelope the proven chronological simulation engine of the EPRI LMSTM IRP model, and provide direct linkages with other EPRI models, including DSManager, EGEAS, and HELM.

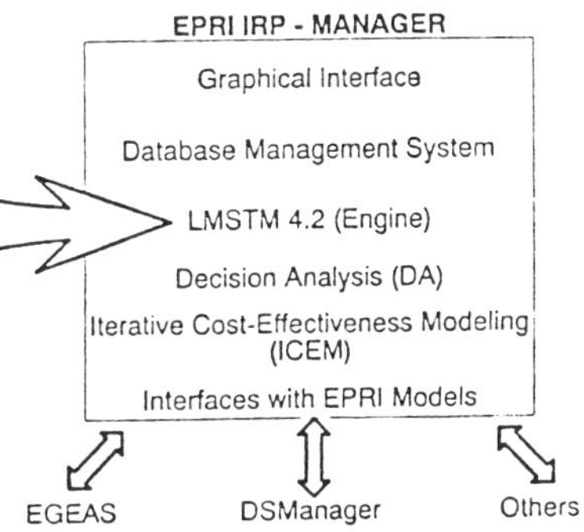

FIGURE 8 IRP-MANAGER

IRP-Manager provides and coordinates an expansive "Tool Box" of integrated resource planning (IRP) capabilities, including:

* Chronological simulation of utility demand, supply, financial and rate strategies;
* Automated resource strategy development with multiple objective functions;
* Decision analysis capabilities, including value of information.

Overview of IRP-Manager's Applications
Planners can develop a range of resource strategies using the ICEM technique incorporated within IRP-Manager. the ICEM module sequentially evaluates and incorporates resource options each year, as chosen by the planners from initial demand and supply screening analysis. The planners can define a single variable or multi-variable decision rule for selection of resource options. Differing resource strategies will result from the application of differing decision criteria or different weighting of similar decision criteria.

Planners can also develop a range of resource strategies by using the IRP-Manager interface with EPRI EGEAS. Planners can then perform a consistent analysis using EGEAS optimization techniques to develop a collection of alternative supply and integrated strategies.

Planners can also perform a detailed evaluation of resource strategies across various scenarios of the future using the IRP-Manager Decision Analysis (DA) Module. this module rapidly constructs a decision analysis scenario "tree" based on user definition of uncertainty variables to be considered, and associated probabilities. IRP-Manager automatically applies data "overlays" to create each end-point scenario of the decision tree, and performs sequential simulation of each scenario.

For each resource plan, IRP-Manager provides a complete forecast of utility system characteristics. This includes chronological marginal energy cost forecasts, electric rates at the system or rate class level, monthly fuel usage, emissions output of utility power plants, capital requirements, and hourly load shape projections including demand-side program impacts.

Key IRP-Manager Features
IRP-Manager combines and enhances the capabilities of several EPRI planning tools, providing accurate, chronological simulation of demand requirements, non-dispatchable and dispatchable DSM programs, and generation resource options. Complete financial reporting is provided for both investor-owned and public-owned utilities, as well as cost-of-service capabilities, regulatory scenario modeling, decision analysis and Iterative Cost-Effectiveness Modeling Module (ICEM).

* Graphical User Interface
IRP-Manager's graphical user interface provides a user-friendly, menu/mouse-driven process of data entry and scenario/project analysis. The GUI menus and screens are highly intuitive and efficient.

The GUI includes instant graphic display of input and output variables. Reports can be designed and graphed on screen; and instantly printed.

IRP-Manager converts historical or forecast EEI format load data to typical day-type format. A typical week can be modeled for each month, if desired.

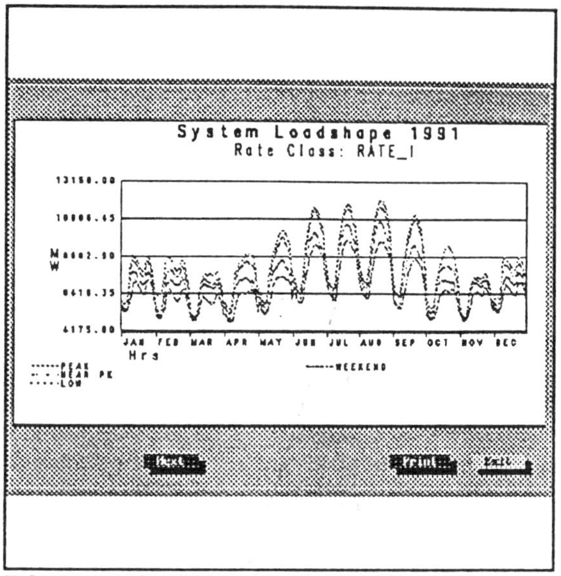

FIGURE 9 HOURLY LOAD SHAPE GRAPHIC

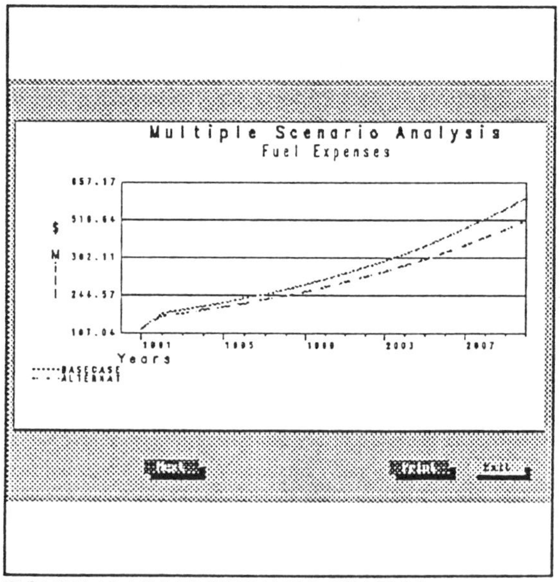

FIGURE 10 MULTI-SCENARIO GRAPHIC

* Database Management System
An efficient and powerful database management system (DBMS) has been developed for IRP-Manager, enabling users to build an extensive database of potential resource options and system characteristics. The framework enables users to concurrently maintain multiple data sets of different vintages as is often necessary for regulatory and internal

planning requirements.

* Scenario Building Evaluation
IRP-Manager includes a Scenario Builder, allowing rapid set-up and organization of data sets for the resource planning scenarios to be evaluated. The Scenario Builder automatically modifies parent "data sets" sequentially and constructs and simulates each scenario defined by the user.

A space-efficient process of applying "overlays" to base scenario data sets dramatically reduces storage requirements in comparison to conventional data management approaches. Scenario-and project-level data archiving further reduces disk storage requirements.

IRP-Manager includes scenario- and project-level validation of completeness and consistency prior to initiation of the simulation process. On-screen documentation included for individual data items, as well as scenarios and projects.

* Decision Analysis (DA) Module
The scenario-building concept has been extended to allow rapid definition and construction of decision analyses. IRP-Manager rapidly constructs a decision analysis "tree", based on user definition of uncertainty variables to be considered, and associated probabilities.

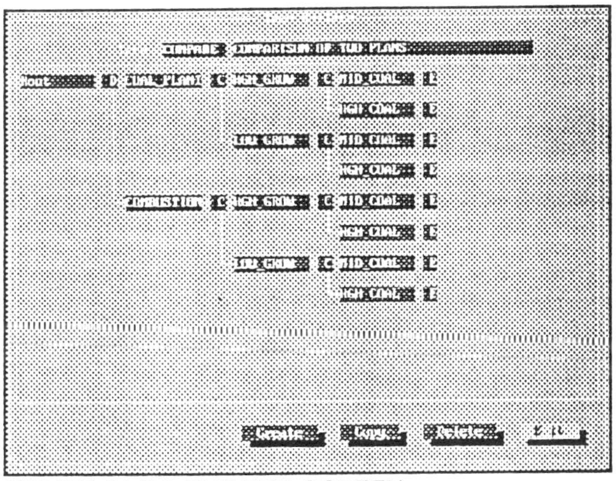

FIGURE 11 DA BUILDER SCREEN

IRP-Manager automatically constructs and applies data overlays to build the end-point scenarios for the decision tree, and verifies consistency and completeness of data. IRP-Manager then sequentially simulates each end-point scenario and stores key results in the database.

End-point scenario results and probability-weighted results are available for any of several key output variables. Both tabular and graphic results are available on-screen.

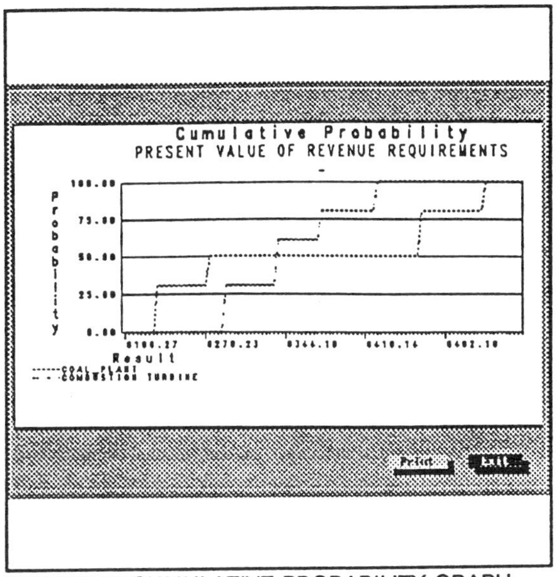

FIGURE 12 CUMULATIVE PROBABILITY GRAPH

* Iterative Cost-Effectiveness Modeling Module
ICEM is a process of evaluating and selecting resource options to obtain a desired combination as defined by a decision rule specified by the user. This process can be applied to develop a range of strategies based on various objective functions such as price growth, benefit/cost tests and emission levels.

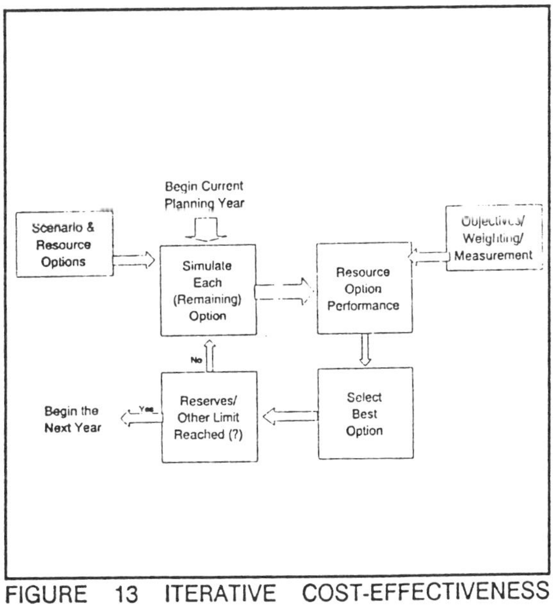

FIGURE 13 ITERATIVE COST-EFFECTIVENESS ANALYSIS

Available time windows are defined for various resource options chosen from earlier demand and supply screening analysis. for each year of the resource evaluation period, each resource block is evaluated separately, and the "best" option is chosen and added to the resource plan, based on the user-specified objective function. The remaining options are each evaluated once again, and additional options iteratively chosen, to the extent that options meet the objective function. This iterative process is repeated for each year, to construct a multi-year resource plan.

ICEM allows a relatively large number of unique demand and supply resource options to be evaluated with acceptable model run times, and with a broad range of objective functions. As such, there is no need for application of oversimplifying "short-cut" methods, or "collapsed" (single or few year) modeling of DSM program implementation in constructing a resource plan at this stage of the planning process.

ICEM as applied within IRP-Manager includes production cost simulation of each resource option for the entire simulation period – up to 45 years. An end-effects adjustment can be applied to estimate economic impact beyond the simulation period. IRP-Manager allows the user to define a single-variable or multi-variable objective function for application within the module. Clear explanation and documentation of the selected options is provided by IRP-Manager.

Coordinated Application of Planning Capabilities
The chart shown below, Figure 14, provides a generalized performance ranking of various planning techniques to each of several IRP modeling considerations. Each planning technique is strong in unique areas of analysis. IRP-Manager provides effective direct application or coordinated application of each of these planning techniques.

Relative Performance of IRP Analytical Technique

Capability to Address	Scenario Analysis	Decision Analysis	Dynamic Programming	ICEM
Possible Futures	M	(H)	L	L
Unique Resource Options	L	L	M	(H)
Combinations of Options	L	L	(H)	M
Multiple Objectives	(H)	(H)	L	M
Uncertainty Thru Time (value of information)	M	M	L	L
Detailed Information	(H)	M	L	L

FIGURE 14 VALUE MATRIX

The IRP-Manager user interface includes a direct database link with EPRI DSManager. DSManager load shape and cost data is not modified or converted in any way for application within IRP-Manager. This unique data management capability provides for much improved coordination of the DSM screening and integrated planning process.

A demand and supply system data export in EPRI EGEAS format is provided by IRP-Manager. The data link between IRP-Manager and EGEAS enhances the productivity of applying both models to achieve truly effective integrated resource planning.

In summary, EPRI DSManager and EPRI IRP-Manager should greatly improve the demand-side and integrated resource planning efforts currently underway at utilities. Both models are available to EPRI members at no charge, and are available to other interested parties upon the payment of a license fee.

Chapter 105

HVAC Retirement: A New Demand-Side Management Program Strategy

J. Stein, C. Robertson, J.S. Wolpert

ABSTRACT

The authors designed a new Demand Side Management (DSM) strategy to achieve comprehensive energy efficiency improvements in commercial buildings. The HVAC Retirement strategy targets buildings with HVAC systems that, on the average, are expected to undergo major maintenance or repair procedures within two years or less. Examples of targeted HVAC systems include rooftop units with compressors 8 years old, or older, or chillers targeted for replacement, compressor rebuilding, or refrigerant conversion.

The HVAC Retirement strategy attempts to influence utility customers to take the following actions: a) Reduce building cooling load by adopting appropriate energy efficiency measures, such as improved lighting systems; and b) Replace existing HVAC systems with new high efficiency systems. Where possible, the new HVAC systems will be accurately sized to reflect the reduced load of the building with retrofit systems. By simultaneously addressing HVAC and other building systems, the HVAC Retirement strategy captures interactive savings that are typically not captured when each system is treated individually.

The paper describes the HVAC Retirement strategy and presents a simplified analysis using building simulation to demonstrate the strategy's cost-effectiveness.

A BRIEF HISTORY OF THE HVAC RETIREMENT STRATEGY

The authors began working together in late 1990, along with their colleague Bruce Jones, as a team focusing on two major interests: (a) how to improve the energy efficiency of buildings by addressing multiple building systems simultaneously; and (b) how utilities could structure their Demand Side Management (DSM) programs to achieve comprehensive, cost-effective energy efficiency resource acquisitions.

Wolpert had already written about a process for designing energy efficient new buildings [1], and Wolpert and Stein, along with others, had monitored the performance of several buildings designed in accordance with that process [2]. Robertson had been involved for many years in the design of energy conservation programs, and was active in the development of comprehensive commercial and industrial DSM programs in New England.

In early 1992, Wolpert and Stein worked on a study of a historic office building renovation that combined retrofitting a number of energy efficiency measures (EEMs) with a new chiller plant [3]. The study concluded that the other EEMs would reduce the building's cooling load, and that as a result, the new chiller plant could be smaller than originally planned. The study further concluded that the capital cost savings achieved by reducing the size of the chiller plant might be enough to fund most of the EEMs that made the reduction possible. As a result of these findings, and others [4], Robertson set about to develop a DSM program that combined lighting and HVAC system improvements in existing buildings.

In the fall of 1992, the authors presented a workshop on HVAC Retirement Program design to a group of utility DSM analysts [5]. In 1993 the authors, with their colleague Bruce Jones, developed a detailed HVAC Retirement Program strategy. They are presently working with several utility collaboratives to develop the implementation details necessary to implement the HVAC Retirement Program.

SECOND-ORDER LOST OPPORTUNITIES

During the summer of 1992, Robertson and Stein developed a set of 16 principles as a guide to evaluate the effectiveness of DSM programs [6]. Two of these principles were fundamental in developing a program design to implement the HVAC Retirement Strategy:

Comprehensiveness of energy conservation measures: The effectiveness of DSM programs can be enhanced by capturing a high percentage of the cost-effective savings available in each building treated by the program. Comprehensive DSM programs achieve this goal, in part, by treating multiple building systems simultaneously.

Resource acquisition priorities: When there is a "window of opportunity" to acquire a particular DSM resource, such that the resource is unlikely to be cost-effective again if it is not acquired during this "window," it is often referred to as a *lost opportunity* resource. *Retrofit* resources are available in existing buildings, and may be acquired at any time with little concern about being lost. Frequently, the first priority in DSM program design is to capture a high percentage of lost opportunity resources. When such is the case, the second priority is to properly combine lost opportunity and retrofit investments.

Reference [6] used the term *second-order lost opportunities* to describe the combination of lost opportunity HVAC improvements with retrofit-like improvements to other building systems. When cost-effective building system retrofits that reduce cooling load (such as high-efficiency lighting upgrades) are performed simultaneously with planned HVAC system improvements, all of the associated equipment in the upgraded HVAC system may be re-engineered to the reduced cooling load. The net effect is to capture the capital cost savings, as well as peak and energy savings that result from installing the downsized high-efficiency HVAC equipment. These interactive savings are substantially greater than if each piece of equipment was changed out individually upon failure.

HVAC RETIREMENT GOALS

The HVAC Retirement strategy was developed in accordance with the two DSM principles above to capture second-order lost opportunities. As such, the authors sought to achieve the following goals:

a. To achieve comprehensive energy efficiency improvements in commercial and industrial air conditioned buildings by replacing aging inefficient equipment.

b. To capture the savings potential that results from interactions between HVAC and other building systems by acquiring cost-effective cooling load reducing measures.

THE HVAC RETIREMENT STRATEGY

The HVAC Retirement treatment begins when a piece of HVAC equipment is identified which on the average is expected to undergo a major maintenance or repair procedure in 2 years or less. Examples of targeted HVAC systems include rooftop units with compressors 8 years old, or older, or chillers targeted for compressor rebuilding or refrigerant conversion. Such maintenance or repairs usually represent a significant cost to the owner, but result in little or no efficiency improvement.

When a candidate building is identified, a comprehensive retrofit is performed on non-HVAC building systems. Usually, the lighting systems are addressed, but other cooling load-reducing measures may be included where appropriate. The HVAC system equipment is then replaced with new high efficiency equipment. Where possible, the new HVAC equipment is re-sized to reflect the reduced load of the building with retrofit systems.

Funding for the HVAC Retirement transaction is structured as follows: The utility pays incentives for improvements to the HVAC system and for other energy efficiency measures. The minimum participant share of the HVAC improvements is the cost the participant would have faced anyway (e.g. emergency compressor change out, chiller compressor rebuild, chiller replacement, etc.). In addition, participants may be required to make a contribution towards the installation of the energy efficiency measures. The utility also funds design assistance, marginal design costs, performance based design incentives, and commissioning.

ESTIMATED SAVINGS USING BUILDING SIMULATION

The authors have developed estimates of the savings that could be achieved by the HVAC Retirement strategy for several utilities in the United States. This paper contains a single analysis conducted for a metropolitan utility on the mid-Atlantic seaboard. The analysis was simplified such that it utilized two micro-AXCESS 10.1 [7] simulation models: one to represent the small/medium buildings sector and the other to represent the large buildings sector.

In addition, the energy efficiency improvements were limited to simple efficiency upgrades which could readily be achieved. Greater cost-effective savings could be acquired by implementing more efficient, but also more technically challenging strategies.

Small/Medium Office Building Model
The small/medium simulation model was designed to represent a 10,400 square foot office building, approximately 10-15 years old, and cooled by seven 5-ton rooftop units. Table 1 contains the basic characteristics of the simulation model.

The Baseline case represented the existing building. The energy efficiencies and power densities of most of the building systems were based on the authors' experience, as the utility had little end-use data available. For example, it was assumed that the existing lighting system largely consisted of four lamp fluorescent fixtures with core-coil ballasts, supplemented by incandescent fixtures. It was also assumed that the existing packaged rooftop units could be characterized by an EER of 6.

The Enhanced case represented the building after a combined high-efficiency lighting retrofit and a high-efficiency packaged roof-top unit retrofit. The lighting retrofit consisted of installing T-8 lamps and electronic ballasts in the four lamp fluorescent fixtures. Additionally, incandescent fixtures were retrofit with 18W modular compact fluorescent units containing a reflector. It was also assumed that the existing packaged rooftop units were replaced with new high efficiency, 10.5 EER units. The Enhanced column is left blank in Table 1 where characteristics were not changed by the HVAC Retirement treatment.

TABLE 1: CHARACTERISTICS OF SMALL OFFICE BUILDING		
Characteristic	Baseline	Enhanced
Gross Floor Area (SqFt)	10,400	
U-Values (Glass/Walls)	1.05/.13	
Shade Coefficient	.75	
Clng SP / Htng SP	73/70	
# People / Density	35/297	
Installed Ltg (W/SqFt)	2.75	1.25
Receptacles (W/SqFt)	1.0	
Total CFM	11544	
Economizer	None	Enthalpy
Cooling Plant (tons)	35	
Cooling EER	6.0	10.5
Heating Fuel	Nat. Gas	

Large Office Building Model
The large building simulation model was designed to represent a 99,225 square foot office building, approximately 10-15 years old, and cooled with a 390-ton central chiller plant. Table 2 contains the basic characteristics of the simulation model.

As with the Small/Medium building, the Baseline case represented the existing building with efficiencies and power densities largely based on the authors' experience of similar buildings in other utility service territories. Again, the existing lighting system largely consisted of four lamp fluorescent fixtures with core-coil ballasts, supplemented by incandescent fixtures. The central chiller was selected to represent an average unit that had degraded in performance somewhat since its installation 10-15 years ago.

Again, the Enhanced case represented the building after a combined high-efficiency lighting and HVAC retrofit. These retrofits included:

Lighting: Four lamp fluorescent fixtures with core-coil ballasts were retrofit with 2 T-8 lamps, electronic ballasts, and reflectors. Incandescent fixtures with 100 W. bulbs were retrofit with 18 W modular compact fluorescent units containing a reflector.

TABLE 2: CHARACTERISTICS OF LARGE OFFICE BUILDING		
Characteristic	Baseline	Enhanced
Gross Floor Area (SqFt)	99,225	
U-Values (Glass/Walls)	.72/.13	
Shade Coefficient	.65	
Clng SP / Htng SP	73/70	
Clng Setup / Htng Stbck	80/63	
# People / Density	495/200	
Installed Ltg (W/SqFt)	2.75	.85
Receptacles (W/SqFt)	1.0	
Cooling Plant (tons)	390	239
Cooling Eff. (kW/ton)	.9	.6
Total CFM	103,132	
Fan kW (Supply/Return)	86.1/29	72.5/24.4
Cooling Aux (Con/Var)	98/54	12/88
Economizer	None	Enthalpy
Heating Fuel	Electric	

HVAC: The 390-ton, .9 kW/ton, existing chiller was replaced with a 239-ton, .6 kW/ton chiller. The auxiliary pumps and fans were re-engineered for the reduced load, which included retrofitting variable speed drives and high efficiency motors.

The Enhanced column is left blank where characteristics were not changed by the HVAC Retirement treatment.

Simulation Results
Tables 3 and 4 contain the results of the building simulation exercise. In the small office building, represented by Table 3, the HVAC Retirement treatment reduced annual electric energy consumption by 42%, and annual peak demand by 43%. Natural gas consumption did increase however, as the building's heating system had to compensate for the reduced heating contribution of the retrofit high efficiency lighting system.

TABLE 3: SMALL OFFICE BUILDING SIMULATION RESULTS

Scenario	Annual Electric Energy (kWh/yr)	Annual Peak Demand (kW)	Annual Natural Gas Energy (therms)
Baseline	269,000	100	2,200
Enhanced	157,000	60	3,500
Savings	112,000	40	-1,300

In the large office building, represented by Table 4, the HVAC Retirement treatment reduced annual electric energy consumption by 57%, and annual peak demand by 50%.

TABLE 4: LARGE OFFICE BUILDING SIMULATION RESULTS

Scenario	Annual Electric Energy (kWh/yr)	Annual Peak Demand (kW)
Baseline	2,994,000	820
Enhanced	1,280,000	410
Savings	1,714,000	410

COST OF HVAC RETIREMENT TREATMENT

In order to analyze the cost-effectiveness of the HVAC Retirement strategy, it was necessary to estimate the cost of the energy efficiency measures. These costs were estimated as follows.

Small Office
Lighting: This measure was designed to test the cost-effectiveness of a lighting retrofit. It was assumed that the retrofit improved the lighting power density from 2.75 W/sq.ft. to 1.25 W/sq. ft., by replacing the existing T-12 lamps and core coil ballasts, with T-8 lamps and electronic ballasts. It was also assumed that any incandescent fixtures would be replaced with compact fluorescent fixtures. Through conversations with leading experts in the field of lighting retrofits, it was determined that the lighting retrofit described could be achieved at a price of 75¢ per square foot. Since it was assumed that the small office was 10,400 sq. ft., it was estimated that this retrofit could be performed for a total cost $7,800.

The retrofit also made it possible to save on the cost of maintaining the lighting system. The greatest savings were made possible by the compact fluorescent downlights, which require replacement far less frequently than did the original incandescent downlights. The cost of the lighting retrofit in the small office including avoided maintenance is shown in Table 5.

TABLE 5: COST OF SMALL OFFICE LIGHTING RETROFIT

Item	Cost
High Efficiency Lighting Retro	$7,800
Avoided O&M	($2,800)
Total	$5,000

HVAC: In the small office, the existing packaged rooftop units, which were assumed to have an EER of 6, were replaced with new high efficiency units, which were assumed to have an EER of 10.5. The cost of purchasing and installing the new units was determined by phone interviews with HVAC contractors. The maintenance savings attributed to installing a new unit were also included in the cost calculations. Such maintenance savings were largely due to avoiding a $1500 compressor replacement within two years of the retrofit. These costs and savings are shown in Table 6 below.

TABLE 6: COST OF SMALL OFFICE HVAC RETROFIT

Item	**Cost**
Retrofit standard 5 ton unit	$4,100
Add economizer	$650
Incremental cost of high EER	$375
Avoided O&M	($1,750)
Total for 1 unit	$3,375
Total for 7 units	$23,625

Large Office
Lighting: This measure tested the impact of improving the lighting power density from 2.75 W/sq.ft. to .85 W/sq. ft. Such a reduction was achieved by retrofitting each

fixture containing 4 T-12 lamps and 2 core-coil ballasts, with 2 T-8 lamps, an electronic ballast, and a reflector. In addition, the incandescent downlights were replaced with compact fluorescent downlights. As with the small office, a credit was given for avoided maintenance costs, largely due to reducing the frequency of downlight replacements and the reduced number of fluorescent lamps requiring replacement. The cost of this retrofit was determined by conversations with lighting experts. A 20% adder was included for the lighting design and construction management that would likely be required to achieve the aggressive savings assumed by this measure. These costs are shown in Table 7.

TABLE 7: COST OF LARGE OFFICE LIGHTING RETROFIT	
Item	Cost
High Efficiency Lighting Retro	$110,000
Design & Constr. Management	$22,000
Avoided Lighting O&M	($55,000)
Total	$77,000

HVAC: This measure included performing a comprehensive HVAC retrofit, combined with taking advantage of downsizing opportunities made possible by the reduced cooling load due to the lighting retrofit. It was assumed that the original chiller was 390 tons and had an efficiency of .9 kW per ton. It was further assumed that the retrofit chiller was 239 tons and had an efficiency of .6 kW/ton. In addition, it was assumed that the HVAC auxiliary pumps and fans would be re-engineered and retrofit with high efficiency motors and variable speed drives to match the reduced load. Furthermore, it was assumed that a water side economizer would be added to the system, and that the system would be recommissioned.

TABLE 8: COST OF COMPREHENSIVE HVAC RETROFIT	
Item	Cost
239 Ton Hi-Efficiency Chiller	$75,600
Installation	$17,500
VSDs	$70,000
Hi-Efficiency Motors	$11,200
Water Side Economizer	$37,800
Engineering Fees	$24,000
Recommissioning	$35,700
Avoided Chiller Rebuilding	($25,600)
Avoided Chiller Maintenance	($13,300)
Total	$232,900

The savings achieved by avoiding a rebuilding to convert the chiller to a non-CFC refrigerant were deducted from the cost, as well as the maintenance savings achieved by the installation of a new chiller. All the costs and savings of this retrofit are summarized in Table 8.

BENEFIT COST ANALYSIS

Table 9 below shows the results of the Benefit-Cost analysis. The analysis method applied was a limited version of the Total Resource Cost (TRC) test, applied at the measure level only. The TRC test is widely used to test the cost-effectiveness of utility DSM programs, because it takes into account the overall costs and benefits experienced by all the utility stakeholders, and as such, approximates the impact of the programs on society as a whole. The TRC test is also often referred to as the Societal test, and may also take into account environmental impacts, although such impacts were not included in this study.

The TRC test compares the total avoided supply costs (societal benefits) to the total costs (both utility and customer) of installing the measures [8]. For the purposes of this study, the avoided supply costs included energy, generation, transmission, and distribution. These avoided supply costs were compared to the measure costs calculated in the previous section. The study made no assumption regarding what portion of the measure costs would be borne by the participant and what portion would be borne by the utility. Although not included for the purposes of this analysis, a complete Total Resource Cost test would also include utility program administration costs.

In Table 9, the Net Retrofit Cost represents the cost of the HVAC Retirement treatment, and was calculated by adding the cost of the lighting retrofit to the cost of the HVAC retrofit. The Increased Natural Gas Cost was only applicable to the small building, and represented the cost of increased natural gas usage due to the lighting retrofit. Total Measure Cost was calculated by adding the Net Retrofit Cost to the Increased Natural Gas Cost. The Measure Benefit was calculated using the utility's avoided cost screening tool.

TABLE 9: BENEFIT-COST ANALYSIS		
Parameter	Small	Large
Net Retrofit Cost	$28,625	$309,900
Inc. Nat. Gas Cost	$7,400	
Total Measure Cost	$36,025	$310,000
Measure Benefit	$123,000	$1,510,000
B/C Ratio	3.4	4.9
Net Benefits	$87,000	$1,200,000

The Benefit/Cost ratio was simply calculated by dividing the Total Measure Cost into the Measure Benefit. As indicated by Table 9, the HVAC Retirement strategy is robustly cost-effective. Both buildings exhibited Benefit/Cost ratios greater than 3.0. Net Benefits equal the Measure Benefit minus the Total Measure Cost. In both the Small and Large buildings, the Net Benefits are more than adequate to support any program administrative costs.

ACKNOWLEDGMENTS
The authors would like to acknowledge Bruce Jones as a member of the team that developed the HVAC Retirement Strategy. Future papers are expected to include more details regarding Mr. Jones' contributions.

REFERENCES

1. Wolpert, J.S. Robbins, C.L. An Energy Responsive Building Design Guideline. Seventh Annual Symposium on Improving Building systems in Hot and Humid Climates. Texas A&M University. October 9-10, 1990.

2. Rozek, L. Stein, J. Wolpert, J.S. CCM3 Monitoring Study. Compaq Computer Corporation. Houston, TX. January 11, 1991.

3. Sieben Energy Associates, Ltd. E-Cube, Inc. Marquette Building Energy-Efficiency Study. John D. and Catherine T. MacArthur Foundation. March 25, 1992. Unpublished.

4. Houghton, D.J. et al. The State of the Art: Space Cooling and Air Handling. E-Source. Boulder, CO. August, 1992.

5. 1992 E-Source Forum. Snowmass, CO. September 29 - October 2, 1992. Proceedings not published.

6. Robertson, C. Stein, J. Comments of Public Interest Intervenors on the 1993-1994 Annual and Long Range Demand Side Management and Integrated Resource Plans of New York Electric Utilities Commercial Program Design Principles. State of New York Public Service Commission. Case No. 28223. Albany, NY. September 14, 1992. pp. II-1 - II-10.

7. micro-AXCESS 10.1 Building Energy Analysis Program. Electric Power Research Institute. Palo Alto, CA.

8. Hirst, E. Sabo, C. Electric-Utility DSM Programs: Terminology and Reporting Formats. Oak Ridge National Laboratory, Oak Ridge, TN. October, 1991.

Chapter 106

Designing Competitive Procurement Systems for Demand-Side Resources

J.C. Dalton, J.E. Higgins

Abstract

Competitive bidding for the procurement of demand-side resources can be an effective means for utilities to obtain additional demand-side resources. While competitive bidding to procure supply-side resources has obtained a significant amount of experience over the last fifteen years, it is important for utilities to realize the dissimilarities between supply and demand-side resources when designing a demand-side resource request for proposals (RFP).

This paper is intended to assist electric utilities in developing a successful demand-side RFP. The paper highlights the importance of identifying and defining the utility's objectives, selecting appropriate evaluation criteria, and developing an evaluation process that will enable the utility to identify and select the best proposals.

Introduction

In the last seven years over thirty electric utilities have issued requests for proposals (RFPs) to procure hundreds of MW of demand-side resources. Electric utilities have chosen to issue RFPs to procure demand-side resources largely in response to regulatory mandates. While supply-side RFPs have provided ample evidence of the significant benefits offered by competitive procurement systems, the experience with demand-side RFPs has been mixed. To a large degree, the differences in utilities' experiences with demand-side and supply-side RFPs reflect the difficulty of developing a demand-side RFP that conforms with the utility's objectives and does not adversely affect its existing investment in demand-side management (DSM) programs. Nonetheless, the experiences of some utilities illustrate that demand-side RFPs can effectively procure demand-side resources and ensure that DSM costs are minimized. By issuing an RFP and properly reviewing all proposals and comparing these to available resources, an electric utility can minimize the risks that it will be denied cost recovery, since the resource essentially has been subjected to a market test and been found to be the resource that best meets the utility's objectives. Furthermore, competitive procurement systems result in a more efficient allocation of development and performance risks by shifting these risks from the utility and ratepayers to energy service companies (Ascots) that are better able to manage these risks. However, an electric utility that chooses to rely on a competitive procurement process to obtain resources exposes itself to new risks since bidders that are not awarded contracts under the RFP can file claims with the public utility commission that their proposals were treated unfairly. Several demand-side bidders have done just this.[1] Therefore, a properly designed RFP and a properly executed evaluation process are crucial for satisfying electric utilities' objectives and minimizing the risks of RFP-related litigation.

Presented below is a discussion of the three major steps in the development of an RFP for demand-side resources: (1) define objectives; (2) determine the evaluation criteria; and (3) establish the evaluation process.

Step 1: Defining Objectives

As with any resource planning effort, the first task in the development of an RFP should be to identify and define the utility's planning objectives. While these objectives can be relatively vague for a supply-side RFP and are more likely to be a restatement of the utility's overall system planning objective (e.g., to minimize power supply costs subject to a specific reliability constraint), the utility's objectives for demand-side RFPs should be expressed in more concrete terms (e.g., to reduce summer peak loads with DSM programs that complement the company's existing DSM programs). Greater specificity is needed for

[1] See, for example, Petition of SESCO, Inc., the Free Lighting Corporation, and BEI Energy Corporation for a Declaratory Ruling, New York Public Service Commission, Case 92-E-0677, p.7

demand-side resources given the more specific nature of DSM resources relative to supply-side resources and the greater variations in the quality of the resource.[2]

Four major objectives should be identified and defined prior to the development of a demand-side bidding program: (1) the utility's load shape objectives; (2) the relationship between the programs bid and the utility's existing DSM programs; (3) the balance between the comprehensiveness of measures and their cost-effectiveness; and (4) the degree to which customers will be encouraged to participate and which customer classes will be targeted.[3]

Identifying and defining the objectives of a utility's demand-side bidding program can be an involved process, unless the utility's management or regulators have clearly articulated the major objectives of the utility's DSM programs. These objectives are likely to be shaped by regulatory considerations, the other DSM programs offered by the utility, the amount of experience the utility has in implementing DSM measures, and the utility's resource requirements.[4] While a utility's load shape objectives are likely to be generally known by the utility's entire demand-side planning staff and thus can be easily identified and defined, greater consideration will typically have to be given to the relationship between the company's existing demand-side programs and the proposals received in response to the RFP. Specifically, the utility must decide whether bidders will be instructed to develop proposals that: (1) complement the company's existing programs and target sectors not fully served by the company's programs; or (2) compete directly with existing company programs. In states where public utility commissions have formally mandated competitive bidding for demand-side resources, guidance is often provided on this point. In most of these jurisdictions, regulators have expressed a preference for direct competition between company programs and ESCo proposals to ensure that the DSM measures offered to customers have the most attractive mix of price and non-price attributes.[5]

[2]Goldman, Charles A. and David R. Wolcott, "Demand-Side Bidding: Assessing Current Experience," *Proceedings of the 1990 ACEEE Summer Study on Energy Efficiency in Buildings.* Vol. 6, p. 11-12.

[3]Finleon, Jane L., Kelly A. Triplett, and Bruce C. Peterson, "Incorporating Lessons Learned into Demand Side Bidding", p. 8.53.

[4]*Ibid.*

[5]Given the diverse mix of program attributes, paticular attention has to be paid to the mix of price and non-price attributes of a propopsal since the lowest cost proposal is not necessarily the most attractive proposal.

The distinction between a complementary and a competitive RFP can be subtle. For example, a recent RFP for demand-side resources that Reed Consulting Group developed with a client had many of the elements of a complementary RFP. NPC preferred proposals that served end-uses that were not fully addressed by its existing programs, yet the only requirement for proposals that served the same end-uses and sectors as its programs was that they offer a different delivery mechanism than NPC's programs (e.g., financing for the installation of measures). The "PowerSavings Partners" RFP issued by Pacific Gas & Electric (PG&E) had many of the elements of a complementary RFP, but also defined competitive programs very narrowly. Bidders were allowed to target the same end-uses and customers as PG&E's programs as long as the bidder's proposal would not make PG&E's programs cost-ineffective. Clearly, many options exist for defining the relationship between a company's programs and bidders' proposals.

Another objective that must be defined as part of the RFP development process is the degree to which the RFP will favor proposals that offer a comprehensive mix of measures versus proposals that focus only on the most cost-effective measures. This issue must be addressed whenever DSM programs are designed. Two considerations for developing the appropriate policy are: (1) whether the electric utility's public utility commission has directed the company to focus primarily on one of these goals; and (2) the company's degree of concern with the rate impacts from DSM programs.

Almost all demand-side RFPs have allowed customers to participate as bidders in the RFP. However, except for the larger energy consumers, which are the most sophisticated customers and typically have energy management specialists on staff, it can be difficult for a customer to participate in a demand-side RFP.[6] Another consideration in the decision whether customers should be encouraged to participate is the potential impact on customer relations if the customer's proposal is not selected.

Whether bidders should be allowed to serve all customer classes is also an issue. Some have argued that

[6]In an effort to encourage customer participation, several utilities have developed different bidding requirements for customers. This is appropriate since a customer bid is obviously able to eliminate marketing risks that are associated with ESCo proposals in which the hosts have not been identified.

residential and small commercial classes are not well-suited for competition since there is not the same degree of competition among service providers that serve these sectors.

Step 2: Determining the Evaluation Criteria

Once the company's objectives have been identified, the next step in the development of the RFP is to determine how these criteria will be reflected in the RFP. In essence, a company can express its objectives in the RFP by using a carrot or a stick. Specifically, the company must decide whether a project attribute is: (1) essential and should be reflected as an eligibility requirement or a minimum threshold that all proposals must satisfy to be considered eligible; or (2) desirable and should be reflected as an evaluation criterion.

Eligibility requirements are standards that all proposals must satisfy if they are to be formally reviewed. Typical eligibility requirements are definition of eligible measures, minimum bid sizes, in-service dates, contract term and eligible customer classes. Minimum threshold requirements, like eligibility requirements, are requirements that all projects must meet, but typically are also part of the evaluation criteria since the proposal is also evaluated to the degree that it exceeds the minimum threshold.

Table 1 provides a list of different evaluation criteria that can be used for demand-side RFPs as well as the rationale for considering these criteria in demand-side RFPs. These evaluation criteria are designed specifically for demand-side RFPs. In general, there is widespread recognition that separate evaluation criteria should be used for demand-side RFPs and supply-side RFPs.[7,8] However, at least one state public utility commission, the Massachusetts Department of Public Utilities, has required that many of the same evaluation criteria be used for both supply and demand-side RFPs and that the weights applied to the criteria categories be the same. Presumably, this was done to ensure that electric utilities treat supply and demand-side resources consistently. However, it is inappropriate to have the same weights for demand-side resources as for supply-side resources, since the two resources have different characteristics that determine their relative risks and attractiveness to electric utilities.[9]

In one demand-side RFP that utilized essentially the same non-price criteria as the supply-side RFP, almost all proposals received the same total non-price score.[10] Thus, to a large extent, proposals were selected only on the basis of price, and no consideration was given to the non-price considerations that are applicable to demand-side proposals. Clearly, such an evaluation system is flawed.

Once these eligibility requirements, minimum thresholds, and evaluation criteria have been identified, it is essential that they be expressed clearly in the RFP so that bidders will be able to craft proposals that satisfy the utility's underlying objectives to the greatest extent possible. More than one RFP has been issued without the utility clearly articulating the evaluation criteria that will be used to select proposals. While being overly prescriptive can stifle innovation, at least one utility, Niagara Mohawk Power Company, has openly expressed its regrets about having issued an RFP that it characterized as too open.[11] One obvious consequence of not clearly articulating or fully identifying the evaluation criteria is that when it comes time to evaluate the proposals it is possible that the RFP will not have requested all the information necessary to perform the evaluation. This can result in the electric utility requiring bidders to respond to a second set of questions, which increases both the time required for the evaluation process and the bidder's degree of frustration.

Step 3: Establishing the Evaluation Process

Once the eligibility requirements and evaluation criteria have been developed, the next step in the development of a demand-side RFP is to establish the process that will be used to evaluate proposals. The most critical issues in establishing the evaluation process are: (1) the degree to which the evaluation process will be self-scoring; (2) the degree to which elements of bidders' proposals will be negotiated; and (3) the price/cost-effectiveness evaluation.

The Degree of Self-Scoring

The advantage of self-scoring, which allows the bidder to determine independently its score for each evaluation criterion, is that it makes bidding "transparent, explicit, and credible to potential bidders."[12] When bidding programs were first established, most regulators believed

[7] Hamilton, William E. and Theresa A. Flaim, "Acquiring DSM Resources Through All-Source Bidding: Lessons Learned," *Proceedings of Conference on DSM Bidding: Status and Results, April, 1992*, pp. 3, 5.

[8] Goldman and Wolcott, *op. cit.*, p. 11-12

[9] Hamilton and Flaim, *op. cit.*, p. 7.

[10] Peters, Jane S., Deborah J. Barry, Marvin J. Horowitz and Fred Gordon, "Evaluation of Orange and Rockland Utilities Competitive Bidding Program for Demand Side Resources," p. 23.

[11] Hamilton and Flaim, *op. cit.*, p. 6.

[12] Peters *et al.*, *op. cit.*, p. 12.

that self-scoring was necessary to ensure that different resources would be evaluated fairly and thus self-scoring systems were generally favored. However, there is currently widespread recognition that full self-scoring severely limits the ability of the utility to exercise its judgment after the evaluation criteria and evaluation process have been established. Unless the utility reserves the right to deviate from the results of the self-scoring system and to select projects that do not have the highest scores, once the self-scoring system has been established and proposals have been received, the utility is bound by the results of its scoring system. This is a major disadvantage of a self-scoring system. If a full self-scoring system is to produce the best results consistently then the utility developing the RFP must be omniscient when developing the scoring system and determining the relative weights of all major evaluation criteria since the relative importance of the evaluation criteria depends on the types of bids received. For example, a utility might receive proposals for 100 MW of commercial lighting that are scored higher than one proposal for 10 MW of thermal energy storage, but given the mix of resources offered, the relative value of the lighting measures drops, and the value of the thermal energy storage project increases.

While the risks of a full self-scoring system can be reduced if the utility reserves the right to reoptimize the project rankings to account for the interactive effects between resources and changed circumstances, the risks of a full self-scoring system cannot be fully mitigated by being able to reoptimize the award group. For example, if a bidder knows that its proposal received a higher score than another proposal (in many instances it is easy to estimate another project's score and some utilities are required by their state public utilities commissions to indicate all bidders' scores) it will be more likely to file a complaint with the state public utility commission calling into question the utility's judgment.[13] This has happened more than once in supply-side RFPs.

Another reason for not using a full self-scoring system is that some evaluation criteria are by nature subjective and thus require the application of judgment. A bidder is unable to determine its score with respect to the criteria because the utility cannot reasonably expect the bidder's judgment of its performance with respect to the criteria to be the same as the utility's. An example of such an evaluation criterion would be the evaluation of a bidder's marketing plan. In general, the evaluation criteria that are used for demand-side RFPs are less suited to full self-scoring systems than are supply-side RFPs.

While it is not in the utility's or ratepayer's best interests to use a full self-scoring system, to the greatest degree possible, the evaluation criteria should be clear, precise, and transparent. Furthermore, when possible, the thresholds that will be used to evaluate a bidder's performance with respect to a specific evaluation criterion should be identified. For example, if the utility has determined the optimal size for a bid, then it should indicate the optimal size for the bid in its evaluation criteria. The only difference between this approach and full self-scoring is that the relative scores associated with the different thresholds are not identified. The one instance in which full-self scoring is likely to be appropriate is when the utility or its subsidiary is participating in the RFP.

Negotiation

Another important decision that must be made in the development of the evaluation process is whether there will be any negotiating of the provisions of the bidder's proposal. While some procurement specialists believe that allowing for negotiation of the elements of the proposal reduces the incentive for the bidder to submit its best proposal, we believe that negotiation between the purchasing utility and the bidder is likely to lead to better proposals and greater ratepayer benefits. This is true for several reasons. First, it is essentially impossible for any competitive procurement system to fully express the purchasing utility's preferences for projects. And as discussed above, it is likely that these preferences will change based on the types of bids received. Thus, a negotiation process provides the purchasing utility with the ability to request revisions to proposals that will better serve the utility and its ratepayers. Second, negotiation provides the bidder and the utility with an opportunity to search for ways in which the both parties can benefit. For example, in one instance, a utility was able to secure through negotiation a price reduction of over ten percent by reducing the rigorousness of the bidder's measurement and verification methods. Clearly, this is a case in which the utility felt that the cost savings to the ratepayer were of greater value than the higher reliability of the savings estimates.

If the evaluation process provides for the negotiation of elements of the proposal, then the RFP should contain a provision that provides that changes will only be allowed, which from the utility's perspective on a net basis result improves the bidder's proposal. This will ensure that bidders do not use the negotiation process to backslide if

[13] See, for example, the complaint filed by DLS Cogen before the Massachusetts Department of Public Utilities, Docket No. 92-153.

they think that they have "over bid" their proposals.

Price/Cost-Effectiveness Evaluation

In almost all RFPs the evaluation criterion that has the largest single weight is price. Typically, price is evaluated on the basis of the following cost-effectiveness tests: (1) total resource cost test; (2) ratepayer impact test; (3) utility cost test; and (4) societal benefits test. Several utilities have evaluated price/cost-effectiveness solely on the basis of the price offered by the bidder and not released avoided costs out of concern that bidders will price their proposals to just undercut avoided costs and thus satisfy the minimum threshold (e.g., a total resource benefit-cost ratio of greater than 1.0).[14] While some utilities have based their evaluation solely on the total cost of the proposal (i.e., the utility payment and the customer contribution), many utilities have also considered in their evaluations the relative contribution of these two elements by taking into account the results of the utility cost test or ratepayer impact test. These two cost-effectiveness tests also allow the utility to assess the impact of the proposal on the utility's rates and the relative cost to the utility and ensure that the bidder does not shift a disproportionate share of the cost of the program to the company. A significant level of host contribution is necessary to ensure that the host has an investment in the program.

There are several other important considerations in the treatment of price for a demand-side RFP: (1) payments should be performance-based; (2) the pricing should be consistent with the utility's load shape objectives; (3) the payment method should be consistent with the measurement and verification methods; (4) the price evaluation methodology should not rely solely on traditional avoided costs, but on the costs of competing resources; (5) the methodology that will be used to calculate the cost-effectiveness of the proposal should be spelled out; (6) the bidder should be required to indicate the formula that will be used to calculate the price to be paid to the bidder each year; and (7) the pricing and payment provisions in the RFP should be developed in view of the degree to which the RFP is designed to encourage bidders to install a comprehensive mix of measures. In some instances these considerations compete, thus it will not be possible to develop a price evaluation system that will completely satisfy all of these considerations.

[14] Boston Edison Company used such a price evaluation system in its draft demand-side RFP filed before the Massachusetts Department of Public Utilties. Under the Boston Edison price evaluation, the bidder's proposed price was adjusted by a factor that considered the relative value of demand reductions at different time periods.

One of the major advantages offered by a competitive bidding system for demand-side services is that it allows the utility and ratepayers to shift many of the performance-related risks to a party that is better able to manage these risks (i.e., ESCos that submit the majority of the bids). If these performance risks are to be truly shifted to the bidder then the utility's payments to the bidder must be performance-based to the degree allowed by the measurement and verification system. However, in practice, it is likely to be difficult to achieve a perfect fit between the pricing and payment mechanism and the measurement and verification plan. For example, a utility may be primarily interested in reducing its system peak loads. This suggests that payments should be made based on the peak load reduction achieved and pricing should be expressed in terms of $/kW. However, given the difficulty and expense of verifying the actual peak load reduction achieved by a measure, for many measures it is likely that the peak load reduction will be measured in terms of the kWh reduced during the peak period. Thus, in this example, it is difficult to tie both the pricing and the measurement and verification methods to the utility's load shape objective of reducing peak load.

The price evaluation methodology should not be based exclusively on traditionally defined avoided costs (i.e., the cost of the utility's next resource and system avoided energy costs), but on the cost of competing resources. For a DSM bidding program this can mean the cost of the utility implementing the program. This is important given the lower costs of DSM resources relative to supply-side resources on many utility systems. For example, if the total cost of a bidder or the utility implementing a DSM measure is 3¢/kWh and the cost of a supply-side resource is 5¢/kWh, then if avoided costs are based on supply-side costs and DSM bidders believe the resource block is larger than the total MW savings likely to be offered by bidders, shrewd DSM bidders will be able to extract an additional profit of almost 2¢/kWh before the utility would select the supply-side resource. Therefore, to increase the competitive pressures on bidders, the costs of bidders' programs should be compared directly to the cost of the utility installing the measure. However, if such a comparison is to be made, it should be made on a comparable basis and fully consider the relative risks of the two programs and relevant non-price considerations.[15]

[15] Central Hudson Gas & Electric Company (Central Hudson) compared the cost of bidders' proposals to its programs in its demand-side RFP and rejected several bidders' proposals because they were more costly than its own programs. In response to a petition by these bidders, the New York Public Service Commission (NYPSC) directed

To enable the bidder to ensure that its pricing proposal conforms with the utility's requirements (e.g., has a total resource benefit-cost ratio of 1.0), the formula that will be used to calculate the proposal's cost-effectiveness ratio should be clearly outlined. The bidder should be able to determine its effective price or the cost-effectiveness ratios of its proposed pricing by following the price/cost-effectiveness evaluation methodology presented in the evaluation sheets. The escalation rates for all escalators and the discount rate that will be used should be included in these evaluation sheets. The bidder should be required to indicate the price that will be paid in each year of the analysis period, and if the price will vary based on specific escalators, the formula to be used to calculate the price should be spelled out.

Finally, the pricing and payment provisions in the RFP should also be developed with consideration of the degree to which the RFP is designed to encourage bidders to install a comprehensive mix of measures. For example, if an RFP which gives credit to bidders offering a comprehensive mix of measures does not have a properly structured evaluation system, then bidders could potentially take credit for offering to install a comprehensive set of measures and yet only install a limited set. To reduce the incentives for bidders to play a bait and switch game with the mix of measures offered, bidders should be required to specify measure-specific pricing, and payment to bidders should be based on this type of pricing. As part of the evaluation process, the utility should also evaluate the measure-specific cost of each measure in the bidder's proposal to ensure that it is competitive with other bidders' pricing for that measure. This will ensure that the utility does not overpay for the measure if the bidder implements only this measure.

Conclusions

Competitive bidding can be a useful tool to procure cost effective and competitive demand-side resources. Relying on the experiences of the over thirty utilities which have issued RFPs for the procurement of demand-side measures provides significant insight into how a utility can design an RFP which will extract the most competitive resources from the marketplace while minimizing the risk of being required to enter into an agreement for the purchase of savings from a DSM measure which does not properly fit into the utility's resource portfolio. Based on this, and our experience in designing and evaluating response to RFPs, the following conclusions can be drawn as to the design and execution of a successful competitive procurement process:

1) The RFP should clearly identify the company's objectives, whether they were developed by the utility or mandated by the state regulatory commission. Clearly identifying the objectives of the RFP will reduce potential frustration on the bidder's part and minimize any possible litigation from bidders who feel that they were not given full consideration in the evaluation.

2) Use the minimum threshold and evaluation criteria as a tool to indicate to the bidder which project attributes are required by all projects and which are regarded as important and desirable to the utility.

3) Develop an evaluation process which allows the utility to use its experience and judgment. Demand-side RFPs solicit a multitude of proposed measures which cannot be adequately compared using a uniform self-scoring process.

4) Pricing of each proposal should be evaluated relative to other proposals and the cost of the Company implementing similar measures, not just the system's avoided supply costs.

5) Payments to the bidder should be linked to the performance of the measure, and be consistent with the measurement and verification methods proposed by the bidder.

---

the Central Hudson in its ongoing rate case to demonstrate "why the cost recovery proposed for its internal DSM programs is reasonable in light of the terms it specified in its RFP for DSM bids." The NYPSC also required Central Hudson to "justify full cost recovery for programs shown to be less cost-effective than bids the company rejected." New York Public Service Commission, Case 92-E-0677, p. 7.

TABLE 1

DEMAND-SIDE RFP EVALUATION CRITERIA

Price/Cost-Effectiveness	Economic & Financial Risk	Project Development Risk	Compatibility	Operating Viability
Total Resource Cost	Financing Plan	Project Team Experience	Hard-to-Reach Sectors	Operating Security
Utility Cost Test	Financial Strength	Host Commitments	Complementary Measure	Maintenance Experience
Ratepayer Impact Test	Amount of Front-Loading	Marketing Plan	Measure Comprehensiveness	
Societal Benefits Test	Breakeven Year	Milestone Schedule	Project Bid Size	
	Measurement & Verification Plan		Acceptance of Standard Contract	
			Measure Dispatchability	
			Project Location	

Review of Non-Price Criteria Listed On Table 1:

Financing Plan: Used to assess the likelihood that the bidder will be able to finance the project.

Financial Resources: Used to determine if the bidder is likely to be able to fulfill its financial obligations contained in its proposal.

Amount of Front Loading: Used to calculate how forecasted payments by the utility compare to the value of the savings.

Break-even Year: Used to determine the first year in which the cumulative value of the savings to the utility exceed the cumulative payments.

Measurement & Verification Plan: Used to assess the method proposed by the bidder to measure and verify the savings attributable to the installation of the project.

Project Team Experience: Used to evaluate the relevant experience the project team in marketing, financing, installing, and maintaining measures.

Host Commitments: Used to determine the degree to which the bidder has already identified and obtained signed commitments from the project that will enable the bidder to fulfill its proposal.

Marketing Plan: Used to evaluate the likelihood of the bidder achieving its savings bid commitment by evaluating the realisticness and reasonableness of the bidders' marketing strategy.

Milestone Schedule: Provides the bidder with a contractual obligation to develop the project in a timely manor to avoid termination of the contract.

Hard-to-Reach Sectors: Encourages bidders to propose measures for customer sectors which the utility's existing programs do not fully serve.

Complementary Measures: Encourages bidders to propose measures which the utility's C&LM programs do not provide.

Project Bid Size: Encourages bidder's to offer a bid size that is consistent with the utility's objective of minimizing contract administration costs yet reducing the risks of over-reliance on any one bidder.

Acceptance of Standard Contract: Encourages bidders to accept the utility's standard contract offer in its entirely and without modification.

Measure Dispatchability: Encourages the bidder to offer measures which can be controlled by the utility and operated at the utility's discretion.

Project Location: Encourages the development of projects in pre-specified areas that have transmission and distribution capacity constraints.

Operating Security: A financial commitment by the bidder to secure the performance of its obligation under the standard contract.

Operating and Maintenance Experience: Used to determine the bidder's experience in operating and maintaining measures thereby increasing the likelihood of the successful operation and maintenance of the project and its ability to continue to deliver savings throughout the contract term.

References

Finelon, Jane L., Kelly A. Triplett and Bruce C. Peterson, "Incorporating Lessons Learned into Demand Side Bidding." Public Service Company of Colorado.

Goldman, Charles A. and John F. Busch, "Review of Utility Experience with Demand-Side Bidding Programs." *Proceedings of 5th National Demand-Side Management Conference.* Electric Power Research Institute, July 1991.

Goldman, Charles A. and David R. Wolcott, "Demand-Side Bidding: Assessing Current Experience." *Proceedings of American Council for an Energy Efficient Economy Summer Study.* American Council for an Energy Efficient Economy, August 1990.

Hamilton, William E. and Theresa A. Flaim, "Acquiring DSM Resources Through All-Source Bidding: Lessons Learned." *Proceedings of Conference on DSM Bidding: Status and Results.* April 1992.

Integrated Resource Management Draft Initial Filing, Volume E: Resource Solicitation Request - RFP. Boston Edison, December 1992. Massachusetts Department of Public Utilities, Docket 92-153.

New York Public Service Commission, Case 92-E-0677.

Peters, Jane S., Deborah J. Barry, Marvin Horowitz and Fred Gordon, "Evaluations of the Orange and Rockland Utilities Competitive Bidding Program for Demand Side Resources." State Energy Research and Development Authority, December 1990.

PowerSaving Partners: Request for Demand-side Management Proposals, Guide to the RFP. Pacific Gas and Electric Company, October 1992.

Wolcott, David R., "Why DSM Bidding Has Failed to Live Up to Expectations." *Proceedings of Infocast Conference on DSM Contracting and Financing.* May 1992

Chapter 107

The Agony and Ecstasy of Implementing DSM Bid Projects

F.G. Mitchell

Utility capacity acquisition via demand-side management (DSM) bidding places supply-side generation and DSM resources on a competitive price basis, and offers unique benefits and risks to the electric utility industry, their customers, and the DSM industry. The purpose of this paper is two-fold:

- To provide a third-party developer's insights into the DSM bid and implementation process, such as participant risks and rewards.
- To identify barriers to utility, DSM service provider, and customer involvement in DSM resource bidding and project development.

This paper offers the perspective of a private DSM developer, Energy Investment, Inc. The firm has substantial "hands-on" experience in successfully competing for and developing conservation-derived capacity under several DSM bid competitions, e.g., Central Maine Power Company, Jersey Central Power & Light Company, and Public Service Electric and Gas Company. Energy Investment is an independent consulting, engineering, and project development firm specializing in the engineering, design, and construction management of conservation and load management projects. Since the company's founding in 1973, we have worked in over 2,000 commercial/industrial, and institutional facilities nationwide. In this capacity, Energy Investment has helped clients develop over $250 million in energy-conserving capital improvements for annual savings of $85 million.

Traditionally, Energy Investment's business has been helping facility clients pursue energy cost control as a matter of good business practice. More recently, with the advent of large scale, utility-sponsored DSM programs, we have been increasingly engaged by utilities to provide program planning, marketing, and engineering services in support of their programs. Energy Investment has assisted more than 50 electric and natural-gas utilities nationwide. Our observations are valued in that they are based on over 20 years of experience working with both facility and utility clients.

DSM BIDDING: DEFINITION AND BACKGROUND

DSM bidding is an outgrowth of the bid process employed by utilities to procure new generation capacity from independent power producers (IPP). As with IPPs, DSM developers compete on a $/kWh basis to deliver measured capacity. The critical difference between the two entities is that DSM developers supply utilities with capacity via "avoided use" by installing energy-efficiency improvements in customer facilities.

This innovative approach to capacity development, like IPP development, offers utilities an opportunity to shift the costs and risks of capacity development to third parties. Two key elements of risk reallocation drive the continued growth of DSM bidding. One, the costs of marketing, engineering, designing, and installing energy-efficiency projects in customer facilities are carried by DSM developers, not ratepayers; DSM developers are compensated only *after* they deliver verified energy savings. Two, measuring and verifying energy savings is becoming increasingly attractive to utilities because more utility commissions are now allowing them to earn preferred rates of return on kW/kWh delivered.

DSM Bidding Innovations

DSM bidding is one approach that allows utilities to (1) offer comprehensive customer services (via third-party developers), and (2) acquire new, low-cost capacity with minimal direct risk. Some DSM bidding process innovations include:

- Allowing private sector DSM developers to price bids against the utility's avoided costs.
- Offering facility customers the opportunity to sign long-term contracts with terms of up to 15 years. This creates the potential for revenue streams of sufficient value to attract the capital/development expertise necessary to finance and develop large-scale conservation projects.
- Allowing developers to find the most efficient methods of marketing, engineering, and installing a broad range of efficiency measures under financial terms that are in the customer's best interest.

Results

To date, more than a dozen utilities nationwide have engaged in DSM bidding. Contracts have been signed and projects are under way for delivery of approximately 400 megawatts of DSM capacity, with costs averaging 85 percent of utility avoided costs.

RISKS/REWARDS OF DEVELOPING DSM BID PROJECTS

While third-party development of DSM capacity offers an innovative way to cost-effectively develop new resources, it carries with it a complex web of risks and rewards. These must be fully understood and fairly allocated among participating utilities, customers, and developers for individual projects to succeed. The following briefly describes the risks and rewards of DSM capacity development for each party.

Utilities

Risks	Rewards
Loss of direct control of DSM projects in customer facilities	No upfront project costs--pay only for savings delivered
Failure to receive delivery of contracted DSM capacity	Management risks transferred to third-party developers
Difficulty measuring energy-efficiency savings	Performance risks transferred to third party
Lack of long-term track record on part of DSM developers	Energy savings guaranteed over long term

Customers

Risks	Rewards
Must sign long-term contracts (5 to 10 years) with developers	Receive turnkey installation of energy-efficiency projects
Must agree to keep energy-efficiency equipment in place for contract term	Developer carries project costs
Must maintain equipment and replace components over contract term	Receive valuable capital improvements that reduce operating costs
Third-party involvement in facility may adversely affect its operations	Retain some or all of energy cost savings generated by project

DSM Developers

Risks	Rewards
Compete with similar utility-sponsored customer incentive programs	DSM resources pricing competitive with supply-side resources
Must market to many customers to develop viable projects	Enter into long-term contracts with stable funding source
Chance of host facilities shutting down/modifying operations, decreasing savings	Develop long-term business relationships with customers
Must guarantee delivery of savings to utility over 5 to 15 years	Successful projects deliver long-term revenues (frequently) tied to utilities' long run avoided costs

BARRIERS TO DEVELOPING DSM CAPACITY

Bidding for DSM-based capacity offers the potential for significant increases in low cost, reliable new capacity. However, there are several key barriers that tend to restrict rapid expansion of this capacity resource, as described below.

Arbitrary Imposition of DSM Price Limits

Utility policies that place arbitrary price limits on DSM capacity (in lieu of an open bid competition with avoided cost as the pricing benchmark) do not allow for fair valuation of DSM delivered capacity. Imposition of DSM price limits at a fraction of avoided cost predetermines a reduced level of delivered capacity.

Complex DSM Project Implementation Requirements

DSM developers should be required to meet minimum performance and delivery guidelines to help ensure quality projects that address a utility's particular power supply needs. Beyond these requirements, DSM developers should be allowed to utilize their talent and expertise in directing marketing strategies, developing project financing arrangements with customers, and determining the most cost-effective application of practical, electrical efficiency upgrades. Overly complex guidelines will only serve to increase the cost of DSM capacity and slow its eventual development.

Limits on Available Manpower and Expertise in the Energy Management Industry

Full and rapid development of DSM bid capacity will be limited by the current available level of manpower and expertise. Successful large-scale DSM projects require high-quality engineering, design, construction management, and monitoring services applied to potentially hundreds of customer facilities throughout a utility's service territory.

The energy management industry, traditionally small-scaled and specialized, will require several years to adapt and expand its organizational resources to meet the increasing demand for large-scale development of DSM capacity.

Limits on Sources of Available DSM Project Financing

Like the construction of a large central power plant by an independent power developer, DSM project developers require access to capital. However, unlike the financing and development of a finite generation plant that has a clear asset value, DSM technologies are diverse, small-scale, and implemented throughout a utility's customer base. In the absence of the financial community's experience and understanding of DSM project financing, capital resources will be somewhat limited. As more DSM projects are developed, however, complex financial issues will be resolved and financing sources will be more readily available.

SUMMARY

DSM bidding has shown that in an open, avoided cost, price-based competition, DSM developers will assume the financial, organizational, and long-term operating risks necessary to develop and deliver large blocks of conservation derived capacity. Future competition from supply-side developers, other DSM developers, and, to some extent, utility-sponsored DSM programs will ensure that the energy management industry prices competitively, delivers high-quality customer services, and maintains successful, reliable DSM projects.

DSM bidding offers a "win-win" situation for all parties involved: utilities receive guaranteed, low cost new capacity with little or no financial/operational risks, utility customers obtain valuable capital efficiency improvements, the energy management industry is given the incentives necessary to grow into viable producers of new capacity, and society benefits from reduced dependence on fossil fuel and increased environmental quality. While DSM bidding is a new concept requiring resolution of important regulatory, legal, and financial issues, it offers the promise of spurring rapid development of high-quality new capacity.

Chapter 108

Development of a DSM Infrastructure: Planning, Implementation, and Verification for Conservation Bidding

R.L. Bandy, J.S. Wolpert

ABSTRACT

Public Service Company of Colorado (PSCo) has undertaken an aggressive program to obtain conservation resources through a competitive bidding process. This program, the largest conservation bidding program in the U.S. to date, has over 50 MW of resource under contract with another 50 entering contract negotiation at the time of this paper (June). As of early 1993 some 12.7 MW had been successfully installed and verified with an estimated 20 MW currently in progress.

This paper follows up on an earlier publication (1) and describes the overall goals and objectives of this program including some of the reasons for the decision to pursue a bid program. This discussion also deals with the overall organization of this program, including the two-stepped approach (two successive 50 MW bids), and will highlight differences in the second phase as a result of lessons learned in the first 50 MW phase.

The discussion then moves on to the infrastructure development process. The specifics of how this was lacking in the early process and how an informal process evaluation allowed for mid-course corrections to be developed. This is followed by a discussion of the specific steps taken to improve the situation as they applied to all the involved parties. The results of this on-going process, including both areas where it has worked well and areas where success has been more limited, will be addressed.

In closing, some specific suggestions will be given for any interested parties who may be considering embarking on DSM bidding programs. In addition, some general observations on infrastructure development and DSM programs in general is offered. Finally, a brief discussion of how this experience is expected to impact PSC's incentive programs, which are presently nearing implementation, is presented.

INTRODUCTION

With 50 MW of demand reduction under contract and another 50 MW that is in contract negotiations, Public Service Company of Colorado (PSC) has one of the largest conservation only demand-side management (DSM) bidding programs in the country. This paper will describe the challenges that were met as a result of implementing these programs and, more particularly, of developing the infrastructure required to successfully run the programs.

GOALS AND OBJECTIVES OF THE BID PROGRAMS

The Colorado Public Utilities Commission (CPUC), in October 1990, authorized a bidding program for 100 MW of demand reduction. It was decided to offer two 50 MW programs, to be implemented sequentially, so that lessons learned from the first program could be integrated into the second program. Goals of the bid programs were:

- Determine which DSM measures were cost-effective, from both a PSC and market perspective.
- Obtain lowest cost DSM measures using competitive bidding as a mechanism.
- Examine a variety of technologies--let the market indicate what it wanted to do. All energy efficiency technologies were allowed, except for cogeneration. This was to be a DSM-only bid program.
- Quick program implementation. Bid programs were deemed to be a good way to come up to speed fast in DSM. The request for proposals (RFP) could be designed and out on the street relatively quickly.
- Serve as a learning tool, not only for PSC, but for customers and energy service contractors (ESCOs) as well.

Barriers
Before the first bid program, DSM was more or less a new concept among the energy community in Colorado. Much education was needed as to why PSC was doing a DSM bid and what was involved in implementing the program. To achieve successful installation of this demand-side resource requires the close cooperation of the utility, its customers, the bidders and the design and installation professionals.

PSC staff had never conducted a full scale DSM bid program before (having previously carried out a 2 MW pilot), and in planning for the first Bid, PSC underestimated time and staff requirements to implement a 50 MW bid program. This led to a series of delays in reviewing the proposals and in negotiating contracts. Lack of performance by a consultant also caused the loss of 3 months in the schedule. PSC staff had to finish the work internally.

PSC's engineering group did not initially know or the level of detail required for site inspections, verification reports or other documentation of DSM programs, being more familiar with standard construction review processes. This required procedures and tracking systems to be established so that the volume of work coming in could be properly accommodated.

Bidders lacked understanding regarding demand reduction, load shape, documentation requirements, and the purpose of the program (some thought it was a fuel switching program). For example, in the first program, bidders were supposed to do all verification. In many cases they did not know what was involved in doing this properly. Some had selected as many verification methods as possible because this maximized their bid score. Some bidders had no intention of carrying out every method or appreciated the fact that not all methods of verification applied equally well to all types of technologies and applications..

Consultants assisting the bidders had similar issues and generally did not understand much about DSM. Typically, the level of rigorous treatment in preparing the proposals was lacking. There tended to be a focus on dollar savings in the studies and not on kW reduction or load shape. In hindsight, this is not surprising, since most of the bidders had experience with performing feasibility studies for marketing purposes, but only limited experience with utility DSM programs (especially bid programs).

Customers did not always jump at the chance to install energy efficient equipment. Customers' concerns are frequently about matters other than energy cost. For many customers, energy cost is less than 5% of their operating budget. As one customer put it, "I'm too busy making a buck to worry about saving a nickel". Customers' lack of familiarity with DSM was also an obstacle. Many times they did not understand how the bid process worked, why it seemed so complicated or how the third party bidders exactly fit in with PSC and the bid program.

Customer Relations
DSM causes a shift in the relationship between utility and customer. The utility goes from a supplier of energy to a customer who is buying demand reduction. Likewise, customers change from buyers of energy to providers of DSM power plants. This shift is not always readily apparent among the customers or the utility field reps who assist them. Many times appeals are made by the reps to make decisions that are friendly to the customers in the name of "customer service", even though these decisions might compromise the integrity of the DSM effort. Holding customers accountable for demand reduction for which they've been paid forces the utility to walk a fine line between its traditional role of energy provider and its new role as efficiency purchaser.

A continued effort must be made to educate all participants that DSM measures must be reliable and verifiable. It is not enough to just get something installed so the customer can get paid. The "DSM power plants" must be as reliable as traditional supply-side resources and have similar high availability on a year-round basis or else be discounted in their value. In reality, since most DSM measures do not operate for all hours of the year, almost all DSM measures are discounted in their value compared to supply side resources. At PSC, this valuing process is referred to as determining the ceiling price of a measure.

Some utility field reps felt left out of the process. In order to keep the competition fair, the reps were not allowed to assist their customers with proposal preparation. Also, some third party bidders were working with customers directly and the field reps did not always know what was going on. This was rectified somewhat in the 2nd bid by implementing a series of technical workshops where potential bidders could come and learn more about the bid preparation process. The workshops also allowed PSC's field reps to come and find out which of their customers might be contemplating a bid.

Communications Issues

Technical Workshops
The first bid program had 2 pre-bid conferences only. The process evaluation that was conducted later indicated that there was a desire on the part of bidders as well as utility field reps for more communication. There needed to be more updates as to what was going on in the process as well as more information about what the process entailed. In addition, many of the proposals received in the first bid were lacking in quality and content. This was partly due to the bidders not having been exposed to a large DSM bid program before, but also to PSC's not defining the requirements with enough detail.

As a result, for the second bid it was decided to conduct technical workshops in addition to the two pre-bid conferences. Three 1/2-day workshops were held, each identical, about three weeks apart. Each workshop began with a general session for all attendees which dealt with issues such as demand reduction, load shape, verification, etc. Breakout sessions were then held on specific technologies such as lighting, heating and cooling conversions, and control measures such as energy management systems and variable speed drives. Although the workshops did not entirely remove the problem of poor quality bids, the level was much improved over those received in the first bid and the bidders seemed to have a much better understanding of the issues.

Newsletters

In order to enhance communication of the bid process for all concerned parties, a newsletter was developed for the second bid that was sent to a database of over 2000 names. The database was assembled from inquiries in the first bid and from lists of local energy decision makers such as consulting engineers, contractors, vendors, and large customers (and was expanded to serve a number of functions, see *Changes from 1st to 2nd Bid Program* section). The newsletters dealt with subjects such as the purpose of the program, updates on the status of the bid process, summaries of the proposals received and the initial award group. Care was taken not to release any information which the bidders had indicated to be confidential. The newsletters also filled a need for a mechanism to update people making inquiries about the program who might not want to bid or who were too late to participate as bidders. In these cases, back issues were mailed out.

Presentations

PSC staff made numerous presentations about the bid programs to local professional groups including AEE, ASHRAE, and others with an interest in energy efficiency. These appearances served to position PSC as being available to the community and allowed a forum for direct answers to questions. This also helped to negate the often leveled criticism of utilities that they are too large and removed from their customers and not accessible enough.

1ST BID VERIFICATION

Verification, for the purposes of PSC's DSM bidding programs, is the process of determining if the bidder has satisfied the goals of their contract and can thus be paid. As a contractual tool it differs from impact evaluation where other broader concerns are addressed (e.g., seasonal load shapes, persistence of savings, free-riders, etc.). In the first bid, verification was the bidder's responsibility with the methodology and some of the requirements outlined in the contract. PSC's role was to review and approve the bidder's required submittals and to carry out pre- and post-installation field inspections.

Both the proposed and accepted measures included a wide range of technologies and of customer and 3rd party bids. As shown by Tables 1 and 2, this mix necessitated PSC being prepared for a corresponding range of approaches to verification.

TABLE 1: 1ST BID - SUMMARY OF BIDS

Quantity	Proposed			Accepted		
	3rd Party	Customer	TOTAL	3rd Party	Customer	TOTAL
Proposals	21	42	63	10	21	31
Measures			128	19	35	54
MW			131	29.77	23.87	53.64

TABLE 2. 1ST BID - BID DETAIL

Technology	Proposed		Accepted				
	MW	# Meas.	MW	# Meas.	Winter	Summer	Annual
Elec. Heat Conversions to Gas/Steam	38.72	17	21.02				
Motor Efficiency	5.44	4	0.22				
Snow Making Efficiency	9.90	2	3.27				
Elec. Cooling Conversions to Gas/Steam	3.86	1	0.50				
Indus. Process Efficiency Improvements	12.48	9	7.17				
Air Conditioning Equipment Efficiency	2.70	3	0.02				
Efficient Lighting	31.97	15	16.89				
Energy Cooperative - Dispatchable	15.00	1	0.00				
Energy Management Systems	7.46	9	4.33				
Other	3.06	2	0.22				
TOTAL	130.59	63	53.64	55	25.00	1.00	28.00
Weighted Average Bid Price, $/kW	397.00		265				

Early Projects

As work proceeded into late 1992, it became increasingly clear that the verification being provided by the bidders was not adequately establishing if the goals were being met. Among the specific shortcomings were lack of detailed engineering calculations, incomplete information on equipment removed and installed, and incomplete and poorly documented metering results. Overall the result was a critical lack of any comprehensive verification report that allowed for an informed review of the projects accomplishments. As this problem came into focus and efforts were made to deal with it, it became apparent that the bidders had not anticipated this level of detail, having successfully generated far simpler engineering reports and studies for customers in the past but not being used to this more intense level of scrutiny, especially as it related to peak demand (as opposed to energy and monthly cost) reductions. Given their expectations, the bidders had not planned for a sufficient level of either staff or effort to deliver this and adjustments had to be made.

From PSC's side, a number of issues began to surface as well. Some of these arose from the fact that during contract development and negotiations PSC's team lacked some technical expertise and accordingly, some issues were left for later determination. In addition, it was initially assumed that both the bidders and PSC's field inspection personnel would be able to proceed satisfactorily with instructions that stated the overall goals and objectives, but without day-to-day management. As work proceeded it became clear that a number of problems were developing including lack of standardized formats for reporting information, missing or incomplete field inspection reports, last minute changes often coupled with requests for expedited inspections and payment (especially as the end of the year approached). Overall this process, coupled with insufficient staffing to provide more detailed management, increasingly lead PSC's project manager to feel trapped in a reactive mode whereby getting things under control would never be achievable.

Early Process Evaluation

After actual installation of some measures had been underway for approximately six months, an assessment of how the process was proceeding was undertaken. This informal process evaluation came about primarily at the request of the PSC's project manager, who was becoming increasingly concerned about a number of issues, including the rising workload. After a brief review period of about one month, a number of important issues were identified. These included:

- having only one person at PSC assigned to this program, without additional support, was insufficient.
- the lack of a structured process to deal with managing the bidders progress was creating numerous difficulties.
- the desired coordination between verification (primarily the bidders responsibility) and evaluation (PSC's area), was not happening.

Overall, the conclusion was that the program was being managed in a "reactive" mode where PSC was not exerting sufficient control over the final outcome. This was resulting in a situation where PSC's objectives were not being well served and where the bidders were unhappy with the lack of clear requirements and the resulting (from their perspective at least) "eleventh hour" changes.

Based on this evaluation, a series of mid-course corrections were developed and implemented over the next two months. The first step in this process was to add an additional person (in the form of a consultant) to share the workload. This enabled PSC to begin developing procedures for managing the bidders in an ongoing fashion, thus avoiding the difficult task of correcting deficiencies at the end of the process.

Short-Term Improvements

As the end of 1992 approached a process was undertaken to address the issues and concerns identified above. A two-stage objective was agreed upon whereby initial efforts would focus on projects targeted for completion by year end (primarily a number of heating conversions). This was to be followed by a process of applying these specific improvements, in a more general form, to the program as a whole.

To aid in the fast-tracking of the remaining 1992 projects the efforts of an outside consultant were added to those of PSC's staff, basically adding an additional full-time equivalent person to the program. Armed with both a direction and staffing, PSC embarked on improving the process. The overall goals were three fold:

- define PSC's requirements, both from it's own staff and from the bidders, for verification.
- communicate these requirements to all parties involved.
- assist all parties in expediting the process so year-end goals could be met.

To aid in achieving these goals a number of specific steps were taken including:

- identify a single point of contact at PSC for bidders.
- identify bidders with sites planned for completion in the remaining 2-1/2 months.
- for these critical path projects, review project documentation including contracts, correspondence, and any site inspection reports
- identify known/potential problem areas based on documentation and internal discussions.
- contact appropriate bidders and arrange progress review meetings.
- identify areas of concern and develop site-by-site plan, with the bidder's project manager, to meet goal of successful year-end completion.
- meet with bidders support staff, including installing contractors, engineering firms, and controls technicians, and assist with process of verifying goals.
- develop preliminary (but usable) forms, instruction sheets, example documents, etc., to aid bidders and PSC's field staff in carrying out needed work.

As a result of these efforts all of the projects targeted for completion in 1992 (approximately 12.7 MW) were successfully carried out.

On-Going Improvements

With the year-end push behind everyone, PSC and the bidders could now work on establishing longer term goals and procedures. This process was aided by the addition of a PSC staff engineer with construction management experience to serve as project manager. Under his guidance PSC's 1st bid team (PSC project manager, a 1/2-time verification consultant, a 1/2-time evaluation consultant, and a full-time equivalent PSC field inspection engineer) was able to work with the remaining bidders to expand on the preliminary procedures and materials developed for the short-term process. This effort was fairly concentrated during the first three months of 1993 with some minor improvements continuing on an on-going basis.

In addition to the focus on the bidder's efforts, two other areas received a fair amount of attention during this period. First, a successful effort went into coordinating the verification (short-term demand reduction focus) and evaluation efforts (longer-term demand and energy focus). This was desirable as many of the requirements and efforts were similar, and although the benefits of this had been recognized, it had not been happening earlier in the program. Second, PSC's field engineering group, responsible for carrying out the pre- and post-installation inspections, received more direction. Foremost in this was a better defining of their responsibilities and the establishment of a clear process and areas of responsibility. Aiding in this, specific improvements included developing inspection process flowcharts, pre-inspection checklists, reporting forms, sample reports, document- tracking forms, and other aids as needed.

PLANNED CHANGES FROM 1ST TO 2ND BID PROGRAM

Having gained invaluable experience through the 1st bid process, the 2nd bid benefited from a greater level of advanced planning. Key steps in this process are detailed in Table 3 as shown.

TABLE 3. OVERVIEW OF BID PROCESS

PSC issued RFP
Pre-bid conferences &
Technical workshops held
Proposal due to PSC from bidders
PSC evaluates and scores proposals
Initial Award Group announced
Contracts developed and negotiated
Contracts signed
Pre-metering if required
Measures installed
Demand reduction verified by PSC
Award Payment
PSC does ongoing monitoring and evaluation

Winter Cap

In the first bid, contracts were signed with bidders for 27 MW of winter-only demand reduction, most of this being electric to gas heating conversions. In an effort to balance this in the second bid, the Quarterly Analysis Committee(QAC) (members include PSC, CPUC staff and Office of Consumer Council(OCC)) requested that a 15 MW cap be instituted for measures with winter impact only. It is a testimony to the cost effectiveness of fuel switching measures that 10 MW of winter-only measures did not make the initial award group due to the 15 MW cap (see subsequent Tables 5,6).

Scoring

The first bid was self-scored, meaning that when a bidder turned in his proposal, he knew what his total score was. It might be adjusted later by PSC in the review process, but he knew the total score at the time the proposal was submitted. The second bid used a combination of self-scoring and subjective scoring by PSC. This was done to reduce "gaming" by some bidders who manipulated their bid using the published self-scoring formula, without intention of carrying out the commitments being made to get the higher score. By limiting the bidders self-score (objective score) to just bid price and measure life, the effects of this "gaming" were minimized. The subjective score included categories such as quality and completeness, technical evaluation, financial capability, marketing plan (for 3rd party bidders), and experience/qualifications.

Deposits

In the first bid, no deposit was required until the time of contract execution. Since the process of reviewing proposals and selecting an award group is lengthy, much can happen in the period of time up until contracts are signed. With no financial stake up until contract signing, third party bidders had an incentive to submit proposals with inflated kW reduction amounts and then hope they could find the customers to work with later. If not, they could always hope to negotiate a lower kW reduction later. They reasoned that at the time of contract signing, the utility would have already made a significant investment in reviewing their proposal and developing the contract and would be hesitant to disqualify the proposal altogether.

In the second bid, a $2/kW deposit was required at the time proposals were submitted. An additional $18/kW was required at the time of executing contracts. If a bidder withdrew a measure or tried to reduce the kW amount, then the $2 deposit was kept. This caused bidders to think more seriously about what they were proposing.

Default Load Shapes

In an attempt to encourage smaller or less sophisticated bidders to participate in the second bid, PSC provided default load shapes for certain common building types and technologies based on computer models (see Table 4).

TABLE 4. DETAIL MEASURES

Building Type/Measure	Lighting	Heat Conv.	Cool Conv.
Offices - Small	x	x	x
" - Medium	x	x	x
" - Large	x	x	x
Schools - Small	x	x	x
" - Large	x	x	x
Residential - Single Family		x	

All bidders had to do was calculate and document their demand reduction goal. Of the 80 measures submitted, 19 used this technique and a surprising number of these were third party bidders (who had been expected to use more generous load shapes resulting from custom calculations). In contrast, for those bidders supplying their own load shape data, the most frequently asked questions by PSC during the proposal review process had to do with load shapes. A common point of confusion was not understanding the difference between average demand reduction for a typical weekday and maximum demand reduction, which might occur as infrequently as once a year.

Verification

Bidders were responsible for verification in the first bid and were able to choose from a menu of verification options. Each of these options had an associated point value for scoring purposes. Some of the bidders were tempted to select more methods than they should have since the more methods they chose, the more points they received. Many of the methods chosen were either inappropriate for the measure involved or the methods chosen were not well understood. For example, "billing analysis" can mean anything from a simple review of utility bills with no adjustments to sophisticated regression methods. Some methods were based solely on engineering calculations which later turned out to be not entirely accurate.

In the second bid, PSC will determine the most appropriate verification method and will pay to conduct it. This will allow PSC to control which verification methods are used and thereby optimize the cost effectiveness of the verification effort. Long term impact evaluation will also benefit as metering and data collection plans can now be coordinated to serve both verification and evaluation. In addition, a much more thorough technical evaluation was conducted on each measure before the Initial Award Group was selected. This served to eliminate measures with high uncertainty of demand reduction, which tend to be harder to verify and therefore more costly.

PCB Disposal

In the first bid, there were no provisions for disposal of PCB ballasts from lighting system retrofits. Neither Colorado nor Federal law requires special handling of PCB ballasts and they may legally be disposed of in landfill sites. However, if there is ever leakage in the landfill there is potential liability under the Federal Superfund law for whoever contributed the ballasts. This liability could run into the millions of dollars. With this in mind, PSC decided for the 2nd Bid that they would secure the services of a recycling and disposal contractor to handle all PCB ballasts removed as part of the program. There is no cost for this service to the bidders. The cost is absorbed as part of the bid program overhead cost. Providing this service also prevents each bidder who is doing lighting measures from having to come up to speed on the intricacies of environmental law regarding PCBs, which is quite complex.

Bidder Assistance

In the first bid, the only assistance provided to potential bidders was two pre-bid conferences which described the bid process and which were held before proposals were due. One of the criticisms of the first bid was that there wasn't enough assistance for bidders, especially customers. In the second bid, a default method for supplying load shapes was provided (see section on *Default Method)* and the RFP gave much more specific guidance on bid preparation. Detailed examples for different technologies and calculation methods were also provided. A series of technical workshops which described some of the most common technologies from the first bid were also conducted. *(see Communications Issues)*.

Database

In planning for a 2nd bid database, service in a number of roles spanning the entire process (in addition to communication), had been planned for. As a result, this database is presently adding PSC's staff and consultants in many tasks including keeping track of PSC/bidder exchanges of information, bidder submittal statistics, measure statistics, and other items as required.

RESULTS TO DATE

As of the end of June of 1993 the 2nd bid process was progressing smoothly. As in the 1st bid, a diverse group of proposals from both customers and 3rd party bidders was received (Tables 5,6).

TABLE 5. 2ND BID - SUMMARY OF BIDS

Quantity	Proposed			Initial Award Group		
	3rd Party	Customer	TOTAL	3rd Party	Customer	TOTAL
Proposals	19.0	16	35	11	13	24
Measures			81	22	23	45
MW	67.7	13.2	80.8	40.85	10.77	51.62

TABLE 6. 2ND BID - BID DETAIL

Technology	Proposed				Initial Award Group		
	MW	# Meas.	MW	# Meas.	Winter	Summer	Annual
Elec. Heat Conversions to Gas/Steam	24.05	22	17.98				
Motor Efficiency	0.71	4	0.00				
Motor Controls - VSD	0.72	5	0.14				
Snow Making Efficiency	2.44	3	0.00				
Elec. Cooling Conversions to Gas/Steam	6.57	6	4.00				
Indus. Process Efficiency Improvements	3.24	3	23.4				
Air Conditioning Equipment Efficiency	0.28	3	0.18				
"Free" Cooling	0.78	2	0.30				
Efficient Lighting	21.54	21	19.3				
Lighting Controls	0.08	2	0.07				
Energy Management Systems	13.99	6	0.00				
Combined measures	3.70	1	3.50				
Refrigeration	1.75	1	1.75				
Thermal Storage	0.20	1	0.00				
Evaporative cooling	1.00	1	1.00				
TOTAL	81.06	81.00	51.64	45	16.99	5.18	29.12
Weighted Average Bid Price, $/kW	228.00		237				

Overall the number of proposals (63 vs. 35) and the total MW's (131 vs. 80.8) dropped somewhat although the average MW/proposal rose from approximately 2.1 to 2.3. In addition, the capacity-weighted cost dropped significantly from $397/MW to $228/MW. These changes, basically carried through into the Initial Award Group, reflect an increasing sophistication on the part of the bidders.

CONCLUSIONS

Public Service Co.'s venture into DSM bidding has proven very successful thus far. A relatively large amount of resource (approximately 90-100 MW) is in various stages of obtainment at a very competitive price (in the range of $235-265/MW). In the concluding paragraphs a number of final points relating to this undertaking are offered for consideration.

Advantages Of Bid Programs

This approach to conservation resources offers a number of advantages including:

- Competitive nature of bidding yields market driven results (lower bid prices). The first bid had a kW weighted average bid price of $265/kW. The second bid Initial Award Group has a weighted average bid price of $237/kW. By any comparison these are extremely low prices for DSM measures.
- A variety of technologies are submitted and installed which gives good experience. See Tables 1,2,5,6 for listing of the technologies in both programs.
- Relatively fast implementation of the program. The first bid RFP was released in December of 1990 and 50 MW of capacity reduction was signed 18 months later. As of June 1993, over 15 MW had been installed and verified.

Disadvantages Of Bid Programs

There are, however, a number of drawbacks to this type of DSM program, among them:

- Bidding requires long term contracts, which some customers are reluctant to sign.
- Not as customer friendly as rebate programs. The bid process is lengthy and expensive to participate in, especially if one's proposal is not successful.
- The potential for conflicts with bidders over competition from other DSM programs exists.
- Third party bidders introduce another layer between the utility and the customer. The

utility may perceive a loss of control over the program with its customers.

Project Planning

Budgeting

The large amounts of incentive payments involved and the bidders' installation schedules makes budgeting both important and difficult. Budgets are usually set six months to a year in advance of expenditures. PSC's bid programs required that all installations be complete within 30 months of contract execution. This means that budgets must be predicted over a 2 1/2 year period. But bidders, especially third parties, may suffer delays in their projects. There could potentially be several million dollars in incentive payments that get delayed from one budget year to the next with little warning. There should be a contingency plan for these situations. If consultants will be used for part of the work, then this too should be allowed for.

Scheduling

Schedules are difficult to adhere to in bid programs due to the high amount of external variables. As an example, it is difficult to control the amount of time required to negotiate contracts. Also, many tasks will not have been performed before thereby making it difficult to accurately estimate the amount of time required. In any event, it is important that the program be broken down into tasks and sub-tasks so that realistic estimates can be made of time requirements. Keep in mind Murphy's law: "Things always take longer than you think". The estimates can always be changed later as better data is available. Bid programs are very complex to administer effectively and if schedules are to be adhered to, it is important that sufficient resources be made available to carry out the task.

Consultants

In dealing with the challenges of resources, one of the assets PSC relied upon was the use of outside consultants. These additional staff were brought in from two firms in the 1st bid where they were utilized in project management, verification, and evaluation work. One of these consulting firms then assisted in development of the RFP and workshops, and both firms plus a third then worked on (to date) bid evaluation and contract development. This use of consultants offers a number of benefits in meeting the demands of resource allocation, including:

- they offer quick access to experience obtained in other parts of the country with a range of DSM and construction programs.
- they can rapidly be brought up to speed and are generally able to quickly change their focus as needed, enabling them to serve in a "peak-shaving" capacity.
- they can be brought on without having to make the long-term commitments to personnel and equipment (computers, printers, etc.) that increased utility staffing requires.

Impact on Other Programs

At PSC, the two bid programs were the first major DSM programs to be launched. As DSM matures at PSC, impact from the bid programs will continue to be felt. Much has been learned about the various technologies that were installed in the programs. Techniques of project management, proposal review, verification, contract negotiation and evaluation have been developed which will be useful in future programs. There has been much progress in educating utility staff and customers as well as the consulting and contracting community about the various aspects of DSM. Finally, the bid programs have allowed PSC to target retrofit of energy efficiency in a cost-effective manner. The next generation of programs will be rebate programs and will be aimed at equipment replacement. Taken together, the bidding and rebate programs will offer a more complete coverage of the market than either could accomplish alone.

REFERENCES

1. Finleon, J.L., and Triplett, K., "Demand-Side Management Bidding - Lessons learned", *Proc. of the 15th World Energy Congress*, Atlanta, Georgia, October 27, 1992.

ACKNOWLEDGMENTS

A large number of people, including PSC/bidder/customer staff and consultants, have worked hard to make this a successful program and are to be complimented. Among those who made day-to-day contributions in parts of either the 1st, 2nd or both bids, include Dave Anderson, Janet Bidwell, Melinda Bise, Meng Chi, Dave Guinn, Dee Liebl, Terry Moore, Bruce Peterson, Jim Pilcher and Kelly Triplet, with PSC; Steve Schiller with Barakat & Chamberlin; Jay Stein with E-Cube; and Rick Brakken and Mark Bowman with Xenergy.

Chapter 109

What Utilities Are Saying about DSM Bidding

W.J. LeBlanc, M.R. McRae, J.S. Peters

ABSTRACT

Competitive DSM bidding programs have been a unique opportunity to test the ability of the marketplace to efficiently and effectively deliver utility services. So far, approximately 15 competitive DSM solicitations have been conducted. Building upon this narrow but valuable base of experience, utilities are becoming more sophisticated in developing their bid solicitations and evaluating bid offers. They are also beginning to consider alternatives to bidding.

This presentation outlines the evolution of competitive DSM bidding programs; it reports on lessons learned from the evaluations of several such programs, as well as from working with utilities on DSM bidding; and it presents insights about what potential bidders can expect in future utility DSM resource solicitations.

INTRODUCTION

Demand side management (DSM) can help utilities meet their resource planning and marketing objectives. Third-party "bidding" is one way to obtain DSM resources. In a DSM bidding program, utilities solicit proposals for projects from customers and energy service companies (ESCOs). Initially some utilities saw the promise of DSM bidding as a way of acquiring DSM resources while minimizing their costs and their risk. What they are learning is that such programs neither eliminate the risk associated with DSM for utilities nor ensure program cost-effectiveness.

Our evaluations of bidding programs and interviews with utility program managers show us that utilities are at least beginning to question the assumption that DSM bidding programs offer a relatively easy way to find new methods of acquiring conservation resources and reduce (or shift) the risk of unrealized savings. Both the nature and the cost of the resources that utilities acquired under DSM bidding have differed from their initial expectations. As a consequence, utilities are becoming more sophisticated in developing their bid solicitations and evaluating bid offers.

Barakat & Chamberlin has reviewed, in depth, DSM bidding solicitations and programs at Bonneville Power Administration, Central Maine Power, Pacific Gas & Electric Company, Public Service Company of Colorado, and Orange and Rockland Utilities, and has surveyed programs at half a dozen other utilities. This paper reports on what utilities have learned from their DSM bidding experience, provides some insights into what bidders may expect in the future, and suggests other ways in which utilities may seek to secure third-party DSM.

EVOLUTION OF COMPETITIVE DSM BIDDING PROGRAMS

DSM bidding evolved for two primary reasons. The simplest reason is that as a resource strategy it is a logical parallel to supply-side bidding; some energy observers see it as a natural outgrowth of the success of competitive procurement on the supply side. The more creative reason is that it appeared to be a promising means of expanding the non-utility market for energy efficiency and perhaps promoting diversity in DSM options and/or service delivery. Bidding is an opportunity for utilities to test the ability of the marketplace to deliver energy services (conservation and others) efficiently and effectively. In California, the three major investor-owned utilities now require DSM bidding with the purpose of stimulating the private DSM marketplace.

Utilities undertaking DSM bidding programs typically have several goals:

- To obtain desired demand and energy reduction
- To pay only for the savings received
- To acquire demand and energy savings at the lowest competition-induced price
- To achieve a greater variety of DSM measures and new, useful approaches to marketing DSM
- To stimulate the private DSM marketplace and expand third-party DSM capabilities
- To reach market niches not easily served by utility DSM programs
- To speed the pace at which customers adopt energy-efficient technologies.

What have utilities learned? What are the implications of these lessons for bidders? What does the future hold for bidding and bidders?

LESSONS LEARNED

Since 1988, when Central Maine Power carried out the industry's first DSM bidding program through an all-source solicitation, over 15 utilities have issued DSM RFPs. These utilities have learned a number of lessons.

Difference Between Supply-Side and Demand-Side Bids

Utilities have learned that DSM solicitations cannot be structured or evaluated the same way as Requests for Proposals (RFPs) for supply-side projects. This is because DSM is both a customer service *and* a power resource. DSM projects must simultaneously provide services that meet customer needs and load shaping to satisfy utility system needs. Success depends more on marketing plans and customer behavior than on construction management techniques or technologies. Unlike generation projects, DSM programs are not individual plants located at remote sites with easily measurable MW and MWh output. Rather DSM programs touch customers directly, are dispersed in terms of both location and end uses, and provide smaller resource increments. Further, their impacts can only be estimated, not measured explicitly.

Importance of Policy Issues

The utilities that are most likely to have the most successful programs and the fewest unhappy surprises are those that already have extensive DSM experience, and those that make critical policy decisions before launching a bidding program. These decisions pertain to market coverage, the types of projects that will be eligible, minimum requirements, the nature of the working relationship with contractors, handling of environmental externalities (air, land, and water impacts), preferred attributes of the bids, acceptable payment methods, bid evaluation philosophies and methods, selection criteria, and acceptable levels of risk.

These decisions define the shape and content of both the RFP submitted to bidders and the utility's internal staff guidelines. They are relevant whether the utility solicits separate block bids, all-source (integrated) bids, or parallel block bids.

Importance of the RFP

The RFP is the heart of the bidding solicitation. It is largely responsible for the quality and success of the solicitation.

The differences between supply- and demand-side projects described above increase the importance of RFP design for DSM solicitations. Because many DSM success factors are not readily quantifiable, it can be difficult to compare and rank alternatives. A successful RFP enables the utility to obtain the information it needs in order to evaluate all proposals systematically and fairly. It is also flexible enough to enable each proposer to describe fully all of the features and benefits its project offers. Not all solicitations accomplish their goals with equal measure and no two have been the same. However, successful solicitations have several features in common (see Exhibit 1).

Bidders must meet the requirements of each RFP criterion in order to be selected. In general, bidders are effective in marketing, fairly good at technology installation, and quick to implement programs. Many, however, lack expertise in areas that utilities regard as critical. Most are related to system planning issues. While utility system planning issues are not a concern of utility customers, ESCOs, as intermediaries, need to understand both marketing and utility needs (see Exhibit 2).

The key stumbling blocks during negotiation are verification and monitoring, estimation of market penetration, and financing:

- **Verification and monitoring.** Utilities need to know with a relatively high level of certainty the timing and magnitude of DSM impacts. This is especially true in a bidding program where ESCOs have pay-for-performance contracts. Ultimately, pay-for-performance contracts require continual monitoring over the life of the project. Sometimes ESCOs see these costs as unnecessary. Yet, the nature of the contractual agreement means the

Exhibit 1
FEATURES OF SUCCESSFUL SOLICITATIONS

- The RFP clearly expresses utility preferences in terms of types of programs, end uses, and target customers;
  - It tells bidders exactly what the utility is looking for in a DSM resource. Exclusions, such as DSM programs that will not be accepted, are explicit.
  - The requirements are easily found and understood. Legal jargon is kept to a minimum. Forms are provided to help the bidder complete all requirements. Sample contracts are included.
  - It includes enough data to allow intelligent bidders to estimate market penetration and the value of the bid to the utility. There is neither too much detail nor too little information.
  - It explains how the bid selection process works (but not necessarily the exact scoring system) and what will help or hurt a bidder's score.
- The selection criteria emphasize realistic marketing plans as well as verifiable savings; and
- The utility staff carrying out the solicitation is composed of a small team that is empowered to make decisions on behalf of the company.

utility must be certain it is paying for reliable and persistent savings.

- **Market penetration estimates.** The estimate of market penetration of any given DSM program is subject to high uncertainty, even though utilities have spent substantial resources to help refine these estimates. ESCOs may find it hard to persuade utilities that their market potential figures are sound.
- **Financing.** ESCOs have widely varying financial capability. Over the long term, ESCOs may have far less financial stability and/or ability to secure appropriate financing than utilities. Utilities typically have a requirement for security, which increases the cost of the DSM. Over time, financing will become less of a negotiation stumbling block than it is today.

Importance of Customer Service Issues

Many of the early proponents of DSM bidding considered only the resource (kW and kWh) aspect of DSM when promoting the concept. This approach is flawed, however: DSM is just as much a customer service as it is an energy resource. Utilities are extremely concerned, as they should be, about how they are portrayed to their customers. Utilities need assurance that ESCOs representing them are providing the same level of superior service to customers that utilities themselves are to provide. At times, the regulatory demands for a "fair" bidding scoring system have overlooked the "softer" yet critical aspect of customer service. The bottom line is: If utilities perceive that they are lowering the service level because of bidding, they will be likely to try to reject the bidding concept and move to alternative procurement methods.

Complexity

Utilities that conducted DSM bidding programs typically thought they were embarking on an innovative means of acquiring DSM resources—but most were not prepared for the vast differences between bidding programs and in-house utility programs. They discovered that bidding programs resurrect and often exacerbate many of the problems they faced in launching a DSM effort in the first place.

In particular, utilities have found that bidding programs require significantly more involvement from utility management and staff than most efficiency-based rebate

Exhibit 2
INFORMATION REQUESTED OF BIDDER

Technical Specifications: Bidders must list all major equipment in the proposed facility, or used in the proposed project, and specifications for the equipment to be used including manufacturer and independent testing laboratory data, and engineering drawings.

Marketing Plan: Bidders must define target markets by geography, building type, end use(s), and other pertinent characteristics. The plan shall specify whether hosts have been selected. If so, site location and plot plans should be included. Marketing methods should be described.

Evaluation Plan: Bidders must include a measurement and verification plan that will demonstrate savings over the life of the contract. Specific methodology for monitoring and verifying the kW reduction achieved should be developed for each significant market sector.

Resource Impacts: Bidders shall describe the timing and level of demand and energy changes due to their programs. They or the utility may specify a range of possible results. The persistence of savings must be presented over the life of the contract. Environmental impacts must also be described. (For example, proper disposal of old ballasts should be outlined, as well as identification and amount of toxic and/or hazardous wastes, if any.) If applicable, minimum and maximum levels of demand reduction attainable under dispatch operations should be included.

Project Schedule: Bidders shall outline a proposed project timetable, including any phasing in of project results. Major milestones should be included, as well as a management plan.

Project and DSM Experience: Bidders should describe their prior experience in DSM implementation, contracting, and other pertinent work; provide references; and cite any problems in past projects.

Project Financing: Bidders must submit requirements for payment and indicate the types of bonds, insurance, and letters of credit they can secure.

programs require. Our research has consistently shown similar patterns of bidding program implementation—and similar difficulties. Not recognizing the programs' complexity at the outset, utilities typically assign the bidding program to a DSM staff person as if it were another equipment efficiency rebate program. Sometime later, they may discover the need to create an interdepartmental management team to provide guidance on the complex bid evaluation and contract negotiation issues that have arisen. In short, they find themselves playing catch-up.

Bidding can be a protracted process. Our survey of more than a dozen utilities that have conducted DSM bidding programs revealed that evaluating proposals took an average of 4 months, and as much as 10 months; negotiating the contract took from 3 to 15 months (average 9); and the length of time required for both evaluation and negotiation ranged from 5 to 15 months (average of 11). Table 1 (based on a report prepared for Bonneville Power Administration) illustrates the time required for each of these categories at several utilities that conducted DSM bidding or integrated bidding programs.

In light of this experience, clearly a bidding program is not the best way to introduce DSM into a utility. The more background a utility has in DSM the better able it is to manage the process and to judge the merits of a third-party proposal.

Negotiation Subtleties

The utilities that have conducted DSM bidding programs have learned the importance of thorough planning and detailed requirements the hard way; and so, in some cases,

Table 1
IMPLEMENTATION TIME REQUIRED BY
COMPETITIVE ACQUISITION PROGRAMS

Utility/Agency	Type of Solicitation	Time for RFP Design	Time for Evaluation	Time for Negotiations
Bonneville Power	All-source	~ 2 years	6 months	DSM: 16+ months (ongoing)
Central Maine Power	DSM	~ 2 weeks	6 months	9 months
Consolidated Edison	All-source	~ 3 years	3 months	DSM: >12 months
Jersey Central Power & Light	All-source	2 years	2 months	3–4 months
Massachusetts Electric Co.	DSM	several weeks	2–3 months	3 months
Niagara Mohawk	All-source	1.5 years	2 months	12 months
Pacific Power & Light	All-source	6 months	8 months	5+ months (ongoing)
Portland General	DSM	9 months	2 months	3 months
PSI Energy	All-source	2 months ("too quickly!")	1.5 months	DSM: 12 months
Public Service Electric & Gas (NJ)	All-source	Not available	3 months	18 months
Public Service of Colorado	DSM	3 months	10 months	5 months
Puget Sound Power & Light	All-source	3–4 months	3 months	7 months
Washington Water Power	All-source	12 months	6.5 months	rejected

Source: Barakat & Chamberlin and Hope Robertson, publisher of *Robertson's Current Competition.*

have bidders. Contract negotiation has frequently become protracted and expensive for utilities and bidders alike (as indicated in Table 1). For example, when detailed verification methods have not been specified at the outset, contract negotiations have frequently become polarized, with the bidder holding on to a minimalist approach to verification while the utility struggles to protect its investment. Naturally relations between utilities and bidders, and between different groups within utilities, are sometimes strained.

Utilities must thoughtfully decide how they wish to conduct negotiations. For example, some utilities approach negotiation as the start of a long partnership. Others prefer an arm's-length negotiation with little interaction in-person. Ultimately, a utility must be comfortable with the load relief promised, the customer service, and the long-term viability of the project. Bidders must believe that they can achieve the contracted DSM level and that they will earn a profit on the project. All players must know their threshold—at what point they can willingly make a deal and at what point they must walk away.

Competition and Inefficiencies

ESCOs sometimes worry that utility programs might reduce the market for ESCO services, which would increase ESCOs' marketing costs and perhaps jeopardize their ability to meet their contracts for delivering energy savings. Utilities, on the other hand, worry that customers may become confused when presented with programs offered by ESCOs as well as those offered directly by the utility. Utilities and ESCOs have yet to develop a successful way of dealing with the overlap of bid programs with programs the utility offers.

WHAT UTILITIES HAVE LOOKED FOR FROM BIDDERS

Utilities have looked for a variety of benefits from bidding. How have they fared?

- **Lower price**. Thus far, bids for single end uses have typically been selected over bids for multiple end uses because, typically, they have been offered at lower cost and have been dominated by commercial lighting. However, although single end-use bids may be cheaper than multiple end-use bids, some utilities think they can serve the single end-use at a lower cost than bidders have proposed. Thus, utilities are increasingly trying to design RFPs that will increase the value to the utility of projects to be selected by making them more comprehensive. This is not an easy task, since the bidding process emphasizes the role of bid price in selecting projects.

- **Expanded delivery and promotion capability**. ESCOs have enabled utilities to reach certain markets more quickly. Bidding may be stimulating an expansion in the number of ESCOs.

- **Technical expertise in DSM**. ESCOs supplement in-house experts and have the benefit of offering turn-key capabilities for design and installation.

- **Stimulation of a broader DSM market**. Utilities' hopes have not been met in this area, apparently because ESCOs view "nontraditional" DSM projects as expensive and not worth considering.

- **Creativity and entrepreneurship**. In this area, also, utility hopes have not been met. Most bids have been standard rather than innovative. Utilities may be in the better position to experiment with new DSM technologies and methods because of their larger and more secure financial base.

- **Reputation**. Reputation is an often-overlooked factor in judging bids. Unfortunately, some utilities have found various third-party providers inferior in their approach to customers and DSM, causing a degradation in service or an excessive increase in costs. We recommend to utilities that the reputation of ESCOs be a strong consideration in selecting winning bidders, and even a threshold value in some cases.

- **Financing.** Utility hopes have not been met here either. ESCO financing has typically been more expensive than utility or customer financing. In many cases, customers contacted by a bidding program contractor opt to self finance their projects or use utility rebates and self financing. Only the most marginally financially viable companies and institutions prefer ESCO financing.

Energy service companies undeniably face a difficult problem in bidding on the basis of technical or market potential for a utility service area before they have had a chance to analyze individual customer facilities. Furthermore, unlike QFs, ESCOs cannot readily transfer developmental experience with DSM resources from one region to another because of the diversity of climate effects and customer types. ESCOs often have the advantage, however, of having very deep experience in a given niche market.

THE FUTURE

Like utility-developed DSM programs, DSM bidding initially was sought for reducing demand and reducing energy consumption. While that will continue, other services may be sought as well. For example, utilities are considering focusing future DSM solicitations to target specific markets and technologies and to get comprehensive savings.

In addition, they may look to bidding as an innovative means of promoting strategic growth or providing value-added services. Not all utilities are focused solely on reducing energy consumption and demand. Bidding could be used to solicit additional off-peak sales, which could have the benefit of increasing system efficiency, lowering rates, and solving minimum load problems.

DSM bidding could also become a vehicle for efficiently delivering value-added services to customers. Programs such as low-income assistance, power quality and reliability enhancements, interruptible rates, appliance repair and replacement, and new construction consulting may be solicited competitively. Electric outage insurance or price stability insurance are other possibilities. Likewise, bidding might be used to create DSM programs that maximize environmental benefits (in which case environmental impacts would become evaluation criteria).

Better Bidding

Despite the difficulties associated with bidding, such as the fear that factors like behavioral inconsistencies, low persistence, and snap-back will interfere with expected savings, bidding continues to evolve. The experience of the utilities with which we have worked demonstrates that there are ways to help DSM bidding programs live up to their promise. Briefly, utilities will benefit both themselves and bidders by (1) creating a structure for distinguishing among proposed DSM projects on the basis of their total value to the utility; (2) specifying the detailed verification methods that they find acceptable; (3) anticipating the time required to evaluate bids, negotiate contracts, verify savings, and administer the contract over its lifetime and including these estimates in its evaluation of program cost-effectiveness; and (4) recognizing that even though they may not have to pay contractors if savings do not materialize, the payment for savings attained will include a risk premium embedded in each bidder's profit margin.

Other Third-Party Alternatives

Even though utilities can take steps to maximize the benefit of bidding programs, bidding is only one possible method by which to involve third parties in demand-side management. Several alternatives are available, including traditional service procurement, which minimizes administrative overlap. For example:

- **Standard offers**. A utility may offer standard contracts for DSM to third parties. The standard offers and the price utilities would pay might vary by specific DSM measure.
- **Franchising**. Utilities might grant franchises to third parties for a specific geographic area, market segment, or end use. Franchises could be awarded using normal procurement methods that make use of specific work statements.
- **Contracting for energy services**. DSM bidding has focused on saving kWs and kWhs, but utilities could contract with ESCOs to deliver energy services for a given price (e.g., cooling, heating, lighting; educational campaigns; or maintenance services).

CONCLUSIONS

DSM bidding has not entirely met the goals of lowering DSM costs, improving third-party DSM infrastructure, and creating broader DSM markets. Utilities are likely to prefer to use third parties in a wide variety of circumstances ranging from turnkey energy service to quick turnaround commodity installations. DSM is ultimately a means of providing superior energy service value. A structure that impedes that goal will be unpopular with utilities. Conversely, enhancing value through competitive means can only be a positive step for the industry.

PD93m062/wee.ppr/7-6-93

REFERENCES

LeBlanc, W. "Stimulating the Private Energy-Efficiency Market: To Bid or Not to Bid?" In *DSM Bidding: Status and Results*. Albany, New York: Conference proceedings, April 1992.

McRae, Marjorie, Pamela Brandis, and Jane S. Peters. "Do You Get What You Pay for in DSM Bidding Programs?" Forthcoming.

Peters, Jane S., Deborah J. Barry, Marvin J. Horowitz, and David R. Wolcott. "A Dual State Bidding Program: Effects of Different Guidelines." In *Energy Program Evaluation: Uses, Methods, and Results*. Chicago: Proceedings of 1991 International Energy Program Evaluation Conference, September 1991. Pp 527–532.

Peters, Jane S., and M. Hossein Haeri. *Evaluation of Central Maine Power's Pilot Power Partners Program (T&C 19.15), Pilot Efficiency Buy-Back Program (T&C 19.13), and Pilot Commercial and Industrial Shared Savings Program (T&C 19.12)*. Augusta, Maine: Central Maine Power Company, December 1989.

Peters, Jane S., Marjorie R. McRae, and Jean Oates. "DSM Bidding as a New Business Venture." Forthcoming.

Peters, Jane S., Marjorie R. McRae, and Kenneth Seiden. *Process Evaluation of Public Service Colorado's 50 MW Bidding Program*. Denver, Colo.: Public Service Company of Colorado, April 1992.

Peters, Jane S., Marjorie R. McRae, and Peggy Seratt. *Resource Acquisition by the Bonneville Power Administration: An Assessment of Current Experience*. Portland, Oreg.: Bonneville Power Administration, May 1993.

Peters, Jane S., Peggy Seratt, Lorna Stucky, and Debora Darden-Butler. *Evaluation of the Orange and Rockland Utilities Competitive Bidding Program for Demand-Side Resources*. Albany, N.Y.: New York State Energy Research and Development Authority, February 1993.

Chapter 110

Healthier Projects Through Measuring and Monitoring

J.A. Polaski

Why measure and monitor energy projects? Your answer will of course depend upon who you are or who you represent within the DSM Industry. One thing is certain, the demand on services from energy measurement professionals has never been greater. Increased pressure for utilities to develop and rely upon DSM as a firm resource shall certainly increase the role of end-use metering to quantify energy savings and long term monitoring to promote persistence of savings.

WHO ARE YOU?

If you are a Regulator, you are looking to end-use metering and verification to assure prudent ratepayer investment in utility DSM Programs. Regulators are most likely to require metering where utilities receive not only lost revenue associated with the sale of fewer kilowatthours, but also receive an incentive or return on investment for dollars spent on promoting and/or installing energy conservation measures. Energy Service Companies (ESCOs) rely upon end-use metering to prove results to their clients as well as sponsoring utilities. Evaluators are called upon by regulators and utilities alike to determine the impacts of utility investment in conservation programs and many now offer end-use sample metering to confirm or question findings of broader evaluating and estimating techniques. Industrial firms have looked to end-use metering to justify budgets and prove benefit of new technologies which are introduced to improve production.

WHY THE RUSH TO METERING?

To embrace some level of end-use metering surely doesn't mean widely accepted evaluation and estimating techniques which have previously stood on their own do not have a place. There are those applications where metering energy savings alone may not be practical or cost-effective (i.e. industrial/commercial envelope upgrades, residential weatherization programs). However, there are applications where recording real-time end-use energy consumption is both practical, cost-effective and just plain smart. Sir Francis Bacon once mused, "*It is pointless to reason to the number of teeth in a horse's mouth. Rather, one should open the horse's mouth and simply count them.*"

DSM Investors (Ratepayers) Understand Counting

Energy savings resulting from lighting and motor retrofits are easily quantified. Savings from drives, chillers and process improvements, while not as easy as counting a horse's teeth, have also been cost-effectively quantified. It is important to remember the DSM industry is ultimately supported by your neighbors. Defending large investments of their cash into your DSM programs is made easier through measurement. For most of us, counting is a lot easier to understand than engineering estimates and statistical analyses. The benefits derived from most retrofits currently financed through DSM Programs can be counted.

DSM Budgets Support the Cost of End-Use Metering

End-use metering is now perceived as cost-effective. When DSM Programs were considered a public service, a "*nice thing for utilities to do*" engineering estimates and impact evaluations were appropriate to the expenditures being made. Early DSM budgets simply could not support the cost of earlier metering equipment. Introduction of less expensive metering alternatives and the increased amount of dollars chasing the "field of DSM dreams" have made those who rely on energy savings resulting from these programs rethink the metering issue. In New Jersey, for example, the DSM budget at just one utility has increased from $29 million per year in 1991 to over $150 million per year.

Increased Ratepayer Investment Fuels End-Use Metering

This level of ratepayer investment in other jurisdictions shall certainly cause ratepayer advocates to revisit end-use metering to confirm utility-purchased energy savings wherever practical. Building DSM portfolios in and of itself does not insure that real, persistent energy savings will come.

"HOW MUCH" IS EASY

We currently have approximately 400 installations in place which monitor savings resulting from light and motor retrofits, chiller and process improvements and residential air-conditioning cycling programs. For the most part, the data we supply to clients is used to quantify energy savings:

- To sell same to a utility
- To aid entry of new energy savings measures into the marketplace
- To justify capital investment in new plant equipment

We currently have one gigabyte (a thousand megabytes) of data regarding energy use. This is enough data to predict how much energy a particular building type will consume. It is ALMOST enough information to predict WHEN a particular building type will consume that same energy.

"WHEN" IS A LITTLE HARDER

Our experience is that while most facility operators have a pretty good grasp of how much energy is consumed in a given month, there are misperceptions regarding where and when the energy is used. This is in part true because we are all creatures of habit and tend to spend more time in the same areas of a building. A facility may also be awake when we are not, thereby excluding the reality of "after hours" energy use from our experience.

We have also found that the coincidental reduction of power associated with a particular project is probably less than expected. This is because every fixture and motor identified to operate at the same time rarely do. See Table 1 below:

Table 1

Point Information #	Gp	Description	KW Saved	Percent Active
1	1	HALLWAY - 2ND FLOOR	4.39	99.9%
2	2	SERVICE CORRIDOR - 2ND FLOOR	5.48	99.9%
3	3	LAB #271 - 2ND FLOOR	9.48	27.3%
4	4	OFFICE #271 (#272 B) - 2ND FLOO	2.02	7.7%
5	3	LAB #268 - 2ND FLOOR	9.48	23.5%
6	3	LAB #264 -2ND FLOOR	9.48	24.2%
7	3	LAB #260 - 2ND FLOOR	9.48	29.6%
8	4	OFFICE #260 (#359 B) - 2ND FLOO	2.02	29.2%
9	4	OFFICE #257 - 2ND FLOOR	2.02	26.7%
10	3	LAB #242 (HIGH CEILING FIXTURE)	9.48	25.3%
11	4	OFFICE #228 - 2ND FLOOR	2.02	23.6%
12	4	OFFICE #229 - 2ND FLOOR	2.02	10.8%
13	4	SECRETARY #230 - 2ND FLOOR	2.02	39.4%
14	4	OFFICE #234 - 2ND FLOOR	2.02	19.4%
15	4	SECRETARY #224 - 2ND FLOOR	2.02	29.5%
16	3	LAB #246 - 2ND FLOOR	9.48	27.0%
17	7	FAN ROOM #172 - 2ND FLOOR	2.89	92.6%
18	4	OFFICE #206 - 2ND FLOOR	2.02	24.5%
19	4	OFFICE #207 - 2ND FLOOR	2.02	19.3%
20	10	CONFERENCE RM # 204 - 2ND FLOOR	2.02	33.3%
21	4	SECRETARY #144 - 2ND FLOOR	2.02	32.0%
22	1	HALLWAY BY LIBRARY - 2ND FLOOR	4.39	31.6%
23	9	LIBRARY - MAIN ROOM - 2ND FLOOR	2.02	30.4%
24	9	LIBRARY - BACK ROOM - 2ND FLOOR	2.02	16.4%
25	3	LAB #186 - 2ND FLOOR	9.48	30.1%
26	4	OFFICE #186 - 2ND FLOOR	2.02	30.1%
27	5	STORAGE #210 - 2ND FLOOR	4.28	38.9%
28	9	KITCHEN #284 - 3RD FLOOR	0.77	34.5%
29	1	HALLWAY 015 OFFICE #288 - 3RD F	4.39	35.2%
30	4	OFFICE #286 - 3RD FLOOR	2.02	16.9%
31	1	CONFERENCE ROOM #280 - 3RD FLOO	2.02	35.1%
32	4	PAULA GREENFIELD'S OFFICE - 3RD	2.02	9.8%
Aggregates			129.31	33.9%

In the example above please note:

- The range of active hours (shown here as percent of 8760 hours) for offices and labs thought to have very similar, if not exact hours, of operation.

- The 24-hour use of service corridors originally thought to operate from 6 in the morning till 6 at night.

The average demand reduction for this lighting project was originally estimated to be 129.31. Annual energy savings were estimated on 8,760 operating hours for hallways and 5,750 operating hours for the rest of the facility. Months of monitoring have for the most part confirmed the 24-hour operation of hallway fixtures. The labs and offices, however, operate for about 2,900 annual hours, significantly reducing the scope of energy savings expected by the facility, the energy service company and the sponsoring utility.

The facility and energy service company had expected to save about 790,000 kilowatthours. Current projections based upon the metered data call for savings of 284,000 kilowatthours. All parties will probably look to Contract language to mitigate their exposure to the shortfalls of this project. Once this occurs the partnership to deliver real energy savings begins to break down. Mitigating the possibility of this circumstance is best down before the project even begins.

BENEFITS OF SHORT TERM METERING

The above lighting example leads to a discussion regarding the importance of short term metering. Portable data loggers to determine hours of operation for various areas of buildings prior to making an investment which relies on run hours for a return on the investment. Where you are required to install metering to monitor the persistence of energy savings, it is advisable to install the metering prior to the retrofit. You can determine the real load factor, (maybe even find the rare constantly loaded motor), run-times and energy consumption of equipment, particular motors and process, prior to investing in the equipment. Trying to make a square peg (your investment) fit into a round hole (unrealistic expectations) often foster customer dissatisfaction, regardless of everyone's good intentions.

LONG TERM MONITORING

Long term monitoring is seen as an ideal means of confirming the persistence of energy savings derived from the installation of energy conservation measures.

Monitoring is also helpful in identifying a program's free riders. Monitoring real-time consumption of residential air-conditioners, for example, will reveal the number and location of air-conditioners which normally operate during utility peak periods. It also identifies those air-conditioners that do not run during these periods, thereby allowing the utility to better invest its cycling technology and administration costs. For example if the customer's energy profile in Table 2 continues through the summer, he should be termed a free-rider. The cycling technology, as well as any incentives for his participation, are best applied to other customers.

Table 2

Points 1 and 2 measure the voltage; points 3 and 4 measure whole house consumption; point 5 measures air-conditioning usage and point 6 measures refrigeration consumption.

Point #	1	2	3	4	5	6
Units	V	V	W	W	W	W
06/10/93 12:52:35	122	121	15	703	0	292
06/10/93 12:57:35	122	121	15	439	0	22
06/10/93 13:02:35	121	121	15	146	0	21
06/10/93 13:07:35	121	121	15	132	0	21
06/10/93 13:12:35	122	121	15	718	0	322
06/10/93 13:17:35	121	120	15	688	0	300
06/10/93 13:22:35	122	121	15	703	0	296
06/10/93 13:27:35	122	121	15	703	0	293
06/10/93 13:32:35	122	121	15	703	0	289
06/10/93 13:37:35	122	121	15	146	0	22
06/10/93 13:42:35	121	121	15	132	0	21
06/10/93 13:47:35	121	121	15	410	0	21
06/10/93 13:52:35	122	121	29	571	0	21
06/10/93 13:57:35	122	121	15	864	0	311
06/10/93 14:02:35	121	120	15	850	0	300
06/10/93 14:07:35	123	121	15	835	0	299
06/10/93 14:12:35	122	121	15	527	0	292
06/10/93 14:17:35	121	120	15	513	0	286
06/10/93 14:22:35	121	120	15	527	0	22
06/10/93 14:27:35	122	121	15	542	0	21
06/10/93 14:32:35	122	121	15	439	0	21
06/10/93 14:37:35	121	121	15	439	0	21
06/10/93 14:42:35	122	121	15	732	0	311
06/10/93 14:47:35	122	121	15	410	0	297
06/10/93 14:52:35	121	120	15	410	0	292
06/10/93 14:57:35	121	121	15	674	0	289
06/10/93 15:02:35	121	120	15	688	0	292
06/10/93 15:07:35	121	121	15	439	0	22
06/10/93 15:12:35	121	120	15	439	0	21

At the end of summer '93, the utility, in addition to quantifying the coincidental peak demand reduction of its metered sample, will also have a much better perspective on the percentage of free-riders in its own program.

HEALTHY PROJECTS CONFIRM THE VALUE OF DSM

Measuring and monitoring energy savings, either alone, or in conjunction with other impact evaluation methods provides valuable information which benefits facility managers, ESCOs or other investing parties, utilizes and evaluators. To date, the majority of this experience has been retroactive. We are all seeing, after the fact, what makes projects healthier. Targeted, cost-effective pre-measurement and short term monitoring allows all participants to see the projects in 20/20 and plan accordingly.